Encyclopedia of Nanotechnology

Bharat Bhushan

Editor

Encyclopedia of Nanotechnology

Volume I

0-9–E

With 1976 Figures and 124 Tables

 Springer

Editor
Professor Bharat Bhushan
Ohio Eminent Scholar and
The Howard D. Winbigler Professor,
Director, Nanoprobe Laboratory for Bio- & Nanotechnology
and Biomimetics (NLB^2)
Ohio State University
201 W. 19th Avenue
Columbus, Ohio, 43210-1142
USA

ISBN 978-90-481-9750-7   ISBN 978-90-481-9751-4 (eBook)
DOI 10.1007/978-90-481-9751-4
ISBN 978-90-481-9752-1 (print and electronic bundle)
Springer Dordrecht Heidelberg New York London

Library of Congress Control Number: 2012940716

Printed on acid-free paper

Springer is part of Springer Science+Business Media (www.springer.com)

Preface

On December 29, 1959, at the California Institute of Technology, Nobel Laureate Richard P. Feynman gave a speech at the Annual meeting of the American Physical Society that has become one of the twentieth-century classic science lectures, titled "There's Plenty of Room at the Bottom." He presented a technological vision of extreme miniaturization in 1959, several years before the word "chip" became part of the lexicon. He spoke about the problem of manipulating and controlling things on a small scale. Extrapolating from known physical laws, Feynman envisioned a technology using the ultimate toolbox of nature, building nanoobjects atom by atom or molecule by molecule. Since the 1980s, many inventions and discoveries in the fabrication of nanoobjects have been testament to his vision. In recognition of this reality, National Science and Technology Council (NSTC) of the White House created the Interagency Working Group on Nanoscience, Engineering and Technology (IWGN) in 1998. In a January 2000 speech at the same institute, former President W. J. Clinton spoke about the exciting promise of "nanotechnology" and the importance of expanding research in nanoscale science and technology, more broadly. Later that month, he announced in his State of the Union Address an ambitious $497 million federal, multiagency National Nanotechnology Initiative (NNI) in the fiscal year 2001 budget, and made the NNI a top science and technology priority. The objective of this initiative was to form a broad-based coalition in which the academe, the private sector, and local, state, and federal governments work together to push the envelope of nanoscience and nanoengineering to reap nanotechnology's potential social and economic benefits.

The funding in the USA has continued to increase. In January 2003, the US senate introduced a bill to establish a National Nanotechnology Program. On December 3, 2003, President George W. Bush signed into law the 21st Century Nanotechnology Research and Development Act. The legislation put into law programs and activities supported by the (NNI). The bill gave nanotechnology a permanent home in the federal government and authorized $3.7 billion to be spent in the 4-year period beginning in October 2005, for nanotechnology initiatives at five federal agencies. The funds have provided grants to researchers, coordinated R&D across five federal agencies (National Science Foundation (NSF), Department of Energy (DOE), NASA, National Institute of Standards and Technology (NIST), and Environmental Protection Agency (EPA)), established interdisciplinary research centers, and accelerated technology transfer into the private sector. In addition, the Department of Defense (DOD), Homeland Security, Agriculture and Justice, as well as the National Institutes of Health (NIH) also fund large R&D activities. They currently account for more than one-third of the federal budget for nanotechnology.

The European Union (EU) made nanosciences and nanotechnologies a priority in the Sixth Framework Program (FP6) in 2002 for the period 2003–2006. They had dedicated small funds in FP4 and FP5 before. FP6 was tailored to help better structure European research and to cope with the strategic objectives set out in Lisbon in 2000. Japan identified nanotechnology as one of its main research priorities in 2001. The funding levels increased sharply from $400 million in 2001 to around $950 million in 2004. In 2003, South Korea embarked upon a 10-year program with around $2 billion of public funding, and Taiwan has committed around $600 million of public funding over 6 years. Singapore and China are also investing on a large scale. Russia is well funded as well.

Nanotechnology literally means any technology done on a nanoscale that has applications in the real world. Nanotechnology encompasses production and application of physical, chemical, and biological systems at scales, ranging from individual atoms or molecules to submicron dimensions, as well as the integration of the resulting nanostructures into larger systems. Nanotechnology is likely to have a profound impact on our economy and society in the early twenty-first century, comparable to that of semiconductor technology, information technology, or cellular and molecular biology. Science and technology research in nanotechnology is leading to break-throughs in areas such as materials and manufacturing, nanoelectronics, medicine and healthcare, energy, biotechnology, information technology, and national security. It is widely felt that nanotechnology will be the next industrial revolution.

There is an increasing need for a multidisciplinary, system-oriented approach toward designing and manufacturing micro/nanodevices which function reliably. This can only be achieved through the cross-fertilization of ideas from different disciplines and the systematic flow of information and people among research groups. Reliability is a critical technology for many micro- and nanosystems and nanostructured materials. The first edition of a broad-based *Handbook of Nanotechnology* from Springer was published in April 2004, the second edition in 2007, and the third edition in 2010. It presents an overview of nanomaterial synthesis, micro/nanofabrication, micro- and nanocomponents and systems, scanning probe microscopy, reliability issues (including nanotribology and nanomechanics) for nanotechnology, and various industrial including biomedical applications.

The field of nanotechnology is getting a strong foothold. It attracts people from various disciplines including science and engineering. Given the explosive growth in nanoscience and nanotechnology, this *Encyclopedia of Nanotechnology* is being launched with essays written by experts in the field from academia and industry.

The objective of this encyclopedia is to introduce a large number of terms, devices, and processes. For each entry, a brief description is provided by experts in the field. The entries have been written by a large number of internationally recognized experts in the field, from academia, national research labs, and industry.

The *Encyclopedia of Nanotechnology* is expected to provide a comprehensive and multidisciplinary reference to the many fields relevant to the general field of nano-technology. The encyclopedia focuses on engineering and applications with some coverage of the science of nanotechnology. It aims to be a comprehensive and genuinely international reference work and is aimed at graduate students, researchers, and practitioners.

The development of the encyclopedia was undertaken by an Editorial Board, who were responsible for the focus and quality of contributions, and an Advisory Board, who were responsible for advising about the selection of topics.

The print version of the encyclopedia contains four volumes with a total of **325** entries and about **3068** pages. The encyclopedia is also available online. Editors expect to update it periodically. The editor-in-chief and all the editors thank a large number of authors for making contributions to this major reference work. We also thank the referees who meticulously read the entries and made their recommendations.

Powell, Ohio
USA
May 2012

Bharat Bhushan

Biography

Bharat Bhushan
Ohio Eminent Scholar
The Howard D. Winbigler Professor, Director
Nanoprobe Laboratory for Bio- & Nanotechnology and Biomimetics (NLB^2)
Ohio State University
201 W. 19th Avenue
Columbus, Ohio, 43210-1142
USA
bhushan.2@osu.edu

Dr. Bharat Bhushan received an M.S. in mechanical engineering from the Massachusetts Institute of Technology in 1971; an M.S. in mechanics and a Ph.D. in mechanical engineering from the University of Colorado at Boulder in 1973 and 1976, respectively; an MBA from Rensselaer Polytechnic Institute at Troy, NY, in 1980; a Doctor Technicae from the University of Trondheim at Trondheim, Norway, in 1990; a Doctor of Technical Sciences from the Warsaw University of Technology at Warsaw, Poland, in 1996; and Doctor Honouris Causa from the National Academy of Sciences at Gomel, Belarus, in 2000 and University of Kragujevac, Serbia, in 2011. He is a registered professional engineer. He is presently an Ohio Eminent Scholar and The Howard D. Winbigler Professor in the College of Engineering and the Director of the Nanoprobe Laboratory for Bio- & Nanotechnology and Biomimetics (NLB^2) at the Ohio State University, Columbus, Ohio. His research interests include fundamental studies with a focus on scanning probe techniques in the interdisciplinary areas of bio/nanotribology, bio/nanomechanics, and bio/nanomaterials characterization and applications to bio/nanotechnology and biomimetics. He is an internationally recognized expert of bio/nanotribology and bio/nanomechanics using scanning probe microscopy and is one of the most prolific authors. He is considered by some a pioneer of the tribology and mechanics of magnetic storage devices. He has authored 8 scientific books, approximately 90 handbook chapters, 700 scientific papers (h-index – 57; ISI Highly Cited in Materials Science, since 2007; ISI Top 5% Cited Authors for Journals in Chemistry since 2011), and 60 technical reports. He has also edited more than 50 books and holds 17 US and foreign patents. He is coeditor of Springer NanoScience and Technology Series and coeditor of Microsystem Technologies. He has given more than 400 invited presentations on six continents and more than 200 keynote/plenary addresses at major international conferences.

Dr. Bhushan is an accomplished organizer. He organized the 1st Symposium on Tribology and Mechanics of Magnetic Storage Systems in 1984 and the 1st International Symposium on Advances in Information Storage Systems in 1990, both of which are now held annually. He is the founder of an ASME Information Storage and Processing Systems Division founded in 1993 and served as the founding chair during the period 1993–1998. His biography has been listed in over two dozen *Who's Who* books including *Who's Who in the World* and has received more than two dozen awards for his contributions to science and technology from professional societies, industry, and US government agencies. He is also the recipient of various international fellowships including the Alexander von Humboldt Research Prize for Senior Scientists, Max Planck Foundation Research Award for Outstanding Foreign Scientists, and the Fulbright Senior Scholar Award. He is a foreign member of the International Academy of Engineering (Russia), Byelorussian Academy of Engineering and Technology, and the Academy of Triboengineering of Ukraine; an honorary member of the Society of Tribologists of Belarus; a fellow of ASME, IEEE, STLE, and the New York Academy of Sciences; and a member of ASEE, Sigma Xi, and Tau Beta Pi.

Dr. Bhushan has previously worked for Mechanical Technology Inc., Latham, NY; SKF Industries Inc., King of Prussia, PA; IBM, Tucson, AZ; and IBM Almaden Research Center, San Jose, CA. He has held visiting professorship at University of California at Berkeley; University of Cambridge, UK; Technical University Vienna, Austria; University of Paris, Orsay; ETH, Zurich; and EPFL, Lausanne. He is currently a visiting professor at KFUPM, Saudi Arabia; Harbin Inst., China; University of Kragujevac, Serbia, and University of Southampton, UK.

Section Editors

Advisory Board

Contributors

Ahmad Nabil Abbas Department of Electrical and Computer Engineering, Florida International University, Miami, FL, USA

Patrick Abgrall Formulaction, L'Union, France

Angelo Accardo Lab. BIONEM, Dipartimento di Medicina Sperimentale e Clinica, Università "Magna Grecia" di Catanzaro, Catanzaro, Italy

Soft Matter Structures Group ID13 – MICROFOCUS Beamline, European Synchrotron Radiation Facility, Grenoble Cedex, France

Wafa Achouak iCEINT, International Consortium for the Environmental Implications of Nanotechnology, Center for the Environmental Implications of NanoTechnology, Aix–En–Provence, Cedex 4, France

Laboratoire d'Ecologie Microbienne de la Rhizosphère et d'Environnements Extrême, UMR 6191 CNRS–CEA–Aix–Marseille Université de la Méditerranée, St Paul lez Durance, France

Ranjeet Agarwala Department of Mechanical & Aerospace Engineering, North Carolina State University, Raleigh, NC, USA

Alessandro Alabastri Nanobiotech Facility, Istituto Italiano di Tecnologia, Genoa, Italy

Muhammad A. Alam School of Electrical and Computer Engineering, Purdue University, West Lafayette, IN, USA

Tuncay Alan Mechanical and Aerospace Engineering Department, Monash University, Victoria, Australia

Antonio Aliano Department of Physics, Politecnico di Torino, Torino, Italy

Wafa' T. Al-Jamal Nanomedicine Laboratory, Centre for Drug Delivery Research, The School of Pharmacy, University of London, London, UK

Paolo Allia Materials Science and Chemical Engineering Department, Politecnico di Torino, Torino, Italy

N. R. Aluru Department of Mechanical Science and Engineering, Beckman Institute for Advanced Science and Technology, University of Illinois at Urbana – Champaign, Urbana, IL, USA

Giampiero Amato Quantum Research Laboratory, Electromagnetism Division, Istituto Nazionale di Ricerca Metrologica, Torino, Italy

Benoy Anand Department of Physics, Sri Sathya Sai Institute of Higher Learning, Prashanthinilayam, Andhra Pradesh, India

Francesco De Angelis Nanobiotech Facility, Istituto Italiano di Tecnologia, Genoa, Italy

José V. Anguita Nano Electronics Center, Advanced Technology Institute, University of Surrey, Guildford, Surrey, UK

Wadih Arap David H. Koch Center, The University of Texas M. D. Anderson Cancer Center, Houston, TX, USA

Departments of Genitourinary Medical Oncology and Cancer Biology, The University of Texas M. D. Anderson Cancer Center, Houston, TX, USA

Walter Arnold Department of Material Science and Technology, Saarland University, Saarbrücken, Germany

Physikalisches Institut, Göttingen University, Göttingen, Germany

Christopher Arntsen Department of Chemistry and Biochemistry, UCLA, Los Angeles, CA, USA

Eduard Arzt INM – Leibniz Institute for New Materials and Saarland University, Saarbrücken, Germany

Burcu Aslan Department of Experimental Therapeutics, M.D. Anderson Cancer Center, The University of Texas, Houston, TX, USA

Elena Astanina University of Torino, Department of Oncological Sciences, Candiolo, Torino, Italy

Orlando Auciello Materials Science Division, Argonne National Laboratory, Argonne, IL, USA

Mélanie Auffan CEREGE, UMR 6635 CNRS/Aix–Marseille Université, Aix–en–Provence, France

iCEINT, International Consortium for the Environmental Implications of Nanotechnology, Center for the Environmental Implications of NanoTechnology, Aix–En–Provence, Cedex 4, France

Thomas Bachmann Institute for Fluid Mechanics and Aerodynamics, Technische Universität Darmstadt, Darmstadt, Germany

Armelle Baeza-Squiban Laboratory of Molecular and Cellular Responses to Xenobiotics, University Paris Diderot - Paris 7, Unit of Functional and Adaptive Biology (BFA) CNRS EAC 4413, Paris cedex 13, France

Darren M. Bagnall Nano Group, Electronics and Computer Science, University of Southampton, Highfield, Southampton, UK

Xuedong Bai Beijing National Laboratory for Condensed Matter Physics, Institute of Physics, Chinese Academy of Sciences, Beijing, P. R. China

Anisullah Baig Department of Electrical and Computer Engineering, University of California, Davis, USA

David J. Bakewell Department of Electrical Engineering and Electronics, University of Liverpool, Liverpool, UK

Mirko Ballarini DIFIS – Dipartimento di Fisica, Politecnico di Torino, Torino, Italy

Belén Ballesteros CIN2 (ICN-CSIC), Catalan Institute of Nanotechnology, Bellaterra, Barcelona, Spain

Antoine Barbier CEA-Saclay, DSM/IRAMIS/SPCSI, Gif-sur-Yvette, France

Robert Barchfeld Department of Electrical and Computer Engineering, University of California, Davis, USA

Yoseph Bar-Cohen Jet Propulsion Laboratory (JPL), California Institute of Technology, Pasadena, CA, USA

W. Jon. P. Barnes Centre for Cell Engineering, University of Glasgow, Glasgow, Scotland, UK

Larry R. Barnett Department of Electrical and Computer Engineering, University of California, Davis, USA

Friedrich G. Barth Life Sciences, Department of Neurobiology, University of Vienna, Vienna, Austria

Michael H. Bartl Department of Chemistry, University of Utah, Salt Lake City, UT, USA

R. Baskaran Department of Electrical Engineering, University of Washington, Seattle, WA, USA

Components Research, Intel Corporation, USA

Werner Baumgartner Department of Cellular Neurobionics, Institut für Biologie II RWTH-Aachen University, Aachen, Germany

Pierre Becker Laboratoire de Biologie des Organismes Marins et Biomimétisme, Université de Mons – UMONS, Mons, Belgium

Novid Beheshti Swedish Biomimetics 3000® Ltd, Birmingham, UK

Rachid Belkhou Synchrotron SOLEIL, L'Orme des Merisiers, Gif-sur-Yvette, France

Laila Benameur Laboratoire de Biogénotoxicologie et Mutagenèse Environnementale (EA 1784/FR CNRS 3098 ECCOREV), Aix-Marseille Université, Marseille Cedex 5, France

Rüdiger Berger Max Planck Institute for Polymer Research, Mainz, Germany

Magnus Berggren Department of Science and Technology, Linköpings University, Norrköping, Sweden

Pierre Berini School of Information Technology and Engineering (SITE), University of Ottawa, Ottawa, ON, Canada

Shekhar Bhansali Department of Electrical and Computer Engineering, Florida International University, Miami, FL, USA

Vikram Bhatia Science and Technology, Corning Incorporated SP-PR-02-1, Corning, NY, USA

Rustom B. Bhiladvala Department of Mechanical Engineering, University of Victoria, Victoria, BC, Canada

Bharat Bhushan Nanoprobe Laboratory for Bio- & Nanotechnology and Biomimetics, The Ohio State University, Columbus, OH, USA

Stefano Bianco Center for Space Human Robotics, Fondazione Istituto Italiano di Tecnologia, Torino, Italy

Ion Bita Qualcomm MEMS Technologies, Inc., San Jose, CA, USA

Martin G. Blaber Department of Chemistry and International Institute for Nanotechnology, Northwestern University, Evanston, IL, USA

Luca Boarino Quantum Research Laboratory, Electromagnetism Division, Istituto Nazionale di Ricerca Metrologica, Torino, Italy

Jorge Boczkowski INSERM U955 Eq04, Créteil, France

University Paris Est Val de Marne (UPEC), Créteil, France

Stuart A. Boden Nano Group, Electronics and Computer Science, University of Southampton, Highfield, Southampton, UK

Peter Bøggild DTU Nanotech – Department of Micro- and Nanotechnology, Technical University of Denmark, Lyngby, Denmark

K. F. Böhringer Department of Electrical Engineering, University of Washington, Seattle, WA, USA

Maarten P. de Boer Department of Mechanical Engineering, Carnegie Mellon University, Pittsburgh, USA

Ardemis A. Boghossian Department of Chemical Engineering, Massachusetts Institute of Technology, Cambridge, MA, USA

Paul W. Bohn Department of Chemical and Biomolecular Engineering, University of Notre Dame, Notre Dame, IN, USA

Department of Chemistry and Biochemistry, University of Notre Dame, Notre Dame, IN, USA

Sonja Boland Laboratory of Molecular and Cellular Responses to Xenobiotics, University Paris Diderot - Paris 7, Unit of Functional and Adaptive Biology (BFA) CNRS EAC 4413, Paris cedex 13, France

Robert D. Bolskar TDA Research, Inc., Wheat Ridge, CO, USA

Alexander Booth Energy and Resources Research Institute, University of Leeds, Leeds, West Yorkshire, UK

Garry J. Bordonaro Cornell NanoScale Science and Technology Facility, Cornell University, Ithaca, NY, USA

Edward Bormashenko Laboratory of Polymers, Applied Physics Department, Ariel University Center of Samaria, Ariel, Israel

Céline Botta CEREGE, UMR 6635, CNRS, Aix-Marseille Université, Aix en Provence, Cedex 04, France

Alain Botta Laboratoire de Biogénotoxicologie et Mutagenèse Environnementale (EA 1784/FR CNRS 3098 ECCOREV), Aix-Marseille Université, Marseille Cedex 5, France

Jean-Yves Bottero CEREGE, UMR 6635 CNRS/Aix–Marseille Université, Aix–en–Provence, France

iCEINT, International Consortium for the Environmental Implications of Nanotechnology, Center for the Environmental Implications of NanoTechnology, Aix–En–Provence, Cedex 4, France

Alessia Bottos University of Torino, Department of Oncological Sciences, Candiolo, Torino, Italy

Floriane Bourdiol EcoLab – Laboratoire d'écologie fonctionnelle et environnement, Université de Toulouse, INP, UPS, Castanet Tolosan, France

CNRS UMR 5245, EcoLab, Castanet Tolosan, France

Institut Carnot Cirimat, Université de Toulouse, UPS, INP, Toulouse cedex 9, France

Olivier Bourgeois Institut Néel CNRS-UJF, Grenoble, France

Herbert Bousack Peter Grünberg Institut, Forschungszentrum Jülich GmbH, Jülich, Germany

Thomas Braschler Laboratory of stem cell dynamics, SV-EPFL, Lausanne, Switzerland

Graham Bratzel Laboratory for Atomistic and Molecular Mechanics, Department of Civil and Environmental Engineering, Massachusetts Institute of Technology, Cambridge, MA, USA

Department of Mechanical Engineering, Massachusetts Institute of Technology, Cambridge, MA, USA

Donald W. Brenner Department of Materials Science and Engineering, North Carolina State University, Raleigh, NC, USA

Victor M. Bright Department of Mechanical Engineering, University of Colorado, Boulder, CO, USA

Lawrence F. Bronk David H. Koch Center, The University of Texas M. D. Anderson Cancer Center, Houston, TX, USA

Joseph J. Brown Department of Mechanical Engineering, University of Colorado, Boulder, CO, USA

Dorothea Brüggemann The Naughton Institute, School of Physics, Trinity College Dublin, CRANN, Dublin, Ireland

Markus J. Buehler Laboratory for Atomistic and Molecular Mechanics, Department of Civil and Environmental Engineering, Massachusetts Institute of Technology, Cambridge, MA, USA

Sven Burger Zuse Institute Berlin, Berlin, Germany

Federico Bussolino University of Torino, Department of Oncological Sciences, Candiolo, Torino, Italy

Donald P. Butler The Department of Electrical Engineering, The University of Texas at Arlington, Arlington, TX, USA

Hans-Jürgen Butt Max Planck Institute for Polymer Research, Mainz, Germany

Javier Calvo Fuentes NANOGAP SUB-NM-POWDER S.A., Milladoiro – Ames (A Coruña), Spain

Mary Cano-Sarabia CIN2 (ICN-CSIC), Catalan Institute of Nanotechnology, Bellaterra, Barcelona, Spain

Andrés Cantarero Materials Science Institute, University of Valencia, Valencia, Spain

Francesca Carpino Department of Chemical and Biomolecular Engineering, University of Notre Dame, Notre Dame, IN, USA

Marie Carrière Laboratoire Lésions des Acides Nucléiques, Commissariat à l'Energie Atomique, SCIB, UMR-E 3 CEA/UJF-Grenoble 1, INAC, Grenoble, Cedex, France

Jérôme Casas Institut de Recherche en Biologie de l'Insecte, IRBI UMR CNRS 6035, Université of Tours, Tours, France

Zeynep Celik-Butler The Department of Electrical Engineering, The University of Texas at Arlington, Arlington, TX, USA

Frederik Ceyssens Department ESAT-MICAS, KULeuven, Leuven, Belgium

Nicolas Chaillet FEMTO-ST/UFC-ENSMM-UTBM-CNRS, Besançon, France

Audrey M. Chamoire Department of Mechanical and Aerospace Engineering, The Ohio State University, E443 Scott Laboratory, Columbus, OH, USA

Peggy Chan Micro/Nanophysics Research Laboratory, RMIT University, Melbourne, VIC, Australia

Munish Chanana Departamento de Química Física, Universidade de Vigo, Vigo, Spain

Pen-Shan Chao Chung-Shan Institute of Science and Technology, Taoyuan, Taiwan ROC

Satish C. Chaparala Science and Technology, Corning Incorporated SP-PR-02-1, Corning, NY, USA

Wei Chen Department of Mechanical and Materials Engineering, Florida International University, Miami, FL, USA

Jian Chen Department of Mechanical and Industrial Engineering, University of Toronto, Toronto, ON, Canada

Department of Chemistry and Biochemistry, University of Wisconsin-Milwaukee, Milwaukee, WI, USA

Angelica Chiodoni Center for Space Human Robotics, Fondazione Istituto Italiano di Tecnologia, Torino, Italy

Alessandro Chiolerio Physics Department, Politecnico di Torino, Torino, Italy

Yoon-Kyoung Cho School of Nano-Bioscience and Chemical Engineering, Ulsan National Institute of Science and Technology (UNIST), Ulsan, Republic of Korea

Jong Hyun Choi School of Mechanical Engineering, Birck Nanotechnology Center, Bindley Bioscience Center, Purdue University, West Lafayette, IN, USA

Michael Chu Advanced Pharmaceutics and Drug Delivery Laboratory, University of Toronto, Toronto, ON, Canada

Han-Sheng Chuang Department of Biomedical Engineering, National Cheng Kung University, Tainan, Taiwan

Medical Device Innovation Center, National Cheng Kung University, Taiwan

Giancarlo Cicero Materials Science and Chemical Engineering Department, Politecnico di Torino, Torino, Italy

C. Cierpka Institute of Fluid Mechanics and Aerodynamics, Universität der Bundeswehr München, Neubiberg, Germany

Dominique Collard LIMMS/CNRS-IIS (UMI 2820), Institute of Industrial Science, The University of Tokyo, Tokyo, Japan

Lucio Colombi Ciacchi Hybrid Materials Interfaces Group, Faculty of Production Engineering and Bremen Center for Computational Materials Science, University of Bremen, Bremen, Germany

Maria Laura Coluccio Nanobiotech Facility, Istituto Italiano di Tecnologia, Genoa, Italy

Lab. BIONEM, Dipartimento di Medicina Sperimentale e Clinica, Università "Magna Grecia" di Catanzaro, Catanzaro, Italy

A. T. Conlisk Department of Mechanical Engineering, The Ohio State University, Columbus, OH, USA

Andrew Copestake Swedish Biomimetics 3000® Ltd, Southampton, UK

Miguel A. Correa-Duarte Departamento de Química Física, Universidade de Vigo, Vigo, Spain

Giovanni Costantini· Department of Chemistry, The University of Warwick, Coventry, UK

Claire Coutris Department of Plant and Environmental Sciences, Norwegian University of Life Sciences, Ås, Norway

Eduardo Cruz-Silva Department of Polymer Science and Engineering, University of Massachusetts Amherst, Amherst, MA, USA

Steven Curley Department of Surgical Oncology, The University of Texas M. D. Anderson Cancer Center, Houston, TX, USA

Department of Mechanical Engineering and Materials Science, Rice University, Houston, TX, USA

Christian Dahmen Division Microrobotics and Control Engineering (AMiR), Department of Computing Science, University of Oldenburg, Oldenburg, Germany

Lu Dai The Key Laboratory of Remodeling-related Cardiovascular Diseases, Capital Medical University, Ministry of Education, Beijing, China

Gobind Das Nanobiotech Facility, Istituto Italiano di Tecnologia, Genoa, Italy

Bakul C. Dave Department of Chemistry and Biochemistry, Southern Illinois University Carbondale, Carbondale, IL, USA

Enrica De Rosa Department of Nanomedicine, The Methodist Hospital Research Institute, Houston, TX, USA

Paolo Decuzzi Dept of Translational Imaging, and Nanomedicine, The Methodist Hospital Research Institute, Houston, TX, USA

Christian L. Degen Department of Physics, ETH Zurich, Zurich, Switzerland

Ada Della Pia Department of Chemistry, The University of Warwick, Coventry, UK

Gregory Denbeaux College of Nanoscale Science and Engineering, University at Albany, Albany, NY, USA

Parag B. Deotare Electrical Engineering, Harvard School of Engineering and Applied Sciences, Cambridge, MA, USA

Emiliano Descrovi DISMIC – Dipartimento di Scienza dei Materiali e Ingegneria Chimica, Politecnico di Torino, Torino, Italy

Joseph M. DeSimone Department of Chemistry, University of North Carolina, Chapel Hill, NC, USA

Department of Pharmacology, Eshelman School of Pharmacy, University of North Carolina, Chapel Hill, NC, USA

Carolina Center of Cancer Nanotechnology Excellence, University of North Carolina, Chapel Hill, NC, USA

Institute for Advanced Materials, University of North Carolina, Chapel Hill, NC, USA

Institute for Nanomedicine, University of North Carolina, Chapel Hill, NC, USA

Lineberger Comprehensive Cancer Center, University of North Carolina, Chapel Hill, NC, USA

Department of Chemical and Biomolecular Engineering, North Carolina State University, Raleigh, NC, USA

Sloan–Kettering Institute for Cancer Research, Memorial Sloan–Kettering Cancer Center, New York, NY, USA

Hans Deyhle Biomaterials Science Center (BMC), University of Basel, Basel, Switzerland

Charles L. Dezelah IV Picosun USA, LLC, Detroit, MI, USA

Nathan Doble The New England College of Optometry, Boston, MA, USA

Mitchel J. Doktycz Biosciences Division, Oak Ridge National Laboratory, Oak Ridge, TN, USA

Center for Nanophase Materials Sciences, Oak Ridge National Laboratory, Oak Ridge, TN, USA

Calvin Domier Department of Electrical and Computer Engineering, University of California, Davis, USA

Lixin Dong Electrical and Computer Engineering, Michigan State University, East Lansing, MI, USA

Avinash M. Dongare Department of Materials Science and Engineering, North Carolina State University, Raleigh, NC, USA

Emmanuel M. Drakakis Department of Bioengineering, The Sir Leon Bagrit Centre, Imperial College London, London, UK

Wouter H. P. Driessen David H. Koch Center, The University of Texas M. D. Anderson Cancer Center, Houston, TX, USA

Carlos Drummond Centre de Recherche Paul Pascal, CNRS–Université Bordeaux 1, Pessac, France

Jie Du Beijing Institute of Heart Lung and Blood Vessel Diseases, Beijing Anzhen Hospital, Beijing, China

Jean-Marie Dupret Laboratory of Molecular and Cellular Responses to Xenobiotics, University Paris Diderot - Paris 7, Unit of Functional and Adaptive Biology (BFA) CNRS EAC 4413, Paris cedex 13, France

Julianna K. Edwards David H. Koch Center, The University of Texas M. D. Anderson Cancer Center, Houston, TX, USA

Volkmar Eichhorn Division Microrobotics and Control Engineering (AMiR), Department of Computing Science, University of Oldenburg, Oldenburg, Germany

Masayoshi Esashi The World Premier International Research Center Initiative for Atom Molecule Materials, Tohoku University, Aramaki, Aoba-ku Sendai, Japan

Mikael Evander Department of Measurement Technology and Industrial Electrical Engineering, Division of Nanobiotechnology, Lund University, Lund, Sweden

Enzo Di Fabrizio Nanobiotech Facility, Istituto Italiano di Tecnologia, Genoa, Italy

Lab. BIONEM, Dipartimento di Medicina Sperimentale e Clinica, Università "Magna Grecia" di Catanzaro, Catanzaro, Italy

Yubo Fan Center for Bioengineering and Informatics, Department of Systems Medicine and Bioengineering, The Methodist Hospital Research Institute, Weill Cornell Medical College, Houston, TX, USA

Zheng Fan Electrical and Computer Engineering, Michigan State University, East Lansing, MI, USA

Sergej Fatikow Division Microrobotics and Control Engineering (AMiR), Department of Computing Science, University of Oldenburg, Oldenburg, Germany

Henry O. Fatoyinbo Centre for Biomedical Engineering, University of Surrey, Guildford, Surrey, UK

Joseph Fernandez-Moure The Methodist Hospital Research Institute, Houston, TX, USA

Mauro Ferrari Department of NanoMedicine, The Methodist Hospital Research Institute, Houston, TX, USA

Benjamin M. Finio School of Engineering and Applied Sciences, Harvard University, Cambridge, MA, USA

Emmanuel Flahaut Institut Carnot Cirimat, Université de Toulouse, UPS, INP, Toulouse cedex 9, France

CNRS, Institut Carnot Cirimat, Toulouse, France

Patrick Flammang Laboratoire de Biologie des Organismes Marins et Biomimétisme, Université de Mons – UMONS, Mons, Belgium

Richard G. Forbes Advanced Technology Institute, Faculty of Engineering and Physical Sciences, University of Surrey, Guildford, UK

Isabelle Fourquaux CMEAB, Centre de Microscopie Electronique Appliquée à la Biologie, Université Paul Sabatier, Faculté de Médecine Rangueil, Toulouse cedex 4, France

Marco Francardi Lab. BIONEM, Dipartimento di Medicina Sperimentale e Clinica, Università "Magna Grecia" di Catanzaro, Catanzaro, Italy

International School for Advanced Studies (SISSA), Edificio Q1 Trieste, Italy

Francesca Frascella DISMIC – Dipartimento di Scienza dei Materiali e Ingegneria Chimica, Politecnico di Torino, Torino, Italy

Roger H. French Department of Materials Science and Engineering, Case Western Reserve University, Cleveland, OH, USA

James Friend Micro/Nanophysics Research Laboratory, RMIT University, Melbourne, VIC, Australia

Hiroyuki Fujita Center for International Research on MicroMechatronics (CIRMM), Institute of Industrial Science, The University of Tokyo, Meguro-ku, Tokyo, Japan

Kenji Fukuzawa Department of Micro System Engineering, Nagoya University, Nagoya, Chikusa-ku, Japan

Diana Gamzina Department of Electrical and Computer Engineering, University of California, Davis, USA

Xuefeng Gao Suzhou Institute of Nano-Tech and Nano-Bionics, Chinese Academy of Sciences, Suzhou, PR China

Pablo García-Sánchez Departamento de Electrónica y Electromagnetismo, Universidad de Sevilla, Sevilla, Spain

Jean-Luc Garden Institut Néel CNRS-UJF, Grenoble, France

Paolo Gasco Nanovector srl, Torino, Italy

Laury Gauthier EcoLab – Laboratoire d'écologie fonctionnelle et environnement, Université de Toulouse, INP, UPS, Castanet Tolosan, France

CNRS UMR 5245, EcoLab, Castanet Tolosan, France

Shady Gawad MEAS Switzerland, Bevaix, Switzerland

Denis Gebauer Department of Chemistry, Physical Chemistry, University of Konstanz, Konstanz, Germany

Ille C. Gebeshuber Institute of Microengineering and Nanoelectronics (IMEN), Universiti Kebangsaan Malaysia, Bangi, Selangor, Malaysia

Institute of Applied Physics, Vienna University of Technology, Vienna, Austria

Francesco Gentile Nanobiotech Facility, Istituto Italiano di Tecnologia, Genoa, Italy

Lab. BIONEM, Dipartimento di Medicina Sperimentale e Clinica, Università "Magna Grecia" di Catanzaro, Catanzaro, Italy

Claudio Gerbaldi Center for Space Human Robotics, Fondazione Istituto Italiano di Tecnologia, Torino, Italy

Amitabha Ghosh Bengal Engineering & Science University, Howrah, India

Ranajay Ghosh Department of Mechanical, Aerospace and Nuclear Engineering, Rensselaer Polytechnic Institute, Troy, NY, USA

Larry R. Gibson II Department of Chemical and Biomolecular Engineering, University of Notre Dame, Notre Dame, IN, USA

Jason P. Gleghorn Department of Chemical and Biological Engineering, Princeton University, Princeton, NJ, USA

Biana Godin Department of Nanomedicine, The Methodist Hospital Research Institute, Houston, TX, USA

Irene González-Valls Laboratory of Nanostructured Materials for Photovoltaic Energy, Escola Tecnica Superior d Enginyeria (ETSE), Centre d'Investigació en Nanociència i Nanotecnología (CIN2, CSIC), Bellaterra (Barcelona), Spain

Ashwini Gopal Department of Biomedical Engineering, The University of Texas at Austin, Austin, TX, USA

Claudia R. Gordijo Advanced Pharmaceutics and Drug Delivery Laboratory, University of Toronto, Toronto, ON, Canada

Yann Le Gorrec FEMTO-ST/UFC-ENSMM-UTBM-CNRS, Besançon, France

Alok Govil Qualcomm MEMS Technologies, Inc., San Jose, CA, USA

Paul Graham Centre for Computational Neuroscience and Robotics, University of Sussex, Brighton, UK

Dmitri K. Gramotnev Nanophotonics Pty Ltd, Brisbane, QLD, Australia

Nicolas G. Green School of Electronics and Computer Science, University of Southampton, Highfield, Southampton, UK

Robert J. Greenberg Second Sight Medical Products (SSMP), Sylmar, CA, USA

Julia R. Greer Division of Engineering and Applied Sciences, California Institute of Technology, Pasadena, CA, USA

Dane A. Grismer Department of Chemical and Biomolecular Engineering, University of Notre Dame, Notre Dame, IN, USA

Petra Gruber Transarch - Biomimetics and Transdisciplinary Architecture, Vienna, Austria

Rina Guadagnini Laboratory of Molecular and Cellular Responses to Xenobiotics, University Paris Diderot - Paris 7, Unit of Functional and Adaptive Biology (BFA) CNRS EAC 4413, Paris cedex 13, France

Vladimir Gubala Biomedical Diagnostics Institute, Dublin City University, Glasnevin, Dublin, Ireland

Pablo Gurman Materials Science Division, Argonne National Laboratory, Argonne, IL, USA

Evgeni Gusev Qualcomm MEMS Technologies, Inc., San Jose, CA, USA

MEMS Research and Innovation Center, Qualcomm MEMS Technologies, Inc., San Jose, CA, USA

Maria Laura Habegger Department of Integrative Biology, University of South Florida, Tampa, FL, USA

Yassine Haddab FEMTO-ST/UFC-ENSMM-UTBM-CNRS, Besançon, France

Neal A. Hall Electrical and Computer Engineering, University of Texas at Austin, Austin, TX, USA

Moon-Ho Ham Department of Chemical Engineering, Massachusetts Institute of Technology, Cambridge, MA, USA

Hee Dong Han Gynecologic Oncology, M.D. Anderson Cancer Center, The University of Texas, Houston, TX, USA

Center for RNA Interference and Non–coding RNA, M.D. Anderson Cancer Center, The University of Texas, Houston, TX, USA

Xiaodong Han Institute of Microstructure and Property of Advanced Materials, Beijing University of Technology, Beijing, People's Republic of China

Aeraj ul Haque Biodetection Technologies Section, Energy Systems Division, Argonne National Laboratory, Lemont, IL, USA

Nadine Harris Department of Chemistry and International Institute for Nanotechnology, Northwestern University, Evanston, IL, USA

Judith A. Harrison Department of Chemistry, United States Naval Academy, Annapolis, MD, USA

Achim Hartschuh Department Chemie and CeNS, Ludwig-Maximilians-Universität München, Munich, Germany

Jian He Department of Physics and Astronomy, Clemson University, Clemson, SC, USA

Martin Hegner The Naughton Institute, School of Physics, Trinity College Dublin, CRANN, Dublin, Ireland

Michael G. Helander Department of Materials Science and Engineering, University of Toronto, Toronto, Ontario, Canada

Michael J. Heller Department of Nanoengineering, University of California San Diego, La Jolla, CA, USA

Department of Bioengineering, University of California San Diego, La Jolla, CA, USA

Simon J. Henley Nano Electronics Center, Advanced Technology Institute, University of Surrey, Guildford, Surrey, UK

Elise Hennebert Laboratoire de Biologie des Organismes Marins et Biomimétisme, Université de Mons – UMONS, Mons, Belgium

Joseph P. Heremans Department of Mechanical and Aerospace Engineering, The Ohio State University, E443 Scott Laboratory, Columbus, OH, USA

Simone Hieber Biomaterials Science Center (BMC), University of Basel, Basel, Switzerland

Dale Hitchcock Department of Physics and Astronomy, Clemson University, Clemson, SC, USA

David Holmes London Centre for Nanotechnology, University College London, London, UK

Hendrik Hölscher Karlsruher Institut für Technologie (KIT), Institut für Mikrostrukturtechnik, Karlsruhe, Germany

J. H. Hoo Department of Electrical Engineering, University of Washington, Seattle, WA, USA

Bart W. Hoogenboom London Centre for Nanotechnology and Department of Physics and Astronomy, University College London, London, UK

Kazunori Hoshino Department of Biomedical Engineering, The University of Texas at Austin, Austin, TX, USA

Larry L. Howell Department of Mechanical Engineering, Brigham Young University, Provo, UT, USA

Hou-Jun Hsu C.C.P. Contact Probes CO., LTD, New Taipei City, Taiwan ROC

Jung-Tang Huang National Taipei University of Technology, Taipei, Taiwan ROC

Michael P. Hughes Centre for Biomedical Engineering, University of Surrey, Guildford, Surrey, UK

Shelby B. Hutchens California Institute of Technology MC 309-81, Pasadena, CA, USA

John W. Hutchinson School of Engineering and Applied Sciences, Harvard University, Cambridge, MA, USA

Gilgueng Hwang Laboratoire de Photonique et de Nanostructures (LPN-CNRS), Site Alcatel de Marcoussis, Route de Nozay, Marcoussis, France

Hyundoo Hwang School of Nano-Bioscience and Chemical Engineering, Ulsan National Institute of Science and Technology (UNIST), Ulsan, Republic of Korea

Barbara Imhof LIQUIFER Systems Group, Austria

Hiromi Inada Hitachi High-Technologies America, Pleasanton, CA, USA

M. Saif Islam Electrical and Computer Engineering, University of California - Davis Integrated Nanodevices & Nanosystems Lab, Davis, CA, USA

Mitsumasa Iwamoto Department of Physical Electronics, Tokyo Institute of Technology, Meguro-ku, Tokyo, Japan

Esmaiel Jabbari Biomimetic Materials and Tissue Engineering Laboratory, Department of Chemical Engineering, Swearingen Engineering Center, Rm 2C11, University of South Carolina, Columbia, SC, USA

Laurent Jalabert LIMMS/CNRS-IIS (UMI 2820), Institute of Industrial Science, The University of Tokyo, Tokyo, Japan

Dongchan Jang Department of Applied Physics and Materials Science, California Institute of Technology, Pasadena, CA, USA

Daniel Jasper Division Microrobotics and Control Engineering (AMiR), Department of Computing Science, University of Oldenburg, Oldenburg, Germany

Debdeep Jena Department of Electrical Engineering, University of Notre Dame, Notre Dame, IN, USA

Taeksoo Ji School of Electronics and Computer Engineering, Chonnam National University, Gwangju, Korea

Lixin Jia The Key Laboratory of Remodeling-related Cardiovascular Diseases, Capital Medical University, Ministry of Education, Beijing, China

Lei Jiang Center of Molecular Sciences, Institute of Chemistry Chinese Academy of Sciences, Beijing, People's Republic of China

Dilip S. Joag Centre for Advanced Studies in Materials Science and Condensed Matter Physics, Department of Physics, University of Pune, Pune, Maharashtra, India

Erik J. Joner Bioforsk Soil and Environment, Ås, Norway

Suhas S. Joshi Department of Mechanical Engineering, Indian Institute of Technology Bombay, Mumbai, Maharastra, India

Gabriela Juarez-Martinez Centeo Biosciences Limited, Dumbarton, UK

Soyoun Jung Samsung Mobile Display Co., LTD, Young-in, Korea

C. J. Kähler Institute of Fluid Mechanics and Aerodynamics, Universität der Bundeswehr München, Neubiberg, Germany

Sergei V. Kalinin Center for Nanophase Materials Sciences, Oak Ridge National Laboratory, Oak Ridge, TN, USA

Ping Kao The Pennsylvania State University, University Park, PA, USA

Swastik Kar Department of Physics, Northeastern University, Boston, MA, USA

Mustafa Karabiyik Department of Electrical and Computer Engineering, Florida International University, Miami, FL, USA

Sinan Karaveli School of Engineering, Brown University, Providence, RI, USA

David Karig Center for Nanophase Materials Sciences, Oak Ridge National Laboratory, Oak Ridge, TN, USA

Michael Karpelson School of Engineering and Applied Sciences, Harvard University, Cambridge, MA, USA

Andreas G. Katsiamis Toumaz Technology Limited, Abingdon, UK

Christine D. Keating Department of Chemistry, Penn State University, University Park, PA, USA

Pamela L. Keating Department of Chemistry, United States Naval Academy, Annapolis, MD, USA

John B. Ketterson Department of Physics and Astronomy, Northwestern University, Evanston, IL, USA

S. M. Khaled Department of Nanomedicine, The Methodist Hospital Research Institute, Houston, TX, USA

Arash Kheyraddini Mousavi Department of Mechanical Engineering, University of New Mexico, Albuquerque, NM, USA

Andrei L. Kholkin Center for Research in Ceramic and Composite Materias (CICECO) & DECV, University of Aveiro, Aveiro, Portugal

Seonghwan Kim Department of Chemical and Materials Engineering, University of Alberta, Edmonton, AB, Canada

Seong H. Kim Department of Chemical Engineering, Pennsylvania State University, University Park, PA, USA

Moon Suk Kim Department of Molecular Science and Technology, Ajou University, Suwon, South Korea

CJ Kim Mechanical and Aerospace Engineering Department, University of California, Los Angeles (UCLA), Los Angeles, CA, USA

Bongsang Kim Advanced MEMS, Sandia National Laboratories, Albuquerque, NM, USA

Pilhan Kim Graduate School of Nanoscience and Technology, Korea Advanced Institute of Science and Technology (KAIST), Daejeon, South Korea

M. Todd Knippenberg Department of Chemistry, High Point University, High Point, NC, USA

Yee Kan Koh Department of Mechanical Engineering, National University of Singapore room: E2 #02-29, Singapore, Singapore

Helmut Kohl Physikalisches Institut, Westfälische Wilhelms-Universität Münster, Wilhelm-Klemm-Straße 10, Münster, Germany

Mathias Kolle Harvard School of Engineering and Applied Sciences, Cambridge, MA, USA

Susan Köppen Hybrid Materials Interfaces Group, Faculty of Production Engineering and Bremen Center for Computational Materials Science, University of Bremen, Bremen, Germany

Kostas Kostarelos Nanomedicine Laboratory, Centre for Drug Delivery Research, The School of Pharmacy, University of London, London, UK

Roman Krahne Nanobiotech Facility, Istituto Italiano di Tecnologia, Genoa, Italy

Gijs Krijnen Transducers Science & Technology group, MESA + Research Institute for Nanotechnology, University of Twente, Enschede, The Netherlands

Rajaram Krishnan Biological Dynamics, Inc., University of California San Diego, San Diego, CA, USA

Florian Krohs Division Microrobotics and Control Engineering (AMiR), Department of Computing Science, University of Oldenburg, Oldenburg, Germany

Elmar Kroner INM – Leibniz Institute for New Materials, Saarbrücken, Germany

Tom N. Krupenkin Department of Mechanical Engineering, The University of Wisconsin-Madison, Madison, WI, USA

Satish Kumar G. W. Woodruff School of Mechanical Engineering, Georgia Institute of Technology, Atlanta, GA, USA

Aloke Kumar Biosciences Division, Oak Ridge National Laboratory, Oak Ridge, TN, USA

Momoko Kumemura LIMMS/CNRS-IIS (UMI 2820), Institute of Industrial Science, The University of Tokyo, Tokyo, Japan

Harry Kwok Department of Electrical and Computer Engineering, University of Victoria, Victoria, Canada

Jae-Sung Kwon School of Mechanical Engineering and Birck Nanotechnology Center, Purdue University, West Lafayette, IN, USA

Jérôme Labille CEREGE, UMR 6635, CNRS, Aix-Marseille Université, Aix en Provence, Cedex 04, France

Jean Christophe Lacroix Interfaces, Traitements, Organisation et Dynamique des Systemes Universite Paris 7-Denis Diderot, Paris Cedex 13, France

Nicolas Lafitte LIMMS/CNRS-IIS (UMI 2820), Institute of Industrial Science, The University of Tokyo, Tokyo, Japan

Jao van de Lagemaat National Renewable Energy Laboratory, Golden, CO, USA

Renewable and Sustainable Energy Institute, Boulder, CO, USA

Akhlesh Lakhtakia Department of Engineering Science and Mechanics, Pennsylvania State University, University Park, PA, USA

Périne Landois Institut Carnot Cirimat, Université de Toulouse, UPS, INP, Toulouse cedex 9, France

Amy Lang Department of Aerospace Engineering & Mechanics, University of Alabama, Tuscaloosa, AL, USA

Sophie Lanone Inserm U955, Équipe 4, Université Paris Est Val de Marne (UPEC), Créteil, France

Hôpital Intercommunal de Créteil, Service de pneumologie et pathologie professionnelle, Créteil, France

Gregory M. Lanza C-TRAIN Labs, Washington University, St. Louis, MO, USA

Lars Uno Larsson Swedish Biomimetics 3000® AB, Stockholm, Sweden

Camille Larue Laboratoire Lésions des Acides Nucléiques, Commissariat à l'Energie Atomique, SCIB, UMR-E 3 CEA/UJF-Grenoble 1, INAC, Grenoble, Cedex, France

Michael J. Laudenslager Department of Materials Science and Engineering, University of Florida, Gainesville, FL, USA

Thomas Laurell Department of Measurement Technology and Industrial Electrical Engineering, Division of Nanobiotechnology, Lund University, Lund, Sweden

M. Laver Laboratory for Neutron Scattering, Paul Scherrer Institut, Villigen, Switzerland

Materials Research Division, Risø DTU, Technical University of Denmark, Roskilde, Denmark

Nano–Science Center, Niels Bohr Institute, University of Copenhagen, Copenhagen, Denmark

Department of Materials Science and Engineering, University of Maryland, Maryland, USA

Falk Lederer Institute of Condensed Matter Theory and Solid State Optics, Abbe Center of Photonics, Friedrich-Schiller-Universität Jena, Jena, Germany

Kuo-Yu Lee National Taipei University of Technology, Taipei, Taiwan ROC

David W. Lee Department of Biological Sciences, Florida International University Modesto Maidique Campus, Miami, FL, USA

Andreas Lenshof Department of Measurement Technology and Industrial Electrical Engineering, Division of Nanobiotechnology, Lund University, Lund, Sweden

Donald J. Leo Mechanical Engineering, Center for Intelligent Material Systems and Structures, Virginia Tech, Virginia Polytechnic Institute and State University, Arlington, VA, USA

Zayd Chad Leseman Department of Mechanical Engineering, University of New Mexico, Albuquerque, NM, USA

Nastassja A. Lewinski Department of Bioengineering, Rice University, Houston, TX, USA

Mo Li Department of Electrical and Computer Engineering, University of Minnesota, Minneapolis, MN, USA

Jason Li Department of Mechanical and Industrial Engineering and Institute of Biomaterials and Biomedical Engineering, University of Toronto, Toronto, ON, Canada

Wenzhi Li Department of Physics, Florida International University, Miami, FL, USA

Chen Li Department of Chemistry and Biochemistry, University of Bern, Bern, Switzerland

Chun Li Department of Experimental Diagnostic Imaging-Unit 59, The University of Texas MD Anderson Cancer Center, Houston, TX, USA

King C. Li Department of Bioengineering, Rice university, Houston, TX, USA

Weicong Li Department of Electrical and Computer Engineering, University of Victoria, Victoria, Canada

Meng Lian Biosciences Division, Oak Ridge National Laboratory, Oak Ridge, TN, USA

Carlo Liberale Nanobiotech Facility, Istituto Italiano di Tecnologia, Genoa, Italy

Ling Lin Beijing National Laboratory for Molecular Sciences (BNLMS), Key Laboratory of Organic Solids, Institute of Chemistry Chinese Academy of Sciences, Beijing, People's Republic of China

Chung-Yi Lin FormFactor, Inc., Taiwan, Hsinchu, Taiwan ROC

Lih Y. Lin Department of Electrical Engineering, University of Washington, Seattle, WA, USA

Mónica Lira-Cantú Laboratory of Nanostructured Materials for Photovoltaic Energy, Escola Tecnica Superior d Enginyeria (ETSE), Centre d'Investigació en Nanociència i Nanotecnología (CIN2, CSIC), Bellaterra (Barcelona), Spain

Shawn Litster Department of Mechanical Engineering, Carnegie Mellon University, Pittsburgh, PA, USA

Ying Liu School of Materials Science and Engineering, Georgia Institute of Technology, Atlanta, GA, USA

Chang Liu Tech Institute, Northwestern University, ME/EECS, Room L288, Evanston, IL, USA

Gang Logan Liu Micro and Nanotechnology Laboratory, Department of Electrical and Computer Engineering, University of Illinois at Urbana-Champaign, Urbana, IL, USA

Matthew T. Lloyd National Renewable Energy Laboratory MS 3211/SERF W100-43, Golden, CO, USA

Sarah B. Lockwood Department of Chemistry and Biochemistry, Southern Illinois University Carbondale, Carbondale, IL, USA

V. J. Logeeswaran Electrical and Computer Engineering, University of California - Davis Integrated Nanodevices & Nanosystems Lab, Davis, CA, USA

Mariangela Lombardi IIT - Italian Institute of Technology @ POLITO - Centre for Space Human Robotics, Torino, Italy

Marko Loncar Electrical Engineering, Harvard School of Engineering and Applied Sciences, Cambridge, MA, USA

Kenneth A. Lopata William R. Wiley Environmental Molecular Sciences Laboratory, Pacific Northwest National Laboratory, Richland, WA, USA

Gabriel Lopez-Berestein Department of Experimental Therapeutics, M.D. Anderson Cancer Center, The University of Texas, Houston, TX, USA

Cancer Biology, M.D. Anderson Cancer Center, The University of Texas, Houston, TX, USA

Center for RNA Interference and Non–coding RNA, M.D. Anderson Cancer Center, The University of Texas, Houston, TX, USA

The Department of Nanomedicine and Bioengineering, UTHealth, Houston, TX, USA

Alejandro Lopez-Bezanilla National Center for Computational Sciences, Oak Ridge National Laboratory, Oak Ridge, TN, USA

Jun Lou Department of Mechanical Engineering and Materials Science, Rice University 223 MEB, Houston, TX, USA

M. Arturo López-Quintela Laboratory of Magnetism and Nanotechnology, Institute for Technological Research, University of Santiago de Compostela, Santiago de Compostela, Spain

Yang Lu Department of Materials Science and Engineering, Massachusetts Institute of Technology, Cambridge, MA, USA

Zheng-Hong Lu Department of Physics, Yunnan University, Yunnan, Kunming, PR China

Department of Materials Science and Engineering, University of Toronto, Toronto, Ontario, Canada

Michael S.-C. Lu Department of Electrical Engineering, Institute of Electronics Engineering, and Institute of NanoEngineering and MicroSystems, National Tsing Hua University, Hsinchu, Taiwan, Republic of China

Wei Lu Department of Mechanical Engineering, University of Michigan, Ann Arbor, MI, USA

Jia Grace Lu Departments of Physics and Electrophysics, University of Southern California Office: SSC 215B, Los Angeles, CA, USA

Vanni Lughi DI3 – Department of Industrial Engineering and Information Technology, University of Trieste, Trieste, Italy

Neville C. Luhmann Jr. Department of Electrical and Computer Engineering, University of California, Davis, USA

Lorenzo Lunelli Biofunctional Surfaces and Interfaces, FBK-CMM Bruno Kessler Foundation and CNR-IBF, Povo, TN, Italy

Richard F. Lyon Google Inc, Santa Clara, CA, USA

Kuo-Sheng Ma IsaCal Technology, Inc., Riverside, CA, USA

Marc Madou Department of Mechanical and Aerospace Engineering & Biomedical Engineering, University of California at Irvine, Irvine, CA, USA

Daniele Malleo Fluxion Biosciences, South San Francisco, CA, USA

Supone Manakasettharn Department of Mechanical Engineering, The University of Wisconsin-Madison, Madison, WI, USA

Richard P. Mann Centre for Interdisciplinary Mathematics, Uppsala University, Uppsala, Sweden

Liberato Manna Nanobiotech Facility, Istituto Italiano di Tecnologia, Genoa, Italy

Shengcheng Mao Institute of Microstructure and Property of Advanced Materials, Beijing University of Technology, Beijing, People's Republic of China

Francelyne Marano Laboratory of Molecular and Cellular Responses to Xenobiotics, University Paris Diderot - Paris 7, Unit of Functional and Adaptive Biology (BFA) CNRS EAC 4413, Paris cedex 13, France

Sylvain Martel NanoRobotics Laboratory, Department of Computer and Software Engineering, and Institute of Biomedical Engineering, École Polytechnique de Montréal (EPM), Montréal, QC, Canada

Pascal Martin University Paris 7-Denis Diderot, ITODYS, Nanoelectrochemistry Group, UMR CNRS 7086, Batiment Lavoisier, Paris cedex 13, France

Paola Martino Politronica inkjet printing technologies S.r.l., Torino, Italy

Armand Masion CEREGE, UMR 6635 CNRS/Aix–Marseille Université, Aix–en–Provence, France

iCEINT, International Consortium for the Environmental Implications of Nanotechnology, Center for the Environmental Implications of NanoTechnology, Aix–En–Provence, Cedex 4, France

Daniel Maspoch CIN2 (ICN-CSIC), Catalan Institute of Nanotechnology, Bellaterra, Barcelona, Spain

Cintia Mateo Departamento de Química Física, Universidade de Vigo, Vigo, Spain

Shinji Matsui Laboratory of Advanced Science and Technology for Industry, University of Hyogo, Hyogo, Japan

Theresa S. Mayer Department of Electrical Engineering and Materials Science and Engineering, The Pennsylvania State University, University Park, PA, USA

Jeffrey S. Mayer Department of Electrical Engineering, Penn State University, University Park, PA, USA

Chimaobi Mbanaso College of Nanoscale Science and Engineering, University at Albany, Albany, NY, USA

Eva McGuire Department of Materials, Imperial College London, London, UK

Andy C. McIntosh Energy and Resources Research Institute, University of Leeds, Leeds, West Yorkshire, UK

Federico Mecarini Nanobiotech Facility, Istituto Italiano di Tecnologia, Genoa, Italy

Ernest Mendoza Centre de Recerca en Nanoenginyeria, Universitat Politècnica de Catalunya, Barcelona, Spain

Christoph Menzel Institute of Condensed Matter Theory and Solid State Optics, Abbe Center of Photonics, Friedrich-Schiller-Universität Jena, Jena, Germany

Timothy J. Merkel Department of Chemistry, University of North Carolina, Chapel Hill, NC, USA

Vincent Meunier Department of Physics, Applied Physics, and Astronomy, Rensselaer Polytechnic Institute, Troy, NY, USA

Paul T. Mikulski Department of Physics, United States Naval Academy, Annapolis, MD, USA

Hwall Min The Pennsylvania State University, University Park, PA, USA

Rodolfo Miranda Dep. Física de la Materia Condensada, Universidad Autónoma de Madrid and Instituto Madrileño de Estudios Avanzados en Nanociencia (IMDEA-Nanociencia), Madrid, Spain

Sushanta K. Mitra Micro and Nano-scale Transport Laboratory, Department of Mechanical Engineering, University of Alberta, Edmonton, AB, Canada

Cristian Mocuta Synchrotron SOLEIL, L'Orme des Merisiers, Gif-sur-Yvette, France

Mohammad R. K. Mofrad Molecular Cell Biomechanics Lab, Department of Bioengineering, University of California, Berkeley, CA, USA

Seyed Moein Moghimi Centre for Pharmaceutical Nanotechnology and Nanotoxicology, Department of Pharmaceutics and Analytical Chemistry, University of Copenhagen, Copenhagen, Denmark

Farghalli A. Mohamed Department of Chemical Engineering and Materials Science, University of California, Irvine The Henry Samueli School of Engineering, Irvine, CA, USA

Nancy A. Monteiro-Riviere Center for Chemical Toxicological Research and Pharmacokinetics, North Carolina State University, Raleigh, NC, USA

Philip Motta Department of Integrative Biology, University of South Florida, Tampa, FL, USA

Florence Mouchet EcoLab – Laboratoire d'écologie fonctionnelle et environnement, Université de Toulouse, INP, UPS, Castanet Tolosan, France

CNRS UMR 5245, EcoLab, Castanet Tolosan, France

Prachya Mruetusatorn Department of Electrical Engineering and Computer Science, University of Tennessee Knoxville, Knoxville, TN, USA

Oakridge National Laboratory, Oak Ridge, TN, USA

Weiqiang Mu Department of Physics and Astronomy, Northwestern University, Evanston, IL, USA

Stefan Mühlig Institute of Condensed Matter Theory and Solid State Optics, Abbe Center of Photonics, Friedrich-Schiller-Universität Jena, Jena, Germany

Partha P. Mukherjee Computer Science and Mathematics Division, Oak Ridge National Laboratory, Oak Ridge, TN, USA

Bert Müller Biomaterials Science Center (BMC), University of Basel, Basel, Switzerland

Claudia Musicanti Nanovector srl, Torino, Italy

Jit Muthuswamy School of Biological and Health Systems Engineering, Arizona State University, Tempe, AZ, USA

Shrikant C. Nagpure Nanoprobe Laboratory for Bio- & Nanotechnology and Biomimetics, The Ohio State University, Columbus, OH, USA

Vishal V. R. Nandigana Department of Mechanical Science and Engineering, Beckman Institute for Advanced Science and Technology, University of Illinois at Urbana – Champaign, Urbana, IL, USA

Avinash P. Nayak Electrical and Computer Engineering, University of California - Davis Integrated Nanodevices & Nanosystems Lab, Davis, CA, USA

Suresh Neethirajan School of Engineering, University of Guelph, Guelph, ON, Canada

Celeste M. Nelson Department of Chemical and Biological Engineering, Princeton University, Princeton, NJ, USA

Department of Molecular Biology, Princeton University, Princeton, NJ, USA

Bradley J. Nelson Institute of Robotics and Intelligent Systems, ETH Zurich, Zurich, Switzerland

Gilbert Daniel Nessim Chemistry department, Bar-Ilan Institute of Nanotechnology and Advanced Materials (BINA), Bar-Ilan University, Ramat Gan, Israel

Daniel Neuhauser Department of Chemistry and Biochemistry, UCLA, Los Angeles, CA, USA

J. Tanner Nevill Fluxion Biosciences, South San Francisco, CA, USA

Nam-Trung Nguyen School of Mechanical and Aerospace Engineering, Nanyang Technological Univeristy, Singapore, Singapore

Hossein Nili School of Electronics and Computer Science, University of Southampton, Highfield, Southampton, UK

Nano Research Group, University of Southampton, Highfield, Southampton, UK

Vincent Niviere GDRI ICEINT: International Center for the Environmental Implications of Nanotechnology, CNRS–CEA, Europôle de l'Arbois BP 80, Aix–en–Provence Cedex 4, France

Laboratoire de Chimie et Biologie des Métaux, UMR 5249, iRTSV–CEA Bat. K', 17 avenue des Martyrs, Grenoble Cedex 9, France

Michael Nosonovsky Department of Mechanical Engineering, University of Wisconsin-Milwaukee, Milwaukee, WI, USA

Thomas Nowotny School of Informatics, University of Sussex, Falmer, Brighton, UK

Seajin Oh Department of Mechanical and Aerospace Engineering & Biomedical Engineering, University of California at Irvine, Irvine, CA, USA

Murat Okandan Advanced MEMS and Novel Silicon Technologies, Sandia National Laboratories, Albuquerque, NM, USA

Brian E. O'Neill Department of Radiology Research, The Methodist Hospital Research Institute, Houston, TX, USA

Takahito Ono Department of Mechanical Systems and Design, Graduate School of Engineering, Tohoku University, Aramaki, Aoba-ku Sendai, Japan

Clifford W. Padgett Chemistry & Physics, Armstrong Atlantic State University, Savannah, GA, USA

Christine Paillès CEREGE, UMR 6635 CNRS/Aix–Marseille Université, Aix–en–Provence, France

iCEINT, International Consortium for the Environmental Implications of Nanotechnology, Center for the Environmental Implications of NanoTechnology, Aix–En–Provence, Cedex 4, France

Nezih Pala Department of Electrical and Computer Engineering, Florida International University, Miami, FL, USA

Manuel L. B. Palacio Nanoprobe Laboratory for Bio- & Nanotechnology and Biomimetics, The Ohio State University, Columbus, OH, USA

Jeong Young Park Graduate School of EEWS (WCU), Korea Advanced Institute of Science and Technology (KAIST), Daejeon, Republic of Korea

K. S. Park Department of Electrical Engineering, University of Washington, Seattle, WA, USA

Kinam Park Departments of Biomedical Engineering and Pharmaceutics, Purdue University, West Lafayette, IN, USA

Woo-Tae Park Seoul National University of Science and Technology, Seoul, Korea

Andrew R. Parker Department of Zoology, The Natural History Museum, London, UK

Green Templeton College, University of Oxford, Oxford, UK

Alessandro Parodi Department of Nanomedicine, The Methodist Hospital Research Institute, Houston, TX, USA

Renata Pasqualini David H. Koch Center, The University of Texas M. D. Anderson Cancer Center, Houston, TX, USA

Laura Pasquardini Biofunctional Surfaces and Interfaces, FBK-CMM Bruno Kessler Foundation, Povo, TN, Italy

Melissa A. Pasquinelli Fiber and Polymer Science, Textile Engineering, Chemistry and Science, North Carolina State University, Raleigh, NC, USA

Siddhartha Pathak California Institute of Technology MC 309-81, Pasadena, CA, USA

Cecilia Pederzolli Biofunctional Surfaces and Interfaces, FBK-CMM Bruno Kessler Foundation, Povo, TN, Italy

Natalia Pelinovskaya CEREGE, UMR 6635, CNRS, Aix-Marseille Université, Aix en Provence, Cedex 04, France

Néstor O. Pérez-Arancibia School of Engineering and Applied Sciences, Harvard University, Cambridge, MA, USA

Dimitrios Peroulis School of Electrical and Computer Engineering, Birck Nanotechnology Center, Purdue University, West Lafayette, IN, USA

Vinh-Nguyen Phan School of Mechanical and Aerospace Engineering, Nanyang Technological University, Singapore, Singapore

Reji Philip Light and Matter Physics Group, Raman Research Institute, Sadashivanagar, Bangalore, India

Andrew Philippides Centre for Computational Neuroscience and Robotics, University of Sussex, Brighton, UK

Gianluca Piazza Department of Electrical and Systems Engineering, University of Pennsylvania, Philadelphia, PA, USA

Remigio Picone Dana-Farber Cancer Institute – Harvard Medical School, David Pellman Lab – Department of Pediatric Oncology, Boston, MA, USA

Ilya V. Pobelov Department of Chemistry and Biochemistry, University of Bern, Bern, Switzerland

Ryan M. Pocratsky Department of Mechanical Engineering, Carnegie Mellon University, Pittsburgh, USA

Ramakrishna Podila Department of Physics and Astronomy, Clemson University, Clemson, SC, USA

Martino Poggio Department of Physics, University of Basel, Basel, Switzerland

R. G. Polcawich US Army Research Laboratory RDRL-SER-L, Adelphi, MD, USA

Jan Pomplun Zuse Institute Berlin, Berlin, Germany

Alexandra Porter Department of Materials, Imperial College London, London, UK

Cristina Potrich Biofunctional Surfaces and Interfaces, FBK-CMM Bruno Kessler Foundation and CNR-IBF, Povo, TN, Italy

Siavash Pourkamali Department of Electrical and Computer Engineering, University of Denver, Denver, CO, USA

Shaurya Prakash Department of Mechanical and Aerospace Engineering, The Ohio State University, Columbus, OH, USA

Luigi Preziosi Dipartimento di Matematica, Politecnico di Torino, Torino, Italy

Luca Primo University of Torino, Department of Clinical and Biological Sciences, Candiolo, Italy

R. M. Proie US Army Research Laboratory RDRL-SER-E, Adelphi, MD, USA

Olivier Proux GDRI ICEINT: International Center for the Environmental Implications of Nanotechnology, CNRS–CEA, Europôle de l'Arbois BP 80, Aix–en–Provence Cedex 4, France

OSUG, Université Joseph Fourier BP 53, Grenoble, France

Pascal Puech CEMES, Toulouse Cedex 4, France

Robert Puers Department ESAT-MICAS, KULeuven, Leuven, Belgium

J. S. Pulskamp US Army Research Laboratory RDRL-SER-L, Adelphi, MD, USA

Yongfen Qi The Key Laboratory of Remodeling-related Cardiovascular Diseases, Capital Medical University, Ministry of Education, Beijing, China

Qiquan Qiao Department of Electrical Engineering and Computer Science, South Dakota State University, Brookings, USA

Aisha Qi Micro/Nanophysics Research Laboratory, RMIT University, Melbourne, VIC, Australia

Yabing Qi Energy Materials and Surface Sciences (EMSS) Unit, Okinawa Institute of Science and Technology, Kunigami-gun, Okinawa, Japan

Hongwei Qu Department of Electrical and Computer Engineering, Oakland University, Rochester, MI, USA

Marzia Quaglio Center for Space Human Robotics, Fondazione Istituto Italiano di Tecnologia, Torino, Italy

Regina Ragan The Henry Samueli School of Engineering, Chemical Engineering and Materials Science University of California, Irvine, Irvine, CA, USA

Melur K. Ramasubramanian Department of Mechanical & Aerospace Engineering, North Carolina State University, Raleigh, NC, USA

Antonio Ramos Departamento de Electrónica y Electromagnetismo, Universidad de Sevilla, Sevilla, Spain

Hyacinthe Randriamahazaka University Paris 7-Denis Diderot, ITODYS, Nanoelectrochemistry Group, UMR CNRS 7086, Batiment Lavoisier, Paris cedex 13, France

Apparao M. Rao Department of Physics and Astronomy, Clemson University, Clemson, SC, USA

Center for Optical Materials Science & Engineering Technologies, Clemson University, Clemson, SC, USA

E. Reina-Romo School of Engineering, University of Seville, Seville, Spain

Stéphane Régnier Institut des Systèmes Intelligents et de Robotique, Université Pierre et Marie Curie, CNRS UMR7222, Paris, France

Philippe Renaud Microsystems Laboratory, Ecole Polytechnique Federale de Lausanne (EPFL), Lausanne, Switzerland

Tian-Ling Ren Institute of Microelectronics, Tsinghua University, Beijing, China

Scott T. Retterer Biosciences Division, Oak Ridge National Laboratory, Oak Ridge, TN, USA

Center for Nanophase Materials Sciences, Oak Ridge National Laboratory, Oak Ridge, TN, USA

Department of Electrical Engineering and Computer Science, University of Tennessee Knoxville, Knoxville, TN, USA

Roberto de la Rica MESA + Institute for Nanotechnology, University of Twente, Enschede, The Netherlands

Agneta Richter-Dahlfors Swedish Medical Nanoscience Center, Department of Neuroscience, Karolinska Institutet, Stockholm, Sweden

Michèle Riesen Department of Genetics Evolution and Environment, Institute of Healthy Ageing, University College London, London, UK

Matteo Rinaldi Department of Electrical and Computer Engineering, Northeastern University, Boston, MA, USA

Aditi Risbud Molecular Foundry, Lawrence Berkeley National Laboratory, Berkeley, CA, USA

José Rivas Laboratory of Magnetism and Nanotechnology, Institute for Technological Research, University of Santiago de Compostela, Santiago de Compostela, Spain

INL – International Iberian Nanotechnology Laboratory, Braga, Portugal

Paola Rivolo Dipartimento di Scienza dei Materiali e Ingegneria Chimica – Politecnico di Torino, Torino, Italy

Stephan Roche CIN2 (ICN–CSIC), Catalan Institute of Nanotechnology, Universidad Autónoma de Barcelona, Bellaterra (Barcelona), Spain

Institució Catalana de Recerca i Estudis Avançats (ICREA), Barcelona, Spain

Carsten Rockstuhl Institute of Condensed Matter Theory and Solid State Optics, Abbe Center of Photonics, Friedrich-Schiller-Universität Jena, Jena, Germany

Fernando Rodrigues-Lima Laboratory of Molecular and Cellular Responses to Xenobiotics, University Paris Diderot - Paris 7, Unit of Functional and Adaptive Biology (BFA) CNRS EAC 4413, Paris cedex 13, France

Brian J. Rodriguez Conway Institute of Biomolecular and Biomedical Research, University College Dublin, Belfield, Dublin 4, Ireland

Jérôme Rose CEREGE UMR 6635– CNRS–Université Paul Cézanne Aix–Marseille III, Aix–Marseille Université, Europôle de l'Arbois BP 80, Aix–en–Provence Cedex 4, France

GDRI ICEINT: International Center for the Environmental Implications of Nanotechnology, CNRS–CEA, Europôle de l'Arbois BP 80, Aix–en–Provence Cedex 4, France

Yitzhak Rosen Superior NanoBiosystems LLC, Washington, DC, USA

M. Rossi Institute of Fluid Mechanics and Aerodynamics, Universität der Bundeswehr München, Neubiberg, Germany

Marina Ruths Department of Chemistry, University of Massachusetts Lowell, Lowell, MA, USA

Kathleen E. Ryan Department of Chemistry, United States Naval Academy, Annapolis, MD, USA

Malgorzata J. Rybak-Smith Department of Pharmacology, University of Oxford, Oxford, UK

V. Sai Muthukumar Department of Physics, Sri Sathya Sai Institute of Higher Learning, Prashanthinilayam, Andhra Pradesh, India

Verónica Salgueirino Departamento de Física Aplicada, Universidade de Vigo, Vigo, Spain

Meghan E. Samberg Center for Chemical Toxicological Research and Pharmacokinetics, North Carolina State University, Raleigh, NC, USA

Florence Sanchez Department of Civil and Environmental Engineering, Vanderbilt University, Nashville, USA

Catherine Santaella iCEINT, International Consortium for the Environmental Implications of Nanotechnology, Center for the Environmental Implications of NanoTechnology, Aix–En–Provence, Cedex 4, France

Laboratoire d'Ecologie Microbienne de la Rhizosphère et d'Environnements Extrême, UMR 6191 CNRS–CEA–Aix–Marseille Université de la Méditerranée, St Paul lez Durance, France

J. A. Sanz-Herrera School of Engineering, University of Seville, Seville, Spain

Stephen Andrew Sarles Mechanical Aerospace and Biomedical Engineering, University of Tennessee, Knoxville, TN, USA

J. David Schall Department of Mechanical Engineering, Oakland University, Rochester, MI, USA

George C. Schatz Department of Chemistry and International Institute for Nanotechnology, Northwestern University, Evanston, IL, USA

André Schirmeisen Institute of Applied Physics, Justus-Liebig-University Giessen, Giessen, Germany

Frank Schmidt Zuse Institute Berlin, Berlin, Germany

Helmut Schmitz Institute of Zoology, University of Bonn Poppelsdorfer Schloss, Bonn, Germany

Scott R. Schricker College of Dentistry, The Ohio State University, Columbus, OH, USA

Georg Schulz Biomaterials Science Center (BMC), University of Basel, Basel, Switzerland

Udo D. Schwarz Department of Mechanical Engineering, Yale University, New Haven, USA

Praveen Kumar Sekhar Electrical Engineering, School of Engineering and Computer Science, Washington State University Vancouver, Vancouver, WA, USA

Rita E. Serda Department of NanoMedicine, The Methodist Hospital Research Institute, Houston, TX, USA

Nika Shakiba Department of Mechanical and Industrial Engineering, University of Toronto, Toronto, ON, Canada

Karthik Shankar Department of Electrical and Computer Engineering, W2-083 ECERF, University of Alberta, Edmonton, AB, Canada

Yunfeng Shi Department of Materials Science and Engineering, MRC RM114, Rensselaer Polytechnic Institute, Troy, NY, USA

Li Shi Department of Mechanical Engineering, The University of Texas at Austin, Austin, TX, USA

Youngmin Shin Department of Electrical and Computer Engineering, University of California, Davis, USA

Wolfgang M. Sigmund Department of Materials Science and Engineering, University of Florida, Gainesville, FL, USA

Department of Energy Engineering, Hanyang University, Seoul, Republic of Korea

S. Ravi P. Silva Nano Electronics Center, Advanced Technology Institute, University of Surrey, Guildford, Surrey, UK

Nipun Sinha Department of Mechanical Engineering and Applied Mechanics, Penn Micro and Nano Systems (PMaNS) Lab, University of Pennsylvania, Philadelphia, PA, USA

S. Siva Sankara Sai Department of Physics, Sri Sathya Sai Institute of Higher Learning, Prashanthinilayam, Andhra Pradesh, India

Dunja Skoko The Naughton Institute, School of Physics, Trinity College Dublin, CRANN, Dublin, Ireland

Craig Snoeyink Birck Nanotechnology Center, Mechanical Engineering, Purdue University, West Lafayette, IN, USA

Konstantin Sobolev Department of Civil Engineering and Mechanics, University of Wisconsin-Milwaukee, Milwaukee, WI, USA

Helmut Soltner Zentralabteilung Technologie, Forschungszentrum Jülich GmbH, Jülich, Germany

Thomas Søndergaard Department of Physics and Nanotechnology, Aalborg University, Aalborg Øst, Denmark

Youngjun Song Department of Electrical and Computer Engineering, University of California San Diego, San Diego, CA, USA

Anil K. Sood Gynecologic Oncology, M.D. Anderson Cancer Center, The University of Texas, Houston, TX, USA

Cancer Biology, M.D. Anderson Cancer Center, The University of Texas, Houston, TX, USA

Center for RNA Interference and Non–coding RNA, M.D. Anderson Cancer Center, The University of Texas, Houston, TX, USA

The Department of Nanomedicine and Bioengineering, UTHealth, Houston, TX, USA

Pratheev S. Sreetharan School of Engineering and Applied Sciences, Harvard University, Cambridge, MA, USA

Bernadeta Srijanto Center for Nanophase Materials Sciences, Oak Ridge National Laboratory, Oak Ridge, TN, USA

Tomasz Stapinski Department of Electronics, AGH University of Science and Technology, Krakow, Poland

Ullrich Steiner Department of Physics, Cavendish Laboratories, University of Cambridge, Cambridge, UK

Michael S. Strano Department of Chemical Engineering, Massachusetts Institute of Technology, Cambridge, MA, USA

Arunkumar Subramanian Department of Mechanical and Nuclear Engineering, Virginia Commonwealth University, Richmond, VA, USA

Maxim Sukharev Department of Applied Sciences and Mathematics, Arizona State University, Mesa, AZ, USA

Bobby G. Sumpter Computer Science and Mathematics Division and Center for Nanophase Materials Sciences, Oak Ridge National Laboratory, Oak Ridge, TN, USA

Yu Sun Department of Mechanical and Industrial Engineering and Institute of Biomaterials and Biomedical Engineering and Department of Electrical and Computer Engineering, University of Toronto, Toronto, ON, Canada

Tao Sun Research Laboratory of Electronics, Department of Electrical Engineering and Computer Science, Massachusetts Institute of Technology, Cambridge, MA, USA

Vishnu-Baba Sundaresan Mechanical and Nuclear Engineering, Virginia Commonwealth University, Richmond, VA, USA

Srinivas Tadigadapa Department of Electrical Engineering, The Pennsylvania State University, University Park, PA, USA

Saikat Talapatra Department of Physics, Southern Illinois University Carbondale, Carbondale, IL, USA

Qingyuan Tan Department of Mechanical and Industrial Engineering, University of Toronto, Toronto, ON, Canada

Hiroto Tanaka School of Engineering and Applied Sciences, Harvard University, Cambridge, MA, USA

Xinyong Tao College of Chemical Engineering and Materials Science, Zhejiang University of Technology, Hangzhou, China

Ennio Tasciotti Department of Nanomedicine, The Methodist Hospital Research Institute, Houston, TX, USA

J. Ashley Taylor Department of Mechanical Engineering, The University of Wisconsin-Madison, Madison, WI, USA

Raviraj Thakur School of Mechanical Engineering and Birck Nanotechnology Center, Purdue University, West Lafayette, IN, USA

Antoine Thill iCEINT, International Consortium for the Environmental Implications of Nanotechnology, Center for the Environmental Implications of NanoTechnology, Aix–En–Provence, Cedex 4, France

Laboratoire Interdisciplinaire sur l'Organisation Nanométrique et Supramoléculaire, UMR 3299 CEA/CNRS SIS2M, Gif–surYvette, France

Alain Thiéry iCEINT, International Consortium for the Environmental Implications of Nanotechnology, Center for the Environmental Implications of NanoTechnology, Aix–En–Provence, Cedex 4, France

IMEP, UMR 6116 CNRS/IRD, Aix–Marseille Université, Marseille, Cedex 03, France

Thomas Thundat Department of Chemical and Materials Engineering, University of Alberta, Edmonton, AB, Canada

Katarzyna Tkacz–Smiech Faculty of Materials Science and Ceramics, AGH University of Science and Technology, Krakow, Poland

Steve To Department of Electrical and Computer Engineering, University of Toronto, Toronto, ON, Canada

Gerard Tobias Institut de Ciència de Materials de Barcelona (ICMAB-CSIC), Bellaterra, Barcelona, Spain

Andrea Toma Nanobiotech Facility, Istituto Italiano di Tecnologia, Genoa, Italy

Katja Tonisch Institut für Mikro- und Nanotechnologien, Technische Universität Ilmenau Fachgebiet Nanotechnologie, Ilmenau, Germany

Elka Touitou Institute of Drug Research, School of Pharmacy, The Hebrew University of Jerusalem, Jerusalem, Israel

Lesa A. Tran Department of Chemistry and the Richard E. Smalley Institute for Nanoscale Science and Technology, Rice University, Houston, TX, USA

Alexander A. Trusov Department of Mechanical and Aerospace Engineering, The University of California, Irvine, CA, USA

Soichiro Tsuda Exploratory Research for Advanced Technology, Japan Science and Technology Agency, Osaka, Japan

Lorenzo Valdevit Department of Mechanical and Aerospace Engineering, University of California, Irvine, CA, USA

Ana Valero Microsystems Laboratory, Ecole Polytechnique Federale de Lausanne (EPFL), Lausanne, Switzerland

Pablo Varona Dpto. de Ingenieria Informatica, Universidad Autónoma de Madrid, Madrid, Spain

Amadeo L. Vázquez de Parga Dep. Física de la Materia Condensada, Universidad Autónoma de Madrid and Instituto Madrileño de Estudios Avanzados en Nanociencia (IMDEA-Nanociencia), Madrid, Spain

K. Venkataramaniah Department of Physics, Sri Sathya Sai Institute of Higher Learning, Prashanthinilayam, Andhra Pradesh, India

Nuria Vergara-Irigaray Department of Genetics Evolution and Environment, Institute of Healthy Ageing, University College London, London, UK

Georgios Veronis Department of Electrical and Computer Engineering and Center for Computation and Technology, Louisiana State University, Baton Rouge, LA, USA

Alexey N. Volkov Department of Materials Science and Engineering, University of Virginia, Charlottesville, VA, USA

Frank Vollmer Laboratory of Biophotonics and Biosensing, Max Planck Institute for the Science of Light, Erlangen, Germany

Fritz Vollrath Department of Zoology, University of Oxford, Oxford, UK

Prashant R. Waghmare Micro and Nano-scale Transport Laboratory, Department of Mechanical Engineering, University of Alberta, Edmonton, AB, Canada

Hermann Wagner Institute for Biology II, RWTH Aachen University, Aachen, Germany

Richard Walker ICON plc, Marlow, Buckinghamshire, UK

Thomas Wandlowski Department of Chemistry and Biochemistry, University of Bern, Bern, Switzerland

Wenlong Wang Beijing National Laboratory for Condensed Matter Physics, Institute of Physics, Chinese Academy of Sciences, Beijing, P. R. China

Feng-Chao Wang State Key Laboratory of Nonlinear Mechanics (LNM), Institute of Mechanics, Chinese Academy of Sciences, Beijing, China

Yu-Feng Wang Institute of Microelectronics, Tsinghua University, Beijing, China

Szu-Wen Wang Chemical Engineering and Materials Science, The Henry Samueli School of Engineering, University of California, Irvine, CA, USA

Chunlei Wang Department of Mechanical and Materials Engineering, Florida International University, Miami, FL, USA

Zhibin Wang Department of Materials Science and Engineering, University of Toronto, Toronto, Ontario, Canada

Enge Wang International Center for Quantum Materials, School of Physics, Peking University, Beijing, China

Guoxing Wang School of Microelectronics, Shanghai Jiao Tong University (SJTU), Minhang, Shanghai P R, China

Zhong Lin Wang School of Materials Science and Engineering, Georgia Institute of Technology, Atlanta, GA, USA

Reinhold Wannemacher Madrid Institute for Advanced Studies, IMDEA Nanociencia, Madrid, Spain

Benjamin L. J. Webb Division of Infection & Immunity, University College London, London, UK

Liu Wei CEREGE UMR 6635-CNRS, Aix-Marseille Université, Europôle de l'Arbois, Aix-en-Provence, France

Michael Weigel-Jech Division Microrobotics and Control Engineering (AMiR), Department of Computing Science, University of Oldenburg, Oldenburg, Germany

Steven T. Wereley School of Mechanical Engineering, Birck Nanotechnology Center, Purdue University, West Lafayette, IN, USA

Tad S. Whiteside Savannah River National Laboratory, Aiken, SC, USA

John P. Whitney School of Engineering and Applied Sciences, Harvard University, Cambridge, MA, USA

Samuel A. Wickline C-TRAIN Labs, Washington University, St. Louis, MO, USA

Mark Wiesner iCEINT, International Consortium for the Environmental Implications of Nanotechnology, Center for the Environmental Implications of NanoTechnology, Aix–En–Provence, Cedex 4, France

CEINT, Center for the Environmental Implications of NanoTechnology, Duke University, Durham, NC, USA

Stuart Williams Department of Mechanical Engineering, University of Louisville, Louisville, KY, USA

Kerry Allan Wilson London Centre For Nanotechnology, University College London, London, UK

Lon J. Wilson Department of Chemistry and the Richard E. Smalley Institute for Nanoscale Science and Technology, Rice University, Houston, TX, USA

Patrick M. Winter Department of Radiology, Imaging Research Center, Cincinnati Children's Hospital Medical Center, Cincinnati, OH, USA

Stephen TC Wong Center for Bioengineering and Informatics, Department of Systems Medicine and Bioengineering, The Methodist Hospital Research Institute, Weill Cornell Medical College, Houston, TX, USA

Robert J. Wood School of Engineering and Applied Sciences, Wyss Institute for Biologically Inspired Engineering, Harvard University, Cambridge, MA, USA

Matthew Wright NanoSight Limited, Amesbury, Wiltshire, UK

Wei Wu Department of Materials Science, Fudan University, Shanghai, China

Xiao Yu Wu Advanced Pharmaceutics and Drug Delivery Laboratory, University of Toronto, Toronto, ON, Canada

H. Xie The State Key Laboratory of Robotics and Systems, Harbin Institute of Technology, Harbin, China

Guoqiang Xie Institute for Materials Research, Tohoku University, Sendai, Japan

Huikai Xie Department of Electrical and Computer Engineering, University of Florida, Gainesville, FL, USA

Tingting Xu Department of Electrical Engineering and Computer Science, South Dakota State University, Brookings, USA

Didi Xu Institute of Robotics and Intelligent Systems, ETH Zurich, Zurich, Switzerland

Christophe Yamahata Microsystem Lab., Ecole Polytechnique Fédérale de Lausanne, Lausanne, Switzerland

Keqin Yang Department of Physics and Astronomy, Clemson University, Clemson, SC, USA

Yuehai Yang Department of Physics, Florida International University, Miami, FL, USA

Yi Yang Institute of Microelectronics, Tsinghua University, Beijing, China

Chun Yang Division of Thermal Fluids Engineering, School of Mechanical and Aerospace Engineering, School of Chemical and Biomedical Engineering (joint appointment), Nanyang Technological University, Singapore

Yoke Khin Yap Department of Physics, Michigan Technological University, Houghton, MI, USA

Leslie Yeo Micro/Nanophysics Research Laboratory, RMIT University, Melbourne, VIC, Australia

Zheng Yin Center for Bioengineering and Informatics, Department of Systems Medicine and Bioengineering, The Methodist Hospital Research Institute, Weill Cornell Medical College, Houston, TX, USA

Yaroslava G. Yingling Materials Science and Engineering, North Carolina State University, Raleigh, NC, USA

Minami Yoda G. W. Woodruff School of Mechanical Engineering, Georgia Institute of Technology, Atlanta, GA, USA

Sang-Hee Yoon Department of Mechanical Engineering, University of California, Berkeley, CA, USA

Molecular Cell Biomechanics Lab, Department of Bioengineering, University of California, Berkeley, CA, USA

Lidan You Department of Mechanical and Industrial Engineering and Institute of Biomaterials and Biomedical Engineering, University of Toronto, Toronto, ON, Canada

Yanlei Yu Department of Materials Science, Fudan University, Shanghai, China

Wei Yu The Key Laboratory of Remodeling-related Cardiovascular Diseases, Capital Medical University, Ministry of Education, Beijing, China

Seok H. Yun Graduate School of Nanoscience and Technology, Korea Advanced Institute of Science and Technology (KAIST), Daejeon, South Korea

Wellman Center for Photomedicine, Department of Dermatology, Harvard Medical School and Massachusetts General Hospital, Boston, MA, USA

The Harvard–Massachusetts Institute of Technology Division of Health Science and Technology, Cambridge, MA, USA

Remo Proietti Zaccaria Nanobiotech Facility, Istituto Italiano di Tecnologia, Genoa, Italy

Li Zhang Institute of Robotics and Intelligent Systems, ETH Zurich, Zurich, Switzerland

Ze Zhang Institute of Microstructure and Property of Advanced Materials, Beijing University of Technology, Beijing, People's Republic of China

State Key Laboratory of Silicon Materials and Department of Materials Science and Engineering, Zhejiang University, Hangzhou, China

Yi Zhang Shanghai Institute of Applied Physics, Chinese Academy of Sciences, Shanghai, China

Jin Z. Zhang Department of Chemistry and Biochemistry, University of California, Santa Cruz, CA, USA

Xiaobin Zhang Department of Materials Science and Engineering, Zhejiang University, Hangzhou, China

John Xiaojing Zhang Department of Biomedical Engineering, The University of Texas at Austin, Austin, TX, USA

Jianqiang Zhao Department of Materials Science, Fudan University, Shanghai, China

Ya-Pu Zhao State Key Laboratory of Nonlinear Mechanics (LNM), Institute of Mechanics, Chinese Academy of Sciences, Beijing, China

Jinfeng Zhao Department of Electrical and Computer Engineering, University of California, Davis, USA

Leonid V. Zhigilei Department of Materials Science and Engineering, University of Virginia, Charlottesville, VA, USA

Ying Zhou Department of Electrical & Computer Engineering, University of Florida, Gainesville, FL, USA

Menghan Zhou Department of Physics and Astronomy, Clemson University, Clemson, SC, USA

Yimei Zhu Center for Functional Nanomaterials, Brookhaven National Lab, Upton, NY, USA

Wenguang Zhu Department of Physics and Astronomy, The University of Tennessee, Knoxville, TN, USA

Rashid Zia School of Engineering, Brown University, Providence, RI, USA

0–9

μMist®

▶ Spray Technologies Inspired by Bombardier Beetle

μVED (Micro Vacuum Electronic Devices)

▶ MEMS Vacuum Electronics

193-nm Lithography

▶ DUV Photolithography and Materials

248-nm Lithography

▶ DUV Photolithography and Materials

3D Micro/Nanomanipulation with Force Spectroscopy

H. Xie[1] and Stéphane Régnier[2]
[1]The State Key Laboratory of Robotics and Systems, Harbin Institute of Technology, Harbin, China
[2]Institut des Systèmes Intelligents et de Robotique, Université Pierre et Marie Curie, CNRS UMR7222, Paris, France

Synonyms

Force spectroscopy; Nano manipulation; Nanorobotics

Definition

A micromanipulation system is a robotic device which is used to physically interact with a sample under a microscope, where a level of precision of movement is necessary that cannot be achieved by the unaided human hand

It is well known that pick-and-place is very important for 3-D microstructure fabrication since it is an indispensable step in the bottom-up building process. The main difficulty in sufficiently completing such pick-and-place manipulation at this scale lies in fabricating a very sharp end-effector that is capable of smoothly releasing microobjects deposited on the substrate. Moreover, this end-effector has to provide enough grasping force to overcome strong adhesion forces [1–3] from the substrate as well as being capable of sensing and controlling interactions with the microobjects. Furthermore, compared with the manipulation of larger microobjects under an optical microscope, visual feedback at several microns more suffers from the shorter depth of focus and the narrower field of view of lenses with high magnifications, although different schemes or algorithms have been introduced on techniques of autofocusing [4, 5] and extending focus depth [7]. Compared with vision-based automated 2-D micromanipulation, automated 3-D micromanipulation at the scale of several microns to submicron scale is more challenging because of optical microscope's resolution limit (typically 200 nm). Moreover, additional manipulation feedback is needed that is beyond the capability of optical vision, such as in the cases of vertical contact detection along the optical axis or manipulation obstructed by opaque components. Therefore, multi-feedback is of vital

B. Bhushan (ed.), *Encyclopedia of Nanotechnology*, DOI 10.1007/978-90-481-9751-4,
© Springer Science+Business Media B.V. 2012

3D Micro/Nanomanipulation with Force Spectroscopy, Fig. 1 (**a**) A photo of the AFM-FRS on the configuration for nanoscale pick-and-place manipulation. (**b**) A SEM image of the cantilever fabricated with a protruding tip

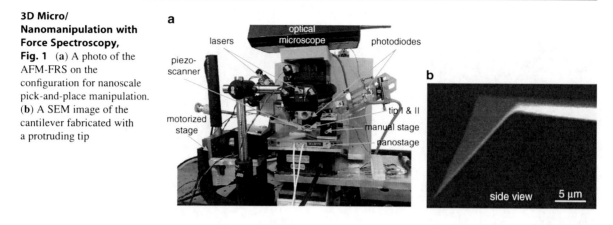

importance to achieve such accurate and stable 3-D micromanipulation at the scale of several microns to submicron scale.

Pick-and-place nanomanipulation is also a promising technique in 3D nanostructure fabrication since it is an indispensable step in the bottom-up building process. It can overcome limitations of bottom-up and top-down methods of nanomanufacturing and further combine advantages of these two methods to build complex 3D nanostructures. In literature, nanostructures have been manipulated, assembled, and characterized by integrating nanomanipulators or nanogrippers into scanning electron microscopes (SEM) and transmission electron microscopes (TEM) [12–16]. Both the SEM and the TEM provide a vacuum environment where the van der Waals force is the main force to be overcome during the manipulation. 3D nanomanipulation could be also achieved with optical tweezers in liquid, where the adhesion forces are greatly reduced [17–19]. However, the pick-and-place nanomanipulation in air is still a great challenge due to the presence of strong adhesion forces, including van der Waals, electrostatic and capillary forces [20]. In this case, the main difficulties in achieving the 3D nanomanipulation are fabricating sharp end-effectors with enough grasping force, as well as capabilities of force sensing while controlling interactions between the nanoobject and the tool or the substrate.

In this entry, a concept is presented to make a prototype of an AFM-based flexible robotic system (FRS) which is equipped with two collaborative cantilevers. By reconfiguring the modular hardware and software of the AFM-FRS, different configurations can be obtained. With different manipulation strategies

and protocols, two of the configurations can be respectively used to complete pick-and-place manipulations from the microscale of several microns to the nanoscale that are still challenges.

This entry is organized as follows. Section I introduces the prototype and experimental setup of the AFM-FRS. In subsection II and III, pick-and-place manipulation of microspheres and nanowires for building 3-D micro/nano structures are performed using the AFM-FRS.

AFM-Based Flexible Robotic System

AFM-FRS Setup

Figure 1a shows the AFM-FRS system setup, which is in the configuration for nanoscale pick-and-place. The system is equipped with an optical microscope and two sets of modules commonly used in a conventional AFM, mainly including two AFM cantilevers (namely, tip I and tip II, ATEC-FM Nanosensors, see Fig. 1b), two sets of nanopositioning devices and optical levers. The motion modules include an open-loop X-Y-Z piezoscanner (PI P-153.10H), an X-Y-Z closed-loop nanostage (MCL Nano-Bio2M on the X- and 7-axis, PI P-732. ZC on the Z-axis), an X-Y-Z motorized stage, and an X-Y-Z manual stage. Detailed specifications of the motion modules are summarized in Table 1. Figure 1c shows an optical microscope image of the collaborating tips in microsphere grasping mode.

A data acquisition (NI 6289) card is used for high-speed (500~800 Hz of sampling frequency for force and 600 kHz for amplitude) capture of the photodiode voltage output to estimate deflections on both tips induced by force loading or resonant oscillation.

**3D Micro/Nanomanipulation with Force Spectroscopy,
Table 1** Specifications of each motion module

Actuator	Travel range	Resolution
X-Y-Z piezoscanner	$10 \times 10 \times 10 (\mu m)$	Sub-nm
X-Y-Z nanostage	$50 \times 50 \times 10 (\mu m)$	0.1 nm
X-Y-Z motorized stage	$25 \times 25 \times 25 (mm)$	50 nm
X-Y-Z manual stage	$5 \times 5 \times 5 (mm)$	0.5 μm

A multi-thread planning and control system based on the C^{++} is developed for AFM image scan and two-tip coordination control during manipulation. This control system enables programming of complex tasks on the highly distributed reconfigurable system.

Force Sensing During Pick-and-Place

Figure 2 shows a schematic diagram of the nanotip gripper for the micro/nanoscale pick-and-place operation that has a clamping angle $\theta \approx 44°$ micro/nanoscale pickup manipulation, for instance, interactive forces applied on tip I include repulsive forces F_{r1}, friction forces F_{f1}, and adhesive forces F_{a1}.

The forces applied on tip I can be resolved into two components on the X-axis and the Z-axis in the defined frame, namely, F_{x1} and F_{z1}, respectively. F_{x1} is the clamping force that holds the micro/nanoobject. F_{z1} is the pickup force that balances adhesion forces from the substrate. To sense the pickup force, it is necessary to know the normal deflection on both cantilevers. The normal deflection ζ_{n1} associated with the normal voltage output of the optical lever on tip I is given by:

$$\zeta_{n1} = \frac{F_{z1} \cos \gamma + F_{x1} \sin \gamma}{k_n} + \frac{F_{z1} \sin \gamma + F_{x1} \cos \gamma}{k_{xz}} \tag{1}$$

where γ is the mounting angle of the cantilever, $k_{xz} = 2lk_n/3h$ is the bending stiffness due to the moment applied on the tip end, where h is the tip height. Assuming the magnitude of F_{z1} and F_{x1} are of the same order, contributions from the F_{x1} to the normal deflection of tip I are relatively very small since $k_{xz} \gg k_n$ and $\gamma = 5°$. Therefore, the normal deflection induced from F_{z1} is only considered in the following calculations of the adhesion force F_a applied on the micro/nanoobject. Thus, F_{z1} can be simplified by estimating the normal voltage output ΔV_{n1} from the tip I:

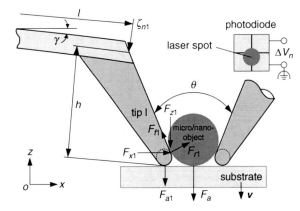

**3D Micro/Nanomanipulation with Force Spectroscopy,
Fig. 2** Schematic diagram of the nanotip gripper and force simulation during the pickup manipulation

$$F_{z1} = \beta_1 \times \Delta V_{n1} \tag{2}$$

where β_1 is the normal force sensitivity of the optical lever. A similar pickup force F_{z2} can be also obtained on Tip II. Before the gripper pulls off the substrate, the adhesion force F_a can be estimated as:

$$F_a = F_{z1} + F_{z2} = \beta_1 \Delta V_{n1} + \beta_2 \Delta V_{n2} - (F_{a1} + F_{a2}) \tag{3}$$

where β_2, ΔV_{n2}, and F_{a2} are respectively the normal force sensitivity, normal voltage output, and adhesive force on tip II. Once the gripper pulls off the substrate, e.g., in the case of nanowire/tube pick-and-place, the adhesion force F_a is estimated as:

$$F_a = F_{z1} + F_{z2} = \beta_1 \Delta V_{n1} + \beta_2 \Delta V_{n2}. \tag{4}$$

Experimental Results

3-D Micromanipulation Robotic System
System Configuration for 3-D Micromanipulation
As the size of microobjects is reduced to several microns or submicrons, problems will arise with these conventional grippers: (1) Sticking phenomena becomes more severe due to the relatively larger contact area between the gripper and the microobject. (2) The tip diameters of the micro-fabricated clamping jaws are comparable in size to the microobjects to be grasped. Conventional grippers are not geometrically

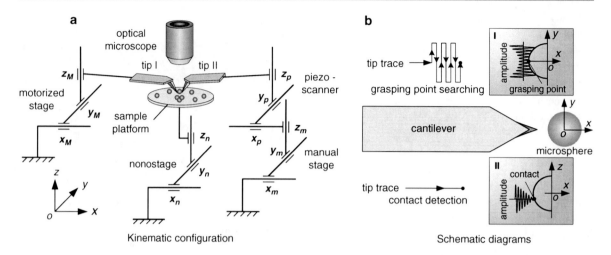

Kinematic configuration

Schematic diagrams

3D Micro/Nanomanipulation with Force Spectroscopy, Fig. 3 A kinematic configuration and schematic diagrams for 3-D micromanipulation at a scale of several micrometers or grasping point searching (*inset* I) and contact detection (*inset* II) with amplitude feedback of the dithering cantilever

sharp enough to pick up microobjects of several micrometers deposited on the substrate. Fortunately, the AFM tip has a very tiny apex (typically ~10 nm in radius) with respect to the size of the microobject to be manipulated. Thus, the nanotip gripper can be used to achieve pick-and-place at the scale of several microns since the contact area of the gripper-microobject is much smaller than the microobject-substrate contact. Moreover, real-time force sensing makes the manipulation more controllable.

As shown in Fig. 3, the system configuration for 3-D micromanipulation is reconfigured as follows:

1. For a large manipulation travel range, the nanostage here is used to support the sample platform and transport the microobject during the manipulation.
2. Tip I, immovable during the pick-and-place micromanipulation, is fixed on the motorized stage for coarse positioning.
3. Tip II is actuated by the piezoscanner for gripper opening and closing operations. The piezoscanner is supported by the manual stage for coarse positioning.

Benefiting from AFM-based accurate and stable amplitude feedback of a dithering cantilever, the grasping state can be successfully achieved by the amplitude feedback, with very weak interaction at the nano-Newton scale, protecting the fragile tips and the microobjects from damage during manipulation. As shown in inset I of Fig. 3, the dithering cantilever with its first resonant mode is used to locate

grasping points and detect contact. When approaching the microsphere with a separation between the tip and the substrate (typically 500 nm), the tip laterally sweeps the microsphere over the lower part of the microsphere. By this means, the grasping point can be accurately found by locating the minimum amplitude response of each single scan.

From the scheme depicted in inset II of Fig. 3, the amplitude feedback is also used for contact detection. Tip-microsphere contact is detected as the amplitude reduces to a steady value close to zero. As shown in Fig. 4, a protocol for pick-and-place microspheres mainly consists of four steps:

- System initialization and task planning: Each axis of the nanostage and the piezoscanner are set in a proper position, supplying the manipulation with enough travel range on each axis. Then the task is planned in Fig. 4a with a global view of the manipulation area that provides coarse positions of the microspheres and tips.
- Making tip I-microsphere in contact: In Fig. 4b, tip I has started to approach the microsphere by moving the nanostage with amplitude feedback to search for the grasping point and detect contact.
- Forming the gripper: Similarly, tip II approaches the microsphere by moving the piezoscanner. Once tip II and the microsphere are in contact, a nanotip gripper is configured in Fig. 4c for a manipulation.
- Pick-and-place micromanipulation: In Fig. 4d, the microsphere is picked up, transported, and released

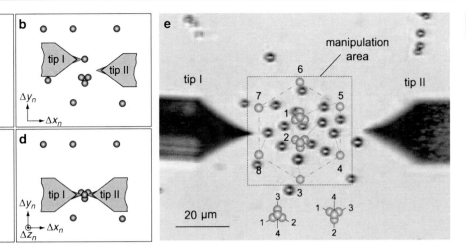

3D Micro/Nanomanipulation with Force Spectroscopy, Fig. 4 Microsphere manipulation protocol. (**a**) Task planning. (**b**) Tip I and the microsphere are in contact. (**c**) The nanotip gripper is formed. (**d**) Pick up and release the microsphere to its target position. (**e**) Task descriptions of the microsphere assembly

by moving the nanostage with a proper displacement on each axis that depends on the diameter of the microsphere and its destination. The whole process of 3-D micromanipulation is monitored by real-time force sensing.

3-D Microsphere Assembly

Nylon microspheres with diameter of about $3 \sim 4 \; \mu m$ were manipulated to build 3-D microstructures in experiments. The microspheres were deposited on a freshly cleaned glass slide and then an area of interest was selected under the optical microscope. Figure 4 shows a plan view of the selected area, in which 14 microspheres separated in a 50 μm square frame are going to be manipulated to build two 3-D micropyramids and a regular 2-D hexagon labeled by assembly sequences from 1 to 8. Each pyramid is constructed from four microspheres with two layers and the assembly sequences are shown in the bottom insets for two different arrangements of the pyramids.

Figure 5 shows a result of grasping point searching, in which the dithering tip II laterally sweeps the microsphere within a range of 1.75 μm on the y-axis and with a free oscillating amplitude of about 285 nm. Ten different distances to the microsphere were tested from 100 to 10 nm with an interval of 10 nm and, consequently, the grasping point is well located with an accuracy of ±10 nm. Figure 5 shows a full force spectroscopy curve during the pick-and-place of a microsphere deposited on a glass slide with an ambient temperature of 20°C and relative humidity of 40%.

In this curve, point A represents the start of the pick-and-place, point B the pull-off location of the microsphere-substrate contact, point C nonlinear force restitution due to the tip-microsphere frictions, and point D the snap-in point between the microsphere and the substrate. The force spectroscopy curve is synthesized from force responses on tip I and tip II.

Figure 6 shows an automated microassembly result consisting of two 3-D micropyramids and a 2-D pattern of a regular hexagon. The whole manipulation process was completed in 11 min, so the average manipulation time for each microsphere is about 47 s, which mainly breaks down into about 20 s for microsphere grasping including the grasping point search and contact detection processes using amplitude feedback, $10\sim35$ s for microsphere release and, the remaining time for transport.

Assembly of the fourth is the key to success in building a micropyramid. During pick-and-place of the fourth microsphere, microscopic vision was firstly used for coarse positioning it target, then the normal force feedback of the gripper was used to detect the vertical contact between the fourth microsphere and other three microspheres on the base. When the contact is established, a small vertical force was applied on the fourth microsphere by moving the nanostage upward and it will adjust to contact with all the base microspheres.

3-D Nanomanipulation Robotic System

System Configuration for 3-D Nanomanipulation
Compared with the 3-D micromanipulation, tip alignment precision is the key factor in succeeding the

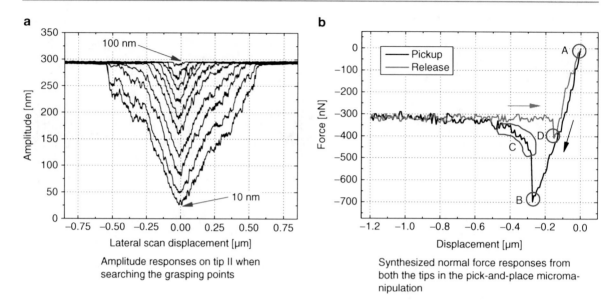

a   Amplitude responses on tip II when searching the grasping points

b   Synthesized normal force responses from both the tips in the pick-and-place micromanipulation

3D Micro/Nanomanipulation with Force Spectroscopy, Fig. 5   Experimental results

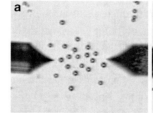

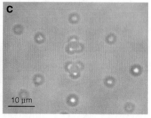

3D Micro/Nanomanipulation with Force Spectroscopy, Fig. 6   A microassembly result

Before the microassembly   Micropyramids were built   Enlarged image of the microassembly result under a magnification of 100×

nanoscale pick-and-place. Therefore, the closed-loop nanostage is considered in this configuration for accurate tip alignment. As shown in Fig. 7, the system configuration for 3-D nanomanipulation is reconfigured as follows:

1. The nanostage is used for image scan with tip I.
2. Nanoobjects are supported and transported by the piezoscanner.
3. Tip I, fixed on the motorized stage for coarse positioning, is immovable during the pick-and-place micromanipulation. Before manipulation, cantilever I acts as an image sensor for nanoobject positioning.
4. For accurate gripper alignment between Tip I and Tip II, Tip II is fixed on the nanostage rather than the piezoscanner. Tip II is supported by the manual stage for coarse positioning.

Nanowires and nanotubes are being intensively investigated. Thus, a protocol is developed here for nanowire or nanotube pick-and-place. However, applications can easily be extended to, for example, pick-and-place of nano-rods or nanoparticles dispersed on a substrate.

Once the manipulation area is selected under the optical microscope, both the tips are aligned as a quasi-gripper above the center of the manipulation area. Each axis of the nanostage and the piezoscanner is initialized at an appropriate position to allow for enough manipulation motion travel.

In this step, tip I is used to fully scan the relevant area obtaining a topographic image that contains nanoobjects to be manipulated and the end of tip II. Figure 7a shows a simulated image that contains the topography of two nanowires and the end of tip II. The image provides the following pick-and-place with

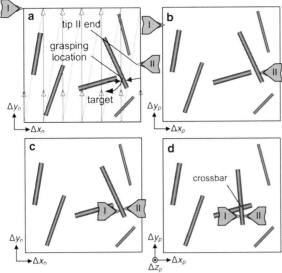

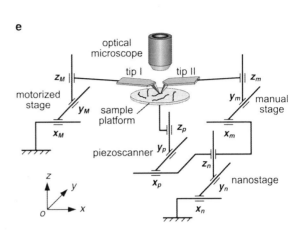

3D Micro/Nanomanipulation with Force Spectroscopy, Fig. 7 A protocol for nanowire pick-and-place (**a**) Image scan for task planning. (**b**) Tip II is in contact with the nanowire. (**c**) Tip I is in contact with the nanowire, forming a nanotip gripper. (**d**) Pick up and release a nanowire to its target position (**e**) A kinematic configuration of the AFM-RFS for nanoscale pick-and-place

relative positions between tip I, tip II and the nanoobjects to be manipulated. However, after a long period image scan, relocating tip II is recommended to eliminate system's thermal drift.

As shown in Fig. 7b, tip II approaches the nanowire to make contact by moving the X-axis of the piezoscanner. A gap (typically $\sim$20 nm above the snap-in boundary) between Tip II and the substrate should be maintained during the approach to enable a negative deflection response in the form of a tiny force applied on tip II, and hence, sensitive detection of the tip-nanowire contact.

Similarly, in Fig. 7c, once tip I is in contact with the nanowire, a nanotip gripper is configured for pick-and-place manipulation of the nanowire.

The nanotip gripper in this step is used to pick up, transport, and release the nanowire to its target position by moving the piezoscanner on the X, Y, or Z-axis. The displacement on each axis depends on the dimensions of the nanowire and the location of the destination. Figure 7d shows a simulated post-manipulation image, in which a nanowire crossbar is built. The complete pick-and-place procedure is monitored by force sensing.

Pick-and-Place Nanomanipulation

In experiments, silicon nanowires (SiNWs) were deposited on a freshly cleaned silicon wafer coated with 300 nm silicon dioxide. AFM images show that the SiNWs have a taper shape and have diameters of 25 nm (top), $\sim$200 nm (root), and lengths of about $4 \sim 7$ µm.

Figure 8 shows an example of the contact detection with tip II: Point A and point C are where the tip contacts with the SiNW and the Si substrate, respectively; Point B and point D are where the tip breaks the contact with the Si substrate and the SiNW, respectively. Figure 8 shows a curve of the peeling force spectroscopy on tip II for the pick-and-place manipulation of the SiNW: point A and point B are where the tip snaps in and pulls off the Si substrate, respectively. The shape of curve of the force responses on tip I are similar except for the force magnitude due to different force sensitivities on each tip and uneven grasping due to asymmetric alignment of the SiNW relative to the grasping direction. This force spectroscopy during the pickup operation shows stable grasping for further SiNW transport.

Figure 9 shows an experimental result of 3-D SiNW manipulation. A prescanned image (9×9 µm) is shown in Fig. 9a, which includes the topographic image of SiNWs and the local image of tip II. A grasping location of the nanowire to be manipulated is marked with a–a, where the SiNW has a height of 160 nm. Figure 9b is a post-manipulation image. It can

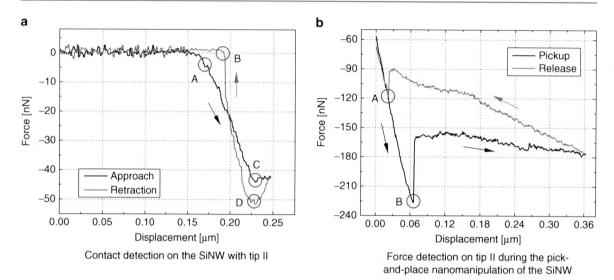

Contact detection on the SiNW with tip II

Force detection on tip II during the pick-and-place nanomanipulation of the SiNW

3D Micro/Nanomanipulation with Force Spectroscopy, Fig. 8 Contact and force detection

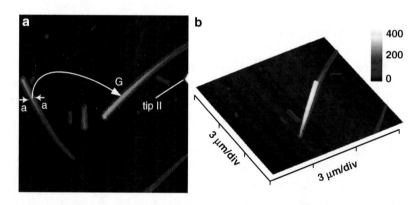

3D Micro/Nanomanipulation with Force Spectroscopy, Fig. 9 Pick-and-place results of the SiNWs. (a) A pre-scanned image. In which a–a and G are the grasping location and the target position, respectively. (b) A post-manipulation image verifies that the manipulated SiNW is piled upon another nanowire deposited on the Si substrate

be seen that the SiNW has been successfully transported and piled onto another SiNW. The manipulation procedure is described as follows. Once the SiNW was reliably grasped, the piezoscanner moved down 560 nm at a velocity of 80 nm/s. In this step, the SiNW was transported a distance of 4.4 μm along the X-axis at a velocity of 120 nm/s and 0.12 μm along the Y-axis at a lower velocity of 3.3 nm/s. In the releasing step, the piezoscanner moved up at a velocity of 100 nm/s. As tip II was slightly bent upward leading to a positive response of 0.015 V, tip I and tip II were separated by moving both the nanostage and the piezoscanner on the X-axis to release the SiNW from the nanotip gripper.

Conclusion and Future Directions

A flexible robotic system developed for multi-scale manipulation and assembly from nanoscale to microscale is presented. This system is based on the principle of atomic force microscopy and comprises two individually functionalized cantilevers. After reconfiguration, the robotic system could be used for pick-and-place manipulation from nanoscale to the scale of several micrometers. Flexibilities and manipulation capabilities of the developed system are validated by pick-and-place manipulation of microspheres and silicon nanowires to build three-dimensional micro/nanostructures in ambient conditions.

Complicated micro/nano manipulation and assembly can be reliably and efficiently performed using the proposed flexible robotic system. 3-D nanomanipulation methods are indispensible for heterogeneous integration of complex nanodevices. Serial nanomanipulation systems could only enable low-volume prototyping applications. However, high-volume and high-speed nanomanipulation systems are indispensable for future nanotechnology products. Therefore, autonomous and massively parallel AFM systems are required.

Cross-References

▶ AFM
▶ Atomic Force Microscopy
▶ Force Modulation in Atomic Force Microscopy
▶ Friction Force Microscopy
▶ Manipulating
▶ Nanogrippers
▶ Nanorobotic Assembly
▶ Nanorobotics
▶ Self-assembly of Nanostructures

References

1. Menciassi, A., Eisinberg, A., Izzo, I., Dario, P.: From "macro" to "micro" manipulation: models and experiments. IEEE-ASME Trans. Mechatro. **9**(2), 311–320 (2004)
2. Lambert, P., Régnier, S.: Surface and contact forces models within the framework of microassembly. J. Micromechatro. **3**(2), 123–157 (2006)
3. Sitti, M.: Microscale and nanoscale robotics systems-characteristics, state of the art, and grand challenges. IEEE Robot. Autom. Mag. **14**(1), 53–60 (2007)
4. Xie, H., Rong, W.B., Sun, L.N.: Construction and evaluation of a wavelet-based focus measure for microscopy. Imaging Microsc. Res. Tech. **70**, 987–995 (2007)
5. Liu, X.Y., Wang, W.H., Sun, Y.: Dynamic evaluation of autofocusing for automated microscopic analysis of blood smear and pap smear. J. Microsc. **227**, 15–223 (2007)
6. Xie, H., Rong, W.B., Sun, L.N., Chen, W.: Image fusion and 3-D surface reconstruction of microparts using complex valued wavelet transforms. In: IEEE International Conference on Image Process, Atlanta, pp. 2137–2140 (2006)
7. Xie, H., Vitard, J., Haliyo, S., Régnier, S.: Optical lever calibration in atomic force microscope with a mechanical lever. Rev. Sci. Instrum. **79**, 096101 (2008)
8. Xie, H., Rakotondrabe, M., Régnier, S.: Characterizing piezoscanner hysteresis and creep using optical levers and a reference nanopositioning stage. Rev. Sci. Instrum. **80**, 046102 (2009)
9. Vögeli, B., von Knel, H.: AFM-study of sticking effects for microparts handling. Wear **238**, 20–24 (2000)
10. Varenberg, M., Etsion, I., Halperin, G.: An improved wedge calibration method for lateral force in atomic force microscopy. Rev. Sci. Instrum. **74**, 3362–3367 (2003)
11. Xie, H., Vitard, J., Haliyo, S., Régnier, S.: Enhanced accuracy of force application for AFM nanomanipulation using nonlinear calibration of optical levers. IEEE Sens. J. **8**(8), 1478–1485 (2008)
12. Fukuda, T., Arai, F., Dong, L.X.: Assembly of nanodevices with carbon nanotubes through nanorobotic manipulations. Proc. IEEE **91**, 1803–1818 (2003)
13. Dong, L.X., Arai, F., Fukuda, T.: Electron-beam-induced deposition with carbon nanotube emitters. Appl. Phys. Lett. **81**, 1919–1921 (2002)
14. Dong, L.X., Arai, F., Fukuda, T.: Destructive constructions of nanostructures with carbon nanotubes through nanorobotic manipulation. IEEE/ASME Trans. Mechatron. **9**, 350–357 (2004)
15. Molhave, K., Wich, T., Kortschack, A., Boggild, P.: Pick-and-place nanomanipulation using microfabricated grippers. Nanotechnology **17**, 2434–2441 (2006)
16. Dong, L.X., Tao, X.Y., Zhang, L., Nelson, B.J., Zhang, X.B.: Nanorobotic spot welding: controlled metal deposition with attogram precision from copper-filled carbon nanotubes. Nano Lett. **7**, 58–63 (2007)
17. Leach, J., Sinclair, G., Jordan, P., Courtial, J., Padgett, M.J., Cooper, J., Laczik, Z.J.: 3D manipulation of particles into crystal structures using holographic optical tweezers. Opt. Exp. **12**, 220–226 (2004)
18. Yu, T., Cheong, F.C., Sow, C.H.: The manipulation and assembly of CuO nanorods with line optical tweezers. Nanotechnology **15**, 1732–1736 (2004)
19. Bosanac, L., Aabo, T., Bendix, P.M., Oddershede, L.B.: Efficient optical trapping and visualization of silver nanoparticles. Nano Lett. **8**, 1486–1491 (2008)
20. Rollot, Y., Régnier, S., Guinot, J.C.: Simulation of micromanipulations: adhesion forces and specific dynamic models. Int. J. Adhes. Adhes. **19**, 35–48 (1999)

A

Ab Initio DFT Simulations of Nanostructures

Antonio Aliano[1] and Giancarlo Cicero[2]
[1]Department of Physics, Politecnico di Torino, Torino, Italy
[2]Materials Science and Chemical Engineering Department, Politecnico di Torino, Torino, Italy

Synonyms

Electronic structure calculations; First principles calculations; Mean field approaches to many-body systems

Definition

Density Functional Theory (DFT) is an exact theory in which the electronic charge density is the basic quantity used to determinate the ground state properties of a many-body quantum system [1, 2]. Such a theory is an alternative formulation of the Schrödinger picture of quantum mechanics and represents a fundamental breakthrough in the application of computer simulation to material science and to nanosystems in particular. In the last decades, DFT-based analysis showed a quantitative predictive role and has been successfully employed in the study of molecules, crystals, and nanostructured materials. DFT calculations permit to investigate phenomena at an atomic scale hardly accessible by other techniques, emerging as a fundamental tool for complementing experimental investigations and for addressing open issues regarding nanomaterials and novel design processes [3, 4]. The DFT methodology is an ab initio (or first principles) approach, since no parameter characterizing the Hamiltonian of the system is tuned to empirical data and simulations results are thus free from empirical hints.

Introduction and DFT Foundations

The chemical and physical properties of materials both at the macro- and nano-scale strongly depend on the fundamental interactions holding together atoms and electrons present in the structure. A complete understanding of the material properties rests upon being able to accurately describe the electronic structure of the system, or, in other words, to solve the Schrödinger equation for the Hamiltonian characterizing the system. Unfortunately, solving this equation is not possible without making some physical assumptions that decrease the mathematical complexity of the problem. One of the most important approximations considered in the quantum treatment of materials consists of the so-called Born-Oppenheimer or adiabatic approximation: considering the three orders of magnitude difference between nuclei and electron masses (and De Broglie wavelengths), only the electrons are regarded as quantum particles and it is accepted that the motion of the nuclei does not influence the electron dynamics (nuclei are considered fixed in space). Within this hypothesis, the description of correlated nuclei and electrons can be decoupled so that the wave function Ψ of an N-electron system depends only on the electron coordinates, namely, $\Psi(r_1, r_2, \ldots, r_N) = \Psi(\{r\})$, and the nuclei positions enter only as parameters. The Hamiltonian operator $\hat{H}$ for

B. Bhushan (ed.), *Encyclopedia of Nanotechnology*, DOI 10.1007/978-90-481-9751-4,
© Springer Science+Business Media B.V. 2012

the electrons at a fixed nuclear configuration can be written in the compact form:

$$\hat{H} = \hat{T} + \hat{V}_{ee} + \hat{V}_{ext} \tag{1}$$

where $\hat{T}$ is the kinetic energy operator, $\hat{V}_{ee}$ the operator corresponding to the electron–electron interactions, and $\hat{V}_{ext}$ the external potential operator typically defined by the Coulomb attractive potential due to nuclei. Despite the simplifications introduced, when the form of the interaction potentials are known, the many-body Schrödinger equation $\hat{H}\Psi = E\Psi$ for the electrons cannot be solved analytically and it is not numerically manageable because of the prohibitive number of degrees of freedom. Density Functional Theory, which was elaborated in 1960s by Hohenberg and Kohn, together with the ansatz made by Kohn and Sham has provided a way to obtain the ground state properties for real systems of many electrons, starting from the general Hamiltonian reported in (1).

Theoretical Background

Density Functional Theory (DFT) takes its historical root in the Thomas-Fermi model [1] and has its solid mathematical justification in the Hohenberg-Kohn theorems [1, 2]. These theorems and their corollaries prove rigorously that the electronic charge density, $\rho(r)$, can be used as the only basic variable to obtain the ground state properties of a system of N interacting electrons moving in an external potential $v_{ext}(r)$. In particular, it can be demonstrated that the total energy of a quantum system is a functional of the electron density $\rho(r)$ only: knowing the electron density explicitly implies having a complete physical description of the system and it is equivalent to know its ground state wave function ($\Psi_0[\rho]$). The energy functional can be written as:

$$E[\rho] = F[\rho] + \int v_{ext}(r)\rho(r)dr$$
$$\text{with } F[\rho] = T[\rho] + V_{ee}[\rho] \tag{2}$$

In (2), the functional $E[\rho]$ is organized in two parts: a term depending on the external potential or, in other words, by the specific material and structure under investigation, and the functional $F[\rho]$ which is the same for all the physical systems (atoms, molecules, solids, or nanostructures) and it is called universal

functional. In quantum mechanics the electron charge density $\rho(r)$ is defined as:

$$\rho(r) = N \int \dots \int |\Psi(\{r\})|^2 dr_2 \dots dr_N \tag{3}$$

and obeys the condition $\int \rho(r)dr = N$ in order to represent the number of electrons per unit volume for a given state (in (3) and throughout the chapter atomic units are used). The second Hohenberg-Kohn theorem states that the exact ground state energy of a system is the global minimum value of the functional $E[\rho]$ and that the density that minimizes this functional corresponds to the exact ground state density ρ_0. In this context, if ρ_0 is unknown, the best estimate of the ground state properties can be obtained following a variational approach by minimizing the energy functional $E[\rho]$ calculated with respect to a trial density. The latter condition can be expressed considering that also the wave function of the system is a functional of the density, and the total energy of the system can be written as:

$$E[\rho] = \langle \Psi[\rho]|\hat{H}|\Psi[\rho]\rangle \tag{4}$$

the ground state charge density ρ_0 results from:

$$\rho_0 = \min_{\rho} E[\rho]. \tag{5}$$

The Hohenberg and Kohn formulation has in principle a very general validity but it is impractical since the exact form of the universal functional $F[\rho]$, is not known. A decisive step toward its practical applications was made possible thanks to the ansatz by Kohn-Sham (KS); according to their hypothesis, the interacting electron systems can be replaced with an auxiliary system of independent electrons characterized by the same ground state charge density, ρ_0, as the original one. For the auxiliary system the energy functional, $E[\rho]$, can be written as:

$$E[\rho] = T_s[\rho] + \frac{1}{2}\int\int\frac{\rho(r)\rho(r')}{|r-r'|}drdr' + \int v_{ext}\rho(r)dr + E_{xc}[\rho] \tag{6}$$

where $T_S[\rho]$ is the kinetic energy term of noninteracting electrons, $1/2 \int\int \rho(r)\rho(r')/|r-r'|drdr'$ is a classical term representing the Coulomb repulsion between two continuous charge distributions (Hartree term), while $E_{xc}[\rho]$ is the "exchange and correlation" term in which all the many-body effects of the original interacting system are included. For the noninteracting electrons, the wave function is expressed by $\Psi = \psi(r_1)\psi(r_2)\ldots\psi(r_N)$ and the charge density $\rho(r)$ becomes

$$\rho(r) = \sum_i f_i |\psi_i|^2, i = 1, \ldots, N \qquad (7)$$

where ψ_i are the single-particle wave functions and f_i are the occupation numbers. Once the density of the auxiliary system is obtained, one knows also the ground state density of the original interacting system. The minimization of (6) with respect to ρ yields a set of N ground state single-particle Schrödinger-like equations that reads:

$$\left(-\frac{1}{2}\nabla^2 + \int \frac{\rho(r')}{|r - r'|}dr' + v_{ext}(r) + \frac{\delta E_{xc}[\rho]}{\delta\rho(r)}\right) \\ \psi_i^{KS} = \varepsilon_i^{KS}\psi_i^{KS}, i = 1, \ldots, N \qquad (8)$$

It is worth to remark that the eigenvalues (ε_i^{KS}), and the wave functions (ψ_i^{KS}) calculated in (8) are referred to a Kohn-Sham system and are not those relative to the real system of interacting electrons, yet, following (7), ψ_i^{KS} are used to obtain the correct density. The main intricacy of (6) and (8) is related to the fact that the explicit dependence of $E_{xc}[\rho]$ on the electron density is not known, thus, even if the KS ansatz has reduced the complexity of the original many-body problem, the correct ground state is still not achievable. Following the work by KS, many approximations to $E_{xc}[\rho]$ have been proposed in literature and these are usually classified as a "Jacob's ladder of DFT" [5] which goes from lower accuracy (lower steps) to the exact unknown form (highest step). The simplest construction of $E_{xc}[\rho]$ is represented by the Local Density Approximation (LDA) in which $E_{xc}[\rho]$ is taken so that $\delta E_{xc}[\rho]/\delta\rho$ is equal to the exchange and correlation energy density of a homogeneous electron gas, $\varepsilon_{xc}^{hom}(\rho(r))$, at the density of the auxiliary system at point r of space. The quantity $\varepsilon_{xc}^{hom}(\rho(r))$ has an analytical expression, that, once substituted in (8) allows finding the eigenfunctions of the auxiliary system. Many approximations other than LDA are available for $E_{xc}[\rho]$; one of the most commonly used consists of

the Generalized Gradient Approximation (GGA) in which the exchange and correlation energy density term has an explicit dependance also on the density gradient $|\nabla\rho(r)|$. Several expressions for such dependence have been proposed in literature. Both LDA and GGA functionals suffer from being approximations of the true energy functional and, for this reason, they fail in describing physical phenomena in which electron correlation has an important role as for example Van der Waals forces or strongly correlated systems such as the d-shells of transition metals. To overcome the limits of KS-DFT formulation, several post-DFT methods have been developed; for example, the LDA + U (or GGA + U) approximation in which an on-site Hubbard-like correction (U) is added to the strongly correlated electrons [6], or the GW approach in which Green functions are used to evaluate self-energy corrections on single particle eigenvalues [7]. Finally, we mention that DFT, which is a ground state theory, has been extended to a time dependent formulation (TD-DFT) to provide a solid methodology to study the excited state properties of a many-body system in the presence of an external time dependent potential [8].

Numerical Details

Once the Exchange and Correlation term is made explicit, the KS wave functions ψ_i^{KS} (and the ground state charge density ρ_0) can be obtained by solving self-consistently the set of nonlinear differential equations (8). In the iterative solution procedure, the wave function is generally represented as a linear combination of basis set functions yielding a matrix representation of the Hamiltonian. This choice transforms the solution of partial integro-differential equations into a diagonalization problem, for which many standard algorithms have been largely tested and developed. Traditionally many DFT codes, implemented to calculate the electronic structure of materials, have been developed to deal with crystalline solids, that is, with periodic arrangements of atoms in a Bravais lattice and consequently yielding $v_{ext}(r)$ appearing in (8) also periodic. In this case, according to the Bloch's theorem [9], the KS eigenfunctions can be expressed as Bloch states denoted by a wave vector $\vec{k}$ within the first Brillouin Zone (BZ). A convenient basis set to expand these wave functions consists of the plane waves (PWs) basis set. Since the system is periodic, not all the PWs appear in the expansion but only those having a wave vector belonging to the reciprocal lattice of the crystal. In principle one would have to include an infinite number of functions to express the solution of the KS eigenvalue equation; in practice it is enough to limit the expansion to a finite subset of the basis set space once the accuracy of the subset has been proved. In the case of nonperiodic systems, like molecules, surfaces, or nanostructures, one can still effectively adopt a finite PWs basis set and use all the established numerical strategies deviced for 3-D periodic structures, if one adopts a supercell approach [2]. With the term "supercell" one indicates a 3-D periodic cell that contains the nonperiodic (or partially periodic) structure and a vacuum region. The cell sizes along the periodic directions correspond to the lattice parameters of the structure in those directions, while in the other directions the cell size has to be large enough to accommodate the finite structure and the vacuum. Periodic boundary conditions are applied to the supercell, in this way the structure is repeated in space even if it is not fully periodic; if the vacuum size is large enough the replicas along the nonperiodic directions can be considered to be isolated. Indeed, the increment of the cell size to create a reliable system increases the computational cost, thus a compromise has to be reached to obtain the desired accuracy. One of the most used methods, employed to decrease the size of the basis set (and thus reducing the computational cost) is represented by the so-called "pseudopotential" approximation [10]. According to this approximation the electrons tightly bound to the nuclei and not involved in bond formation (core electrons) are "frozen" into the core and their effect on the other electrons (valence electrons) is included in an effective way into the external potential ($v_{ext}(r)$). The pseudopotential represents an electrostatic potential that contains the effect of the bare ionic charge screened by the core electrons and it is characterized by pseudo-wavefunctions that, in the core region, are smoother than the wave functions of the bare nuclear potential and, for this reason, require less PWs to be expanded.

Once the solutions of the KS equations are obtained, it is possible to calculate any physical quantity of interest for the investigated system. For the case of periodic structures many properties such as the counting of electrons in bands, total energy, density, etc., are obtained by summing over all the states labeled by $\vec{k}$ in the BZ. For a general function $F_i(\vec{k})$ associated with a certain physical quantity, the average value $\overline{F}_i$ is defined as:

$$\overline{F}_i = \frac{1}{\Omega_{BZ}} \int_{BZ} F_i(\vec{k})d\vec{k} \tag{9}$$

where i denotes discrete states for each $\vec{k}$, Ω_{BZ} is the BZ volume, and the integral is extended over the entire BZ. In computer codes, accurate integration is performed by summation over discrete set of points in the BZ. One of the methods that has been largely used to sample the BZ is the one proposed by Monkhorst and Pack [2]. This method consists of constructing a uniform grid of points indicated by three integer numbers (n_1, n_2, n_3) that define the mesh discretization along the reciprocal space axis. If these points are chosen carefully and the grid is fine enough, the error in the estimate of $\overline{F}_i$ is small and the integral is well approximated by the summation. Finally once the ground state density of the KS system is known, according to the Hellmann-Feynman theorem [2], it is possible to calculate analytically the forces acting on each atom of the structure. Thus, if the structure is not at its equilibrium geometry, atoms can be moved and systems relaxed following the calculated forces until a selected threshold value is reached.

DFT Applications to Nanostructures

In the following the application of DFT to the investigation of the structural and electronic properties of a particular type of nanostructures, namely, one-dimensional nanowires (NWs) is presented. Understanding how the properties of a material change when it is reduced to a nanostructure is important for a wide range of scientific and technological applications as well as for basic science. The reduction to nanometers in the size of a semiconductor [11] demands theoretical approaches based on an atomistic description of a material; first principles–based approaches have demonstrated to be capable of exploring the electronic structure of bulk materials and small nanostructures with highly predictive character reaching a degree of detail that can hardly be achieved with experimental techniques. In particular, DFT simulations have played a major role in the study of electron confinement, and in determining how the band gap, the optical properties, and the electronic transport change at the nanoscale. In the last years, many different materials have been studied ranging from low (e.g., Ge, Si, InN) to wide band gap (e.g., ZnO, SiC) semiconductor, in the form of both simple and core/shell wires. Hereinafter the results for simple nanowires are discussed for a particular type of a technologically important material: Indium Nitride (InN). More in detail the effects of size reduction on the energy gap of the material are presented and the changes induced by surface termination on the electronic properties of the nanostructure are analyzed.

DFT Simulations of InN Nanowires

Group-III nitrides represent a material class with promising electronic and optical properties [12]. Among these, InN exhibits the narrowest band gap of about 0.67 eV at low temperatures, the lowest effective electron mass, and the highest peak drift velocity and electron mobility. InN nanowires have been proposed for applications in new generation integrated systems like quantum wire transistors with the aim of reducing the power consumption in large-scale integrated circuits. Due to their high surface to volume ratio, nanowires are also potential candidates for sensor applications and they have been proposed as key elements in new generation solar energy harvesting devices such as dye sensitized or other hybrid solar cells. A crucial role for applications of InN nanowires as novel nanoelectronic devices is played by a quasi-two-dimensional electron accumulation layer around the wire. To fully exploit the potential of InN nanowires, their physical properties have to be understood in depth.

Here the LDA and LDA + U results are discussed for wurtzite InN NWs [13] of increasing diameter with both clean and hydrogenated surfaces of three sizes (see Fig. 1): NW1 has one atom ring and a diameter d ≈ 4.1 Å, NW2 has two rings and a diameter d ≈ 10.6 Å, while NW3 has three rings and a diameter d ≈ 17.6 Å. By construction, these NWs have hexagonal cross-section and present nonpolar (1$\bar{1}$00) facets, in

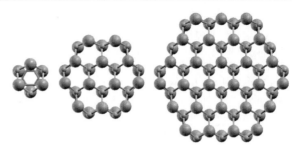

Ab Initio DFT Simulations of Nanostructures, Fig. 1 Ball and stick representation of the investigated InN nanowires: *left* – 1 ring NW (d ≈ 4.1 Å), middle – 2 rings NW (d ≈ 10.6 Å), *right* – 3 rings NW (d ≈ 17.6 Å). Large balls represent In atoms, small balls represent N atoms

agreement with experimental observation. Typically for a DFT simulation, due to the numerical approximations adopted to solve the KS equations presented previously, accuracy of the results has to be checked carefully once the following computational conditions have been chosen:

- Pseudopotentials needed to describe the external potential felt by the valence electrons.
- Size of the plane waves basis set employed to expand the one-electron wave function. This is expressed by an energy cutoff that corresponds to the highest kinetic energy (lowest wave vector) of the PWs included in the expansion.
- Convergence criteria on the total energy of the system and on the forces acting on the atoms in the case of structural relaxation based on Hellman-Feynman forces.
- Accuracy of the Monkhorst-Pack grid used to sample the Brillouin Zone of the periodic structure.
- Vacuum size in the supercell, in the case of structures that are nonperiodic in one or more dimensions (e.g., surfaces, nanowires, quantum dots, etc.).

To simulate InN 1-D nanostructures previously mentioned, the supercell approach was used: in this case the nanostructures are periodic only along the NW axis (z), thus large supercells have to be employed to accommodate enough vacuum space surrounding the wire (i.e., ∼ 10 Å) in the directions perpendicular to the NW axis. The integration over the Brillouin Zone was performed using a (1 × 1 × 6) k-point grid. All the structures were fully relaxed until the forces acting on the atoms were less than 10^{-3} Ry/bohr. The lattice parameter along the direction of the NW axis can be optimized by varying the cell size along that direction and relaxing all the atomic positions. The minimum energy structure is then obtained by fitting the total energy values with a polynomial fit. It has been found that in nanowires two main relaxation mechanisms occur: the lattice parameter along the NW axis changes with respect to the bulk value and the atoms at the NW surfaces relax.

The lattice constant along the NW axis increases decreasing the NW diameter and, except for the smallest diameter NW1, the behavior appears to be linear. While the NWs elongate along their axis, the lattice contracts in the perpendicular directions. This behavior has been observed also for other more ionic compound such as ZnO [14]. After relaxation, changes of InN bonds occur at the nanowire surface: the InN bond length becomes 2.05 Å (shorter than InN length in bulk of ~ 0.1 Å) and it tilts by about 9.5°, with the In atoms relaxing inward. Upon hydrogenation the In–N bond length recovers its bulk value (~ 2.18 Å), but the In–N bond tilting is now opposite to the clean case ($\sim 5 - 6°$), due to the moving out of the In atom.

In Fig. 2, the band gap dependence on the inverse of the NW diameter (1/d) for NWs with clean and hydrogenated surfaces is reported. It can be noted that in both LDA and LDA + U approaches the dependence of quantum confinements is almost linear and that the inclusion of the Hubbard (U) correction determines a rigid shift of the band gap (about 0.2 eV for NWs with clean surfaces and 0.35 eV for NWs with

hydrogenated surfaces) that depends only slightly on the NW diameter. The larger energy gaps observed in the case of nanowires with hydrogenated surfaces imply that quantum confinement effects are more noticeable in passivated NWs and that one might expect to see spectroscopic differences in experiments performed in different environments (e.g., solution vs ultra high vacuum). LDA and LDA + U predict similar confinement effects also in terms of charge density distribution when comparing nanowires with clean and hydrogenated surfaces. In the latter case, due to

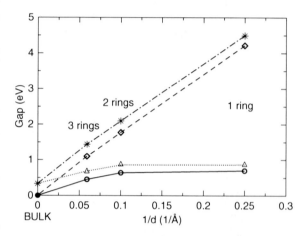

Ab Initio DFT Simulations of Nanostructures, Fig. 2 Band gap versus inverse diameter (1/d) of InN nanowires with clean surface indicated by *solid* (LDA) and *dotted* (LDA + U) lines, and for hydrogenated nanowires indicated by *dashed* (LDA) and *dashed-dotted* (LDA + U) lines

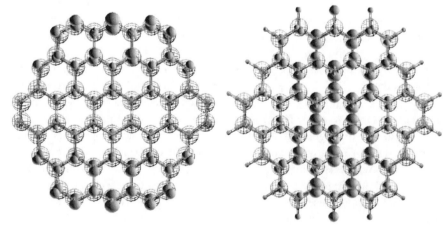

Ab Initio DFT Simulations of Nanostructures, Fig. 3 Electronic charge density isosurface (denoted by shaded *yellow-green* area) for the highest occupied state: *left* panel – NW3, *right* panel – NW3 + H. Sketched globes represent atoms

surface dangling bonds saturation, the electronic charge density accumulates in the center of the wire, while in nanowires with clean surfaces the density is more delocalized on the surface atoms. This effect is visible in Fig. 3 where the electronic charge density isosurface of the highest occupied state for NW3 is shown for clean and hydrogenated surfaces.

Conclusion

In conclusion, first principles DFT electronic structure calculations can be effectively used to systematically investigate the relative stability and electronic properties of nanostructures with clean, reconstructued, or passivated surfaces. In the case of Nanowires one can also consider the effects of the growth direction and of the diameter. Electronic properties of NWs, including band gaps, band structures, and effective masses, are usually found to depend sensitively on all the NW structural parameters. In many cases, the size dependence of the band gap depends on the growth direction of the NW, and the band gaps for a given size also depend on surface structure; moreover NWs with clean surfaces have HOMO orbitals localized on the facets and exhibit smaller band gaps. DFT calculations demonstrate the extremely rich nature of the electronic properties of nanostructures and NWs in particular and have a unique role in describing confinement effects in these small dimension systems. By changing the growth direction and NW diameter, or tuning the surface structure via chemical or physical methods, it is possible to engineer NWs with a wide range of technologically important electronic properties.

Cross-References

- ▶ Hybrid Solar Cells
- ▶ Nanomaterials for Excitonic Solar Cells
- ▶ Nanostructured Materials for Sensing
- ▶ Optical and Electronic Properties
- ▶ Surface Electronic Structure
- ▶ Theory of Optical Metamaterials

References

1. Parr, R.G., Yang, W.: Density-Functional Theory of Atoms and Molecules. Oxford University Press, Oxford, UK (1989)
2. Martin, R.M.: Electronic Structure-Basic Theory and Practical Methods. Cambridge University Press, Cambridge, UK (2008)
3. Hafner, J., Wolverton, C., Ceder, G.: Towars computational meterial design: the impact of density functional theory on material research. MRS Bull. **31**, 659 (2006)
4. Marzari, N.: Realistic modeling of nanostructures using density functional theory. MRS Bull. **31**, 681 (2006)
5. Perdew, J.P., Shmidt, K.: Density Functional Theory and Its Application to Materials, p. 1. American Institute of Physics, New York (2001)
6. Anisimovy, V.I., Aryasetiawanz, F., Lichtenstein, A.I.: First-principles calculations of the electronic structure and spectra of strongly correlated systems: the LDA + U method. J. Phys. Condens. Matter **9**, 768 (1997)
7. Onida, G., Reining, L., Rubio, A.: Electronic excitations: density-functional versus manybody Green's-function approaches. Rev. Mod. Phys. **74**, 601 (2002)
8. Marques, M.A.L., Ullrich, C.A., Nogueira, F., Rubio, A., Burke, K., Gross, E.K.U.: Time-Dependent Density Functional Theory. Springer, Berlin/Heidelberg (2006)
9. Ashcroft, N.W., Mermin, N.D.: Solid State Physics. Saunders College Publishing, Philadelphia (1976)
10. Bachelet, G.B., Hamann, D.R., Schlüter, M.: Pseudopotentials that work: from H to Pu. Phys. Rev. B **26**, 4199 (1982)
11. Law, M., Goldberger, J., Yang, P.: Semiconductor mamowires and nanotubes. Annu. Rev. Mater. Res. **34**, 83 (2004)
12. Wu, J.: When group-III nitrides go infrared: new properties and perspectives. Appl. Phys. Rev., J. Appl. Phys. **106**, 011101 (2009)
13. Terentjevs, A., Catellani, A., Prendergast, D., Cicero, G.: Importance of on-site corrections to the electronic and structural properties of InN in crystalline solid, nonpolar surface, and nanowire forms. Phys. Rev. B **82**, 165307 (2010)
14. Cicero, G., Ferretti, A., Catellani, A.: Surface induced polarity inversion in ZnO nanowires. Phys. Rev. B **80**, 201304(R) (2009)

AC Electrokinetics

- ▶ Dielectrophoresis

AC Electrokinetics of Colloidal Particles

- ▶ AC Electrokinetics of Nanoparticles

AC Electrokinetics of Nanoparticles

Hossein Nili[1,2] and Nicolas G. Green[2]

[1]School of Electronics and Computer Science, University of Southampton, Highfield, Southampton, UK

[2]Nano Research Group, University of Southampton, Highfield, Southampton, UK

Synonyms

AC electrokinetics of colloidal particles; AC electrokinetics of sub-micrometer particles

Definition

AC electrokinetics is the name given to a group of techniques that utilize alternating (AC) electric fields to move dielectric particles in suspension. AC electrokinetics of nanoparticles refers to methods of exerting electrical force and/or torque on particles of nanometer dimensions, examples of which are viruses, macromolecules, and colloidal particles.

Introduction

Dielectrics do not bear a net charge, but rather polarize when subjected to electric fields. When dielectric particles in suspension are subjected to electric fields, polarization results in charge accumulation at the particle/electrolyte interface. Electrode polarization may also occur, giving rise to the buildup of a double layer of ions and counterions at the electrode/electrolyte interface. Particle polarization is represented by effective dipole and higher-order moments. AC electrokinetic forces and torques on particles result from interactions of an applied electric field with the effective moments. Electric field-induced fluid motion is also studied under AC electrokinetics due to the important effect it can have on particle behavior. Electric field interactions with the double layer and gradients in fluid permittivity and conductivity (resulting from localized heating of the medium) are the main factors causing fluid motion under AC electric fields.

AC electrokinetic techniques are particularly fitted to lab-on-a-chip applications where multiple processes involving manipulation, separation, or characterization of biological particles, mostly dielectrics, are integrated onto a single chip. The direction and magnitude of AC electrokinetic forces and torques can be easily controlled by varying electric field frequency and geometry. The techniques are advantageous over alternative means of moving particles in their noninvasive nature and easy integration onto micro-devices, not relying on any moving parts.

It was long believed that for sub-micrometer particles, AC electrokinetic forces would be overwhelmed by thermal effects such as Brownian motion. The strong electric fields required to exert sufficient electrical force to dominate particle behavior at the nanometer scale were believed to generate excessive heat giving rise to strong fluid motion that would hinder AC electrokinetic interactions. Thanks to fabrication techniques that realized micro- and nano-electrode geometries, electric fields of sufficient strength to overcome Brownian motion of nanoparticles could be generated with the application of modest voltages, avoiding excessive heating of the suspension.

AC Electric Field Interactions with Nanoparticle Suspensions

When an electric field is applied to dielectric particles in suspension, surface charge accumulates at interfaces between the dielectrics due to the differences in electrical properties. Since the polarizabilities of each dielectric are frequency dependent, the magnitude of the surface charge is also frequency dependent and the total (complex) permittivity of the system exhibits dispersions solely due to the polarization of the interfaces. This is referred to as the Maxwell-Wagner interfacial polarization. For a spherical particle of radius R, the effective dipole moment is given by:

$$p = 4\pi\varepsilon_m R^3 [K(\omega)]E \tag{1}$$

where ω is the frequency of the applied electric field E, and $K(\omega)$, known as the Clausius-Mossotti factor, describes a relaxation in the polarizability of the particle with a relaxation time:

$$\tau_{MW} = \frac{\varepsilon_p + 2\varepsilon_m}{\sigma_p + 2\sigma_m} \tag{2}$$

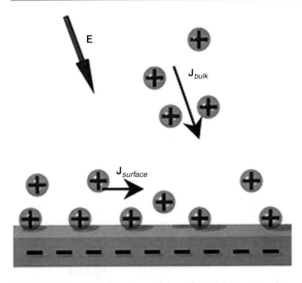

AC Electrokinetics of Nanoparticles, Fig. 1 Schematic of the mechanism with which surface conductance affects interfacial polarization of nanoparticles in suspension. The AC electric field produces both a bulk flow of ions and a surface flow around the particle

The angular frequency $\omega_{MW} = 2\pi f_{MW} = 1/\tau_{MW}$ is often referred to as the Maxwell-Wagner relaxation frequency.

During the first half of the twentieth century, dielectric spectroscopic measurements of suspensions of nanoparticles identified that the Maxwell-Wagner relaxation frequency for charged nanoparticles was higher than expected. Later, O'Konski [1] showed that the dielectric properties of nanoparticles in this frequency regime were dominated by surface conductance effects. The high value of particle conductivity determined from dielectric measurements were explained by the inclusion of a surface conductance component in the derivation of the particle's dipole moment. The model developed by O'Konski is shown in Fig. 1. He assumed that the flux due to the transport of charge carriers, associated with the fixed charge on the particle surface, could be added to the flux due to the transport of bulk charge to and from the surface. With this assumption, he derived the potential around the spherical particle and derived the equation for the dipole moment of a dielectric sphere (1) but with the particle conductivity given by the sum of the bulk conductivity of the particle and a surface conductivity term:

$\sigma_p = \sigma_{p,bulk} + \sigma_{p,surface}$. The surface conductivity of the particle is given by:

$$\sigma_{p,surface} = \frac{2K_s}{R} \tag{3}$$

where K_s is the surface conductance.

AC Electrokinetic Techniques

Dielectrophoresis (DEP): Figure 2 shows the underlying principle of dielectrophoresis, as the most widely used of AC electrokinetic techniques. The gradient in electric field strength gives rise to unequal forces experienced by polarization charges at the particle/electrolyte interface, hence a net force on the particle moving it toward or away from regions of high field intensity depending on whether the particle is more or less polarizable than the suspending medium. When particle dimensions are smaller than a characteristic length scale of electric field nonuniformity, higher-order terms can be neglected and the time-averaged dielectrophoretic force on a spherical particle of radius R suspended in a fluid of dielectric constant $\in_m$ is given by [2]:

$$\langle F_{DEP} \rangle = \pi \in_m R^3 Re[K(\omega)]\nabla|E|^2 \tag{4}$$

It is understood from (4) that the dielectrophoretic force scales with particle volume and also inversely with the cube of a characteristic dimension of the electrode geometry. It is due to the latter proportionality that micro- and nano-electrode geometries are capable of exerting electric fields of sufficient strength to move particles as small in volume as nanoparticles, overcoming Brownian and field-induced fluid motion.

Although first observations of the effect date back to earlier times, Pohl was the first to use the term dielectrophoresis for the force exerted by a nonuniform electric field on a dielectric [3]. He observed coagulation of carbon particles from a polymer solution upon application of a DC or AC electric field. Pohl used the term "dielectrophoresis" to distinguish the effect from electrophoresis, which describes motion of charged particles under DC (and not AC) electric fields.

Since its advent, dielectrophoresis has been used in a broad range of applications. Separation, as a fundamental part of many processes in micro-total-analysis

AC Electrokinetics of Nanoparticles, Fig. 2 Dielectrophoresis – Interaction of a nonuniform electric field with a dielectric particle in suspension gives rise to unequal forces experienced by polarization forces at the particle/electrolyte interface. The result is a net dielectrophoretic force that moves the particle (**a**) toward or (**b**) away from regions of high electric field intensity depending on whether the particle is more or less polarizable than its suspending medium

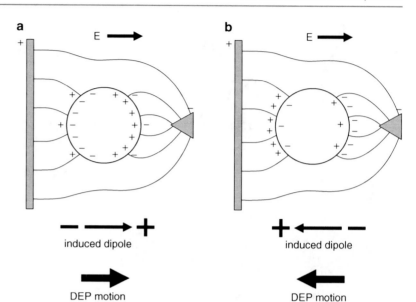

systems (μTAS), has been diversely and controllably accomplished using dielectrophoresis. The dependence of the dielectric properties of particles on their structure and composition has broadened the applicability of DEP-based separation techniques. In electrode geometries with well-defined regions of electric field maxima and minima, particles of different properties can be separated into subpopulations. Figure 3 shows the separation of 280 nm tobacco mosaic viruses (TMV) from 250 nm herpes simplex viruses (HSV) in a polynomial electrode configuration [4]. As the TMV particles are more polarizable than the suspending medium, they gather at electrode edges where the electric field is strongest, while HSV particles – less polarizable than the fluid at the applied field frequency – are concentrated at the electric field minimum in the center of the electrode geometry.

Dielectrophoresis has been used extensively for manipulation and characterization of nanoparticles. One of the most notable application areas is diagnostics and healthcare, toward which a huge amount of effort has been directed involving sub-micrometer biological particles such as viruses, DNA, and chromosomes [5]. Dielectrophoresis has also been used for separation of metallic from semiconducting carbon nanotubes [6], assembly of nanoparticles into microwires [7], and three-dimensional focusing of nanoparticles in microfluidic channels [8]. With the emergence of nano-electrode geometries, there is huge prospect for further applications involving

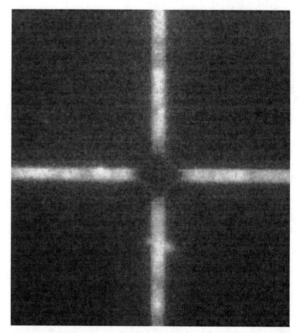

AC Electrokinetics of Nanoparticles, Fig. 3 Dielectrophoretic separation of nanoparticles – Applying a 6 MHz AC electric field leads to the collection of 280 nm tobacco mosaic viruses (TMV) at the electrode edges where the field is strongest, and the concentration of 250 nm herpes simplex viruses (HSV) at the center of the polynomial electrode geometry where there is a well-defined field minimum [4]

dielectrophoresis of nanoparticles including the fabrication of a new generation of electronic devices and sensors [5].

AC Electrokinetics of Nanoparticles, Fig. 4 Principle of electro-orientation – In a uniform field **E** (indicated by the *vector* and the *dotted field lines*) the two charges experience equal and opposite forces resulting in a torque about the center point of the dipole

Electro-orientation: As shown in Fig. 4, when a dipole sits in a uniform field, each charge on the dipole experiences an equal and opposite force tending to align the dipole with the electric field. The effect can also be observed in nonuniform and rotating electric fields and is the basis of a phenomenon known as electro-orientation.

A dielectric particle of no net charge in a uniform electric field will be subject to no net force but will experience a torque given by:

$$\Gamma = p \times E \qquad (5)$$

This torque always tends to align the dipole with the electric field. However, small dipoles, such as those of nanoparticles, will not completely align with the field due to the dominating effect of Brownian motion. In typical dielectric spectroscopy measurements, the electric fields are only of sufficient magnitude to align individual molecules by fractions of degrees, but the total effect of alignments of many molecules gives rise to a large net or average alignment [9].

In AC electric fields, electro-orientation is frequency-dependent with nonspherical particles orienting along different axes in different frequency ranges. The set of cross-over frequencies is referred to

as the orientation spectrum. The orientation spectra of elongated biological particles such as erythrocytes and bacteria have been used to study the dielectric properties of these particles [10].

Electro-rotation (ROT): Dielectric particles subjected to rotating electric fields experience a torque based on a principle which is somewhat different from that of electro-orientation. As mentioned previously, applying an electric field to a dipole will result in the exertion of a (electro-orientational) torque that tends to align the dipole with the electric field. Alignment of the dipole with the electric field vector will not be immediate as it takes a finite amount of time for polarization charges to move toward the particle/electrolyte interface and form a dipole. In a rotating electric field, where the field vector changes direction, this time delay gives rise to a (electro-rotational) torque, as shown in Fig. 5. The figure also shows an example electro-rotation setup, where voltages of 90° phase differences are applied to successive electrodes encircling the particle to generate a rotating electric field.

The first order electro-rotational torque on a spherical particle of radius R is given by [11]:

$$\Gamma_{ROT} = -4\pi \in_m R^3 Im[K(\omega)]|E|^2 \qquad (6)$$

where $Im[K(\omega)]$ denotes the imaginary part of the Clausius-Mossotti factor. The particle will rotate with or counter to the applied field depending on whether $Im[K(\omega)]$ is negative or positive, respectively. Accounting for viscous drag force from the suspending fluid, the rotation rate of the particle is given by [11]:

$$\Omega = -\frac{\in_m Im[K(\omega)]|E|^2}{2\eta} \qquad (7)$$

where η denotes fluid viscosity. Variations with frequency of the rotation rate are referred to as electro-rotation spectra.

Two important distinctions can be readily identified from a comparison of (4) and (6) for dielectrophoretic force and electro-rotational torque. Firstly, the ROT torque is proportional to the square of the electric field magnitude while the DEP force is a function of the gradient of the square of the field magnitude – and is therefore zero in uniform electric fields. Secondly, the electro-rotational torque depends on the imaginary rather than the real part of the Clausius-Mossotti

AC Electrokinetics of Nanoparticles,
Fig. 5 Electro-rotation – (a) A schematic diagram of an electro-rotation setup. Four signals, successively 90° out of phase are applied to four electrodes encircling the particle. (b) Schematic diagram showing how the induced dipole moment of a particle lags behind a rotating applied electric field

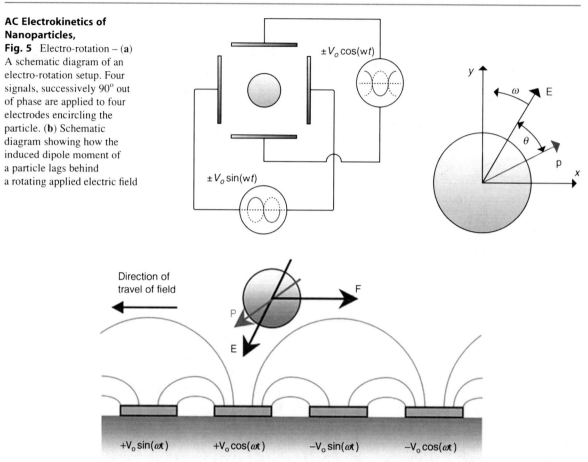

AC Electrokinetics of Nanoparticles, Fig. 6 Traveling-wave dielectrophoresis – Schematic diagram of a linear traveling wave dielectrophoresis array and the consecutive phase-shifted signals required to generate the traveling electric field. Also shown are the approximate field lines for time $t = 0$, the electric field and the dipole moment induced in the particle together with the force on the particle

factor. As a result, particles may experience both dielectrophoresis and electro-rotation with the relationship between the two determined by the dielectric properties of the particles and their suspending media.

Initially observed as a nuisance in cell electrofusion procedures, electro-rotation has since developed into a useful means of measuring dielectric properties and membrane electrical parameters of biological particles. Rotational spectra are analyzed to determine fundamental cell properties and to monitor changes in these properties upon different types of treatments [10]. Among the major works done on electro-rotation of nanoparticles are investigations by Washizu and coworkers into the torque-speed behavior of the flaggellar motor mechanism of bacterial cells [12].

Traveling-wave dielectrophoresis (twDEP): Traveling-wave DEP can be considered as the linear analog of electro-rotation where voltages of 90° phase difference are applied to successive electrodes that are laid out as tracks, rather than being arranged in a circle. This generates an electric field wave which travels along the electrodes. The dipole induced in the particle moves with the electric field but lags behind the field, as in electro-rotation. As shown in Fig. 6, the result is the induction of a force, rather than a torque, given – for a spherical particle of radius R – by [13]:

$$F_{twDEP} = -\frac{4\pi \in_m R^3 Im[K(\omega)]|E|^2}{\lambda} \tag{8}$$

where λ is the wavelength of the traveling wave. The negative sign in (8) indicates that, as shown in Fig. 6, the twDEP force propels particles in the opposite direction to the moving field vector. For a finite twDEP force to be exerted on a particle, two criteria need to be met: (a) the particle must experience a DEP force that levitates it above the electrode array, and (b) some loss mechanism needs to be present for the imaginary part of the Clausius-Mossotti factor to be nonzero.

Masuda et al. [14] were the first to describe the principles of traveling-wave DEP. Since then, twDEP has been used for characterizing and separating cells and microorganisms [15]. Nanoparticle manipulation and separation has also been accomplished using the traveling electric field above interdigitated electrodes [16].

AC Electrokinetic Fluid Motion

AC Electro-osmosis (ACEO): AC electric fields applied to a suspension of dielectric particles can give rise to fluid motion in an effect known as AC electro-osmosis [17]. The effect arises from interaction of the electric field with charges at the double layer and is only observed if the field is nonuniform.

The mechanism for AC electroosmosis is shown in Fig. 7. Applying equal and opposite voltages to successive electrodes gives rise to the electric field E with tangential component E_t outside the double layer and an induced charge on each electrode. The induced charge experiences a force F_q due to the action of the tangential field, resulting in fluid flow. Figure 7a shows the system for one half-cycle of an AC field. In the other half-cycle, the sign of the potential, that of the induced charge, and the direction of the tangential field are all opposite. As a result, the direction of the force vector remains the same giving a nonzero time-averaged force and steady-state fluid flow occurs, as shown in Fig. 7b.

Electrothermal fluid flow: At higher frequencies (>100 kHz), AC electroosmotic flow is negligible and the dominant fluid flow is due to electrothermal effects. This type of flow requires temperature gradients in the fluid and these can be generated both by internal and/or external sources. The internal source is Joule heating, where the electric field causes power dissipation in the fluid, and the corresponding temperature rise diffuses through the system. This gives rise to gradients in the conductivity and permittivity; the electric field acts on these gradients to give a body force on the fluid and consequently a flow. In this case, the velocity of the flow is proportional to the temperature rise in the fluid, which is in turn

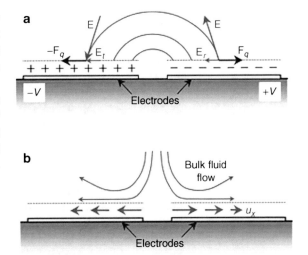

AC Electrokinetics of Nanoparticles, Fig. 7 AC electroosmosis – (**a**) Schematic diagram outlining the mechanism of AC electroosmosis. (**b**) The interaction of the tangential field at the surface with the charge in the double layer gives rise to a surface fluid velocity and a resulting bulk flow

proportional to the conductivity of the electrolyte and the magnitude of the electric field. As a result, this type of fluid flow occurs mainly at high electrolyte conductivities.

Apart from the conductivity dependence of the electrothermal effect, the magnitude of the flow is also frequency dependent. The ratio of low- to high-frequency limiting values is approximately 10:1 and the flow changes magnitude and direction at a frequency of the order of the charge relaxation frequency of the medium. The change in the velocity of the fluid as a function of the two parameters, frequency of the applied field and conductivity of the electrolyte, is complicated and is summarized in Fig. 8. In these plots, the logarithm of the magnitude of the flow velocity is indicated by a gray scale as a function of the conductivity of the electrolyte and frequency of the field, for two applied voltages. Both figures demonstrate the regions in which strong fluid flow occurs and also which regions might be "safe" from fluid motion.

AC Electrokinetic Forces on Nanoparticles

As described through the different mechanisms responsible, particle behavior under AC electric fields is determined by a variety of factors. The total force on any particle is given by the sum of many forces

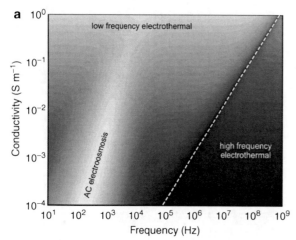

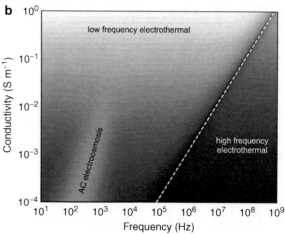

AC Electrokinetics of Nanoparticles, Fig. 8 Schematic frequency/conductivity maps of the two types of electric field-driven fluid flow: AC electroosmosis and electrothermal. The *white dotted line* indicates the frequency at which electrothermal flow changes direction. The velocities are plotted on a log *gray scale* and were determined for the same position (10 μm from the edge of the electrode) and applied signals: (**a**) 1 V and (**b**) 10 V on each electrode. Note that the ratio of AC electroosmotic flow is much less at 10 V

including Brownian, dielectrophoretic, and hydrodynamic forces, the latter arising from fluid motion. These forces can be comparable to, or in certain circumstances much stronger than, the dielectrophoretic force exerted on the particle. At certain combinations of medium conductivity, electric field frequency, and strength, the hydrodynamic forces can dominate particle behavior. Under other experimental conditions these forces are negligible and as a result small changes in the dielectric properties of the particles can be detected and measured by DEP.

Ramos and coworkers have conducted order-of-magnitude calculations for the forces on nanoparticles in AC electric fields [18]. It has been shown that for low-conductivity suspensions, Joule heating has little effect on system temperature and is therefore negligible. With ± 10 V applied to a micro-electrode system with a medium conductivity of 10mS/m, temperature rises of the order of only 1°C have been observed. In contrast, electrothermal forces have been shown to be of sufficient strength to compete with the dielectrophoretic force. However, it is important to note that the DEP force varies much more rapidly than electrothermal forces upon proximity with electrode edges. Applying ± 5 V to a system of parallel finger micro-electrodes and a 10 ms/m-conductivity suspension of 282 nm particles has been shown to generate a fluid velocity of 5 μm/s against a DEP-induced velocity of 1.8 μm/s. Near electrode edges, DEP-induced and electrothermal velocities have

been reported to be 200 and 50 μm/s, respectively. Changes in fluid density arising from field-induced temperature gradients have been shown to be on the order of 0.01% per degree. As a result, natural convection has been found to be very small compared to electrothermal and DEP forces.

Brownian motion was for long considered the biggest obstacle in AC electrokinetic motion of sub-micrometer particles. Pohl had estimated that electric field strengths required to overcome Brownian motion of nanoparticles would (a) not be realizable, and (b) lead to excessive heating, and hence electrothermal motion, of the fluid, thereby hindering DEP motion of particles [1]. Experimental findings have shown that Pohl had largely overestimated the electric field strength required to overcome thermal effects, the reason being his comparison of thermal energy with dielectrophoretic *potential* rather than *force*, the latter being the gradient of the former [19]. It has been shown that DEP forces required to overcome Brownian motion of nanoparticles are in the sub-picoNewton range and can be easily generated using micro- and nano-electrode geometries by application of voltages modest enough to avoid excessive heating of the medium. The observable deterministic force required to move a 282 nm particle has been found to be of the order of 0.01pN over a 1 s time frame of observation, requiring an electric field on the order of 100 kV/m, which can be generated by applying voltages on the order of 1 V across a 10 μm gap.

In summary, although hydrodynamic effects can complicate AC electrokinetic motion of nanoparticles, it has been shown that the forces can be predicted and therefore controlled. Consequently, it is expected that recent technological advances in AC electrokinetics could be applied, together with field-induced fluid motion, to develop new methods for the characterization, manipulation, and separation of nanoparticles.

References

1. O'Konski, C.T.: Electric properties of macromolecules V: theory of ionic polarization in polyelectrolytes. J. Phys. Chem. **64**, 605–619 (1960)
2. Pohl, H.A.: Dielectrophoresis: the behavior of neutral matter in nonuniform electric fields. Cambridge University Press, Cambridge (1978)
3. Pohl, H.A.: The motion and precipitation of suspensoids in divergent electric fields. J. Appl. Phys. **22**, 869–871 (1951)
4. Morgan, H., Hughes, M.P., Green, N.G.: Separation of sub-micron bioparticles by dielectrophoresis. Biophys. J. **77**, 516–525 (1999)
5. Pethig, R.: Review article – dielectrophoresis: status of the theory, technology, and applications. Biomicrofluidics **4**, 022811 (2010)
6. Krupke, R., Hennrich, F., von Lohneysen, H., Kappes, M.M.: Separation of metallic from semiconducting single-walled carbon nanotubes. Science **301**, 344–347 (2003)
7. Hermanson, K.D., Lumsdon, S.O., Williams, J.P., Kaler, E.W., Velev, O.D.: Dielectrophoretic assembly of electrically functional microwires from nanoparticle suspensions. Science **238**, 1082–1086 (2001)
8. Morgan, H., Holmes, D., Green, N.G.: 3D focusing of nanoparticles in microfluidic channels. IEE Proc. Nanobiotechnol. **150**, 76–81 (2003)
9. Morgan, H., Green, N.G.: AC electrokinetics of colloids and nanoparticles. Research Studies Press, Hertfordshire (2003)
10. Jones, T.B.: Electromechanics of Particles. Cambridge University Press, Cambridge (1995)
11. Arnold, W.M., Zimmermann, U.: Electro-rotation – development of a technique for dielectric measurements on individual cells and particles. J. Electrostat. **21**, 151–191 (1988)
12. Washizu, M., Shikida, M., Aizawa, S., Hotani, H.: Orientation and transformation of flagella in electrostatic field. IEEE Trans. IAS **28**, 1194–1202 (1992)
13. Hughes, M.P.: AC electrokinetics: applications for nanotechnology. Nanotechnology **11**, 124–132 (2000)
14. Masuda, S., Washizu, M., Iwadare, M.: Separation of small particles suspended in liquid by nonuniform traveling field. IEEE Trans. Ind. Appl. **23**, 474–480 (1987)
15. Hagedorn, R., Fuhr, G., Muller, T., Schnelle, T., Schnakenberg, U., Wagner, B.: Design of asynchronous dielectric micromotors. J. Electrostat. **33**, 159–85 (1994)
16. Li, W.H., Du, H., Chen, D.F., Shu, C.: Analysis of dielectrophoretic electrode arrays for nanoparticle manipulation. Comp. Mater. Sci. **30**, 320–325 (2004)
17. Ramos, A., Morgan, H., Green, N.G., Castellanos, A.: AC electric-field-induced fluid flow in microelectrodes. J. Colloid Interface Sci. **217**, 420–422 (1999)
18. Ramos, A., Morgan, H., Green, N.G., Castellanos, A.: AC electrokinetics: a review of forces in microelectrode structures. J. Phys. D: Appl. Phys. **31**, 2338–2353 (1998)
19. Green, N.G.: Dielectrophoresis of sub-micrometre particles. Thesis, University of Glasgow, Glasgow, UK (1998)

AC Electrokinetics of Sub-micrometer Particles

▶ AC Electrokinetics of Nanoparticles

AC Electroosmosis: Basics and Lab-on-a-Chip Applications

Pablo García-Sánchez and Antonio Ramos
Departamento de Electrónica y Electromagnetismo, Universidad de Sevilla, Sevilla, Spain

Synonyms

Induced-charge electroosmosis

Definition

Microelectrode structures subjected to AC voltages can generate flow of aqueous solutions by the influence of the AC field on the charges induced by itself at the electrode–electrolyte interface, i.e., induced charges in the electrical double layer (EDL). The phenomenon is called *AC Electroosmosis* (ACEO) in analogy with the "classical" electroosmosis, where fluid flow is generated by the action of an applied electric field on the mobile charge of a given solid–liquid interface [1]. The main difference is that the applied field in ACEO is responsible for both inducing the charge and pulling on it. The term Induced-Charge Electrokinetics (ICEK) has been proposed to refer to all phenomena where the electric field acts on the electrical double layer induced by itself [2].

Basic Mechanism

Possibly, the simplest system for the study of AC Electroosmosis consists of a couple of coplanar electrodes covered by an aqueous electrolyte. The electrodes are subjected to a harmonic potential difference of amplitude V_0 and frequency ω, $V(t) = V_0 \cos(\omega t)$. Figure 1 shows a sketch of the physical system at a given time of the AC signal cycle. The applied field attracts counterions at the solid–liquid interfaces with a certain delay and, since the electric field is nonuniform, there appears an electrical force pulling the liquid outward $F = qE_t$ (E_t is the component of the electric field tangential to the electrode surfaces). Note that in the other half cycle, both the tangential field and the induced charge change sign and, therefore, the force direction remains the same and the time-averaged force is nonzero.

The fluid velocity generated by this mechanism is frequency dependent. At low frequencies, the electrical charges have sufficient time to completely screen the electric field at the electrodes, the electric field in the electrolyte bulk vanishes, and so does the induced velocity. In this view, the electrodes are supposed to be perfectly polarisable, i.e., no Faradaic currents occur. At high frequencies, the electrical charges have no time to follow the electric field, the induced charge at the electrode tends to zero, and, therefore, the electrical force is negligible. The theoretical expectation for the velocity versus frequency plot is a bell-shaped curve, in accordance with the experimental observations (Fig. 2).

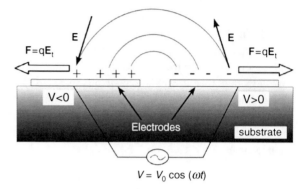

AC Electroosmosis: Basics and Lab-on-a-Chip Applications, Fig. 1 Basics of the ACEO mechanism. Electrical charge is induced at the electrode–electrolyte interface. The tangential component of the electric field acts on this charge giving rise to a surface fluid velocity

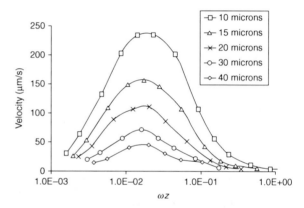

AC Electroosmosis: Basics and Lab-on-a-Chip Applications, Fig. 2 Fluid velocity at several positions on the electrode surface. The velocity is plotted against the product of the angular frequency and the distance of measurement from the center of the two electrodes (Reprinted with permission from [1], © 1999 Elsevier)

Simple Model for Small Voltages

The electroosmotic velocity generated at metal surfaces can be computed from the Helmholtz–Smoluchowski formula, $v_{HS} = \varepsilon \zeta E_t / \eta$, where ε is the dielectric constant of water, ζ is the zeta potential of the solid–liquid interface, and η is the liquid viscosity. For small voltage amplitudes, the relation between ζ and the charge in the EDL is linear and the formula reads $v_{HS} = \sigma_q \lambda_D E_t / \eta$, with λ_d the Debye length and σ_q the charge per unit area in the EDL.

In ACEO, σ_q is the induced charge in the EDL by the AC potential. A constant (non-oscillating) term in the electrode–electrolyte surface charge is not considered here although it might exist, the so-called intrinsic

surface charge. However, in this linear model, the time-average velocity due to that charge is zero. In the system of symmetric electrodes, σ_q can be computed from the RC-circuit model shown in Fig. 3. In this model, the electrical current in the electrolyte is discretized in current tubes of semicircular shape and width Δz. The resistance of one of these tubes is $R = \pi z / \sigma d \Delta z$, where z is the distance to the interelectrode gap center, σ the electrolyte conductivity, and d the depth of the electrodes. The electrical current in the tubes charges the electrical double layer, represented in this model by a distributed capacitor with a capacitance per surface area that, at low

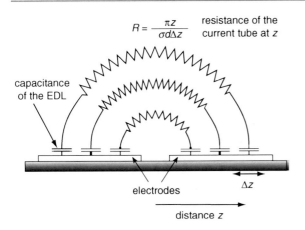

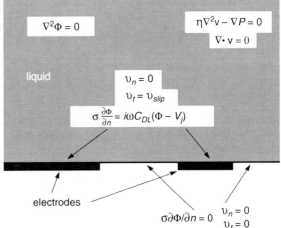

AC Electroosmosis: Basics and Lab-on-a-Chip Applications, Fig. 3 RC-circuit model for computing the charge density in the electrical double layer as a function of the position on the electrodes. The resistance of the liquid is modeled by the resistors. The capacitance of the electrical double layer is represented by the capacitors

AC Electroosmosis: Basics and Lab-on-a-Chip Applications, Fig. 4 Equations and boundary conditions for computing the fluid velocity generated by ACEO. Indexes n and t indicate normal and tangential component, respectively. The electric potential is obtained from Laplace equation with specific boundary conditions at the electrodes. From the solution of the potential at the level of the electrodes, the slip velocity is computed with (3) and introduced as boundary condition in solving Stokes equation for the fluid velocity

amplitudes, can be estimated from the Debye–Hückel theory as $\varepsilon d\Delta z/\lambda_D$. The value of σ_q at any position can be computed from $\sigma_q = \varepsilon V_d/\lambda_D$, where V_d is the voltage drop across the EDL and it is a function of the position z:

$$V_d(z) = \frac{V_0}{1 + j\omega\pi\varepsilon z/\sigma\lambda_D} \tag{1}$$

and the tangential electric field is obtained from $E_t = -\partial V_d/\partial z$. The time-average of the Helmholtz–Smoluchowski formula reads:

$$\langle v_{HS} \rangle = \frac{1}{2}\,\text{Re}\left[\frac{\sigma_q E_t^* \lambda_D}{\eta}\right] = \frac{\varepsilon V_0^2 \Omega^2}{8\eta z(1 + \Omega^2)^2} \tag{2}$$

where * indicates the complex conjugate and the nondimensional frequency Ω is written as $\Omega = \omega\pi\varepsilon z/2\sigma\lambda_D$. Maximum velocity is predicted for $\Omega = 1$, i.e., $\omega_{max} = 2\sigma\lambda_D/\pi\varepsilon z$. The frequency for maximum velocity is dependent on position on the electrode z, and the electroosmotic velocity vanishes for either $\Omega \ll 1$ or $\Omega \gg 1$, as expected. The model is in qualitative agreement with experimental measurements [3].

A rigorous study of ACEO flows for small potentials can be found in ref. [4], where the electrokinetic equations are solved in the limit of thin electrical double layer, i.e., the Debye length is much smaller

than the size of the electrodes. In this limit, the fluid velocity in the liquid can be obtained from Stokes equation with boundary condition of slip velocity on the electrodes, the electroosmotic velocity. For electrode j, subjected to a harmonic potential $V(t) = \text{Re}[V_j \exp(i\omega t)]$, the slip velocity is computed according to:

$$v_{\text{slip}} = -\frac{\varepsilon}{4\eta}\Lambda\frac{\partial|\Phi - V_j|^2}{\partial x} \tag{3}$$

where Φ is the electric potential phasor evaluated at the level of the electrodes (the electric potential in the liquid is written as $\phi(t) = \text{Re}[\Phi\exp(i\omega t)]$). The parameter Λ appears to account for the effect of a compact layer on the electrodes, acting as a capacitor in series with the Debye layer and, as a consequence, diminishing the voltage drop across the latter [5], $\Lambda = C_s/(C_s + C_d) \leq 1$, where C_s and C_d are the capacitances of the compact and Debye layers, respectively.

Thus, it is first required to solve the electric potential in the bulk of the liquid, which is solution of Laplace equation with specific boundary conditions, as shown in Fig. 4. Charge accumulation at the

AC Electroosmosis: Basics and Lab-on-a-Chip Applications, Fig. 5 Side view of the streamlines in ACEO with two coplanar electrodes. (*Left*) Experiments with ac signals at 100 Hz. Fluorescent particles are used as flow tracers. (*Right*) Numerical Simulation (Reprinted with permission from [5], © 2002 APS)

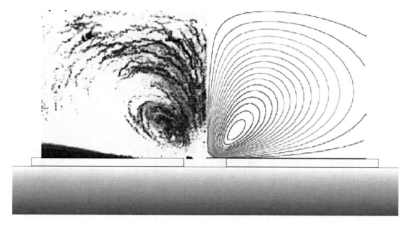

electrode–electrolyte interface yields the following boundary condition:

$$\sigma \frac{\partial \Phi}{\partial x} = i\omega C_{DL}(\Phi - V_j) \tag{4}$$

where σ is the liquid conductivity and the total capacitance of the double layer is given by $C_{DL} = C_s C_d / (C_s + C_d)$. Zero normal current, $\partial \Phi / \partial n = 0$, is imposed at the other boundaries of the domain. Figure 5 shows a comparison of experimental streamlines obtained using fluorescent particles as tracers and streamlines computed with this model [5].

Limiting Effects

The simple model presented in the previous section is rigorous for voltages of the order of $k_B T / e \sim 25 \, \text{mV}$ or smaller. For these voltages, the Debye–Hückel approximation is valid, and the capacitance of the EDL per unit area is given by ε / λ_D. This does not hold true at higher voltages, and analytical models become more complicated, especially when the finite size effect of ions must be considered [6].

Another simplification is that the electrodes are considered to be perfectly polarizable, meaning that there is not charge transfer from the electrode to the electrolyte or vice versa. However, Faradaic currents appear for increasing voltage, and it has been found to have an important influence in the flow [7].

Finally, there is one more limitation not related to the amplitude of the applied voltage. The simple model does not consider the possible existence of an oxide layer on top of the electrode, the so-called

compact layer. Very good fit to experimental results were obtained when this layer was taken into account [4, 8].

In general, ACEO flow of electrolytes with high ionic strength (greater or equal to 100 mM) has not been reported. It seems that there are fundamental limitations around this ionic strength related to steric effects at high electrolyte concentrations as well as problems with the generation of Faradaic reactions.

Applications for the Lab-on-a-Chip

Current clean-room technologies allow for the fabrication of complex microelectrode structures and their integration in microfluidic devices. Therefore, AC electrokinetic phenomena represent an opportunity for performing standard operations in Lab-on-a-Chip systems. In particular, AC Electroosmosis has been demonstrated to be useful for the following applications:

ACEO Pumping and Mixing

The example of two coplanar electrodes of the same size is perhaps the easiest system to illustrate the mechanism of ACEO. As a consequence of the symmetry of the system, no net pumping of fluid is produced. The flow rolls on top of the two electrodes are identical and cancel each other. However, as predicted in [9], any electrode array with some asymmetry will lead to net pumping of the electrolyte.

Two different electrode arrays have been broadly explored for achieving net pumping of fluid, see Fig. 6. When pairs of asymmetric electrodes are used, the flow roll on top of the larger electrode dominates and net

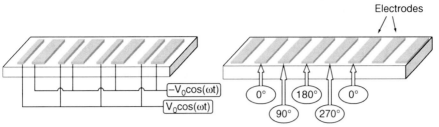

AC Electroosmosis: Basics and Lab-on-a-Chip Applications, Fig. 6 Microelectrode structures for pumping of electrolytes. (*Left*) Array of asymmetric couples of electrodes subjected to a harmonic potential. (*Right*) Array of equal size electrodes subjected to a four-phase traveling-wave potential

flow is created (from left to right in the figure) [10–12]. Reference [13] demonstrates a microfluidic pump consisting of a long serpentine microchannel lined with 3D stepped electrode arrays, allowing to achieve high pressures with voltages around 1 V_{rms}. Because the ACEO pump employs low voltage, it has potential applications in portable and/or implantable biomedical microfluidic devices. A different strategy to break the symmetry is the application of a traveling-wave electric potential [14, 15]. In experiments, this is implemented by using an array of equal-size electrodes and applying a sinusoidal potential to each electrode but with a phase lag of 90° between neighbors. Net flow is achieved in the same direction of the traveling wave (left to right in the figure). A common feature of both arrays is that the flow direction is reversed for sufficiently high voltages [16]. This observation cannot be explained by the standard ACEO model, and limiting effects like ion crowding and Faradaic currents should be accounted for.

The predominant laminar character of the flow in microfluidics makes mixing a slow process. References [17, 18] show how AC Electroosmosis can be exploited for generating micro-vortices locally and enhancing the mixing.

Transport and Aggregation of Particles and Molecules

AC Electroosmotic flows have also been used in combination with other forces (dielectrophoresis in most cases) for concentration of bio-particles. The flow is used to bring and concentrate the particles on top of the electrodes while the other forces are used for holding them [19, 20]. For example, reference [21] demonstrates ACEO for concentrating DNA molecules;

the generated bulk flow was able to transport DNA molecules from a large effective region to the electrode surface.

The detection times for heterogeneous immunoassays can be very long if analytes have to travel only by diffusion to the device surface, where they then bind or react with surface-bound receptors. Reference [22] demonstrates that the fluid flow generated by ACEO can enhance the analyte transport and reduce the detection time.

Promising results have recently been obtained by using transparent semiconductors that become conductors when light is projected on them. In this way, light patterns created at will on the substrate will function as electrodes and, for example, allow for dynamic and in situ concentrations of particles [23, 24]. Reference [23] showed the light-patterning of 31,000 microfluidic ACEO vortices on a featureless photoconductive surface which were able to concentrate and transport micro- and nanoscale particles.

Cross-References

▶ Induced-Charge Electroosmosis

References

1. Ramos, A., Morgan, H., Green, N.G., Castellanos, A: AC electric-field-induced fluid flow in microelectrodes. J. Colloid. Interface. Sci. **217**, 420–422 (1999)
2. Bazant, M.Z., Squires, T.M: Induced-charge electrokinetic phenomena: theory and microfluidic applications. Phys. Rev. Lett. **92**, 066101 (2004)
3. Green, N.G., Ramos, A., González, A., Morgan, H., Castellanos, A: Fluid flow induced by nonuniform ac

electric fields in electrolytes on microelectrodes. I. Experimental measurements. Phys. Rev. E. **61**, 4011–4018 (2000)

4. González, A., Ramos, A., Green, N.G., Castellanos, A., Morgan, H: Fluid flow induced by nonuniform ac electric fields in electrolytes on microelectrodes. II. A linear double-layer analysis. Phys. Rev. E. **61**, 4019–4028 (2000)

5. Green, N.G., Ramos, A., González, A., Morgan, H., Castellanos, A: Fluid flow induced by nonuniform ac electric fields in electrolytes on microelectrodes. III. Observations of streamlines and numerical simulation. Phys. Rev. E. **66**, 026305 (2002)

6. Bazant, M.Z., Kilic, M.S., Storey, B.D., Ajdari, A: Towards an understanding of induced-charge electrokinetics at large applied voltages in concentrated solutions. Adv. Colloid. Interface. Sci. **152**, 48–88 (2009)

7. González, A., Ramos, A., García-Sánchez, P., Castellanos, A: Effect of the combined action of Faradaic currents and mobility differences in ac electro-osmosis. Phys. Rev. E. **81**, 016320 (2010)

8. Pascall, A.J., Squires, T.M: Induced charge electro-osmosis over controllably contaminated electrodes. Phys. Rev. Lett. **104**, 088301 (2010)

9. Ajdari, A: Pumping liquids using asymmetric electrode arrays. Phys. Rev. E. **61**, R45–R48 (2000)

10. Brown, A.B.D., Smith, C.G., Rennie, A.R: Pumping of water with ac electric fields applied to asymmetric pairs of microelectrodes. Phys. Rev. E. **63**, 016305 (2000)

11. Ramos, A., González, A., Castellanos, A., Green, N.G., Morgan, H: Pumping of liquids with ac voltages applied to asymmetric pairs of microelectrodes. Phys. Rev. E. **67**, 056302 (2003)

12. Studer, V., Pépin, A., Chen, Y., Ajdari, A: An integrated AC electrokinetic pump in a microfluidic loop for fast and tunable flow control. Analyst **129**, 944–949 (2004)

13. Huang, C., Bazant, M.Z., Thorsen, T: Ultrafast high-pressure AC electro-osmotic pumps for portable biomedical microfluidics. Lab. Chip. **10**, 80–85 (2010)

14. Cahill, B.P., Heyderman, L.J., Gobrecht, J., Stemmer, A: Electro-osmotic streaming on application of traveling-wave electric fields. Phys. Rev. E. **70**, 036305 (2004)

15. Ramos, A., Morgan, H., Green, N.G., González, A., Castellanos, A: Pumping of liquids with traveling-wave electroosmosis. J. Appl. Phys. **97**, 084906 (2005)

16. García-Sánchez, P., Ramos, A., Green, N.G., Morgan, H: Experiments on AC electrokinetic pumping of liquids using arrays of microelectrodes. IEEE Trans. Dielectr. Electr. Insul. **13**, 670–677 (2006)

17. Wang, S.H., Chen, H.P., Lee, C.Y., Yu, C.C., Chang, H.C: AC electro-osmotic mixing induced by non-contact external electrodes. Biosens. Bioelectron. **22**, 563–567 (2006)

18. Harnett, C.K., Templeton, J., Dunphy-Guzman, K.A., Sensousy, Y.M., Kanouff, M.P: Model based design of a microfluidic mixer driven by induced charge electroomosis. Lab. Chip. **8**, 565–572 (2008)

19. Wu, J., Ben, Y., Battigelli, D., Chang, H.C: Long-range AC electrosmotic trapping and detection of bioparticles. Ind. Eng. Chem. Res. **44**(8), 2815–2822 (2005)

20. Wong, P.K., Chen, C.Y., Wang, T.H., Ho, C.M: Electrokinetic bioprocessor for concentrating cells ad Molecules. Anal. Chem. **76**, 6908–6914 (2004)

21. Lei, K.F., Cheng, H., Choy, K.Y., Chow, L: Electrokinetic DNA concentration in microsystems. Sensor. Actuat. A-Phys. **156**, 381–387 (2009)

22. Hart, R., Lec, R., Noh, H: Enhancement of heterogeneous immunoassays using AC electroomosis. Sensor. Actuat. B-Chem. **147**, 366–375 (2010)

23. Chiou, P.Y., Ohta, A.T., Jamshidi, A., Hsu, H.Y., Wu, M.C: Light-actuated AC electroosmosis for nanoparticle manipulation. J. Microelectromech. S. **17**, 525–531 (2008)

24. Hwang, H., Park., J.K: Rapid and selective concentration of microparticles in an optoelectrofluidic platform. Lab. Chip. **9**, 199–206 (2009)

ac-Calorimetry

▶ Nanocalorimetry

Accumulator

▶ Nanomaterials for Electrical Energy Storage Devices

Acoustic Contrast Factor

Andreas Lenshof and Thomas Laurell
Department of Measurement Technology and Industrial Electrical Engineering, Division of Nanobiotechnology, Lund University, Lund, Sweden

Definition

The acoustic contrast factor (Φ) governs whether an object will be affected by the acoustic radiation force in an acoustic standing wave field [1]. It is comprised of the density and speed of the sound of the object (ρ_p and c_p) and the surrounding medium (ρ_0 and c_0), as seen in the equation below. A positive contrast factor means that the object will move to a pressure node, while a negative factor will move the object toward a pressure antinode:

$$\Phi = \frac{\rho_p + \frac{2}{3}(\rho_p - \rho_0)}{2\rho_p + \rho_0} - \frac{1}{3}\frac{\rho_0 c_0^2}{\rho_p c_p^2}$$

Cross-References

▶ Acoustic Trapping
▶ Acoustophoresis
▶ Integrated Micro-acoustic Devices

References

1. Gorkov L.P.: On the forces acting on a small particle in an acoustical field in an ideal fluid. Soviet Physics Doklady (**6**), 9, 773–775 (1962)

Acoustic Nanoparticle Synthesis for Applications in Nanomedicine

Aisha Qi, Peggy Chan, Leslie Yeo and James Friend
Micro/Nanophysics Research Laboratory,
RMIT University, Melbourne, VIC, Australia

Synonyms

Drug delivery and encapsulation; Nanocarriers; Nanomedicine; Polymer nanocapsules; Sound propagation in fluids; Ultrasonic atomization

Definition

Sound wave propagation arising from the high-frequency acoustic irradiation of a fluid can generate considerable stresses at the free surface of the fluid leading toward its destabilization and subsequent breakup. If the fluid comprises a polymer solution, the evaporation of the solvent from the aerosols that are generated as a consequence of the atomization process leaves behind a solidified polymer core with submicron dimensions. Here, any discussion of nanoparticle synthesis due to chemical reactions driven by sound waves, i.e., sonochemistry, is omitted and the discourse is limited to the physical synthesis of such nanoparticles due to the atomization of polymer solutions into a gas, typically dry air.

Overview

The synthesis of functional nanoparticles is an area of tremendous interest, particularly from the standpoint of many industry applications from catalysis, optics, and electronics to biomolecular sensing, regenerative medicine, and pharmaceutical science. In the latter, there are several challenges that current drug delivery technologies and systems face. For example, the low solubility of many drugs in biological environments prevents their absorption and distribution in vivo [1, 2]. While drug solubility limitations can be circumvented through chemical structure modification or the introduction of surfactant, the former often requires expensive and complicated procedures whereas the latter may not only increase the dosage volume but is also associated with toxicity problems [1]. In addition, the delivery of most drugs cannot be localized to target a site of interest, leading to undesirable side effects or immune responses when taken up by uninfected tissues and organs. Further, drugs are often susceptible to decomposition, enzymatic degradation, aggregation, or denaturation, thus either reducing their shelf life or their efficacy in vivo.

Nanoparticle-based delivery systems, however, offer the possibility of addressing some of these issues. For example, the drug dissolution rate can be increased by reducing the particle size [1, 2]. Tumor tissues are characterized by leaky vasculature and poor lymphatic clearance; due to their subcellular dimensions, nanoparticles are able to preferentially accumulate in tumor tissue through an enhanced permeation and retention (EPR) effect [3]. As an attractive alternative to conventional invasive surgical procedures, nanoparticles have also been reported to be able to cross the blood–brain barrier to be administrated into the central nervous system [1].

The encapsulation of drugs within the polymer nanoparticle not only acts as a protective layer surrounding the drug from hostile in vivo environments but also enables the drug to diffuse out slowly over extended periods. Such controlled release as well as direct local targeting of diseased regions can also be further tuned by judicious choice of the polymeric excipient chemistry, comonomer ratios, and their degradability in certain regions. Further site-specific targeting can be achieved by functionalizing the nanoparticle surface with tissue-specific or cell-specific ligands to increase uptake into receptor expressing cells [3]. Alternatively, it is straightforward to chemically modify the nanoparticle surface to actively target a diseased site by attaching a surface-bound ligand, e.g., monoclonal antibodies or polymeric conjugates that specifically bind to target cells [1].

There are numerous routes for nanoparticle synthesis, all of which can be broadly delineated into three general approaches. Mechanical milling techniques are widely used commercially but are typically confined to industrial applications such as ceramics and paints due to contamination issues. While the level of impurities can be reduced, for example, by carrying out the process in vacuum, this is expensive and the powders are often polydispersed. Wet chemical deposition and precipitation techniques involve phase separation and include crystallization and sol-gel processing. Polymeric nanoparticles can be synthesized using emulsion (solvent extraction/evaporation) and self- or directed-assembly methods. These are usually batch operations and mass production can lead to scale up in complexity and cost. Moreover, while these methods allow for size and size distribution reproducibility, controlling these parameters is difficult. Gas phase evaporation and condensation techniques, on the other hand, encompass a range of methods that include combustion flame pyrolysis, plasma chemical vapor deposition, and laser ablation as well as spray drying. As with all aerosol processing strategies, spray drying is typically straightforward, fast, and allows high throughput, producing dried nanoparticles by atomizing a solution into micron or submicron-sized aerosols which then rapidly pass through a drying configuration. By removing the solvent either through evaporation or extraction using a hardening agent [4–6], the material of interest, e.g., proteins, peptides, and a wide range of polymers, can solidify into dried particles usually with desired dimensions and morphologies governed by the parent aerosol [7].

Other aerosol processing techniques include hydrodynamic flow focusing and electrospraying. Hydrodynamic flow focusing generates droplets by forcing liquid through an orifice using direct pressure or a fast moving air stream [8], although this typically results in an uneven and large particle size distribution. As such, it is typically used by the food industry, for example, in the manufacture of milk powder, in which stringent particle size requirements are not essential. Electrospraying, on the other hand, employs large voltages to stretch and pinch off aerosol droplets from a liquid meniscus at the tip of a metal capillary [9]. While this produces aerosol droplets and polymeric nanoparticles with a uniform size distribution, the throughput is generally low and the large electric fields required not only poses safety hazards but could also result in molecular lysis, although these could potentially be circumvented using high-frequency AC fields [10].

New technologies for nanoparticle synthesis and drug encapsulation, however, have not kept up with the rapid advances achieved in nanomedicine, particularly, from the standpoint of drug discovery and formulation. The rest of this entry is concerned with atomization processes via acoustic means for polymeric nanoparticle synthesis and recent advances in this field that allow it to be exploited as a powerful tool for drug delivery.

Ultrasonic Atomization

Ultrasonic atomization generally refers to a method for aerosol production induced by irradiating a fluid in order to destabilize its interface with acoustic energy at driving frequencies between 20 kHz to a few MHz [11], although novel technologies such as surface acoustic waves (SAWs) allow operation at higher frequencies above 10 MHz [12]. The source of the acoustic energy is usually a piezoelectric transducer driven by an alternating electrical signal. Ultrasonic atomization therefore typically encompasses both indirect and direct vibration-induced atomization [11, 13], ultrasonic microjetting [14], and, more recently, SAW atomization [15]. Figure 1 illustrates these various conceptual strategies that are generally classified as ultrasonic atomization methods, the difference between these arising through the mechanism by which the acoustic energy is introduced or the way in which the aerosol droplets are produced. For example, *bulk* vibration energy is transferred indirectly from the piezoelectric transducer to the fluid drop to be atomized through a metallic horn (Fig. 1a) or directly to the sessile fluid drop placed on the piezoelectric

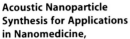

Acoustic Nanoparticle Synthesis for Applications in Nanomedicine,
Fig. 1 Different conceptual configurations for ultrasonic atomization. (**a**) Indirect bulk vibration driven by a piezoelectric transducer via a metallic horn on which the working fluid drop is deposited. (**b**) Direct bulk vibration transmitted to the fluid drop placed on the piezoelectric transducer. (**c**) Use of micron-sized nozzles and orifices to drive ultrasonic microjets. (**d**) Transmission of surface vibration energy in the form of SAWs into a fluid drop to drive interfacial destabilization

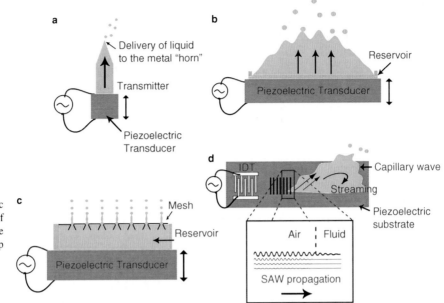

substrate (Fig. 1b). Nozzles and orifice plates from which fluid jets emanate that subsequently break up into aerosol droplets can also be used in an extension of the indirect method to provide finer control over the final droplet size (Fig. 1c). Alternatively, a nozzle can assume the place of the horn in the direct method in Fig. 1a, with the fluid issuing through the needle instead of placing the fluid on the horn. Nozzles or orifices however tend to clog over time and require constant cleaning; the pressure drop is also significantly increased, and hence more power is typically required to drive the atomization process. Alternatively, the *surface* vibration in the form of SAWs can couple acoustic energy into a fluid drop sitting on the piezoelectric substrate to effect interfacial destabilization (Fig. 1d). The operating principle and governing mechanisms that underpin ultrasonic atomization (cavitation or capillary wave destabilization) are discussed in detail in Yeo et al. [12].

The average diameter D of the aerosol droplets produced depends on the most unstable wavelength λ of the capillary wave instability that is induced, specified by the Kelvin equation, which essentially arises from a dominant force balance between the stabilizing capillary stresses and the destabilizing stress imposed by the vibration:

$$D \approx C_1 \lambda \approx C_2 \left(\frac{2\pi\gamma}{\rho f_c^2} \right)^{\frac{1}{3}}, \qquad (1)$$

where γ and ρ are the surface tension and density of the fluid, respectively, and f_c is the capillary wave frequency. C_1 and C_2 are empirically determined coefficients commonly used to fit the experimental data to the predictions, and vary widely in the literature, although they are typically of order unity. It should be noted that Kelvin's theory does not provide a way in which the capillary wave frequency is related to the forcing frequency due to the acoustic vibration, and the usual assumption made (see, for example, [15]) is that the capillary waves are excited at a subharmonic frequency that is one half of the forcing frequency, i.e., $f_c = f/2$. Recent studies have however shown that this may not always be true and that the capillary wave frequency can be predicted from a dominant balance between capillary and viscous stresses, at least for a sessile drop or thick liquid film [12, 15]:

$$f_c \sim \frac{\gamma}{\mu L}, \qquad (2)$$

where μ is the viscosity of the fluid and L is the characteristic length scale of the drop or film. A correction factor $(H/L)^2$, wherein H is the characteristic height scale, can be included to account for the geometry of the parent drop, in particular, the axial capillary stress, when substituting Eq. 2 into Eq. 1 [15].

SAWs, generated by applying an oscillating electrical signal to the interdigital transducer electrodes patterned onto a piezoelectric substrate (Fig. 1d), the gap and width of which specifies the SAW wavelength and hence the resonant frequency (typically 10–100 MHz), have also been shown as an effective mechanism for atomizing drops. Essentially, SAWs are nanometer order amplitude ultrasonic waves that are confined to and propagate along the substrate surface in the form of a Rayleigh wave. Owing to high MHz order frequencies and substrate displacement velocities, on the order of 1 m/s in the vertical direction perpendicular to the substrate, substrate accelerations on the order of 10^7 m^2/s arise. This, together with the strong acoustic streaming that is induced within the fluid drop placed on the substrate, leads to fast destabilization and breakup of the drop free surface to produce 1–10 μm dimension aerosol drops [15], as illustrated in Fig. 1d. One advantage of the SAW atomization over its ultrasonic counterparts is the efficient energy transfer mechanism from the substrate to the liquid – unlike bulk vibration, most of the acoustic energy of the SAW is localized within a region on the surface of the substrate about 3–4 wavelengths thick (a SAW wavelength is typically on the order of 100 μm) and is transferred to the liquid drop to drive the atomization. Consequently, it is possible to atomize fluids using the SAW at input powers of around 1 W, which is one to two orders of magnitude smaller than that required using other ultrasonic methods. Another advantage of the SAW is the high frequencies (>10 MHz) that can be assessed. The timescale associated with the period (inverse frequency) of the oscillating acoustic and electromechanical field at these frequencies is much shorter than the hydrodynamic timescale $\mu L/\gamma \sim 10^{-4}$ s as well as the 10^{-5}–10^{-6} s relaxation timescales (inverse of the strain rate) associated with shear-induced molecular lysis [12]. In addition, it is not possible to induce cavitation, which is known to cause molecular lysis, at the low powers

and high frequencies associated with SAW atomization [15].

Synthesis of Polymeric Microparticles and Nanoparticles

Early work to demonstrate the possibility of synthesizing polymeric particles, albeit with micron dimensions, was reported by Tsai et al. [16] and Berkland et al. [4]. The apparatus in both studies is similar, based on concept known as two-fluid atomization. A piezoelectric transducer is used to vibrate a nozzle or orifice from which a jet comprising the working fluid issues and which subsequently suffers from Rayleigh-Plateau instabilities to break up into individual aerosol droplets. In a manner similar to the Kelvin equation given in Eq. 1, the droplet diameter is controlled by the most unstable wavelength of the axisymmetric instability, which can be predicted from a dominant force balance between the inertia imposed on the jet and the capillary stresses that stabilize it, and is a function of the diameter of the undistorted jet, which is slightly larger than the nozzle or orifice diameter. In both cases, an external annular stream surrounding the nozzle through which the liquid jet issues is employed. The annular sheath fluid consists of air in the former and an immiscible carrier liquid that does not dissolve the polymer solution in the latter. In both cases, the sheath fluid, whether air or a second immiscible liquid, is flowed much faster than the jetting fluid although the underlying reason provided for its necessity differs. In the former, the airflow is suggested to vibrate in resonance with the vibrating nozzle due to the coaxial annular arrangement, resulting in a magnification of the amplitude of the capillary waves on the liquid jet. In the latter, the interfacial shear imposed by the carrier fluid surrounding the jet is suggested to aid the primary breakup of the jet away from its parent fluid at the orifice, allowing the production of droplets one order of magnitude smaller. 40 μm diameter xanthan gum particles were synthesized in the former when driven at a fundamental resonant frequency of 54 kHz through a 0.93 mm nozzle whereas 5–500 μm diameter poly(D,L-lactic-co-glycolic acid) (PLGA) particles (Fig. 2) were synthesized in the latter when driven at frequencies between 19 and 70 kHz through 60 and

100 μm diameter orifices. The carrier fluid, poly(vinyl alcohol), in which the PLGA particles were collected and solidified within, however, remained on the surface of the particles, which could potentially affect physical and cellular uptake [3].

Forde et al. [17] subsequently showed the possibility of synthesizing poly(ε-caprolactone) (PCL)

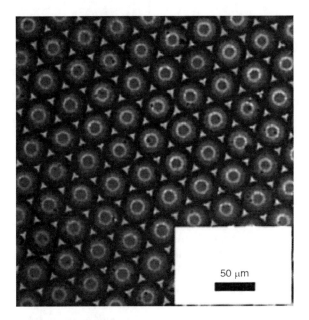

Acoustic Nanoparticle Synthesis for Applications in Nanomedicine, Fig. 2 45 μm PLGA microparticles synthesized by ultrasonically atomizing the polymer solution through a 60 μm orifice which is indirectly vibrated using a piezoelectric transducer at frequencies between 19 and 70 kHz (Reprinted with permission from Berkland et al. J. Control. Release 73, 59–74 (2001). Copyright (2001) Elsevier)

*nano*particles around 200 nm in dimension by atomizing a PCL/acetone solution using the direct method shown in Fig. 1b involving a piston-vibrated hard lead zirconate titanate piezoelectric disk between 1 and 5 MHz, as illustrated in Fig. 3. The nanoparticle diameter D_p can be estimated from volume considerations [18]:

$$D_p \approx D \left(\frac{C}{\rho} \right)^{\frac{1}{3}}, \tag{3}$$

wherein C denotes the initial polymer concentration, such that it is possible to tailor the nanoparticle size by tuning the aerosol size through its physical properties as well as the geometry of the parent drop, as suggested by Eq. 2.

Polymeric nanoparticles have also been produced using SAW atomization. Friend et al. [19] demonstrated the synthesis of 150–200 nm clusters of 5–10 nm PCL nanoparticle aggregates, as shown in Fig. 4, the cluster size being weakly dependent on several parameters such as the surfactant used and the drying length. The grape-bunch-like cluster morphology was attributed to nonuniform solvent evaporation during in-flight drying of the droplets that drive a thermodynamic instability. This results in spinodal decomposition and phase separation, which gives rise to separate regions that are polymer-rich and solvent-rich. The polymer-rich regions then solidify more rapidly, creating nucleation sites that lead to the formation of a cluster of PCL molecules until a critical nucleation size is attained, estimated from classical nucleation theory to be around 10 nm and consistent

Acoustic Nanoparticle Synthesis for Applications in Nanomedicine,

Fig. 3 Schematic illustration of the direct ultrasonic atomization setup of Forde et al. in which a sessile drop is vibrated at 1.645 MHz or 5.345 MHz in a piston-like manner using a hard PZT disk (Reprinted with permission from Forde et al. Appl. Phys. Lett. 89, 064105 (2006), Copyright (2006) American Institute of Physics)

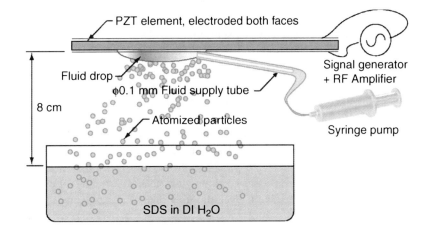

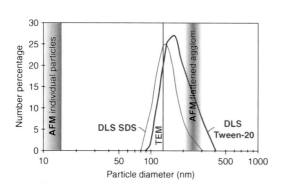

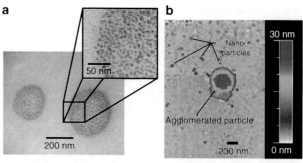

Acoustic Nanoparticle Synthesis for Applications in Nanomedicine, Fig. 4 The *left panel* shows dynamic light scattering measurements of the PCL nanoparticles produced using SAW atomization. The right panel shows (**a**) transmission electron microscopy and (**b**) atomic force microscopy images showing 150–200 nm clusters of 5–10 nm nanoparticles (Reprinted with permission from Friend et al. Nanotechnology 19, 145301 (2008). Copyright (2008) Institute of Physics)

with the 5–10 nm nanoparticles observed. In addition to PCL, 1–10 μm dimension protein (bovine serum albumin and insulin) aerosols have been generated using the SAW, which evaporate in flight to form 50 to 100 nm nanoparticles [18].

Drug Encapsulation

Early work on drug encapsulation via ultrasonic atomization was carried out by Felder et al. [5], who showed the possibility of encapsulating protein and peptides into poly(lactic acid) (PLA) and PLGA directly using ultrasonic atomization. The polymer solution was sonicated with a protein (bovine serum albumin; BSA) or peptide solution (2–10% w/w concentration) to create a stable water-in-oil emulsion, which was then atomized into a beaker in which hardening agent (octamethylcyclotetrasil, hexane, isopropyl myristate, or water) was agitated to yield 10–100 μm order microparticles. This solidification method relies on *solvent extraction* and hence polymer desolvation within the hardening agent as opposed to *solvent evaporation* inflight used in the other methods reported here. While smaller nanoparticles on the order of 100 nm diameter were also reported, it is not clear whether these contained any encapsulated material, as the authors only reported size distribution data for the naked PLA and PLGA particles. It would, however, be consistent with the other studies reported below that encapsulated particles are generally of micron-order dimension due to the size of the peptides and proteins within. The frequency at which the ultrasonic atomization was carried out was also unspecified. Reported encapsulation efficiencies, obtained by dissolving the polymer particles in the original solvent (or an acetonitrile/chloroform mixture) to release the encapsulated material which is then recovered on a 200 nm cellulose filter followed by elution with phosphate buffered saline and quantitative characterization using a spectrofluorometer (protein) or through a HPLC assay (peptides), were typically low, between 10% and 35%. This was attributed to comparable extraction rates between that of the aqueous phase in which the protein or peptide resides with that of the polymer solvent through diffusion and partitioning. Consequently, the protein or peptide is removed during washing. When a different peptide which is only weakly soluble in water was encapsulated, the efficiency was observed to increase to 63% and 93%. As such, it is instructive to note here that encapsulation efficiencies are therefore likely to be higher if solvent evaporation is used as the polymer desolvation method instead of solvent extraction, since the solvent is much more likely to be removed rapidly through evaporation while in flight at a rate much faster than that of water, therefore resulting in more efficient encapsulation of the aqueous phase and the

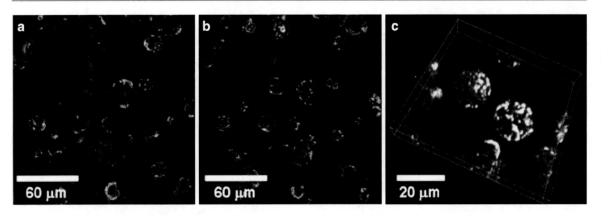

Acoustic Nanoparticle Synthesis for Applications in Nanomedicine, Fig. 5 Confocal microscope images of the cross-sectional slice (*top, bottom, and center*) across a PCL microparticle showing the encapsulation of fluorescent biotin within (Reprinted with permission from Alvarez et al. Biomicrofluidics 3, 014102 (2009). Copyright (2009) American Institute of Physics)

proteins or peptides within the fast solidifying polymer shell.

BSA was also encapsulated in PLGA using 100 kHz ultrasonic atomization of a similar water-in-oil emulsion in a study by Freitas et al. [6]. The solvent was evaporated and the solidified particles were collected and further desolvated in an agitated aqueous solution and subsequently recovered using a filter. To allow for aseptic encapsulation conditions, the conventional spray dryer unit was shortened by running the unit under near-vacuum conditions instead of using hot air, and the cyclone unit that is usually present in spray dryers was replaced by an aqueous collection bath; in that way, the entire unit can be placed within a laminar flow chamber. Again, the particles were in the micron dimension (13–24 μm mean diameter). As expected due to the use of solvent evaporation over solvent extraction, the BSA encapsulation efficiencies were higher, around 50–60% compared to 10–35%, which the authors claim could be improved if BSA loss in the aqueous collection solution was replaced by a fluid which is a nonsolvent for BSA. Alternatively, they suggest reducing the polymer concentration.

Plasmid DNA (pDNA) and poly(ethyleneimine) (PEI) complexes were encapsulated within 10–20 μm diameter PLGA microparticles using a 40 kHz ultrasonic atomizer [20]. Solvent extraction via a poly(vinyl alcohol) hardening agent was used in this case. Encapsulation efficiencies in a wide range between

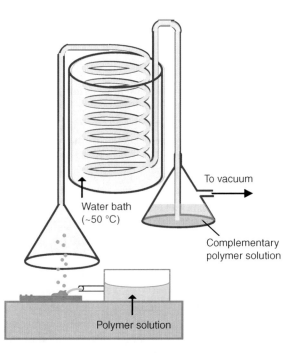

Acoustic Nanoparticle Synthesis for Applications in Nanomedicine, Fig. 6 Schematic illustration of the SAW atomization setup involving a single atomization–evaporation–resuspension step used to deposit a single polymer layer and to resuspend it in a complementary polymer solution. The step is repeated to deposit subsequent layers for as many layers as required

20% and 90% were reported, with lower polymer concentrations and higher pDNA-PEI volume fractions giving higher efficiencies; contrary to prior claims,

Acoustic Nanoparticle Synthesis for Applications in Nanomedicine,
Fig. 7 Dynamic light scattering measurements of (**a**) single-layer chitosan nanoparticles, (**b**) bilayer chitosan and CMC nanoparticles, and (**c**) pDNA encapsulated chitosan nanoparticles

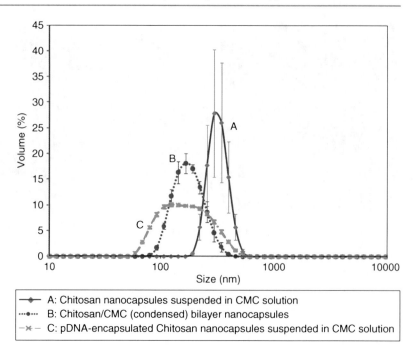

A: Chitosan nanocapsules suspended in CMC solution
B: Chitosan/CMC (condensed) bilayer nanocapsules
C: pDNA-encapsulated Chitosan nanocapsules suspended in CMC solution

the authors however did not find significant dependence of the encapsulation efficiency on the N/P ratio (N referring to the nitrogen content and P referring to the DNA phosphate content of the pDNA-PEI complex). These values were however obtained by subtracting the mass of pDNA remaining in the hardening agent, measured using a spectrofluorometer, from that in the feedstock, which does not take into account loss of pDNA in the environment during atomization, the amount of unatomized pDNA left on the atomizer as well as the pDNA that resides on the surface of the particles instead of being encapsulated within. The authors also investigated the post-atomization structural integrity of the pDNA using gel electrophoresis due to the susceptibility of pDNA to shear degradation, and found 80% retention of the pDNA in supercoiled conformation. This was attributed to cationic complexation between pDNA and PEI, which reduces its size and hence the possibility of shear-induced degradation (when naked pDNA was assessed, only 8% of this remained supercoiled). Nevertheless, PEI has been known to induce cytotoxic effects and hence its use may be limited for gene transfection and delivery. An alternative mitigation strategy involving multilayer polymer nanoparticles will be discussed in the next section.

Alvarez et al. [18] later demonstrated the encapsulation of BSA in PCL particles using SAW

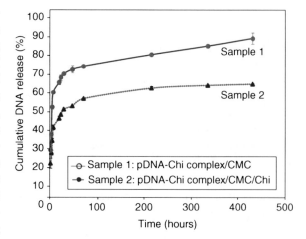

Acoustic Nanoparticle Synthesis for Applications in Nanomedicine, Fig. 8 In vitro release profile showing the diffusion of pDNA out of chitosan/CMC bilayer nanoparticles (sample 1) and chitosan/CMC/chitosan trilayer nanoparticles (sample 2) acquired through fluorescence absorbance measurements of fluorescently labeled pDNA. The lines are added to aid visualization

atomization. With 10 MHz SAWs, the particle sizes had a mean of 23 μm whereas this decreased to around 6 μm with 20 MHz SAWs. Proof of encapsulation was acquired through confocal image slices of the particles showing the fluorescently tagged BSA appearing not only at the top and bottom of the cross-sectional slices across the particle but also in the particle center,

Acoustic Nanoparticle Synthesis for Applications in Nanomedicine, Fig. 9 Confocal microscopy images of (**a**) COS-7 cells and (**b**) human mesenchymal progenitor cells (MPCs) transfected with pDNA encoded with a yellow fluorescent protein (the expression is depicted in the images in *green*) encapsulated in PEI/CMC bilayer nanoparticles

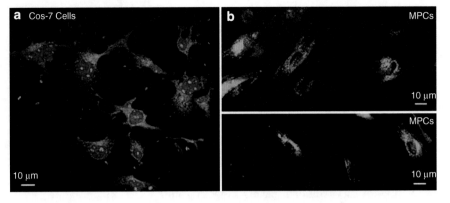

thus verifying that the BSA was entrapped within and not simply bound to the surface of the particle (Fig. 5). The encapsulation efficiency value reported was slightly different compared to other studies in that 54% of BSA from the feedstock was found to be encapsulated with its chemical structure intact (as opposed to the total BSA content encapsulated reported in previous studies). Thus, the total encapsulation efficiency value could be significantly higher (if a comparison were to be made with the values in Felder et al. [5] and Freitas et al. [6]). As discussed above, the high-frequency operation of the SAW limits the amount of shear and cavitation damage caused to the molecules, and constitutes a considerable advantage of using SAWs over conventional ultrasonic atomization.

Multilayer Nanoparticle Synthesis and Encapsulation

Multilayered polymer particles offer significant advantages over a particle comprising a single layer. Layers of different polymers with varying chemistry and hence degradation profiles offer the possibility of tuning the drug release over time and in different physiological regions, therefore affording tremendous opportunities for targeted and controlled release delivery. Very recently, SAW atomization has been exploited to demonstrate the potential for rapidly synthesizing nanoparticles with alternating layers of complementary polymers of opposing charge [21]. This is done through a variation of the usual procedure of atomizing an initial polymer solution followed by evaporated-assisted solidification to produce a single-layer polymer nanoparticle. In addition, however, the

solidified polymer particle is collected in a solution in which the second polymer is dissolved (Fig. 6), which is then re-atomized and dried to deposit the second polymer layer over the initial polymer core. By collecting the two-polymer-layer nanoparticle in a solution comprising the polymer to form the third layer and re-atomizing, a further layer can be deposited. The atomization–evaporation–resuspension procedure is then repeated for as many times as the number of layers desired. The requirement of the polymers comprising alternating layers is that each successive polymer must be complementary to the previous polymer, i.e., they must have opposing charges. In addition, the polymer making up the subsequent layer must be soluble in a solvent that cannot dissolve the polymer making up the previous layer. Up to eight layers of alternating chitosan (or PEI) and CMC layers were synthesized; proof of the deposition of successive layers was provided by visual inspection (atomic force microscopy), charge characterization (reversal of the zeta-potential after the deposition of each successive layer), Fourier transform infrared spectrometry showing ionic complexation between the deposited layers, and fluorescence measurements of labeled polymers. In addition, pDNA was also encapsulated within the multilayer polymeric nanoparticles to demonstrate the therapeutic capability of the nanoparticles in particular for gene therapy.

Figure 7 shows the size distribution of negatively charged chitosan and positively charged CMC polymer bilayer particles. The decrease in size upon deposition of the second CMC layer over the chitosan core or the encapsulated pDNA can be attributed to ionic complexation that tends to compact the particle size. Even with encapsulation, however, the size of the particles remains in the 100–200 nm range, which is a significant advance over the micron-sized particles obtained whenever a therapeutic molecule is encapsulated (see previous section). In targeted cancer therapy, for example, the mean vascular pore size of most human tumors is around 400 nm [1], and hence extravasation is likely to be more effective with the multilayer nanoparticle drug carriers synthesized through this technique. Figure 8 shows in vitro release profiles of the pDNA, showing the possibility for slowing and hence controlling the release with the deposition of an additional layer. Good in vitro DNA transfection is also demonstrated in COS-7 and human mesenchymal progenitor cells, as seen in Fig. 9.

Cross-References

▶ Nanoencapsulation
▶ Nanomedicine
▶ Nanoparticles
▶ Polymer Coatings

References

1. Kumar, M.N.V.R. (ed.): Handbook of Particulate Drug Delivery. American Scientific, California (2008)
2. Mehnert, W., Mäder, K.M.: Solid lipid nanoparticles: production, characterization and applications. Adv. Drug. Deliv. Rev. **47**, 165–196 (2001)
3. Panyam, J., Labhasetwar, V.: Biodegradable nanoparticles for drug and gene delivery to cells and tissue. Adv. Drug. Deliv. Rev. **55**, 329–347 (2003)
4. Berkland, C., Kim, K., Pack, D.W.: Fabrication of PLG microspheres with precisely controlled and monodisperse size distributions. J. Control. Release **73**, 59–74 (2001)
5. Felder, C., Blanco-Prieto, M., Heizmann, J., Merkle, H., Gander, B.: Ultrasonic atomization and subsequent polymer desolvation for peptide and protein microencapsulation into biodegradable polyesters. J Microencapsul. **20**, 553–567 (2003)
6. Freitas, S., Merkle, H., Gander, G.: Ultrasonic atomisation into reduced pressure atmosphere – envisaging aseptic spray-drying for microencapsulation. J. Control. Release **95**, 185–195 (2004)
7. Alvarez, M., Friend, J., Yeo, L.Y.: Rapid generation of protein aerosols and nanoparticles via surface acoustic wave atomization. Nanotechnology **19**, 455103 (2008)
8. Gañán-Calvo, A.M.: Enhanced liquid atomization: from flow-focusing to flow-blurring. Appl. Phys. Lett. **86**, 214101 (2005)
9. Grace, J., Marijnissen, J.: A review of liquid atomization by electrical means. J. Aerosol. Sci. **25**, 1005–1019 (1994)
10. Yeo, L.Y., Gagnon, Z., Chang, H.-C.: AC electrospray biomaterials synthesis. Biomaterials **26**, 6122–6128 (2005)
11. Friend, J., Yeo, L.Y.: Microscale acoustofluidics: microfluidics driven via acoustics and ultrasonics. Rev. Mod. Phys. **83**, 647–704 (2011)
12. Yeo, L.Y., Friend, J.R., McIntosh, M.P., Meeusen, E.N.T., Morton, D.A.V.: Ultrasonic nebulization platforms for pulmonary drug delivery. Expert. Opin. Drug. Deliv. **7**, 663–679 (2010)
13. James, A.J., Vukasinovic, B., Smith, M.K., Glezer, A.: Vibration-induced drop atomization and bursting. J. Fluid. Mech. **476**, 1–28 (2003)
14. Meacham, J.M., Varady, M.J., Degertekin, F.L., Fedorov, A.G.: Droplet formation and ejection from a micromachined ultrasonic droplet generator: visualization and scaling. Phys. Fluids **17**, 100605 (2005)
15. Qi, A., Yeo, L., Friend, J.: Interfacial destabilization and atomization driven by surface acoustic waves. Phys. Fluids **20**, 074103 (2008)
16. Tsai, S.C., Luu, P., Song, Y.L., Tsai, C.S., Lin, H.M.: Ultrasound-enhanced atomization of polymer solutions and applications to nanoparticles synthesis. Trans. Ultrason. Symp. **1**, 687–690 (2000)
17. Forde, G., Friend, J., Williamson, T.: Straightforward biodegradable nanoparticle generation through megahertz-order ultrasonic atomization. Appl. Phys. Lett. **89**, 064105 (2006)
18. Alvarez, M., Yeo, L.Y., Friend, J.R., Jamriska, M.: Rapid production of protein-loaded biodegradable microparticles using surface acoustic waves. Biomicrofluidics **3**, 014102 (2009)
19. Friend, J.R., Yeo, L.Y., Arifin, D.R., Mechler, A.: Evaporative self-assembly assisted synthesis of polymeric nanoparticles by surface acoustic wave atomization. Nanotechnology **19**, 145301 (2008)
20. Ho, J., Wang, H., Forde, G.: Process considerations related to the microencapsulation of plasmid DNA via ultrasonic atomization. Biotechnol. Bioeng. **101**, 172–181 (2008)
21. Qi, A., Chan, P., Ho, J., Rajapaksa, A., Friend, J., Yeo, L.: Template-free synthesis and encapsulation technique for layer-by-layer polymer nanocarrier fabrication. ACS Nano, doi:10.1021/nn202833n

Acoustic Particle Agglomeration

▶ Acoustic Trapping

Acoustic Separation

▶ Acoustophoresis

Acoustic Trapping

Mikael Evander and Thomas Laurell
Department of Measurement Technology and
Industrial Electrical Engineering, Division of
Nanobiotechnology, Lund University, Lund, Sweden

Synonyms

Acoustic tweezers; Acoustic particle agglomeration

Definition

Acoustic trapping is the immobilization of particles
and cells in the node of an ultrasonic standing wave
field.

Overview

Acoustic trapping has been shown to be a gentle way of
performing non-contact immobilization of cells and
particles in microfluidic systems. Localized ultrasonic
standing waves are created in microfluidic channels,
cavities, or other small, confined spaces creating
pressure nodes that attract and hold particles and
cells. Commonly, structures in the range of a couple
of 100 μm are used, corresponding to acoustic frequencies in the MHz-range.

Background

Some 30 years ago, NASA and ESA started to
develop containerless processing and one of several
possible techniques was acoustic trapping/levitation
[1]. These open-air levitators, still in use today, are
rather large instruments that create a standing wave in
air and make it possible to position and levitate liquid
droplets or small and light solid matter. The use of
radiation forces on particles and cells in liquid-
suspensions have been a more common approach
than the open-air levitators however. Baker showed
in 1972 that a band-formation owing to the radiation
forces could be observed when subjecting erythrocytes
to a 1 MHz standing wave in polystyrene containers
[2]. Since then, the technique has been refined and
successfully incorporated into the lab-on-a-chip field
as one of several methods to control cells and particles
in laminar flows.

Theory

In order to create a standing wave, the geometry of the
channel/cavity must match the actuation frequency of
the transducer. To create a single pressure node, the
length of the resonance cavity should be $\lambda/2$, where λ is
the wavelength of the ultrasound in the media. To
increase the trapping capacity, this can then be scaled
so that the number of pressure nodes where objects can
be trapped is equal to $n*\lambda/2$. For water-based buffers,
the sound velocity is typically 1,500 m/s. So in order to
create a standing wave at 2 MHz, the resonance cavity
should be 375 μm (v/2f).

There are several acoustic forces acting on the
objects in an acoustic standing wave, the *primary radiation force*, *the lateral radiation force*, and
interparticular secondary forces. The most dominant
force is the *primary radiation force* (Eq. 1) that is
proportional to the particle volume (V_c) and is thus
strongly size dependent [3]. The force also depends
on the acoustic wavelength (λ), the acoustic pressure
amplitude (p_0), and the particles' position in the acoustic wave (x). *The acoustic contrast factor* (Eq. 2) is
dependent on the density and compressibility of both
the particle (ρ_c, β_c) and the medium (ρ_w, β_w).

$$F_r = -\left(\frac{\pi p_0^2 V_c \beta_w}{2\lambda}\right) \cdot \phi(\beta, \rho) \cdot \sin(2kx) \quad (1)$$

$$\phi(\beta, \rho) = \frac{5\rho_c - 2\rho_w}{2\rho_c + \rho_w} - \frac{\beta_c}{\beta_w} \quad (2)$$

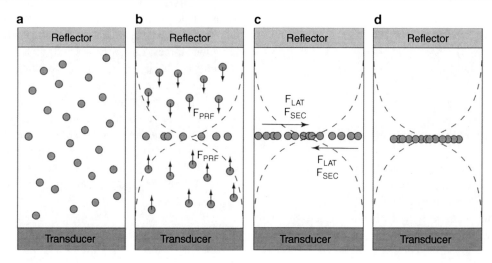

Acoustic Trapping, Fig. 1 A schematic image demonstrating how different acoustic forces act on particles in a standing wave. (**a**) Before activating the ultrasound, particles will be evenly distributed in the channel. (**b**) When activating the ultrasound, the particles will be moved into the pressure node by the primary radiation force. (**c**) Lateral forces will then focus the particles to the center of the sound field. Once the distance between the particles is short enough, secondary interparticular forces will create an attractive force between the particles and pull the cluster firmer together. (**d**) The end result is a centered, levitated cluster of particles situated in the middle of the resonance cavity (Reprinted from [5] with permission from the author)

The acoustic contrast factor can change the direction of the force so that, depending on the material parameters, an object may be pushed toward the pressure node or the pressure antinode. For most cells and particles, the force will be directed toward the pressure node however. *Acoustophoresis* uses this factor to enable separation between different cell types that may differ in density and/or compressibility.

While the primary radiation force is the main force at play in acoustophoresis, *the lateral force* is what in most trapping designs keeps the particles and cells locked in the standing wave. The lateral force arises from the fact that the wave front of the sound is divergent and has reducing amplitude at the edges of the sound beam. The pressure difference between the center of the sound beam and the edges will create a component that draws objects into the center of the beams and locks them in place. By looking at a dense particle with a radius R positioned in a standing wave with a constant amplitude gradient, Gröschl estimated the lateral force component to [4]:

$$F_{Lat} = \pi \rho \omega^2 R^2 \hat{u}_0 \hat{u}_m \qquad (3)$$

As can be seen in Eq. 3, the lateral force depends on the medium density (ρ), the angular frequency of the ultrasound (ω), the square of the particle radius, and the difference in displacement amplitude between the center ($\hat{u}_0$) and the edge of the particle ($\hat{u}_0 + \hat{u}_m$). Comparing Eqs. 3 and 1 will show that the lateral force is not as size dependent as the primary radiation force since it only depends on the square of the radius. Also, the greater the amplitude difference ($\hat{u}_m$) between the center of the particle and the edge of the particle, the greater the trapping force. So in order to create a strong acoustic trap, a large lateral pressure gradient is needed.

There are also *secondary interparticular forces* that arise when the incident sound wave is scattered on the objects in the standing wave. These forces only act on short distances but help to form a cluster of the objects and keep them together (Fig. 1).

System Design

An acoustic trap is usually designed to either use a localized pressure field or create a localized resonance. An example of a localized pressure field would be a small transducer that couples sound to a fluid only in a small portion of a channel, also called a layered

Acoustic Trapping,
Fig. 2 (a) Shows an acoustic trapping approach using a small transducer that creates a very localized pressure field in a channel. The transducer is embedded into the bottom of the fluidic channel and creates a large pressure gradient over a small area; (b) is an example of an acoustic trapping cavity. Here the entire chip is actuated and a standing wave is created only where the geometry matches the ultrasound

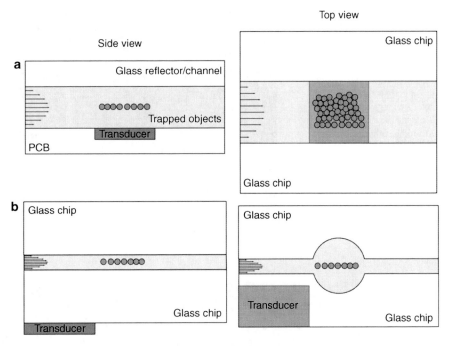

resonator. The alternative approach is to actuate the entire chip and design a resonance cavity on the chip that matches the actuation frequency. The standing wave will then only exist where the geometry matches the standing wave criteria. Another alternative that has not been as common as the other two designs is the use of focused transducers. By focusing the ultrasound into a very narrow beam and then reflecting the wave in the focal plane, a hemispheric standing wave that can trap and hold objects can be created [6] (Fig. 2).

Designs Using Local Resonance Cavities

Designs that use a transducer coupled directly or through a matching layer to the fluid typically have a larger trapping capacity than the cavity approach. However, since the cavity trap is usually created using photolithographic techniques it is easy to array and to integrate with other sample processing steps on chip. An example of a cavity array was presented by Vanherberghen et al. where a 10×10 array was used for aggregating cells for isolated cell studies using microscopy [7] (Fig. 3).

Designs Using a Localized Transducer

Spengler and Coakley presented a trapping system that used a layered design in 2000 [8]. By using a PTFE spacer and a glass reflector, they were able to study cell aggregates in a standing wave at 1.93 MHz. This system was later redesigned to a stainless steel system using a matching layer to couple the ultrasound into the resonator and used for studies of cell membrane spreading [9].

An alternative to using a matching layer is to integrate the transducer in the bottom of a microfluidic channel. A system based around a miniature transducer that was embedded in a printed circuit board was presented by Evander et al. in 2007 [10]. A glass lid with a microfluidic channel was placed on top of the printed circuit board and a standing wave was formed above the transducer in the channel. To test the system, yeast cells were grown in the standing wave and an online viability assay was performed on trapped neural stem cells (see Fig. 4).

A further development of the trapping system was presented by Hammarström et al. [11]. Instead of etched microfluidic channels, a square borosilicate capillary was coupled to an external miniature transducer through a thin glycerol layer. Using

Acoustic Trapping,
Fig. 3   The 10 × 10 ultrasonic
cavity array presented by
Vanherberghen et al. [7].
In (**a**), human B cells are
shown in the cavities without
ultrasound and in (**b**) the cells
have been aggregated
using standing waves
(Reproduced by permission
of The Royal Society of
Chemistry: http://dx.doi.org/
10.1039/c004707d/)

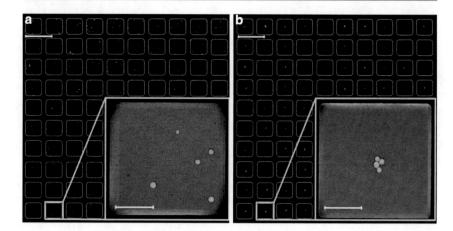

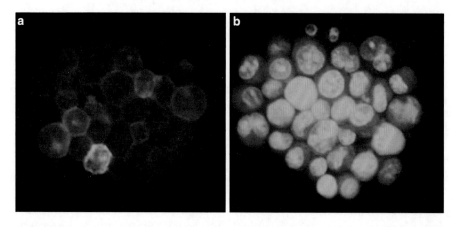

Acoustic Trapping, Fig. 4 Neural stem cells, HiB5-GFP trapped in an acoustic standing wave. In (**a**) the cells have just been trapped in the standing wave and in (**b**) the cells have been levitated for 15 min after which they were perfused with

Acridine Orange to test for viability. The increase in fluorescence indicates that the cells are still viable (Reprinted and modified with permission from [10]. Copyright 2007 American Chemical Society)

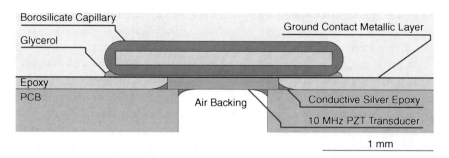

Acoustic Trapping, Fig. 5 The trapping platform presented in [11] uses a square borosilicate capillary as microfluidic channel and resonance cavity. A system with an external transducer and off-the-shelf capillaries has a lower production cost and is

also compatible with more sensitive assays where disposable channels must be used (Reproduced by permission of The Royal Society of Chemistry: http://dx.doi.org/10.1039/c004504g)

commercially available capillaries both lowers the production costs and allows for disposable channels for more sensitive analysis.

Cell Viability

Several studies have shown that ultrasonic levitation does not seem to have any negative effects on live cells. Hultström et al. cultured cells that had been levitated for over an hour and studied the doubling times of the cells [12]. No direct or delayed damage could be detected on the cells. Bazou et al. performed a study on mouse embryonic stem cells and confirmed that they could see no changes in gene expression after ultrasonic treatments up to 1 h and the cells maintained their pluripotency [13].

Future Directions of the Field

The main research focus has so far been on technology development but as the robustness of the systems is steadily increasing, more application-driven development can be expected. Further work aimed at integrating the trapping systems with other lab-on-a-chip systems can also be expected as well as an ambition to increase the particle size range that can be manipulated, in particular toward smaller objects such as bacteria.

Cross-References

▶ Acoustic Contrast Factor
▶ Acoustic Tweezers
▶ Acoustophoresis
▶ Integrated Micro-acoustic Devices

References

1. Lierke, E.G.: Akustische Positionierung – Ein umfassender Überblick über Grundlagen und Anwendungen. Acustica 82(2), 220–237 (1996)
2. Baker, N.V.: Segregation and sedimentation of red blood-cells in ultrasonic standing waves. Nature 239(5372), 398–399 (1972)
3. Gorkov, L.P.: On the forces acting on a small particle in an acoustic field in an ideal fluid. Sov. Phys. Doklady 6(9), 773–775 (1962)
4. Groschl, M.: Ultrasonic separation of suspended particles – part I: fundamentals. Acustica 84(3), 432–447 (1998)
5. Evander, M.: Cell and particle traping in microfluidic systems using ultrasonic standing waves. Dissertation, Department of Electrical Measurements and Industrial Engineering and Automation, Lund University, Lund (2008)
6. Wiklund, M., Nilsson, S., Hertz, H.M.: Ultrasonic trapping in capillaries for trace-amount biomedical analysis. J. Appl. Phys. 90(1), 421–426 (2001)
7. Vanherberghen, B., et al.: Ultrasound-controlled cell aggregation in a multi-well chip. Lab Chip 10(20), 2727–2732 (2010)
8. Spengler, J.F., et al.: Observation of yeast cell movement and aggregation in a small-scale MHz-ultrasonic standing wave field. Bioseparation 9(6), 329–341 (2000)
9. Coakley, W.T., et al.: Cell-cell contact and membrane spreading in an ultrasound trap. Colloids Surf. B Biointerfaces 34(4), 221–230 (2004)
10. Evander, M., et al.: Noninvasive acoustic cell trapping in a microfluidic perfusion system for online bioassays. Anal. Chem. 79(7), 2984–2991 (2007)
11. Hammarstrom, B., et al.: Non-contact acoustic cell trapping in disposable glass capillaries. Lab Chip 10(17), 2251–2257 (2010)
12. Hultstrom, J., et al.: Proliferation and viability of adherent cells manipulated by standing-wave ultrasound in a microfluidic chip. Ultrasound Med. Biol. 33(1), 145–151 (2007)
13. Bazou, D., et al.: Gene expression analysis of mouse embryonic stem cells following levitation in an ultrasound standing wave trap. Ultrasound Med. Biol. 37(2), 321–330 (2011)

Acoustic Tweezers

▶ Acoustic Trapping

Acoustophoresis

Andreas Lenshof and Thomas Laurell
Department of Measurement Technology and
Industrial Electrical Engineering, Division
of Nanobiotechnology, Lund University, Lund,
Sweden

Synonyms

Acoustic separation; Free flow aoustophoresis (FFA)

Definition

"Acoustophoresis" means migration with sound, i.e., "phoresis" – migration and "acousto" – sound waves are the executors of the movement. In related concepts, electric forces move particles in electrophoresis and magnetic forces in magnetophoresis [1]. Acoustophoresis is a noncontact and label-free mode of manipulating particles and cell populations and allows for implementation of several separation modes [2]. The technology is currently finding increased applications in bioanalytical and clinical applications of cell handling and manipulation.

Theory

Particles in suspension exposed to an acoustic standing wave field will be affected by an acoustic radiation force [3]. The force will cause the particle to move in the sound field if the acoustic properties of the particle differ from the surrounding medium. The magnitude of the movement depends on factors, such as the size of the particle, the acoustic energy density, and the frequency of the sound wave. The direction of the particle movement is dependent on the density and speed of sound of the particle as well as the liquid media. In an acoustic standing wave generated in a liquid (e.g., water)-filled channel with wall boundaries of dense materials such as metal, silicon, or glass, a standing wave pressure maxima will form at the walls/fluid interface. If the width of the channel is matched to half a wavelength, a pressure node will form in the center of the channel.

The most predominant acoustic force acting on microparticles in an acoustic standing wave is the primary axial acoustic radiation force (PRF), Eq. 1. The magnitude of the PRF is dependent on the acoustic energy density, E_{ac}, and the radius, a, of the particle. The direction of the particle movement depends on the acoustic contrast factor Φ, which comprises the inherent physical properties of the cell or particle, such as the density, ρ_p, and speed of sound, c_p, relative to the properties of the surrounding medium, ρ_0 and c_0. The sign of the acoustic contrast factor defines the direction of the movement.

$$F_y^{\text{rad}} = 4\pi k a^3 E_{ac}\Phi\sin(2ky) \tag{1}$$

$$\Phi = \frac{\rho_p + \frac{2}{3}(\rho_p - \rho_0)}{2\rho_p + \rho_0} - \frac{1}{3}\frac{\rho_0 c_0^2}{\rho_p c_p^2} \tag{2}$$

As shown in Fig. 1, most rigid cells and particles (blue particles) have a positive contrast factor and are moved to the pressure node located in the center of the channel. Liquid vesicles or air bubbles will move to the antinodes at the wall (yellow particles). If the size of the flow channel is in micro-domain, the flow conditions is generally laminar, and the particles passing through the standing wave field will be moved to their nodal position and remain in that position throughout the flow channel even after exiting the sound field. By terminating the flow channel with a trifurcation it is possible to separate and/or concentrate the particles from the medium, see Fig. 1D.

Beside the primary axial radiation force, also secondary forces act on particles in a standing wave field [4]. These interparticle forces can mostly be negligible as they are only effective when particles are very close to each other. The secondary forces assist in particles forming clusters which are less common in continuous flow systems, but are useful in stagnant systems which rely on agglomeration and sedimentation to clarify medium [5].

When designing flow channels with dimensions that match a half wavelength in the MHz frequency regime, conventional microfabrication offers simple means of fabricating acoustophoresis chips typically in glass or silicon or other high Young's modulus materials.

There are several ways of actuating an acoustophoresis chip. The most straightforward approach is to set up the standing wave between a coupling layer bonded to the transducer and the opposing channel wall, commonly referred to as a layered acoustic resonator, Fig. 2a. Another way is to place the transducer underneath the entire separation system or at any location on the chip surface where space is available. This causes the whole acoustophoresis chip to be actuated, not just the separation channel. By matching the actuation frequency to the channel dimension, it is possible to obtain a standing wave horizontally or vertically or in both directions in the microchannel, Fig. 2b. Two-dimensional focusing can be obtained either using two transducers of different frequencies or using a single transducer if the cavity has height-to-width proportions, Fig. 2c, d.

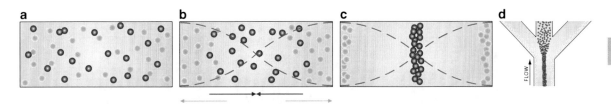

Acoustophoresis, Fig. 1 Illustration of particles with different sign of the acoustic contrast factor being exposed to an acoustic stranding wave field. (**a**) No ultrasound active. (**b**) The radiation force move the particles with positive contrast factor (dark grey) to the pressure node located in the center of the channel while the particles with negative contrast factor (light grey) move to the anti nodes at the channel side walls. (**c**) Particles at equilibrium states after acoustic exposure. **d**) The laminar flow enables separation of the two particle types using a trifurcation at the end of the channel

Acoustophoresis, Fig. 2 Two ways of actuation of the resonator system. (**a**) Layered resonator where the standing wave is generated in the direction of the primary direction of actuation and (**b**) transversal resonance. The resonance is created perpendicular to the primary direction of actuation. (**c**) Two-dimensional focusing using transducers of two different frequencies; (**d**) two-dimensional focusing using a single transducer

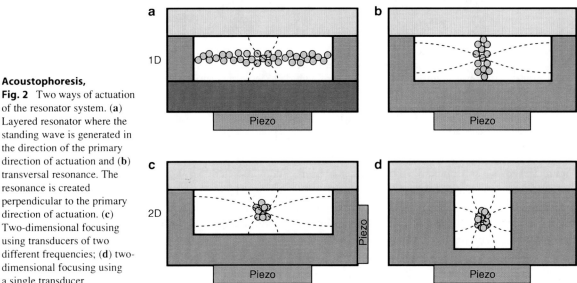

Applications

Acoustophoretic Enrichment and Depletion of Cells and Particles

Acoustophoretic enrichment/concentration of a sample is the most straightforward operation as most cells and particles easily focus in pressure nodes and thus deplete the surrounding medium of solid matter. By using flow splitters such as a trifurcation, the particle dense fraction in the central pressure node can be collected via the middle outlet of the trifurcation, Fig. 3a. Likewise, cell and particle depletion from a fluid is also obtained with the chip configuration of Fig. 3a if the fluid of interest is collected from the side branches.

However, like many microfluidic devices presented in the literature, acoustophoretic systems display problems in handling fluids of high concentrations, such as whole blood. Plasmapheresis, i.e., the removal of the cellular content from the blood plasma, is a quite common microfluidic procedure where free plasma is desired for further diagnostic purposes. Commonly these devices are forced to work with dilute blood samples, which deteriorate the plasma composition from an analytical point of view. The acoustical chips presented earlier are no exception and the plasma separation efficiency is known to decrease significantly when the hematocrit is increased above 5–10%).

Means to handle samples with higher amounts of cellular content/hematocrit can be realized by a sequential acoustophoretic enrichment and partial removal of enriched cells from the standing wave node region [6]. This requires a microchannel configuration that extends over a longer distance such that the duration in the acoustic field is sufficiently long to focus cells between each point of cell removal.

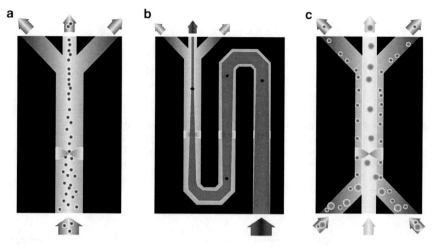

Acoustophoresis, Fig. 3 (**a**) An acoustophoretic concentrator. The particles are concentrated in the central pressure node, while particle-depleted fluid exits to the sides. (**b**) Depletion of high cell or particle contents. Acoustically concentrated particles leave outlets in the bottom of the chip (*black squares*) thus lowering the particle load until eventually all particles can be removed through the central duct of the trifurcation outlet. This mode of operation is suitable for plasmapheresis of whole blood. (**c**) Wash or fractionation of cells. Sample enters the separation channel from the side branches and becomes laminated against the side walls. Clean buffer enters through the central inlet. The acoustic radiation force moves the large cells into the clean buffer and thus completes a wash in form of a medium switch

Figure 3b shows a schematic of a chip surface conservative meander style layout, where focused blood cells are partially removed at each outlet in the center of the channel bottom indicated as dark squares. At the end of the meander structure, the concentrations of blood cells are sufficiently low for all remaining cells to be removed via the central outlet and the clean plasma fraction can be extracted from the side branches of the trifurcation.

Acoustophoretic Cell Washing

Microchip acoustophoresis can be configured to perform buffer exchange operations by adding two extra inlets on each side of an acoustophoresis channel. The sample solution is supplied via the side inlets while a clean buffer is provided via the central channel inlet, Fig. 3c. Cells/particles are acoustophoretically moved from their original sample stream along the channel side wall into the central, particle-free medium stream thus exchanging the medium the particles are suspended in [7]. The typical buffer exchange efficiency displays performance data equal or better than conventional centrifugation steps and allows continuous flow-based operation, where several sequential wash steps can be implemented depending on the requirements on the wash efficiency [8]. Acoustophoretic cell and particle washing has been demonstrated for red blood cell washing in glass chips [9] and in washing of microbeads used to selectively affinity capture bacteriophages in a large phage library [10] as well as extraction of phosphopeptides from protein digests using metal ion affinity-specific microbeads [8] and more recently also as a means of performing stain and wash unit operations in flow cytometry applications [11].

Acoustophoretic Separation/Fractionation

Acoustophoretic separation can be performed in two modes where:
1. The species to be separated display different signs of the acoustic contras factor and hence gather in the acoustic pressure node and antinode, respectively, so-called binary separation.
2. The rate of acoustophoretic transport of a cell or particle into a pressure node determines the lateral position in the flow stream of a given cell or particle type as they leave the separation zone and thus the acoustophoretic mobility determines the outcome of the separation, so-called free flow acoustophoresis – FFA.

An early clinical application of microchip-integrated acoustophoretic binary separation targeted blood wash in open-heart surgery. Patient bloodshed during the surgery is contaminated by lipid microemboli from tissue undergoing surgery and has to be removed prior to autologous retransfusion to the patient. Jönsson et al. [12] demonstrated that acoustophoresis can remove the lipid microemboli from blood, utilizing the fact that erythrocytes display a positive acoustic contrast factor ($\sim +0.05$) while lipid particles have negative contrast factor (~ -0.18). Hence, the blood cells and the lipid vesicles will move to different lateral positions in the acoustic standing wave field and can thus be separated from each other as they pass through the acoustic separation channel, Fig. 1d, following the principles of binary separation.

Binary separation has also been demonstrated as an efficient sample preprocessing step in milk quality control where the lipid emulsion was removed by binary acoustophoresis, enabling online Fourier transform infrared spectroscopic analysis of lactose and protein content of the milk [13].

When performing separation based on free flow acoustophoresis (FFA) the major factor that influences separation outcome is the primary axial radiation force which is dependent on the size, density and speed of sound of the particle, see Eq. 1. This commonly means that larger particles will experience a larger acoustic force than smaller-sized particles and thus size fractionation can be performed by acoustophoresis. Petersson et al. demonstrated successful FFA of suspensions with mixed particle sizes by tuning the acoustic force and the flow rate such that the largest particles reached the pressure node shortly before entering the center outlet and hence, laterally from the center to the side wall a gradient of particle sizes were located, which could be routed to individuals outlets of the acoustophoresis chip [14].

By manipulating the density of the medium through the addition of a buffer that alters the fluid density, it is also possible to improve separation conditions between species that display similar acoustophoretic mobility. Buffer density manipulation in combination with the FFA-fractionation device demonstrated different separation profiles of fractionation of erythrocytes, platelets, and leucocytes [14].

Free flow acoustophoresis has recently also been demonstrated in the preparation of peripheral blood progenitor cells from apheresis product. Platelets are an unwanted contaminant when the apheresis instrument continuously separates the buffy coat from the whole blood as they compromise the subsequent immunomagnetic extraction of stem cells from the apheresis product. Acoustophoresis can be used to solve this problem using the same chip design as in Fig. 3c where sample enters the channel at the side inlets and clean buffer enters centrally [15]. The acoustophoretic force on the platelets is very low and thus proceeds with the laminar flow along the channel side walls while the larger leukocytes and the few remaining erythrocytes present in the apheresis product are focused in the center of the channel and exited through the central outlet as a platelet-depleted fraction [15].

Summary

Microchip acoustophoresis offers simple means to noncontact handling/processing of cells in continuous flow mode. Acoustophoresis is to a large extent independent of surface charge, ionic strength, and pH variations (within physiological conditions). It is also a non-perturbing technique in the sense that cells are not experiencing any major stress when undergoing acoustophoresis as seen in several viability studies [15, 16].

Cross-References

▶ Acoustic Contrast Factor
▶ Acoustic Trapping
▶ Acoustic Tweezers
▶ Integrated Micro-acoustic Devices

References

1. Lenshof, A., Laurell, T.: Continuous separation of cells and particles in microfluidic systems. Chem. Soc. Rev. **39**(3), 1203–1217 (2010)
2. Laurell, T., Petersson, F., Nilsson, A.: Chip integrated strategies for acoustic separation and manipulation of cells and particles. Chem. Soc. Rev. **36**(3), 492–506 (2007)
3. Gorkov, L.P.: On the forces acting on a small particle in an acoustical field in an ideal fluid. Soviet Physics Doklady **6**(9), 773–775 (1962)
4. Groschl, M.: Ultrasonic separation of suspended particles – part I: fundamentals. Acustica **84**(3), 432–447 (1998)
5. Trampler, F., et al.: Acoustic cell filter for high-density perfusion culture of hybridoma cells. Biotechnology **12**(3), 281–284 (1994)

6. Lenshof, A., et al.: Acoustic whole blood plasmapheresis chip for prostate specific antigen microarray diagnostics. Anal. Chem. **81**(15), 6030–6037 (2009)
7. Petersson, F., et al.: Carrier medium exchange through ultrasonic particle switching in microfluidic channels. Anal. Chem. **77**(5), 1216–1221 (2005)
8. Augustsson, P., et al.: Decomplexing biofluids using microchip based acoustophoresis. Lab Chip **9**(6), 810–818 (2009)
9. Evander, M., et al.: Acoustophoresis in wet-etched glass chips. Anal. Chem. **80**(13), 5178–5185 (2008)
10. Persson, J., et al.: Acoustic microfluidic chip technology to facilitate automation of phage display selection. FEBS J. **275**(22), 5657–5666 (2008)
11. Lenshof, A., Warner, B., Laurell, T.: Acoustophoretic pretreatment of cell lysate prior to FACS analysis. In: Micro Total Analysis Systems 2010, Groningen (2010)
12. Jonsson, H., et al.: Particle separation using ultrasound can radically reduce embolic load to brain after cardiac surgery. Ann. Thorac. Surg. **78**(5), 1572–1578 (2004)
13. Grenvall, C., et al.: Harmonic microchip acoustophoresis: A route to online raw milk sample precondition in protein and lipid content quality control. Anal. Chem. **81**(15), 6195–6200 (2009)
14. Petersson, F., et al.: Free flow acoustophoresis: Microfluidic-based mode of particle and cell separation. Anal. Chem. **79**(14), 5117–5123 (2007)
15. Dykes J., et al.: Efficient removal of platelets from peripheral blood progenitor cell products using a novel microchip based acoustophoretic platform. PLoS ONE, 6, e23074, (2011)
16. Hultstrom, J., et al.: Proliferation and viability of adherent cells manipulated by standing-wave ultrasound in a microfluidic chip. Ultrasound Med. Biol. **33**(1), 145–151 (2007)

Active Carbon Nanotube-Polymer Composites

Jian Chen
Department of Mechanical and Industrial Engineering,
University of Toronto, Toronto, ON, Canada
Department of Chemistry and Biochemistry,
University of Wisconsin-Milwaukee, Milwaukee,
WI, USA

Synonyms

Smart carbon nanotube-polymer composites

Definition

Active carbon nanotube-polymer composites are carbon nanotube-polymer composites that display active material functions such as actuation and sensing.

Introduction

Carbon nanotubes (CNTs) represent a rare class of materials, which exhibit a number of outstanding properties in a single material system, such as high aspect ratio, small diameter, light weight, high mechanical strength, high electrical and thermal conductivities, and unique optical and optoelectronic properties. CNTs are recognized as the ultimate carbon fibers for high-performance, multifunctional polymer composites, where an addition of only a small amount of CNTs, if engineered appropriately, could lead to simultaneously enhanced mechanical strength and electrical conductivity [1, 2]. While most efforts in the field of CNT-polymer composites have been focused on passive material properties such as mechanical, electrical, and thermal properties, there is growing interest in harnessing active material functions such as actuation, sensing, and power generation in designed CNT-polymer composites. The synergy between CNTs and the polymer matrix has been judiciously exploited to create highly desirable active material functions. In this entry, recent progress in active CNT-polymer composites is briefly highlighted with a focus on smart materials and infrared (IR) sensors.

Shape-Memory CNT-Polymer Composites

Shape-memory polymers (SMPs) are polymeric smart materials that can memorize a temporary shape and are able to return from a temporary shape to their permanent shape upon exposure to an external stimulus such as heat [3]. Compared with shape-memory alloys and ceramics, SMPs offer a number of distinctive advantages, which include high recoverable strain (up to 400%), low density, ease of processing and the ability to tailor the recovery temperature, programmable and controllable recovery behavior, and low cost. Such advantages could enable a broad spectrum of demanding applications including deployable space structures, morphing wings, information storage, smart textiles, biomedical devices, and drug delivery.

Although SMPs show promising shape-memory effects, several major issues remain to be addressed: (1) slow recovery speed; (2) low recovery stress; (3) lack of remote control. Vaia and coworkers demonstrate that the addition of multiwalled CNTs

(MWNTs) into a thermoplastic polyurethane matrix (Morthane) could address these issues simultaneously (Fig. 1) [4]. The low recovery speed (up to several minutes) of thermal-responsive SMPs originates from their intrinsically low thermal conductivity (< 0.3 Wm^{-1} K^{-1}) [3]. Therefore, the heat transfer from an external heating source to the core of a SMP sample will take considerable amount of time. Vaia and coworkers show that non-radiative decay of IR photons absorbed by the nanotubes raises the internal temperature, melting strain-induced polymer crystallites (which act as physical cross-links that secure the deformed shape), and remotely trigger the release of the stored strain energy. The 1 wt% MWNT-Morthane nanocomposite displays a rapid shape-recovery within 5 s upon exposure to an IR light (Fig. 1b). Comparable effects occur for electrically induced actuation associated with Joule heating of the matrix when a current is passed through the conductive percolative network of the nanotubes within the resin in a 16.7 wt% MWNT-Morthane nanocomposite (Fig. 1d).

The low recovery stress of thermal-responsive SMPs originates from their intrinsically low modulus [3]. CNTs have proven to enhance the mechanical properties, particularly modulus, of various polymers. For example, 8.5 wt% MWNTs could increase the room temperature rubbery modulus of Morthane by a factor of 5 [4]. In fact, incorporation of 5 wt% of MWNTs into polyurethane SMPs results in an increase of the recovery stress by $\sim 230\%$ [4].

CNT-Polymer Composite Actuators

Polymeric actuators are polymers that deform in response to an external stimulus such as heat, light, or electric field, and they normally recover their original shapes after the stimulus is terminated [5]. Potential applications of polymeric actuators include artificial muscles, mini- and microrobots, "smart skins," pumps, and valves in microfluidic systems for drug delivery, ventricular assist devices for failing hearts, and sensors for mechanical strain, humidity, and gases. Researchers have recently demonstrated that the synergistic combination of CNTs and a suitable polymer matrix could lead to promising CNT-polymer composites with either new or significantly enhanced actuation properties.

Nematic liquid crystalline elastomers (LCEs) have fascinated scientists and engineers continually since 1981 when they were first synthesized by Finkelmann and coworkers [6]. LCEs bring together, as no other polymer, three important features: orientational order exhibited by the mesogenic units in amorphous soft materials; topological constraints via cross-links; and responsive molecular shape due to the strong coupling between the orientational order and the mechanical strain [7]. Acting together, they create many new physical phenomena that can lead to many potential applications. Nematic LCEs can dramatically and reversibly elongate or contract in response to temperature changes; however, this type of LCE actuator tends to have a slow response time due to the low thermal conductivity of LCEs and cannot be actuated remotely, thereby limiting its potential applications.

Terentjev and coworkers have studied the rich photomechanical behavior of three types of MWNT-elastomer nanocomposites irradiated with near-IR light, including polydimethylsiloxane, styrene–isoprene–styrene, and a nematic LCE with a polysiloxane backbone [8, 9]. They show that these composite materials have the novel ability to change their actuation direction, from expansive to contractive response, as greater imposed strain is applied to the sample.

Unlike MWNTs, which have featureless visible/near-IR absorption, single-walled CNTs (SWNTs) show strong and specific absorptions in the visible/near-IR region owing to bandgap transitions [10]. Recent advances in nanotube separation make it possible to obtain enriched single-chirality SWNTs with strong and narrow optical absorptions, which may ultimately offer unique opportunities for the wavelength-selective IR actuation of LCE nanocomposites. The effective use of SWNTs in polymer-composite applications strongly depends on the ability to disperse them uniformly in a polymer matrix without destroying their integrity. Pristine SWNTs are incompatible with most solvents and polymers, which lead to poor dispersion of the nanotubes in solvents and polymer matrices. Chen and coworkers have recently developed a versatile, nondamaging chemistry platform that enables them to modify specific CNT surface properties, while preserving the CNT's intrinsic properties [11]. They have discovered that rigid, conjugated macromolecules, for example, poly(p-phenyleneethynylene) (PPE), can be used to noncovalently functionalize and

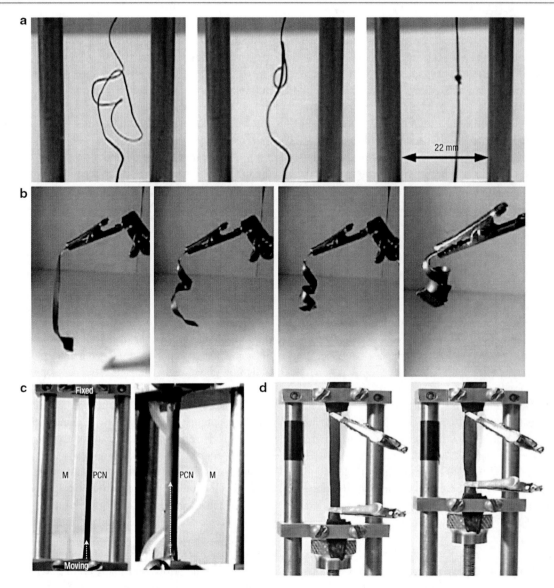

Active Carbon Nanotube-Polymer Composites, Fig. 1
(**a**) Stretched (800%) Morthane ribbon containing 1 wt% MWNTs, tied into a loose knot and heated at 55°C. The knot closes on strain recovery. (**b**) Strain recovery and curling of the 1 wt% MWNT-Morthane nanocomposite ribbon upon IR irradiation within 5 s. (**c**) Comparison of the stress recovery before (*left*) and after (*right*) remote actuation by IR irradiation.

Neat Morthane (M) bends and does not recover. In contrast, a 1 wt% MWNT-Morthane nanocomposite (PCN) contracts on exposure to IR irradiation (*arrow* indicates moving direction). (**d**) Electrically stimulated stress recovery of a 16.7 wt% MWNT-Morthane nanocomposite (Reprinted by permission from Macmillan Publishers Ltd [4])

solubilize CNTs and disperse CNTs homogeneously in polymer matrices [2, 11].

Chen and coworkers have recently reported the reversible IR actuation behavior of a new type of nematic SWNT-LCE nanocomposite using PPE-functionalized SWNTs (PPE-SWNTs) as a filler in a LCE matrix with side-on mesogenic units [12]. The SWNT-LCE nanocomposites have been prepared through a two-stage photopolymerization process coupled with a "hotdrawing" technique, which enable the fabrication of relatively thick films ($\sim$ 160–260 µm). The excellent dispersion of

Active Carbon Nanotube-Polymer Composites,
Fig. 2 IR actuation of a 0.2 wt % SWNT-LCE nanocomposite film: (**a**) before the IR was turned on; (**b**) after the IR was turned on for 4 s; (**c**) after the IR was turned on for 10 s; (**d**) the film returned to its original length with the IR beam turned off (Reprinted by permission from Wiley-VCH Verlag GmbH & Co. KGaA [12])

PPE-SWNTs in the LCE matrix has allowed them to observe a significant and reversible IR-induced strain ($\sim$ 30%) at very low SWNT loading levels (0.1–0.2 wt %, Fig. 2). The effects of various parameters have been investigated, including the degree of pre-alignment, the loading level of the SWNTs, and the curing time. In SWNT-LCE composites, semiconducting SWNTs can efficiently absorb and transform IR light into thermal energy, thereby serving as numerous nanoscale heaters being uniformly embedded in the LCE matrix. The absorbed thermal energy then induces the LCE nematic-isotropic phase transition, which leads to the shape change of the nanocomposite film. The IR strain response of SWNT-LCE nanocomposites increases with an increasing SWNT loading level and a higher degree of hot-drawing, and decreases with longer photocuring [12].

Electric field driven pure nematic LCE actuators have not been developed because the characteristic rotation energy density is too low compared with the characteristic resistance energy density from the rubbery elastic network due to the low anisotropic dielectric permittivity of the LCEs, even under a very high electric field [13]. However, adding CNTs with high anisotropic

polarizability into the LCEs, aligned along the uniaxial director of a monodomain nematic elastomer network, the CNT-LCE nanocomposites below the conductivity percolation threshold can achieve an effective dielectric anisotropy many orders of magnitude higher than the LCEs alone [13]. Therefore, introducing highly polarizable anisotropic CNTs into the LCE will lead to a significantly enhanced electromechanical effect and make electrical-field driven LCE actuators possible. Terentjev and coworkers have demonstrated, for the first time, a large electromechanical response (uniaxial stress of $\sim$1 kPa in response to a constant field of $\sim$ 1 MV/m) in nematic LCEs filled with a very low ($\sim$0.0085 wt%) concentration of MWNTs, pre-aligned along the nematic director at preparation [13]. However, the response stress in the isostrain conditions shows a linear dependence on the applied field E, instead of E^4 that would be expected from a CNT rotation-induced electromechanical response, a puzzling finding to the researchers as stated in the original paper [13].

Park and coworkers have investigated the electro-mechanical properties of a 0.05 wt% SWNT/LaRC-EAP (Langley Research Center-ElectroActive Polyimide) composite, which forms an intrinsic

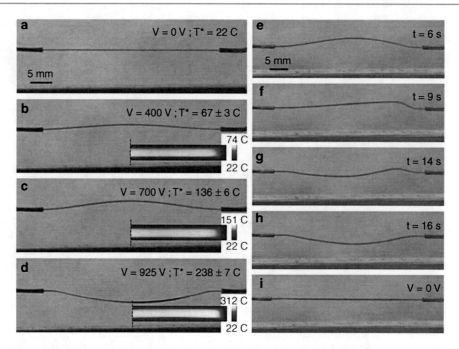

Active Carbon Nanotube-Polymer Composites, Fig. 3
(**a–d**) Deflection of a 0.5 vol% SWNT-CP2 beam (L = 4 cm; d = 140 μm; w = 1.9 mm) as a function of applied voltage and temperature. For T* < T_g (~220°C), thermal expansion induces a positive deflection in the beam due to a buckling instability (**b**, **c**). At T* ≈ T_g, mechanical softening promotes a deflection inversion as gravity becomes the dominant force acting on the beam (**d**). Sample temperature was determined at each voltage using the thermal images shown inset; (**e–i**) Deflection of a 0.5 vol% SWNT-CP2 beam as a function of time at V_{th} (voltage where T* is first observed to exceed T_g). The sample initially buckles as a result of thermal expansion (**e**); however, as the sample temperature continues to increase, the sample kinks (**f**, **g**), and then eventually sags under its own weight at T_g (**h**). Once the voltage is removed, the beam reversibly actuates back to its original form (**i**) (Reprinted by permission from Wiley-VCH Verlag GmbH & Co. KGaA [15])

unimorph during the fabrication process to actuate without the need for additional inactive layers [14]. The 0.05 wt% SWNT/LaRC-EAP exhibits a large strain (~ 2.6%) at a low driving voltage (<1 MV/m) while possessing excellent mechanical and thermal properties. The out-of-plane strain is proportional to the square of the electric field E, which suggests that the actuation mechanism of the SWNT/LaRC-EAP composite is primarily due to electrostriction [14].

The addition of CNTs also enables electrically stimulated thermo-mechanical actuation via electrical Joule heating when the loading level of CNTs is above its electrical conductivity percolation threshold. Vaia and coworkers have studied the electrothermal properties of SWNT-polyimide (CP2) nanocomposites (Fig. 3) [15]. They show that the reversible softening associated with temperature transitions through T_g can provide substantial strain increases ($\Delta S \approx 4\%$) over relatively small changes in applied electric field ($\Delta E \approx 0.01$ MV/m).

CNT-Polymer Composite IR Sensors

Organic electronic materials offer ease of materials processing and integration, low cost, physical flexibility, and large device area as compared to traditional inorganic semiconductors. Optoelectronic materials that are responsive at wavelengths in the near-infrared (NIR) region (e.g., 800–2,000 nm) are highly desirable for various demanding applications such as telecommunication, thermal imaging, remote sensing, thermal photovoltaics and solar cells [16]. SWNTs have strong absorptions in the NIR region owing to the first optical transition (S_{11}) of semiconducting nanotubes [10]. The inverse diameter dependence of S_{11} optical transition energy enables the wavelength-tuning of SWNT's NIR absorptions, which, coupled with the rapid progress in nanotube separation, will ultimately allow the development of SWNT IR sensors tailored to specific regions of the NIR spectrum. There are a number of reports on the IR photoelectrical property

Active Carbon Nanotube-Polymer Composites, Fig. 4 (a) Schematic experimental setup. IR light covers all sample area. (b) Relative conductivity (σ/σ_{dark}) response of SWNT-PC nanocomposites to the on/off IR illumination (power intensity: 7 mW/mm^2)

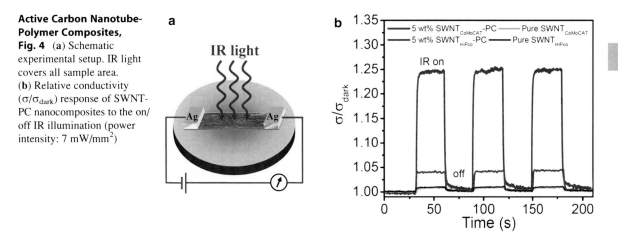

of SWNT films. Levitsky and coworker have demonstrated that the arc-produced SWNT (SWNT$_{arc}$) film is capable of generating a very weak photocurrent upon continuous-wave IR illumination (12 mW/mm^2) in the air at room temperature, and the current increase upon IR illumination is only about 0.2% [17]. In addition, they have observed a dark current drift due to oxygen adsorption and a relatively slow rise/decay ($\sim$ 4–5 s relaxation time) of the photocurrent in response to the on/off illumination. Haddon and coworkers have reported that the IR photoresponse in the electrical conductivity of a SWNT$_{arc}$ film is dramatically enhanced when the nanotube film is suspended in vacuum at low temperature (e.g., 50 K) [18].

Chen and coworkers have discovered that the IR photoresponse in the electrical conductivity of SWNTs is dramatically enhanced by embedding SWNTs in an insulating polymer matrix such as polycarbonate (PC) in the air at room temperature (Fig. 4) [19, 20]. In contrast to the gradual photoresponse and weak conductivity change (1.10%) observed in a HiPco$^{\circledR}$-produced SWNT (SWNT$_{HiPco}$) film in the air at room temperature, the 5 wt% SWNT$_{HiPco}$-PC nanocomposite under the same IR illumination (power intensity: 7 mW/mm^2) shows a sharp photoresponse and strong conductivity change (4.26%) [19]. The dark current drift owing to oxygen adsorption is also minimized by embedding nanotubes in the polymer matrix. They have shown that semiconducting SWNTs are critical to the IR photoresponse of nanotube materials [19, 20]. They have also demonstrated that both SWNT types and nanotube-matrix polymer-nanotube junctions have profound impact on

the IR photoelectrical property of SWNT-polymer composites [20]. Composite IR sensors based on CoMoCAT$^{\circledR}$-produced SWNTs (SWNTs$_{CoMoCAT}$) significantly outperform those based on SWNTs$_{HiPco}$. The 5 wt% SWNT$_{CoMoCAT}$-PC nanocomposite demonstrates a very strong conductivity change of 23.45% upon the IR illumination (7 mW/mm^2) in the air at room temperature [20], which is 5.5 times of that (4.26%) observed in the 5 wt% SWNT$_{HiPco}$-PC composite film [19], nearly 27 times of that (0.88%) observed in the pure SWNT$_{CoMoCAT}$ film [20], 21 times of that (1.10%) observed in the pure SWNT$_{HiPco}$ film [19], and 117 times of that (0.2%) observed in the pure SWNT$_{arc}$ film in the air at room temperature (12 mW/mm^2) [17]. In addition, the 5 wt% SWNT$_{CoMoCAT}$-PC composite film shows a detectable IR photoresponse at a light intensity as low as 23.4 μW/mm^2 [20].

Acknowledgments JC is greatly indebted to his students and collaborators and his colleagues in this field, whose names are cited in the references. Financial support from the National Science Foundation (DMI-06200338, CMMI-0625245, and CMMI-0856162), UWM start-up fund, UWM Research Growth Initiative award, and the Lynde and Harry Bradley Foundation is gratefully acknowledged.

Cross-References

▶ Carbon Nanotubes
▶ Functionalization of Carbon Nanotubes
▶ Insect Infrared Sensors

- ▶ Light-Element Nanotubes and Related Structures
- ▶ Nanorobotics
- ▶ Nanostructured Materials for Sensing
- ▶ Organic Actuators
- ▶ Synthesis of Carbon Nanotubes
- ▶ Thermal Actuators

References

1. Winey, K.I., Kashiwagi, T., Mu, M.: Improving electrical conductivity and thermal properties of polymers by the addition of carbon nanotubes as fillers. MRS Bull. **32**, 348–353 (2007)
2. Chen, J., Ramasubramaniam, R., Xue, C., Liu, H.Y.: A versatile, molecular engineering approach to simultaneously enhanced, multifunctional carbon nanotube-polymer composites. Adv. Funct. Mater. **16**, 114–119 (2006)
3. Leng, J., Lu, H., Liu, Y., Huang, W.M., Du, S.: Shape-memory polymers—a class of novel smart materials. MRS Bull. **34**, 848–855 (2009)
4. Koerner, H., Price, G., Pearce, N.A., Alexander, M., Vaia, R.A.: Remotely actuated polymer nanocomposites—stress-recovery of carbon-nanotube-filled thermoplastic elastomers. Nat. Mater. **3**, 115–120 (2004)
5. Madden, J.D., Vandesteeg, N.A., Anquetil, P.A., Madden, P.G.A., Takshi, A., Pytel, R.Z., Lafontaine, S.R., Wieringa, P.A., Hunter, I.W.: Artificial muscle technology: physical principles and naval prospects. IEEE J. Ocean. Eng. **29**, 706–728 (2004)
6. Finkelmann, H., Kock, H.-J., Rehage, G.: Investigation on liquid crystalline siloxanes: 3. liquid crystalline elastomers. Makromol. Chem. Rapid Commun. **2**, 317–322 (1981)
7. Warner, M., Terentjev, E.M.: Liquid crystal elastomers. Oxford University Press, Oxford (2003)
8. Ahir, S.V., Terentjev, E.M.: Photomechanical actuation in polymer-nanotube composites. Nat. Mater. **4**, 491–495 (2005)
9. Ahir, S.V., Squires, A.M., Tajbakhsh, A.R., Terentjev, E.M.: Infrared actuation in aligned polymer-nanotube composites. Phys. Rev. B **73**, 085420 (2006)
10. Hamon, M.A., Itkis, M.E., Niyogi, S., Alvaraez, T., Kuper, C., Menon, M., Haddon, R.C.: Effect of rehybridization on the electronic structure of single-walled carbon nanotubes. J. Am. Chem. Soc. **123**, 11292–11293 (2001)
11. Chen, J., Liu, H., Weimer, W.A., Halls, M.D., Waldeck, D.H., Walker, G.C.: Noncovalent engineering of carbon nanotube surfaces by rigid, functional conjugated polymers. J. Am. Chem. Soc. **124**, 9034–9035 (2002)
12. Yang, L., Setyowati, K., Li, A., Gong, S., Chen, J.: Reversible infrared actuation of carbon nanotube-liquid crystalline elastomer nanocomposites. Adv. Mater. **20**, 2271–2275 (2008)
13. Courty, S., Mine, J., Tajbakhsh, A.R., Terentjev, E.M.: Nematic elastomers with aligned carbon nanotubes: new electromechanical actuators. Europhys. Lett. **64**, 654–660 (2003)
14. Park, C., Kang, J.H., Harrison, J.S., Costen, R.C., Lowther, S.E.: Actuating single wall carbon nanotube-polymer composites: intrinsic unimorphs. Adv. Mater. **20**, 2074–2079 (2008)
15. Sellinger, A.T., Wang, D.H., Tan, L.-S., Vaia, R.A.: Electrothermal polymer nanocomposite actuators. Adv. Mater. **22**, 3430–3435 (2010)
16. McDonald, S.A., Konstantatos, G., Zhang, S., Cyr, P.W., Klem, E.J.D., Levina, L., Sargent, E.H.: Solution-processed PbS quantum dot infrared photodetectors and photovoltaics. Nat. Mater. **4**, 138–142 (2005)
17. Levitsky, I.A., Euler, W.B.: Photoconductivity of single-wall carbon nanotubes under continuous-wave near-infrared illumination. Appl. Phys. Lett. **83**, 1857–1859 (2010)
18. Itkis, M.E., Borondics, F., Yu, A., Haddon, R.C.: Bolometric infrared photoresponse of suspended single-walled carbon nanotube films. Science **312**, 413–416 (2006)
19. Pradhan, B., Setyowati, K., Liu, H., Waldeck, D.H., Chen, J.: Carbon nanotube-polymer nanocomposite infrared sensor. Nano Lett. **8**, 1142–1146 (2008)
20. Pradhan, B., Kohlmeyer, R.R., Setyowati, K., Owen, H.A., Chen, J.: Advanced carbon nanotube/polymer composite infrared sensors. Carbon **47**, 1686–1692 (2009)

Active Nanoantenna System

- ▶ Active Plasmonic Devices

Active Plasmonic Devices

Jean Christophe Lacroix[1], Pascal Martin[2] and Hyacinthe Randriamahazaka[2]
[1]Interfaces, Traitements, Organisation et Dynamique des Systemes Universite Paris 7-Denis Diderot, Paris Cedex 13, France
[2]University Paris 7-Denis Diderot, ITODYS, Nanoelectrochemistry Group, UMR CNRS 7086, Batiment Lavoisier, Paris cedex 13, France

Synonyms

Active nanoantenna system

Definition

Active plasmonic devices combine plasmonic systems and a physical or chemical control input.

Main Text

Published papers using the term "active plasmonics" and citations of those papers are shown in Fig. 1a, b. The first papers explicitly devoted to this new field were published in 2004 and 2005, even though many previous publications had already pointed out the need for such devices. The growth of these two curves clearly demonstrates that breakthroughs have been achieved in the last 5 years and that many new devices will briefly be demonstrated.

Active plasmonic systems generally use two components: one of them supporting *Localized Surface Plasmons* (LSP) or *Surface Plasmon Polaritons* (SPPs) [1]. One needs to distinguish two types of approaches. In the first approach, the plasmonic component acts as an input in order to control or tune the properties of the second component. Such active plasmonic systems are widely studied but do not belong to the field of active plasmonic devices as described in this Essay. Surface Enhanced Raman Spectroscopy (SERS) is an example of such approaches using an extreme light concentration in order to develop a highly sensitive analysis that can now even detect single molecules [2]. In the second approach, the second component is used in order to control, switch, or tune the properties of the plasmonic component, and the plasmonic response to the external stimulus is the output of the device. This review will be devoted only to this second approach, even though some examples of plasmonic systems using the first approach will be given when they are of interest in the design of active devices.

Surface Plasmons: Qualitative Description

In order to understand the evolution depicted in Fig. 1 and the tremendous research and technological opportunities of this new field, one has first to have in mind why in the last decade passive plasmonic devices have been studied so much and what kinds are of interest.

Briefly, they can be divided in two classes.

- Those using SPPs, i.e., electromagnetic waves coupled to the coherent oscillation of free-charge carriers at the interface between a dielectric and a conductor (Fig. 2). SPPs propagate along subwavelength nanostructured metals or semiconductors and are often referred to as low-dimensional waves due to the strong confinement of the electromagnetic field on the surface [3].

- Those using LSPs, some metal NanoParticles (NPs) exhibit coherent but confined oscillations of the quasi-free electrons in the conduction bands (Fig. 3). When the characteristic frequency of these oscillations coincides with that of the light excitation, the response of the metallic NPs becomes resonant and strong absorption in the visible and near-infrared range occurs (so-called localized surface plasmon resonance, LSPR). The frequency of LSPR strongly depends on the size, shape, spacing of the particles, and on the dielectric constants of the substrate and surrounding medium [4].

Because of surface plasmons capability to enhance electric fields in the very close vicinity of NP structures or to localize and guide light in metallic structures, plasmonics allows the manipulation of the flow of light and its interaction with matter at the nanoscale. In this sense, NPs work in a similar way to that of antennas in radio and telecommunication systems, but at optical frequencies, i.e., at frequencies corresponding to typical electronic excitations in matter. Plasmonics also offer an opportunity to merge photonics and electronics at nanoscale dimensions to obtain unusual optical properties and opportunities for unprecedented levels of synergy between optical and electronic functions. Figure 4a depicts the main passive devices already demonstrated, including light sources, filters, waveguides, lenses, plasmonic antennas, and polarizers [5].

Nevertheless, in order to become a high valuable technology, plasmonics needs switches, modulators, and routers. Indeed, such active plasmonic systems are needed to achieve functional Surface Plasmons (SP) circuits that will, for instance, first convert light into SPs using nanoantennas; SPP would then propagate along SP waveguides and be processed by logic elements before being converted back to light using nanoscopic sources. Active plasmonic devices can thus be the basic building blocks for an alternative technology capable of writing, reading, storing, and processing information at the nanoscale. Figure 4b show the various types of physical and chemical external stimuli already demonstrated to control passive plasmonic devices. Currently, their number is expanding.

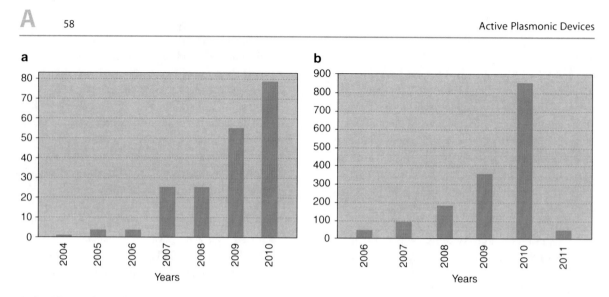

Active Plasmonic Devices, Fig. 1   Histogram of (**a**) published papers and (**b**) number of citations of papers including terms "active plasmonics"

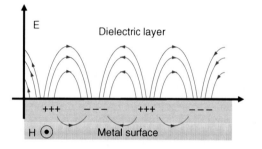

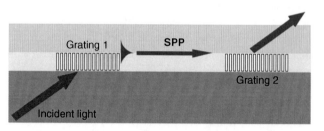

Active Plasmonic Devices, Fig. 2   SPs propagating along the interface between metal and dielectric materials. These waves have a combined electromagnetic and surface charge character.

Schema of a plasmonic signal, coupled to and from the waveguide by gratings on a metallic layer

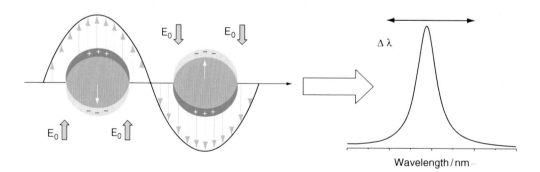

Active Plasmonic Devices, Fig. 3   Schema of the localized oscillations of electrons exposed to an electric field. The electrons are shifted, and the polarization charges on opposite surface elements apply a restoring force on the electrons. The

extinction spectra of NPs shows the localized surface plasmons resonance peak which shift with the shape, size, and spacing of particles

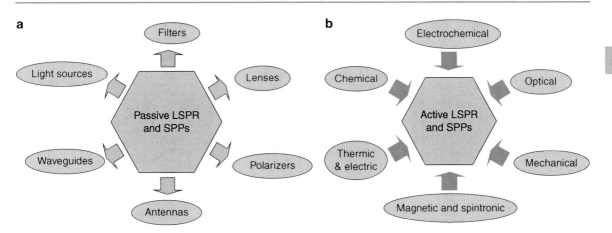

Active Plasmonic Devices, Fig. 4 (**a**) Major plasmonic devices. (**b**) Overview of major routes for the development of active plasmonics

Many recent reviews [1] have been devoted to passive plasmonics, and this contribution will only describe active plasmonic devices, and few aspects of theoretical background will be used to define and understand strategies to reversibly switch or modulate plasmons.

Controlling Localized Surface Plasmons

The frequency of LSPR depends mainly on the size, shape, and spacing of the particle and on the dielectric constants of the substrate and surrounding medium. Adding to a LSPR-based plasmonic system, an external stimulus that will tune the size, the shape, the spacing of the particle or the dielectric functions of the surrounding medium can therefore lead to active plasmonic devices. Some strategies will easily allow reversibility; others based on tuning the size and shape of the NPs will be less successful for reversible control. Of particular interest is the recent discovery of mixed exciton-plasmon states in metal-molecule complexes, which makes them promising candidates for active molecular plasmonic devices.

Thermally Driven Active Plasmonic Devices
The LSPR is not tuned by changing the temperature of a substrate. A thermoresponsive layer surrounding NPs could be use to design *thermally driven active plasmonic devices*. Most examples use thermoresponsive polymers, inorganic layers, and liquid crystals in which the thermal effect induces

a variation of the spacing between the NPs or of the dielectric function of the surrounding medium. For all these systems, a temperature-driven phase transition is observed and induces a plasmon resonance shift.

Several systems based on the thermally induced and reversible collapse of poly(N-isopropylacrylamide), pNIPAM have been reported (Fig. 5) [6]. Inorganic composites were also used. For example, Metal–VO$_2$ nanocomposites have shown LSPR shifts of about 50 nm for a temperature variation of 95°C.

Other thermoresponsive active plasmonic devices have been developed using liquid crystals (LCs). LCs are unique materials that show such characteristic properties as self-organization, fluidity with long-range order, cooperative motion and alignment change with various external stimuli (temperature, electric fields, etc.) at surfaces and interfaces. Thereby, the modulation of alignment of LCs gives rise to a change in optical properties, which forms the basis of LC displays (LCDs) [7]. The LSPR in anisotropic media such as liquid crystals are strongly dependent on the local ordering of the media, and that induced orientational transitions of the liquid crystal may be used to tune the optical properties of metallic NPs.

Mechanical Input
An alternative approach is to fabricate plasmonic structures on mechanically tunable substrates. In most of the cases, small metallic nanospheres have been deposited on stretchable elastomeric films like polydimethylsiloxane (PDMS). As the film is stretched

Active Plasmonic Devices, Fig. 5 TEM image of a single hybrid particle based on Gold nanorod-loaded poly-(NIPAM-co-AAA) microgel with high surface coverage. UV-Vis spectra of hybrid sample at 15 °C (swollen state), at 42 °C, and at 60 °C (collapsed state) (Reprinted with permission from Karg et al. [6])

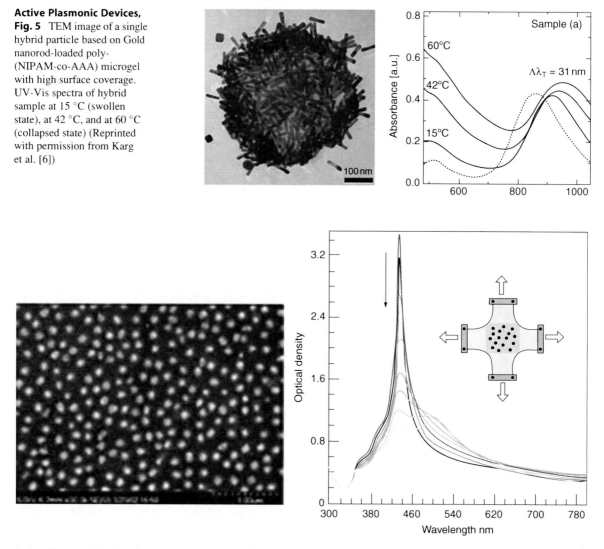

Active Plasmonic Devices, Fig. 6 SEM image of the 2D array of Ag nanoparticles into PDMS. Extinction spectra of the array as a function of the stretching force. *Red curve* represents the film stretched 50% from its original, nonstretched state (*black curve*) (Reprinted with permission from Journal of American Chemical Society, Malynych and Chumanov [8])

to different degrees, the intensity of plasmon resonance rapidly decreases, and a new band appears at higher wavelength (Fig. 6).

Chemically Driven Plasmonic Devices

LSPR modulation using molecular recognition is a mature technology that is widely used to design sensors and biosensors. The LSPR of isolated NP shows limited sensitivity of absorption spectra (for small colloidal 5–20 nm nanoparticles), to molecular or biomolecular recognition events due to the small changes of the refractive index of the surrounding

medium induced by the molecular recognition events. Additionally, the LSPR wavelength and the UV-Vis absorption of the molecules do not match in most cases. However, LSPRs are highly sensitive to interparticle spacing, originated from the near-field plasmon coupling of nanoparticles. The strength of the effect falls rapidly with distance; at separations above 2.5 times the particle size, the nanoparticles behave essentially as isolated clusters. In particular, particle aggregation results in a pronounced red shift of the LSPR peaks, and most of the applications for sensors are based on this distance variation, resulting

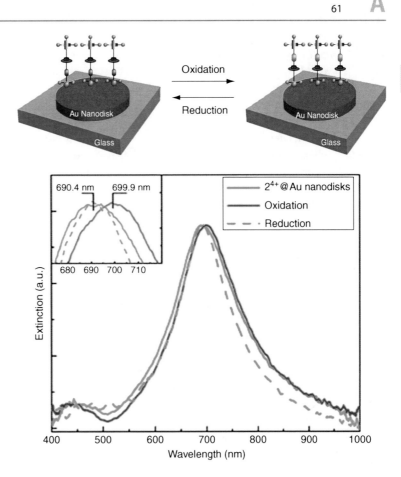

Active Plasmonic Devices,
Fig. 7 Schematic of working principle of the molecular-machine-based active plasmonics. Extinction spectra recorded from Au nanodisks in air during the redox process (Reprinted with permission from Zheng et al. [10])

in the change of their plasmon coupling. Recent reviews [9] have described the large body of research on plasmon-based sensors. Indeed, even though such devices can be considered as active plasmonic devices, their purpose is not to control LSPR but to use LSPR as a transduction of the molecular events occurring at the NP surface. This part will only focus on some recent systems illustrating the various chemical stimuli which tuned LSPR.

As first examples, the pH of a solution could act as a chemical input. In a typical configuration, metal nanoparticles are immobilized as a monolayer on a glass substrate, while a polymer material is either deposited on the top of the nanoparticles monolayer or between nanoparticles. PH-induced conformational changes will alter the interparticle distance and, hence, the strength of plasmon coupling. The variation of the nanoparticles–island distance was limited to 1–20 nm, which corresponds to the optimal distance between metal nanoparticles for electromagnetic coupling. Reversible operation was achieved by returning the pH and optical properties of the system to the initial state.

Electroactive molecular machines with grafted NPs have also been used but, in this case, the input was a chemical chosen to reduce or oxidize the electroactive machine. Zheng et al. recently showed that a gold nanodisc array, coated with bistable, redox-controllable [2]rotaxane molecules, when exposed to chemical oxidants and reductants, undergoes reversible switching of its plasmonic properties (Fig. 7) [10]. These observations suggest that the nanoscale movements within surface-bound "molecular machines" can be used as the active components in plasmonic devices. One has, nevertheless, to note that the observed LSPR shifts were rather small (10 nm) due to the small change of monolayer's effective refractive index and the very thin molecular film.

In these previous examples, the molecules used did not absorb light near the nanoantenna extinction. New phenomena occur when LSPR and molecular absorption coincide. When the peak position of the NPs LSPR did not overlap with the molecular resonance, a small shift (20 nm for example) was observed upon monolayer deposition, whereas a higher shift is observed when the LSPR peak partially overlaps with the molecular resonance [11]. The oscillatory dependence of the peak shift on wavelength was shown to track with the wavelength dependence of the real part of the refractive index of the monolayer, as determined by a Kramers–Koenig transformation of the molecular resonance absorption spectrum.

The influence of surrounding adsorbates is important even for monolayers as explained above or at low-coverage adsorption. In such cases, it is difficult to estimate the dielectric constant and an atomistic description is needed [12]. Thus, the influence of molecular adsorbates on the absorption spectra of spherical NPs was investigated using arbitrary coupling to the surface and changing the molecular excitation characteristics through its HOMO–LUMO gap. For the weak adsorption regime, the adsorbates influenced the resonance width; the resonance energy, however, is increased by less than 10 nm. For strong coupling to the surface, and when the excitation of the adsorbates becomes lower in energy than that of the plasmon, it was shown that the peak splits into two excitations. The higher-energy broad peak was shown to be located at the particle surface and was, therefore, attributed to plasmonic excitation. The lower-energy excitation was located over the adsorbates and was attributed to an in-phase oscillation of molecular dipoles on opposite sites of the NP. This new excitation is similar to a completely symmetric molecular exciton, in which molecular dipole oscillations couple in-phase. However, the electrostatic coupling between molecular dipoles occurs through the metallic particle, allowing very significant interaction at large distances.

Strong coupling of chromophore and plasmon resonances has been exploited for molecular sensors. Zhao et al. showed an extreme LSPR shift for isolated NPs due to small changes in the resonances of a chromophore that are induced by binding an additional analyte [13]. In this case, the binding of substrate or inhibitor molecules to a cytochrome caused its absorption band to shift to shorter or longer wavelengths, respectively, and to be *off-resonance* with the plasmon, which was at the origin of a huge LSPR shift for isolated NP (Fig. 8).

The creation of such responsive plasmonic devices by coupling a chromophore to the plasmon through absorption overlap with the wavelength of the LSPR is an important new research trend for applications in molecular sensing and active plasmonics. Excitons and plasmons are the two main excited states of matter. Understanding their interactions is likely to become a very important research field. It yields to mixed exciton-plasmon states, so-called hybridized states. Indeed, when excitons of molecules, such as in J-aggregates, resonate with the plasmon modes of metal NPs, new peaks appear in the extinction spectra. These peaks are governed by the strength of the coupling between excitons and plasmons, and strategies based on tuning this strength, through some physical or chemical input, have recently been reported and considerably widen the opportunities to reach active molecular plasmonic devices. Change of the angle of incident light or change of the geometry of the NPs was used to control the coupling strength. For strong coupling, the LSPR splits into two bands upon adsorption of the J-aggregates. Reversible tuning of the coupling strength was not obtained since tuning was generated using the flexible geometry or changing the incident angle. Based on these recent results, the reversible tuning of exciton–plasmon coupling using other stimuli will be an easy and flourishing means to develop active plasmonic devices.

Electrochemically Driven Active Plasmonic Devices

Nanoelectrochemistry is rapidly expanding its frontiers toward many new aspects including redox-gated molecular junctions and photovoltaic integrated systems. In this context, electrochemical switching is a useful tool to reversibly control the properties of metallic NPs.

Electrochemistry on nanostructured electrodes supporting LSPR, used as the working electrode, has been widely investigated. Sweeping or changing the potential of such an electrode in an electrolyte usually yields very small LSPR wavelength shifts. Ung et al. were the first to monitor the position of the surface plasmon band of silver colloids as a function of the applied potential [14]. The SP shift amounted to 10 nm/V in the potential regime dominated by double-layer charging. With various electro-inactive organic shells a higher blue shift could be obtained. The LSPR shift was attributed to the increased of metal plasma frequency due to the excess stored electrons.

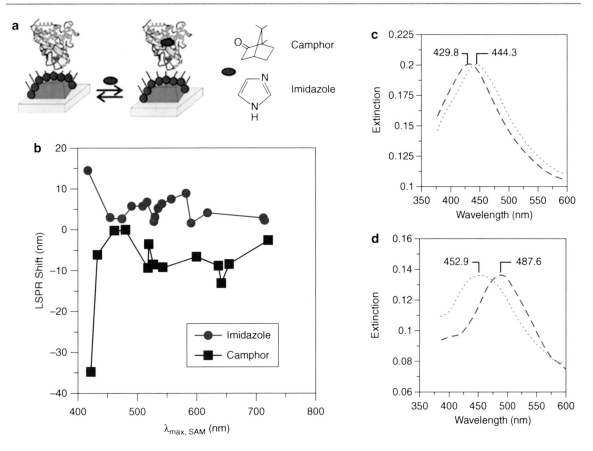

Active Plasmonic Devices, Fig. 8 Schematic representation of small molecule binding to receptors on functionalized Ag nanoparticles. (*below*) The wavelength-dependent LSPR shift induced by inhibitor (*red line with dots*) and substrate (*blue line with squares*). LSPR spectra of NPs with cytochrome before (*dashed line*) and after (*dotted line*) substrate (*up*) or inhibitor (*down*) binding (Reprinted with permission from Zhao et al. [13])

Changing the spacing of the particle through electrochemistry has not yet been widely used, probably because reversibility is not easy to obtain. Electrochemical etching induces LSPR wavelength modulation but is unlikely to be used for reversible active plasmonic devices. On the contrary, reversible control through electrochemistry of the surrounding dielectric medium is easy to obtain. Deposition on the NPs of an electroactive thin film or of an electroactive monolayer makes it possible to trigger LSPR through the oxidoreduction properties of the deposited layer and thus through the applied potential to the electrode. For example, Ag NPs, arranged in closely spaced two-dimensional arrays, were coated by a thin film of tungsten oxide [15]. WO_3 is a well-known electro-active electrochromic material that undergoes a reversible redox process to form reduced tungsten oxide. Reduced and oxidized forms of WO_3 have different electrical conductivities and thus different dielectric functions. By switching the externally applied potential, the optical extinction signal was reversibly modulated and a blue shift was observed.

Following that, Lacroix et al. reported the use of electroactive conductive polymer (CP) thin layers deposited on gold gratings [16]. In a first study, they used polyaniline (PANI) because such films exhibit ultrafast switching (around 1 μs, RC limited) between their reduced non-conductive state and their oxidized conductive state. Blue shift was observed from $\lambda_{LSP} = 633$–571 nm, below that of λ_{LSP} observed in air (593 nm), accompanied by a strong damping of LSPR. This reversible effect was observed repeatedly

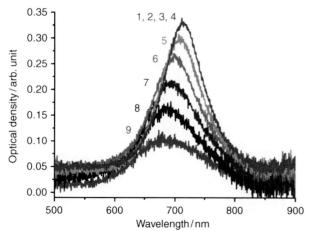 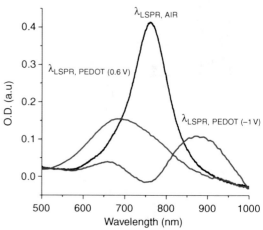

Active Plasmonic Devices, Fig. 9 (*left*) Extinction spectra of prolate gold grating under x-polarized light and polarization from −500 to 500 mV. (Reprinted with permission from Leroux et al. [17]). (*Right*) Extinction spectra of prolate gold particle grating (X polarized light) in air (*black curve*); overcoated with a PEDOT/SDS film in its oxidized state (*blue curve*); overcoated with a PEDOT/SDS film in its reduced state (*red curve*) (Reprinted with permission from Stockhausen et al. [18])

for more than 1 h. The interpretation lies in the switching of the dielectric constant ε of PANI. When PANI is in its reduced state, ε is real, whereas in its oxidized state, ε is complex ($\varepsilon = \varepsilon_{Re} + i\ \varepsilon_{Im}$). Moreover, oxidized metallic PANI samples follow a Drude-like behavior [1] with a plasma frequency $\omega_p = 2$ eV yielding a simplified expression of ε_{Re} given by $\varepsilon_{Re} = \varepsilon_{\infty} - (\omega_p/\omega)$ [2]. The imaginary part of the dielectric constant can be expressed as $\varepsilon_{Im} = (2\sigma\lambda/c)$, where λ is the vacuum wavelength of incident light, σ the wavelength-dependent conductivity, and c the speed of light. The electrochemical switching of PANI thus generates a large change in both the real and the imaginary part of the dielectric function of the material around the nanoparticles. The large blue shift can be attributed to a decrease in the real part of the dielectric constant at wavelength around 632.8 nm, whereas switching in the imaginary part of the dielectric constant (from zero to a finite value) induces quenching and damping of the LSP. In a second step, they showed that the LSPR in such a PANI/gold NP hybrid material exhibits a progressive damping as a function of the applied potential, as evidenced in Fig. 9 (left). These results clearly show that the observed modulations of the LSPR are closely related to the charge-carrier density injected into the conductive polymer through electrochemical doping.

More recently, poly(3,4-ethylenedioxythiophene) (PEDOT) was used to shift the LSPR of gold NP arrays [18]. Figure 9 shows the giant red shift from 685 nm to

877 nm under polarization at −1.0 V. The reversible shift between reduced and oxidized states is 192 nm. Such a giant plasmon resonance shift, covering almost half of the visible spectrum, is, to the best of our knowledge, unique.

Bringing together two systems used in two emerging scientific areas with important applications, i.e., plastic electronics and plasmonics, can thus yield new active plasmonic devices. In this context, it is important to evaluate the potentialities and the flexibility of such devices. CPs can be tailored at will in order to change their switching potential or the variation of their dielectric constant upon switching which will give electrochemically driven plasmonic devices with tunable optical properties. Switching speed is another key point for photonics application and in the present case, switching times of the LSPR in the millisecond range have been observed. Electrode size reduction allows switching times in the microsecond range, and it can thus be predicted that switching frequencies in the MHz range could be easily obtained. Moreover, the external stimuli inducing CP switch is not restricted to the electrochemical input. CPs can also be doped using light, and all-optical solid-state devices using CPs for controlling plasmonic systems are likely to be demonstrated in the near future. Finally, other plasmonic systems, such as waveguides or nanoscopic light sources, can also be combined with plastic electronic components providing a wide range of active plasmonic components for solid-state nanophotonics.

Electric Field-Driven Active Iasmonic Devices

Another input for controlling LSPR is the application of an external electric field. The applied voltage can then be used to exert direct control over an electrooptically active dielectric medium, changing its refractive index and thus the SPR modes at the metal/dielectric interface. Using an electric field as an input could lead to faster switches than thermally, mechanically, chemically, or electrochemically driven systems, and is especially significant because both plasmonic and electric signals can be guided in the same metallic circuitry.

Coupling LSPR devices and organic field-effect transistors has not yet been reported but is likely to be developed in the next couple of years. On the contrary, many studies have used liquid crystals (LC) or ferroelectric films. In all these cases, shifts have been observed, and the spectral position of the surface plasmon resonances depends on the applied voltage. The variations of the local density of the LC molecules and/or changes in the refractive index in the active films provide a sufficient effect to modulate the LSPR of gold nanostructures. The switching time for the system was in the order of ms.

Optically Driven Active Plasmonic Devices

Light-driven plasmonic devices have several potential advantages: they could operate faster, light can be used for inducing (writing) as well as detecting (reading) the behavior of plasmonic devices; and they can be integrated in future all-optical plasmonic circuits. Efficient methods to achieve optically driven plasmonic devices have been developed and are based on tuning the spacing between particles or the dielectric constant of the surrounding medium.

In this context, it is worth mentioning that there are a number of reports on photosensitive Au NPs with surface coating with small molecules bearing a photochromic moiety. In the case of colloidal metallic NPs, azobenzene trans-cis photoisomerization was used to induce NP aggregation, resulting in the change of the interparticle distance and a shift of the surface plasmon resonance. Sidhaye et al. reported the red shift (close assembly of particles) and the blue shift (loose assembly of the gold nanoparticles) of the surface plasmon peak position ($\Delta\lambda = 80$ nm) induced by the reversible manipulation of interparticle spacing in gold nanoparticle networks connected with azobenzene derivatives [19]. Isomerization yields and switching

speeds in such devices remain clear limitations, even though lowering the surface coverage of azobenzene provided more free space for isomerization to take place and enhanced the LSPR shift. Reversibility was clearly demonstrated.

Finally, several studies described the use of specific photoresponsive liquid crystals due to their large and controllable birefringence [7]. The surrounding media was homogeneous mixtures of a liquid crystal and a photoswitchable molecule such as azobenzene. Upon photo-irradiation, the LCs undergo a phase transition and a change in their refractive index, leading to a LSPR shift of few of tens nm (for Gold arrays). Reversibility was demonstrated, and a response time in the second range was observed.

Magnetic and Magneto-optic Driven Active Plasmonic Devices

The study of the excitation of SPs on ferromagnetic materials has recently attracted much interest not only from a fundamental point of view but also because of the possibility of realizing spin-plasmonic devices, bringing together the rapidly developing plasmon technology and the field of spintronics. Active spin-plasmonic devices have been developed for controlling SPP and will be briefly reviewed in the next part of this essay. To the best of our knowledge, the only study, combining LSPR and magneto-active materials, have been recently reported by Du et al., using spherical and elliptical nanocylinders [20]. However, they focused on how plasmons can enhance the magneto-optical properties of a thin magnetic metal layer (Co, Ni or Fe). Active control of the LSPR using magnetic layers remains to be demonstrated.

Controlling Surface Plasmon Polaritons

Surface plasmon polaritons (SPPs) are electromagnetic waves coupled to the coherent oscillation of free charge carriers at the interface between a dielectric and a conductor. The propagation length of SPPs depends on the metal, even though resistive heating losses in metals can severely limit the performance of devices. Several passive plasmonic devices, such as waveguides, and nanoscopic light sources have been demonstrated. In the latter system, the light could be transmitted through holes with lateral dimensions smaller than half the incident wavelength, while classical theory predicts orders-of-magnitude less transmission. (extraordinary or enhanced optical transmission [EOT]) [3]. For a better efficiency, the metal should be corrugated with a periodic structure such as an array of holes or a single hole surrounded by grooves.

In order to fully exploit the possibilities of such SPP-based plasmonic devices, it is necessary to achieve active control. Since SPP fields are strongly confined, they are again very sensitive to refractive index changes within a few tens of nanometers of the surface. As a consequence, many strategies used to control SPPs are similar to those used for LSPR tuning.

Most systems are solid-state devices avoiding the use of liquid, and when molecules are used, they are deposited as thin films and not used in solution which excludes, to some extent, PH or redox input. A notable exception lies in Surface Plasmon Resonance (SPR) spectroscopy which is very sensitive and widely used to design sensors and biosensors [9]. Readers are directed toward other reviews for many examples of such system. One of the considered configuration replaces the "dielectric" layer in the structure with an active thin film which can be artificially doped by suitable inputs and allows the dynamic modulation of the SPP wavevector [5]. One of the first examples was demonstrated by Krasavin and Zheludev in [21] and will be described in the optical input section.

Inorganic semiconducting materials supporting SPP are also used, whereas metal NP remains the main basic blocks for LSPR devices. Indeed, as the carrier densities in semiconductors are much lower than those in metals, the plasma frequency is much smaller, being typically at mid- or far-infrared frequencies. The main advantage of semiconductors is that their carrier density and mobility, and consequently the SPPs, can be easily controlled by many external stimuli (thermal, electrical, and optical excitation).

Thermally Driven Active Plasmonic Devices

Nikolajsen's group fabricated and characterized thermo-optic Mach–Zender interferometric modulators (MZIMs) and directional-coupler switches (DCSs) at telecom wavelengths [22]. MZIMs and DCSs operation utilize the SPP waveguiding along thin gold stripes embedded in polymer and heated by electrical signal currents. The operation of a thermo-optic MZIM is based on changing the SPP propagation constant in a heated arm resulting in the phase difference of two SPP modes that interfere in the output Y-junction. By sandwiched, the stripes between a thick layer of benzocyclobutene, they observed the modulation of the SPP propagation (extinction ratio of 35 dB for a 8 mW electrical power). This achieved driving power was considerably lower than that of conventional thermo-optic MZIMs.

In the field of EOT, ordered arrays of nano-apertures have been considered as field enhancers and controllers for near-field electromagnetic waves. The EOT transmission takes place in a double-layer structure (a metal-oxide layer). The characteristics of the transmission spectra are fixed by the dielectric constants of the materials that bound the array and by the hole's depth and pitch. Using a specific oxide layer in which semiconductor-metal transition is observed; the transmission of light incident on a metallic nanoholes array is modulated by switching, at transition temperature, the oxide layer between metallic and semiconducting phases with distinctly different dielectric functions.

Electric Field-Driven Active SPP Devices

Replacing the "dielectric" layer in the structure with an electro active thin film (for example barium titanate or GaAs layer) allows the dynamic modulation of the SPP wavevector. Under an electric field, the orientation of ferroelectric domains is affected by film stresses. Thus, the SPP waves are modulated by the switch of the in-plane to out-of-plane domains.

Coupling field-effect transistor with plasmonic devices supporting SPP has not yet been widely developed. In 2009, a Metal-Oxide-Semiconductor (MOS) field-effect plasmonic modulator has nevertheless been demonstrated with a device operating at 1.55 μm [23]. A four layer metal-MOS–metal (Ag-SiO$_2$-Si-Ag) waveguide structure supporting both photonic and plasmonic modes was fabricated. Its transmission coefficient is determined by interference between the two modes. When the MOS is driven into accumulation under a bias of 0.75 V, the Si index changes and the photonic mode is cut off, which leads

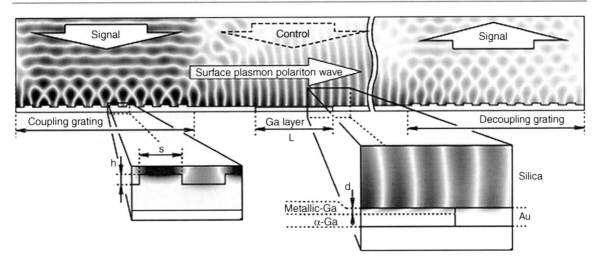

Active Plasmonic Devices, Fig. 10 Representation of gold waveguide containing a Ga switching layer (Reprinted with permission from Krasavin and Zheludev [21])

to transmission modulation. Switching time of 10 ns has been demonstrated (limited by the experimental apparatus), but gigahertz frequencies could be reachable.

Optically Field-Driven Active SPP Devices

Due to the optical properties of dielectric layer, the switching may be achieved by external optical excitation through a phase transition in the layer section. For example, Krasavin et al. investigated the control of a dielectric waveguides containing a gallium section a few microns long (Fig. 10).

In such a waveguide, the SPP waves propagated at the interface between the metal film and silica substrate, through the gold and gallium sections, but the transmitted SPP wave was attenuated due to the mismatch of dielectric characteristics. The transition of the gallium layer from the semiconductor solid phase to the metallic liquid phase changes the local refractive index and provides a high-contrast modulation of the SPP propagation. This system could also be thermally driven. A fluence about 10 mJ/cm [2] was needed to induce the Ga-phase transition. Since this first work, many studies have been done in order to implement such system in different active plasmonic devices.

The use of photochromic (PC) molecules was developed by several groups. Among many interesting examples is the use of spyropyran, which are mixed with PMMA and deposited on top of the aluminum layer. Spiropyrans are PC molecules that can undergo a ring-opening reaction and can reversibly be switched between transparent and absorbing states using a free space optical pump. In the transparent (signal "on") state, the SPPs freely propagate through the molecular layer, and in the absorbing (signal "off") state, the SPPs are strongly attenuated (26% of the original intensity). Upon exposure with 532 nm green light, the signals go back up to 90%. Therefore the switch was very reproducible but slow, and the incomplete recovery of the signal was attributed to degradation.

With sub-microsecond or nanosecond response times at best, these techniques are likely to be too slow for future applications in data transport and processing fields. Few groups are working to develop some solutions to control or modulate a femtosecond optical frequency plasmon pulses by direct ultrafast optical excitation of the metal. MacDonald's group have discovered a nonlinear interaction between a propagating SPP and light that takes place in the skin layer of the metal surface along which the plasmon wave is propagating [24]. A femtosecond optical pulse incident on the metal surface disturbs the equilibrium in the energy–momentum distribution of electrons, thereby influencing SPP propagation along the surface (Fig. 11).

Active Plasmonic Devices,
Fig. 11 Schematic
representation of SPP
modulation by ultrafast pulse
in a Al-silica waveguide
(Reprinted with permission
from MacDonald et al. [24])

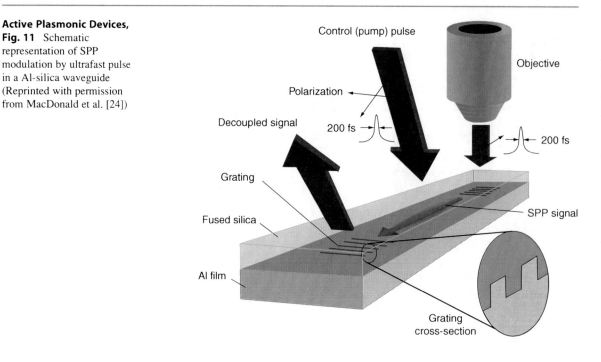

The transient effect of control pulse excitation on the propagation of the SPP signal was monitored by varying the time delay between the SPP excitation and optical pump pulses. Two components were observed in the transient pump-probe interaction. The "fast" component was only seen where the pump polarization has a component parallel to the direction of SPP propagation with a magnitude of around 7.5% for pump fluence of 10 mJ cm^{-2}. On the contrary, the "slow" component was a characteristic decay time of about 60 ps and reaches about 4% of magnitude at the same fluence. The ultrafast femtosecond switching times and modest switching energy reported here open the gates to the exploration of what can ultimately be achieved in nonlinear plasmonics and active plasmonic switching, in particular for the purposes of high-bandwidth interconnects, plasmon–polariton modulation, all-plasmonic switching.

Spin Coupling SPP

A first proposal for a spin-plasmonic device was reported by Chau et al. [25]. The properties of a gigahertz magnetoplasmon optical modulator based on a bismuth-substituted yttrium iron garnet system are studied, showing the possibility of achieving a high-speed magneto-optic modulation of improved efficiency compared to bulk devices. This system is characterized by relevant magneto-optical properties

(with Faraday rotations as high as 1°/μm). However, the high absorptions of this material in the visible range of wavelengths represent one of the major limitations in its application in devices. In 2007, the first experimental evidence of a spin-dependent transport in spin-plasmonic media was given by the same group, showing a magnetic manipulation of near-field-mediated light transport via the electron spin. They demonstrated that ferromagnetic particles coated with nonmagnetic metal nanolayers exhibit an enhanced magnetic field controlled attenuation of the electromagnetic field propagated through the sample. The mechanism is related to dynamic, electromagnetically induced electron spin accumulation in the non-magnet. Since 2007, many studies have been done based on core shell metal/magnetic material/metal or magnetic material/metal/magnetic material nanosandwichs system demonstrating a strong coupling between the plasmon resonance and the magneto-optics activity.

Future Directions for Research

Active plasmonic is still in its infancy, even though controlling LSPR and SPP by various external stimuli has been widely successful. Many new devices are likely to be fabricated shortly. Strategies are similar, but some of them have only been used in order to control one kind of plasmonic device. Switching speed measurements in LSPR-based systems are still lacking, whereas SPP control does not yet use plastic electronic components. It can be anticipated that works on LSPR and SPP control will feed each other, and that active molecular plasmonic devices will lead to a wealth of new important results.

Cross-References

► Biosensors
► Gold Nanorods
► Nanoparticles
► Nanophotonic Structures for Biosensing
► Nanopolaritonics
► Nanostructures for Photonics
► Optical properties of Metal Nanoparticles
► Plasmon Resonance Energy Transfer Nanospectroscopy
► Surface Plasmon-Polariton-Based Detectors

References

1. Brongersma, M.L., Kik, P.G.: Surface Plasmon Nanophotonics. Springer, New York (2007)
2. Baia, M., Astilean, S., Iliescu, T.: New developments in SERS-active substrates. In: Baia, M., Astilean, S., Iliescu, T. (eds.) Raman and SERS Investigations of Pharmaceuticals, pp. 187–205. Springer, Berlin (2008)
3. Barnes, W.L., Dereux, A., Ebbesen, T.W.: Surface plasmon subwavelength optics. Nature **424**, 824–830 (2003)
4. Jain, P.K., Huang, X., El-Sayed, I.H., El-Sayed, M.A.: Noble metals on the nanoscale: Optical and photothermal properties and some applications in imaging, sensing, biology, and medicine. Acc. Chem. Res. **41**, 1578–1586 (2008)
5. MacDonald, K.F., Zheludev, N.I.: Active plasmonics: Current status. Laser Photon. Rev. **4**, 562–567 (2010)
6. Karg, M., Pastoriza-Santos, I., Perez-Juste, J., Hellweg, T., Liz-Marzan, L.M.: Nanorod-coated PNIPAM microgels: Thermoresponsive optical properties. Small **3**, 1222–1229 (2007)
7. Ikeda, T.: Photomodulation of liquid crystal orientations for photonic applications. J. Mater. Chem. **13**, 2037–2057 (2003)
8. Malynych, S., Chumanov, G.: J. Am. Chem. Soc. **125**(10), 2896–2898 (2003)
9. Homola, J.: Surface plasmon resonance based sensors. Springer, Berlin (2006)
10. Zheng, Y.B., Yang, Y.W., Jensen, L., Fang, L., Juluri, B.K., Flood, A.H., Weiss, P.S., Stoddart, J.F., Huang, T.J.: Active molecular plasmonics: Controlling plasmon resonances with molecular switches. Nano Lett. **9**, 819–825 (2009)
11. Haes, A.J., Zou, S., Zhao, J., Schatz, G.C., Van Duyne, R.P.: Localized surface plasmon resonance spectroscopy near molecular resonances. J. Am. Chem. Soc. **128**, 10905–10914 (2006)
12. Negre, C.F.A., Sánchez, C.G.: Effect of molecular adsorbates on the plasmon resonance of metallic nanoparticles. Chem. Phys. Lett. **494**, 255–259 (2010)
13. Zhao, J., Das, A., Schatz, G.C., Sligar, S.G., Van Duyne, R.P.: Resonance localized surface plasmon spectroscopy: Sensing substrate and inhibitor binding to cytochrome P450. J. Phys. Chem. C **112**, 13084–13088 (2008)
14. Ung, T., Giersig, M., Dunstan, D., Mulvaney, P.: Spectroelectrochemistry of colloidal silver. Langmuir **13**, 1773–1782 (1997)
15. Wang, Z.C., Chumanov, G.: WO3 sol-gel modified Ag nanoparticle arrays for electrochemical modulation of surface plasmon resonance. Adv. Mater. **15**, 1285–1289 (2003)
16. Leroux, Y.R., Lacroix, J.C., Chane-Ching, K.I., Fave, C., Felidj, N., Levi, G., Aubard, J., Krenn, J.R., Hohenau, A.: Conducting polymer electrochemical switching as an easy means for designing active plasmonic devices. J. Am. Chem. Soc. **127**, 16022–16023 (2005)
17. Leroux, Y., et al.: ACS Nano **2**(4), 728–732 (2008)
18. Stockhausen, V., Martin, P., Ghilane, J., Leroux, Y., Randriamahazaka, H., Grand, J., Felidj, N., Lacroix, J.C.: Giant plasmon resonance shift using poly (3,4-ethylenedioxythiophene) electrochemical switching. J. Am. Chem. Soc. **132**, 10224–10226 (2010)
19. Sidhaye, D.S., Kashyap, S., Sastry, M., Hotha, S., Prasad, B.L.V.: Gold nanoparticle networks with photoresponsive interparticle spacings. Langmuir **21**, 7979–7984 (2005)
20. Du, G.X., Mori, T., Saito, S., Takahashi, M.: Shape-enhanced magneto-optical activity: Degree of freedom for active plasmonics. Phys. Rev. B Condens. Matter **82**, 4 (2010)
21. Krasavin, A.V., Zheludev, N.I.: Active plasmonics: Controlling signals in Au/Ga waveguide using nanoscale structural transformations. Appl. Phys. Lett. **84**(8), 1416–1418 (2004)
22. Nikolajsen, T., Leosson, K., Bozhevolnyi, S.I.: Surface plasmon polariton based modulators and switches operating at telecom wavelengths. Appl. Phys. Lett. **85**, 2 (2004)
23. Dionne, J.A., Diest, K., Sweatlock, L.A., Atwater, H.A.: PlasMOStor: A metal-oxide-Si field effect plasmonic modulator. Nano Lett. **9**, 897–902 (2009)
24. MacDonald, K.F., Samson, Z.L., Stockman, M.I., Zheludev, N.I.: Ultrafast active plasmonics. Nat. Photon. **3**, 55–58 (2009)
25. Chau, K.J., Irvine, S.E., Elezzabi, A.Y.: A gigahertz surface magneto-plasmon optical modulator. IEEE J. Quant. Electron. **40**, 571–579 (2004)

Adhesion

► Nanotribology

Adhesion in Wet Environments: Frogs

W. Jon. P. Barnes
Centre for Cell Engineering, University of Glasgow, Glasgow, Scotland, UK

Synonyms

Wet adhesion in tree frogs

Definition

Mechanisms of adhesion in climbing frogs.

Overview

Biomimetics of Animal Adhesion

Mankind's understanding of the adhesive mechanisms of climbing animals has increased rapidly in recent years, in no small way due to advances in materials science providing both the tools and the theoretical background. This research has shown that, as a result of millions of years of evolution, the adhesive mechanisms of climbing animals are highly dynamic, showing many features that are the envy of materials scientists. Unlike most man-made adhesives, they (1) cope well with rough and antiadhesive substrates, (2) have self-cleaning mechanisms and so recover quickly following contamination, (3) can control attachment so that it only occurs when required, and (4) possess mechanisms for effortless detachment. Finally (5), since animals attach and detach their adhesive pads every time they take a step, sticking ability is not lost with repeated application, this being mainly due to their resistance to both wear and contamination. To date, most research effort has been directed at investigating the adhesive mechanisms of geckos (▶ Gecko Adhesion) and insects [1, 2], but tree and torrent frogs, adapted to living in habitats with a high rainfall, offer solutions to different problems, namely, adhesion under wet conditions.

Climbing and Adhesion

It goes without saying that climbing animals, in addition to adaptations for climbing, need mechanisms for maintaining a grip and thus avoiding falling. Taking, e.g., an animal on an inclined tree trunk, the force of gravity will cause this animal to fall unless it can generate a reaction force between itself and the trunk that is equal and opposite to the force generated by the gravitational pull. Such forces can be generated in a number of ways, namely, mechanical interlocking, friction, and bonding. *Interlocking* involves interlocking the surface of the animal with that of the support. This is how claws, present in many climbing animals (e.g., squirrels and members of the cat family), work. Claws may either catch on preexisting surface irregularities or be pushed into the surface if it is soft enough. Many animals without claws (e.g., monkeys and humans) utilize a simple *friction* grip. It only works, however, when hands or feet can span almost the entire branch diameter, i.e., a vertical wall cannot be climbed by friction alone. Many specialist climbers, however, possess specialized adhesive pads on their feet and adhere by *bonding* (▶ Bioadhesion). Such bonds need to be reversible; otherwise, they would only be good for permanent attachment. The main mechanisms of bonding are dry and wet adhesion. In *dry adhesion*, the adhesive pads are able to get so close to the substrate that adhesion is primarily by van der Waals forces. Such forces can hold solids together and only operate at distances of a few nanometers. In *wet adhesion*, on the other hand, there is a fluid joint, secreted liquid between the pad and substrate, and adhesion is thought to be mainly by capillarity and viscosity. It is generally accepted that dry adhesion is found in lizards such as geckos, while wet adhesion is found in insects, amphibians, and among mammals, in arboreal possums and at least one bat species. However, it is probable (though unproven) that capillarity plays a minor role in animals that mainly adhere by way of van der Waals forces, and van der Waals forces play a minor role in animals that mainly adhere by wet adhesion.

Two rather different types of adhesive pads – smooth and hairy – have evolved that allow animals to climb and adhere to vertical and overhanging surfaces. Hairy adhesive pads are found in many insects, spiders, and lizards. In geckos, they consist of millions of tiny branching hairs (setae), each branch ending in a flattened spatula no more than 200 nm across. This allows the tips of the hairs to form extremely close contact to rough surfaces, enabling intermolecular van der Waals forces (as described above) to provide strong adhesion [3]. In contrast, the smooth adhesive pads of tree frogs, arboreal salamanders, and insects such as ants secrete a fluid so that contact is maintained by wet adhesion [4]. Interestingly, the hairy pads of insects also secrete tiny amounts of fluid, so that, in contrast to geckos, they mainly use wet adhesion. Until recently, spiders were thought to use dry adhesion, but there is new evidence of a fluid secretion from the adhesive setae, so they may, like insects, use wet adhesion.

Adhesion in Tree and Torrent/Stream Frogs

Among the Amphibia, adaptations for climbing are found in Neotropical salamanders of the genus *Bolitoglossa* and at least ten families of frogs. Where the habitat is vegetation (herbs, shrubs, trees), the frogs are known as tree frogs, while those found clambering on wet rock in the vicinity of streams and waterfalls are known as stream and torrent frogs, respectively. All are characterized by the presence of adhesive pads on their digits. In spite of the fact that these pads have evolved independently in at least seven different families, their structures are very similar. The development of adhesive pads is thus an excellent example of convergent evolution. This suggests that there is an optimal design for a toe pad, in frogs at least, and this is important for biomimetics. Most frog adhesion research has been carried out on tree frogs of the families Hylidae and Rhacophoridae, while work on torrent frogs of the family Ranidae is still at an early stage.

Physical Principles

Although the dominant force in tree frog adhesion is believed to be capillarity, rate-dependent viscous forces such as Stefan adhesion are also considered to play a role. Additionally, since the thickness of the fluid layer was found to be less than 35 nm in some areas of the contact zone of tree frogs [5], it is probable that there is direct contact between nanostructural features of the pad surface and the substrate. Thus van der Waals forces may also be involved, but to what extent is unknown. The physical principles underlying capillarity and viscous forces are outlined below.

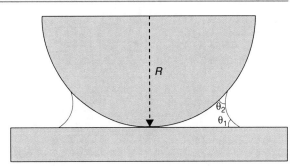

Adhesion in Wet Environments: Frogs, Fig. 1 Capillarity model of sphere on a plane surface connected by a drop of fluid

Wet Adhesion – Capillary Forces

A common model that has been used for quantifying capillary forces consists of a sphere on a plane surface, connected by a drop of liquid (Fig. 1). In this model, the surface tension of the meniscus results in a pressure difference, where the pressure inside the liquid is lower than it is outside, so long as the meniscus is concave. This pressure difference, the Laplace pressure, will resist the separation of the two surfaces. The attractive Laplace force for a macroscopic, perfectly smooth, and homogeneous sphere in contact with a plane is given by:

$$F_L = -2\pi R\gamma(\cos\theta_1 + \cos\theta_2), \quad (1)$$

where R is the radius of the sphere, γ is the surface tension of the liquid, and θ_1 and θ_2 are the contact angles of the liquid with the plane surface and sphere, respectively [6]. If the liquid completely wets both surfaces (i.e., $\theta_1 = \theta_2 = 0$), Eq. 1 simplifies to:

$$F_L = -4\pi R\gamma. \quad (2)$$

An alternative model examines capillary forces between two rigid plates, separated by a thin layer of fluid (Fig. 2). This model predicts substantial forces if the radius of the area of contact (r) is large and the separation distance of the plates (h) is small, according to the equation:

$$F_L = \pi r^2 \gamma[r^{-1} - (\cos\theta_1 + \cos\theta_2)h^{-1}]. \quad (3)$$

In a large flat meniscus where r $\gg$ h and $\theta_1 = \theta_2 = 0$, Eq. 3 approximates to Eq. 4, namely:

$$F_L \approx -2\pi r^2 \gamma/h \quad (4)$$

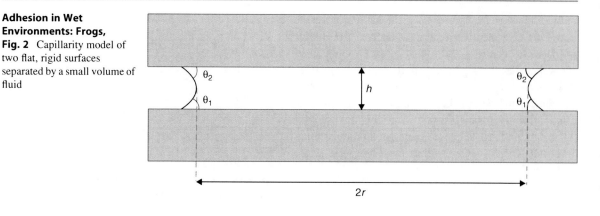

Adhesion in Wet Environments: Frogs, Fig. 2 Capillarity model of two flat, rigid surfaces separated by a small volume of fluid

A second component of the adhesive force is the tensile force of the meniscus. In the rigid plate model, this surface tension force is:

$$F_T = -2\pi r \gamma . \sin \theta \qquad (5)$$

The total meniscus force (F_M) is the sum of F_L and F_T:

$$F_M = F_L + F_T \qquad (6)$$

F_T is negligible in any single large meniscus (e.g., ones with a radius of 0.5–3 mm as occur in tree frog toe pads), but would probably dominate adhesive forces in a fly's adhesive pad where you have large numbers of microscopic menisci (radii of ca. 1 µm).

In a recent study [7], Eq. 2 has been extended from hard, undeformable surfaces and spheres to soft, elastic materials, such as the toe pad of a tree frog. The mathematical equations relate the capillary attraction between the bodies to their elastic repulsion. Although they do not lend themselves to any easy calculation of the adhesive force, they are of interest as they predict, for the sphere on a plane surface model, that F_M scaling will gradually change from length scaling to area scaling with increasing R, this change occurring more rapidly for materials with a lower effective elastic modulus (E_{eff}). For instance, in Eq. 7 below, which deals with the Laplace pressure component of adhesion, the first term is proportional to R and is identical to the force with a hard sphere (Eq. 2), while the second term is proportional to R^2. It is negligible for high Young's moduli, but becomes significant for soft materials:

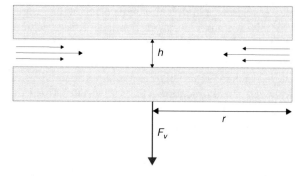

Adhesion in Wet Environments: Frogs, Fig. 3 Stefan adhesion model of two flat rigid plates, fully immersed in a fluid, subject to an external force

$$F_L = -4\pi R\gamma - \left(\frac{\pi\gamma}{2r'}\right)^3 . \frac{2R^2}{3(E_{eff})^2}, \qquad (7)$$

where r' is the radius of curvature of the meniscus.

Wet Adhesion – Viscous Forces

The second component of wet adhesion is provided by time-dependent viscous forces (such as Stefan adhesion). Consider two rigid plates fully submersed in a fluid (Fig. 3). Separating the plates involves fluid flowing into the gap between them so that separation is resisted by a viscous force until the fluid movements are complete. This force will be greater for more viscous fluids and for smaller values of h. As Eq. 8 indicates, Stefan adhesion (F_V) scales with area squared:

$$F_V = -\frac{dh}{dt} . \frac{3\pi\eta r^4}{2h^3}, \qquad (8)$$

Adhesion in Wet Environments: Frogs, Fig. 4 (**a**) White's tree frog, *Litoria caerulea*; (**b–d**) scanning electron microscopy of toe pad epithelium; (**b**) low-power micrograph of whole pad: (**c**) micrograph showing a mucus pore and (largely) hexagonal epithelial cells separated from each other at their distal ends by channels; (**d**) higher-power micrograph indicating the presence of nanostructuring on the "flat" surface of the epithelial cells (From [15]

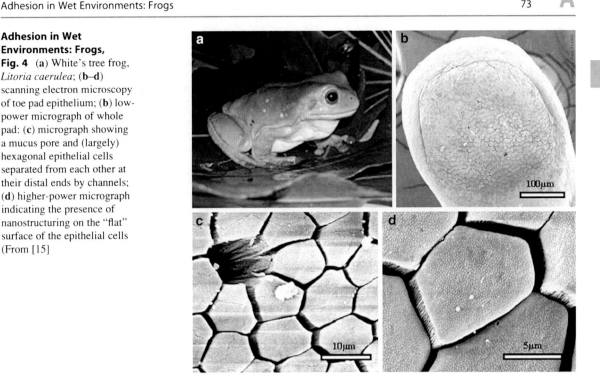

where η is the viscosity of the liquid and t is the time needed to separate the two plates. For completeness, the equation for the hydrodynamic force between a sphere and a plate (Eq. 9) is also included:

$$F_V = -6\pi\eta R \frac{dh}{dt}\left(1+\frac{R}{h}\right). \tag{9}$$

When the sphere touches the plate, as in Fig. 1, h = 0.

Key Research Findings

Toe Pad Micro- and Nanostructure

In both tree and torrent frogs, soft domed pads occur on the ventral surface of the tip of each digit, the specialized pad epithelium being delineated from normal skin by distinct grooves (Fig. 4). There are also much smaller adhesive areas located elsewhere on the ventral surface of the feet, in particular, the subarticular tubercles located more proximally on the digits. Tree frog adhesive pads have a stratified columnar epithelium, the cells being separated from each other at their apices. Scanning electron microscopic studies (▶ Scanning electron microscopy) show that most of these cells are hexagonal, but as Fig. 4 shows, some are

pentagonal and a few heptagonal. Pores of mucous glands open into the channels between the cells. The toe pad epithelium thus consists of an array of flat-topped cells separated by mucous-filled grooves [8–12]. In functional terms, having the cells separated at their tips enables the pads to conform to the shape of surface irregularities to which the frog is adhering. The mucous glands are necessary to produce the watery secretion that forms an essential part of the adhesive mechanism of the pad. The hexagonal array of channels that surround each epithelial cell presumably functions to spread mucus evenly over the pad surface and, under wet conditions (most tree frog species live in rainforests), remove surplus water. Finally, the presence of grooves could aid adhesion by reduction of crack propagation (peeling). Pull-off stress would be spread between a larger number of hexagons rather than being concentrated at the edge of the contact zone. Such features have been incorporated into bio-inspired artificially patterned surfaces to increase their adhesion [13]. In torrent frogs, the epithelial cells have become elongated along the pad's proximal-distal axis, which results in the channels between them being straighter and shorter in this direction, presumably an adaptation for efficient drainage of excess water.

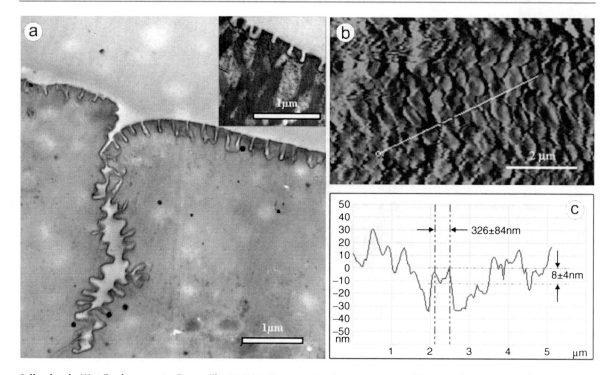

Adhesion in Wet Environments: Frogs, Fig. 5 White's tree frog, *Litoria caerulea*. (**a**) Transmission electron micrograph of a toe pad showing a section of one of the channels that separate adjacent epithelial cells and the surface nanostructures, which are themselves separated from each other by narrow channels. The inset shows similar nanostructures on a toe pad of the hylid tree frog, *Scinax ruber*. Here, the nanostructures are associated with filaments running at right angles to the cell surface. (**b**) Atomic force microscope deflection image showing the shape of the dense array of peg-like nanopillars that constitutes the adhesive surface of the pad. (**c**) Height profile of line in (**b**). The two crosses delineate a columnar nanopillar that lay precisely on the profile line, showing that these peg-like structures have a small dimple at their center. The figures are means ± standard deviations of nanopillar width and dimple depth. (NB: The height of the nanopillars cannot be measured by the atomic force microscope as the cantilever cannot reach the base of the channels between them) (Redrawn from [15])

At the nanometer scale, the pad surface is not smooth but is covered by a dense array of nanopillars, 300–500 nm in diameter and 200–300 nm in height. Under the atomic force microscope, they can be clearly seen to have small dimples at their tips (Fig. 5). In many species, these nanostructures are the ends of filament bundles lying within the cytoplasm of the epithelial cells that run at right angles to their external surface. Similarly oriented fibers are found in the smooth adhesive pads of insects. In tree frogs, they form the cytoskeleton of the pads. The functions of the nanopillars that constitute the so-called flat surface of these cells remain a matter for speculation. The following is a list of obvious possibilities, none of which are mutually exclusive. (1) Like the epithelial cells and the channels that surround them, the nanopillar array may allow close conformation to surface irregularities but on a much smaller length scale (nanometers rather than micrometers) than applies to whole epithelial cells. (2) The narrow channels between them could serve to absorb excess water (like a sponge), much as sipes (fine-scale grooves) do on a wet-weather car tire [14]. This would allow rapid optimization of the thickness of the intervening fluid layer (as thin as possible, but without any air pockets). (3) The nanopillars could be very important in the generation of friction forces in that their tips will be in direct contact with the surface (see later). (4) It cannot be excluded that the dimples give rise to a suction effect, but the existence of channels in the dimple wall make this unlikely. Indeed, it is more likely that the channels allow the escape of fluid from the dimples when the pad is pressed against a smooth surface, thus minimizing fluid layer thickness and increasing close contact of nanopillars with the surface.

Neotropical salamanders of the genus *Bolitoglossa* are also arboreal. Some species have toe pads, whereas in others, digits are much reduced and there is extensive digital webbing. In these cases, the whole foot appears to function as a single smooth pad. In smaller species, adhesion is thought to be by wet adhesion, but in some larger species, there is circumstantial evidence that suction may play a role. In most species, the surfaces of the cells are smooth, although in *Bolitoglossa odonelli*, the surface is covered by nanopillars that appear very similar to those found in tree frogs.

Physical Properties of Toe Pads

The elastic modulus (or in this case, the effective elastic modulus, as the Poisson's ratio for toe pads is not known) provides a quantitative measure of material stiffness, an important parameter for adhesion. Stiffness is estimated from force-distance curves, resulting from indentation of the toe pads, using equations based on, for instance, the Johnson Kendall Roberts (JKR) theory. Such equations will also allow calculation of the work of adhesion and the pull-off force. Microindentation usually involves specialized microindenters, while at the nanolevel, either a nanoindenter or atomic force microscope can be used.

Use of the atomic force microscope (▶ Atomic force microscopy) in nanoindenter mode indicates that the outer keratinized layer of tree frog toe pads, which is about 15 μm thick, has an effective elastic modulus (E_{eff}) of 5–15 MPa, equivalent to that of silicone rubber [15]. However, it gives no information on the physical properties of deeper structures. To measure the stiffness and elasticity of the pad as a whole, microindentation is used. It demonstrates that tree frog toe pads are among the softest of biological tissues and that they show a gradient of stiffness, being stiffest on the outside. Estimates of E_{eff} lie in the range 4–25 KPa [16]. Although not as soft as adipose tissue ($E_{eff} \approx$ 3 KPa), values are much lower than for aorta ($E_{eff} \approx$ 500 KPa), cartilage ($E_{eff} \approx$ 20 MPa), or the adhesive pads of locusts ($E_{eff} =$ 250–750 KPa). At the opposite extreme, tooth enamel has an E_{eff} value of 60,000 MPa [17–19]. Functionally, the stiffer outer surface provides a degree of resistance against wear, while the softer interior provides an increased ability to form intimate contact with various surface profiles. Since the epithelium as a whole is a living, growing tissue, the long-term solution to the wear problem is a simple one; the outer cell layer is shed at intervals, presumably after the cell layer below it has become keratinized.

A dense network of blood capillaries lies below the pad epithelium. This may well have a mechanical function, acting as a shock absorber. This could be particularly important when a tree frog lands on a vertical surface after a jump. Having soft inelastic material beneath the pad allows the pad to mold to the surface the frog has landed on without any tendency for elastic recoil.

Viscosity and Thickness of Fluid Layer

Microrheometry of toe pad mucus has been undertaken using a laser tweezer technique ([5]; ▶ Optical Tweezers). Latex beads (3 μm in size) were mixed with droplets of toe pad mucus (ca. 5 μL in volume) on a coverslip on a motorized stage of a microscope equipped with a laser tweezer. The laser tweezer trapped individual beads in its focus. The stage was then oscillated sinusoidally at a number of different frequencies. The fluid movement exerts a drag on the beads, which is dependent on the velocity and also on the viscosity of the fluid. Graphs were plotted of the amplitude of the bead's lateral displacement against oscillation frequency. The viscosity of the mucus can be estimated from a comparison of the linear regression slope obtained for frog mucus with that obtained for beads in pure water. Such measurements (Fig. 6) show that the viscosity of toe pad mucus is approximately 1.5 MPa s (i.e., 1.5 times the viscosity of water). No evidence for non-Newtonian properties was found, but it is possible that the situation changes for more concentrated mucus.

The thickness of the mucus layer beneath the frog's toe pad has been investigated by a technique called interference reflection microscopy. This enables the calculation of the fluid layer thickness from an optical interference pattern of reflected wave fronts [5]. Monochromatic light was reflected both from the surface of the pad and from the top surface of the glass plate on which the frog was sitting (a large coverslip). This leads to a series of interference fringes reflecting the thickness of the fluid layer. Different interference fringes were distinguished by comparing images obtained by light of different wavelengths (Fig. 7).

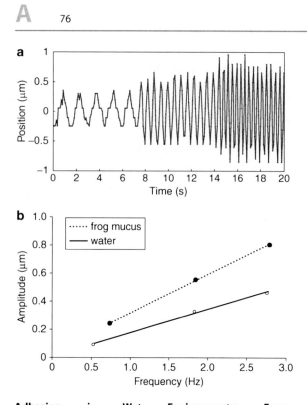

Adhesion in Wet Environments: Frogs, Fig. 6 Microrheology of mucus from White's tree frog, *Litoria caerulea*, using a laser tweezer technique. (**a**) Amplitude of bead displacement elicited by sinusoidal movement of the microscope stage (stimulus amplitude ≈10μm) at three different frequencies; (**b**) relationship of bead displacement amplitude to the stimulus frequency, measured for both toe pad mucus and pure water. The difference between the slopes of the two lines indicates that this sample of toe pad mucus was 1.65 times more viscous than water (Redrawn from [5])

Examination of images such as this shows that most of the toe pad epithelial cells had dark zones in the center, which correspond to zero-order interference minima (i.e., to a mucus film thickness of less than 100 nm). Further analysis of the intensity of these central areas in relation to interference maxima and minima leads to the conclusion that the film thickness ranges from 0 to 35 nm and is indistinguishable from zero in many epithelial cells (median 6.0 nm). Given that there will be fluid in the narrow channels between the nanopillars, it seems likely that the tops of many nanopillars will be in direct contact with the substrate in an attached toe pad. Additionally, in some species, an increase in the proportion of the area in closest contact was observed as frogs were tilted from the horizontal toward the vertical.

Adhesion and Friction Forces

Tree frogs are good climbers, many tropical species being found high in the canopy of rain forests. When objects such as twigs are available, they are grasped and climbing is by reaching and pulling as in human climbing. Toe pads come into their own on smooth substrates such as leaves. Unlike geckos, tree frogs cannot run across a ceiling, but they can climb vertical and overhanging surfaces, and a small tree frog can hang on to an inverted glass plate using just its toe pads. At rest, thigh and belly skin aid adhesion.

Quantitative measurements of the adhesive (normal) and friction (parallel) forces that can be generated by tree frogs have been made by a simple procedure first developed by Emerson and Diehl [10]. Immediately after being weighed, each frog was placed "head-up" on a smooth surface on a rotation platform that was rotated slowly from 0° (horizontal) through 90° (vertical) to 180° (upside-down). The angles at which the frog first slipped on the surface (slip angle) and falls from the surface (fall angle) were recorded. These slip and fall angles are used to calculate maximum shear and adhesive forces, respectively, by simple trigonometry. If the total toe pad area is also measured, these values can be converted to the force per unit area that toe pads can generate. In all species studied, toe pad adhesive force per unit area on a smooth, high-surface energy surface (glass) lies in the range 0.7–2.0 mN mm^{-2} (0.7–2.0 KPa).

Frogs are good models for allometric studies, as general body shape is maintained from metamorphosis to full adult size, i.e., frogs grow approximately isometrically. By plotting mass, toe pad area and adhesive force against frog length on log-log coordinates, both between and within species, it can be shown that adhesive force scales with toe pad area or length squared ($f_a \propto a$ or $f_a \propto l^2$) (Fig. 8) [10, 12, 20, 21]. This presents a problem for larger species since mass is proportional to the cube of linear dimensions, while adhesive force, depending upon area, would only be proportional to the square of the linear dimension. A within-species study of hylid tree frogs in Trinidad [21] showed that toe pad area increases as length squared as expected, but mass increases by a factor rather less than the expected length cubed. Also, toe pad efficiency (force per unit area) increases with age, perhaps associated with the increased structural complexity of adult toe pads. However, in the majority of the species studied, adult

Adhesion in Wet Environments: Frogs, Fig. 7 In vivo analysis of White's tree frog, *Litoria caerulea* toe pad contact with glass using interference reflection microscopy. (**a**, **b**) Images of the same area of the contact zone using different wavelengths (**a**, 436 nm; **b**, 546 nm) using an illuminating numerical aperture of 0.27. *Asterisks* indicate an epithelial cell that is not in close contact (fluid film thickness >40 nm). (**c**) Intensity profile along the *arrow* shown in **a** and **b**. *Arrows* indicate that the position of the second order "*green*" minimum coincides with the second order "*blue*" maximum. From this, it can be deduced that the central dark areas of most epithelial cells are zero order minima. (**d**) Reconstruction of the fluid film thickness along the *arrow* show in **a** and **b** (Redrawn from [5])

frogs did not completely overcome the allometry problem, for they fell from the rotation platform at smaller angles than younger frogs. In ecological terms, this disadvantage is less than one might expect, in that frogs can use their digits to grasp objects such as twigs, as described above. Indeed, there is a reasonable correlation between adult size and degree of arboreality in that the largest species are often found high in the canopy, while smaller species are most commonly found in shrubs no more than a meter or so above the ground. Finally, the way in which forces scale with size can provide insights into the underlying mechanisms of adhesion. Classic formulae for wet adhesion, based on flat metal disks (e.g., Eq. 4), show capillarity forces scaling with toe pad area and have

therefore been used as evidence for the importance of capillarity in tree frog adhesion. However, at the small values of h measured by interference reflection microscopy, they predict adhesive forces far greater than toe pads produce, even when the wedge of fluid around the pad perimeter is taken into account. Indeed, the softness of the pad material makes this model inappropriate for tree frogs. Equation 2 predicts forces in the right range but scaling with length and not area. Peeling models, applicable to the forces that prevent peeling in adhesive tape, are also worthy of consideration because of the way in which toe pads detach (see below). However, they also have force scaling with length, the forces being concentrated in a very narrow zone along the peeling line and thus dependent

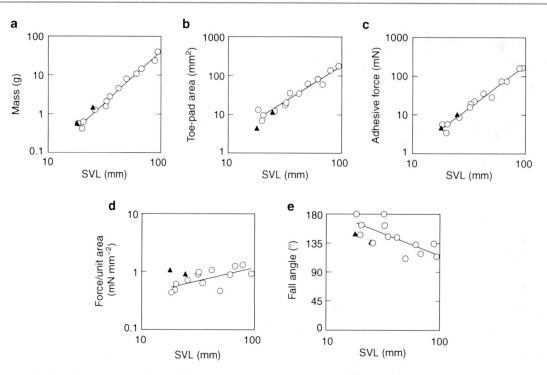

Adhesion in Wet Environments: Frogs, Fig. 8 Allometric relationships of 13 hylid species (open circles) and two nonhylids (filled triangles) plotted against snout-vent length (SVL). (**a, b**) Morphometric relationships between frog mass, toe pad area, and length (log-log plots); mass $\propto$ SVL$^{2.65}$, significantly different from expected value of 3.0 (p < 0.01), and area $\propto$ SVL$^{1.85}$, not significantly different from expected value

of 2.0; (**c, d**) Adhesive force and force per unit area plotted against length (log-log plots); force $\propto$ SVL$^{2.24}$, not significantly different from 2.0, the value expected from area scaling, and unit force $\propto$ SVL$^{0.46}$, a significant positive relationship (p < 0.05); (**e**) fall angle against length (log-linear plot). The regression line has a slope of −68, a significant negative relationship (p < 0.01) (From [20])

on the width of the tape. Possibly, the solution lies in the recent analysis of Butt et al. [7] for adhesion using the model illustrated in Fig. 1. This analysis predicts a change from length to area scaling with increasing R and decreasing E_{eff}.

A common misconception is that the subdivision of the surface by the channels between the cells increases adhesion according to the principle of contact splitting. This theory states that adhesive force is proportional to the length of the contact; therefore, by splitting up the contact zone into many small areas of contact, the total adhesive force can be increased in direct proportion to the density of these small areas [22]. This principle clearly applies to wet adhesion. However, as the major force component of wet adhesion is capillarity, it is necessary that an air-water interface (meniscus) should surround each small area of contact. This is true of the hairy pads of insects, but is not applicable to tree frogs, where the meniscus surrounds the whole toe pad, not the individual epithelial cells.

In recent years, the development of a technique to measure adhesion and friction forces from single toe pads (Fig. 9) has resulted in a number of interesting findings. The frog is held in a container with holes in its base through which individual digits can protrude. Toe pad forces are measured by a two-axis force plate, manipulated so that it makes contact with the protruding toe. As well as recording forces, the setup has a feedback facility so that, for instance, load can be kept constant while recording friction forces. This technique has provided good evidence for the presence of static friction in tree frog toe pads [5]. This was demonstrated both by the build-up of friction force at the onset of sliding and also from the presence of a remaining shear force 2 min after the sliding motion had stopped (Fig. 10). Since static friction is expected to be very small in a completely fluid-filled joint, this is strongly indicative of direct contact of the nanopillars with the substrate, in accord with the findings of interference reflection microscopy

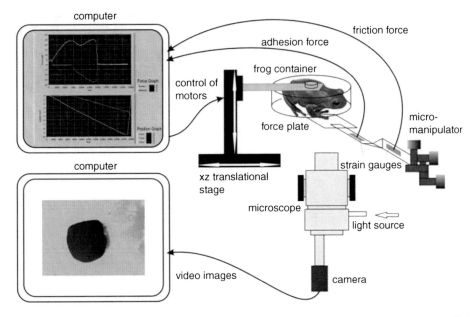

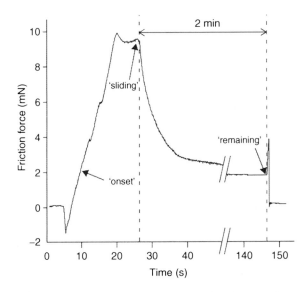

Adhesion in Wet Environments: Frogs, Fig. 9 Experimental setup for measuring adhesion and friction forces from single toe pads of tree frogs. Frogs are held stationary in a small chamber (foam padding not shown), with a toe pad protruding through a hole in its base. Using a micromanipulator, a toe pad is brought into contact with a glass force plate attached to a 2D bending beam force transducer for measuring friction (shear) and adhesion (normal) forces. A computer-controlled XZ translational stage moves the frog in relation to the bending beam. Normal forces can be controlled by a feedback mechanism when required. Contact area is imaged from below using reflected light (Developed from a setup for measuring adhesion and friction forces from insects devised by Drs Drechsler and Federle at the University of Würzburg, Germany)

Adhesion in Wet Environments: Frogs, Fig. 10 Shear force measurement in single toe pads of White's tree frog, *Litoria caerulea*. The recording shows the result of a 20-s sliding movement toward the body (500 μm s^{-1}), followed by a 2-min period when no movement occurred. The arrows indicate the force when the pad began to slide ('onset'), the force when sliding ended ('sliding'), and the force when the experiment was terminated after the 2-min rest ('remaining') (Redrawn from [5])

(see above). By making measurements from pads of different sizes, one can also do scaling experiments with single pads. Such experiments demonstrate that friction forces are independent of load but correlate well with the area of contact. This is contrary to what one would expect of classic (Coulomb) friction but in line with rubber friction. This conclusion fits well with the results of indentation experiments (see above), since the low elastic modulus of tree frog toe pads allows close contact between pad and substrate. Adhesion forces of single pads scale with pad area when pads are pulled off the force plate at a low angle (20–50°), in line with whole animal measurements. However, they scale with length when the pull-off is at right angles to the surface. This has significance for toe pad detachment mechanisms (see below). Finally, such single pad experiments provide evidence for the dominant role of capillarity in tree frog adhesion since toe pad adhesive forces fall to low levels when the meniscus surrounding the pad is removed by submerging the whole pad in a drop of water. Preliminary results from studies on torrent frogs

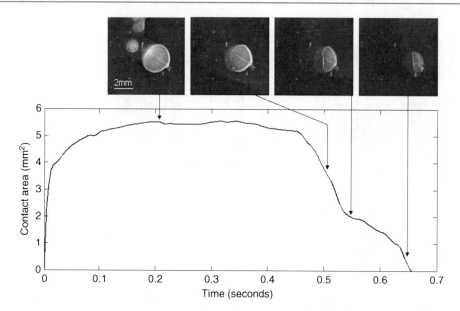

Adhesion in Wet Environments: Frogs, Fig. 11 Toe pad detachment by peeling during walking on a level surface in the tree frog, *Litoria caerulea*. The graph shows the pad contact area during a complete step, while the images are single frames from a 250-frames-per-sec video, taken by a camera on the underside of the glass plate on which the frog walked. The glass plate is illuminated from the side by arrays of light-emitting diodes that direct the light within the thickness of the plate, the light only escaping (and therefore visible to the camera) when an object, here the toe pad, contacts the glass (Unpublished data kindly provided by Diana Samuel, University of Glasgow)

show that they adhere particularly well in wet and flooded conditions, but the underlying mechanisms remain to be determined.

Toe Pad Detachment

During locomotion, toe pads are detached by peeling, a mechanism that reduces the forces required to detach their feet from the substrate every time they wish to jump or take a step to very low levels [11]. During forward walking, climbing or jumping, toe pads are peeled from the rear forward, i.e., in a proximal to distal direction. When frogs are induced to walk backward, peeling occurs in the opposite direction, from the front of the pad rearward in a distal to proximal direction. Use of a force platform to measure directly the forces generated by the feet during locomotion shows that during forward locomotion, this peeling is not accompanied by any detectable detachment forces. Such forces of detachment are seen, however, during backward walking and when belly skin inadvertently comes into contact with the force platform. That peeling occurs automatically during forward locomotion is supported both by observations of peeling in single toe pads of anesthetized frogs and by the inability of frogs to adhere to vertical surfaces in a head-down orientation. Indeed, frogs on a rotating vertical surface have been observed to adjust their orientations back toward the vertical whenever their deviation from the vertical reached $85° \pm 21°$. During forward locomotion, peeling seems to occur as a natural consequence of the way in which toes are lifted off surfaces from the rear forward, while during backward locomotion, it is an active process involving the distal tendons of the toes. The process is seen most clearly when the pad is viewed from below using high-speed video imaging (Fig. 11).

Behavioral Strategies of Adhering Frogs

Interestingly, peeling has consequences for frog behavior. This is because pads can produce strong adhesion and friction when the angle of force application is low but peel easily off the surface when the pull-off angle approaches $90°$. On vertical and overhanging surfaces, frogs must therefore adjust their posture so that gravity acts on the pads at angles where they can withstand the force and not fall from the surface. On a rotation platform, the first response to the changing angle occurs at angles between $60°$ and $100°$, heavier frogs tending to respond first. Any

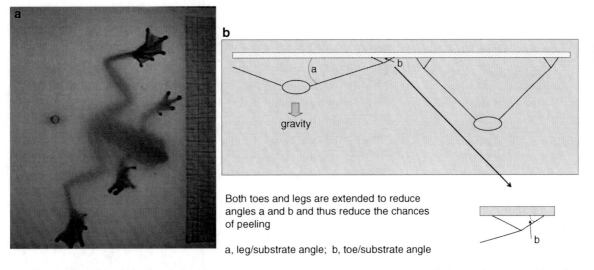

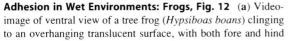

Adhesion in Wet Environments: Frogs, Fig. 12 (**a**) Video-image of ventral view of a tree frog (*Hypsiboas boans*) clinging to an overhanging translucent surface, with both fore and hind limbs stretched out sideways. Only areas in contact with surface are in sharp focus. Scale: large squares are 10 mm across (From [30]). (**b**) Explanatory model

frog not already in a head-up position will turn to face head-up so that gravity will not be acting to pull the pads off the surface. At still higher angles, both tree and torrent frogs spread their legs out sideways (Fig. 12). This is an adaptation to prevent peeling since, as the figure shows, spread-out limbs have a much smaller pad-substrate angle than limbs held close to the body. Gravity, of course, will tend to pull the limbs inward, but this is resisted by pad friction forces. In this way, friction and adhesion both play roles in helping frogs to maintain their grip under challenging conditions. The term "frictional adhesion" has been coined to describe adhesion in geckos [23]. It is unclear whether this term should also be applied to tree frog adhesion, but there is certainly interplay between friction and adhesion forces in frog toe pads too.

Biomimetic Advances

As far as tree frog adhesion is concerned, the process of micro- or nanofabrication of toe pad replicas, where the effect of the individual design parameters can be analyzed and quantified precisely, is in its infancy. An early attempt at copying the hexagonal cell structure of tree frog toe pads in polydimethylsiloxane (PDMS) is illustrated in Fig. 13a, whereas a recent mimic of the structure of a tree frog's toe pad epithelium that replicates both nano- and microstructures, from del Campo's research group at the Max Planck Institute for Polymer Research in Mainz, Germany, is shown in Fig. 13b, c.

However, the role of surface patterning in increasing adhesion is well established. Adhesive tape detaches by the spreading of cracks into the adhesive from the point of peeling. When all the energy can be concentrated at a single crack, then peeling occurs readily, but micropatterning can increase the force required to produce peeling by up to a factor of 3 by the principle of contact separation. Cracks form wherever there is a groove in the pattern so that the energy is spread between many cracks, resulting in an increased force being required to produce separation [13]. Recently, this principle has been taken a step further, and the role of subsurface structures (air- or oil-filled microchannels) has been investigated. These have crack-arresting properties just like the transverse grooves of a patterned surface, but the effect is much more dramatic, for adhesion can be increased up to 30-fold. The adhesive itself remains elastic, which renders it reusable with no reduction in adhesive efficiency [24].

Turning from patterned dry adhesion to wet adhesion, all published work is insect- rather than frog-inspired. Of particular interest is a study of the frictional properties of a structure like that in Fig. 13a. Although the addition of fluid lowered friction as would be expected, the micropatterned surface was able to generate significantly higher friction forces

Adhesion in Wet Environments: Frogs, Fig. 13 (a) Polydimethylsiloxane (PDMS) replica of a tree frog toe pad, copying the hexagonal pattern of epithelial cells (Image: Centre for Cell Engineering, University of Glasgow); (b, c) hierarchical PDMS replica of both micro- and nanostructuring on a tree frog toe pad. The PDMS was made hydrophilic by treatment in a plasma chamber and has an elastic modulus in the range 1–5 MPa (Image: Max Planck Institute for Polymer Research, Mainz)

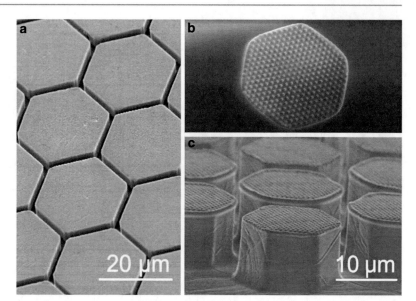

during sliding than an otherwise identical smooth one [25]. In another study, the same authors studied surfaces with a mushroom-shaped fibrillar structure not unlike the shape of the nanopillars seen on tree frog toe pads. Surfaces with such structures adhered over 20 times better than flat ones underwater, possibly due to a suction effect [26].

Examples of Possible Applications

Since adhesive tapes inspired by the dry adhesive mechanisms of gecko toe pad setae are now reaching a stage where commercialization is imminent, it is appropriate to consider whether the rather different wet adhesion mechanism of tree and torrent frogs might also have biomimetic relevance. Improved wet weather tires [14, 27, 28], nonslip footwear, plasters for surgery able to adhere to tissue, and holding devices for neurosurgery or MEMS devices are obvious examples of the many uses to which these toe pad analogues might be applied. However, as biomimetics (▶ biomimetics) has as much to do with inspiration as copying, a beetle-inspired device providing reversible adhesion based on capillarity is worthy of mention as it uses wet adhesion [29]. It generates an adhesive force through the formation of a large number of liquid bridges between two plates, which can be quickly made or broken by a low-voltage pulse that drives electroosmotic flow. There is good potential for the development of a grab-and-release device that could carry significant loads.

Cross-References

▶ Atomic Force Microscopy
▶ Bioadhesion
▶ Biomimetics
▶ Gecko Adhesion
▶ Optical Tweezers
▶ Scanning Electron Microscopy

References

1. Gorb, S.N.: Uncovering insect stickiness: structure and properties of hairy attachment devices. Am. Entomol. **51**, 31–35 (2005)
2. Gorb, S.N.: Smooth attachment devices in insects. In: Casas, J., Simpson, S.J. (eds.) Advances in Insect Physiology: Insect Mechanics and Control. Adv. Insect Physiol. **34**, 81–116 (2008)
3. Autumn, K., Sitti, M., Liang, Y.C.A., Peattie, A.M., Hansen, W.R., Sponberg, S., Kenny, T.W., Fearing, R., Israelachvili, J.N., Full, R.J.: Evidence for van der Waals adhesion in gecko setae. Proc. Natl. Acad. Sci. USA **99**, 12252–12256 (2002)
4. Barnes, W.J.P.: Functional morphology and design constraints of smooth adhesive pads. Mater. Res. Soc. Bull. **32**, 479–485 (2007)
5. Federle, W., Barnes, W.J.P., Baumgartner, W., Drechsler, P., Smith, J.M.: Wet but not slippery: boundary friction in tree frog adhesive toe pads. J. R. Soc. Interface **3**, 689–697 (2006)
6. Bhushan, B.: Introduction to Tribology. Wiley, New York (2002)
7. Butt, H.-J., Barnes, W.J.P., del Campo, A., Kappl, M.: Capillary forces between soft, elastic spheres. Soft Matter. **6**, 5930–5936 (2010)

8. Ernst, V.: The digital pads of the tree frog, *Hyla cinerea*. 1. The epidermis. Tissue Cell **5**, 83–96 (1973)
9. Green, D.M.: Treefrog toe pads: comparative surface morphology using scanning electron microscopy. Can. J. Zool. **57**, 2033–2046 (1979)
10. Emerson, S.B., Diehl, D.: Toe pad morphology and mechanisms of sticking in frogs. Biol. J. Linn. Soc. **13**, 199–216 (1980)
11. Hanna, G., Barnes, W.J.P.: Adhesion and detachment of the toe pads of tree frogs. J. Exp. Biol. **155**, 103–125 (1991)
12. Smith, J.M., Barnes, W.J.P., Downie, J.R., Ruxton, G.D.: Structural correlates of increased adhesive efficiency with adult size in the toe pads of hylid tree frogs. J. Comp. Physiol. A **192**, 1193–1204 (2006)
13. Ghatak, A., Mahadevan, L., Chung, J.Y., Chaudhury, M.K., Shenoy, V.: Peeling from a biomimetically patterned thin elastic film. Proc. Roy. Soc. Lond. A **460**, 2725–2735 (2004)
14. Persson, B.N.J.: Wet adhesion with application to tree frog adhesive toe pads and tires. J. Phys. Cond. Matter **19**, 376110 (16 pp) (2007)
15. Scholz, I., Barnes, W.J.P., Smith, J.M., Baumgartner, W.: Ultrastructure and physical properties of an adhesive surface, the toe pad epithelium of the tree frog, *Litoria caerulea* White. J. Exp. Biol. **212**, 155–162 (2009)
16. Barnes, W.J.P., Perez Goodwyn, P., Nokhbatolfoghahai, M., Gorb, S.N.: Elastic modulus of tree frog adhesive toe pads. J. Comp. Physiol. A **197**, 969-978 (2011)
17. Samani, A., Zubovitz, J., Plewer, D.: Elastic moduli of normal and pathological human breast tissue: an inversion-technique-based investigation of 169 samples. Phys. Med. Biol. **52**, 1565–1576 (2007)
18. Vogel, S.: Comparative Biomechanics: Life's Physical World. Princeton University Press, Princeton (2003)
19. Perez Goodwyn, P., Peressadko, A., Schwarz, H., Kastne, V., Gorb, S.: Material structure, stiffness, and adhesion: why attachment pads of the grasshopper (*Tettigonia viridissima*) adhere more strongly than those of the locust (*Locusta migratoria*) (Insecta: Orthoptera). J. Comp. Physiol. A **192**, 1233–1243 (2006)
20. Barnes, W.J.P., Oines, C., Smith, J.M.: Whole animal measurements of shear and adhesive forces in adult tree frogs: insights into underlying mechanisms of adhesion obtained from studying the effects of size and scale. J. Comp. Physiol. A **192**, 1179–1191 (2006)
21. Smith, J.M., Barnes, W.J.P., Downie, J.R., Ruxton, G.D.: Adhesion and allometry from metamorphosis to maturation in hylid tree frogs – a sticky problem. J. Zool. **270**, 372–383 (2006)
22. Arzt, E., Gorb, S., Spolenak, R.: From micro to nano contacts in biological attachment devices. Proc. Natl. Acad. Sci. USA **100**, 10603–10606 (2003)
23. Autumn, K., Dittmore, A., Santos, D., Spenko, M., Cutkosky, M.: Frictional adhesion: a new angle on gecko attachment. J. Exp. Biol. **209**, 3569–3579 (2006)
24. Majumder, A., Ghatak, A., Sharmer, A.: Microfluidic adhesion induced by subsurface microstructures. Science **318**, 258–261 (2007)
25. Varenberg, M., Gorb, S.N.: Hexagonal surface micropattern for dry and wet friction. Adv. Mater. **21**, 483–486 (2009)
26. Varenberg, M., Gorb, S.N.: A beetle-inspired solution for underwater adhesion. J. R. Soc. Interface **5**, 383–385 (2008)
27. Barnes, W.J.P.: Tree frogs and tire technology. Tire Technol. Int. March 42–47 (1999)
28. Barnes, W.J.P., Smith, J., Oines, C., Mundl, R.: Bionics and wet grip. Tire Technol. Int. Dec. 56–60 (2002)
29. Vogel, M.J., Steen, P.H.: Capillary-based switchable adhesion, beetle inspired. Proc. Natl. Acad. Sci. USA **107**, 3377–3381 (2010)
30. Barnes, W.J.P., Pearman, J., Platter, J.: Application of peeling theory to tree frog adhesion, a biological system with biomimetic implications. Eur. Acad. Sci. E-Newsletter Sci. Technol. **1**(1), 1–2 (2008)

Adsorbate Adhesion

▶ Disjoining Pressure and Capillary Adhesion

Aerosol-Assisted Chemical Vapor Deposition (AACVD)

▶ Chemical Vapor Deposition (CVD)

AFM

▶ Robot-Based Automation on the Nanoscale

AFM Force Sensors

▶ AFM Probes

AFM in Liquids

Bart W. Hoogenboom
London Centre for Nanotechnology and Department of Physics and Astronomy, University College London, London, UK

Synonyms

Atomic force microscopy in liquids; Scanning force microscopy in liquids

Definition

Atomic force microscopy (AFM) in liquids is the application of AFM in liquid environment, i.e., in which both the surface under investigation and the scanning probe are immersed in liquid.

Overview and Definitions

AFM is a microscopy technique that can provide three-dimensional images of virtually any surface at nanometer-scale resolution. It relies on the force between a sharp probe and the surface, which is detected while scanning the probe over the sample. Unlike many other microscopy techniques at such a resolution, it can readily be applied in liquid environment.

An atomic force microscope consists of a sharp probe ("tip") mounted on a microfabricated cantilever beam and a mechanism ("scanner") to scan the tip over the surface at subnanometer resolution [1], see Fig. 1. Typically, an optical detection scheme is used to detect the deflection of the cantilever. Via the spring constant of the cantilever, the cantilever deflection can be translated to a force between tip and sample. For a rectangular lever, the spring constant is given by

$$k = \frac{Et^3 w}{4l^3},\tag{1}$$

where E is the Young's modulus of the cantilever material (typically silicon or siliconnitride), and t, w, and l are its thickness, width, and length, respectively, which are in the micron range. Most cantilevers for AFM have spring constants between 0.01 and 100 N/m. In its most common mode of operation, the deflection of the cantilever (and thus the tip–sample force) is kept constant by adjusting the vertical position of the sample with respect to the tip.

In most instruments, the cantilever deflection is recorded via the position of a laser beam that is deflected from the cantilever. Images of the surfaces are acquired by line-by-line scanning the surface and tracing its surface contours on a false-color scale. For AFM in liquid, the surface and the cantilever are immersed in liquid. Typically, the minimum liquid volume is some tens of microliters, which can be either contained in a closed liquid cell, or in the shape of

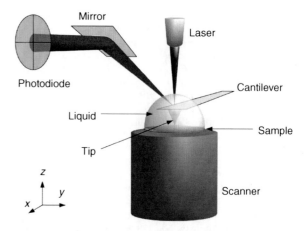

AFM in Liquids, Fig. 1 Schematic of AFM in liquid. A sharp tip is scanned across the sample surface while the tip–sample interaction is monitored via the deflection of the cantilever to which the tip is attached. The sample is mounted on a (usually piezoelectric) scanner for three-dimensional positioning with sub-nanometer accuracy. The sample, the tip, and the cantilever are immersed in liquid. The bending of the cantilever is usually detected via a laser beam deflected on a position-sensitive detector (4-quadrant photodiode)

a droplet that is formed by capillary forces between the sample surface and the cantilever holder. In principle, the liquid can be of arbitrary nature, though the optical deflection detection in most microscopes limit their use to liquids that are transparent for (near infra-) red light.

There are many possible imaging modes of AFM and their names are rather confusing. They are divided into static mode and a variety of dynamic modes in which the cantilever is oscillating [2]. The latter are of particular importance when the lateral (drag) forces on the sample need to be minimized, e.g., for imaging DNA molecules that are loosely bound to a flat surface. The most common modes of operation in liquid are static or contact mode and amplitude-modulation or tapping mode. In contact mode, the bending or deflection of the cantilever is used to detect the tip–sample force. In tapping mode, the cantilever is vertically oscillated above the sample, and the tip–sample interactions detected via the change (usually decrease) of amplitude of the cantilever oscillation. In variations along the same theme, the cantilever may be oscillated and tip–sample interactions are detected via changes in phase or resonance frequency of the cantilever [3]. An overview of the most relevant AFM modes in liquids is given in Table 1. Confusion may arise because the

AFM in Liquids, Table 1 Different modes of operation in AFM, with the parameter that is used to control the tip–sample distance while scanning over the surface

Mode of operation	Other common names	Control parameter
Static	Contact mode	Static deflection
Amplitude modulation (AM)	Tapping mode	Oscillation amplitude
	Intermittent contact AFM AC mode	
Frequency modulation (FM)	Non-contact (NC) AFM	Resonance frequency

commonly used terms contact, intermittent-contact, and non-contact often – but not always – refer to a mode of operation (as referred to in Table 1), and not necessarily to the actual contact (or lack of it) between the tip and the sample. As a particular example, FM ("non-contact") AFM in liquids is usually performed in the range of repulsive tip–sample interactions, with the tip in actual intermittent contact with the sample. In addition, if the amplitude is used to trace and control the tip–sample distance (i.e., tapping mode), the phase of the cantilever can be recorded simultaneously for phase imaging, which provides an additional means to measure tip–sample interactions. Or, in FM AFM, a frequency shift is used to trace elastic tip–sample interactions and control the tip–sample distance, while the oscillation amplitude can be kept constant by adjusting the cantilever drive signal. This cantilever drive signal then provides a simultaneous measure of the dissipative tip–sample interactions.

Physical and Chemical Principles

The spatial resolution of AFM critically depends on the sharpness of the tip and on the range of the tip–sample interactions. Ideally, only the very (nanometer-scale) end of the tip interacts with the surface, thus avoiding convolution effects due to interactions between the sample and the (micron-scale) bulk of the tip. The presence of long-range tip–sample forces is therefore decremental to the spatial resolution. On the other hand, the shorter the range of the tip–sample interaction, the more likely the tip is to enter in hard contact with the sample, which may damage the tip and/or the sample.

By immersing the tip and sample in liquid, it is not only possible to access scientifically and technologically interesting solid–liquid interfaces (including biological samples), but also to tune the tip–sample interactions by varying the ingredients of the liquid. In particular, the electrostatic sample interaction between the tip and a flat sample can be approximated by [4]:

$$F_{el} = g_\kappa \sigma_t \sigma_s e^{-\kappa z} + g_{2\kappa} \left(\sigma_t^2 + \sigma_s^2 \right) e^{-2\kappa z}, \quad (2)$$

where g_κ and $g_{2\kappa}$ are constants that depend on the tip geometry and the dielectric properties of the medium, σ_t and σ_s are the surface charges of the tip and the sample, respectively, z is the tip–sample distance, and κ^{-1} the Debye screening length [5],

$$\kappa^{-1} = \sqrt{\frac{\varepsilon_0 \varepsilon_b k_B T}{2e^2 I}}. \quad (3)$$

In the latter, ε_0 is the permittivity of vacuum (8.854×10^{-23} F/m), ε_b the permittivity of the bulk solution (or the dielectric medium), k_B is the Boltzmann's constant (1.38×10^{-23} J/K), and T is the temperature in Kelvin. The parameter I is the ionic strength of the medium,

$$I = \frac{1}{2} \sum_i z_i^2 c_i. \quad (4)$$

where z_i is the valence of ion i and c_i is its concentration. The Debye length is historically written as an inverse length (i.e., κ has units of m^{-1}). In typical biological buffers, the Debye length is between 1 and 10 nm.

In a chemically reactive medium, σ_t and σ_s are generally not zero. For typical tip terminations such as silicon oxide or silicon nitride in water, $\sigma_t < 0$ at neutral pH. As can be seen from Eq. 2, the strength of the electrostatic tip–sample interaction depends on σ_t and σ_s. These can be changed by adding or removing solutes that react with or bind to the tip and sample surfaces. In aqueous solutions, this is most readily achieved by changing the pH (i.e., adding H_3O^+ or OH^-). Moreover, the dielectric properties – and in particular the ionic strength (Eq. 4) – of the medium can be tuned to control the range of the electrostatic tip–sample interaction. This provides a significant advantage over AFM in gaseous mediums and vacuum environment, and is one of the reasons why the first true atomic resolution (see below) AFM images were

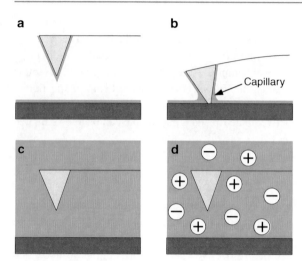

AFM in Liquids, Fig. 2 (a) Capillary forces for AFM in air arise due to the thin water layers that are usually present at both tip and sample under ambient conditions. (**b**) At small tip–sample distances, they pull the tip toward the surface to create a larger contact area and thus limit the lateral resolution. (**c**) In liquid environment, these capillary forces are absent. (**d**) Moreover, long-range electrostatic forces can be screened by ions in the liquid

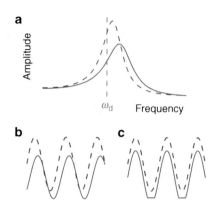

AFM in Liquids, Fig. 3 (a) Schematic resonance curve of a freely oscillating cantilever (*dashed, blue curve*), based on a model of a simple harmonic oscillator. In the presence of a (repulsive) elastic tip–sample force as well as dissipative tip–sample interactions, the resonance shifts to higher frequencies, broadens and decreases in amplitude (*red curve*). At a fixed driving frequency ω_d, such as in tapping mode, the effects of elastic or dissipative interactions will cause the amplitude to decrease or increase, depending on the choice of ω_d. (**b**) Sinusoidal oscillation of a cantilever above the surface (*dashed, blue*). Close to the sample, the oscillation amplitude decreases and changes phase (*red*). (**c**) For heavily damped cantilever oscillations ($Q \sim 1$), however, the harmonic-oscillator analysis is not valid any more. The tip–sample interaction will only affect the bottom part of the oscillation, where the tip is closest to the sample

obtained, in contact mode, in aqueous environment rather than in vacuum or air [6].

Compared to AFM in air, AFM in liquids has the additional advantage of preventing the capillary forces that arise from the humid coverage of both sample and tip under ambient conditions, as illustrated in Fig. 2. The capillary forces create a strong pull on the tip toward the sample at distances of many nanometers. On close approach, this results in a tip–sample contact area that covers many atoms at the surface, thus prohibiting high resolution. Under such conditions, AFM is likely to result in one of its most extensively documented artifacts: The apparent "atomic resolution" that results from the periodic interaction averaged over many atoms of tip and surface. This represents the atomic periodicity of the surface rather than individual atomic-scale features such as atomic defects (which are often used to demonstrate "true atomic resolution").

Apart from electrostatic and capillary forces, AFM in liquids is also (and not exclusively) sensitive to atomic bonding forces and hard-core repulsion, van der Waals interactions, dissolution/solvation forces, and general dissipative interactions.

The latter are particularly important in the dynamic modes of AFM operation. To illustrate this, it is helpful to depict the cantilever resonance for different elastic and dissipative interactions, see Fig. 3.

A repulsive (elastic) tip–sample interaction increases the effective spring constant of the cantilever and thus increases its resonance frequency, shifting the whole resonance curve to the right. Any dissipative interaction reduces the sharpness of the resonance curve and the maximum of the curve. The quality factor Q is the most common measure for this sharpness, defined as $\omega_0/\Delta\omega$, where ω_0 is the natural frequency of the resonator and $\Delta\omega$ the width of the resonance curve at $1/\sqrt{2} \times$ its maximum amplitude. In tapping mode, the driving frequency is usually set just below the resonance frequency, such that both elastic and dissipative interactions cause the amplitude to decrease when the tip comes into contact with the sample. In FM AFM, the shift in the resonance frequency provides a direct measure of the elastic interaction, and the conversion to a force is straightforward when the amplitude of the oscillation is (kept) constant [7]. The dissipated energy can be deduced from the energy that is required to maintain a fixed amplitude. Thus, tapping mode is sensitive to a mixture of elastic and dissipative tip–sample interactions, whereas FM AFM can track elastic and dissipative interactions separately (and simultaneously).

This analysis relies on the cantilever behaving as a simple harmonic oscillator, which is appropriate for $Q \gg 1$, typical for operation in vacuum ($Q \stackrel{>}{\sim} 1000$) and air ($Q \stackrel{>}{\sim} 100$). However, in liquid, viscous damping reduces Q to $\stackrel{<}{\sim} 10$. In particular for softer cantilevers $k \stackrel{<}{\sim} 1 \mathrm{N/m}$, $Q \approx 1$, and the contact between tip and sample will prevent the cantilever from following a harmonic, sinusoidal oscillation. The cantilever oscillation will simply be topped off at the bottom of the oscillation, at the closest tip–sample approach. In this case, significant information on the tip–sample interaction will be contained in higher harmonics of the oscillation. The interpretation of cantilever oscillation in terms of tip–sample forces thus becomes considerably more complicated.

The low quality factor has another undesirable side effect for AFM operation that involves oscillating the cantilever. Cantilevers are usually driven by a small piezoelectric element that drives not only the cantilever itself, but also its (macroscopic) support chip and often the whole cantilever holder. In vacuum and air, the high Q singles out the cantilever resonance from other mechanical resonances in the instrument. In liquid, this is not the case any more, and a typical excitation spectrum contains a so-called forest of peaks, i.e., a convolution of the broad cantilever resonance with many sharp(er) mechanical resonances from the microscope, as illustrated in Fig. 4. Again, this complicates the interpretation of the cantilever behavior in terms of a simple harmonic oscillator. It also makes the choice of optimum driving frequency much less straightforward (usually, this is done by trial and error). Moreover, since this excitation spectrum depends on the macroscopic geometry of the fluid cell, it is much more dependent on drift than the cantilever alone. For these reasons, there are a number of alternative methods to only drive the microscopic cantilever (and no macroscopic parts), such as magnetic and optical actuation. In the former, the cantilever is coated with a magnetic material and driven by an AC magnetic field; in the latter, an AC-modulated actuation laser locally heats the metal-coated surface of cantilever, which acts as a bimetal and thus transduces

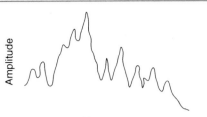

AFM in Liquids, Fig. 4 Excitation spectrum such as can be obtained for a cantilever in water that is actuated by piezoelectrically driving the cantilever holder rather than the cantilever alone. Note the contrast with the ideal resonance shape in Fig. 3a

laser power to cantilever deflection. These and other methods have the disadvantage of adding complexity to the instrument and eventually to the microfabricated cantilever itself.

These concerns can partly be addressed by so-called Q-control or self-oscillatory methods [8], in which the signal from the cantilever oscillation itself is phase-shifted, amplified, and added to the actuation signal. This positive feedback is very familiar in electronic oscillatory circuits. It causes any change to the oscillation to ring for a longer time, stretching its effect over many oscillations. As a result the effective Q of the cantilever oscillation is enhanced, and the cantilever is forced to follow a perfectly harmonic, sinusoidal oscillation. As for problems related to the forest of peaks in nonideal cantilever excitation, these are not resolved by Q control, since the positive feedback does not distinguish between the cantilever resonance and other mechanical resonances. Q control or self-oscillatory methods can be applied to each mode of operation (see Table 1), but for tapping mode has the side effect of slowing down the measurement: Changes in amplitude occur at a timescale of $\sim Q/\omega_0$, unlike changes in phase or resonance frequency, which are instantaneous. In FM AFM, the cantilever can even entirely be driven by its phase-shifted and amplified thermal noise.

Applications

AFM has been used to probe a vast amount of interfaces between (hard and soft) condensed matter and liquids. It is of particular importance for the study of surfaces that are instable or show distinctly different behavior when taken out of the liquid, and for the study of dynamic processes that depend on an exchange of material (ions, macromolecules) between the surface and the liquid. AFM in liquids has been important for the study of, among others, crystal growth/dissolution, polymer science, and the behavior of liquids in confined geometries. The vast majority of its applications, however, lie in the life sciences [9]: For most biomolecular structures, aqueous solutions represent their natural environment. Their structures and functions are strongly dependent on the presence of water, a number of ions, and other macromolecules. The structure and function of biological samples can thus be studied at (sub)nanometer resolution while they are still "alive," and for varying liquid contents. The following represents some illustrative examples of the use of AFM in aqueous solutions on easily accessible samples, with a strong focus on biological applications.

Up to now, highest resolution AFM images in liquid have been obtained on atomically flat and inorganic surfaces, with calcite and muscovite mica as most common examples. On such surfaces, the lateral resolution can be less than an Ångström and the vertical resolution a few picometers. Though initially such resolution was only obtained in contact mode [6], more recently FM AFM has become popular as a method for atomic-resolution imaging that is significantly less susceptible to thermal drift. One of the key steps toward high-resolution imaging in liquids was the development of low-noise deflection sensors [10, 11]. This has brought the measurement noise down to the thermal noise of the cantilevers, even for stiff and heavily damped cantilevers in liquid. As a result, surfaces can now be stably measured with Ångström amplitudes of cantilever oscillation. This enhances the sensitivity to short-range forces and yields images such as depicted in Fig. 5. Because it is flat, hydrophilic, and easily cleaved, mica is one of the substrates of choice for adsorbing molecules in AFM experiments in aqueous solutions.

AFM has been successful in imaging DNA as well as DNA–protein complexes. DNA is routinely adsorbed on mica, where it is generally assumed to adopt a two-dimensional projection of its original three-dimensional configuration [13]. The rather flat geometry required for high-resolution AFM is a natural one for proteins that are embedded in a lipid membrane. In particular when two-dimensional crystals are available, such as for bacteriorhodopsin,

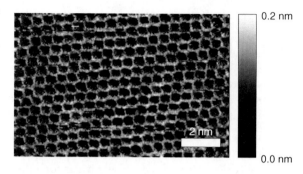

AFM in Liquids, Fig. 5 Atomic-resolution image of mica in (aquaeous) buffer solution, obtained by FM AFM (Reproduced from [12], with permission)

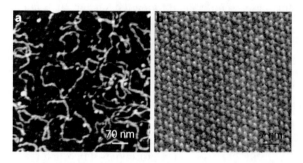

AFM in Liquids, Fig. 6 (a) Tapping-mode AFM image of DNA adsorbed on mica. Image courtesy Elliot Menter. (b) Two-dimensional crystal of the membrane protein bacteriorhodopsin, obtained by FM AFM on the extracellular side of the protein. Image courtesy Carl Leung

high-resolution images can readily be obtained in the various modes of AFM operation [14]. DNA and bacteriorhodopsin represent some of the most often imaged biomolecules by AFM in liquid (Fig. 6).

On a completely different scale, AFM can image objects as large as whole cells (Fig. 7). Due to thermal fluctuations, vibrations, and the overall softness of the cell, the spatial resolution is considerably lower than that obtained on flat surfaces or molecules directly adsorbed on a hard substrate. As, for cell imaging, alone, the advantage of AFM over optical techniques is therefore limited.

AFM can also be used to deliberately probe sample properties other than structure [15]. Single molecules that are tethered between the tip and the substrate can be deliberately stretched, unfolded, and allowed to refold at controlled load force [16], similar to optical-tweezers experiments. Using the same force

AFM in Liquids, Fig. 7 $100 \times 100 \times 3 \ \mu m^3$ contact-mode AFM topograph of an osteoblast (bone cell) adsorbed on a glass coverslip. Image courtesy Guillaume Charras

spectroscopy techniques, force–distance curves can be obtained on whole cells, yielding information about their local elasticity, or – when using chemically modified tips – local binding sites for specific ligands [17].

Finally, fast AFM methods [18] have improved the image rate from the typical frame per minute up to many frames per second. This gives direct access to the kinetics of molecular-scale processes at nanometer resolution.

Cross-References

▶ AFM
▶ AFM Probes
▶ AFM, Non-Contact Mode
▶ AFM, Tapping Mode
▶ Atomic Force Microscopy
▶ Force Modulation in Atomic Force Microscopy
▶ Friction Force Microscopy
▶ Imaging Human Body Down to Molecular Level
▶ Kelvin Probe Force Microscopy
▶ Magnetic Resonance Force Microscopy
▶ Mechanical Properties of Hierarchical Protein Materials
▶ Nanomedicine
▶ Nanotechnology
▶ Nanotribology
▶ Optical Tweezers
▶ Optomechanical Resonators
▶ Surface Forces Apparatus

References

1. Sarid, D.: Scanning Force Microscopy with Applications to Electric, Magnetic, and Atomic Forces, 2nd edn. Oxford University Press, Oxford (1994)
2. Garcia, R., Perez, R.: Dynamic atomic force microscopy methods. Surf. Sci. Rep. **47**, 197–301 (2002)
3. Giessibl, F.J.: Advances in atomic force microscopy. Rev. Mod. Phys. **75**, 949–983 (2003)
4. Butt, H.-J.: Electrostatic interaction in atomic force microscopy. Biophys. J. **60**, 777–785 (1991)
5. Israelachvili, J.N.: Intramolecular and Surface Forces, 3rd edn. Academic, Oxford (2011)
6. Ohnesorge, F., Binnig, G.: True atomic resolution by atomic force microscopy through repulsive and attractive forces. Science **260**, 1451–1456 (1993)
7. Sader, J.E., Jarvis, S.P.: Accurate formulas for interaction force and energy in frequency modulation force spectroscopy. Appl. Phys. Lett. **84**, 1801–1803 (2004)
8. Bhushan, B., Fuchs, H., Hosaka, S. (eds.): Applied Scanning Probe Methods, vol. I. Springer, Berlin/Heidelberg (2004)
9. Morris, V.J., Gunnig, A.P., Kirby, A.R.: Atomic Force Microscopy for Biologists. World Scientific, River Edge (1999)
10. Hoogenboom, B.W., Frederix, P.L.T.M., Yang, J.L., Martin, S., Pellmont, Y., Steinacher, M., Zäch, S., Langenbach, E., Heimbeck, H.-J., Engel, A., Hug, H.J.: A Fabry-Perot interferometer for micrometer-sized cantilevers. Appl. Phys. Lett. **86**, 074101 (2005)
11. Fukuma, T., Kimura, M., Kobayashi, K., Matsushige, K., Yamada, H.: Development of low noise cantilever deflection sensor for multienvironment frequency-modulation atomic force microscopy. Rev. Sci. Instrum. **76**, 053704 (2005)
12. Khan, Z., Leung, C., Tahir, B., Hoogenboom, B.W.: Digitally tunable, wide-band amplitude, phase, and frequency detection for atomic-resolution scanning force microscopy. Rev. Sci. Instrum. **81**, 073704 (2010)
13. Hansma, H.G.: Surface biology of DNA by atomic force microscopy. Annu. Rev. Phys. Chem. **52**, 71–92 (2001)
14. Müller, D.J., Engel, A.: Atomic force microscopy and spectroscopy of native membrane proteins. Nat. Protoc. **2**, 2191–2197 (2007)
15. Müller, D.J., Dufrêne, Y.F.: Atomic force microscopy as a multifunctional molecular toolbox in nanobiotechnology. Nat. Nanotechnol. **5**, 261–269 (2008)
16. Fisher, T.E., Marszalek, P.E., Fernandez, J.M.: Stretching single molecules into novel conformations using the atomic force microscope. Nat. Struct. Biol. **7**, 719–724 (2000)
17. Hinterdorfer, P., Dufrêne, Y.F.: Detection and localization of single molecular recognition events using atomic force microscopy. Nat. Method **3**, 347–355 (2006)
18. Yamamoto, D., Uchihashi, T., Kodera, N., Yamashita, H., Nishikori, S., Ogura, T., Shibata, M., Ando, T.: High-speed atomic force microscopy techniques for observing dynamic biomolecular processes. In: Walter, N.G. (ed.) Methods in Enzymology, vol. 475, pp. 541–564. Academic, Burlington (2010)

AFM Probes

Kenji Fukuzawa
Department of Micro System Engineering, Nagoya
University, Nagoya, Chikusa-ku, Japan

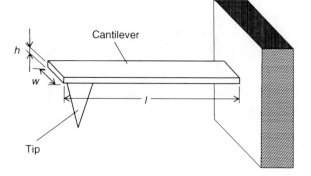

AFM Probes, Fig. 1 Schematic of a microcantilever probe

Synonyms

AFM force sensors; AFM tips

Definition

AFM probes are transducers that convert the interaction force with a sample surface into a deformation or a change of the vibrational state of the probe. Most probes consist of a sharp microtip and a force transducer. The former determines the lateral resolution of the AFM and the latter provides the force sensitivity.

Overview

A typical AFM probe consists of a sharp tip and a microcantilever, which plays the role of a force transducer. In this case, the interaction force between the tip and sample deflects the cantilever. The deflection can be detected by a displacement measurement method, such as an optical lever or interference techniques. Assuming that the deflection and the spring constant of the cantilever are Δz and k, respectively, the interaction force F is given by

$$F = k\Delta z. \qquad (1)$$

In contact AFM mode, the probe height is controlled so that the probe deflection is constant, which means that the interaction force does not change during the probe scanning. Mapping the controlled height can thus provide the topography of the sample.

In the early days, a metal foil with a glued microparticle or sharpened metal wire was used as an AFM probe. Here, the particle or wire acted as a tip [1]. Improving the force sensitivity and reproducibility was not easy with those probes. Then, microfabricated cantilever probes were introduced [2]. These probes are made of silicon nitride or silicon and are produced by microfabrication techniques such as anisotropic chemical etching. Microfabricated probes enabled to improve the force sensitivity and expand the freedom of the probe design. They were also suitable for mass production. Therefore, microfabricated probes are the most common at present and various types of them are commercially available nowadays. Undoubtedly, the microfabricated probes have been playing an important role in AFM's becoming one of the most widely used methods in nanotechnology.

Probe Design

Since the specifications of the tip are rather limited by the fabrication method, the main part of the probe design is the force transducer, such as the rectangular micro cantilever sketched in Fig. 1. In addition to rectangular cantilevers, V-shaped cantilevers are widely used, especially in contact AFM mode. The force sensitivity of the probe should be high enough to detect weak interaction forces. From Eq. 1, the cantilever should convert a small force F into a large displacement Δz, which requires a small spring constant $k = F/\Delta z$. This means that higher sensitivity requires softer cantilevers. In addition to the force sensitivity, the robustness against mechanical disturbances from the environment should be considered. If the vibrations induced by the disturbances exceed the deflection due to the interaction force, the AFM signal is buried in the noise. The vibrations due to disturbances are amplified when the disturbance frequency is around the resonance frequency (natural frequency) of the probe. Since the disturbance frequencies are generally low, the resonance frequency should be set at a high value, typically larger than about 10 kHz. If the force transducer of the probe is simplified as a point mass and spring system, the resonance frequency is given by

$$f_r = \frac{1}{2\pi}\sqrt{\frac{k}{m}}, \tag{2}$$

where m is the mass of the transducer. The sensitivity needs low values of k whereas the robustness against the environmental disturbances requires high values of f_r. This means that low values of m are required. Therefore, the force transducer of the probe should be soft and small. If a rectangle cantilever is chosen as a force transducer, it has to be thin and small. Microfabricated cantilevers meet this demand and, for this reason, they became standard probes for AFM.

The design details of a rectangular cantilever are considered below. If the length, width, and thickness of the cantilever are l, w, and h, respectively, the spring constant k is given by

$$k = \frac{Ewh^3}{4l^3}, \tag{3}$$

where E is the Young's modulus of the cantilever. It should be noted that the spring constant is proportional to $(h/l)^3$. This means that longer and thinner cantilevers can provide higher force sensitivity. The force F to be measured depends on the interaction between the tip and sample and the deflection Δz depends on the sensitivity of the displacement measurement method. Therefore, the spring constant k should be determined by considering the typical values of F and Δz for the targeted samples and the measurement system. On the other hand, the resonance frequency of a rectangular cantilever is given by

$$f_r = \frac{\lambda^2}{4\pi}\frac{h}{l^2}\sqrt{\frac{E}{3\rho}}, \tag{4}$$

where ρ is the density of the cantilever and λ is a constant that depends on the vibrational mode. For the first mode λ is about 1.875. It should be noted that the cantilever width is not included in Eq. 4. Since f_r is proportional to h/l^2, higher resonance frequencies f_r require shorter and thicker cantilevers. However, as described above, high sensitivity requires

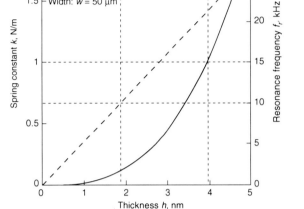

AFM Probes, Fig. 2 Thickness dependencies of the spring constant and resonance frequency of a microcantilever

opposite conditions. Thus, a balance between the spring constant and the resonance frequency has to be found in designing the probe. An example for contact mode cantilevers is shown in Fig. 2, where the thickness dependencies of the spring constant and resonance frequency are plotted using Eqs. 3 and 4 for a length $l = 500$ μm and a width $w = 50$ μm, respectively. If a spring constant of less than 1 N/m and a resonance frequency larger than 10 kHz are selected, the former curve requires a thickness of more than 1.9 μm and the latter curve requires a thickness of less than 4.0 μm. Therefore, a thickness in the overlapping range should be selected. A probe with these dimensions is not easy to fabricate by conventional machining and requires micromachining techniques.

Measurements of the spring constant k and resonance frequency f_r are important for converting the measured deflection signal into a force signal. The resonance frequency f_r can be measured by varying the drive frequency as it is routinely done in commercial AFM equipments. However, the accurate measurement of the spring constant k is not easy. An approximate value can be calculated from the dimensions of the cantilever using Eq. 3. An experimental method using the thermal fluctuation was presented in Ref. [3].

AFM Probes,
Fig. 3 Fabrication process of
a microfabricated probe: (**a**)
silicon substrate, (**b**) formation
of pit for a microtip, (**c**)
deposition of silicon nitride
layer, (**d**) patterning of
cantilever, and (**e**) removal of
substrate

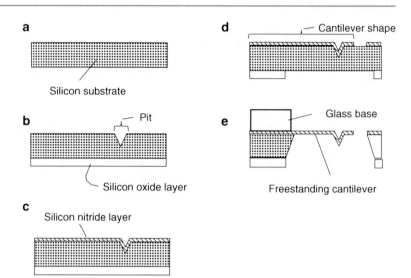

AFM Probes,
Fig. 3 Fabrication process of a microfabricated probe: (**a**) silicon substrate, (**b**) formation of pit for a microtip, (**c**) deposition of silicon nitride layer, (**d**) patterning of cantilever, and (**e**) removal of substrate

Many types of AFMs using a vibrating cantilever have been developed, such as cyclic contact mode and non-contact mode AFMs. The probe vibrations can be induced using a quartz-oscillator. In this case, the resonance properties, especially the quality factor Q, are very important. The quality factor Q is defined as $Q = 1/2\zeta$, where ζ is the damping coefficient. A probe with a high Q is able to resonate effectively and its vibration is hard to damp. The minimum detectable force gradient in the frequency-modulation detection, which is widely used in highly sensitive AFMs such as non-contact AFMs and magnetic force microscopy, is given by

$$\left(\frac{\partial F}{\partial z}\right)_{\min} = \frac{2}{A}\sqrt{\frac{kk_{\mathrm{B}}TB}{\omega_{\mathrm{r}}Q}}, \tag{5}$$

where B, ω_{r}, and A are the measurement band width, resonance angular frequency, and vibration amplitude, respectively [4]. Higher force sensitivity requires a higher Q. Since the vibration damping of the probe is caused by the internal friction and by friction between the probe and environmental gas such as air, the quality factor Q usually depends on the probe structure and environment. Typical values of Q of silicon cantilevers are on the order of 10–100 in air whereas they can increase up to more than 10,000 in vacuum. A method to increase Q electronically was presented, where the deflection signal is fed back to the drive signal after the phase of the deflection signal is shifted by 90° [5].

Probe Fabrication

At present, most of the probes are fabricated by microfabrication techniques and various types of probes have been designed. Moreover, a considerable number of them are commercially available because microfabricated probes are suitable for mass production. The fabrication methods have also been improved, corresponding to a general improvement of the probe quality. Here, the fabrication methods for two typical types of probes are outlined. We will first address the probes made of silicon nitride. These probes are widely used in contact AFM mode. The fabrication process is shown in Fig. 3 [6]. In this method, a small pit that is formed in a silicon substrate by anisotropic wet etching is used as a mold. The silicon oxide layer is formed as an etching mask for the later substrate removal (Figs. 3a and b). A silicon nitride film is deposited onto the substrate and the cantilever is patterned by lithography. At this point, the cantilever with tip is shaped (Figs. 3c and d). By attachment of the glass base by anodic bonding and removal of the silicon substrate by wet etching, a freestanding cantilever is obtained (Fig. 3e). Another type of probes are silicon cantilevers [7]. Silicon cantilevers are the most widely used probes in AFM and they are operated in various modes. By using the undercut etching of the underneath of a circular mask, a tip is formed. Then the cantilever is patterned. The substrate silicon is removed by wet etching with protection of the fabricated tip and cantilever. Sharpening the tip end can be achieved by thermal oxidation followed by oxide removal [6].

Future Directions for Research

In addition to standard silicon or silicon nitride cantilever probes, various types of probes have been recently introduced. In a piezoresistive probe a piezoresistive strain sensor is embedded with a silicon cantilever [8]. The embedded sensor detects the cantilever deflection and makes the detection system such as an optical lever unnecessary. This not only makes the setup compact but also provides easier operation in the special case of liquid or vacuum environments. Along the growth of microfabricated cantilever probes, probes that use a quartz-oscillator as a force transducer have also been developed. Various types of oscillators such as tuning fork or linear-extension resonators are used [9]. In many probes, a microtip is manually glued to the quartz-oscillator although the oscillator is microfabricated. Quartz-oscillator probes have advantages of high Q and large spring constant. The latter allows to avoid the jump-in of the probe. In addition, quartz-oscillator probes provide self-sensing as piezoresistive probes.

As stated above, the improvement of the probes is essential for the development of AFM. Recent advances in micro/nanomachining are promoting this improvement, which is expected to open up new applications of AFM.

Cross-References

► Atomic Force Microscopy
► Friction Force Microscopy
► MEMS
► Micromachining
► NEMS
► Scanning Probe Microscopy

References

1. Binnig, C., Quate, C.F., Gerber, Ch: Atomic force microscope. Phys. Rev. Lett. **56**(9), 930–933 (1986)
2. Binnig, G., Gerber, C., Stoll, E., Albrecht, T.R., Quate, C.F.: Atomic resolution with atomic force microscope. Europhys. Lett. **3**(12), 1281–1286 (1987)
3. Hutter, J.L., Bechhoefer, J.: Calibration of atomic-force microscope tips. Rev. Sci. Instrum. **64**(7), 1868–1873 (1993)
4. Albrecht, T.R., Grutter, P., Horne, D., Rugar, D.: Frequency modulation detection using high-Q cantilevers for enhanced force microscope sensitivity. J. Appl. Phys. **69**(2), 668–673 (1991)
5. Anczykowski, B., Cleveland, J.P., Krüger, D., Elings, V., Fuchs, H.: Analysis of the interaction mechanisms in dynamic mode SFM by means of experimental data and computer simulation. Appl. Phys. **A66**, S885–S889 (1998)
6. Albrecht, T.R., Akamine, S., Carver, T.E., Quate, C.F.: Microfabrication of cantilever styli for the atomic force microscope. J. Vac. Sci. Technol. **A8**, 3386–3396 (1990)
7. Wolter, O., Bayer, Th, Greschner, J.: Micromachined silicon sensors for scanning force microscopy. J. Vac. Sci. Technol. **B9**, 1353–1357 (1991)
8. Tortonese, M., Barrett, R.C., Quate, C.F.: Atomic resolution with an atomic force microscope using piezoresistive detection. Appl. Phys. Lett. **62**(8), 834–836 (1993)
9. Giessibl, F.J.: Atomic resolution on Si(111)-(7 × 7) by noncontact atomic force microscopy with a force sensor based on a quartz tuning fork. Appl. Phys. Lett. **76**(11), 1470–1472 (2000)

AFM Tips

► AFM Probes

AFM, Non-contact Mode

Hendrik Hölscher
Karlsruher Institut für Technologie (KIT), Institut für Mikrostrukturtechnik, Karlsruhe, Germany

Definition

The *noncontact atomic force microscopy* (NC-AFM) is a specific AFM technique primarily developed for

AFM, Non-contact Mode,
Fig. 1 The schematic set-up
of a dynamic force microscope
based on the frequency
modulation technique often
used in UHV. A significant
feature is the positive feedback
of the self-driven cantilever.
The detector signal is
amplified and phase shifted
before it is used to drive the
piezo. The measured quantity
is the frequency shift due to
the tip-sample interaction,
which serves as the feedback
signal for the cantilever-
sample distance

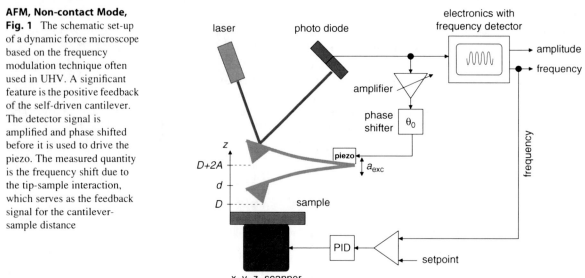

Overview

To obtain high resolution images with an atomic force microscope it is most important to prepare clean sample surfaces free from unwanted adsorbates. Therefore, these experiments are usually performed in ultra-high vacuum with pressures below 1×10^{-10} mbar. As a consequence most dynamic force microscope experiments in vacuum utilize the so-called *frequency modulation* (FM) detection scheme introduced by Albrecht et al. [1]. In this mode the cantilever is self-oscillated, in contrast to the AM- or tapping-mode (▶ AFM, Tapping Mode). The FM-technique enables

application in vacuum where standard AFM cantilevers made from silicon or silicon nitride exhibit very high quality factors Q, what makes the response of the system slow if driven in AM or tapping mode (▶ AFM, Tapping Mode). The technique to oscillate the cantilever also in high Q environments is called *frequency-modulation* (FM) mode. In contrast to the tapping or AM mode typically applied in air or liquids, this approach features a so-called self-driven oscillator, which uses the cantilever deflection itself as drive signal, thus ensuring that the cantilever instantaneously adapts to changes in the resonance frequency. The NC-AFM technique is the method of choice to obtain true atomic resolution on non-conduction surfaces with an atomic force microscope.

the imaging of single point defects on clean sample surfaces in vacuum and its resolution is comparable with the scanning tunneling microscope, while not restricted to conducting surfaces [2–6]. In the years after the invention of the FM-technique the term *non-contact atomic force microscopy* (NC-AFM) was established, because it is commonly believed that a repulsive, destructive contact between tip and sample is prevented by this technique.

Set-Up of FM-AFM

In vacuum applications, the Q-factor of silicon cantilevers is in the range of 10,000–30 000. High Q-factors, however, limit the acquisition time (bandwidth) of a dynamic force microscopy, since the oscillation amplitude of the cantilever needs a long time to adjust. This problem is avoided by the FM-detection scheme based on the specific features of a self-driven oscillator.

The basic set-up of a dynamic force microscope utilizing this driving mechanism is schematically shown in Fig. 1. The movement of microfabricated cantilevers is typically measured with the laser beam deflection method or an interferometer. A self-detecting sensor like a tuning fork do not need an additional detection sensor. In any case the amplitude signal fed back into an amplifier with an *automatic gain control* (AGC) and is subsequently used to excite the piezo oscillating the cantilever. The time delay between the excitation signal and cantilever deflection is adjusted by a time ("phase") shifter to a value of $\approx 90°$, since this ensures an oscillation at resonance. Two different modes have been established: The *constant amplitude*-mode [1], where the oscillation amplitude A is kept at a constant value by the AGC, and the *constant excitation mode* [7], where the excitation amplitude is kept constant. In the following, however, only the constant amplitude mode is discussed.

The key feature of the described set-up is the positive feedback-loop which oscillates the cantilever always at its resonance frequency f [8]. The reason for this behavior is that the cantilever serves as the frequency determining element. This is in contrast to an external driving of the cantilever in tapping mode by a frequency generator (▶ AFM, Tapping Mode). If the cantilever oscillates near the sample surface, the tip-sample interaction alters its resonant frequency, which is then different from the eigenfrequency f_0 of the free cantilever. The actual value of the resonant frequency depends on the nearest tip-sample distance and the oscillation amplitude. The measured quantity is the *frequency shift* Δf, which is defined as the difference between both frequencies ($\Delta f := f - f_0$). For imaging the frequency shift Δf is used to control the cantilever sample distance. Thus, the frequency shift is constant and the acquired data represents planes of constant Δf, which can be related to the surface topography in many cases.

Origin of the Frequency Shift

Before presenting experimental results obtained in vacuum the origin of the frequency shift is analyzed in mode detail. A good insight into the cantilever dynamics is given by looking at the tip potential displayed in Fig. 2. If the cantilever is far away from the sample surface, the tip moves in a symmetric parabolic potential (dotted line), and its oscillation is harmonic. In such a case, the tip motion is sinusoidal and the resonance frequency is given by the eigenfrequency f_0 of the cantilever. If, however, the cantilever approaches the sample surface, the potential – which determines the tip oscillation – is modified to an effective potential V_{eff} (solid line) given by the sum of the parabolic potential and the tip-sample interaction potential V_{ts} (dashed line). This effective potential differs from the original parabolic potential and shows an asymmetric shape.

As a result of this modification of the tip potential the oscillation becomes anharmonic, and the resonance

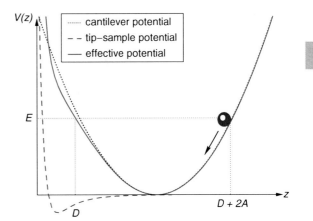

AFM, Non-contact Mode, Fig. 2 The frequency shift in dynamic force microscopy is caused by the tip-sample interaction potential (*dashed line*), which alters the harmonic cantilever potential (*dotted line*). Therefore, the tip moves in an anharmonic and asymmetric effective potential (*solid line*)

frequency of the cantilever depends now on the oscillation amplitude A. Since the effective potential experienced by the tip changes also with the nearest distance D, the frequency shift is a functional of both parameters ($\Rightarrow \Delta f := \Delta f(D, A)$).

Figure 3 displays some experimental frequency shift versus distance curves for different oscillation amplitudes [9]. The obtained experimental frequency shift vs. distance curves show a behavior expected from the simple model explained above. All curves show a similar overall shape, but differ in magnitude in dependence of the oscillation amplitude and the nearest tip-sample distance. During the approach of the cantilever towards the sample surface, the frequency shift decreases and reaches a minimum. With a further reduction of the nearest tip-sample distance, the frequency shift increases again and becomes positive. For smaller oscillation amplitudes, the minimum of the $\Delta f(z)$-curves is deeper and the slope after the minimum is steeper than for larger amplitudes, i.e., the overall effect is larger for smaller amplitudes.

This can be explained by the simple potential model as well: A decrease of the amplitude A for a fixed nearest distance D moves the minimum of the effective potential closer to the sample surface. Therefore, the relative perturbation of the harmonic cantilever potential increases, which increases also the absolute value of the frequency shift.

AFM, Non-contact Mode, Fig. 3 (a) Experimental frequency shift versus distance curves acquired with a silicon cantilever ($c_z = 38$ N/m; $f_0 = 171$ kHz) and a graphite sample for different amplitudes (54–180 Å) in UHV at low temperature ($T = 80$ K). (b) Transformation of all frequency shift curves shown in (a) to one universal curves using Eq. 6. The normalized frequency shift $\gamma\,(D)$ is nearly identical for all amplitudes. (c) The tip-sample force calculated with the experimental data shown in (a) and (b) using the formula Eq. 7

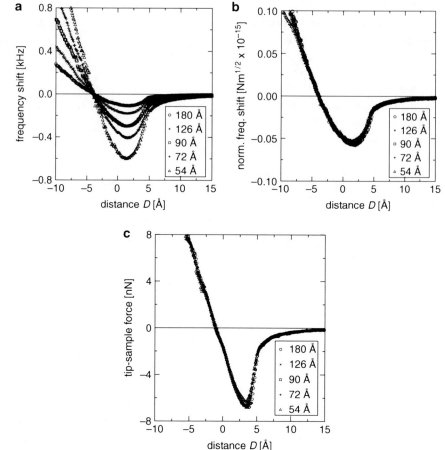

Theory of FM-AFM

As already described in the previous subsection it is a specific feature of the FM-modulation technique that the cantilever is "self-driven" by a positive feedback loop. Due to this experimental set-up, the corresponding equation of motion is different from the case of the externally driven cantilever discussed for the tapping-mode. The external driving term has to be replaced in order to describe the self-driving mechanism correctly. Therefore, the equation of motion is given by

$$m\ddot{z}(t) + \frac{2\pi f_0 m}{Q}\dot{z}(t) + c_z(z(t) - d)$$
$$+ \underbrace{gc_z(z(t - t_0) - d)}_{driving} \qquad (1)$$
$$= F_{ts}[z(t), \dot{z}(t)].$$

where $z := z(t)$ represents the position of the tip at the time t; c_z, m, and Q are the spring constant, the effective mass, and the quality factor of the cantilever, respectively. $F_{ts} = - (\partial\,V_{ts})/(\partial z)$ is the tip-sample interaction force. The last term on the left describes the active feedback of the system by the amplification of the displacement signal by the *gain factor g* measured at the retarded time $t - t_0$.

The frequency shift can be calculated from the above equation of motion with the ansatz

$$z(t) = d + A\cos(2\pi ft) \qquad (2)$$

describing the stationary solutions of Eq. 1. Again, it is assumed that the cantilever oscillations are more or less sinusoidal and develop the tip-sample force F_{ts}

into a Fourier-series. As a result the following equation for the frequency shift is obtained

$$\Delta f = \frac{1}{\pi c_z A^2} \int_{d-A}^{d+A} (F_\downarrow + F_\uparrow) \frac{z-d}{\sqrt{A^2 - (z-d)^2}} dz \quad (3)$$

and the energy dissipation

$$\Delta E = \left(g - \frac{1}{Q} \frac{f}{f_0} \right) \pi c_z A^2. \quad (4)$$

Since the amplitudes in FM-AFM are often considerably larger than the distance range of the tip-sample interaction, apply the "large amplitude approximation" [10, 11] can be applied. This yields the formula

$$\Delta f = \frac{1}{\sqrt{2\pi}} \frac{f_0}{c_z A^{3/2}} \int_{D}^{D+2A} \frac{F_{ts}(z)}{\sqrt{z-D}} dz \quad (5)$$

It is interesting to note that the integral in this equation is virtually independent of the oscillation amplitude. The experimental parameters (c_z, f_0, and A) appear as pre-factors. Consequently, it is possible to define the *normalized frequency shift* [10]

$$\gamma(z) := \frac{c_z A^{3/2}}{f_0} \Delta f(z) \quad (6)$$

This is a very useful quantity to compare experiments obtained with different amplitudes and cantilevers. The validity of Eq. 6 is nicely demonstrated by the application of this equation to the frequency shift curves already presented in Fig. 3a. As shown in Fig. 3b all curves obtained for different amplitudes result into one universal γ-curve, which depends only on the actual tip-sample distance D.

These equations help to calculate the frequency shift for a given tip-sample interaction law. The inverse problem, however, is even more interesting: *How can the tip-sample interaction be determined from frequency shift data?* Several solutions to this question have been presented by various authors and have lead to the *dynamic force spectroscopy* (DFS) technique, which is a direct extension of the FM-AFM mode [12].

Here the approach of Dürig [11] is presented, which is based on the inversion of the integral Eq. 5. It can be transformed to

$$F_{ts}(D) = \sqrt{2} \frac{c_z A^{3/2}}{f_0} \frac{\partial}{\partial D} \int_{D}^{\infty} \frac{\Delta f(z)}{\sqrt{z-D}} dz, \quad (7)$$

which allows a direct calculation of the tip-sample interaction force from the frequency shift versus distance curves.

An application of this formula to the experimental frequency shift curves already presented in Fig. 3a is shown in Fig. 3c. The obtained force curves are nearly identical although obtained with different oscillation amplitudes. Since the tip-sample interactions can be measured with high resolution, dynamic force spectroscopy opens a direct way to compare experiments with theoretical models and predictions.

Applications of FM-AFM

The excitement about the NC-AFM technique in ultrahigh vacuum was driven by the first results of Giessibl [13] who achieved to image the true atomic structure of the Si(111)-7 × 7-surface with this technique in 1995. In the same year Sugawara et al. [14] observed the motion of single atomic defects on InP with true atomic resolution. However, imaging on conducting or semi-conducting surfaces is also possible with the scanning tunneling microscope (STM) and these first NC-AFM images provided no new information on surface properties. The true potential of NC-AFM lies in the imaging of non-conducting surface with atomic precision. A long-standing question about the surface reconstruction of the technological relevant material Aluminium oxide could be answered by Barth et al. [15], who imaged the atomic structure of the high temperature phase of α-Al$_2$O$_3$(0001).

The high resolution capabilities of non-contact atomic force microscopy are nicely demonstrated by the images shown in Fig. 4. Allers et al. [16] resolved atomic steps and defects with atomic resolution on Nickel oxide. Today such a resolution is routinely obtained by various research groups (for an overview see, e.g., Refs. [2–6]). Recent efforts have also been concentrated on the analysis of functional organic molecules, since in the field of nanoelectronics it is anticipated that in particular organic molecules will

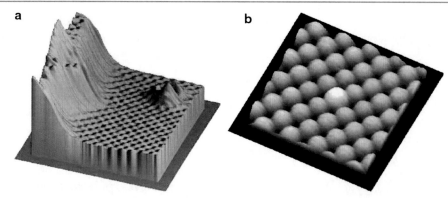

AFM, Non-contact Mode, Fig. 4 Imaging of a NiO(001) sample surface with a non-contact AFM. (**a**) Surface step and an atomic defect. The lateral distance between two atoms is 4.17 Å.

(**b**) A dopant atom is imaged as a light protrusion about 0.1 Å higher as the other atoms (Images courtesy of W. Allers and S. Langkat, University of Hamburg; used with kind permission)

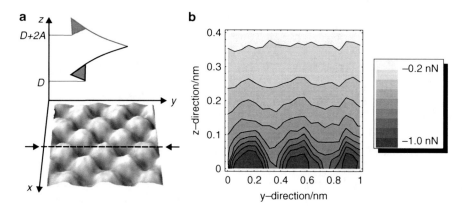

AFM, Non-contact Mode, Fig. 5 (**a**) Principle of 3D-force spectroscopy. The cantilever oscillates near the sample surface and measure the frequency shift in a x-y-z-box. The three dimensional surface shows the topography of the sample

(image size: 10 Å × 10 Å) obtained immediately before the recording of the spectroscopy field. (**b**) The reconstructed force field of NiO(001) shows atomic resolution. The data is taken along the line shown in (**a**)

play an important role as the fundamental building blocks of nanoscale electronic device elements.

The concept of dynamic force spectroscopy can be also extended to 3D-force spectroscopy by mapping the complete force field above the sample surface [12]. Figure 5a shows a schematic of the measurement principle [17]. Frequency shift vs. distance curves are recorded on a matrix of points perpendicular to the sample surface. Using Eq. 7 the complete three-dimensional force field between tip and sample can be recovered with atomic resolution. Figure 5b shows a cut through the force field as a two-dimensional map.

If the NC-AFM is capable of measuring forces between single atoms with sub-nN precision, why should it not be possible to also exert forces with this technique? In fact, the new and exciting field of

nanomanipulation would be driven to a whole new dimension, if defined forces can be reliably applied to single atoms or molecules. In this respect, Loppacher et al. [18] achieved to push on different parts of an isolated Cu-TBBP molecule, which is known to possess four rotatable legs. They measured the force-distance curves while one of the legs was pushed by the AFM tip and turned by 90°, thus being able to measure the energy which was dissipated during "switching" this molecule between different conformational states. The manipulation of single Sn-atoms with the NC-AFM was nicely demonstrated by Sugimoto et al. who manipulated single Sn-atoms on the Ge(111)-c(2 × 8) semiconductor surface (Fig. 6). By pushing single Sn-atoms from one lattice site to the other they finally succeeded to write the letter "Sn" with single atoms.

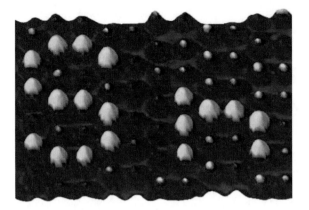

AFM, Non-contact Mode, Fig. 6 Final topographic NC-AFM image of the process of rearranging single Sn-atoms on a Ge (111)-c(2 × 8) semiconductor surface at room temperature (Reproduced from [19])

Cross-References

▶ AFM, Tapping mode
▶ Friction Force Microscopy
▶ Kelvin Probe Force Microscopy
▶ Magnetic Resonance Force Microscopy
▶ Piezoresponse Force Microscopy and Spectroscopy

References

1. Albrecht, T.R., Grütter, P., Horne, D., Rugar, D.: Frequency modulation detection using high-Q cantilevers for enhanced force microscope sensitivity. J. Appl. Phys. **69**, 668 (1991)
2. Morita, S., Wiesendanger, R., Meyer, E. (eds.): Noncontact Atomic Force Microscopy. Springer, Berlin (2002)
3. Garcia, R., Pérez, R.: Dynamic atomic force microscopy methods. Surf. Sci. Rep. **47**, 197 (2002)
4. Giessibl, F.-J.: Advances in atomic force microscopy. Rev. Mod. Phys. **75**, 949 (2003)
5. Meyer, E., Hug, H.J., Bennewitz, R.: Scanning Probe Microscopy - The Lab on a Tip. Springer, Berlin (2004)
6. Morita, S., Giessibl, F.J., Wiesendanger, R. (eds.): Noncontact Atomic Force Microscopy, vol. 2. Springer, Berlin (2009)
7. Ueyama, H., Sugawara, Y., Morita, S.: Stable operation mode for dynamic noncontact atomic force microscopy. Appl. Phys. A **66**, S295 (1998)
8. Hölscher, H., Gotsmann, B., Allers, W., Schwarz, U.D., Fuchs, H., Wiesendanger, R.: Comment on "Damping mechanism in dynamic force microscopy". Phys. Rev. Lett. **88**, 019601 (2002)
9. Hölscher, H., Schwarz, A., Allers, W., Schwarz, U.D., Wiesendanger, R.: Quantitative analysis of dynamic force spectroscopy data on graphite(0001) in the contact and non-contact regime. Phys. Rev. B **61**, 12678 (2000)
10. Giessibl, F.J.: Forces and frequency shifts in atomic-resolution dynamic-force microscopy. Phys. Rev. B **56**, 16010 (1997)
11. Dürig, U.: Relations between interaction force and frequency shift in large-amplitude dynamic force microscopy. Appl. Phys. Lett. **75**, 433 (1999)
12. Baykara, M.Z., Schwendemann, T.C., Altman, E.I., Schwarz, U.D.: Three-dimensional atomic force microscopy taking surface imaging to the next level. Adv. Mater. **22**, 2838 (2010)
13. Giessibl, F.-J.: Atomic resolution of the silicon (111)-(7 × 7) surface by atomic force microscopy. Science **267**, 68 (1995)
14. Sugawara, Y., Otha, M., Ueyama, H., Morita, S.: Defect motion on an InP(110) surface observed with noncontact atomic force microscopy. Science **270**, 1646 (1995)
15. Barth, C., Reichling, M.: Imaging the atomic arrangement on the high-temperature reconstructed α-Al$_2$O$_3$(0001) surface. Nature **414**, 54 (2001)
16. Allers, W., Langkat, S., Wiesendanger, R.: Dynamic low-temperature scanning force microscopy on nickel oxide (001). Appl. Phys. A **72**, S27 (2001)
17. Hölscher, H., Langkat, S.M., Schwarz, A., Wiesendanger, R.: Measurement of threedimensional force fields with atomic resolution using dynamic force spectroscopy. Appl. Phys. Lett. **81**, 4428 (2002)
18. Loppacher, C., Guggisberg, M., Pfeiffer, O., Meyer, E., Bammerlin, M., Luthi, R., Schlittler, R., Gimzewski, J.K., Tang, H., Joachim, C.: Direct determination of the energy required to operate a single molecule switch. Phys. Rev. Lett. **90**, 066107 (2003)
19. Sugimoto, Y., Abe, M., Hirayama, S., Oyabu, N., Custance, O., Morita, S.: Atom inlays performed at room temperature using atomic force microscopy. Nat. Mater. **4**, 156 (2005)

AFM, Tapping Mode

Hendrik Hölscher
Karlsruher Institut für Technologie (KIT), Institut für Mikrostrukturtechnik, Karlsruhe, Germany

Definition

The *tapping* or AM-mode is the most common dynamic mode used in atomic force microscopy. In dynamic mode AFM the cantilever is oscillated with (or near) its resonance frequency near the sample surface. Using a feedback electronic the cantilever sample distance is controlled by keeping either the amplitude or the phase of the oscillating cantilever constant. Since lateral tip–sample forces are avoided by this technique the resolution is typically higher

AFM, Tapping Mode,
Fig. 1 Setup of a dynamic
force microscope operated in
the tapping (or AM-) mode.
A laser beam is deflected by
the back side of the cantilever,
and its deflection is detected
by a split photo-diode. The
cantilever vibration is caused
by an external frequency
generator driving an excitation
piezo. A lock-in amplifier is
used to compare the cantilever
driving with its oscillation.
The amplitude signal is held
constant by a feedback loop
controlling the cantilever
sample distance

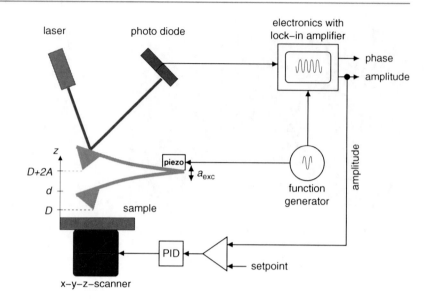

compared to the classical contact mode AFM where tip
and sample are in direct mechanical contact.

Overview

Since its introduction in 1986 [1], the *atomic force
microscope* became a standard tool in nanotechnology.
In early experimental setups, a sharp tip located at or
near the end of a microstructured cantilever profiled
the sample surface in direct mechanical contact
(*contact mode*) to measure the force acting between
tip and sample. Maps of constant tip–sample interaction force, which are usually regarded as representing
the sample's "topography," were then recovered by
keeping the deflection of the cantilever constant
(▶ Friction Force Microscopy). This is achieved by
means of a feedback loop that continuously adjusts
the z-position of the sample during the scan process
so that the output of the deflection sensor remains
unchanged at a preselected set-point.

Despite the widespread success of contact mode
AFM in various applications, the resolution was
frequently found to be limited (in particular for soft
samples) by lateral forces acting between tip and
sample. In order to avoid this effect, it is advantageous
to vibrate the cantilever in vertical direction near the
sample surface. AFM imaging with an oscillating
cantilever is often denoted as *dynamic force
microscopy* (DFM) [2].

The cantilever dynamics are governed by the tip–
sample interaction as well as by its driving method.
For instance, the historically oldest scheme of cantilever
excitation in DFM imaging is the external driving of the
cantilever at a fixed excitation frequency chosen to be
exactly at, or very close to, the cantilever's first resonance. For this driving mechanism, different detection
schemes measuring either the change of the oscillation
amplitude or the phase shift between driving signal and
resulting cantilever motion were proposed. Over the
years, the *amplitude modulation* (AM) or *tapping*
mode, where the actual value of the oscillation amplitude is employed as a measure of the tip–sample distance, has been established as the most widely applied
technique for ambient conditions and liquids.

Experimental Setup
A schematic of the experimental setup of an atomic
force microscope driven in amplitude modulation
mode is shown in Fig. 1. In this mode the cantilever
is vibrated close to its resonant frequency near the
sample surface. Due to the tip–sample interaction
the resonant frequency (and consequently also
amplitude A and phase ϕ) of the cantilever changes
with the cantilever sample distance d. Therefore, the
amplitude as well as the phase can be used as feedback
channels. A certain set-point for example, the
amplitude is given, and the feedback loop will adjust
the tip–sample distance such that the amplitude
remains constant. The cantilever sample distance is

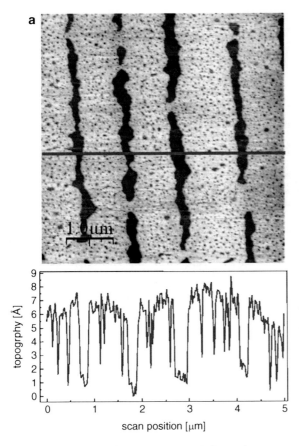

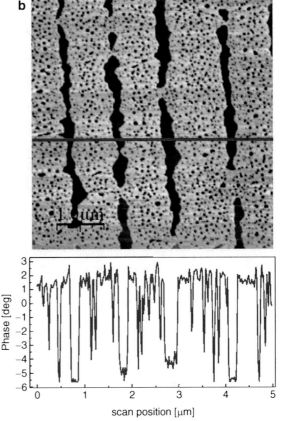

AFM, Tapping Mode, Fig. 2 (**a**) A dynamic force microscopy image of a monomolecular DPPC (L-α-dipalmitoyl-phophatidycholine) film adsorbed on mica. (**b**) The phase contrast is different between substrate and DPPC film. Equation 6 explains how this contrast is related to the energy dissipation caused by the tip–sample contact interaction

recorded as a function of the lateral position of the tip with respect to the sample and the scanned height essentially represents the surface topography.

The deflection of the cantilever is typically measured with the laser beam deflection method. During operation in conventional tapping mode, the cantilever is driven with a fixed frequency and a constant excitation amplitude using an external function generator, while the resulting oscillation amplitude and/or the phase shift are detected by a lock-in amplifier. The function generator supplies not only the signal for the dither piezo; its signal serves simultaneously as a reference for the lock-in amplifier.

This setup is mostly used in air and in liquids. A typical image obtained with this experimental setup in ambient conditions is shown in Fig. 2. The phase between excitation and oscillation can be acquired as an additional channel and gives information about the different material properties of DPPC and the mica substrate. As shown at the end of the next section, the phase signal is closely related to the energy dissipated in the tip–sample contact.

Due to its technical relevance the investigation of polymers has been the focus of many studies and high-resolution imaging has been extensively performed in the area of material science. Using specific tips with additionally grown sharp spikes Klinov et al. [3] obtained molecular resolution on a polydiacetylene crystal. Imaging in liquids opens up the avenue for the investigation of biological samples in their natural environment. For example, Möller et al. [4] have obtained high-resolution images

AFM, Tapping Mode,
Fig. 3 Topography of DNA
adsorbed on mica imaged in
buffer solution by tapping
mode AFM. The *right* graph
shows a single scan line
obtained at the position
marked by an arrow in the *left*
image

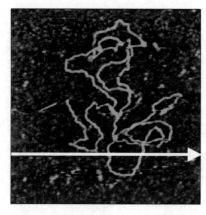
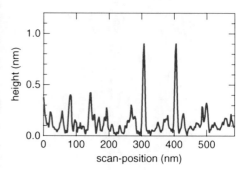

of the topography of hexagonally packed intermediate
(HPI) layer of *Deinococcus radiodurans* with tapping-
mode AFM. A typical example for the imaging of
DNA in liquid solution is shown in Fig. 3.

Theory of Tapping-Mode AFM

Many features observed in tapping-mode AFM can be
described by a simple spring-mass model which
includes the tip–sample interaction force.

$$m\ddot{z}(t) + \frac{2\pi f_0 m}{Q_0}\dot{z}(t) + c_z(z(t) - d)$$
$$= \underbrace{a_{\rm d}cz\cos(2\pi f_{\rm d}t)}_{\text{external driving force}} + \underbrace{F_{\rm ts}[z(t),\dot{z}(t)]}_{\text{tip–sample force}}. \tag{1}$$

Here, $z(t)$ is the position of the tip at the time t; c_z, m,
and $f_0 = \sqrt{(c_z/m)}/(2\pi)$ are the spring constant, the
effective mass, and the eigenfrequency of the cantilever, respectively. The quality factor Q_0 combines the
intrinsic damping of the cantilever and all influences
from surrounding media, such as air or liquid. The
equilibrium position of the tip is denoted as d. The
first term on the right-hand side of the equation
represents the external driving force of the cantilever
by the frequency generator. It is modulated with the
constant excitation amplitude $a_{\rm d}$ at a fixed frequency
$f_{\rm d}$. The (nonlinear) tip–sample interaction force $F_{\rm ts}$ is
introduced by the second term.

The actual tip–sample force is unknown in nearly
all cases. However, in order to understand the most
important effects it is often sufficient to apply the
DMT-M theory [5]. In this approach it is assumed the

tip is nearly spherical and that the noncontact forces
are given by the long-range van-der-Waals forces
while the contact forces are described by the
well-known Hertz model. The resulting overall force
law is given by

$$F_{\rm DMT-M}(z) = \begin{cases} -\dfrac{A_H R}{6z^2} & \text{for} \quad z \geq z_0, \\ \dfrac{4}{3}E^*\sqrt{R}(z_0 - z)^{3/2} & \text{for} \quad z < z_0, \end{cases} \tag{2}$$

where A_H is the Hamaker constant, E^* the effective
modulus, and R the tip radius. At z_0 tip and sample
come in contact. Figure 4 displays the resulting tip–
sample force curve for this model.

For the analysis of dynamic force microscopy,
experiments only on the steady states with sinusoidal
cantilever oscillation are of interest. Consequently, the
steady-state solution is given by the ansatz

$$z(t \gg 0) = d + A\cos(2\pi f_{\rm d}t + \phi), \tag{3}$$

where ϕ is the phase difference between the excitation
and the oscillation of the cantilever.

Here, the situation where the driving frequency is
set *exactly* to the eigenfrequency of the cantilever
($f_{\rm d} = f_0$) is analyzed. With this choice, which is also
very common in actual DFM experiments, defined
imaging conditions are given leading to a handy
formula relationship between the free oscillation
amplitude A_0, the actual amplitude A, and the equilibrium tip position d [6].

AFM, Tapping Mode,
Fig. 4 Tip–sample model
force after the DMT-M model
for air Eq. 2 using the typical
parameters: $A_H = 0.2$ aJ,
$R = 10$ nm, $z_0 = 0.3$ nm, and
$E^* = 1$ GPa. The *dashed line*
marks the position z_0 where
the tip touches the surface

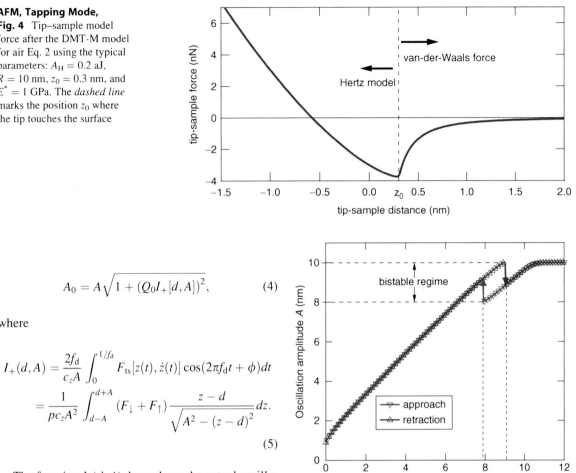

where

$$A_0 = A\sqrt{1 + (Q_0 I_+[d, A])^2}, \qquad (4)$$

where

$$I_+(d, A) = \frac{2f_d}{c_z A} \int_0^{1/f_d} F_{ts}[z(t), \dot{z}(t)] \cos(2\pi f_d t + \phi)dt$$

$$= \frac{1}{p c_z A^2} \int_{d-A}^{d+A} (F_\downarrow + F_\uparrow) \frac{z - d}{\sqrt{A^2 - (z - d)^2}} dz. \qquad (5)$$

The function $I_+(d, A)$ depends on the actual oscillation amplitude A and cantilever–sample distance d and it is a weighted average of the tip–sample forces during approach and retraction ($F_\downarrow + F_\uparrow$).

The solution of this equation allows us to study amplitude vs. distance curves as shown in Fig. 5 for a Q-factor of 300. Most noticeable, the tapping mode curve exhibits jumps between unstable branches, which occur at different locations for approach and retraction. The resulting bistable regime then causes a hysteresis between approach and retraction and divides the tip–sample interaction into two regimes. In order to identify the forces acting between tip and sample in these two regimes, the oscillation amplitude is plotted as a function of the nearest tip–sample distance in Fig. 6. In addition, the bottom graph depicts the corresponding tip–sample force (cf. Fig. 4). The origin of the nearest tip–sample position D is defined by this force curve. Since both, the amplitude curves and the tip–sample force curve, are plotted as a function of the nearest tip–sample position, it is

AFM, Tapping Mode, Fig. 5 Amplitude vs. distance curve for tapping-mode AFM assuming the parameters $A_0 = 10$ nm, $f_0 = 300$ kHz, and the tip–sample interaction force shown in Fig. 4. The overall amplitude decreases the sample surface distance, but instabilities (indicated by *arrows*) occur during approach and retraction

possible to identify the resulting maximum tip–sample interaction force for a given oscillation amplitude.

During the approach of the vibrating cantilever toward the sample surface, there is discontinuity in the nearest tip–sample position D. This gap corresponds to the bi-stability and the resulting jumps in the amplitude vs. distance curve. After the jump from the attractive to the repulsive regime has occurred, the amplitude decreases continuously. The nearest tip–sample position, however, does not reduce accordingly, remaining roughly between -0.8 and -1.5 nm. As a result, larger A/A_0 ratios do not necessarily result into lower tip–sample interactions, which is important to keep in

AFM, Tapping Mode,
Fig. 6 A comparison between the maximum tip–sample forces (tip–sample forces acting at the point of closest tip–sample approach/nearest tip–sample position D) experienced by conventional "tapping mode" AM-AFM assuming the same parameters as in Fig. 5. The *upper* graph shows the nearest tip–sample position D vs. the actual oscillation amplitude A for tapping mode. The graph at the *bottom* reveals the force regimes sensed by the tip. The maximal tip–sample forces in tapping mode are on the repulsive (tapping regime) as well as attractive (bistable tapping regime) part of the tip–sample force curve

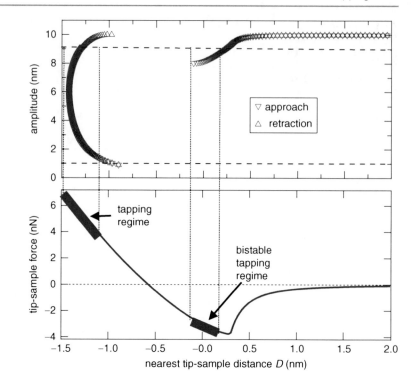

mind while adjusting imaging parameters in tapping mode. For practical applications, it is reasonable to assume that the set-point of the amplitude used for imaging has been set to a value between 90% and 10% of the free oscillation amplitude. With this condition, one can identify the accessible imaging regimes indicated by the horizontal (dashed) lines and the corresponding vertical (dotted) lines. Therefore, in tapping mode, two imaging regimes are typically accessible: the *tapping regime* (left) and the *bistable tapping regime* (middle). The bistable imaging state is only accessible during approach. Imaging in this regime is indeed possible with the limitation that the oscillating cantilever might jump into the repulsive regime [7, 8].

In the above paragraphs, the influence of the tip–sample interaction on the cantilever oscillation was analyzed, the maximum tip–sample interaction forces based on the assumption of a specific model force was calculated, and subsequently possible routes for image optimization were discussed. However, in practical imaging, the tip–sample interaction is not a priori known. Therefore, several authors [9–14] suggested solutions to this inversion problem, but some of them need further technical equipment. However, most commercial systems give access to

amplitude and phase vs. distance data which already allows the reconstruction of the tip–sample force curve applying numerical procedures [11, 12, 14].

The energy dissipation caused by the tip–sample interaction can be easily calculated using the conservation of energy principle [15].

$$\Delta E = \left(\frac{1}{Q_0} \frac{f_d}{f_0} + \frac{a_d}{A} \sin \phi \right) \pi c_z A^2. \qquad (6)$$

Since all parameters – except the phase ϕ – are constant during scanning, this formula shows that the energy dissipation is roughly proportional to $\sin \phi$ as already mentioned in the caption of Fig. 2.

Cross-References

- AFM, Non-contact Mode
- Friction Force Microscopy
- Kelvin Probe Force Microscopy
- Magnetic Resonance Force Microscopy
- Piezoresponse Force Microscopy and Spectroscopy

References

1. Binnig, G., Quate, C.F., Gerber, Ch: Atomic force microscopy. Phys. Rev. Lett. **56**, 930–933 (1986)
2. Garcia, R., Pérez, R.: Dynamic atomic force microscopy methods. Surf. Sci. Rep. **47**, 197–301 (2002)
3. Klinov, D., Maganov, S.: True molecular resolution in tapping-mode atomic force microscopy with high-resolution probes. Appl. Phys. Lett. **84**, 2697–2698 (2004)
4. Möller, C., Allen, M., Elings, V., Engel, A., Müller, D.J.: Tapping-mode atomic force microscopy produces faithful high-resolution images of protein surfaces. Biophys. J. **77**, 1150–1158 (1999)
5. Schwarz, U.D.: A generalized analytical model for the elastic deformation of an adhesive contact between a sphere and a flat surface. J. Coll. Interf. Sci. **261**, 99–106 (2003)
6. Hölscher, H., Schwarz, U.D.: Theory of amplitude modulation atomic force microscopy with and without Q-control. Int. J. Nonlinear Mech. **42**, 608–625 (2007)
7. Paulo, A.S., García, R.: High-resolution imaging of antibodies by tapping-mode atomic force microscopy: Attractive and repulsive tip-sample interaction regimes. Biophys. J. **78**, 1599–1605 (2000)
8. Stark, R.W., Schitter, G., Stemmer, A.: Tuning the interaction forces in tapping mode atomic force microscopy. Phys. Rev. B **68**, 085401 (2003)
9. Stark, M., Stark, R.W., Heckl, W.M., Guckenberger, R.: Inverting dynamic force microscopy: From signals to time-resolved interaction forces. PNAS **99**, 8473–8478 (2002)
10. Legleiter, J., Park, M., Cusick, B., Kowalewski, T.: Scanning probe acceleration microscopy (SPAM) in fluids: Mapping mechanical properties of surfaces at the nanoscale. Proc. Natl. Acad. Sci. U.S.A. **103**, 4813–4818 (2006)
11. Lee, M., Jhe, W.: General solution of amplitude-modulation atomic force microscopy. Phys. Rev. Lett. **97**, 036104 (2006)
12. Hölscher, H.: Quantitative measurement of tip-sample interactions in amplitude modulation atomic force microscopy. Appl. Phys. Lett. **89**, 123109 (2006)
13. Sahin, O., Maganov, S., Chanmin, Su, Quate, C., Solgaard, O.: An atomic force microscope tip designed to measure time-varying nanomechanical forces. Nat. Nanotechnol. **2**, 507–514 (2007)
14. Shuiqing, Hu, Raman, A.: Inverting amplitude and phase to reconstruct tip-sample interaction forces in tapping mode atomic force microscopy. Nanotechnology **19**, 375704 (2008)
15. Cleveland, J.P., Anczykowski, B., Schmid, A.E., Elings, V.B.: Energy dissipation in tapping-mode atomic force microscopy. Appl. Phys. Lett. **72**, 2613 (1998)

Aging

► Fate of Manufactured Nanoparticles in Aqueous Environment

ALD

► Atomic Layer Deposition

Alkanethiols

► BioPatterning

AlN

► Piezoelectric MEMS Switches

Alteration

► Fate of Manufactured Nanoparticles in Aqueous Environment

Aluminum Nitride

► Piezoelectric MEMS Switches

Amphibian Larvae

► Ecotoxicology of Carbon Nanotubes Toward Amphibian Larvae

Anamorphic Particle Tracking

► Astigmatic Micro Particle Imaging

Angiogenesis

Alessia Bottos[1], Elena Astanina[1], Luca Primo[2] and Federico Bussolino[1]
[1]University of Torino, Department of Oncological Sciences, Candiolo, Torino, Italy
[2]University of Torino, Department of Clinical and Biological Sciences, Candiolo, Italy

Definition

Originally, the term "angiogenesis" was strictly confined to describe the sprouting of blood vessels from preexisting capillaries or post-capillary venules. Presently, angiogenesis refers to the molecular and cellular mechanisms leading to the formation and remodeling of vascular tree occurring during embryo development and in adult life. However, this term often includes the concept of vasculogenesis. Vasculogenesis indicates the earlier events of the formation of the cardiovascular system and is characterized by the assembly of a primitive vascular plexus constituted by angioblasts, which are believed to arise from the hemangioblast, a common precursor for endothelial and hematopoietic lineages.

There are two other general concepts that strictly parallel angiogenesis. The first is arteriogenesis, which describes the growth of functional collateral arteries from preexisting arterio-arteriolar anastomoses in response to ischemic injuries. The second is lymphangiogenesis, which illustrates the formation of the lymphatic system. This forms a unidirectional network of blind-ended terminal lymphatic vessels that collect the protein-rich fluid that exudes from blood capillaries and then drain through a conduit system of collecting vessels, lymph nodes, lymphatic trunks and ducts into the venous circulation.

Introduction

The notion that blood circulation is indispensable for bringing nutrients to the organs and for organism growth has been known for millennia. However, the connected concept – that growth of blood vessels may promote disease – is more recent [1, 2].

The perception that tumor growth is linked to an increased vascularization took place in the mid-twentieth century. In 1971, Judah Folkman gave the first formal demonstration that tumors release soluble molecules able to recruit from surrounding tissue capillaries, which enable cancer to growth over few millions of cells, infiltrate the normal parenchyma, and colonize distant organs. These observations opened an extraordinary research area that is far to be completely exploited and is at the frontier of the molecular medicine. The greatest achievement of these studies has been made over the past years in targeting angiogenesis for human therapy. In particular, in the 2004 vascular-endothelial growth factor (VEGF)-A removal by the humanized monoclonal antibody bevacizumab and by the specific aptamer pegaptanib was, respectively, approved for the treatment of metastatic colorectal cancer and exudative retinopathy [3]. Currently more than 50 anti-angiogenic compounds are in clinical trials. However, clinical experience has shown that the antivascular effect of bevacizumab as well as of other anti-angiogenic compounds are often transient or even negligible [4, 5], indicating that new knowledge is required to better understand the complexity of vascular formation and to properly select responder patients.

Nevertheless the impact of angiogenesis in medicine overcomes oncology and is extended to other relevant pathologies including ischemic diseases, chronic inflammatory diseases, metabolic disorders, obesity, some mendelian inherited diseases, reproductive disorders, and bone diseases (Table 1).

The Morpho-functional Vascular Unit

Arteries, veins, and capillaries share a common structure (tunica intima) characterized by endothelial cells (EC) that line vascular lumens and are surrounded by the basement membrane (BM) constituted by extracellular matrix proteins and complex sugars (i.e., heparan sulfates, proteoglycans). The external face of tunica intima is covered by vascular smooth muscle cells and pericytes (mural cells) that form a multilayer wall of different size (thinner in the veins and ticker in the arteries), named tunica media. In capillaries, pericytes do not form a continuous layer but they are scattered and embedded in the BM. Mural cells exert key roles in multiple microvascular processes, including:

Angiogenesis, Table 1 Human diseases characterized by abnormal vascularization

Organs	Diseases
Whole body	Cancer, autoimmune diseases, ischemic diseases, diabetes, infections
Vessels	Vascular malformations, cavernous hemangiomas, hemangiomas, atherosclerosis, hereditary hemorrhagic telangiectasia, Di George syndrome, glomerulonephritis, transplant arteriopathy
Adipose tissue	Obesity
Skin	Psoriasis, warts, blistering disease, allergic dermatitis, scar keloid, pyogenic granulomas
Eye	Persistent hyperplastic vitreous syndrome, Norrin disease, exudative retinopathy, retinopathy of prematurity
Lung	Primary pulmonary hypertension, asthma, nasal polyps
Intestines	Inflammatory bowel disease, ascites, periodontal diseases
Reproductive system	Endometriosis, ovarian cysts, ovarian hyperstimulation
Bone, joints	Arthritis, synovitis, osteomyelitis, osteophyte formation

(1) endothelial cell proliferation and differentiation, (2) contractility and tone, (3) stabilization and permeability, and (4) morphogenesis and remodeling during disease onset [6, 7].

EC thickness varies from less than 0.1 μm in capillaries and veins to 1 μm in the aorta and in arteries they are aligned in the direction of blood flow and this remodeling occurs in response to hemodynamic shear stress. On the contrary in vasculature characterized by a slow blood stream, endothelial shape may be irregular, rounded, or elliptical. Endothelium may be continuous or discontinuous (Fig. 1). Continuous endothelium, in turn, is fenestrated or nonfenestrated. Nonfenestrated continuous endothelium is found in arteries, veins, and capillaries of the brain, skin, heart, and lung. Fenestrated continuous endothelium occurs in tissues that are characterized by increased filtration or increased transendothelial transport, including exocrine and endocrine glands, gastrointestinal mucosa, and kidney. Fenestrae are transcellular pores (~70 nm in diameter) that develop through the whole thickness of the cell and have a thin 5–6 nm nonmembranous diaphragm across their aperture.

Discontinuous endothelium characterizes capillaries of sinusoidal vascular beds (i.e., liver and bone marrow). It is characterized by large fenestrations (100–200 nm in diameter) that lack a diaphragm and contain gaps (or large holes) within individual cells.

The regulation of these kinds of EC-EC interactions occurs at two types of intercellular junctions: the tight junctions (also named *zona occludens*), which are mainly located at the apical region of the intercellular cleft and the adherens junctions (*zona adherens*). Zonula occluden-1, occludin, claudin, and JAM are the major elements of tight junctions, while vascular-endothelial cadherin is a specific constituent of adherens junctions and its clustering at cell-cell contacts promotes the formation of multimolecular complexes that comprise signaling, regulatory, and scaffold proteins.

In physiologic and pathological conditions, the above-described structural features of endothelium contribute to maintain the fluid and the cellular homeostasis between blood and tissues. Actually, the vascular system needs to be sufficiently permeable to allow the ready exchange of small molecules (gases, nutrients, waste products) with the tissues. Plasma proteins also need to cross the normal vascular barrier, at least in small amounts. Albumin, for example, transports fatty acids and vitamins and immunoglobulin antibodies are required for host defense. The passage of small molecules across the endothelial layer is almost passive and has to take into account three parameters: hydraulic conductivity, reflection coefficient, and diffusion. Diffusion is the most important of these for the exchange of small molecules and is driven by the molecular concentration gradient across vascular endothelium (Fick equation). On the other hand, filtration is much more important than diffusion for the flux of large molecules such as plasma proteins and is determined by the Starling equation.

On structural point of view, endothelial permeability is mediated by the so-called transcellular and paracellular pathways – that is, solutes and cells can pass through (transcellular) or between (paracellular) EC. While the paracellular route restricts the passage of solutes larger than 3 nm in radius, transcellular vesicle trafficking selectively transports macromolecules such as albumin across the endothelium. Transcellular passage (transcytosis) is an energy-dependent mechanism that requires either cell fenestration or a complex system of transport vesicles, which includes caveolae and organelles called vesiculo-vacuolar organelles (VVO). These organelles, or similar structures, can also fuse and appear

Angiogenesis, Fig. 1 Morphology of capillaries. Capillaries in different anatomical districts have different morphology. Continuous nonfenestrated EC is mainly found in skin, heart, muscle, brain, and lung. It regulates the passage of molecules through transendothelial channels or by caveolae-mediated transcytosis. Continuous fenestrated ECs characterize glomerulus, endocrine glands, and intestines and demonstrate greater permeability to water and small molecules but do not allow the passage of larger macromolecules by using diaphragms as molecular filters. Discontinuous endothelium is present in liver, bone marrow, and spleen and has fenestrae (without diaphragms), gaps, and partially structured BM. The intercellular cleft represents tight junctions or adherens junctions

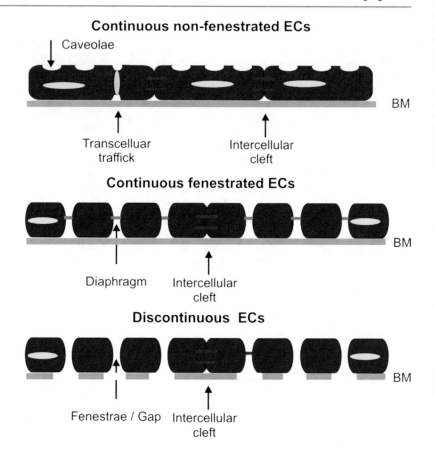

as channels that traverse single cells and allow the passage of leukocytes and solutes through the endothelium. Caveolae are 70-nm membrane-bound, flask-shaped vesicles that usually open to the luminal or abluminal side of the EC membrane. They are enriched in cholesterol and some, but not all, caveolae possess a thin diaphragm similar to that found in fenestrae. The density of caveolae is far greater in capillary endothelium (up to 10,000 per cell) compared with other vessels and their number is highest in continuous nonfenestrated endothelium. A notable exception is the blood-brain barrier, where caveolae are rare, which reflects the protective role of this vascular bed for the brain.

VVOs are comprised of hundreds of cytoplasmic vesicles and vacuoles that together form an organelle that traverses venular endothelial cytoplasm from lumen to albumen. VVOs often extend to inter-endothelial cell interfaces and their individual vesicles commonly open to the inter-endothelial cell cleft. The vesicles and vacuoles comprising VVOs are linked to each other and to the luminal and abluminal plasma

membranes by stomata that are normally closed by thin diaphragms that appear similar to those found in caveolae and fenestrae.

The paracellular pathway, by contrast, is mediated by the coordinated opening and closure tight junctions and adherens junctions, and is mainly operative during pathological conditions, in which this barrier is altered by the release of permeability factors (acute and chronic inflammatory injury) or by an abnormal and chaotic vascularization (i.e., tumor angiogenesis).

The Morpho-functional Vascular Unit in Pathological Settings

The formation of vascular bed depends from well-regulated circuits that involve soluble molecules with inducing and inhibiting activity, mechanical forces, different cell types and features of the extracellular matrix. The final result of these dynamic interactions is the ordered distribution of capillaries (the intercapillary distance ranged from ~100 to

~150 μm) allowing the appropriate cell oxygenation on the basis of local pO_2 and O_2 coefficient diffusion [6–8].

In pathological settings (i.e., tumor, chronic inflamed tissues, wound healing, postischemic revascularization) this homeostatic balance is disturbed leading to the formation of vessels with functional and structural defects.

In pathological tissues and in particular in tumors, vessels are dilated, tortuous, and show a chaotic spatial distribution. Normal vascular tree is characterized by vessels with decreasing diameter and by dichotomous branching, but tumor vessels are unorganized with trifurcations and irregular diameters. Endothelial phenotype is also modified in tumors. Large discontinuities characterize inter-endothelial junctions, as well as the increased numbers of fenestrae and VVOs and alterations or lack of BM. Similarly, mural cells are poorly distributed and show defects in association with vessel wall. The molecular and cellular mechanisms causing these abnormal vascular architectures are largely unknown, but the imbalance of pro- and anti-angiogenic associated with mechanicals stresses is believed to be instrumental in achieving these features.

These structural alterations are accountable for some features of tumor microenvironment. In normal tissues permeation of hydrophilic solutes is progressively restricted with increasing molecular size, as by a sieve, with many openings 8 nm and a few 40–60 nm wide. In solid tumors vascular permeability is generally higher allowing ($\geq$80–90 nm) or facilitating ($\geq$10 nm) the extravasation of molecules with sizes that in general are retained. The blood flow is irregular because the flow resistance, which depends from vascular geometry and blood viscosity, is increased. Focal interruption of vessel wall may also compromise the downstream blood flow. As a result, overall perfusion rates (blood flow rate per unit volume) in tumors are lower than in healthy organs. Blood flux is erratically distributed and fluctuates with time causing a jeopardized perfusion.

The increased permeability abolishes the differences between hydrostatic and colloid osmotic pressures that characterize normal tissues. As a result, tumors show interstitial hypertension and reduction of transmural pressure gradient and convection across tumor vessel walls. Consequently, the increased permeability connected with the interstitial hypertension

reduces pressure difference between up and down stream of blood vessels and leads to blood flow stasis.

The final consequences of these functional abnormalities takes place at metabolic level leading to hypoxia and acidosis. Tumor vessels fail to adequately deliver oxygen and nutrient as well as to remove wastes. The imbalance of vascular network development and tumor cell proliferation results in the formation of hypovascular regions in tumors. Since tissue diffusion limit of oxygen is 80–200 μm, the regions far from blood vessels become hypoxic. Extracellular acidosis is a consequence of hypoxia and Warburg effect (cancer cells produce energy through glycolysis also in the presence of adequate amount of oxygen). When pO_2 is low, the metabolic process (glycolysis followed by oxidative phosphorylation) sustaining oxidative demolition of glucose and ATP production decreases its efficiency. Actually, in the absence of oxygen, the end product of glycolysis, pyruvic acid, does not proceed toward oxidative phosphorylation but it is reduced to lactic acid. This metabolite, together carbonic acid generated by growing cells, accounts for the reduction of extracellular pH in tumors.

The above-described metabolic and functional characteristics render the microenvironment hostile, leading to a genetic selection that favors the most aggressive and malignant cancer cells. Finally, the combination of acidosis (which may modify the polarity of hydrophilic compound) with the blood stasis and the increased interstitial pressure hinders the delivery and efficacy of therapeutic agents to tumors.

The Mechanisms of Angiogenesis

Mesenchymal cells differentiate in angioblasts, which undergo vasculogenesis where they aggregate and form a primitive vascular network firstly located into extraembryonic tissues and then invading the embryo. As vessels begin to be remodeled, they undergo localized proliferation and regression, as well as programmed branching and migration into different regions of the body. They need to be specified into different calibers and types of vessel, including division into arteries, veins, and lymphatics, with further subdivision into large vessels, venules, arterioles, and capillaries. In addition, they need to recruit supporting mural cells and create the optimal extracellular matrix to ensure the stability of the vessels formed. Pulsatile and shear stress forces mediated by heart beats are instrumental for capillary branching and pruning. The establishment of final vascular architecture is characterized by four different mechanisms, which are also present in adult life: (1) formation of primitive and immature plexus, (2) stabilization of primitive plexus, (3) remodeling toward the final architecture, and (4) vascular specialization. Molecules involved in these steps are summarized in Table 2 [9–12].

Formation of primitive and immature vascular plexus – During embryo development, the nascent capillary network is generated by combining vasculogenesis and angiogenesis. Vasculogenesis describes the *de novo* formation of vascular plexus. Mesenchymal cells are stimulated by soluble molecules (i.e., Indian hedgehog, basic fibroblast growth factor, bone morphogenic protein 4) released by endodermal cells to differentiate in vascular-endothelial growth receptor-2 (VEGFR-2)-positive cells. In extra-embryonic yolk sac, these cells form hemangioblast, the common precursor of vascular and hematopoietic system. Primitive endothelial and hematopoietic cells coalesce to form blood islands. The outer cells of the blood islands are endothelial, whereas the inner cells give rise to hematopoietic progenitors. The endothelial-lined blood islands fuse to generate *de novo* a primary capillary plexus. This process is named vasculogenesis. In intra-embryonic tissues, VEGFR-2-positive mesenchymal cells differentiate into angioblasts which directly aggregate into the dorsal aorta or cardinal vein, without a plexus intermediate. These large vessels then undergo connection with the primitive vascular network.

Besides vasculogenesis, sprouting and intussusceptive angiogenesis take place (Fig. 2). Sprouting angiogenesis if facilitated by hypoxia, which upregulates the transcription of a large set of pro-angiogenic genes. EC loose the cell-cell contacts and vessels become leaky. The BM is digested by the release of specific proteases (metalloproteases) and a specific suppression of protease inhibitors allowing EC migration and proliferation. Intriguingly, BM cleavage can produce multiple anti-angiogenic molecules that counteract and balance the effect of angiogenic inducers thus allowing an accurate vascularization. The sprouts reorganize internally to form a vascular lumen and are finally connected to other capillary segments. VEGF signaling is critically important in this process, as it promotes endothelial cell proliferation and modulates migration.

Angiopoietin-2, in the presence of VEGF, facilitates sprouting angiogenesis because it contributes in dismantling cell contacts. The pivotal effect of VEGF in sprouting angiogenesis is regulated by Notch receptors and their Delta-like-4 (DLL4) and Jagged-1 ligands to allow a specific focalization of capillary area from which a new capillary stems. The fine-tuning activity between VEGF and DLL4 allows the selection of the tip EC, which is devoid of proliferative activity and drives the sprout by emitting filopodia. Initially, Dll4 and Notch signaling are thought to be balanced in EC until presumptive tip cells eventually increase Dll4 expression in response to VEGF signaling. Consequently, Notch is upregulated in neighboring cells, which inhibits the expression of VEGF receptors and the tip cell differentiation. Thus, Dll4-expressing tip cells react strongest to VEGF signaling and acquire a motile, invasive, and sprouting phenotype whereas tip cell behavior is suppressed in Jag1-expressing (stalk) cells, which form the base of the emerging sprout. Reduced levels of DLL4 or blocking of Notch signaling enhance the formation of tip cells, resulting in dramatically increased sprouting and chaotic angiogenesis. The principle in which a differentiating cell represses the differentiation of adjacent cells is called lateral inhibition (Fig. 3).

More recently ephrins and wnt pathway has been demonstrated to modulate tip cell dynamics. Although little is known about the lumenization of blood vessels, there is strong evidence for the involvement of pinocytosis and vacuole formation. High-resolution time-lapse imaging has established that the lumen in these ECs is formed by intracellular and, subsequently, intercellular fusion of large vacuoles.

Intussusception (growth within itself) is an alternative to the sprouting mode of angiogenesis. The protrusion of opposing microvascular walls into the capillary lumen creates a contact zone between ECs. The endothelial bilayer is perforated, intercellular contacts are reorganized, and a transluminal pillar with an interstitial core is formed, which is soon invaded by myofibroblasts and pericytes leading to its rapid enlargement by the deposition of collagen fibrils. When a primitive capillary plexus is generated by vasculogenesis or sprouting, intussusception is triggered and is responsible for rapid vascular growth and optimal remodeling.

Stabilization of primitive plexus – Blood perfusion of nascent capillaries requires their maturation. This

Angiogenesis, Table 2 Most important molecules involved in angiogenesis

Angiogenic inducers (ligand/receptor)

VEGF-A/VEGFR-1	Negative regulation of embryonic angiogenesis. Later, VEGFR-1 acts as decoy receptor. Recruitment of monocytes and macrophages
VEGF-A/VEGFR-2	Regulation of proliferation, permeability, migration, and survival of ECs. ECM remodeling. EC differentiation and specialization (i.e., fenestrae formation). The activity of VEGFR-2 is modulated by association with co-receptors (neuropilins, VE-cadherin, $\alpha v \beta 3$ integrin). VEGF-A isoforms contribute to vascular patterning
Placental growth factor/VEGFR-1	Regulation of pathological angiogenesis (i.e., tumor, collateral formation, wound healing). Contribution to hematopoiesis. Recruitment of inflammatory cells
VEGF-B/VEGFR-1	Cardiac vascularization
VEGF-C or VEGF-D/VEGFR-2 and/or VEGFR-3	Control of lymphangiogenesis in physiologic and pathological settings. Modulators of vascular angiogenesis
Notch/Dll4 or Jagged	Control of EC tip formation in sprouting angiogenesis. Differentiation and recruitment of mural cells. Venous-arterial differentiation
Angiopoietins/Tie-2	Regulation of the relationships between mural cells and EC. Regulation of vessel stabilization. Accessory role in lymphangiogenesis. Activation of inflammatory cells
TGFβ/ALK or endoglin	Production of ECM and proteases. Venous-arterial differentiation. Mural cell differentiation. Control of EC proliferation and migration
PDGFs/PDGF receptors	Differentiation, survival, and recruitment of mural cells
Ephrin B2/EphB4	Venous-arterial differentiation. Vascular branching
Slits/Robo	Formation of primitive plexus. Vascular stabilization. Control of EC motility
Netrins/UNC5B	Negative control of sprouting angiogenesis. Regulation of vascular branching
Semaphorins/plexins-Neuropilins	Regulation of vascular morphogenesis. Inhibition of EC adhesion and migration. Anti-angiogenic effect in tumor angiogenesis
Hepatocyte growth factor/Met	Regulation of pathological angiogenesis. Collateral formation
Chemokines/Chemokine receptors	Members that contain the "ELR" motif are potent promoters of angiogenesis via binding and activating CXCR2 on ECs. In contrast, members, in general, those lack the ELR motif (ELR-) are potent inhibitors of angiogenesis, and bind to CXCR3 on ECs

Endogenous angiogenic inhibitors

Tumstatin, arresten, and canstatin	They are cryptic angiostatic peptides released by collagen type IV during ECM proteolysis in sprouting angiogenesis
Endostatin	A cryptic angiostatic peptide released by collagen type XVIII during ECM proteolysis in sprouting angiogenesis
Angiostatin	A plasminogen fragment produced during sprouting angiogenesis. It Inhibits EC migration and proliferation and induces apoptosis
Vasostatin	Vasostatin is an N-terminal fragment of the calreticulin. It inhibits angiogenesis presumably by binding to laminin and blocking EC adhesion and proliferation
Thrombospondin	It inhibits EC proliferation and migration by interacting with CD36 molecule
2-methoxyestradiol	It inhibits angiogenesis by reducing EC proliferation and inducing EC apoptosis
Histidine-rich glycoprotein	It inhibits tumor angiogenesis by preventing release of angiogenic factors from the matrix and inhibiting growth factor-induced EC migration
Pigment epithelial-derived factor	It is released by retinal pigment epithelium and causes EC apoptosis
Maspin	It is a serpin inhibitor of serine-proteases with anti-angiogenic activity in tumor
PEX	It is a fragment of metalloproteinase-2 that prevents its collagenolytic activity during sprouting angiogenesis
Fibulin 5	It is a ECM protein that antagonizes VEGF activities and enhances that of thrombospondin
Interferons (α,β,γ)	They inhibit angiogenesis by multiple and direct effects on EC or by modulating innate immunity

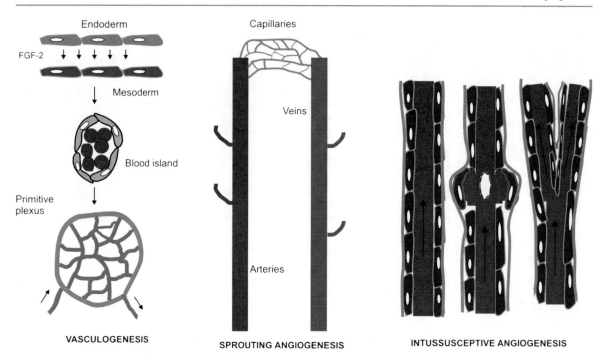

Angiogenesis, Fig. 2 Vasculature assembly. *Vasculogenesis* starts when mesenchymal cells stimulated by soluble molecule released by endoderm differentiate into EC precursors (hemangioblasts and angioblasts) and form the blood islands (*left*). Fusion of blood islands leads to formation of primary capillary. *Sprouting angiogenesis.* Blood circulation is established and primary plexi are remodeled into a hierarchical network of arteries, capillaries, and veins. Sprouting angiogenesis starts when a gradient of VEGF triggers the detachment of EC-EC contacts leading EC to proliferate and migrate.

The nascent capillaries undergo maturation characterized by reorganization of ECM and recruitment of mural cells. This step parallels the selection of the proper trajectory for the best blood supply. *Intussusception.* The division of vessels through the insertion of tissue pillars is an alternative mechanism that leads to the expansion of blood vessels in several organs. Little is known about the function or regulation of intussusceptive growth but the process seems to involve EC growth, motility, and remodeling of ECM

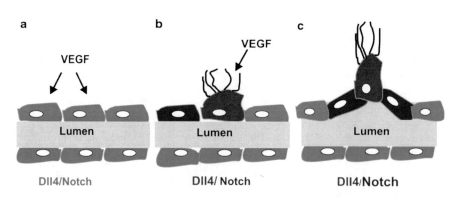

Angiogenesis, Fig. 3 Tip cell selection in sprouting angiogenesis. (**a**) At the beginning of VEGF stimulation ECs most probably contain equal amount of Dll4 and its receptor Notch. (**b**) Tip selection occurs by increasing Dll4 expression in a specific EC (*red* cell) stimulated by VEGF. As consequence, Notch is upregulated in neighboring cells (*blue*), which inhibit the

expression of VEGF receptors. (**c**) This modulation allows tip cells reacting strongest to VEGF signaling and acquiring sprouting phenotype whereas tip cell behavior is suppressed in Jag1-expressing (stalk) cells (*blue*), which form the base of the emerging sprout step is mainly characterized by recruitment of mural cells and remodeling of extracellular matrix (ECM). Mural cells are recruited by molecules released by EC and in particular by platelet-derived growth factors (PDGF) and sphingosine-1 phosphate, which, respectively, bind the tyrosine kinase PDGF receptor and the seven spanning domains endothelial differentiation sphingolipid G-protein-coupled receptor-1. Compelling support for this hypothesis comes from mouse genetic models in which the modification of the expression of these molecules impair the correct relationships between mural cells and EC and alter production of ECM. Mural and stroma cells produce angiopoietin-1 (Ang-1), which binds the tyrosine kinase receptor, Tie2 on endothelial cells, resulting in enhanced EC-mural cell interactions leading the vessels leak-resistant. A homolog, Ang-2, appears to antagonize Ang-1, relaxing these interactions and triggering BM degradation in preparation for angiogenic sprout. Transforming-growth factor β promotes the differentiation of mesenchymal cells into mural cells and contributes to ECM production. Mutations in endoglin and activin-like kinase, two receptors of transforming-growth factor β, cause hereditary hemorrhagic telangiectasia, a disease that is characterized by capillary dilatation, microaneurysms, and bleeding.

Interactions between ECs and their surrounding BM and ECM are critical for vascular maturation and occur through specialized adhesive receptors named integrins. The importance of these connections are underscored by the vascular defects seen in mice deficient in some types of integrins and various BM constituents such as type IV collagen, fibronectin, and laminins. Interestingly, the adhesion of EC to BM through integrins mediates anti-apoptotic signals that are instrumental to avoid vessel regression. Actually, anti-integrin molecules, including specific antibodies, have anti-angiogenic effects. Besides an adhesive role, integrins act as co-receptors of several tyrosine kinase receptors involved in angiogenesis modulating their activities. Following BM assembly, vessels are stabilized by cross-linking of ECM components and associated proteins by transglutaminase-2-mediated lysine bonds. The ECM cross-linking activity leads to inhibition of angiogenesis by increasing ECM stiffness and deposition.

Remodeling toward the final architecture – The transition from the plexus to the vascular tree is characterized by regression and pruning of some vascular segments, new branching, remodeling, and navigation to build the most appropriate network to distribute oxygen, nutrients, and wash catabolites off. This step is regulated by the combined activities of physical forces, axon guidance cues, and ECM features. Using in vivo molecular imaging associated with genetic and vascular grafting studies it has been established that physical cues drive guidance of lumenized vessel sprouts. ECs are subjected to tension created by cell-cell and cell-ECM interactions or by blood flow (shear stress) and pressure (circumferential wall stress). Interestingly, the absence of blood flow or sufficient tensions promotes vascular regression. Physical stimuli must be sensed by cells and transmitted through intracellular transduction pathways to the nucleus, resulting in gene expression. However, how cells sense biomechanical stimuli is still under debate. Several hypotheses experimentally supported have been proposed. They include the capacity of heterotypic (integrin-mediated cell-ECM) and homotypic (VE-cadherin mediated EC-EC interaction) adhesive receptors to link the internal cytoskeleton and provide mechanical coupling across the cell surface, including ion-channel opening [13].

As above mentioned, proteases released by angiogenic EC modify the structure of ECM during the formation of primitive plexus. Furthermore, this mechanism occurs also during the remodeling phase leading to pro- or anti-angiogenic activities. This stimulatory effect may be due to decreasing the density of extracellular matrix proteins, and/or by exposing cryptic binding sites within matrix molecules that promote migration. In contrast, the action of proteases can be anti-angiogenic, due to their ability to generate fragments of matrix molecules that have pro-apoptotic properties not possessed by the intact molecule. Therefore the characteristics of surrounding ECM may favor the progression or the regression of primitive capillary network. On the other hand ECM elements, and in particular fibronectin, collagens, laminins, and heparan sulfates, participate in vascular stabilization and patterning as demonstrated by specific vascular defects (increase in lumen size, edema, leakage, vessel rupture) in mouse genetic models.

ECM also affects the availability of angiogenic regulators. For instance, different isoforms of VEGF-A, characterized by increased affinity to ECM, form specific gradients required for vessel patterning. Similarly, fibroblast growth factor-2 may be sequestered by ECM and released to cell surface by proteolysis. Generally, ECM and proteolytic activities adapt the strength and the duration of angiogenic signals and train ECs to initiate fine-tuned and specific signal genetic programs regulating vascular patterning.

During the remodeling step, vessels select specific trajectories leading to an optimal spatial distribution instrumental for the molecular exchanger in the tissues. This strategy is shared with the nervous system, in which axon guidance molecules are responsible for guiding the navigating axon toward its target area. Indeed ECs and neurons share some fundamental mechanisms during the formation of their networks. On an anatomical point of view, in several tissues (skin, muscles, intestine) nerves and larger blood vessel align into two parallel structures and genetic studies indicate that peripheral nerves in the dermis provide a spatial template for the growth and differentiation of arteries. During axon navigation, growth cones project numerous filopodia as shown in endothelial tip cells that actively extend and retract in response to extracellular cues. Four ligand-receptor pairs of guidance cues are known to play attractive of repulsive effects in nervous and vascular systems: ephrins/Eph, semaphorins/plexins-neuropilins, slit/Robo, and netrins/UNC-DCC. Many studies performed in genetic modified mice or zebrafish clearly indicate that these molecules mainly act in defining the final shape and architecture of vessels in body tissues.

Vascular specialization – The paradigmatic example of this concept is the differentiation of arteries and veins. During vascular development cardiac work presses high-pressure blood flow through large diameter arterial vessels, which acquire an extensive supporting system of mural cells. Blood comes back to the heart with low pressure through veins, which develop *venous valves* to prevent blood from flowing back. Therefore hemodynamic forces have a crucial role in defining arterial and venous circulation. These mechanisms are strictly intermingled with early genetic determinant. The landmark of arterial–venous cell fate is the expression of ephrinB2 and EphB4, respectively, in plasma membrane of arterial and venous ECs. The interactions between these two markers are essential for proper vascular development. However, the expression of these molecules is differently controlled. VEGF upregulates the Notch pathway that through hesr1/hesr2 transcription factors leads to arterial specification via EphrinB2 activation. In veins,

COUP-TFII is required for strong expression of EphB4 and venous differentiation. Therefore it seems that molecular differences between arteries and veins are predetermined but the final fate is contributed by epigenetic and physical cues indicating a prominent plasticity of the system.

The vascular architecture has definite features in different organs to respond to specific needs in terms of nutrient requirements and transport [14]. For instance, the filtration barrier of glomerular endothelial cells in kidney results from the presence of continuous fenestrations allowing bulk fluid and anionically charged molecules to be filtered and facilitates movement of cationically charged molecules. The sinusoidal hepatic system is characterized by discontinuous endothelium with sieve-like pores ($\sim$ 100 nm diameter) and nondiaphragmatic fenestrae, which regulate the permeability and the blood flow in hepatic sinusoids. Endocrine organs have fenestrated capillaries that contain pores sometimes lined with thin diaphragms. These structures are thought to allow increased transport of long-range hormones between extravascular cells and blood. In contrast to the fenestrated vascular beds present in high-transport organs, brain capillaries are linked by extensive tight junctions that restrict transendothelial movement.

These differences are furtherer envisaged by specific patterns of gene expression orchestrated by mechanisms largely unexplored. It is emerging the existence of angiogenic inducers specific for each organ. They include members of Wnt family and endocrine gland-VEGF.

The Mechanisms of Pathological Angiogenesis

In adult life, physiologic angiogenesis employs the same mechanisms illustrated above and occurs in endometrium during female cycle, in *corpus luteum* during reproductive cycle, in hypertrophic glands (i.e., breast during lactation), and in exercise-trained skeletal muscles. Several diseases are connected with and sustained by an abnormal angiogenesis. In some pathological settings, including tumors and chronic inflammatory diseases, abnormal angiogenesis results from an altered homeostasis between angiogenic inducers and inhibitors driven and sustained by specific pathogenic events. On the contrary, ischemic

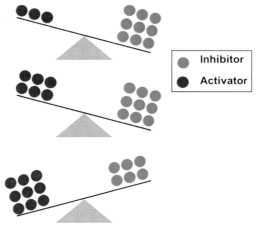

Physiologic angiogenesis
during development and in
selected conditions in adult life

Physiologic conditions in adults:
it's difficult to lunch angiogenesis

Heart disease and other
pathologies associated with
an impaired neovascularization
capacity

Cancer, Inflammation,
Retinopathies, Obesity

● Inhibitor
● Activator

Angiogenesis, Fig. 4 Homeostatic hypothesis in pathological angiogenesis. During embryonic development and in some process in adult life (i.e., female cycle, reproduction, lactation, muscle exercise), tissues express balanced amounts of activators and inhibitors of angiogenesis, which allow the formation of regular vessels. In other adult tissues, angiogenesis is repressed by an increased amount of inhibitors. Vascular regeneration may occur in ischemic tissues but usually it is insufficient to recover tissue oxygenation. On the other hand, tumors, obesity, chronic inflammatory diseases, and retinopathies are characterized by a prominent but chaotic angiogenesis. These different scenarios are associated with unbalanced ratios between activators and inhibitors

diseases are characterized by attempts to revascularize the necrotic area through angiogenesis and arteriogenesis (Fig. 4) [2, 15, 16].

Tumor angiogenesis – Oncogene-mediated cell transformation is necessary but not sufficient to promote metastatic and aggressive cancer. Tumor cell proliferation alone, in the absence of angiogenesis, can give rise to dormant, microscopic tumors of ~1 mm^3 or less, but these in situ cancers are harmless to the host. Oncogene somatic mutations deeply modify the transcription profile of transformed cells, including the upregulation of VEGF and other growth factors often associated with the down-modulation of inhibitors of vessel formation (i.e., thrombospondin). Therefore the accumulation of genetic mutation along the tumor span-life progressively modifies the angiogenic homeostasis leading to the formation of a chaotic vasculature (see above). However the process of tumor angiogenesis is not limited to the altered production of factors in tumor cells and other mechanisms may occur. The angiogenic process may be partially sustained by EC precursors mobilized by bone marrow and migrated to the angiogenic site. Bone marrow participates to the angiogenic process by producing myeloid cells that differently modulate angiogenesis. For example, macrophages have long been characterized as a highly plastic cell type capable of tumor-suppressive or tumor-promoting effects, depending on the polarization state (M1 vs. M2). Tie2-expressing monocytes have also been reported to promote tumor growth through secretion of angiogenic factors, as also demonstrated for neutrophils and platelets. Interestingly the accumulation of CD11b$^+$Gr1$^+$ cells in mice has been associated with the refractoriness to anti-VEGF therapies.

At least in their early phase, some tumors do not mediate sprouting angiogenesis but they utilize preexisting vessels. For instance, brain tumors are highly vascularized and vessels have a similar phenotype than normal brain. Vessels are surrounded, co-opted by tumor cells, and no sprouts are observed (co-option mechanism). Later, when the tumor mass collapses normal capillaries, hypoxia takes places and activates a VEGF-mediated sprouting angiogenesis.

In some tumors and in particular in colon-rectal cancers, the capillary wall is often constituted by EC and tumor cells (mosaic vessel). This scenario may be induced by the endothelial migration during rapid vessel growth without sufficient proliferation to complete the endothelial lining, leaving cancer cells exposed to the lumen. Second, ECs could be shed from the vessel lining, leaving exposed tumor cells. Third, migrating tumor cells could invade the vessel wall, displacing endothelial cells from the lining.

Finally vascular mimicry describes the potential of tumor cells (i.e., in melanomas) or cancer stem cells (at least in glioblastoma) to differentiate in EC with a mixed phenotype and be able to form functional vessels. Despite a significant body of literature, vascular mimicry is not yet a universally accepted mechanism, as the concept has been challenged by several investigators.

Arteriogenesis – The development of an arterial circulation circumventing the occlusion of a large artery from preexistent small arterioles is named arteriogenesis and occurs exclusively in adult injured tissues. It shares some features with angiogenesis, but the pathways leading to it are different. Furthermore arteriogenesis is potentially able to fully replace an occluded artery whereas angiogenesis cannot. As demonstrated in embryonic arterial differentiation, a pivotal role in arteriogenesis is played by physical forces and in particular by shear stress. It is proportional to blood flow velocity and inversely related to the cube of the collateral vessel radius Therefore, since growth increases the collateral vessel radius, shear stress falls quickly. This may be the reason why arteriogenesis arrests without a completed vascularization of ischemic tissue. Collateral vessels of an occluded artery show activated ECs, able to release molecules involved in the recruitment of circulating monocytes, in vascular proliferation and remodeling (proteases, Notch ligands, integrins, VEGF, angiopoietins, fibroblast growth factors) and in control of vascular tone (nitric oxide). Furthermore these collaterals are leaky. A similar activation is observed in smooth muscle cells, which undergo a massive proliferation. Therefore it seems to be conceivable to hypothesize that high fluid shear stress starts at least two signaling mechanisms: one that attracts bone marrow-derived cells for remodeling and another that causes endothelial and smooth muscle cells to enter the cell cycle, leading to proliferation.

Acknowledgments Some of the concepts here reported originate from experimental papers supported by "Associazione Italiana per la Ricerca sul Cancro" (AIRC).

Cross-Reference

▶ Biosensors
▶ Cell Adhesion
▶ Intravital Microscopy Analysis
▶ Nanomedicine
▶ Nanotechnology in Cardiovascular Diseases
▶ Synthetic Biology

References

1. Carmeliet, P.: Angiogenesis in health and disease. Nat. Med. **9**, 653–660 (2003)
2. Folkman, J.: Angiogenesis: an organizing principle for drug discovery? Nat. Rev. Drug Discov. **6**, 273–286 (2007)
3. Ferrara, N., Kerbel, R.S.: Angiogenesis as a therapeutic target. Nature **438**, 967–974 (2005)
4. Bergers, G., Hanahan, D.: Modes of resistance to anti-angiogenic therapy. Nat. Rev. Cancer **8**, 592–603 (2008)
5. Jain, R.K.: Lessons from multidisciplinary translational trials on anti-angiogenic therapy of cancer. Nat. Rev. Cancer **8**, 309–316 (2008)
6. Aird, W.C.: Phenotypic heterogeneity of the endothelium: II. Representative vascular beds. Circ. Res. **100**, 174–190 (2007)
7. Aird, W.C.: Phenotypic heterogeneity of the endothelium: I. Structure, function, and mechanisms. Circ. Res. **100**, 158–173 (2007)
8. Fukumura, D., Jain, R.K.: Tumor microvasculature and microenvironment: targets for anti-angiogenesis and normalization. Microvasc. Res. **74**, 72–84 (2007)
9. Jain, R.K.: Molecular regulation of vessel maturation. Nat. Med. **9**, 685–693 (2003)
10. Serini, G., Napione, L., Arese, M., Bussolino, F.: Besides adhesion: new perspectives of integrin functions in angiogenesis. Cardiovasc. Res. **78**, 213–222 (2008)
11. Roca, C., Adams, R.H.: Regulation of vascular morphogenesis by Notch signaling. Genes Dev. **21**, 2511–2524 (2007)
12. Coultas, L., Chawengsaksophak, K., Rossant, J.: Endothelial cells and VEGF in vascular development. Nature **438**, 937–945 (2005)
13. Hoffman, B.D., Grashoff, C., Schwartz, M.A.: Dynamic molecular processes mediate cellular mechanotransduction. Nature **475**(7356), 316–23 (2011)
14. Aitsebaomo, J., Portbury, A.L., Schisler, J.C., Patterson, C.: Brothers and sisters: molecular insights into arterial-venous heterogeneity. Circ. Res. **103**, 929–939 (2008)
15. Schaper, W.: Collateral circulation: past and present. Basic Res. Cardiol. **104**, 5–21 (2009)
16. Coffelt, S.B., Lewis, C.E., Naldini, L., Brown, J.M., Ferrara, N., De Palma, M.: Elusive identities and overlapping phenotypes of proangiogenic myeloid cells in tumors. Am. J. Pathol. **176**, 1564–1576 (2010)

Animal Coloration

▶ Structural Color in Animals

Animal Reflectors and Antireflectors

▶ Biomimetics of Optical Nanostructures

Anodic Arc Deposition

▶ Physical Vapor Deposition

Antifogging Properties in Mosquito Eyes

Xuefeng Gao
Suzhou Institute of Nano-Tech and Nano-Bionics,
Chinese Academy of Sciences, Suzhou, PR China

Definition

Fogging occurs when vapor condenses into droplets on condensation nuclei (e.g., dust) or exposed objects (e.g., glass and plastics) with diameters larger than 190 nm, namely, half of the shortest wavelength of visible light. These droplets can firmly stick to the surface of optical transparent materials so as to reduce the visibility by light scattering and reflection, which are undesired. Antifogging is to avoid condensed water droplets on a surface. Antifogging properties in mosquito eyes present a novel surface nonstick superhydrophobicity for micrometer-size water droplets [1].

Overview

Surface wettability is a very common but important surface property closely related to the chemical

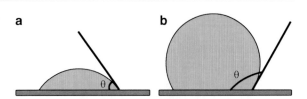

Antifogging Properties in Mosquito Eyes, Fig. 1 Contact angle and wetting states. (**a**) A hydrophilic surface with a water contact angle less than 90°. (**b**) A hydrophobic surface with a water contact angle higher than 90°

composition and microstructures of material surfaces. According to the affinity of water to material chemicals, the wettability may be simply divided into the hydrophobicity and hydrophilicity. The intrinsic contact angle θ of a water droplet on a smooth solid surface is given by the classical Young's equation:

$$\cos \theta = \frac{\gamma_{SV} - \gamma_{SL}}{\gamma_{LV}} \tag{1}$$

where γ_{SL}, γ_{SV}, γ_{LV} are the interfacial free energies per unit area of the solid–liquid, solid–gas, and liquid–gas interfaces, respectively. The higher surface energy of solid materials means a lower contact angle with water, and vice versa. Generally, one defines that a surface with $\theta < 90°$ is called a hydrophilic surface; conversely, a surface with $\theta > 90°$ is called the hydrophobic (Fig. 1). Note that glass and PMMA are two kinds of most widely used optical transparent materials, both of which are hydrophilic with water contact angles of about 30° and 68°, respectively.

Water vapor easily condenses in the form of tiny droplets on any flat surface made of hydrophobic and hydrophilic materials once its temperature is lower than the dew point. Strictly speaking, the water condensed on exposed objects such as plastic and glass should be called dew although the formation of fog and dew virtually follows the same thermodynamic principles. However, for material sciences and engineering, the measures of preventing the moisture condensation on optical transparent materials are customarily called antifogging while those aiming at other opaque materials such as metals and paints are called antidewing. It is known that the ordinary wettability makes the condensed water droplets firmly stick to the surfaces once contacting, which generates light scattering and reflection so as to greatly blur the visual field. It often occurs that the eyeglasses fog up when one comes inside on a cold day. This is because the eyeglasses have been cooled outside down to a temperature that is below the dew point and the moisture in the inside air will come in contact with the cooled eyeglasses and condense into tiny droplets. Similar phenomena can frequently occur on the aluminum heat exchange fins in air conditioners, resulting in higher energy consumption and mildew wind. Besides, other side effects caused by moisture condensation are often neglected, for example, the corrosion of metal, the swelling and peeling of paints, and the attachment of dust particles.

Up to now, most antifogging coatings reported are highly hydrophilic, aiming at protecting the optical transparency. A known example is the use of photocatalytic TiO_2 nanoparticle coatings that become superhydrophilic under UV irradiation [2]. Besides, a three-dimensional capillary effect was adopted to achieve superhydrophilic properties by constructing porous nanostructures from layer-by-layer assembled nanoparticles [3]. The key to these two wet-style antifogging strategies is to make these condensed droplets rapidly spread into a uniform thin film, which can avoid the light scattering and reflection. However, many other problems caused by adsorption of moisture are hard to solve by the superhydrophilic approach due to its inherently wet nature. Thus, a dry-style antifogging technique that can prevent the condensed drops from sticking to the material surface is highly desired, which may be developed into multifunctional coatings such as antifogging, antirust, and antifouling.

Basic Principles

Fogging of material surfaces generally forms via the impact of fog drops suspending in air or the direct condensation of moisture. The antifogging property can be achieved by changing the water affinity of the material surfaces toward the two extremities, as shown in Fig. 2. One is the superhydrophilic approach, with a water contact angle close to 0° [2, 3]. The other is the nonstick superhydrophobic approach, with a contact angle close to 180° and extremely low adhesion for micrometer-size water droplets [1]. The first strategy is now well understood. Thus, we mainly intend to introduce the possibility of the superhydrophobic strategy in this text.

With the advance of nanofabrication techniques, one can easily construct micro- and nanostructures on

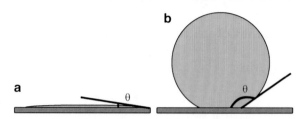

Antifogging Properties in Mosquito Eyes, Fig. 2 (a) A superhydrophilic surface with a water contact angle close to zero. (**b**) A superhydrophobic surface with a water contact angle larger than 150°

the surfaces of different materials, which provide the possibility to study and modulate surface wettability toward the nonstick superhydrophobicity. The real material surfaces are actually not ideally smooth at atomic and molecular scale, which means that the classic Young's equation is not suitable in analyzing their water contact angles. Thus, Wenzel [4] proposed a theoretical model describing the apparent contact angle θ_W on a rough surface and modified Young's equation as follows:

$$\cos \theta_W = \frac{r(\gamma_{SV} - \gamma_{SL})}{\gamma_{LV}} = r \cos \theta \qquad (2)$$

where r is a roughness factor, defined as the ratio of the actual area of a rough surface to the geometric projected area. Apparently, this factor is always larger than unity and the hydrophobicity of a rough surface can be augmented by the increase of the solid–liquid contact area. However, such mechanism brings about very strong water adhesion, as shown in Fig. 3a. Thus, Wenzel model may be effective in the case of superhydrophobic surfaces with microscopic large interspaces or hollows enabling the water penetration but inapplicable to the development of nonstick and antifogging materials.

Air is an effective hydrophobic medium with a water contact angle ~180° [5]. As the air phase can be stably trapped into the microscopic hollows, a rough hydrophobic surface may be considered a composite surface composed of air and solid. Accordingly, the contact angle θ_C of a water droplet on the composite surface can be given by Cassie and Baxter's equation [5] as follows:

$$\cos \theta_C = f \cos \theta + (1 - f) \cos 180°$$
$$= f(\cos \theta + 1) - 1 \qquad (3)$$

a   **b**

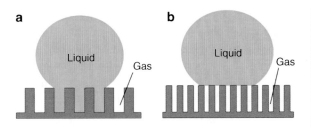

Antifogging Properties in Mosquito Eyes, Fig. 3 Two types of superhydrophobic states. (**a**) Wenzel's model with water trapping into the intervals of structures on surface. (**b**) Cassie's model without any water trapping

Antifogging Properties in Mosquito Eyes, Fig. 4 A photograph of antifogging mosquito eyes (*C. pipiens*). Even though they are exposed to moisture, the surface of the eyes remains dry and clear while the surrounding hairs nucleate many drops

where f and $(1-f)$ represent the area fraction of solid and trapped air, respectively; θ is the intrinsic contact angle of solid materials. Clearly, the hydrophobicity of a rough surface can be amplified by the decrease of the solid–liquid contact area (Fig. 3b). It has been reported that the surface adhesion of a superhydrophobic surface is governed by van der Waals attraction at the solid–liquid interfaces and the capillary force caused by the curving of the liquid–air interfaces [6]. Thus, it is theoretically possible that the nonstick superhydrophobic property can be obtained by fine surface nanoengineering for the antifogging purpose.

Key Research Findings

Lotus leaves are a sort of classic examples of natural nonstick superhydrophobic with a water contact angle $>160°$ and a sliding angle $<5°$, exhibiting remarkable raindrop self-cleaning ability. Recent bionic research has clearly revealed that such property is attributed to the cooperative effects of the hierarchical micropapillae and branch-like nanostructures and the low-surface-energy wax [7, 8]. Up to now, much efforts have been made toward the bio-mimetic fabrication of the lotus-like nonstick superhydrophobic materials, which was designed for various applications such as self-cleaning and drag-reduction [9]. The emergence of such nonstick superhydrophobic technique was expected to be able to effectively solve the fogging problem. However, there are no reports on man-made dry-style antifogging coatings. Recently, researchers from General Motors have reported that the surface of lotus leaves becomes wet with moisture because the micrometer-size fog drops can easily trap in the interspaces among micropapillae [10]. Thus, the lotus-like microstructures are unsuitable for the

creation of superhydrophobic antifogging coatings, and a new inspiration from nature is desired for solving this problem.

Recently, Gao et al. reported a novel dry-style antifogging strategy, inspired by the eyes of mosquitoes, *Culex pipiens*, which have ideal superhydrophobic properties (Fig. 4), providing an effective protective mechanism for maintaining clear vision in a humid habitat. As shown in Fig. 5, scanning electronic microscopic (SEM) observation reveals that the mosquito eye is a compound structure composed of hundreds of micro-hemispheres, which act as individual sensory units called ommatidia. These ommatidia are uniform with a diameter of ca. 26 µm and organize in a hexagonal close-packing (hcp) arrangement. The surface of each ommatidium was covered with numerous nanoscale nipples. These uniform nanonipples, with diameters of (101.1 ± 7.6) nm and interparticle spacings of (47.6 ± 8.5) nm, are organized in an approximately hexagonal non-close-packed (ncp) array.

According to Cassie's principle for surface wettability, such hierarchical micro- and nanostructures may be considered as heterogeneous curved surfaces composed of air and solids. Similar to the surfaces of lotus leaves [7, 8], air may be trapped in the spaces between the nanonipples and micro-hemispheres to form a stable air cushion that acts as an effective water barrier, greatly reducing the contact of tiny fog drops with the eye surface. Chandler and co-workers have reported that it is theoretically difficult for a hydrogen-bonded network of water molecules to

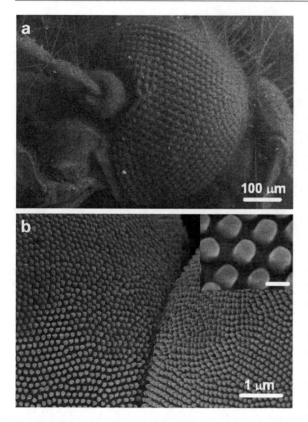

Antifogging Properties in Mosquito Eyes, Fig. 5 SEM images of (**a**) a single mosquito eye (*C. pipiens*) and (**b**) two neighboring ommatidia covered by hexagonally non-close-packed nanonipples. The inset is a high-resolution image. The scale bar: 100 nm (Adapted from [1]. Copyright © 2007 Wiley-VCH)

invade the voids in nanomaterials such as parallel hydrophobic plates with a critical separation of ca. 100 nm at room temperature and atmospheric pressure [11]. As a result, air is firmly trapped in the small voids between these neighboring nanonipples, which possess an average interparticle spacing of 47.6 nm. In this case, water surface tension is strong enough to cause the micrometer-sized fog drops to contract, assuming a perfect spherical shape due to their negligible weight. Moreover, the loose arrangement of nanonipples makes an extremely discrete and nonplanar triple-phase (liquid–air–solid) contact line, which is energetically favorable for driving the spherical fog drops effortlessly from the surface [12–15]. Thus, the combination of the ncp nipples at the nanoscale and hcp ommatidia at the microscale on the surface of mosquito eyes may induce ideal antifogging properties via a superhydrophobic approach.

Although mosquito eyes and lotus leaves are both superhydrophobic, the difference in their surface microstructures gives rise to different biological behavior: the former is antifogging for microscale fog drops while the latter is self-cleaning for millimeter-scale raindrops. It has been reported that the arrangement of lotus papillae at the microscale is random and diffuse with a spacing larger than fog drop diameters. Accordingly, the tiny fog drops are easily trapped in the spaces between the papillae so that the lotus leaf surface may become wet during a long exposure to moisture. In comparison, the mosquito ommatidia at the microscale assume a compact hcp arrangement with triangular voids less than 3 μm, which effectively prevent the plunge of fog drops, while the hexagonal ncp nanonipples trap a stable air cushion, thus preventing microscale fog drops from condensing on the corneal surface.

It is noteworthy that similar nanonipple array structures have also been found on other insects, such as moth eyes and cicada wings [16–19]. Detailed microscopic investigations of these nanonipples indicate that they have in common a characteristic tapered configuration, with periods in the range of 180–250 nm [16–18]. Interestingly, such nanostructures can have other biological functions such as anti-adhesive and broadband antireflective properties [16–19]. Inspired by these, one can develop advanced (multi-)functional interfacial materials. For example, the bio-inspired self-cleaning and zero-reflective coatings can greatly enhance the photoelectric conversion efficiency of solar cells. Many efforts have been made in fabricating nanotaper (e.g., nanonipple and nanocone) arrays by self-masked and masked reactive ion etching techniques [20]. However, the dry-style antifogging properties of these nanostructures have not been reported so far.

Future Directions for Research

The superhydrophobic antifogging technique can keep material surfaces dry, and thus is very attractive as comparing with other antifogging techniques. Nowadays, how to quantitatively measure the interaction of nanostructure surfaces with moisture and condensed drops remains a great challenge. In the near future, one should develop a widely accepted method. There are no successful reports on man-made dry-style antifogging coatings based on the current nonstick superhydrophobic techniques. Accordingly, continuous efforts should be made in designing and fabricating the nonstick superhydrophobic surfaces for microscale water drops. At the same time, their potential physical mechanism is under scrutiny, which is helpful to further guide the experimental research. Finally, when the dry-style antifogging technique is developed for practical applications, it becomes increasingly important to build up low-cost and large-scale nanomanufacture techniques and endow the robust nanofilms with excellent anti-wear, anti-aging properties as well as durability, in order to ensure functioning over longer operation periods.

Cross-References

► Lotus Effect
► Moth-eye Antireflective Structures
► Nanoscale Properties of Solid–Liquid Interfaces
► Nanostructured Functionalized Surfaces
► Nanostructures for Surface Functionalization and Surface Properties
► Nanotechnology

References

1. Gao, X., Yan, X., Yao, X., Xu, L., Zhang, K., Zhang, J., Yang, B., Jiang, L.: The dry-style antifogging properties of mosquito compound eyes and artificial analogues prepared by soft lithography. Adv. Mater. **19**, 2213–2217 (2007)
2. Wang, R., Hashimoto, K., Fujishima, A., Chikuni, M., Kojima, E., Kitamura, A., Shimohigoshi, M., Watanabe, T.: Light-induced amphiphilic surfaces. Nature **388**, 431–432 (1997)
3. Lee, D., Rubner, M.F., Cohen, R.E.: All-nanoparticle thin-film coatings. Nano Lett. **6**, 2305–2312 (2006)
4. Wenzel, R.N.: Resistance of solid surfaces to wetting by water. Ind. Eng. Chem. **28**, 988–994 (1936)
5. Cassie, A.B.D., Baxter, S.: Wettability of porous surfaces. Trans Faraday Soc **40**, 546–551 (1944)
6. Lai, Y., Gao, X., Zhuang, H., Huang, J., Lin, C., Jiang, L.: Designing superhydrophobic porous nanostructures with tunable water adhesion. Adv. Mater. **21**, 3799–3803 (2009)
7. Barthlott, W., Neinhuis, C.: Purity of the sacred lotus, or escape from contamination in biological surfaces. Planta **202**, 1–8 (1997)
8. Feng, L., Li, S., Li, Y., Li, H., Zhang, L., Zhai, J., Song, Y., Liu, B., Jiang, L., Zhu, D.: Super-hydrophobic surfaces: from natural to artificial. Adv. Mater. **14**, 1857–1860 (2002)
9. Bhushan, B., Jung, Y.C.: Natural and biomimetic artificial surfaces for superhydrophobicity, self-cleaning, low adhesion and drag reduction. Prog Mater Sci **56**, 1–108 (2011)
10. Cheng, Y.T., Rodak, D.E., Angelopoulos, A., Gacek, T.: Microscopic observations of condensation of water on lotus leaves. Appl. Phys. Lett. **87**, 194112 (2005)
11. Lum, K., Chandler, D., Weeks, J.D.: Hydrophobicity at small and large length scales. J. Phys. Chem. B **103**, 4570–4577 (1999)
12. Chen, W., Fadeev, A.Y., Hsieh, M.C., Öner, D., Youngblood, J., McCarthy, T.J.: Ultrahydrophobic and ultralyophobic surfaces: some comments and examples. Langmuir **15**, 3395–3399 (1999)
13. Yoshimitsu, Z., Nakajima, A., Watanabe, T., Hashimoto, K.: Effects of surface structure on the hydrophobicity and sliding behavior of water droplets. Langmuir **18**, 5818–5822 (2002)
14. Gao, X., Yao, X., Jiang, L.: Effects of rugged nanoprotrusions on the surface hydrophobicity and water adhesion of anisotropic micropatterns. Langmuir **23**, 4886–4891 (2007)
15. Zheng, Y., Gao, X., Jiang, L.: Directional adhesion of superhydrophobic butterfly wings. Soft Matter **3**, 178–182 (2007)
16. Bernhard, C.G., Gemne, G., Sällström, J.: Comparative ultrastructure of corneal surface topography in insects with aspects on phylogenesis and function. Z Vergl Physiol **67**, 1–25 (1970)
17. Parker, A.R., Hegedus, Z., Watts, R.A.: Solar-absorber antireflector on the eye of an eocene fly (45 Ma). Proc. R. Soc. Lond. B **265**, 811–815 (1998)
18. Sun, T.L., Feng, L., Gao, X.F., Jiang, L.: Bioinspired surface with special wettability. Acc. Chem. Res. **38**, 644–652 (2005)
19. Peisker, H., Gorb, S.N.: Always on the bright side of life: anti-adhesive properties of insect ommatidia grating. J. Exp. Biol. **213**, 3457–3462 (2010)
20. Min, W., Jiang, B., Jiang, P.: Bioinspired self-cleaning antireflection coatings. Adv. Mater. **20**, 3914–3918 (2008)

Applications of Nanofluidics

Vinh-Nguyen Phan[1], Nam-Trung Nguyen[1], Chun Yang[2] and Patrick Abgrall[3]
[1]School of Mechanical and Aerospace Engineering, Nanyang Technological University, Singapore, Singapore
[2]Division of Thermal Fluids Engineering, School of Mechanical and Aerospace Engineering, School of Chemical and Biomedical Engineering (joint appointment), Nanyang Technological University, Singapore
[3]Formulaction, L'Union, France

Definition

Nanofluidics studies the properties and transport phenomena of fluid in nanoscale, which typically ranges from 1 nm to 100 nm. The applications of nanofluidics employ those properties and transport phenomena either for real life needs or to support other fields of research.

Overview

The transport of fluid confined in nanoscale structures (1–100 nm) possesses many interesting unique characteristics so that it promises various applications in science and technology. In history, people have applied the principle of transport in nanoscale to perform chromatography with porous materials. In chromatography, a mobile phase, usually liquid or gas, is driven through a static phase, or matrix, usually in solid. Due to difference in affinity of the particles in the mobile phase to the surface of the matrix, different kinds of particles will move to a different distance. By observing the pattern that the particles form, one can point out the composition of the particles in the mobile phase. Because the surface affinity plays a key role in chromatography, it is preferable for the matrix to have large area-to-volume ratio. Natural nanoporous materials satisfy well this criterion. The driving force in chromatography may be of various kinds, such as pressure driven, capillary, and electrophoresis. In techniques such as paper chromatography, the capillary is preferred because of its simplicity. However, in nanoporous material, where viscous force becomes more important, the capillary and pressure-driven chromatography turn out to be less effective. In such case, the electrophoresis is more favorable. Electrophoresis phenomenon describes the movement of particles submerged in fluids under the influence of the electrostatic field. The movement happens because of the electric charge associated with the intersurface of fluid and solid. Due to the chemical affinity, defects in the crystalline structure of the solid surface, and the adsorption, the ions of the fluid attach to the solid surface. This process is generally selective. Therefore, the intersurface between the fluid and the solid gains a net charge. The thickness of this charge layer is usually represented by the Debye's length $\kappa^{-1} = \sqrt{\varepsilon\varepsilon_0 kT \left/ e^2 \sum_i z_i^2 n_i\right.}$. For large object, the effect of this charge layer is negligible. However, for small particles, where the surface-over-volume ratio is high,

the charged surface becomes more important. When an external electric field is applied, the charged particles move.

Chromatography

Gel electrophoresis is traditionally used to analyze the biomolecules such as DNA, RNA, or proteins. The gel is a porous matrix, usually made from polymer. Several wells are punched in the gel to hold the analyte. When the external electric field is applied, the analyte moves across the gel. Because of the difference in size, mass, shape, and charge of the molecules, different kinds of the molecules migrate to different distances. The molecules are then stained and then observed to obtain the information about the molecular distribution. For small molecules such as proteins or shorter DNA, the polyacrylamide gel is preferred. For the molecules whose size is about 100 bases, the carbohydrate-based matrix agarose is used. However, conventional gel electrophoresis shows to be less effective for large molecules.

With the introduction of new devices, which is better controlled fabricated, the applications of nanofluidics are widely expanded. General perspectives of these new devices and their application were given by Abgrall and Nguyen [1]. Random patterns of the porous material can be substituted by more regular fabricated pillar arrays. Bakajin et al. introduced an improving technique to separate DNA strands using switching electric fields [2] along different orientations. The DNA strands react with the electric field by first realigning their orientation and then migrating along the direction of the electric field. Longer strands take more time to realign; therefore, they have less time to migrate in a cycle. The shorter strands therefore migrate through longer distances than the long ones. The authors demonstrated the separation of 100-kbp DNA in 10 s.

However, in the pillars array design, the large molecules tend to clog the array and easily break up. Baba et al. introduced a new structure to solve the problem [3]. The structure consists of parallel wide channels connected with narrower channels run perpendicular from the wide channels. Electrophoresis is used to drive molecules along the wide channels. Molecules smaller than the narrow channel can enter the narrow channel by Brownian motion while the large molecules remain in the wide channels. As a result, the large molecules migrate in faster than the small molecules. Han et al. performed a series of studies in DNA separation in entropic trap [4]. The entropic trap is a pattern of alternative deep (1.5–3-μm) and shallow (75–100-nm) regions. The molecules were driven across the pattern by electrophoresis technique. The system showed effective separation of long chain DNA. The depths value can be changed to optimize for different size range of molecules. The devices showed to be effective in separating long DNA (5–200 kbp) within 30 min. This principle is also applied to separate protein or smaller DNA in similar devices [1].

Chromatography methods can also be performed in well-defined microchannels or nanochannels rather than in porous material. Various studies have been carried out to establish the relation between the species retention and coefficient of dispersion as functions of Debye's length and the species charge [1]. The channel size is crucial in electrophoresis separation effect. Pennathur and Santiago [5] introduced an investigation on the electrokinetic transport in nanochannels. The investigation described a complete dependence of the effective mobility of species in electrophoretic motion on the bulk electrolyte mobility values, zeta potential, ion valence, and the background electrolyte concentration. From this theory, the authors suggested a method called electrokinetic separation by ion valence. This method determines the ion valence and the ion mobility by comparing transport motion of species in micro- and nanochannels.

Beside electrophoresis, pressure driven flow can also be used in DNA chromatography, as demonstrated by Wang et al. [6]. The DNA sample was injected into the 500-nm channel. The sample was pushed along the channel with pressure. The longer molecules migrate faster than the shorter ones. Real demonstration with DNA sample from Arabidopsis showed that this method is effective. The method, therefore, also promises a great potential to be used for protein and other biomolecules.

Other concept such as shear-driven separation and anisotropic continuous flow separation are also used to separate DNA. Shear-driven separation employs the relative movement of two parallel substrates separated by a spacer. This method does not require a high pressure as in pressure-driven separation. It also does not use electric field as in electrokinetic-

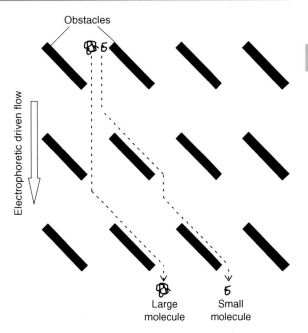

Applications of Nanofluidics, Fig. 1 Array of oblong obstacle used to sieve biomolecules. The smaller molecules suffer more from diffusion and deflect further from the electrophoretic direction

driven method; therefore, it is suitable for samples which are sensitive to electric field. It is also faster than electrokinetic-driven method to reach the optimal velocity for separation. Experiments with dyes showed effective separation within only 0.2 s and over a distance of 1.8 mm [7].

Continuous flow separation technique offers advantage such as real-time monitoring or integration of signal over time. Duke and Austin [8] introduced a sieving mechanism for biological molecules. A stream of molecules are driven by electrophoresis along a sieve consisting of oblong obstacle in oblique direction (Fig. 1). Beside the electrophoretic movement, the molecules also move orthogonally depending on their diffusion coefficient. Therefore, the molecules with different sizes follow different trajectories. This principle was tapped to sieve the DNA molecules of different sizes [1]. Huang et al. also used an array of pillars for sieving, but instead of natural diffusion process, a weaker electric field pulse is used to sieve the molecules [9]. The sieving action is done by alternating strong and weak electric field pulses in different directions. Because of the apparent similarity between the sieving phenomenon and optical dispersion, the author proposed the term "DNA prism" for the device. Using this DNA prism, separation of DNA with sizes in the range of 61–209 kbp was completed within 15 s. The principle can be applied with different shape and arrangement of the obstacles to optimize the sieving effect. Fu et al. [10] introduced a device with parallel deep channels connected by perpendicular nanochannels. The macromolecules such as DNA and proteins are driven by diagonal electric field. One component of the electric field drives the molecules along the deep channels while the other component leads the molecules to jump from one deep channel to another through the nanochannels. The molecules jump more or less frequently depending on their size. As a result, channels with different depths contain molecules of different size ranges.

Separation effect such as in chromatography is usually used for analysis purpose. However, it is also useful in preparation of solution, thanks to the electrokinetic mechanism. Kuo et al. demonstrated the delivery of analyte from one microchannel (injection) to another (collection) [11]. The rate of delivery is controlled by bias voltage across two microchannels and the type of analyte. Different configuration of voltages (specified, ground, or float) at four ends of the injection and collection channels shows different levels of control. Karnik et al. [12] realized the nanofluidic transistor circuit to control the transport of protein between microchannels by connecting nanochannels, serving as the gate. The electric field across the gate is controlled with electrode. The gate then can be switched on or off to control the delivery. Microfabricated porous membrane and nanochannels were used as electrical current passages incorporated in the microchannel system; preconcentration with a factor of up to millions can be reached [1].

Single Molecules Analysis

The size of the nanoscale structures allows them to capture less and less molecules, approaching the goal of handling individual macromolecules for analysis. With recent advances in nanotechnology, the single molecule sensing technique has improved significantly. Conventionally, molecules such as DNA are handled in porous material. The properties of the molecules are studied statistically on a large population, usually obtained after amplification processes, such as

Polymerase Chain Reaction (PCR). The well-defined nanostructure allows the study of the motion of individual DNA. The obtained information is then more specific than that with statistical approach.

When driven into the nanochannels, the DNA molecules separate from one another. Florescent-labeled technique was used to capture the individual images of each molecules. By correlating the photon burst size obtained with the length of the DNA, the length of each molecule can be measured. This technique also can be used to count the number of protein bound to DNA using integrated photon molecular counting method. It was shown that the minimum channel size required to stretch a DNA molecule not only depends on the length of the DNA but also on the ionic strength of the buffer [1]. In a study performed by Krishnan et al. [13], the DNA molecules introduced by capillary filling in a nanochannel were observed to extend at the edge of the channel. This phenomenon promises a simple way to stretch the DNA molecules. However, the physics of this phenomenon is still unknown. Zevenbergen et al. [14] introduced a device to perform redox reaction of molecules in a nanodevice. The solution is filled in a nanochannel with two electrodes on top and at the bottom. The potential of each electrode can be adjusted individually with respect to a reference Ag/AgCl electrode immersed in the solution. The product generated at one electrode becomes the reactant at the other electrode and vice versa. The process is called redox cycling. The redox cycling enables one molecule to contribute many electrons to the induced current, making it possible to detect a small number of molecules by measuring the induced current. The amplification factor up to 400 and a resolution down to 70 molecules were detected.

Figure 2 explains the principle of reading information of individual DNA strand using its translocation through a nanopore. The DNA moves through the nanopore under an applied electric field. During the movement, the DNA partially blocks the pore. The ionic movement along the pore reduces, leading to a drop in the electric current. If the molecules move across the pore individually, each down peak of the current associates with one translocation. With the nanopore small enough, the DNA passes across the nanopore linearly. Then, the DNA length can be deduced from the time of the translocation. However, the traveling speed depends both on the composition (A, C, G, or T) and the orientation (3′ or 5′) of the DNA [1]. Therefore, measuring the length of DNA is not

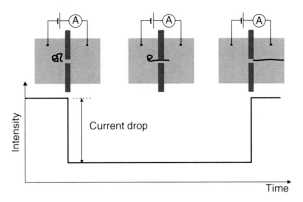

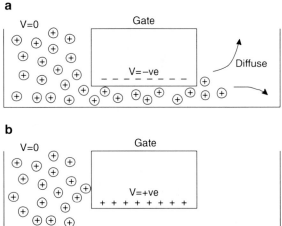

Applications of Nanofluidics, Fig. 2 Translocation of DNA molecules through a nanopore, causing a drop in electric current intensity

Applications of Nanofluidics, Fig. 3 Working principle of field-effect nanofluidic transistor. (**a**) A negative gate potential attracts cations in the nanochannel, making them possible to pass through the channel. (**b**) A positive gate potential repulses the cations from the nanochannels, blocking the flow of cations through the channel

a simple problem, unless the interaction between the nanopore and the molecules is well understood. To analyze such an interaction, Keyser et al. [15] placed a double-stranded RNA molecule attached to a bead by optical tweezers near the nanopore. The strand was pulled into the pore by electric field, but the optical restoring prevents the complete translocation of the molecules. The optical restoring force causes a dislocation of the beads which can be measured. The translocation also caused a drop in electric current. Correlation between the dislocation of the bead and the electric current drops reveals the interaction between the molecules and the nanopore. The information of such interactions provided tools for further studies in single molecule DNA sequencing.

Nanofluidic Electronics

Applications of nanofluidics require the fluid flow controlling in nanoscale, where most of conventional fluid controlling methods are impossible to implement. Electrokinetic properties of the fluid in nanoscale inspire the use of electric field to control the fluid, similar to its use in controlling the electric current in diodes and transistors; this idea leads to the concept of nanofluidic electronics [1]. In a nanochannel, it is easier to create a strong electric field with a relatively small value of electric potential. The electric field in the nanochannel attracts or repulses the counterions or coions. Therefore, the flow of counterions and coions through the channels can be switched on or off. This principle is illustrated in Fig. 3. Gajar et al. [16]

introduced the concept of ionic liquid-channel field-effect transistor, which uses the electric field at the gate to control the concentration ratio of cation and anion. For demonstration with solution of KCl in glycerol in 88-nm-thick channel, when the gate voltage changed from 0 to 25 V, the anion/cation ratio increases by 17.2, the conductance increases by 5.2. This device is not meant to substitute traditional transistor in electronic circuit, but to inspire the possibility that the flow of chemical can be logically controlled, which gives great advantage in process automation. The same principle was also employed by other authors to control the transport of proteins or in the field-effect transistor biosensors [1]. The transistor principle can also be realized with carbon nanotube instead of nanochannels [17]. Especially, surface treatment of the nanotube can modify the polarity of the transistor; therefore, many types of transistor structures can be made in a similar way. Beside the transistor, the idea of nanofluidic diodes was also demonstrated with both nanochannels and nanotubes [1].

Future Directions

Apart from these broadly developed applications, nanofluidics suggests many other applications. The ability to control the diffusivity makes it greatly

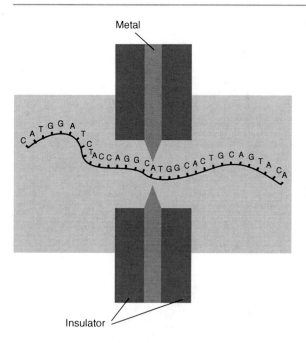

Metal

Insulator

Applications of Nanofluidics, Fig. 4 Reading of DNA sequence as the DNA moves through a nanopore. Preliminary theoretical principles for such an application have been already sketched out

advantageous in drug delivery in vivo [1]. The diffusivity of the chemicals can be controlled by changing the size and shape of the porous pore. Therefore, it is possible to determine the lag time after which the drug will release, as well as the drug releasing rate. Gersborg-Hansen et al. [18] integrated the microfluidic channels in a resonator for dye laser. The laser wave length can be alternated by changing the dye or solvent. Theoretical models have been developed to directly read the sequence of a DNA molecule as it moves linearly through the nanopore like the magnetic tape reading in a tape deck [1], as illustrated in Fig. 4. The nanostructures for energy emerge as a promising field, as the need for new source of energy is increasing. It was showed that conversion of hydrostatic to electrical energy has a maximum efficiency of 12% with aqueous-based solutions of lithium ions [19]. In practice, a lower efficiency was achieved in preliminary experiments but the conditions to improve the efficiency were also sketched out [1]. Current prototypes of fuel cell are mainly based on proton exchange across the membranes. However, because of the enrichment of counterions in a nanochannel, Liu et al. [20] proposed the use of fabricated nanochannels to substitute the porous membranes in fuel cell.

Cross-References

▶ Nanostructure Field Effect Transistor Biosensors
▶ Nanostructures for Energy
▶ Nanotechnology Applications in Polymerase Chain Reaction (PCR)
▶ Optical Tweezers

References

1. Abgrall, P., Nguyen, N.T.: Nanofluidic devices and their applications. Anal. Chem. **80**(7), 2326–2341 (2008)
2. Bakajin, O., Duke, T.A.J., Tegenfeldt, J., Chou, C.F., Chan, S.S., Austin, R.H., Cox, E.C.: Separation of 100-kilobase DNA molecules in 10 seconds. Anal. Chem. **73**(24), 6053–6056 (2001)
3. Baba, M., Sano, T., Iguchi, N., Lida, K., Sakamoto, T., Kawaura, H.: DNA size separation using artificially nanostructured matrix. Appl. Phys. Lett. **83**(7), 1468–1470 (2003)
4. Han, J., Craighead, H.G.: Characterization and optimization of an entropic trap for DNA separation. Anal. Chem. **74**(2), 394–401 (2002)
5. Pennathur, S., Santiago, J.G.: Electrokinetic transport in nanochannels. 1. Theory. Anal. Chem. **77**(21), 6772–6781 (2005)
6. Wang, X., Wang, S., Veerappan, V., Chang, K.B., Nguyen, H., Gendhar, B., Allen, R.D., Liu, S.: Bare nanocapillary for DNA separation and genotyping analysis in gel-free solutions without application of external electric field. Anal. Chem. **80**(14), 5583–5589 (2008)
7. Clicq, D., Vankrunkelsven, S., Ranson, W., De Tandt, C., Baron, G.V., Desmet, G.: High-resolution liquid chromatographic separations in 400 nm deep micro-machined silicon channels and fluorescence charge-coupled device camera detection under stopped-flow conditions. Anal. Chim. Acta **507**(1), 79–86 (2004)
8. Duke, T.A.J., Austin, R.H.: Microfabricated sieve for the continuous sorting of macromolecules. Phys. Rev. Lett. **80**(7), 1552–1555 (1998)
9. Huang, L.R., Tegenfeld, J.O., Kraeft, J.J., Sturm, J.C., Austin, R.H., Cox, E.C.: A DNA prism for high-speed continuous fractionation of large DNA molecules. Nat. Biotechnol. **20**(10), 1048–1051 (2002)
10. Fu, J., Schoch, R.B., Stevens, A.L., Tannenbaum, S.R., Han, J.: A patterned anisotropic nanofluidic sieving structure for continuous-flow separation of DNA and proteins. Nat. Nanotechnol. **2**(2), 121–128 (2007)
11. Kuo, T.C., Cannon Jr., D.M., Chen, Y., Tulock, J.J., Shannon, M.A., Sweedler, J.V., Bohn, P.W.: Gateable nanofluidic interconnects for multilayered microfluidic separation systems. Anal. Chem. **75**(8), 1861–1867 (2003)
12. Karnik, R., Castelino, K., Majumdar, A.: Field-effect control of protein transport in a nanofluidic transistor circuit. Appl. Phys. Lett. **88**(12), 123114 (2006)

13. Krishnan, M., Mönch, I., Schwille, P.: Spontaneous stretching of DNA in a two-dimensional nanoslit. Nano Lett. **7**(5), 1270–1275 (2007)

14. Zevenbergen, M.A.G., Krapf, D., Zuiddam, M.R., Lemay, S.G.: Mesoscopic concentration fluctuations in a fluidic nanocavity detected by redox cycling. Nano Lett. **7**(2), 384–388 (2007)

15. Keyser, U.F., Koeleman, B.N., Van Dorp, S., Krapf, D., Smeets, R.M.M., Lemay, S.G., Dekker, N.H., Dekker, C.: Direct force measurements on DNA in a solid-state nanopore. Nat. Phys. **2**(7), 473–477 (2006)

16. Gajar, S.A., Geis, M.W.: Ionic liquid-channel field-effect transistor. J. Electrochem. Soc. **139**(10), 2833–2840 (1992)

17. Fan, R., Yue, M., Karnik, R., Majumdar, A., Yang, P.: Polarity switching and transient responses in single nanotube nanofluidic transistors. Phys. Rev. Lett. **95**(8), 1–4 (2005)

18. Gersborg-Hansen, M., Balslev, S., Mortensen, N.A., Kristensen, A.: A coupled cavity micro-fluidic dye ring laser. Microelectron. Eng. **78–79**(1–4), 185–189 (2005)

19. Van Der Heyden, F.H.J., Bonthuis, D.J., Stein, D., Meyer, C., Dekker, C.: Electrokinetic energy conversion efficiency in nanofluidic channels. Nano Lett. **6**(10), 2232–2237 (2006)

20. Liu, S., Pu, Q., Gao, L., Korzeniewski, C., Matzke, C.: From nanochannel-induced proton conduction enhancement to a nanochannel-based fuel cell. Nano Lett. **5**(7), 1389–1393 (2005)

Arc Discharge

▶ Physical Vapor Deposition

Arthropod Strain Sensors

Friedrich G. Barth
Life Sciences, Department of Neurobiology,
University of Vienna, Vienna, Austria

Synonyms

Biological sensors; Biosensors; Cuticle; Exoskeleton; Fiber reinforced composite; Insects; Mechanoreceptors; Sense organ; Sensors; Spiders

Definition

Arthropods have highly refined sensors which provide them with a detailed picture of the mechanical events going on in their exoskeleton. These exoskeletal strain sensors are innervated holes embedded in the stiff cuticle that locally increase compliance when the exoskeleton is loaded by internal or external forces. Their deformation in the nanometer range suffices to set off nervous impulses. Arthropod strain sensors are called slit sensilla in arachnids and campaniform sensilla in insects. Their function is particularly closely linked to locomotion.

Introduction

Arthropods are well informed about the mechanical events in their outside and inside worlds. They have numerous highly developed sense organs which respond to a large variety of mechanical events such as tactile, acoustic, vibratory, and airflow stimuli. Whereas their proprioreceptive sensors tell them about self-generated mechanical stimuli like the position and movements of joints and hemolymph pressure, exteroreceptive sensors keep them informed about the external world. In both cases it is in particular changing conditions as opposed to stationary ones which they respond to and use for the guidance of their behavior.

This chapter deals with a sense alien to us humans and specifically linked to the cuticular *exoskeleton*, a structure most typical of arthropods and of dominant importance in regard to their behavior and way of life. The exoskeletons of insects and spiders and related animals not only provide mechanical support and protection and attachment sites for the musculature. They also advertise the animal to mates and conspecifics or conceal it from predators if adequately shaped and colored. In addition to all that, the exoskeleton carries tens or even hundreds of thousands of sensory organs, called sensilla, and has a particular sensory function not known in vertebrates. Embedded into their exoskeleton, insects and arachnids have *strain sensors* which may be considered analogous to technical strain gauges (reviews [1, 2]). These biological strain gauges form tiny holes in the cuticular exoskeleton which represent areas of increased mechanical compliance and locally amplified deformation. When the exoskeleton is loaded by internal or external forces, the resulting minute strains in the cuticle deform the sensors. This deformation in turn elicits nervous responses by straining the dendritic membrane of sensory cells attached to them [1, 3]. Apart from actually measuring strains, the biological strain sensors

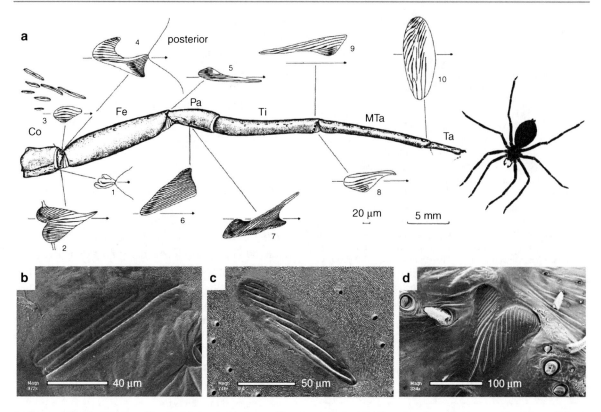

Arthropod Strain Sensors, Fig. 1 (a) Arachnid slit sensilla. Posterior aspect of a spider leg (*Cupiennius salei*) showing the occurrence of slit sense organs. Enlarged drawings of lyriform organs with details of slit arrangements (from [7], with permission) (b) Scanning electron micrographs of three different lyriform organs.(courtesy R. Müllan) *Co* coxa, *Fe* femur, *Pa* patella, *Ti* tibia, *MTa* metatarsus, *Ta* tarsus

have control functions as parts of reflex and more complex nervous circuits which help the animal to detect substrate engagement, to temporally and spatially control its movements and posture, to adapt them to different environments, and to avoid overloading of the exoskeleton [2, 4, 5]. In night-active spiders it was shown that exoskeletal strain sensors even help the animal to remember motion sequences which it can later use to find back to a starting point where they might have lost a prey animal or where they usually return to as their home site or retreat. The zoologists refer to such a remarkable way of orientation as kinesthetic or idiothetic orientation [1, 6].

The largest *number* of strain sensors so far known in an arthropod is that found in the spider *Cupiennius salei* [1, 7]. The presence of some 3,300 strain sensors suggests that there is much detail in the picture the spider receives about the spatiotemporal patterns of the mechanical events going on in its exoskeleton. The vast majority of the sensors (86%) are embedded in the stiff sclerotized cuticle of the walking legs and pedipalps whose mechanical properties are close to that of bone with a Young's modulus of around 18 GPa [8]. The *slit sense organs or slit sensilla*, as the strain sensors are called in spiders and their arachnid relations because of their slit-like form, have aspect ratios (length *l*/width *b*) of up to more than 100, being from 8 to 200 μm long and 1 to 2 μm wide. They come in different configurations as individual slits, loose groups of slits, and so-called lyriform organs (Figs. 1 and 2). The latter are the most intriguing type of slit sensilla. Lyriform organs, of which *Cupiennius* has 144 (the majority on the legs, Figs. 1a, 2a), consist of arrays of up to about 30 slits of different length closely arranged in parallel. Some of these organs show bizarre outlines which substantially impact their mechanical behavior and as a consequence the information made available to the central nervous system.

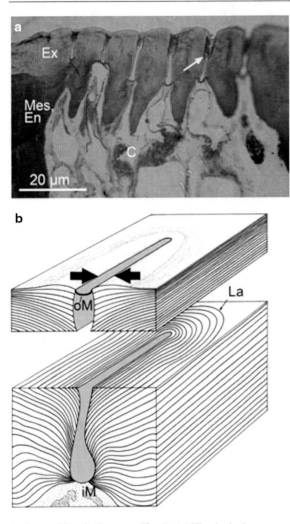

Arthropod Strain Sensors, Fig. 2 (a) Histological cross section through the slits of a lyriform organ. The outer and the inner covering membranes are clearly seen. *Ex* stiff, mechanically dominating exocuticle; *Mes, En* meso- and endocuticle, respectively; *c* cellular components. (**b**) Arrangement of cuticular laminae (*La*) around a slit according to 3D reconstruction. *oM* outer membrane, *iM* inner membrane which one of the two sensory cells penetrates to end at *oM*, whereas the dendrite of the other sensory cell ends at *iM* (**a** and **b** courtesy R. Müllan)

In insects, the equivalent (analogous) strain sensors are called *campaniform sensilla* (Fig. 3). Their strain sensors usually form oval, more rarely round, holes in the cuticle which are covered by a dome-shaped cuticular membrane to which the dendrite of a mechanosensory cell attaches. The organs are called campaniform sensilla because the dendrite and the covering membrane remind of a bell and its clapper. In different campaniform sensilla the aspect ratios range from 1 (round) to rarely more than 3 [1]. The most complete mapping of campaniform sensilla is available for a fly (*Calliphora vicina*), where a total of about 1,200 such campaniform sensilla were found [9, 10]. Similar to arachnid slit sensilla campaniform sensilla occur as single sensilla, loose groups, and what Grünert and Gnatzy [10] referred to as fields. In *Calliphora vicina* there were 86 single campaniform sensilla, 52 groups consisting of 3–32 sensilla with a total of 350 sensilla, and 12 fields with a total of 730 sensilla. As much as 55% (about 730) of all campaniform sensilla were located on the halteres (see Fig. 3a and below) arranged in 10 of the 12 sensilla fields, about 250 sensilla were on the wings and ca. 210 on the walking legs underlining their central role in the coordination of locomotion.

Like in the case of technical strain gauges, the *mechanical parameter measured* depends to a large extent on the sensor's location and arrangement in the exoskeleton. In arthropods, strain sensors tell the animals about sources of load such as muscle contraction, hemolymph pressure, vibration of the substrate they are sitting on, wing deformation during flight due to aerodynamic forces and – in the case of dipteran flies – the Coriolis forces generated during flight (see section "Sensor Arrays"). A special case are insect strain sensors closely associated with the base of wind-detecting sensory hairs and responding when these are strongly stimulated.

After a short description of some functional principles of individual arthropod strain sensors, the main focus of this chapter will be on the intriguing questions of what the mechanical implications of the strange patterns of sensor arrangement in the lyriform organs may be and how nature managed to successfully maintain the mechanical intactness of the exoskeleton using holes as sensory structures (which an engineer always tends to avoid because of stress concentration problems leading to cracks and fracture). As experimental analyses and Finite Element modeling show, quantitative predictions can be made in regard to the mechanical effects of a number of morphological features which are indeed seen to vary in nature. These predictions will be a valuable basis for the design of bio-inspired synthetic sensors used for technical applications.

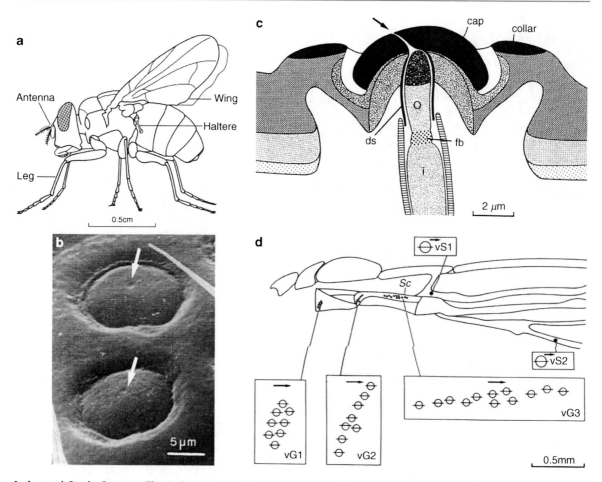

Arthropod Strain Sensors, Fig. 3 Insect campaniform sensilla. (**a**) The fly *Calliphora vicina*. Note halteres which are modified wings. (**b**) Scanning electron micrograph of two campaniform sensilla on the fly leg. (**c**) Schematic drawing of section through stimulus conducting structures of a campaniform sensillum. The different components seen in transmission electron micrographs are shaded differently. *o, i* outer and inner segment of the sensory cell's dendrite, with a so-called fibrillar body (fb) in between; *ds* cuticular dendritic sheath. (**d**) Arrangement of campaniform sensilla on the ventral aspect of the fly's wing base. (**a** and **d** from [9]; **b** and **c** from [10], with permission)

Form and Function

Individual Sensors

According to experimental studies the adequate stimulus of slit sensilla is compressive loading [1]. As demonstrated by Finite Element analyses [11] and earlier pioneering experiments with polymer models [1, 12], several different geometrical parameters affect stimulus transformation and thus the mechanical sensitivity of a slit. One of these parameters is (a) *orientation* in regard to load direction. Single slits are highly directional and best suited for detecting compressive stresses acting normal to their long axis. The same applies to oval campaniform sensilla responding to compression at right angles to their long axes [13]. The other parameter to which slit compression is most sensitive is (b) *slit length l* which indeed varies between about 8 and 200 µm in *Cupiennius* and is the most appropriate parameter to adjust slit compression. Slit length linearly scales with slit face displacement, that is, magnitude of stimulation. A slit's directionality increases with aspect ratio and slit length as well. Remarkably, the (c) *aspect ratio* (*l/b*) of slits plays a minor role in this context when assuming values between 20 and 100. However, when comparing a slit of aspect ratio 1 (reminding of some insect campaniform sensilla) with a reference slit of aspect ratio 100 (well within the range of arachnid slit sensilla) the deformation in the former exceeds that of the latter by 50%. This is mainly due to the area covered by such round sensilla, which is about 80 times larger than in case of a sensor with an aspect ratio of 100. Similarly, (d) capped rectangular slits are compressed more than elliptic holes of identical aspect ratio due to their larger area. Apart from straight slits there are (e) C- and S-shaped slit geometries which, as a rule of thumb, are less mechanically sensitive than the straight slit under normal compressive loading. If one had to adjust/modify the mechanical sensitivity in a synthetic slit sense organ the primary parameter to adjust is slit length. Since slit deformation, which ultimately induces strains in the dendritic membrane and thus nervous activity, is largest in the middle along the slit's long axis mechanical sensitivity could also be adjusted (reduced) by moving the (f) dendrite attachment site toward the slit's ends.

Sensor Arrays

By forming arrays of slits in the lyriform organs, the measurement capabilities at a particular location of the exoskeleton can be substantially increased. What is it that a lyriform organ can do more and better than a single slit?

Due to limited spatial resolution, it is quite difficult to reveal the mechanical tricks underlying the various arrangement patters found in spider lyriform organs by direct observation. Modeling again has helped to better understand the principles [11, 12, 14]. Whereas the analytical approach of Kachanov, developed within fracture mechanics to calculate stress intensity factors at the tips of interacting cracks, can be applied to single slits and loose groups of slits (where the distance S between slits is larger than about ½ slit length l), there are difficulties with lyriform organs where typically S < 0.05 l [15]. A more efficient way to study the mechanical properties of a large variety of slit shapes, positions, and orientations in lyriform organs is finite element analysis. Thus the effects of lateral spacing (S), longitudinal shift (λ), and length gradation (Δl) on slit deformation could be studied under uniaxial normal load (in plane, normal to long axis of longest slit in the array) (Fig. 4). In short, some basic results are the following.

(1) For a group of parallel slits of equal length and as closely spaced as in real lyriform organs the inner slits are much less compressed than the peripheral

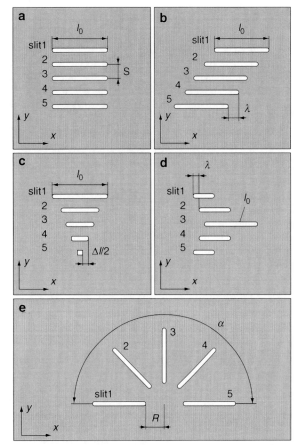

Arthropod Strain Sensors, Fig. 4 Some of the patterns of planar models studied by FE analysis. (**a**) Non-staggered array of five slits, (**b**) oblique bar arrangement, (**c**) triangular arrangement, (**d**) heart shaped arrangement, and (**e**) fan like arrangement. l_0 length of longest slit in array; S distance to neighboring slit; λ longitudinal shift between neighboring slits. (From [11], with permission)

ones. The reason for this is the shielding effect of the outer slits which are deformed much more. (2) Slit compression may be larger by up to several hundred percent than that of a single slit if there is a longitudinal shift between slits. A longitudinal shift between neighboring slits is typical of natural lyriform organs; actually no case without it is known. (3) By arranging slits with different length in a group, mechanical sensitivity can be fractionated and the range of absolute and directional mechanical sensitivity of the entire organ extended. Each of the individual slits in one lyriform organ is innervated independently. Details are found in the papers by Hößl et al. [11, 14].

Arthropod Strain Sensors, Fig. 5 Deformed configurations of three finite element models subjected to uniaxial compressive load at right angle to slit long axes. The displacements are scaled up. (**a**) Oblique bar array with similar mid slit compression for all of the slits. (**b**) Triangular formation with linear gradation of slit length; note considerable shielding effect. (**c**) Heart-shaped array, where maximum slit deformation is not in the middle along the slits. Orange diamonds mark position of maximum compression. (From [11], with permission)

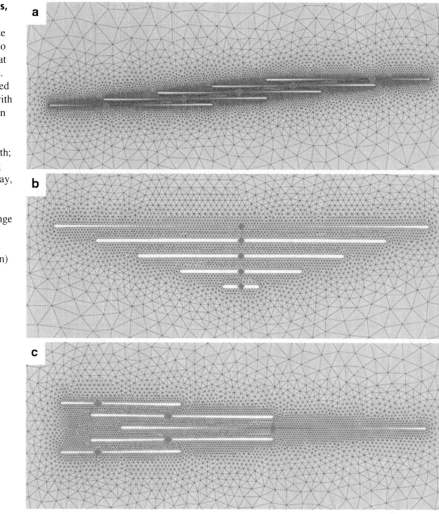

Considering different typical outlines of lyriform slit arrays (represented by five slits in the FE models) provides some answers to the question initially asked: What can a lyriform array do more or better than a single slit? A few examples shall illustrate this (Figs. 4 and 5). (1) In the "oblique bar" array the slits have the same length l but are longitudinally shifted by λ. The level of slit face displacement is similar in all slits and considerably larger than that of a single slit as long as $0.25 \leq \lambda/l \leq 2.5$. One might speculate that by such an arrangement and the central nervous convergence of the corresponding nervous information the signal-to-noise ratio (an ever-present problem in both biological and technical sensing) may be enhanced. (2) Another typical array has a "triangular outline" with slits of decreasing length where the shorter slits are strongly shielded at length gradations $\Delta l/l_0 \geq 0.1$ (Δl, length difference of neighboring slits; l_0, length of longest slit) and will respond to relatively large strains (loads, respectively) only. The widely differing mechanical response in a triangular array suggests that it serves to maximize the range of signal magnitude the organ can cope with and exhibits a pronounced range fractionation. (iii) A third type of array is referred to as "heart-shaped" (Figs. 4 and 5). Similar to fan-like configurations with slit axes diverging in different directions, which are indeed known to occur in nature, one achieves sensitivity over a wide range of loading angles with heart-shaped arrays. This implies that even with particular types of parallel slit arrangements the range of effective load directions can be considerably enlarged. It also implies that such an array fractionates the stimulus both in regard to magnitude and direction. As Fig. 6 demonstrates, small differences in the shape and arrangement of slits

Arthropod Strain Sensors, Fig. 6 FE simulation of the directional sensitivities of the seven slits making up a lyriform organ on the spider leg of the oblique bar type. (**a, b**): Von Mises equivalent stresses (*MPa*) under loads normal to the long axis of slit 1, the longest slit in the array. (**a', b'**): Directional mechanical sensitivity to uniaxial compressive loads in the simple (**a**) and more detailed (**b**) model formed according to scanning electron micrographs of the original organ. The mechanical sensitivity is given as the ratio between the deformation at the position of the dendrite (D_d) and the deformation in the middle of an isolated single slit (D_{sc}). Note the even fractionation of sensitivity among the slits in **b'**. (From [14], with permission)

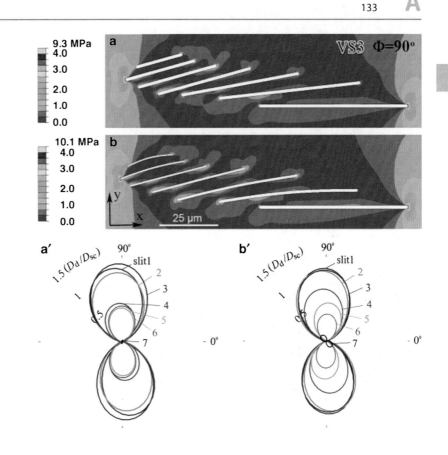

in a lyriform array may affect the deformation pattern substantially.

The most spectacular arrays of insect campaniform sensilla are those at the base of the halteres of flies (Diptera) [9]. The halteres are club-shaped structures derived from the second pair of wings which oscillate during flight and work like gyroscopes, stimulated by Coriolis forces due to body rotations and providing inertial sensory information for flight control with very high temporal resolution [16] (Fig. 3a).

Vincent et al. [17] used finite element modeling to see stress concentration levels in order to analyze the reduction of the material strength by the "campaniform" holes. Their model was a multiple hole arrangement derived from a sensilla field on the base of the fly haltere and consisting of 4 rows of 5 sensilla each (aspect ratio of sensilla 6.5). They found the maximum von Mises stress to be about 10 times higher than in case of the plate with no hole (stress concentration factor 11.1), whereas in case of the single hole it was only 4. Strain amplification in the multiple hole model was predicted to be ×8.75 parallel to the loading. The effect of the aspect ratio,

orientation of the holes, and the influence of other holes in the vicinity (distance between holes) were not studied.

Naturally Occurring Loads and Strains

An intriguing question is that of naturally occurring strains in the arthropod exoskeleton. It is these strains which must have shaped the evolution of the sensors receiving them. Again it is large spiders (*Cupiennius salei*; *Aphonopelma* spec.) where the corresponding quantitative measurements were done in some detail on the legs (where most of the sensilla are found) in living animals. Here the focus will be on just a few values. More details are found in the original literature [1, 8].

The three main sources of load are (1) *hemolymph pressure* which spiders use to extend their legs at joints lacking extensor muscles, (2) contraction of the *musculature* which is concentrated proximally in the legs so that the forces of inertia are kept low in these long and slender structures, and (3) *external sources* such as gravity and substrate vibration. Under the conditions of slow walk at speeds of 1–10 cm/s the strains found at the site of lyriform organs varied between −13 and −20 µε in a particular leg but differed in time course at different positions, relative to the walking pattern. Whereas one organ was stimulated during the stance or load bearing phase signaling muscle force (up to 200 mN) during joint flexion (strain due to resisted muscle action) another one responded to the increased hemolymph pressure (up to 5.3 kPa) during leg extension (swing phase). Since in areas without slit sensilla the strains occurring simultaneously may be significantly higher the conclusion is that lyriform organs are not necessarily located at sites of particularly high strain. In the given case the cyclical change of leg loading and thus the sensory timing of leg movements and the phases of walking seem to be a biologically more relevant aspect.

During situations like rapid starts and stops and jumps one would expect particularly high strain values. At a walking speed of 30 cm/s strains as high as −360 µε were indeed measured in the cuticle of the joint region during the stance phase. During starts and stops and jumps the high accelerations imposed loads similar to those found during rapid walks. Clearly at least many of the lyriform organs are in regions with strongly varying strains and high peak loadings. According to both the model studies referred to above and electrophysiological analyses (see below), the lyriform organs are well adapted to monitor such variable load conditions and to avoid dangerous peaks by eliciting synergic reflexes [4].

Physiological Properties

For a lyriform organ on the spider leg tibia identified as HS-8 [7], which belongs to the "triangular" type (Fig. 3), the available analyses include FE-modeling, studies of organ deformation in the living animal using white light interferometry [14] and electrophysiological experiments [18]. By a comparison of all these data it turns out that the values for slit compression predicted by the FE analysis are very close to those obtained by direct measurement. At the electrophysiologically determined thresholds, where action potentials are elicited in the sensory cells, slit face displacements are as small as 1.7 nm. Clearly,

physiologically relevant measurements down to the nanometer level are possible even in a material with a modulus of elasticity close to that of bone. The sequence of slits in the organ in regard to the magnitude of their deformation (and therefore sensitivity) follows slit length and is the same both in the FE models and in the interferometric measurements. Electrophysiologically determined threshold values differed by more than 40 dB (×100) among the slits of organ HS-8 and the linear parts of their working ranges complemented each other, thereby considerably increasing the linear range and range of high increment sensitivity of the entire lyriform organ as compared to a single slit [18].

Readers interested in the cellular mechanisms of mechanotransduction in slit sensilla are referred to reviews by French and Torkkeli [1, 3]. Zill et al. [2] reviewed the role of insect campaniform sensilla in load sensing and the control of posture and locomotion in some detail.

The Problem of Embedding Holes and Synthetic Sensors

Considering the basically simple design of arthropod strain sensors and the enormous range of their properties achieved by adjusting just a few parameters (whose relevance is understood quantitatively at least in slit sensilla), they are indeed of interest as models for the design of bio-inspired synthetic strain sensors for technical applications. Bio-inspired arrays would allow a highly resolved analysis of strains at small areas of particular interest and help the process of economizing (which is what natural evolution does) by saving material, weight, and energy. A network of sensors could help the decentralized steering of a robot and its feedback control in adaptive locomotion in particular in uneven terrain. As embedded sensors [19] they would not cause a problem by being attached to the outside of a piece of material under load and thereby changing its mechanical properties in particular if the material studied is very thin. There are attempts to design high-performance MEMS sensors inspired by fly campaniform sensilla but the gap between the refinement and performance of the natural sensor and the engineered one is still considerable [17, 20].

In engineering one would normally try to avoid micro-holes like those of arthropod strain sensors because they are sources of stress concentration and the initiation of cracks. Obviously nature has found ways to effectively use holes for sensory purposes without endangering the intactness and provoking fracture of the material they are embedded in. There are several reasons why in nature the intactness of the material obviously is not a problem. (1) As already said the strain sensors not only measure strains but are also part of feedback loops and nervous circuits which can be used to avoid dangerous loads. (2) The resistance to tensile and pressure forces is increased close to the slits by an increased thickness of the stiff exocuticle around the opening (Fig. 2a). (3) The arthropod cuticle is a fiber-reinforced laminate material with the fibers arranged in layers parallel to the surface. Near the slits the fibers deviate from their arrangement parallel to the surface and follow a course reminiscent of the stress lines near a notch subjected to tensile force or pressure. They curve inward like the lines of greatest principal stress conforming to the contours of the slit. Thus the load-bearing capacity of the cuticle is largest in the direction of the greatest principal stress (Fig. 2b). (4) The sensory holes are not holes drilled into a homogeneous material but shaped within a fibrous laminate composite material while the exoskeleton is laid down. In line with this the chitinous microfibers are not interrupted by the hole but molded around it (Fig. 2b). (5) The ends of the notches are not sharp-edged but rounded (Fig. 2b).

The stress concentration factors SCF (local maximum equivalent stress/far-field stress) seen in the models of slit sense organs (which do not fully describe the natural case because they are homogeneous, in-plane isotropic plates of constant thickness) are about 22 at the ends of an isolated slit of aspect ratio 100 and only 4.2 when the aspect ratio is reduced to 2.5. For ellipses with aspect ratios of 2.5 as found in insect campaniform sensilla the SCF is about 6.5. For the lyriform groups highest SCFs were in the range between about 15 and 25 [14, 17].

Cross-References

▶ Biomimetics
▶ Biosensors
▶ Finite Element Methods for Computational Nano-optics
▶ Insect Infrared Sensors
▶ Nanostructured Materials for Sensing
▶ Organic Sensors

References

1. Barth, F.G.: A Spider's World. Senses and Behavior. Springer, Berlin/Heidelberg (2002)
2. Zill, S., Schmitz, J., Büschges, A.: Load sensing and control of posture and locomotion. Arthropod Struct. Dev. **33**, 273–286 (2004)
3. French, A.S., Torkkeli, P.H.: Mechanotransduction in spider slit sensilla. Can. J. Physiol. Pharmacol. **82**, 541–548 (2004)
4. Seyfarth, E.-A.: Spider proprioception: receptors, reflexes and control of locomotion. In: Barth, F.G. (ed.) Neurobiology of Arachnids, pp. 230–248. Springer, Berlin/Heidelberg (1985)
5. Zill, S., Keller, B., Chaudry, S., Duke, E., Neff, D., Quinn, R., Flannigan, C.: Detecting substrate engagement: responses of tarsal campaniform sensilla in cockroaches. J. Comp. Physiol. A **196**, 407–420 (2010)
6. Seyfarth, E.-A., Barth, F.G.: Compound slit sense organs on the spider leg: mechanoreceptors involved in kinaesthetic orientation. J. Comp. Physiol. **78**, 176–191 (1972)
7. Barth, F.G., Libera, W.: Ein Atlas der Spaltsinnesorgane von *Cupiennius salei* keys. Chelicerata (Araneae). Z. Morph. Tiere **68**, 343–369 (1970)
8. Blickhan, R., Barth, F.G.: Strains in the exoskeleton of spiders. J. Comp. Physiol. A **157**, 115–147 (1985)
9. Gnatzy, W., Grünert, U., Bender, M.: Campaniform sensilla of *Calliphora vicina* (Insecta, Diptera) I. Topography. Zoomorphology **106**, 312–319 (1987)
10. Grünert, U., Gnatzy, W.: Campaniform sensilla of *Calliphora vicina* (Insecta, Diptera). II. Typology. Zoomorphology **106**, 320–328 (1987)
11. Hößl, B., Böhm, H.J., Rammerstorfer, F.G., Barth, F.G.: Finite element modeling of arachnid slit sensilla – I. The mechanical significance of different slit arrays. J. Comp. Physiol. A **193**, 445–459 (2007)
12. Barth, F.G., Ficker, E., Federle, H.U.: Model studies on the mechanical significance of grouping in spider slit sensilla (Chelicerata, Araneida). Zoomorphology **104**, 204–215 (1984)
13. Spinola, S.M., Chapman, K.M.: Proprioceptive indentation of the campaniform sensilla of cockroach legs. J. Comp. Physiol. **96**, 257–272 (1975)
14. Hößl, B., Böhm, H.J., Schaber, C.F., Rammerstorfer, F.G., Barth, F.G.: Finite element modelling of arachnid slit sensilla: II. Actual lyriform organs and the face deformations of the individual slits. J. Comp. Physiol. A **193**, 881–894 (2009)
15. Hößl, B., Böhm, H.J., Rammerstorfer, F.G., Müllan, R., Barth, F.G.: Studying the deformation of arachnid slit sensilla with a fracture mechanical approach. J. Biomech. **39**, 1761–1768 (2006)
16. Daniel, T.L., Dieugonne, A., Fox, J., Myhrvold, C., Sane, S., Wark, B.: Inertial guidance systems in insects: from the neurobiology to the structural mechanics of biological gyroscopes. J. Inst. Navig. **55**, 235–240 (2008)

17. Vincent, J.F.V., Clift, S.E., Menon, C.: Biomimetics of campaniform sensilla: measuring strain from the deformation of holes. J. Bionic Eng. **4**(2), 63–76 (2007)
18. Bohnenberger, J.: Matched transfer characteristics of single units in a compound slit sense organ. J. Comp. Physiol. **142**, 391–401 (1981)
19. Calvert, P.: Embedded mechanical sensors in artificial and biological systems. In: Barth, F.G., Humphrey, J.A.C., Secomb, T.W. (eds.) Sensors and Sensing in Biology and Engineering, pp. 359–378. Springer, Wien/New York (2003)
20. Wicaksono, D.H.B., Vincent, J.F.V., Pandraud, G., Craciun, G., French, P.J.: Biomimetic strain-sensing microstructure for improved strain sensor: fabrication results and optical characterization. J. Micromech. Microeng. **15**, S72–S81 (2005)

Artificial Muscles

▶ Biomimetic Muscles and Actuators Using Electroactive Polymers (EAP)

Artificial Retina: Focus on Clinical and Fabrication Considerations

Pablo Gurman[1], Yitzhak Rosen[2] and Orlando Auciello[1]
[1]Materials Science Division, Argonne National Laboratory, Argonne, IL, USA
[2]Superior NanoBiosystems LLC, Washington, DC, USA

List of Acronyms

DOE   Department of energy
EP   Electric potentials
MD   Macular degeneration
OCT   Optical coherence tomography
RP   Retinitis pigmentosa
VPU   Video processing unit

Synonyms

Artificial vision; Bionic eye; Electronic visual prosthesis; Epiretinal implant; Retina implant; Retina silicon chip; Sub-retinal implant; Visual prosthesis

Definition

A device intended to restore the vision in patients suffering from vision loss caused by retinal disorders by replacing or improving the natural retina function or the vision pathway.

Visual Prosthesis for Retinal Disorders

Visual prostheses are currently under development by the scientific and industrial communities aimed to restore the vision for people suffering from retinal disorders such as retinitis pigmentosa (RP) and age-related macular degeneration (AMD). Several types of visual prosthesis have been developed to date by different groups using different approaches. All approaches focus on developing a so called "artificial retina."

The artificial retina includes a device intended to mimic the function of the photoreceptors in the natural retina. The natural retina is made of several layers. In retinal disorders, such as RP and AMD, the photoreceptor layer is compromised. The photoreceptor layer comprises photoreceptor cells that are responsible for capturing the incoming light and transforming it into electrical pulses that will travel to the brain where the image will be created. The artificial retina approach is based on the finding that, although the photoreceptor layer is dysfunctional, the rest of the retina (bipolar and ganglion cells) and the visual pathway is still capable of transmitting the visual information, as electrical pulses, to the brain.

This entry focuses on the Department of Energy's (DOE) artificial retina because it has been successfully developed and commercialized, although other prototypes using different approaches are being developed by other groups worldwide [1].

DOE Epiretinal Device Function

As with the natural retina, an artificial retina should be able to transform the light (a flow of photons) into an electrical pulse (a flow of electrons). This is performed by a camera mounted on a glass connected to a video processing unit (VPU) that is located on the belt of the patient. The camera receives the visual information and transmits this information to a VPU that converts this information into a digital format in the form of pixels that are sent to the internal unit. The internal unit uses a stimulator system to generate electrical pulses that activate the neuron cells through a microelectrode array placed in contact with the bodies of these neurons (the group of neurons that are thought to be activated are the ganglion cells). These cells trigger an electrical wave called action potential that transports the visual information codified in electric pulses through the optic nerve to the occipital cortex of the brain where the image percept is created.

Main Components of the DOE Artificial Retina

The device can be divided in two parts, an external component and an internal component:

External Components
Camera: A CCD camera, located on glasses used by the patient, collects the visual information in the form of pixels sending this information to a microprocessor located in the patient belt.

Microprocessor: A microprocessor is placed on the patient's belt in order to capture the visual information (pixels) and transforms it into electronic signals. This part of the device is located externally allowing adjustments of the visual information to send the most accurate data to the internal unit. In this manner, the ophthalmologist could evaluate the patient's vision and modify it accordingly.

Battery: A battery is placed externally to provide energy to the external microprocessor, but also to send energy wirelessly to the internal unit. Having an external battery is an important requirement since a battery located inside would have more constraints such as being of small size, and its replacement could demand frequent surgeries plus the fact that a battery inside the eye would require stringent bio-isolation from the eye to prevent eye rejection.

Internal Components
Transmitting Unit: Using a special algorithm, the internal unit codifies the electronic signals sent by the microprocessor and send electrical pulses through a stimulator to the microelectrode array located at the retina surface via wire connection. The power to supply this unit is provided wireless by the external component.

Microelectrode array: Microelectrodes are the interface between the device and the visual system that are responsible for stimulating the neurons that will transport the visual stimulus to the brain. Microelectrodes are micrometer size structures ranging from flat circular platinum discs to elongated sharp pyramidal structures, in the size ranging between 250 and 500 µ in diameter, responsible for transmitting the electrical pulses coming from the microstimulators to the neurons, which become activated and transport the electrical pulses to the brain. To date, microelectrode arrays have been developed, featuring 16, 64, and 200 microelectrodes. The 64 microelectrode array is currently the first artificial retina implanted commercially in blind people, first in Europe, where Second Sight has obtained approval for commercial implantation of the epiretinal device. Several kinds of microelectrodes have been developed for retina prosthesis. Desired features for retina microelectrodes are low impedance which means easy transfer of electrical pulses to the neurons, biocompatibility, or an acceptable tissue response to the implant, long lifetime to avoid additional surgeries, and flexibility (that means, adaptation of the implant to the curvature of the eyeball without hindering the electrode-neuron distance that could promote heating effects if they are too close or could not activate the neurons if they are too separate). Miniaturization of electrodes is highly desirable and currently under development as the larger number of electrodes allows to achieving higher resolution of the image. The problem, which a tough one, is that impedance increases as electrode size becomes smaller. As impedance increase, voltage should increase to deliver the same amount of charge to the neurons, resulting in an increase in energy consumption, and concurrently an increase in heat dissipation, which could damage the surrounding tissue.

Wireless Transfer of Information and Energy between External and Internal Units: In the current epiretinal devices, both information and energy communication between the external and internal units transmitted via electromagnetic signals wirelessly and coupled to the microchip via inductively coupling coils. Inductively coupling coils are conductive wires built up as coils that work based on the physical phenomena of electromagnetic induction by which an electric current flowing in one coil induces a voltage on the other coil by the electromagnetic field created

Artificial Retina: Focus on Clinical and Fabrication Considerations,
Fig. 1 Photograph of the internal unit with the microelectrode array and the stimulator circuit (With permission from Springer, from Ref [4])

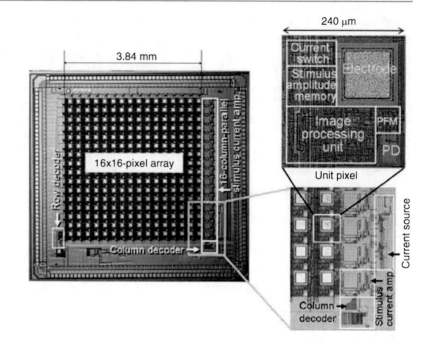

in the former by the current flow. One of the coils is located in the external component while the other is located in the internal component (Fig. 1).

Materials Used in the DOE Artificial Retina

Silicon: Silicon is the current material used in most of the microelectronic components and microfabrication processes. In the artificial retina, microprocessors are necessary for translating the visual stimulus into an electric signal. Also, the internal unit has a microchip stimulator to deliver the electric signal to the microelectrodes.

Diamond: In the DOE artificial retina prototype, a novel diamond material known as Ultrananocrystalline Diamond (UNCD) is being investigated in thin film form to provide a hermetic/bioinert/biocompatible coating to encapsulate the silicon microchip to enable implantation inside the eye. The UNCD film provides a coating extremely resistant to the corrosive environment of the eye saline, which otherwise would attack the silicon material underneath dissolving it [5].

Polymers: Some polymers including polyimide, parylene, and silicone have been tested to provide a flexible platform to deposit microelectrodes. In this manner, it is possible to adjust the microelectrode array to the curvature of the eye reducing the distance between the device and the neurons which results in lower voltage thresholds to activate the neurons decreasing power consumption.

Metals

Gold: gold is being used as one of the noble metals for wires to connect the microelectrode array to the pads where the current is delivered from the internal electronic unit that drives the energy provided wireless by the external battery.

Platinum: Platinum is being used as the material for microelectrodes. Key properties that make platinum an optimum candidate as microelectrode material includes its capacity to transfer electrical charge to the neurons (low impedance), its biocompatibility, and its durability. A potential problem of Pt, which has not been investigated in detail, is the catalytic properties of Pt. As the size of the Pt electrodes becomes smaller, the catalytic properties of Pt becomes bigger. Under the liquid environment of the eye, this could increase the risk of unwanted electrochemical reactions.

Indium Tin Oxide: material used for leads
Silicon Nitride: material used as insulator layer
Silicon Oxide: material used as insulator layer

Fabrication Process of the DOE Artificial Retina

Due to the tiny and delicate space that occupies the retina inside the eyeball, miniaturization technologies are deemed extremely necessary to create complex devices with integrated functionality. Some of the miniaturization techniques currently used to develop an artificial retina are:

Photolithography: Photolithography is a technique developed to imprint micro and nano patterns using a photosensitive polymer that when exposed to UV light it could reproduce a pattern imprinted in a mask over a substrate like silicon.

Reactive Ion Etching (RIE): Once a photolithography pattern is created, in order to etch away the parts that do not correspond to the pattern, an etching technique is needed. RIE is a technique that uses high-energy ions, radicals, and electrons created in plasma and accelerated toward the substrate to enhance chemical reactions of the reactive atoms in the system to produce directional (anisotropic) etching of the substrate.

Electroplating: Electroplating is a technique that allows growing metals to create microstructures such as the platinum discs microelectrode array.

Thin Film Deposition (Sputtering, Physical Evaporation, Chemical Vapor Deposition): Sputtering, physical evaporation, and chemical vapor deposition techniques are used to produce films with thickness from a few nanometers ($1 \text{ nm} = 10^{-7} \text{ cm}$) to a few microns ($1 \text{ μm} = 10^{-4} \text{ cm}$). For the components of the artificial retina, different type of films are used for different purposes. In the hermetic encapsulation of the microchip to provide bioinertness (UNCD films), in the metal contacts to allow current passage to the microelectrodes (platinum films), and insulation layers to minimize or eliminate leakage currents that could promote unwanted electrochemical reactions within the biological media of the eye.

Most of these techniques are carried out in special facilities known as clean room designed to keep the air free from dust particles, preserving the surface of the retina components free from contaminants (dust and other species).

Laboratory Studies for Assessing Retina Structure and Function

In Vitro and In Vivo Testing

In vitro and vivo testing of biomedical devices becomes of critical importance to accomplish the goal of developing a safe and effective commercial device. In vitro systems could provide valuable information regarding electrical requirements that are needed to activate ganglion cells and thus the visual pathway. Electrophysiological techniques are used to study and assess the response of the artificial retina components to electrical stimulus and are some of the laboratory techniques to perform analysis of the artificial retina performance. Electrophysiological studies are based on recording electric potentials of the retina neurons. Electric potentials (EP) are electric waves that travel through the neurons carrying information. In this case, visual information is codified by frequency and amplitude changes of the potentials. In retina diseases such as RP and AMD, it is found that electrical potentials are altered and therefore become an important tool to perform a diagnosis of retina disease. In the same manner, the EPs allow an electrical characterization of artificial retina performance by analyzing the effect of microelectrodes stimulation on the activation of retina neurons. The technique developed to record EPs on retina cells is called *Electroretinogram. In order to perform an Electroretinogram* it is necessary to mount electrodes in the ocular surface to be able to record electric potentials in the neurons of the retina. A more invasive technique, called intraocular electroretinogram, consists of placing an electrode inside the eye, near the retina surface. In spite of providing useful information during the stage of animal testing, the clinical utility of Electroretinogram remains controversial.

In addition to the microelectrode system, the experimental setup to perform these types of experiments includes isolating the retina from the eyeball of an animal (i.e., salamander). The retina tissue is stimulated by a microelectrode array displayed in such a way to contact the retina neurons. In this manner, it is possible to record the electrical activity of a group of neurons (i.e., bipolar cells, ganglion cells) after an electric pulse is applied through a microelectrode array. Before performing studies in human subjects, it is necessary to evaluate the safety and effectiveness of the artificial retina in an *animal model*. An animal model mimics the human disease in an animal. Animal models of RP and AMD have been developed to assess the safety and effectiveness of an artificial retina [6].

Clinical and Surgical Considerations

Device Implantation

The eye is divided in two main compartments, the anterior compartment composed of the cornea, iris, and lens, and the posterior segment composed of the sclera, choroid, retina vitreous, and optic nerve. In order to implant the artificial retina, it becomes mandatory to access the posterior compartment of the eye, which is filled with a gelatinous substance known as vitreous. Being able to place the device on the retina surface (epiretinal) requires to remove the vitreous out of the posterior chamber while simultaneously keeping the pressure of the eye constant to avoid the collapse of the eyeball. The procedure is called vitreoretinal surgery and is done by performing three incisions on the sclera, one for introducing an illuminating system, another one for introducing a cannula to insert the device and cut the vitreous (it is not possible to suction it because it could track the retina and produce a retinal detachment, therefore it should be cut first and then aspirated), and the last incision for infusing liquid to keep the intraocular pressure constant while the small pieces of vitreous are aspirated. The device is fixed to the surface of the inner part of the retina using a tack that keeps the device in place avoiding any displacement [7].

Clinical Assessment of a Patient with an Artificial Retina

Vision Before Surgery

Patients with RP initially suffer from night blindness. As the disease progress, patients start to relate symptoms associated with peripheral loss of rods such as

inability to see pedestrians while they are driving. In addition, a difficulty in seeing the light impedes the patient from reading. At more advances stages, the patient loses independency because they are only able to see a small center region of the visual field (tunnel vision) and they are not able to walk by themselves. The patient with MD on the other hand consults for central vision impairment such as vision blurring, or a scotoma, which is a patch in the center of the visual field. At later stages, in both diseases, blindness could be the final outcome. Therefore, an artificial retina could restore partial vision for these patients enhancing their quality of life significantly. Computer simulations indicate that in order for a blind person to recover basic visual capabilities such as face recognition, reading, and independent mobility, the artificial retina should have at least 1,000 electrodes in order to create an image of 1,000 pixels. Current technologies have not overcome this challenge yet because miniaturization of electrodes is difficult. However, improvement of everyday life activities performance has been observed in patients that were implanted with the device with 60 electrodes [1].

Clinical Studies to Assess Retina Structure and Function

OCT (Optical Coherence Tomography): OCT is a clinical technique that allows the measurement of the thickness of the retina. When an artificial retina is implanted, OCT brings about useful information about distance between microelectrodes and the retina surface which is an important parameter, since in order to obtain a neural response the microelectrode array should be placed in close proximity to the group of neurons. In addition, if the microelectrode array is placed too far apart from the retina neurons the system consumes more power and dissipates more heat, increasing the chances of tissue damage.

Fundus Photo: Fundus refers to the inside back part of the eye where the retina is located. A special tool called ophthalmoscope allows the physician to take pictures of this part of the eye obtaining valuable information of the status of the retina tissue affected by RP or AMD.

Fluorescein Angiogram

This technique allows visualization of retina vessels by injecting a fluorophore inside the artery vessels. It is a useful technique for diagnosis of RP.

Clinical Performance After Artificial Retina Implantation

After implantation with an artificial retina containing 16 microelectrodes, patients were capable of discerning light and dark patterns (called phospenes) and localizing the spatial origin of the phospenes (which microelectrode was stimulated), whereas with the new generation of 64 microelectrodes the patients were capable of performing spatial motor task where visual input results are of critical importance to perform motor activities. In order to assess visual performance in these patients, several specific test including psychophysical studies, specific questionnaires, and subjective perceptions are being used, and new strategies to objectively assess visual function in patients implanted with an artificial retina are being pursued.

Cross-References

► CMOS MEMS Fabrication Technologies
► Epiretinal Prosthesis
► MEMS Neural Probes
► MEMS on Flexible Substrates

References

1. Artificial Retina Project, DOE website. http://artificialretina.energy.gov/gpra2010.shtml. Accessed 1 Nov 2011
2. Auner, G.W., You, R., Siy, P., McAllister, J.P., Takluder, M., Abrams, G.W.: Development of a high frequency microarray implant for retina stimulation. In: Artificial Sight: Basic Research, Biomedical Engineering, and Clinical Advances. Springer, New York (2007)
3. Liu, W., Mohanasankar, S., Wang, G., Mingcui, Z., Weiland,J.D., Humayun, M.S.: Challenges in realizing a chronic high-resolution retinal prostheses. In: Artificial Sight: Basic Research, Biomedical Egineering, and Clinical Advances. Springer, New York (2007)
4. Ohta, J., Tokuda, T., Kagawa, K., Terasawa, Y., Ozawa, M., Fujikado, T., Yasuo, T.: Large scale integration base stimulus electrodes for retinal prosthesis. In: Artificial Sight: Basic Research, Biomedical Engineering, and Clinical Advances. Springer, New York (2007)
5. Auciello, O., Sumant, A.V.: Status review of the science and technology of ultrananocristalline diamond (UNCD™) films and application to multifunctional devices. Diamond Relat. Mater. **19**, 699–718 (2010)
6. Fletcher, E.L., Jobling, A.I., Vessey, K.A., Luu, C., Guymer, R.H., Baird, P.N.: Animal models of retinal disease. Prog. Mol. Biol. Transl. Sci. **100**, 211–286 (2011)
7. Hameri, H., Weiland, J.D., Humayun, M.S.: Biological considerations for an intraocular retinal prosthesis. In: Artificial Sight: Basic Research, Biomedical Engineering, and Clinical Advances. Springer, New York (2007)

Artificial Synapse

► Dynamic Clamp

Artificial Vision

► Artificial Retina: Focus on Clinical and Fabrication Considerations

Astigmatic Micro Particle Imaging

C. J. Kähler, C. Cierpka and M. Rossi
Institute of Fluid Mechanics and Aerodynamics,
Universität der Bundeswehr München, Neubiberg,
Germany

Synonyms

Anamorphic particle tracking; Wavefront deformation particle tracking

Definition

Astigmatism particle imaging is an optical method to determine the instantaneous volumetric position of spherical particles in a volume of transparent medium using only a single camera view. The method is mainly applied in microfluidics to measure the velocity and trajectories of tracer particles that follow the fluid motion. However, recently it could be shown that this measurement principle without bias errors due to the depth of correlation and spatial averaging is well suited to reconstruct the arbitrary-shaped interface of mixing layers in a micromixer with extended high precision.

Optical Principle

Astigmatic imaging is a way to break the axis symmetry of an optical system thus allowing for the depth coding of particle positions in 2D images. Among the manifold of 3D3C methods recently developed, it can be classied as a method that makes use of the particle defocus without using an extra aperture [13]. The word astigmatism is based on the Greek language description for not (a-) point-shaped (-stigma). An optical system that features astigmatism has two focal planes and the astigmatic image of a point source is thus an ellipse as shown in Fig. 1. Astigmatism is commonly used for technical applications, e.g., auto-focus systems such as CD, DVD, or Blu-ray players to measure the distance between the rotating disk and the reader head.

One approach producing an astigmatic image aberration is the use of a tilt angle between the measurement plane or volume and the camera, which results in off-axis imaging. This method is often not applicable due to the fixed optical axis in microscopes and requires complex calibration procedures. However, another more elegant method is the use of cylindrical lenses. Already in 1994, Kao and Verkman [7] applied this technique to measure the position of fluorescent particles in living cells. The basic principle and the implementation to a standard epi-fluorescence microscope are shown in Fig. 2. The only difference from a standard micro particle image velocimetry (μPIV) system [12] is the cylindrical lens, which is installed directly in front of the camera chip as shown in Fig. 2. As for standard μPIV, the transparent flow of interest is usually seeded with monodisperse spherical fluorescent tracer particles that follow the flow faithfully. The whole microfluidic channel is then illuminated twice by a laser or another bright light source (continuous or pulsed). To homogenize the beam profile often a diffuser plate is used. A dichroic beam splitter is applied to allow only short wavelength light pass through (indicated in green in Fig. 2), if a white light source is used. This short wavelength light is then absorbed by the tracer particles' fluorescent dye and longer wavelength light (indicated in red) is emitted. The dichroic filter is now used to reject direct reflections of the microfluidic device and only allow the emitted red light to contribute to the signal. A combination of a field lens and the cylindrical lens is used to image the particles on the digital CCD

(charge coupled device), CMOS (complementary metal oxide semiconductor), or intensified camera. On the right side of Fig. 2, the imaging of the particles is schematically shown. The curvature of the cylindrical lens acts in one direction only. Therefore, two distinct focal planes are produced. In the left (red dotted) schematic, the cylindrical lens causes a shortening of the distance between the focal plane and the objective lens. Particles that are close to this focal plane show a sharp and small extension in the x-direction (a_x) and a larger stretched one in the y-direction (a_y), thus producing an oblate-shaped particle image. On the right side of the schematic (blue dashed) the cylindrical lens is rotated by $90°$. The focal plane is now further away from the objective lens. a_y is thus smaller than a_x, resulting in a prolate-shaped particle image. The particle image width and height can now be related to the depth position z by a proper calibration.

Data Evaluation and Calibration

Since the volumetric particle position is used later to estimate the velocity or fluid interfaces, the accuracy of the method relies on the precise determination of the particle's position. For the in-plane, or x- and y-direction, an in-plane calibration by means of a grid, or a known object, can be performed. The in-plane position can later be determined with sub-pixel accuracy by either a two-dimensional Gauss fit or via wavelet analysis, where the latter method is much faster. The width and height of the particle images are identified by either a two-dimensional Gauss fit or the autocorrelation function of the particle. These algorithms have shown accurate results, even for high noise levels for particle images that extend at least three pixels in both directions [2]. Since the technique relies on the exact determination of the particle images' shapes, at least an 8-bit or better sampling is necessary to yield reliable results. Once a_x and a_y are known, a relationship with the depth distance is necessary. If the particles' diameter distribution is very coarse, the difference $a_x - a_y$ [1] or the ratio a_x/a_y [2] can be used as an independent criteria. The trust region is then limited to the linear region between the two focal planes. For a narrow particle diameter distribution, a direct fit of a_x and a_y can be used to substantially extend the measurement depth. A model of the particle image diameter, as a

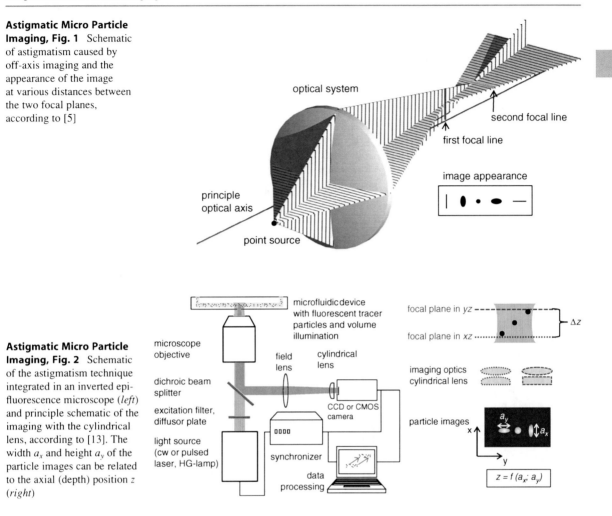

Astigmatic Micro Particle Imaging, Fig. 2 Schematic of the astigmatism technique integrated in an inverted epifluorescence microscope (*left*) and principle schematic of the imaging with the cylindrical lens, according to [13]. The width a_x and height a_y of the particle images can be related to the axial (depth) position z (*right*)

function of the distance z from the in-focus plane, can be applied for this purpose [9]. Under the assumption that the geometric image of the particles and the effect of diffraction and defocusing can be approximated by a Gaussian function, and that the working distance of the lens is significantly larger than z, the particle image diameter can be estimated by the following equation [14]:

$$a(z) = M\sqrt{d_p^2 + 1.49\lambda^2\left(\frac{n_0^2}{NA^2} - 1\right) + 4z^2\left(\frac{n_0^2}{NA^2} - 1\right)^{-1}}$$

(1)

where d_p denotes the particle diameter, λ is the wavelength of emitted light, n_0 the refractive index of the immersion medium of the lens, and M and NA are the magnification and numerical aperture of the lens,

respectively. Equation 1 represents an arc of a hyperbola. The function can be used to extend the measurement depth beyond the two focal planes and the fit is applied to the measured data itself:

$$a_x(z) = \sqrt{c_1^2(z - F_{xz})^2 + c_2^2} + c_3$$
$$a_y(z) = \sqrt{c_4^2(z - F_{yz})^2 + c_5^2} + c_6$$

(2)

F_{xz} and F_{yz} denote the positions of the respective infocus planes for the x- and y-directions, and $c_{1...6}$ are parameters for the fit function. Particles that are beyond the in-focus planes can be taken into account by this fit and thus the measurement volume is not restricted to the linear region between the in-focus planes. The distance between F_{xz} and F_{yz} in physical space can be determined automatically by scanning a grid reticule through the volume. The axial position of F_{xz} and F_{yz} can be measured by a fit of a focus function (e.g., rms of the intensity) with sub-micron resolution. If additional image aberrations are present and the focal planes are not completely flat, these aberrations will also be estimated by the procedure and the final results can be corrected for this systematic error. The validity of the model was tested on simulated and real data [3]. The advantages of the intrinsic calibration are:

- The measurement volume depth is not limited by the distance between the two in-focus planes.
- The optical path through the different media does not change from calibration to experiment.
- All data points are taken into account, and thus the calibration is statistically relevant to a high degree.

Nevertheless, the distance between the two in-focus planes Δz is the quantity that primarily defines the depth of the measurement volume. It can be adjusted by the focal length of the cylindrical lens or the distance between the imaging lens of the CCD and the cylindrical lens [1]. The specific measurement volume is finally dependent on the specific optical arrangement of the whole microscope and the magnification [1]. Increasing the magnification gives lower Δz for the same cylindrical lens, whereas decreasing the focal length of the cylindrical lens increases the measurement volume by approximately $1/f_{cyl}$, [3].

Applications

In general, the astigmatic imaging approach based on cylindrical lenses is very easy to apply and allows for the extension of existing 2D measurement systems to fully 3D measurements without changing the illumination light path. Knowing the particle positions in the volume at two different time instants t and $t + \Delta t$ allows for an estimation of the first-order approximation of the particles' velocity. To find matching particles, several elaborate tracking schemes can be used [10]. Despite the necessity for lower seeding concentration to avoid particle image overlapping, the major benefits of a 3D tracking scheme opposed to correlation based methods are:

- Out-of-focus particles do not bias the velocity estimate by the depth of correlation effect [9].

- In-plane gradients and out-of-plane gradients are not averaged over interrogation volumes [6].
- The spatial resolution can be increased by taking a larger number of images.

The tracking approach was applied in recent investigations to measure flow velocities in micron-sized channels [1, 2] and close to the electrode during electrochemical metal deposition [15]. A comparison with standard and stereoscopic μPIV showed a similar uncertainty for the in-plane velocity components but an almost two times lower uncertainty for the out-of-plane component using A-μPTV [4]. Using time-resolved image sampling, the temporal and spatial evolution of the particle trajectories in an electrothermally generated micro vortex was measured in a micro volume of 50 μm in depth without traversing [8]. Selected trajectories are shown in a volumetric representation on the left side of Fig. 3. The three-dimensional structure of the vortex and the mechanisms of particle trapping inside the vortex could be studied and the results were used to validate the numerical boundary conditions for a simulation of such a scenario. Since the particle trajectories were sampled with sufficient temporal resolution, Lagrangian accelerations were also measured as shown on the right side of Fig. 3.

Astigmatic particle imaging can also be used to estimate the boundary of two mixing fluid regions of arbitrary shape or to reconstruct the three-dimensional shape of particle-seeded stream tubes. The volumetric topology of the interface can be estimated from the measured 3D positions of the particles. Algorithms based on Delaunay triangulation and numerical diffusion have been examined to perform the surface reconstruction [11]. The accuracy and resolution of the reconstruction is limited by the maximum particle density achievable and the error of the particle position determination. However, for steady flows (as typical in microfluidics) any desired accuracy can be achieved for a sufficient large number of samples. A classic example in which this approach is useful is the characterization of passive microfluidic mixers, where the topology of the mutual interface between the two mixing fluid streams plays an important role. In this case, one of the two mixing fluids is seeded with a homogeneous distribution of tracer particles. A typical result for the case of a micromixer, based on a curved microchannel, is shown in Fig. 4.

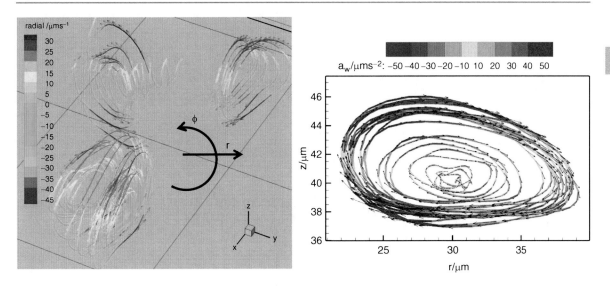

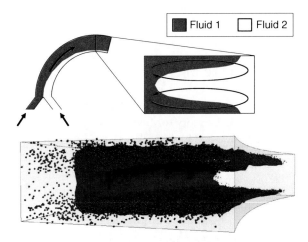

Astigmatic Micro Particle Imaging, Fig. 3 Selected trajectories in an electro-thermally driven microvortex, color coded with the radial velocity (*left*). Representation of one trajectory in the *rz*-plane, color coded with the axial acceleration (*right*)

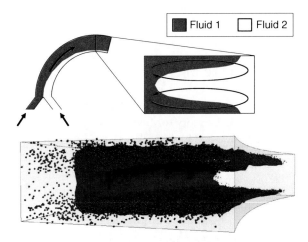

Astigmatic Micro Particle Imaging, Fig. 4 Reconstruction of the mutual interface between two mixing fluids in a microfluidic mixer using the volumetric distribution of tracer particles. The flow is laminar and the convective mixing is induced by secondary flows established as a consequence of the channel curvature

The cross-section of the microchannel was $200 \times 70~\mu m^2$ and the average particle density was set to 0.023 particles/μm^3, resulting in a resolution of 2.5 μm in the optical-axis direction with a maximum estimated error of 5.5 μm [11].

Conclusion and Future Aspects

It is obvious that astigmatic particle imaging is a technique that provides the instantaneous volumetric position of particles in a transparent medium with high precision. The displacement between two successive recordings can be used to estimate the volumetric three-component velocity field. In addition, the Lagrangian velocity and acceleration of the particle can be measured for sufficiently high sampling rates without systematic errors due to the depth of correlation and spatial averaging as typical for μPIV. The positions of the particles can also be used to estimate their volumetric distribution and to analyze the mixing of fluid streams [11], particle agglomeration, trapping, deposition, or other process engineering relevant phenomena. Since not only the position but also the intensity of the particle images can be measured, other possible extensions of the technique are the measurement of temperature, pH-value, or even pressure by particles that show a sensitivity in the fluorescence signal to these quantities.

Cross-References

▶ Micro/Nano Flow Characterization Techniques

References

1. Chen, S., Angarita-Jaimes, N., Angarita-Jaimes, D., Pelc, B., Greenaway, A.H., Towers, C.E., Lin, D., Towers, P.D.: Wavefront sensing for three-component three-dimensional flow velocimetry in microfluidics. Exp. Fluids **47**, 849–863 (2009). doi:10.1007/s00348-009-0737-z

2. Cierpka, C., Segura, R., Hain, R., Kähler, C.J.: A simple single camera 3C3D velocity measurement technique without errors due to depth of correlation and spatial averaging for micro fluidics. Meas. Sci. Technol. **21**, 045401 (2010). doi:10.1088/0957-0233/21/4/045401

3. Cierpka, C., Rossi, M., Segura, R., Kähler, C.J.: On the calibration of astigmatism particle tracking velocimetry for microflows. Meas. Sci. Technol. **22**, 015401 (2011). doi:10.1088/0957-0233/22/1/015401

4. Cierpka, C., Rossi, M., Segura, R., Mastrangelo, F., Kähler, C.J.: A comparative analysis of the uncertainty of astigmatism-μPTV, stereo-μPIV, and μPIV. Exp. Fluids (2011). doi:10.1007/s00348-011-1075-5. open access

5. Hain, R., Kähler, C.J., Radespiel, R.: Principles of a Volumetric Velocity Measurement Technique Based on Optical Aberration. Notes on Numerical Fluid Mechanics and Multidisciplinary Design, vol. 106, pp. 1–10. Springer, Berlin/Heidelberg (2009). doi:10.1007/978-3-642-01106-1-1

6. Kähler, C.J., Scharnowski, S.: On the resolution limit of digital particle image velocimetry. In: 9th international symposium on particle image velocimetry – PIV11, Kobe (2011)

7. Kao, H.P., Verkman, A.S.: Tracking of single fluorescent particles in three dimensions: Use of cylindrical optics to encode particle postition. Biophys. J. **67**, 1291–1300 (1994). doi:10.1016/S0006-3495(94)80601-0

8. Kumar, A., Cierpka, C., Williams, S.J., Kähler, C.J., Wereley, S.T.: 3D3C velocimetry measurements of an electrothermal microvortex using wave-front deformation PTV and a single camera. Micro. Nano. **10**, 355–365 (2012). doi:10.1007/s10404-010-0674-4

9. Olsen, M.G., Adrian, R.J.: Out-of-focus effects on particle image visibility and correlation in microscopic particle image velocimetry. Exp. Fluids **29**, S166–S174 (2000). doi:10.1007/s003480070018

10. Oulette, N.T., Xu, H., Bodenschatz, E.: A quantitative study of three-dimensional Langrangian particle tracking algorithms. Exp. Fluids **40**, 301–313 (2006). doi:10.1007/s00348-005-0068-7

11. Rossi, M., Cierpka, C., Segura, R., Kähler, C.J.: Volumetric reconstruction of the 3D boundary of stream tubes with general topology using tracer particles. Meas. Sci. Technol. **22**, 105405 (2011). doi:10.1088/09570233/22/10/105405. open access

12. Santiago, J.G., Wereley, S.T., Meinhart, C.D., Beebe, D.J., Adrian, R.J.: A particle image velocimetry system for microfluidics. Exp. Fluids **25**, 316–319 (1998). doi:10.1007/s003480050235

13. Cierpka, C., Kähler, C.J.: Particle imaging techniques for volumetric three-component (3D3C) velocity measurements in microuidics. J Vis, (2011). doi:10.1007/s12650-011-0107-9. open access

14. Rossi, M., Segura, R., Cierpka, C., Kähler, C.J.: On the effect of particle image intensity and image preprocessing on depth of correlation in micro-PIV. Exp. Fluids (2011). doi:10.1007/s00348-011-1194-z. open access

15. Tschulik, K., Cierpka, C., Uhlemann, M., Gebert, A., Kähler, C. J., Schultz, L.: In-situ analysis of three-dimensional electrolyte convection evolving during the electrodeposition of copper in magnetic gradient fields. Anal. Chem. **83**, 3275–3281 (2011). doi:10.1021/ac102763m

Atmospheric Pressure Chemical Vapor Deposition (APCVD)

► Chemical Vapor Deposition (CVD)

Atomic Cluster

► Synthesis of Subnanometric Metal Nanoparticles

Atomic Force Microscopy

Bharat Bhushan and Manuel L. B. Palacio
Nanoprobe Laboratory for Bio- & Nanotechnology and Biomimetics, The Ohio State University, Columbus, OH, USA

Synonyms

Scanning probe microscopy

Definition

Atomic force microscopy (AFM) is a scanning probe technique used to measure very small forces (as low as <1nN) and produce very high resolution, three-dimensional images of sample surfaces. The AFM is capable of investigating surfaces of both conductors and insulators on the atomic scale. Aside from topographic imaging, the AFM is used in a wide variety of applications, including, but not limited to, the measurement of mechanical properties, chemical forces, electrical properties, among others.

Overview

Gerd Binnig and his colleagues developed the AFM in 1985. It is capable of investigating surfaces of scientific and engineering interest on an atomic scale [1]. As shown in Table 1, it belongs to a general class of related instrumental techniques that are used in determining surface nanotopography, nanomechanical, and nanotribological properties, such as the surface force apparatus (SFA), the scanning tunneling microscopes (STM), and the friction force microscope (FFM). The SFA was developed in 1968 and is commonly employed to study both static and dynamic properties of molecularly thin films sandwiched between two molecularly smooth surfaces. The STM, developed in 1981, allows imaging of electrically conducting surfaces with atomic resolution, and has been used for imaging of clean surfaces. AFM/FFM techniques have the advantage of imaging all types of sample surfaces, regardless of its roughness or electrical properties, unlike the SFA or STM.

The AFM relies on a scanning technique to produce very high resolution, three-dimensional images of sample surfaces. It measures ultrasmall forces (less than 1 nN) present between the AFM tip surface mounted on a flexible cantilever beam and a sample surface. These small forces are obtained by measuring the motion of a very flexible cantilever beam having an ultrasmall mass, by a variety of measurement techniques including optical deflection, optical interference, capacitance, and tunneling current. The deflection can be measured to within 0.02 nm, so for a typical cantilever spring constant of 10 N/m, a force as low as 0.2 nN can be detected. To put these numbers in perspective, individual atoms and human hair are typically a fraction of a nanometer and about 75 µm in diameter, respectively, and a drop of water and an eyelash have a weight of about 10 and 100 nN, respectively.

In the operation of high-resolution AFM, the sample is generally scanned rather than the tip because any cantilever movement would add vibrations. AFMs are available for measurement of large samples, where the tip is scanned and the sample is stationary. To obtain atomic resolution with the AFM, the spring constant of the cantilever should be weaker than the equivalent spring between atoms. A cantilever beam with a spring constant of about 1 N/m or lower is desirable. For high lateral resolution, tips should be as sharp as possible. Tips with a radius ranging from 5 to 50 nm are commonly available. Aside from topographic imaging, the AFM is also widely used for the characterization of surface/near-surface properties such as friction, adhesion, scratch, wear, fabrication/machining, electrical properties (such as surface potential, resistance, capacitance, tunneling current, electric field gradient distribution), local deformation, nanoindentation, local surface elasticity, viscoelasticity, and boundary lubrication.

Various measurement techniques using atomic force microscopy are described next.

Description of Various Measurement Techniques

Surface Roughness and Friction Force Measurements

The AFM is widely used for surface height imaging and roughness characterization down to the nanoscale. Commercial AFM/FFM is routinely used for simultaneous measurements of surface roughness and friction force [2–4]. These instruments are available for measurement of small samples and large samples. In a small sample AFM shown in Fig. 1a, the sample, generally no larger than 10 mm × 10 mm, is mounted on a piezoelectric crystal in the form of a cylindrical tube (referred to as a PZT tube scanner) which consists of separate electrodes to scan the sample precisely in the X–Y plane in a raster pattern and to move the sample in the vertical (Z) direction. A sharp tip at the free end of a flexible cantilever is brought in contact with the sample. Normal and frictional forces being applied at the tip-sample interface are measured using a laser beam deflection technique. A laser beam from a diode laser is directed by a prism onto the back of a cantilever near its free end, tilted downward at about 10° with respect to the horizontal plane. The reflected beam from the vertex of the cantilever is directed through a mirror onto a quad photodetector (split photodetector with four quadrants). The differential signal from the top and bottom photodiodes provides the AFM signal which is a sensitive measure of the cantilever vertical deflection. Topographic features of the

Atomic Force Microscopy, Table 1 Comparison of typical operating parameters in SFA, STM, and AFM/FFM used for micro/nanotribological studies

Operating parameter	SFA	STM[a]	AFM/FFM
Radius of mating surface/tip	~10 mm[b]	5–100 nm	5–100 nm
Radius of contact area	10–40 μm	N/a	0.05–0.5 nm
Normal load	10–100 mN	N/a	<0.1–500 nN
Sliding velocity	0.001–100 μm/s	0.02 – 200 μm/s (scan size ~1 nm × 1 nm–125 μm × 125 μm; scan rate <1–122 Hz)	0.02–200 μm/s (scan size ~1 nm × 1 nm–125 μm × 125 μm; scan rate <1–122 Hz)
Sample limitations	Typically atomically–smooth, optically transparent mica; opaque ceramic, smooth surfaces can also be used	Electrically conducting samples	None of the above

[a]Can be used for atomic-scale imaging
[b]Since stresses scale with the inverse of tip radius, SFA can provide very low stress measurement capabilities

sample cause the tip to deflect in the vertical direction as the sample is scanned under the tip. This tip deflection will change the direction of the reflected laser beam, changing the intensity difference between the top and bottom sets of photodetectors (AFM signal). In the AFM operating mode called the contact mode, for topographic imaging or for any other operation in which the applied normal force is to be kept constant, a feedback circuit is used to modulate the voltage applied to the PZT scanner to adjust the height of the PZT, so that the cantilever vertical deflection (given by the intensity difference between the top and bottom detector) will remain constant during scanning. The PZT height variation is thus a direct measure of the surface roughness of the sample.

In a large sample AFM, both force sensors using optical deflection method and scanning unit are mounted on the microscope head, Fig. 1b. Because of vibrations added by cantilever movement, lateral resolution of this design can be somewhat poorer than the design in Fig. 1a in which the sample is scanned instead of cantilever beam. The advantage of the large sample AFM is that large samples can be measured readily.

Aside from contact mode, most AFMs are equipped with dynamic operational mode capabilities, such as the amplitude modulation mode and the frequency modulated mode (also referred to as the self-excitation mode), which also can be used for surface roughness measurements. It should be noted that these dynamic modes could be implemented as an intermittent contact or noncontact imaging experiments. The amplitude modulation mode, also referred to as "tapping mode"

will be described further below, as it is the most widely used among the dynamic modes of AFM operation.

During a surface scan in tapping mode, the cantilever/tip assembly with a normal stiffness of 20–100 N/m (tapping mode etched Si probe or TESP) is sinusoidally vibrated at its resonance frequency (350–400 kHz) by a piezo mounted above it, and the oscillating tip slightly taps the surface. The piezo is adjusted using feedback control in the Z-direction to maintain a constant (20–100 nm) oscillating amplitude (setpoint) and constant average normal force, Fig. 2 [2–4]. The feedback signal to the Z-direction sample piezo (to keep the setpoint constant) is a measure of surface roughness. The cantilever/tip assembly is vibrated at some amplitude, here referred to as the free amplitude, before the tip engages the sample. The tip engages the sample at some setpoint, which may be thought of as the amplitude of the cantilever as influenced by contact with the sample. The setpoint is defined as a ratio of the vibration amplitude after engagement to the vibration amplitude in free air before engagement. A lower setpoint gives a reduced amplitude and closer mean tip-to-sample distance. The amplitude should be kept large enough so that the tip does not get stuck to the sample because of adhesive attractions. Also, the oscillating amplitude applies less average (normal) load as compared to contact mode and reduces sample damage. The tapping mode is used in topography measurements to minimize effects of friction and other lateral forces, and is well suited for topographic imaging of soft surfaces.

For measurement of friction force at the tip surface during sliding, left hand, and right hand sets of

Atomic Force Microscopy,
Fig. 1 Schematics (**a**) of
a commercial small sample
atomic force microscope/
friction force microscope
(AFM/FFM) and (**b**) of a large
sample AFM/FFM

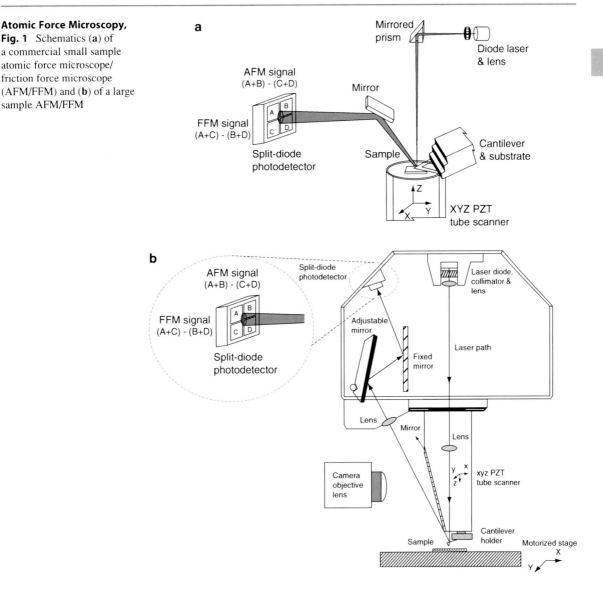

quadrants of the photodetector are used. In the so-called friction mode, the sample is scanned back and forth in a direction orthogonal to the long axis of the cantilever beam. A friction force between the sample and the tip will produce a twisting of the cantilever. As a result, the laser beam will be reflected out of the plane defined by the incident beam and the beam reflected vertically from an untwisted cantilever. This produces an intensity difference of the laser beam received in the left hand and right hand sets of quadrants of the photodetector. The intensity difference between the two sets of detectors (FFM signal) is directly related to the degree of twisting and hence to the magnitude of the

friction force. One problem associated with this method is that any misalignment between the laser beam and the photodetector axis would introduce error in the measurement. However, by following the procedures developed by Ruan and Bhushan [5], in which the average FFM signal for the sample scanned in two opposite directions is subtracted from the friction profiles of each of the two scans, the misalignment effect is eliminated. This method provides three-dimensional maps of friction force. By following the friction force, calibration procedures developed by Ruan and Bhushan [5], voltages corresponding to friction forces can be converted to force units [6]. The

Tapping mode imaging

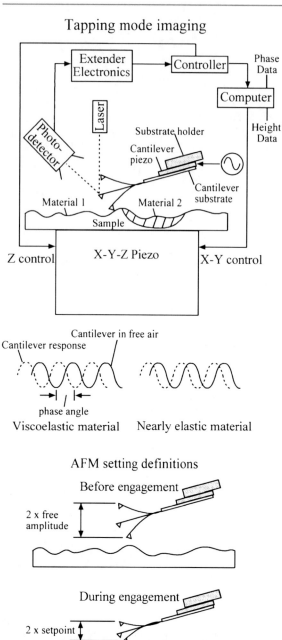

Atomic Force Microscopy, Fig. 2 Schematic of tapping mode used to obtain height and phase data and definitions of free amplitude and setpoint. During scanning, the cantilever is vibrated at its resonance frequency and the sample X-Y-Z piezo is adjusted by feedback control in the Z-direction to maintain a constant setpoint. The computer records height (which is a measure of surface roughness) and phase angle (which is a function of the viscoelastic properties of the sample) data

coefficient of friction is obtained from the slope of friction force data measured as a function of normal loads typically ranging from 10 to 150 nN. This approach eliminates any contributions due to the adhesive forces [2–4]. For calculation of the coefficient of friction based on a single-point measurement, friction force should be divided by the sum of applied normal load and intrinsic adhesive force. Furthermore it should be pointed out that for a single asperity contact, the coefficient of friction is not independent of load.

Surface roughness measurements in the contact mode are typically made using a sharp, microfabricated square-pyramidal Si_3N_4 tip with a radius of 30–50 nm on a triangular cantilever beam (Fig. 3a) with normal stiffness on the order of 0.06–0.58 N/m with a normal resonant frequency of 13–40 kHz (silicon nitride probe or NP) at a normal load of about 1–50 nN, and friction measurements are carried out in the load range of 1–100 nN. Surface roughness measurements in the tapping mode utilize a stiff cantilever with high resonance frequency; typically a square-pyramidal etched single-crystal silicon tip, with a tip radius of 5–10 nm, integrated with a stiff rectangular silicon cantilever beam (Fig. 3a) with a normal stiffness on the order of 17–60 N/m and a normal resonance frequency of 250–400 kHz (TESP), is used. Multiwalled carbon nanotube tips having a small diameter (few nm) and a length of about 1 μm (high aspect ratio) attached on the single-crystal silicon, square-pyramidal tips are used for high-resolution imaging of surfaces and of deep trenches in the tapping mode (noncontact mode) (Fig. 3b) [3, 4]. The MWNT tips are hydrophobic. To study the effect of the radius of a single asperity (tip) on adhesion and friction, microspheres of silica with radii ranging from about 4 to 15 μm are attached at the end of cantilever beams. Optical micrographs of two of the microspheres at the ends of triangular cantilever beams are shown in Fig. 3c.

The tip is scanned in such a way that its trajectory on the sample forms a triangular pattern, Fig. 4. Scanning speeds in the fast and slow scan directions depend on the scan area and scan frequency. Scan sizes ranging from less than 1 nm × 1 nm to 125 μm × 125 μm and scan rates from less than 0.5–122 Hz typically can be used. Higher scan rates are used for smaller scan lengths. For example, scan rates in the fast and slow scan directions for an area of 10 μm × 10 μm scanned at 0.5 Hz are 10 μm/s and 20 nm/s, respectively.

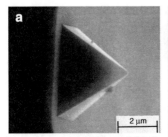

Square pyramidal silicon nitride tip

Square pyramidal single-crystal
silicon tip

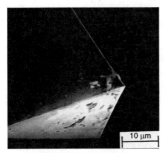

Three-sided pyramidal
natural diamond tip

Atomic Force Microscopy, Fig. 3 (**a**) SEM micrographs of a square-pyramidal plasma-enhanced chemical vapor deposition (PECVD) Si_3N_4 tip with a triangular cantilever beam, a square-pyramidal etched single-crystal silicon tip with a rectangular silicon cantilever beam, and a three-sided pyramidal natural diamond tip with a square stainless steel cantilever beam, (**b**) SEM micrograph of a multiwalled carbon nanotube (MWNT) physically attached on the single-crystal silicon and square-pyramidal tip, and (**c**) optical micrographs of commercial Si_3N_4 tip and two modified tips showing SiO_2 spheres mounted over the sharp tip, at the end of the triangular Si_3N_4 cantilever beams (radii of the tips are given in the figure)

In an AFM measurement during surface imaging, the tip comes in intimate contact with the sample surface and leads to surface deformation with finite tip-sample contact area (typically few atoms). The finite size of contact area prevents the imaging of individual point defects, and only the periodicity of the atomic lattice can be imaged. Figure 5 shows the topography and friction force images of a freshly cleaved surface of highly oriented pyrolytic graphite (HOPG) [2–4]. The atomic-scale friction force of HOPG exhibited the same periodicity as the corresponding topography.

Adhesion Measurements

Adhesive force measurements are performed in the so-called force calibration mode. In this mode, force-distance curves are obtained. An example is shown in Fig. 6 for the contact between a Si_3N_4 tip and a single-crystal silicon surface. The horizontal axis gives the distance the piezo (and hence the sample) travels, and the vertical axis gives the tip deflection. As the piezo extends, it approaches the tip, which is at this point in free air and hence shows no deflection. This is indicated by the flat portion of the curve. As the tip approaches the sample within a few nanometers

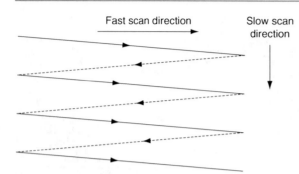

Fast scan direction   Slow scan
                           direction

Atomic Force Microscopy, Fig. 4 Schematic of triangular pattern trajectory of the tip as the sample (or the tip) is scanned in two dimensions. During scanning, data are recorded only during scans along the solid scan lines

(point A), an attractive force exists between the atoms of the tip surface and the atoms of the sample surface. The tip is pulled toward the sample and contact occurs at point B on the graph. From this point on, the tip is in contact with the surface and as the piezo further extends, the tip gets further deflected. This is represented by the sloped portion of the curve. As the piezo retracts, the tip goes beyond the zero deflection (flat) line because of attractive forces (van der Waals forces and long-range meniscus forces), into the adhesive regime. At point C in the graph, the tip snaps free of the adhesive forces and is again in free air. The horizontal distance between points B and C along the retrace line gives the distance moved by the tip in the adhesive regime. This distance multiplied by the stiffness of the cantilever gives the adhesive force. Incidentally, the horizontal shift between the loading and unloading curves results from the hysteresis in the PZT tube [2–4].

Scratching, Wear, and Fabrication/Machining

For microscale scratching, microscale wear, nanofabrication/nanomachining, and nanoindentation hardness measurements, an extremely hard tip is required. A three-sided pyramidal single-crystal natural diamond tip with an apex angle of $80°$ and a radius of about 100 nm mounted on a stainless steel cantilever beam with normal stiffness of about 25 N/m is used at relatively higher loads (1–150 µN), Fig. 3a. For scratching and wear studies, the sample is generally scanned in a direction orthogonal to the long axis of the cantilever beam (typically at a rate of 0.5–50 Hz) so that friction can be measured during scratching and wear. The tip is mounted on the cantilever such that

one of its edges is orthogonal to the long axis of the beam; therefore, wear during scanning along the beam axis is higher (about 2–3 x) than that during scanning orthogonal to the beam axis. For wear studies, an area on the order of 2 µm × 2 µm is scanned at various normal loads (ranging from 1 to 100 µN) for a selected number of cycles [2, 4].

Scratching can also be performed at ramped loads and the coefficient of friction can be measured during scratching [2–4]. A linear increase in the normal load approximated by a large number of normal load increments of small magnitude is applied using a software interface (such as the lithography module in Nanoscope) that allows the user to generate controlled movement of the tip with respect to the sample. The friction signal is tapped out of the AFM and is recorded on a computer. A scratch length on the order of 25 µm and a velocity on the order of 0.5 µm/s are used and the number of loading steps is usually taken to be 50.

Nanofabrication/nanomachining is conducted by scratching the sample surface with a diamond tip at specified locations and scratching angles. The normal load used for scratching (writing) is on the order of 1–100 µN with a writing speed on the order of 0.1–200 µm/s [2–4, 7].

Electrical Properties Measurements

The use of a conductive tip in AFM experiments enables the mapping of surface or subsurface electrical properties of materials. Various AFM-based electrical characterization methods have been developed, such as scanning Kelvin probe force microscopy, scanning impedance microscopy, scanning capacitance microscopy, piezoresponse force microscopy, conductive AFM, scanning spreading resistance microscopy, and scanning gate microscopy. As an example, scanning Kelvin probe force microscopy, which is used for the measurement of the surface potential, is described below.

Mapping of the surface potential is made in the so-called lift mode (Fig. 7). These measurements are made simultaneously with the topography scan in the tapping mode, using an electrically conducting (nickel-coated single-crystal silicon) tip. After each line of the topography scan is completed, the feedback loop controlling the vertical piezo is turned off, and the tip is lifted from the surface and traced over the same topography at a constant distance of 100 nm. During the lift mode, a DC bias potential and an oscillating

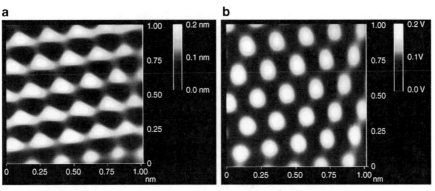

Atomic Force Microscopy, Fig. 5 (**a**) Gray-scale plots of surface topography and friction force maps (2D spectrum filtered), measured simultaneously, of a 1 nm × 1 nm area of freshly cleaved HOPG, showing the atomic-scale variation of topography and friction and (**b**) schematic of superimposed topography and friction maps from (**a**); the symbols correspond to maxima. Note the spatial shift between the two plots (Bhushan 2010, 2011).

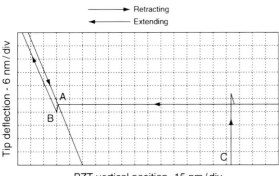

Atomic Force Microscopy, Fig. 6 Typical force-distance curve for a contact between Si_3N_4 tip and single-crystal silicon surface in measurements made in the ambient environment. Snap-in occurs at point A; contact between the tip and silicon occurs at point B; tip breaks free of adhesive forces at point C as the sample moves away from the tip

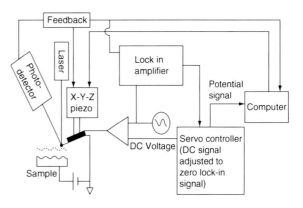

Atomic Force Microscopy, Fig. 7 Schematic of lift mode used to make surface potential measurement. The topography is collected in tapping mode in the primary scan. The cantilever piezo is deactivated. Using topography information of the primary scan, the cantilever is scanned across the surface at a constant height above the sample. An oscillating voltage at the resonant frequency is applied to the tip and a feedback loop adjusts the DC bias of the tip to maintain the cantilever amplitude at zero. The output of the feedback loop is recorded by the computer and becomes the surface potential map

potential (3–7 V) is applied to the tip. The frequency of oscillation is chosen to be equal to the resonance frequency of the cantilever (∼80 kHz). When a DC bias potential equal to the negative value of surface potential of the sample (on the order of ± 2 V) is applied to the tip, it does not vibrate. During scanning, a difference between the DC bias potential applied to the tip and the potential of the surface will create DC electric fields that interact with the oscillating charges (as a result of the AC potential), causing the cantilever to oscillate at its resonance frequency, as in tapping mode. However, a feedback loop is used to adjust the DC bias on the tip to exactly nullify the electric field, and thus the vibrations of the cantilever. The required bias voltage follows the localized potential of the surface [8].

In Situ Characterization of Local Deformation Studies

In situ characterization of local deformation of materials can be carried out by performing tensile, bending, or compression experiments inside an AFM and by

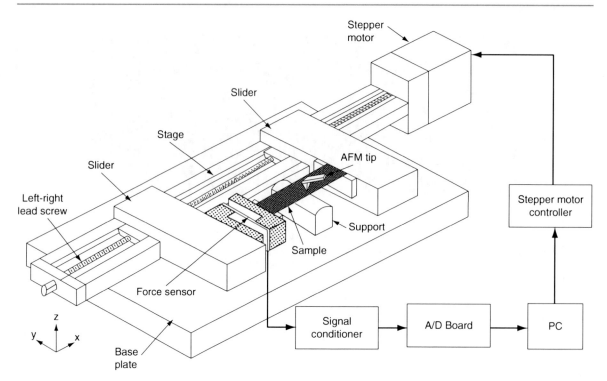

Atomic Force Microscopy, Fig. 8 Schematic of the tensile stage to conduct in situ tensile testing of the polymeric films in AFM

observing nanoscale changes during the deformation experiment [2–4]. In these experiments, small deformation stages are used to deform the samples inside an AFM. In tensile testing of the polymeric films carried out by Tambe and Bhushan [9], a tensile stage was used (Fig. 8). The stage with a left-right combination lead screw (that helps to move the slider in the opposite direction) was used to stretch the sample to minimize the movement of the scanning area, which was kept close to the center of the tensile specimen. One end of the sample was mounted on the slider via a force sensor to monitor the tensile load. The samples were stretched for various strains using a stepper motor and the same control area at different strains was imaged. In order to better locate the control area for imaging, a set of four markers was created at the corners of a 30 μm × 30 μm square at the center of the sample by scratching the sample with a sharp silicon tip. The scratching depth was controlled such that it did not affect cracking behavior of the coating. A minimum displacement of 1.6 μm could be obtained. This corresponded to a strain increment of $8 \times 10^{-3}\%$ for a sample length of 38 mm. The maximum travel was about 100 mm. The resolution of the force sensor was 10 mN with a capacity of 45 N. During stretching, a stress-strain curve was obtained during the experiment to study any correlation between the degree of plastic strain and propensity of cracks.

Nanoindentation Measurements

For nanoindentation hardness measurements the scan size is set to zero, and then a normal load is applied to make the indents using the diamond tip. During this procedure, the tip is continuously pressed against the sample surface for about 2 s at various indentation loads. The sample surface is scanned before and after the scratching, wear, or indentation to obtain the initial and the final surface topography, at a low normal load on the order of 100–500 nN using the same diamond tip. An area larger than the indentation region is scanned to observe the indentation marks. Nanohardness is calculated by dividing the indentation load by the projected residual area of the indents [2–4].

Direct imaging of the indent allows one to quantify piling up of ductile material around the indenter. However, it becomes difficult to identify the boundary of the indentation mark with great accuracy. This makes the direct measurement of contact area somewhat inaccurate. A technique with the dual capability of depth-sensing as well as in-situ imaging, which is most

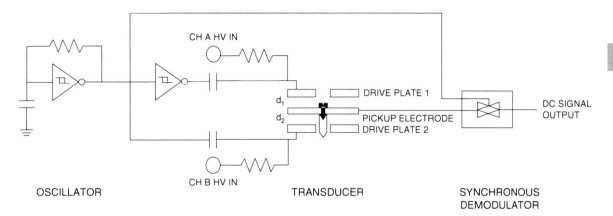

Atomic Force Microscopy, Fig. 9 Schematic of a nano/picoindentation system with three-plate transducer with electrostatic actuation hardware and capacitance sensor [10]

appropriate in nanomechanical property studies, is used for accurate measurement of hardness with shallow depths [2–4, 10]. This nano/picoindentation system is used to make load-displacement measurement and subsequently carry out in-situ imaging of the indent, if required. The indentation system, shown in Fig. 9, consists of a three-plate transducer with electrostatic actuation hardware used for direct application of a normal load and a capacitive sensor used for measurement of vertical displacement. The AFM head is replaced with this transducer assembly while the specimen is mounted on the PZT scanner, which remains stationary during indentation experiments. The transducer consists of a three (Be-Cu) plate capacitive structure, and the tip is mounted on the center plate. The upper and lower plates serve as drive electrodes, and the load is applied by applying appropriate voltage to the drive electrodes. Vertical displacement of the tip (indentation depth) is measured by measuring the displacement of the center plate relative to the two outer electrodes using capacitance technique. Indent area and consequently hardness value can be obtained from the load-displacement data. The Young's modulus of elasticity is obtained from the slope of the unloading curve.

Localized Surface Elasticity and Viscoelasticity Mapping

Localized Surface Elasticity Indentation experiments provide a single-point measurement of the Young's modulus of elasticity calculated from the slope of the indentation curve during unloading. Localized surface elasticity maps can be obtained using dynamic force microscopy, in which an oscillating tip is scanned over the sample surface in contact under steady and oscillating load. Lower frequency operation mode in the kHz range, such as force modulation mode [11] or pulsed force mode [12], are well suited for soft samples such as polymers. However, if the tip-sample contact stiffness becomes significantly higher than the cantilever stiffness, the sensitivity of these techniques strongly decreases. In this case, the sensitivity of the measurement of stiff materials can be improved by using high-frequency operation modes in the MHz range with a lateral motion, such as acoustic (ultrasonic) force microscopy, referred to as atomic force acoustic microscopy (AFAM) or contact resonance spectroscopy [13]. Inclusion of vibration frequencies other than only the first cantilever flexural or torsional resonance frequency also allows additional information to be obtained.

In negative lift mode force modulation technique, during primary scanning height data is recorded in tapping mode as described earlier. During interleave scanning, the entire cantilever/tip assembly is moved up and down at the force modulation holder's bimorph resonance frequency (about 24 kHz) at some amplitude, here referred to as the force modulation amplitude, and the Z-direction feedback control for the sample X-Y-Z piezo is deactivated, Fig. 10a [11]. During this scanning, height information from the primary scan is used to maintain a constant lift scan height. This eliminates the influence of height on the measured signals during the interleave scan. Lift scan height is the mean tip-to-sample distance between the tip and sample during the interleave scan. The lift scan

Atomic Force Microscopy, Fig. 10 (**a**) Schematic of force modulation mode used to obtain amplitude (stiffness) and definitions of force modulation amplitude and lift scan height. During primary scanning, height data is recorded in tapping mode. During interleave scanning, the entire cantilever/tip assembly is vibrated at the bimorph's resonance frequency and the Z-direction feedback control for the sample X-Y-Z piezo is deactivated. During this scanning, height information from the primary scan is used to maintain a constant lift scan height. The computer records amplitude (which is a function of material stiffness) during the interleave scan, and (**b**) Schematic of an AFM incorporating shear wave transducer which generates in-plane lateral sample surface vibrations. Because of the forces between the tip and the surface, torsional vibrations of the cantilever are excited [14]. The shift in contact resonance frequency is a measure of contact stiffness

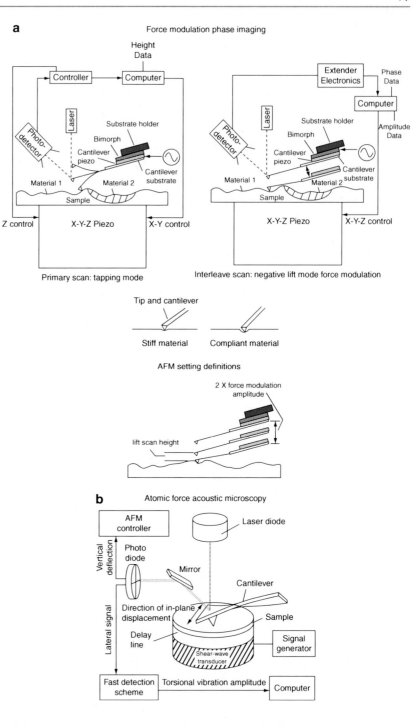

height is set such that the tip is in constant contact with the sample, that is, a constant static load is applied. (A higher lift scan height gives a closer mean tip-to-sample distance.) In addition, the tip motion caused by the bimorph vibration results in a modulating periodic force. The sample surface resists the oscillations of the tip to a greater or lesser extent depending upon the sample's stiffness. The computer records amplitude (which is a function of elastic stiffness of the material). Contact analyses can be used to obtain a quantitative measure of localized elasticity of soft surfaces [3, 4]. Etched single-crystal silicon cantilevers with integrated tips (DI force modulation etched Si probe or FESP) with a radius of 25–50 nm, a stiffness of 1–5 N/m, and a resonant frequency of 60–100 kHz, are commonly used for the measurements. The scanning rate could be set between 0.1 and 2 Hz along the fast axis.

In the AFAM technique [13], the cantilever/tip assembly is moved either in the normal or lateral mode, and the contact stiffness is evaluated by comparing the resonance frequency of the cantilever in contact with the sample surface to those of the free vibrations of the cantilever. Several free resonance frequencies are measured. Based on the shift of the measured frequencies, the contact stiffness is determined by solving the characteristic equation for the tip vibrating in contact with the sample surface. The elastic modulus is calculated from contact stiffness using Hertz analysis for a spherical tip indenting a plane. Contact stiffness is equal to 8 multiplied with the contact radius and the reduced shear modulus in shear mode.

In the lateral mode using the AFAM technique, the sample is glued on cylindrical pieces of aluminum, which serves as ultrasonic delay lines coupled to an ultrasonic shear wave transducer, Fig. 10b [14]. The transducer is driven with frequency sweeps to generate in-plane lateral sample surface vibrations. These couple to the cantilever via the tip-sample contact. To measure torsional vibrations of the cantilever at frequencies up to 3 MHz, the original electronic circuit of the lateral channel of the AFM (using a lowpass filter with limited bandwidth to few hundred kHz) was replaced by a high-speed scheme which bypasses the lowpass filter. The high- frequency signal was fed to a lock-in amplifier, digitized using a fast A/D card, and fed into a broadband amplifier, followed by an rms-to-dc converter and read by a computer. Etched single-crystal silicon cantilevers (normal stiffness of 3.8–40 N/m) integrated tips are used.

Viscoelastic Mapping Another form of dynamic force microscopy, phase contrast microscopy, is used to detect the contrast in viscoelastic (viscous energy dissipation) properties of the different materials across the surface [15–18]. In these techniques, both deflection amplitude and phase angle contrasts are measured, which are measures of the relative stiffness and viscoelastic properties, respectively. Two phase measurement techniques – tapping mode and torsional resonance (TR) mode – have been developed, and are described below.

In the tapping mode (TM) technique, as described earlier, the cantilever/tip assembly is sinusoidally vibrated at its resonant frequency, and the sample X-Y-Z piezo is adjusted using feedback control in the Z-direction to maintain a constant setpoint, Fig. 3 [15]. The feedback signal to the Z-direction sample piezo (to keep the setpoint constant) is a measure of surface roughness. The extender electronics is used to measure the phase angle lag between the cantilever piezo drive signal and the cantilever response during sample engagement. As illustrated in Fig. 3, the phase angle lag (at least partially) is a function of the viscoelastic properties of the sample material. A range of tapping amplitudes and setpoints is used for measurements. Commercially etched single-crystal silicon tip (TESP) used for tapping mode, with a radius of 5–10 nm, a stiffness of 20–100 N/m, and a resonant frequency of 350–400 kHz, is normally used. Scanning is typically set between 0.1 and 2 Hz along the fast axis.

In the torsional mode (TR mode), a tip is vibrated in the torsional mode at high frequency at the resonance frequency of the cantilever beam. An etched single-crystal silicon cantilever with integrated tip (FESP) with a radius of about 5–10 nm, normal stiffness of 1–5 N/m, torsional stiffness of about 30 times normal stiffness and torsional resonant frequency of 800 kHz is typically used. A major difference between the TM and TR modes is the directionality of the applied oscillation – a normal (compressive) amplitude exerted for the TM and a torsional amplitude for the TR mode. The TR mode is expected to provide good contrast in the tribological and mechanical properties of the near-surface region as compared to the TM. Two of the reasons are as follows. (1) In the TM, the interaction is dominated by the vertical properties of the sample, so the tip spends a small fraction of its time in the near field interaction with the sample. Furthermore, the distance between the tip and the sample changes during the measurements, which changes interaction time and forces, and affects measured data. In the TR mode, the distance remains nearly constant. (2) The lateral stiffness of a cantilever is typically about two orders of magnitude larger than the normal (flexural) stiffness. Therefore, in the TM, if the sample is relatively rigid, much of the deformation occurs in the cantilever beam, whereas in the TR mode, much of the deformation occurs in the sample. A few comments on the special applications of the TR mode are made next. Since most of the deformation occurs in the sample, the TR mode can be used to measure stiff and hard samples. Furthermore, properties of thin films can be measured more readily with the TR mode. For both the TM and TR modes, if the cantilever is driven to vibrate at frequencies above resonance, it would have less motion (high apparent stiffness), leading to higher sample deformation and better contrast. It should be further noted that the TM exerts a compressive force, whereas the TR mode exerts torsional force, therefore normal and shear properties are measured in the TM and TR modes, respectively.

In the TR mode, the torsional vibration of the cantilever beam is achieved using a specially designed cantilever holder. It is equipped with a piezo system mounted in a cantilever holder, in which two piezos vibrate out-of-phase with respect to each other. A tuning process prior to scanning is used to select the torsional vibration frequency. The piezo system excites torsional vibration at the cantilever's resonance frequency. The torsional vibration amplitude of the tip (TR amplitude) is detected by the lateral segments of the split-diode photodetector, Fig. 11 [18]. The TR mode measures surface roughness and phase angle as follows. During the measurement, the cantilever/tip assembly is first vibrated at its resonance at some amplitude dependent upon the excitation voltage, before the tip engages the sample. Next, the tip engages the sample at some setpoint. A feedback system coupled to a piezo stage is used to keep a constant TR amplitude during scanning. This is done by controlling the vertical position of the sample using a piezo moving in the Z direction, which changes the degree of tip interaction. The displacement of the sample Z piezo gives a roughness image of the sample. A phase angle image can be obtained by measuring the phase angle of the cantilever vibration response in the torsional mode during engagement with respect to the cantilever vibration response in free air before engagement. The control feedback of the TR mode is similar to that of tapping, except that the torsional resonance amplitude replaces flexural resonance amplitude [18].

Chen and Bhushan [16] used a variation to the approach just described (referred to as mode I here). They performed measurements at constant normal cantilever deflection (constant load) (mode II) instead

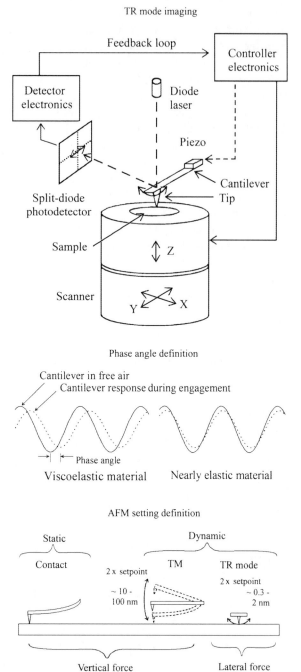

Atomic Force Microscopy, Fig. 11 Schematic of torsional resonance mode shown at the *top*. Two examples of the phase angle response are shown in the *middle*. One is for materials exhibiting viscoelastic (*left*) and the other nearly elastic properties (*right*). Three AFM settings are compared at the *bottom*: contact, tapping mode (TM), and TR modes. The TR mode is a dynamic approach with a laterally vibrating cantilever tip that can interact with the surface more intensively than other modes. Therefore, more detailed near-surface information is available

Atomic Force Microscopy, Table 2 Summary of various operating modes of AFM for surface roughness, stiffness, phase angle, and friction

Operating mode	Direction of cantilever vibration	Vibration frequency of cantilever (kHz)	Vibration amplitude (nm)	Feedback control	Data obtained
Contact	N/a			Constant normal load	Surface height, friction
Tapping	Vertical	350–400	10–100	Setpoint (constant tip amplitude)	Surface height, phase angle (normal viscoelasticity)
Force Modulation	Vertical	10–20 (bimorph)	10–100	Constant normal load	Surface height, amplitude (normal stiffness)
Lateral	Lateral (AFAM)	100–3,000 (sample)	∼5 (sample)	Constant normal load	Shift in contact resonance (normal stiffness, friction)
TR mode I	Torsional	∼800	0.3–2	Setpoint (constant tip amplitude)	Surface height, phase angle (lateral viscoelasticity)
TR mode II	Torsional	∼800	0.3–2	Constant normal load	Surface height, amplitude and phase angle (lateral stiffness and lateral viscoelasticity)
TR mode III	Torsional	>800 in contact	0.3–2	Constant normal load	Shift in contact resonance (friction)

of using constant setpoint in the mode I approach. Their approach overcomes the meniscus adhesion problem present in mode I, and reveals true surface properties.

Song and Bhushan [19] presented a forced torsional vibration model for a tip-cantilever assembly under viscoelastic tip-sample interaction. This model provides a relationship of torsional amplitude and phase shift with lateral contact stiffness and viscosity, which can be used to extract in-plane interfacial mechanical properties.

Various operating modes of AFM used for surface roughness, localized surface elasticity, and viscoelastic mapping, and friction force measurements are summarized in Table 2.

Boundary Lubrication Measurements

AFM can be used to study nanoscale boundary lubrication properties, mainly adhesion, friction, and wear (or durability). Adhesive forces at the nanoscale can be measured in the force calibration mode, as previously described. The adhesive forces can also be calculated from the horizontal intercept of friction versus normal load curves at a zero value of friction force [20]. For friction measurements, the unlubricated and lubricated surfaces (such as silicon wafers, bulk metal or metallic films, ceramics and polymer films, among others) are

typically scanned using a Si_3N_4 or Si tip over scan length of 2–10 μm with the normal load ranging from 5 to 150 nN. Velocity effects on friction are studied by changing the scan frequency from 0.1 to 60 Hz. As an example, if the scan length is maintained at 2 μm, the sliding velocity then varies from 0.4 to 240 μm/s. To study the durability properties, the friction force and coefficient of friction are monitored during scanning at an arbitrary nanoscale normal load (on the order of 10–100 nN) and a predefined scanning speed for a desired number of cycles [20].

Cross-References

▶ AFM, Non-Contact Mode
▶ AFM, Tapping Mode
▶ AFM in Liquids
▶ AFM Probes
▶ Robot-Based Automation on the Nanoscale

References

1. Binnig, G., Quate, C.F., Gerber, Ch.: Atomic force microscopy. Phys. Rev. Lett. **56**, 930–933 (1986)
2. Bhushan, B.: Handbook of Micro/Nanotribology, 2nd edn. CRC Press, Boca Raton (1999)
3. Bhushan, B.: Springer Handbook of Nanotechology, 3rd edn. Springer, Heidelberg (2010)

4. Bhushan, B.: Nanotribology and Nanomechanics I – Measurement Techniques and Nanomechanics, II – Nanotribology, Biomimetics, and Industrial Applications, 3rd edn. Springer, Heidelberg (2011)

5. Ruan, J., Bhushan, B.: Atomic-scale friction measurements using friction force microscopy: part I – general principles and new measurement techniques. ASME J. Tribol. 116, 378–388 (1994)

6. Palacio, M., Bhushan, B.: Normal and lateral force calibration techniques for AFM cantilevers. Crit. Rev. Solid State Mater. Sci. 35, 73–104 (2010)

7. Bhushan, B., Israelachvili, J.N., Landman, U.: Nanotribology: friction, wear and lubrication at the atomic scale. Nature 374, 607–616 (1995)

8. DeVecchio, D., Bhushan, B.: Use of a nanoscale kelvin probe for detecting wear precursors. Rev. Sci. Instrum. 69, 3618–3624 (1998)

9. Tambe, N., Bhushan, B.: In situ study of nano-cracking of multilayered magnetic tapes under monotonic and fatigue loading using an AFM. Ultramicroscopy 100, 359–373 (2004)

10. Bhushan, B., Kulkarni, A.V., Bonin, W., Wyrobek, J.T.: Nano/picoindentation measurement using a capacitance transducer system in atomic force microscopy. Philos. Mag. 74, 1117–1128 (1996)

11. Maivald, P., Butt, H.J., Gould, S.A.C., Prater, C.B., Drake, B., Gurley, J.A., Elings, V.B., Hansma, P.K.: Using force modulation to image surface elasticities with the atomic force microscope. Nanotechnology 2, 103–106 (1991)

12. Krotil, H.U., Stifter, T., Waschipky, H., Weishaupt, K., Hild, S., Marti, O.: Pulse force mode: a new method for the investigation of surface properties. Surf. Interface Anal. 27, 336–340 (1999)

13. Rabe, U., Janser, K., Arnold, W.: Vibrations of free and surface-coupled atomic force microscope cantilevers: theory and experiment. Rev. Sci. Instrum. 67, 3281–3293 (1996)

14. Reinstaedtler, M., Rabe, U., Scherer, V., Hartmann, U., Goldade, A., Bhushan, B., Arnold, W.: On the nanoscale measurement of friction using atomic-force microscope cantilever torsional resonances. Appl. Phys. Lett. 82, 2604–2606 (2003)

15. Bhushan, B., Qi, J.: Phase contrast imaging of nanocomposites and molecularly-thick lubricant films in magnetic media. Nanotechnology 14, 886–895 (2003)

16. Chen, N., Bhushan, B.: Morphological, nanomechanical and cellular structural characterization of human hair and conditioner distribution using torsional resonance mode in an AFM. J. Micros. 220, 96–112 (2005)

17. Garcia, R., Tamayo, J., Calleja, M., Garcia, F.: Phase contrast in tapping-mode scanning force microscopy. Appl. Phys. A 66, S309–S312 (1998)

18. Kasai, T., Bhushan, B., Huang, L., Su, C.: Topography and phase imaging using the torsional resonance mode. Nanotechnology 15, 731–742 (2004)

19. Song, Y., Bhushan, B.: Atomic force microscopy dynamic modes: modeling and applications. J. Phys. Condens. Matter 20, 225012 (2008)

20. Liu, H., Bhushan, B.: Nanotribological characterization of molecularly-thick lubricant films for applications to MEMS/NEMS by AFM. Ultramicroscopy 97, 321–340 (2003)

Further Reading

Bhushan, B.: Adhesion and stiction: mechanisms, measurement techniques, and methods for reduction (invited). J. Vac. Sci. Technol. B 21, 2262–2296 (2003)

Bhushan, B.: Scanning Probe Microscopy in Nanoscience and Nanotechnology. Springer, Heidelberg (2010)

Bhushan, B.: Scanning Probe Microscopy in Nanoscience and Nanotechnology, vol. 2. Springer, Heidelberg (2011)

Bhushan, B., Fuchs, H.: Applied Scanning Probe Methods II-IV. Springer, Heidelberg (2006)

Bhushan, B., Fuchs, H.: Applied Scanning Probe Methods VII. Springer, Heidelberg (2007)

Bhushan, B., Fuchs, H.: Applied Scanning Probe Methods XI-XIII. Springer, Heidelberg (2009)

Bhushan, B., Kawata, S.: Applied Scanning Probe Methods VI. Springer, Heidelberg (2007)

Bhushan, B., Fuchs, H., Hosaka, S.: Applied Scanning Probe Methods. Springer, Heidelberg (2004)

Bhushan, B., Fuchs, H., Kawata, S.: Applied Scanning Probe Methods V. Springer, Heidelberg (2007)

Bhushan, B., Fuchs, H., Tomitori, M.: Applied Scanning Probe Methods VIII-X. Springer, Heidelberg (2008)

Fuchs, H., Bhushan, B.: Biosystems – Investigated by Scanning Probe Microscopy. Springer, Heidelberg (2010)

Magonov, S.N., Whangbo, M.H.: Surface Analysis with STM and AFM: Experimental and Theoretical Aspects of Image Analysis. VCH, Weinheim (1996)

Meyer, E., Hug, H.J., Bennewitz, R.: Scanning Probe Microscopy: The Lab on a Tip. Springer, Heidelberg (2004)

Morita, S., Wiesendanger, R., Meyer, E.: Noncontact Atomic Force Microscopy. Springer, Heidelberg (2002)

Quate, C.F.: (1994), "The AFM as a tool for surface imaging, In: Duke, C.B. (ed.) Surface Science: The First Thirty Years. North-Holland, Amsterdam (1994)

Atomic Force Microscopy in Liquids

▶ AFM in Liquids

Atomic Layer Chemical Vapor Deposition

▶ Atomic Layer Deposition

Atomic Layer Chemical Vapor Deposition (ALCVD)

▶ Chemical Vapor Deposition (CVD)

Atomic Layer CVD (AL-CVD)

▶ Atomic Layer Deposition

Atomic Layer Deposition

Charles L. Dezelah IV
Picosun USA, LLC, Detroit, MI, USA

Synonyms

ALD; Atomic layer chemical vapor deposition; Atomic layer CVD (AL-CVD); Atomic layer epitaxy (ALE); Molecular layer deposition (MLD)

Definition

Atomic layer deposition is a method of thin film growth related to chemical vapor deposition whereby two or more vapor phase precursor chemicals are individually and sequentially introduced to a substrate surface via discrete precursor pulses that are separated by purge steps. Each precursor pulse must yield saturative, self-limiting growth of a monolayer or sub-monolayer of the desired material if an ALD mechanism is operant. The resultant film exhibits excellent thickness uniformity and conformality, even on highly complex and demanding nanoscale substrate features, with the process affording an excellent degree of thickness control at the sub-nanometer level.

Introduction

Atomic layer deposition (ALD) is a technique for the deposition of thin films of solid state or polymeric materials onto solid substrates from vapor phase precursor chemicals. While it is closely related to the chemical vapor deposition (CVD) technique and is similar to it in important ways, it also has substantial differences that distinguish it from CVD. Unlike CVD, ALD achieves film growth via the introduction of two or more precursor chemicals separately in discrete

pulses into a reaction chamber. As a result, gas-phase reactions are avoided and only surface chemisorbed species are present at the time of exposure to the subsequent precursor. Growth with each exposure is self-limited and saturative, with each deposition cycle giving controlled deposition and layer-by-layer growth. As a result ALD is able to coat difficult substrate features and has been identified with its ability to overcome technological problems in the microelectronics industry, as well as potential for contributing to new areas in nanotechnology. ALD was first developed in Finland in the mid-1970s by Tuomo Suntola for thin film electroluminescent display applications, but experienced an enormous resurgence in the 2000s, mostly due to the need to replace SiO_2 gate dielectrics with high permittivity alternatives to allow for future scaling of capacitor structures in microelectronics [1]. ALD continues to find new possibilities for future applications as the need for high-quality coatings on ever-smaller nano-sized features expands.

Key Features of an ALD Process

A number of key features are principal to the understanding of ALD [1–4]. Foremost is the concept of the ALD growth cycle. An ALD process is necessarily comprised of a sequence of pulses by which doses of precursor are delivered to a reaction chamber (see Fig. 1). In each pulse, a precursor is supplied in a dose sufficient to ensure reaction with every available surface site and over a time period that is long enough for the precursor vapor to effectively diffuse into substrate features and for reactions to occur on the surfaces that are present. The reaction between the precursor and the substrate must be self-limiting; the presence of excess precursor beyond that which is necessary for the reaction of all surface sites must not lead to additional film growth (Fig. 2). This behavior is often described as "saturative" growth and is the defining feature of ALD. A dose that has less than the required precursor amount to achieve this type of growth is considered "sub-saturative" and the resulting film is likely to not have the full characteristics expected of ALD. Additionally, it is crucial that no significant self-reaction or decomposition of the precursor occurs, as this may lead to an uncontrolled component of film growth.

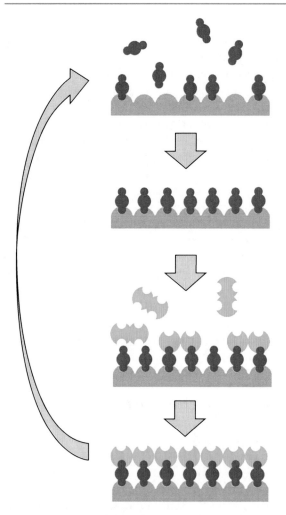

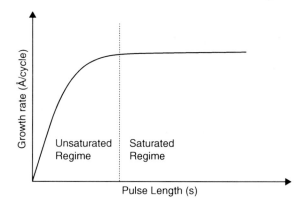

Atomic Layer Deposition, Fig. 2 ALD saturation curve. Each exposure of the substrate to precursor vapor leads to self-limiting growth, which is the defining feature of ALD. Any dose of sufficient size (here indicated by pulse length, which is one means of determining dose) leads to saturative growth whereby additional exposure does not lead to additional growth

Atomic Layer Deposition, Fig. 1 A depiction of the ALD deposition cycle. A precursor is alternately exposed to precursors that undergo self-limiting reactions with the surface. Repetition of the deposition cycle leads to layer-by-layer growth mechanism inherent to ALD (Credit: T. Suntola, Picosun Oy)

without wasting precursor. The latter method, however, often exhibits shorter deposition cycles, faster deposition times, and more effective purging. The purge step must be long enough and effective enough to ensure that gas-phase reactions between precursors are avoided. The failure to avoid gas phase mixing of the precursors can result in uncontrolled poor quality dual-source type CVD film growth and the formation of particulates in the deposition chamber, both of which can have deleterious effects on the intended film function.

Following the above described pulse and purge steps, the substrate is exposed to a second precursor chemical chosen to react rapidly and completely with the first precursor, which is now already reacted with the surface, to achieve the desired material. This is followed again by a purging step to remove excess precursor vapor and any gas-phase reaction by-products, thus completing the second "half-reaction" of the ALD cycle. By the repetition of the ALD cycle, the layer-by-layer growth of the thin film is achieved. A constant growth rate per deposition cycle is a feature of every ALD process, and as a result, the thickness of deposited film varies linearly with the number of cycle repetitions. The number of necessary cycles for a given film thickness for a particular process operated at a specific temperature is easily determined and is the basis for the degree of thickness control inherent in ALD (see Fig. 3).

The influence of deposition temperature on the growth rate is another fundamental ALD concept. Depending on the mechanistic phenomena taking

Upon the conclusion of a precursor pulse, the deposition chamber is purged, typically by the combined treatment of the deposition chamber by an inert gas flow and evacuation. Depending on the design of the ALD system being used, the application of inert gas flow and vacuum may be stopped and resumed to create a purge phase, or may be continuous, with precursor vapor introduced to the inert gas flow intermittently, so as to make the inert gas flow function alternately as a carrier gas and purge gas. The former of these approaches may be advantageous in that the pulse period can be lengthened to increase the residence time and effectively incubate the substrate

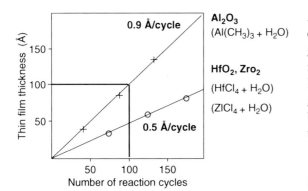

Atomic Layer Deposition, Fig. 3 ALD exhibits excellent control of film thickness. By selecting a number of deposition cycles appropriate to the desired thickness, precise control can be achieved. For a given process at a given temperature, the growth per cycle is constant. As a result, film thickness increases linearly with the applied number of deposition cycles

place during the deposition of the film, the growth rate may change considerably, sometimes with loss of the self-limiting growth necessary for ALD. The common modes of growth behavior are outlined in Fig. 4. The concept of an ALD window, a temperature range over which the growth rate does not change with temperature, is important to ALD. It is a common misunderstanding that an ALD window is a required attribute of every ALD process and that ALD only exists within an ALD window. This is not true, however, and in fact many well-known ALD processes totally lack an ALD window. These processes instead will display a modest increase or decrease in growth rate over a temperature range where self-limiting growth is observed. Nevertheless, an ALD window is considered an attractive feature, as it affords a greater degree of reproducibility by mitigating the influence of small variations in temperature. Regardless of the presence or absence of an ALD window, there is frequently a more dramatic change in growth rate as the temperature moves to a regime where the growth is no longer self-limiting. At low temperatures, a process may undergo a sizable decrease in growth rate as the degree of thermal activation becomes insufficient to sustain effective surface reactions. Alternatively, an increase in growth behavior may be observed as a result of precursor condensation and/or physisorption leading to the association of multiple monolayers of precursor with the substrate surface, all of which become involved with film growth upon exposure to the following precursor. At high temperatures a similar departure from normal

ALD conditions can occur as well. Most often an excessive temperature will lead to a large increase in growth rate as precursor decomposition or self-reaction becomes a significant mode of film growth. High temperatures can also lead to a substantial drop in growth rate if precursor desorption from the surface becomes favorable. Both of these situations are likely to lead to loss of ALD growth.

Precursor Considerations

A number of factors are vital to the consideration of which precursor should be used for an ALD process [1–3]. Of primary importance is volatility. If a precursor compound cannot be effectively vaporized and delivered to the substrate surface in the gas phase, it will be impossible for the ALD process to proceed. A vapor pressure curve can be a very helpful tool in determining whether the volatility of a particular compound is suitable for effective delivery. Many precursors require heating to generate adequate vapor pressure for its delivery to the ALD deposition chamber. The exact temperature required depends considerably on the configuration of the source delivery system used and the minimum vapor pressure requirement afforded by its design. A few precursors possess high enough vapor pressure at room temperature to ensure adequate dosing for ALD. These include some well-known examples such as trimethylaluminum (TMA), $Al(CH_3)_3$; titanium tetrachloride, $TiCl_4$; and diethylzinc, $Zn(CH_2CH_3)_2$. These provide the advantage not only of simplicity of delivery, but also in allowing for low-temperature depositions of the corresponding metal oxides. Because the deposition temperature in a typical ALD reactor must exceed the temperature of the accompanying source delivery systems, low volatilization temperatures can help facilitate lower deposition temperatures. A further necessity of such low-temperature depositions – and for that matter, all ALD deposition – is sufficient reactivity. In order for surface reactions to proceed via an ALD mechanism, both precursors must react rapidly and completely with the chemisorbed species left on the surface from the previous reaction. Insufficient reactivity may result in reduced growth rates, increased impurity content due to the inclusion of unreacted surface groups, and compromised film conformality and thickness uniformity. A third essential feature of

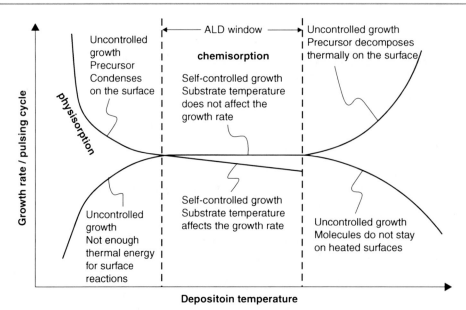

Atomic Layer Deposition, Fig. 4 Depiction of how growth rate per cycle varies with temperature. Some ALD processes have an "ALD window" or temperature range over which the growth rate is constant. At high or low temperatures, a number of phenomena can occur leading to a large increase or decrease in growth rate and the loss of ALD behavior

an ALD precursor is thermal stability, that is, the inability of the precursor to undergo decomposition or self-reaction at a sufficient rate at the temperatures to which it is exposed to noticeably affect the quality of film growth. The lack of optimal thermal stability may result in uncontrolled single source CVD-like growth and loss of ALD-type growth, oftentimes along with the loss of associated advantages of ALD. In addition to the three critical features of an ALD precursor, a number of other factors should be considered, such as availability, cost, ease of handling, the presence of undesired elements, the steric bulk of precursor molecules and its impact on film growth rate, reactivity and volatility of surface reaction by-products and their effects on the film and substrate, as well as environmental health and safety concerns.

Precursor Types

A variety of precursors exists for ALD and can be classified according to their key features [2, 3]. These features, as well as any associated advantages or disadvantages must be considered carefully when determining the best process for a given application. Metal-containing precursors (i.e., for the delivery of the metal atoms to the growing film surface) can be divided into three categories: inorganic, metalorganic, and organometallic. Inorganic precursors are defined as those that do not contain carbon whatsoever. Typically inorganic precursors are restricted to metal halides, such as $TiCl_4$ for the growth of TiO_2 or TiN, but other examples exist as well, that is, the growth of HfO_2 from hafnium nitrate, $Hf(NO_3)_4$. Metalorganic precursors are those that contain one or more metal atoms and also contain at least one carbon-containing ligand bound to the metal through a heteroatom such as nitrogen or oxygen. Figure 5 shows common examples of metalorganic precursors. Classes of metalorganic precursors include alkoxides, amides, and amidinates. Such precursors are noted for their high volatility and good reactivity, although many suffer from suboptimal thermal stability, and exhibit a loss of ALD growth when the thermal budget has been exceeded. Due to this combination of characteristics, metalorganic precursors can be expected oftentimes to give ALD growth at the lower end of the range typical for ALD processes. Organometallic precursors are those that contain direct metal-carbon bond. Figure 5 shows common examples of organometallic precursors. These precursors primarily consist of metal alkyls, such as TMA, or cyclopentadienyl (Cp) compounds, such as dichlorohafnocene, $HfCp_2Cl_2$, or compounds containing both ligand types, such as

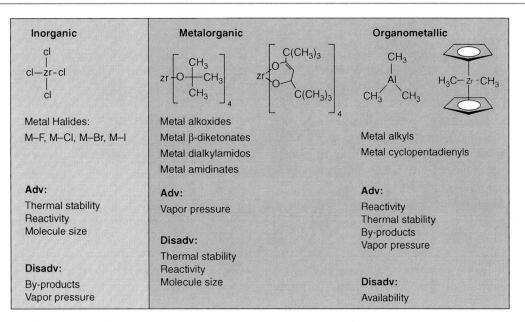

Atomic Layer Deposition, Fig. 5 An overview of ALD precursor types and their advantages or disadvantages

dimethylhafnocene, HfCp2(CH$_3$)$_2$. Organometallic compounds typically have fairly good to excellent volatility, good to excellent reactivity, and excellent thermal stability. Their primary drawback is the limited availability of these compounds for many metals. For some organometallic compounds, high cost can also be an additional disadvantageous factor.

Besides metal-containing precursors, one must give proper consideration to the non-metal precursor as well. A good non-metal precursor must abide by all of the conditions and requirements for a good metal-containing precursor. Although many of the most used non-metal precursors are compressed gases where volatility is not an issue of concern, some non-metal precursors are liquids, or perhaps even solids. In this case, good volatility is essential. Likewise, good thermal stability and reactivity are necessary, but due to the nature of the simple and robust non-metal precursors commonly employed, only the second of these properties is usually a constraint. The reactivity of the non-metal precursor must be appropriately matched to the metal-containing precursor, and ideally should react in a vigorous and complete way that precludes the incorporation of impurities, as well as having oxidative or reductive properties adequate for achieving the desired oxidation state of material. Typically metal oxide processes employ either H$_2$O or ozone. The use of H$_2$O commonly affords the metal oxide product by means of

protonolysis and loss of the ligands bound to the metal of the adsorbed metal-containing precursor, along with replacement of the ligands by a hydroxyl group. The hydroxyl groups will then dominate the surface, making it amenable to reaction with the subsequent reintroduction of the metal-containing precursor. Ozone operates by a different mechanism, however, where the ligands undergo oxidation to CO$_2$ and H$_2$O, the latter of which is presumed to regenerate a hydroxyl-terminated surface. Ozone also brings with it a much greater oxidative power than water and its use typically assures that the resulting material will be in the highest accessible oxidation state. There are other precursors that have seen use for oxide growth, such as H$_2$O$_2$ and N$_2$O, but these are only used rarely. Oxygen (O$_2$) is typically not used in the growth of oxides, as its oxidative power and capability of promoting oxide growth is low at the temperatures used for ALD. Oxygen has been used extensively, however, in the growth of ALD films of the noble metals, where the combustion of organic ligands and the resulting reduction of the metal atoms are featured. The deposition of nitrides and sulfides is achieved most frequently using ammonia and hydrogen sulfide, respectively, which are analogous to water in that they are "hydrides" of their respective non-metals and undergo protonolysis and ligand replacement reactions to achieve film growth.

Materials Grown by ALD

An ever-increasing variety of materials are available for growth by ALD. Comprehensive review articles exist that list and describe all known processes in detail [1–5]. It is beyond the scope of this entry to exhaustively list these processes, describe, and give references for them. Instead, a general overview of the materials according to type will be given, and the reader is encouraged to consult the recommended reviews for leading articles.

The large majority of ALD grown materials are binary, that is, comprised of one metallic and one non-metallic element. Metal oxides are easily the most common binary materials for which ALD processes are known, and encompass a considerable portion of the periodic table. The oxides of the group 3 metals (Sc, Y, and lanthanide elements), group 4 metals (Ti, Zr, Hf), as well several main group metals (Al, Mg, Zn) are particularly well represented, with many published studies and multiple processes developed from a variety of ligand types. Given the choice of available processes for these materials, ALD users can often select the process that best matches their needs and can choose optimal volatility, reactivity, thermal stability from among the available precursors. Beyond these well-established processes are numerous others for which at least preliminary process studies have been published. These include nearly all of the first row transition metal oxides, a few second and third row transition metal oxides (e.g., Nb_2O_5, Ta_2O_5, WO_3, RuO_2, RhO_2, IrO_2), as well as additional main group oxides beyond those mentioned earlier (e.g., B_2O_3, Ga_2O_3, In_2O_3, SiO_2, GeO_2). A complete listing of known processes is given in Table 1. Besides binary metal oxide processes, there are numerous examples of ternary (two metals and oxygen) and quaternary (three metals and oxygen) oxides that have been investigated. The metal titanates are the most frequently reported materials of this type, and include examples such as $SrTiO_3$, $BaTiO_3$, $PbTiO_3$, and Bi_xTi_yO. Additionally, lanthanide-based ternary and quaternary materials are also well-precedented, with accounts of the following in the literature: $YScO_3$, $LaAlO_3$, $NdAlO_3$, $GdScO_3$, $LaScO_3$, $LaLuO_3$, $Er_3Ga_5O_{13}$, $LaCoO_3$, $LaNiO_3$, $LaMnO_3$, and $La_{1-x}Ca_xMnO_3$. Besides stoichiometric multi-element oxide films, doping of films by ALD is a frequently used approach. In fact due to its ability to precisely control film composition by setting a pulse sequence designed for the repeated introduction of a dopant-containing precursor, ALD is specially suited to the formation of thin films of such materials. Most of the work done to date has involved the doping of zinc, tin, or indium oxide with a single element, namely, ZnO:M, where M = B, Al, or Ga; SnO:M, where M = Sb; and In2O3:M, where M = Sn, F, or Zr.

After oxides, metal pnictides represent the second most explored group of ALD materials. Of these, nitrides have received by far the most attention. Metal nitride deposition by ALD generally falls into two categories: early transition metal nitrides and main group element nitrides. Many examples from the former group have metallic properties and have potential applications as an adhesion and barrier layers in microelectronics, thus creating a motivation for their study. Early transition metal nitrides include TiN, Ti-Si-N, Ti-Al-N, TaN, NbN, MoN, WN_x, and WN_xC_y.

Atomic Layer Deposition, Table 1 An overview of known ALD materials. Based on information in Ref. [1]

Types	Materials
Metal oxides	Al_2O_3, TiO_2, ZrO_2, HfO_2, Ta_2O_5, Nb_2O_5, Sc_2O_3, Y_2O_3, MgO, B_2O_3, SiO_2, GeO_2, La_2O_3, CeO_2, PrO_x, Nd_2O_3, Sm_2O_3, EuO_x, Gd_2O_3, Dy_2O_3, Ho_2O_3, Er_2O_3, Tm_2O_3, Yb_2O_3, Lu_2O_3, $SrTiO_3$, $BaTiO_3$, $PbTiO_3$, $PbZrO_3$, Bi_xTi_yO, Bi_xSi_yO, $SrTa_2O_6$, $SrBi_2Ta_2O_9$, $YScO_3$, $LaAlO_3$, $NdAlO_3$, $GdScO_3$, $LaScO_3$, $LaLuO_3$, $Er_3Ga_5O_{13}$, In_2O_3, In_2O_3:Sn, In_2O_3:F, In_2O_3:Zr, SnO_2, SnO_2:Sb, ZnO, ZnO:Al, ZnO:B, ZnO:Ga, RuO_2, RhO_2, IrO_2, Ga_2O_3, V_2O_5, WO_3, W_2O_3, NiO, FeO_x, CrO_x, CoO_x, MnO_x, $LaCoO_3$, $LaNiO_3$, $LaMnO_3$, $La_{1-x}Ca_xMnO_3$
Metal nitrides	BN, AlN, GaN, InN, SiN_x, Ta_3N_5, Cu_3N, Zr_3N_4, Hf_3N_4, TiN, Ti-Si-N, Ti-Al-N, TaN, NbN, MoN, WN_x, WN_xC_y
Metal chalcogenides	ZnS, ZnSe, ZnTe, CaS, SrS, BaS, CdS, CdTe, MnTe, HgTe, La_2S_3, PbS, In_2S_3, Cu_xS, $CuGaS_2$, Y_2O_2S, WS_2, TiS_2, ZnS:M (M = Mn, Tb, Tm); CaS:M (M = Eu, Ce, Tb,Pb); SrS: M (M = Ce, Tb, Pb)
Metal pnictides	GaAs, AlAs, AlP, InP, GaP, InAs
Elements	Ru, Pt, Ir, Pd, Rh, Ag, W, Cu, Co, Fe, Ni, Mo, Ta, Ti, Al, Si, Ge
Others	CaF_2, SrF_2, MgF_2, LaF_3, ZnF_2, SiC, TiC_x, TaC_x, WC_x, $Ca_x(PO_4)_y$, $CaCO_3$, $Ge_2Sb_2Te_5$

The remaining metal nitrides are semiconductor materials where the metallic element is from either group 13 or 14: BN, AlN, GaN, InN, and SiN_x. Other metal pnictides rather predictably fall into the category of III–V compound semiconductor materials: AlP, GaP, InP, AlAs, GaAs, and InAs.

In addition to the above mentioned material types that have been deposited by ALD, metal chalcogenides have been demonstrated for a number of metals. Upon examination, some trends become apparent: most of the grown films contain either Group 2 or Group 12 metals. Zinc is particularly well studied, as ZnS, ZnSe, ZnTe, and ZnS:M (M = Mn, Tb, Tm) are all known. Work on the Group 2 metals, however, has been thus far restricted to sulfides and doped-variants thereof, such as CaS, SrS, BaS, CaS:M, and SrS:M, where M = Ce, Tb, Pb. Other cases of metal chalcogenides are shown in Table 1.

The ALD of pure elements, particularly metallic elements has been of increasing interest in recent years. Most of the attention has focused on processes for the "noble" metals, which are late transition metals which have a relatively unstable oxide and readily exist in the zero-valent metallic state. Although several notable noble metals have yet to be reported for ALD, elements such as Ru, Rh, Ir, Pd, and Pt have well-behaved processes. The noble metal ALD work has largely centered on the counterintuitive reduction of organometallic compounds using oxygen as the second precursor compound. Studies have indicated that the organic ligands undergo a combustion-like reaction in the presence of oxygen affording the reduction of the metal center and the production of CO_2 as a by-product. Other metal ALD processes have been based on the use of traditional reducing agents to ensure the formation of metallic species, including hydrogen, formaldehyde (as formalin), silanes, diborane, or the vapor of reductive metals such as zinc. Using such an approach, additional metals have been deposited, such as Cu and W, but with varying degrees of success. Recent advances have led to new processes that have started to make transition metal films feasible. Remote plasma ALD and radical-enhanced ALD are now actively being used to access metal films that were previously difficult or unknown by traditional thermal ALD methods. These include metals such as Ag, Ni, and others. This continues to be an active area of research and with time the number of ALD films in this category is likely to expand.

Molecular Layer Deposition

One variant of ALD that has gained considerable attention in recent years is what has become known as molecular layer deposition (MLD). MLD is essentially an ALD process where one or more of the sequential self-limiting surface reactions involve a molecular species that is deposited and itself becomes a layer in the resulting film [4, 6]. MLD precursors are typically organic molecules that possess two or more reactive groups capable of undergoing ALD reactions. At least two such groups are needed in an MLD precursor, so that reaction can be achieved with the existing surface while generating a new surface that bears groups that are amenable to reaction with the precursor molecules of the subsequent precursor pulse. A precursor that contains only one such reactive group can react with the surface, but it will effectively terminate film growth due to the lack of ability to carry out subsequent reactions. Films grown by MLD generally fall into two categories: organic polymers and hybrid organic–inorganic materials.

To grow organic polymer films by MLD, many of the same principle apply as for ALD. The precursors must be sufficiently volatile for gas phase delivery. There must be no significant self-reaction or decomposition that leads to uncontrolled growth behavior – a fact that can become an even larger constraint for MLD than for ALD due to the nature of the precursors. Finally, the organic precursors must react with each other rapidly and completely within the time frame of an MLD growth cycle. This restricts MLD to those organic precursors that exhibit highly reactive functional groups. A common motif has been the reaction of a difunctional precursor molecule with two heteroatoms bearing protonizable hydrogen atoms (e.g., a diamine) with a second precursor that contains two functional groups that react vigorously with such functional groups (e.g., a diacid chloride or a dianhydride). An example of the diacid chloride and diamine MLD reaction is terephthaloyl chloride ($ClCOC_6H_4COCl$) and p-phenylenediamine ($NH_2C_6H_4NH_2$) to form the aromatic polyamide poly(p-phenylene terephthalamide) or PPTA. Similarly, films can be grown by pyromellitic dianhydride and the use of an amine such as ethylenediamine or p-phenylenediamine, to form the corresponding polyimide layers.

The MLD of hybrid organic–inorganic materials has been of increasing interest to researchers [4].

The approach for growing hybrid films is similar to that of other MLD films with one main difference: the precursor containing two or more groups with protonizable hydrogen atoms is reacted with a metal-containing compound having ligands that vigorously react with these functional groups. The obvious choice would be the use of TMA, which possesses methyl groups that are highly and irreversibly reactive toward such hydrogen atoms, and as has been described earlier, is amenable toward ALD growth. The reaction of TMA or other aluminum alkyl compounds with organic diols, such as ethylene glycol, lead to the formation of "alucones," which are the most well-studied class of MLD-grown hybrid inorganic–organic films. Alucone MLD processes have been described as robust and exhibiting the linear growth with number of cycles expected of an ALD process [4]. By extension, other highly reactive metal alkyls should exhibit similar ability to grow MLD organic–inorganic hybrids. "Zincones" have been formed by the MLD reaction of diethylzinc with organic diols with similar growth characteristics to those observed for the alucones [4]. The number and variety of available hybrid materials should be expected to increase rapidly in coming years as additional research is performed.

A number of obstacles exist for MLD processes [4, 6]. The vapor pressure requirements for an MLD precursor can substantially limit the available choice of organic precursor compounds that are suitable. Furthermore, the organic compounds typically under consideration nearly always have thermal stability limitations, which put an upper bound on the temperatures that can be used for volatilization, as well as putting an effective upper limit on the deposition temperature. Another problematic issue is the porous structure of the MLD-deposited films and the potential it creates for diffusion of precursor vapor into these small narrow features – and worse – the extra time required for the unreacted precursor to diffuse out during the purge step of the MLD process. This creates the possibility for gas-phase reactions that can lead to uncontrolled CVD and loss of the self-limiting growth mechanism. A further problem in MLD is the potential reaction of both functional groups of a precursor molecule with available surface reactive sites which can act to terminate or poison growth and can perhaps alter the film characteristics. Double reactions of this type can be reduced or eliminated by selecting precursors with inherent conformational hindrances that greatly reduce the probability of this reaction type. Alternatively, the use of heterobifunctional precursors, where two chemically different reactive groups are present, has been used to limit the incidence of double reactions with the surface.

Plasma-Assisted ALD

Plasma-assisted ALD is an area that is currently seeing a trend toward increasing importance [1, 4, 7]. The central concept of plasma-assisted ALD differs from standard "thermal" ALD in that the reactivity of a precursor is increased by the application of plasma to generate radicals or other energetic species that can help promote reactions that might not otherwise progress by means of thermal activation alone. This advantage is commonly used to accomplish two ends: the reduction in deposition temperature of processes for which a thermal route is known, and the ability to access difficult materials for which a thermal process may not be feasible or practical. In considering the former of these, the plasma-assisted process using TMA and O_2 plasma has been used to grow Al_2O_3 at low temperature, including down to room temperature, and has been shown to have improved dielectric properties relative to thermally grown Al_2O_3 ALD films [4]. An example of the latter type is TaN, which has not been grown in high quality by thermal only methods, but has been achieved using an amido-/imido-based precursor and hydrogen radicals generated in a plasma source. The deposition of metal and metal nitride films is a common goal of plasma-assisted ALD processes, as these processes often require or benefit considerably from the advantages provided by plasma. Incidentally, it should be noted that plasma is always employed only for the generation of energetic non-metal precursor species; it is not used with metal sources.

Plasma sources can be divided into two categories: direct plasma and remote plasma. Direct plasma sources are those where the substrate is directly exposed to the plasma, typically with the substrate resting on one of the plasma source electrodes [1]. Direct plasma is less commonly used in ALD due to drawbacks such as potential substrate damage. In remote plasma ALD, the substrate is situated some distance downstream from the plasma, with the energetic reactive species generated therein conveyed toward the substrate surface by the existing gas flow.

Atomic Layer Deposition, Fig. 6 ALD growth of Al_2O_3 in bottom of a 40:1 aspect ratio trench. ALD excels at coating challenging high aspect ratio features with uniform films and excellent conformality (Image: Picosun Oy proprietary)

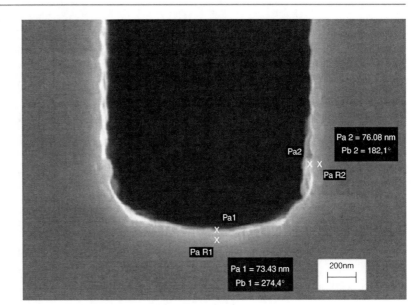

The precursor pulse may contain a mixture of radicals and ions generated in the plasma, or in the case of a system that can be considered to be purely "radical-enhanced ALD," only radicals will be present as the reactive species.

Overview of Key Application Areas

Because of its ability to coat surfaces with exceptional control of thickness, excellent uniformity, and outstanding conformality and step-coverage on complex and high aspect ratio surfaces (Fig. 6), ALD has found great potential for a range of applications [1, 8]. Foremost among these application areas is in microelectronics, where these three advantages are critical to the continued miniaturization of device features and the increase in performance demanded by Moore's Law. Accordingly the ability to make coatings of advanced functional materials on increasingly challenging structures has proved valuable to the success of ALD, and microelectronics has been the major impetus for ALD technology development for the past 10 years. Applications have generally fallen into three areas: (1) high-k dielectric gate oxide materials in metal-oxide field effect transistors (MOSFET), (2) capacitor dielectrics in dynamic random access memory (DRAM), and (3) interconnect materials [1]. The driving force behind the use of high-k dielectrics in MOSFETs has

been the need to replace SiO_2-based gate materials. The continual shrinkage of MOSFET features has led to a requirement that gate dielectric layers are only a few monolayers thick, which is a constraint that has allowed for tunneling and high leakage currents [1]. Alternate materials with high permittivity values that can avoid the scaling problems inherent in SiO_2 have been a major focal point of ALD technology development. Hafnium oxide and zirconium oxide based materials have received the most attention, with phases containing silicon and nitrogen (HfSiON) considered best solution for the near-term future [1, 8]. Active research continues on other materials that could become even better substitutes. DRAM production has also benefitted from the application of ALD. As with much of microelectronics, miniaturization has created a demand for DRAM features that are increasingly difficult to coat with the desired materials. Maintaining the required storage cell capacitance on a continually downsized surface area and increasingly complex-shaped capacitor structure is a challenge to which ALD is particularly well suited. Capacitor dielectrics such as Hf-Al-O and nanolaminates of ZrO_2-Al_2O_3-ZrO_2 have seen recent use, however, it should be expected that the requirements of the 35 nm node will necessitate the use of materials with a permittivity exceeding current benchmarks [1]. ALD is well suited to the deposition of such nanolaminate structures as shown in Fig. 7. Interconnects are another

Atomic Layer Deposition, Fig. 7 Nanolaminate thin film comprised of alternating layer of Al_2O_3 and TiO_2. Because of the stepwise methodology of ALD thin film growth, the deposition of nanolaminates and multi-component composite materials is straightforward (Image: Picosun Oy proprietary)

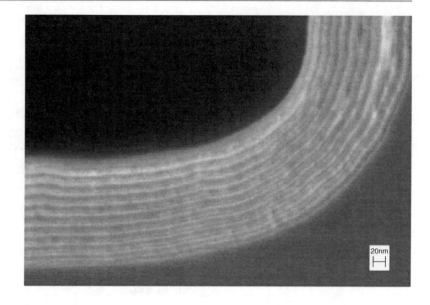

area where the application of ALD for microelectronics manufacture has been employed [1, 8]. Ruthenium and other noble metals have been grown by ALD and have been considered for seed layers for copper electroplating that can also act as an effective copper diffusion barrier. Early transition metal nitrides have been evaluated also as copper diffusion barriers, but obstacles to their use remain. Tungsten processes based on the reduction of WF_6 have been used as seed layers for CVD tungsten contact plugs. Copper deposition for the formation of metallization seed layers is an attractive application area that continues to see active development and may eventually become a manufacturable process.

Optical applications represent a notable field where ALD has been used. ALD has the advantages of high uniformity, the ability to tailor refractive index, and film stress by combining several materials into one structure, all of which are appealing for optical devices [1, 8]. For planar optics, however, it seems unlikely that ALD will be helpful enough to overcome the traditional methods that have higher throughput. Optical designs that implement three dimensions into the structure, however, are more likely to gain from the use of ALD. For example, pillar structures of a selected arrangement have been conformally coated to produce microlens arrays, with integral interference filters incorporated via the deposition of an interference film stack [1]. Also ALD might be applicable to the fabrication of inverse opal photonic

crystals, Fresnel zone plates for X-ray microscopy, and nanograting structures for polarizers and wave plates [1].

A variety of other application areas have also been explored. Due to the recent rapid miniaturization of magnetic read/write heads for inputting data onto hard disks, the deposition of conformal dielectrics by ALD has been implemented as insulating gap layers, adhesion layers between the substrate and film stack, or as encapsulation layers [1, 8]. Thin film electroluminescent displays, which were the original application for which ALD was developed, continue to be a notable application area due to the necessity for defect-free films of uniform thickness. Doped metal sulfides, such as ZnS:Mn, are commonly used as the luminescent layer, whereas Al_2O_3 can be used as a passivation or protective layer [1]. Microelectromechanical systems (MEMS) are yet another area where ALD is experiencing a surge of interest. Because ALD can grow highly conformal and defect-free films on complex geometries, it can be used in MEMS components for wear-resistant, lubricating, anti-stiction, or charge dissipating coatings [1, 9]. An emergent area of ALD research is in what could be described as nanotechnology: the coating of nano-sized features, such as those inside of porous materials, on the surface of nanowires, or onto nanotubes. Although the applications of this area are at this point largely prospective, the possibilities that could be realized are difficult to ignore. This topic has been the subject of a recent review to which the reader is referred for a detailed description [10].

Concluding Remarks

It is the purpose of this entry to give an overview of the basic principles, capabilities, and applications of ALD, while referring the reader to several recent comprehensive review articles that treat specific areas in greater detail where relevant. In describing the unique qualities of ALD, it is apparent that the strong advantages of precise thickness control, thickness uniformity, and exceptional conformality lend themselves to a range of possible applications. Despite the fact that ALD is an inherently slow deposition process, averaging around 1.0 Å/deposition cycle, the rapid miniaturization and increase in structural complexity of many devices has mitigated much of downside associated with its slowness. Correspondingly, these same developments have highlighted the need for ALD, as the limitations of traditional film growth methods have become an impediment to their future use in situations where challenging substrate features are involved. Additionally, in some areas, such as nanotechnology, ALD is likely to become a "critical enabler" that allows for unusual structures to be produced that would be unattainable by other methods. As such, the future for ALD is likely to yield new processes, novel research, and innovative use for an increasing number of applications.

Cross-References

▶ Chemical Vapor Deposition (CVD)
▶ Physical Vapor Deposition

References

1. Ritala, M., Niinistö, J.: Atomic layer deposition. In: Jones, A.C., Hitchman, M.L. (eds.) Chemical Vapor Deposition: Precursors, Processes and Application, pp. 158–206. Royal Society of Chemistry, Cambridge (2009)
2. Ritala, M., Leskelä, M.: Atomic layer deposition. In: Nalwa, H.S. (ed.) Handbook of Thin Film Materials, vol. 1, pp. 103–159. Academic, New York (2002)
3. Puurunen, R.L.: Surface chemistry of atomic layer deposition: a case study for the trimethylaluminum/water process. J. Appl. Phys. **97**, 121301 (2005)
4. George, S.M.: Atomic layer deposition: an overview. Chem. Rev. **110**, 111–131 (2010)
5. Kim, H.: Atomic layer deposition of metal and nitride thin films: current research efforts and applications for semiconductor device processing. J. Vac. Sci. Technol. B **21**, 2231–2261 (2003)
6. George, S.M., Yoon, B., Dameron, A.A.: Surface chemistry for molecular layer deposition of organic and hybrid organic-inorganic polymers. Acc. Chem. Res. **42**, 498–508 (2009)
7. Knoops, H.C.M., Langereis, E., van de Sanden, M.C.M., Kessels, W.M.M.: Conformality of plasma-assisted ALD: physical processes and modeling. J. Electrochem. Soc. **157**, G241–G249 (2010)
8. Ritala, M., Niinistö, J.: Industrial applications of atomic layer deposition. ESC Trans. **902**, 2564–2579 (2009)
9. Stoldt, C.R., Bright, V.M.: Ultra-thin film encapsulation processes for micro-electro-mechanical devices and system. J. Phys. D Appl. Phys. **39**, R-163–R-170 (2006)
10. Knez, M., Nielsch, K., Niinistö, L.: Synthesis and surface engineering of complex nanostructures by atomic layer deposition. Adv. Mater. **19**, 3425–3438 (2007)

Atomic Layer Deposition (ALD)

▶ Chemical Vapor Deposition (CVD)

Atomic Layer Epitaxial (ALE)

▶ Chemical Vapor Deposition (CVD)

Atomic Layer Epitaxy (ALE)

▶ Atomic Layer Deposition

Attocalorimetry

▶ Nanocalorimetry

Automatic Data Analysis Workflow for RNAi

▶ Computational Systems Bioinformatics for RNAi

B

Bacterial Electrical Conduction

▶ Micro/Nano Transport in Microbial Energy Harvesting

Band Alignment

▶ Electrode–Organic Interface Physics

Basic MEMS Actuators

Arash Kheyraddini Mousavi and Zayd Chad Leseman
Department of Mechanical Engineering,
University of New Mexico, Albuquerque, NM, USA

Synonyms

Microactuators; MicroElectroMechanical Systems; Micromachines

Definition

MEMS actuators are a type of MicroElectro-Mechanical System (MEMS) that convert energy into motion. MEMS are systems that integrate mechanical and electrical components with dimensions on the order of micrometers. Therefore, the typical motions achieved by MEMS actuators are on the order of micrometers as well.

Overview

MEMS are systems that integrate mechanical and electrical components with dimensions on the order of micrometers. Though the concept of MEMS has existed since the 1960s, due to the advent of microfabrication techniques for miniaturizing electronic components, the term MEMS was not coined until 1986. Professors Jacobsen and Wood from the University of Utah devised this terminology in the course of writing a proposal to the Defense Advanced Research Projects Agency (DARPA) [1]. The term was then disseminated via a National Science Foundation (NSF) report, the Utah-held IEEE MEMS workshop in 1989, the IEEE/ASME Journal of MEMS, and subsequent DARPA MEMS funding solicitations. Since its inception, this term has gained wide acceptance as a catchall for microdevices in general.

MEMS actuators are a specific class of MEMS that convert energy into motion. The mechanisms by which energy is converted into motion are typically physical or chemical. Motion of the MEMS actuator can be used for positioning, open and closing valves, characterization of energy conversion processes, switching, and material characterization at the micro/nanoscale. A commercial application for positioning with MEMS actuators is that of Texas Instruments Digital Micromirror Device (DMD), wherein micromirrors are positioned (rotated) in order to direct light to create images in projectors for display applications. Opening and closing of (micro) valves is important for microfluidics and lab-on-a-chip applications. MEMS actuators can be configured to be energy storage devices or to detect quantities pertaining to energy conversion. Radio Frequency (RF) MEMS make use

B. Bhushan (ed.), *Encyclopedia of Nanotechnology*, DOI 10.1007/978-90-481-9751-4,
© Springer Science+Business Media B.V. 2012

of MEMS actuators to create resonators for use in filters, reference oscillators, switches, switched capacitors, and varactors. Finally, MEMS actuators are commonly used to study material responses at the micro and nanoscales. They are used to apply mechanical forces to materials in order to characterize their material properties mechanically.

Fabrication of MEMS actuators typically occurs via top-down fabrication methods. These methods begin with a larger piece of material, typically Si, and then shape the Si into the form of an actuator. Most MEMS actuators are extrusions of a 2-D pattern that are transferred into the Si via a photolithographic process. Three common methods by which MEMS actuators are fabricated are: Surface Micromachining, Silicon on Insulator (SOI) Surface Micromachining, and Single Crystal Reactive Etching and Metallization (SCREAM). Other processes exist, but many are a hybrid processes based in one of the prior mentioned processes.

This discussion of Basic MEMS Actuators will be broken into four parts. First, there will be a description of energy conversion mechanisms that generate forces; a specific discussion follows on comb capacitors and how they convert (electrical) energy into a force. Second, a discussion will ensue on how MEMS actuators convert the force into motion via a compliant structure. Third, methods for quantification of the motion will be discussed with specific emphasis on an integrated method for measurement of displacements. Finally, a brief description of the three main processes for fabrication of MEMS actuators will be given.

A MEMS actuator is depicted in Fig. 1. It uses comb capacitors for energy conversion into motion by deflecting a fixed-fixed beam flexure. Displacements of this actuator are measured using a vernier. All components mentioned will be discussed in detail here.

Energy Conversion into Forces and Motion

MEMS actuators convert different forms of energy into force and motion. For example, two electrically isolated parallel plates with an applied potential difference will develop an electrostatic force between the plates. If those plates are on flexible springs then they will move toward one another. This type of energy conversion into motion is termed electrostatic

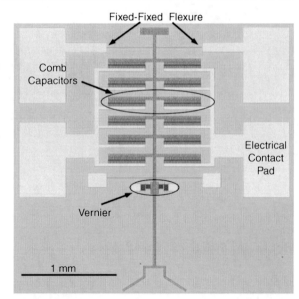

Basic MEMS Actuators, Fig. 1 Basic MEMS Actuator that utilizes comb capacitors to generate force, a fixed-fixed flexure to guide the motion of the applied force, and a vernier to measure displacement

actuation. Other types of actuators also use electrostatic actuation such as the nanotractor or inchworm actuator [2], scratch drive actuators [3], and the shuffle motor [4]. Beyond electrostatic actuators, many other actuators have been developed that use fluidic forces [5], magnetic forces [6], radiation pressure [7], piezoelectrics [8], shape-memory alloys [9], and thermal expansion [10].

Of particular importance for MEMS actuators is the way with which these forces scale with length; they scale differently due to the different physics involved. Table 1 displays a list of common forces and quantities that are of interest for MEMS actuators and gives how they scale with length.

Table 1 displays the way with which each force scales with length. Variables containing a geometric length pertinent to the physical system are in parentheses and moved to the right-hand-side of the equation. It is these quantities that are used to determine a quantities' sensitivity to length scale. Note each entry in Table 1 is highly dependent on the configuration of the system. In addition to common forces generated in MEMS actuators, this table also includes gravity, van der Waals forces, inertia, and the mass moment of inertia for reference.

Basic MEMS Actuators, Table 1 Sensitivity of physical quantities to length scale

Physical quantities	Examples	Governing equation	Sensitivity to length scale
van der Waals [11] $\breve{A} :=$ *Hamaker constant* $A :=$ *Area* $d :=$ *Distance between objects* $r :=$ *Radius* $r_i :=$ *Radius of object "i"* $l =$ *Facing lengths of two objects*	*Case 1:* Two spheres	$F = \dfrac{\breve{A}}{6}\left(\dfrac{r_1 r_2}{r_1 + r_2}\dfrac{1}{d^2}\right)$	l^{-1}
	Case 2: A sphere and a surface	$F = \dfrac{A}{6}\left(\dfrac{r}{d^2}\right)$	l^{-1}
	Case 3: Two cylinders	$F = -\dfrac{A}{8\sqrt{2}}\left(\sqrt{\dfrac{r_1 r_2}{r_1 + r_2}}\dfrac{l}{\sqrt{d^5}}\right)$	l^{-1}
	Case 4: Two crossed cylinders	$F = \dfrac{A}{6}\left(\dfrac{\sqrt{r_1 r_2}}{d^2}\right)$	l^{-1}
	Case 5: Two Surfaces	$F = \dfrac{A}{6\pi}\left(\dfrac{1}{d^3}\right)$	l^{-3}
Viscous forces $\mu :=$ *Dynamic viscosity* $V_0 :=$ *Relative velocity*	*Case 1:* Two infinite plates	$F = \mu V_0\left(\dfrac{1}{d}\right)$	l^{-1}
	Case 2: Two finite plates	$F = \mu V_0\left(\dfrac{A}{d}\right)$	l^{1}
Electrostatic force $\in_r :=$ *Relative static permittivity* $\in_0 :=$ *Vacuum permittivity* $V_e :=$ *Electrical potential* $h :=$ *out of plane thickness*	*Case 1:* Finite parallel plates at distance d	$F = \dfrac{\in_0 \in_r V_e^2}{2}\left(\dfrac{A}{d^2}\right)$	l^{0}
	Case 2: Comb drive [12]	$F = \dfrac{\in_0 \in_r V_e^2}{2}\left(\dfrac{h}{d}\right)$	l^{0}
Thermal expansion $E_y :=$ *Young modulus of elasticity* $\alpha_T :=$ *Therma expansion coefficient* $\Delta T :=$ *Temperature change*	*Case 1:* Constrained column	$F = E_y \alpha_T \Delta T A$	l^{2}
Magnetic forces [13] $\mu_0 :=$ *Vacuum permeability* $d :=$ *Distance between wires* $l :=$ *Length along wire* $I_i :=$ *Current in wire "i"* $A_0 :=$ *Cross sectional area* $A_S :=$ *Surface area* $\dot{Q}_S :=$ *Surface heat flow rate* $I_e :=$ *Electrical current*	*Case 1:* Constant current density with the boundary condition $\dfrac{I_e}{A_o} =$ cons.	$F = \dfrac{\mu_0}{2\pi}\dfrac{l}{d}\,I_1 I_2$	l^{4}
	Case 2: Constant heat flow through the surface of the wire with the boundary condition $\dfrac{\dot{Q}_S}{A_S} =$ cons.	$F = \dfrac{\mu_0}{2\pi}\dfrac{l}{d}\,I_1 I_2$	l^{3}
	Case 3: Constant temperature rise of wire with the boundary condition $\Delta T =$ cons.	$F = \dfrac{\mu_0}{2\pi}\dfrac{l}{d}\,I_1 I_2$	l^{2}
Piezoelectric force [14] $\in :=$ *Mechanical strain* $E_e :=$ *Electrical field* $e_p :=$ *Piezoelectric constant* $\dfrac{\text{Charge density}}{\text{Applied strain}}$ $E_y^E :=$ *Young modulus at constant E_e*	*Case 1:* 1-D unconstrained actuation	$F = -e_p V_e\left(\dfrac{A_0}{d}\right) + E_y^E \in (A_0)$	$l^{0}\,\&\,l^{2}$
Drag force $\rho :=$ *Density* $C_d :=$ *Drag coefficient* $A_p :=$ *Projected are normal to flow*	*Case 1:* Infinite cylinder ($C_d = .47$) *Case 2:* Flat plate perpendicular to flow ($C_d = 1.28$)	$F = \dfrac{1}{2}\rho V_0^2 C_d (A_p)$	l^{2}
Surface tension force $\gamma :=$ *Surface tension* $p :=$ *Perimeter*		$F = \gamma(p)$	l^{1}
Inertia and weight $g :=$ *Gravity* $\alpha :=$ *Acceleration* $\forall :=$ *Volume*	*Case 1:* Weight	$F = \rho g(\forall)$	l^{3}
	Case 2: Inertia force	$F = \rho a(\forall)$	l^{3}

(continued)

Basic MEMS Actuators, Table 1 (continued)

Physical quantities	Examples	Governing equation	Sensitivity to length scale
Mass moment of inertia $\rho_l :=$ Mass per unit length $\rho_A :=$ Mass per unit area $\rho_V :=$ Mass per unit volume $r :=$ Radius	*Case 1:* Sphere	$\bar{I} = \dfrac{8\pi}{15}\rho_V(r^5)$	l^5
	Case 2: Thin circular disk	$\bar{I} = \dfrac{\pi}{2}\rho_A(r^4)$	l^4
	Case 3: Slender bar	$\bar{I} = \dfrac{1}{12}\rho_l(l^3)$	l^3
Shape-memory alloy [9] $v :=$ Poisson's ratio $h_f :=$ Shape memory alloy film thickness $h_s :=$ Substrate thickness $R_i :=$ Initial radius of curvature $R_i :=$ Radius of curvature of the substrate with SAM film	*Case 1:* Force caused by shape memory alloy on a substrate	$F = -\dfrac{E_y}{1-v}\left(\dfrac{A_0 h_s^2(r_1 - r_2)}{6h_f r_1 r_2}\right)$	l^2

Interpretation of Table 1 is accomplished by comparing the exponents of the fourth column. These exponents are used to compare the importance of a given energy conversion process at different length scales. l^0 indicates that the force generating mechanism is invariant with length scale. l raised to a positive power means that the property grows with increasing scale, while a negative power implies that the property grows with decreasing scale.

Energy conversion mechanisms with negative exponents will behave more favorably at smaller length scales. Of the mechanisms listed, the easiest to employ and control are thermal (l^2) and the electrostatic force generated by comb capacitors (l^0). Further discussion on thermal actuators is contained in the thermal actuators entry in this encyclopedia. Comb versus parallel plate capacitors are explained in more detail in the subsequent section. When compared to how inertia scales (l^3), both thermal and electrostatic actuation schemes behave more favorably at smaller length scales. The mass moment of inertia is also displayed in Table 1. The exponent of five (sphere) implies that it diminishes at shorter length scales more rapidly than any of the other forces. Because inertial force and mass moment of inertia are relatively small, compared to common forces used in MEMS actuators, their dynamic responses are much faster than their macroscopic counterparts.

Electrostatic and thermal actuators scale favorably with decreasing length scales and are more easily implemented than most other methods. Therefore, they have considerably more popularity than other energy conversion methods. In the remainder of this encyclopedia entry, MEMS actuators that use electrostatic forces, in particular capacitance due to comb capacitors, will be the topic of discussion. Thermal actuators are more thoroughly discussed in their entry in this encyclopedia.

Comb Capacitors

In order to generate the force applied to the MEMS actuator, one of the more common methods is to use capacitors. The common methods are parallel plate capacitors and comb capacitors. Parallel plate capacitors have a capacitance equal to:

$$C_{pp} = \frac{\varepsilon_0 A}{d} \tag{1}$$

where ε_0 is the permittivity of the material between the plates (usually air or vacuum), A is the area of parallel plate capacitor, and d is the gap between the parallel plates. The electrostatic energy is $\frac{1}{2}C_{pp}V^2$, where V is the voltage between the parallel plate capacitors. Taking its derivative with respective to the gap, the force is calculated to be

$$F_{pp} = -\frac{\varepsilon_0 A V^2}{2d^2} \tag{2}$$

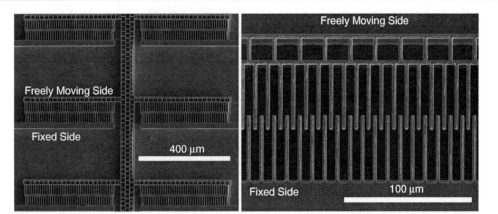

Basic MEMS Actuators, Fig. 2 Arrays of comb capacitors arranged symmetrically about the midline of a MEMS actuator (*left*). Closer view of a set of comb capacitors displaying the comb nature of the structure (*right*)

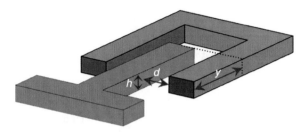

Basic MEMS Actuators, Fig. 3 A single unit of a comb drive actuator

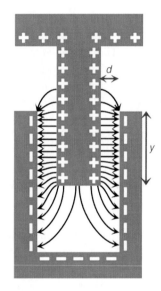

Note that F_{pp} is always attractive and is nonlinear in not only the voltage, but also in the gap between the parallel plates. This geometric nonlinearity makes control of actuators difficult and led to the development of comb capacitors [15].

SEM images of a comb drive array are shown in Fig. 2. Comb capacitors are arrays of interdigitated fingers. One side of the array is rigidly attached to the substrate while the other is allowed to move via a flexure structure which will be discussed in the next section.

Comb capacitor drives typically have hundreds to thousands of combs to generate a desired force. For analysis purposes, Fig. 3 shows one unit of a comb drive. When connected to an external source, opposite charges gather on the surfaces of the comb fingers which generate an electric field; see Fig. 4. Note that along the vertical line of symmetry of Fig. 4 that the horizontal components to the right and left are equal and opposite. Thus the net force due to the horizontal

Basic MEMS Actuators, Fig. 4 Electric field of a comb capacitor

components is zero. The vertical components of the fringing field at the tip of the comb finger, however, are not balanced and a net attractive force results. The capacitance of an array of comb capacitors is found to be:

$$C_{cc} = 2n\frac{\varepsilon_0 hy}{d} \tag{3}$$

where n is the number of combs in the array, h is the depth of the comb, y is the overlap between the moving and fixed comb, and d is the gap between the moving

Basic MEMS Actuators,
Fig. 5 Typical configurations
for springs in MEMS
actuators: (**a**) fixed-fixed or
fixed-guided beam
configuration and (**b**) folded-
beam flexure configuration

and fixed comb. The attractive force generated by the comb capacitors when a voltage is applied is:

$$F_{cc} = \frac{n\varepsilon_0 h}{d} V^2 \qquad (4)$$

The parameters in Eq. 4 are constant with the exception of the voltage. This type of relationship is more desirable than the relationship in Eq. 2, especially for dynamic excitation. Equations 3 and 4 are valid when the initial engagement of the comb is $>4d$ until the tip of the moving comb is within a distance $4d$ of the fixed side of the comb drive.

Spring Configurations for Basic MEMS Actuators

The force created by the conversion of energy is transmitted to a compliant structure (spring) via a stiff structure that traditionally comprises a majority of the mass – the proof mass. Though many different configurations have been used for the compliant structure, the two most common structures used are shown in Fig. 5. Figure 5a is a fixed-fixed (clamped-clamped) beam configuration while Fig. 5b is referred to as the folded-beam flexure. Note that Fig. 5a is also referred to as a fixed-guided configuration. 'F' in each of the figures represents the applied force onto the structure by a means of energy conversion, as was described in the previous section.

One-dimensional motion is often desired. Therefore, one direction is designed to be considerably more compliant than the two orthogonal directions. This typically leads to beams that have slenderness ratios greater than 100. Additionally, structures must be designed such that the guided motion is parallel to

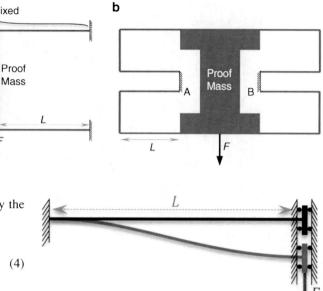

Basic MEMS Actuators, Fig. 6 The fixed-guided beam fixed at one end (*left*) and guided on the other (*right*)

the force. This requires a configuration that is symmetric about the axis of motion. Thus only even numbers of beams can be connected to the proof mass. An additional concern is that asymmetric loading may cause the structure to rotate if there are only two connections. Therefore, many designs will have four or more connections at the extremities of the actuator to avoid rotations.

Analysis of a Fixed-Guided Beam Flexure

The linear response of fixed-guided configuration shown in Fig. 6 can be studied using the moment curvature relationship:

$$\frac{d^2 v}{dx^2} = \frac{M}{EI} \qquad (5)$$

where E is the elastic modulus, v is the deflection, and I is the second moment of inertia. The second moment of inertia for many MEMS structures is that of a rectangular cross section ($\frac{hw^3}{12}$ where h and w are the height and width of the structure, respectively). The rectangular cross section is a result of the fabrication techniques used to make these devices (see Fabrication Methods Section). Solving the ordinary differential equation, Eq. 5, one finds that:

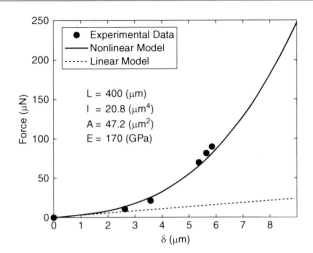

Basic MEMS Actuators, Fig. 7 Force versus displacement response for a fixed-guided beam that compares the linear and nonlinear models developed. Parameters used are for the actuator in Fig. 1. Experimental data is given as found using the method described in [17]

$$F = \frac{12\,EI}{L^3}\,\delta \qquad (6)$$

where δ is the maximum deflection of the structure. Note that this solution is for a single fixed-guided beam (one half of a fixed-fixed beam). In order to find the solution for multiple fixed-guided beams as in Fig. 5a, springs constants are added in parallel, as appropriate. For example, for the structure shown in Fig. 5a the spring constant would be $48\,EI/L^3$.

For small deflections, $\delta < w/4$, this relationship is sufficient for matching the true center line displacement of the fixed-guided beam with better than 5% accuracy. However for larger deflections, stretching of the beam must be considered. Stretching of the beam induces an axial force F_x, which in turn results in an applied moment. This changes the moment in the ordinary differential equation to be solved. In order to solve this new equation, small displacement assumptions must be used in conjunction with Hooke's Law. These assumptions result in the following equations:

$$\delta = 2\left(\frac{2I}{A}\right)^{\frac{1}{2}}(u - \tanh u)\left(\frac{3}{2} - \frac{1}{2}\tanh^2 u - \frac{3}{2}\frac{\tanh u}{u}\right)^{-\frac{1}{2}} \qquad (7)$$

$$F = \frac{8EI}{L^3}\left(\frac{2I}{A}\right)^{\frac{1}{2}}u^3\left(\frac{3}{2} - \frac{1}{2}\tanh^2 u - \frac{3}{2}\frac{\tanh u}{u}\right)^{-\frac{1}{2}} \qquad (8)$$

The parameter $u = \sqrt{\frac{F_x L^2}{4EI}}$ in Eqs. 7 and 8. No analytical solution is possible, because F_x is not known a priori. In practice, δ is commonly found in experiments. With δ, Eq. 7 can be solved for u. Next, F is found by inserting the previously determined value of u into Eq. 8. A more detailed solution is located in Frisch-Fay [16].

Figure 7 plots both the linear and nonlinear solution for the device shown in Fig. 1 which is schematically depicted in Fig. 5a. In addition to the derived solutions, experimental data is included on the plot. Data is taken by hanging calibrated weights from the device as is detailed in [17]. The width of the fixed-guided beams is 2 µm. Note that the nonlinear model diverges at about one-quarter of the value, that is, 500 nm.

Analysis of a Folded-Beam Flexure

The folded-beam flexure, Fig. 5b, is a variation of the fixed-guided beam configuration. Because the beams are folded, this structure experiences less axial stretch than the fixed-guided structure of Fig. 5a. However, for small deformations the behavior is approximately the same and Eq. 6 can be used. Because there is less axial stretching in the folded-beam flexure, the linear range is much larger for folded-beam flexures than for fixed-guided beam flexures. In order to find the solution for folded-beam flexure as in Fig. 5b, springs constants are added in series and parallel, as appropriate. For example, for the structure shown in Fig. 5b the spring constant would be $24EI/L^3$.

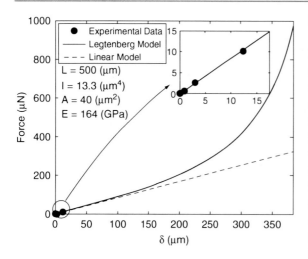

Basic MEMS Actuators, Fig. 8 Force versus displacement response for a folded-beam flexure that compares the linear and nonlinear models developed. Parameters used are for the actuator in Fig. 15. Experimental data is given as found using the method described in [17]

For even larger deflections, elastica theory must be applied. Along the y-direction, the displacement and force are [18]:

$$\frac{L}{2}\sqrt{\frac{F_y}{EI}} = \int_{\phi_1}^{\frac{\pi}{2}} \frac{d\phi}{\sqrt{1 - p^2\sin^2\phi}} \quad (9)$$

$$\delta = \sqrt{\frac{EI}{F_y}} \int_{\phi_1}^{\frac{\pi}{2}} \frac{2p^2\sin^2\phi - 1}{\sqrt{1 - p^2\sin^2\phi}} d\phi \quad (10)$$

where:

$$\phi_1 = \sin^{-1}\left(\frac{1}{\sqrt{2p^2}}\right) \quad (11)$$

The solution method involves choosing p, and finding ϕ_1 from Eq. 11. Then, the elliptic integrals of Eqs. 9 and 10, each of which depends only on p, are calculated.

Figure 8 plots both the linear and nonlinear solution for the device shown in Fig. 15. In addition to the derived solutions experimental data is included on the plot. Data is taken by hanging calibrated weights from the device as is detailed in [17]. The width of the folded-beam flexure's beams is 2 μm. Note that the nonlinear model diverges at almost 55 times the beam's thickness!

Motion Quantification

Many different schemes have been proposed for the quantification of the displacement of the MEMS actuators. The most common types include measuring capacitance, resistance, or interpreting an optical signal. Capacitance measurements are especially common when the method of actuation is by capacitors as well. By utilizing a high frequency voltage on the capacitors, one can measure the impedance. Equation 1 is for the capacitance due to parallel plate capacitors and Eq. 3 is for comb capacitors. Both equations contain a term that relates the displacement to the capacitance of the MEMS actuator. Piezoresistance is also a common method because of the ease of measurement. Limitations of either capacitance or piezoresistance measurements include the sensitivity of the instrument and the ability to isolate the actuator from noise. Optical measurements can be broken into two main categories, inteferometric and vernier. Inteferometric techniques can make very sensitive measurements of displacement and (depending on the technique) may be limited only by thermal vibrations. A vernier is a simple method and can be co-fabricated into the device; see Fig. 9.

In order to measure the displacements of the springs, two sets of scales are incorporated in the design of the MEMS actuator. These are a fixed "main scale" that is attached to the substrate, and a moving "vernier scale" that is integrated into the MEMS actuator, Figs. 9 and 10.

The vernier is an auxiliary scale, whose graduations are of different spacing from those of the main scale, but that bear a simple relation to them. The vernier scale of Fig. 10 has ten divisions that correspond in length to nine divisions on the main scale. Each vernier division is therefore shorter than a main-scale division by 1/10 of a main-scale division. Main-scale divisions are 5 μm apart. The Least Count (LC), which is defined as the smallest value that can be read on the vernier scale, is therefore:

$$LC = 5 \times \frac{1}{10} = 0.5 \,\mu\text{m} \quad (12)$$

For example, in Fig. 11, the zero of the vernier scale is moved past two complete divisions of the

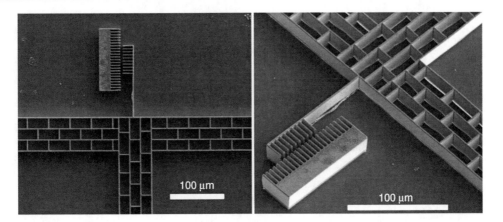

Basic MEMS Actuators, Fig. 9 SEM micrographs of a vernier for the measurement of displacement of a Basic MEMS Actuator

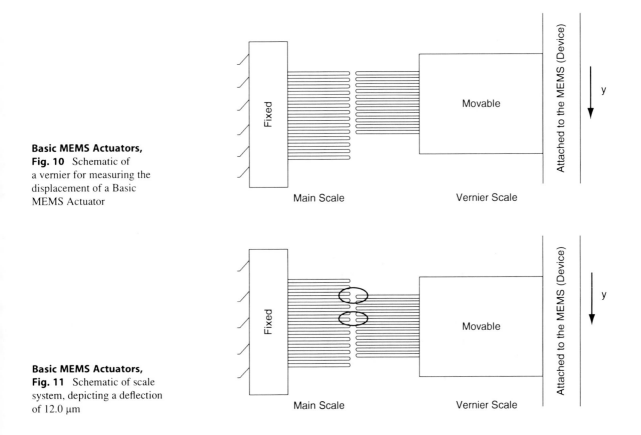

Basic MEMS Actuators, Fig. 10 Schematic of a vernier for measuring the displacement of a Basic MEMS Actuator

Basic MEMS Actuators, Fig. 11 Schematic of scale system, depicting a deflection of 12.0 μm

main scale, and the fifth division of the vernier scale is aligned perfectly to the same fifth main-scale reading.

Therefore the reading from the above setting would be:

Main-scale divisions = 2

Main-scale spacing = 5 μm

Total main-scale reading = 2 × 5 μm = 10 μm

Vernier-scale division that coincides with main scale = 4

Least count of the vernier scale = 0.5 μm

Total vernier-scale reading = 4 × 0.5 μm = 2.0 μm

Total = main-scale reading + vernier-scale reading

Total = 10 + 2.0 μm = 12.0 μm

Basic MEMS Actuators,
Fig. 12 Surface
micromachining of
a freestanding
microcantilever:
(**a**) deposition of a sacrificial
SiO$_2$ layer on Si followed by
photolithographic patterning,
(**b**) removal of photo resist
after etching, (**c**) poly-Si
deposition,
(**d**) photolithographic
patterning and etching of Si to
expose the sacrificial oxide
layer, (**e**) microcantilever
before release, and (**f**) etching
the sacrificial oxide layer thus
releasing the microcantilever

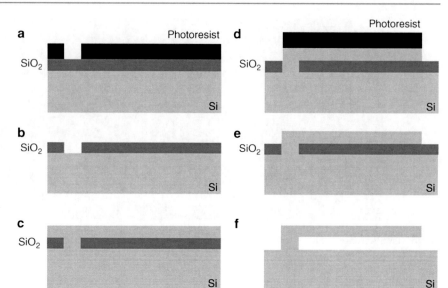

Fabrication Methods

As mentioned previously, there are three main methods for fabrication of MEMS actuators: Surface Micromachining, SOI, and SCREAM [14]. It is important to emphasize that many other processes exist and new processes are being developed constantly. Modifications are typically introduced to accommodate for the actuators' ultimate application or for the type of energy conversion method that the actuator will be utilizing.

Surface micromachining is a process built by alternating the deposition and patterning of sacrificial layers and structural layers of material. Typically, the sacrificial layers are a type of silicate glass (Phospho-Silicate-Glass (PSG) or SiO$_2$). This layer is commonly referred to as the sacrificial oxide (sac-ox). The structural layers are often made of polycrystalline Si (poly-Si). After the deposition of each layer, that layer is patterned using a photolithographic process to define either holes through the sacrificial layer or physical features of the device in the case of a structural layer.

Figure 12 details a simple surface micromachining process for a cantilever beam. Each step (a–f) is a cross-sectional view for the structure. This is a common method of representing a microfabricated structure's process flow. This method typically revolves around an important feature of the final structure. Note that due to the nature of this process, numerous alternating layers of sac-ox and structural layers can be built up to form devices with multiple layers of functionality. One of the more common processes used

for surface-micromachined MEMS is the SUMMiT process [19]. At the writing of this entry, the current process is entitled SUMMiT^{TM} V, where V stand for five structural layers of poly-Si. An example of a SUMMiT surface-micromachined device is shown in Fig. 13. A device is displayed with a pair of orthogonal comb capacitors that turn gears and raise a micromirror out-of-plane.

SOI surface-micromachined devices are similar to the surface micromachining process previously described except that the process begins with an SOI wafer. SOI wafers are comprised of three layers: a handle layer, a buried oxide layer (BOX), and a device layer. The handle layer acts as a substrate for the BOX and device layer. BOX layers are the sacrificial layer to be removed by an oxide etch at the end of the process, whereas the device layer is the structural layer of Si for the actual MEMS. The device layer is a piece of single crystal Si, not poly-Si as in standard surface micromachining. Due to the nature of the SOI wafer, SOI surface micromachining is used to make single structural layer devices. The general process flow for an SOI surface-micromachined process is shown in Fig. 14. The freestanding beam in the Fig. 14d would be for any part of a MEMS actuator that is capable of movement. These freestanding beams however must be anchored to the substrate. This accomplished by having structures with larger cross sections that are not fully undercut by the HF etching process. The left and right side of the device layer in Fig. 14d are a partial view of cross section

Basic MEMS Actuators,
Fig. 13 A surface-
micromachined set of Basic
MEMS Actuators with comb
capacitors and folded-beam
flexure used to drive a gear
train to raise a micromirror.
Image courtesy of Sandia
National Laboraotries,
SUMMiT™ Technologies,
www.mems.sandia.gov

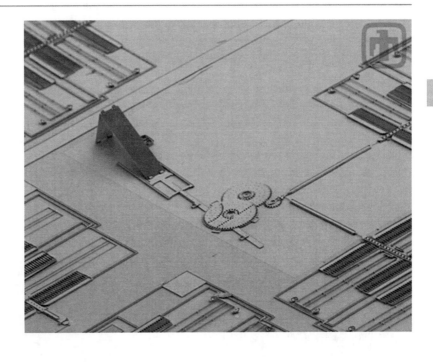

Basic MEMS Actuators,
Fig. 14 Processing steps for
the SOI surface
micromachining process:
(**a**) SOI wafer anatomy,
(**b**) photolithographic
patterning of device layer,
(**c**) deep reactive ion etching of
device layer, and (**d**) release of
device by oxide etch

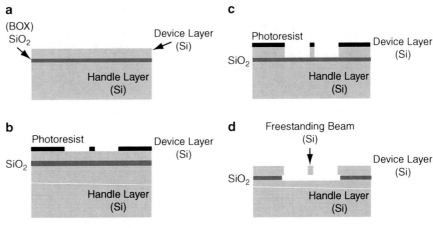

where the SiO_2 has not been fully undercut, thereby allowing for a physical connection between the device layer and handle layer. Note that the connections of the device to ground can be isolated electrically from one another. This is an important feature for placing biases on comb structures such as the one shown in Fig. 15. Although SOI surface micromachining and standard surface micromachining are similar, the device layer in the SOI process can have a broader range of thickness varying from 100's of nm to several 10's of µm. This is important because (1) layer thickness increases the actuator force and (2) a thick layer can make the out-of-plane stiffness of folded-beam suspension much greater, thereby minimizing unintended displacements.

The SCREAM process is a bulk processing method that utilizes a bulk piece of single crystal Si – a Si wafer. Typically, this process utilizes one mask (one photolithography step) as does SOI surface micromachining. In order to create freestanding structures, SCREAM utilizes the selectivity of etchants to certain deposited films and isotropic versus anisotropic etching.

Figure 16 demonstrates the SCREAM process [20]. A Si wafer is the base substrate. SiO_2 is deposited on the surface of the Si. Photoresist is then patterned to define the 2D layout of the microactuator (Fig. 16a).

Narrow features, approximately 1 μm, will be released (freed), while wider structures, approximately 100 μm or larger, will remain fixed to the substrate during the release process. After patterning, the Si is etched anisotropically (vertically) typically via a deep reactive ion etch (DRIE) (Fig. 16b). The photoresist is then removed and a conformal layer of SiO₂ is deposited (Fig. 16c). This step is important to the process in that it places an electrical insulator on the sidewalls of the structures. The SiO₂ on the floor of the wafer is then back etched to expose the Si. A subsequent Si etch then isotropically removes Si from the floor (Fig. 16d). This etch is allowed to continue to the point where the two etch fronts from either side of the narrower structures

are allowed to merge, thereby releasing the narrow structures (Fig. 16e). The wider adjacent structures are also undercut, but not released. In the final step, a metal is deposited (Fig. 16f). Metal is allowed to coat the top horizontal surface and vertical surfaces of the device. Depending on the layout of the device, the islands on the left and right of the freestanding beam can be electrically isolated from the beam and from one another, connect to one or the other, or connect to both. The isotropic etch of Si in step e) allowed for

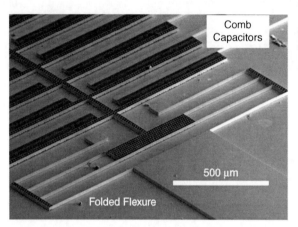

Basic MEMS Actuators, Fig. 15 An SOI surface-micromachined device with comb capacitors and folded-beam flexures

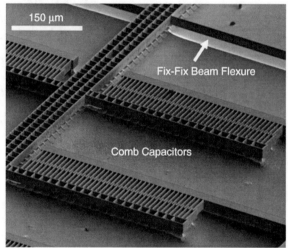

Basic MEMS Actuators, Fig. 17 Example of a Basic MEMS Actuator fabricated using the SCREAM process with comb capacitors and fixed-fixed beams. Note the scalloped floor beneath the actuators. This is indicative of isotropic Si etch in step e) in Fig. 16

Basic MEMS Actuators, Fig. 16 The SCREAM process: (**a**) Deposition of SiO₂ on a silicon wafer followed by photolithographic patterning, (**b**) deep reactive ion etching (DRIE) of Si, (**c**) conformal deposition of SiO₂, (**d**) etching of SiO₂ to expose the floors adjacent to the structure to be released, (**e**) isotropic etching of Si to release the freestanding structure, and (**f**) metallization

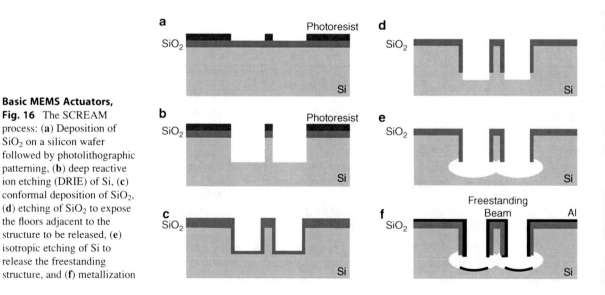

the SiO$_2$ layer to shield the metal (Al) from coming into contact with the Si. If the metal of each structure had touched the Si then the entire device would be shorted and no potential could be set up for actuation of the device. Figure 17 is an SEM micrograph of a MEMS actuator fabricated with the SCREAM process.

Summary

This encyclopedia entry covered the basics of MEMS actuators. The discussion was broken into four parts covering: energy conversion mechanisms for generation of force, compliant structures, motion quantification, and fabrication techniques. The section covering energy conversion mechanisms discussed the common mechanisms used in MEMS actuators and went into detail on electrostatic actuation. In particular, the salient features of parallel plate and comb capacitors were presented. Converted forces are used to move MEMS actuators to perform a host of applications. This motion occurs in controlled manner via a compliant structure. Two of the more common compliant structures used for MEMS actuators are fixed-guided beams and the folded flexure structure. Both the linear and nonlinear force versus displacement relationships are given for both of these common structures. Motion of the actuator is quantified in many different ways. Use of a simple co-fabricated vernier is detailed in this entry. Finally, the three main processes for fabrication of MEMS actuators were detailed. These included: Surface Micromachining, SOI, and the SCREAM processes.

Cross-References

▶ Micropumps
▶ Thermal Actuators

References

1. Wood, J.: Personal communication (2011)
2. de Boer, M.P., Luck, D.L., Ashurst, W.R., Maboudian, R., Corwin A.D., Walraven, J.A., Redmond, J.M.: High-performance surface-micromachined inchworm actuator. J. Microelectromech. Syst. 13(1), 63–74 (2004)
3. Akiyama, T., Collard, D., Fujita, H.: Scratch drive actuator with mechanical links for self-assembly of three-dimensional MEMS. J. Microelectromech. Syst. 6(1) (1997)
4. Deladi, S., Sarajlic, E., Kuijpers, A.A., Krijnen, G.J.M., Elwenspoek, M.C.: Bidirectional electrostatic linear shuffle motor with two degrees of freedom. In: Proceedings of the IEEE MEMS Conference (2005)
5. Seidemann, V., Bütefisch, S., Büttgenbach, S.: Fabrication and investigation of in-plane compliant SU8 structures for MEMS and their application to micro valves and micro grippers. Sens. Actuator A: Phys. 97–98, 457–461 (2002)
6. Basantkumar, R.R., Hills Stadler, B.J., Robbins, W.P., Summers, E.M.: Integration of thin-film galfenol with MEMS cantilevers for magnetic actuation. IEEE Trans. Magn. 42(10) (2006)
7. Sulfridge, M., Saif, T., Miller, N., O'Hara, K.: Optical actuation of a bistable MEMS. J. Microelectromech. Syst. 11(5) (2002)
8. Kommepalli, H.K.R., Yu, H.G., Muhlstein, C.L., Trolier-McKinstry, S., Rahn, C.D., Tadigadapa, S.A.: Design, fabrication, and performance of a piezoelectric uniflex microactuator. J. Microelectromech. Syst. 18(3) (2009)
9. Gill, J.J., Ho, K., Carman, G.P.: Three-dimensional thin-film shape memory alloy microactuator with two-way effect. J. Microelectromech. Syst. 11(1) (2002)
10. Que, L., Park, J.-S., Gianchandani, Y.B.: Bent-beam electrothermal actuators – part I: single beam and cascaded devices. J. Microelectromech. Syst. 10(2) (2001)
11. Israelachvili, J.N.: Intermolecular and Surface Forces: With Applications to Colloidal and Biological Systems (Colloid Science), 2nd edn. Academic, London (2006)
12. Naraghi, M., Chasiotis, I.: Optimization of comb-driven devices for mechanical testing of polymeric nanofibers subjected to large deformations. J. Microelectromech. Syst. 18(5) (2009)
13. Trimmer, W.S.N.: Micro robots and micro mechanical systems. Sens. Actuator. 19, 267–287 (1989)
14. Madou, M.J.: Fundamentals of Microfabrication: The Science of Miniaturization, 2nd edn. CRC Press, Boca Raton (2002)
15. Tang, W.C., Nguyen, T.-C.H., Howe, R.T.: Laterally driven polysilicon resonant microstructures. Sens. Actuator. 20(1–2), 25–32 (1989)
16. 16.Frisch-Fay, R.: Flexible bars. Butterworths Scientific, Washington, DC (1962)
17. Abbas, K., Leseman, Z.C., Mackin, T.J.: A traceable calibration procedure for MEMS-based load cells. Int. J. Mech. Mater. Des. 4(4), 383–389 (2008)
18. Legtenberg, R., Groeneveld, A.W., Elwenspoek, M.: Comb-drive actuators for large displacements. J. Micromech. Microeng. 6, 320–329 (1996)
19. Sniegowski, J.J., de Boer, M.P.: IC-compatible polysilicon surface micromachining. Annu. Rev. Mater. Sci. 30, 299–333 (2000)
20. Shaw, K.A., Zhang, Z.L., MacDonald, N.C.: SCREAM I: a single mask, single-crystal silicon, reactive ion etching process for microelectromechanical structures. Sens. Actuator A Phys. A40(1), 63–70 (1994)

Batteries

▶ Nanomaterials for Electrical Energy Storage Devices

Bendable Electronics

▶ Flexible Electronics

Bending Strength

▶ Nanomechanical Properties of Nanostructures

Bilayer Graphene

▶ Graphene

Bioaccessibility/Bioavailability

▶ Exposure and Toxicity of Metal and Oxide Nanoparticles to Earthworms

Bioadhesion

Manuel L. B. Palacio and Bharat Bhushan
Nanoprobe Laboratory for Bio- & Nanotechnology and Biomimetics, The Ohio State University, Columbus, OH, USA

Synonyms

Cell adhesion; Mucoadhesion

Definition

Bioadhesion refers to the phenomenon where natural and synthetic materials adhere to biological surfaces. This adherence may or may not be associated with the use of a bioadhesive. Bioadhesion also refers to the incorporation of the biomaterial in the body, which is manifested by the formation of a biofilm on the biomaterial. A specific type of bioadhesion is referred to as mucoadhesion, where a mucous gel layer is formed on the surface of the biological surface during the adhesion process.

Occurrence

Depending on the chemical composition of the biomaterial and the function of the biological surface involved, bioadhesion may be desirable or undesirable. A few examples of the occurrence of bioadhesion interactions are described as follows.

Synthetic and Natural Biomaterial Interactions

The implantation of biomaterials to the body (in tissue engineering, orthopedics, or dental prostheses, among others) is an example of a situation where bioadhesion is desired. In this case, the bioadhesion between the biomaterial and the target organs lead to the formation of a biofilm (protein layer), which, in turn, facilitates the integration of the implant to the body. Studying the fundamental mechanisms that influence protein adhesion on biomaterials is of interest in the context of the "Vroman effect," which involves a series of adsorption and displacement of proteins on biomaterial surfaces. The proteins on the biomaterial surface interact with cells and influence their behavior. The formation of protein-containing adhesive complexes known as focal adhesions (or cell-matrix adhesions) determines cell migration, cell cycle progression, and eventual differentiation. The proteins involved in focal adhesions are sensitive to the properties of the underlying substrate, such as chemical composition and hydrophobicity [1, 2].

On the other hand, biofilm formation may not be considered as beneficial in some cases. For ocular (contact) lens applications, the formation of a biofilm on the surface of the polymeric lens leads to the accumulation of inflammatory cells, and is therefore undesired [3].

An example of bioadhesion between natural biomaterial surfaces involves lectins, which are proteins or glycoprotein complexes that can bind to polysaccharides (sugars). Lectins are found in both animals and plants. One of the common animal lectins is the calcium-dependent binding lectins (also referred to as "C-type"). One of these C-type lectins, the asialoglycoprotein lectins, is specific to liver cells and is involved in animal biological function [4]. Meanwhile, plant lectins are abundant and can be found in common plants such as tomato, soybeans, and kidney beans. Plant lectins are commonly regarded as toxins when they bind to animal cells. [4].

Mucoadhesion

The adherence of materials onto a mucous gel layer of a biological surface is referred to as mucoadhesion. A well-known example of mucoadhesion is the attachment of mussels to underwater surfaces (such as corals or ship surfaces). The animal's foot extends from its shell to secrete a pad of the mussel adhesive protein (MAP) on the underwater surface. The foot then retreats to the shell interior, leaving a thread that secures the mussel to the pad. As shown in Fig. 1, this process is repeated multiple times [5]. The MAP has been used as the model for the development of mucoadhesive materials.

Another important example of mucoadhesion pertains to drug-delivery systems, where the drug is carried in a bioadhesive system, which then adheres to the mucous gel layer located on the surface of the biological membrane. Examples of polymers used as bioadhesives include chitosan, poly(vinyl alcohol), poly(vinyl pyrrolidone), agarose, as well as the cellulose and poly(acrylic acid)-type polymers [4]. The mucous gel layer is composed mostly of water (approx. 95% by weight), with the rest being the glycoprotein class known as mucins. The mucosal routes of drug delivery include the eye, mouth, nose, vagina, rectum, and the gastrointestinal (GI) tract [6].

Medical Adhesives

Adhesives are used to close skin incisions, seal and strengthen tissues, stop the leakage of blood and other body fluids, and serve as an adjunct to staples or sutures. These adhesives can be used in various surgical operations such as cardiovascular, pulmonary, and gastrointestinal surgeries, among others. Medical adhesives could either be synthetic (such as

Bioadhesion, Fig. 1 Adhesion of a mussel on a glass slide, where multiple threads composed of the mussel adhesive protein were secreted by the animal to attach itself to the glass surface [5]

cyanoacrylates) or naturally derived (such as fibrin sealants and bovine serum albumin-glutaraldehyde mixtures) [7]. In the case of cyanoacrylates, bonding occurs due to polymerization of the adhesive in the presence of hydroxyl groups in the environment. For the bovine serum albumin-glutaraldehyde glues, bonding takes place between the available lysine groups in the albumin and tissue surfaces and the aldehyde group in glutaraldehyde.

Dental Prostheses

Resin-based composites and polyacid-modified resin-based composites (compomers) are used to restore teeth damaged by cavities, fractures, erosion or aging. Moreover, adhesives are also used to repair fractured porcelain and composite restoratives. These adhesives are applied either to the enamel or to the dentin layer of the tooth requiring the restoration. Acrylic-based resins are used to restore human enamel. The enamel is pre-etched with acid (usually phosphoric acid) in order to create pores and increase surface wettability. The compositions of the enamel and dentin layers are widely different. Relative to enamel, dentin has much higher organic content, is less homogeneous, and more difficult to bond to restorative materials. As a consequence, the composition of adhesives intended for dentin is different from enamel adhesives. Dentin adhesives contain acrylic-based resins, such as bisphenol A-glycidyl methacrylate (BisGMA) and hydroxyethyl methacrylate (HEMA), but other components are present, such as a primer and a bonding agent, which are necessary to enhance bonding to dentin [8].

Key Research Findings

Bioadhesion Mechanisms
Interface Energetics
In systems exhibiting bioadhesion, the liquid environment influences the spreading of one material phase over another. Consider the case of a polymer surface adhering to the mucous gel layer on a biological membrane immersed in liquid medium (e.g., drug-delivery systems). Figure 2 is a schematic illustrating the interfacial energy components that should be considered in evaluating the thermodynamic work of adhesion of this system. During adhesion, a unit interface between the polymer and the liquid and between the mucus and the liquid vanishes, while an interface between the polymer and mucus forms. The thermodynamic work of adhesion (W_{PM}^{adh}) between the polymer and mucus is then defined as follows [6, 9]:

$$W_{PM}^{adh} = \gamma_{PM} - (\gamma_{PL} + \gamma_{ML}) \tag{1}$$

where γ represent surface energies and the subscripts P, M, and L are for polymer, mucus, and liquid, respectively. A positive value for W_{PM}^{adh} is regarded to be a necessary condition to achieve successful bonding between the two surfaces.

Chemical Interactions
Various interactions between two chemically active surfaces (i.e., not inert) facilitate the bioadhesion process. Strong adhesion can occur if the two surfaces are capable of forming either covalent, ionic or metallic bonds. At the same time, weaker interaction forces, such as polar (dipole–dipole), hydrogen bonding, or van der Waals interactions (induced-dipoles), also aid in bonding the two surfaces [4, 9].

The case of mussel adhesion on underwater surfaces is an example of how the chemical composition of the contacting surfaces affects the adhesion mechanism. The initial interaction between the mussel and the underwater surface involves the removal of weak boundary layers (mostly water). If the underwater surface is nonpolar, the water boundary layer interacts through weak dispersive forces. Since the mussel adhesive protein (MAP) is larger than a water molecule, the protein experiences greater dispersive interactions with the nonpolar surface relative to water, leading to the displacement of the water boundary layer, and

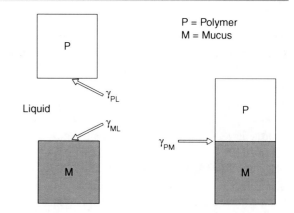

Bioadhesion, Fig. 2 Schematic of the relevant interfacial energies (γ) that determine the formation of a bioadhesive bond, as illustrated for the adhesion between a polymer (P) and the mucus layer (M) on the biological membrane in the presence of a liquid medium (L) (adapted from [6])

adhesion of the protein. In the case of polar underwater surfaces, the water boundary layer cannot be easily displaced. In this case, the MAP uses its hydrophilic amino acid side chains, which contain aminoalkyl, hydroxyalkyl, and phenolic groups (such as 3,4-dihydroxyphenylalanine or Dopa), all of which are capable of forming strong hydrogen bonds. Hence, the MAP is able to displace water and the mussel is able to adhere to polar underwater surfaces as well [10]. The unique chemical composition and properties of the mussel adhesive proteins have led to the development of synthetic analogs for potential use as mucoadhesives for drug-delivery systems.

Mechanical Effects
While interfacial contact and chemical bonding interactions are needed for the initial stages of bioadhesion, the interpenetration or interdiffusion between the molecules of the two contacting surfaces will maintain the adhesive bond. If a polymer is one of the contacting systems involved in the bioadhesion process, the interpenetration process involves the mobility of the individual chains and its entanglement in the opposing biological membrane. A related concept is the swelling capacity of the polymer, which is the ratio of the wet to dry weights. A high swelling capacity for a given polymer implies that it has greater chain mobility, and a higher tendency toward interpenetration [6].

For bioadhesion in dental prostheses, mechanical interlocking or interpenetration is the primary adhesive mechanism. When bonding to enamel, surface etching is performed to create micropores on the surface that will increase the penetration of the restorative material. Dentin, on the other hand, contains tubules that radiate from the pulp. These tubular structures facilitate the penetration of resin monomers into the dentin and their retention once the resin has been polymerized [11].

Interface Optimization

The contact between the two surfaces in contact during the bioadhesion process may be optimized using various physical, chemical, and mechanical methods.

Physical Processes

Plasma processing is a widely used technique for surface cleaning and modification of the surface chemical composition. The plasma, which is generated by applying an electric field to a low-pressure gas in vacuum, produces reactive species such as ions and free radicals. For surface cleaning purposes, the plasma breaks most organic bonds, such as C–H, C–C, C=C, C–O and C–N. In addition, if oxygen plasma is used, another cleaning mode is possible. The activated oxygen species ($O\bullet$, O^+, O^-, O_2^+, O_2^-, O_3, among others) can combine with organic compounds to form H_2O, CO, CO_2, and low molecular weight hydrocarbons, thereby removing organic contaminants on the surface [12].

Surface modification through plasma processing can render the surface more hydrophilic or hydrophobic. If the gas used is oxygen or carbon dioxide, polar groups are introduced to the surface (such as –C–OH, –C=O and –COOH), enhancing surface hydrophilicity. The presence of more polar surface groups is valuable in enhancing adhesion, e.g., between proteins and polymeric biomaterials used in tissue engineering applications. On the other hand, the use of fluorinated gases such as CF_4 in the plasma results in the introduction of a low surface energy polytetrafluoroethylene (PTFE)-like structure, enhancing surface hydrophobicity. This approach has been used to prevent the undesirable protein adhesion and the ensuing generation of inflammatory cells on poly (methyl methacrylate) contact lenses [3].

Chemical Functionalization

Specific functional groups can be attached to surfaces using chemical synthesis techniques to construct model systems for use in adhesion studies. An example is shown in Fig. 3, illustrating how a silica surface is chemically functionalized with the protein streptavidin [13]. In this illustration, the functionalization process consists of multiple steps, starting with the addition of the aminosilane 3-aminopropyltriethoxysilane (APTES) in order to introduce amine groups to the silica surface. This is then followed by reacting the surface with sulfo-*N*-hydroxysuccin-imido-biotin (abbreviated as biotin), which forms a covalent bond with the amine group of the APTES. In the final step, the protein streptavidin will bind to biotin. Streptavidin has a high affinity to biotin, and this interaction is regarded as one of the strongest non-covalent bonds known.

Mechanical Texturing

Surface modification through mechanical texturing is a simple and cost-effective method to roughen biomaterials and create anchor points for enhanced adhesion. In this method, abrasive particles (such as silicon carbide or alumina) impinge on the biomaterial, and the resulting impact increases its surface roughness [14]. This method can be used on orthopedic implant materials such as titanium or cobalt/chrome.

Characterization Methods

Nanoscale

Fundamental studies of bioadhesion at the nanoscale are conducted with atomic force microscopy (AFM). Aside from its high-resolution surface imaging capability, the AFM can measure forces in the pico- to nanonewton range, allowing for the detection of single molecule interactions [15, 16]. In AFM, force-distance curves for the contact between the probe and the sample surface can be obtained. An example is shown in Fig. 4, which corresponds to the interaction between the protein streptavidin and its ligand biotin while immersed in phosphate buffered saline (PBS) [17]. In this example, the AFM probe consisted of a cantilever with a glass microsphere functionalized with biotin on its end, where the biotin is covalently bonded to bovine serum albumin (BSA). The streptavidin was deposited onto a mica surface. The force-distance curve consists of two segments, namely, the advancing curve, which shows how the probe approaches the surface, and the retracting curve, which illustrates how the probe detaches from the surface. In the retracting force curve, a distinct snap-off point is observed, which corresponds to the force necessary to separate the tip from the sample surface. This is the measured adhesive force.

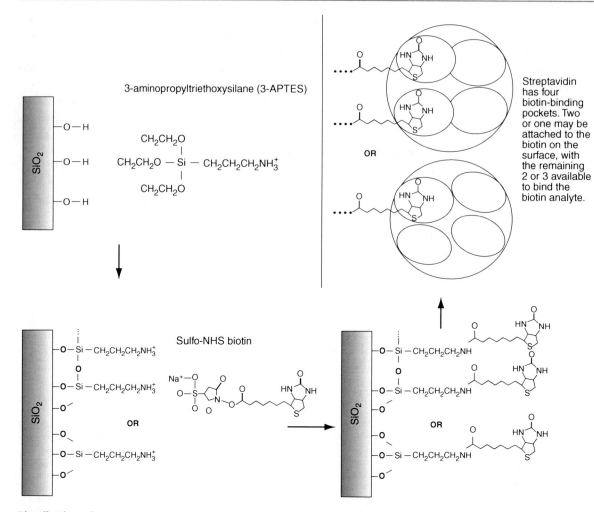

Bioadhesion, Fig. 3 Schematic showing the steps in the functionalization of silica surface with the protein streptavidin (adapted from [13])

In Fig. 4, the quantity τ corresponds to the rupture period, the time over which the bond will break as related by the equation [17]:

$$F = [U - kT \ln(\tau/\tau_0)]/\lambda \qquad (2)$$

where F is the adhesive force, U is the bond energy, kT is the thermal energy, τ_0 is the reciprocal of the natural frequency of oscillation, and λ is a parameter assumed to be equal to the effective rupture length.

The force resolution of the AFM enables it to detect differences on the net charge of biomolecules brought about by a change in the surrounding ionic environment. In another AFM study, Bhushan et al. [18] were able to demonstrate that the adhesion between streptavidin and biotin increases with the pH of the buffer solution, as shown in Fig. 5. This is related to the charge interactions between the two biomolecules. Streptavidin has an isolectric point of pH 5.5, such that it has a net positive charge at pH 4.4 and net negative charge at pH 7.4 and 9.1. The net negative charge in streptavidin facilitates adhesive interactions with the biotin molecule.

Force-distance curves can also be taken over a predefined area divided into an array of points, where the average adhesive force can be determined. This method is useful when analyzing the adhesive properties of next-generation biomaterials such as block copolymers, which have heterogeneous surface properties. Their morphology is a function of the

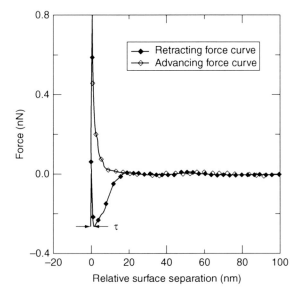

Bioadhesion, Fig. 4 Atomic force microscopy (AFM) force-distance curve between biotinylated bovine serum albumin and streptavidin surface imaged in phosphate buffered saline (PBS) medium (adapted from [17])

composition and molecular weight of the individual blocks, as well as the spatial relationship of the blocks, for instance A-B block copolymers (or diblock) will have a different morphology than A-B-A block copolymers (or triblock), and so on. This variation in the morphology translates to observable differences in the adhesive force. Figure 6 is a summary of the average adhesive force for the interactions between three proteins, namely, fibronectin, BSA, and collagen, on block copolymers composed of poly(methyl methacrylate) (PMMA) and poly(acrylic acid) (PAA), but with different block arrangements (random, diblock, and triblock) [2].

The data in Fig. 6 shows the variation of the adhesive force at two ionic environments, pH 7.4 (PBS) and pH 6.2, where it is seen that the liquid environment affects the adhesive force. At both buffer media, the highest average adhesive force was observed in the triblock copolymer (PAA-b-PMMA-b-PAA, where "b" denotes block copolymer), and the lowest force was obtained from the PMMA reference. The higher average adhesive force in the triblock copolymer surface can be due to greater ordering on its surface, which would expose more PAA-rich areas than it would for the diblock or random copolymer.

The average adhesive forces measured at pH 6.2 are higher throughout the entire series, relative to the PBS medium (pH 7.4) data. This is attributed to higher repulsion at pH 7.4 (or conversely, increased attractive forces at the lower pH). Chemical interactions between the protein and the polymer surface are influenced by the surface charges of the protein and the polymer itself. The amine groups of the proteins are protonated at both pH 7.4 and 6.2. But these proteins carry a net negative charge when immersed in either the pH 7.4 or 6.2 buffer media, since the isoelectric points of fibronectin, BSA, and collagen are all lower than 6.2 [2]. There are more negative charges on the protein surface at pH 7.4 than at pH 6.2. Also, the acrylic acid chains are ionized (to acrylates) during the adhesive force mapping experiment as it is immersed in aqueous medium. The presence of negative charges on both the tip and sample surface leads to repulsive interactions. As illustrated in Fig. 7, the reduced repulsion between the acrylate groups in the acrylic acid and the smaller amount of surface negative charges in fibronectin, BSA and collagen at the lower pH increases the measured adhesion between the two surfaces [2]. At the higher pH (7.4), the proteins carry a greater net negative charge. Since the acrylic acid in the block copolymer is present in its ionized form, the repulsion from the negative charges present on both the protein and the polymer surface is more significant at pH 7.4 than at pH 6.2. This accounts for the higher adhesive force measured at pH 6.2. The results demonstrate how the AFM is able to measure the effect of the ionic environment on bioadhesion.

Macroscale

Standard ASTM characterization procedures, such as the peel test (ASTM D903-98), shear strength (ASTM D-3654), and shear adhesion failure temperature (SAFT, ASTM D4498-00), are conducted on interfaces where an adhesive is applied to the biomaterial. These tests are applicable to systems such as fibrin glues used in surgical operations and pressure-sensitive adhesives (PSA) intended for transdermal drug-delivery applications. Figure 8 shows examples of macroscale adhesion data on pressure-sensitive adhesives. The adhesive system investigated was poly(dimethylsiloxane) (PDMS). It was loaded with an organoclay based on montmorillonite (MMT) to form a composite material. The addition of the

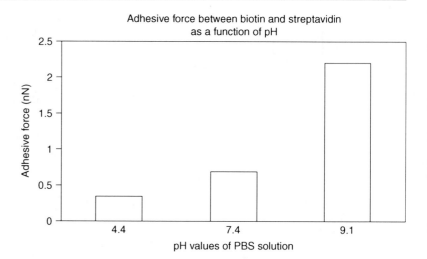

Bioadhesion, Fig. 5 Effect of pH on the adhesion between streptavidin and biotin as measured by AFM (adapted from [18])

Bioadhesion, Fig. 6 AFM data on the average adhesive force for the interaction between block copolymers containing PMMA/PAA with the proteins fibronectin, BSA, and collagen (attached to functionalized AFM tips) in pH 7.4 (PBS) and pH 6.2 buffer media, with data for PMMA as a reference [2]

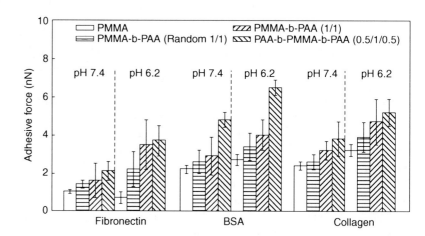

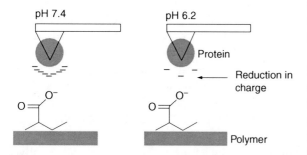

Bioadhesion, Fig. 7 Schematic illustrating the effect of pH on protein-block copolymer interactions (focusing on the effect of PAA, which is ionized at pH 7.4 and 6.2), showing negative charge reduction at lower pH conditions, which leads to decreased repulsion [2]

organoclay was shown to improve the shear strength and the SAFT, with a minimal reduction in the peel strength [19].

The flow-through method, which measures the flow rate needed to remove a bioadhesive-coated sphere, is a common macroscale experiment suited for the characterization of mucoadhesion of potential drug-delivery systems. A widely used biophysical assay involves the monitoring of molecular weight change through the variation in the sedimentation coefficient as measured by the analytical ultracentrifuge [20]. The Wilhelmy plate experiment is another technique for evaluating bioadhesion at the macroscale. Instead of having a liquid medium such as water (as used in

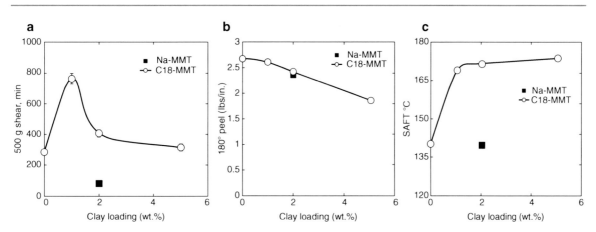

Bioadhesion, Fig. 8 Macroscale testing of bioadhesion. (**a**) Shear strength versus clay loading in pressure-sensitive adhesive (PSA) loaded with an organosilicate nanocomposite (C18-MMT); (**b**) 180° peel strength versus clay loading in the PSA; (**c**) shear adhesion failure temperature (SAFT) versus clay loading. Na-MMT denotes the pure clay used as the reference material [19]

surface tension measurements), it is replaced with either natural or synthetic mucus and the plate is coated with the polymer of interest [6, 14]. This method is useful for determining interfacial forces between mucus and polymer, as well as for the investigation of changes in bioadhesion properties over time.

Cross-References

▶ Bioadhesives
▶ BioPatterning

References

1. Keselowsky, B.G., Collard, D.M., Garcia, A.J.: Surface chemistry modulates focal adhesion composition and signaling through changes in integrin binding. Biomaterials **25**, 5947–5954 (2004)
2. Palacio, M.L.B., Schricker, S.R., Bhushan, B.: Bioadhesion of various proteins on random, diblock and triblock copolymer surfaces and the effect of pH conditions. J. R. Soc. Interface **8**, 630–640 (2011)
3. Legeay, G., Poncin-Epaillard, F.: Surface engineering by coating of hydrophilic layers: bioadhesion and biocontamination. In: Possat, W. (ed.) Adhesion: Current Research and Applications. Wiley, Weinheim (2005)
4. Yang, X., Robinson, J.R.: Bioadhesion in mucosal drug delivery. In: Okano, T. (ed.) Biorelated Polymers and Gels. Academic, San Diego (1994)
5. Burkett, J.R., Wojtas, J.L., Cloud, J.L., Wilker, J.J.: A method for measuring the adhesion strength of marine mussels. J. Adhes. **85**, 601–615 (2009)
6. Buckton, G.: Interfacial phenomena in drug delivery and targeting. Harwood Academic, Chur (1995)
7. Spotnitz, W.D., Burks, S.: Hemostats, sealants, and adhesives: components of the surgical toolbox. Transfusion **48**, 1502–1516 (2008)
8. van Noort, R.: Introduction to Dental Materials, 3rd edn. Elsevier, London (2007)
9. Bhushan, B.: Introduction to Tribology. Wiley, New York (2002)
10. Waite, J.H.: Nature's underwater adhesive specialist. Int. J. Adh. Adhes. **7**, 9–14 (1987)
11. Asmussen, E., Hansen, E.K., Peutzfeldt, A.: Influence of the solubility parameter of intermediary resin on the effectiveness of the Gluma bonding system. J. Dent. Res. **70**, 1290–1293 (1991)
12. Hozbor, M.A., Hansen, W.P., McPherson, M.: Plasma cleaning of metal surfaces. Precis. Clean. **2**, 46 (1994)
13. Bhushan, B., Tokachichu, D.R., Keener, M.T., Lee, S.C.: Morphology and adhesion of biomolecules on silicon based surfaces. Acta Biomater. **1**, 327–341 (2005)
14. Bhushan, B., Gupta, B.K.: Handbook of Tribology: Materials, Coatings and Surface Treatments. McGraw-Hill, New York (1991)
15. Bhushan, B.: Springer Handbook of Nanotechology, 3rd edn. Springer, Heidelberg (2010)
16. Bhushan, B.: Nanotribology and Nanomechanics, 3rd edn. Springer, Heidelberg (2011)
17. Lee, G.U., Kidwell, D.A., Colton, R.J.: Sensing discrete streptavidin-biotin interactions with atomic force microscopy. Langmuir **10**, 354–357 (1994)
18. Bhushan, B., Tokachichu, D.R., Keener, M.T., Lee, S.C.: Nanoscale adhesion, friction and wear studies of biomolecules on silicon based surfaces. Acta Biomater. **2**, 39–49 (2006)
19. Shaikh, S., Birdi, A., Qutubuddin, S., Lakatosh, E., Baskaran, H.: Controlled release in transdermal pressure sensitive adhesives using organosilicate nanocomposites. Ann. Biomed. Eng. **35**, 2130–2137 (2007)
20. Harding, S.E.: Trends in mucoadhesive analysis. Trends Food Sci. Technol. **17**, 255–262 (2006)

Bioadhesives

Scott R. Schricker
College of Dentistry, The Ohio State University,
Columbus, OH, USA

Synonyms

Clinical adhesives; Medical adhesives

Definition

Materials that are designed to bond interfaces for clinical applications.

Introduction

Bioadhesives can be broadly defined as materials that bond interfaces in biological or clinical situations. There are many well-known examples of biological adhesion including mussels or barnacles sticking to rocks, gecko lizards that walk on ceilings, and spider webs. Some examples of bioadhesion have detrimental consequences: Barnacles adhering to ship hulls is a significant problem for maritime operations and bacteria will form adhesive biofilms that resist treatment resulting in significant harm to human health. These topics have been widely covered in the literature and many excellent reviews on these topics have been written.

The main subject area of this entry is ▶ bioadhesives that are used in clinical situations to improve human health. While the proteins involved in mussel adhesion have been studied for clinical applications, most of the adhesives used clinically are synthetically derived. A notable exception to this is fibrin glue which will be discussed in detail. One of the reasons for this is that synthetic adhesives can offer a level of consistency and reproducibility that is not always possible from biologically derived materials. This entry will cover synthetic adhesives, fibrin glue and discuss some recent developments in nanostructured adhesives.

Applications and Clinical Problems

Bioadhesives are used in many situations: to adhere biological interfaces, a biological interface to a material or two material interfaces. Many types of materials are used as bioadhesives, and they can have a wide range of mechanical and physical properties depending on the application. For instance, methacrylate adhesives for dental and orthopedic applications are permanent and nondegradable, whereas most soft tissue adhesives such as cyanoacrylates and fibrin glue are temporary and degradable, while adhesives for maxillofacial prostheses and dentures are temporary and the adhesives must allow for strong bonding until removal for daily cleaning is desired. In addition, bioadhesives must be resistant to biological attack such as hydrolysis, pH, temperature cycles, and enzymatic attack.

Adhesion is a multifaceted phenomenon involving mechanical locking and chemical interactions. For most hard tissue applications, mechanical locking is the dominant interaction that provides interfacial stability. The classic example is a dental amalgam that is bonded to the tooth almost exclusively by mechanical forces. Proper tooth preparation allows for a macroscale mechanical locking and the malleable nature of the uncured amalgam allows for microscale mechanical locking with the tooth surface. The ability for an uncured adhesive to flow in-between the crevices and ridges of biological surface is important to establish a stable interface. Low viscosity is an important requirement for uncured bioadhesives and is the reason that most bioadhesives are lower-molecular-weight chemicals that are cured in vivo.

Chemical interactions play a significant role in determining the interfacial strength of a bioadhesive. The chemical composition of a bioadhesive will determine its ability to wet the surface of the substrate as well as its ability to form bonds with the substrate. Establishing a good mechanical interface requires the adhesive to wet the substrate and poor wetting can result in voids/defects at the interface. As most biological surfaces are hydrophilic, low contact angle adhesives will flow into the crevices and ridges better than a hydrophobic adhesive. This will allow for good mechanical locking on the surface. Chemical interactions such as hydrogen bonding, polar interactions, ionic bonding, and covalent bonding will also play a role in determining the bond strength of a bioadhesive.

Recent Trends

Adhesives in Dentistry

Bioadhesives play a very important role in many aspects of clinical dentistry, binding restorations to tooth structure, repair of restorations, binding of orthodontic appliances (braces) to teeth, and retention of dentures and maxillofacial prostheses. In restorative dentistry, it is estimated that 70% of procedures are for failed restorations. Most of these are directly or indirectly due to poor interfacial bonding between the restorative and the tooth structure. For this reason, a great deal of research has focused on improving the adhesive interface between a restorative and the tooth structure. As this constitutes the vast majority of dental adhesion research, this will be the focus of the following discussion. It is important to note that the adhesives described can also be used for orthodontic applications and the repair of restorations or dental appliances. The chemistry and principles of adhesion are the same and often only minor changes in formulation are made in order to adapt an adhesive to a different dental application.

Most of the adhesives used in clinical dentistry are based on methacrylate polymers and oligomers. A generic outline of a restorative procedure will illustrate how a dental adhesive is used. The clinician will grind away the carries (tooth decay) and expose the underlying dentin. The exposed dentin is coated with ground dentin, known as the "smear layer" that will prevent good bonding. The smear layer is etched away with a strong acid, normally phosphoric acid. The dental adhesive is then placed on the dentin and polymerized with visible light. An uncured composite restorative is then placed on the adhesive and polymerized by visible light. Current dental adhesives incorporate phosphorous-based acid groups that can simultaneously etch the smear layer and bond to the dentin, thus eliminating a step. A visible light–cured adhesive is used in this example; however, if the restoration is preformed to fit the defect, such as crown, precured composite, or a veneer, then a chemically cured adhesive is used. The light cannot penetrate to the adhesive layer and the adhesive must be polymerized while in contact with the cured restorative to achieve the full adhesive potential.

Methacrylate systems offer many advantages over other curing chemistries and meet most of the requirements for clinical dentistry. As previously mentioned, it is desirable to start with lower-molecular-weight precursors and cure them in vivo. Methacrylate monomers can be cured by visible light within a minute, a longer timeframe is not efficient from a practitioner's standpoint and not well tolerated by the patient. Functional groups are well tolerated in methacrylate polymerizations and hydroxyl-, carboxylic-, and phosphorous-based acid groups can be incorporated as part of dental adhesives. Current formulations of methacrylate monomers and oligomers are stable and have a good shelf life, curing on command with visible light. In addition, most methacrylate polymers and monomers are well tolerated from a biological standpoint. While some concerns about these systems will be addressed later, most methacrylate monomers are considered nontoxic and acceptable for human use.

Early generation dental adhesives are characterized by the use of hydrophilic monomers such as triethylene glycol dimethacrylate (TEGDMA) and 2-hydroxyethyl-methacrylate (HEMA). Often a hydrophobic monomer such as (BisGMA) or fillers are added to enhance the mechanical strength of the adhesive. Subsequent generations have incorporated ionic groups such as carboxylic acid, phosphinic acid, and phosphonic acid to enhance bonding to the tooth structure. These acidic groups bind calcium and adhere to calcium-rich surfaces. Enamel and dentin have high levels of the calcium-based mineral hydroxyapatite, 40–50% and over 90% respectively. So these acidic groups contribute significantly to the bonding to the tooth surface. The varied chemistry of adhesive monomers is due to the desire to control properties such as viscosity, mechanical strength, and bonding to the organic components of the tooth structure. The structures and formulations of many commercial and experimental adhesives are found in the literature, and examples of adhesive components are shown in Fig. 1 [1]. The monomers and initiators are often dissolved in a volatile organic solvent such as ethanol or acetone or a solvent-water combination. The solvent aids in allowing the monomers to penetrate into the dentin to achieve maximum bonding.

Dental adhesives and their monomers are considered bioacceptable and are approved for clinical use. However, recent studies in cell culture and animal models have raised some concerns about the unpolymerized monomers and breakdown products that can elute from the adhesive. One of the concerns involves these chemicals acting as endocrine

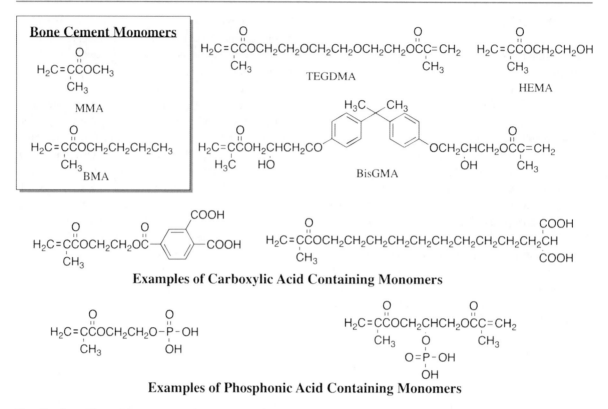

Bioadhesives, Fig. 1 Monomers used in dental adhesives and bone cements

disruptors. Bisphenol A, a known endocrine disruptor, is a precursor to BisGMA and is often detectable in BisGMA-based materials in trace amounts. Additionally, some of the visible light curable initiators such as camphorquinone have been implicated as endocrine disruptors. Unpolymerized methacrylate monomers, in particular TEGDMA and HEMA, are known to induce apoptosis in cell culture. This suggests that as unpolymerized monomer leaches from the adhesive, it will be toxic to the surrounding tissue. In addition, some of these same monomers have been shown to cause genetic damage in cell lines [2–4]. While these are areas of concern, no human studies have demonstrated any significant negative impact on human health. However, as the definition of biocompatibility evolves, such concerns may need to be addressed.

Bone Cements

The chemistry and use of bone cements are similar to dental adhesives. Bone cements are primarily used to affix implants into live bone. Bone implants are normally made from titanium, ceramic, or derived from a human source. None of these implants are malleable and will not form a mechanical bond to the bone. Stress will be placed on the implant bone interface soon after placement and the bone will not regrow quickly enough to establish a good interface to handle the stress. Unpolymerized methacrylate bone cements will mold to the implant and bone upon implantation. Polymerization will harden the cement and form a mechanical lock.

Bone cements are formed by mixing liquid and solid components which initiates a polymerization reaction. The liquid is a mixture of methacrylate monomers, typically methyl methacrylate (MMA) and butyl methacrylate (BMA) along with a polymerization accelerator. The solid component is a fine powder of polymerized methyl methacrylate polymers or copolymers. The most common powder composition is a mixture of methyl methacrylate homopolymer and a copolymer of methylmethacrylate and styrene. The powder also contains a polymerization initiator and upon mixing, the accelerator and initiator start the curing process. In addition, the powder may contain fillers to improve mechanical properties or antibiotics to prevent infection. Upon mixing, the cement has a doughy consistency allowing the clinician to place and mold the cement. The set cement is an interpenetrating network of the powder and polymerized monomers. This provides good mechanical properties that are needed to withstand the stresses across the interface.

The biological properties of the bone cement are considered to be acceptable. The major drawback is that the polymerization is exothermic and the heat generated can cause tissue damage. There are some concerns about the unpolymerized monomers being cytotoxic, but the long history and extensive use of these materials provides good evidence that they are safe [5, 6].

Surgical and Wound Healing Applications

Clinical procedures often require the adhesion of tissue to close a wound or seal an incision after surgery. The applications can range from closing a wound on the skin to closing the small or large intestine after abdominal surgery. Suturing or stitching is a common technique to close wounds that is both labor intesive and requires skill and training. Surgical adhesives offer a quick, relatively simple alternative to suturing that is highly effective and in some cases superior to suturing. In some cases, sealing the wound immediately is important: For instance, athletes are typically not allowed to compete while they are bleeding. Adhesives offer a quick and convenient route to closing a wound or stopping bleeding.

Cyanoacrylates have been widely used as tissue adhesives particularly for external applications. The chemical structures of the common alkyl derivatives and the polymerization process are shown in Fig. 2. Cyanoacrylates are commonly known as "superglue" and are characterized by a rapid set and a strong adhesive bond. The double bond is electron deficient due to the attached cyano and ester groups. The result is that weak nucleophiles such as water can initiate the polymerization. Hydroxyl, amine, and thiol groups that are often found on tissue can also initiate polymerization. Upon contact with blood or tissue, polymerization will start. This is one of its attractive features; no initiator is needed and the rapid polymerization will stop bleeding and quickly seal a wound.

The degradation chemistry of cyanoacrylates is the biggest barrier that prevents broader clinical use. Hydrolysis of polycyanoacrylates will result in the release of formaldehyde and cyanoacetate.

Both by-products are considered toxic and as a result, internal use is limited. The chemistry of the cyanoacrylates can be manipulated to reduce this toxicity. As the alkyl group on the cyanoacrylates becomes longer, the polymer is more hydrophobic and the rate of degradation is slower. While the by-products are still toxic, the overall effects are diminished because the rate of release is slow. Toxicity is a combination of the agent and concentration, so slower release will result in diminished toxicity to the tissue. Another result of a longer alkyl chain is that the polymerization rate is reduced. So the handling properties and degradation rate can be manipulated by mixing different ratios of cyanoacrylate monomers.

In addition to external wound closure, there are several clinical applications for cyanoacrylates. A cyanoacrylate-based product known as InteguSeal is used to seal surgical sites to prevent infections. Upon curing, a barrier is formed that prevents bacteria from migrating to the infection site. Typically, wound sites are treated with iodine solution to reduce infection and the application of cyanoacrylate after the iodine solution forms a barrier that prevents the iodine sterilization solution from evaporating and increases it effectiveness. In one clinical study, the rates of infection dropped to 53% compared to 68.7% with iodine alone [7]. Cyanoacrylates have also been used in ophthalmic and cardiovascular procedures, and it is being explored as a sealant for gastrointestinal surgery.

Fibrin Glue

Fibrin glue is a surgical sealant that is based on the chemistry and components of the blood-clotting cascade. The entire clotting cascade is very complex and is controlled by multiple feedback mechanisms. Figure 3 shows the last few biochemical steps in clotting that are relevant to fibrin glue. In commercial formulations, fibrinogen and factor XIII are isolated from whole blood as are the thrombin components. In most cases, the fibrinogen is isolated from human blood and the thrombin can be isolated from a human or bovine source. Batches of the components are often pooled from multiple sources to reduce variability and make a more consistent product. There are other factors involved in the setting process that are isolated with the fibrinogen and thrombin; they are usually referred to as fibrinogen and thrombin components. A solution of the fibrinogen is mixed with a thrombin

Bioadhesives,
Fig. 2 Structure and
polymerization of
cyanoacrylates

Structure of Cyanoacrylates

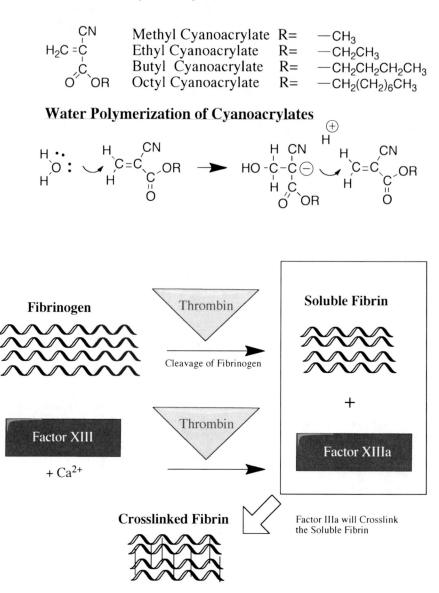

Methyl Cyanoacrylate	R=	$-CH_3$
Ethyl Cyanoacrylate	R=	$-CH_2CH_3$
Butyl Cyanoacrylate	R=	$-CH_2CH_2CH_2CH_3$
Octyl Cyanoacrylate	R=	$-CH_2(CH_2)_6CH_3$

Water Polymerization of Cyanoacrylates

Bioadhesives,
Fig. 3 Overview of the final
steps of the clotting process
and the basic steps involved
the formation of fibrin glue.
Many more steps and factors
are involved in the clotting
process, but this is relevant to
fibrin glue

solution containing calcium salts to initiate the biochemical cascade and thus form the fibrin glue.

Fibrin glue was developed from the observation that fibrinogen itself could be used to stop bleeding and help to close wounds. As it became feasible to isolate more components of blood, thrombin was added to enhance the effectiveness of the glue. Because the components of the glue are naturally derived, it is well tolerated by the body. There is no inflammatory response, the fibrin glue can be resorbed by the body and the breakdown components are nontoxic. The major drawback is the potential of contamination from the biological source. Viruses and other pathogens can be transmitted along with fibrinogen and thrombin components. While no evidence of serious contamination has been documented and many decontamination protocols are utilized, the potential for contamination remains. Prions represent an area of potential concern and also highlight the unknown factors that may be responsible for disease transmission.

Until a few decades ago, prions were not known as infectious agents, but their ability to cause disease is well documented. Bovine spongiform encephalopathy (BSE) could potentially be transmitted from bovine sources, as could the human variant Creutzfeldt–Jakob disease. The fact that sources are often pooled to produce the components can increase the risk of transmission.

Fibrin glue was approved for clinical use in the United States in the late 1990s and has been approved for use in Europe and elsewhere for almost two decades prior to that. Many surgical applications use fibrin glue such as cardiovascular surgery, neurosurgery, reconstructive surgery, and gastrointestinal surgery. Though it is beyond the scope of this entry, it is worth noting that fibrin glue is considered useful as a drug delivery carrier and a tissue engineering scaffold. In addition to previously stated convenience of tissue adhesives, there is documented clinical benefit to the use of fibrin glue. One of the complications of lung surgery is the leakage of air from a closure site on the lung. In a clinical study, one patient group had the surgical site closed with a suture and in the second group, the site was closed with a suture plus fibrin glue. The percentage of patients without air leakage for the first group was 34% while fibrin glue group was 61%. In another example, fibrin glue was compared with other agents, based on collagen, cellulose, or gelatin, to stop emergency bleeding following surgery. After 5 min, 92.6% of the fibrin glue patients had stopped bleeding compared to 12.4% of patients using the other methods [8, 9].

Bioadhesives Derived from Nanotechnology

Ophthalmological Applications

Dendrimers have been utilized to develop bioadhesives for ophthalmologic applications and represent one of the few examples of nanotechnology applied to adhesives for clinical applications. Dendrimers are a class of highly branched polymers that can adopt a spherical conformation at higher generations. Dendrimers are characterized by being unimolecular, having a well-defined architecture, and having a highly controlled synthesis. The unimolecular and defined architecture allows for structure-property relationships to be well defined. So, by changing the chemistry of the dendrimers, the resulting architecture will be modified and lead to different properties. Due to its spherical nature, the terminal groups will

determine the interactions of the dendrimer with solvent, other dendrimers, surfaces, and biological moieties. Complete or partial modification of these terminal groups will have a significant impact on dendrimer properties. The core group and branching units will determine the arrangement and architecture of the terminal groups as well as the relative stiffness or the dendrimer itself. While these principles are generally true, dendrimers are not rigid spheres but rather molecules that are flexible and can react to its external environment. So the branching units can potentially contribute to the solubility of a dendrimer, for example.

The advantageous properties of dendrimers have led Grinstaff and coworkers to develop ophthalmologic sealants with them. The dendrimers were based on polyester chemistry that could hydrolytically degrade over time. The terminal groups were functionalized with methacrylate groups that allowed them to be visible light cured. Animal models have demonstrated that these materials serve well as adhesives for eye surgery. The investigators note little inflammation and high success rate comparable to standard suturing technique. The dendrimer adhesives also provide advantages over suturing. Suturing can further damage the delicate tissue of the eye and the visible light–cured adhesive is more rapid than suturing. Additionally, the Grinstaff group has developed dendrimers that have peptides as terminal groups instead of methacrylates. The peptides can chemically link to the tissue of the eye and avoids any potential damage from the heat and light of methacrylate curing [10, 11].

Nanostructured Biological Adhesives

The application of nanotechnology to bioadhesives is a growing but relatively new area. Other than the dendrimer-based adhesives mentioned previously, nanotechnology has not made significant inroads to clinical bioadhesives. One aspect of bioadhesion that has incorporated nanotechnology is the creation of surfaces that can control cell and protein adhesion. It has been recognized that micro- and nanotopography can affect cell and protein adhesion in vitro. While much of the nanotopography has focused on surface roughness, there is a growing body of literature that is examining the role of well-defined nanostructured surfaces in controlling cell and protein adhesion. In particular, self-assembled monolayers [12] and

Bioadhesives, Table 1 Summary of the measured adhesive force

Material	Adhesive force (nN)					
	Fibronectin (pH 7.4)	Fibronectin (pH 6.2)	BSA (pH 7.4)	BSA (pH 6.2)	Collagen (pH 7.4)	Collagen (pH 6.2)
PMMA	1.0 ± 0.1	0.7 ± 0.3	2.2 ± 0.2	2.7 ± 0.3	2.4 ± 0.2	3.2 ± 0.2
PMMA-co-PAA (1/1)[a]	1.4 ± 0.2	2.2 ± 0.9	2.6 ± 0.6	3.4 ± 0.7	2.6 ± 0.4	3.9 ± 0.8
PMMA-b-PAA (1/1)	1.6 ± 0.9	3.5 ± 1.3	2.9 ± 1.0	4.0 ± 0.8	3.2 ± 0.5	4.7 ± 1.1
PAA-b-PMMA-b-PAA (0.5/1/0.5)	2.1 ± 0.5	3.7 ± 0.8	4.8 ± 0.4	6.5 ± 0.4	3.8 ± 0.9	5.2 ± 0.7
PMMA-co-PHEMA (1/1)[a]	1.6 ± 0.3	3.4 ± 0.4	1.6 ± 0.4	4.2 ± 0.6	3.6 ± 0.8	5.1 ± 0.6
PMMA-b-PHEMA (1/1)	3.0 ± 0.2	4.4 ± 0.9	3.8 ± 0.4	3.7 ± 0.7	3.9 ± 0.7	5.0 ± 0.6
PMMA-b-PHEMA-b-PMMA (0.5/1/0.5)	4.2 ± 0.3	4.6 ± 1.0	3.3 ± 0.2	3.7 ± 0.4	4.0 ± 0.8	4.5 ± 1.0

[a]Random copolymer

nanopatterned surfaces [13, 14] have been used to control cell and protein adhesion.

Block copolymers provide a platform for creating a variety of nanostructures. While their use for many different applications has been well established, recent work has demonstrated that block copolymer morphology can be used to control protein and cell adhesion on a surface. Work by Bhushan and Schricker [15, 16] has demonstrated that varying surface nanomorphology through block copolymer design can affect protein adhesion. As shown in Table 1, varying the arrangement of the blocks while keeping the composition constant can affect protein adhesion. The protein adhesion is measured by covalently attaching a protein to an AFM tip and measuring the amount of force required to pull the tip away from the surface. For the polymethyl methacrylate (PMMA) and polyacrylic acid (PAA) block copolymers, an effect in the adhesion of fibronectin and bovine serum albumin (BSA) is observed at low pH but not neutral. However, for the PMMA and poly(2-hydroxy-ethyl-methacrylate) (PHEMA), the effect is observed at neutral pH but not at low pH.

Work by Cooper-White et al. [17, 18] has also demonstrated the use of block copolymers in controlling cellular behavior. Block copolymers of polystyrene and polyethylene oxide have been utilized to effect cellular and protein adhesion. In particular, as the domain sizes of the block changed, the cellular adhesion was affected.

Cross-References

▶ Bioadhesion

References

1. Van Landuyt, K.L., et al.: Systematic review of the chemical composition of contemporary dental adhesives. Biomaterials **28**, 3757–3785 (2007)
2. Janke, V., von Neuhoff, N., Schlegelberger, B., Leyhausen, G., Geurtsen, W.: TEGDMA causes apoptosis in primary human gingival fibroblasts. J. Dent. Res. **82**, 814–818 (2003)
3. Wada, H., Tarumi, H., Imazato, S., Narimatsu, M., Ebisu, S.: In vitro estrogenicity of resin composites. J. Dent. Res. **83**, 222–226 (2004)
4. Schweikl, H., Spagnuolo, G., Schmalz, G.: Genetic and cellular toxicology of dental resin monomers. J. Dent. Res. **85**, 870–877 (2006)
5. Kuehn, K.D., Ege, W., Gopp, U.: Acrylic bone cements: composition and properties. Orthop. Clin. North Am. **36**, 17–28 (2005)
6. Smith, D.C.: The genesis and evolution of acrylic bone cement. Orthop. Clin. North Am. **36**, 1–10 (2005)
7. Wilson, S.E.: Microbial sealing: a new approach to reducing contamination. J. Hosp. Infect. **70**(Suppl 2), 11–14 (2008)
8. Jackson, M.R.: Fibrin sealants in surgical practice: an overview. Am. J. Surg. **182**, 1S–7S (2001)
9. Spotnitz, W.D.: Fibrin sealant: past, present, and future: a brief review. World J. Surg. **34**, 632–634 (2010)
10. Grinstaff, M.W.: Designing hydrogel adhesives for corneal wound repair. Biomaterials **28**, 5205–5214 (2007)
11. Mintzer, M.A., Grinstaff, M.W.: Biomedical applications of dendrimers: a tutorial. Chem. Soc. Rev. **40**, 173–190 (2010)
12. Raynor, J.E., Capadona, J.R., Collard, D.M., Petrie, T.A., Garcia, A.J.: Polymer brushes and self-assembled monolayers: versatile platforms to control cell adhesion to biomaterials (Review). Biointerphases **4**, FA3–FA16 (2009)
13. Lu, J., Rao, M.P., MacDonald, N.C., Khang, D., Webster, T.J.: Improved endothelial cell adhesion and proliferation on patterned titanium surfaces with rationally designed, micrometer to nanometer features. Acta Biomater. **4**, 192–201 (2008)
14. Dalby, M.J., Gadegaard, N., Wilkinson, C.D.: The response of fibroblasts to hexagonal nanotopography fabricated by electron beam lithography. J. Biomed. Mater. Res. A **84**, 973–979 (2008)

15. Schricker, S., Palacio, M., Thirumamagal, B.T., Bhushan, B.: Synthesis and morphological characterization of block copolymers for improved biomaterials. Ultramicroscopy **110**, 639–649 (2010)
16. Palacio, M., Schricker, S., Bhushan, B.: Bioadhesion of various proteins on random, diblock, and triblock copolymer surfaces and the effect of pH conditions. J.R. Soc. Interface **8**, 630–640 (2011)
17. George, P.A., Donose, B.C., Cooper-White, J.J.: Self-assembling polystyrene-block-poly(ethylene oxide) copolymer surface coatings: resistance to protein and cell adhesion. Biomaterials **30**, 2449–2456 (2009)
18. George, P.A., Doran, M.R., Croll, T.I., Munro, T.P., Cooper-White, J.J.: Nanoscale presentation of cell adhesive molecules via block copolymer self-assembly. Biomaterials **30**, 4732–4737 (2009)

Biocalorimetry

▶ Nanocalorimetry

Bioderived Smart Materials

Vishnu-Baba Sundaresan[1], Stephen Andrew Sarles[2] and Donald J. Leo[3]
[1]Mechanical and Nuclear Engineering, Virginia Commonwealth University, Richmond, VA, USA
[2]Mechanical Aerospace and Biomedical Engineering, University of Tennessee, Knoxville, TN, USA
[3]Mechanical Engineering, Center for Intelligent Material Systems and Structures, Virginia Tech, Virginia Polytechnic Institute and State University, Arlington, VA, USA

Definition

Bioderived smart materials are a category of ionic active materials that are fabricated from biological macromolecules and utilize the ion transport properties of cell membranes to couple multiple physical domains (chemical, electrical, mechanical, and optical). Owing to this coupling exhibited by biological macromolecules, bioderived smart materials are used as actuators, sensors, and energy harvesting applications.

Cell Membrane Proteins: Transporter Macromolecules

The cell membranes of plant and animal cells are host to cholesterol and a variety of proteins that provide structural rigidity and serve as anchor points for cell attachment. In addition, some proteins serve as transporters and allow the cell to exchange ions and neutral molecules between the cell cytoplasm and its surroundings. The transport of species through a protein transporter in the cell membrane is driven by biological processes that have inspired the development of novel sensing, actuation, and energy harvesting concepts. This entry will provide the technical background on biological processes that have inspired the development of *bioderived smart materials* and discuss the system level concepts, fabrication, and characterization of these material systems. Specific emphasis will be given to novel actuation systems, sensing platforms, and recent advances in fabrication of these materials systems.

Transporter proteins are biological macromolecules embedded in cell membranes that convert electrical, mechanical, or chemical stimuli into ion transport and regulate chemoelectrical potentials across the membrane. It is this stimuli response of protein transporters that makes it viable for the development of a bioderived smart material using biological ion transport processes. The transport of ions through the transporter protein occurs via one of the following processes:

1. Simple diffusion
2. Facilitated diffusion
   a. Voltage-gated diffusion
   b. Ligand-gated diffusion
   c. Mechanically gated diffusion
3. Active transport

These ion transport processes are illustrated in Fig. 1 and are used to classify protein transporters found in cell membranes. In simple diffusion, the protein transporter behaves like an open channel and allows diffusion of a species along a concentration gradient across the membrane. In facilitated diffusion, the protein transporter uses a gating signal such as light, electrical field, mechanical stretch (or) an analyte to open and conduct an ion or neutral molecule across the membrane. In active transport, the protein transporter uses the energy from biosynthetic molecules such as adenosine triphosphate (ATP) to transport

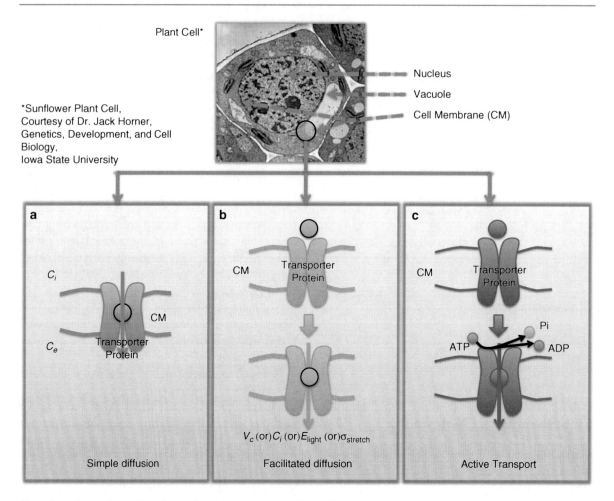

Bioderived Smart Materials, Fig. 1 Protein transporters in plant cell membrane

ions across the membrane. Since active transporters expend chemical energy available in triphosphates, ions are transported across the membrane against an existing concentration gradient. In addition to ion transport, active transporters establish the necessary chemoelectrical gradients required for facilitated diffusion across the membrane and provide the driving force for long-range ion and solute transport in plant and animal cells.

The mechanism for ion transport through the protein transporter is explained by its molecular composition and structural arrangement. The structural composition of the protein transporter is unique to a plant (or) animal cell membrane and hence transporters are unique in their ion transport function. The diversity of plant and animal cell membranes and the uniqueness of protein transporters found in the cell membranes lead to a large number of protein transporters that can be used as an active component in a bioderived smart material. Protein transporters ion proteins are formed from amino acids that are connected by a polypeptide bond. There are some 20 amino acids that form the majority of the protein transporters and have clearly identified C-C backbone and polar, nonpolar, and aromatic side chains. The linear network of polypeptide bonds has a well-defined two-dimensional structure and is woven (folded) into a three-dimensional structure in the cell membrane. The arrangement of amino acid side chains in the two-dimensional and three-dimensional structures contained within the protein transporters leads to hydrogen bonding, van der Waals, and steric interactions in response to external stimuli. This response in structural changes for an applied optical, electrical,

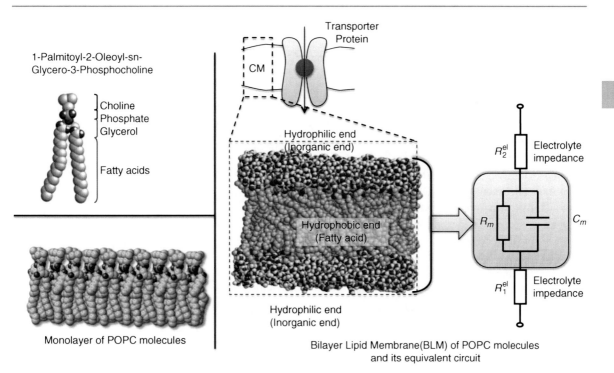

Bioderived Smart Materials, Fig. 2 l-palmitoyl-2-oleoyl-sn-glycero-3-phosphocholine molecule, monolayer, bilayer, and equivalent electrical circuit

mechanical, or chemical stimuli results in the formation of a conductive pathway for a specific ion to migrate across the membrane and results in an ionic current through the protein transporter and the cell membrane carrying the protein transporter. As a consequence of the ionic current, the protein transporter maintains and regulates the electrical potential across the membrane in response to the applied stimulus.

Bilayer Lipid Membranes: A Breadboard for Bioderived Smart Material Concepts

The fabrication of a bioderived smart material requires a membrane that can serve as a host to transporter proteins and exhibit coupling between multiple physical domains (chemical, electrical, mechanical, and optical). This membrane should have the structural properties of a cell membrane and assume the ion transport function imparted by the transporter protein. In addition, this membrane should be sufficiently modular to accommodate multiple proteins and provide a composite ion transport function. The current state

of the art in bioderived smart materials uses the same macromolecules that form the cell membranes of plant and animal cells. These molecules, referred to as glycerophospholipids (also referred to as phospholipids or lipids), are extracted and purified from cell membranes and are amphiphilic in nature. Phospholipids comprise of an organic alkyl end group with one or few double bonds and a polar end group connected to each other via a glycerol-phosphate group. An example phospholipid molecule 1-*p*almitoyl-2-*o*leoyl-sn-glycero-3-*p*hospho*c*holine (POPC) with an organic palmitoyl- oleoyl- and polar -choline end group is shown in Fig. 2. These molecules, owing to their amphiphilic nature assemble into a bilayer formed from stacked monolayers with the hydrophilic ends of the monolayer facing outside as shown in Fig. 2. This stacked arrangement of hydrophilic-hydrophobic-hydrophilic groups in the bilayer membrane offers high impedance for transport of ions and neutral molecules through the membrane. The bilayer lipid membrane (BLM) measures 6–10 nm in height, spans $10–20/\mu m^2$ area, and serves as the host to various transporter proteins and forms the basic structure of a bioderived smart material. Since various transporter proteins with different ion transport functions can be incorporated in the BLM to provide a composite ionic function, the BLM becomes the breadboard for various smart material concepts. Owing to their biological origin, the BLM with the protein transporter is referred to in this entry as the bioderived membrane.

In order to quantify the ionic current through the bioderived membrane in the presence of the stimulus, it is required to establish a measure of baseline ion transport (current) through the BLM and charge separation (electrical potential) across the membrane without the protein transporter. The conductance of the membrane is represented using a frequency-dependent complex impedance function and the membrane is modeled using electrical equivalents as shown in Fig. 2. The complex impedance is measured using electrical impedance spectroscopy (EIS) and fitted to an equivalent circuit. In the absence of protein transporters in the BLM, the membrane has a high electrical impedance ($\sim O$ 0.1–1 $G\Omega.cm^2$) and a capacitance of $1/\mu F.cm^{-2}$ [1, 2]. In the presence of protein transporters, the membrane assumes the transport properties of the protein and has markedly different conductance states. The impedance of the membrane with the protein in the presence of the stimuli measured using EIS and the conductance states through cyclic voltammetry (CV) and chronoamperometry (CA) establish the incorporation and functioning of the protein transporter in the membrane.

Concepts for Ionic Active Materials Using Bioderived Membranes

The design of smart materials system using protein transporters employs one or several of the previously discussed ion transport processes. In order to understand the energy conversion processes in a protein transporter and its performance as an ionic active material, it is necessary to quantify the work performed by the bioderived membrane for an applied input energy. The electrical work done by bioderived membranes from ion transport is

$$\Delta U_E = \int_0^q E dQ \qquad (1)$$

where E is the electrical field established by the protein and dQ is the charge displacement through the membrane. The electrical field established across the membrane is given by $E = V/t_m$ where V is the transmembrane potential and t_m is the thickness of the membrane. The charge displacement dQ is due to the ionic current i_c through the protein transporter embedded in the membrane and can be obtained from electrochemical measurements on the membrane. The ionic current results in altering the concentration of the ion across the membrane that sets up an additional transmembrane potential V_c give by

$$V_c^\infty = \frac{RT}{zF} \ln\left(\frac{C_e^\infty}{C_i^\infty}\right) \qquad (2)$$

where C_e^∞, C_i^∞ are the concentration of the ion on either sides of the membrane at the end of the process and z is the valency on the transported ion. Similarly, the chemical energy gradient across the membrane due to a concentration gradient is given by

$$\Delta\mu = RT \ln\left(\frac{y_e^\infty}{y_i^\infty}\right) \qquad (3)$$

where y_e^∞, y_e^∞ are the mole fractions of species y on either side of the membrane at steady state. The interconversion between chemical potential and electrical potential across the membrane shown in Eqs. 1, 2, and 3 occur via ionic currents through the protein transporter.

The ionic currents through the bioderived membrane can be written for various transport process from *phenomenological equations* as a function of one or more of the following as shown in Sundaresan et al. [3]

1. Applied transmembrane potential(V)
2. Chemical energy gradient($\Delta\mu$)
3. Chemical energy released from biochemical reaction($\Delta\mu$)

The diffusion ionic current through the membrane of area A and permeability constant P_m in the presence of concentration gradients C_e, C_i, and transmembrane potential V is given by Goldman-Hodgkin-Katz equation as

$$i_{\text{diff}} = -AP_mF\left(\frac{zFV}{RT}\right)\left[\frac{[C]_i - [C]_e e^{-\frac{zFV}{RT}}}{1 - e^{-\frac{zFV}{RT}}}\right]. \qquad (4)$$

The ionic current through the protein transporter via voltage-gated diffusion i_{vg} is given by

$$i_{vg} = 2\psi zekT \sqrt{[C]_e[C]_i} \frac{A_p}{\in d} \sinh\left(\frac{e(V - Vc)}{2kT}\right), \quad (5)$$

where ψ is the voltage gating coefficient, z is the valency of the transported species, e is the charge on the ion, A_p is the area of the ion conducting channel, and d is the length of the pore in the protein transporter. The ionic current through the protein transporter via active transport i_{pump} is given by

$$i_{pump} = Me\lambda \tanh\left(\frac{F[-V - V_{ATP} + V_c]}{2kT}\right), \quad (6)$$

where M is the pump density in the membrane, z is the charge of the pumped ion, λ is the sum of the forward and backward reaction rates in the pump, V_c is the electrochemical gradient for the pumped ion given by Eq. 2, and k is Boltzmann constant.

The choice of transporter protein in bioderived materials depends on the nature of the stimulus to be detected for sensing, stimulus that is available for actuation and ambient energy source available for harvesting. This choice is also dependent on the ability to extract a protein transporter from a cell membrane and the ability to reconstitute the cell membrane-like structure on a synthetic platform. Thus, a bioderived membrane measuring 6–10 nm in thickness with finite number of transporter proteins supported or suspended on a solid substrate and membrane area of $\sim 10~\mu m^2$ serves as the fundamental unit in a bioderived smart material. In an actuator, sensor, or energy harvesting device, a large number of such membranes with protein transporters are assembled to work in parallel as a smart material. An overview of concepts that use the bioderived membrane as the active component in sensors, actuators, and energy harvesting devices is shown in Fig. 3. The ionic currents through the membrane can be obtained from the ionic current through the transporter from expressions similar to Eqs. 4–6 for the design of smart material systems.

Sensor: The bioderived membrane is used as the sensing element by monitoring the electrical response of the membrane in response to an applied stimulus. The bioderived membrane is formed with electrodes on either sides of the membrane on a synthetic substrate and packaged into a system for monitoring electrical current, transmembrane voltage, or electrical impedance. With the choice of an appropriate protein transporter, the sensor is tailored to respond to chemical analyte, bioelectrical sensors, mechanical strain, light, and temperature.

Actuator: A bioderived membrane is used to regulate the ionic concentration between two closed volumes in response to an applied stimulus. The ionic currents through the bioderived membrane sets up ionic and osmotic gradients across the membrane. This concentration gradient becomes an intermediate stimulus for osmotic regulation and movement of the solvent molecules (water) across the membrane. The bulk movement of water molecules leads to volumetric expansion of a closed volume that will lead to bulk stress in the material. The bulk stress generated in a closed volume encompassing the bioderived membrane is placed in a system that will lead to the development of a stack- or a bimorph-type actuator.

Energy Harvesting: The definition of an energy harvesting device has evolved over the last decade and is now defined as a system that converts ambient energy into electrical energy that can be stored locally and used later for powering electronic devices. In this context, a bioderived membrane generates a chemoelectrical potential from ion transport by varying the ionic concentration as shown in Eq. 2. The concentration gradient and the resulting chemoelectrical potential can be generated by active transport using incident light, consuming triphosphates (adenosine, guanosine), etc. Due to the nature of the ionic transport processes and the resulting gradient across the membrane, protein transporters behave like constant current power sources.

Fabrication Methods and Recent Advances

The methods to form BLM utilize the amphiphilic nature of the phospholipid molecule to produce a highly organized structure in the presence of an aqueous medium. In the past five decades, many techniques have emerged to assemble phospholipids into vesicles and planar BLM [1, 4–8]. Vesicles are hollow-spherical structures that are useful as carrier vehicles for chemical payloads. The planar configuration of the lipid membrane (planar BLM) permits access to both

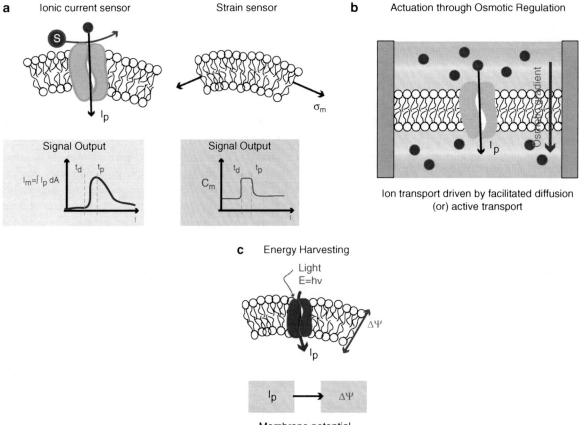

a Ionic current sensor   Strain sensor   **b** Actuation through Osmotic Regulation

Signal Output   Signal Output

Ion transport driven by facilitated diffusion
(or) active transport

c Energy Harvesting

Membrane potential
established by ion transport

Bioderived Smart Materials, Fig. 3 Concepts for (**a**) sensor elements, (**b**) actuation units, and (**c**) energy harvesting cells using bioderived membranes

sides of the membrane and is preferred for the investigation of ion transport through transporter proteins. The bioderived smart materials typically use a planar BLM supported on a solid substrate (supported BLM) or suspended across the pores of a nanofabricated porous substrate (suspended BLM).

The tools and methods to fabricate a synthetic biological membrane trace its origin to the pioneering research by Mueller and Rudin and coworkers [1, 4] and Tien, Ottova, and coworkers [9]. The initial studies focused on forming planar BLMs from egg phospholipids across a 200-μm pore in a polymeric membrane with voltage-gated ion channels [4]. These initial experimental investigations and subsequent advances by Tien, Ottova, and coworkers were motivated by the need to create in vitro test platforms for analyzing the physiological and immunological function of cell membranes [10–12]. Recently, the experimental

methods to form a BLM with proteins are motivated by the engineering applications of bioderived membranes. The anticipated applications for bioderived membranes have significantly influenced the fabrication methods and material choice in the past decade.

A consistent challenge in transitioning supported and suspended lipid bilayers into robust material platforms is the fragility and limited shelf life of the thin membrane. Various approaches have been investigated to stabilize lipid bilayers, including suspending bilayers across nanopores, tethering bilayers to solid surfaces, and sandwiching suspended membranes between water-swollen hydrogels. Among them, this entry will discuss a novel platform for reconstituting phospholipids into a BLM in a curable polymerizable structure. The method, referred to as droplet interface bilayer (DIB), utilizes the affinity of the hydrophobic tails to organic solvents (or) oils. The DIB technique

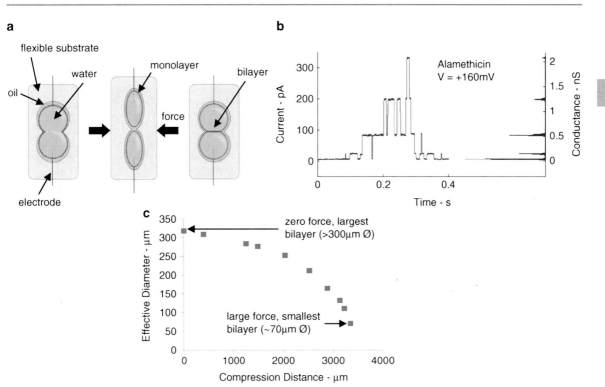

Bioderived Smart Materials, Fig. 4 (**a**) The regulated attachment method uses a flexible substrate and applied force to enable bilayer formation. (**b**) Membranes produced via the RAM enable single-channel measurements. (**c**) Modulating the applied force allows for reversible control on the size (area) of the bilayer

first demonstrated by Funakoshi et al. [13] and then refined and expanded by Bayley et al. [14–16] showed that the stability of the bilayer can be improved by minimizing the direct interaction of phospholipids with a solid substrate. This leads to a BLM with improved longevity (DIBs lasted for days to weeks) and increased resistance to rupture. In this approach, a liquid-supported interface bilayer is formed by connecting two lipid-encased water droplets submerged in oil as shown in Fig. 4. The monolayers that make up the bilayer self-assemble at the oil-water interface surrounding each droplet and spontaneously "zip" together when the droplets are placed into contact. The formation of a lipid bilayer at this interface occurs by thinning when depletion flocculation removes excess oil from between the droplets. Using this principle, the amount of contact between two droplets can be modulated by the choice of oil.

Aside from advantages of simplicity and stability, the DIB embodiment affords the ability to form a biomolecular network of lipid bilayers by connecting more than two droplets. The ability of the DIB

membrane to accommodate proteins makes it feasible to demonstrate selective transport across multiple interfaces formed from a network of droplets. The primary disadvantage of this method, though, is the need to dispense and position individual water droplets a task that becomes increasingly difficult for small droplets (<100 μm). To remedy this limitation and to provide added support to the droplets themselves, Sarles and Leo recently developed an alternative technique called the regulated attachment method (RAM) for interface bilayer formation within a flexible, nonwetting solid substrate [17, 18]. Instead of positioning droplets individually, the RAM uses mechanical force applied to the flexible substrate to control the attachment of lipid-coated aqueous volumes contained in neighboring compartments. The applied force regulates the dimensions of the aperture separating the compartment as shown in Fig. 4a. With the initial design, increasing the force closes the aperture and separates the volumes. Relaxing this force, then, opens the aperture and allows the volumes to come into contact. The results of this method proved that, like DIBs, interface bilayers formed using RAM lead to stable membranes that exhibit high electrical resistances ($>10G\Omega$) necessary for measuring single-channel and multiple-channel ion currents through proteins such as alamethicin (Fig. 4b). Further, the RAM enables the size of the bilayer (i.e., the area of contact) to be reversibly modulated after membrane thinning. It has been shown through experiments that the effective diameter of the bilayer formed from RAM technique could be varied by a factor of 5X by merely modulating compressive force (Fig. 4c). The flexibility of the regulated attachment method (RAM) provides additional advantages over earlier methods to form lipid bilayers, including the DIB approach. First, this method provides control of the size of the bilayer, independent of the sizes or shapes of the aqueous volumes. Sarles and Leo later utilized this advantage in forming interface bilayers between nonspherical hydrogels [19]. Second, supporting the aqueous volumes within a flexible substrate helps to reduce relative motion between the connected volumes under vibration and shock, making for a more portable system. With this added support, additional contents in the form of discrete aqueous droplets can be added to an existing lipid-coated volume via a microchannel or syringe without causing the membrane to rupture. This capability enables biomolecules (or other species) to be added to the membrane at a designated time after bilayer formation.

Bioderived Microhydraulic Actuators

Motion in biomolecules, especially protein transporters, is typically associated with conformational change in proteins and is of the order of few angstroms of displacement and/or angular rotation of the subunits in the protein transporter [20]. There are numerous challenges in coupling this mechanical motion resulting from three-dimensional conformational change of proteins to an external system and hence cannot be used for performing mechanical work. Another interesting example of chemomechanical actuation is the demonstration of kinesin motility on microtubules [21]. In this system, a kinesin-functionalized nanoparticle traverses along the length of a microtubule resulting from the rotation of kinesin dimer from ATP hydrolysis. Various research groups have used nanobeads functionalized with kinesin as

platform for demonstrating this novel actuation concept. While this concept is significantly advanced than molecular motors, coupling this linear motion to generate force and volumetric strain has challenges in interfacing the bimolecular motion with a structure. In order to demonstrate a system that uses biomolecules to generate volumetric strain similar to ferroelectrics, electroactive polymers, a membrane-based hydraulic actuation concept was developed by Sundaresan and Leo [3].

The membrane-based microhydraulic actuator uses a BLM with ion transporters as shown in Fig. 5(a–b). The chemomechanical actuator uses osmotic regulation between two chambers resulting from ion and sucrose transport through the bioderived membrane. The bioderived membrane is suspended across the pores of a microporous substrate and separates the contents of the two chambers and is reconstituted with a sucrose transporter protein (SUT4) extracted from *Arabidopsis thaliana*. The SUT4 protein, grown in and extracted from yeast cell membrane, is a cotransporter that transports proton and sucrose from the side of the membrane with higher proton and sucrose concentration. On applying a buffered proton gradient, the cotransporter balances the pH gradient with sucrose concentration gradient. The sucrose concentration gradient generates an osmotic gradient across the membrane and produces water transport through the membrane. The additional volume of fluid from osmotic regulation, demonstrated in Sundaresan et al. [22], builds pressure in the chamber, balances the osmotic gradient established by sucrose transport, and deforms a flexible wall in the chamber.

The mechanics of ion transport leading to an osmotic gradient due to a cotransporter is shown using phenomenological equations. The forces in the system during proton-sucrose cotransport through the membrane for actuation are chemical potentials ($\Delta\mu$) due to

1. Concentration gradient of proton ($\Delta\mu_p$)
2. Concentration gradient of sucrose ($\Delta\mu_s$) – cotransported specie
3. Osmotic gradient due to concentration gradients of charged species and nonelectrolytes

The proton and sucrose fluxes (ϕ_p, ϕ_s) through the membrane due to the applied concentration gradients across a membrane can be represented by phenomenological equations for ion transport formulated by Katachalsky et al. [23, 24].

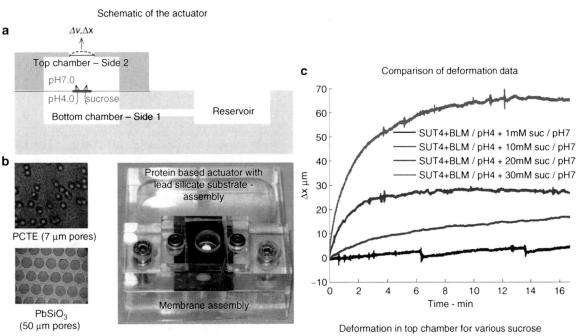

Bottom chamber – Side 1

Bioderived Smart Materials, Fig. 5 (**a–b**) Schematic of a membrane-based chemomechanical actuator. (**c**) Deformation of cover plate in the actuator for various initial sucrose concentrations

$$\phi_s = L_{s,s}\,\Delta\mu_s + L_{s,p}\,\Delta\mu_p$$
$$\phi_p = L_{p,s}\,\Delta\mu_s + L_{p,p}\Delta\mu_p. \tag{7}$$

Assuming unity values for activity coefficients, the chemical potentials that contribute to the flux are [25]

$$\Delta\mu_p = RT\ln\left(\frac{c_1^p}{c_2^p}\right)$$
$$\Delta\mu_s = RT\ln\left(\frac{c_1^s}{c_2^s}\right). \tag{8}$$

The coefficients in the flux equations in Eq. 7 represent the coupling between proton and sucrose transport through the membrane. The coefficient $L_{s,s}$ represents the coupling between sucrose chemical potential and sucrose flux. The coefficient $L_{s,p}$ relates sucrose flux to the chemical potential due to proton and similarly, $L_{p,s}$ relates the proton flux to sucrose chemical potential. The coefficients $L_{s,p}$, $L_{p,s}$ are assumed equal due to the Onsager symmetry relationship [26]. The chemical potentials due to sucrose and pH

gradients are assumed to balance each other as shown in Sundaresan and Leo [22] and the equilibrium concentration of sucrose is obtained for various pH and sucrose gradients. For an infinite source of sucrose on side 1 of the membrane and zero initial concentration of sucrose on side 2 of the membrane, it is assumed that the concentration on side 1 of the membrane remains unchanged from the initial concentration. The concentration of sucrose on side 2 balanced by a buffered pH gradient at equilibrium is given by the expression

$$c_2^s = c_1^s 10^{\Delta pH}, \tag{9}$$

where $\Delta pH = pH_2 - pH_1$. In the presence of a limited source of sucrose on side 1 of the membrane as in the actuator, sucrose is transported through the membrane until the concentration gradient of sucrose balances the applied proton gradient. The balancing concentration as a function of the volume ratios. The number of moles of the sucrose on side 1 and side 2 at equilibrium is denoted by $n_1^s|_{eq}$ and $n_2^s|_{eq}$, respectively. The volume of the chamber on side 1 and side 2 of the membrane is represented by V_1 and V_2 and the volume ratio by

$V_r = V_1/V_2$. At equilibrium condition the number of moles in the system is constrained by

$$n_1^s\big|_{eq} + n_2^s\big|_{eq} = V_1 c_1^s \tag{10}$$

The equilibrium concentration on side 2 of the membrane from Eq. 9 is rewritten in terms of the number of moles of the sucrose on both the sides of the membrane,

$$n_2^s\big|_{eq} = n_1^s\big|_{eq} \frac{V_r}{10^{\Delta pH}} \tag{11}$$

The concentration on side 1 at equilibrium condition is obtained in terms of the initial sucrose concentration on side 1 of the membrane by substituting the mass balance expression in Eq. 10 into Eq. 11

$$c_1^s\big|_{eq} = c_1^s \frac{V_r}{V_r + 10^{\Delta pH}}. \tag{12}$$

Similarly, the concentration of sucrose at equilibrium on side 2 of the membrane is

$$c_2^s\big|_{eq} = c_1^s \frac{10^{\Delta pH} V_r}{V_r + 10^{\Delta pH}}. \tag{13}$$

The equilibrium concentration of sucrose on side 2 of the membrane $(c_2^s\big|_{eq})$ due to an imposed initial sucrose concentration (c_1^s) is computed for a volume ratio $V_r = 100$ in Table 1. It is observed from the table that the proton gradient concentrates sucrose on side 2 of the membrane. From the applied proton gradient and different initial concentration of sucrose on side 1 of the membrane, the membrane with cotransporter will develop different osmotic pressures on side 2 of the membrane within a finite duration. This analysis shows the feasibility to use concentration of sucrose as a control variable in the actuator with a cotransporter for modulating the force and deformation. A prototype chemomechanical actuator similar to the schematic in Fig. 5a was fabricated to demonstrate the concept of microhydraulic actuation using bioderived membranes. The porous glass plate is attached to a 500-μm Kapton film and sandwiched in between the two chambers with a rubber gasket. This serves as the supporting substrate for forming suspended BLM with SUT4 transporters. A clamping plate screwed

Bioderived Smart Materials, Table 1 Equilibrium concentration of sucrose on side 2 of the membrane in the actuator for pH4/pH7 applied across the membrane assembly

c_1^s mM (Initial)	1	10	15	20
$c_2^s\big|_{eq}$ mM	90	900	1,360	1,810

onto the top chamber holds the PET (polyethylene terepthalate) coverplate that deforms out of the chamber due to increasing pressure from osmotic regulation across the two chambers. The volumes of the chambers are designed to be 0.54 ml and 50 ml so that the larger chamber resembles a reservoir. The procedure to assemble the actuator follows the description in Sundaresan and Leo [22]. The BLM is formed from POPS and POPE lipids by painting method and the proteins are reconstituted by vesicle fusion method. In this method, 10 μl of lipids (POPS:POPE mixed in 3:1 w/w ratio and dissolved in n-decane at 40 mg/ml) is painted on the surface and allowed to dry for 10 min under a stream of nitrogen. This is followed by the addition of 10 μl of SUT4 transporters in liposomes suspended in pH7.0 medium. The porous substrate with added components are allowed to stand in air for 15 min and assembled into the actuator.

The assembled prototype actuator is characterized by applying different concentrations of sucrose in the bottom chamber and measuring the deformation of the coverplate. As a control study, different baseline tests with a key ingredient left out is tested. The different baseline tests performed on the actuator are:
1. BLM with pH7.0 buffer on both sides of the membrane
2. BLM with pH4.0/pH7.0 gradient without sucrose dissolved in the bottom chamber
3. BLM with pH4.0/pH7.0 gradient with 5 mM sucrose dissolved in the bottom chamber

The results from baseline tests on the prototype shown in Sundaresan and Leo [22] demonstrate that there is no deformation in the coverplate for the case with pH7.0 buffer on both the sides of the membrane. Similarly, the coverplate does not show an appreciable deformation for a pH gradient applied to the membrane without sucrose. In the third baseline, pH gradient is applied across a BLM without SUT4 and 5 mM sucrose is added to the bottom chamber. It is observed that the coverplate deforms into the top chamber and is attributed to the migration of water molecules from the top chamber to the bottom chamber driven by the osmotic gradient.

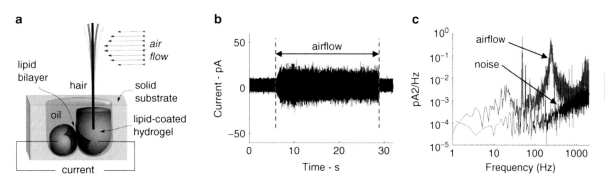

Bioderived Smart Materials, Fig. 6 A membrane-based hair cell (**a**) uses the mechanoelectrical properties of a lipid bilayer to transduce airflow into current (**b**). (**c**) The frequency content of the measured signal correlates to the mechanical vibrations of the hair

The next set of experiments are performed on an actuator assembled with pH4.0/pH7.0 gradient applied across the BLM with SUT4 and different concentrations of sucrose. Representative results from these experiments are shown in Fig. 5c. From the experimental results in Fig. 5c, it is observed that the deformation in the actuator and rate of deformation increases with the initial sucrose concentration on side 1 of the membrane. The performance metrics of this actuator of this bioderived membrane-based actuator discussed in Sundaresan and Leo [22] compares well with soft polymeric electrochemomechanical actuators [27]. Thus, this demonstration follows the theoretical discussion that varying the sucrose concentration across the SUT4 cotransporter results in an osmotic gradient and leads to volumetric strain in the flexible wall of the actuator.

Hair Cell Sensors

Sensory hairs, or hair cells, are one of the most common forms of transducers found in nature. In general, a hair cell is type of sensory receptor that uses protruding structures, called cilia, to probe the surrounding environment. Sharing this basic trait, natural hair cells display tremendous diversity in their morphologies, mechanical properties, and physiological functions. As a result, animals use hair cells to detect sound, pressure, flow, vibration, chemical species, and even position (inertial tilt). Motivated by the diversity of stimuli-responsive behavior of natural hair cells, several groups have used synthetic active materials to engineer hair cell-inspired sensors. Most notably, Liu's research group [28] pioneered multiple

generations of artificial cilia that used microfabricated cilia mounted on strain gauges or polymeric force-sensitive resistors (FSRs) to create an electrical response in response to flow or touch. However, living hair cells such as mammalian outer hair cells (OHCs) rely heavily on soft biological materials, namely, cell membranes and stretch-activated ion channels to sense deflections the cilia.

Building on recent advances to assemble and encapsulate an artificial cell membrane within a flexible substrate, Sarles et al. [29] demonstrated that a gel-supported lipid bilayer could be used to build a new type of membrane-based hair cell-inspired sensor. As shown in Fig. 6a, the hair cell consists of synthetic cilium (hair) that features an electrolyte-swollen polymeric gel at lower end and which is held vertically within one compartment of a nonswelling solid substrate. A second, lipid-encased aqueous volume contained in the neighboring compartment connects to the gel at the base of the hair to form a lipid bilayer. Both compartments also contain immiscible oil needed for promoting phospholipid self-assembly at the oil-water interfaces. The performance of the hair cell sensor was characterized by measuring the current through the membrane (using conductive probes inserted into each aqueous volume) in response to the applied airflow across the hair. The results of this initial study demonstrated that the membrane-based hair cell generates an increase in the amplitude of the current through the bilayer in response to flow-induced vibration of the synthetic hair (Fig. 6b). Also, the power spectral density revealed that the frequency content of the sensing current (Fig. 6c) differs from that of the background noise evidence that the mechanical vibrations of the hair contribute to the measured current. Unlike living hair cells that use stretch-activated ion channels to provide both static and oscillatory sensing, this system only contained an Ac response (i.e., static deflection of the hair did not result in a change in the current). Yet, these initial responses did not explain the source of the current. By systematically varying the applied potential across the bilayer and holding the area of the membrane constant during the experiment, the authors discovered that the sensing current increased linearly with respect to the voltage. This result provided proof that a change in the electrical capacitance of the membrane, caused by bending in the bilayer, is the source of the measured current. In this work, the sensing currents ranged from 10 to 100 pA, depending on the speed of airflow, the length of the hair, and the transmembrane potential. Current efforts to increase the sensitivity of the sensor, provide static and directional responses, and fully encapsulate the liquid contents are underway.

Concluding Remarks

This entry presents a novel framework to develop smart materials using biomolecular components extracted from cell membranes. The bioderived smart materials use ion and fluid transport through protein transporters embedded in an impervious membrane and couple one or more of the following – chemical, electrochemical gradients across the membrane, mechanical stretch in the membrane and light energy incident on the membrane. This entry discusses a unique bioderived actuator that uses sucrose transport through a proton-sucrose transporter to generate volumetric strain. It is shown that the blocked force generated by these materials resemble soft polymeric electrochemomechanical actuators. We also present a novel electromechanical sensing concept using the BLM and demonstrate the potential of this device as a mechanical flow sensor. In the bioderived smart material systems described in this entry, two methods to fabricate the bioderived membrane are presented. Among the two methods, the first method utilizes the self-assembly of phospholipids in a pore to form a BLM with proteins. The second method, inspired by DIB, demonstrates an innovative method to fabricate a durable bilayer membrane, demonstrates the force and flow sensing, and provides a rugged

framework for future development in this field. The framework for building bioderived smart materials and their application as sensors, chemomechanical actuators, and energy harvesting devices discussed in this entry benefits from a wide variety of transporters that can be extracted from cell membranes and offers innovative approaches to couple various forms of energy.

Cross-References

▶ Biomimetic Flow Sensors
▶ Biomimetics
▶ Biosensors
▶ Molecular Dynamics Simulations of Nano-Bio Materials
▶ Organic Actuators
▶ Organic Bioelectronics
▶ Synthetic Biology

References

1. Montal, M., Mueller, P.: Formation of bimolecular membranes from lipid monolayers and a study of their electrical properties. Proc. Nat. Acad. Sci. USA 69(12), 3561–3566 (1972)
2. Ottova, A.L., Ti Tien, H.: Self-assembled bilayer lipid membranes: from mimicking biomembranes to practical applications. Bioelectrochem. Bioenerg. 42(2), 141–152 (1997)
3. Sundaresan, V.B., Homison, C., Weiland, L.M., Leo, D.J.: Biological transport processes for micro-hydraulic actuation. Sens. Actuators B Chem. 123, 685–695 (2007)
4. Mueller, P., Rudin, D.O., Tien, H.T., Wescott, W.C.: Reconstitution of excitable cell membrane structure in vitro. Circulation 26(5), 1167–1171 (1962)
5. Needham, D., Haydon, D.A.: Tensions and free energies of formation of "solventless" lipid bilayers measurement of high contact angles. Biophys. J. 41(3), 251–257 (1983)
6. Poulin, P., Bibette, J.: Adhesion of water droplets in organic solvent. Langmuir 14(22), 6341–6343 (1998)
7. Haydon, D.: Properties of lipid bilayers at a water-water interface. J. Am. Oil Chem. Soc. 45(4), 230–240 (1968)
8. Fettiplace, R., Andrews, D.M., Haydon, D.A.: The thickness, composition and structure of some lipid bilayers and natural membranes. J. Membr. Biol. 5(3), 277–296 (1971)
9. Tien, H.T., Wurster, S.H., Ottova, A.L.: Electrochemistry of supported bilayer lipid membranes: background and techniques for biosensor development. Bioelectrochem. Bioenerg. 42, 77–94 (1997)
10. Tien, H.T.: Self-assembled lipid bilayers as a smart material for nanotechnology. Mater. Sci. Eng. C. 3, 7–12 (1995)

11. White, S.H.: Analysis of the torus surrounding planar lipid bilayer membranes. Biophys. J. **12**(4), 432–445 (1972)

12. Singer, S.J., Nicolson, G.L.: The fluid mosaic model of the structure of cell membranes. Science **175**(4023), 720–731 (1972)

13. Funakoshi, K., Suzuki, H., Takeuchi, S.: Lipid bilayer formation by contacting monolayers in a microfluidic device for membrane protein analysis. Anal. Chem. **78**(24), 8169–8174 (2006)

14. Holden, M.A., Needham, D., Bayley, H.: Functional bionetworks from nanoliter water droplets. J. Am. Chem. Soc. **129**(27), 8650–8655 (2007)

15. Hwang, W.L., Holden, M.A., White, S., Bayley, H.: Electrical behavior of droplet interface bilayer networks: Experimental analysis and modeling. J. Am. Chem. Soc. **129**(38), 11854–11864 (2007)

16. Bayley, H., Cronin, B., Heron, A., Holden, M.A., Hwang, W.L., Syeda, R., Thompson, J., Wallace, M.: Droplet interface bilayers. Mol. Biosyst. **4**(12), 1191–1208 (2008)

17. Sarles, S.A., Leo, D.J.: Regulated attachment method for reconstituting lipid bilayers of prescribed size within flexible substrates. Anal. Chem. **82**(3), 959–966 (2010)

18. Sarles, S.A., Leo, D.J.: Membrane-based biomolecular smart materials. Smart Mater. Struct. **20**(9), 094018 (2011)

19. Sarles, S.A., Stiltner, L.J., Williams, C.B., Leo, D.J.: Bilayer formation between lipid-encased hydrogels contained in solid substrates. ACS Appl. Mater. Interfaces **2**(12), 3654–3663 (2010)

20. Oster, G., Wang, H.: Rotary protein motors. Trends Cell Biol. **13**(3), 114–121 (2003)

21. Hess, H., Bachand, G.D., Vogel, V.: Powering nanodevices with biomolecular motors. Chem. A Eur. J. **10**(9), 2110–2116 (2004)

22. Sundaresan, V.B., Leo, D.J.: Modeling and characterization of a chemomechanical actuator using protein transporters. Sens. Actuators B Chem. **131**(2), 384–393 (2008)

23. Kedem, O., Katchalsky, A.: A physical interpretation of the phenomenological coefficients of membrane permeability. J. Gen. Physiol. **45**(1), 143–179 (1961). doi:10.1085/jgp.45.1.143. arXiv: http://www.jgp.org/cgi/reprint/45/1/143.pdf

24. Schultz, S.G., Curran, P.F.: Coupled transport of sodium and organic solutes. Physiol. Rev. **50**(4), 637–718 (1970). arXiv: http://physrev.physiology.org/cgi/reprint/50/4/637.pdf

25. Gyftopoulos, E.P., Berreta, G.P.: Thermodynamics: foundations and applications. Macmillan, New York (1996)

26. Onsager, L.: Reciprocal relations in irreversible processes. II. Phys. Rev. **38**(12), 2265–2279 (1931). doi:10.1103/PhysRev.38.2265

27. Akle, B., Bennett, M., Leo, D.J.: Development of a novel electrochemically active membrane and "smart" material based vibration sensor/damper. Sens. Actuators A Phys. **126**(1), 173–181 (2006)

28. Liu, C.: Micromachined biomimetic artificial haircell sensors. Bioinspir. Biomim. **2**(4), S162 (2007)

29. Sarles, S.A., Madden, J.D.W., Leo, D.J.: Hair cell inspired mechanotransduction with a gel-supported, artificial lipid membrane. Soft Matter **7**(10), 4644–4653 (2011)

Bio-FET

▶ Nanostructure Field Effect Transistor Biosensors

Biofilms in Microfluidic Devices

Suresh Neethirajan[1], David Karig[2], Aloke Kumar[3], Partha P. Mukherjee[4], Scott T. Retterer[3] and Mitchel J. Doktycz[2,3]

[1]School of Engineering, University of Guelph, Guelph, ON, Canada

[2]Center for Nanophase Materials Sciences, Oak Ridge National Laboratory, Oak Ridge, TN, USA

[3]Biosciences Division, Oak Ridge National Laboratory, Oak Ridge, TN, USA

[4]Computer Science and Mathematics Division, Oak Ridge National Laboratory, Oak Ridge, TN, USA

Synonyms

Floccules; Microbial aggregations

Definition

Biofilms are aggregations of microbes that are encased by extracellular polymeric substances (EPS) and adhere to surfaces or interfaces. Biofilms exist in a very wide diversity of environments, and microfluidic devices are being increasingly utilized to study and understand their formation and properties.

Overview

Microbes often form aggregates on interfaces, and due to a production of EPS, the aggregates become encased in a matrix [1]. Though microbes in a biofilm are physiologically distinct from bacteria growing in a free swimming state (planktonic bacteria), biofilm growth is a complex process that is typically initiated by planktonic bacteria themselves. Biofilm growth is initiated with bacterial adhesion to a surface, followed by events such as growth, EPS secretion, and morphological and physiological changes. Microbial biofilms are excellent examples of multi-scale phenomena. Cell-to-cell communication, which plays a role in biofilm formation, is molecular in nature, but occurs over a scale of several cells. Adhesion events occur at the nanometer scale and are mediated by pili or flagella, while the cells themselves are typically micron-sized. Finally the biofilms themselves typically range between 10 and 1,000 μm in thickness. These length scales are compatible with the scale of microfluidic devices, thus making such tools useful for exploring the spatiotemporal properties of biofilms. Moreover, microfluidic devices are often optimized for online optical monitoring and/or incorporation of sensors. These factors make microfluidic devices appropriate for studying biofilms. For example, the effect of miniscule changes in molecular cues (such as nanomolar concentrations) can be characterized and studied easily with microfluidic devices. Another advantage of microfluidics is that they enable the precise control of the microenvironment, thereby allowing biofilms to be subjected to controlled external stimuli. Thus, when molecular cues or signals are externally applied, minute responses in the biofilm can be effectively studied. The use of microfluidics offers distinct advantages for fundamental studies regarding the nature, properties, and evolution of microbial biofilms. Beyond being a platform for such studies, microfluidics is being increasingly applied toward miniaturized device creation. In this entry, some of the basic methodologies involved in biofilm studies in microfluidic devices are first discussed, followed by a discussion on some of the key findings reported in this area.

Methodology

Microfluidics

Microfluidics is the precise control and manipulation of fluids contained to miniaturized channels (typical length scale <100 μm). In the microfluidics regime, fluid flow is laminar, and flow is dominated by Stokes drag and surface tension effects. The study of biofilms in microfluidic devices typically requires device design and fabrication, controlled microbial growth, and analysis. Microfluidic device fabrication by itself is a significantly evolved science. There are several different techniques for device fabrication including, but not limited to, various micromachining processes

and polymer-based soft-lithography. Complex structures such as micropumps, microvalves and mixers can also be incorporated on microfluidic devices with the help of different microfabrication techniques. The techniques of microfabrication are beyond the scope of this article and interested readers can refer to one or more manuscripts on this topic [2, 3].

Biofilms

A wide range of microbial species produce biofilms. Once the appropriate device is fabricated, controlled biofilm growth in the device may be desirable. For controlled in situ biofilm growth in a microfluidic device, a dilute microbial culture may be introduced into the microfluidic device. The diluted samples are made from liquid microbial cultures once the culture achieves a specified optical density (OD). A fluid port usually provides a means for the introduction of the inoculum into the device. Depending on the microorganism and its microenvironment and the specific experimental need, proper biofilm growth can require anywhere between a few hours to several days. During this interval, it might be desirable to control environmental conditions such as temperature and humidity. In such circumstances, the device is typically housed in an incubator. Depending on the organism, aerobic or anaerobic environment may be necessary. For microfluidic devices made from PDMS or similar permeable polymers, maintaining an aerobic environment is usually not an issue. On the other hand, to maintain anaerobic conditions typically necessitates more complex device fabrication techniques. Analysis of biofilm formation may be done by various forms of microscopy. Microfluidic devices usually lend themselves to optical microscopy with considerable ease – one of the reasons why it is popular as a diagnostic setup. Use of thin-walled chambers and optically clear materials in microfluidic devices allows probing through optical means and use of high-magnification lenses. Confocal laser scanning microscopy (CLSM) has been demonstrated as a tool for integrating microfluidics in studying the evolution and structural heterogeneity of biofilms [4]. Other forms of microscopy may also be employed such as scanning electron microscopy (SEM) and atomic force microscopy (AFM). Besides microscopy, micro- and nanosensors can also be easily incorporated into a microfluidic device for in situ monitoring. For example, micro-electrodes fabricated on a glass substrate can be incorporated in a microfluidic device for electrochemical impedance spectroscopy studies.

Key Findings

Microfluidic devices, in conjunction with microscopy and other analytical techniques, enable flexible and novel approaches for probing the multiple determinants of biofilm formation. Two factors shaping the dynamics of biofilm formation are fluid dynamics and cell phenotype. Fluid dynamics determine shear forces that govern cell attachment and detachment rates, which can alter whether or not a biofilm will form, biofilm depth, biofilm density, and surface coverage. On the other hand, cell phenotype dictates important processes such as EPS production, growth rate, and flocculation, all of which can also have profound effects on biofilm structure and function. Cell phenotype, in turn, is a function of both environmental cues and cell-cell communication. Accordingly, recent studies have harnessed microfluidic technology to probe the effects of fluidic dynamics, cell phenotype, and cell-cell communication. Mathematical modeling efforts have integrated these findings and guided further investigations.

Influence of Fluid Dynamics on Biofilms
Several studies have examined the effects of hydrodynamics on biofilm development. For example, microfluidic devices were used by Lee at al. [5] to study the influences of hydrodynamics of local microenvironments of *Staphylococcus epidermis* biofilm formation (Fig. 1). They observed that at high flow velocity, the cells formed an elongated biofilm morphology, and at low fluid velocity clump-like multilayered biofilms were produced. The results of this study indicate that microfluidic devices with embedded microvalves can perhaps be used for screening the effects of therapeutic reagents, and as novel tools for developing predictable in vitro models of biofilm-related infections. Also, Rusconi et al. [6] studied suspended filamentous biofilms in a microfluidic device. Experiments with several bacterial strains under high flow inside the miniaturized channels demonstrated the link between the extracellular matrix and the development of biofilm structures. Secondary vertical motion from the numerical simulations of the flow

in the curved channels of the microfluidic device proved that the hydrodynamic forces were key in influencing the formation of suspended biofilm filamentous structures. Another study by Richter et al. [7] examined the influence of shear stress on the growth and structure of fungal biofilms using microfluidic systems. Electrode structures were incorporated into a microfluidic device for performing cellular dielectric spectroscopy, enabling real-time, noninvasive quantification of cell morphology changes. Increase in shear stress caused a significant change in the biofilm formation patterns, and the addition of amphotericin B resulted in distinct dynamic behavior of the biofilm.

Effect of Cell Phenotype on Attachment
In addition to hydrodynamics, cell phenotype plays an important role in biofilm formation. For example, after initial attachment, pili, flagella, and adhesins can help cells adhere to surfaces. Characterizing these determinants of cell attachment is critical for understanding the early stages of biofilm development. Using a microfluidic device, De La Fuente et al. recently examined the roles of different *Xylella fastidiosa* pili in determining cell adhesion at different flow rates [8]. To do so, they compared different genetic variants: wild-type cells, cells with only type I pili, cells with only type IV pili, and cells with no pili. They subjected these variants to different drag forces by exposing the cells to different fluid flow rates. The results enabled quantification of the role of the different pili in attachment, and the adhesion force values were within the range of adhesion forces determined by AFM and by laser tweezers for other microbes.

Cellular-Communication Inside Microfluidic Systems
Several cell processes relevant to biofilm formation, including dispersion, EPS secretion, and lipid secretion, are often regulated by cell-cell communication. Thus, several studies have focused on studying intercellular communication in biofilm contexts. Specifically, many microbes communicate through the process of quorum sensing. Quorum sensing (QS) enables a group of cells to measure their local population density through the synthesis of and response to small signal molecules that can pass from cell to cell. While quorum sensing regulates several microbial behaviors that influence biofilm formation, biofilm

Biofilms in Microfluidic Devices, Fig. 1 Microfluidic devices can be engineered to produce well-defined flow structures and shear rates. Such devices can be used to investigate hydrodynamic influences on biofilms (From Lee et al. [5]. Reproduced with kind permission from Springer Science and Business Media)

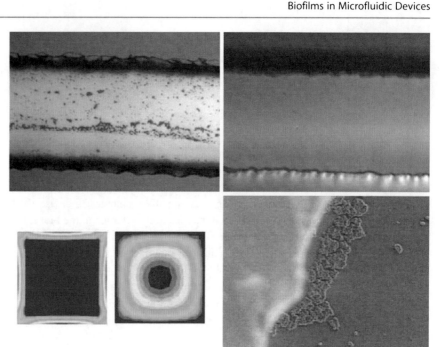

characteristics in turn strongly impact the efficiency of quorum sensing and signal transport, thus setting up an interplay between quorum sensing and biofilm formation. Consequently, microfluidic studies have been used to examine quorum sensing in a biofilm context.

Timp et al. [9] studied genetically engineered cells in microfluidic devices to explore the fundamental principles of quorum sensing in biofilms. Engineered cells were flowed through a microfluidic channel into a region where optical traps were used to position cells into defined patterns. The patterned cells were then enclosed in a hydrogel to mimic a biofilm. "Transmitter" cells synthesized an N-Acyl Homoserine Lactone (AHL) communication signal which was detected by "receiver" cells that fluoresced in response. Activation of the receiver cells by the transmitters was found to be heavily dependent on the hydrodynamics around the biofilm mimic. In particular, low flow rates corresponding to diffusion-dominated transport yielded efficient communication between transmitters and receivers, but this communication began to break down at high flow rates corresponding to convection-dominated transport. These results indicate that the density of bacteria necessary to constitute a "quorum" depends on the hydrodynamic properties of the environment. A similar conclusion was reached by Connell et al. [10] who studied QS in fabricated picoliter-scale microcavities. Specifically, the authors

found that they could achieve a sixfold increase in QS-dependent gene expression by bacteria trapped within their microcavities when they reduced the external flow rate from 250 to 5 μL min^{-1}.

Another interesting aspect of QS highlighted by the Connell study was the importance of cell density, rather than total cell count as a factor governing QS behavior. In particular, because the authors were able to capture small numbers of cells in small volumes, they were able to control cell density independent of cell population size. As a result, they were able to show that as few as 150 cells are capable of exhibiting QS behavior, provided that the cells are constrained such that cell density remains relatively high. This is important, since it demonstrates the potential for small bacterial communities to exhibit some of the properties (e.g., antibiotic resistance), which are typical of biofilms. In a related study, Boedicker et al. used microfluidics to examine the QS behavior of small numbers of bacteria trapped in poly(dimethylsiloxane) (PDMS) wells [11]. In this study, the authors were able to trap as few as one to two cells in each well, and while the majority of cells needed several rounds of cell division to initiate QS, on occasion the authors found single cells expressing QS controlled genes. Again, this result showed that it is bacterial density, rather than bacterial cell numbers, that is of fundamental importance to the onset of QS behavior.

Another aspect of QS and microbial aggregation was investigated by Park et al. [12], who devised a microfluidic device resembling a maze. They investigated the growth of *Escherichia coli* cells in the device. They found that the device geometry affected aggregation of the cells. Their results led them to conclude that self-attraction in microbes could allow them to easily exceed critical QS densities.

Collectively, these studies exemplify the flexibility offered by microfluidic systems to precisely define the cell's microenvironment, including cell position, confinement geometry, and fluid flow rate. These studies have begun to shed light on fundamental questions such as when quorum sensing is truly used to sense local population as opposed to merely sensing environmental conditions such as the local diffusion rate or degree of confinement. Thus, microfluidics is enabling the study of biological principles that would otherwise not be possible using conventional approaches in biology. We next focus on how data from these unique experimental studies can be used to validate mathematical models of biofilm formation and lead to new insights.

Mathematical Biofilm Modeling and Microfluidics

Thus far several determinants of the process of biofilm formation including fluid dynamics, cell phenotype, and cell-cell communication have been discussed. The complex nature of biofilm formation makes mathematical modeling challenging, yet critical for developing a fundamental quantitative understanding of the underpinnings of the process. Proper mathematical models take into account the multistaged process with its broad range of time and length scales. Mathematical development of biofilms has been long pursued with substantial developments having taken place in the past two decades. Klapper and Dockery present a comprehensive review of mathematical model development for biofilms [13]. Today, rapid developments in the biofilm community are pushing the need for better and comprehensive models; microfluidics presents itself as an invaluable aid to the modeling community. Microfluidics not only affords researchers to explore in real time biofilm growth and dynamics, but it also allows for a controlled microenvironment that can be adjusted to suit the need of the user. A controlled microfluidic environment amenable to probing by various sensors can allow researchers to pursue valuable validation experiments for proposed models. In one such study, Janakiraman et al. [14] used a microfluidic device to study biofilm development of *Pseudomonas aeruginosa* PA14, where mass and species transport effected biofilm development and vice versa. Their use of a microfluidic device allowed assessment of biofilm growth variation with shear rates, finally leading to model validation. The validated model, which took into account mass transport and its effect on quorum sensing, allowed valuable conclusions such as that flow rate could be used to turn on and turn off quorum sensing within the biofilm. Experimentally validated models like the one presented by Janakiraman et al. [14] allow a greater understanding of biofilm kinetics. In another study, Volfson et al. [15] used microfluidic devices to investigate spatial ordering and self-organization of microbes in a microfluidic device. The authors were investigating the role of contact biomechanics in the formation of dense colonies and toward this end they used a controlled microfluidic device to monitor the two-dimensional growth of motile microbes. The experimental investigations were combined with discrete element simulations (DES), and the authors showed that biomechanical interactions could lead to highly ordered structures in a microbial colony. Such investigations can help in the complete understanding of the role of various environmental and self-generated factors that play a role in biofilm formation. Deeper quantitative understanding of biofilm formation, resulting from the interplay between modeling and experimental validation with microfluidics, will drive progress in various applications.

Other Uses of Lab-On-A-Chip Technologies

Lab-on-a-chip (LoC) technologies refer to a suite of technologies that have evolved primarily over the last decade, where complex operations such as cell culture and sensing are integrated in a miniaturized platform. Apart from sensors, LoC technologies allow microfluidic systems to be interfaced with other systems such as micro-electro-mechanical (MEMs) and opto-electric systems. MEMs devices in LoC systems have been harnessed for various studies on microbial biofilms. Several different MEMs devices have been used to pursue biofilm-related studies, thus providing key insights into fundamental phenomena such as cellular self-organization in biofilms and the role of motility [16]. Purely electrical systems have also been incorporated into LoC systems and have been employed to investigate physiological heterogeneity within microbial biofilms [17]. Emerging lab-on-a-chip technologies can also benefit biofilm research. For example, novel opto-electric techniques [18] might find a use in biofilm engineering. Rapid electro kinetic patterning (REP) is an opto-electric technique, which can be used to manipulate micro- and nanosized objects noninvasively in a microfluidic device. REP has already been used to capture an aggregation of *Shewanella oneidensis* MR-1 in a microfluidic device and place it at user-defined intervals using an infrared laser and electric fields [18]. Such an approach could be used to trigger and organize the formation of beneficial biofilms.

Other than such appeal, LoC technologies are also being employed to fabricate miniaturized devices for applications such as current production and miniaturization of assays for clinical testing.

Future Perspectives

Microfluidic systems integrated with mechanical and/or electrical transducers and sensor components open up new ways to understand biofilm formation and bacterial surface interactions. Biofilms subjected to external stimuli through molecular cues during the growth and establishment stage can be characterized for their mechanical properties more accurately using microfluidic systems. Qualitative assessment of temporal and spatial patterns of biofilm formation against antimicrobial actions (chemical method) or electric field interference [19] or a physical disruption can be effectively studied using microfluidic systems.

Acknowledgments A. Kumar performed the work as a Eugene P. Wigner Fellow at the Oak Ridge National Laboratory (ORNL). The authors acknowledge research support from the US Department of Energy (US DOE) Office of Biological and Environmental Sciences. ORNL is managed by UT-Battelle, LLC, for the US DOE under Contract no. DEAC05-00OR22725.

Cross-References

References

1. Costerton, J.W., Lewandowski, Z., Caldwell, D.E., Korber, D.R., Lappinscott, H.M.: Microbial biofilms. Annu. Rev. Microbiol. **49**, 711–745 (1995)
2. Becker, H., Gartner, C.: Polymer microfabrication methods for microfluidic analytical applications. Electrophoresis **21**(1), 12–26 (2000)
3. Madou, M.: Fundamentals of Microfabrication: The Science of Miniaturization, 2nd edn. CRC Press, Boca Raton (2002)
4. Yawata, Y., Toda, K., Setoyama, E., Fukuda, J., Suzuki, H., Uchiyama, H., Nomura, N.: Bacterial growth monitoring in a microfluidic device by confocal reflection microscopy. J. Biosci. Bioeng. **110**(1), 130–133 (2010)
5. Lee, J.-H., Kaplan, J., Lee, W.: Microfluidic devices for studying growth and detachment of *Staphylococcus epidermidis* biofilms. Biomed. Microdevices **10**(4), 489–498 (2008)
6. Rusconi, R., Lecuyer, S., Guglielmini, L., Stone, H.A.: Laminar flow around corners triggers the formation of biofilm streamers. J. R. Soc. Interface **7**(50), 1293–1299 (2010)
7. Richter, L., Stepper, C., Mak, A., Reinthaler, A., Heer, R., Kast, M., Bruckl, H., Ertl, P.: Development of a microfluidic biochip for online monitoring of fungal biofilm dynamics. Lab Chip **7**(12), 1723–1731 (2007)
8. De La Fuente, L., Montanes, E., Meng, Y., Li, Y., Burr, T.J., Hoch, H.C., Wu, M.: Assessing adhesion forces of type I and type IV pili of Xylella fastidiosa bacteria by use of a microfluidic flow chamber. Appl. Environ. Microbiol. **73**(8), 2690–2696 (2007)
9. Timp, W., Mirsaidov, U., Matsudaira, P., Timp, G.: Jamming prokaryotic cell-to-cell communications in a model biofilm. Lab Chip **9**(7), 925–934 (2009)
10. Connell, J.L., Wessel, A.K., Parsek, M.R., Ellington, A.D., Whiteley, M., Shear, J.B.: Probing prokaryotic social behaviors with bacterial "Lobster Traps". MBio **1**(4), e00202–e00210 (2010)
11. Boedicker, J.Q., Vincent, M.E., Ismagilov, R.F.: Microfluidic confinement of single cells of bacteria in small volumes initiates high-density behavior of quorum sensing and growth and reveals its variability. Angew. Chem. Int. Ed. Engl **48**(32), 5908–5911 (2009)
12. Park, S., Wolanin, P.M., Yuzbashyan, E.A., Silberzan, P., Stock, J.B., Austin, R.H.: Motion to form a quorum. Science **301**(5630), 188 (2003)
13. Klapper, I., Dockery, J.: Mathematical description of microbial biofilms. SIAM Rev. **52**(2), 221–265 (2010)
14. Janakiraman, V., Englert, D., Jayaraman, A., Baskaran, H.: Modeling growth and auorum sensing in biofilms grown in microfluidic chambers. Ann. Biomed. Eng. **37**(6), 1206–1216 (2009)
15. Volfson, D., Cookson, S., Hasty, J., Tsimring, L.S.: Biomechanical ordering of dense cell populations. Proc. Natl. Acad. Sci. U. S. A. **105**(40), 15346–15351 (2008)
16. Ingham, C.J., Vlieg, J.: MEMS and the microbe. Lab Chip **8**(10), 1604–1616 (2008)
17. Stewart, P.S., Franklin, M.J.: Physiological heterogeneity in biofilms. Nat. Rev. Microbiol. **6**(3), 199–210 (2008)
18. Kumar, A., Williams, S.J., Chuang, H.-S., Green, N.G., Wereley, S.T.: Hybrid opto-electric manipulation in

microfluidics-opportunities and challenges. Lab Chip **11**(13), 2135–2148 (2011)

19. Kumar, A., Mortensen, N.P., Mukherjee, P.P., Retterer, S.T., Doktycz, M.J.: Electric field induced bacterial flocculation of enteroaggregative Escherichia coli 042. Appl. Phys. Lett. **98**(25), 253701–253703 (2011)

Biognosis

▶ Biomimetics

Bio-inspired CMOS Cochlea

Andreas G. Katsiamis[1], Emmanuel M. Drakakis[2] and Richard F. Lyon[3]
[1]Toumaz Technology Limited, Abingdon, UK
[2]Department of Bioengineering, The Sir Leon Bagrit Centre, Imperial College London, London, UK
[3]Google Inc, Santa Clara, CA, USA

Synonyms

Bionic ear; Cochlea implant; Gammatone filters; Log-domain; Low-power

Definition

This chapter deals with the design and performance evaluation of a new analogue CMOS cochlea channel of increased biorealism. The design implements a recently proposed transfer function [12], namely the One-Zero Gammatone Filter (or OZGF), which provides a robust foundation for modeling a variety of auditory data such as realistic passband asymmetry, linear low-frequency tail, and level-dependent gain. Moreover, the OZGF is attractive because it can be implemented efficiently in any technological medium – analogue or digital – using standard building blocks. The channel was synthesized using novel, low-power, Class-AB, log-domain, biquadratic filters employing MOS transistors operating in their weak inversion regime. Furthermore, the chapter details the design of a new low-power automatic gain-control circuit that adapts the gain of the channel according to the input signal strength, thereby extending significantly its input dynamic range. The performance of a fourth-order OZGF channel (equivalent to an eighth-order cascaded filter structure) was evaluated through both detailed simulations and measurements from a fabricated chip using the commercially available 0.35 μm AMS CMOS process. The whole system is tuned at 3 kHz, dissipates a mere 4.46 μW of static power, accommodates 124 dB (at <5% THD) of input dynamic range at the center frequency, and is set to provide up to 70 dB of amplification for small signals.

Introduction

The first generations of high-performance cochlea designs were synthesized using g_m-C filters and relied on power-hungry linearization techniques and/or the compressive action of the AGC for extending the input DR. However, recent advances in the field of analogue filter design have led to the development of inherently compressive systems that operate internally in the nonlinear domain while preserving overall input–output linearity. The application of *companding* in filter design resulted in the successful realization of topologies that were able to attain a wider input DR with a lower power-supply requirement compared to traditional g_m-C filters employing linearized transconductors. These companding filters or processors belong to the more general class of ELIN (Externally-Linear–Internally-Nonlinear) systems [21], and their systematic synthesis is articulated in the pioneering works of Frey [7] and Tsividis [20]. It is worth noting that the need for inherently compressive filters emerged very early in the development process of micropower, high DR cochlea designs and, in fact, a bit earlier than the first 1993 Log-domain paper by Frey [6].

Since Frey's [6] paper, Log-domain circuits progressed significantly with several contributions aiming at increasing the input DR and lowering the quiescent power dissipation. The two most thoroughly studied techniques are: (a) The use of two Class-A filters in a pseudo-differential Class-AB arrangement [8] that increases the DR without spending too much power and (b) the use of an AGC scheme that dynamically changes certain biasing levels of the filter (according to a particular measure of input signal

Bio-inspired CMOS Cochlea, Fig. 1 Block diagram of the active fourth-order OZGF channel

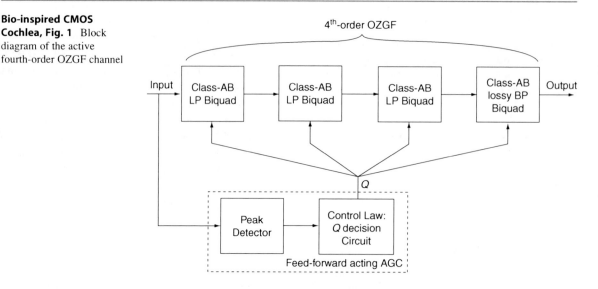

strength) in order to optimize its output SNR and power dissipation – this technique is otherwise known as *syllabic companding* [9] and constitutes indeed a parallel engineering strategy to the one performed by the real cochlea to widen its input DR and maintain signal integrity. It is interesting to note that several mechanisms within the cochlea may include an adjustable "DC bias" in the BM caused by hair-cell interactions that affect the operating point on their detection nonlinearity and their sensitivity [15].

In the following sections, the proposed pseudo-differential, Class-AB, Log-domain implementation of a fourth-order (i.e., a cascade of four, second-order transfer functions) OZGF with the ability to automatically adapt its gain by means of a low-power, AGC circuit acting *in feed-forward* is discussed. The reason a feed-forward scheme was employed (like the one demonstrated in [17]), as opposed to a feedback (like in [5]), was in order to avoid any latency issues and to obtain faster adaptation times.

The AGC senses the strength of the input signal and according to a particular control law sets the Q values of the individual filter stages of the cascade comprising the channel. The action of the AGC helps to accommodate a wider input DR because when the input is small (say below or close to the noise floor), the OZGF adapts its gain and provides significant amplification for the output to be well above noise levels. On the other hand, when the input is large, the OZGF provides low or no amplification so that acceptable distortion levels are maintained at the output. Since, in effect, the AGC is changing the state variables or pole positions

of the individual filters, the whole OZGF channel falls under the category of a *high-order, syllabically companding, ELIN, auditory (cochlea) amplifier/processor*. Figure 1 depicts the high-level block diagram of our *active* fourth-order OZGF channel architecture.

Clarification: Please note that the term "active" we refer to the channel together with its feed-forward AGC mechanism to distinguish it from the case where the AGC is disconnected (i.e., the OZGF gain is set manually off-chip); the latter case will be simply referred to as "open loop." It should be clarified here that strictly speaking, the term "feed-forward" always refers to an open-loop controller since it does not employ feedback to determine if its input has achieved the desired goal. This means that the system does not observe the output of the processes that it is controlling. Under this strict definition, the implemented OZGF is considered open-loop *irrespective of whether the AGC is active or not*, but during the course of this chapter, we shall use the terms "active" and "open loop" to refer to an OZGF channel "with-" and "without-" AGC, respectively.

The Circuits

This section deals with the design of the circuits comprising the fourth-order OZGF channel with its AGC. All device sizes, biasing currents, capacitors, and power supplies are displayed in Table 1 in section Active OZGF Channel Measured Results.

Bio-inspired CMOS Cochlea, Table 1 Electrical and device parameters

Topology	Biquads	GMS	LPF	E-cell[a]	OTA[a]
$(W/L)_{PMOS}$	300 μm/1.5 μm				20 μm/1.5 μm
$(W/L)_{NMOS}$	60 μm/8 μm				10 μm/1.5 μm
I_o	20 nA				
I_Q	2 nA–24 nA (i.e., Q between 0.834–10)				
I_Z	2 nA (i.e., $0.1 I_o$ for a–20 dB DC gain)				
I_{LPF}	20 nA				
$I_{o\_control}$	42 nA				
I_{tail}	24 nA				
V_{DD}	1.8 V				
C	20 pF				
C_{LPF}	80 pF				

[a]The reported dimensions correspond to the minimum device sizes of Fig. 13.

Pseudo-differential Class-AB Log-Domain Biquads

Class-AB designs are offered as the most efficient solution for balancing DR and low quiescent power-consumption performance requirements. This is because for small signals, a Class-AB design operates with the quality of a Class-A system, whereas for large signals with the efficiency of a Class-B system.

In Log-domain, one way to design a Class-AB filter is by connecting two Class-A filters in a pseudo-differential arrangement with a signal conditioner at the input. The signal conditioner (a geometric mean splitter (GMS)) ensures that a bidirectional input signal is split into two complementary, unidirectional, positive signals, which are then processed separately by the two Class-A filters. The respective unidirectional processed outputs are subsequently subtracted to form the total bidirectional output, which is a linearly filtered version of the input (see Fig. 2). However, this is not enough to guarantee Class-AB operation unless the following two rules (implied by the vertical, two-sided arrows in Fig. 2) are obeyed [8]:

1. Firstly, a nonzero DC-operating-point solution must always exist for all the state variables (these are currents in Log-domain filter implementations) for any strictly positive, static (i.e., DC) values of the complementary inputs. This can be achieved by ensuring that the derivatives of the state-variables can become equal to zero while the state-variables themselves remain strictly positive.

2. Secondly, all state-variables must remain strictly positive provided that the complementary inputs

remain strictly positive and bounded for all time. This can be achieved by enforcing the derivative of a state-variable to be strictly positive in the limit as the variable approaches zero; hence, that variable can never reach zero and thus will stay strictly positive for all time.

In short, the above two rules state that in a transistor-level implementation of a Class-AB Log-domain topology, all devices should carry nonzero and strictly positive currents at all times.

One possible Class-AB-compatible states-space (SS) description for a differential biquad is given below [8]:

$$
\begin{aligned}
u &= u_u - u_l \\
\dot{x}_{u1} &= \omega_o \left(-\frac{1}{Q} x_{u1} + x_{l2} + u_u - \frac{x_{u1} x_{l1}}{g} \right) \\
\dot{x}_{u2} &= \omega_o \left(x_{u1} - \frac{x_{u2} x_{l2}}{g} \right) \\
\dot{x}_{l1} &= \omega_o \left(-\frac{1}{Q} x_{l1} + x_{u2} + u_l - \frac{x_{u1} x_{l1}}{g} \right) \\
\dot{x}_{l2} &= \omega_o \left(x_{l1} - \frac{x_{u2} x_{l2}}{g} \right) \\
y_{BP} &= x_{u1} - x_{l1}; \; y_{LP} = x_{u2} - x_{l2}
\end{aligned}
\tag{1}
$$

where u is the differential input, y_{LP}, y_{BP} are the LP and BP differential outputs, respectively, and g is a positive factor that depends on implementation and will be defined later on. The subscripts u and l denote "upper" and "lower" referring to the two individual Class-A filter branches. The interested reader could verify that the derivatives $\dot{x}_{uj}, \dot{x}_{lj} (j = 1, 2)$ always attain a positive value whenever their respective state-variables tend to zero, or in other words to guarantee the existence of a DC operating point (condition 1). Also, nonlinear cross-coupling terms of the form $x_{uj} x_{lj} (j = 1, 2)$ have been added to ensure that the derivatives can reach a zero value while the state-variables themselves remain strictly positive. Since the outputs are formed differentially, the nonlinear cross-coupling terms cancel out ensuring an overall linear characteristic.

Even though there exist various ways to synthesize Log-domain circuits, in the next section, we describe the transistor-level Log-domain implementation of (1) using the Bernoulli Cell (BC) formalism.

Bio-inspired CMOS Cochlea, Fig. 2 Block diagram of a general pseudo-differential Class-AB architecture. The two vertical *arrows* between the two Class-A filters indicate coupling that ensures correct Class-AB operation

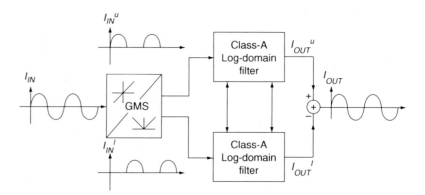

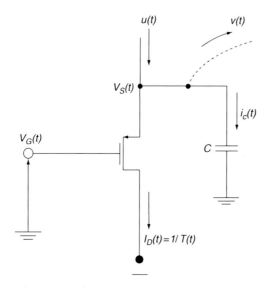

Bio-inspired CMOS Cochlea, Fig. 3 The Bernoulli cell

$$I_D(t) = I_S \exp\left(\frac{V_S(t) - V_G(t)}{nV_T}\right) = \frac{1}{T(t)} \quad (2)$$

$$\dot{T}(t) - \left[\frac{\dot{V}_G(t)}{nV_T} - \frac{u(t) - v(t)}{nCV_T}\right]T(t) - \frac{1}{nCV_T} = 0 \quad (3)$$

The synthesis of the Log-domain biquad will now be described in more detail. The transistor-level synthesis procedure is based on the direct comparison of the required dynamics of the original prototype system (i.e., (1)) with those codified by the LDSS equations; the scope of this comparison is to identify the necessary time-domain relations (implemented by means of static translinear loops (TL)) that modify the LDSS dynamics in such a way so that they become identical to the desired ones [3]. Having in mind a MOS-WI implementation, the two generic LDSS equations describing a Class-A 2nd-order system (i.e., the differential equations describing two interconnected, logarithmically driven BCs [3]) are:

$$nCV_T\dot{w}_1 + (u_1 - v_1)w_1 = I_{IN}$$
$$nCV_T\dot{w}_2 + (u_2 - v_2)w_2 = w_1 \quad (4)$$

with the w state variables being defined as:

$$w_1 = T_1 I_{IN} \text{(dimensionless)}$$
$$w_2 = T_2 T_1 I_{IN} = T_2 w_1 \text{(units of amperes}^{-1}) \quad (5)$$

and with C, n, and V_T denoting the grounded integrating capacitor, subthreshold slope parameter (usually between 1 and 2), and thermal voltage (25 mV at 300 K), respectively. At this point, the designer needs to decide on what kind of TL arrangement he/she wants to use for the interconnection of the BC. In this

The BC was introduced in [3] as a low-level operator for synthesizing Log-domain filters. It consists of an exponential transconductor and a grounded capacitor and, through a nonlinear change of its state-variable, see (2), linearizes and electronically solves a nonlinear differential equation of the well-known Bernoulli form, see (3). A cascade of BC is known as the "Bernoulli backbone" and implements a generic set of differential equations termed the Log-Domain State-Space or LDSS. Such a "backbone" is depicted in the Log-domain biquad implementation of Fig. 6 which realizes the SS equations shown in (1). In Figs. 4, 5, and 6, the BCs are denoted by circles, each consisting of a PMOS-WI transistor and a capacitor connected between its source terminal and ground (Fig. 3).

Bio-inspired CMOS Cochlea, Fig. 4 Two interconnected logarithmically driven BCs representing the differential equations described in (4). The *green* TL is used to "sense" the w_1-state variable, whereas the *red* TL the w_2-state variable

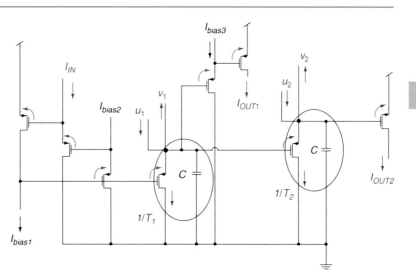

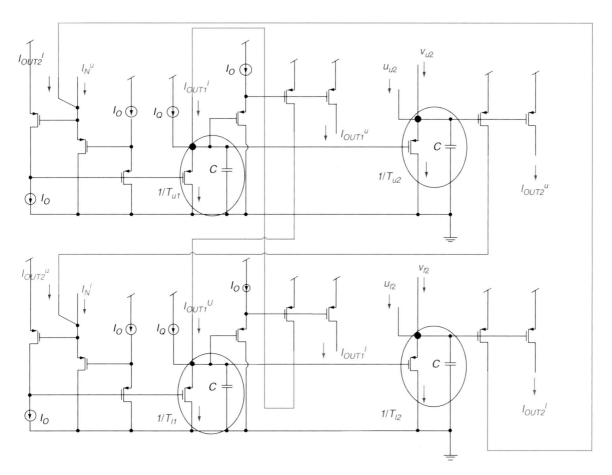

Bio-inspired CMOS Cochlea, Fig. 5 Circuit implementation of the *upper* and *lower* u_1 and v_1 currents

implementation, a used the "stacked" TL was used (as opposed to the "alternating" one), shown in Fig. 4 below. Note that the first BC is logarithmically driven because its gate voltage is essentially a level-shifted voltage arising from logarithmically compressing the input current I_{IN}.

Bio-inspired CMOS Cochlea, Fig. 6 The pseudo-differential Class-AB Log-domain Biquad. The design implements the state-space equations shown in (10)–(14). The feeding of the currents $I_{OUT1,2}^{u}$ ($I_{OUT1,2}^{l}$) to the lower (upper) topology (implementing the linear and nonlinear cross-coupling terms to ensure correct Class-AB operation) is shown in colored connections. All biasing current sources, as well as the subtraction of the I_{OUT} currents to form the outputs, were facilitated by means of both PMOS and NMOS cascode current mirrors not shown for clarity

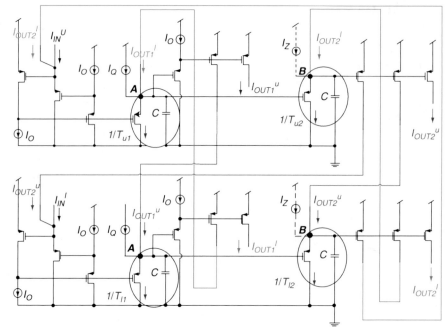

With Fig. 4 as a reference, the w state variables can be "sensed" through two TL and extracted as the two output currents I_{OUT1} and I_{OUT2}. Note that even though w_l is dimensionless and w_2 has dimensions of amperes^{-1}, both I_{OUT1} and I_{OUT2} have dimensions of amperes:

$$I_{OUT1} = \left(\frac{I_{bias1}I_{bias2}}{I_{bias3}}\right)T_1 I_{IN} = \left(\frac{I_{bias1}I_{bias2}}{I_{bias3}}\right)w_1$$
$$I_{OUT2} = T_2(I_{bias1}I_{bias2})T_1 I_{IN} = T_2(I_{bias1}I_{bias2})w_1 \qquad (6)$$
$$= (I_{bias1}I_{bias2})w2$$

In the subsequent implementation $I_{bias1} = I_{bias2} = I_{bias3} = I_o$, thus (6) becomes:

$$I_{OUT1} = I_o w_1$$
$$I_{OUT2} = I_o^2 w_2 \qquad (7)$$

By,
1. Splitting the input I_{IN} into two positive complementary inputs I_{IN}^{u} and I_{IN}^{l} (we will describe how to split I_{IN} later in section The Input Signal Conditioner – The Geometric Mean Splitter (GMS)
2. Duplicating the circuit topology in Fig. 4

The LDSS equations in (4) are now described in differential form as:

$$nCV_T\dot{w}_{u1} + (u_{u1} - v_{u1})w_{u1} = I_{IN}^{u}$$
$$nCV_T\dot{w}_{u2} + (u_{u2} - v_{u2})w_{u2} = w_{u1}$$
$$nCV_T\dot{w}_{l1} + (u_{l1} - v_{l1})w_{l1} = I_{IN}^{l} \qquad (8)$$
$$nCV_T\dot{w}_{l2} + (u_{l2} - v_{l2})w_{l2} = w_{l1}$$

where the subscripts u, l again indicate "upper" and "lower." Now, one needs to compare (8) with (1) and define the u, v currents via TL so that the two sets of relations become identical. As an example consider the relation of (1) and rearrange it slightly to:

$$\dot{x}_{u1} + \left(\frac{\omega_o}{Q} + \omega_o\frac{x_{l1}}{g}\right)x_{u1} = \omega_o x_{l2} + \omega_o u_u \qquad (9)$$

At this point, one needs to define the u, v current of the first relation in (8) (i.e., the u, v currents of the first BC) so that it "matches" that of (9). This comparison results in:

$$nCV_T\dot{w}_{u1} + (I_Q + I_o w_{l1})w_{u1} = I_o^2 w_{l2} + I_{IN}^{u} \qquad (10)$$

where I_Q, I_o will be defined later on. Please check the dimensional consistency of (10), given (5). Consequently, the same comparison is performed for the lower (complementary) part of the circuit.

$$nCV_T\dot{w}_{l1} + (I_Q + I_o w_{u1})w_{l1} = I_o^2 w_{u2} + I_{IN}^{l} \qquad (11)$$

Bio-inspired CMOS Cochlea, Fig. 7 The fourth-order OZGF pseudo-differential Class-AB Log-domain channel architecture

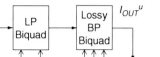

The above comparison has defined the w state variables in (6) now to be:

$$
\begin{aligned}
w_{u1} &= T_1 I_{IN} = \left[T_{u1}\left(I_{IN}{}^u + I_o{}^2 w_{l2} \right) \right] \\
&= \left[T_{u1}\left(I_{IN}{}^u + I_{OUT2}{}^l \right) \right] \\
w_{l1} &= T_1 I_{IN} = \left[T_{l1}\left(I_{IN}{}^l + I_o{}^2 w_{u2} \right) \right] \\
&= \left[T_{l1}\left(I_{IN}{}^l + I_{OUT2}{}^u \right) \right]
\end{aligned}
\tag{12}
$$

The circuit implementation of (10)–(12) is shown in Fig. 5. Note the feeding of the "sensed" w state variables (i.e., $I_{OUT1,\,2}{}^u$) from the upper to the lower portion of the circuit (purple-colored connections) and vice-versa (green-colored connections) which implement the nonlinear cross-coupling terms that guarantee true Class-AB operation.

Similarly, by performing the same comparison for the other two sets of relations, the u, v currents for the second BC are defined.

$$
\begin{aligned}
nCV_T \dot{w}_{u2} + \left(I_o{}^2 w_{l2} \right) w_{u2} &= w_{u1} \\
nCV_T \dot{w}_{l2} + \left(I_o{}^2 w_{u2} \right) w_{l2} &= w_{l1}
\end{aligned}
\tag{13}
$$

With the w state variables defined as:

$$
\begin{aligned}
w_{u2} &= T_2 T_1 I_{IN} = \left[T_{u2} T_{u1}\left(I_{IN}{}^u + I_o{}^2 w_{l2} \right) \right] \\
&= \left[T_{u2} T_{u1}\left(I_{IN}{}^u + I_{OUT2}{}^l \right) \right] = T_{u2} w_{u1} \\
w_{l2} &= T_2 T_1 I_{IN} = \left[T_{u2} T_{u1}\left(I_{IN}{}^u + I_o{}^2 w_{u2} \right) \right] \\
&= \left[T_{l2} T_{l1}\left(I_{IN}{}^l + I_{OUT2}{}^u \right) \right] = T_{l2} w_{l1}
\end{aligned}
\tag{14}
$$

In relations (10)–(14), $I_{IN}{}^{u,l}$ denote the two upper and lower complementary inputs (generated from the input signal conditioner described next), $1/T_{u,lj}(j = 1, 2)$ are the upper and lower BC drain currents, and I_Q, I_o

are biasing currents that control the quality factor Q and pole frequency ω_o, respectively. Finally, the positive factor g appearing in (1) was set equal to the biasing current I_o. The final circuit implementation is shown in Fig. 5. Finally, the outputs are formed by taking the difference of the respective "sensed" w state variables (i.e., $I_{OUT1}{}^u - I_{OUT1}{}^l$ and $I_{OUT2}{}^u - I_{OUT2}{}^l$). The LP and BP transfer functions are given by (16) and (17), whereas (18) describes the two-pole, one-zero transfer function required for the final OZGF channel. The transfer function in (18) is obtained by adding the biasing current I_Z at points B of Fig. 6. Note also that the addition of the biasing current I_Z alters (13) in the following way:

$$
\begin{aligned}
nCV_T \dot{w}_{u2} + \left(I_o{}^2 w_{l2} + I_Z \right) w_{u2} &= w_{u1} \\
nCV_T \dot{w}_{l2} + \left(I_o{}^2 w_{u2} + I_Z \right) w_{l2} &= w_{l1}
\end{aligned}
\tag{15}
$$

By inspecting (16)–(18): $\omega_o = I_o/nCV_T$, $Q = I_o/I_Q$ and the zero position is controlled by the ratio I_o/I_Z. The final fourth-order OZGF pseudo-differential Class-AB channel architecture is depicted in Fig. 7.

$$
\begin{aligned}
\frac{I_{OUT_{LP}}}{I_{IN}} &= \frac{I_{OUT2}{}^u - I_{OUT2}{}^l}{I_{IN}{}^u - I_{IN}{}^l} \\
&= \frac{\left(I_o/nCV_T \right)^2}{s^2 + \frac{(I_o/nCV_T)}{(I_o/I_Q)} s + \left(I_o/nCV_T \right)^2}
\end{aligned}
\tag{16}
$$

$$
\begin{aligned}
\frac{I_{OUT_{BP}}}{I_{IN}} &= \frac{I_{OUT1}{}^u - I_{OUT1}{}^l}{I_{IN}{}^u - I_{IN}{}^l} \\
&= \frac{\left(I_o/nCV_T \right)s}{s^2 + \frac{(I_o/nCV_T)}{(I_o/I_Q)} s + \left(I_o/nCV_T \right)^2}
\end{aligned}
\tag{17}
$$

$$\frac{I_{OUT_{2P1Z}}}{I_{IN}} = \frac{I_{OUT1}{}^{u} - I_{OUT1}{}^{l}}{I_{IN}{}^{u} - I_{IN}{}^{l}} = \frac{\left(\frac{I_Q}{nCV_T}\right)\left[s + \frac{I_Z}{nCV_T}\right]}{s^2 + \frac{a \times I_Q}{nCV_T}s + \frac{\beta \times I_o{}^2}{(nCV_T)^2}}$$

$$\text{with } a = \left(1 + \frac{I_Z}{I_Q}\right) \text{and } \beta = \left(1 + \frac{I_Q I_Z}{I_o{}^2}\right)$$

$$(18)$$

Clarification: Note that (18) has not the same denominator as (16) or (17). If one cascades (18) to a cascade of identical (16), then he/she will obtain a quasi-OZGF response which is not consistent with the transfer function discussed in [12] The choice for this discrepancy was deliberate for the following two reasons:

1. By adding a DC current I_Z at point B in Fig. 6, a real zero in the transfer function is realized with the simplest possible way without adding extra circuitry. If one opts for (18) to have the same denominator as (16) or (17), then the various loops would have to be closed through exponential transconductors (E-cells) which would possibly degrade the performance of the resulting topology.

2. Since in this implementation I_Z is always fixed and 10 times smaller than I_o (in order to realize the -20 dB tail at DC), both a, β in (18) approach to unity when I_Q is maximum (i.e., for low gains – recall $Q = I_o/I_Q$). In that case, (18) approaches to a lossy BP transfer function with the same denominator as (16) or (17). For smaller I_Q values (i.e., for higher gains) the variation in gain and pole frequency due to a and β increases but still does not affect the overall response noticeably due to the fact that in our fourth-order OZGF implementation, (18) is preceded by three identical LP biquads. This fact will become more apparent later on when we will discuss the measured frequency response of our OZGF implementation.

In other words, we chose to deviate from an ideal OZGF response due to the fact that it was easier to design (18) without sacrificing significantly any performance or modeling (like peak gain, CF, etc.) specifications.

The Input Signal Conditioner – The Geometric Mean Splitter (GMS)

For the input signal conditioner, we chose to implement the GMS for three reasons:

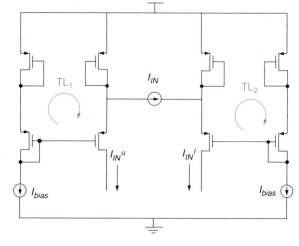

Bio-inspired CMOS Cochlea, Fig. 8 PMOS-WI implementation of the geometric mean splitter

1. It exhibits a good frequency response since it contains well-defined low-impedance points.
2. It can be realized with few noncritical components.
3. And for its low DC levels, ensuring lower static power consumption and noise, relative to other splitters employing, for example, the harmonic mean law.

Figure 8 depicts the PMOS-WI GMS implementation used in this work. From the two (left and right) TL, one may deduce that:

$$I_{IN}{}^{u} \times I_{IN}{}^{l} = I_{bias}{}^2 (= I_o{}^2) \tag{19}$$

It also holds that : $I_{IN} = I_{IN}{}^{u} - I_{IN}{}^{l}$ (20)

Moreover, the large-signal expressions for the two positive, complementary, upper and lower, unidirectional currents are given by:

$$I_{IN}{}^{u} = \frac{I_{IN} + \sqrt{I_{IN}{}^2 + 4I_o{}^2}}{2} \tag{21}$$

$$I_{IN}{}^{l} = \frac{-I_{IN} + \sqrt{I_{IN}{}^2 + 4I_o{}^2}}{2} \tag{22}$$

Relations (19) and (20) together describe the splitting action and the law with which the two positive complementary inputs are defined. Figure 9 illustrates typical current waveforms generated by the GMS used in this work. The biasing current I_{bias} could in principle

Bio-inspired CMOS Cochlea, Fig. 9 Indicative current waveforms generated by the GMS. The upper (*solid*) and lower (*dotted*) complementary inputs resemble half-wave rectified signals

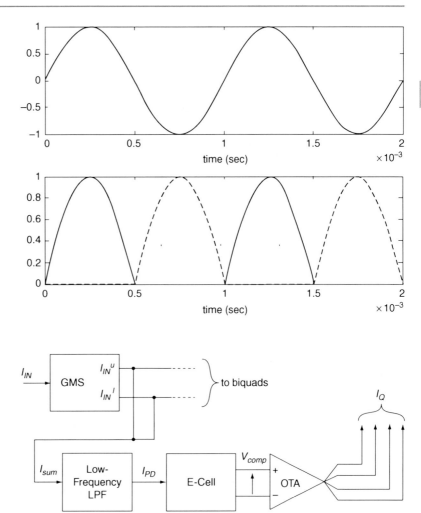

Bio-inspired CMOS Cochlea, Fig. 10 Block diagram describing the processing stages of the AGC

be set arbitrarily small to reduce the noise and static power but, in the subsequent OZGF, implementation was set equal to I_o in order to ensure similar biasing conditions (and hence impedance levels) for all the transistors comprising the OZGF channel and reduce the overall number of external pin connections.

Finally, it should be noted that one of the potential disadvantages of the GMS topology of Fig. 8 is the implementation of the floating current source I_{IN}. In practice, this current source was realized via two grounded AC-current sources at an 180° phase difference with each other. The relative matching between the two sources will play a significant role in the linearity performance of the OZGF channel since the performance of the whole system cannot exceed that of its input stage. This point will be addressed in more detail in a later section.

The Automatic Gain-Control Circuits

The AGC should be designed in such a way so that a potentially wide DR at the input of the OZGF channel (greater than six orders of magnitude in this implementation) can be compressed to a small current range (typically within one order of magnitude) for the biasing current I_Q that sets the Q of the biquads. The proposed AGC acts in feed-forward and consists of a cascade of four circuit blocks: a GMS, a low-frequency LP filter (LPF), an exponential transconductor (or E-cell) operating as a logarithmic transimpedance amplifier, and a wide linear-range OTA.

With the aid of Fig. 10, the processing stages can be summarized as follows:

- Stage 1: The GMS ensures that for small signals, the OZGF gain does not exceed a certain value or shoot to instability (i.e., it sets the highest OZGF peak gain for zero input) and also full-wave rectifies the bidirectional input I_{IN} (I_{sum}).

- Stage 2: The low-frequency current-input current-output LPF gives a quasi-DC output (I_{PD}), whose value corresponds to the peak value of the full-wave rectified current-input I_{sum}. In other words, the GMS together with the LPF implement a peak detector (i.e., full-wave rectification plus averaging).
- Stage 3: The E-cell compresses the, still-wide, current range at the output of the LPF to a differential voltage range (V_{comp}) that can be easily accommodated by the OTA that follows. The compressive I-V transfer characteristic is electronically controlled by a single biasing current $I_{o\_control}$.
- Stage 4: The OTA converts V_{comp} to a DC current range, corresponding to the Q range of the OZGF. The tail current of the OTA (I_{tail}) together with $I_{o\_control}$ of the E-cell are the two main parameters that control the law with which the Q values are determined (i.e., which input strength value corresponds to which Q).

In the subsections that follow, the AGC circuits are described in detail.

AGC Stages 1 and 2: The GMS and Low-Frequency LPF
By adding the large-signal expressions of the two splitted complementary inputs (i.e., (21) and (22)) and by defining the peak values of the input as $\hat{I}_{IN} = m \times I_o$, with m being the modulation index, we arrive at the following result:

$$\hat{I}_{sum} = \frac{\hat{I}_{IN} + \sqrt{\hat{I}_{IN}^2 + 4I_o^2}}{2}$$
$$+ \frac{-\hat{I}_{IN} + \sqrt{\hat{I}_{IN}^2 + 4I_o^2}}{2}$$
$$= \sqrt{\hat{I}_{IN}^2 + 4I_o^2} = I_o\sqrt{m^2 + 4} \qquad (23)$$

with m indicating how many times larger or smaller is the input zero-to-peak amplitude relative to the biasing current I_o. The above relation states that for small m and less than 1 (i.e., when the biquads operate in their Class-A mode), $\hat{I}_{sum} \approx 2I_o$, whereas for $m \gg 1$, $\hat{I}_{sum} \approx mI_o = \hat{I}_{IN}$. In other words, for zero or small signals, the GMS outputs a current which has a minimum peak value close to $2I_o$, whereas for larger signals, it

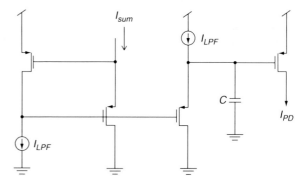

Bio-inspired CMOS Cochlea, Fig. 11 The low-frequency LPF topology – A first-order Class-A Log-domain integrator

follows the input peak almost exactly. Thus, by setting the control law to map the value $2I_o$ to the minimum I_Q value (corresponding to the maximum Q, since $Q = I_o/I_Q$), the OZGF channel will always have a bounded peak gain according to the required specifications and will never shoot up to large unwanted gain values or instability. That is indeed a "hidden" and very useful operation of the GMS because it can rectify the input and set the highest bound for the OZGF peak gain in a simple and elegant manner without needing to resort to additional circuitry.

The current I_{sum} is consequently filtered by the low-frequency LPF to extract its DC component I_{PD}. However, to accurately obtain the peak value, I_{sum} needs to be multiplied by an appropriate scale factor (the crest factor). This multiplication factor, which is waveform dependent and was set here to $\pi/2$, was implemented by means of scaled current mirrors in the GMS. Since I_{sum} is a rectified positive current, the LPF does not have to operate in Class-AB mode. The LPF topology used in this work and depicted in Fig. 11 is a simple Class-A, 1st-order, Log-domain integrator which implements the transfer function in (24). All symbols have their usual meaning.

$$H_{LPF}(s) = \frac{\omega'_o}{s + \omega'_o} = \frac{(I_{LPF}/nC_{LPF}V_T)}{s + (I_{LPF}/nC_{LPF}V_T)} \qquad (24)$$

Clarification: It is known from physiology that the IHC perform half-wave rectification to extract the energy of the BM vibrations. So why did we opt for a full-wave rectification in this implementation since clearly this choice is inconsistent with biology? The reason is that IHC and OHC are two separate sensors in the real cochlea. The IHC do not control the gain of the cochlea as OHC do, but through the auditory nerves create action potentials to convey information to the brain. For this reason, both the phase and frequency information of the signal they detect is important. *Half-wave rectification enhances the zero-crossings of the signal while preserving its frequency information, whereas full-wave does not (it doubles the frequency).* In this implementation, the silicon neurons that convert the output of the OZGF channel into spikes were not designed; only a circuit that models (at least qualitatively) the function of OHC for controlling the channel's gain. The proposed AGC operates virtually at DC and so both frequency and phase information is unimportant at this point. The choice for a full-wave rectification was made since the capacitor value used in the low-frequency LPF for smoothing the full-wave rectified signal can me made smaller than in the half-wave case, thereby saving area. That is one example of a choice that is not neuromorphic in its strict sense but, given the particular chip's area requirements, serves as a viable alternative.

AGC Stages 3 and 4: The E-cell and Wide Linear-Range OTA

The E-cell, shown in Fig. 12, is an exponential transconductor operating as a nonlinear transimpedance amplifier. It consists of two low-impedance points (the drain-gates of devices Q_1 and Q_3), where the compressed differential voltage V_{comp} is generated when the current I_{PD} is sourced from the diode-connected drain of device Q_1. Device Q_2 serves as a degeneration (tail) resistor which, if appropriately sized, gives additional headroom for accommodating the upper biasing current source $I_{o\_control}$ (implemented by means of cascoded mirrors not shown for simplicity). The reason for using an E-cell compared to a standard diode to perform the logarithmic compression is attributed to the fact that with an E-cell, the compressive *I-V* transfer characteristic can be electronically tuned by $I_{o\_control}$ as described by (25). From the TL Q_1Q_3 and by assuming that $I_{Do1} = I_{Do3}$ (i.e., matched devices):

$$V_+ - nV_T \ln\left(\frac{I_{PD}}{I_{Do1}}\right) + nV_T \ln\left(\frac{I_{o\_control}}{I_{Do3}}\right) - V_- = 0$$

$$\Rightarrow V_+ - V_- = V_{comp} = nV_T \ln\left(\frac{I_{PD}}{I_{o\_control}}\right)$$

(25)

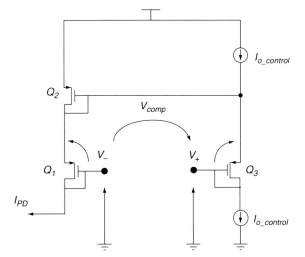

Bio-inspired CMOS Cochlea, Fig. 12 The exponential transconductor – E-cell used for AGC action

The compressed differential voltage V_{comp} is subsequently applied to an OTA whose linear range was widened by using two standard techniques:

1. Plurality of cross-coupled, source-coupled pairs [14, 19].
2. Source degeneration. More specifically, four OTAs, each with an appropriate offset voltage and tail current, were degenerated and connected in tandem in order to realize an overall flatter input–output slope. This was achieved by choosing the offset voltages and tail currents in such a way so that the higher-order even and odd derivatives of the total (combined) transconductance become zero.

The offset voltages were realized by scaling asymmetrically the dimensions of the devices that comprise the branches of each OTA pair (see Fig. 13). In [19], for maximally flat transconductance characteristics, these aspect ratios were calculated to be 13.4, 2.04, 1, 1 and with tail current multiplication factors of 1.83, 1, 1, 1.83. However, since accurate linearity is not a crucial performance requirement for the AGC, the device aspect ratios and tail current multiplication factors were set to 12, 2, 1, 1 and 2, 1, 1, 2, respectively, in order to simplify our layout matching efforts during fabrication phase. The scaling factor of 4.8, which all the tail currents are divided with, was chosen empirically so that the upper and lower saturating levels of the *V-I* tanh function coincide with the actual I_{tail} value. Figure 14 describes qualitatively the input to every AGC stage output transfer characteristics. The

Bio-inspired CMOS Cochlea, Fig. 13 The E-cell (dotted green lines) connected together with a wide linear-range OTA employing asymmetric cross-coupled, source-coupled pairs. The devices aspect ratios are indicated on the schematic

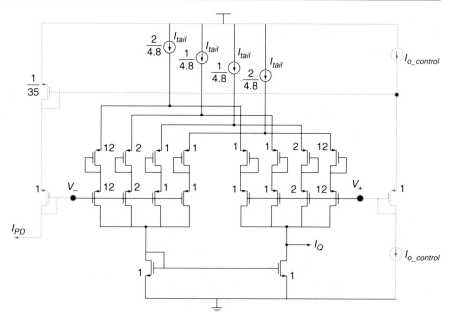

AGC Simulation and Measured Results

The simulation results presented in this section were obtained from Cadence IC Design Framework® and with the parameters presented in Table 1. Figure 15 shows the simulated V-I and transconductance transfer characteristics of the wide linear-range OTA for an I_{tail} of 24 nA. Strictly speaking, the achieved linear input range is around 200mV$_{peak}$. However, since a) the OTA operates in quasi-DC and b) the overall AGC characteristic must be compressive, the practically usable range is around 400mV$_{peak}$. It should be noted that since $I_Q^{max} \equiv I_{tail}$, the OTA sets the lowest bound of the OZGF peak gain according to $Q_{min} = I_o/I_{tail}$. The control law was calibrated (via the currents $I_{o\_control}$ and I_{tail}) to map the minimum signal at the input of the OZGF (which results in a value of $2I_o \times (\pi/2)$ at the output of the LPF of the AGC) to a maximum biquad Q of 10. This corresponds to a value of 2 nA for the respective I_Q currents and a nominal OZGF peak gain of ~80 dB. On the other

x-axes were exaggerated to indicate the various different ranges of operation. In essence, the AGC is able to compress a wide input range of 120 + dB to a much smaller range (~20 dB) for the Q values setting effectively the peak gain of the OZGF response according to input level.

hand, for a maximum allowable input of 10 μA (i.e., for $m = 500$), the AGC gives an I_Q equal to 20 nA corresponding to a Q of 1 (see Fig. 16). Since I_{tail} sets the upper level of the V-I transfer characteristic of the OTA, for very large input signals ($m > 500$) the maximum I_Q value saturates at 24 nA, corresponding to a lowest Q of 0.834 and a passive OZGF peak gain of 1.43 dB. Indicative waveforms at the outputs of each AGC stage are shown in Fig. 17.

The law with which the Q values are determined was obtained by performing a DC-sweep analysis on the whole AGC system. Figure 17 depicts I_Q versus I_{IN} curves with $I_{o\_control}$ and I_{tail} as the two implicit parameters. It can be observed that the overall input–output AGC transfer characteristic is logarithmically compressive in nature as suggested by (25) with $I_{o\_control}$ controlling the "vertical shift" of the characteristic and with I_{tail} controlling the actual level of compression as expected from (25) and OTA operation. Also observe that a linear change in I_{tail} results in a logarithmic change on the compression level, whereas a linear change in $I_{o\_control}$ results in a linear vertical shift of the whole characteristic. In conclusion, the above two electronically controllable degrees of freedom give a certain level of versatility to the AGC system because a variety of input DR values can be accommodated and mapped to a specific Q range according to the particular physiological (modelling) or design (performance) specifications (Figs. 18–20).

Bio-inspired CMOS Cochlea, Fig. 14 Input to every AGC stage output transfer characteristics. Note that the actual I_Q expression is an approximate one since a closed form expression for the transconductance of the wide linear-range OTA cannot be obtained

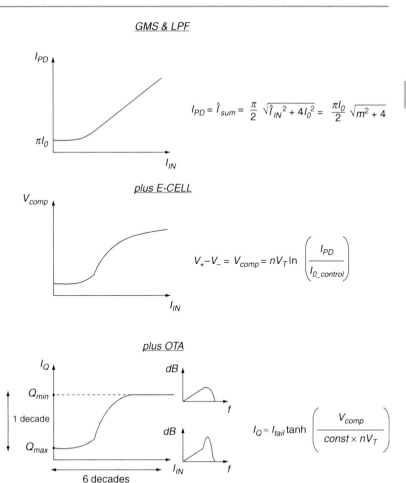

GMS & LPF

$$I_{PD} = \hat{I}_{sum} = \frac{\pi}{2}\sqrt{\hat{I}_{IN}^2 + 4I_0^2} = \frac{\pi I_0}{2}\sqrt{m^2 + 4}$$

plus E-CELL

$$V_+ - V_- = V_{comp} = nV_T \ln\left(\frac{I_{PD}}{I_{0\_control}}\right)$$

plus OTA

$$I_Q \approx I_{tail}\tanh\left(\frac{V_{comp}}{const \times nV_T}\right)$$

Active OZGF Channel–Measured Results

The whole channel, together with its AGC, was integrated in the standard 0.35 μm AMS 2P/4 M CMOS process. The OZGF was tuned so that its nominal center frequency (CF) falls at 3.3 kHz for a Q of 1. This corresponds to $I_o = I_Q = 20$ nA.

Measurement Setup: In section The Input Signal Conditioner – The Geometric Mean Splitter (GMS) it was mentioned that one of the disadvantages of the particular GMS topology was the need to accurately realize the floating current source I_{IN} in order to minimize distortion. In this case, this was achieved by using two 6221 precision AC/DC Keithley current sources. These were programmed using MatLab™ (via their GPIB interface) and triggered externally to output two AC-current waveforms at an exact 180° phase difference with each other. Moreover, by fine-tuning their relative amplitudes and phases, we could

partly overcome any mismatches between the two Class-A paths and, consequently, further optimize the linearity performance of the OZGF channel. The OZGF output current was measured via standard I-to-V converters (utilizing 100fA-leakage current AD549 Opamps) and through the 1MΩ input resistance of a Stanford Research 1mHz–100 kHz spectrum analyzer (the SRT785).

Table 1 shows the various dimensions of the devices comprising the channel and AGC. It should be emphasized that this choice occurred after careful deliberation and optimization. For the case of the Log-domain biquad (where high DR performance is of primary interest), a large W (such as 300 μm) shifts the $\ln I_{DS} - V_{GS}$ characteristic vertically so that the upper limit of the WI region extends to the μA-range, _while the leakage currents are still maintained below noise levels_. This is an important design choice to ensure that the full WI range, offered by the particular

Bio-inspired CMOS Cochlea, Fig. 15 *V-I-* (*upper*) and transconductance- (*lower*) simulated transfer characteristics of the wide linear-range OTA for $I_{tail} = 24$ nA

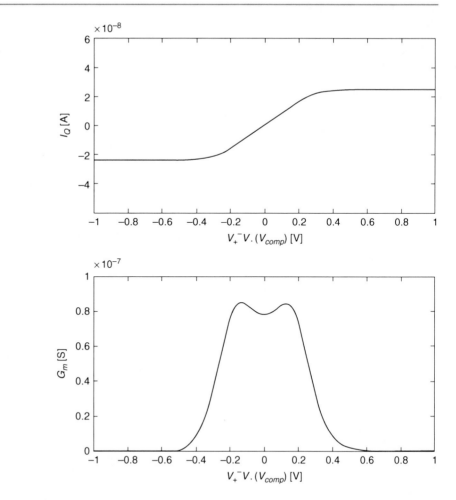

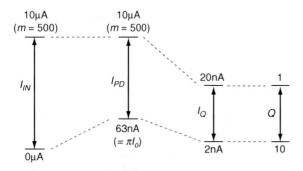

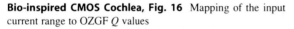

Bio-inspired CMOS Cochlea, Fig. 16 Mapping of the input current range to OZGF Q values

CMOS process, is exploited. In addition, the choice for L is equally important: For best THD performance, L should be made minimum since it affects the linearity of the WI slope, (i.e., the range of the flat portion of the $d(\ln I_{DS})/dV_{GS}$ versus V_{GS} curve) but a small L will give rise to channel-length modulation effects [1].

A larger L will definitely make the respective device's output impedance larger, but overall output THD will be reduced. For this reason, it was found that $L = 1.5$ µm (i.e., around 4 times the feature size) was a good compromise between driving capability and linearity. In addition, the device areas satisfied a 3σ-V_T mismatch of about ± 2 mV which is somewhere in between moderate to precise matching.

Frequency Response: Figure 21 shows the Q-tunability of the active fourth-order OZGF frequency response. The current I_Q was changed automatically via the action of the AGC by changing accordingly the strength of the input signal from the spectrum analyzer. Observe also that a linear change in I_Q corresponds to a logarithmic change in the peak gain (or Q) due to a) the compressive law of the AGC and b) due to the fact that $Q = I_o/I_Q$; this is indeed another form of compressive behavior which is embedded in the frequency response and stems directly from the

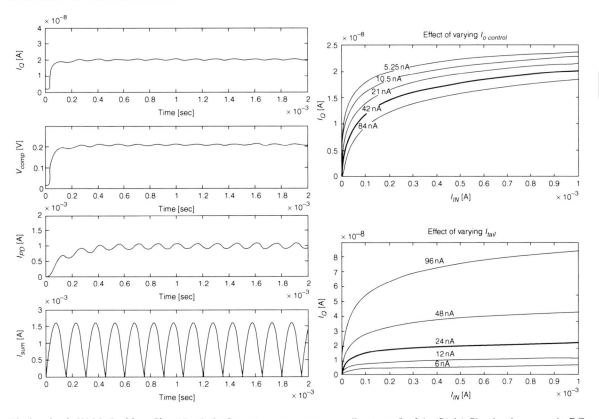

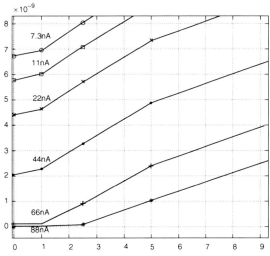

Bio-inspired CMOS Cochlea, Fig. 17 (*Left*) Current waveforms from the output of each AGC stage for an input signal of $m = 500$. Observe that the peak of I_{sum} is at $\sim$16 μA instead of 10 μA (due to scaling by $\pi/2$), whereas the LP-filtered version of I_{sum}, I_{PD} is at 10 μA as expected. The I_Q value is at 20 nA corresponding to a Q of 1. (*Right*) Simulated parametric DC response plots of the Q-control law with varying $I_{o\_control}$ (*upper*) and I_{tail} (*lower*). The bold curves are the ones corresponding to the values shown in

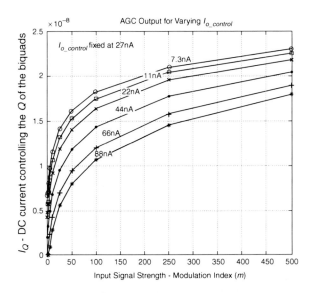

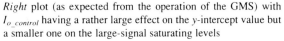

Bio-inspired CMOS Cochlea, Fig. 18 (*Left*) Measured parametric DC response plots of the Q-control law with varying $I_{o\_control}$. Note the characteristics for small-signals level of the Right plot (as expected from the operation of the GMS) with $I_{o\_control}$ having a rather large effect on the y-intercept value but a smaller one on the large-signal saturating levels

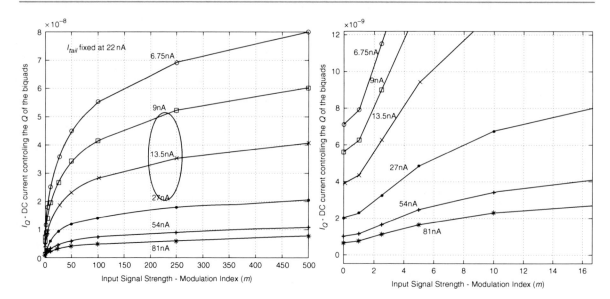

Bio-inspired CMOS Cochlea, Fig. 19 (*Left*) Measured parametric DC response plots of the Q-control law with varying I_{tail}. Note that the characteristics level of for small signals (as expected from the operation of the GMS) with I_{tail} having a small effect on the y-intercept value (*Right*) but a much larger one on the large-signal saturating levels

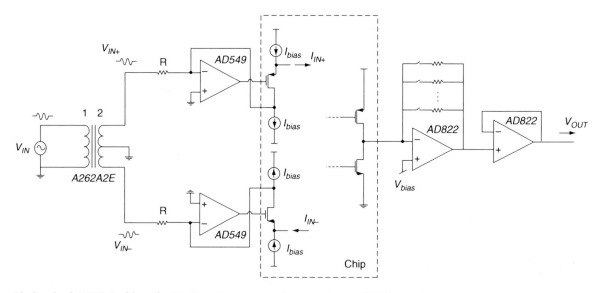

Bio-inspired CMOS Cochlea, Fig. 20   Experimental setup for measuring the OZGF channel

design equations of the Log-domain biquads. Figure 22 shows tuning of the OZGF's low-frequency tail by varying the current I_Z (note the quite small gain and pole frequency variation due to a and β terms in (18) – see again discussion at the end of section Pseudo-differential Class-AB Log-domain Biquads), whereas Fig. 23 shows the two orders of magnitude ω_o-tunability when changing the biasing current I_o to 2, 20 and 200 nA, successively. For each ω_o, two

indicative Q setups are shown to demonstrate gain adaptation within the whole audio spectrum. Observe the quite stable characteristics over frequency. Also note that the low-ω_o, high-Q response is affected by input signal noise because of the low biasing current levels of the particular configuration. Finally, it should be emphasized that all low-to-moderate Q responses were obtained with a large m ($\gg 1$) in order to validate the OZGF's Class-AB operation.

Bio-inspired CMOS Cochlea,
Fig. 21 *Q*-tunability of the
active fourth-order OZGF
frequency response. The
maximum peak gain was
measured at 70 dB

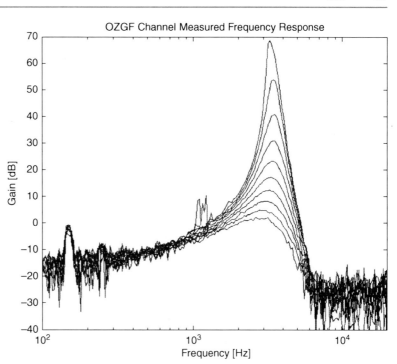

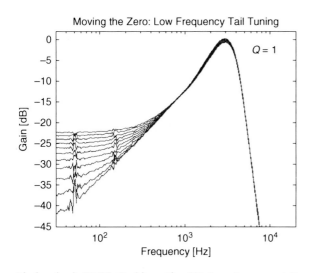

Bio-inspired CMOS Cochlea, Fig. 22 Low-frequency tail
tuning by means of varying the biasing current I_Z

Bio-inspired CMOS Cochlea, Fig. 23 Gain adaptation and
ω_o-tunability (CF from 330 to 33 kHz)

Total Harmonic Distortion (THD): The THD is
a common measure for assessing the linearity performance of amplifier systems. However, when talking
about audio-auditory processors, its value must be
appropriately judged since the "shape" of a signal's
spectrum is something that affects one's hearing perception. In other words, a signal with a certain THD
value might sound "aesthetically more pleasing" to our

ears than another of equal or lower THD value. In
commercial and academic cochlea designs,
a common upper limit for THD is 4–5% (e.g.,
Sarpeshkar's neuromorphic cochlea: <5% [17, 18],
MED-EL's state-of-the-art DUET EAS hearing system: <5%@500 Hz, Phonak Super-Front PPC-4
<7%@500 Hz, etc.). This was also justified experi-
mentally from a recent (and one of the first)

Bio-inspired CMOS
Cochlea, Fig. 24 Measured
single-tone linearity of the
fourth-order ($Q = 1$) OZGF
for various passband
frequencies

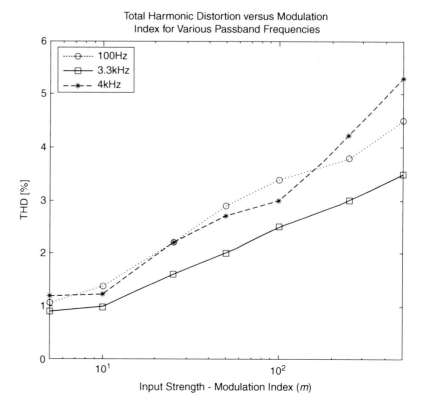

Log-domain implementations of a CMOS cochlea, where the effect of distortion started to become evident to the ear after 5% THD [10]. Finally, note that the 10%@1 kHz THD value represents a desirable amount of allowable distortion for typical low-cost wide-band audio material such as FM broadcasts, CD-media or cassette tape media. For these reasons, it was decided that 5% is a reasonable upper limit of THD for our OZGF cochlea channel.

Due to their differential nature, CMOS-WI Class-AB Log-domain filters exhibit mainly odd-harmonic distortion content. Their THD (for frequencies near the pole frequency) usually reaches the 1% value quite early (for m values as low as 1, i.e., at the border between Class-A and Class-AB operation) but grows rather slowly thereafter; this behavior renders Log-domain filters as adequate candidates for applications where moderate THD values can be tolerated. Figure 24 shows THD versus input strength (m) for various tones near-and-at the CF and deep in the passband for the open-loop $Q = 1$ OZGF response. At the CF, the THD for an input of 10 μA ($m = 500$) is at 3.5%. Figure 25 shows how the active OZGF adapts

automatically its peak gain with input level together with the corresponding output THD. For a 14pA input tone, the OZGF provides a peak gain of 70 dB at 0.4% output THD, whereas for an input tone of 20 μA, the peak gain is at −0.85 dB with an output THD of 5%.

Two-Tone Intermodulation Tests: Figure 26 shows both even- and odd-order measured intermodulation distortion products (IMD) for 2 t placed at a frequency ±2% away from the (CF) 3 kHz of the open-loop $Q = 1$ OZGF response. The IMD_2 starts at around −34 dB for small-signals and increases gradually to higher dB values with $f_2 - f_1$ increasing faster than $f_1 - f_2$. The IMD_3 is less attractive with a much unpredictable behavior toward smaller signals. These results unfortunately do not compare favorably to IMD figures reported from similar (biquadratic and not) prior CMOS-WI Log-domain efforts [11, 16] – even though those were (a) only IMD_2 (a quite unreasonable test for differential filters), (b) obtained at ±10% of a given frequency, and (c) for lower-order filters with some gain. It is believed that this set of results can be improved considerably in future implementations; nonetheless, one should appreciate that the real

Bio-inspired CMOS Cochlea, Fig. 25 Measured peak gain at CF and corresponding output THD versus I_{IN}

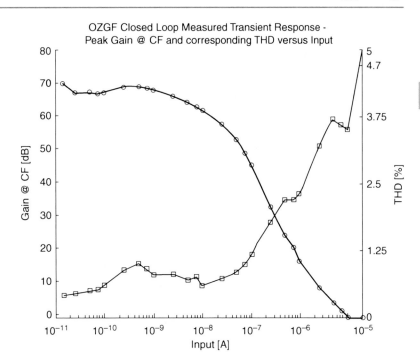

Bio-inspired CMOS Cochlea, Fig. 26 Measured two-tone third- and second-order intermodulation products of the fourth-order ($Q = 1$) OZGF for two frequencies at $\pm 2\%$ away from 3 kHz

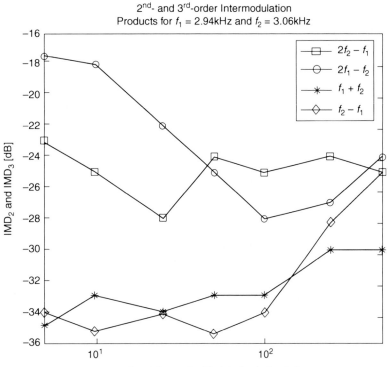

**Bio-inspired CMOS
Cochlea, Fig. 27** Simulated
two-tone third- and second-
order intermodulation
products of the second-
generation, fourth-order
($Q = 1$) OZGF for two
frequencies at $\pm 2\%$ away
from 3 kHz

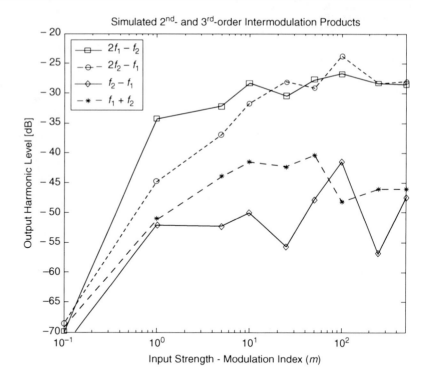

cochlea exhibits considerable intermodulation distortion, with both its $IMD_{2,\,3}$ levels around -20 dB for a large range of input intensities. In any case, we have managed recently to identify some potential optimization steps that may lead to an improved overall intermodulation linearity performance. A second-generation, fourth-order OZGF channel (one where its lossy BP stage is at the front of the cascade and not at the end) exhibited in simulation IMD_3 levels below -35 dB for small-signals, which also stayed below -25 dB across the whole input range (see Fig. 27).

Mismatch: One potential disadvantage of all pseudo-differential Class-AB filters of the type shown in Fig. 6 is that any potential mismatch between the two signal paths translates to distortion at the output. Therefore, considerable attention was given during layout in order to try and accurately match the two Class-A branches. Specifically, the two (upper and lower) TL comprising each Class-A biquad were nested together in such a way so that every upper transistor was interdigitized with its corresponding lower counterpart. Common centroid arrangements along both axes of symmetry were also employed. Figure 28 shows $Q = 1$ (upper plot) and $Q = 10$

(lower plot) measured OZGF frequency responses across 19 chips. Observe that in both cases the relative variation between responses does not exceed 6 dB.

Input Dynamic Range (DR): In this work, by input DR, we imply the ratio of the maximum input signal for a given allowable distortion at the output (measured in % of THD) over the noise floor for zero input. The measured input DR was calculated by taking into consideration the compressive action of the AGC by dividing the maximum signal at the input of the passive ($Q = 1$) OZGF response over the measured noise floor (integrated over the 3 dB bandwidth) of the fully active ($Q = 10$) OZGF response. The maximum input DR at CF was found to be 124 dB at $<5\%$ THD, although at different frequencies, this figure changed since the distortion is not uniform across the whole bandwidth of the OZGF. More specifically, we observed an abrupt increase in THD around 1/3 of the CF; a problem partially attributed to the fact that at 1/3 of the CF the input tone experiences a gain of ~ -10 dB, whereas the 3 rd-harmonic (which is generated and propagated across the four Log-domain biquads) a gain of ~ 2 dB. *In other words, the fundamental gets suppressed, whereas the distortion gets amplified.* Moving deeper in the passband, one would expect

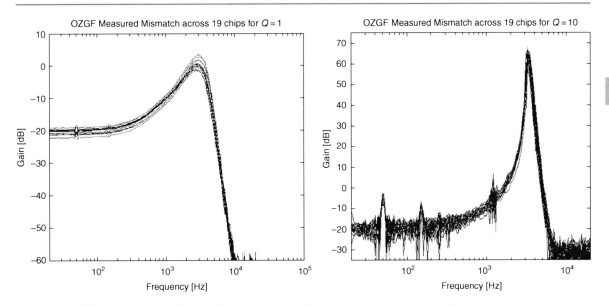

Bio-inspired CMOS Cochlea, Fig. 28 OZGF mismatch across 19 chips for both extreme cases of Q

that this particular problem would become more severe since both the third and fifth harmonics would get amplified while the fundamental would get suppressed even further. However, this is not entirely the case because for low frequencies, the whole structure operates virtually at DC, the capacitors draw smaller currents and, hence, the exerted nonlinearities are weaker.

It is really interesting to notice that the aforementioned phenomenon (that of "amplified distortion" as opposed to the more common "distorted amplification") is also observed in the biological cochlea. Recent measurements in the basal and apical turns of the guinea-pig cochlea [2, 13] revealed that the real cochlea exhibits even-order harmonic content with maximum second-order harmonic distortion levels of 4% and 28% near and one octave below the CF, respectively. Since our implementation contains odd-harmonic content, we expect the maximum distortion to occur at 1/3 of the CF, which indeed was the case. Figure 29 illustrates large-signal transient waveforms at 1/3 of the CF and CF, respectively. Similar distortion occurred in the design of [17], where THD values of as high as 40% were observed around one octave below the CF.

Figure 30 shows simulated results from the improved second-generation, fourth-order OZGF channel operating in open loop and with a $Q = 1$. The plot depicts THD values as a function of frequency with varying modulation index. Observe that for small m, the responses resemble HP filters, and BP filters for large m. The "sharpness" of these BP responses seem to increase with m, and their peak occurs approximately at 1/3 of the CF (around 1 kHz). Observe that for a wide range of input frequencies and strengths, the THD stays below 7%.

Out-of-Band Interferer: A well-known problem of all ELIN topologies is that their noise floor is signal-dependent, resulting in a constant output SNR of about 60 dB for large m values. This observation was experimentally verified as well in our OZGF implementation. One manifestation of this problem is the ability of the filter to maintain good signal integrity at the presence of an out-of-bound interferer. Figure 31 shows the measured PSD of an 1 µA ($m = 50$) in-band signal at CF. An out-of-band interferer was placed at 20 kHz and for two distinct amplitudes, 1 µA and 10 µA. It was observed that the noise PSD increased from around −60 dB for the case of an absent interferer to about −40 dB for the case of the 10 µA interferer; that is about 10 dB per decade of interferer amplitude increase as predicted in [4] and experimentally verified in [16].

In conclusion, Table 2 summarizes the measured performance of the active fourth-order OZGF cochlea channel, whereas Fig. 31: Masking effect of an out-of-band interferer.

Figure 32 shows a chip micrograph of the channel's layout.

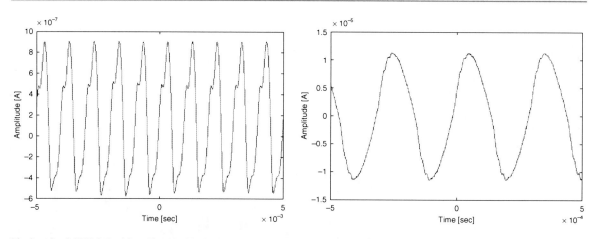

Bio-inspired CMOS Cochlea, Fig. 29   Typical large-signal output waveforms from the OZGF channel at 1/3 of the CF (*upper*) and at the CF (*lower*), respectively. The THD in the *upper plot* is in excess of 20%, whereas the distortion in the *lower plot* is less than 4%

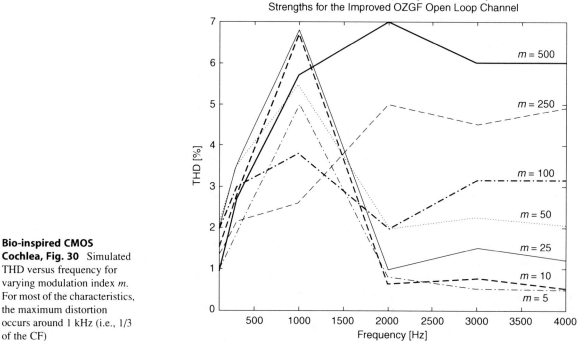

Bio-inspired CMOS Cochlea, Fig. 30 Simulated THD versus frequency for varying modulation index m. For most of the characteristics, the maximum distortion occurs around 1 kHz (i.e., 1/3 of the CF)

Summary, Discussion, and Conclusion

The architectural choices and measured results of this cochlea channel implementation can be summarized in the following points:

1. Instead of opting for a filter-cascade implementation, which is prone to noise and offset accumulation, gain sensitivity, and yield, this design is based on separate filter-bank channels of short-cascades. In this way, it was possible to model the propagation of distortion products; conserve on computation; and obtain realistic roll-off slopes, amplitude, and group-delay responses without the associate disadvantages of long-cascade models.

2. The real cochlea exhibits 120 dB of input dynamic range at 3 kHz at <5% THD. In this work, the

Bio-inspired CMOS Cochlea, Fig. 31 Masking effect of an out-of-band interferer

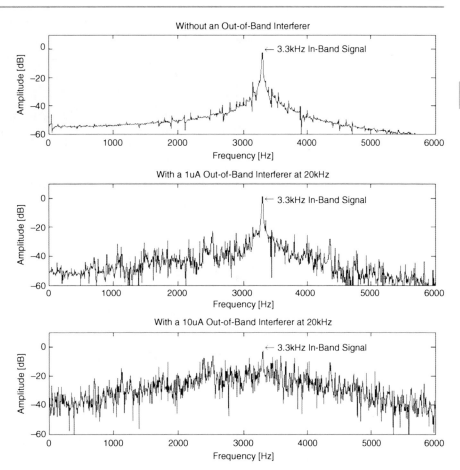

B

Power consumption	3.4 µW (no AGC), 4.46 µW (with AGC)
Noise floor	67.5pA ($Q = 1$), 14pA ($Q = 10$)
Centre frequency	3.3 kHz ($Q = 1$)–3.7 kHz ($Q = 10$)
Peak gain	70 dB with <1% THD
Input DR at CF	124 dB with <5% THD (with AGC)
SNR at CF	~60 dB for $m > 10$
Total on-chip capacitance	400 pF
Chip area	1.5 mm × 1.5 mm (2.25 mm^2)

measured results at the exact same frequency revealed an input DR of 124 dB at <5% THD.

3. The BM in the real cochlea has an area of approximately 70mm^2. This is equivalent to having a filter bank employing approximately thirty fourth-order OZGF channels.

4. A linear change in Q results in a logarithmic change in the peak gain (and in the temporal resolution) of our Log-domain OZGF response (see Fig. 21). Contrary to prior cochlea implementations, this is compatible with biology.

5. The implemented fourth-order OZGF frequency response achieves 70 dB of amplification for small-signals (even though in practice, we could get gains as high as 100 dB) while maintaining a linear low-frequency tail (20 dB/Dec) with steep high-frequency roll-off slopes (−160 dB/Dec).

6. The AGC dissipates 1 µW and compresses the six orders of magnitude of input signal intensity into one decade of Q-value tuning range. Moreover, it sets the lower and upper bounds of the peak-gain

Bio-inspired CMOS Cochlea, Fig. 32 Chip micrograph of the fourth-order OZGF channel with the AGC

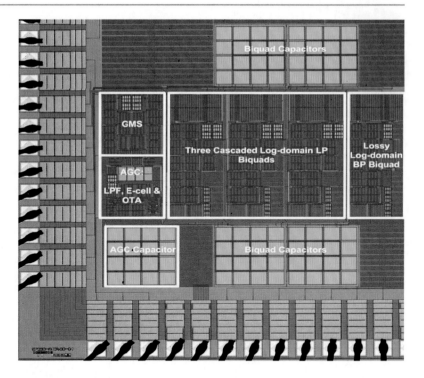

response avoiding excessive distortion or instability. In addition, asymmetric attack and release time constant programmability can be easily incorporated by modifying the LPF like in [22]. Lastly, in this implementation, the AGC acts in feed-forward and takes as input the global input. The downside of such a scheme is that the AGC does not contain any frequency information. The AGC can become frequency-dependent by modifying the architecture to accept as an input the output from the last BP OZGF stage or any other stage in between (in that case the AGC regulation will be closed-loop).

All the circuits presented in this work may serve as basic building blocks for the realization of new biorealistic cochlea processors designed in any of the two commonly used architectures; filter bank, or filter-cascade. Nonetheless, the fourth-order OZGF channel presented here is a good compromise between biorealism, circuit complexity, and yield and may be suitable in filter-bank applications were biological fidelity and/or high DR performance is of primary interest. The choice for using CMOS-WI pseudo-differential Class-AB Log-domain biquads for low-power, high DR operation has proven successful at

the expense of an increased chip area since the capacitor and transistor count increases by a factor of 2.

References

1. Binkley, D.M., Hopper, C.E., Tucker, S.D., Moss, B.C., Rochelle, J.M., Foty, D.P.: A CAD methodology for optimizing transistor current and sizing in analog CMOS design. IEEE Trans. Comput. Aided Des. Integr. Circ. Syst. **22**, 225–237 (2003)
2. Cooper, N.P., Rhode, W.S.: Nonlinear mechanics at the apex of the guinea-pig cochlea. Hear. Res. **82**, 225–243 (1995)
3. Drakakis, E.M., Payne, A.J., Toumazou, C.: "Log-domain state-space": a systematic transistor-level approach for log-domain filtering. IEEE Trans. Circ. Syst. II Analog Digit. Signal Process. **46**, 290–305 (1999)
4. Enz, C., Punzenberger, M., Python, D.: Low-voltage log-domain signal processing in CMOS and BiCMOS. IEEE Trans. Circ. Syst. II Analog Digit. Signal Process. **46**, 279–289 (1999)
5. Fragniere, E., van Schaik, A., Vittoz, E.A.: Design of an analogue VLSI model of an active cochlea. Analog Integr. Circ. Signal Process. **13**, 19–35 (1997)
6. Frey, D.R.: Log-domain filtering: an approach to current-mode filtering. IEE Proc. G: Circ. Devices Syst. **140**, 406–416 (1993)

7. Frey, D.R.: Exponential state space filters: a generic current mode-design strategy. IEEE Trans. Circ. Syst. I: Fundamental Theory Appl. **43**, 34–42 (1996)
8. Frey, D.R., Tola, A.T.: A state-space formulation for externally linear class AB dynamical circuits. IEEE Trans. Circ. Syst. II Analog Digit. Signal Process. **46**, 306–314 (1999)
9. Frey, D.R., Tsividis, Y.P.: Syllabically companding log domain filter using dynamic biasing. Electron. Lett **33**, 1506–1507 (1997)
10. Georgiou J. Personal Communication with Dr Julius Georgiou. 2007.
11. Grech, I., Micallef, J., Vladimirova, T.: Low-power log-domain CMOS filter bank for 2-D sound source localization. Analog Integr. Circ. Signal Process. **36**, 99–117 (2003)
12. Katsiamis, A.G., Drakakis, E.M., Lyon, R.F.: Practical gammatone-like filters for auditory processing. EURASIP J Audio, Speech, and Music Process. 15 pp (2007)
13. Khanna, S.M., Hao, L.F.: Nonlinearity in the apical turn of living guinea pig cochlea. Hear. Res. **135**, 89–104 (1999)
14. Kimura, K.: The ultra-multi-tanh technique for bipolar linear transconductance amplifiers. IEEE Trans. Circ. Syst. I: Fundamental Theory Appl. **44**, 288–302 (1997)
15. Lyon, R.: A computational model of filtering, detection, and compression in the cochlea. In: Acoustics, speech, and signal processing, IEEE International Conference on ICASSP'82, vol. 7, pp. 1282–1285 (1982)
16. Python, D., Enz, C.C.: A micropower class-AB CMOS log-domain filter for DECT applications. IEEE J. Solid-State Circ. **36**, 1067–1075 (2001)
17. Sarpeshkar, R., Lyon, R.F., Mead, C.: A low-power wide-dynamic-range analog VLSI cochlea. Analog Integr. Circ. Signal Process. **16**, 245–274 (1998)
18. Sarpeshkar, R., Salthouse, C., Ji-Jon, S., Baker, M.W., Zhak, S.M., Lu, T.K.T., Turicchia, L., Balster, S.: An ultra-low-power programmable analog bionic ear processor. IEEE Trans. Biomed. Eng. **52**, 711–727 (2005)
19. Tanimoto, H., Koyama, M., Yoshida, Y.: Realization of a 1-V active filter using a linearization technique employing plurality of emitter-coupled pairs. IEEE J. Solid-State Circ. **26**, 937–945 (1991)
20. Tsividis, Y.: On linear integrators and differentiators using instantaneous companding. IEEE Trans. Circ. Syst. II Analog Digit. Signal Process. **42**, 561–564 (1995)
21. Tsividis, Y.: Externally linear, time-invariant systems and their application to companding signal processors. IEEE Trans. Circ. Syst. II Analog Digit. Signal Process. **44**, 65–85 (1997)
22. Zhak, S.M., Baker, M.W., Sarpeshkar, R.: A low-power wide dynamic range envelope detector. IEEE J. Solid-State Circ. **38**, 1750–1753 (2003)

Bioinspired Microneedles

▶ Biomimetic Mosquito-Like Microneedles

Bioinspired Synthesis of Nanomaterials

Roberto de la Rica
MESA + Institute for Nanotechnology, University of Twente, Enschede, The Netherlands

Synonyms

Biomimetic synthesis; Biomimetic synthesis of nanomaterials

Definition

Synthetic routes that mimic natural processes for the fabrication of nanomaterials in mild conditions and with improved control over key features such as size, shape, and hierarchical organization of the end product. These bioinspired processes frequently utilize biocatalytic templates, such as DNA, peptides, and proteins as well as cell bioreactors, whose performance can be enhanced via molecular biology tools.

The transition from the microscale to the nanoscale has been a great endeavor for mankind, which has opened a new horizon in technological applications due to the superior features of nanomaterials compared to their bulk counterparts. Although major breakthroughs in nanotechnology have been accomplished, some areas of this field remain challenging for the widespread application of these materials in everyday life. For example, the organization of nanomaterials to yield hierarchical structures with defined shapes and sizes could generate materials with novel functions derived from the spatial arrangement of the nanometric building blocks. This feature is also essential for the easy integration of nanomaterials with complex circuits and microprocessors, a crucial step for electronic applications such as sensors and solar cells. The fabrication in mild conditions such as at room temperature and in aqueous solution would greatly help disseminating the utilization of nanomaterials in real applications, since less energy would be required for the fabrication and less toxic waste generated during the manufacture process, therefore reducing costs and environmental impact related to their synthesis. To address these issues, many scientists have searched for inspiration in biological systems. Living organisms have evolved the capability of growing inorganic nanomaterials with an exquisite control over the shape, size, crystal phase, and three-dimensional arrangement. These biological processes are orchestrated by a battery of biomolecules that act as biocatalytic templates for the synthetic process. Moreover, these biological phenomena take place under mild conditions, that is, at room temperature and in aqueous solution, which are not only advantageous from an economic and ecological standpoint, but can also result in the fabrication of materials with improved properties such as defect-free crystals. Usually, a bioinspired route begins from the observation of a natural process and its replication ex vivo, for example, by recreating the fabrication in the presence of biomolecules found to be crucial in the biological synthesis. Subsequently, the synthesis can be refined by generating alterations on the biocatalytic templates via expression of mutant DNA sequences in microorganisms. This approach is particularly advantageous because it allows one to create a vast array of catalytic candidates, whose appropriateness for the fabrication of the target material can be tested with high-throughput methodologies. Furthermore, once a particular template has been recognized, its production can be easily scaled up by selecting the microorganism encoding the biomolecule and making it grow in adequate conditions. Subsequently, the target biomolecule can be isolated by well-known purification protocols. The utilization of biomolecules is not only advantageous for the synthesis of nanomaterials in benign conditions and with desired shapes and sizes; the intrinsic recognition capability of certain biomolecules can be also exploited to target them at particular sites on substrates for their integration with circuits and microprocessors, which is essential for the implementation of nanomaterials in real applications. Finally, bioinspired approaches have been described that mimic the biological mechanism of formation of inorganic materials rather than utilizing biomolecules as biocatalytic templates. Among these, biomimetic approaches for the fabrication of perfectly aligned arrays of nanomaterials, also called mesocrystals, are drawing much attention since the resulting superstructures could show novel physical properties derived from the intrinsic order of the collectives.

In the following sections the relevance of several biocatalytic templates such as DNA, peptides, and proteins will be outlined and illustrated with key examples. The utilization of cells for the synthesis of nanomaterials as well as other biomimetic approaches for the fabrication of mesocrystals will also be covered, followed by conclusions and future directions in the field.

DNA

DNA molecules have been extensively used as templates for the synthesis of nanomaterials due to their inherent biorecognition capabilities and excellent chemical properties for the deposition of metallic and semiconducting materials [1]. The recognition of DNA strands bearing complementary sequences has proved to be extremely useful for the programmable assembly of DNA templates on substrates, which is an important step for the integration of nanomaterials with common technologies. Moreover, the negatively charged phosphate groups of DNA molecules can concentrate metal ion precursors in the vicinity of the template for the selective deposition of materials. In certain cases, the coordination of these metal ions to the bases of DNA can result in the formation of nanocrystals with unique properties. An early example on the advantages of using DNA molecules as templates for the fabrication of nanomaterials can be found in the work of Ben-Yoseph et al. [2]. By immobilizing oligonucleotides bearing a complementary sequence on two adjacent gold electrodes, the positioning of the DNA template was programmed to happen at the gap between the electrodes. Subsequently, silver ions were concentrated in the vicinity of the DNA bridge via electrostatic interactions, and reduction with hydroquinone yielded silver nanowires of 12 μm length that bridged the electrodes. However, the metal nanowires prepared this way showed poor conductivity mainly due to the presence of granules and defects as a consequence of the fast growth kinetics of metallization in the aforementioned conditions. More recently it was demonstrated that the conductivity of the metal coatings on DNA templates could be greatly improved by first depositing a metal oxide such as PdO and subsequently reducing the intermediate with hydrogen gas [3]. The resulting Pd nanowires are continuous and show higher conductivities (Fig. 1).

Short DNA sequences have proven to be also very useful in the formation of extremely small metal

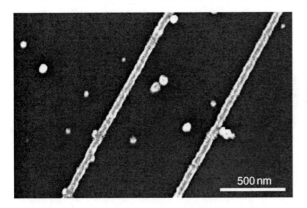

Bioinspired Synthesis of Nanomaterials, Fig. 1 Pd nanowires templated by DNA (Reprinted from [3], copyright Wiley-VCH Verlag GmbH & Co. KGaA)

nanoparticles, the so-called metal nanoclusters. When the particle size is further reduced and approaches the Fermi wavelength of electrons the continuous density of states breaks up into discrete energy levels leading to the observation of dramatically different optical, electrical, and chemical properties compared to larger nanoparticles. For example, Ag nanoclusters are highly fluorescent, and their reduced dimensions compared to quantum dots make them ideal candidates for bioimaging provided that they are soluble in aqueous solution. By using cytosine-rich DNA sequence as templates, Dickson et al. demonstrated the formation of Ag nanoclusters with 2–4 atoms [4]. Contrary to former approaches where the interaction between the negative charged phosphate backbone and metal ions was the driving force for the selective growth of the material, the sequence of the DNA templates plays a crucial role in the formation of metal nanoclusters. The utilization of biotechnology tools can greatly speed up the screening of DNA sequences that are useful for the fabrication of a particular material. For example, short DNA sequences can be chemically synthesized in a random fashion and spotted on a glass substrate to generate DNA microarrays. After incubation with silver and reduction, the spots containing DNA sequences capable of growing silver nanoclusters can be detected by measuring the fluorescence of the whole substrate with a scanner. By this methodology, blue-, green-, yellow-, and red-emitting Ag nanoclusters could be synthesized from a DNA library [5].

Peptides

Peptides have proved to be extremely useful tools for the fabrication of nanomaterials [6]. Compared to proteins, they possess shorter sequences, which facilitate their fabrication via synthetic methods. This feature also makes their 3D structure simpler and more predictable when synthesizing de novo sequences. Moreover, when carefully designed, peptides can assemble to yield supramolecular architectures with novel functions. A well-established area in nanomaterials fabrication synthesis is the application of peptides for growing nanocrystals with a particular crystalline structure, shape, and size. The selection of peptide sequences for the fabrication of nanocrystals is usually performed via molecular biology tools such as phage display. In this approach, random DNA sequences encoding peptides are inserted in bacteriophages. When the viruses replicate, they express the peptides on their capsids. Subsequently, the affinity of the peptide for a particular material is tested by applying stringent selection protocols. The peptides with the highest affinity for the target material are tested for the fabrication of nanocrystals in different conditions. Although the mechanism behind this bioinspired approach is not clear yet, it has been demonstrated that the amino acid composition and the 3D arrangement of chemical moieties dictated by the conformation of the peptides are crucial for the fabrication of nanocrystals. For example, Matsui et al. applied peptides with the sequence His-Gly-Gly-Gly-His-Gly-His-Gly-Gly-Gly-His-Gly for the fabrication of Cu nanocrystals on peptide nanotubes [7]. They found that the peptides could undergo a conformational change depending on the pH of the growing solution that could control the size and monodispersity of the resulting nanocrystals (Fig. 2). The same group demonstrated that peptides with the sequence Asn-Pro-Ser-Ser-Leu-Phe-Arg-Tyr-Leu-Pro-Ser-Asp could catalyze the growth of hexagonal Ag nanocrystals. The selectivity for growing a particular shape was provided by the high affinity of the peptide for the face (111) of Ag crystals.

Certain peptide sequences have been reported to self-assemble and yield complex architectures with catalytic functions. For example, peptides derived from naturally occurring leucine-zipper motifs have been described to fold in a coil-coil conformation that can assemble into templates for crystal growth. By

Bioinspired Synthesis of Nanomaterials,
Fig. 2 Peptide-templated Cu nanocrystals. The conformation of the peptide dictated by the pH of the growing solution controls the size and shape of the nanocrystals; (**a**) pH 6 and (**b**) pH 8 (Reprinted from [7], copyright The National Academy of Sciences of the USA)

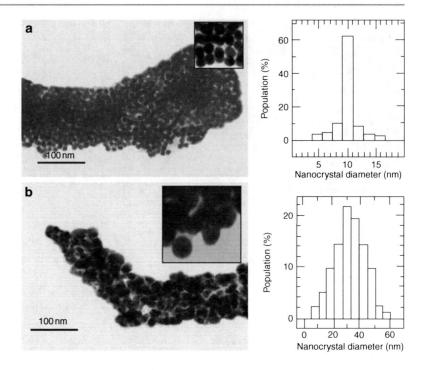

changing key amino acids of the primary sequence for thiol-rich cysteine units, the supramolecular peptide aggregates can be used as templates for the fabrication of monodisperse Ag nanoparticles [8]. In another example, the self-assembly of bolaamphiphile peptides resulted in the formation of peptide "nanodoughnuts" that could catalyze the fabrication of Ga_2O_3 [9]. Chemical residues with affinity for the metal ions preconcentrated the precursors inside the cavity of the templates and generated a gallium hydroxide intermediate. The higher hydrophobicity inside the template played a crucial role in the dehydratation of the intermediate to yield crystalline Ga_2O_3, a wide bandgap semiconductor of great interest in electronics. These examples highlight the ability of peptide assemblies to work as catalytic templates for crystal growth.

Proteins

When compared to peptides, proteins show a more complex tertiary structure and can perform highly specialized tasks such as catalyzing specific chemical reactions [6]. Therefore, the adaptation of a protein for the synthesis of a nanomaterial can be a very advantageous approach provided that it can be easily purified from natural sources in large amounts. Although the

higher degree of complexity seen in proteins makes it more challenging to modify their structure to gain a new function, recent advances in molecular biology have enabled the fabrication of mutants with improved features for the fabrication of nanomaterials. The rich structural variety of proteins can be exploited for the fabrication of nanomaterials with desired shape and dimensions when these proteins act as template for the synthetic process. For example, the protein ferritin is formed by 24 subunits that generate a cavity with a diameter of 7 nm. In cells, it contains iron crystallites and hence serves as an important intracellular iron storage unit. In vitro, iron-free ferritin (apoferritin) can be loaded with metal ions to generate nanoparticles of materials such as iron and cobalt oxide, which grow inside the cavity with uniform size and shape. This feature is especially advantageous compared to classical methods for nanoparticle synthesis, since in the absence of a template it is difficult to predict the final shape and size of the nanocrystals. For instance, in the fabrication of nanoparticle-based floating gates it is crucial that all the building blocks have a precise size, shape, and surface density. These requirements could be met by synthesizing CoO_3 nanoparticles in apoferritin and immobilizing the protein-covered nanocrystals on the device via programmed electrostatic interactions [10]. The resulting floating gate memory showed excellent charge capacity, long charge retention, and stress resistance. Moreover, it has been demonstrated that the inner cavity of the cage can be modified via genetic engineering protocols to display peptide sequences that have high affinity for a certain metal ion precursor, which paves the way for the utilization of ferritin in the fabrication of nanocrystals of diverse technologically relevant materials.

While protein cages have proved to be extremely useful for the fabrication of nanocrystals, nanostructures with a higher aspect ratio such as nanowires are required for a wide variety of nanotechnology applications. To address this issue, a fibrillar protein such as collagen could be used as a template for the deposition of metallic or semiconducting materials. However, naturally occurring collagen may not be suitable for this application because the relatively harsh conditions for metallization can affect the conformation of the protein, therefore rendering nanomaterials of polydisperse size. This issue could be overcome by using a collagen-like triple helix protein that is genetically engineered to possess improved stability and rigidity, as well as uniform size. When used as a template for the electroless deposition of gold, the mutant protein could template the fabrication of gold nanowires of monodisperse size with a length of 40 nm and a width of 4 nm [11]. The fabrication of this protein via expression in a host microorganism such as Escherichia Coli allows one to obtain large quantities of the biomolecular template by simply growing the bacteria and applying well-established protein purification protocols.

An emerging area in bionanotechnology is the adaptation of proteins with biocatalytic activity such as enzymes to the synthesis of nanomaterials because the mild conditions required for the biological catalysis make the manufacturing process more environmentally benign, and in some cases, with less side products. In a pioneering work by Samuelson et al., the enzyme peroxidase was utilized for the synthesis of the conductive polymer polyaniline [12]. In this approach, peroxidase catalyzed the oxidation of the monomer to trigger the polymerization in the presence of a polyelectrolyte template to yield water-soluble, high molecular weight polyaniline with minimal parasitic branching. The approach is particularly attractive in that it is simple (one step), uses very mild conditions, and requires minimal separation and purification. Another area that has benefited greatly from the use of enzymes for the fabrication of nanostructures is the room-temperature synthesis of oxide semiconductors. Oxide semiconductors such as ZnO are wide bandgap semiconductors with a myriad of applications in optics, electronics, solar cells, and catalysis. However, the high temperatures required for the fabrication of these materials with conventional methods increase the cost associated to the manufacturing process, which may hamper the widespread application of these materials. This issue could be solved by using the enzyme urease as biocatalytic template for the synthesis of ZnO nanoshells. In this approach, urease converted urea into ammonia and carbon dioxide, and the resulting fine-tuning of the pH at the enzyme-solution interphase could grow highly crystalline ZnO as nanoshells around the protein [13]. When immobilized on a substrate, the enzymatic production of ammonia could be harnessed to produce amorphous intermediates that could be used as ink to be patterned with an AFM tip for a pen in a procedure similar to dip-pen nanolithography [14]. By this technique called Biomimetic Crystallization Nanolithography, ZnO nanocrystals could be patterned on the surface with desired shapes and sizes, which is a crucial step for the integration of nanomaterials with electrical components of circuits and microprocessors (Fig. 3). Moreover, urease could be also applied for the fabrication of other relevant semiconductors such as metal sulfides by harnessing the production of S^{2-} via hydrolysis of thiorea. By this approach, the size, shape, and crystallinity of Ag_2S as a test material could be controlled to be as desired by fine-tuning the activity of the enzyme via inhibition with the metal ion precursor [15].

Cells

The utilization of cells such as bacteria for the fabrication of nanomaterials can be very advantageous since the fast growth kinetics of these microorganisms in adequate conditions allows one to obtain large amounts of the target material with minimal resources. A well-known example of the application of cells in nanomaterials synthesis is the fabrication of magnetic nanoparticles by magnetotactic bacteria [16]. These microorganisms generate intracellular magnetic particles of iron oxide that allow them to migrate along

Bioinspired Synthesis of Nanomaterials, Fig. 3 ZnO nanopatterns via biomimetic crystallization nanolithography with different tip velocity during the patterning step; (**a**) $v = 100$ nm/s; (**b**) $v = 500$ nm/s (Reprinted from [14], copyright Wiley-VCH Verlag GmbH & Co. KGaA, Weinheim)

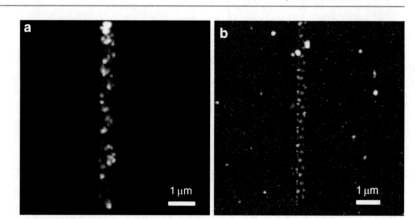

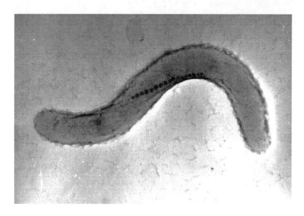

Bioinspired Synthesis of Nanomaterials, Fig. 4 Image of *Magnetospirillum magneticum* strain AMB-1 showing the magnetic nanoparticles (Reprinted from [16], copyright the Royal Society)

points for their functionalization with a wide variety of molecules such as enzymes and antibodies. Particularly, the amino groups of the phospholipids present at the membrane can be cross-linked to antibodies and oligonucleotides with well-known covalent chemistries, which confer the particles biomolecular recognition capabilities that are useful for medical and diagnostic applications. Alternatively, the protein of interest can be generated during the biological synthesis of bacterial magnetic particles by fusing its gene to proteins known to be present on the membrane of the nanoparticles so that when the bacteria express the resulting fusion protein, it is automatically incorporated on the surface of the nanoparticles during the biosynthesis.

Mesocrystals

Biomineralizing organisms such as sea urchins and marine stars have evolved the capability to generate nanocrystals and arrange them into highly ordered mesoscopic arrangements, also known as mesocrystals [17]. The perfect alignment of the constituent nanocrystals results in similar scattering pattern and behavior in polarized light to those of a single crystal, although a close look into the ultrastructure reveals the presence of the small constituent nanoparticles. This phenomenon is responsible for the improved mechanical properties of the exoskeleton of crustaceans, and the replication ex vivo for the fabrication of assemblies of perfectly aligned nanocrystals of functional materials is extremely interesting to study the physical properties of nanoparticle collectives. Several mechanisms have pointed out to yield three-dimensional

oxygen gradients by acting as compass needles. Different bacterial strains have been reported to grow these bacterial magnetic particles in different shapes, which can be very advantageous for obtaining materials with unusual nanostructures (Fig. 4). For example, the *M. magneticum* strains AMB-1 and MGT-1 synthesize cubo-octahedral nanoparticles, whereas *Desulfovibrio magneticus* RS-1 produces irregular bullet-shaped particles. Bacterial magnetic particles also have the great advantage of being synthesized surrounded by an organic membrane, whose main components are phospholipids and proteins. These biomolecules confer better solubility to the particles, which disperse well in aqueous solvents without further surface treatment. Moreover, the biomolecules present at the surface of bacterial magnetic particles show chemical moieties that can be used as anchoring

Bioinspired Synthesis of Nanomaterials,
Fig. 5 Scheme of the different mechanisms to grow nanoparticle superstructures or mesocrystals (Reprinted from [17], copyright Wiley-VCH Verlag GmbH & Co. KGaA, Weinheim)

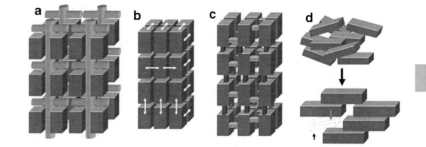

nanoparticle superstructures, as depicted in Fig. 5. One possibility is that crystal growth takes places inside an organic matrix such as a biopolymer, which guides the growth and alignment of the constituent nanocrystals (Fig. 5a). The presence of certain chemical groups with affinity for cationic precursors is important to initiate the nucleation of the nanoparticles inside the matrix, and in some cases, to stabilize a particular face of the crystal and direct the growth in a particular direction. Another possibility is the presence of physical fields as the driving force for nanoparticle aggregation, for example, the spontaneous alignment of dipoles (Fig. 5b). The growth of mesocrystals can also be permitted by the existence of so-called "mineral bridges," that is, the nanoparticles are hold together by another inorganic structure in a "brick and mortar" fashion (Fig. 5c). Finally, when growing in a constraint reaction vessel, nanoparticles can be forced to interact and align with each other to reduce the entropy of the system (Fig. 5d). Although these basic rules for the formation of mesocrystals have been successfully replicated ex vivo by the correct choice of crystallization environment such as the presence of a polymer matrix or addition of soluble additives, the exact mechanism of this nanoparticle-mediated crystallization route remains still elusive. For example, nanospheres formed by ZnO nanocrystals oriented along the [001] direction could be obtained by using the polymer poly(vinylpyrrolidone) in a mixture of N,N-dimethylformamide (DMF) and water as a guiding matrix. The coordination of Zn^{2+} to active groups in the polymer as well as the stabilization of particular crystal planes play a crucial role for obtaining nanocrystal-based superstructures. The resulting porous arrangements of aligned nanocrystals are promising for photonic applications. Moreover, it is possible to fabricate composite thin films via self-assembly of the freestanding spheres by simply coating with the as-prepared suspensions, which facilitates the integration of these hybrid materials with planar devices. Recently, the biomimetic conditions for the fabrication of mesocrystals could be replicated by a novel approach in which a nanopore building block was self-assembled to yield a highly porous matrix. The macrocycle cucurbit[7]uril aggregated in the presence of Ag^+, and the ion-impregnated nanopore assemblies guided the growth of perfectly aligned Ag_2S quantum dots after the addition of S^{2-} to yield a nanoparticle superstructure with a pitch of 0.8 nm. This example represents an exquisite control over the crystallization of this material due to the extremely low solubility of Ag_2S in water, which usually results in the formation of particles of uncontrolled size and poor crystallinity. The key step of the process was to confine the metal ion precursor inside the porous matrix, which controls mass transport via diffusion of the anions through the pores of the assemblies [19].

Conclusions and Outlook

The utilization of bioinspired approaches in the fabrication of nanomaterials is extremely useful to control key parameters such as size monodispersity, shape, and three-dimensional arrangement. These processes usually take place under mild biological conditions, which is advantageous for reducing costs and environmental impact of the manufacturing process. The utilization of microorganisms as bioreactors, whether to generate the biocatalytic template or to synthesize the target nanomaterial, facilitates the large-scale production due to the rapid growth kinetics of cells in adequate media. Moreover, in certain cases, the intrinsic recognition capabilities of the biomolecule templates can be exploited to locate the biomaterials at target places on substrates, which facilitate their integration with common technologies. While these features have been demonstrated in the last years, challenges in the field still remain. From a fundamental standpoint, it would be desirable to unravel the exact mechanism of formation of nanomaterials in biological conditions. A deeper understanding of the process would allow the formulation of rules for bionspired synthesis so that the outcome of the manufacturing process is more predictable, and hence the fabrication of the nanomaterial requires less effort. For example, discerning the relationship between sequence and conformation in peptide and DNA templates for the fabrication of nanomaterials with specific shape, size, and polymorph could help designing de novo sequences for the synthesis of another material by applying lessons learnt from previous experiences. Another important step toward the widespread utilization of bioinspired routes would be to demonstrate the fabrication of real devices with the resulting nanomaterials. Although in the last years several key applications have been demonstrated such as the fabrication of floating memories [10] and the design of ultra-sensitive sensors for the detection of heavy metal ions via target-specific bioinspired crystallization [18], the integration of nanomaterials obtained by bioinspired routes is still challenging in certain fields. For example, a common problem is that these nanomaterials are usually covered by the biological template. While this can be advantageous for their programmed assembly via biorecognition, the presence of an insulating organic layer can hamper certain applications such as in the fabrication of electronic components of circuits, in which the conductance of the nanomaterials and the electric contact with the circuitry is crucial for the performance of the device. Nevertheless, biocatalytic templates are extremely useful components of the nanotechnology toolbox that show great potential to change classical routes for the synthesis of nanomaterials to be more environmentally benign and that offer improved control over fey features such as size, shape and hierarchical organization.

Cross-References

▶ Dip-Pen Nanolithography

References

1. Becerril, H.A., Woolley, A.T.: DNA-templated nanofabrication. Chem. Soc. Rev. **38**, 329–337 (2009)
2. Braun, E., Eichen, Y., Sivan, U., Ben-Yoseph, G.: DNA-templated assembly and electrode attachment of a conducting silver wire. Nature **391**, 775–778 (1998)
3. Nguyen, K., Monteverde, M., Filoramo, A., Goux-Capes, L., Lyonnais, S., Jegou, P., Viel, P., Goffman, M., Bourgoin, J. P.: Synthesis of thin and highly conductive DNA-based palladium nanowires. Adv. Mater. **20**, 1099–1104 (2008)
4. Petty, J.T., Zheng, J., Hud, N.V., Dickson, R.M.: DNA-templated Ag nanocluster formation. J. Am. Chem. Soc. **126**, 5207–5212 (2004)
5. Richards, C.I., Choi, S., Hsiang, J.C., Antoku, Y., Vosch, T., Bongiorno, A., Tzeng, Y.L., Dickson, R.M.: Oligonucleotide-stabilized Ag nanocluster fluorophores. J. Am. Chem. Soc. **130**, 5038–5039 (2008)
6. de la Rica, R., Matsui, H.: Applications of peptide and protein-based materials in bionanotechnology. Chem. Soc. Rev. **39**, 3499–3509 (2010)
7. Banerjee, I.A., Yu, L., Matsui, H.: Cu nanocrystal growth on peptide nanotubes by biomineralization: size control of Cu nanocrystals by tuning peptide conformation. Proc. Natl Acad. Sci. U.S.A. **100**, 14678–14682 (2003)
8. Ryadnov, M.G.: A self-assembling peptide polynanoreactor. Angew. Chem. Int. Ed. **46**, 969–972 (2007)
9. Lee, S.Y., Gao, X., Matsui, H.: Biomimetic and aggregation-driven crystallization route for room-temperature material synthesis: growth of β-Ga_2O_3 nanoparticles on peptide assemblies as nanoreactors. J. Am. Chem. Soc. **129**, 2954–2958 (2007)
10. Yamashita, I., Iwahori, K., Kumagai, S.: Ferritin in the field of nanodevices. Biochim. Biophys. Acta **1800**, 846–857 (2010)
11. Bai, H., Xu, K., Xu, Y., Matsui, H.: Fabrication of Au nanowires of uniform length and diameter using a monodisperse and rigid biomolecular template: collagen-like triple helix. Angew. Chem. Int. Ed. **46**, 3319–3322 (2007)
12. Liu, W., Kumar, J., Tripathy, S., Senecal, K.J., Samuelson, L.: Enzymatically synthesized conducting polyaniline. J. Am. Chem. Soc. **121**, 71–78 (1999)
13. de la Rica, R., Matsui, H.: Urease as a nanoreactor for growing crystalline ZnO nanoshells at room temperature. Angew. Chem. Int. Ed. **47**, 5415–5417 (2008)
14. de la Rica, R., Fabijanic, K., Matsui, H.: Biomimetic crystallization nanolithography: simultaneous nanopatterning and crystallization. Angew. Chem. Int. Ed. **49**, 1447–1450 (2010)
15. Pejoux, C., de la Rica, R., Matsui, H.: Biomimetic crystallization of sulfide semiconductor nanoparticles in aqueous solution. Small **6**, 999–1002 (2010)
16. Arakaki, A., Nakazawa, H., Nemoto, M., Mori, T., Matsunaga, T.: Formation of magnetite by bacteria and its application. J. R. Soc. Interface **5**, 977–999 (2008)
17. Song, R.Q., Colfen, H.: Mesocrystals – ordered nanoparticle superstructures. Adv. Mater. **22**, 1301–1330 (2010)

18. de la Rica, R., Mendoza, E., Matsui, H.: Bioinspired target-specific crystallization on peptide nanotubes for ultrasensitive Pb ion detection. Small **6**, 1753–1756 (2010)
19. de la Rica, R., Velders, A. H.: Biomimetic crystallization of Ag_2S nanoclusters in nanopore assemblies. J. Am. Chem. Soc. **133**, 2875–2877 (2011)

Biological Breadboard Platform for Studies of Cellular Dynamics

Sang-Hee Yoon[1,2] and Mohammad R. K. Mofrad[2]
[1]Department of Mechanical Engineering, University of California, Berkeley, CA, USA
[2]Molecular Cell Biomechanics Lab, Department of Bioengineering, University of California, Berkeley, CA, USA

Synonyms

Biological platform; Cell manipulation platform

Definition

Biological breadboards (BBBs) are platforms that carry out the spatiotemporal manipulation of cell behavior related to cell adhesion and detachment (e.g., cell positioning, cell patterning, and cell motility control), thus offering a new, versatile tool for studies of cellular dynamics. The platforms are composed of an array of identical gold electrodes patterned on a Pyrex glass substrate. The gold electrodes are functionalized with arginine-glycine-aspartic acid (RGD)-terminated thiol to create a cytophilic surface, and the Pyrex glass substrate is modified with polyethylene glycol (PEG) to achieve a cytophobic one. In the BBBs, cell adhesion and detachment are controlled by the reductive desorption of a gold-alkanethiol self-assembled monolayer (SAM) with activation potential of -0.9 to -1.8 V.

Overview

Cell adhesion and detachment processes are mediated by complex biomolecules from both sides of the cell-matrix interface, fitting together like pieces of a three-dimensional puzzle. Only biomolecules with the right shape can be involved in cell adhesion and detachment, and even small changes in the shape may significantly affect cell behavior. When cells adhere to extracellular matrix (ECM) components, integrins are activated, the activated integrins bind target ligands, and the bound integrins cluster together by changing their conformation [1]. The cytoplasmic domain of the clustered integrins interacts with focal adhesion (FA) proteins (e.g., talin, focal adhesion kinase, vinculin, paxillin, etc.) to form FAs, and then binds actin filaments [2]. Likewise, the cell detachment or de-adhesion is also resulted from an orchestrated process involving the molecular machinery composed of a host of extracellular, transmembrane, and cytoplasmic proteins.

Cell adhesion and detachment have profound effects on the biological behavior of anchorage-dependent cells. For example, cell adhesion and detachment are controlling parameters in a variety of biological phenomena (e.g., embryonic development, cancer metastasis, and wound healing), and any abnormality in cell adhesion and detachment could lead to diverse pathophysiological consequences [1]. The quantitative characterization of the dynamic nature of cell adhesion and detachment is therefore essential for understanding a variety of pathophysiological phenomena.

Despite significant progresses over the past decade in characterizing biomolecules and signaling pathways for cell adhesion and detachment, the biophysical (and dynamic) details of cell adhesion and detachment still remain elusive and challenging. Thus, various experimental methods which characterize cell adhesion or detachment have been developed. Previous techniques for the characterization of cell adhesion can be classified into photolithography [3], e-beam lithography [4], dip-pen lithography [5], nanoimprint lithography [6], microcontact printing [7], elastomeric stencil [8], ink-jet printing [9], optical tweezer [10], electrophoresis [11], and switchable surface [12]. Although the conventional lithography techniques (i.e., photolithography, e-beam lithography, dip-pen lithography, and nanoimprint lithography), microcontact printing, elastomer stencil, and ink-jet printing technologies characterized cell adhesion using specific proteins, they altered cell adhesion only at the beginning of microfabrication and were not involved in the characterization of cell detachment. The optical tweezer and electrophoresis had a possibility of protein denaturation and cell electrolysis, respectively. The switchable surface showed its questionable biocompatibility. Meanwhile, the existing techniques for the characterization of cell detachment can be categorized into hydrodynamic shear force assay [13, 14], centrifugal assay [15], and micropipette aspiration [16]. The hydrodynamic shear force assay was designed to characterize cell detachment with a parallel, rotational, or radial flow. Although this method successfully quantified cell detachment with a wide range of shear forces, it had several inherent drawbacks: cell rupture during experiments and considerable local deformation of cells. Moreover, it provided a relatively low centrifugal force for cell detachment, thus restricting its application to cells which were either cultured for short term (less than 1 h) or coupled with weak cell-to-substrate interactions. The micropipette aspiration also resulted in rapid changes cytoskeletal remodeling during its experiments due to the direct contact between cell and micropipette. Recently, electrochemical techniques have been developed to solve the limitations of the previous approaches such as protein denaturation, cell electrolysis, local deformation, and direct contact. The reductive desorption of a gold-alkanethiol SAM, combining with microcontact printing technique, was used to quantify cell adhesion [17]. The success of this method was, however, restricted to only cell adhesion characterization at a cellular level and was involved in relatively complex preparation processes. The electrochemical reaction of polyelectrolytes was also proposed to study cell adhesion as well as cell detachment [18]. However, the multiple deposition of polyelectrolytes on a substrate and even cells of interest accompanied disturbed results due to unwanted changes in the cells. The shortcomings of previous approaches, described above, motivate us to develop a more promising platform for a quantitative analysis of cellular dynamics.

Biological breadboards (BBBs) are developed for spatiotemporal manipulation of cell adhesion and detachment at cellular and subcellular levels. The BBBs, inspired by electrical breadboards, have the following features in the characterization of cellular dynamics: addressability, multifunctionality, reusability, and ease of use. The BBBs have addressability in manipulating cell adhesion and detachment due to their exquisite structure, that is, an array of independently operated gold electrodes patterned on a Pyrex

glass substrate. The BBBs are used to perform a variety of biological characterizations (e.g., cell positioning, cell patterning, and cell motility control) related to cell adhesion and detachment; the electrochemistry involved in the BBBs, reductive desorption of a gold-thiol SAM, makes the BBBs reusable. The BBBs are easily incorporated with other conventional instruments such as optical, fluorescent, and confocal microscopes. Moreover, the BBBs provide cells with a microenvironment (RGD peptide) that is as similar as possible to in vivo microenvironment. The platforms therefore make it possible to characterize the adhesion and detachment of *living* and *intact* cells, thus enabling quantitative characterization of cellular dynamics.

Structure and Working Principle

The BBBs consist of an array of identical gold electrode micropatterned on a Pyrex substrate which secures a high-degree-of-freedom in a programmable manipulation of cell adhesion and detachment, as shown in Fig. 1a. The gold and Pyrex surfaces are modified with RGD-terminated thiol (RTT) and PEG, respectively. The RTT functionalization on gold electrodes is to make a cell-adhesive surface by tethering an RGD peptide to the electrode via thiol compound, following the spontaneous chemisorption:

$$R - S - H + Au \rightarrow R - S - Au + \frac{1}{2}H_2, \quad (1)$$

where R is a substituent [19]. The RGD-ligand provides a binding site to integrins in a cell adhesion process. The PEG treatment on Pyrex substrate is made to achieve a cell-resistive surface where hydrated neutral PEG chains sterically repulse cells.

A spatiotemporal manipulation of cell adhesion is implemented by selectively detaching the RGD from gold electrodes with negative potential of about -0.9 to -1.6 V, following the electrochemical reaction [20]

$$R - S - Au + H^+ + e^- \rightarrow R - S - H + Au. \quad (2)$$

After the surface modifications with PEG and RTT, the RTT on a target area is detached by activating a corresponding electrode with potential, followed by sonication in cell culture media for 3 min and cell loading. The loaded cell adheres only to the inactivated (still, RTT-functionalized) gold electrode (Fig. 1b), because the cell has substantially more affinity for cell adhesion to the RGD (of the inactivated electrode) than to gold (of the activated electrode). The spatiotemporal manipulation of cell detachment is the same as that of cell adhesion except the sequential order of electrode activation and cell loading. For the spatiotemporal manipulation of cell detachment, the surface of the BBBs is also modified with PEG and RTT, followed by cell loading. The loaded cell grafts and stretches to the RGD on a gold electrode. On cell detachment manipulation, the cell or a part of the cell is detached from the BBB by activating a target electrode and then spontaneously retracts because a chemical bonding between gold and thiol is broken by activation potential, as shown in Fig. 1c. When the detached part of the cell senses no external mechanical anchorage (focal adhesions), it begins to experience the liquefaction (gel–sol transition) of the cytoskeleton which accompanies cellular retraction by changing the length of actin filaments. This is how the BBBs manipulate cell adhesion and detachment in a spatiotemporal way.

Materials and Methods

Microfabrication Process

The BBBs were fabricated on a 4-in. Pyrex glass wafer with a thickness of 500 μm. After cleaning it with a piranha solution of 1:1 (v:v) 96% sulfuric acid (H_2SO_4) and 30% hydrogen peroxide (H_2O_2) for 10 min, 1 μm-thick LOR resist (LOR 10A, MicroChem Corp.) was spin-coated at 4,000 rpm for 40 s, followed by soft baking at 170°C for 5 min. A 2 μm-thick positive photoresist (S1818, Rohm and Haas Corp.) was spin-coated on the LOR resist at 4,000 rpm for 40 s for double-layer resist stack, followed by soft baking at 110°C for 1 min. An optical lithography was done to pattern the double-layer resist stack before e-beam evaporation process, as shown in Fig. 2a, left. Next was a deposition of 5 nm-thick chromium (Cr) adhesion layer and 100 nm-thick gold (Au) layer on the wafer, as shown in Fig. 2a, center. Because the optical transparency in a visual light range is essential to make the BBBs incorporate with other biological instruments (e.g., inverted optical and fluorescent microscopes), the Au layer thickness was reduced to 30 nm for biological experiments (see Fig. 2d, right). Next, the Cr/Au-deposited wafer was immersed in an organic

solvent mixture (BAKER PRS-3000 Stripper, Mallinckrodt Baker, Inc.) at 80°C for 4 h to lift off the double-layer resist stack, thus achieving two kinds of BBBs, as shown in Fig. 2a, right. One is for cell adhesion and detachment manipulations at a cellular level, each electrode of which is 500 μm in length and 500 μm in width, as shown in Fig. 2b. The other is for cell adhesion and detachment manipulations at a subcellular level, each gold line of which is 10 μm in width and 3 μm in gap between two neighboring gold lines, as shown in Fig. 2c. The microfabricated BBBs were wire-bonded in a chip carrier and then were assembled with a cell-culture-well made of polystyrene, as shown in Fig. 2d.

PEG Treatment on Pyrex Surface

Before PEG treatment, the microfabricated BBBs were cleaned with an oxygen plasma chamber (PM-100 Plasma Treatment System, March Plasma Systems, Inc.) at 100 W for 30 s. The BBBs were then incubated with 2 mL m-PEG silane (2% v/v, Gelest, Inc.) and 1 mL hydrochloric acid (HCl, 1% v/v, Fisher Scientific) dissolved in 97 mL anhydrous toluene (Fisher Scientific) for 2 h, as shown in Fig. 2e, left. This process was carried out in a glove box under a nitrogen purge to avoid atmospheric moisture. The incubated BBBs were sequentially rinsed in fresh toluene and ethanol, dried with nitrogen, and cured at 120°C for 2 h. The surface-modified BBBs were stored in a vacuum desiccator until the next surface modification, RTT functionalization on gold surface.

RTT Functionalization on Gold Surface

The gold electrodes of the BBBs were functionalized with RTT whose solution was synthesized by chemically combining *cyclo* (Arg-Gly-Asp-D-Phe-Lys) ($C_{27}H_{41}N_9O_7$, Peptides International, Inc.) with dithiobis(succinimidylundecanoate) ($C_{30}H_{48}N_2O_8S_2$, Dojindo Molecular Technologies, Inc.) as follows. The 3.02 mg *cyclo* (Arg-Gly-Asp-D-Phe-Lys) was dissolved in 5 mL dimethoxysulfoxide (DMSO, Sigma-Aldrich) to get 1 mM aliquot and stored at −20°C. This reaction was made in a glove box under a nitrogen purge to protect the RGD peptide from exposure to atmospheric moisture. The 3.14 mg dithiobis(succinimidylundecanoate) was also dissolved in 5 mL DMSO, and then stored at −20°C. This preparation was also done in moisture-free environment. Before a gold surface functionalization, both aliquots were warmed to room temperature in a desiccator. The 5 mL RGD peptide aliquot was mixed with 50 μL triethylamine (1% v/v, Fisher

Scientific) for 5 min to make all primary amines of a lysine amino acid unprotonated. The 5 mL the dithiobis(succinimidylundecanoate) was added to the 5 mL RGD peptide aliquot, and then mixed well using

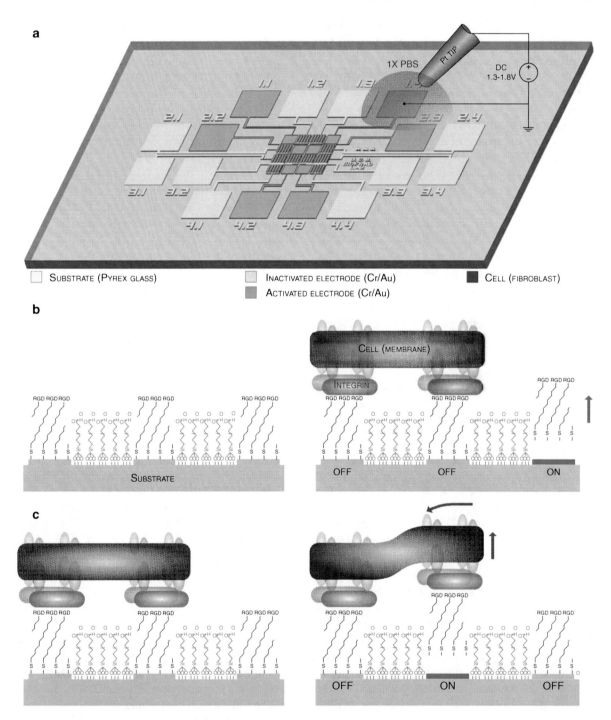

Biological Breadboard Platform for Studies of Cellular Dynamics, Fig. 1 (continued)

a vortex mixer for 4 h to synthesize the RTT solution. For the gold surface functionalization, the PEG-treated BBBs were incubated with the solution for 1 h at room temperature to promote a spontaneous chemisorption between thiol and gold, followed by sonification in DMSO for 3 min, rinses in ethanol and phosphate-buffered saline (PBS, Sigma-Aldrich) to eliminate all unbound RTTs from a gold surface, as shown in Fig. 2e, right. The thiol contacted with gold made a SAM and tethered an RGD peptide to a gold surface.

Contact Angle Measurement

The contact angles of PEG-treated Pyrex surface and RTT-functionalized gold surface were measured with a contact angle measurement system (KRÜSS582, KRÜSS), goniometer. A sessile drop mode was used to estimate the wetting properties of above two solid surfaces. The contact angles were averaged from 10 measurements. The contact angle of PEG-treated Pyrex surface was compared to that of pure Pyrex surface, and the contact angle of RTT-functionalized gold surface was compared to those of bare gold surface and thiol-treated gold surface.

X-Ray Photoelectron Spectroscopy (XPS) Sample Preparation and Characterization

An XPS survey scan was used to confirm the existence of RGD peptide linked to a gold surface via thiol after RTT functionalization. An XPS sample was prepared on a 4-in. silicon wafer e-beam evaporated with 5 nm Cr adhesion layer and 50 nm Au layer. This wafer was immersed for 2 h in the prepared RTT solution for the RTT functionalization on gold surface. A bare gold sample without RTT functionalization was run as a control experiment. The XPS analysis was carried out with a customized ESCA (Omicron

NanoTechnology) at 1×10^{-8} Torr, and all measured spectra were referenced to the position of the Au $4f$ peaks. The scans were collected over a range of 20 eV around the peak of interest with pass energy of 23.5 eV.

Potentiodynamic Electrochemical Characterization of Reductive Desorption of Gold-Thiol SAM

A silicon wafer e-beam evaporated with 5 nm-thick Cr and 50 nm-thick Au was functionalized with RTT to prepare a cyclic voltammetry (CV) sample. This RTT-functionalized gold electrode was used as a working electrode while platinum and Ag/AgCl electrodes were used as counter and reference electrodes, respectively. A voltage supplied by a DC power source (B&K Precision Corporation, Yorba Linda, CA) was applied between the gold-thiol SAM (or Ag/AgCl electrode) and the platinum electrode. The CV was carried out in the Dubecco's phosphate buffered saline (DPBS (pH 7.4), Sigma-Aldrich, St. Louis, MO) solution with an EG&G potentiostat model 362 (AMETEK Princeton Applied Research, Oak Ridge, TN). A scan started cathodically from 0 to -2 V, then anodically back to 0 V at a scan rate of 50 mV/s.

Cell Culture

NIH 3T3 mouse embryonic fibroblast cell (NIH 3T3 fibroblast) was cultured in a Dulbecco's modified eagle medium (DMEM, GIBCO™) supplemented with 10% fetal bovine serum (FBS, GIBCO™) and 1% Penicillin-Streptomycin (GIBCO™) at 37°C in humidified 5% CO_2 atmosphere. The cell was passaged every 4 days as follows. The cell was washed once in $1 \times$ PBS and trypsinized with Trypsin-EDTA solution 0.5%. After centrifuging the cell, it was inoculated into a new Petri dish. The NIH 3T3 fibroblasts with a passage number of 5–20 were used in the

---

Biological Breadboard Platform for Studies of Cellular Dynamics, Fig. 1 Biological breadboards for spatiotemporal manipulation of cell adhesion and cell detachment at cellular and subcellular levels. (**a**) Schematic of the BBBs consisting of a gold electrode array patterned on a Pyrex surface. The addressability, multifunctionality, and reusability of the BBBs are due to their exquisite structure and working principle (electrochemistry). (**b**) Spatiotemporal manipulation of cell adhesion. Before cell adhesion (*left*), RGD peptides are bound to all gold electrodes functionalized with RTT, whereas no cell adhesion site is formed on the Pyrex surface treated with PEG. On cell adhesion (*right*), the RGD peptides on a target electrode (third electrode

from *left*) are detached from the target electrode with activation potential, followed by cell loading. The loaded cell adheres only to inactivated electrodes (first and second electrodes from *left*) through RGD binding to integrin. (**c**) Spatiotemporal manipulation of cell detachment. Before cell detachment (*left*), a cell adheres to the gold electrodes functionalized with RTT through RGD binding to integrin. On cell detachment (*right*), the cell or a part of the cell adhered to a target electrode (second electrode from *left*) is detached by activating the target electrode with activating potential which yields the reductive desorption of a gold-thiol SAM. The detached cell (or part of the cell) retracts spontaneously

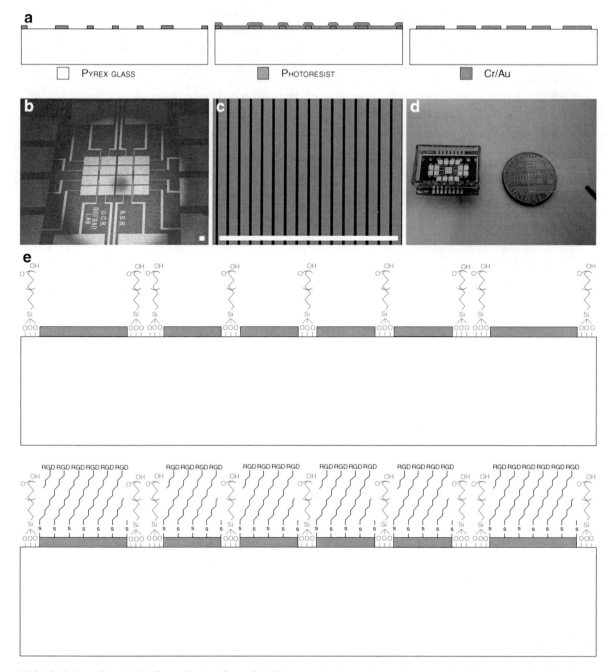

Biological Breadboard Platform for Studies of Cellular Dynamics, Fig. 2 Microfabrication and surface modification of the BBBs. (a) The BBBs are fabricated by patterning an array of gold electrodes on Pyrex substrate through patterning of photoresist by lithography (*left*), deposition of Cr and Au layers by e-beam evaporation (*center*), and patterning of the deposited Cr/Au layer by lift-off (*right*). (b) BBB for cell adhesion and detachment manipulation at a cellular level, each electrode of which is 500 μm in length and 500 μm in width. (c) BBB for cell adhesion and detachment manipulation at a subcellular level, each electrode of which is 10 μm in width and 3 μm in gap. (d) Photograph of the BBB before (*right*) and after (*left*) assembly, showing its transparency in the visual spectrum. (e) Surface modification process. The BBBs are incubated with a PEG solution to make a Pyrex surface cell-resistive (*left*), and then they are incubated with a synthesized RTT solution to make gold electrodes cell-adhesive (*right*). Scale bars of (b) and (c) are 100 μm experiments. Before each experiment, the surface-modified BBB was sterilized with 70% ethanol, washed twice with $1 \times$ PBS, and placed into a Petri dish containing 5 mL cell culture medium with the cell suspension of about 1×10^6 cells/mL. For the subcellular detachment experiments, the cell concentration was changed into 1×10^4 cells/mL. After 1 h, unadhered NIH 3T3 fibroblasts were removed by additional wash in $1 \times$ PBS, followed by cell culture medium replacement. All experiments were carried out after 24 h of cell loading in a self-designed chamber with humidified 5% CO_2 atmosphere and at 37°C.

Key Research Findings

Verification of Surface Modifications

The surface modifications were verified by two methods of contact angle measurement and XPS survey. The contact angle of the PEG-treated Pyrex surface measured by a goniometer was $61.5 \pm 3.8°$ (averaged from 10 measurements), whereas that of the untreated Pyrex surface was $25.7 \pm 1.5°$, as shown in Fig. 3a. This shows that a PEG-treated surface has strong hydrophobicity, and consequently prevents cell adhesion and protein fouling. An NIH 3T3 fibroblast load test was also made, as shown in Fig. 3b. The images obtained after 24 h of cell loading show the Pyrex surface is modified into cell-resistive through PEG treatment. To characterize the RTT functionalization, a contact angle was measured for bare gold, thiol-treated gold, and RTT-functionalized gold, each of which was $67.3° \pm 2.5°$, $53.3° \pm 1.3°$, and $24.6° \pm 2.8°$, respectively, as shown in Fig. 3c. This confirms the RTT functionalization is made as designed. The above experimental results verify that our surface modifications are effective to achieve cell-resistive and cell-adhesive surfaces in the BBBs.

After RTT functionalization on gold surface, the existence of RGD, linked to a gold electrode via thiol, was also demonstrated by an XPS survey scan, a measured XPS survey spectrum of RGD/thiol/Au interface, as shown in Fig. 3d. The peaks of Au $4s$, Au $4p$, Au $4d$, and Au $4f$ indicate the presence of e-beam evaporated gold (Au(111)); the peaks of S $2p_{1/2}$ and S $2p_{3/2}$ (right inset) show sulfur from thiol is in existence on the surface; and the peaks of

C $1s$, O $1s$, O KLL, and N $1s$ (left inset) demonstrate there are carbon, oxygen, and nitrogen from amine functional group ($-NH_2$) and carboxylic acid functional group ($-COOH$) of an RGD peptide. For reference, hydrogen was not detected due to XPS working principle. This XPS survey spectrum verifies that the RTT functionalization on the gold electrode is well made as designed.

Reductive Desorption of Gold-Thiol SAM

The rapid desorption of a gold-thiol SAM under activation potential was also characterized with cyclic voltammetry using the three-electrode system where the gold electrode (of the BBBs), a platinum electrode, and an Ag/AgCl electrode work as working, counter, and reference electrodes, respectively, as shown in Fig. 3e. The measured CV shows the current measured at the working electrode as a function of the applied voltage with respect to the Ag/AgCl electrode, as shown in Fig. 3f. At a potential range of 0 V to -0.9 V (section a), the measured current was negligible, which means there is no reductive desorption of the SAM and consequently the SAM impedes electron transfer across an electrolyte-electrode interface. When the applied voltage increased (section c), the reductive desorption of the SAM started and finished at a potential of -0.9 V (point b) and -1.55 V (point d), respectively, breaking the chemical binding between thiol and gold. The reductive desorption of the SAM got maximized at a peak of -1.4 V. After finishing the rapid desorption of the gold-thiol SAM (sections e and f), the gold electrode worked as a resistor. This electrochemical result shows the activation potential needs to be larger than -0.9 V. This CV brings forward an indisputable evidence of the reductive desorption of the gold-thiol SAM, and also demonstrates the negative bias potential is required to detach the RGD peptide from the gold electrode.

Cell Adhesion Manipulation

The adhesion of an adherent cell (NIH 3T3 fibroblast) was spatiotemporally manipulated with the surface-modified BBB composed of two-by-one gold electrodes where a left electrode was activated with activation potential of -1.2 V but a right one was inactivated. The BBB was then sonicated in cell culture media and NIH 3T3 fibroblasts were loaded. An

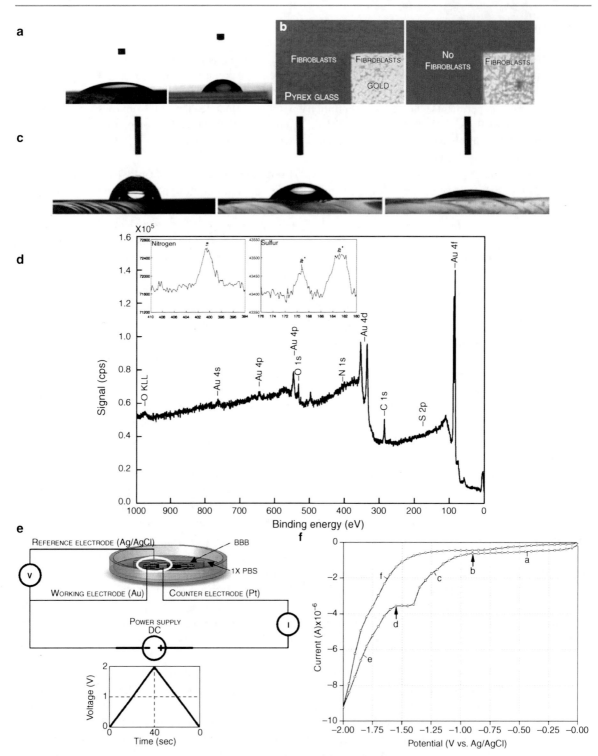

Biological Breadboard Platform for Studies of Cellular Dynamics, Fig. 3 (continued)

adherent cell in general has much higher affinity for RGD peptide than for gold when the cell adheres to the substrate. The loaded cell therefore adheres only to the right electrode, as shown in optical and immunofluorescent images (Fig. 4a, b). This result demonstrates the BBB's addressability in cell adhesion manipulation. Next, the dependence of cell adhesion at a single cell level on the size and geometric shape of a gold electrode was studied to determine the geometry of the gold electrode for subcellular detachment. Forty types of gold-electrode-arrays were prepared, as shown in Fig. 4c. Each array was designed to have 25 identical gold surfaces whose size was 9, 25, 64, 100, 225, 400, 625, or 900 μm^2 and their shape was an n-sided regular polygon ($n = 3, 4, 5, 6, \infty$ (circle)). NIH 3T3 fibroblasts were loaded into the surface-modified assays at a cell suspension concentration of 1×10^6 cells/mL. As an index for quantifying cell adhesion at a single cell level, a CA-ratio which is the ratio of the number of gold electrodes with cell adhesion to the total number of gold electrodes was measured as a function of the size and geometric shape of a gold surface, as shown in Fig. 4d. The measured CA-ratio provides the following facts. First, the CA-ratio in a single cell level is proportional to the size of a gold electrode, and the minimum size of a gold electrode for single cell adhesion is the diameter of a cell in a floating state (10 μm for NIH 3T3 fibroblast). Secondly, a cell wants to make its adhesion on the circumferential zone of a gold electrode rather than the central zone.

Cell Detachment Manipulation at a Cellular Level

The cell detachment dynamics at a cellular level was characterized by disconnecting an entire living cell from a substrate with the BBBs. The cell detachment experiments were carried out for three cases: two fibroblasts (see Fig. 5a), sparse fibroblasts with about 25% confluence (see Fig. 5b), and confluent fibroblasts with 100% confluence (see Fig. 5c). The experiments at cellular and subcellular levels were performed at the same experimental conditions and apparatus as those of cell adhesion experiment. As an evaluation index for cell detachment at a cellular level, a time required to detach 95% of the adhered fibroblasts (CD-time) was measured. The three cases, respectively, had the CD-times of 45.2 ± 6.8 s, 36.7 ± 8.7 s, and 21.1 ± 3.5 s (averaged from at least ten measurements) at activation potential of -1.5 V, showing more confluent cells have shorter CD-time, as shown in Fig. 5d. This is because all cells in confluent culture are connected to each other through cell-to-cell interaction. The detachment of one cell allows the neighboring cells to be easily detached by providing an external vertical force to them. The detached cell seems to continue to communicate with the neighboring cells through cell-to-cell interaction. The effect of activation potential on cell detachment was also investigated by counting the ratio of detached cells to total cells (CD-ratio) as a function of activation time and potential, as shown in Fig. 5e. The CD-ratio gets decreased as activation potential gets increased due to the reductive desorption of gold-thiol SAM which gets faster as negative potential gets increased (see Fig. 3f). The CD-ratio is monotonically increasing with two inflection points, s-shape curve, which clearly demonstrates there is a large deviation in integrin binding to extracellular matrix or other cells related to cell-to-cell interaction. A programmable cell patterning of NIH 3T3 fibroblasts was also made with a single four-by-four to demonstrate the BBBs' addressability and reusability in cell detachment manipulation at a cellular level, as shown in Fig. 5f. The cells were sequentially patterned in the shape of "C," "A," and "L" by using a single device. The BBB

Biological Breadboard Platform for Studies of Cellular Dynamics, Fig. 3 Characterization of two surface modifications and potentiodynamic electrochemical characterization of reductive desorption of gold-thiol SAM. (**a**) Contact angle of Pyrex surface before (*left*) and after (*right*) PEG treatment. The contact angle is changed from $25.7° \pm 1.5°$ to $61.5° \pm 3.8°$ by PEG treatment on the Pyrex surface. (**b**) Cell adhesion on the Pyrex surface before (*left*) and after (*right*) PEG treatment, showing the Pyrex surface is modified to be cell-resistive through PEG treatment. (**c**) Contact angles of bare gold ($67.3° \pm 2.5°$, *left*), thiol-treat one ($53.3° \pm 1.3°$, *middle*), and RTT-functionalized one ($24.6° \pm 2.8°$, *right*), confirming the surface modification by RTT. (**d**) XPS survey spectrum of RTT-functionalized gold surface. Detected are the gold peak from gold element, sulfur peak from thiol, nitrogen peak from amine group of RGD peptide, and carbon and oxygen peaks from carboxylic acid group of RGD peptide. (**e**) Experimental setup for CV measurement where a gold electrode, a platinum electrode, and an Ag/AgCl electrode work as working, counter, and reference electrodes, respectively. (**f**) Measured CV showing that the reductive desorption of the gold-thiol SAM starts and finishes at -0.9 and -1.55 V, respectively, and gets maximized at -1.4 V

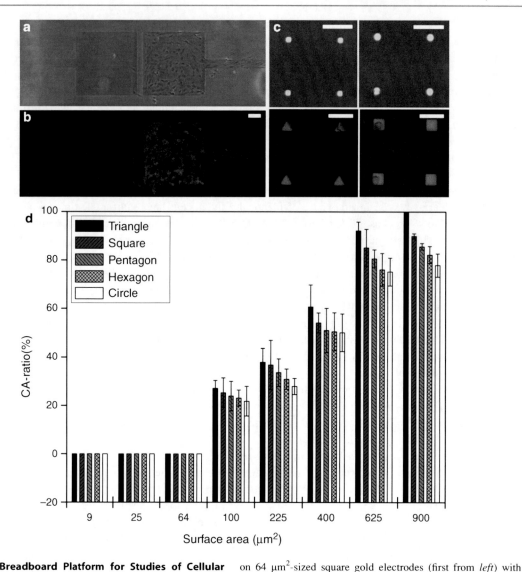

Biological Breadboard Platform for Studies of Cellular Dynamics, Fig. 4 Spatiotemporal manipulation of cell adhesion using the BBBs. (**a**) Optical image. The spatiotemporal manipulation of cell adhesion is demonstrated with a two-by-one BBB where a left electrode is activated but a right one is inactivated. (**b**) Immunofluorescent image. NIH 3T3 fibroblasts are stained for actin with rhodamine phalloidin (*red*) and for cell nucleus with DAPI (*blue*). (**c**) Cell adhesion as a function of gold electrodes shape and size. No cell adhesion is made on 64 μm^2-sized square gold electrodes (first from *left*) with the cell suspension of 1×10^6 cells/mL; cell adhesion is made on 25% of 100 μm^2-sized regular hexagonal gold electrodes (second); cell adhesion is made on 25% of 225 μm^2-sized equilateral triangular gold electrodes (third); and cell adhesion is made on 50% of 400 μm^2-sized square gold electrodes (fourth). (**d**) CA-ratio as a function of the size and geometric shape of gold electrodes. Scale bar of (**b**) is 100 μm, and those of (**c**) are 50 μm

was reused through simplified organic cleaning process after each usage, experimentally verifying the addressability, multifunctionality, and reusability of the BBBs at the same time. The BBBs are reusable due to their reductive desorption of gold-thiol SAM which yields original gold again.

Cell Detachment Manipulation at a Subcellular Level

The cell detachment at a subcellular level was also characterized by disconnecting a part of an NIH 3T3 fibroblast from a substrate using the BBBs which consists of gold electrode lines with 10 μm-width and

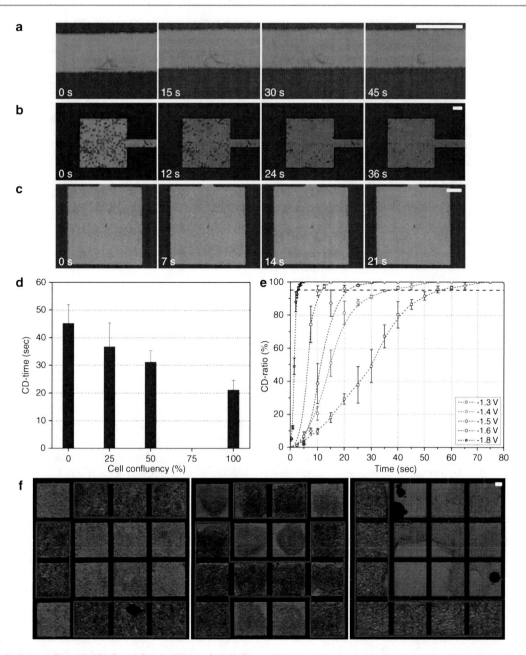

Biological Breadboard Platform for Studies of Cellular Dynamics, Fig. 5 Spatiotemporal manipulation of cell detachment at a cellular level using the BBBs. (**a**) Optical sequential images showing the spatiotemporal manipulation of cell detachment for two cells case. Two cells are detached from a gold electrode activated with negative potential of −1.2 V. The average CD-time for two cells is 45.2 ± 6.8 s. (**b**) Cell detachment manipulation of 25% confluent cells whose average CD-time is 36.7 ± 8.7 s. (**c**) Cell detachment manipulation of 100% confluent cells whose average CD-time is 21.1 ± 3.5 s. (**d**) Measured CD-time as a function of cell confluency, showing

CD-time is inversely proportional to cell confluency. (**e**) Measured CD-ratio as a function of activation time and potential for 100% confluent cells. The CD-ratio is monotonically increasing with two inflection points (s-shaped curve), indicating there is a large deviation in integrin binding to ECM or other cells related to cell-to-cell interaction. (**f**) Programmable cell patterning (cell detachment manipulation) using a single four-by-four BBB. The cells are patterned into "C," "A," and "L" shapes, verifying the addressability and reusability of the BBBs. Scale bars are 100 μm

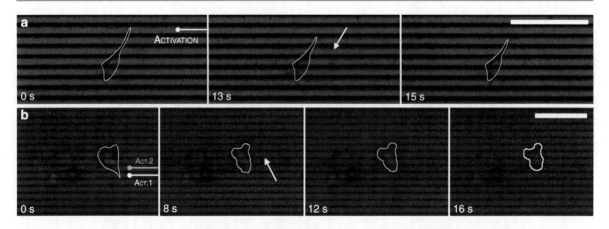

Biological Breadboard Platform for Studies of Cellular Dynamics, Fig. 6 Spatiotemporal manipulation of cell detachment at a subcellular level using the BBBs. (**a**) A part of NIH 3T3 fibroblast is detached by a single activation. This subcellular detachment is accompanied by the spontaneous retraction of the detached part of a cell. (**b**) Sequential subcellular detachment manipulation. A part of NIH 3T3 fibroblast is sequentially detached by a series of activations (Act. 1 and Act. 2). Scale bars are 100 μm

3 μm-gap (see Fig. 2d). As shown in the optical sequential images (see Fig. 6a), the subcellular detachment was accomplished by retracting the detached cytoskeleton for 16 s, which is faster retraction than cell detachment at a cellular level. This is because a single cell which has no constraint in stretching itself with focal adhesions is always under higher strain than confluent cells. In the next subcellular detachment experiment, a part of NIH 3T3 fibroblast was sequentially detached with a series of activations where the first activation (activation 1) was followed by the second one (activation 2) after 16 s. The optical sequential images reveal that repetitive activations within dozens of seconds do not have any influence on the cell's viability and a subcellular detachment amount is gradually increasing, as shown in Fig. 6b. These subcellular detachment experiments open a possibility for spatiotemporal manipulation of cell motility by disconnecting a part of the cell from a substrate. For example, when an adherent cell is exposed to subcellular detachment at 3 o'clock direction, it will migrate into 9 o'clock direction.

In our biological breadboard platforms, the spatiotemporal manipulation of cell adhesion and detachment is made by deliberately detaching an RGD peptide from a gold electrode, leading to adhere and detach an adherent cell at cellular and subcellular levels. It is therefore necessary to check whether a detached cell is still alive rather than electrocuted. If a detached cell is electrocuted, this technology cannot be used for spatiotemporal manipulation of cell motility by disconnecting a part of the cell from a substrate, and furthermore for cell adhesion and detachment manipulation for a better understanding of cellular dynamics. The viability of the detached cell was therefore characterized. In this experiment, NIH 3T3 fibroblasts bound to the BBBs were detached by activation potential of -1.5 V. The detached cells collected with a centrifuge, and then it was loaded into a plastic Petri dish. Their viability was observed after 24 h of cell loading. This experiment shows 80% of the detached cells are still alive. Compared to a general cell passage at which about 95% of cells are viable, this result is comparably excellent and lays a technical foundation for spatiotemporal manipulation of cell adhesion and detachment using the BBBs.

Future Directions

Although we have developed the biological breadboards for spatiotemporal manipulation of cell adhesion and detachment at cellular and subcellular levels, the platforms are still at their earliest stage, that is, there is no commercially available one. Four exquisite features of the BBBs over the previous cell manipulation methods, however, make them a promising platform for studies of cellular dynamics: (1) addressability in the spatiotemporal manipulation of cell adhesion and cell detachment; (2) multifunctionality in the characterization of cell biomechanics such as cell positioning, cell patterning, and cell motility control; (3) reusability due to the electrochemical reaction of a gold-thiol SAM; and (4) no need for special accessories in incorporating the BBBs with other conventional instruments. Extrapolation of this technology to other adherent cells might help us to investigate a critical cellular function and behavior, thereby leading to a better understanding of cellular dynamics. Ongoing work is focusing on more in-depth control of cell motility by developing a large-scale assay to shed light on the dynamics of cell motility. Combined with molecular dynamics models, the proposed devices for programmable subcellular adhesion and detachment will offer a new platform for studies of molecular biomechanics of the cell, especially mechanotransduction at focal adhesions.

Cross-References

▶ Cell Adhesion

References

1. Geiger, B., Bershadsky, A., Pankov, R., Yamada, K.M.: Transmembrane crosstalk between the extracellular matrix and the cytoskeleton crosstalk. Nat. Rev. Mol. Cell Biol. **2**, 793–805 (2001)
2. Galbraith, C.G., Yamada, K.M., Sheetz, M.P.: The relationship between force and focal complex development. J. Cell Biol. **159**, 695–705 (2002)
3. Clark, P., Connolly, P., Curtis, A.S., Dow, J.A., Wilkinson, C.D.: Topographical control of cell behavior. I. Simple step cues. Development **99**, 439–448 (1987)
4. Lussi, J.W., Tang, C., Kuenzi, P.-A., Staufer, U., Csucs, G., Vörös, J., Danuser, G., Hubbell, J.A., Textor, M.: Selective molecular assembly patterning at the nanoscale: a novel platform for producing protein patterns by electron-beam lithography on SiO_2/indium tin oxide-coated glass substrates. Nanotechnology **16**, 1781–1786 (2005)
5. Lee, K.-B., Park, S.J., Mirkin, C.A., Smith, J.C., Mrksich, M.: Protein nanoarrays generated by dip-pen nanolithography. Science **295**, 1702–1705 (2002)
6. Hoff, J.D., Cheng, L.-J., Meyhöfer, E., Guo, L.J., Hunt, A.J.: Nanoscale protein patterning by imprint lithography. Nano Lett. **4**, 853–857 (2004)
7. Chen, C.S., Mrksich, M., Huang, S., Whitesides, G.M., Ingber, D.E.: Geometric control of cell life and death. Science **276**, 1425–1428 (1997)
8. Folch, A., Jo, B.H., Hurtado, O., Beebe, D.J., Toner, M.: Microfabricated elastomeric stencils for micropatterning cell cultures. J. Biomed. Mater. Res. A **52**, 346–353 (2000)
9. Roth, E.A., Xu, T., Das, M., Gregory, C., Hickman, J.J., Boland, T.: Inkjet printing for high-throughput cell patterning. Biomaterials **25**, 3707–3715 (2004)
10. Birkbeck, A.L., Flynn, R.A., Ozkan, M., Song, D., Gross, M., Esener, S.C.: VCSEL arrays as micromanipulators in chip-based biosystems. Biomed. Microdevices **5**, 47–54 (2003)
11. Rosenthal, A., Voldman, J.: Dielectrophoretic traps for single-particle patterning. Biophys. J. **88**, 2193–2205 (2005)
12. Lahann, J., Mitragotri, S., Tran, T.-N., Kaido, H., Sundaram, J., Choi, I.S., Hoffer, S., Somorjai, G.A., Langer, R.: A reversibly switching surface. Science **299**, 371–374 (2003)
13. van Kooten, T.G., Schakenraad, J.M., van der Mei, H.C., Dekker, A., Kirkpatrick, C.J., Busscher, H.J.: Fluid shear induced endothelial cell detachment from glass-influence of adhesion time and shear stress. Med. Eng. Phys. **16**, 506–512 (1994)
14. Kuo, S.C., Hammer, D.A., Lauffenburger, D.A.: Simulation of detachment of specifically bound particles from surfaces by shear flow. Biophys. J. **73**, 517–531 (1997)
15. Lotz, M.M., Burdsal, C.A., Erickson, H.P., McClay, D.R.: Cell adhesion to fibronectin and tenascin: quantitative measurements of initial binding and subsequent strengthening response. J. Cell Biol. **109**, 1795–1805 (1989)
16. Shao, J.-Y., Hochmuth, R.M.: Micropipette suction for measuring piconewton forces of adhesion and tether formation from neutrophil membranes. Biophys. J. **71**, 2892–2901 (1996)
17. Inaba, R., Khademhosseini, A., Suzuki, H., Fukuda, J.: Electrochemical desorption of self-assembled monolayers for engineering cellular tissues. Biomaterials **30**, 3573–3579 (2009)
18. Guillaume-Gentil, O., Gabi, M., Zenobi-Wong, M., Vörös, J.: Electrochemically switchable platform for the micropatterning and release of heterotypic cell sheets. Biomed. Microdevices **13**, 221–230 (2011)
19. Karp, G.: Cell and Molecular Biology: Concepts and Experiments. Wiley, New York (2005)
20. Dalton, B.A., Walboomers, X.F., Dziegielewski, M., Evans, M.D., Taylor, S., Jansen, J.A., Steele, J.G.: Modulation of epithelial tissue and cell migration by microgrooves. J. Biomed. Mater. Res. A **56**, 195–207 (2001)

Biological Nano-crystallization

▶ Macromolecular Crystallization Using Nano-volumes

Biological Photonic Structures

▶ Structural Color in Animals

Biological Platform

▶ Biological Breadboard Platform for Studies of Cellular Dynamics

Biological Sensors

▶ Arthropod Strain Sensors

Biological Structural Color

▶ Structural Color in Animals

BioMEMS/NEMS

▶ Nanotechnology

Biomimetic Antireflective Surfaces Arrays

▶ Moth-Eye Antireflective Structures

Biomimetic Energy Conversion Devices

▶ Self-repairing Photoelectrochemical Complexes Based on Nanoscale Synthetic and Biological Components

Biomimetic Flow Sensors

Jérôme Casas[1], Chang Liu[2] and Gijs Krijnen[3]
[1]Institut de Recherche en Biologie de l'Insecte, IRBI UMR CNRS 6035, Université of Tours, Tours, France
[2]Tech Institute, Northwestern University, ME/EECS, Room L288, Evanston, IL, USA
[3]Transducers Science & Technology group, MESA + Research Institute for Nanotechnology, University of Twente, Enschede, The Netherlands

Definition

Biomimetic flow sensors are biologically inspired devices that measure the speed and direction of fluids.

Introduction

This survey starts by describing the role and functioning of airflow-sensing hairs in arthropods and in fishes, carries on with the biomimetic MEMS implementations, both for air and water flow sensing, and ends up with some perspectives for bio-inspired micro-technologies based on the latest understanding of the biological sensors.

Inspiring Biological Systems

Flow sensing systems have been mainly studied on crickets and cockroaches, and to a lesser degree on spiders. The fact that the latter are predators of the former seems to matter little so far, and this entry will concentrate on the cricket for comprehensiveness. Flow sensors in crustaceans living in water have been studied to an even lesser degree but in a similar fashion, without any relationship with biomimetics, so that these works are not mentioned here.

Arthropod Hairs

The filiform hairs of many insects, spiders, and other invertebrates are among the most delicate and sensitive flow sensing cells: they measure displacements on the order of a fraction of the hydrogen atomic diameter (sensitivity ca. 10^{-10} m $= 1$ Å) and react to flow speeds down to 30 μm/s. If one considers the energy needed to elicit a neuronal spike, one finds that they react with a thousandth of the energy contained in a photon, so that they surpass photoreceptors. In fact, these mechanoreceptors work at the thermal noise level [19]. These hairs pick up air motion, implying that they are measuring both the direction and speed of air particles, in contrast to pressure receivers, i.e., ears. Many insects have ears, so that the extra information available in the flow field must be useful. The biomechanics of the filiform hairs have been studied with care since several decades by several groups worldwide, based on the analogy with a single degree of freedom inverted pendulum without bending (see the latest review of hair biomechanics in [8]).

Among insects, mainly cockroaches and crickets have been studied, because their airflow-sensitive hairs are put on two antenna-like appendages, the cerci (cercus in singular, see Fig. 1). Insect hairs have usually a high aspect ratio, with a length of a few hundreds of microns up to 2 mm, and with a diameter of less than a dozen of microns (Fig. 1). Their tapered shape have been found to have an influence on the drag forces. Hairs and sockets are ellipsoidal in cross section, which leads to a movement in a preferred direction. The hairs of spiders, called trichobothria, are often curved, are innervated by several sensory cells and have an ultrastructure which is favorable to drag while minimizing weight (a feathery structure which decreases the mass and hence their inertia considerably, while increasing the friction in air). The studied airflow-sensing trichobothria are located on the legs. These two aspects regarding the exact form of cricket and spider hairs show that natural selection is acting on the tiniest details of biological organization. The base of the hairs is much more complex, and its mechanics poorly understood. In crickets, only a single sensory cell is below the hair shaft, while spiders have several (Fig. 1). At the base of the socket, another mechanoreceptor type, the campaniform sensilla, reacts to deformation of the cuticula produced by the movement of the socket. Crickets possess often two sensilla around each hair. Thus the two sensors types act as a coupled system, extending thereby the range of forces it can measure.

Arthropods are very often quite hairy, and the high density of flow sensing hairs implies that they interact with each other in order to produce a high-resolution map of flow characteristics, acting like a flow camera (Fig. 1). Up to now, only the hydrodynamic type of interaction has been studied, called viscous coupling. It was found to be highly dependent on the geometrical arrangement of hairs, of their respective lengths and preferential planes of movement, as well as on the frequency content of the input signal [4]. Hairs often interact over long distances, up to 50 times their radius, and usually negatively. Short hairs in particular "suffer" substantially from the presence of longer hairs nearby, due to a shadowing effect. Positive interactions, where the flow velocity at one hair is increased by the presence of nearby hairs, have been however observed in real animals and reproduced computationally. The biological implications of these interactions have only very recently been addressed, and hint toward a coding of incoming signals which relies strongly on the specific sequence of hairs been triggered [16]. In other words, the signature of the incoming signal is mapped into a given sequence of recruited hairs, which in turn produces a typical sequence of action potentials.

Single hairs, or groups of hairs, are not placed at random on the body. For the same reasons that the exact shape of hairs and their relative position within a group have been most likely molded by natural selection, the position of the hairs on the sensory organ must be shaped by selection. This aspect of mechanosensory research is however badly neglected. As for the positions along the cercus, the presence of a potential acoustic fovea (i.e., a location with particularly high acuity) at the base of the cercus has been hinted already twice, not only due to the highest hair density in this region, but also because it corresponds to the region with the largest flow velocities, the cercus being the largest there. Putting hairs radially around the cercus enables crickets also to pick up transversal flows, whose peak flow velocities are larger than in the situation where the hairs would be placed on a flat surface; in the later case, hairs are submitted to longitudinal flow with lower peak velocities. In summary, where you put your sensors relative to your body geometry matters a lot.

Biomimetic Flow Sensors, Fig. 1 Habitus of a wood cricket (**a**), with the two cerci highlighted, on which a large number of filiform hairs, intermixed with spines and chemoreceptive hairs, can be observed (**b**). The socket of a hair is a complex system made of several membranous stoppers. It deflection is sensed through another type of mechanosensor, the campaniform sensilla, which is a strain detector (**c**). Both act as a coupled system, the cuticula being deformed once the hair touches the border of the socket. Each sensor is innervated by a single hair cell (**d**)

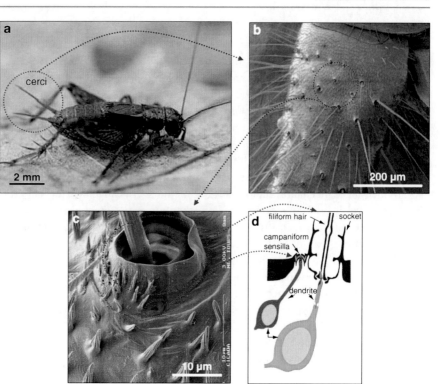

An action potential triggered by a moving hair ends up directly in the terminal abdominal ganglion (TAG), a local neuronal processing unit. Information from all the hairs, as well as from other sensors, converges there and is processed by interneurons. The compression and convergence of information at this stage is huge: About 1,500 afferent neurons of hairs are connected to only some 20 interneurons [10]. The fact that invertebrates possess few large, singly identifiable neurons enabling comprehensive mapping and repeated recordings of activities is a unique asset which explains the interest in such exotic systems. Information coming from the central brain as well as from the higher ganglia also descends into the TAG. Once processed, the combined information moves up quickly toward higher neuronal centers, in particular the ganglia in which the hindleg movements are being decided. This local feedback loop, with little input from the main brain, enables the animal to process vital information and act accordingly very quickly. As so often with invertebrates, what can be processed locally should be done so, a distributed processing scheme which explains why biomimetics has so much to gain from this group of animals.

The last level of integration is behavior, and flow sensing is known to be of importance in predator and prey perception, sexual selection, and most likely other context, such as noticing its own speed and movement. Predator avoidance is obviously a major selection force, where speed is of paramount importance. Jumping or running away is the behavior which is elicited using appropriate stimuli. The cricket possesses in the TAG an internal map of the direction of the stimuli from the outside world and the geometric computation of the direction of incoming flow by the cercus is one of the nicest case studies of spatial representation [10]. Computing the speed of an approaching predator is also carried out by the TAG, and has been only recently established using appropriate stimuli. Where to jump is a different question, in which directing stimuli and other conditions intervene.

Natural selection therefore acts along the full chain of information transfer, from acquisition and processing, up to actuation. This is important to restate in a biomimetic context, as the extreme sensitivity on the biomechanical side of the hair shaft, which has been the exclusive focus of the engineer attention, could be otherwise lost into an inefficient sequence of information transfer. As of today, one has however very little information about the constraints acting on the different parts of the chain, and hence no idea about their optimization levels.

Fish Neuromasts

Hair cells of the lateral-line system of fish are able to detect water displacement of the order of ca. 20 nm [2]. This ability is used in various ecological contexts, such as prey and predator localization, schooling, and most likely a host of other behaviors. Fish can also detect obstacle-created flow distortions of a flow field they have generated. The blind cave fish, *Astyanax fasciatus*, is an interesting model because it is believed that this species, having lost sight and having to maneuver in darkness, possesses particularly sensitive hair cells. The spatial discrimination ability, through self-induced water movements when approaching objects, of the blind cave fishes is about 1 mm^2. The hairs of fish are located within specialized units, called neuromasts. These neuromasts, which are localized in many different parts of the body and which do not display a strong correlation with the flow structure of their environments, are located either just on the skin (called superficial neuromasts), or in specialized fluid-filled canals which produce a reticulated pattern on the head and along the body sides. A single neuromast consists of sensory cells that project into a jelly-like material, the cupula, in which several hairs (or cilia) of different lengths are embedded. The cupula is therefore the medium which connects the hairs and the external fluid.

An Engineering Perspective on Biomimetic Hairs

Hair-based flow sensing systems of arthropods and fish contain various levels, ranging from the mechanical structure, to the generation of action potentials in neurons, and signal transport, collecting and processing signals from many (hair-) sensors in the various neuronal centers. Taking inspiration from such complex systems is a non-trivial endeavor. The mechanics of these sensory systems are relatively easy to mimic in an engineering context, at least in comparison to the complexity of the neural system which requires

delicate materials (e.g., membranes with ion-channels) and fluids with ion concentrations robustly and actively maintained by the organism. Consequently, up to date most of the bio-inspired flow sensors actually are defined merely by the fact that they use a hair, if needed covered by a cupula resembling cover, to capture viscous drag forces. Having no neural system in an engineering context implies that signal transduction, i.e., from the mechanical to the electrical domain, needs to be done by rather "un-bio-inspired" methods, e.g., using capacitive or piezoresistive interrogations methods. The next level up, namely the arrangement of the hair sensors into an array like structure (a cercal canopy for crickets and a lateral line for fish), is again open to engineering approaches. Mimicking the entire hair-based sensory system requires the ability to recreate dense hair-sensor arrays, including the associated electronic interfacing, asking for engineering solutions to the multiplexing problems. On the signal processing level, the biomimetic content can be high again, e.g., by exploitation of artificial neural networks to extract features and events from the multitude of array signals. The insight in the neuronal architecture of the animals needed for such exercise is however at best sketchy, with the exception of the architecture behind the directional acuity of crickets.

In view of the overall shape and sizes of the hairs in combination with the high densities and large numbers of hair sensors on the hair canopies, it is obvious that an artificial version of this kind of sensor arrays cannot be assembled but should be made monolithically. As it turns out, the range of dimensions is well attainable by micromachining techniques as used in the field of micro-electro-mechanical systems (MEMS). However, the materials generally used in MEMS are rather stiff compared to the materials found in nature, calling for some creative engineering.

Various types of hair-based flow sensors, for operation in air as well as in water, have been developed using both conventional machining and micromachining technologies. They are based on a variety of materials (including semiconductor, oxide, and polymers) and transduction principles (including electrostatic sensing, piezoresistive sensing, thermal sensing, etc.). However, artificial hair-cell (AHC) sensors represent unique challenges to microengineering design and fabrication. The high aspect ratio, vertical, hair must be fabricated using processing steps that are amenable to engineering, manufacturing and eventually scaled production. The overall process must be scalable to large areas as the AHC sensors are often used in arrays. Ideally, the sensors should be integrated with highly sensitive transduction mechanisms to couple mechanical input to electrical output. Further, it is desirable that the AHC sensors be made on a substrate that also houses the integrated electronics elements for signal conditioning and amplification.

For reasons outlined above, most of the hair-based flow sensors reported in literature are made using MEMS fabrication technology. Ironically, the fabrication of the relative large hair structures poses considerable difficulties in this technology. Two basic types of artificial hairs can be distinguished, namely, hairs fabricated in the wafer plane and hairs fabricated perpendicular to the wafer plane. The first is straightforward since surface micromachining techniques can be used. However, surface-micromachined hairs cannot easily be combined into high-density arrays. An overview of the various reports in literature on ciliar inspired actuators and sensors is found in [22].

Ozaki et al. [17] were probably the first to provide flow sensors inspired by insects. Their piezoresistance-based sensors had either 400–800 μm long hairs in plane, being cantilevers exposed to flow after removing large part of the silicon substrate, or consisted of wires manually glued to a cross-shaped piezoresistive structure delivering two DOF sensitivity. Characterization by continuous flows in the m/s range proofed the functionality and directivity of the sensors. A comparable in-plane hair fabrication process was used in [12], but incorporated a stress gradient in the artificial hairs to make them curve out of plane. Readout is obtained by tracking the flow-induced vibrations of a set of dissimilar hairs resonating at different frequencies when exposed to specific flows. Argyrakis et al. [1] also use in-plane cantilevers with piezoresistive readout connected to an artificial neuronal circuit for spike generation. An alternative approach was proposed in [15] where the actual hair consisted of an optical fiber with reduced optical transmission when displaced by viscous drag. The sensor is meant for DC-flow measurements and achieves high volumetric sensitivity (μL/min range). Xueb et al. [20] more or less reverted to the structure of Ozaki using a cross-shaped frame with piezoresistors with a plastic hair of about 5 mm length. A nice twist on the hair

sensor is presented in [14] by exploiting piezoelectric transduction by polyvinylidene fluoride in thermo-direct written clamped–clamped beam-type structures. Recently, interesting results have been obtained using a straightforward cantilever design in combination with appropriate electronic feedback which helped to incorporate adaptation of the sensor allowing for sharp filtering and gain [13].

Hair-Sensor Systems for Operation in Air: A Case Study

Over the course of several years, the possibilities to fabricate hair-based flow sensors and sensor arrays inspired by the hair sensors and canopies as found on the cerci of crickets have been explored. In several sensor iterations, understanding optimization criteria alongside with viable and robust fabrication technology and suitable interfacing was targeted, not only for single hair sensors but also for arrays.

The scaling of the hair sensors is clearly confined by various requirements. First of all the sensors are supposed to capture a sufficient amount of drag-force. Evidently, the longer and thicker the hairs, the larger the drag torque is (see Fig. 2). Even more so since air flowing over a substrate forms a boundary layer in which the flow-velocity transitions from zero, at the substrate, to the maximum (far-field) value. This boundary layer depends on viscosity and becomes thinner with increasing frequency. Next, for given drag torque the angular rotation will increase when the rotational stiffness decreases. Hence, a small rotational stiffness will be beneficial for the sensitivity and flow-velocity threshold. However, animals live in dynamic environments in which the amount of information derived from the flow sensors not only depends on the sensors sensitivity but also on the bandwidth in which they are sufficiently sensitive. Combining these observations one finds, unfortunately, that this bandwidth goes down with the square-root of the ratio of the spring stiffness and moment of inertia. In order for the animals to capture a large amount of drag torque while still maintaining a low spring-stiffness at a usable large bandwidth, it turns out that long and thin hairs are beneficial. Parametric studies on the drag-force as modeled by Stokes' expressions showed that the dependence on hair diameter is far less than the dependence on hair length. Based on the mechanical

Biomimetic Flow Sensors,
Fig. 2 Viscous drag torque on
a cylinder of length L_h,
diameter of 50 μm for various
frequencies with flow
amplitude of 10 mm/s

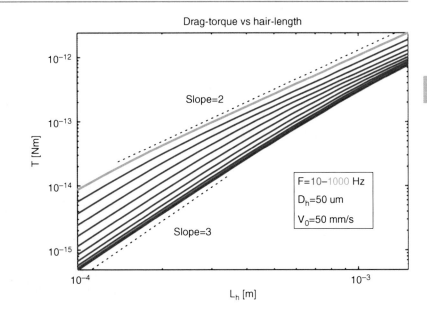

responsivity and available bandwidth, one can propose
a figure of merit (FoM) given by:

$$\frac{L^{1/2}}{\rho^{1/2}S^{1/2}D^{2/3}} \qquad (1)$$

where L and D are the hair length and diameter, S is the
rotational stiffness, and ρ is the specific mass density
of the material from which the hairs are formed. The
FoM is the product of the bandwidth of the sensors
(expressed by the resonance frequency, i.e., square
root of the rotational spring-stiffness S over the
moment of inertia J) times their mechanical
responsivity (i.e., the magnitude of angular rotation α
per unit of flow-velocity V). Clearly it pays off to have
long and thin hairs mounted on flexible suspensions.

The conditions under which hair sensors have to
operate are such that the Reynolds (R) and Strouhal
(St) numbers are relatively low. For a hair diameter of
25–50 μm, an air-oscillation frequency of 250 Hz and
a flow-velocity amplitude of 10 mm/s, R varies
between 0.008 and 0.016 and St between 1.96 and
3.92 (for the flow around the hairs). The rather small
Reynolds numbers and the large hair-length to hair-
diameter ratio allow using the Stokes expressions for
the drag torque exerted by the airflow on the hairs.
Furthermore, these hair sensors are mounted on flat
substrates allowing the use of the Stokes expressions
for a harmonic flow along the hairs. These expressions

predict a viscous flow over an infinite substrate to be
harmonic in time with zero flow velocity and 45° phase
advance at the substrate interface and a boundary
layer thickness (δ_b) proportional to the inverse of
$\beta = (\omega/2v)^{0.5}$ where v is the kinematic viscosity
($1.79 \cdot 10^{-5}$ m²/s for air at room temperature) and ω is
the radial frequency. The Stokes expressions can be
usefully employed for this situation [7]. As an example, for a harmonic airflow of 100 Hz, the boundary
layer is roughly 0.5 mm. Note that under most conditions the artificial hairs can be assumed infinitely stiff
and the rotation angles are rather small (in the order
of 1–100 mrad amplitude per m/s flow-velocity
amplitude).

On a basal level the simple principle of a hair sitting
on a torsional mount allowing for drag-induced rotation has been the starting point for these artificial hair
sensors (see Fig. 3). In the implementation, the hairs
are made of SU-8. Using two deposition-exposure
cycles and subsequent development, these hairs can
be made up to 1 mm long with diameters of about
50 μm for the bottom part and 25 μm for the top part.
This configuration hardly reduces the drag torque but
decreases the moment of inertia of the artificial hairs
by about 65%. The rotational freedom comes from two
torsion beams. These beams as well as two membranes
connected to the beams are made of silicon nitride. On
the membranes an Al layer is deposited which forms
a capacitor together with the underlying highly doped

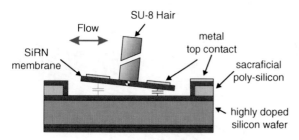

Biomimetic Flow Sensors, Fig. 3 Schematic of artificial hair-based flow sensor with differential capacitive layout

silicon substrate. On drag-induced tilting of hair and membranes, the capacitances change in opposite fashion allowing for a differential capacitive read-out scheme. The hairs sit on the membranes as well. Rotational freedom is obtained when the structure is released by sacrificial etching of the poly-silicon layer (0.5–1.0 μm thick) that sits between the membrane and a protective silicon-nitride layer on the silicon substrate.

The choice for capacitive transduction is motivated by the fact that differential capacitive readout can be very sensitive and allows for rejection of various common mode signals (e.g., those that originate from flow in the direction perpendicular to the plane in which the hair is supposed to tilt). Moreover, contrary to e.g., thermal readout, it requires little power and does not thermally pollute the air around the sensors. This may not be important for a single hair but since the interest is in larger arrays, thermal effects could be far from negligible. Additionally capacitive readout allows for interrogation of many hair sensors in an array by frequency division multiplexing. There are also some clear drawbacks: the electronics required for capacitive readout are far more complex, certainly when compared to piezoresistive transduction and the method asks for rather complex mechanical structures when applied to sensors for use in liquids, especially in water. In the latter situation, the volume between the electrodes is ideally sealed in order to prevent liquid to get between since this could lead to electrolysis and large viscous damping. Figure 4 shows an exploded view of these capacitive, artificial hair-based flow sensors. Measurements of these sensors are shown in Fig. 5. Due to a combination of the boundary layer effect (high pass) and the second-order mechanics (i.e., its behavior is described by a second-order ordinary differential equation acting as low pass), the sensors show a band-pass characteristic.

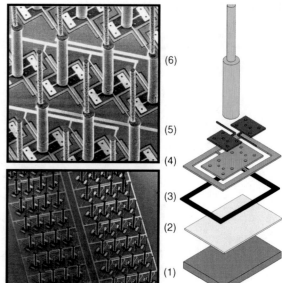

Biomimetic Flow Sensors, Fig. 4 *Right*: Exploded design view of a single hair sensor showing: (*1*) highly conductive silicon bulk (*bottom* electrode), (*2*) 200 nm thick SiRN layer for insulation and etch-stop, (*3*) Poly-silicon layer after final sacrificial etching, (*4*) 1 μm thick SiRN layer patterned into membranes, (*5*) 100 nm thick aluminum for *top* electrodes and (*6*) two 450 μm thick SU-8 layers patterned into a long hair. *Left*: SEM images of microfabricated hair sensor arrays with zoom of the two-stage artificial hairs

The beauty of the neural system lies in its robustness as well as its largely localized generation of signals. In contrast, the capacitive measurement method employed in these sensors is plagued by parasitic capacitances that can hardly be avoided and have a far less localized nature. Changes in temperature, humidity, wiring geometry, etc. indeed cause fluctuations of the parasitics which are large relative to the small capacitive changes that the hair sensors induce. Therefore, control of the parasitics is of paramount importance for a proper device functioning. A reduction in parasitics can be obtained using much smaller electrode areas as well as exploiting shielding. Both have been shown to be feasible using silicon-on-insulator (SOI) technology. Results of the latest generation of sensors are shown in Fig. 6.

Successive iterations led to airflow sensors delivering sensitivities of about 400 μm/s (300 Hz bandwidth) with near-perfect figure of eight directivity. In order to do this, segmented cylinders, each up to 1 mm long, were used. Large arrays of sensors (over 100) can

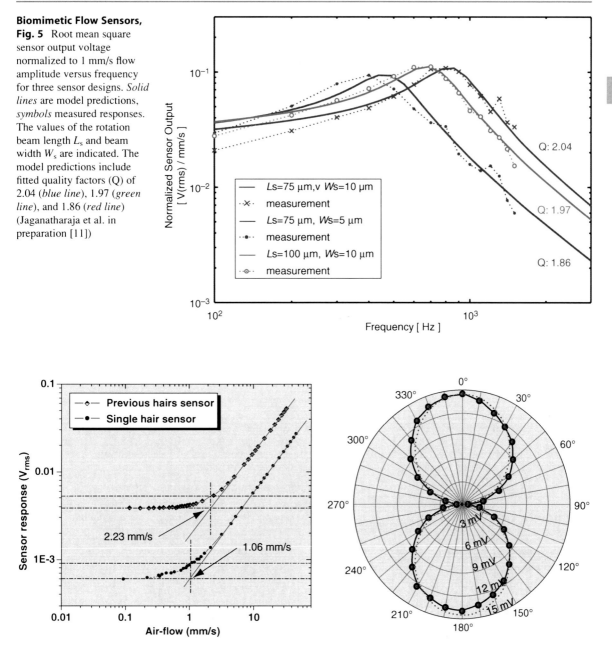

Biomimetic Flow Sensors, Fig. 5 Root mean square sensor output voltage normalized to 1 mm/s flow amplitude versus frequency for three sensor designs. *Solid lines* are model predictions, *symbols* measured responses. The values of the rotation beam length L_s and beam width W_s are indicated. The model predictions include fitted quality factors (Q) of 2.04 (*blue line*), 1.97 (*green line*), and 1.86 (*red line*) (Jaganatharaja et al. in preparation [11])

Biomimetic Flow Sensors, Fig. 6 *Left*: Output voltage of two generations of sensors measured in a 3 KHz bandwidth. The single hairs sensor is fabricated using a silicon on insulator substrate with a technology slightly different from the one indicated in Fig. 3. Measurements were done as a function of flow-amplitude for a 250 Hz sinusoidal flow. *Red lines* are asymptotic lines to determine the threshold flow. *Right*: directivity of the single hair sensor showing an almost ideal figure of eight (*red dashed lines*)

be fabricated and, utilizing frequency division multiplexing, these sensors can be simultaneously and individually interrogated (see Fig. 7). This allows one to carry out measurements of spatiotemporal flow-patterns rather than single point measurements. Hence

the aim is to develop a system delivering the functionality of a "flow camera" with sensor densities of about 25–100 sensors per mm^2 [6]. This development has largely benefitted from insights in the cricket cercal system.

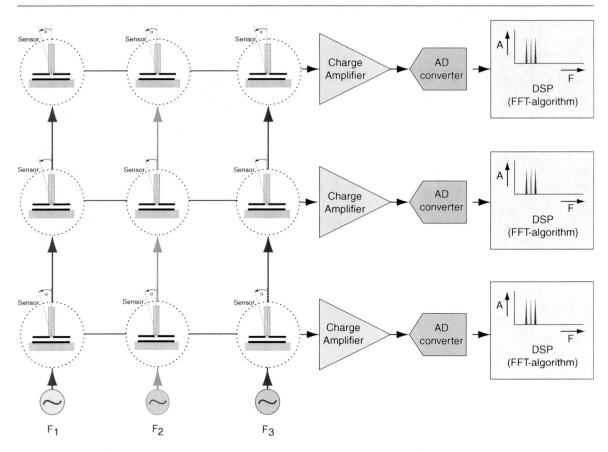

Biomimetic Flow Sensors, Fig. 7   Schematic of frequency division multiplexing as applied to hair-sensor array interfacing

Biomimetic Hairs in Water

Though functionally not necessarily different, hair-based flow sensors in water are confronted with some different scaling than sensors meant for air. The differences arise due to the fact that water has a higher mass density (about 1,000 times) and a smaller kinematic viscosity (about 20 times smaller). The first means that the effects of the moment of inertia of the hairs are smaller whereas the latter translates into boundary layers about 4.5 times smaller in water than in air. Hence, hairs for use in water can be both shorter and larger in diameter.

Hair-based sensing in water may exploit arbitrary transduction principles. However, the practical problems associated with capacitive readout caused by large viscous damping and necessary prevention of electrolysis may call for rather complicated technologies [9], as indicated in Fig. 8. This may be one of the

reasons that up to now not much work has been presented on capacitive aquatic hair-based sensors.

The piezoresistance-based artificial hair cell device, shown schematically in Fig. 9, consists of a cilium located at the distal end of a paddle-shaped silicon cantilever [5]. Doped silicon strain gauges are located at the base of the cantilever. The cilium is made of photodefinable SU-8 epoxy and is considered rigid. Lateral force along the on-axis acting on the cilium will create a bending moment (M), which is translated to the silicon beam through the stiff joint. The torque introduces a longitudinal strain and can be detected by piezoresistors at the base. The relation between the induced strain (ε) and the moment is given by

$$\varepsilon = \frac{6M}{Ewt^2} \qquad (2)$$

Biomimetic Flow Sensors, Fig. 8 Exploded view of a capacitive aquatic hair-based flow sensor. The complex technology allows for sealed cavities (etched from the back side) safeguarding low damping and preventing electrolysis

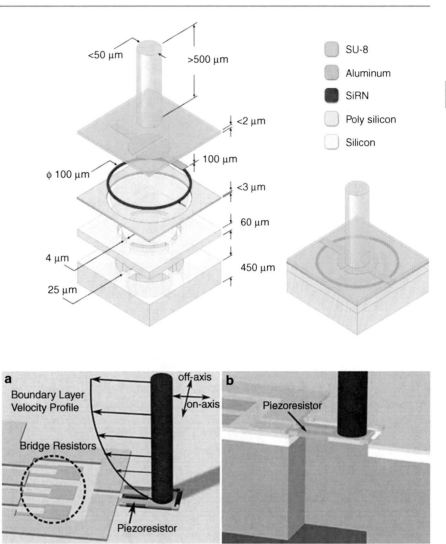

Biomimetic Flow Sensors, Fig. 9 (a) Schematic drawing of an individual AHC sensor with superimposed boundary layer flow velocity profile. (b) Cross-sectional perspective view of the AHC

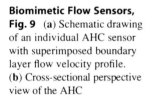

where E is the Young's modulus of silicon, w the cantilever width, and t the cantilever thickness. When used as a flow sensor, flow passing along the cilium introduces a bending moment (M) due to frictional and pressure drag. The device is primarily designed for underwater applications that require sensing of low-velocity and low-frequency flows. Structural and hydrodynamic model analyses suggest that the sensor's sensitivity is mainly determined by the following parameters: cantilever width (w), cantilever thickness (t), cilium height (h), cilium diameter (d), and sensor's distance form the leading edge (x).

For real-life underwater applications, good sensitivity in the sub-1 mm/s flow velocity range is desirable. Based on the existing signal conditioning circuitry

capability, one can assume the minimum reliable voltage reading at 1,000 times signal gain as 1 mV. Taking 70 as a reasonable estimate for the gauge factor value, 1 mV output at 1,000 times signal gain translates into approximately 0.06 micro-strain using the quarter bridge model. In order to create and detect 0.06 micro-strain under 1 mm/s flow, the values of w, h, and d need to be carefully chosen. Figure 10 is a 3-D map with x, y, and z-axis being the three pertinent design variables, w, h, and d, respectively. A point within the plotted plane indicates a cantilever design that will satisfy the sensitivity requirement. The choice of w, h, and d are further narrowed down according to processing capabilities and frequency response considerations.

Biomimetic Flow Sensors,
Fig. 10 3-D map showing the
possible combinations of
cilium height, diameter, and
cantilever width that satisfy
the design requirements

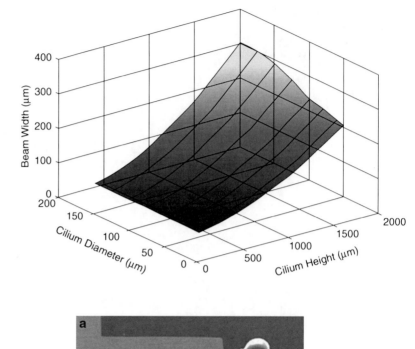

Biomimetic Flow Sensors,
Fig. 11 Scanning electron
micrograph of (a) an
individual and of (b) an array
of AHC sensors

The cantilever thickness is chosen as 2 μm; this correlates to the thickness of the single crystal silicon epitaxial layer on the silicon-on-insulator (SOI) wafer used. For the analysis at hand, the sensor's distance from the leading edge (x) is assumed to be 2 mm, set by the die size. The devices are fabricated on SOI wafers with a 2 μm thick epitaxial silicon layer on top, 2 μm thick oxide, and 300 μm thick handle wafer. SU-8 epoxy is chosen for its ability to form rigid high aspect ratio structures. Figure 11 shows an SEM of an actual device, and an array of AHC sensors. A series of

mechanical and electrical experiments were performed to characterize the sensor performance. The steady-state response to flow in water is presented in Fig. 12.

The sensitivity of artificial hair sensors can be tailored up to some extent. However, as indicated by the FoM above there is a trade-off between bandwidth and sensitivity. There seem to be some indications that nature does not only, or not per se, maximize the mechanical response (e.g., the rotational displacement of the hairs) but also optimizes the energy captured from the environmental flow. The allometric scaling of

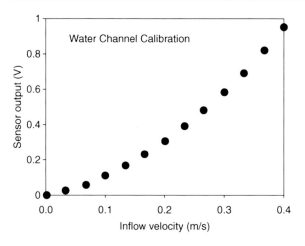

Biomimetic Flow Sensors, Fig. 12 Experimental data of underwater steady-state flow velocity calibration of a single hair

cricket hairs seems to reflect the principle of mechanical impedance matching [19]. One may reason that capturing the maximum amount of energy will lead to the best signal-to-noise ratio whereas a high sensitivity may be obtained by proper adaptations, e.g., the lever ratio in the cricket hair-sensor system. Such systems are found more often in nature; for example, the middle ear of the mammalian hearing system is merely an impedance transformer. It offsets the impedance difference by a factor of a few tens in order to capture maximum energy from air-borne sound waves and convert it to traveling waves in the cochlea despite a factor of a few thousand difference in acoustic impedance. Up to date little has been done in artificial hair sensors to address this aspect. Nevertheless some studies have shown the possibility to improve sensor performance by properly covering the sensory hairs by hydrogel structures, as discussed above. The high water content, reminiscent of the cupular material found in fish neuromast, alleviates the mechanical impedance difference, which together with the increased hydrodynamic surface, leads to a 40-fold improvement in responsivity [18, 21].

Perspectives

The performance of single flow detecting biomimetic MEMS hairs has steadily improved over the last two decades and the threshold flow velocity is now only about one order of magnitude higher than for the most spectacular biological sensors. This is quite an achievement, given that the signal transduction aspect of it is not carried out in a biomimetic approach. Some clever ideas might however originate from biology in this respect too. While active sensing, now found not only in humans and vertebrates but also in insects, has never been reported in these hairs yet, stochastic resonance has been already observed. This ability of biological systems to harness background or within-cell noise in order to improve signal capture needs to be seriously considered. Another as yet untapped possibility for increasing sensitivity and pattern recognition lies in the relative positioning of MEMS hairs, exploiting positive and negative viscous coupling [4]. Major challenges exist on the road until such sensors will be ready to use. They are brittle and break easily, in contrast to their natural counter parts which sustain very large deformations without harm. Furthermore, stiction problems (static friction) at the MEMS hair basis are occurring as soon as the flow velocity is too high, asking for incorporation of anti-stiction measures such as bumps and adhesion-reducing coatings. Nature is here again surprisingly resilient to extremes. Beyond potential technological applications, these hairs can now finally be turned into excellent physical models to tackle biological questions too difficult to address directly on biological materials: for example, the hydrodynamical interactions of tandem hairs moving within the same plane is nearly impossible to measure on real crickets, while MEMS of different lengths can be arranged at will.

Hair-based flow sensing is one of the few areas in which applied mathematics has been the common language of groups as diverse as field ecologists, materials scientists, aerodynamicists, theoretical physicists, computer scientists, and neurobiologists [3]. The mathematical toolbox, and its wide acceptance throughout the scientific community, is an unusual asset in biomimetics, as is the MEMS fabrication. Hair-based flow sensing is therefore *uniquely* suited for interactions among biologists and engineers working toward bio-inspired technologies.

Acknowledgments Without the many discussions we have had with our colleagues in the Cicada and Cilia projects this work would never have had the depth and breadth we have attained. These projects were the Cricket Inspired perCeption and Autonomous Decision Automata (CICADA) project (IST-2001-34718) and the Customized Intelligent Life Inspired Arrays (CILIA) project (FP6-IST-016039). Both projects were funded by the European Community under the Information Society Technologies (IST)

Program, Future and Emergent Technologies (FET), Lifelike Perception Systems action. We also like to acknowledge the contributions by our MSc and PhD students, PostDocs and technicians without which this paper would not exist.

Cross-References

▶ Bio-inspired CMOS Cochlea
▶ Biosensors
▶ Integrated Micro-acoustic Devices
▶ Micro/Nano Flow Characterization Techniques
▶ Nanomechanical Resonant Sensors and Fluid Interactions

References

1. Argyrakis, P., Hamilton, A., Webb, B., Zhang, Y., Gonos, T., Cheung, R.: Fabrication and characterization of a wind sensor for integration with a neuron circuit. Microelectron. Eng. **84**, 1749–1753 (2007)
2. Bleckmann, H., Zelick, R.: Lateral line system of fish. Integr. Zool. **4**, 13–25 (2009)
3. Casas, J., Dangles, O.: Physical ecology of fluid flow sensing in arthropods. Annu. Rev. Entomol. **55**, 505–520 (2010)
4. Casas, J., Steinmann, T., Krijnen, G.: Why do insects have such a high density of flow-sensing hairs? Insights from the hydromechanics of biomimetic MEMS sensors. J. R. Soc. Interface **7**, 1487–1495 (2010)
5. Chen, J., Fan, Z., Zou, J., Engel, J., Liu, C.: Two dimensional micromachined flow sensor array for fluid mechanics studies. ASCE J. Aerosp. Eng. **16**, 85–97 (2003)
6. Dagamseh, A., Bruinink, C., Kolster, M., Wiegerink, R., Lammerink, T., Krijnen, G.: Array of biomimetic hair sensor dedicated for flow pattern recognition. Proc. DTIP 2010, 5–7 May 2010, Seville, ISBN: 978-2-35500-011-9
7. Dijkstra, M., van Baar, J., Wiegerink, R., Lammerink, T., de Boer, J., Krijnen, G.: Artificial sensory hairs based on the flow sensitive receptor hairs of crickets. J. Micromech. Microeng. **15**, S132–S138 (2005)
8. Humphrey, J.A., Barth, F.G.: Medium flow-sensing hairs: biomechanics and models. In: Casas, J., Simpson, S.J. (eds.) Insect Mechanics and Control. Advances in Insect Physiology, vol. 34, pp. 1–80. Elsevier, Amsterdam (2008)
9. Izadi, N., de Boer, M.J., Berenschot, J.W., Krijnen, G.J.M.: Fabrication of superficial neuromast inspired capacitive flow sensors. J. Micromech. Microeng. **20**, 085041 (2010)
10. Jacobs, G.A., Miller, J.P., Aldworth, Z.: Computational mechanisms of mechanosensory processing in the cricket. J. Exp. Biol. **211**, 1819–1828 (2008)
11. Jaganatharaja, R.K.: Cricket inspired flow-sensor arrays. PhD thesis, University of Twente., 2011, ISBN 978-90-365-3215-0, http://dx.doi.org/10.3990/1.9789036532150
12. Kao, I., Kumar, A., Binder, J.: Smart MEMS flow sensor: theoretical analysis and experimental characterization. IEEE Sens. J. **7**(5), 713–722 (2007)
13. Kim, H., Song, T., Kang-Hun, A.: Sharply tuned small force measurement with a biomimetic sensor. Appl. Phys. Lett. **98**, 013704 (2011)
14. Li, F., Liu, W., Stefanini, C., Dario, P.: A novel bioinspired PVDF micro/nano hair receptor for a robot sensing system. Sensors **10**, 994–1011 (2010)
15. Lien, V., Vollmer, F.: Microfluidic flow rate detection based on integrated optical fiber cantilever. Lab on a Chip **7**, 1352–1357 (2007)
16. Mulder-Rosi, J., Cummins, G.I., Miller, J.P.: The cricket cercal system implements delay-line processing. J. Neurophysiol. **103**(4), 1823–1832 (2010)
17. Ozaki, Y., Ohyama, T., Yasuda T., Shimoyama I.: An air flow sensor modeled on wind receptor hair of insects. Presented at International Conference on MEMS, Miyazaki, 2000
18. Peleshanko, S., Julian, M.D., Ornatska, M., McConney, M.E., LeMieux, M.C., Chen, N., Tucker, C., Yang, Y., Liu, C., Humphrey, J.A.C., Tsukruk, V.V.: Hydrogel-encapsulated microfabricated haircells mimicking fish cupula neuromast. Adv. Mater. (19), 2903–2909 (2007).
19. Shimozawa, T., Murakami, J., Kumagai, T.: Cricket wind receptors: thermal noise for the highest sensitivity known. In: Barth, F.G., Humphrey, J.A., Secomb, T.W. (eds.) Sensors and Sensing in Biology and Engineering, pp. 145–159. Springer, Berlin (2003)
20. Xueb, C., Chena, S., Zhanga, W., Zhanga, B.: Design, fabrication, and preliminary characterization of a novel MEMS bionic vector hydrophone. Microelectron. J. **38**, 1021–1026 (2007)
21. Yang, Y.C., Nguyen, N., Chen, N.N., Lockwood, M., Tucker, C., Hu, H., Bleckmann, H., Liu, C., Jones, D.L.: Artificial lateral line with biomimetic neuromasts to emulate fish sensing. Bioinspir. Biomim. **5**(1) (2010)
22. Zhou, Z., Liu, Zhi-wen: Biomimetic cilia based on MEMS technology. J. Bion. Eng. **5**, 358–365 (2008)

Biomimetic Infrared Detector

▶ Insect Infrared Sensors

Biomimetic Mosquito-Like Microneedles

Melur K. Ramasubramanian and Ranjeet Agarwala
Department of Mechanical & Aerospace Engineering,
North Carolina State University, Raleigh, NC, USA

Synonyms

Bioinspired microneedles; Biomimicked microneedles; Bionic microneedles; Mosquito fascicle inspired microneedles; Nature inspired microneedle

Definition

Microneedles have evolved over recent years as a replacement for traditional hypodermic needles to minimize pain during transdermal drug delivery (TDD) or during blood draw for clinical analysis. Female mosquitoes have offered scientists an excellent insight into how an organism has evolved to drive a flexible microneedle into human skin and draw blood without causing pain to the host. Microneedle designs inspired by the anatomy of a mosquito fascicle and the mechanics of feeding by female mosquitoes have been developed by several researchers.

A microneedle is able to penetrate the skin painlessly when the critical load applied by it is greater than the load required to puncture the skin and the microneedle does not buckle. Once inserted, the microneedle draws blood or delivers therapeutic agents quickly into the blood stream.

Microneedles

Microneedles are used widely as a substitute for hypodermic needles for relatively painless blood draw, drug delivery, and for precise delivery of reagents in chemical and biological systems. Contrary to hypodermic needles, microneedles do not penetrate the skin far enough to contact, pinch, or pierce nerve endings. Khumpuang et al. [1] reported that penetration between 150 and 1,500 μm is painless due to the relative absence of nerve endings at these depths. This is further illustrated by viewing the human skin anatomy as depicted in Fig. 1 [2].

As shown in Fig. 1 [2], the human skin comprises three predominant layers. The outermost skin layer (living epidermis and stratum corneum) is approximately 150 μm thick and is devoid of blood vessels and nerve endings. The next skin layer (dermis) contains blood capillaries and is about 1,100 μm thick. Although the dermis has some sensory perception, it is mostly painless to penetration. The next layer of human skin (hypodermis or subcutaneous layer) consists of fat cells, large blood vessels, and nerve endings. Hence, penetration into this layer causes pain. However, with hypodermic needles, penetration into the subcutaneous layer is necessary to deliver drug into the larger blood vessels. Generally, a microneedle penetration is limited to the dermis layer and is effective in accessing blood capillaries for drug delivery or blood draw while minimizing if not eliminating pain in the process. Microneedles diameters generally range between 40 and 100 μm and can be solid or hollow, whereas the smallest size hypodermic needles are about 320 μm in outer diameter and 160 μm in inner diameter.

Researchers in the past have successfully created stainless steel and titanium microneedles claiming them to be painless [3–5]. Many microneedle models designed by researchers have been fabricated using silicon. Roxhed et al. [2] designed and built silicon microneedles that had side openings and ultrasharp tips. Needle tip radius has a direct bearing on the stress needed to puncture the skin. Their needles were cylindrical and tapered in shape, 400 μm long with a base diameter of 108 μm, and a tip radius below 100 nm. The bore of the needle was chosen to be eccentrically elliptic with major and minor axes dimensions of 40 and 62 μm, respectively. The side opening of the bore enables the formation of ultrasharp tips. Their needles exhibited insertion forces of 10 mN which was much smaller than previously reported data. The theoretical pressure required to penetrate human skin is given as 3.183×10^6 Pa. Silicon microneedles are generally very short, most less than 500 μm in length, and are generally painless when pushed against the skin.

Oka et al. [6] fabricated and tested silicon microneedles that were 40 μm in diameter, wall thickness of 1.6 μm, and about 1,000 μm long. They reported that their microneedle failed due to brittle fracture during insertion. Silicon microneedles which are brittle can fracture during skin penetration can be carried by blood to other parts of the body, potentially causing injury. Hence, efforts to make polymeric microneedles to prevent brittle fracture were attempted. Davis et al. [7] designed microneedles that were 720 μm long with a tip radius and thickness ranging from 30 to 80 μm and 5 to 58 μm, respectively. Their polymeric needles exhibited insertion forces between 0.1 and 3.0 N. Chaudhri et al. [8] fabricated high aspect ratio hollow micro needles made of SU-8 polymer based on the anatomy of a mosquito fascicle. Microneedles of about 1,600 μm in length, 100 μm in inner diameter, and about 15 μm in wall thickness were produced. The fabrication process was delineated but no results from performance studies were presented.

Reducing the diameter and wall thickness of hollow microneedles for painless insertion necessitates the use

Biomimetic Mosquito-Like Microneedles,
Fig. 1 Schematic representation of the anatomy of human skin [2]

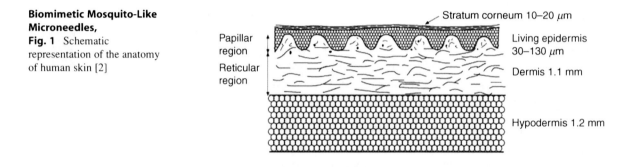

of high stiffness materials from a mechanics point of view. Although silicon microneedles are stiff, they are susceptible to fracture during penetration. On the other hand polymeric needles, although very tough, are not stiff enough to penetrate the skin painlessly at comparable dimensions without buckling. In nature, however, a mosquito uses its fascicle (the blood sucking tube) made of natural polymers and successfully penetrates the skin to draw blood. By studying the anatomy of a mosquito, the process a mosquito uses to insert its fascicle into the skin, and the process of blood draw, one can develop novel ideas leading to scientific breakthroughs in microneedle design, fabrication, and analysis.

Bioinspiration and Biomimetics

Many of nature's organisms are involved in human skin penetration and blood extraction process as part of their life cycle. Meyers et al. [9] describe sharp cutting materials from plants, insects, fish, and mammals such as razor grass, bees, mosquitoes, piranhas, shark, rabbits, and rats. They report various biomimetic devices such as scissors inspired by piranha teeth, shredder cutting blades inspired by rabbit's incisors, and syringes inspired by a mosquito.

Why Mosquito?

Daniel et al. [10] compared blood feeding strategies of three common insects, mosquitoes (*Aedes aegypti*), bedbugs (*Cimex lectularis*), and body louse (*Pediculus humanus*) representing a wide range of sizes and feeding rates. Maximization of nutrient intake and minimization of blood feeding time seems to be the strategy used by all these insects given their

anatomy. The results of their simulation model depicted that the mosquito fed at lower rates compared to that of a bedbug at a pressure drop that was one order of magnitude lower. The body louse on the other hand fed at one tenth the rate of a mosquito, but the pressure drop required was five times higher. The summary of their results showed that the tube radius was a critical variable that affected the feeding rate at a given pressure drop. Insects with larger food canal radius fed at faster rates at any applied pressure drop. The food canal radii for a bedbug, a mosquito, and a body louse were 8, 11, and 3.6 μm, respectively. Therefore, mosquitoes with their large syringe-like tubular food canal anatomy are an ideal model for bioinspiration and biomimetics for developing blood draw and drug delivery devices.

Figure 2 [11] depicts a mosquito that has just inserted its fascicle into the skin of a human host and has started to draw blood from the host.

Mosquito Anatomy

In order to understand how a mosquito draws blood from the host, it will be prudent to first introduce the anatomy of mosquito mouthparts, a compound structure, known as the proboscis.

The proboscis is 1,500–2,000 μm in length. The outer sheath is known as the labium that has a hairy structure at the end and has an inner diameter of 40 μm. The labium has a transverse slit for buckling out of the way during the insertion process. Inside the labium is the actual needle known as the fascicle with an outer diameter of 40 μm and inner diameter of 20 μm for blood draw. The fascicle of the mosquito which serves as a conduit for blood flow is generally retracted into the labium. It comprises of a labrum tube which is the main path for blood flow and hypopharynx which has

Biomimetic Mosquito-Like Microneedles,
Fig. 2 Mosquito drawing blood from a human host [11]

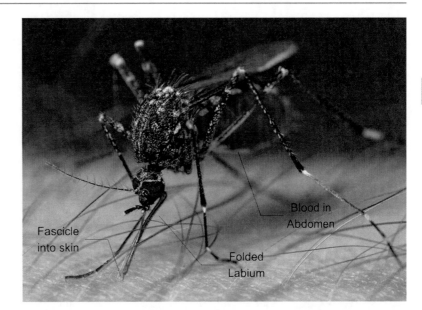

Biomimetic Mosquito-Like Microneedles, Fig. 3 (a) Mosquito proboscis attached to the head (b) Tip of the fascicle shown protruding from the labium [12]

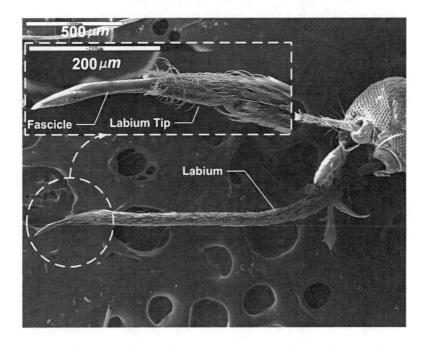

a salivary duct. The fascicle has a maxilla with saw toothed tips. Figure 3 [12] depicts the tip of the fascicle of a mosquito protruding out of the labium. A fascicle extracted from the labium is shown in Fig. 4 [12]. The fascicle has a remarkable resemblance to that of a hypodermic needle. The fascicle's lateral dimension as shown in Fig. 4, decreases from about 10 µm to less than 1 µm at the tip in the last 50 µm length segment creating a very sharp tip.

Mosquito-Inspired Microneedles

Many researchers have studied the mosquito anatomy and have designed microneedles by mimicking its morphology. Xin et al. [13] observed the surface morphology of mosquito and cicada fascicles using scanning electron microscopy. They drew inspiration from the fascicle and its jagged shape to design a biomimetic painless needle. The jagged shape of the microneedle

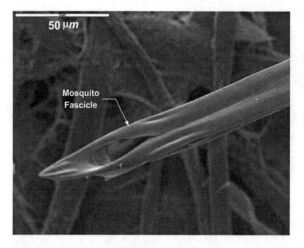

Biomimetic Mosquito-Like Microneedles, Fig. 4 A Mosquito fascicle tip extracted from the labium [12]

design enabled drag reduction. They proposed that their needle design was able to pierce the skin without much resistance and the surface of their needle design exhibited smaller contact area and friction. Chakraborty et al. [14] researched a female mosquito's blood extraction activity to develop and conduct a fluid mechanics analyzis of a microneedle integrated with a painless blood extraction system. They used sputtering deposition method to fabricate titanium microneedles. They reported that their theoretical transport analysis agreed well with experimental observations in terms of flow rates.

Tsuchiya et al. [4] designed a titanium microneedle inspired by the anatomy of a female mosquito fascicle. They describe a self-monitoring of blood glucose device (SMBG) triggered by a shape memory alloy (SMA) actuator. The system comprised a SMA-based indentation system, microelectric pump to extract blood, and a biosensor to evaluate the level of glucose. They evaluated the performance of the insertion system which required 0.1 N force to indent the skin and the pumping system extracted 2 µl/min of blood. Kosoglu et al. [15] developed pliant microneedles for transdermal lighting. These microneedles were inspired by the anatomy of a mosquito fascicle. The usage of these microcneedles as transdermal light probes or optical fibers has the potential to illuminate tissue at a greater depth for surgery or therapeutic procedures like treatment of cancer. They used a melt-drawing process to make various tip sized silica probes with an average diameter of 73 µm. They also

experimentally investigated the needle performance in piercing porcine skins. They observed that sharper probe tips had a higher success rate of penetration compared to probes with more blunt tips.

Kong et al. [16, 17] studied the structure of Aedes albopictus mosquito fascicle using scanning electron microscope (SEM). They observed and documented the entire process from skin penetration to blood feeding. They experimentally measured the insertion force using a digital balance, a glass box, and a screen cloth. The mosquito was housed within the glass box and was free to move within it. The screen cloth connected to the bottom of the glass box deflected once the mosquito extended its fascicle to penetrate and draw blood from the volunteer's hands. The force causing the deflection of the skin cloth was recorded using a digital balance connected to the system. An average force value of 1.8 µN was required for the skin penetration, which is remarkably low. Tsuchiya et al. [3] designed and fabricated a painless needle inspired by the anatomy of a female mosquito for blood extraction. The microneedles were made using sputtering deposition method wherein a rotary substrate was used to deposit the microneedle material, subsequently etched by a chemical solution, and finally heat treated to produce the microneedle. They also reported that the titanium microneedle could be inserted into the skin without buckling.

Aoyagi et al. [18] were able to create high aspect ratio microneedles using biodegradable polymers like PMMA, polyimide, SU-8, Parylene, and polycarbonate. They successfully created 1,000 µm long microneedles that were 10 µm in diameter. The microneedle was able to collect blood plasma with high reliability using capillary suction. Aoyagi et al. [19] used mosquito-inspired microneedles to investigate the resistance force during insertion into silicone rubber. They used a static and a vibrating needle for their experiment. Chichkov [20] proposed a two-photon-polymerization process to create a hollow microneedle and other medical devices using this process. The process used a hybrid material called ormocers; needles were able to penetrate porcine tissue surfaces without fracturing. McAllister et al. [21] used microfabrication techniques to produce microneedles motivated by the anatomy of female mosquitoes. They fabricated microhypodermic needles using silicon and assembled them with heat-controlled bubble pumps. These microneedles allowed flow of water through their bores under acceptable pressures and were able to permeate skin with a greater degree of ease. Suzuki et al. [22] devised a microglucose measuring system comprising a microneedle based on the feeding anatomy of a mosquito. Their system comprised a microneedle, a micro glucose sensor, and a sampling mechanism. Yoshida et al. [23] developed a microneedle design that mimicked a mosquito fascicle. They used human hair as a starting point for manufacturing microneedles. They drilled a hollow bore into the center of a human hair using a microdrilling machine. They created a pen-shaped micro needle which was 1.1 mm long, 80 μm base diameter and, 40 μm tip diameter. Finally, they bleached the drilled human hair using a diluted bleach solution.

Prow et al. [24] researched and developed the Nanopatch (NP). The NP design was inspired by mosquito anatomy. The NP consisted of projections of closely packed microneedles that are smaller than other available microneedles (less than 100 μm). The NP was coated with antigen, adjuvants, and DNA payloads which were transdermally delivered to mice in the lab. The NP exhibited good results with regards to a painless and efficient vaccine delivery system. Izumi et al. [25] used isotropic dry etching process to create long silicon microneedles with three-dimensional sharp tips. Their design did not impose any limit on either the length or the shape of the microneedle. They were able to fabricate straight, jagged, and harpoon like microneedles that mimicked the anatomy of a mosquito fascicle. They also reported that their microneedles with 3-D tips penetrated artificial skins relatively easier compared to prior microneedle designs.

Roxhed [5] investigated a patch-based microneedle drug delivery system. These microneedles were inspired by mosquitoes and interfaced with small dispenser mechanisms. The dispenser mechanisms were triggered by thermal actuators and contained highly expandable microspheres. The microspheres when actuated expanded into a liquid reservoir and dispensed liquid through the reservoir outlets. These microneedles were made of monocrystalline silicon and were chemically treated by a deep reactive ion etching process. The microneedles were hollow with opening along the side to have better penetration ability. The patch was tested for insulin delivery and exhibited promising results with comparable performance to other drug delivery devices.

Mosquito Fascicle Insertion Dynamics

Almost all prior work to produce biomimetic microneedles were based on morphology as discussed in the earlier section. Very few published work exist to assess the significance of the dynamics of a mosquito penetration process in the design of bioinspired microneedles. It can be easily shown that if the morphology is reproduced without considering the material, it is possible to build a needle with high stiffness materials such as metal and silicon and penetrate the skin. If a polymeric needle is produced based on the morphology alone, the needle will not be able to penetrate the skin without bucking for comparable dimensions and material properties to that of a mosquito fascicle. Hence, studying the dynamics of the process of insertion of a fascicle into the skin by a mosquito is important. Ramasubramanian et al. [12] observed in detail how a mosquito (Aedes Aegypti) uses its fascicle to penetrate the skin, and have carried out detailed anatomical studies, and subsequently modeled the penetration process and provided insights into mosquito skin penetration process.

As observed in their research, the mosquito contacts the host to commence the feeding process. It hovers on the host from one spot to another until it is able to thrust its proboscis into the skin without bending it. Once the proboscis is thrust into the skin, the fascicle affixes itself into the epidermis of the skin with its tip and commences probing with irregular frequency. The mosquito keeps on applying axial force for some time by keeping its anatomy poised to prevent proboscis buckling and completes the penetration process by oscillating its head. The mosquito head is observed to oscillate at frequencies near 15 Hz at the commencement of the penetration process. The progressive insertion of the fascicle results in its shortening of the free length making it increasingly stable and consequently the mosquito head oscillates at relatively lower frequency, nearly 6 Hz. The labium with its axial slit is detached from the fascicle near the mouth and forms a loop that looks like a question mark (see Fig. 2). When the mosquito reaches a capillary in the dermis layer, it starts to draw blood until it is full. Figure 2 shows blood in the fascicle and collected in the abdomen.

Study of the penetration mechanics of mosquito fascicle into the skin is paramount as it is antecedent to the feeding behavior of the mosquito. Understanding the forces involved during the penetration process provides critical insights in the process of fascicle penetration as the mosquito tries to prevent the onset of buckling by keeping its posture perched appropriately. One of the movements is the forward tilt of the mosquito head between 10° and 20° with respect to vertical as shown in Fig. 2, which results in the application of a follower force which is a constituent of axial thrust. Such a feat may translate into considerable enhancement of buckling load once the fascicle is thrust into the skin. This effect of increasing the buckling load is further compounded by the labium with the labium providing lateral support during penetration. In Fig. 2, it can be seen that substantial part of the fascicle is supported by the labium and at the tip only a small part of the fascicle is visible before entering the skin. The penetration process and the mechanism of buckling force enhancement by follower force and lateral support has been described effectively by a mathematical model, quantifying their relative importance to the penetration process [12]. The fascicle (an inspiration for microneedle design described earlier) is subjected to an Euler load (P_e) and a nonconservative load known as Beck Load (P_b) supported laterally by the labium as shown in Fig. 5 [12]. The labium is treated as an elastic foundation which resists sideways motion of the fascicle. Here q is the distributed force per unit length and β is the foundation modulus (force per unit deflection per unit length).

The equation of motion as obtained is given by,

$$\frac{\partial^4 y}{\partial x^4} + \left(\frac{P_e + P_b}{EI}\right)\frac{\partial^2 y}{\partial x^2} + \frac{m}{EI}\frac{\partial^2 y}{\partial t^2} + \frac{\beta}{EI}y = 0 \quad (1)$$

With appropriate boundary conditions, namely, $y(0,t) = 0$; $y'(0,t) = 0$; $y''(l,t) = 0$;   and $y'''(l,t) + \left(\frac{P_e}{EI}\right)y'(l,t) = 0$ where, m is mass per unit length, E is the Young's Modulus and I is the moment of inertia, the general solution for the differential equation can be obtained [12]. As depicted in Fig. 6, the distance labeled "A" illustrates the effect of nonconservative force as an increase in Euler load from 1 to 2 is attained solely by applying a Beck load of around 0.5 without substantial increment of lateral

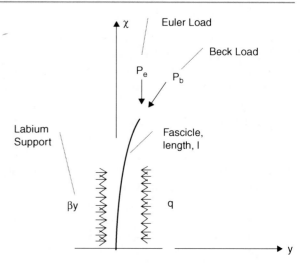

Biomimetic Mosquito-Like Microneedles, Fig. 5 Schematic representation of fascicle-skin interaction with a mosquito applying the force components at the upper end [12]

support. The distance labeled "B" illustrates the effect of foundation stiffness as an increase of Euler load from 1 to 6 is attained by increasing the foundation stiffness value from 0 to 100 for a value of nonconservative load of 0.5, representing the maximum enhancement with both factors. Figure 6 further illustrates that in the absence of nonconservative load, an increment of Euler load by a factor of nearly 5 is still attainable for foundation stiffness values of 100. It is also observed that in the absence of foundation stiffness, the Euler load can be increased by a factor of 3.5 for a Beck load of 0.7 before the onset of flutter instability thus representing the theoretical upper bound. Therefore, the lateral support plays an important role in helping the mosquito reach a force above the limit to penetrate the skin by increasing the critical buckling load by a factor of almost 5. From the mathematical model and numerical solution, it is evident that the lateral support along with Beck load has the effect of increasing the Euler buckling load by a factor of 6 or approximately 18 mN while the lateral support by itself increases the Euler buckling load by a factor of 5. Previous research has shown that the lowest penetration force exhibited by sharpest microneedles is around 10 mN and hence it can be seen that the presence of lateral support would suffice in aiding the microneedle to penetrate painlessly. Taking these findings into account, a more realistic biomimetic mosquito-like microneedle can be developed.

Biomimetic Mosquito-Like Microneedles,
Fig. 6 Critical load pairs plot showing the effect of lateral support by the labium and the application of Beck load [12]

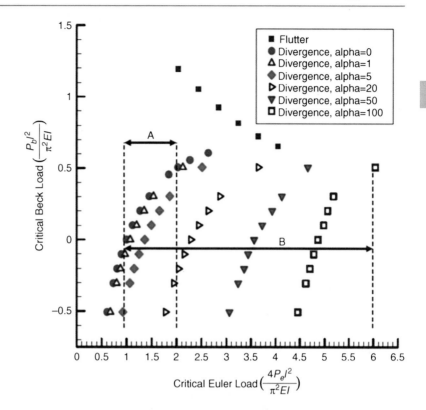

Future Directions

Current research in this area includes constructing carbon nanotube (CNT) based reinforced polymeric microneedles with high axial stiffness along with a lateral support system. Research to construct a synthetic mosquito fascicle based on CNT yarns developed by Zhao et al. [26] is in progress. As depicted in Fig. 7 [26], CNT yarns are spun from CNT arrays. Yarns with diameters of around 40 μm coated with polymers increases the axial stiffness. The coated yarns will be supported laterally and tested for buckling behavior while penetrating porcine skin during the application of an axial force. Successful construction of a polymeric microneedle with a lateral support system should be completed in the near future.

Conclusions

A mosquito is a sophisticated biological machinery that drives a flexible microneedle painlessly into the skin and draw blood successfully. Female mosquitoes of the type *Aedes aegypti* have offered scientists an excellent peek into how a nature's organism is

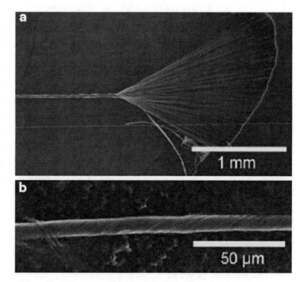

Biomimetic Mosquito-Like Microneedles, Fig. 7 Ultralong carbon nanotubes spun into micro-yarns [26]

involved in human skin penetration and blood extraction. Microneedle designs inspired by the feeding morphology and mechanics of female mosquito offer an excellent tool for researchers and medical practitioners. Fascicles of female mosquitoes of the type *Aedes aegypti* are one of nature's pliant polymeric microneedles that penetrate human skin effectively to draw blood painlessly by remaining within the bounds of epidermis and dermis, suitably maneuvring its anatomy to avoid fascicle breakage. Significant progress has been made to develop microneedles based on mosquito morphology. Further progress is possible when the dynamics and the importance of lateral support are incorporated in the design.

Cross-References

▶ Biomimicry
▶ Bionics
▶ Nanomechanics
▶ Nanomedicine

References

1. Khumpuang, S., Horde, M., Fujioko, K., Sugiyama, S.: Geometrical strengthening and tip-sharpening of a microneedle array fabricated by X-ray lithography. Microsyst. Technol. **13**, 209–214 (2007)
2. Roxhed, N., Gasser, T.C., Griss, P., Holzapfel, G.A., Stemme, G.: Penetration-enhanced ultrasharp microneedles and prediction on skin interaction for efficient transdermal drug delivery. J. Electromech. Syst. **16**(6), 1429–1440 (2007)
3. Tsuchiya, K., Isobata, K., Sato, M., Uetsuji, Y., Nakamachi, E., Kajiwara, K., Kimura, M.: Design of painless microneedle for blood extraction system. Proc. SPIE. **67990Q**, 1–10 (2007)
4. Tsuchiya, K., Nakanishi, N., Uetsuji, Y., Nakamachi, E.: Development of blood extraction system for health monitoring system. Biomed. Microdevices **7**(4), 347–353 (2005)
5. Oki, A., Takai, M., Ogawa, H., Takamura, Y., Fukasawa, T., Kikuchi, J., Ito, Y., Ichiki, T., Horiike, Y.: Healthcare chip for checking health condition from analysis of trace blood collected by painless needle. J. Appl. Phys. **42**, 3722–3727 (2003)
6. Oka, K., Aoyagi, S., Arai, Y., Isono, Y., Hashiguchi, G., Fujita, H.: Fabrication of a micro needle for trace blood test. Sens. Actuator. A **97–98**, 478–485 (2002)
7. Davis, P.S., Landis, B.J., Adams, Z.H., Allen, M.G., Prausnitz, M.R.: Insertion of microneedles into skin: measurement and prediction of insertion and needle fracture force. J. Biomech. **37**, 1155–1163 (2004)
8. Chaudhri, B.P., Ceyssens, F., Moor, P.D., Hoof, C.V., Puers, R.: A high aspect ratio SU-8 fabrication technique for hollow microneedles for transdermal drug delivery and blood extraction. J. Micromech. Microeng. **20**(6), 064006 (2010)
9. Meyers, M.A., Lin, A.Y.M., Lin, Y.S., Olevsky, E.A., Georgalis, S.: The cutting edge: sharp biological materials. J. Miner. Metals Mater. Soc. **60**(3), 19–24 (2008)
10. Daniel, T.L., Kingsolver, J.G.: Feeding strategy and the mechanics of blood sucking in insects. J. Theor. Biol. **105**, 661–672 (1983)
11. Harrison, J.J.: Wikipedia, http://www.wikipedia.com, (2009)
12. Ramasubramanian, M.K., Barham, O.M., Swaminathan, V.: Mechanics of a mosquito bite with applications to microneedle design. Bioinsp. Biomim. **3**, 046001 (2008) (10 pp)
13. Xin, Q.I., Ying-Chun, Q.I., Yan, L.I., Qian, C.: The insect fascicle morphology research and bionic needle pierced mechanical mechanism analysis. Adv. Nat. Sci. **3**(2), 251–257 (2010)
14. Chakraborty, S., Tsuchiya, K.: Development and fluidic simulation of microneedles for painless pathological interfacing with living systems. J. Appl. Phys. **103**(11), 114701–114701-9 (2008)
15. Kosoglu, M.A., Hood, R.L., Chen, Y., Xu, Y., Rylander, M.N., Rylander, C.G.: Fiber optic microneedles for transdermal light delivery: ex vivo porcine skin penetration experiments. J. Biomech. Eng. **132**(9), 091014 (2010)
16. Kong, X.Q., Wu, C.W.: Mosquito proboscis: an elegant biomicroelectromechanical system. Am. Phys. Soc. E **82**, 011910 (2010)
17. Kong, X.Q., Wu, C.W.: Measurement and prediction of insertion force for the mosquito fascicle penetrating into human skin. J. Bionic Eng. **6**(2), 143–152 (2009)
18. Aoyagi, S., Izumi, H., Isono, Y., Fukuda, M., Ogawa, H.: Laser fabrication of high aspect ratio thin holes on biodegradable polymer and its application to a microneedle. Sens. Actuator. **139**, 293–302 (2007)
19. Aoyagi, S., Izumi, H., Fukuda, M.: Biodegradable polymer needle with various tip angles and consideration on insertion mechanism of mosquito's proboscis. Sens. Actuator. A **143**, 20–28 (2007)
20. Chichkov, B.: Two-photon polymerization enhances rapid prototyping of medical devices. Int Soc Opt. Eng., 7 (2007)
21. McAllister, D.V., Allen, M.G., Prausnitz, M.R.: Microfabricated microneedles for gene and drug delivery. Adv. Drug Deliv. Rev. **56**, 581–587 (2004)
22. Suzuki, H., Tokuda, T., Miyagishi, T., Yoshida, H., Honda, N.: A disposable on-line microsystem for continuous sampling and monitoring of glucose. Sens. Actuator. B **97**, 90–97 (2004)
23. Yoshida, Y., Takei, T.: Fabrication of a microneedle using human hair. Japanese J. Appl. Phys. **48**, 098007 (2009) 2 pp
24. Prow, T.W., Chen, X., Crichton, M., Tiwari, Y., Gradassi, F., Raphelli, K., Mahony, D., Fernando, G., Roberts, M.S., Kendall, M.A.F.: Targeted epidermal delivery of vaccines from coated micro-nanoprojection patches. Nanosci. Nanotechnol. **10287125**, 125–128 (2008)
25. Izumi, H., Aoyagi, S.: Novel fabrication method for long silicon microneedles with three-dimensional sharp tips and complicated shank shapes by isotropic dry etching. Trans. Electr. Electr. Eng. **2**, 328–334 (2006)
26. Zhao, H., Zhang, Y., Bradford, P.D., Zhou, Q., Jia, Q., Yuan, F.G., Zhu, Y.: Carbon nanotube yarn strain sensors. Nanotechnology **21**(5), 305502 (2010)

Biomimetic Muscles and Actuators

▶ Biomimetic Muscles and Actuators Using Electroactive Polymers (EAP)

Biomimetic Muscles and Actuators Using Electroactive Polymers (EAP)

Yoseph Bar-Cohen
Jet Propulsion Laboratory (JPL), California Institute of Technology, Pasadena, CA, USA

Synonyms

Artificial muscles; Biomimetic muscles and actuators; Electroresponsive polymers

Definition

Electroactive polymers (EAP) are also known as artificial muscles. They have the closest functional response similarity to biological muscles and therefore they gained the name artificial muscles. As such they are being considered for use to mimic the mechanical performance of biological systems. Actuation via the use of electrical energy is an attractive activation method for causing elastic deformation in polymers and it offers great convenience and practicality.

Introduction

The ability to copy, adapt, and/or be inspired by nature is critically dependant on the availability of effective capabilities and tools. In most biological systems that are larger than bacteria, muscles are their actuators and they are able to lift loads as large as hundreds of kilograms with response time of milliseconds. Biological muscles are driven by a complex mechanism that is very difficult to mimic. Electroactive polymers (EAP) offer the closest functional similarity to biological muscles and therefore they gained the name artificial muscles. There are many polymers that vary their shape or size when subjected to electric, chemical,

pneumatic, optical, or magnetic stimulation. However, electrical excitation is one of the most attractive activation methods for causing elastic deformation in polymers. The convenience and practicality of the electrical stimulation and the recent response improvements have made EAP materials the most preferred among the responsive polymers [1].

Polymers that respond to electrical stimulation were known for over 100 years; however, they received little attention since they responded with quite a small strain at the level of fraction of a percent. In the last 20 years, a series of EAP materials have emerged that respond with a very significant shape change. The impressive advances in improving their performance are attracting the attention of engineers and scientists from many different disciplines. Using the developed materials as actuators, many novel mechanisms and devices were already demonstrated including robot fish, fish-shaped blimp propelled by wagging its body and tail, refreshable braille displays, haptic interfaces, miniature gripper, catheter steering element, artificial eye lid, and dust-wiper.

The Available EAP Materials

The first milestone in developing EAP materials was marked by Roentgen [2] when he subjected to an electric field a rubber-band with fixed end and a mass attached to the free end. Other important steps in the development of these materials include the discoveries of the electrets by Eguchi [3] and the Polyvinylidene Fluoride (PVDF) by Kawai [4]. However, up until the 1990s, the generated strain by the known EAP materials was at the level of tenth of a percent [1]. Most of the EAP materials known today that generate large strain have emerged around 1990. One of the notable ones is the polyacrylate dielectric elastomer (VHB^{TM} 4910 from 3M^{TM}) that was reported to reach strain levels of about 380% with a relatively short response time (<0.1 s) [5]. To simplify the categorization of the many EAP material types, the author divided them into two groups according to their characteristics, including the *ionic* and *electronic* (also known as the *field activated*) [1].

- *Ionic EAP*: As actuators, they consist of a polymer film with electrolyte and it is coated by two electrodes. Their activation mechanism involves mobility of ions under electrical excitation [6]. These polymers have the advantages of generating large bending displacement under activation by a relatively low voltage (1–2 V). Their disadvantages are the need to maintain electrolytes wetness, low energy conversion efficiency (~1%), and difficulties sustaining constant displacement under activation of a DC voltage (except for conducting polymers). Since the strain in conducting polymers is proportional to charge and since they behave electrically as capacitors, they maintain a constant position under constant voltage [7]. Moreover, due to the related macroscopic motion of ions that is involved with their activation, these polymers respond relatively slow in the range of tens to a fraction of a second. Examples of ionic EAP materials include the ionomeric polymer-metal composites (IPMC), conducting polymers, carbon nanotubes, and ionic polymer gels.

Among the materials in this group, IPMC is one of the most studied EAP materials. The base polymer provides channels for mobility of positive ions in a fixed network of negative ions on interconnected clusters [6,8]. In 1992, three different groups of researchers have reported independently the development of IPMC-based EAP, and they include: Oguro et al. [9] in Japan, as well as Shahinpoor [10] and Sadeghipour et al. [11] in the United States. The attractive characteristic of IPMC is the significant bending in response to a relatively low electrical voltage at the level of 1 V (Fig. 1). Generally, because of the fact that ions need to physically travel though the material, the response of IPMC is relatively slow (<10 Hz).

- *Electronic (also known as field-activated) EAP*: These materials are activated by Coulomb force that causes dimension change either due to direct thickness reduction (e.g., dielectric elastomers) or due to internal polarization with molecular alignment (e.g., ferroelectric EAP [12]). Generally, high electric field (>10-V/μm) is required, which may be close to the electric breakdown level that is the result of the low dielectric constant. To reduce the required high voltage two alternative approaches are used: (a) creating a stack of thin multilayers and (b) forming a composite using filler material with a high dielectric constant. Photographs of such a composite ferroelectric EAP in passive and activated states are shown in Fig. 2. The EAP material

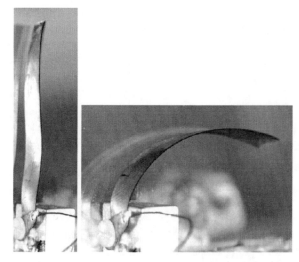

Biomimetic Muscles and Actuators Using Electroactive Polymers (EAP), Fig. 1 IPMC in passive (*left*) and activated states (*right*)

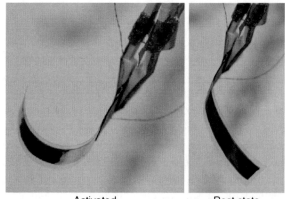

Activated   Rest state

Biomimetic Muscles and Actuators Using Electroactive Polymers (EAP), Fig. 2 Photographs of a composite ferroelectric EAP in passive (*right*) and activated states (*left*). This EAP material was provided to the author as a courtesy of Qiming Zhang, Penn State University

with the filler requires about an order of magnitude less voltage for its activation.

Since the materials in this group of EAP are activated by the electric field, the response is quite fast and can reach milliseconds. The advantages of the materials in this group include holding the generated displacement under a DC voltage making them attractive for robotic applications. Also, these EAP materials have a greater mechanical energy density than the ionic EAP and they can be activated in air with no major constraints. Their main disadvantage is the required high activation field that can be close to the electric breakdown level. Examples of these materials are dielectric elastomers, piezoelectric, and ferroelectric.

Currently, the dielectric elastomer EAP is considered to have the highest technology readiness with the least issues associated with the understanding of its operation. Generally, this EAP type is based on the use of polymers with low elastic stiffness and high dielectric breakdown strength that are subjected to an electrostatic field and they generate a large actuation strain. An actuator that is made of dielectric elastomer EAP can be represented by a parallel plate capacitor [5]. The electrodes have to be highly compliant in order to avoid impeding the generated strain and conductive carbon grease that are widely used for this purpose. The grease provides the electrodes with the required resilience and the carbon provides the required high electrical conductivity. The required very high flexibility of the conductive carbon grease and high electrical conductivity make it best fit for the task. Application of an electric field results in a strain that is proportional to the square of the electric field and to the dielectric constant while inversely proportional to the elastic modulus. The first observation of the fact that dielectric elastomers sustain large strain (23% in silicone films) when subjected to high electric field was reported in a 1992–1993 study by Pelrine and his coinvestigators [13,14].

Since EAP materials are still relatively new it is necessary to establish the related scientific and engineering foundations, including improvement of the understanding of the basic principles that drive them [1]. These require developing effective computational chemistry models, comprehensive material science, electromechanical analytical tools, and material processing techniques. In order to maximize the actuation capability and operation durability, effective processing techniques are being developed for their fabrication, shaping, and electroding. Methods of reliably characterizing the response of EAP materials are being developed and efforts are underway to create databases with documented material properties (http://www.actuatorweb.org/). Bringing these materials to

the level of making daily used products will necessitate finding a niche application that addresses critical needs.

Applications of EAP

Significant efforts are underway to develop practical EAP actuators and commercial products are starting to emerge. Eamax, Japan, is credited for developing the first EAP-related commercial product, and at the end of 2002 it announced the marketing of a fish robot that is actuated by IPMC [1]. Mechanisms and devices applicable to many fields are being considered or developed including zoom lens of cellular phones, valves, energy harvesting, pumps, and many others. EAP actuators offer many important capabilities for devices with biomimetic characteristics [1,15]. One may produce such devices as artificial bugs that may walk, swim, hop, crawl, and dig while reconfiguring themselves as needed. Mimicking nature would immensely expand the collection and functionality of robots allowing performance of tasks that are impossible with existing capabilities [15–17]. For example, the author and his coinvestigators constructed a miniature gripper with four IPMC-based fingers as bending actuators [1]. This gripper was demonstrated to grab rocks very similar to the human hand using hooks at the bottom emulating fingernails. Other recently reported robotic applications include a blimp with steering fins as well as a fish-shaped blimp propelled by wagging its body and tail (developed by EMPA, Switzerland).

Space applications are among the most demanding in terms of the harshness of the operating conditions (extreme temperatures, high pressure, or vacuum) and they require very high reliability and durability. Today's available materials are not applicable to handle the required actuators that need to operate down to as low as $-200°C$ as on Europa and Titan or $+460°C$ as on Venus. Another challenge is the need for large-scale EAP in the form of films, fibers, and others. The required dimensions can be as large as several meters or kilometers to produce large gossamer structures such as antennas, solar sails, and various large optical components. Making biomimetic capability using EAP material will potentially allow NASA and other space agencies to conduct missions in other planets using humanlike robots that emulate human operation before sending real humans.

Addressing the Challenges to EAP

In an effort to promote worldwide development toward the realization of the potential of EAP materials, the author posed in 1999 an arm-wrestling challenge (http://ndeaa.jpl.nasa.gov/nasa-nde/lommas/eap/EAP-armwrestling.htm). The challenge is to have an EAP-activated robotic arm win against a human in a wrestling match. The icon of this challenge is illustrated graphically in Fig. 3. The emphasis on arm wrestling with a human was chosen in order to use the intuitive comparison of our muscles as a baseline for gauging the advances in the capability of EAP actuators. Success will allow applying EAP materials to improve many aspects of our life including the development of products with unmatched capabilities and dexterity. The first arm-wrestling match was held in San Diego, California, USA, on March 7, 2005, as part of the EAP-in-Action Session of the SPIE's EAPAD Conference. Three robotic arms participated in the contest against a 17-year-old high school female student and the student won against all these arms (see Fig. 4).

The second Artificial Muscles Armwrestling Contest was held on February 27, 2006. Rather than wrestling with a human opponent the performance of the arms was measured and compared to the performance of the student (http://eap.jpl.nasa.gov). The measuring fixture (developed jointly by individuals from UCLA and the author's group at JPL) was strapped to the contest table and the EAP-actuated arms were tested for speed and pulling force. Each competing arm pulled on the fixture cable that has a force gauge on its other end. To simulate wrestling action, the pulling on the cable lifted a 0.5-kg weight and the speed of lifting was measured from the travel time of the weight from the bottom to the top section of the fixture. To establish a baseline for performance comparison, the capability of the student (the same student who wrestled in 2005) was measured first and then the three participating robotic arms were tested. The results have shown two orders of magnitude lower performance of the arms compared to the student.

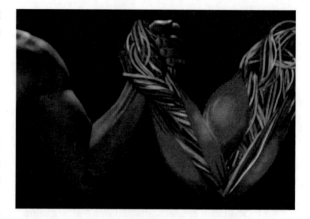

Biomimetic Muscles and Actuators Using Electroactive Polymers (EAP), Fig. 3 The icon of the grand challenge for the development of EAP-actuated robotics

Biomimetic Muscles and Actuators Using Electroactive Polymers (EAP), Fig. 4 An EAP-driven arm made by students from Virginia Tech and the human opponent, 17-year-old student

Challenges, Trend, and Potential Development

Since the early 1990s, new EAP materials have been developed that generate large strains making them highly attractive for biomimetic and many other applications. Their operational similarity to biological muscles, including resilience, damage tolerance, and ability to induce large actuation strains (stretching, contracting, or bending) makes them unique compared to other electroactive materials. Researchers are increasingly making improvements in the various related areas including a better understanding of the operation mechanism of the various EAP material types. The processes of synthesizing, fabricating, electroding, shaping, and handling are being refined to maximize the actuation capability and durability. Methods of reliably characterizing the response of these materials are being developed and efforts are being made to establish databases with documented material properties to support engineers that are considering the use of these materials. An initiative in this area has been taken by the University of British Columbia and they formed a Web database for viewing, comparing, and submitting EAP properties [18].

Applying EAP materials as actuators of manipulation, mobility, and robotic devices involves multidiscipline including materials, chemistry, electromechanics, computers, and electronics. Even though the actuation force of the existing materials requires further improvement, there has already been a series of reported successes in the development of mechanisms that are driven by EAP actuators. However, seeing EAP replace existing actuators in commercial devices and engineering mechanisms may be a difficult challenge and would require identifying niche applications where EAP materials would not need to compete with existing technologies. It is quite encouraging to see the growing number of researchers and engineers who are pursuing careers in EAP-related disciplines. Hopefully, the growth in the research and development activity will lead to making these materials becoming the actuators of choice.

Acknowledgment Some of the research reported in this entry was conducted at the Jet Propulsion Laboratory (JPL), California Institute of Technology, under a contract with National Aeronautics and Space Administration (NASA).

Cross-References

▶ Bioinspired Synthesis of Nanomaterials
▶ Biomimetic Flow Sensors
▶ Biomimetic Mosquito-like Microneedles
▶ Biomimetic Synthesis of Nanomaterials
▶ Biomimetics
▶ Biomimetics of Marine Adhesives
▶ Biosensors

References

1. Bar-Cohen, Y.: Electroactive polymer (EAP) actuators as artificial muscles – reality, potential and challenges, vol. PM136, 2nd edn, pp. 1–765. SPIE Press, Bellingham, Washington (2004). ISBN 0-8194-5297-1

2. Roentgen, W.C.: About the changes in shape and volume of dielectrics caused by electricity, section III. In: Wiedemann, G. (ed.) Annual physics and chemistry series, vol. 11, pp. 771–786. John Ambrosius Barth Publisher, Leipzig, German (1880). In German

3. Eguchi, M.: On the permanent electret. Philos. Mag. **4**(9), 178 (1925)

4. Kawai, H.: Piezoelectricity of poly(vinylidene fluoride). Jpn. J. Appl. Phys. **8**, 975–976 (1969)

5. Pelrine, R., Kornbluh, R., Pei, Q., Joseph, J.: High-speed electrically actuated elastomers with strain greater than 100%. Science **287**(5454), 836–839 (2000)

6. Park, I.S., Jung, K., Kim, D.S.M., Kim, K.J.: Physical principles of ionic polymer-metal composites as electroactive actuators and sensors, special issue dedicated to EAP. Mater. Res. Soc. MRS Bull. **33**(3), 190–195 (2008)

7. Madden, J.D.W., Madden, P.G., Hunter, I.W.: Conducting polymer actuators as engineering materials. In: Bar-Cohen, Y. (ed.) Proceeding of the SPIE smart structures and materials 2002: electroactive polymer actuators and devices (EAPAD), pp. 176–190. SPIE Press, Bellingham, Washington (2002). doi:10.1117/12.475163

8. Nemat-Nasser, S., Thomas, C.W.: Ionomeric polymer-metal composites, Chapter 6. In: Bar-Cohen, Y. (ed.) Electroative polymer (EAP) actuators as artificial muscles - reality, potential and challenges, 2nd edn, vol. PM136, pp. 171–230. SPIE Press, Bellingham, Washington (2004). ISBN 0-8194-5297-1

9. Oguro, K., Kawami, Y., Takenaka, H.: Bending of an ion-conducting polymer film-electrode composite by an electric stimulus at low voltage. Trans. J. Micromachine Soc. **5**, 27–30 (1992)

10. Shahinpoor, M.: Conceptual design, kinematics and dynamics of swimming robotic structures using ionic polymeric gel muscles. Smart Mater. Struct. **1**(1), 91–94 (1992)

11. Sadeghipour, K., Salomon, R., Neogi, S.: Development of a novel electrochemically active membrane and 'smart' material based vibration sensor/damper. J. Smart Mater. Struct. **1**(1), 172–179 (1992)

12. Cheng, Z., Zhang, Q.: Field-activated electroactive polymers, special issue dedicated to EAP. Mater. Res. Soc. MRS Bull. **33**(3), 190–195 (2008)

13. Pelrine, R., Joseph, J.: FY 1992 Final report on artificial muscles for small robots, ITAD-3393-FR-93-063, SRI International, Menlo Park, California, submitted to Micro Machine Center, MBR99 Bldg. 6 F 67, Kanda-sakumagashi, Chiyoda-ku, Tokyo, 101–0026 Japan (1993)

14. Pelrine, R., Joseph, J.: FY 1993 final report on artificial muscles for small robots, ITAD-4570-FR-94-076, SRI International, Menlo Park, California, submitted to Micro Machine Center, MBR99 Bldg. 6 F 67, Kanda-sakumagashi, Chiyoda-ku, Tokyo, 101–0026 Japan (1994)

15. Bar-Cohen, Y. (ed.): Biomimetics – biologically inspired technologies, pp. 1–527. CRC Press, Boca Raton (2005). ISBN 0849331633
16. Bar-Cohen, Y., Breazeal, C. (eds.): Biologically-inspired intelligent robots, vol. PM122, pp. 1–393. SPIE Press, Bellingham, Washington (2003). ISBN 0-8194-4872-9
17. Bar-Cohen, Y., Hanson, D.: The coming robot revolution – expectations and fears about emerging intelligent, human-like machines. Springer, New York (2009). ISBN 978-0-387-85348-2
18. Madden, J.D.W.: Actuator selection tool, http://www.actuatorweb.org/. Accessed 12 Jan 2010

Biomimetic Synthesis

▶ Bioinspired Synthesis of Nanomaterials

Biomimetic Synthesis of Nanomaterials

▶ Bioinspired Synthesis of Nanomaterials

Biomimetics

Bharat Bhushan
Nanoprobe Laboratory for Bio- & Nanotechnology and Biomimetics, The Ohio State University, Columbus, OH, USA

Synonyms

Bionics; Biognosis; Biomimicry; Lesson from Nature; Nanotechnology

Definition

The emerging field of biomimetics allows one to mimic biology or nature to develop nanomaterials, nanodevices, and processes. Properties of biological materials and surfaces result from a complex interplay between surface morphology and physical and chemical properties. Hierarchical structures with dimensions of features ranging from macroscale to the nanoscale are extremely common in nature to provide properties of interest. Molecular scale devices, superhydrophobicity, self-cleaning, drag reduction in fluid flow, energy conversion and conservation, high adhesion, reversible adhesion, aerodynamic lift, materials and fibers with high mechanical strength, biological self-assembly, anti-reflection, structural coloration, thermal insulation, self-healing, and sensory aid mechanisms are some of the examples found in nature which are of commercial interest.

Introduction

Nature has gone through evolution over the 3.8 billion years, since life is estimated to have appeared on earth [33]. Nature has evolved objects with high performance using commonly found materials. These function on the macroscale to nanoscale. The understanding of the functions provided by objects and processes found in nature can guide researchers to imitate and produce nanomaterials, nanodevices, and processes. Biologically inspired design or adaptation or derivation from nature is referred to as "biomimetics." It means mimicking biology or nature. Biomimetics is derived from the Greek word biomimesis. The word was coined by polymath Otto Schmitt in 1957, who in his doctoral research, developed a physical device that mimicked the electrical action of a nerve. Other words used include bionics (coined in 1960 by Jack Steele of Wright-Patterson Air Force Base in Dayton, Ohio), biomimicry, and biognosis. The field of biomimetics is highly interdisciplinary. It involves the understanding of biological functions, structures, and principles of various objects found in nature by biologists, physicists, chemists, and material scientists and the design and fabrication of various materials and devices of commercial interest by engineers, material scientists, chemists, and others. The word biomimetics first appeared in Webster's dictionary in 1974 and is defined as "the study of the formation, structure or function of biologically produced substances and materials (as enzymes or silk) and biological mechanisms and processes (as protein synthesis or photosynthesis) especially for the purpose of synthesizing similar products by artificial mechanisms which mimic natural ones."

Biological materials are highly organized from the molecular to the nano-, micro-, and macroscales,

Biomimetics, Fig. 1 An overview of various objects from nature and their selected functions [13]

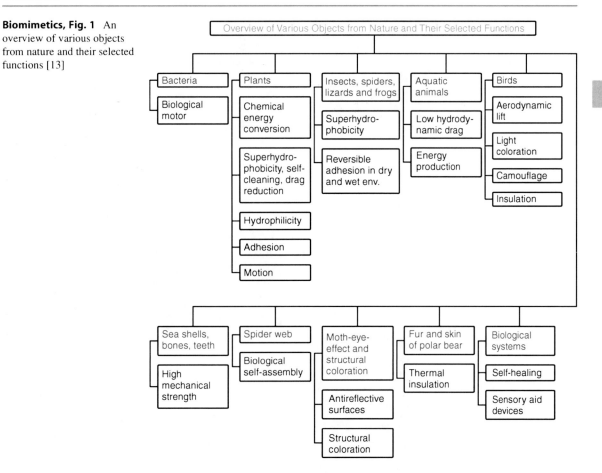

often in a hierarchical manner with intricate nanoarchitecture that ultimately makes up a myriad of different functional elements [1]. Nature uses commonly found materials. Properties of materials and surfaces result from a complex interplay between surface structure and morphology and physical and chemical properties. Many materials, surfaces, and devices provide multifunctionality. Molecular scale devices, superhydrophobicity, self-cleaning, drag reduction in fluid flow, energy conversion and conservation, reversible adhesion, aerodynamic lift, materials and fibers with high mechanical strength, biological self-assembly, anti-reflection, structural coloration, thermal insulation, self-healing, and sensory aid mechanisms are some of the examples found in nature which are of commercial interest.

Various features found in nature objects are on the nanoscale. The major emphasis on nanoscience and nanotechnology since early 1990s has provided a significant impetus in mimicking nature using nanofabrication techniques for commercial applications [14]. Biomimetics has spurred interest across many disciplines.

It is estimated that the 100 largest biomimetic products had generated about US $1.5 billion over 2005–2008. The annual sales are expected to continue to increase dramatically.

Lessons from Nature and Applications

There are a large number of objects, including bacteria, plants, land and aquatic animals, and sea shells, with properties of commercial interest. Figure 1 provides an overview of various objects from nature and their selected functions. Figure 2 shows a montage of some examples from nature. These serve as the inspiration for various technological developments [13].

Some leaves of water repellent plants such as *Nelumbo nucifera* (Lotus) and *Colocasia esculenta* are known to be superhydrophobic and self-cleaning due to hierarchical roughness (microbumps

Biomimetics,
Fig. 2 Montage of some
examples from nature. (**a**)
Lotus effect [18], (**b**) glands of
carnivorous plant secrete
adhesive to trap insects [40],
(**c**) pond skater walking on
water [26], (**d**) gecko foot
exhibiting reversible adhesion
[27], (**e**) scale structure of
shark reducing drag [49], (**f**)
wings of a bird in landing
approach, (**g**) spiderweb made
of silk material [5], and (**h**)
antireflective moth's eye [29]

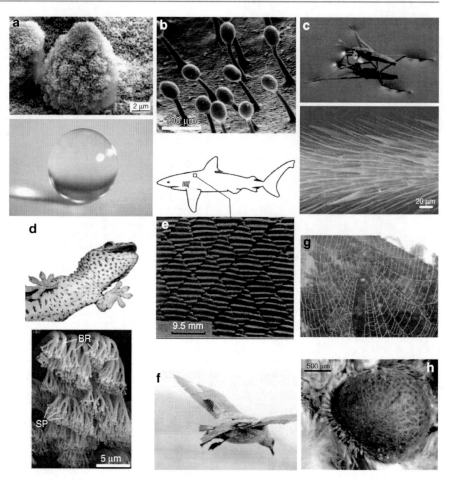

superimposed with nanostructure) and the presence of
a hydrophobic coating [16, 38, 39, 41, 45–47]. Roughness-induced superhydrophobic and self-cleaning surfaces are of interest in various applications, including
self-cleaning windows, windshields, and exterior
paints for buildings, boats, ships, and aircrafts, uten-
sils, roof tiles, textiles, solar panels, and applications
requiring antifouling and a reduction of drag in fluid
flow, for example, in micro/nanofluidics, boats, ships,
and aircrafts. Superhydrophobic surfaces can also be
used for energy conversion and conservation.
Nonwetting surfaces also reduce stiction at
a contacting interface in machinery [9, 10]. A model
surface for superhydrophobicity and self-cleaning is
provided by the leaves of the Lotus plant (*Nelumbo
nucifera*) (Fig. 2a) [6, 15, 38, 39, 41, 44, 52]. The so-
called papillose epidermal cells form asperities or
papillae and provide roughness on the microscale.
The surface of the leaves is usually covered with
a range of waxes made from a mixture of long-chain

hydrocarbon compounds that have a strong phobia of
being wet. Sub-micron sized asperities composed of
the 3-D epicuticular waxes are superimposed over
microscale roughness, creating a hierarchical structure. The wax asperities consist of different morphologies, like tubules on Lotus or platelets on *Colocasia*
[38, 39, 41]. The water droplets on these surfaces
readily sit on the apex of nanostructures because air
bubbles fill in the valleys of the structure under the
droplet. Therefore, these leaves exhibit considerable
superhydrophobicity. The water droplets on the leaves
remove any contaminant particles from their surfaces
when they roll off, leading to self-cleaning. A contact
angle of 164° and a contact angle hysteresis of 3° have
been reported for the Lotus leaf [18, 40].

Two strategies used for catching insects by plants
for digestion are having sticky surfaces or sliding
structures. As an example, for catching insects by
using sticky surfaces, the glands of the carnivorous
plants of the genus *Pinguicula* (butter-worts) and

Drosera (sundew), shown in Fig. 2b, secrete adhesives and enzymes to trap and digest small insects like mosquitoes and fruit flies [39]. In *Pinguicula*, the stalked glands are the sticky ones which trap the insects by secreting an adhesive solution. The shorter ones are those which secrete digestive enzymes, including protease and phosphotase, and later resorb the digested material. In *Drosera*, the stalked glands can effectively enclose small flies by bringing numerous glands in contact with the prey.

Water striders (Gerris remigis) are insects which live on the surfaces of ponds, slow streams, and quiet waters. A water strider has the ability to stand and walk upon water surface without getting wet, Fig. 2c. Even the impact of rain droplets with a size greater than the strider's size does not make it immerse into water. Gao and Jiang [26] showed that the special hierarchical structure of strider legs, which are covered by large numbers of oriented tiny hairs (microsetae) with fine nanogrooves and covered with cuticle wax, makes the leg surfaces superhydrophobic and is responsible for the water resistance and enables them to stand and walk quickly on the water surface. They measured the contact angle of the insect's legs with water to be about 167°. Scanning electron microscope (SEM) micrographs revealed numerous oriented setae on the legs. The setae are needle-shaped hairs with diameters ranging from 3 μm down to several 100 nm. Most setae are roughly 50 μm in length and arranged at an inclined angle of about 20° from the surface of the leg. Many elaborate nanoscale grooves were found on each microseta, and these form a unique hierarchical structure. This hierarchical micro- and nanostructuring on the leg's surface seems to be responsible for its water resistance [45, 46] and the strong supporting force. Gao and Jiang [26] reported that a leg does not pierce the water surface until a dimple of 4.4 mm depth is formed. They found that the maximal supporting force of a single leg is 1.52 mN, or about 15 times the total body weight of the insect. The corresponding volume of water ejected is roughly 300 times that of the leg itself.

Leg attachment pads of several animals, including many insects (e.g., beetles and flies), spiders, and lizards (e.g., geckos) are capable of attaching to a variety of surfaces and are used for locomotion even on vertical walls or across ceilings [4, 11, 12, 30]. Biological evolution over a long period of time has led to the optimization of their leg attachment systems. This dynamic attachment ability is referred to as reversible adhesion or smart adhesion [11, 12].

Attachment systems in various creatures such as insects, spiders, and lizards have similar structures. As the size (mass) of the creature increases, the radius of the terminal attachment elements decreases [3, 23]. This allows a greater number of setae to be packed into an area, hence increasing the linear dimension of contact and the adhesion strength. Based on surface energy approach, it has been reported that adhesion force is proportional to a linear dimension of the contact [9, 10]. Therefore, it increases with the division of contacts. The density of the terminal attachment elements strongly increases with increasing body mass. Flies and beetles have the largest attachment pads and the lowest density of terminal attachment elements. Spiders have highly refined attachment elements that cover their legs. Geckos have both the highest body mass and greatest density of terminal elements (spatula). Spiders and geckos can generate high dry adhesion, whereas beetles and flies increase adhesion by secreting liquids at the contacting interface.

Gecko is the largest animal that can produce high (dry) adhesion to support its weight with a high factor of safety. The gecko skin is comprised of a complex hierarchical structure of lamellae, setae, branches, and spatula [4, 11]. As shown in Fig. 2d, the gecko consists of an intricate hierarchy of structures beginning with lamellae, soft ridges that are 1–2 mm in length, that are located on the attachment pads (toes) that compress easily so that contact can be made with rough bumpy surfaces. Tiny curved hairs known as setae extend from the lamellae with a density of approximately 14,000 per square millimeter. These setae are typically 30–130 μm in length and 5–10 μm in diameter and are composed primarily of β-keratin with some α-keratin component. At the end of each seta, 100–1,000 spatulae with a diameter of 0.1–0.2 μm branch out and form the points of contact with the surface. The tips of the spatula are approximately 0.2–0.3 μm in width, 0.5 μm in length, and 0.01 μm in thickness and get their name from their resemblance to a spatula.

The attachment pads on two feet of the Tokay gecko have an area of about 220 mm^2. About three million setae on their toes can produce a clinging ability of about 20 N (vertical force required to pull a lizard down a nearly vertical (85°) surface) and allow them to climb vertical surfaces at speeds of over 1 m/s with the capability to attach or detach their toes in milliseconds. It should be noted that a three-level hierarchical structure allows adaptability to surfaces with different magnitudes of roughness. The gecko uses peeling action to unstick itself [11, 12].

Replication of the structure of gecko feet would enable the development of a superadhesive polymer tape capable of clean, dry adhesion which is reversible (e.g., [11, 12, 17, 28, 32]). (It should be noted that common man-made adhesives such as tape or glue involve the use of wet adhesives that permanently attach two surfaces.) The reusable gecko-inspired adhesives have the potential for use in everyday objects such as tapes, fasteners, toys, and in high technology such as microelectronic and space applications. Replication of the dynamic climbing and peeling ability of geckos could find use in the treads of wall-climbing robots [21].

Many aquatic animals can move in water at high speeds with a low energy input. Drag is a major hindrance to movement. Most shark species move through water with high efficiency and maintain buoyancy. Through its ingenious design, their skin turns out to be an essential aid in this behavior by reducing drag by 5 – 10% and auto-cleaning ecto-parasites from their surface [7, 8]. The very small individual tooth-like scales of shark skin, called dermal denticles (little skin teeth), are ribbed with longitudinal grooves (aligned parallel to the local flow direction of the water) which result in water moving very efficiently over their surface. An example of scale structure on the right front of a Galapagos shark (*C. Galapagensis*) is shown in Fig. 2e. The detailed structure varies from one location to another for a given shark. The scales are present over most of the shark's body. These are V-shaped, about 200–500 μm in height, and regularly spaced (100–300 μm) [22].

Due to the relatively high Reynolds number of a swimming shark, the turbulent flow occurs. The skin drag (wall shear stress) is not generally affected by the surface roughness. Longitudinal scales on the surface result in lower wall shear stresses than that on a smooth surface and control boundary layer separation. The longitudinal scales result in water moving more efficiently over their surface than it would were shark scales completely oriented differently. Over smooth surfaces, fast-moving water begins to break up into turbulent vortices, or eddies, in part because the water flowing at the surface of an object moves slower than water flowing further away from the object with so-called low boundary slip. This difference in water speed causes the faster water to get "tripped up" by the adjacent layer of slower water flowing around an object, just as upstream swirls form along riverbanks. The grooves in a shark's scales simultaneously reduce eddy formation in a surprising number of ways: (1) the grooves reinforce the direction of flow by channeling it, (2) they speed up the slower water at the shark's surface (as the same volume of water going through a narrower channel increases in speed), reducing the difference in speed of this surface flow and the water just beyond the shark's surface, (3) conversely, they pull faster water toward the shark's surface so that it mixes with the slower water, reducing this speed differential, and finally, (4) they divide up the sheet of water flowing over the shark's surface so that any turbulence created results in smaller, rather than larger, vortices [2].

It is also reported that longitudinal scales influence the fluid flow in the transverse direction by limiting the degree of momentum transfer. It is the difference in the protrusion height in the longitudinal and transverse directions which govern how much the scales impede the transverse flow. Bechert et al. [7] reported that, based on their experimental data, thin, vertical scales result in the low transverse flow and low drag. They also reported that the ratio of the scale height to tip-to-tip spacing of 0.5 is the optimum value for low drag.

In addition to reduction of drag, the shark skin surface prevents marine organisms from being able to adhere to ("foul") it. It is not because of the Lotus effect, the shark skin is hydrophilic and wets with water. There are three factors which appear to keep the surface clean: (1) the accelerated water flow at a shark's surface reduces the contact time of fouling organisms, (2) the roughened nanotexture of shark skin reduces the available surface area for adhering organisms, and (3) the dermal scales themselves perpetually realign or flex in response to changes in internal and external pressure as the shark moves through water, creating a "moving target" for fouling organisms [2].

Speedo created the wholebody swimsuit called Fastskin bodysuit (TYR Trace Rise) in 2006 for elite swimming. The suit is made of polyurethane woven fabric with a texture based on shark scales. In the 2008 Summer Olympics, two thirds of the swimmers wore Speedo swim suits, and a large number of world records were broken. Boat, ship, and aircraft manufacturers are trying to mimic shark skin to reduce friction drag and minimize the attachment of organisms on their bodies. One can create riblets on the surface by painting or attaching a film (3 M). Skin friction contributes about half of the total drag between a solid surface and air in an aircraft. The transparent sheets with ribbed structure in the longitudinal direction have been used on the commercial Airbus 340 aircrafts. It is expected that riblet film on the body of the aircraft can reduce drag on the order of 10% [24].

Mucus on the skin of aquatic animals, including sharks, acts as an osmotic barrier against the salinity of seawater and protects the creature from parasites and infections [8]. The mucus also operates as a drag-reducing agent on some fast predatory fishes, which allows the fish to attack more easily [35]. The artificial derivatives of fish mucus, that is, polymer additives for liquids, are used in drag reduction technology, for example, to propel crude oil in the Alaska pipeline [42].

Bird feathers perform multiple functions – make the body water repellant, create wing and tail for aerodynamic lift during flying, provide coloration for appearance as well as camouflage, and provide an insulating layer to keep the body warm. Many bird feathers exhibit hydrophobicity (an apparent contact angle of 100–140°). Bormashenko et al. [19] studied Feral Rock pigeon feathers. The morphology consists of a network formed by barbs and barbules made of keratin. It is the morphology which plays an important role in hydrophobicity. Birds consist of several consecutive rows of covering feathers on their wings which are flexible, Fig. 2f. These movable flaps develop the lift. When a bird lands, a few feathers are deployed in front of the leading edges of the wings which help to reduce the drag on the wings. Self-activated movable flaps (artificial bird feathers) have been shown to provide an increase in lift in flight experiments [8]. Birds serve as the inspiration for aircrafts and early development of wing design [36]. However, aircrafts do not flap their wings like birds to simultaneously produce lift and thrust. This is impractical in aircrafts due to limitations of scaling phenomena and the high speeds.

Spiders produce a variety of proteins, among which are major ampullate silks (MAS). MAS fibers are used by spiders as a scaffold upon which they attach other silk fibers during the formation of the web. The spider generates the silk fiber, and at the same time it is hanging on it. It has a sufficient supply of raw material for its silk to span great distances [5, 37]. Spider web is a structure built of a one-dimensional fiber, Fig. 2g. The fiber is very strong and continuous and is insoluble in water. The web can hold a significant amount of water droplets, and it is resistant to rain, wind, and sunlight [5, 50]. Spider silk is three times stronger than steel, having a tensile strength of about 1.2 GPa [51]. Some spider silks have high stiffness with a tensile modulus of about 10 GPa, while others are elastomeric with a modulus of about 1 GPa and extension to rupture of 200%. The combination of strength and extensibility is primarily derived from the domains of crystalline β-sheets and flexible helixes within the polypeptide chain, imparting a toughness that is greater than bone, Kevlar, and high strength steel. The web is designed to catch insects (food for the spider) that cross the net and get stuck to its stickiness and complex structure.

Optical reflection and anti-reflection is achieved in nature by using nanoscale architecture [48]. The eyes of moths are antireflective to visible light. The so-called moth-eye effect was discovered in the 1960s as a result of the study of insect eyes. For nocturnal insects, it is important not to reflect the light, since the reflection makes the insect vulnerable to predators. The eyes of moths consist of hundreds of hexagonally organized nanoscopic pillars, each about 200 nm in diameter and height, that result in a very low reflectance for visible light, Fig. 2h [29, 43]. These nanostructures' optical surfaces make the eye surface nearly antireflective in any direction.

Light reflection is avoided by a continuously increasing refractive index of the optical medium. The little protuberances upon the cornea surface increase the refractive index. These protuberances are very small microtrichia (about 200 nm in diameter). For an increase in transmission and reduced reflection, a continuous matching of the refraction index at the boundary of the adjacent materials (cornea and air) is required. If the periodicity of the surface pattern is smaller than the light wavelength, the light is not reflected [31]. If this condition is satisfied, it may be assumed that at any depth the effective refraction index is the mean of that of air and the bulk material, weighted in proportion to the amount of material present at that depth [45]. For a moth eye surface with the height of the protuberances of h and the spacing of d, it is expected that the reflectance is very low for wavelengths less than about 2.5 h and greater than d at normal incidence, and for wavelengths greater than 2d for oblique incidence. For protuberances with 220 nm depth and the same spacing (typical values for the moth eye), a very low reflectance is expected for the wavelengths between 440 and 550 nm [53].

This moth eye effect should not be confused with reduction of the specular reflectance by roughening of a surface. Roughness merely redistributes the reflected light as diffuse scattering. In the case of the moth eye there is no increase in diffuse scattering, the transmitted wavefront is not degraded, and the reduction in reflection gives rise to a corresponding increase in transmission [53].

Attempts are being made to incorporate microscopic corrugations in solar panels to reduce light reflection. Attempts are also being made to produce a glare-free computer screen by creating facets on a photosensitive lacquer using lasers. Some 25,000 dots of texture per square mm can essentially eliminate the glare on the screen [43]. Hadobás et al. [34] prepared patterned silicon surfaces with 300 nm periodicity and depth up to 190 nm. They found a significant reduction in reflectivity, partially due to the moth-eye effect. Gao et al. [54] used epoxy and resin to replicate the antireflective surface of a cicada's eye. It is also possible to create transparent surfaces using the moth-eye effect [20].

Hierarchical Organization in Biomaterials

Nature develops biological objects by means of growth or biologically controlled self-assembly adapting to the environmental condition and by using the most commonly found materials. Biological materials are developed by using the recipes contained in the genetic code. As a result, biological materials and tissues are created by hierarchical structuring at all levels in order to adapt form and structure to the function, which have the capability of adaptation to changing conditions and self-healing [25, 45]. The genetic algorithm interacts with the environmental condition, which provides flexibility. For example, a tree branch can grow differently in the direction of the wind and in the opposite direction. The only way to provide this adaptive self-assembly is a hierarchical self-organization of the material. Hierarchical structuring allows the adaptation and optimization of the material at each level.

It is apparent that nature uses hierarchical structures, consisting of nanostructures in many cases, to achieve the required performance [45–47]. Understanding the role of hierarchical structure and development of low cost and flexible fabrication techniques would facilitate commercial applications.

Outlook

The emerging field of biomimetics is already gaining a foothold in the scientific and technical arena. It is clear that nature has evolved and optimized a large number of materials and structured surfaces with rather unique characteristics. As the underlying mechanisms are understood, industry can begin to exploit them for commercial applications.

The commercial applications include new nanomaterials, nanodevices, and processes. As for devices, these include superhydrophobic self-cleaning and/or low drag surfaces, surfaces for energy conversion and conservation, superadhesives, robotics, objects which provide aerodynamic lift, materials and fibers with high mechanical strength, anti-reflective surfaces and surfaces with hues, artificial furs and textiles, various biomedical devices and implants, self-healing materials, and sensory aid devices, to name a few.

Cross-References

▶ Gecko Feet
▶ Lotus Effect
▶ Rose Petal Effect
▶ Shark Skin Effect

References

1. Alberts, B., Johnson, A., Lewis, J., Raff, M., Roberts, K., Walter, P. (eds.): Molecular Biology of the Cell. Garland Science, New York (2008)
2. Anonymous, www.biomimicryinstitute.org.
3. Arzt, E., Gorb, S., Spolenak, R.: From micro to nano contacts in biological attachment devices. Proc. Natl Acad. Sci. U.S.A. **100**, 10603–10606 (2003)
4. Autumn, K., Liang, Y.A., Hsieh, S.T., Zesch, W., Chan, W. P., Kenny, T.W., Fearing, R., Full, R.J.: Adhesive force of a single Gecko foot-hair. Nature **405**, 681–685 (2000)

5. Bar-Cohen, Y. (ed.): Biomimetics: Biologically Inspired Technologies. Taylor & Francis, Boca Raton (2006)
6. Barthlott, W., Neinhuis, C.: Purity of the sacred lotus, of escape from contamination in biological surfaces. Planta **202**, 1–8 (1997)
7. Bechert, D.W., Bruse, M., Hage, W., van der Hoeven, J.G. T., Hoppe, G.: Experiments on drag-reducing surfaces and their optimization with an adjustable geometry. J. Fluid Mech. **338**, 59–87 (1997)
8. Bechert, D.W., Bruse, M., Hage, W., Meyer, R.: Fluid mechanics of biological surfaces and their technological application. Naturwissenschaften **87**, 157–171 (2000)
9. Bhushan, B.: Principles and Applications of Tribology. Wiley, New York (1999)
10. Bhushan, B.: Introduction to Tribology. Wiley, New York (2002)
11. Bhushan, B.: Adhesion of multi-level hierarchical attachment systems in Gecko feet, (invited). J. Adhes. Sci. Technol. **21**, 1213–1258 (2007)
12. Bhushan, B. (ed.): Nanotribology and Nanomechanics: An Introduction, 2nd edn. Springer, Heidelberg (2008)
13. Bhushan, B.: Biomimetics: lessons from nature; an overview. Phil. Trans. R. Soc. A **367**, 1445–1486 (2009)
14. Bhushan, B. (ed.): Springer Handbook of Nanotechnology, 3rd edn. Springer, Heidelberg (2010)
15. Bhushan, B., Jung, Y.C.: Micro and nanoscale characterization of hydrophobic and hydrophilic leave surface. Nanotechnology **17**, 2758–2772 (2006)
16. Bhushan, B., Jung, Y.C.: Wetting, adhesion and friction of superhydrophobic and hydrophilic leaves and fabricated micro/nanopatterned surfaces. J. Phys Conden. Matter **20**, 225010 (2008)
17. Bhushan, B., Sayer, R.A.: Surface characterization and friction of a bio-Inspired reversible adhesive tape. Microsyst. Technol. **13**, 71–78 (2007)
18. Bhushan, B., Jung, Y.C., Koch, K.: Micro-, Nano-, and Hierarchical structures for superhydrophobicity, self-cleaning, and low adhesion. Phil. Trans. R. Soc. A **367**, 1631–1672 (2009)
19. Bormashenko, E., Bormashenko, Y., Stein, T., Whyman, G., Bormashenko, E.: Why do pigeon feathers repel water? Hydrophobicity of pennae, cassie-baxter wetting hypothesis and cassie-wenzel capillarity-induced wetting transition. J. Colloid Interf. Sci. **311**, 212–216 (2007)
20. Clapham, P.B., Hutley, M.C.: Reduction of length reflection by moth eye principle. Nature **244**, 281–282 (1973)
21. Cutkosky, M.R., Kim, S.: Design and fabrication of multi-materials structures for bio-inspired robots. Philos. Trans. R. Soc. A **367**, 1799–1813 (2009)
22. Dean, B., Bhushan, B.: Shark skin surfaces for fluid drag reduction in turbulent flow: a review. Phil. Trans. R. Soc. A **368**, 4775–4806 (2010)
23. Federle, W.: Why are so many adhesive pads hairy? J. Exp. Biol. **209**, 2611–2621 (2006)
24. Fish, F.E.: Limits of nature and advances of technology: what does biomimetics have to offer to aquatic robots? ABBI **3**, 49–60 (2006)
25. Fratzl, P., Weinkamer, R.: Nature's hierarchical materials. Prog. Mat. Sci. **52**, 1263–1334 (2007)
26. Gao, X.F., Jiang, L.: Biophysics: water-repellent legs of water striders. Nature **432**, 36 (2004)
27. Gao, H., Wang, X., Yao, H., Gorb, S., Arzt, E.: Mechanics of hierarchical adhesion structures of geckos. Mech. Mater. **37**, 275–285 (2005)
28. Geim, A.K., Dubonos, S.V., Grigorieva, I.V., Novoselov, K. S., Zhukov, A.A., Shapoval, S.Y.: Microfabricated adhesive mimicking Gecko Foot-hair. Nat. Mater. **2**, 461–463 (2003)
29. Genzer, J., Efimenko, K.: Recent developments in superhydrophobic surfaces and their relevance to marine fouling: a review. Biofouling **22**, 339–360 (2006)
30. Gorb, S.: Attachment Devices of Insect Cuticle. Kluwer, Dordrecht (2001)
31. Gorb, S.: Functional surfaces in biology: mechanisms and applications. In: Bar-Cohen, Y. (ed.) Biomimetics: Biologically Inspired Technologies, pp. 381–397. Taylor & Francis, Boca Raton (2006)
32. Gorb, S., Varenberg, M., Peressadko, A., Tuma, J.: Biomimetic mushroom-shaped fibrillar adhesive microstructure. J. R. Soc. Interf. **4**, 271–275 (2007)
33. Gordon, J.E.: The New Science of Strong Materials, or Why You Don't Fall Through the Floor, 2nd edn. Pelican-Penguin, London (1976)
34. Hadobás, K., Kirsch, S., Carl, A., Acet, M., Wasserman, E.F.: Reflection properties of nanostructure-arrayed silicon surfaces. Nanotechnology **11**, 161–164 (2000)
35. Hoyt, J.W.: Hydrodynamic drag reduction due to fish slimes. In: Wu, T.Y.T., Brokaw, C.J., Brennen, C. (eds.) Swimming and Flying in Nature, vol. 2. Plenum, New York (1975)
36. Jakab, P.L.: Vision of a Flying Machine. Smithsonian Institution Press, Washington D.C. (1990)
37. Jin, H.J., Kaplan, D.L.: Mechanism of silk processing in insects and spiders. Nature **424**, 1057–1061 (2003)
38. Koch, K., Bhushan, B., Barthlott, W.: Diversity of structure, morphology, and wetting of plant surfaces (invited). Soft Matter **4**, 1943–1963 (2008)
39. Koch, K., Bhushan, B., Barthlott, W.: Multifunctional surface structures of plants: an inspiration for biomimetics (invited). Prog. Mater. Sci. **54**, 137–178 (2009)
40. Koch, K., Bhushan, B., Jung, Y.C., Barthlott, W.: Fabrication of artificial lotus leaves and significance of hierarchical structure for superhydrophobicity and low adhesion. Soft Matter **5**, 1386–1393 (2009)
41. Koch, K., Bhushan, B., Barthlott, W.: Multifunctional plant surfaces, smart materials. In: Bhushan, B. (ed.) Springer Handbook of Nanotechnology, 3rd edn. Springer, Heidelberg (2010)
42. Motier, J.F., Carrier, A.M.: Recent studies on polymer drag reduction in commercial Pipelines. In: Sellin, R.H.J., Moses, R.T. (eds.) Drag Reduction in Fluid Flows: Techniques for Friction Control. Ellis Horwood, Chichester (1989)
43. Mueller, T.: Biomimetics design by natures. Natl Geogr. **2008**, 68–90 (2008)
44. Neinhuis, C., Barthlott, W.: Characterization and distribution of water-repellent, self-cleaning plant surfaces. Ann. Bot. **79**, 667–677 (1997)
45. Nosonovsky, M., Bhushan, B.: Multiscale Dissipative Mechanisms and Hierarchical Surfaces: Friction, Superhydrophobicity, and Biomimetics. Springer, Heidelberg (2008)
46. Nosonovsky, M., Bhushan, B.: Roughness-induced superhydrophobicity: a way to design non-adhesive surfaces. J. Phys Conden. Matter **20**, 225009 (2008)

47. Nosonovsky, M., Bhushan, B.: Biologically-inspired surfaces: broadening the scope of roughness. Adv. Func. Mater. **18**, 843–855 (2008)
48. Parker, A.R.: Natural photonics for industrial applications. Phil. Trans. R. Soc. A **367**, 1759–1782 (2009)
49. Reif, W.E.: Squamation and ecology of sharks. Cour. Forschungsinstitut Senckenberg **78**, 1–255 (1985) (Frankfurt am Main)
50. Sarikaya, M., Aksay, I.A. (eds.): Biomimetic Design and Processing of Materials. American Institute of Physics, Woodbury, New York (1995)
51. Vogel, S.: Comparative Biomechanics: Life's Physical World. Princeton Univ. Press, Princeton (2003)
52. Wagner, P., Furstner, R., Barthlott, W., Neinhuis, C.: Quantitative assessment to the structural basis of water repellency in natural and technical surfaces. J. Exper. Bot. **54**, 1295–1303 (2003)
53. Wilson, S.J., Hutley, M.C.: The optical properties of the 'moth-eye' antireflective surfaces. J. Mod. Opt. **29**, 993–1009 (1982)
54. Gao, H., Liu, Z., Zhang, J., Zhang, G., Xie, G.: Precise replication of antireflective nanostructures from biotemplates. Appl. Phys. Lett. **90**, 123115 (2007)

Biomimetics of Marine Adhesives

Pierre Becker, Elise Hennebert and Patrick Flammang
Laboratoire de Biologie des Organismes Marins et Biomimétisme, Université de Mons – UMONS, Mons, Belgium

Synonyms

Marine bio-inspired adhesives

Definition

A large diversity of marine organisms possesses adhesive mechanisms allowing strong attachment to various substrata in a wet and saline environment. This remarkable capacity raised the interest of scientists for the development of bio-inspired underwater adhesives for various applications, notably in the biomedical field. Model organisms that have been used for the development of such biomimetic adhesives include mussels, barnacles, tubeworms, and algae. All attach permanently by means of solid cements mostly made up of specialized proteins. Synthesis of biomimetic marine adhesives is performed either by the recombinant DNA technology (i.e., the production of adhesive proteins by transformed host cells such as bacteria or yeasts) or by the chemical synthesis of polymers incorporating functional groups copying key amino acids from the natural cement proteins. Other remarkable properties of natural cements, such as the capacity to form complex coacervates, to self-assemble and to cross-link, have been mimicked for the underwater delivery and in situ curing of the biomimetic adhesives.

Introduction

Of all biological phenomena that have been investigated with a view to biomimetics, adhesion in nature has perhaps received the most interest. Indeed, biological adhesives often offer impressive performance in their natural context and, therewith, the potential to inspire novel, superior industrial adhesives for an increasing variety of high-tech applications [1, 2]. Yet, because of the complexity of biological adhesion and the multidisciplinarity needed to tackle it, there has been little visible progress in the development of bio-inspired adhesives and many technological challenges remain, such as the development of adhesives which can function underwater or in aqueous environments. The marine environment is a place of choice for the search of inspiration for such adhesives. Indeed, numerous invertebrates like molluscs, worms, or sea stars produce adhesive secretions that are able to form strong attachments even when totally immersed [1, 2]. These adhesive secretions contain specialized adhesive proteins sometimes associated with other macromolecules such as polysaccharides.

Three types of adhesions can be recognized in marine organisms [3]. Permanent adhesion is characteristic of algae and animals that attach strongly to a substratum with a solid glue possessing high adhesive and cohesive strength (e.g., mussel, barnacles, tube worms, tunicates [1, 2]). Nonpermanent adhesion occurs in invertebrates that attach firmly, but only temporarily, to the substratum through viscoelastic secretions and therefore retain the capacity to move (e.g., sea anemones, gastropods, cephalopods, sea stars, sea urchins [1, 2]). Finally, instantaneous adhesion comprises adhesive systems that rely on single-use organs or cells and are used in functions requiring a very fast formation of adhesive bonds like prey capture or defense reactions (e.g., comb jellies, sea cucumbers [1, 2]). All these systems allow underwater attachment but differ by their structure, functioning, and the physicochemical characteristics of their adhesives. Among them, however, a more intense research effort has been devoted to the elucidation of the mechanisms allowing permanent adhesion of sessile invertebrates and algae [4]. These organisms are indeed involved in the economically important problem of fouling in the marine environment. Because biomimetic efforts are only possible when composition and key molecular components of biological adhesives are understood, so far only a very limited number of organisms have been used for the development of bio-inspired adhesives. The best-characterized marine bioadhesive is that from the mussel and it has inspired most of the biomimetic adhesives currently available [5]. Recently, however, other organisms such as tubeworms, brown algae, and barnacles have also been used as models. The following paragraphs will focus on these four organisms, from the features of their adhesives to the imaginative applications that have been developed for their biomimetic counterparts.

Mussels

Natural Adhesive

To attach themselves to the substratum, mussels produce a byssus (Fig. 1), which consists of a bundle of proteinaceous threads connected proximally to the base of the animal's foot, within the shell, and terminating distally with a flattened plaque which mediates adhesion to the substratum [5, 6]. These plaques are formed by the auto-assembly of secretory products originating from four distinct glands enclosed in the mussel foot. These products comprise a collagenous substance, a mucous material, a mixture of polyphenolic proteins (known as foot proteins 2 to 6, abbreviated as fp-2-6) and an accessory protein (fp-1). Only the different foot proteins (fp-1-6), also known as mussel adhesive proteins (MAPs), will be considered here.

The Mussel Adhesive Proteins (MAPs) Since the characterization of fp-1 in the early 1980s, MAPs have been the subject of a very large number of studies leading to a detailed knowledge on their structures, functions, and interactions within the byssal attachment plaque. Proteins fp-2 and fp-4 form the central core of the plaque; fp-3, fp-5, and fp-6 are located at

Biomimetics of Marine Adhesives, Fig. 1 A mussel of the species *Mytilus edulis* attached on a glass substratum by the mean of its byssus. The byssus consists of a bundle of threads (*T*), each terminating with a flattened plaque (*P*) which mediates adhesion to the surface

the interface between the plaque and the substratum (primer layer); and fp-1 forms a hard cuticle protecting the core from hydrolysis, abrasion, and microbial attack [5, 6].

The 6 foot proteins (fps) constituting the attachment plaques have been mostly characterized from mussels of the genus *Mytilus*. Their characteristics are summarized in Table 1. Most MAPs exhibit repeated sequence motifs, whose number and amino acid composition vary according to the species considered. Moreover, all the proteins identified in the plaque share a common distinctive feature: the presence of 3,4-dihydroxyphenylalanine (DOPA), a residue formed by the posttranslational hydroxylation of tyrosine [5, 6] (Fig. 2). This modified amino acid fulfills two important roles in the attachment plaque (Fig. 2): it is involved in the formation of cross-links between the different fps (cohesion) and it mediates physicochemical interactions with the surface (adhesion) [5]. It is generally accepted that cross-linking reactions are related to the oxidation of DOPA to DOPA-quinone, a reaction catalyzed by a catecholoxidase in the byssus. DOPA-quinones may also result from redox reactions involving transition metal ions or may form spontaneously at alkaline pH. Once formed, DOPA-quinone is capable of participating in a number of different reaction pathways leading to intermolecular cross-link formation [6]. DOPA also confers to proteins a capacity for intermolecular metal complexation

Biomimetics of Marine Adhesives, Table 1 Characteristics of the proteins constituting the attachment plaques in the byssus of mussels from the genus *Mytilus*

Protein	Localization in the plaque	Mass (kDa)	Repeated unit (frequency)	DOPA (mol%)	Feature
fp1	Cuticle	110	AKPSYPPTYK (80)	10–15	Hyp, DiHyp
fp2	Core	47	EGF-like motif (11)	2–3	Cys
fp3	Primer layer	6	None	20–25	Arg-OH
fp4	Core	93	HVHTHRVLHK (36, in N-term half)	<2	His
			DDHVNDIAQTA (16, in C-term half)		
fp5	Primer layer	9.5	None	27	p-Ser
fp6	Primer layer	12	None	4	p-Ser, Cys

Arg-OH 4-hydroxyarginine, *Cys* cysteine, *DiHyp* 3,4 dihydroxyproline, *His* histidine, *Hyp* 4-hydroxyproline, *p-Ser* O-phosphoserine

through the formation of tris- and/or bis-catechol-metal complexes. On the other hand, DOPA is also involved in surface coupling, either through hydrogen bonds or by forming complexes with metal ions and metal oxides present in mineral surfaces [5, 6]. These complexes possess some of the highest known stability constants of metal-ligand chelates. Some of the MAPs also contain other posttranslationally modified amino acids such as O-phosphoserine (fp-5 and fp-6), which could mediate adhesion to calcareous substrata, or 4-hydroxyproline (fp-1), 3,4 dihydroxyproline (fp-1) and 4-hydroxyarginine (fp-3) which could make hydrogen bonds with surfaces [5, 6].

MAP Extraction and Purification The extraction and purification of MAPs directly from mussel tissues for purposes is feasible but is economically and ecologically problematic. Indeed, several thousands mussels are needed to obtain 1 g of proteins [4, 7, 8]. Harvesting the natural adhesive is therefore an unrealistic solution for most industrial applications. A commercial MAP extract is however commercially available under the name Cell-Tak™ (BD Biosciences, Franklin Lakes, NJ USA) and consists of a mixture of fp-1 and fp-2. It is sold as an adhesive for the in vitro immobilization of cells and tissue sections on various surfaces, including plastic, glass, metal, and biological materials [4]. In vivo transplantation experiments (e.g., corneal transplantation in rabbit) also showed that Cell-Tak appeared to be nontoxic and well tolerated by biological systems. In the laboratory, the adhesive properties of an extracted MAP mixture similar to Cell-Tak (i.e., containing ~80% fp-1 and ~20% fp-2) were investigated by bonding together strips of porcine tissues, either skin or small intestinal submucosa [8]. With skin, a satisfactory joint strength of about 1 MPa, similar to the commercial fibrin glue Tisseel™ (Baxter, Deerfield, IL USA) was obtained, but only for long curing time exceeding 12 h. With small intestine, an adhesive strength of almost 0.5 MPa was reported and the curing was reduced to 1 h by the addition of oxidizing transition metal ions such as Fe^{3+} and V^{5+}. It is usually admitted that adhesives suitable for medical applications should possess an adhesive and cohesive strength ranging from 0.01 to 6 MPa, depending on the particular bonding site (e.g., relatively unstressed soft tissues versus bones or tendons) [2]. In addition, the properties of the adhesive should be adapted to the "dry" (topical tissues such as teeth or skin) or "wet" (internal organs bathed in body fluids) state of the bonding site.

Production of Mussel Mimetic Adhesives and Their Applications

To bypass the problem of obtaining MAPs directly from mussels, biomimetic adhesives have been developed. These bio-inspired molecules are produced either in the form of recombinant preparations of the adhesive proteins or in the form of chemically synthesized polymers incorporating DOPA or catechol groups.

Production of Recombinant MAPs Recombinant DNA technology has been used in order to obtain larger quantities of mussel adhesive proteins [6–9]. Attempts to produce recombinant forms of MAPs started in the early 1990s with the expression, production, and purification of complete fp-1 and of synthetic fp-1 protein analogs consisting of 6–20 repeats of the

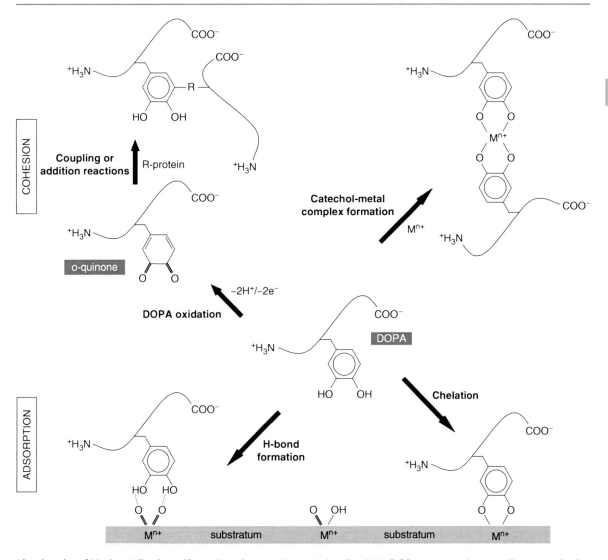

Biomimetics of Marine Adhesives, Fig. 2 Dual functionality of peptidyl DOPA groups in mussel adhesive proteins. The catechol functions of DOPA-containing proteins contribute to adhesive adsorption through hydrogen bonding of the phenolic OH groups to the oxygen atoms of the surface (*lower left*) or by forming mono-bidentate complexes with metal ions at the surface of mineral, metal oxide or metal hydroxide substrata (*lower right*). Peptidyl DOPA groups also contribute to adhesive cohesion by forming bis- or tris-catechol-metal complexes (*upper right*) or by forming intermolecular cross-links. Cross-linking follows the oxidation of DOPA to an *o*-quinone that reacts with amino acid side chains (R) of other proteins, in which R can belong to cysteine, histidine, lysine, and other DOPA residues (*upper left*)

consensus fp-1 decapeptide (Ala-Lys-Pro-Ser-Tyr-Pro-Pro-Thr-Tyr-Lys) in the yeast or in the bacterium *Escherichia coli* [6, 9]. However, these first attempts failed to obtain large amounts of recombinant proteins (Table 2) for different reasons including the highly biased amino acid composition of fp-1 and differences in codon usage between mussels and the heterologous expression systems [6, 9]. Bacterial expression of recombinant fp-1 was enhanced by the fusion of fp-1 with an *E. coli* signal peptide (OmpASP), which boosts secretion by directing the expressed protein to the periplasm. Another problem with the heterologous expression of fp-1 and MAPs in general is that the recombinant proteins lack the posttranslational modifications such as the hydroxylation of proline, arginine, or most importantly, tyrosine residues. The production of functional recombinant MAPs therefore requires an additional in vitro modification step: the enzyme-

Biomimetics of Marine Adhesives, Table 2 Comparison of several recombinant mussel adhesive proteins

Recombinant protein	Host	Production yield (mg/L) after purification	Solubility (g/L)
fp-1	*Saccharomyces cerevisiae* (Y)	nd	nd
	Escherichia coli (B)	4–10	nd
fp-2	*Saccharomyces cerevisiae* (Y)	nd	nd
fp-3	*Kluyveromyces lactis* (Y)	1	nd
	Escherichia coli (B)	0.8–3	1
fp-5	*Escherichia coli* (B)	2.8	1
fp-131	*Escherichia coli* (B)	nd	nd
fp-151	*Escherichia coli* (B)	1,000	330
fp-353	*Escherichia coli* (B)	39	90

B bacterium, *Y* yeast, *nd* not determined

catalyzed modification of tyrosine residues into DOPA [6]. This is usually done using a commercially available mushroom tyrosinase.

Fp-3 and fp-5 are other interesting adhesive proteins as they have a high DOPA content and are present at the interface between the byssus attachment plaques and the substratum. Both were cloned in expression vectors containing a histidine tag sequence which allows purification of the protein using metal affinity chromatography [9]. The vectors were expressed in *E. coli* with subsequent in vitro treatment by mushroom tyrosinase for DOPA formation. Recombinant fp-5 presented a low purification yield (Table 2) but remarkable adsorption and adhesive abilities on various surfaces [6, 9]. Indeed adsorption and adhesion forces were comparable to – and sometimes exceeded – those of Cell-Tak. A better purification yield was achieved for recombinant fp-3 (Table 2). The adhesion capabilities of this protein were, however, lower than those of recombinant fp-5, but still comparable to those of Cell-Tak. Recombinant fp-5 was also tested successfully as a cell-adhesion material for in vitro cell culture [6, 9].

The successful production of recombinant MAPs and their good performance in microscale adhesion tests were encouraging. However, macroscale testing and large-scale applications were still prevented by poor production due to post-induction bacterial cell growth inhibition and by the low solubility of purified proteins in aqueous buffers (Table 2). To overcome these limitations, recombinant hybrid adhesive proteins were designed, the so-called fp-353, fp-151, and fp-131 [7, 9]. The former is a fusion protein with fp-3 at each terminus of fp-5. Because fp-353 formed inclusion bodies, host cell growth inhibition did not occur. In addition, the solubility of fp-353 was better than that of fp-3 or fp-5 alone, permitting the preparation of a viscous concentrated glue solution for large-scale adhesion strength measurements. The fp-151 and fp-131 hybrids resulted from the fusion of six fp-1 decapeptide repeats at each terminus of fp-5 and fp-3, respectively [6, 7, 9]. Recombinant fp-151 displayed the highest production yield reaching 1 g/L of batch-type flask culture (Table 2). In fed-batch-type bioreactor cultures, similar high production levels were maintained through a co-expression of fp-151 with bacterial hemoglobin that facilitates the utilization of cellular oxygen [6]. Of all recombinant MAPs, fp-151 possesses the highest solubility (up to 330 g/L), allowing the concentration of the protein solution into a sufficiently viscous liquid for practical adhesive application. Additionally, fp-151 showed a better adsorption and a similar adhesion force, compared to recombinant fp-5, and its adhesive strength (~0.8 MPa on cowhide, ~1.1 MPa on aluminum, and ~1.8 MPa on poly(methyl methacrylate [Plexiglas])) is always largely in excess to that of fibrin glue. The recombinant fusion proteins, marketed in milligram quantities (Kollodis, Inc), are less expensive than extracted MAPs (Cell-Tak) [7].

Fp-151 proved to be biocompatible for the in vitro culture of various cell types including both anchorage-dependent and anchorage-independent cells [6, 9]. In this context, fp-151 was also fused to a RGD peptide, a sequence identified as a cell-adhesion recognition motif by integrins, and used in culture plates coating. The resulting hybrid, fp-151-RGD, maintained the high production yield of fp-151, but also presented superior spreading and cell-adhesion abilities compared to the commercially produced cell-adhesion materials poly-L-lysine (PLL) and Cell-Tak, and this regardless of mammalian cell line used [6]. The excellent adhesion and spreading abilities of fp-151-RGD might be due to the fact that it combines three types of cell-binding mechanisms: DOPA adhesion of Cell-Tak, cationic binding force of PLL, and RGD sequence-mediated adhesion of fibronectin. These characteristics make the two hybrid proteins fp-151 and fp-151-RGD suitable for use as cell-adhesion material in cell culture or tissue engineering. In a totally different context, fp-151 has been proposed as a potential gene delivery material in view of its similar amino acid composition to histone proteins, which are known as effective mediators of transfection. The fusion protein displayed comparable transfection efficiency in human and mouse cells compared to the widely used transfection agent Lipofectamine™ (Invitrogen).

Production of MAP Mimetic Adhesive Polymers Another way to mimic mussel adhesive proteins is by the chemical synthesis of important functional parts of these adhesives. This has been done either by peptide synthesis or by the functionalization of various other polymers with MAP reactive groups, mostly DOPA or catechol groups.

Various synthetic polypeptides inspired by MAPs were studied for adhesion strength [7, 8, 10]. This approach was used to experimentally identify the exact functions of the amino acids that are active in the chemistry of the adhesive proteins. The importance of particular MAP over-expressed amino acids, mainly Lys and Tyr/DOPA, was therefore tested through different polytripeptides and polydipeptides such as poly (X-Tyr-Lys), poly(Gly-Lys), poly(Tyr-Lys), and poly (Gly-Tyr), each peptide being subsequently incubated with tyrosinase for the enzymatic hydroxylation of tyrosines into DOPA. The adhesive strength of poly (X-Tyr-Lys) peptides, irrespective of whether X was Gly, Ala, Pro, Ser, Leu, Ile, or Phe, measured on pig skin (i.e., 0.01 MPa) was higher than that of poly (Gly-Lys) peptides but similar to the one of poly (Tyr-Lys) peptides and of a poly(fp-1 decapeptide) [10]. Tyrosine (DOPA) and lysine thus appears as key amino acids for adhesion efficiency, with little importance of the other residues and of the primary structure of the MAPs. A poly(Gly-Tyr-Lys) peptide

was also used as a surgical glue to close a skin incision in a living pig. Good incision adhesion and reduced immunological response after 1 week were observed.

Peptide synthesis also allows the direct incorporation of DOPA into the polymers, thus avoiding the complications associated with the oxidation of precursor tyrosine residues [7, 8, 10]. Water soluble copolypeptides containing DOPA and Lys residues were prepared by polymerization of α-amino acid N-carboxyanhydride monomers. Using lap shear tensile adhesive measurements, these copolymers were found to form moisture-resistant adhesive bonds to a variety of substrata after cross-linking with different oxidizing agents. Adhesion was strongest to metals and polar surfaces such as glass (3–5 MPa) [10]. Adhesive strength of about 0.2 MPa was also measured on porcine skin and bone in vitro. Furthermore, outcomes of endothelial cell cultures demonstrated that the copolypeptide presented a good cell affinity, which would provide basic data for its application in the biomaterial field [10].

DOPA being considered as a key component of MAPs for adhesion, functionalization of synthetic polymers other than polypeptides with catechol groups has emerged as a promising strategy for the development of new biomimetic adhesives [1, 7, 11]. Following this approach, simplified polymer mimics of MAPs have been designed in which a polystyrene backbone is used to take the place of the protein polyamide chain. Hardening of this poly[(3,4-dihydroxystyrene)-*co*-styrene] by treatments with various oxidizing agents including dichromate and Fe^{3+} yielded adhesive strength of up to 1.2 MPa on aluminum [7]. Linear and branched poly(ethyleneglycol) (PEG) molecules were also chemically coupled to one to four DOPA endgroups. PEG has a low toxicity and is considered to be nonallergenic, therefore allowing its utilization as a convenient platform in medicine [1]. The addition of oxidizing agents resulted in the polymerization of PEG-DOPA *via* DOPA-quinone cross-links, forming polymer networks and rapid gelation. Although these gels were not adhesive, it has been shown that unoxidized four-armed PEG polymers functionalized with a single DOPA residue at the extremity of each arm (PEG-DOPA$_4$) adsorbed strongly on mucin, this mucoadsorption being largely due to the presence of DOPA [11]. A tissue adhesive was investigated by using liposomes in order to compartmentalize an oxidizing reagent (NaIO$_4$) in a solution containing

PEG-DOPA$_4$ [11]. While sodium periodate is sequestered in these vesicles at room temperature, it is released in the polymer solution when liposomes melted at wound site (37°C). Oxidation of PEG-DOPA$_4$ resulted in hydrogel formation with interesting adhesive properties. Indeed, lap shear strength between two porcine dermal tissues was five times higher than that of fibrin glue [11]. A similar strategy was used for efficient immobilization of transplanted pancreatic islets in mice, with minimal inflammatory response. Polymers made up of PEG-DOPA coupled to polycaprolactone (PCL) were also used to coat the biologic meshes used for hernia repair.

Paradoxically, PEG-DOPA polymers can also be used for antifouling applications. In this last case, DOPA or DOPA-containing peptides are conjugated to PEG polymers to allow their adsorption to Au and Ti surfaces [1]. The surfaces modified with these polymers were shown to be resistant to cell attachment and protein adsorption suggesting that they might be used to design cell and protein-resistant surfaces for medical implants, anti-icing coating on aircraft wings and nonfouling marine structures (Kensey Nash Corporation, Exton, PA USA).

The use of toxic oxidizing reagents for solidification of hydrogels represents a medical concern for in vivo applications [1, 2, 7, 11]. To avoid such reagents, DOPA was then conjugated to poly(ethyleneoxide)-poly(propyleneoxide)-poly(ethyleneoxide) (PEO-PPO-PEO) block copolymers. These DOPA-modified block copolymers were soluble in cold water but self-aggregated into micelles at a higher temperature that was dependent on the block copolymer concentration. For instance, a 20 wt% solution formed gels when heated to body temperature. Photopolymerization is another approach to achieve oxidation-free cross-linking and monomers combining DOPA and a UV (or visible light)-polymerizable methacrylate group were therefore copolymerized with PEG diacrylate. In both heat and light-triggered polymerization, gelation did not result from DOPA-quinones cross-links. DOPA groups were thus available for adhesion but the adhesive strength of these hydrogels was not investigated. Self-assembly of amphiphilic triblock copolymers is also an alternative strategy to obtain oxidation-free polymerization. Triblock copolymers were made up of hydrophobic endblocks consisting of poly(methylmethacrylate) (PMMA) and of a poly(methacrylic acid) (PMAA) water-soluble midblock. DOPA was incorporated into the hydrophilic PMAA. Hydrogels were then obtained by exposing triblock copolymers solution in DMSO to water vapor. As water diffused into the solution, the hydrophobic endblocks formed aggregates that were bridged by water-soluble midblocks. Underwater adhesive properties of a DOPA-modified PMMA-PMAA-PMMA membrane on titanium and pig skin were assessed and showed strong adhesion in both cases.

The strategy centered on the functionality of the catechol group reaches its highest point with the report of a method to form multifunctional polymer coatings through simple dip-coating of objects in an aqueous solution of dopamine, a DOPA analog [12]. Indeed, dopamine self-polymerization forms thin, surface-adherent polydopamine films onto a wide range of inorganic and organic materials including metals, oxides, polymers, semiconductors, and ceramics. Secondary reactions can then be used to create a variety of ad-layers [12, 13]. On metals, these polydopamine films can provide corrosion protection properties. They can also promote cell adhesion on any type of material surfaces including the well-known anti-adhesive substrate, poly(tetrafluoroethylene), and therefore convert a variety of bioinert substrates into bioactive ones [12, 13]. Another important application of this method is the improvement of interfacial properties in composites. The macroscale properties of polymer-matrix composites depend immensely on the quality of the interaction between the reinforcement phase and the bulk polymer, and polydopamine films deposited on the reinforcement phase have been demonstrated to increase interfacial shear stress. For solar cell applications, binding photosensitizer onto semiconductor surfaces, especially TiO$_2$ surfaces, via catechol groups is a very promising method for creating stable interfaces [13].

Tubeworms

Some polychaete worms of the family Sabellariidae are tube-dwelling and live in the intertidal zone. They are commonly called honeycomb worms or sandcastle worms because they are gregarious and the tubes of all individuals are closely imbricated to form large reef-like mounds (Fig. 3). To build the tube in which they live, they collect particles such as sand grains or shell fragments with their tentacles from the water column and sea bottom. These particles are then conveyed to the building organ which is a crescent-shaped structure near the mouth. There, the particles are

Biomimetics of Marine Adhesives, Fig. 3 A polychaete of the species *Sabellaria alveolata* (**a**) and a reef fragment made of the imbricated tubes built by individuals of this species (**b**). *BO* building organ, *Te* tentacles, *Th* thorax

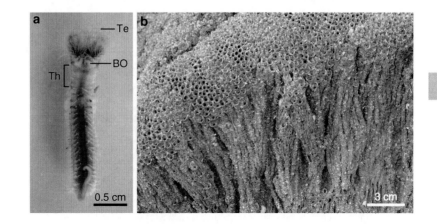

dabbed with spots of cement secreted by two types of unicellular glands located in the worm's thorax, and then added to the end of the preexisting tube by the building organ [2, 7, 14].

Cement composition has been investigated in the species *Phragmatopoma californica*. The adhesive consists mostly of three proteins (known as *Phragmatopoma* cement proteins 1–3, abbreviated as Pc-1, 2, and 3) and large amounts of Mg^{2+} and Ca^{2+} ions [7, 14]. The three cement proteins have highly repetitive primary structures with limited amino acid diversity (Table 3). As it is the case for plaque proteins in mussels, Pc-1 and 3 are present in the form of several variants. Pc-1 and 2 contain basic residues with amine side chains and DOPA residues, some of which are halogenated into 2-chloro-DOPA residues [14, 15]. DOPA residues presumably play the same functions as in MAPs (Fig. 2), contributing to both adsorptive interactions with the surface (adhesion) and cross-link formation (cohesion). Regarding the chloro-DOPA residues, they could be involved in the protection of the cement from microbial fouling and degradation. Pc-3 is particularly rich in serine (72.9 mol%) and careful calculations indicated that up to 90% of these serine residues are posttranslationally phosphorylated [14, 15]. Pc-3 is therefore a remarkably acidic protein (pI $\sim$ 1). Phosphorylation is thought to impart a potential for both cohesive (by Ca^{2+} or Mg^{2+} bridging) and adhesive contributions to the cement [14].

The co-occurrence of the positively charged Pc-1 and Pc-2 proteins and the negatively charged Pc-3 protein in the tube worm cement led to the hypothesis that a phenomenon called complex coacervation could play a role in the condensation of the adhesive in the form of a dense water-immiscible fluid [14, 15].

Biomimetics of Marine Adhesives, Table 3 Characteristics of the cement proteins of the tubeworm *Phragmatopoma californica*

Protein	Mass (kDa)	Repeated unit (frequency)	Features
Pc-1	18	VGGYGYGGKK (15)	DOPA, basic
Pc-2	21	HPAVHKALGGYG (8)	DOPA, basic
Pc-3	10–52[a]	Poly(S)	p-Ser (about 80 mol%), highly acidic
			p-Ser definition

p-Ser O-phosphoserine

[a]The mass of Pc-3 varies according to the variant considered

Complex coacervation is the spontaneous separation of an aqueous solution of two oppositely charged polyelectrolytes into two immiscible aqueous phases, a dilute equilibrium phase and a denser solute-rich phase (Fig. 4). Coacervation occurs when the charges of the polyelectrolytes are balanced. This phenomenon is therefore pH dependent, occurring to the maximum extent at the pH where the solution is electrically neutral; ionic strength dependent, since shielding of charges can change the charge balance of the system [15]; and dependent on the ratio of polyelectrolyte concentrations. Whether complex coacervation is really involved in the formation of the natural cement or not is still under debate. However, the ideal material and rheological properties of complex coacervates provided a valuable blueprint for the synthesis of a biomimetic, water-borne, underwater adhesive [15]. Based on the tubeworm model, poly(meth)acrylate polymers with phosphate, amine, and catechol side chains in the same molar ratios as the natural adhesive proteins Pc1-3 were synthesized [7, 15]. Analogs of Pc-1 and 2 were created by copolymerization of

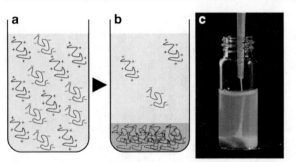

Biomimetics of Marine Adhesives, Fig. 4 Complex coacervation is the associative phase separation in a solution of positively and negatively charged macroions. From a stable colloidal solution of polyelectrolyte complexes (**a**), a trigger such as a pH change leads to the formation of two immiscible aqueous phases, a dilute equilibrium phase and a denser solute-rich phase (**b**). The latter, called the coacervate phase, is an isotropic liquid containing amorphous associative particles that move freely relative to one another ((**c**) Courtesy of R. Stewart, University of Utah, USA)

N-(3-aminopropyl)methacrylamide hydrochloride (APMA) and acrylamide. Mimics of Pc-3 were synthesized by free radical copolymerization of monoacryloxyethyl phosphate (MAEP), acrylamide, and dopamine methacrylate (DMA), the latter being a DOPA analog monomer [15]. Both copolymers were mixed, with divalent cations, in similar proportions as the natural proteins. Around neutral pH, they condensed into a liquid complex coacervate (Fig. 4c) while at pH 10, they formed a hydrogel through cross-links between dopaquinone and amine side chains [7, 15]. The underwater bond strength of the biomimetic adhesive on aluminum was ~0.8 MPa, about twice the value estimated for the natural glue [15]. Moreover, a biodegradable, temperature-triggered version of the adhesive coacervate was designed by replacing the APMA/acrylamide copolymer by an amine-modified collagen hydrolysate. This adhesive coacervate was used to repair rat calvarial bone defect and was capable of maintaining three-dimensional bone alignment in freely moving rats over a 12-week indwelling period [15]. Histological evaluation demonstrated that the adhesive was gradually resorbed and replaced by new bone with an inflammatory response commensurate with normal wound healing. This noncytotoxic degradable coacervate therefore appears to be suitable for use in the reconstruction of craniofacial features [15].

The tubeworm cement-inspired complex coacervates prompted efforts to form complex coacervates with other biomimetic adhesives such as the mussel hybrid fusion proteins fp-151 and fp-131 [15]. The MAP hybrids are insoluble at physiological pH so coacervation was done at pH 3.8 with hyaluronic acid (HA) at HA:fp-151 (or fp-131) ratios of 3:7. Importantly, it was found that the highly condensed complex coacervates significantly increased the bulk adhesive strength of MAPs in both dry and wet environments (up to ~4 MPa on aluminum) [15]. In addition, oil droplets were successfully incorporated in the coacervate, forming a microencapsulation system that could be useful in the development of self-adhesive microencapsulated drug carriers, for use in biotechnological and biomedical applications. Coacervates were also prepared using HA and fp-151-RGD [15]. The low interfacial energy of the coacervate was exploited to coat titanium (Ti), a metal widely used in implant materials. The coacervate effectively distributed both HA and fp-151-RGD over the Ti surfaces and enhanced osteoblast proliferation.

Brown Algae

Brown algae live firmly attached on substrata in the subtidal and intertidal rocky shores where they are often exposed to high gradients of turbulence [1]. They have therefore developed adhesive strategies to attach themselves strongly and durably throughout their life cycle, from the microscopic reproductive cell stages to the large thalli of mature algae (Fig. 5a). Enzymatic extractions and cytochemical studies indicated that the adhesive mucilage produced by these algae contain polysaccharides and glycoproteins [1, 16]. Phenolic polymers also play a role in brown algae adhesion. They appear to be secreted a few hours after fertilization, allowing initial substratum adhesion and, after germination, are localized at the site of adhesive mucilage formation, the rhizoid tip in the case of *Fucus*. These polyphenols are composed of phloroglucinol units linked by carbon–carbon and ether bonds (Fig. 5b). They are reminiscent of the DOPA-containing proteins of mussels and tubeworms in that they can be oxidized and form cross-links. The proposed mechanism of algal adhesion indeed postulates that the polyphenols are oxidized by an extracellular vanadium bromoperoxidase to enable their cross-linking with the polysaccharide components of the adhesive such as alginate [16].

The adhesive of the species *Fucus serratus* inspired a biomimetic glue composed of synthetic

Biomimetics of Marine Adhesives, Fig. 5 The brown alga *Fucus serratus* ((a) Courtesy of H. Bianco-Peled, Technion – Israel Institute of Technology) and the chemical structure of brown algal polyphenol (b) and phloroglucinol monomer (c)

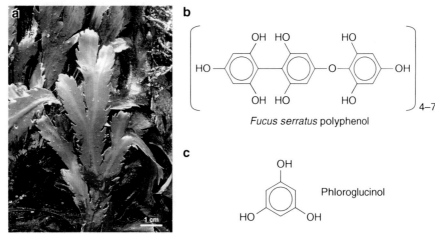

Fucus serratus polyphenol

Phloroglucinol

phloroglucinol units (Fig. 5c), alginate, and calcium for medical applications [17]. This synthetic adhesive was shown to be capable of adhering to a variety of both hydrophilic and hydrophobic surfaces, with strength similar to that of the algal-born natural adhesive (0.017–0.025 MPa). It also adhered well to porcine tissue and was shown to be safe for living cells. The biomimetic adhesive was also tested on porcine muscle tissue and presented a tensile strength similar to that of the commercially available fibrin glue Tisseel™. It is marketed as a vascular sealant for surgical adjunctive leakage control (SEAlantis Ltd., Haifa, Israel).

Barnacles

Barnacles are sessile crustaceans that attach firmly and massively not only to a large variety of underwater natural substrata such as rocks (Fig. 6), but also to man-made substrata such as ship hulls therefore causing major economical losses [2]. In these organisms, attachment is mediated by the release of a permanent adhesive called cement [1, 2, 18]. In acorn barnacles, this cement is produced by large isolated secretory cells (the cement cells) joined together by ducts which open onto the base of the animal [18]. The cement is composed of approximately 90% proteins with the remainder as carbohydrates (1%), lipids (1%), and inorganic ash (4%, of which calcium accounts for 30%). More than ten proteins have been found in the cement (cement proteins abbreviated as cp) [1, 19], of which six have been purified and characterized, originally from the species *Megabalanus rosa* and later from other species from the genera *Amphibalanus*,

Biomimetics of Marine Adhesives, Fig. 6 Group of barnacles of the species *Elminius modestus* attached on a rock (Picture courtesy of N. Aldred, Newcastle University, UK)

Balanus, Fistulobalanus and *Semibalanus* [1]. Their features are summarized in Table 4. Among these proteins, three (cp-19k, cp-20k, and cp-68k) have a surface coupling function, two (cp-52k and cp-100k) have a bulk function, and the last one (cp-16k) is an enzyme whose possible function is the protection of the cement from microbial degradation [18, 19]. Evidence to date suggests that barnacles display a contrasting adhesive system compared to mussels and tubeworms, two animals characterized by their reliance on posttranslational modification of adhesive proteins, especially the formation of DOPA residues. On the other hand, little or no posttranslational modifications have been found so far in individual barnacle cement proteins [18]. Their strong attachment to surfaces does not involve therefore the

Biomimetics of Marine Adhesives, Table 4 Characteristics of the cement proteins of the barnacle *Megabalanus rosa*

Protein[a]	Repeated unit	Feature	Proposed function
cp-16k	None	Similar to lysozyme	Antimicrobial
cp-19k	None	Bias toward Ser, Thr, Gly, Ala, Lys, Val (these residues amount for 67% of the total amino acids)	Surface coupling
cp-20k	Six Cys-rich repeated sequences	Cys, charged amino acids	Surface coupling
cp-52k	Repetitive sequence	Hydrophobic	Cement structural framework
cp-68k	None	Bias toward Ser, Thr, Ala, Gly (these residues amount for 60% of the total amino acids); glycosylated	Surface coupling
cp-100k	None	Hydrophobic	Cement structural framework

Ala alanine, *Cys* cysteine, *Gly* glycine, *Lys* lysine, *Ser* serine, *Thr* threonine, *Val* valine
[a]Denotes cement protein with its apparent molecular weight (kDa)

"DOPA-system." Only one of the cement proteins, cp-68k, appears to be posttranslationally modified. This protein is indeed glycosylated, although the glycosyl moiety is limited in amount.

As in mussels, the first attempts to exploit the potentialities of barnacle glue were undertaken in the early 1980s. Native cement was shown to be effective in tests of rabbit bone repair. However, the difficulty to collect sufficient quantities of adhesive material prevented further use of this material in concrete medical procedures [7]. Recently, the proteins cp-19k and cp-20k, have been produced in bacteria in the form of recombinant proteins at the laboratory scale. Both recombinant cps were demonstrated to present underwater irreversible adsorption activity to a variety of surface materials, including positively charged, negatively charged, and hydrophobic ones [19].

The protein cp-20k was used for the design of self-assembling peptides [18]. This protein is characterized by its high content of charged amino acids and cysteines, and by the presence of six repetitive sequences deduced from the alignment of cysteines (Cys) [1]. A single repetitive sequence, containing four Cys, has been chemically synthesized to obtain a water-soluble peptide solution. The addition of salt at a final concentration of 1 M triggered an irreversible self-assembly into a macroscopic membrane made up of interwoven nanofilaments [18]. Treatment with dithiothreitol (DTT), a reducing agent, and the replacement of cysteines by serines in the peptide did not affect self-assembly. This means that the latter is not dependent on the formation of disulfide bonds but rather on

noncovalent molecular interactions. However, extended incubation of the peptide solution in buffers at alkaline pH caused a conformational change and the formation of a three-dimensional mesh-like structure stabilized by disulfide bonds. It was proposed that this self-assembling peptide may present a novel source of inspiration for the design of new materials [18].

Perspectives

From the biologist's perspective, marine adhesives may serve one or more of the following functions: (a) the temporary or permanent attachment of an organism to a surface, including dynamic attachment during locomotion and maintenance of position; (b) the collection or capture of food items; or (c) the building of tubes or burrows. The evolutionary background and biology of the species, on one hand, and environmental constraints on the other hand, both influence the specific composition and properties of adhesive secretions in a particular organism. The diversity of marine adhesives is therefore huge and daunting for researchers but, in return, it will also allow a great deal of flexibility in applications. To date, only a very limited number of marine organisms have been used as models for the development of biomimetic adhesives and most of this diversity remains unexploited. The study of new model organisms representative of the three different types of adhesion is therefore required in order to extract essential structural, mechanical, and chemical principles from their adhesives. Simplification of natural systems while retaining the desired efficacy is, indeed, one of the foundation stones of biomimetics. As quoted by Herbert Waite [2], it is no longer absurd to predict that a compilation of bio-inspired adhesive designs will someday fill a textbook with recipes for a whole range of applications, from technical to medical.

Cross-References

▶ Adhesion in Wet Environments: Frogs
▶ Bioadhesion
▶ Bioadhesives
▶ Bioinspired Synthesis of Nanomaterials
▶ Biomimetics
▶ Gecko Adhesion

References

1. Smith, A.M., Callow, J.A.: Biological Adhesives. Springer, Heidelberg (2006)
2. von Byern, J., Grunwald, I.: Biological Adhesive Systems – From Nature to Technical and Medical Application. Springer, Wien (2010)
3. Flammang, P., Santos, R., Haesaerts, D.: Echinoderm adhesive secretions: from experimental characterization to biotechnological applications. In: Matranga, V. (ed.) Marine Molecular Biotechnology: Echinodermata, pp. 201–220. Springer, Berlin (2005)
4. Taylor, S.W., Waite, J.H.: Marine adhesives: from molecular dissection to application. In: McGrath, K., Kaplan, D. (eds.) Protein-Based Materials, pp. 217–248. Birkhäuser, Boston (1997)
5. Waite, J.H., Andersen, N.H., Jewhurst, S., Sun, C.: Mussel adhesion: finding the tricks worth mimicking. J. Adhes. 81, 297–317 (2005)
6. Silverman, H.G., Roberto, F.F.: Understanding marine mussel adhesion. Mar. Biotechnol. 9, 661–681 (2007)
7. Stewart, R.J.: Protein-based underwater adhesives and the prospects for their biotechnological production. Appl. Microbiol. Biotechnol. 89, 27–33 (2011)
8. Wilker, J.J.: Marine bioinorganic materials: mussels pumping iron. Curr. Opin. Chem. Biol. 14, 276–283 (2010)
9. Cha, H.J., Hwang, D.S., Lim, S.: Development of bioadhesives from marine mussels. Biotechnol. J. 3, 1–8 (2008)
10. Deming, T.J.: Synthetic polypeptides for biomedical applications. Prog. Polym. Sci. 32, 858–875 (2007)
11. Lee, B.P., Dalsin, J.L., Messersmith, P.B.: Synthesis and gelation of DOPA-modified poly(ethylene glycol) hydrogels. Biomacromolecules 3, 1038–1047 (2002)
12. Lee, H., Dellatore, S.M., Miller, W.M., Messersmith, P.B.: Mussel-inspired surface chemistry for multifunctional coatings. Science 318, 426–430 (2007)
13. Ye, Q., Zhou, F., Liu, W.: Bioinspired catecholic chemistry for surface modification. Chem. Soc. Rev. 40, 4244–4258 (2011)
14. Zhao, H., Sun, C., Stewart, R.J., Waite, J.H.: Cement proteins of the tube-building polychaete Phragmatopoma californica. J. Biol. Chem. 280, 42938–42944 (2005)
15. Stewart, R.J., Wang, C.S., Shao, H.: Complex coacervates as a foundation for synthetic underwater adhesives. Adv. Colloid Interface Sci. 167, 85–93 (2011)
16. Vreeland, V., Waite, J.H., Epstein, L.: Polyphenols and oxidases in substratum adhesion by marine algae and mussels. J. Phycol. 34, 1–8 (1998)
17. Bitton, R., Bianco-Peled, H.: Novel biomimetic adhesives based on algae glue. Macromol. Biosci. 8, 393–400 (2008)
18. Kamino, K.: Underwater adhesive of marine organisms as the vital link between biological science and material science. Mar. Biotechnol. 10, 111–121 (2008)
19. Kamino, K.: Molecular design of banacle cement in comparison with those of mussel and tubeworm. J. Adhes. 86, 96–110 (2010)

Biomimetics of Optical Nanostructures

Andrew R. Parker
Department of Zoology, The Natural History Museum, London, UK
Green Templeton College, University of Oxford, Oxford, UK

Synonyms

Animal reflectors and antireflectors; Photonics in nature; Structural color in nature; Submicron structures in nature

Definition

Animals and plants boast a range of submicron photonic devices comparable to the portfolio of physicists. While physicists design their photonic devices to suit the human eye or specific detectors, the devices of animals and plants have evolved to target different but equally precise detectors – eyes – and under specific conditions. Optical biomimetics involves the characterization of natural photonic devices and the manufacture of artificial analogues. It exploits modern, submicron fabrication techniques and methods of cell culture. The ultimate goal of optical biomimetics is to contribute to commercial products or to provide some degree of inspiration for industry.

Concept

Three centuries of research, beginning with Hooke and Newton, have revealed a diversity of optical devices at the nanoscale (or at least the submicron scale) in nature [1]. These include structures that cause random scattering, 2D diffraction gratings, 1D multilayer reflectors, and 3D liquid crystals (Fig. 1a–d). In 2001, the first photonic crystal was identified as such in animals [2], and since then the scientific effort in this subject has accelerated. Now a variety of 2D and 3D photonic crystals in nature are known (e.g., Fig. 1e, f), including some designs not encountered previously in physics.

Biomimetics is the extraction of good design from nature. Some optical biomimetic successes have resulted from the use of conventional (and constantly advancing) engineering methods to make direct analogues of the reflectors and antireflectors found in nature. However, recent collaborations between biologists, physicists, engineers, chemists, and material scientists have ventured beyond merely mimicking in the laboratory what happens in nature, leading to a thriving new area of research involving biomimetics via cell culture. Here, the nano-engineering efficiency of living cells is harnessed, and nanostructures such as diatom "shells" can be made for commercial applications via culturing the cells themselves.

Procedures

Engineering of Antireflectors

Some insects benefit from antireflective surfaces, either on their eyes to see under low-light conditions or on their wings to reduce surface reflections in transparent (camouflaged) areas. Antireflective surfaces, therefore, occur on the corneas of moth and butterfly eyes [3] and on the transparent wings of hawkmoths [4]. These consist of nodules, with rounded tips, arranged in a hexagonal array with a periodicity of around 240 nm (Fig. 2b). Effectively they introduce a gradual refractive index profile at an interface between chitin (a polysaccharide, often embedded in a proteinaceous matrix; r.i. 1.54) and air, and hence reduce reflectivity by a factor of 10.

This "moth-eye structure" was first reproduced at its correct scale by crossing three gratings at 120° using lithographic techniques, and employed as antireflective surfaces on glass windows in Scandinavia [5]. Here, plastic sheets bearing the antireflector were attached to each interior surface of triple glazed windows using refractive-index-matching glue to provide a significant difference in reflectivity. Today the moth-eye structure can be made extremely accurately using e-beam etching [6], and is also employed commercially on solid plastic and other lenses.

A different form of antireflective device, in the form of a sinusoidal grating of 250 nm periodicity, was discovered on the cornea of a 45 million-year-old fly preserved in amber [7] (Fig. 2a). This is particularly useful where light is incident at a range of angles (within a single plane, perpendicular to the grating grooves), as demonstrated by a model made in photoresist using lithographic methods [7]. Consequently, it has been employed on the surfaces of solar panels, providing a 10% increase in energy capture through reducing the reflected portion of sunlight [8]. Again, this device is embossed onto plastic sheets using holographic techniques.

Engineering of Iridescent Devices

Many birds, insects (particularly butterflies and beetles), fishes, and lesser-known marine animals display iridescent (changing color with angle) and/or "metallic" colored effects resulting from photonic nanostructures. These appear comparatively brighter than the effects of pigments and often function in animals to attract the attention of a potential mate or to startle a predator. An obvious application for such visually attractive and optically sophisticated devices is within the anticounterfeiting industry. For secrecy reasons, work in this area cannot be described, although devices are sought at different levels of sophistication, from effects that are discernable by the eye to fine-scale optical characteristics (e.g., polarization and angular properties) that can be read only by specialized detectors. However, new research aims to exploit these devices in the cosmetics, paint, printing/ink, and clothing industries. They are even being tested in art to provide a sophisticated color change effect.

Original work on exploiting nature's reflectors involved copying the design but not the size, where reflectors were scaled-up to target longer wavelengths. For example, rapid prototyping was employed to manufacture a microwave analogue of a *Morpho* butterfly scale that is suitable for reflection in the

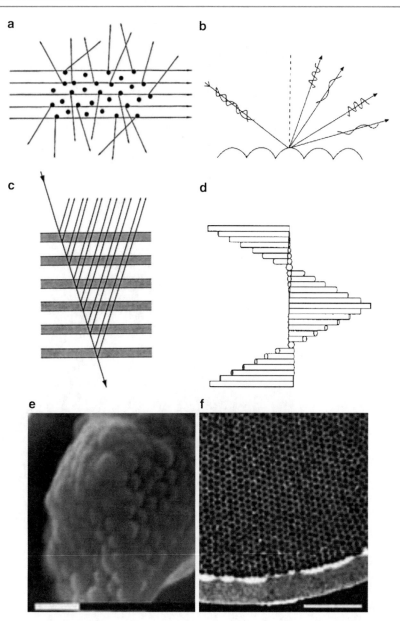

Biomimetics of Optical Nanostructures, Fig. 1 *Summary of the main types of optical reflectors found in nature*; (**a–d**) where a light ray is (generally) reflected only "once" within the system (i.e., they adhere to the single scattering, or First Born, approximation), and E and F where each light ray is (generally) reflected multiple times within the system. (**a**) An irregular array of elements that scatter incident light into random directions. The scattered (or reflected) rays do not interfere. (**b**) A diffraction grating, a surface structure, from which incident light is diffracted into specific angular directions resulting in a spatial separation of its angular wavelength/color components. Each corrugation is about 500-nm wide. Diffracted rays interfere either constructively or destructively. (**c**) A multilayer reflector, composed of thin (ca. 100-nm thick) layers of alternating refractive index, where

light rays reflected from each interface in the system interfere either constructively or destructively. (**d**) A "liquid crystal" composed of nano-fibers arranged in layers, where the nano-fibers of one layer lie parallel to each other yet are orientated slightly differently to those of adjacent layers. Hence, spiral patterns can be distinguished within the structure. The height of the section shown here – one "period" of the system – is around 200 nm. (**e**) Scanning electron micrograph of the "opal" structure – a close-packed array of submicron spheres (a "3D photonic crystal") – found within a single scale of the weevil *Metapocyrtus* sp.; scale bar = 1 μm. (**f**) Transmission electron micrograph of a section through a hair (neuroseta) of the sea mouse *Aphrodita* sp. (Polychaeta), showing a cross section through a stack of submicron tubes (a "2D photonic crystal"); scale bar = 5 μm

Biomimetics of Optical Nanostructures,
Fig. 2 Scanning electron micrographs of antireflective surfaces. (**a**) Fly-eye antireflector (ridges on four facets) on a 45-million-year-old dolichopodid fly's eye (Micrograph by P. Mierzejewski, reproduced with permission) and (**b**) moth-eye antireflective surfaces. (**c**) Moth-eye mimic fabricated using ion-beam etching. Micrograph by S.A. Boden and D.M. Bagnall, reproduced with permission. Scale bars = 3 μm (**a**), 1 μm (**b**), 2 μm (**c**)

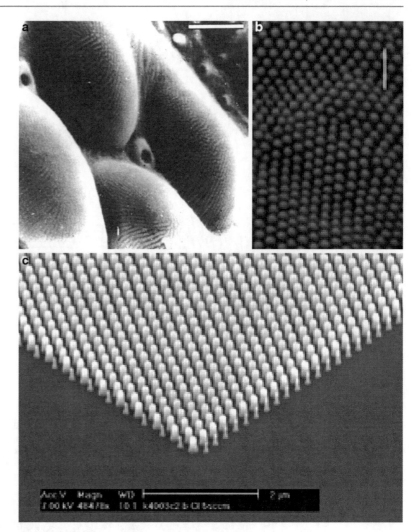

10–30 GHz region. Here layer thicknesses would be in the order of 1 mm rather than 100 nm as in the butterfly, but the device could be employed as an antenna with broad radiation characteristics, or as an antireflection coating for radar. However, today techniques are available to manufacture nature's reflectors at their true size.

Nanostructures causing iridescence include photonic crystal fibers, opal and inverse opal, and unusually sculpted 3D architectures. Photonic crystals are ordered, often complex, sub-wavelength (nano) lattices that can control the propagation of light at the single wave scale in the manner that atomic crystals control electrons [9]. Examples include opal (a hexagonal or cubic close-packed array of 250-nm spheres) and inverse opal (a hexagonal array of similar sized holes in a solid matrix). Hummingbird feather

barbs contain ultrathin layers with variations in porosity that cause their iridescent effects, due to the alternating nanoporous/fully dense ultrastructure [10]. Such layers have been mimicked using aqueous-based layering techniques [10]. The greatest diversity of 3D architectures can be found in butterfly scales, which can include micro-ribs with nano-ridges, concave multilayered pits, blazed gratings, and randomly punctate nano-layers [11, 12]. The cuticle of many beetles contains structurally chiral films that produce iridescent effects with circular or elliptical polarization properties [13]. These have been replicated in titania for specialized coatings [13], where a mimetic sample can be compared with the model beetle and an accurate variation in spectra with angle is observed (Fig. 3). The titania mimic can be nanoengineered for a wide range of resonant wavelengths; the lowest so far is

Biomimetics of Optical Nanostructures, Fig. 3 (a) A Manuka (scarab) beetle with (b) titania mimetic films of slightly different pitches. (c) Scanning electron micrograph of the chiral reflector in the beetle's cuticle. (d) Scanning electron micrograph of the titania mimetic film (Images by L. DeSilva and I. Hodgkinson, reproduced with permission)

a pitch of 60 nm for a circular Bragg resonance at 220 nm in a Sc_2O_3 film (Ian Hodgkinson, personal communation).

Biomimetic work on the photonic crystal fibers of the *Aphrodita* sea mouse is underway. The sea mouse contains spines (tubes) with walls packed with smaller tubes of 500 nm, with varying internal diameters (400–50 nm). These provide a band gap in the red region, and are to be manufactured via an extrusion technique. Larger glass tubes packed together in the proportion of the spine's nanotubes will be heated and pulled through a drawing tower until they reach the correct dimensions. The sea mouse fiber mimics will be tested for standard PCF applications (e.g., in telecommunications) but also for anticounterfeiting structures readable by a detector.

Analogues of the famous blue *Morpho* butterfly (Fig. 4a) scales have been manufactured [14, 15]. Originally, corners were cut. Where the *Morpho* wing contained two layers of scales – one to generate color (a quarter-wave stack) and another above it to scatter the light – the model copied only the principle [14]. The substrate was roughened at the nanoscale, and coated with 80-nm thick layers alternating in refractive

index [14]. Therefore, the device retained a quarter-wave stack centered in the blue region, but incorporated a degree of randomness to generate scattering. The engineered device closely matched the butterfly wing – the color observed changed only slightly with changing angle over 180°, an effect difficult to achieve and useful for a broad-angle optical filter without dyes.

A new approach to making the 2D "Christmas tree" structure (a vertical, elongated ridge with several layers of 70-nm-thick side branches; Fig. 4b) has been achieved using focused-ion-beam chemical-vapor-deposition (FIB-CVD) [15]. By combining the lateral growth mode with ion-beam scanning, the Christmas tree structures were made accurately (Fig. 4c). However, this method is not ideal for low-cost mass production of 2D and 3D nanostructures, and, therefore, the ion-beam-etched Christmas trees are currently limited to high cost items including nano- or micron-sized filters (such as "pixels" in a display screen or a filter). Recently further corners have been cut in manufacturing the complex nanostructures found in many butterfly scales, involving the replication of the scales in ZnO, using the scales themselves as templates [16] (Fig. 4d, e).

Biomimetics of Optical Nanostructures, Fig. 4 (**a**) A *Morpho* butterfly with (**b**) a scanning electron micrograph of the structure causing the blue reflector in its scales. (**c**) A scanning electron micrograph of the FIB-CVD fabricated mimic. A Ga^+ ion beam (beam diameter 7 nm at 0.4 pA; 30 kV), held perpendicular to the surface was used to etch a precursor of phenanthrene ($C_{14}H_{10}$). Both give a wavelength peak at around 440 nm and at the same angle (30°). (**d**) Scanning electron micrograph of the base of a scale of the butterfly *Ideopsis similes*. (**e**) Scanning electron micrograph of a ZnO replica of the same part of the scale in (**d**). (**a**)–(**c**) by K. Watanabe, and (**d**) and (**e**) by W. Zhang, all reproduced with permission of the authors. Scale bars: (**b**) and (**c**) = 100 nm, (**d**) = 5 μm, (**e**) = 2 μm

Cell Culture

Sometimes nature's optical nanostructures have such an elaborate architecture at such a small (nano)scale that we simply cannot copy them using current engineering techniques. Additionally, sometimes they can be made as individual reflectors (as for the *Morpho* structure) but the effort is so great that commercial-scale manufacture would never be cost effective.

An alternative approach to making nature's reflectors is to exploit an aspect other than design – that the animals or plants can make them efficiently. Therefore, we can let nature manufacture the devices for us via cell culture techniques. Animal cells are in

the order of 10 μm in size and plant cells up to about 100 μm, and hence suitable for nanostructure production. The success of cell culture depends on the species and on type of cell from that species. Insect cells, for instance, can be cultured at room temperature, whereas an incubator is required for mammalian cells. Cell culture is not a straightforward method, however, since a culture medium must be established to which the cells adhere, before they can be induced to develop to the stage where they make their photonic devices.

Current work in this area centers on butterfly scales. The cells that make the scales are identified in chrysalises, dissected and plated out. Then the individual cells are separated, kept alive in culture, and prompted to manufacture scales through the addition of growth hormones. Currently, blue *Morpho* butterfly scales have been cultured in the lab that have identical optical and structural characteristics to natural scales. The cultured scales could be embedded in a polymer or mixed into a paint, where they may float to the surface and self-align. Further work, however, is required to increase the level of scale production and to harvest the scales from laboratory equipment in appropriate ways. A far simpler task emerges where the iridescent organism is single celled.

Diatoms and Coccolithophores

Diatoms are unicellular photosynthetic microorganisms. The cell wall is called the frustule and is made of the polysaccharide pectin impregnated with silica. The frustule contains pores (Fig. 5a–c) and slits which give the protoplasm access to the external environment. There are more than 100,000 different species of diatoms, generally 20–200 μm in diameter or length, but some can be up to 2-mm long. Diatoms have been proposed to build photonic devices directly in 3D [17]. The biological function of the optical property (Fig. 5d) is at present unknown, but may affect light collection by the diatom. This type of photonic device can be made in silicon using a deep photochemical etching technique (initially developed by Lehmann [18]) (e.g., Fig. 5e). However, there is a new potential here since diatoms carry the added advantage of exponential growth in numbers – each individual can give rise to 100 million descendents in a month.

Unlike most manufacturing processes, diatoms achieve a high degree of complexity and hierarchical structure under mild physiological conditions. Importantly, the size of the pores does not scale with the size of the cell, thus maintaining the pattern. Fuhrmann et al. [17] showed that the presence of these pores in the silica cell wall of the diatom *Coscinodiscus granii* means that the frustule can be regarded as a photonic crystal slab waveguide. Furthermore, they present models to show that light may be coupled into the waveguide and give photonic resonances in the visible spectral range.

The silica surface of the diatom is amenable to simple chemical functionalization (e.g., Fig. 6a–c). An interesting example of this uses a DNA-modified diatom template for the control of nanoparticle

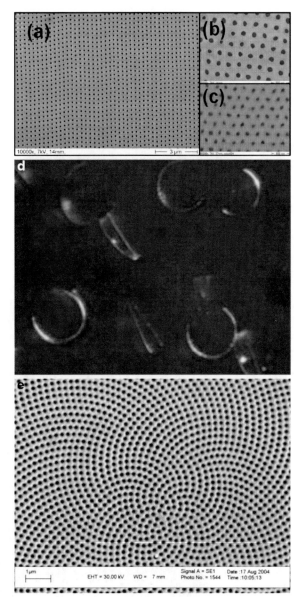

Biomimetics of Optical Nanostructures, Fig. 5 (a–c) Scanning electron micrographs of the intercalary band of the frustule from two species of diatoms, showing the square array of pores from *C. granii* ((**a**) and (**b**)) and the hexagonal arrays of pores from *C. wailesii* (**c**). These periodic arrays are proposed to act as photonic crystal waveguides. (**d**) Iridescence of the *C. granii* girdle bands. (**e**) Southampton University mimic of a diatom frustule (patented for photonic crystal applications); scanning electron micrograph (By G. Parker, reproduced with permission)

assembly [21]. Gold particles were coated with DNA complementary to that bound to the surface of the diatom. Subsequently, the gold particles were bound to the diatom surface via the sequence specific DNA

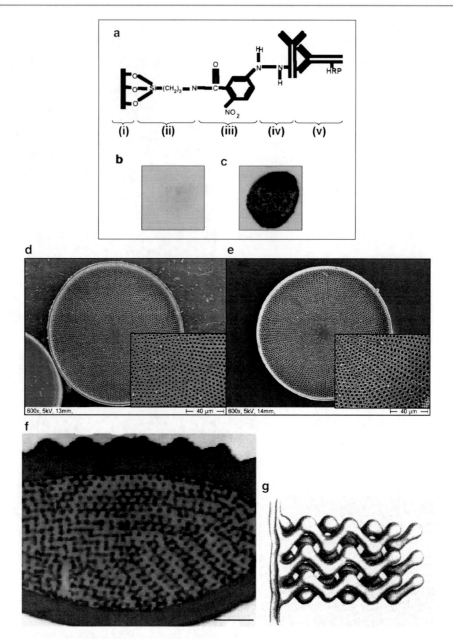

Biomimetics of Optical Nanostructures, Fig. 6 Modification of natural photonic devices. (**a**)–(**c**) Diatom surface modification. The surface of the diatom was silanized, then treated with a heterobifunctional cross-linker, followed by attachment of an antibody via a primary amine group. (**a**) (i) Diatom exterior surface (ii) APS (3-aminopropyl trimethoxy silane) (iii) ANB-NOS (*N*-5-azido-2-nitrobenzoyloxysuccinimide) (iv) primary antibody (v) secondary antibody with HRP (horseradish peroxide) conjugate. Immunoblots of diatoms show negative (**b**) and positive (**c**) chemiluminescence (of HRP-antibody conjugate), where the diatoms were all treated with primary and secondary antibodies, and with (**b**) no surface modification and (**c**) with surface modification. This demonstrates that diatoms can be employed as a solid support for the attachment of bioactive molecules (e.g., in Lab-on-a-Chip technologies) when their surfaces have been suitably modified. (**d**), (**e**) Scanning electron micrographs showing the pore pattern of the diatom *C. wailesii* (**d**) and after growth in the presence of nickel sulfate (**e**). Note the enlargement of pores, and hence change in optical properties, in (**e**). (**f**) "Photonic crystal" of the weevil *Metapocyrtus* sp., section through a scale, transmission electron micrograph; scale bar: 1 μm (see Parker [19]). (**g**) A comparatively enlarged diagrammatic example of cell membrane architecture: tubular cristae in mitochondria from the chloride cell of sardine larvae (From Threadgold [20]). Evidence suggests that preexisting internal cell structures play a role in the manufacture of natural nanostructures; if these can be altered so can the nanostructure made by the cell interaction. Using this method up to seven layers were added showing how a hierarchical structure could be built onto the template.

Porous silicon is known to luminesce in the visible region of the spectrum when irradiated with ultraviolet light [22]. This photoluminescence (PL) emission from the silica skeleton of diatoms was exploited by De Stefano [23] in the production of an optical gas sensor. It was shown that the PL of *Thalassiosira rotula* is strongly dependent on the surrounding environment. Both the optical intensity and peaks are affected by gases and organic vapors. Depending on the electronegativity and polarizing ability, some substances quench the luminescence, while others effectively enhance it. In the presence of the gaseous substances NO_2, acetone, and ethanol, the photoluminescence was quenched. This was because these substances attract electrons from the silica skeleton of the diatoms and hence quench the PL. Nucleophiles, such as xylene and pyridine, which donate electrons, had the opposite effect, and increased PL intensity almost ten times. Both quenching and enhancements were reversible as soon as the atmosphere was replaced by air.

The silica inherent to diatoms does not provide the optimum chemistry/refractive index for many applications. Sandhage et al. [24] have devised an inorganic molecular conversion reaction that preserves the size, shape, and morphology of the diatom while changing its composition. They perfected a gas/silica displacement reaction to convert biologically derived silica structures such as frustules into new compositions. Magnesium was shown to convert SiO_2 diatoms by a vapor phase reaction at 900°C to MgO of identical shape and structure, with a liquid Mg_2Si by-product. Similarly, when diatoms were exposed to titanium fluoride gas, the titanium displaced the silicon, yielding a diatom structure made up entirely of titanium dioxide, a material used in some commercial solar cells.

An alternative route to silica replacement hijacks the native route for silica deposition in vivo. Rorrer et al. [25] sought to incorporate elements such as germanium into the frustule; a semiconductor material that has interesting properties that could be of value in optoelectronics, photonics, thin film displays, solar cells, and a wide range of electronic devices. Using a two-stage cultivation process, the photosynthetic marine diatom *Nitzschia frustulum* was shown to assimilate soluble germanium and fabricate Si-Ge oxide nanostructured composite materials.

Porous glasses impregnated with organic dye molecules are promising solid media for tunable lasers and nonlinear optical devices, luminescent solar concentrators, gas sensors, and active waveguides. Biogenic porous silica has an open sponge-like structure and its surface is naturally OH-terminated. Hildebrand and Palenik [26] have shown that Rhodamine B and 6G are able to stain diatom silica in vivo, and determined that the dye treatment could survive the harsh acid treatment needed to remove the surface organic layer from the silica frustule.

Now attention is beginning to turn additionally to coccolithophores – single-celled marine algae, also abundant in marine environments. Here, the cell secretes calcitic photonic crystal frustules which, like diatoms, can take a diversity of forms, including complex 3D architectures at the nano- and microscales.

Iridoviruses

Viruses are infectious particles made up of the viral genome packaged inside a protein capsid. The iridovirus family comprises a diverse array of large (120–300 nm in diameter) viruses with icosahedral symmetry. The viruses replicate in the cytoplasm of insect cells. Within the infected cell, the virus particles produce a paracrystalline array that causes Bragg refraction of light. This property has largely been considered aesthetic to date but the research group of Vernon Ward (New Zealand), in collaboration with the Biomaterials laboratory at Wright-Patterson Air Force base, is using iridoviruses to create biophotonic crystals. These can be used for the control of light, with this laboratory undertaking large-scale virus production and purification as well as targeting manipulation of the surface of iridoviruses for altered crystal properties. These can provide a structural platform for a broad range of optical technologies, ranging from sensors to waveguides.

Virus nanoparticles, specifically *Chilo* and *Wiseana* invertebrate iridovirus, have been used as building blocks for iridescent nanoparticle assemblies. Here, virus particles were assembled in vitro, yielding films and monoliths with optical iridescence arising from multiple Bragg scattering from close-packed crystalline structures of the iridovirus. Bulk viral assemblies were prepared by centrifugation followed by the addition of glutaraldehyde, a cross-linking agent. Long-range assemblies were prepared by employing a cell design that forced virus assembly within a confined geometry followed by cross-linking. In addition, virus particles were used as core substrates in the fabrication of metallodielectric nanostructures. These comprise a dielectric core surrounded by a metallic shell. More specifically, a gold shell was assembled around the viral core by attaching small gold nanoparticles to the virus surface using inherent chemical functionality of the protein capsid [27]. These gold nanoparticles then acted as nucleation sites for electroless deposition of gold ions from solution. Such nano-shells could be manufactured in large quantities, and provide cores with a narrower size distribution and smaller diameters (below 80 nm) than currently used for silica. These investigations demonstrated that direct harvesting of biological structures, rather than biochemical modification of protein sequences, is a viable route to create unique, optically active materials.

Future Research

Where cell culture is concerned, it is enough to know that cells *do* make optical nanostructures, which can be farmed appropriately. However, in the future, an alternative may be to emulate the natural engineering processes ourselves, through reacting the same concentrations of chemicals under the same environmental conditions, and possibly substituting analogous nano- or macro-machinery.

To date, the process best studied is the silica cell wall formation in diatoms. The valves are formed by the controlled precipitation of silica within a specialized membrane vesicle called the silica deposition vesicle (SDV). Once inside the SDV, silicic acid is converted into silica particles, each measuring approximately 50 nm in diameter. These then aggregate to form larger blocks of material. Silica deposition is molded into a pattern by the presence of organelles such as mitochondria spaced at regular intervals along the cytoplasmic side of the SDV [28]. These organelles are thought to physically restrict the targeting of silica from the cytoplasm, to ensure laying down of a correctly patterned structure. This process is very fast, presumably due to optimal reaction conditions

for the synthesis of amorphous solid silica. Tight structural control results in the final species-specific, intricate exoskeleton morphology.

The mechanism whereby diatoms use intracellular components to dictate the final pattern of the frustule may provide a route for directed evolution. Alterations in the cytoplasmic morphology of *Skeletonema costatum* have been observed in cells grown in sublethal concentrations of mercury and zinc [29], resulting in swollen organelles, dilated membranes, and vacuolated cytoplasm. Frustule abnormalities have also been reported in *Nitzschia liebethrutti* grown in the presence of mercury and tin [30]. Both metals resulted in a reduction in the length to width ratios of the diatoms, fused pores, and a reduction in the number of pores per frustule. These abnormalities were thought to arise from enzyme disruption either at the silica deposition site or at the nuclear level. *C. wailesii* was grown in sublethal concentrations of nickel and observed an increase in the size of the pores (Fig. 6d, e), and a change in the phospholuminescent properties of the frustule. Here, the diatom can be "made to measure" for distinct applications such as stimuli-specific sensors.

Further, *trans*-Golgi-derived vesicles are known to manufacture the coccolithophore 3D "photonic crystals" [31]. So the organelles within the cell appear to have exact control of (photonic) crystal growth ($CaCO_2$ in the coccolithophores) and packing (SiO_3 in the diatoms) [32, 33]. Indeed, Ghiradella [9] suggested that the employment of preexisting, intracellular structures lay behind the development of some butterfly scales and Overton [34] reported the action of microtubules and microfibrils during butterfly scale morphogenesis. Further evidence has been found to suggest that these mechanisms, involving the use of molds and nano-machinery (e.g., Fig. 6f, g), reoccur with unrelated species, indicating that the basic "eukaryote" (containing a nucleus) cell can make complex photonic nanostructures with minimal genetic mutation [19]. The ultimate goal in the field of optical biomimetics, therefore, could be to replicate such machinery and provide conditions under which, if the correct ingredients are supplied, the optical nanostructures will self-assemble with precision.

For further information on the evolution of optical devices in nature, including those found in fossils, or when they first appeared on earth, see references [35, 36].

Cross-References

▶ Bioinspired Synthesis of Nanomaterials
▶ Biomimetic Synthesis of Nanomaterials
▶ Moth-eye Antireflective Structures
▶ Nanostructures for Photonics

References

1. Parker, A.R.: 515 Million years of structural colour. J. Opt. A **2**, R15–R28 (2000)
2. Parker, A.R., McPhedran, R.C., McKenzie, D.R., Botten, L.C., Nicorovici, N.-A.P.: Aphrodite's iridescence. Nature **409**, 36–37 (2001)
3. Miller, W.H., Moller, A.R., Bernhard, C.G.: The corneal nipple array. In: Bernhard, C.G. (ed.) The Functional Organisation of the Compound Eye, pp. 21–33. Pergamon Press, Oxford (1966)
4. Yoshida, A., Motoyama, M., Kosaku, A., Miyamoto, K.: Antireflective nanoprotuberance array in the transparent wing of a hawkmoth Cephanodes hylas. Zool. Sci. **14**, 737–741 (1997)
5. Gale, M.: Diffraction, beauty and commerce. Phys. World **2**, 24–28 (1989)
6. Boden, S.A., Bagnall, D.M.: Biomimetic subwavelength surfaces for near-zero reflection sunrise to sunset. *Proc. 4th World Conference on Photovoltaic Energy, Conversion, Hawaii* (2006)
7. Parker, A.R., Hegedus, Z., Watts, R.A.: Solar-absorber type antireflector on the eye of an Eocene fly (45Ma). Proc. R. Soc. Lond. B **265**, 811–815 (1998)
8. Beale, B.: Fly eye on the prize. Bull.**1**, 46 (1999)
9. Yablonovitch, E.: Liquid versus photonic crystals. Nature **401**, 539–541 (1999)
10. Cohen, R.E., Zhai, L., Nolte, A., Rubner, M.F.: pH gated porosity transitions of polyelectrolyte multilayers in confined geometries and their applications as tunable Bragg reflectors. Macromol **37**, 6113 (2004)
11. Ghiradella, H.: Structure and development of iridescent butterfly scales: lattices and laminae. J. Morph. **202**, 69–88 (1989)
12. Berthier, S.: Les coulers des papillons ou l'imperative beauté. Proprietes optiques des ailes de papillons. Springer, Paris (2005). 142pp
13. DeSilva, L., Hodgkinson, I., Murray, P., Wu, Q., Arnold, M., Leader, J., Mcnaughton, A.: Natural and nanoengineered chiral reflectors: structural colour of manuka beetles and titania coatings. Electromagnetics **25**, 391–408 (2005)
14. Kinoshita, S., Yoshioka, S., Fujii, Y., Okamoto, N.: Photophysics of structural color in the Morpho butterflies. Forma **17**, 103 (2002)
15. Watanabe, K., Hoshino, T., Kanda, K., Haruyama, Y., Matsui, S.: Brilliant blue observation from a *Morpho*-butterfly-scale quasi-structure. Jap. J. Appl. Phys. **44**, L48–L50 (2005)
16. Zhang, W., Zhang, D., Fan, T., Ding, J., Gu, J., Guo, Q., Ogawa, H.: Biomimetic zinc oxide replica with structural color using butterfly (*Ideopsis similis*) wings as templates. Bioinspir. Biomim. **1**, 89–95 (2006)
17. Fuhrmann, T., Lanwehr, S., El Rharbi-Kucki, M., Sumper, M.: Diatoms as living photonic crystals. Appl. Phys. B **78**, 257–260 (2004)
18. Lehmann, V.: On the origin of electrochemical oscillations at silicon electrodes. J. Electrochem. Soc. **143**, 1313 (1993)
19. Parker, A.R.: Conservative photonic crystals imply indirect transcription from genotype to phenotype. Rec. Res. Dev. Entomol. **5**, 1–10 (2006)
20. Threadgold, L.T.: The ultrastructure of the animal cell. Pregamon Press, Oxford (1967). 313 pp
21. Rosi, N.L., Thaxton, C.S., Mirkin, C.A.: Control of nanoparticle assembly by using DNA-modified diatom templates. Angew. Chem. Int. Ed. **43**, 5500–5503 (2004)
22. Cullis, A.G., Canham, L.T., Calcott, P.D.J.: The structural and luminescence properties of porous silicon. J. Appl. Phys. **82**, 909–965 (1997)
23. De Stefano, L., Rendina, I., De Stefano, M., Bismuto, A., Maddalena, P.: Marine diatoms as optical chemical sensors. Appl. Phys. Let. **87**, 233902 (2005)
24. Sandhage, K.H., Dickerson, M.B., Huseman, P.M., Caranna, M.A., Clifton, J.D., Bull, T.A., Heibel, T.J., Overton, W.R., Schoenwaelder, M.E.A.: Novel, bioclastic route to self-assembled, 3D, chemically tailored meso/nanostructures: Shape-preserving reactive conversion of biosilica (diatom) microshells. Adv. Mater. **14**, 429–433 (2002)
25. Rorrer, G.L., Chang, C.H., Liu, S.H., Jeffryes, C., Jiao, J., Hedberg, J.A.: Biosynthesis of silicon-germanium oxide nanocomposites by the marine diatom Nitzschia frustulum. J. Nanosci. Nanotechnol. **5**, 41–49 (2004)
26. Hildebrand, M., Palenik, B.: Grant report Investigation into the Optical Properties of Nanostructured Silica from Diatoms (2003). (See also Hildebrand, M.: The prospects of manipulating diatom silica. J. Nanosci. Nanotechnol. **5**: 146–157 (2004))
27. Radloff, C., Vaia, R.A., Brunton, J., Bouwer, G.T., Ward, V.K.: Metal nanoshell assembly on a virus bioscaffold. Nano Lett. **5**, 1187–1191 (2005)
28. Schmid, A.M.M.: Aspects of morphogenesis and function of diatom cell walls with implications for taxonomy. Protoplasma **181**, 43–60 (1994)
29. Smith, M.A.: The effect of heavy metals on the cytoplasmic fine structure of Skeletonema costatum (Bacillariophyta). Protoplasma **116**, 14–23 (1983)
30. Saboski, E.: Effects of mercury and tin on frustular ultrastructure of the marine diatom Nitzschia liebethrutti. Water Air Soil Pollut. **8**, 461–466 (1977)
31. Corstjens, P.L.A.M., Gonzales, E.L.: Effects of nitrogen and phosphorus availability on the expression of the coccolith-vesicle v-ATPase (subunit C) of Pleurochrysis (Haptophyta). J. Phycol. **40**, 82–87 (2004)
32. Klaveness, D., Paasche, E.: Physiology of coccolithophorids. In: Levandowsky, M., Hutner, S.H. (eds.) Biochemistry and Physiology of Protozoa, vol. 1, 2nd edn. Academic, New York (1979)

33. Klaveness, D., Guillard, R.R.L.: The requirement for silicon in *Synura petersenii* (Chrysophyceae). J. Phycol. **11**, 349–355 (1975)
34. Overton, J.: Microtubules and microfibrils in morphogenesis of the scale cells of *Ephestia kuhniella*. J. Cell Biol. **29**, 293–305 (1966)
35. Parker, A.R.: In the Blink of an Eye. Simon & Schuster/Perseus Press, London/Cambridge (2003). 316pp
36. Parker, A.R.: A geological history of reflecting optics. J. R. Soc. Lond. Interface **2**, 1–17 (2005)

Biomimicked Microneedles

▶ Biomimetic Mosquito-Like Microneedles

Biomimicry

▶ Biomimetics

Biomolecular Mechanics

▶ Mechanical Properties of Hierarchical Protein Materials

Bionic Ear

▶ Bio-inspired CMOS Cochlea

Bionic Eye

▶ Artificial Retina: Focus on Clinical and Fabrication Considerations

Bionic Microneedles

▶ Biomimetic Mosquito-Like Microneedles

Bionics

▶ Biomimetics

Bio-optics

▶ Structural Color in Animals

BioPatterning

Remigio Picone
Dana-Farber Cancer Institute – Harvard Medical School, David Pellman Lab – Department of Pediatric Oncology, Boston, MA, USA

Synonyms

Alkanethiols; Bioadhesion; Cell adhesion; Cell morphology; Cell patterning; Micro-patterning; Molecular patterning; Nano-patterning; Photolithography; Self-assembled monolayers; Siloxanes; Soft lithography

Definition

BioPatterning techniques allow the spatial organization of cell adhesive and non-adhesive chemistries on surfaces for biological applications.

Overview

Patterning techniques allow the spatial organization of chemistries on surfaces. The need to fabricate surface patterns of materials emerged with the miniaturization of electrical components in the first integrated circuits. In particular, photolithography, the technology that allowed the generation of the initial template patterns of electrical circuits, has been successfully used to create surface patterns for a wide range of applications in various fields. Photolithography techniques are used to create geometrical relief patterns onto substrates. This is done in three main steps (Fig. 1a). First, a thin

BioPatterning,
Fig. 1 Photolithographic master fabrication (**a**) and surface patterning (**b**) by using photolithography

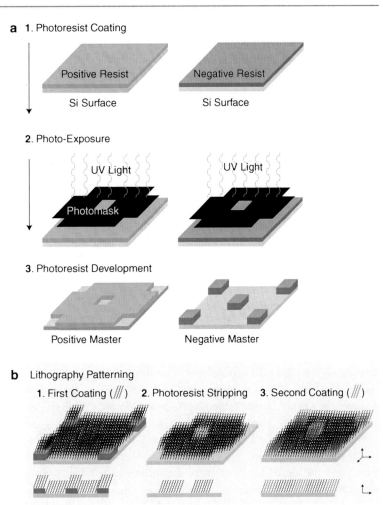

a 1. Photoresist Coating

Positive Resist   Negative Resist

Si Surface   Si Surface

2. Photo-Exposure

UV Light   UV Light

Photomask

3. Photoresist Development

Positive Master   Negative Master

b Lithography Patterning

1. First Coating (⫽)   2. Photoresist Stripping   3. Second Coating (⫽)

layer of photosensitive chemical material, also called the photoresist, is laid on top of a surface (Fig. 1a: 1). Second, an opaque plate, called the photomask, with transparencies or holes of defined shapes, is put into close contact with the pre-coated photoresist surface. Then, light is shone, usually UV light, above the photomask, allowing the light to access and modify the photoresist at areas just underneath the transparent photomask regions (Fig. 1a: 2). Third, immersion of the entire surface in an etchant solution removes the photo-exposed regions. As a result, the photo exposure and development steps produce a three-dimensional surface with the non-etched photoresist areas in relief (Fig. 1a: 3). An inverse three-dimensional pattern can be obtained by using a negative photoresist. In this case, the non-illuminated photoresist areas will be etched, while the illuminated ones will form the pattern relief regions (Fig. 1a: 3). Higher numerical aperture

lenses, new short-wavelength light sources, and development of new photoresist chemistry have helped to achieve fabrication of fine patterns over large areas. The photoresist pattern regions, once created by photomask photo exposure, can be used to spatially control distribution of various materials.

Photolithography patterning techniques can be dividend in two main groups: hard and soft lithography. Hard lithography involves the combination of lithography with the use of etchants to remove the unprotected parts left from the photomask. Soft lithography involves the use of elastomeric materials. The advantage of techniques derived from hard lithography is that they build upon well established technology, but a drawback for biological applications is that chemicals that are sensitive to etchants and lights, such as certain biological agents, cannot be used. Furthermore, it involves the use of special tools and skills that are not always accessible to biologists. The soft-lithography techniques are generally more affordable for biologists, and potentially suitable for patterning a wide variety of chemicals and materials. The main disadvantage is that they are not as accurate and reproducible as hard-lithography techniques.

Hard Lithography

Hard-lithography techniques involve the combination of surface chemistry, etchant agents, and photolithographic techniques. One of the first examples of the application of these techniques in biology is from Kleinfeld and collaborators. To study developmental properties of neural systems, they fabricated a two-dimensional in vitro model that mimicked neuronal architectures [1]. In this study, in order to fabricate the two-dimensional pattern, they took advantage of the adhesive properties of either silicon or silicon dioxide (silica) that binds covalently to small organic molecules with silane coupling agents. To do this, a silicon or silicon dioxide (silica) surface was coated with a photoresist, exposed through a lithographic mask containing the desired pattern developed to etch the resist from the exposed areas. This formed an image of the desired pattern into the photoresist. Open areas in the resist pattern allowed alkane chains to be bound to the underlying surface. These areas formed cell adhesion-resistant pattern regions. The resist pattern was subsequently removed and the previously protected areas were bound with amine derivatives. The regions with amines exposed were shown to promote cell adhesion.

Self assembly is the key mechanism used by nature for the creation of complex structures such as DNA, proteins, lipid membranes, and many other cellular structures. Research advances in the self-assembled monolayer (SAM) formation were crucial for developing a better surface chemistry control. Nuzzo and collaborators showed that organosulfur compounds adsorbed on gold surfaces form SAMs, and by changing their functional groups it was possible to control the structural and chemical properties of the exposed surface [2].

The SAM-forming molecules consist of amphiphilic molecules in which head domains strongly bind to a surface substrate, and tail chains, typically alkyl chains, form a highly ordered monolayer oriented away from the surface with the functional groups exposed (Fig. 2a, b). The functional groups can be changed to control the surface chemistry properties of substrates. In particular, they can expose the ability to either attract or repel water (hydrophilicity or hydrophobicity), reject proteins, or capture particular chemical compounds.

Alkanethiols and siloxanes are probably the most commonly used SAM-forming molecules. The sulfur atom in the head of alkanethiols has a strong binding with the gold, or other metals, surface, and the alkyl tail chain orients away from the surface at a specific angle and exposes a functional group on the surface forming an ordered two-dimensional layer (Fig. 2c). Similar to the alkanethiols, the heads of the siloxane molecules strongly bind to the surface and the alkyl tails orient away from the surface. An advantage of using the siloxanes is that their thriclorosilane head coordinates strongly with hydroxylated glass, or silicon, making it suitable for biological applications as no layer of an evaporized metal is needed (Fig. 2d). One disadvantage with respect to the alkanethiols is that the attached siloxane can be easily altered by hydrolysis generating an unstable and not completely ordered surface monolayer.

The combination of SAMs with photolithography has been used to precisely characterize the spatial organization of surfaces chemistries. Similar to the method described above, UV light and a photomask were used to selectively remove pre-coated photoresist areas from the gold surface (Fig. 1a). Thus, a first layer of SAM molecules can be strongly chemisorbed onto the revealed gold areas (Fig. 1b: 1). Successively, by using etchants, the remaining photoresist areas can be removed (stripping) and a second layer of SAM molecules can be chemisorbed onto them (Fig. 1b: 2, 3). Once the different SAMs are patterned, their functional groups can be used to confer specific surface properties to the different patterned regions. This is especially useful for biological application in which, by controlling cell adhesion surface properties, many important biological processes can be studied.

In particular, the use of specific amine derivative molecules as functional group has been found to promote cell adhesion [1]. In contrast, negatively charged functional groups show a poor surface adherence for cultures of different cell types [1].

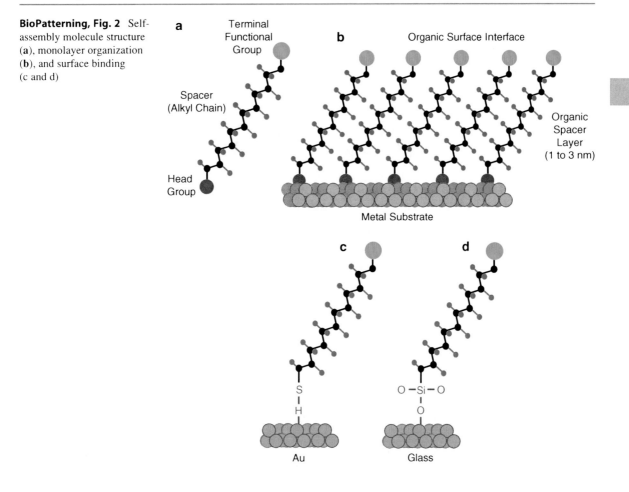

BioPatterning, Fig. 2 Self-assembly molecule structure (**a**), monolayer organization (**b**), and surface binding (c and d)

Although the use of SAMs in combination with photolithography permit a precise spatial and chemical control of patterning on different surfaces, these techniques have the significant drawbacks of requiring a long series of procedural steps, high cost tools, and the access to a clean room, all of which make these techniques inconvenient for biologists.

Soft Lithography

More suitable techniques for biological applications were developed by Kumar and Whitesides [3]. They called this set of techniques soft lithography, as each of them required the used of elastomeric surfaces ("soft material") with defined pattern features in relief for transferring molecules onto a surface. Soft-lithographic techniques are simple to be applied,

not expensive, and do not require the access to a clean room.

Micro-contact printing (μCP) was the first of the soft-lithography techniques to be developed. In this case, Kumar and Whitesides managed to print submicron patterns of SAMs on gold by using a polydimethylsiloxane (PDMS) stamp [3]. PDMS is a suitable material for biological applications as it is biocompatible, permeable to gases, and can be used for culturing cells and tissues.

The way that μCP works is simple and requires a few procedural steps. First, the PDMS stamp is fabricated by replica molding of a pre-fabricated master surface (Fig. 3a). The master surface is usually built by using the photolithography method described above (Fig. 1a). It can also be fabricated by using a laser/ion beam machine, plasma etching, rapid prototyping, or even the soft-lithography techniques themselves. The

BioPatterning,
Fig. 3 Fabrication of an
elastomeric stamp by replica
molding (**a**) and micro-contact
printing of a SAM multi-
molecules patterning (**b**–**d**)

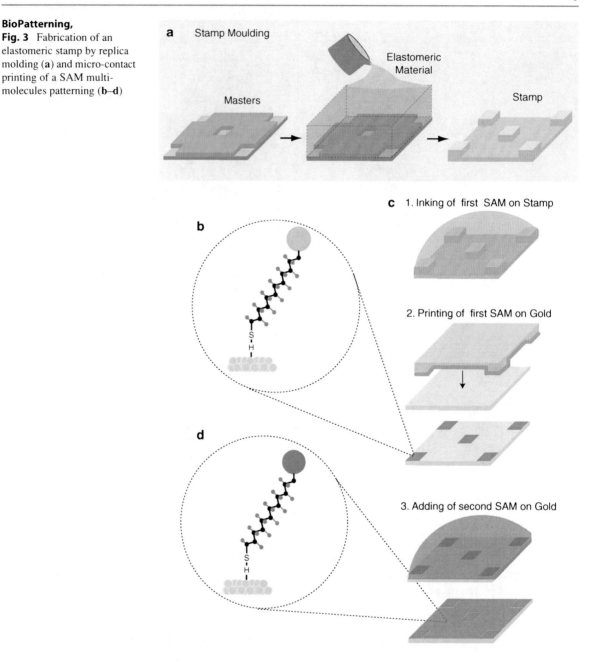

use of a specific technology for producing the master surface depends on different factors, such as pattern resolution and accuracy, size of the pattern area, and cost. Once fabricated, the stamp surface can be inked with a molecule of interest, gently dried by an inert gas stream, and finally, put in contact with the surface (Fig. 3c: 1, 2). Then, the stamp is left on the surface for a few seconds, and subsequently lifted up.

Successively, the non-patterned surface regions that were not in contact with the stamp can be filled by immersing the entire substrate in a second molecule solution (Fig. 3c: 3). Alternatively, the patterned surface can be micro-contact printed once again.

One major drawback associated with μCP is the difficulty of controlling the amount of molecules transferred, first on the stamp by inking, and successively on the surface by conformal contact. Then, it is important that the elastomeric stamp surface has properties making it more favorable for the molecules to transfer than to remain on the stamp.

Tan and collaborators studied the transfer efficiency of proteins printed onto characterized SAMs substrates composed of different quantities of methyl-terminated alkanethiol mixed with the hydrophilic hydroxyl, carboxylic acid, or poly(ethylene glycol)-terminated alkanethiols [4]. Interestingly, as the percentage of hydrophilic alkanethiols increased, the transfer of proteins onto the surface improved. On the contrary, increased amount of hydrophobic methyl-terminated thiol showed higher resistance to the protein transfer. Finally, increased stamp hydrophobicity also improved the protein transfer onto the substrate. These results showed that the relative stamp and substrate hydrophobicities are key parameters for μCP transfer of proteins on surfaces.

Other soft-lithography techniques are: MicroTransfer Molding, MicroMolding In Capillaries, Lift-Up, Put-Down, REplica Molding, Solvent-Assisted MIcroMolding, and Soft Embossing. All these techniques require the use of a patterned elastomeric stamp, and can be considered as variations of the μCP technique. Qui and collaborators give a wide overview of these techniques [5].

Molecular Patterning in Biology

Since the introduction of soft lithography, many studies demonstrated the importance of the surface patterning for the investigation of fundamental biological processes.

One of the first studies that opened up the use of soft lithography in biology was conducted by Singhvi and collaborators [6]. In their work, soft lithography was used to print gold surfaces with specific patterns of alkanethiol SEMs to create extracellular matrix (ECM) regions of defined shape and size for cell attachment [6]. With this technique, it was possible to attach cells in predetermined locations, spaced by defined distances, and to impose patterned shapes to the cells. This method allowed the analysis of individual cells, and was shown to be experimentally simple, and useful for studying fundamental biological problems.

Chen and collaborators, by using μCP, forced cells to assume different sizes and shapes. Interestingly, they found that cell growth and apoptosis is directly dependent on the cell morphology confinement [7]. This study revealed a fundamental mechanism that can have important implications for the biological processes involved in the regulation of organism development. They patterned a first layer of SAM by printing hexadecanethiol [$HS(CH_2)_{15}CH_3$] on gold-coated substrates, and backfilling the unpatterned areas by immersing the surface in a solution of tri(ethylene glycol)-terminated alkanethiol [$HS(CH_2)_{11}(OCH_2CH_2)_3OH$ in ethanol]. Then, the SAM surface was patterned on the stamped hexadecanethiol regions by immersing the entire substrate in the protein solution of interest.

To study cell migration, Jiang and collaborators confined cells to specific shapes, and subsequently, released the cell from the confined shapes by applying a pulse of voltage, which induces the electrochemical desorption of SAMs [8]. With this method, it was possible to investigate how predefined cell morphologies can influence the motility of cells. Interestingly, they demonstrated that an asymmetric cell shape is a fundamental pre-requisite for promoting the migration of cells onto a surface. In this work, the cell adhesive and non-adhesive regions were produced respectively by using μCP, and in particular printing $HS(CH_2)_{17}CH_3$ on pre-coated gold surfaces, and successive substrate immersion in a ethanol solution bath of tri(ethylene glycol)-terminated alkanethiol [$HS(CH_2)_{11}(OCH_2CH_2)_3OH$].

Thery and collaborators used μCP to study mechanisms controlling the axis orientation during cell [9]. By using defined patterned shapes of cell adhesion proteins, they were able to decouple adhesion from overall cell shape. Interestingly, this work revealed that cell contractility, through specific adhesion patterns, decides the axis of cell division beforehand, and thus determines the future cell orientation.

By using patterned lines of cell adhesion proteins, cell length homeostasis was studied [10]. Noticeably, in this work the length of cells on patterned lines was found to be independent of the pattern line width, and dependent directly on microtubule organization and dynamics. This mechanism was demonstrated to be important in in vivo tissues like the Zebrafish neural tube, in which the length homeostasis of single cells plays an important role in the structural integrity of tissues.

DNA microarray technologies have been extensively used in biology to investigate expression or mutation of thousands of genes simultaneously. A common way to fabricate DNA arrays is to spot, by using microactuated nozzles or metal pins, fluids containing the desired DNA fragment onto a microscope slide [11]. A drawback of this technique is that the DNA liquid spots must be deposited drop by drop on the surface, creating spots inhomogeneity and slowing down the fabrication process. Lange and collaborators used μCP to pattern DNA microarrays, and demonstrated that this approach can potentially overcome drawbacks of the traditional DNA microarray method [12]. In this study, a PDMS stamp was inked with a DNA solution and transferred onto a glass surface. To control DNA transfer from the elastomeric stamp to the surface, they regulated the surface electrical charge by adjusting the pre-coated (aminopropyl)trimethoxysilane with specific Ph value ranges. Microfluidic channels were used to Ink the different DNA probes onto the stamp surfaces as described by Delamarche and collaborators [13].

Future Directions

Photolithography is normally used to give photoresist layers a three-dimensional topography. Alternative approach uses the photomask and light to directly control the spatial organization of chemistries on substrates. Dulcey and collaborators, by modulating the UV light through a photomask, were able to control the density of patterned molecules on different surfaces [14]. The main disadvantage of this photolithography patterning technique is that only a fraction of chemicals can presently be patterned by light.

Doyle and collaborators developed a patterning technique based on molecular photolytic ablation termed microphotopatterning, and used it to investigate the role of ECM topography in cell migration [15]. They used polyvinyl alcohol for its high hydrophilicity and ability to prevent protein adsorption and cell adhesion. Because polyvinyl alcohol is susceptible to photolytic ablation, a two-photon confocal microscope was used to photo-ablate precise patterns in the polyvinyl alcohol film. After local polyvinyl alcohol ablation, ECM proteins were adsorbed to the etched

surfaces. The protein adsorption was dependent on the extent of ablation from the total amount of light energy focused on a given region. This method permits rapid molecular patterning without the need of photolithographic tools.

An alternative method for creating patterns of molecules on surfaces involves the use of microfluidic channels. Delamarche and collaborators, by using parallel laminar streams in a microfluidic channel, created adjacent linear patterns of different molecules [13]. An advantage of this technique is that the volume of the channels is extremely small, and so, very little reagent is needed. A drawback is that only linear pattern shapes can be created.

Although surface patterning for biological applications does not require a high degree of resolution, the control of surface patterning at the nanometer scale level could afford the creation of nano-structures with topographical features comparable to the size of the ECM, and thus, to facilitate and enhance the study of fundamental cellular processes, such as stem cell differentiation, cell migration, and morphology. One of the main limitations of the photolithographic patterning techniques is the minimum patterning feature obtainable. In fact, the resolution limit of light diffraction at the opaque area edges of the mask, and the photoresist thickness, represent one of the main limitations of photolithography for the fabrication of pattern features at the nanometer scale.

Micromachining and microwriting are the first techniques that afforded the patterning of chemistries on a surface without the use of photolithographic tools [16]. Lopez and collaborators, by using a scalpel, or carbon fibers, were able to scratch patterns across a pre-coated photoresist layer on a gold substrate [17]. They called this technique micromachining. By using this technique they managed to functionalize the exposed bare gold simply by immersing the entire surface in a solution of SAM-forming molecules. One limitation of this method is the sensitiveness of many materials, which can be easily damaged by the physical deformation produced by the micromachining tip. Moreover, the serial process makes this technique unsuitable for patterning large surface areas.

Microwriting technologies use a micropen to write chemistries on substrates [17]. This technique suffered from similar limitations as the micromachining, with the difference that the adjacent micro-written spots can potentially produce nonuniformities in the patterned regions [16]. An alternative possibility is to micro-write a pattern of molecules on the photoresist surface functioning as photo mask layer, and so, to use the above described photolithography techniques to create the final surface pattern. In this case, the light wave-length represents a limitation to the pattern resolution and the serial process, and the number of procedural steps makes this technique laborious.

One of the most promising techniques to pattern features at the nanoscale level is dip-pen nanolithography (DPN) [18]. DPN uses the capillary action between an atomic force microscope (AFM) tip and the surface allowing the patterning of 50 nm fea-ture size. Another interesting advantage is the possi-bility of creating multi-molecule patterns [19]. Similar to the micromachining and microwriting, in this case, the serial process represents the main disadvantage.

A scanning probe has been used by Armin Knoll and colleagues to pattern resist materials to a resolution of 15 nm in three dimensions [20]. Their method involves applying a controlled amount of heat and pressure to an organic resist by using a scanning probe tip, so overcoming the weak hydrogen bonds that bind the resist molecules and causing the resist to evaporate. The "hot tip" technology allows a resolu-tion of 15 nm to be achieved at speeds comparable to electron beam lithography. Three-dimensional pat-terns can be successively scanned over the same area of resist, removing varying depths with each scan. The authors used this approach to construct a 25-nm-high replica of the Swiss mountain, the Matterhorn.

Molecular patterning techniques have proven to be invaluable in the study of many biological processes and have led to new findings that were otherwise impossible with traditional techniques. Many of these technologies, if routinely used, could potentially be very helpful in enhancing a better understanding of fundamental biological processes, and, at same time, lead to innovative applications in many scientific, and non-scientific, fields. Therefore, in the future, it will be crucial to make these technologies simpler, more affordable, and accessible to a wider public.

Acknowledgments The author acknowledges Jennifer Rohn for reading the manuscript and helpful advices, David Holmes from London Centre of Nanotechnology who has made this study possible, and support from Engineering and Physical Sciences Research Council PhD + fellowship and University College London.

Cross-References

▶ Bioadhesion
▶ Bioadhesives
▶ Biosensors
▶ Cell Adhesion
▶ Chemical Milling and Photochemical milling
▶ Dip-Pen Nanolithography
▶ Dry Etching
▶ DUV Photolithography and Materials
▶ Electron Beam Lithography (EBL)
▶ Microcontact Printing
▶ Microfludic Whole-cell Biosensor
▶ Nanoimprint Lithography
▶ Nanoscale Printing
▶ Self-assembled Monolayers
▶ Self-assembly
▶ SU-8 Photoresist
▶ Wet Etching

References

1. Kleinfeld, D, Kahler, K H, Hockberger, P E: Controlled outgrowth of dissociated neurons on patterned substrates. J. Neurosci. **8**, 4098–4120 (1988)
2. Nuzzo, R G, Zegarski, B G, Dubois, L H: Fundamental studies of the chemisorption of organosulfur compounds on gold(111). Implications for molecular self-assembly on gold surfaces. J. Am. Chem. Soc. **109**, 733–740 (1987)
3. Kumar, A, Whitesides, G M: Patterned condensation figures as optical diffraction gratings. Science **263**, 60–62 (1994)
4. Tan, J L, Tien, J, Chen, C S: Microcontact printing of pro-teins on mixed self-assembled monolayers. Langmuir **18**, 519–523 (2001)
5. Qin, D, Xia, Y, Whitesides, G M: Soft lithography for micro- and nanoscale patterning. Nat. Protoc. **5**, 491–502 (2010)
6. Singhvi, R, Kumar, A, Lopez, G P, Stephanopoulos, G N, Wang, D I, Whitesides, G M, Ingber, D E: Engineering cell shape and function. Science **264**, 696–698 (1994)
7. Chen, C S, Mrksich, M, Huang, S, Whitesides, G M, Ingber, D E: Geometric control of cell life and death. Science **276**, 1425–1428 (1997)
8. Jiang, X, Bruzewicz, D A, Wong, A P, Piel, M, Whitesides, G M: Directing cell migration with asymmetric micropatterns. PNAS **102**, 975–978 (2004)
9. Théry, M, Racine, V, Piel, M, Pépin, A, Dimitrov, A, Chen, Y, Sibarita, J, Bornens, M: Anisotropy of cell adhesive microenvironment governs cell internal organization and orientation of polarity. Proc. Natl Acad. Sci. U.S.A. **103**, 19771–19776 (2006)
10. Picone, R, Ren, X, Ivanovitch, K D, Clarke, J D W, McKendry, R A, Baum, B: A polarised population of

dynamic microtubules mediates homeostatic length control in animal cells. PLoS Biol. **8**, e1000542 (2010)

11. Schober, A, Günther, R, Schwienhorst, A, Döring, M, Lindemann, B F: Accurate high-speed liquid handling of very small biological samples. Biotechniques **15**, 324–329 (1993)

12. Lange, S A, Benes, V, Kern, D P, Hörber, J K H, Bernard, A: Microcontact printing of dna molecules. Anal. Chem. **76**, 1641–1647 (2004)

13. Delamarche, E, Bernard, A, Schmid, H, Michel, B, Biebuyck, H: Patterned delivery of immunoglobulins to surfaces using microfluidic networks. Science **276**, 779–781 (1997)

14. Dulcey, C S, Georger, J H J, Krauthamer, V, Stenger, D A, Fare, T L, Calvert, J M: Deep uv photochemistry of chemisorbed monolayers: patterned coplanar molecular assemblies. Science **252**, 551–554 (1991)

15. Doyle, A D, Wang, F W, Matsumoto, K, Yamada, K M: One-dimensional topography underlies three-dimensional fibrillar cell migration. J. Cell Biol. **184**, 481–490 (2009)

16. López, G P, Biebuyck, H A, Frisbie, C D, Whitesides, G M: Imaging of features on surfaces by condensation figures. Science **260**, 647–649 (1993)

17. Lopez, G, Biebuyck, H, Harter, R, Kumar, A, Whitesides, G: Fabrication and imaging of two-dimensional patterns of proteins adsorbed on self-assembled monolayers by scanning electron microscopy. J. Am. Chem. Soc. **115**, 10774 (1993)

18. Hong, S, Zhu, J, Mirkin, C: Multiple ink nanolithography: toward a multiple-pen nano-plotter. Science **286**, 523–525 (1999)

19. Piner, R, Zhu, J, Xu, F, Hong, S, Mirkin, C: "dip-pen" nanolithography. Science **283**, 661–663 (1999)

20. Pires, D, Hedrick, J L, De Silva, A, Frommer, J, Gotsmann, B, Wolf, H, Despont, M, Duerig, U, Knoll, A W: Nanoscale three-dimensional patterning of molecular resists by scanning probes. Science **328**, 732–735 (2010)

Bio-photonics

▶ Structural Color in Animals

Biopolymer

▶ Spider Silk

Bioprobes

▶ Biosensors

Biornametics – Architecture Defined by Natural Patterns

Ille C. Gebeshuber[1,2], Petra Gruber[3] and Barbara Imhof[4]
[1]Institute of Microengineering and Nanoelectronics (IMEN), Universiti Kebangsaan Malaysia, Bangi, Selangor, Malaysia
[2]Institute of Applied Physics, Vienna University of Technology, Vienna, Austria
[3]Transarch - Biomimetics and Transdisciplinary Architecture, Vienna, Austria
[4]LIQUIFER Systems Group, Austria

Biornametics – Architecture defined by Natural Patterns – is an emerging contemporary design practice that explores a new methodology to interconnect scientific evidence with creative design in the field of architecture. The word biornametics is generated from "ornament," referring to the famous Austrian architect Adolf Loos, and "biomimetics" (▶ Biomimetics).

Role models from nature, static and dynamic patterns (e.g., nanostructured surfaces or materials with functional hierarchy from the nano- to the macroscale) are investigated and the findings are applied to design strategies. The emergence of patterns in nature at all scales of existence of organisms as one of the most important signs of life – order – is not arbitrary, but highly interconnected with boundary conditions, functional requirements, systems requirements, material, and structure. The three main areas of investigation for role models in biornametics are, firstly, surface patterns, nanosurfaces, and nanostructured materials, secondly, shape, growth, and deployable structures, and thirdly, adaptation and reorganization. Biological building strategies rely basically on repetition, variation, and self-similarity. Often simple building blocks are arranged with molecular-precision and thus achieve diverse and highly specialized material properties.

The research performed in biornametics aims at understanding the functionality of these natural patterns by extracting the principles found in current nanotechnology research, and transferring these principles to an architectural interpretation. Colors are just one very important example. In contrast to pigment colors, structural colors that are found on some butterfly wings, beetles, and even plants ("▶ Nanostructures

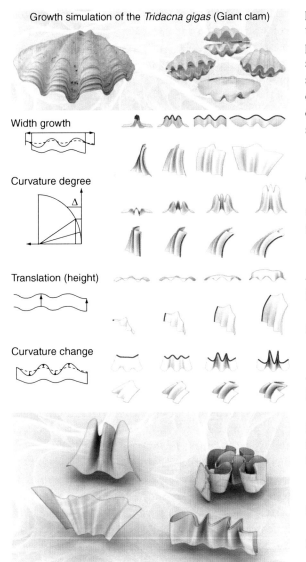

Growth simulation of the *Tridacna gigas* (Giant clam)

Width growth

Curvature degree

Translation (height)

Curvature change

Biornametics – Architecture Defined by Natural Patterns, Fig. 1 Ornaments for architecture inspired by a growth simulation of the giant clam, starting from the initial nanocrystalline composite, followed by the formation of a two-dimensional disc protrusion and the formation of crystalline curls

for Coloration (Organisms other than Animals)") are primarily determined by the geometry of the underlying material. Interesting is also the generation of these surfaces and materials, as well as multifunctional properties such as durability, degradation, or self-repair. Further examples include plant–environment interactions such as the pitcher plant that lures animals onto a super-sliding surface, or lotus leaf self-cleaning

properties and nanostructured composite materials with high toughness such as the abalone shell. The patterns found do not only fulfill their purpose but are surprisingly elegant and appeal to the aesthetic dimension of the human perception (see Fig. 1). The transfer of surface patterning to architectural elements may deliver added or integrated functionality or reinterpret specific functions on another scale.

Cross-References

▶ Biomimetics
▶ Nanostructures for Coloration (Organisms other than Animals)

Biosensing

▶ Biosensors

Biosensors

Henry O. Fatoyinbo and Michael P. Hughes
Centre for Biomedical Engineering,
University of Surrey, Guildford, Surrey, UK

Synonyms

Bioprobes; Biosensing; Nanobiosensors

Definition

A biosensor is a system or device which incorporates a biologically active material in intimate contact with an appropriate transduction element for the purpose of detecting the concentration or activity of chemical species in any type of sample.

Overview

The important components of a biosensor are (1) a bioreceptor (e.g., enzymes, antibody, microorganism, or cells); (2) a transducer of the physicochemical signal, and (3) a signal processor to interpret the information that has been converted. Depending on the type of physicochemical signal produced, i.e., electrochemical, mass, thermal, or optical, the transduction element can come in various forms, such as electrodes for amperometric and potentiometric detection, field effect transistors (FET), piezoelectric quartz (TSM, QCM), thermistors, and optical fibers/optoelectronic systems. Biosensors have found many applications in a range of industries including defense, medicine, food, pharmaceutical, and healthcare owing to their rapid response time in detection, coupled with the emergence of complementary technologies. As fabrication technologies become more sophisticated, research and development has continually reduced the size of biosensing devices, giving rise to submicron detection systems known collectively as *nanobiosensors*.

One of the first descriptions of a man-made biosensor was the *enzyme electrode*, coined by Clark and Lyons in 1962 [1]. In this system an oxidoreductase enzyme was held next to a platinum electrode in a membrane sandwich. Holding the platinum anode to a positive potential (0.6 V) relative to an Ag/AgCl reference electrode, an enzymatic catalyzed oxidation reaction of glucose produced hydrogen peroxide detected by the platinum electrode. It was not until 1967 that Updike and Hicks gave the necessary details required to build a functional enzyme electrode for glucose monitoring [2], paving the way for the first commercial biosensor for measurement of glucose in whole blood to come on to the market in 1974. The biological entity (i.e., enzyme) catalyzing this reaction produced a physicochemical signal, which was converted to an electrical signal, similar in process to the different sensory systems found in the human body.

Biosensors which incorporate an electrochemical (EC) transduction element monitor an electro-active species which is produced or consumed by the action of the biological element. These include amperometric, potentiometric, conductimetric, and field effect transistor (FET)-based detection [3, 4]. Enzyme-based electrodes are the most commonly used biosensors associated with electrochemical transduction elements (i.e., amperometric and potentiometric). The basic principle behind amperometric detection (e.g., glucose electrode) works by holding the voltage potential between a working and reference electrode constant, while measuring the variation in current in relation to the concentration of the active species. In contrast, potentiometric detection (e.g., pH electrode) is based on holding the current constant or at zero and measuring the variation in voltage as a logarithmic proportion of the reactive species. Oxidoreductases and hydrolases are commonly used enzymes in these systems, and can be immobilized on to the electrode in a variety of ways. To improve these sensor characteristics, high porosity membrane structures (e.g., polycarbonate) allow a more controlled substrate flux to the immobilized enzyme layer, situated just above the detection electrode. Both potentiometric and amperometric biosensing systems receive considerable amounts of attention as they have the advantages of being low cost, single-use devices with a high degree of reproducibility. The associated instrumentation is also very inexpensive and compact, lending these biosensors to on-site measurements in fields as varied as environmental monitoring to bioprocess monitoring.

Classification of Bioreceptors

Enzymes and Antibodies

The vast majority of biochemical reactions taking place in a cell are permitted through the catalytic ability of enzymes. Enzymes are globular proteins; their primary structure consists of sequences of L-amino acids linked via condensation reactions to form polypeptide chains. The basic generic structure of an amino acid can be seen in Fig. 1, where the letter R denotes the side chain of the peptide. These peptides can be classified as polar neutral (cysteine and tyrosine being the most hydrophilic amino acids), nonpolar neutral (hydrophobic), basic (negatively charged), or acidic (positively charged) depending on the attached groups. There are 20 different types of L-amino acids, and with variations in size from 62 to 2,500 amino acid residues an enormous combination of polypeptides can be created. The primary structured polypeptide is the starting point for the final globular protein. As the amino acids begin to interact with each other through ionic interactions, van der Waals forces, hydrogen bonding, and hydrophobic bonding, complex folding arrangements begin to occur. This secondary structure, commonly termed α-helix or β-pleated sheet, may contain regions of regular folding patterns, stabilized by weak interactions of neighboring carboxylic (-COOH) and amine (-NH$_2$) side groups. Neighboring

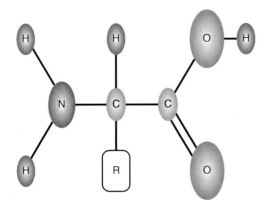

Biosensors, Fig. 1 Generic structure of an amino acid residue. The R-side group denotes the amino group

Biosensors, Table 1 Class of enzymes and examples within each class

Enzyme definition	Examples
Oxidoreductases: catalyze oxidation and reduction reactions	*Glucose oxidase, Alcohol dehydrogenase, Lactate dehydrogenase, Diacetyl reductase*
Transferases: transfer functional groups	*Thiaminase, DNA methyltransferase, Choline acetyltransferase*
Hydrolases: catalyze the hydrolysis of bonds	*Phospholipase, Lipoprotein lipase, Phosphatase, Dexyribonuclease, RNase, Lysozyme*
Lyases: cleaves bonds, not via hydrolysis or oxidation	*Carbonic anhydrase, Tryptophan synthase, Adenylate cyclise*
Isomerases: catalyzes isomerization change of molecules	*Protein disulfide isomerise, Topoisomerase, Phosphoglucomutase*
Ligases: links two molecules via covalent bonds	*DNA ligase, CTP synthase, Acetyl-CoA carboxylase*

interactions of the R-side groups determine the 3D conformational shape of the protein, referred to as the tertiary structure. Quaternary structures are formed when separate macromolecules (e.g., cofactors) get linked in a specific order forming enzymes of high stereospecificity, regiospecificity, and chemoselectivity.

Enzymes are classified into six major groups: (1) oxidoreductases, (2) transferases, (3) hydrolases, (4) lyases, (5) isomerases, and (6) ligases (Table 1). They are able to retain their catalytic activity within a limited range of parameters (i.e., pH, temperature, fluid forces, and chemical agents), but are well suited for amperometric and potentiometric biosensors due to their high turnover numbers, defined by the net number of substrate molecules reacted per catalyst active site per unit time. The dependence of the speed of reaction to the concentration of substrate can be described using the equation derived by Michaelis–Menten in Eq. 1, where V is the reaction rate, $[S]$ is the substrate concentration, and K_m is the Michaelis–Menten constant at which the substrate concentration gives the reaction rate which is half of V_{max} (maximum rate at substrate saturation). Although enzymes are most widely used in biosensing systems, they have a high cost factor in terms of production and purification as they have to be derived from biological sources:

$$V = \frac{V_{max}[S]}{K_m + [S]} \qquad (1)$$

Antibodies, also known as immunoglobulins (Ig), are a class of globular proteins ($\sim$150 kDa) which bind to specific antigenic structures forming an antibody-antigen complex, and they have widespread applications in clinical diagnostics and biotechnology. Their basic structure comprises of four polypeptide chains arranged in a Y configuration (Fig. 2). There are two heavy chains and two light chains connected by disulfide bonds. The base of the Y-structure is known as the Fc (fragment crystallizable) region and is found in the constant domain containing the heavy chains which define the class or isotype of antibody in mammals (i.e., IgA, IgM, IgG, IgD, and IgE). The Fab (fragment-antigen binding) region is found in the variable domain and contains the light chains and heavy chains which make up the amino terminal–shaped antigen binding site, known as the paratope, the most important region for antigen binding on the antibody. The two binding sites on the arms of the antibody can be extremely variable, allowing millions of different antibodies with slightly varying paratopes to exist. This lends antibodies to be widely applied in immunoassays for molecule detection across cells, large proteins, and even amino acids.

Nucleic Acids

Nucleic acids are a group of linear polymers made up of nucleotides. Nucleotides comprise a purine or pyrimidine nucleobase, a pentose sugar, and a phosphate group. The five known nucleotides or

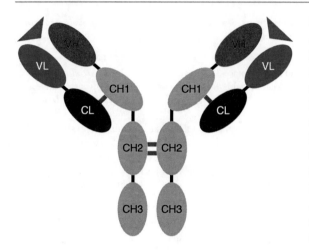

Biosensors, Fig. 2 Representative structure of an IgG antibody consisting of four polypeptide chains, interlinked with disulfide bridges (CH2–CH2; CH1–CL). CH1–CH3 represents the constant heavy chains; CL represents the constant light chains; VL and VH are the variable light and heavy molecules, respectively, which form the paratope regions for antigenic binding (*triangle*)

bases are (1) Adenine (A), (2) Guanine (G), (3) Cytosine (C), (4) Thymine (T), and (5) Uracil (U). The base pairing (b.p) of these nucleotides in the biopolymer molecule DNA takes the form of A-T and G-C. Deoxyribonucleic acid (DNA) and ribonucleic acid (RNA) are double-stranded helix and single-stranded polynucleotides, respectively, though this is more of a generalization rather than a rule. DNA and RNA have very similar chemical structures, apart from two distinct differences: (1) DNA contains 2-deoxyribose, a sugar lacking one oxygen atom compared to the ribose sugar of RNA; (2) RNA uses the nucleobase Uracil as a base, while DNA uses Thymine. The main role of DNA is the long-term storage of the genetic code of living organisms, containing instructions for the construction of macromolecules needed in cells such as proteins and RNA. The genetic information is coded into groups of three adjacent nucleotides termed codons, which represent the 20 different amino acids, the start codon, and the stop codon encoded by RNA. RNA can be considered as the workhorse for protein synthesis, copying the genetic information through transcription (messenger RNA) and translation (transfer RNA and ribosomal RNA) providing information to synthesize all the proteins needed by cells in the body. Development of microarrays has led to a new generation of biochips capable of multiplexing for rapid drug discovery and environmental analysis as will be discussed later.

Cells and Microorganisms

A vast range of microbes (i.e., yeast, algae, bacteria) have been utilized in biosensor systems, having the advantage of being massively produced through cell culturing. They are classified into two major groups based on the sensing technique employed: electrochemical microbial biosensors and optical microbial biosensors [5]. The basic principle is that certain strains of microorganisms preferentially metabolize certain compounds, with the metabolic waste being excreted detected by a transduction system. Although microbial biosensors are less sensitive than enzymatic biosensors, and the overall detection period can be longer, they are well suited to environmental analysis (e.g., Biological Oxygen Demand (BOD)) where the global effect of some toxin is detected rather than a specific toxin or compound. An interesting and growing area of research which uses this principle is that of *synthetic biology*, where a commonly used microbial chassis (e.g., *E. coli*, *S. cerevisiae*) is transfected with a gene that is turned on in the presence of a specific substrate. This has rapid sensing applications in healthcare monitoring of viruses and toxins.

Artificial Bioreceptors

The application of artificial bioreceptors is based on the naturally occurring structures of catalytic antibodies (abzymes), where these antibodies have been produced artificially by directing them against an analogue compound of the transition state for a given reaction. These abzymes have the same specific recognition of natural antibodies and the catalytic activity of enzymes reacting to substrates (antigen) to form an end product with comparable kinetics.

Molecular-imprinted polymers (MIPS) are another form of artificial bioreceptors. They are created by mixing functionalized monomers with an analogue or target molecule which binds to the monomers through hydrogen bonds, hydrophobic bonds, or ionic interactions, thus creating a template around which the polymer will form [6]. Through polymerization reactions (i.e., UV irradiation) in the presence of a cross-linking agent a 3D rigid structure is created with complementary sites for the target molecule. Removal of the template target leaves a molecularly imprinted polymer designed for specific recognition of target molecules. Although this method for creating receptors is useful for hard to obtain natural bioreceptors, there are issues regarding the polymerization stage which can often lead to wrong structure formations.

Unlike enzymes, antibodies and MIPS form ligand-receptor complexes which do not produce physicochemical signal, so quantitative or qualitative detection must be made via indirect or direct methods. Indirect methods use a secondary biological system label (bioconjugate) which emits a physicochemical signal, such as a light-emitting probe (e.g., Horseradish peroxidase) or a radioactive element, commonly used in immunosensor applications. Direct methods do not require labeling, and involve measurement changes in mass, electrical properties, or optical properties as the ligand-receptor complex is formed. The non-catalytic activity of these receptors means that once the complex has been formed reuse of the receptor in the biosensor can only be achieved through device washing to dissociate the complex.

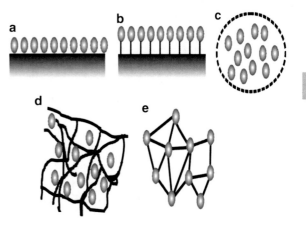

Biosensors, Fig. 3 Representation of techniques employed for bioreceptor (*sphere*) immobilization on and within substrates for biosensor. (**a**) Adsorption; (**b**) covalent attachment to functionalized surfaces; (**c**) encapsulation in a semipermeable membrane; (**d**) inclusion in gels; (**e**) cross-linking using cross-linking agents, i.e., glutaraldehyde

Immobilization Techniques

Attachment of bioreceptors to the surface or in very close proximity of a signal transduction element is usually a prerequisite for biosensors to operate efficiently. A range of immobilization methods have been developed for immobilizing cells and nanoscale bioparticles (enzymes, antibodies, etc.) on and within a variety of substrates including microscope glass slides, micro- and nano-electrodes (i.e., gold, metal oxides, platinum, silver, etc.), SiO_2, polymers, and hydrogels. The technique employed is very much dependent on the bioreceptor and the substrate to which it has to be immobilized for subsequent signal transduction. Originally developed for enzyme immobilization, techniques can be classified into five broad areas which are adsorption, inclusion, covalent coupling on activated surfaces, cross-linking, and confinement (Fig. 3).

Inclusion uses polymerization reactions to embed or encapsulate the bioreceptor within a cross-linked system such as a hydrogel. Mainly used to immobilize whole cells or subcellular fractions, it provides a homogeneous distribution of bioreceptor within the gel but the reagents used for cross-linking (i.e., photo-initiators) can produce free radicals on UV-activated polymerization which can be detrimental to the biological structures present. A well-developed technology, widely used in fiber-optic biosensors is that of the

sol-gel technology. Initially centered on metalloproteins it has been extended to encapsulate proteins, enzymes, antibodies, and whole cells with the biomolecules retaining their structural integrity and biological function [7].

Adsorption processes occur due to simple interactions of the bioreceptor and the substrate. The substrate can be either charged, leading to ionic bonds being created, or uncharged leading to weaker bonds (hydrogen, hydrophilic, and van der Waals effects). Due to varying environmental factors which may occur at the interface, desorption of the bioreceptors is common when change in pH or ionic strength is present, making this technique less widely used for immobilization.

Entrapment involves the bioreceptor remaining in solution, compartmentalized by a semipermeable membrane, allowing small molecules through. This technique is particularly useful in hydrogel biosensing applications where control is of huge importance such as in drug delivery systems which rely upon the slow release of drugs into a system based on the target analyte concentration.

Cross-linking is a technique whereby a bifunctional agent such as glutaraldehyde encourages attachment of polypeptide structures to each other or on to substrate surfaces to make the complex more stable. This technique is particularly useful in the processes involving adsorption or inclusion.

Covalent binding of proteins on activated surfaces can be achieved through the covalent bonds developed between functional groups not required in the biorecognition process. The surface activation process prior to immobilization can be achieved either via a direct method whereby amine (-NH$_2$), carboxylic (-COOH), hydroxyl (-OH), or sulfhydryl (-SH) groups on the substrate surface are functionalized through a range of reagents before covalent binding can occur with the functional group of the bioreceptor. Another common method is silanization of glass surfaces with an organosilane containing an organic functional group at one end and an alkoxysilyl group at the other end of the molecule.

Carboxylic groups can be activated using the acyl nitride method or carbodiimide method. The acyl nitride method transforms the carboxylic acid group to hydrazides through esterification by methanol in acidic solution, then treatment with hydrazine. A final treatment step with sodium nitrite gives acyl nitride which forms an amide bond with the amine groups of the bioreceptor. The carbodiimide method reacts in acidic solution with the carboxylic group forming a key intermediate, O-acylisourea a carboxylic ester, subsequently forming an amide bond with the amine group of the bioreceptor. Hydroxyl groups are activated using cyanogen bromide forming an intermediate which reacts with the amine group of the bioreceptor. Amine groups can be activated using diazonium salts, while sulfhydryl groups are activated through the formation of disulfide bridges between thiol groups of the bioreceptor and the substrate, after surface treatment with 2,2′-dipyridyldisulfide. Although most of these techniques are straightforward to perform, ready to use activated surfaces are commercially available (e.g., Sigma Aldrich). As nanostructures, nanoparticles (NP), and thin films become increasingly incorporated into biosensors, variations to these immobilization techniques for biomolecule attachment have been applied such as layer by layer (LbL) and Langmuir–Blodgett (LB) films for better molecular architectures and ultimately enhanced performances in biosensing systems [8].

Detection Technologies for Nanobiosensors

A wide range of biosensor technologies have been developed since the electrode enzyme of 1962. In recent decades nanomaterials have made significant headway in to biosensor systems for improving the limit of detection (LOD) in a host of complex sample solutions. These new nanobiosensors (or bionanosensors) typically use traditional detection techniques with nanotechnology to develop a host of enhanced bioelectronic devices for a variety of biomedical research and industrial applications. Along with optical, electrochemical, and mass transduction systems, biosensors have also utilized thermistors and magnetic beads as transduction elements. The state of art for current nanotechnologies involved in biosensing is extremely large and intertwined between detection technologies; hence for simplification the following section serves as an illustrative description of novel, and by no means exhaustive, approaches in the most common biosensing techniques.

Electrochemical-Based Detection
Carbon Nanotube Biosensors

An emerging building block of novel nanostructures and devices used in electrochemical biosensing applications are *carbon nanotubes* (CNT). Synthesized in various forms (e.g., single walled (SWCNTs), multi-walled (MWCNTs)), CNTs possess a large surface per unit mass and excellent mechanical and electrical properties making them extremely powerful in electronic detection of biomolecules, viruses, and cells. Functionalization of CNTs with different chemical groups has been achieved using both covalent and non-covalent methods [9]. At low potentials, various analytes can be oxidized with little surface fouling, making them ideal for high selectivity and reusability. Commonly used is the –COOH group for covalent bonding on CNTs via oxidizing treatment, which may also create other functional groups (e.g., hydroxyl, carbonyl, etc.) at defect sites of the graphene sheet. These other functional groups can be easily converted to carboxylic acids through perchloric acid (HClO$_4$) treatment or potassium permanganate (KMnO$_4$), which sets the surface for immobilization of bioreceptors through the carbodiimide method. CNT electrochemical (CNT-EC) biosensing has been reported for detecting glucose and neurotransmitters and a range of other biological entities, including nucleic acids and viruses. Immobilization of single-stranded DNA (ssDNA) molecules, mainly through covalent attachment, on CNTs has led to the development of electrochemical DNA biosensors which have been used to sense nucleic acids, genes, and DNA molecules. In addition, the enhanced amplification in signal transduction has seen their use in detecting DNA hybridization reactions. There are a host of doped and nanocomposite CNT electrodes developed for a variety of DNA detection applications, with sensing techniques that include impedance spectroscopy, surface conductance, and differential pulse voltammetry. Studies into the characteristics of immobilized ssDNA and hybridization of DNA have also been conducted using techniques such as cyclic voltammetry (CV) and electrochemical impedance spectroscopy (EIS).

Since about 85% of the biosensor market, worth approximately US$ 8.5 billion, belongs to glucose sensors, a large amount of novel biosensing ideas are based around blood glucose monitoring [9]. Strategies based on the presence or absence of enzymes as the bioreceptor element has been applied to CNTs for glucose monitoring. The enzyme glucose oxidase (GOD), which catalyzes the oxidation of glucose, has been immobilized on an Ag-NP/CNT/Chitosan composite giving a rapid and steady response to the substrate. This was also demonstrated on MWCNTs/ZnO-NPs, where GOD was electrostatically bound to ZnO-NPs based on isoelectric point differences. Incorporation of GOD into a colloidal Au-CNT composite showed that the apparent Michaelis–Menten constant showed a high affinity of the enzyme to the substrate under the microenvironment of the AuNPs. MWCNTs modified with ferrocene electrodes have been immobilized with GOD, and gave an improved response to glucose oxidation as the ferrocene acted as an electron transfer mediator while the MWCNT acted as the conductor. Though doped-CNT electrodes can promote electron transfer in redox reactions interference could result as endogenous O_2, unless extracted, may react with GOD producing H_2O_2, thus underestimating the glucose level. CNT-based biosensors not employing enzymes for glucose monitoring tend to be low in cost and can be achieved through, for instance, the direct growth of CuNPs on CNTs through self-assembly or via the use of immobilized microbes on CNT-modified carbon paste electrodes [10].

CNT-EC biosensors largely tend to be based upon amperometric sensing, with the CNT either surface modified with an oxide or chemically doped with a variety of NPs for label-free detection of biomolecules such as immunoglobulins and insulin. A wide range of CNT-bioreceptor bioconjugate-based biosensors have also recently been developed for real-time detection with detection limits down to a zeptomole (10^{-21}) based on double-step amplification. For example, immunoaffinity reactions (i.e., formation of antibody-antigen complexes) have been detected on CNTs as a function of mechanical and electrical properties of the CNT; CNT-FET (field effect transistors) have been used to detect mismatches of dsDNA sequences through decreases in conductance; viral antibodies have been immobilized on to CNTs through the use of polymers to detect viruses; and metal-NPs have been used for DNA attachment to CNTs for enhanced detection of sequence-specific target DNA and human hepatitis B virus ssDNA with LOD as low as 0.035pM [10].

Conducting Polymer Biosensors

Another class of nanomaterial which has recently received widespread interest for biosensor applications are a class of functional polymers known as *conducting polymers* (CP). Their attractiveness as electrochemical transducers in sensing stems from their highly conjugated polymer chain composed of alternating single and double carbon–carbon bonds, which can be doped or de-doped in a controllable manner. When used as nanostructures, these unique properties are further enhanced with large surface area, nanoscale size, and the exhibition of quantum effects making response times significantly faster than bulk polymers. The primary principle behind CPs as sensing devices is based on the "before and after" sensing of a target molecule on the surface on the CP, meaning that the sensitivity of the CP is highly dependent on its surface area where target molecules have the ability to increase or decrease mobile charge carriers, thus affecting the conductivity of the CP. Figure 4 shows the typical molecular structures of CPs for biosensing applications. Techniques used to synthesize CP nanostructures include hard- and soft-template methods and electrospinning. In the hard-template method, which is most commonly used, a template membrane is required to guide the growth of the nanostructures within the pores leading to a completely controlled nano-morphology dictated by pores and channels, which can give CP nanowires of 3 nm in diameter. As the template needs removing post-process, the technique has some limitations in that the nanostructures can be destroyed or deformed. Various preparation protocols using the hard-template

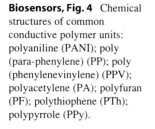

Biosensors, Fig. 4 Chemical structures of common conductive polymer units: polyaniline (PANI); poly (para-phenylene) (PP); poly (phenylenevinylene) (PPV); polyacetylene (PA); polyfuran (PF); polythiophene (PTh); polypyrrole (PPy).

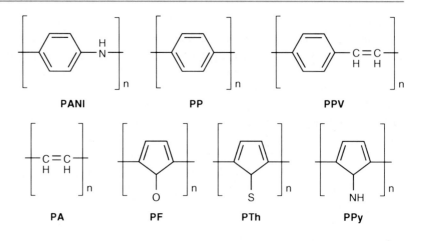

method have been described using either chemical or electrochemical polymerization, but the important variable in these processes is the actual template which can take the form of porous alumina membranes (Al_2O_3), polycarbonate membranes, or nanofibers [11]. In contrast, the soft-template method sometimes referred to as a self-assembled method requires no template. It is a cheap and fairly simple technique based on non-covalent interactions driving the self-assembly (e.g., π–π stacking interactions, hydrogen bonding, and dispersive forces), and can use oligomers, DNA, surfactants, colloidal particles, or colloids to synthesize CP nanostructures, also via both electrochemical and chemical methods [11].

Conducting polymers are extremely versatile when used in biosensing devices and have been used as immobilization matrices, artificial bioreceptors, and transducers for electrical charge from redox reactions [12]. Based on conductometric detection, a change of electrical conductivity of the CP nanostructure upon interaction with target analytes can be measured. The basis of potentiometric detection is the detection of changes to the chemical potential of the CP due to internal shifts in the anion equilibrium. Using these two detection techniques, CP-based biosensors have been developed based on nanothin films and nanostructures. Nanothin CP films, produced by electrochemical polymerization, spin coating, or layer by layer techniques, tend to have bio-active molecules embedded allowing target analytes adsorbing on to the film diffuse into the film for interaction with the receptor. As examples, an ultra thin film of PPy-GOD composite was fabricated for potentiometric sensing of

glucose, showing a high sensitivity to glucose at low current. An organic thin film transistor (OTFT) for glucose monitoring had a channel of immobilized GOD on poly (3,4-ethylenedioxythiophene)-polystyrene sulfonate (PEDOT-PSS) CP film, giving rapid response times (~20 s) and high selectivity between drain current and glucose concentration in test conditions. Using layer by layer self-assembly, an ion-sensitive field effect transistor (ISFET) was fabricated with polyaniline (PANI) as the thin film channel giving glucose sensitivity of <0.8 μA/mM, further indicating the potential of nanothin film CP as low cost biosensor materials. Other defining factors which influence the performance of these nanothin films include the thickness of the film, the roughness of the film, and the nanoporosity of the film. CP nanostructures include CP-nanoparticles (CP-NP) which tend to be dispersed on electrode surfaces to increase area/volume ratio for favorable adsorption of biomolecules; nanoarrays of nanotubes/nanowires in highly ordered alignment have shown increased sensitivity in hybridization, mass transport, and kinetic studies; and individually addressable 1D nanostructures (nanowires/nanotubes) which have shown extremely powerful transduction capabilities due to the high contact area of the analyte on the nanostructure which produces polarons and bipolarons along the backbone of the CP [11]. This effectively affects the electrical conductivity of the CP nanotube or nanowire which subsequently results in an enhanced and amplified signal based on nanosurface interactions.

Nanobiosensing has evolved even further through the development of ultrasensitive, electrical biosensors known as *nanowire biosensors* (inorganic or semiconductor), which tend to be comparable in size of the biological or chemical species being sensed. This enables excellent primary transduction of signals to interfacing macroscopic instrumentation, through direct integration of nanodevices (e.g., nano-FET) with miniaturized systems

Optical-Based Biosensors

Variations in optical phenomena based on the resultant interactions of target analyte and bioreceptor form the basis of optical transducer technologies. Fiber-optic technology has been a key driving force in the development of many of these optical techniques for biosensor applications. As such the term *optrodes* has come to define devices which incorporate optic fibers as transduction elements in sensors. A fiber-optic biosensor is a device in which a reaction phase is associated with a transduction system made from optical fibers or bundles of optical fibers directly coupled to some form of light detector. The reaction phase (e.g., sol-gel film) contains immobilized bioreceptors which generates an optical signal upon biorecognition with target analytes, transmitting the signal through the optic fiber to the detector.

Covering different regions of the electromagnetic spectrum (UV, visible, IR, NIR), optical sensors are based on various optical principles (absorbance, fluorescence, luminescence, reflectance), which allow the measurement of a range of properties including light intensity, refractive index, scattering, diffraction, and polarization. Nanotechnology has played an important part in enhancing optical sensing techniques as will be highlighted.

Surface-Plasmon Resonance (SPR)

The most common technology used in optical biosensing is surface-plasmon resonance (SPR). Belonging to the class of refractometric sensing it exploits *surface-plasmon polaritons* (electromagnetic waves) to probe interactions of analytes in solution forming complexes with bioreceptors immobilized on the SPR sensor surface through adsorption. This highly sensitive method is based on the collective oscillatory excitement of electrons propagating between a metal film and dielectric interface, leading to a total absorption of light at a particular angle of incidence (Fig. 5). Surface plasmons are optically excited if the projection of the incident light wave vector along the x-axis

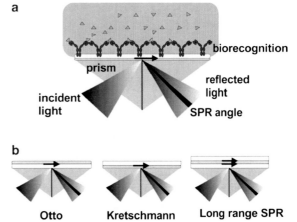

Biosensors, Fig. 5 (a) Typical setup of a SPR biosensor, with the black arrow showing the direction of the SPP wave at the interface of the metallic surface and the dielectric (*white*); (b) Typical configurations used in SPR biosensors

matches the propagation constant of the surface plasmon (k_{sp}), described as:

$$k_{sp} = \frac{\omega}{c} \left[\frac{\varepsilon_m \varepsilon_d}{\varepsilon_m + \varepsilon_d} \right]^{0.5} \qquad (2)$$

where ε_m and ε_d are the dielectric constants of metal and dielectric, respectively, ω is the angular frequency of light, and c is the speed of light in vacuum.

A change to k_{sp} resulting from a change to the refractive index of the sensed medium, i.e., the dielectric, subsequently alters the resonance conditions of the surface plasmons interacting with the optical wave. Three relevant characteristics are critical to sensor design and performance, which are:

1. *Enhancement of electric field*: at the resonance angle, the intensity of the electric field at metal-dielectric interface is significantly enhanced due to the smaller complex permittivity in the dielectric compared to the metal.

2. *Propagation length*: the propagation constant, k_{sp}, contains complex permittivity quantities for both metal and dielectric (Eq. 2) resulting in a complex wave vector. The imaginary component of the propagation constant (k_{sp}'') represents the attenuation by metal absorption and radiative losses, implying the field intensity decays by a characteristic length of $1/2k_{sp}''$. The surface plasmon decay length (δ_{sp}) is

thus defined in Eq. 3, where λ_0 is the optical wavelength:

$$\delta_{sp} = \lambda_0 \frac{(\varepsilon'_m)^2}{2\pi\varepsilon''_m} \sqrt{\left(\frac{\varepsilon'_m + \varepsilon_d}{\varepsilon'_m \varepsilon_d}\right)^3} \qquad (3)$$

3. *Penetration depth*: the other component of the parallel wave vector is the perpendicular wave vector to the interface which is an imaginary amplitude wave vector that decays in the z-direction. The penetration depth (δ_i), a couple of hundred nanometers in water, is defined as the distance in the perpendicular direction to the interface where the electromagnetic field inside the medium (i) falls to $1/e$ ($\sim$37%) and can be expressed by Eq. 4:

$$\delta_i = \frac{\lambda_0}{2\pi} \sqrt{\left|\frac{\varepsilon'_m + \varepsilon_d}{\varepsilon_i^2}\right|} \qquad (4)$$

Optical excitation methods for surface plasmons include prism, grating, and waveguide coupling. The characteristic measurement of the resonance has lead to SPR sensors to be categorized into angular (monochromatic light), wavelength (polychromatic light), intensity, or phase modulation [13].

SPR-based biosensors work on the basis of an alteration to the refractive index near the surface of the functionalized sensor when a biomolecular interaction occurs. The performance of the SPR-based biosensor is highly dependent on the intrinsic optical performance of SPR sensor (sensitivity and resolution) and the characteristics of the functionalized surface. Advances in SPR sensor performance has led to increased sensitivity of systems containing immobilized biorecognition elements, with commercial systems having refractive index resolution (LOD equivalent) as low as 1×10^{-7} RIU (Refractive Index Unit). Researchers worldwide are continuing to construct novel coupling approaches, mainly based on Kretschmann's configuration (as depicted in Fig. 5b), for greater sensitivity such as the multilayer structure long range surface plasmons (LRSPs) which offer greater optical field enhancement, lower propagation loss, and longer field penetration (for virus and bacteria detection) over conventional surface plasmons. Optical fiber-based sensors and integrated optical waveguide SPR sensors offer the development of miniaturized systems for in vivo applications and multichannel sensing on a single chip [13].

Functionalization of the metallic surface (typically Au film) is commonly achieved through covalent attachment as this creates a stable and strong binding to the surface and allows regeneration of the sensor interface after removing analytes from the surface but not the attached bioreceptor. The functional self-assembled monolayers (SAMs) of thiolated organic compounds is a routine approach for covalent attachment of thiol groups of the bioreceptors, while polymeric layers such as carboxymethylated dextran (CM-dextran) thin films offer advantages of reduced binding interference, reduction in nonspecific binding by acting as a buffer, and increased attachment points over SAMs. Although gold thin films are commonly used substrates for SPR biosensors because of their stability in aqueous environments and easy functionalization through alkanethiol SAMs, they have been found not to be the best candidates for high sensitivity SPR sensing. Indium tin oxide (ITO) has been reported to be a better candidate, though excitation and detection would occur in the infrared range. For conventional visible range excitation and detection, silver/gold layers have been used and lamellar overcoat structures, where a thin layer of a dielectric (oxide based) is deposited over silver metal thin film providing efficient protection from the aqueous solution which could damage the silver film [14]. There has also been a large focus on polymeric thin-film electro-deposition of oligomeric and conducting polymers on to gold surfaces for SPR biosensing. They have been used as they possess properties (hydrophobicity, pH, surface charge) which reduce nonspecific adsorption, which in turn increases the sensitivity and specificity of a biosensor. The conductive polymer polypyrrole has been intensively used in biosensors due to its biocompatibility, stability, and conductivity. Entrapment of bioreceptors in polypyrrole has been shown to be less attractive as the hydrophobicity of the substrate makes target molecule diffusion difficult. A more promising approach is the functionalization of the conducting polymer through covalent binding of the bioreceptor on the carboxyl groups. This method of functionalization has become a routine technique, based on electrospotting, for DNA array SPR chip fabrication.

The most common bioreceptor used on SPR-based biosensors are antibodies and commercial SPR systems (BIAcore^{TM}, real-time biospecific interaction analysis) have been developed for pharmaceutical drug discovery, antibody characterization, proteomics, immunogenicity, and much more in life science research. Essentially, SPR-based biosensing is a label-free detection method, but for signal amplification and enhanced surface chemistry new approaches which involve nanomaterials (e.g., CNTs, NP, nanowires) have been developed [13]. The effect of nanostructures on surface plasmons has led to a growing area of photonic studies known as nanoplasmonics. Metal-NPs are incorporated into SPR-based biosensors due to the lack of small molecule detection with conventional systems. As low molecular weight compounds produce a negligible change in the refractive index of the sensing surface, metal-NPs overcome this due to their localized SPR (LSPR) frequency (refer to ▸ Optical Properties of Metal Nanoparticles). This occurs as a result of the incident light interacting with particles much smaller than the incident wavelength, and the presence of metal-NPs is characterized by intense adsorption and light scattering. The enhanced electric field has shown to give an order of magnitude increase in sensitivity (5×10^{-8} RIU) and has enhanced other spectroscopic methods such as surface-enhanced Raman spectroscopy (SERS). For instance, SPR signal output has been increased through mass or enzymatic processes at the bioreceptor-ligand complex. Immobilized peptide nucleic acids (PNA) for point mutation detection were hybridized with DNA-modified nanoparticles. The LOD for this ultrasensitive mismatch was as low as 1fM using this technique [15]. A similar approach was applied to Poly-T modified nanoparticles, which bind to surface-immobilized Poly-A tails forming a ternary surface complex, with LOD for 19–23 mer as low as 10fM.

Experimental measurements based on labeling may be problematic due to metal-NP influences on binding kinetics and biomolecular interactions. A different approach toward incorporating nanotechnology into SPR-based systems focuses on the sensing matrices, which alleviate problems associated with biomolecular labeling. SPR-based immunosensors had immobilized AuNPs encapsulated in dendrimers on SAM-modified gold surfaces providing an enhanced surface density for analyzing insulin, while the sensitivity of SPR-based hormone assays were significantly enhanced using AuNPs [14]. A sensing medium which had gold embedded in a hydrogel and functionalized for small molecule detection swelled upon biorecognition, which increased the distance between AuNPs in the hydrogel. A shift in the SPR curve to higher angles was a result of the intercorrelation of both SPR and LSPR.

Molecular Beacons and Quantum Dots

Luminescence is the emission of electromagnetic radiation in the ultraviolet, visible, and infrared spectra from atoms or molecules as a result of the transition of electronically excited states to a lower energy state, usually the ground state. There are various forms of luminescence (e.g., phosphorescence, bioluminescence, chemiluminescence, and fluorescence) which have been applied to biosensors and fiber-optic biosensing for a diverse range of activities ranging from glucose monitoring and genosensors to environmental monitoring and biopharmaceutical process monitoring.

Fluorescence is a widely used optical technique in biosensing, and there are three types of sensing employed. First is direct sensing when a specific target molecule is detected before and after a reaction or change takes place. The second form involves an indirect detection by the addition of a dye (e.g., FITC, GFP) which will tag the presence of the target molecule. The final form is known as fluorescence resonance energy transfer (FRET), which generates a unique fluorescence signal. Developed by Förster and Weber, FRET requires two probes: a donor (fluorophore) and an acceptor (chromophore) [10, 16]. Irradiating the donor with light of an appropriate wavelength produces an oscillating dipole which resonates with the dipole of the acceptor in close proximity. This dipole–dipole interaction is a radiationless transfer of energy between the donor and acceptor. As such if the acceptor is another fluorophore with an excitation wavelength that overlaps with emission wavelength of the donor, the transfer of energy from the donor to the acceptor will stimulate fluorescence of the acceptor. The two major strategies used to develop FRET biosensors are: (1) two chain probes in which the fluorophores are on two different molecules resulting in intermolecular FRET when two molecules come into close proximity to each other, or (2) FRET due to intramolecular conformational change based on single-chain probes with fluorophores attached to different regions of a single molecule.

Optical DNA nanobiosensors have been developed through the use of molecular beacons (MB) and

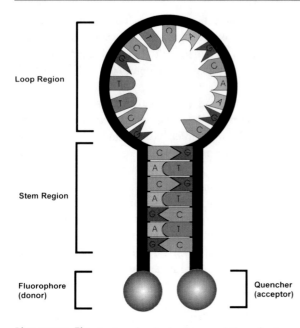

Biosensors, Fig. 6 A molecular beacon consisting of a loop region, stem region, and a fluorophore/quencher pair separated by ~10 nm, inhibiting fluorescence

quantum dots (QD) in optical nano-imaging systems. MBs are single-stranded nucleic acid probes consisting of a stem, a loop, and a fluorophore/quencher pair (Fig. 6). The close proximity (~10 nm) of the donor fluorophore and acceptor quencher in the hairpin, made possible by the "locking" mechanism of the stem region (~4–7 nucleobase pairs), effectively turns off the fluorescence via FRET. They are widely used in fluorometric analysis of nucleic acids. A conformational change of the MB occurs upon hybridization with the complementary target (ssDNA or RNA) which increases the physical separation distance between the fluorophore and quencher, thus providing a fluorescence signal (Fig. 7). Different color fluorophores can be used simultaneously for detection of multiple targets in the same solution. This label-free detection tool is sensitive and its importance in biosensor technology offers high selectivity with real-time monitoring of hybridization for analysis of gene polymorphism in genetically modified organisms, polymerase chain reaction (PCR), and ligase reactions. They have been used to detect bacteria, viruses, and other microorganisms in a variety of conditions; studies of genotyping, genetic disorders associated with point mutations, and drug discovery have been conducted with various aspects of MBs. Other

innovative approaches for using MBs as nanobiosensors have included visualizing the transport and distribution of mRNA; incorporating MBs in flow cytometry studies for oxidative stress analysis; detecting mutations in the *rpoB* of *Mycobacterium tuberculosis* (TB) via MB-qPCR as a screening tool for multidrug-resistant tuberculosis; real-time detection of dsDNA; and linking MBs to the fuse silica surface for drug delivery. These applications highlight the importance of MBs as a powerful tool in intracellular monitoring of nucleic acids and in the ongoing development of highly selective optical biosensors.

QDs are nanoscale semiconductor structures grown by epitaxial techniques in nanocrystals by chemical methods. They consist of a colloidal core (e.g., cadmium selenide, cadmium sulfide, and indium arsenide) covered with surface coatings which decrease leaching of the metal from the core (Fig. 8). Based on the reduced size of QDs, they possess discrete electronic energy which gives rise to their unique optical properties and their use as a fluorescent probe for biomolecular imaging or even as a donor in molecular beacons [17]. In comparison to organic dyes, and fluorescent proteins, QDs are controllable in many ways including their size-oriented optical and electronic properties, tunable light emission, resistance against photobleaching, and simultaneous excitation of multiple fluorescence colors [10]. More recently, development of heavy metal-free QDs with similar optical properties has overcome issues regarding cytotoxicity presented by heavy metals such as cadmium. Their photo-stable fluorescence and water solubility upon coating with amphiphilic polymers has seen QDs being used in biodetection systems, biosensing and high-throughput bio-analytical systems, quantitative cellular investigative studies, and nanobiomedicine.

Surface immobilization of antibodies, DNA, proteins, biopolymers, peptides, and nucleic acid on QDs has significantly advanced bio-imaging and biosensing over the past decade. Bioreceptor attachment via crosslink molecules with diverse reactive groups, e.g., –SH, NH_2, or –COOH, is attained through covalent or electrostatic interactions. A broad absorption spectra with high extinction coefficients and a full width at half maximum (FWHM) of 20–30 nm is afforded by QDs. Emissions of QDs can be quenched by gold nanoparticles (AuNPs), and this has been exploited in fluorescent assays for DNA detection where QDs and AuNPs were used as a FRET donor-acceptor couple

Biosensors,
Fig. 7 Hybridization reaction of a molecular beacon with a target sequence oligomer. The increased separation distance between the fluorophore and the quencher allows fluorescence to occur

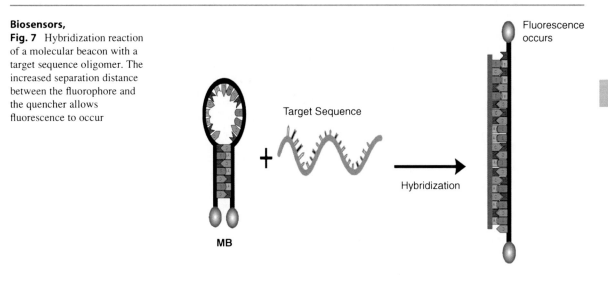

Target Sequence

Hybridization

MB

Fluorescence occurs

bioconjugated receptors

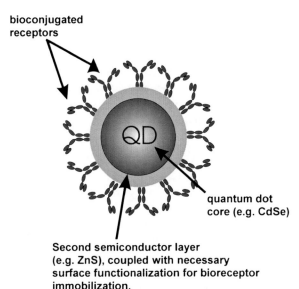

quantum dot core (e.g. CdSe)

Second semiconductor layer (e.g. ZnS), coupled with necessary surface functionalization for bioreceptor immobilization.

Biosensors, Fig. 8 Quantum dot for nanobiosensing applications conjugated with antibodies on the outer surface

based on cadmium selenide QDs bioconjugated with a short DNA strand and hybridized with a AuNP-DNA strand. The development of a single-QD-based nanobiosensor for detection of multiple DNA of HIV-1 and HIV-2 has been developed [10] using the nanobiosensor as a local nanoconcentrator, capable of significantly amplifying both fluorescence signal and FRET signals, with potential use for point of care diagnostics and high-throughput screening. QDs have also been applied in chip-based DNA microarrays for analyzing single nucleotide polymorphism and multi-marker discovery, with anticipated LODs as low as

2nM concentrations of DNA-NP probes. Interestingly, the use of QDs has been discovered in gene technology, where QD-FRET has been utilized in labeling and recognizing target sequences of DNA with red, blue, and green QDs in different combinations. Discrimination and detection of a single base pair mismatch has been reported using a QD-DNA nanobiosensor, potentially leading to very advanced applications in clinical diagnostics of DNA mutants with drug resistance. Chemiluminescence and bioluminescence resonant energy transfer (CRET and BRET, respectively) generate an excited state donor through chemical and biochemical reactions and are believed to be more advantageous than FRET due to lower background from optical excitations [17].

Mass Sensitive Biosensors
Piezoelectric crystals, more commonly referred to as quartz crystal microbalances (QCM) are mass sensitive devices which transform a mass loading at the gold electrode sensor surface, due to adsorbing analytes, into a change in the intrinsic resonance frequency of the QCM. When a low voltage AC signal is applied to thin layer electrodes sandwiching a quartz crystal, the piezoelectric nature of the crystal produces a shear (tangential) deformation causing both surfaces to move in parallel, but in opposite directions, thereby producing acoustic waves which propagate through the bulk of the crystal perpendicular to the surface (Fig. 9). The resonance frequency of the crystal depends on the shear acoustic wave velocity and the thickness of the crystal substrate. The fundamental resonance

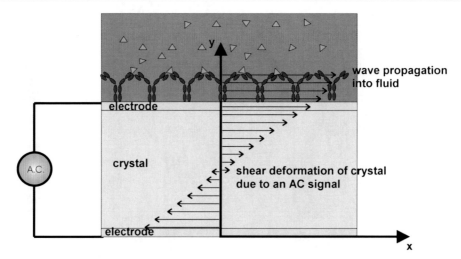

Biosensors, Fig. 9 Illustration of a quartz crystal microbalance biosensor depicting the shear acoustic wave propagating in the bulk of the crystal and into the fluid in contact with one surface. Typical decay lengths of the wave into the fluid are of the order of nanometers, making ligand-complex interactions with surface-immobilized nanoscale bioreceptors and thin film deposition of materials on the surface detectable due to a change in resonance frequency

frequency and odd overtones are the only harmonics which can be excited as predicted using the shear acoustic wave (v) propagation equation. Also, as v is very dependent on the properties of the propagating material, i.e., Young's Modulus and density, thickness shear mode (TSM) resonators are well suited for determination of material-specific parameters, which are difficult to obtain using other biosensing techniques such as SPR.

Able to operate in three modes (i.e., flexural, extensional, and thickness shear), thickness shear mode is the most sensitive with the resonant frequency change being a measure of the thin layer mass attached to the crystal surface. It is also the only mode which can be electrically excited with a low voltage, as the other modes have a coupling factor of zero. In a vacuum, the relation between mass and resonant frequency is considered to be proportional to the added thin layer solid mass on the device surface as described by the Sauerbrey equation (Eq. 5), where f_0 is the resonance frequency of the unperturbed quartz resonator, F_q is the frequency constant of the crystal, ρ_q is the quartz density, and A_{el} is the electrode area:

$$\Delta f_m = -\frac{f_0^2}{F_q \rho_q} \frac{\Delta m_s}{A_{el}} \qquad (5)$$

To take into consideration a QCM in contact with fluids, Kanazawa and Gordon proposed a theoretical

relationship which factors in the material properties of both quartz and a Newtonian fluid as defined in Eq. 6, where μ_q is the shear module in the x-direction, ρ_l is the density of the liquid, and η is the liquid viscosity:

$$\Delta f_l = -\sqrt{f_0^3 \frac{\eta_l \rho_l}{\pi \mu_q \rho_q}} \qquad (6)$$

Suitable oscillator circuits have been designed for the application of these sensors to biological detection in liquid samples, combined with modified surface chemistries of the sensing surface for the immobilization of bioreceptors. QCMs typically operate at frequencies between 1 MHz and 10 MHz, with a frequency-determination limit of 0.1 Hz and a LOD of mass bound to the electrode surface of 10^{-10}–10^{-11} g, subject to calibration of the instrumentation. QCM-based biosensors have been used extensively by researchers for detection of micron-sized bacteria such as *Escherichia coli* and to study the mechanisms involved in surface adhesion and formation of microbial biofilms [18]. The sensitivity of QCMs has been shown to increase with the incorporation of NPs.

It is appropriate to mention here that DNA biosensors, generically known as genosensors, have also been applied to QCMs. These genosensors monitor and detect in real-time DNA hybridization events, through immobilized macromolecules. An AuNP-amplified

QCM DNA biosensor for *E. coli* detection had AuNPs immobilized on the surface of the QCM, increasing the surface area of the sensor and allowing more ssDNA probes to be absorbed on to the surface which then captured the biotinylated *E. coli* DNA. Amplification has been enhanced further using outer-avidin-coated AuNPs giving an LOD as low as 2×10^3 CFU/mL. A similar technique has recently been applied for the specific and rapid detection of avian influenza virus H5N1. A magnetic nanobead amplification-based QCM immunosensor was fabricated, whereby poly-clonal antibodies against the AI H5N1 virus surface antigen hemagglutinin were immobilized on the gold surface through SAMs of 16-mercaptohexadecanoic acid. Capture of the virus on the immobilized anti-bodies resulted in a decrease in resonant frequency of the QCM, which was further amplified by the magnetic nanobeads coated with anti-H5 antibodies bound between antibody and antigen.

The real-time detection of lysozymes on a QCM nanosensor has been achieved using immobilized molecular-imprinted polymers (MIPs) giving a detection limit of 1.2 ng/mL and a high selectivity and sensitivity in a concentration range of 0.2–1,500 µg/mL. They include the use of QCMs in bacteria detection, hemostasis studies; immunoassay studies; carbohydrate binding; oligonucleotide interactions; protein, enzyme kinetic, and cellular studies; and small molecule interactions via MIP and small molecule surface immobilization [18]. The vast majority of these applications had the gold surface of the QCM chemically functionalized through SAMs of both natural and artificial bioreceptor elements which aided in the detection of specific entities, thereby increasing the limit of detection and quantifi-cation significantly in comparison to some optical techniques.

Microarray Biosensors

Microarrays, also known as biochips, are 2D or 3D molecular receptor arrays which allow for the simulta-neous detection of a large number of substances through the interaction of target analytes with sub-strate-bound bioreceptors. There are several categories of microarrays (e.g., DNA, proteins, aptamers, whole-cell) which typically comprise several hundred to sev-eral thousand "spots" or "zones" on which the bioreceptor(s) can be immobilized. The applications of these microarray biosensors are far-reaching and have been used in gene expression, protein-expression profiling, point mutation analysis, immunodiagnostics, drug screening, cell analysis, and DNA sequencing to name a few. A brief description of the most common microarray technologies to date is provided.

DNA microarrays consist of immobilized cDNA ($<$ 5,000 b.p) or oligonucleotides ($<$ 80 b.p) which interrogate target analytes (e.g., DNA) in sample solu-tions. In principle, the assays developed largely depend on the specific molecular recognition between comple-mentary sequences of nucleobases. Depending on the method of detection or "microarray reading," bioreceptors have usually been immobilized on glass/silicon substrates, metal-coated (i.e., ITO, gold) glass substrates, or polymeric substrates. Companies (i.e., Affymetrix) have developed precision equipment for high-density "spotting" of bioreceptors on functionalized substrates for rapid screening studies, along with appropriate microarray reading software. The most widely used detection method for DNA microarrays is fluorescence detection. Target mole-cules are tagged with a fluorophore and upon hybridi-zation with the bioprobe, the spot or zone will fluoresce. Other forms of detection for DNA microarrays include chemiluminescence/electroche-miluminescence. DNA microarray technologies have been used to map the genome sequence of multiple organisms including humans and yeast.

Protein microarrays can generally be classed as either analytical/diagnostic or functional microarrays. They were developed upon completion of genome mapping, leading to the prediction of open reading frames (ORF) from genomes making it possible to express the full or partial proteome with large-scale cloning and gene expression methods. An analytical form of these microarrays is the antibody microarray which forms complexes with the target protein or anti-gen, and offers tremendous potential in clinical research and diagnosis. The functional form of protein microarrays which possess high-density proteins or peptides can represent complete or partial proteome of an organism. They are mainly used in studies involving protein–protein, protein–DNA, and pro-tein–small molecule interactions, also offering wide-ranging applications in clinical cancer studies, viral infections, and autoimmune disease studies [19].

A modified analogue to DNA is peptide nucleic acids (PNA), which possesses a neutral peptide backbone (polyamide) instead of a negatively charged phosphate-sugar backbone. In comparison to DNA, PNA exhibits high chemical stability, resistance to enzymatic degradation, the ability to recognize specific nucleic acid sequences, strand invasion capabilities, high thermal stability, and ionic strength and the ability to hybridize with nucleic acids. Since their discovery in 1991, researchers have shown PNA-based biosensors to be extremely versatile in a range of systems including electrochemical, mass, and SPR-based on its unique biophysical properties. For example, μ-patterning of DNA on to microarrays has been achieved using electrohydrodynamics (EHD). DNA was found to be affected by thermal and electrical properties upon immobilization, but the use of PNA in the patterning process did not affect its function or structural integrity.

Future Trends

Biosensors have undergone a rapid transformation over the past half century. It continually remains in a state of flux with innovative applications of nanotechnology for even faster, cheaper, and smaller biosensing systems for areas as diverse as personalized healthcare, point of care (POC) diagnostics, and rapid biodetection of pathogens in bioterrorism. These are being realized as microfabrication and miniaturization of electronic and transduction elements become incorporated into systems known collectively as lab-on-a-chip (LOC) systems or micro-total analysis systems (μ-TAS). These systems rely on the complete processing and transportation of biological samples on a single microchip through micro- and nanofluidic channels, from sample preparation to sample analysis through advanced biosensing techniques. Novel material properties, the development and integration of particle manipulation techniques, and increasing processing power will drive the next generation of biosensing nanosystems.

Graphene, a carbon allotrope isolated from graphite, has been reported as a new alternative to CNTs and metal alloy NPs ever since its discovery in 2004 [20]. Advances in graphene-based biosensors show significant advantages over CNT and metal-NPs. Purported problems associated with CNTs and metal-NPs, such as inconsistent signal amplification and metal impurities interfering with redox reaction with the

biomolecules are thought to be avoided with the use of graphene in electrochemical nanobiosensor research. There is still a need of bioconjugation protocol development for bioreceptor immobilization on graphene before the uptake of electrochemical graphene-based biosensors becomes more commonplace.

Particle manipulation through AC electrokinetic phenomena, including electrowetting deposition (EWOD), *dielectrophoresis* (DEP), and *AC electroosmosis* (ACEO) have shown tremendous potential in bionanotechnology. They have shown simplicity in integration with various biosensing transducers, and coupled with high specificity and sensitivity for particle manipulation in a range of media solutions, the rate of detection in biosensor systems is significantly greater than conventional methods (i.e., adsorption and electrostatic attraction) which rely heavily on bringing the target analyte in contact to the bioreceptor for high-quality sensing.

Cross-References

▶ AC Electrokinetics of Nanoparticles
▶ Arthropod Strain Sensors
▶ Carbon Nanotubes
▶ Dielectrophoresis
▶ Electrospinning
▶ Graphene
▶ Nanoparticles
▶ Nanostructured Functionalized Surfaces
▶ Optical Properties of Metal Nanoparticles
▶ Quantum-Dot Toxicity
▶ Self-assembly of Nanostructures
▶ Sol-Gel Method
▶ Surface Plasmon-Polariton-Based Detectors
▶ Synthetic Biology

References

1. Clark, L.C., Jr., Lyons, C.: Electrode systems for continuous monitoring in cardio vascular surgery. Ann. N.Y Acad. Sci. **102**, 29–45 (1962)
2. Updike, S.J., Hicks, G.P.: The enzyme electrode. Nature **214**, 986–988 (1967)
3. Blum, L., Marquette, C.: Biosensors. From the glucose electrode to the biochip. In: Boisseau, P., Houdy, P.,

Lahmani, M. (eds.) NanoScience – Nanobiotechnology and Nanobiology, pp. 871–909. Springer, New York (2010)

4. Kress-Rogers, E.: Handbook of Biosensors and Electronic Noses. Taylor and Francis/CRC Press, Boca Raton (1996)

5. Su, L., Jia, W., Hou, C., Lei, Y.: Microbial biosensors: a review. Biosens. Bioelectron. **26**, 1788–1799 (2011)

6. Haupt, K.: Molecularly imprinted polymers in analytical chemistry. Analyst **126**, 747–756 (2001)

7. Jerónimo, P.C.A., Araújo, A.N., Montenegro, M.C.B.S.M.: Optical sensors and biosensors based on sol-gel films. Talanta **72**, 13–27 (2007)

8. Siqueira, J.R., Jr., Caseli, L., Crespilho, F.N., Zucolotto, V., Oliveira, O.N., Jr.: Immobilization of biomolecules on nanostructured films for biosensing. Biosens. Bioelectron. **25**, 1254–1263 (2010)

9. Vashist, S.K., Zheng, D., Al-Rubeaan, K., Luong, J.H.T., Sheu, F.-S.: Advances in carbon nanotube based electrochemical sensors for bioanalytical applications. Biotech. Adv. **29**, 169–188 (2011)

10. Dolatabadi, J.E.N., Maschinchian, O., Ayoubi, B., Jamali, A.A., Mobed, A., Losic, D., Omidi, Y., Guardia, M.: Optical and electrochemical DNA nanobiosensors. Trends. Anal. Chem. **30**(3), 459–472 (2011)

11. Xia, L., Wei, Z., Wan, M.: Conducting polymer nanostructures and their application in biosensors. J. Colloid Interf. Sci. **341**, 1–11 (2010)

12. Teles, F.R.R., Fonseca, L.P.: Applications of polymers for biomolecule immobilization in electrochemical biosensors. Mat. Sci. Eng. C-Bio. S **28**, 1530–1543 (2008)

13. Abbas, A., Linman, M.J., Cheng, Q.: New trends in instrumental design for surface plasmon resonance-based biosensors. Biosens. Bioelectron. **26**, 1815–1824 (2011)

14. Wijaya, E., Lenaerts, C., Maricot, S., Hastanin, J., Habraken, S., Vilcot, J.-P., Boukherroub, R., Szunerits, S.: Surface plasmon resonance-based biosensors: from the development of SPR structures to novel surface functionalization strategies. Curr. Opin. Solid. St. Mat. **15**(5), 208–224 (2011)

15. Scarano, S., Mascini, M., Turner, A.P.F., Minunni, M.: Surface plasmon resonance imaging for affinity based biosensors. Biosens. Bioelectron. **25**, 957–966 (2010)

16. Ibraheem, A., Campbell, R.E.: Designs and applications of fluorescent protein based biosensors. Curr. Opin. Chem. Biol. **14**, 30–36 (2010)

17. Algar, W.R., Tavares, A.J., Krull, U.J.: Beyond labels: a review of the applications of quantum dots as integrated components of assays, bioprobes, and biosensors utilizing optical transduction. Anal. Chim. Acta **673**, 1–25 (2010)

18. Becker, B., Cooper, M.A.: A survey of the 2006–2009 quartz crystal microbalance biosensor literature. J. Mol. Recognit. **24**(5), 754–787 (2011)

19. Chen, H., Jiang, C.J., Yu, C., Zhang, S., Liu, B., Kong, J.: Protein chips and nanomaterials for application in tumor marker immunoassays. Biosens. Bioelectron. **24**, 3399–3411 (2009)

20. Kuila, T., Bose, S., Khanra, P., Mishra, A.K., Kim, N.H., Lee, J.H.: Recent advances in graphene-based biosensors. Biosens. Bioelectron. **26**(12), 4637–4648 (2011)

Bond-Order Potential

▶ Reactive Empirical Bond-Order Potentials

Bone Remodeling

J. A. Sanz-Herrera and E. Reina-Romo
School of Engineering, University of Seville, Seville, Spain

Synonyms

Bone turnover

Definition

Bone remodeling is a dynamic physiological process in which the combined effect of bone formation and bone resorption occurs at a specific location of the bone architecture to enable bones to adapt to mechanical stresses, to repair its microstructure and thus maintain the mechanical integrity of the skeleton, and to maintain mineral homeostasis. Bone remodeling consists of two main subprocesses, bone resorption by osteclasts followed by bone formation by osteoblasts without causing large changes in bone quantity, geometry, or size. An imbalance in the regulation of bone remodeling's two subprocesses results in many metabolic bone diseases, such as osteoporosis.

Overview

Bone is a metabolically active tissue capable of adapting its structure to mechanical stimuli and repairing structural damage. It has the capability of forming new osseous tissue at locations that are damaged or missing, such as in fracture healing, distraction osteogenesis or bone implants interface, or removing bone matrix in case of disuse. (Distraction osteogenesis: surgical process in which new bone formation is induced through gradual traction in an osteotomy site to reconstruct skeletal deformities or lengthen bones.) This calcified, vascular, connective tissue has many functions, including support, protection, mineral storage, blood cell production, hematopoiesis, and locomotion through the attachment of muscles.

There are three basic mechanisms involved in the turnover and development of bone: the longitudinal growth, modeling, and remodeling. The former, the longitudinal growth, ceases with closure of the epiphyseal plates after the growing period. The second, modeling, serves to alter the bone that is present and to determine its geometry and size [1]. Bone modeling at any surface involves an uncoupled activity of osteoclast activation (A) followed either by formation (A-F) or resorption (A-R). (Osteoclast: specialized cells capable of bone resorption.) The latter, bone remodeling, starts with fetal osteogenesis and continues throughout life. It involves the coupled action of the osteoclastic and osteoblastic activities at localized areas on the bone surface (A-R-F) and differs from bone modeling, in which these processes occur independently at distinct anatomical locations (Osteoblast: bone forming cells). By the process of remodeling, bone is continuously resorbed and rebuilt at about 1–2 million microscopic sites per adult skeleton by teams of cells or remodeling units. These remodeling units are structures which persist for a variable but unknown period of time, traveling through the bone structure to carry out the functions of bone turnover in an ordered and predictable manner. Osteoclast–osteoblast interaction is needed for coupling bone resorption and formation both temporarily and spatially [1, 2]. Temporarily, since in each remodeling cycle, there occurs a succession of cellular events in the same place but at different times. Spatially, since in each remodeling unit, the same events are occurring in different places but at the same time.

Biological Principles

Bone Structure
At the structural level, mature bone can be classified into compact/cortical bone and cancellous/trabecular bone. *Cancellous bone* has large, open spaces surrounded by thin, interconnected plates of bone. The large spaces contain red bone marrow where hematopoiesis occurs and the plates of bone are trabeculae composed of several layers of lamellae. By contrast, *compact bone* is much denser than cancellous bone, with spaces reduced in size and lamellar

organization more precise and thicker. The typical range of apparent density of cancellous bone is 1.0–1.4 g/cm^3 compared to about 1.8–2.0 g/cm^3 for compact bone [3]. (Apparent density: mass per unit volume of a region of bulk bone, which includes Haversian canals, marrow spaces, and other soft tissue spaces.) Cancellous bone can be found in the epiphysis or heads of long bones while cortical bone is found in the diaphysis or shaft of the bone and the outer layer of the trabecular bone. In the entire skeleton around 80% of the total skeletal mass is composed of cortical bone and only 20% is cancellous bone and the volume fraction (including internal pores like lacunae and canaliculi) is 0.8 mm^3/mm^3 and 0.2 mm^3/mm^3, respectively.

At the tissue level, bone may be divided into two wide categories:

- *Woven bone/primary bone*: This type of bone is laid down rapidly as a disorganized arrangement of collagen fibers and osteocytes. In adult skeleton, this young regenerated bone can be found in a fracture callus, during distraction osteogenesis and in areas undergoing active endochondral ossification.

- *Lamellar bone/secondary bone*. This mature bone can exist in cortical and trabecular bone. Unlike woven bone, this tissue is well organized and regular.

The Remodeling Cells
The remodeling process requires the synchronized activities of molecular mechanisms and multiple cellular participants, such as osteoblasts, osteoclasts and osteocytes to ensure that bone formation and bone resorption occur sequentially.

- *Osteoclasts*: Are multinucleated cells derived from hematopoietic cells formed by the fusion of mononuclear progenitors of the monocyte/macrophage family. They are the principal resorptive cells of bone by secretion of acid and proteases.

- *Osteoblasts*: Are mononuclear bone cells of mesenchymal origin that have several important roles in bone remodeling: expression of osteoclastogenic factors, production of bone matrix proteins (osteoid) and bone mineralization. It represents a heterogeneous family of cells, including immature osteoblast lineage cells, and differentiating and mature matrix-producing osteoblasts.

- *Osteocytes*: Terminally differentiated osteoblasts embedded in the mineralized collagen matrix that

Bone Remodeling,
Fig. 1 Cross-sectional diagram of a BMU in cortical bone. Osteoclasts resorb the bone matrix at the front while osteoblasts lay down osteoid toward the back to refill the cavity. The central capillary provides a supply of precursor cells, as well as various nutrients [5]

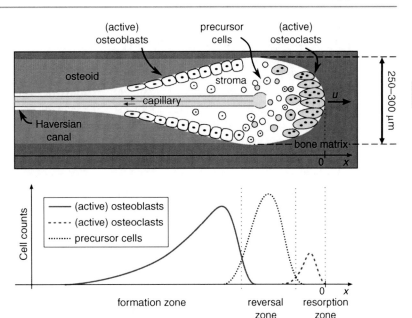

The Remodeling Unit

are likely to determine which bone surface osteoclasts will resorb. These cells are the most abundant cells in bone, accounting for 90–95% of all bone cells. They may act as sensors for the chemical and mechanical signals, such as microcracks and loss of mechanical loading. Osteocytes communicate with each other and with cells on the bone surface, via the dendritic processes encased in canaliculi (Canaliculi: microscopic canals between the various lacunae of ossified bone).

The Remodeling Unit

The remodeling process is not performed individually by each cell but by groups of cells acting as organized units. Clusters of osteoclasts and osteoblasts are arranged within temporary anatomical structures known as "basic multicellular units" (BMUs). A BMU consists of around 20 osteoclasts and several hundred osteoblasts. Active BMUs consists of a leading front or "head" of bone resorbing osteoclasts and a tail portion occupied by osteoblasts (Fig. 1). The "head" contains a capillary bud to supply nutrients and probably to supply the progenitor cells for osteoclasts and osteoblasts. BMUs are not permanent structures since they form in response to a specific signal, perform their function and die, leaving a few residual lining cells and osteocytes. The number of BMUs working in a specified volume of tissue depends on the "activation frequency" and the "sigma period"

[4]. The former refers to the frequency with which new remodeling cycles are created per unit cross-sectional area of arbitrary thickness and thus is a measure of the rate of bone turnover. The latter refers to the speed with which the BMU travels through the tissue space. It quantifies the time of completion of one remodeling cycle at a given location in the bone. Sigma periods are frequently subdivided into resorption and formation periods. Formation is much slower than resorption. In humans, the total remodeling cycle is about 4 months, about 3 weeks for resorption, and about 3 months for rebuilding the lost bone.

The Remodeling Cycle

There are three principal stages in the BMU lifetime: activation of osteoclasts, resorption of the old bone, and formation of new bone matrix (A-R-F) (Fig. 1). The spatial and temporal arrangements of the cells within the BMUs ensure a coordination between the different phases of the remodeling process. The A-R-F process may be further divided into six separated phases: activation, resorption, reversal, formation, mineralization, and quiescence (see Fig. 1):

- *Activation phase*: Activation takes place when an activation signal or stimulus, including chemical, mechanical, and electrical signals, converts a previously inactive bone surface into a remodeling one. It is important to distinguish between origination and activation. The former occurs only once for each BMU at a quiescent surface while the latter is a continuous process that occurs at the cutting edge of the BMU. As the lifespan of the individual cells in a BMU is much shorter than that of the BMU, activation involves gathering of the cells that will form the BMU. In both cortical and cancellous bones, osteoclasts participate in activation or recruitment. In addition, bone lining cells of osteoblastic nature are available on the cancellous surfaces. Systemic hormones, growth factors, and interleukins play a role at this step, helping to recruit new osteoclasts by enlarging the precursor pool.

- *Resorption phase*: Bone resorption is a multistep process initiated by the proliferation of immature osteoclast precursors, the commitment of these cells to the osteoclast phenotype, and finally the degradation of the organic and inorganic phases of bone by the mature resorptive cells. It involves the following steps:
  - Attachment of the osteoclasts to the target matrix: Newly created osteoclasts attach themselves to a bone surface and form a ruffled membrane.
  - Acidification of resorption lacunae. Osteoclasts acidify the extracellular microenvironment. For instance, they secrete both hydrochloric acid, to resorb hydroxyapatite, and proteases like cathepsin K to degrade collagen and other bone matrix proteins.
  - Bone demineralization. The acidic milieu mobilizes the mineral phase of bone and provides an optimal environment for organic matrix degradation.
  - Dissolution of the bone tissue forming a resorption lacuna.
  - The products of bone degradation are endocytosed by the osteoclast, transported, and finally released to the extracellular space.
  - Detachment of osteoclasts and movement to a new site of bone degradation.
- *Reversal phase*: The transition from osteoclastic to osteoblastic activity takes several days (around 9 days) and results in a cylindrical space lying between the resorptive region and the refilling region [1]. The length of this region may vary considerably, depending on the lag between the resorptive and formative phases.

- *Formation phase*: In this step, the osteoblasts initially synthesize the organic matrix called osteoid to restore the previously resorbed bone. The apposition rate of matrix and of mineral is more rapid initially, with an initial rate of 1–2 µm/day which gradually decreases to zero as the cavity is filled. As an individual ages, the amount of bone deposited in the resorption lacunae by osteoblasts is less than that previously removed by osteoclasts. This negative balance of osteoblastic relative to osteoclastic activity is responsible for the loss of bone that occurs with age and that when pronounced leads to osteoporosis.
- *Mineralization phase*: Following deposition of the unmineralized bone matrix, called osteoid, mineral is deposited within and between the collagen fibers. This process is delayed by a period of time known as the mineralization lag time, which is normally about 15 days. Once begun, approximately 60% of the mineralization of the osteoid occurs during the first few days [4]. This is called primary mineralization. The remainder of the mineral is added at a decreasing rate for around 6 months during the secondary mineralization.
- *Quiescence*: Once the resorption and formation processes are completed, the osteoclasts disappear, and the osteoblasts become osteocytes or Haversian canal lining cells, or they disappear.

Cortical Versus Trabecular Remodeling

Bone remodeling has been well described in cortical [1] and cancellous bone [6]. Current evidence indicates that the differences in the remodeling activities of these two types of bone reside in the different mechanical environments of the bone cells in cortical or cancellous bone.

For instance, in *cortical remodeling*, since BMUs become isolated within the cortex and osteoclasts and osteoblasts need nourishment, the tunnel cannot be entirely refilled during the formation phase. Each BMU leaves a central passageway in the bone to support the metabolism of the BMU and bone matrix osteocytes, and to carry calcium and phosphorous to and from the bone when necessary. This passageway is called Haversian canal and is typically 40–50 µm in diameter in humans and contains two capillaries: a "supply" and a "return" vessel. In human adults, osteonal BMUs replace about 5% of compact bone each year [3].

In *trabecular and endocortical surfaces* since BMUs work on trabeculae surfaces, the cells do not dig and refill tunnels but trenches (half-tunnels). Because a BMU is about the same diameter as a trabecula, the remodeling unit in cancellous bone is equivalent to half of a cortical remodeling unit. BMUs replace human adult trabeculae bone at a higher rate than cortical bone, about 25% each year [3] since the surface/volume relationships are much greater in the cancellous bone. Another difference is that cancellous bone BMUs proceed along a surface lined with lining cells, which could participate in remodeling, whereas there are no lining bone cells within the cortex. Cancellous bone has also more ready access to marrow cells than cortical bone.

Key Research Findings

Bone remodeling is the final phenomenological consequence, observable at the bone microarchitectural level, of a cascade of biological events which take place at different spatial scales ranging from molecular to cell/tissue levels. The nature of the phenomena involved in the process of bone remodeling is diverse including molecular mechanisms due to genetic factors or interaction between cells, where mechanics plays an important role. The importance of the former (genetics) versus the latter (mechanics) has been widely discussed during the last decades. Specifically, a number of cytokines, receptors, and proteins have been reported to have an important role in bone remodeling. On the other hand, it has been evidenced that mechanical factors such as the interstitial fluid flow, dynamic loading, or the presence of microcracks influence bone remodeling. Both mechanics and genetics are coupled effects, such that mechanics or genetics themselves cannot explain the phenomenon of bone remodeling [7]. In fact, it may be assumed that mechanical forces acting at the tissue level induce a cascade of molecular mechanisms activated at a lower observation (protein) scale. Then, molecular mechanisms provide a cue transcripted at the cell level which develops a specific function. This hypothesis is termed *mechanotransduction*.

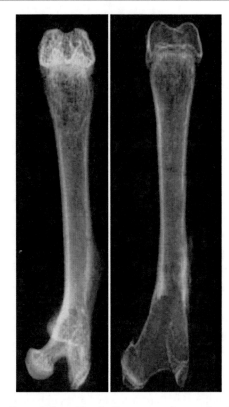

Bone Remodeling, Fig. 2 Two mouse femurs showing the isolated effect of genetics (*right*) and the combined effect of genetics and mechanics (*left*) during growth (From Chalmers and Ray [8])

The effect of genetics in contrast to mechanics was highlighted in an animal experiment presented in Ref. [8]. The experiment analyzed bone growth rather than bone remodeling and proceeded as follows. The cartilaginous primordia of the femur (the cartilage anlage where femur develops from) of a developing mouse was isolated to analyze the effect of the mechanical loading during growth. It was carried out transplanting this tissue into the mouse's spleen site, where mechanical loads are absent, during growth. As a result, the transplanted femur in the spleen, where no mechanical loading was developed, was compared after growth to a normal physiological one, which was subjected to a normal mechanical environment. Figure 2 shows the appearance of both femurs with and without a normal loading environment during growth. It can be observed the lack of bone tissue of the unloaded femur in terms of lower bone mass distribution and organization, thinner cortical wall, undeveloped femoral neck, and curvature of the femoral diaphysis [9]. Even though bone growth during development is a different process than bone remodeling, this experiment illustrates the importance of both genetics and mechanics, and genetics itself, in bone tissue function and form.

In the following sections, some key research findings related to the cellular and molecular mechanisms as well as the mechanical factors involved in bone remodeling are discussed.

Cellular and Molecular Mechanisms Involved in Bone Remodeling

Bone remodeling is the coordinated action of BMUs, as previously introduced. First, a signal, probably coming from a mechanical cue or hormone, activates bone cells in the so-called activation phase. It is thought that osteocytes are the cells involved in the transduction of the above-mentioned mechanical stimulus into a biological signal. In this phase, PTH, a calciotropic hormone secreted by the parathyroid glands, binds to the seven-transmembrane G-protein-coupled receptor (called the PTH receptor) of osteoblastic cells. It activates protein kinase A, protein kinase C, and calcium intracellular signaling pathways of osteoblastic cells, which induces a wave of transcriptional response that produces a secretion of molecules that recruit osteoclast precursors, induce osteoclast differentiation and activation, and establish *bone resorption* [10]. Then, in response to PTH, osteoblasts produce the chemokine MCP-1 (monocyte chemoattractant protein-1), which is a chemoattractant for osteoblasts precursors [10]. In addition, during the resorption phase, cytokines CSF-1, RANKL, and OPG are modulated in response to PTH to activate and maintain osteoclasts activity. Specifically, these cytokines promote osteoclasts proliferation, spreading, motility, cytoskeletal organization, and coordinate differentiation from osteoclasts precursors to multinucleated osteoclasts. Then, bone organic matrix is degraded and dissolved by matrix metalloproteinases and a collection of collagenolytic enzymes with a low pH [10]. At the *reversal phase*, a cell from the osteoblast lineage removes collagen remnants from the bone surface and prepares it for subsequent osteoblast-mediated bone formation. In the *formation phase*, mechanical strain on bone and PTH signaling, via PTH receptors on osteocytes, activate a signaling pathway which determines the early stage of bone deposition. Moreover, mesenchymal stem cells or early osteoblast progenitors differentiate

and secrete molecules that ultimately form replacement bone [10]. The molecules include collagen I proteins and non-collagenous ones, such as proteoclycans, glycosylated proteins, and lipids. During the *mineralization phase*, once an equal quantity of resorbed bone has been formed, an activation signal informs the remodeling cycle to cease.

Many of the above summarized processes which take place during different phases of bone remodeling are under research and some of them are still poorly understood. However, it has been briefly evidenced that many times a mechanical signal is involved in the related molecular and cellular mechanisms present in bone remodeling. In the next section, a discussion of the effects of mechanical signals on bone remodeling is introduced from a cell/tissue perspective.

Mechanics in Bone Remodeling

The important role that different mechanical signals have in the process of bone remodeling has been evidenced in recent years. It is thought that mechanical forces, transferred by muscle and bones, interact at the protein level allowing a transcription into biochemical signals which influences cell function. Some of the mechanisms of the transcription of a mechanical cue into a biochemical signal involved at the molecular level in bone remodeling have been exposed in the previous section. This process, termed mechanotransduction, remains to be elucidated.

A phenomenological consideration of the effect of mechanics in bone remodeling aims to establish a cause-effect statement of the problem. In this sense, "cause" would be a certain mechanical factor and "effect" would be its specific consequence on the continuous formation/resorption of bone, i.e., bone remodeling, observable at the macroscopic level of the bone architecture. In bone remodeling, osteocytes have been traditionally considered as a candidate for bone mechanosensing [7]. Cowin and coworkers have reported that osteocytes sense a range of strain levels exerted at different frequencies at the bone tissue, by means of the fluid flow pressure in the lacunar-canalicular porosity system [11]. Specifically, it has been suggested that strain, shear stress, or fluid flow give rise to changes in the ion channel conductance in the osteocytic membrane, which releases intracellular Ca^{2+} that regulates the opening and closing of membrane ion channels in the communicating junctions. Therefore, the gap junctions modulate the intracellular potential and current that passes through the network of interconnected osteocytes and surface osteoblasts which induce their function [11].

As discussed above, both mechanical load (strain), frequency and fluid flow within the bone pores are essential stimulus for bone adaptation. On one hand, it has been shown that application of load frequency stimulates bone formation when load is applied over 0.5 Hz. Specifically, Hsieh and Turner [12] reported more bone formation in a rat forelimb when application load was 10 cycle/s versus 1 cycle/s. Moreover, small loads were applied to the sheep hindlimbs at 30 cycle/s [13] showing encouraging results. As a conclusion, experimental results suggest that bone formation is proportional to the product of frequency times strain amplitude [11, 14]. It has been proved that a certain strain level at a given frequency may lay down within the physiological regime which stimulates osteocytes' mechanosensory system [11]. On the other hand, interstitial fluid flow circulates through the bone pores, canaliculi, and canals associated to the different porosity levels of bone. Interstitial fluid flow has been postulated as another important stimulus for bone formation due to the transport of nutrients to individual cells buried within the matrix. Moreover, the indirect shear stress and hydrostatic pressure produced by the fluid circulation may induce the necessary stimulus to activate the process of bone remodeling. Different sources to induce fluid movement are gravity forces, capillary forces (i.e., vasculature), hydrostatic pressure-induced flow of blood through vasculature, or forces exerted by muscles and bones and transmitted to the bone fluid due to the poroelastic architecture of the bone tissue.

A consequence of the effect of mechanics on bone formation and remodeling is the loss of bone mass in the absence of activity or bone disuse. It has been widely reported that astronauts, under the effect of microgravity and hence low mechanical stimulation of bones, undergo a rate of bone mass loss of the skeleton approximately of 1% per month. In this context, LeBlanc et al. [15] reported the measurements made from 1990 to 1995 on 18 cosmonauts involving 12 space missions measured before and after missions lasting from 4 to 14.4 months. Eighteen crew members received a lumbar spine and hip scan before and after flight; 17 crew members received an additional whole body scan. Results, shown in Table 1, were expressed

Bone Remodeling, Table 1 Bone mineral density and body composition changes ($p < 0.01$) significantly different from baseline, after 4–14.4 months of space flight

Variable	N	%/Month change	SD
Lumbar spine	18	−1.06	0.63
Femoral neck	18	−1.15	0.84
Trochanter	18	−1.56	0.99
Total body	17	−0.35	0.25
Pelvis	17	−1.35	0.54
Arm	17	−0.04	0.88
Leg	16	−0.34	0.33

Adapted from [15]
N number of samples, *SD* standard deviation

as percent change from baseline per month of flight in order to account for the different flight times.

Moreover, another evidence of the importance of the mechanical forces in bone remodeling is the fact that the skull may actually gain bone due to the increased fluid pressure owing to the caudal shift of fluids in microgravity [7].

Another interesting mechanical cue for bone remodeling is the presence of microcracks along the bone tissue. Microcracks appear as a consequence of accumulated microdamage due to repetitive loading. Microdamage is even produced at physiological levels of strain and has been reported to be a mechanical signal to initiate the process of tissue repair during bone remodeling [16]. On the other hand, under severe or extreme amplitude of the loading, for example athletes during competition, microcracks may eventually produce failure or breaking of the bone organ.

Future Directions

Understanding bone remodeling as well as the implicated mechanisms in such a process may provide insight in bone function and related bone diseases. Specifically, once the mechanical and molecular mechanisms involved are elucidated, one may control bone formation or resorption by the design of drugs, biomaterials or actuators which interact with the bone tissue. To this end, how forces are transferred from muscles and skeleton to the protein level, and how these forces induce a cascade of biochemical signals,

Bone Remodeling,
Fig. 3 Simulation in silico of
bone remodeling of a white
New Zealand rabbit femur:
macroscopic bone density
[g/cc] distribution. (**a**, **b**, **d**, **f**)
Simulated until convergence.
(**c**, **e**, **g**) Actual tomographies
for different sections
(From [22])

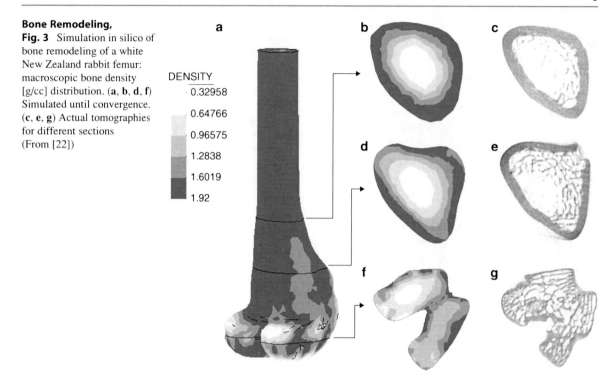

DENSITY

0.32958
0.64766
0.96575
1.2838
1.6019
1.92

i.e., mechanotransduction, is a keystone which remains to be investigated.

A proper knowledge of bone remodeling mechanisms may add some light for the explanation of the process, and hopefully may provide a clinical solution for diseases involving abnormalities in bone remodeling such as osteoporosis and osteosclerosis. In one hand, osteoporosis, the most common bone disease, is a remodeling disease in which patients have a low bone mass condition with a high risk of fracture. This disease reflects an imbalance in the skeletal turnover so that bone resorption exceeds bone formation. On the other hand, patients with osteosclerosis have an increased bone mass due to stimulated osteoblast activity. These bone diseases and their consequences are major public health problems. Their study may benefit advances in the field of bone remodeling such as bone healing, distraction osteogenesis, or osteointegration processes around implants, among many other orthopedic protocols and techniques. Mathematical models have been developed for the analysis of bone remodeling. Currently, this is an intense area of research carried out by engineers, mathematicians, and physicists. The main objective of mathematical modeling is to predict

information for a number of applications with a clinical or research interest to reduce animal experimentation, reduce cost and time, improve protocols, and aid the clinician to take decisions related to diseases in bone remodeling. Specially, many models have focused on the analysis of mechanics in bone remodeling from a continuum approach. The input of these models is normally a so-called mechanical stimulus, namely, fluid flow velocity, shear stress, pressure, or a certain combination of the components of the stress or strain tensors. Normally, models assume a higher amount of deposited bone in places where the mechanical stimulus is higher. Continuum models may be classified into phenomenological and mechanistic ones. In one hand, phenomenological models aim at predicting how macroscopic forces induce bone function and form during bone remodeling [17–21]. Many phenomenological parameters have to be calibrated in this kind of models whereas results should be analyzed in qualitative terms. Figure 3 shows a numerical simulation of bone remodeling and how it compares to actual clinical results [22]. Mechanistic models establish a link between mechanical forces and the implicated biological functions to explain certain features of bone remodeling [23–25].

Mechanistic approaches are usually more complex since they analyze the coupled effect of mechanics and biology in bone remodeling. In the recent years, the term in silico approach has been coined to classify mathematical models into the different scientific methodologies present in the clinical practice such as bone remodeling.

Recently, the use of biomaterials to enhance or improve the function of a damaged tissue, or to promote tissue growth, has been proposed. This new discipline is named *Tissue Engineering* and offers many applications in the case of bone tissue. Specifically, new biomaterials (ceramic-based, bioactive glasses, polymers) are used as a scaffold which interacts with the bone tissue, with the purpose that cells colonize the scaffold biomaterial, attach, and develop their specific function finally leading to the growth of new bone tissue. Several requirements must be fulfilled by the scaffold biomaterial to develop its proper function. First, it must be a biocompatible biomaterial which interacts with the bone tissue. Moreover, it must be strong enough to support physiological loads at its specific location allowing cell attachment and fluid circulation in the interior. Mechanics is also an important factor in bone tissue engineering, as the signaling pathways are similar to the bone tissue but here the extracellular matrix is the biomaterial itself. The gold standard in bone tissue engineering is to promote the growth of functional bone tissue in circumstances where nature has lost its ability to develop new tissue under a disease or age-related complications.

References

1. Frost, H.M.: Tetracycline-based histological analysis of bone remodeling. Calcif. Tissue Res. **3**, 211–237 (1969)
2. Parfitt, A.M.: The actions of parathyroid hormone on bone: relation to bone remodeling and turnover, calcium homeostasis, and metabolic bone disease. Part I of IV parts: mechanisms of calcium transfer between blood and bone and their cellular basis: morphological and kinetic approaches to bone turnover. Metabol. Clin. Exp. **25**, 809–844 (1976)
3. Martin, R.B., Burr, D.B., Sharkey, N.A.: *Skeletal Tissue Mechanics* (Springer, New York, 1998)
4. Parfitt, A.M.: The physiologic and clinical significance of bone histomorphometric data, in *Bone Histomorphometry: Techniques and Interpretations*, ed. by R.R. Recker (CRC Press, Boca Raton, 1983)
5. Buenzli, P.R., Pivonka, P., Smith, D.W.: Spatio-temporal structure of cell distribution in cortical bone multicellular units: a mathematical model. Bone **48**, 918–926 (2011)
6. Eriksen, E.F., Gundersen, H.J., Melsen, F., Mosekilde, L.: Reconstruction of the formative site in iliac trabecular bone in 20 normal individuals employing a kinetic model for matrix and mineral apposition. Metab. Bone Dis. Relat. Res. **5**, 243–252 (1984)
7. Cowin, S.C.: Tissue growth and remodeling. Annu. Rev. Biomed. Eng. **6**, 77–107 (2004)
8. Chalmers, J., Ray, R.D.: The growth of transplanted foetal bones in different immunological environments. J. Bone Joint Surg. **44B**, 149–164 (1962)
9. Goodship, A.E., Cunningham, J.L.: Pathophysiology of functional adaptation of bone in remodeling and repair in-vivo, in *Bone Mechanics Handbook*, ed. by S.C. Cowin (CRC Press, Boca Raton, 2001), pp. 26-1–26-31
10. Raggatt, L.J., Partridge, N.C.: Cellular and molecular mechanisms of bone remodeling. J. Biol. Chem. **286**, 25103–25108 (2010)
11. Weinbaum, S., Cowin, S.C., Zeng, Y.: A model for the excitation of osteocytes by mechanical loading-induced bone fluid shear stresses. J. Biomech. **27**, 339–360 (1994)
12. Hsieh, Y.F., Turner, C.H.: Effects of loading frequency on mechanically induced bone formation. J. Bone Miner. Res. **16**, 918–924 (2001)
13. Rubin, C., Turner, A.S., Bain, S., Mallinckrodt, C., McLeod, K.: Anabolism. Low mechanical signals strengthen long bones. Nature **412**, 603–604 (2001)
14. Robling, A.G., Castillo, A.B., Turner, C.H.: Biomechanical and molecular regulation of bone remodeling. Annu. Rev. Biomed. Eng. **8**, 455–498 (2006)
15. LeBlanc, A., Schneider, V., Shackelford, L., West, S., Oganov, V., Bakulin, A., Voronin, L.: Bone mineral and lean tissue loss after long duration space flight. J. Musculoskelet. Neuronal. Interact. **1**, 157–160 (2000)
16. Bentolila, V., Boyce, T.M., Fyhrie, D.P., Drumb, R., Skerry, T.M., Schaffler, M.B.: Intracortical remodeling in adult rat long bones after fatigue loading. Bone **23**, 271–281 (1998)
17. Pauwels, F.: Eine neue theorie über den einflub mechanischer reize auf die differenzierung der stützgewebe. Z. Anat. Entwicklungsgeschichte **121**, 478–515 (1960)
18. Cowin, S.C., Hegedus, D.H.: Bone remodeling I: a theory of adaptive elasticity. J. Elast. **6**, 313–326 (1976)
19. Huiskes, R., Weinans, H., Grootenboer, H.J., Dalstra, M., Fudala, B., Sloof, T.J.: Adaptive bone-remodeling theory applied to prosthetic-design analysis. J. Biomech. **20**, 1135–1150 (1987)
20. Beaupré, G.S., Orr, T.E., Carter, D.R.: An approach for time-dependent bone modeling and remodeling: theoretical development. J. Orthop. Res. **8**, 651–661 (1990)
21. Doblaré, M., García, J.M.: Application of an anisotropic bone-remodeling model based on a damage-repair theory to the analysis of the proximal femur before and after total hip replacement. J. Biomech. **34**, 1157–1170 (2001)
22. Sanz-Herrera, J.A., García-Aznar, J.M., Doblaré, M.: Micro–macro numerical modeling of bone regeneration in tissue engineering. Comput. Method Appl. Mech. Eng. **197**, 3092–3107 (2008)
23. Hernandez, C.J., Beaupré, G.S., Carter, D.R.: A model of mechanobiologic and metabolic influences on bone adaptation. J. Rehabil. Res. Dev. **37**, 235–244 (2000)

24. Huiskes, R., Ruimerman, R., van Lenthe, G.H., Janssen, J.D.: Effects of mechanical forces on maintenance and adaptation of form in trabecular bone. Nature **405**, 704–706 (2000)
25. Hazelwood, S.J., Martin, R.B., Rashid, M.M., Rodrigo, J.J.: A mechanistic model for internal bone remodeling exhibits different dynamic responses in disuse and overload. J. Biomech. **34**, 299–308 (2001)

Bone Turnover

▶ Bone Remodeling

Boron- and/or Nitrogen-Doped Carbon Nanotubes

▶ Light-Element Nanotubes and Related Structures

Boron Nitride Nanotubes (BNNTs)

▶ Chemical Vapor Deposition (CVD)
▶ Physical Vapor Deposition

Bottom-Up Nanofabrication

▶ Electric Field–Directed Assembly of Bioderivatized Nanoparticles

Boundary Lubrication

Bharat Bhushan
Nanoprobe Laboratory for Bio- & Nanotechnology and Biomimetics, The Ohio State University, Columbus, OH, USA

Synonyms

Greases; Perfluoropolyethers; Petroleum lubricants; Synthetic lubricants

Definition

Lubricants are commonly used for reduction in friction and wear of interfaces. In some applications, the solid surfaces are so close together that some asperities come in contact and others are mitigated by a thin film of lubricant, Fig. 1. Under these conditions, the lubricant viscosity is relatively unimportant and the physical and chemical interactions of the lubricant with the solid bodies controls friction and wear. Even a monolayer of adsorbed molecules may provide some protection against wear. Lubrication in some situations can be achieved by the use of multimolecular lubricant films. Monolayer lubrication is referred to as boundary lubrication and multimolecular lubrication is referred to as mixed lubrication. Boundary lubrication usually occurs under high-load and low-speed conditions in machine components such as bearings, gears, cam and tappet interfaces, and piston ring and liner interfaces. Boundary lubrication forms a last line of defense. In many cases, it is the regime which controls the component life [1, 2, 4–7, 11, 16, 19, 22–24].

Various lubricants and greases are used for lubrication of machine components operating in various lubrication regimes [3–11, 14, 15, 17, 18, 24, 25]. Additives are commonly used to provide the desirable properties and interaction with the interface.

Overview

For the case of two contacting bodies coated with a continuous solid monolayer of lubricant with a load too small to cause plastic deformation, the interface is in equilibrium under load for some time since the films prevent contact between the substrates. The films do not allow any state of lower free surface energy than the initial state. The films will thus lubricate over a considerable sliding distance if the bodies are subjected to low-speed sliding, although they will eventually be worn away. Alternatively, if the temperature is raised slightly above the melting point of the films, the admolecules will acquire some mobility and no state involving more than one complete monolayer can be stable. Activation energy for migrating away from the loaded region is provided thermally – and mechanically also, if a low-speed sliding is imposed. The two layers initially present thus penetrate each other and adsorbed molecules tend to move away

Boundary Lubrication,
Fig. 1 Schematic of two
surfaces separated by
a boundary layer of lubricant

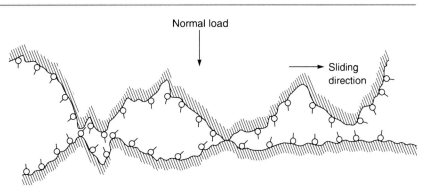

from the loaded interface. An equilibrium between a partial monolayer at the loaded interface and a surrounding vapor, if it exists, is generally metastable since the state for which the bodies are in direct contact usually has lower energy than any lubricated one.

The boundary films are formed by physical adsorption, chemical adsorption, and chemical reaction; for typical examples, see Fig. 2. The physisorbed film can be of either monomolecular (typically <3 nm) or polymolecular thickness. The chemisorbed films are monomolecular, but stoichiometric films formed by chemical reaction can have a large film thickness. In general, the stability and durability of surface films decrease in the following order: chemical reaction films, chemisorbed films, and physisorbed films.

A good boundary lubricant should have a high degree of interaction between its molecules and the sliding surface. As a general rule, liquids are good lubricants when they are polar and thus able to grip solid surfaces (or be adsorbed) [4–8]. Polar lubricants contain reactive functional groups with low ionization potential or groups having high polarizability. Boundary-lubrication properties of lubricants are also dependent upon the molecular conformation and lubricant spreading. Examples of nonpolar and polar molecules are shown in Fig. 3. In the case of Z-DOL, a hydrogen atom, covalently bonded with oxygen atom in the O–H bond, exposes a bare proton on the end of the bond. This proton can be easily attracted to the negative charge of other molecules because the proton is not shielded by electrons, and this is responsible for the polarity of the O–H ends. Likewise, the lone pairs of electrons in the oxygen and fluorine atoms in both molecules are unshielded, and can be attracted to positive charges of other molecules, and thus exhibit electronegativity which is responsible for some

polarity. The CF_3 end in Z-15 is symmetric, and its polarity is low.

In addition to the polarity of liquids, the shape of their molecules governs the effectiveness, which determines whether they can form a dense, thick layer on the solid surface. Ring molecules or branch chain molecules tend to be poorer than straight chain molecules because there is no way in which they can achieve a high packing density. Straight chain molecules with one polar end, such as alcohols and soaps of fatty acids, are highly desirable because they enable a thick film to be formed with the polar end tightly held on the surface and the rest of the molecule normal to the surface. If the sliding surface has to operate under humid conditions, the lubricant should be hydrophobic (i.e., it should not absorb water or be displaced by water).

The most readily observed cause of breakdown of thin-film lubrication is the melting of a solid film and degradation of liquid films, but some degree of lubrication may persist to a higher temperature. Sliding speed and load influence the performance of multilayers.

The properties of the solid surface that are desirable for good lubrication are discussed. A solid should have a high surface energy so that there will be a strong tendency for molecules to adsorb on the surface. Consequently, metals tend to be the easiest surfaces to be lubricated. The solid surface should have a high wetting (or low contact angle) so that the liquid lubricant wets the solid easily. For better lubrication, the surface should be reactive to the lubricant under test conditions so that durable, chemically reacted films can form. Another property of solid surfaces is hydrophobicity. The surfaces should be highly functional with polar groups and dangling bonds (unpaired electrons) so that

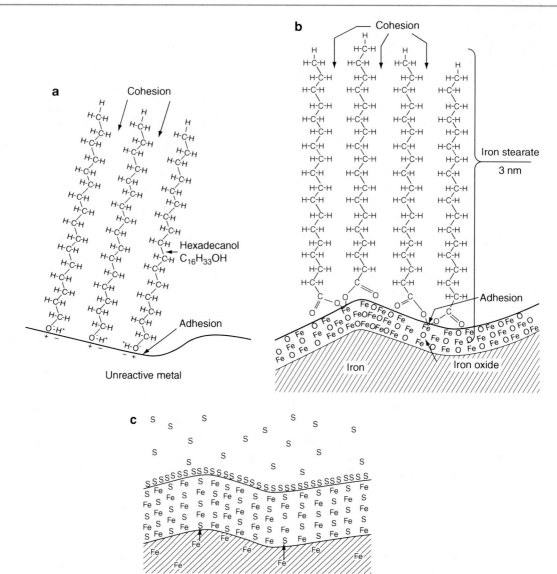

Boundary Lubrication, Fig. 2 (**a**) Schematic diagram representing the physisorption with preferred orientation of three polar molecules of hexadecanol to a metal surface; (**b**) schematic diagram representing the chemisorption of stearic acid on an iron surface to form a monolayer of iron stearate, a soap; (**c**) schematic representation of an inorganic film formed by chemical reaction of sulfur with iron to form iron sulfide (Ku 1970)

Boundary Lubrication, Fig. 3 Structures of nonpolar and polar (−OH) organic lubricant molecules

Nonpolar molecule

Fomblin Z-15 $CF_3-O-(CF_2-CF_2-O)_m-(CF_2-O)_n-CF_3$ $(m/n \sim 2/3)$

Polar molecule

Fomblin Z-DOL $HO-CH_2-CF_2-O-(CF_2-CF_2-O)_m-(CF_2-O)_n-CF_2-CH_2-OH$

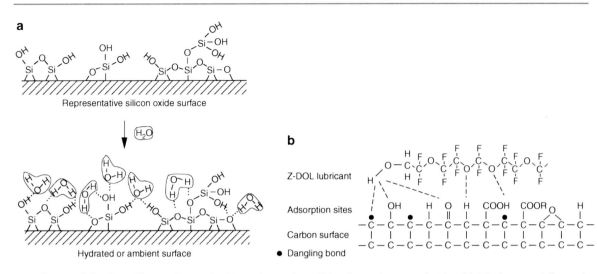

Boundary Lubrication, Fig. 4 Schematic illustrations of (a) a hydrophilic silicon-oxide surface before and after adsorption of water molecules; hydrogen bonding occurs between the solid surface and water molecules. (b) A hydrogenated diamond-like carbon surface with adsorbed polar perfluoropolyether (Z-DOL) lubricant molecules; • represents dangling bonds

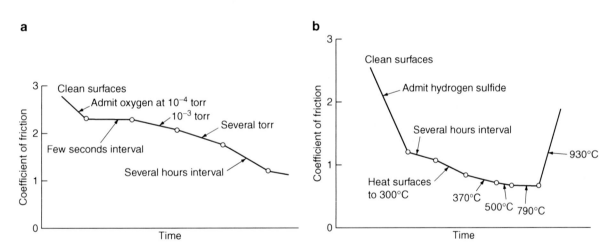

Boundary Lubrication, Fig. 5 (a) Effect of oxygen on the coefficient of friction of outgassed iron surfaces and (b) effect of hydrogen sulfide on the coefficient of friction of outgassed iron surfaces (Bowden 1951)

they can react with lubricant molecules and adsorb them. Examples of hydrophilic silicon oxide surface and its reactivity to ambient water and amorphous carbon surface with polar groups and dangling bonds which promote adsorption of perfluoropolyether molecules are shown in Fig. 4 [8]. Additives to the lubricants can also enhance the formation of chemically reacted films.

Data showing the effect of environment and types of lubricants and their interaction with solid surfaces on boundary lubrication behavior are now presented.

Effect of Adsorbed Gases

Boundary films occur on almost all surfaces because they reduce the surface energy and are thus thermodynamically favored. Normally, air covers any surface with an oxidized film plus adsorbed moisture and organic material. Inadvertent lubrication by air is the most common boundary lubrication. Figure 5a shows the reduction in the coefficient of friction that is obtained by the adsorption and/or chemical reaction of oxygen on clean iron surfaces outgassed in a vacuum (roughly 10^{-6} torr or mmHg). The coefficient of friction is markedly reduced by admission of oxygen gas, though the oxygen pressure is very low (roughly 10^{-4} torr). As oxygen pressure is allowed to increase, the friction is reduced still more. Finally, if the surfaces are allowed to stand for some period of time, the adsorbed oxygen film becomes more complete and the friction drops still further. Note that the seizure of clean metals is prevented by even a trace of oxygen, as obtained at 10^{-4} torr [12].

Figure 5b shows the effect of the addition of hydrogen sulfide on the coefficient of friction of outgassed (clean) iron surfaces; the friction reduced abruptly and appreciably. It is necessary to heat the surface over 790°C before the decomposition of the film takes place and friction rises. It is probable that hydrogen sulfide reacts with the clean iron surfaces to form an iron sulfide (FeS) film.

Effect of Monolayers and Multilayers

It is possible to show by use of monomolecular layers and multimolecular layers that a very thin film of lubricant at the surface can be effective in reducing friction. In studying the effects of monolayers or multilayers, it is convenient to use the well-known Langmuir–Blodgett (L–B) technique. This technique involves floating an insoluble monolayer on the surface of water and then transferring it from the surface of the water to the surface of the solid (by successive dippings) to which the monolayer or multilayer is to be applied. This technique is convenient for deposition of films of known and controllable thicknesses.

Bowden and Tabor [13] deposited the films of a long-chain fatty acid (stearic acid) on a stainless-steel surface. The lubricated surface was slid against an unlubricated surface at 10 mm/s, and the coefficient of friction was recorded from the beginning of sliding. Data shown in Fig. 6 for a monolayer and with multilayers of 3, 9, or 53 films show that the greater the number of films, the longer it takes to wear off or displace this protective film and, consequently, the longer the time in which the film is an effective boundary lubricant. The films deposited by the L–B technique are not entirely equivalent to the type of protective film developed from lubricants in practice with respect to either molecular packing or composition. The stearic acid films, however, were close packed and regularly oriented with the polar group in the water surface.

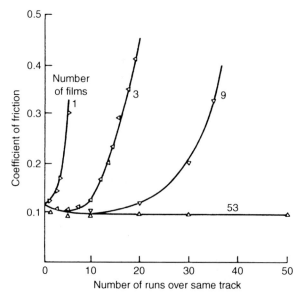

Boundary Lubrication, Fig. 6 Wear behavior of a number of stearic acid films deposited on stainless-steel sliding against unlubricated stainless-steel surface (Bowden and Tabor 1950)

Boundary Lubrication, Table 1 Shear stress as a function of number of boundary layers trapped between two mica surfaces for octamethylcyclotetrasiloxane (OMCTS)[a] and cyclohexane[b]

| Number of layers | Shear stress (MPa) | |
	OMCTS	Cyclohexane
1	8.0 ± 0.5	$2.3 \pm 0.6 \times 10$
2	6.0 ± 1.0	1.0 ± 0.2
3	3.0 ± 1.0	$4.3 \pm 1.5 \times 10^{-1}$
4	Not measured	$2.0 \pm 1.0 \times 10^{-2}$

[a]Molecular diameter ~ 0.85 nm.
[b]Molecular diameter ~ 0.5 nm.

Israelachvili et al. [21] have elegantly measured the frictional force or shear stress (frictional force divided by the apparent area of contact) required to sustain sliding (shearing) of two molecularly smooth mica surfaces with various molecular layers(s) of a liquid film in between. The two liquids used were octamethylcyclotetrasiloxane (OMCTS) and cyclohexane, and have mean molecular diameters of 0.85 and 0.5 nm, respectively. Measurements were made at sliding velocities ranging from 0.25 to 2 μm/s after steady-state sliding was attained. They found that the shear stress depended on the number of boundary liquid layers, Table 1; in cyclohexane, for example, shear stress fell by about an order of magnitude per additional layer. In other words, the friction is

Boundary Lubrication, Table 2 Efficiency of lubrication by paraffin oils or 1% lauric acid in paraffin oil compared with reactivity of metal to lauric acid (Bowden and Tabor 1950)

Metal	Coefficient of friction at 20°C			Transition temperature (°C)	%Acid reactive[a]
	Clean	Paraffin oil	1% Lauric acid in paraffin oil		
Unreactive					
Platinum	1.2	0.28	0.25	20	None
Silver	1.4	0.8	0.7	20	None
Nickel	0.7	0.3	0.28	20	None
Chromium	0.4	0.3	0.3	20	None
Glass	0.9	0.4	0.4	20	None
Reactive					
Zinc	0.6	0.2	0.04	94	10
Cadmium	0.5	0.45	0.05	103	9.3
Copper	1.4	0.3	0.08	97	4.6
Magnesium	0.6	0.5	0.08	80	Trace
Iron	1.0	0.3	0.2		Trace

[a]Estimated amount of acid involved in the reaction assuming formation of a normal salt.

quantized depending on the number of molecular layers separating the surfaces. By extrapolation, one may infer that when seven to ten layers are present, the shear stress of the liquid film would have fallen to the value expected for bulk continuum Newtonian flow. It is noteworthy that this is about the same number of layers as when the forces across a thin film and the whole concept of viscosity begin to be described by continuum theories [20].

Effect of Chemical Films

The addition of a small trace of a fatty acid (polar lubricant) to a nonpolar mineral oil or to a pure hydrocarbon can bring about a considerable reduction in the friction and wear of chemically reactive surfaces. Typical results taken from [13] are given in Table 2. In these experiments, friction was measured using identical materials sliding against each other and lubricated with a (nonpolar) paraffin oil or with 1% (polar) lauric acid added to the paraffinic oil. They found that with unreactive metals (such as Pt, Ag, Ni, and Cr) and glass-the fatty acid is no more effective than a paraffin oil. In contrast, the results for lubrication of reactive metals (such as Zn, Cd, Cu, Mg, and Fe) show that very effective lubrication can be obtained with a 1% solution of lauric acid in paraffin oil. Bowden and Tabor [13] have further shown that a 0.01% solution of lauric acid in paraffin oil reduces the coefficient of friction of chemically reactive cadium surfaces from 0.45 to 0.10. Even smaller concentrations (0.001%) of lauric acid can reduce friction slowly with time (after

a few hours). In the case of the less-reactive metals, such as iron, which cannot be lubricated by a 1% solution of fatty acid, they are well lubricated by a more concentrated solution. These results indicate the strong effects of films formed by chemical reaction on friction and wear.

Friction and wear measurements on monolayers of various substances show that whereas a single monolayer of a given polar compound on a reactive surface provides low friction and wear, on unreactive surfaces it may be completely ineffective, and as many as ten or more layers may be needed. Bowden and Tabor [13] deposited multilayers of stearic acid and metal soaps (believed to be responsible for good lubrication behavior when stearic acid is used as a lubricant on a metal surface) on unreactive and reactive metal surfaces and made the friction measurements. Table 3 shows the number of layers required for effective lubrication. It is found that a larger number of films are needed for unreactive surfaces (see also Ling et al. [23]).

Parraffins, alcohols, ketones, and amides become ineffective lubricants at the bulk melting point of the lubricant. When the melting occurs, the adhesion between the molecules in the boundary film is diminished and breakdown of the film takes place. The increased metallic contact through the lubricant film leads to increased friction and wear. With saturated fatty acids on reactive metals, however, the breakdown does not occur at their melting points but at considerably higher temperatures. This is shown in Fig. 7 for a series of fatty acids on steel surfaces, and it is seen

Boundary Lubrication, Table 3 Lubrication of various metal surfaces by layers of stearic acid and metal stearates deposited by the Langmuir–Blodgett technique (Bowden and Tabor 1950)

Metal	Number of layers for effective lubrication	
	Stearic acid	Metal soap (Cu or Ag stearates)
Unreactive		
Platinum	>10	7–9
Silver	7	3
Nickel	3	3
Reactive		
Copper	3	3
Stainless steel	3	1

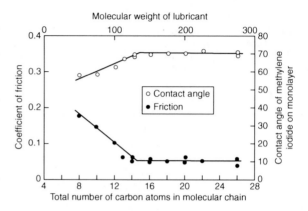

Boundary Lubrication, Fig. 8 Effect of chain length (or molecular weight) on coefficient of friction (of stainless-steel sliding on glass lubricated with a monolayer of fatty acid) and contact angle (of methyl iodide on condensed monolayers of fatty acids on glass) (Zisman 1959)

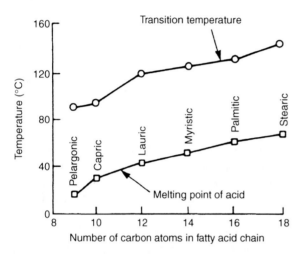

Boundary Lubrication, Fig. 7 Breakdown or transition temperature of fatty acids on steel surfaces and their melting points as a function of chain length (Bowden and Tabor 1950)

Effect of Chain Length (or Molecular Weight)

The effect of chain length of the carbon atoms of paraffins (nonpolar), alcohols (nonpolar), and fatty acids (polar) on the coefficient of friction was studied by Bowden and Tabor [13] and Zisman [27]. Figure 8 shows the coefficient of friction of a stainless-steel surface sliding against a glass surface lubricated with a monolayer of fatty acid. It is seen that there is a steady decrease in friction with increasing chain length. At a sufficiently long chain length, the coefficient of friction reaches a lower limit of above 0.07. Similar trends were found for other lubricant films (paraffins and alcohols). Zisman [27] also found that the contact angle of methylene iodide on a monolayer rises to a high constant value for polar, long-chain molecules of paraffins, alcohols, and fatty acids, indicating an increase in packing to an optimum condition with an increase in the number of carbon atoms in the chain. Owens [26] has confirmed that in polymer lubrication a complete surface coverage occurs with fatty acids having long chain lengths.

Zisman [27] and Owens [26] have shown that the durability of the lubricant film also increases with an increase in the film chain length. These results suggest that monolayers having a chain length below 12 carbon atoms behave as liquids (poor durability), those with chain lengths of 12–15 carbon atoms behave like a plastic solid (medium durability), whereas those with chain lengths above 15 carbon atoms behave like a crystalline solid (high durability).

that breakdown (transition temperature) occurs at 50–70°C above the melting point. The transition temperature corresponds approximately to the melting point of the metallic soaps formed by chemical reaction [13]. The actual value of the breakdown temperature depends on the nature of the metals and on the load and speed of sliding. The esters of saturated fatty acids also behave like acids, except that the difference between the transition and melting temperatures (T_t-T_m) decreases with increasing ester group length, approaching zero at 26 carbon atoms.

Thus, lubrication is affected not by the fatty acid itself but by the metallic soap formed as a result of the chemical reaction between the metal and the fatty acid. Also, T_t-T_m is a measure of the strength of adsorption that is due to a dipole–metal interaction.

Boundary Lubrication, Table 4 Types of liquid lubricants (oils)

Natural organics	Synthetic organics
Animal fat	Synthetic hydrocarbons (polybutene)
Shark oil	Chlorinated hydrocarbons
Whale oil	Chlorofluorocarbons
Vegetable oils	Esters
Mineral (petroleum)	Organic acid
oils	Fatty acid
Paraffinic	Dibasic acid (di)
Naphthenic	Neopentyl polyol
Aromatic	Polyglycol ethers
	Fluoro
	Phosphate
	Silicate
	Disiloxane
	Silicones
	Dimethyl
	Phenyl methyl
	Chlorophenyl methyl
	Alkyl methyl
	Fluoro
	Silanes
	Polyphenyl ethers
	Perfluoropolyethers

Principal Classes of Lubricants

Liquid lubricants (oils) include natural organics consisting of animal fat, vegetable oils, mineral (or petroleum) fractions, synthetic organics, and mixtures of two or more of these materials. Various additives are used to improve the specific properties [4–7, 9, 11, 14, 15, 17, 18, 24]. A partial list of the lubricant types is shown in Table 4.

Closure

Boundary lubrication is accomplished by mono- or multimolecular films. The films are so thin that their behavior is controlled by interaction with the substrates. Molecular structure of the lubricants and functionality of the substrate affect the type and degree of bonding of the lubricant to the substrate.

Liquid lubricants include mineral (or petroleum) and synthetic organics. Various additives are used to improve specific properties. Mineral oils are excellent boundary lubricants and by far the most used lubricants. Synthetic lubricants can be used at greater extremes of environment, including temperature,

humidity, and vapor pressure. Mineral oils are typically used up to a maximum temperature of about 130°C and some synthetic oils up to about 370°C. However, synthetic lubricants are more expensive than mineral oils. There are several properties of lubricants which are important for lubrication; their relative importance depends upon the industrial application. Additives are commonly used to modify friction and wear of lubricants and greases. These are classified as friction modifier, antiwear, and extreme pressure.

Greases are used where circulating liquid lubricant cannot be contained because of space and cost and where cooling by the oil is not required or the application of a liquid lubricant is not feasible.

Cross-References

► Nanotechnology
► Nanotribology
► Reliability of Nanostructures

References

1. Anon.: Limits of lubrication, Special issue of Tribol. Lett. **3**, 1 (1997)
2. Beerbower, A.: Boundary lubrication, Report No. AD-747336, U.S. Department of Commerce, Office of the Chief of Research and Development. Department of the Army, Washington, DC (1972)
3. Bhushan, B.: Tribology and Mechanics of magnetic storage devices, 2nd edn. Springer, New York (1996)
4. Bhushan, B.: Principles and Applications of Tribology. Wiley, New York (1999)
5. Bhushan, B.: Modern Tribology Handbook, Vol. 1 – Principles of Tribology; Vol. 2 – Materials, Coatings, and Industrial Applications. CRC Press, Boca Raton (2001)
6. Bhushan, B.: Introduction to Tribology. Wiley, New York (2002)
7. Bhushan, B.: Nanotribology and Nanomechanics I – Measurement Techniques and Nanomechanics, II – Nanotribology, Biomimetics, and Industrial Applications, 3rd edn. Springer, Heidelberg (2011)
8. Bhushan, B., Zhao, Z.: Macro- and microscale tribological studies of molecularly-thick boundary layers of perfluoropolyether lubricants for magnetic thin-film rigid disks. J. Info. Storage Proc. Syst. **1**, 1–21 (1999)
9. Bisson, E.E., Anderson, W.J.: Advanced Bearing Technology, SP-38. NASA, Washington, DC (1964)
10. Boner, C.J.: Modern Lubricating Greases. Scientific Publications, Broseley (1976)
11. Booser, E.R.: CRC Handbook of Lubrication. Theory and Design, vol. 2. CRC Press, Boca Raton (1984)

12. Bowden, F.P.: The influence of surface films on the friction, adhesion and surface damage of solid, the fundamental aspects of lubrication. Ann. NY Acad. Sci., **53**, Art 4, June 27, 753–994 (1951)
13. Bowden, F.P., Tabor, D.: Friction and Lubrication of Solids, vol. I. Clarendon, Oxford (1950)
14. Braithwaite, E.R.: Lubrication and Lubricants. Elsevier, Amsterdam (1967)
15. Evans, G.G., Galvin, V.M., Robertson, W.S., Walker, W.F.: Lubrication in Practice. Macmillan, Basingstoke (1972)
16. Godfrey, D.: Boundary lubrication. In: Ku, P.M. (ed.) Interdisciplinary Approach to Friction and Wear, vol. SP-181, pp. 335–384. NASA, Washington, DC (1968)
17. Gunderson, R.C., Hart, A.W.: Synthetic Lubricants. Reinhold, New York (1962)
18. Gunther, R.C.: Lubrication. Bailey Brothers and Swinfen, Folkestone (1971)
19. Iliuc, I.: Tribology of Thin Films. Elsevier, New York (1980)
20. Israelachvili, J.N.: Intermolecular and Surface Forces, 2nd edn. Academic, San Diego (1992)
21. Israelachvili, J.N., McGuiggan, P.M., Homola, A.M.: Dynamic Properties of Molecularly Thin Liquid Films. Science **240**, 189–191 (1988)
22. Ku, P.M.: Interdisciplinary Approach to the Lubrication of Concentrated Contacts, vol. SP-237. NASA, Washington, DC (1970)
23. Ling, F.F., Klaus, E.E., Fein, R.S.: Boundary Lubrication – An Appraisal of World Literature. ASME, New York (1969)
24. Loomis, W.R.: New Directions in Lubrication, Materials, Wear, and Surface Interactions – Tribology in the 80's. Noyes, Park Ridge (1985)
25. McConnell, B.D.: Assessment of Lubricant Technology. ASME, New York (1972)
26. Owens, D.K.: Friction of Polymers I. Lubrication. J. Appl. Poly. Sci. **8**, 1465–1475 (1964)
27. Zisman, W.A.: Durability and wettability properties of monomolecular films on solids. In: Davies, R. (ed.) Friction and Wear, pp. 110–148. Elsevier, Amsterdam (1959)

Brain Implants

▶ MEMS Neural Probes

Buckminsterfullerene

▶ Fullerenes for Drug Delivery

Bulk Acoustic Wave MEMS Resonators

▶ Laterally Vibrating Piezoelectric Resonators

C

C$_{60}$

▶ Fullerenes for Drug Delivery

Cancer Modeling

▶ Models for Tumor Growth

Capacitive MEMS Switches

Dimitrios Peroulis
School of Electrical and Computer Engineering,
Birck Nanotechnology Center, Purdue University,
West Lafayette, IN, USA

Synonyms

Electrostatic RF MEMS switches; Micromechanical switches; RF MEMS switches

Definition

Capacitive micro-electro-mechanical systems (MEMS) switches are a special type of micromachined switches that control radio frequency (RF) signal paths in microwave and millimeter-wave circuits through mechanical motion and contact.

Overview

Capacitive and direct current (dc)-contact MEMS switches are among the most important micromachined devices for high-frequency applications due to their near-ideal RF performance. Dc-contact switches function similarly to conventional relays: micromachined beams or plates move under the influence of an appropriately applied force (e.g., electrostatic force) to open or close a metal-to-metal contact. While micromachined beams or plates are also utilized in capacitive switches, these switches rely on metal-to-dielectric contacts to implement their on and off states. Capacitive switches are particularly attractive for demanding high-frequency communications, electronic warfare, and radar systems due to their ultralow loss (< 0.1–0.2 dB up to 40 GHz), high isolation (>20–50 dB for frequencies beyond 10 GHz), very high linearity (>66 dBm third-order intercept point), and near-zero power consumption ($\sim$tens of nJ per switching cycle and zero quiescent power for electrostatically actuated switches). When compared to solid-state switches, capacitive switches are relatively slow devices with speeds ranging in the tens to hundreds of microseconds range. This speed is primarily limited by switch inertia and squeeze film damping. Their relatively large lateral dimensions of tens or hundreds of μm allow capacitive switches to handle several hundred mW of RF power. Long-term operation, however, can only be achieved if they are hermetically sealed in order to avoid contamination- and humidity-induced failure. Hermetically sealed capacitive switches have successfully switched over

B. Bhushan (ed.), *Encyclopedia of Nanotechnology*, DOI 10.1007/978-90-481-9751-4,
© Springer Science+Business Media B.V. 2012

Capacitive MEMS Switches,
Fig. 1 (a) Side-view and (b) top-view schematics of a typical shunt capacitive MEMS switch. Both the up and down states are shown

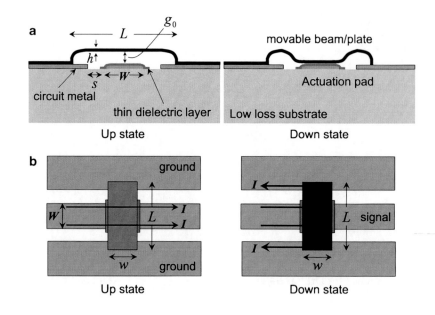

100 billion cycles at room temperature and under low RF power conditions (20 dBm). Despite the aforementioned RF advantages, capacitive switches are currently not available commercially and are not widely utilized in defense or communication systems. This is primarily due to the facts that (1) high-yield manufacturing processes are not widely available yet and (2) their main failure modes such as dielectric charging, dc/RF gas discharge and metal creep and the physics behind them have not been adequately understood and addressed today.

Switch Structure and Actuation Mechanisms

Figure 1 shows a typical capacitive MEMS switch [1]. This is a shunt switch configuration, and is the dominant capacitive switch configuration in the literature today. The signal travels down the center conductor, and if the switch closes, will return along the outside conductors. It is, however, possible to design geometries for series configurations. Their characteristics, nevertheless, are very similar to the ones found in shunt switches. Consequently, this entry focuses primarily on shunt capacitive switches.

The switch in Fig. 1 consists of a movable beam or plate that is anchored to the switch substrate. While typically the term "plate" may better characterize the geometry of Fig. 1 for typical lateral switch dimensions, it is common in the RF MEMS literature to refer

to this geometry as a fixed-fixed beam. Hence the term "beam" is adopted here to describe such capacitive switch geometries. Cantilever beams are also possible particularly for series switch configurations. The fixed-fixed beam anchors are typically metallic and are connected to the RF line. For example, they can be connected to the ground planes of a coplanar waveguide line as shown in Fig. 1b. The movable beam is typically composed of a thin-film (thickness h of 0.5–2 μm) metal such as gold, aluminum, nickel, or molybdenum. It is also possible that the beam is comprised of multiple layers including thin-film dielectrics such as silicon nitride or silicon dioxide and metals. One or more metallic pads are placed underneath the beam. In the simplest case, a single metallic pad is placed under the beam as shown in Fig. 1b. In this case, this pad is the center conductor of the coplanar waveguide line. This pad is usually covered by a thin (0.1–0.3 μm) dielectric layer such as silicon nitride, silicon dioxide, or more recently amorphous [2] or ultrananocrystalline [3] diamond.

The beam's length and width are determined by the required down-state switch capacitance as explained in the following section. This results in length and width in the tens to hundreds of μm for 5–40 GHz capacitive switches. The thickness of the dielectric layer also impacts the down-state capacitance. While high capacitance is in general required, thicknesses lower than 0.1 μm are hard to achieve in practice due to the need to handle high electric fields (dc and/or RF) across this

Capacitive MEMS Switches, Table 1 Main advantages and drawbacks of common actuation schemes for capacitive RF MEMS switches

Actuation mechanism	Advantages	Drawbacks
Electrostatic	Zero quiescent power consumption, easy biasing circuit, fast transient response (tens of microseconds)	Need to generate high voltage, high voltage may lead to charging and breakdown issues
Magnetostatic	Low voltage, high contact pressure, potentially low power with latching mechanism	Low quiescent power consumption requires latching, slower than electrostatic due to increased switch size
Electrothermal	Low voltage, high contact pressure, size comparable to electrostatic schemes	High quiescent power consumption (mW), slow response time (tens to hundreds of milliseconds)
Piezoelectric	Same as electrostatic by with low voltage	Difficult to achieve high-quality piezoelectric layer and integrate it with RF circuit at low temperatures

dielectric. The gap (g_0) between the bottom surface of the movable beam and the top surface of the dielectric layer is determined by the need to minimize the up-state capacitance. A low up-state capacitance is necessary for low insertion loss. Capacitive switches in the 5–40 GHz range typically have gaps in the 2–5 μm range.

The switch has two states of operation. In the up state, the RF signal goes through the signal line almost unaffected by the movable beam. The up state is also called "zero biased" or "on state". In the down state the RF signal does not go through the RF line because it is reflected (see following section). The down state is also called the "biased" or "off state." An actuation mechanism is required to move the switch beam between these two states. Several possible actuation schemes exist, including electrostatic, electrothermal, magnetostatic, and piezoelectric. Table 1 summarizes the main advantages and drawbacks of each actuation scheme. The vast majority of reported capacitive switches are electrostatically actuated. They use the same actuation principle as the one originally proposed when the first capacitive switch was invented and reduced to practice in 1994 [4, 5]. In this original scheme a dc actuation voltage is applied between the

movable beam and the actuation pad underneath it (Fig. 1). The beam is attracted due to the generated electrostatic field and collapses on the switch dielectric layer. Despite the need to generate a high dc voltage (30–100 V), which can be readily accomplished using a dc-dc converter, electrostatic switches exhibit the most desirable electromechanical characteristics, including the fastest possible response, zero quiescent power consumption, and the easiest possible biasing circuits. The resulting high fields though may lead to (gas and solid) dielectric charging and their associated reliability issues. More detailed discussion can be found in the last section.

RF Performance

Figure 2 shows a simple but physically meaningful and accurate lumped-element equivalent circuit of the switch shown in Fig. 1. It also includes typical values for all the equivalent circuit components [1]. The parameters α and β in this figure represent the attenuation constant and propagation constant of the transmission line respectively. The up- and down-state capacitance values are the most critical ones in this equivalent circuit. The up-state capacitance can often be accurately calculated by a typical quasi-static expression

$$C_{UP} = C_{pp} + C_{ff} = \frac{\varepsilon_0 A}{g_0 + \frac{t_d}{\varepsilon_r}} + C_{ff}$$

where C_{pp} and C_{ff} are the parallel-plate and fringing-field capacitances, respectively, A is the RF area of the switch ($A = Ww$ in Fig. 1), g_0 is the initial switch height, t_d is the dielectric layer thickness, and ε_r is this layer's dielectric constant. For typical switch geometries, the fringing-field capacitance could reach 25–50% of the parallel plate capacitance. If improved accuracy is needed, a full-wave simulation is performed to estimate the switch up-state capacitance. The up-state capacitance must be sufficiently small to minimize the up-state insertion loss. Assuming a well-designed switch where the contributions from L and R_s can be ignored in the up state, the switch up-state reflection coefficient can be calculated from

$$S_{11} = \frac{-j\omega C_{UP} Z_0}{2 + j\omega C_{UP} Z_0}$$

Capacitive MEMS Switches,
Fig. 2 Lumped-element
equivalent circuit of the
capacitive switch shown in
Fig. 1. Typical values are
provided for the lumped
components of this circuit

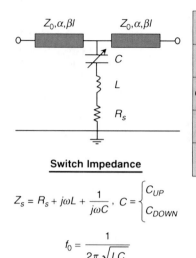

Switch Impedance

$$Z_s = R_s + j\omega L + \frac{1}{j\omega C}, \quad C = \begin{cases} C_{UP} \\ C_{DOWN} \end{cases}$$

$$f_0 = \frac{1}{2\pi\sqrt{LC}}$$

	Typical Values	
	X-band	**K-band**
C_{UP}/C_{DOWN}	0.1/6 pF	0.04/3 pF
L	4 – 80 pH	6 – 50 pH
R_s	0.1 – 0.3 Ω	0.1 – 0.3 Ω

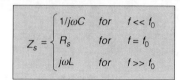

where Z_0 is the characteristic impedance of the transmission line (typically 50 Ω). For example, an up-state capacitance of 70 fF results in $S_{11} < -10$ dB up to approximately 30 GHz. The up-state switch ohmic loss is the other critical up-state characteristic. The total ohmic loss of the switch can be calculated as

$$Loss = 1 - |S_{11}|^2 - |S_{21}|^2$$

This depends on (a) the attenuation α (dB/cm) of the transmission line underneath the movable beam and (b) on the switch series resistance R_s. Well-designed capacitive switches can exhibit a total loss of less than 0.1 dB up to 40 GHz. Figure 3 shows numerical values for this loss for typical switch characteristics.

The switch down-state capacitance is more complicated to calculate because the switch beam may not be perfectly flat against the dielectric layer. Even in good designs this may not be possible due to the roughness of the layers involved. A model that is often used to capture the nonideal down-state switch capacitance is [1]

$$C_{DN} = \frac{\varepsilon_0 A}{2}\left(\frac{1}{r + \frac{t_d}{\varepsilon_r}} + \frac{\varepsilon_r}{t_d}\right)$$

where r is the roughness amplitude. The down-state fringing-field capacitance is not included in this equation because it is typically not significant (<5% of the parallel-plate capacitance) due to the small dielectric layer thickness. As shown by the equation above, the experimentally achieved down-state capacitance can

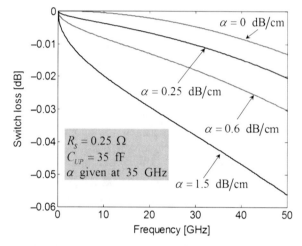

Capacitive MEMS Switches, Fig. 3 Simulated ohmic loss for a typical shunt capacitive switch. The attenuation a (dB/cm) depends on the transmission line characteristics

vary greatly depending on the true contact area and the dielectric layer characteristics including its roughness. In practice, it is difficult to avoid a 30–50% degradation of the down-state capacitance compared to the theoretical parallel-plate value. A high down-state capacitance (2–5 pF) is typically required in order to achieve an acceptable isolation (>20–50 dB) level at the desired frequency. One way to achieve this is to decrease the dielectric layer thickness. However, this dielectric layer needs to sustain very high electric fields (50–150 V/μm) across its thickness. Given typical fabrication process limitations that prohibit high-temperature growth processes for the dielectric layer

Capacitive MEMS Switches,
Fig. 4 Simulated and
measured down-state
scattering parameters of
a capacitive MEMS switch
(After [6] with permission)

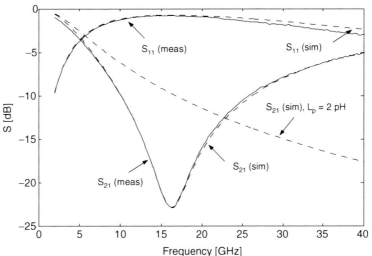

(please see fabrication section), dielectric layers thinner than 0.1–0.2 μm become impractical. A second way to increase the down-state capacitance is to increase the dielectric constant. For instance, barium-strontium-titanate (BST) or strontium-titanate-oxide (STO) films with dielectric constants up to 400 can be employed. Besides additional fabrication complexities, these films have not been thoroughly studied in MEMS switches and may exhibit unacceptably high dielectric charging. Most switches typically employ some form of silicon dioxide or silicon nitride dielectric with dielectric constant in the 3.9–7.5 range. For switches with very low inductances ($L < 1–2$ pH), the switch isolation can be approximately calculated as

$$S_{21} = \frac{2}{2 + j\omega C_{DN} Z_0}$$

A more accurate calculation reveals that the switch inductance and series resistance also determine the total down-state isolation. In particular, its inductance cancels the down-state capacitance at the switch resonant frequency

$$f_0 = \frac{1}{2\pi\sqrt{LC_{DN}}}$$

This frequency can be adjusted by controlling the switch physical geometries. Switch inductances in the range of 1–100 pH can be readily achieved [1]. However, higher switch inductance typically results in a higher switch series resistance. Typical switch

resistance values range in the 0.1–2 Ω range. The series resistance is the primary limiting factor of the switch isolation at its resonant frequency. At that frequency the switch isolation can be calculated as

$$S_{21} = \frac{2R_s}{2R_s + Z_0} \approx \frac{2R_s}{Z_0} \text{ at } f = f_0$$

Figure 4 shows measured and simulated results of a typical capacitive switch [6] with $C_{DN} = 1.1$ pF, $L = 87$ pH, $R_s = 1.95$ Ω. This figure also shows the expected performance when the inductance is reduced to $L = 2$ pH.

Electromechanical Considerations: Static Behavior

Figure 5 shows a simple but physically meaningful one-dimensional electromechanical model of the switch geometry of Fig. 1. The beam is modeled as a spring-mass system with a spring constant k. This spring constant depends on (a) the beam geometry, (b) the electrostatic force distribution on the beam, and (c) the residual stress of its structural film. The residual stress σ (MPa) is due to the fabrication process and depends on the exact deposition conditions. Typically a tensile stress ($\sigma > 0$) is needed in order to avoid buckling. The spring constant can be expressed as

$$k = k_1 + k_2$$

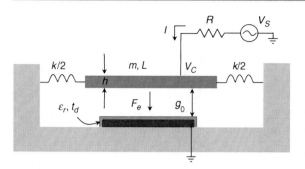

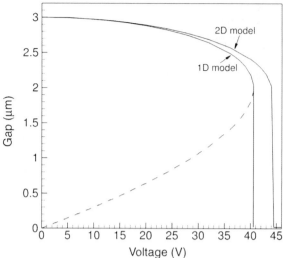

Capacitive MEMS Switches, Fig. 5 One-dimensional electromechanical model of a capacitive RF MEMS switch

Capacitive MEMS Switches, Fig. 6 Simulated gap-voltage relationship for a capacitive MEMS switch with the following characteristics: $L = 300$ μm, $w = 100$ μm, $W = 120$ μm, $h = 1$ μm, $g_0 = 3$ μm, E (Young's modulus) = 79 GPa, $\sigma = 10$ MPa. These results have been obtained with the PRISM center online simulation tool in MEMShub [13]

where k_1 depends on the first two factors and k_2 depends on the residual stress. The exact values can be calculated based on the specific switch design. For example, for the fixed-fixed beam of Fig. 1 and assuming that $W = L/3$ and that the electrostatic attractive force is uniformly distributed along the beam section directly above the coplanar waveguide center conductor, the spring constant can be calculated as [1]

$$k = k_1 + k_2$$
$$= 32Ew\left(\frac{h}{L}\right)^3\left(\frac{27}{49}\right) + 8\sigma(1 - v)w\left(\frac{h}{L}\right)\left(\frac{3}{5}\right)$$

where v is the beam's Poisson's ratio. For usual beam geometries and fabrication processes with residual stress in the order of 10–50 MPa, the second term dominates the spring constant. Typical spring constant values range from 10 to 50 N/m in order to provide sufficiently high restoring force and avoid stiction issues. While low spring-constant designs have been successfully demonstrated [7], special care needs to be taken in avoiding stiction and self-actuation due to high RF power [8]. It is also important to mention that the above spring constant calculations are based on small-deflection theory. A nonlinear spring constant may need to be derived if this condition is not satisfied.

The electrostatic force F_e on the switch beam can be calculated as

$$F_e = \frac{\partial W_e}{\partial g} = \frac{1}{2}V^2\frac{\partial C(g)}{\partial g} \approx -\frac{1}{2}\frac{\varepsilon_0 AV^2}{g^2}$$

where W_e is the stored electrostatic energy, g is the switch gap between the beam and the actuation pad, and V is the applied electrostatic voltage. The last

approximation is based on assuming a parallel-plate capacitance approximation and by ignoring the dielectric contribution. The static switch gap can be calculated by taking into account the static equilibrium of the forces applied on the beam

$$\frac{1}{2}\frac{\varepsilon_0 AV^2}{g^2} = k(g_0 - g)$$

The above equation can be solved for the applied voltage as

$$V = \sqrt{\frac{2k}{\varepsilon_0 A}g^2(g_0 - g)}$$

This equation is plotted in Fig. 6 for typical switch parameters. As Fig. 6 shows, there are two possible gaps for any given actuation voltage. This is not observed in practice and is a result of the unstable behavior of the beam. In particular, for small actuation voltages, the electrostatic force is increased proportionally to $\frac{1}{g^2}$. However, the restoring force is only increased proportionally to g. Hence, there is a critical gap beyond which the restoring force cannot hold the beam and the beam collapses on the dielectric surface. This critical gap can be found from the

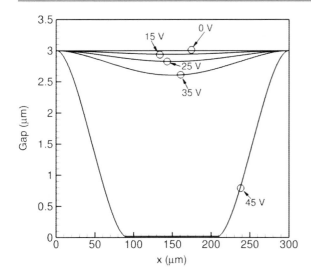

Capacitive MEMS Switches, Fig. 7 Simulated shape of the movable beam of a capacitive MEMS switch for different bias voltages. The characteristics of the switch geometry are the same as in Fig. 6. These results have been obtained with the PRISM center online simulation tool in MEMShub [13]

previous equation by taking its derivative and setting it up to zero

$$\frac{\partial V}{\partial g} = 0 \xrightarrow{yields} g_c = \frac{2}{3}g_0$$

The voltage, therefore, required for actuating the beam, called the pull-in or pull-down voltage V_p, is given by

$$V_p = V(g_c) = \sqrt{\frac{8kg_0^3}{27\varepsilon_0 A}}$$

Figure 6 also shows the voltage-gap relationship as obtained by a two-dimensional beam model [13]. The gap plotted is between the center of the beam and the actuation pad. This curve is slightly different because the beam is not deformed as a perfectly flat object as assumed by the one-dimensional model. Figure 7 shows the actual deformed shape of the movable beam for voltages up to the pull-down voltage. Actuation voltages in the range of 30–100 V are typical in RF MEMS switches. Notice, however, that once the beam is actuated, the voltage required to hold the beam down (hold-down voltage V_h) is much lower because the gap between the beam and the actuation pad is much lower

than g_0. It is hard to analytically calculate the hold-down voltage because it depends on many fabrication-dependent conditions such as the adhesion force between the beam and the dielectric layer. Typical hold-down voltages are in the range of 5–15 V [1]. Consequently, the gap-voltage relationship is strongly hysteretic. It is also worth mentioning that nonideal conditions such as an initial beam curvature and nonlinear bending are not included in this model. These effects can be captured by more complicated nonlinear beam models such as the ones presented in [9, 10].

Electromechanical Considerations: Dynamic Behavior

The dynamic behavior of the capacitive MEMS switch of Fig. 1 can be approximately captured by a one-dimensional model as described by the following equation

$$mg''(t) + bg'(t) + kg(t) = F$$

where m is the switch mass, b is the damping coefficient, and f is the externally applied force. RF MEMS switches are typically packaged in an environment of 1 atm in order to avoid excessive ringing due to an underdamped response. As a result, the switch damping is dominated by squeeze-film damping as the gas under the beam is displaced during the switch motion. Due to the small gap g_0, it is in general difficult to accurately calculate the damping coefficient particularly in the near-contact region. This is further complicated by the possible existence of holes in the beam that aid its fabrication and substantially improve its switching speed. While there are several published approximations that can yield a reasonable approximation to the damping coefficient [1], accurate macro-models based primarily on rarefied gas dynamics only recently started becoming available [11, 12]. An equivalent way to characterize the switch damping is by its mechanical quality factor defined as

$$Q = \frac{k}{\omega_0 b}$$

where $\omega_0 = \sqrt{\frac{k}{m}}$ is the switch mechanical frequency. Typical switches show mechanical frequencies in the 20–100 kHz range and quality factors in the 0.5–2

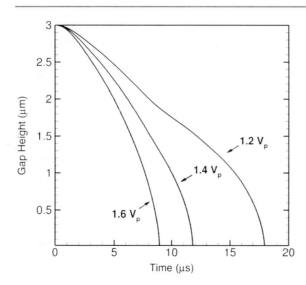

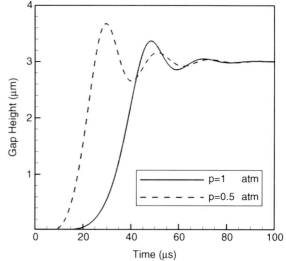

Capacitive MEMS Switches, Fig. 8 Simulated switching time (pull-down) of a capacitive MEMS switch for different bias voltages. The characteristics of the switch geometry are the same as in Fig. 6. These results have been obtained with the PRISM center online simulation tool in MEMShub [13]

Capacitive MEMS Switches, Fig. 9 Simulated switching time (release) of a capacitive MEMS switch for different pressure levels. The characteristics of the switch geometry are the same as in Fig. 6. These results have been obtained with the PRISM center online simulation tool in MEMShub [13]

range. Figures 8 (switch closure) and 9 (switch opening) illustrate dynamic responses as calculated by two-dimensional beam models that accurately capture squeeze-film damping [13]. Notice that the displacement at the center of the beam is plotted in these graphs.

The switching speed can also be estimated based on the simple one-dimensional model. While it is difficult, in general, to derive an exact analytical solution, this model can provide reasonable approximations for common cases. For example, for $Q > 2$, the closing time can be estimated by [1]

$$t_s \approx 3.67 \frac{V_p}{V_s \omega_0}$$

where V_s is the applied voltage. Other limiting cases can be found in [1]. In general, closing times in the 5–50 μs range can be achieved. A similar range is typically possible for the release times.

The one-dimensional model can also be used to estimate the velocity and acceleration of the switch. Switching velocity in the 1–10 m/s range can be observed in the near-contact region. Due to its small mass, the switch acceleration can exceed 10^6 m/s^2 in the same region.

Fabrication Methods

Capacitive switches can be fabricated with conventional micromachining processes and require a small number of masks. Figure 10 illustrates the masks of a typical fabrication process.

- Step (a): The first mask defines the circuit metal of Fig. 1 after this metal layer is deposited on the substrate through evaporation, sputtering, or electroplating. Gold, aluminum, and copper are common metal choices. A smooth metal surface is particularly important directly underneath the switch beam in order to minimize local electric field enhancement.
- Step (b): The second mask defines the dielectric layer to cover the portion of the metal that will be under the beam. Unless a high melting temperature metal has been deposited (e.g., tungsten), a relatively low-temperature process is required for the deposition and patterning of the dielectric layer in order to avoid damaging the circuit metal. Plasma-enhanced chemical vapor deposition (PECVD) is the most common process for depositing silicon nitride/oxide films. This is usually followed by a reactive ion etching (RIE) step that helps etching the unwanted dielectric layer parts.

Capacitive MEMS Switches,
Fig. 10 Simplified typical
fabrication process for
a capacitive MEMS switch

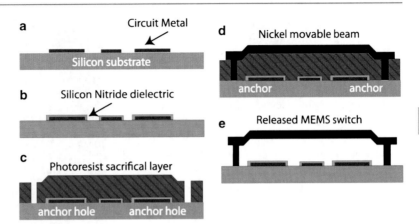

- Step (c): The third mask is used to define the sacrificial layer of the switch. The sacrificial layer is the layer upon which the beam will be deposited, and this material needs to be removed at the end of the process to release the beam. The beam anchor points are defined in this step by selectively etching the sacrificial layer. This can be a dry- or a wet-etch step. More complicated processes may involve an additional planarization step before the next mask. The choice of the sacrificial layer material is critical as it controls many important parameters of the switch design, including the residual stress of the beam. Common material choices include photoresists, other photoconductive polymers, and polyimides. There is no sufficient understanding in the open literature of the exact processes that are involved in controlling the beam layer residual stress in the presence of a sacrificial layer. Important parameters though include, among others, the atomic structures of each film and the deposition temperatures of each film.
- Step (d): The fourth mask defines the actual beam layer. A variety of processes can be utilized including evaporation, sputtering, and electroplating of the beam layer(s). This step is also critical in determining the final residual stress of the beam.
- Step (e): The beam is finally released by etching (wet or dry) the sacrificial layer and by drying (if needed) the switch wafer. Etching of the sacrificial layer can also influence the beam residual stress particularly if a high-temperature process is necessary. If wafer drying is needed, this needs to be done carefully in order to avoid damage to the beam or causing stiction to the substrate. Stiction may occur if drying involves removing a liquid with high

surface tension (e.g., water) underneath the beam. Such a liquid will pull the beam down as it evaporates. Special drying processes and equipment based on supercritical carbon dioxide have been successfully developed [14] and followed by many MEMS researchers.

The last step in the fabrication process is packaging, which is discussed in the following section.

Packaging

Hermetic packaging is required for capacitive RF MEMS switches to avoid any contamination- or humidity-induced early failure. While conventional hermetic packages exist, they are not well suited for capacitive RF MEMS switches or circuits. First, if switches need to be inserted in conventional hermetic packages, these switches will have to be diced first since several thousands of them can be simultaneously fabricated on a wafer. Dicing released switches is particularly dangerous for the switches and may considerably reduce the process yield. Second, conventional hermetic packages are expensive (~tens of dollars/package) and not well-suited for cost-driven consumer applications. Third, they typically exhibit a relatively high insertion loss, which is often much higher than the switch itself (e.g., a DC-40 GHz package could exhibit 0.6 dB at 20 GHz [1]).

As a result, it is important to follow a cost-effective on-wafer hermetic packaging scheme. In this case, a wafer-scale package is first completed and then dicing follows. A wide variety of approaches have been developed so far to accomplish this. These approaches can be divided into three main categories:

- Two-wafer hermetic packages completed by fusion, glass-frit, thermocompression, eutectic, or anodic bonding. These packages are created by bonding two wafers together using one of the aforementioned approaches. The main advantage of these techniques is that they result in excellent hermetic bonds. Their main drawback is that they may require high temperatures (300–1,000°C depending on the technique) with the exception of low-temperature eutectic bonds (e.g., indium–gold bonds). In addition, these techniques tend to be relatively expensive since packaging cost usually accounts for 60–80% of the total cost.

- Two-wafer quasi-hermetic packages completed by low-temperature polymer or solder-bump bonding. This technique is similar to the previous one, except that sealing is achieved by low-temperature bonding (room temperature − 150°C). A wide variety of polymers can be used for this. While low temperatures can be achieved, packages fabricated with such bonding typically exhibit very low but nonzero leak rates [1].

- Hermetic packages fabricated by on-wafer micro-encapsulation using micromachining techniques [15]. These techniques do not require a second wafer cap. Instead, every switch or switching circuit is encapsulated in a tiny package on its own wafer by a microfabricated technique. Typical temperatures in this process are in the 200–250°C range. These techniques are ideally suited for the small size and high RF bandwidth of MEMS devices and typically result in low-cost fabrication. Particular attention needs to be paid though to ensure compatibility between the switch and package fabrication processes. Figure 11 shows a switch packaged with this technique.

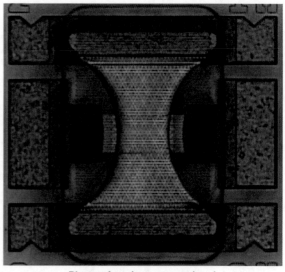

Photo of a microencapsulated RF MEMS capacitive switch.

Capacitive MEMS Switches, Fig. 11 The packaged memtronics switch (After [15] with permission)

pursuing this particularly due to the large cell phone market size.

The most important of the high-frequency circuit applications (> 10 GHz) are high-isolation switching packets, true-time delay networks and phase shifters, reconfigurable impedance tuners for amplifiers and antennas, high-quality-factor reconfigurable filters, and tunable oscillators. Many of these circuits exploit the near-ideal RF performance of capacitive switches. Consequently, optimal performance can be usually achieved by employing a circuit- or sub-system-level package instead of a device-level package. Examples of several of these circuits can be found in [1].

Circuits and Applications

Capacitive MEMS switches may be employed in a number of circuits mostly for communication, radar, and electronic warfare systems [1]. Variable capacitors and impedance tuners for cell phones and other radios in mobile form factors constitute the most important applications in the commercial sector. MEMS variable capacitors (varactors) can be formed by connecting in parallel several capacitive MEMS switches and selectively activating them. Several companies including WiSpry and Cavendish Kinetics are

Failure Mechanisms and Reliability

Capacitive MEMS switches suffer from high electric dc and/or RF fields through narrow gaps and dielectric layers. These fields can readily reach 5–50 V/μm in the up state and may increase further during actuation. Such fields may cause field emission and ionize the gas in the switch gap [16]. The long-term effects of field emission and gas discharge are not known at this point. In addition, when such fields are applied across a thin-film solid dielectric, charges may get trapped in the dielectric layer leading to dielectric charging.

A number of studies have been completed (see for example [17, 18]) focusing mostly on charges trapped in the bulk of the dielectric. However, surface charging of the solid dielectric that is also influenced by gas ionization is potentially more detrimental to the switch performance and is not well understood today. Charging phenomena are the leading cause of failure in capacitive switches today. Long-term drift of actuation voltage, stiction, and breakdown can be observed as a result of these charging issues.

Besides solid and gas dielectric charging, metal creep is another potential failure mechanism. Creep may be developed in the movable beam material if a switch is subjected to a constant stress. For example, if a switch is left in its down state for a long time (typically tens to thousands of hours), the beam material may creep resulting in a temporary or permanent change of the switch spring constant. Several recent papers show it is potentially an area of concern for capacitive MEMS switches [19, 20]. Creep at high temperature may be even a more significant area of concern.

Other possible failure modes include

- Beam buckling due to high temperatures. This may be caused during release process, normal high-temperature operation, or due to high RF currents through the movable beam under high RF power conditions.
- Self-actuation of the switch movable beam due to high RF power [8]. A high RF voltage may result in self-actuation because the attractive electrostatic force is proportional to the square of the switch voltage. This limits the switch power handling.
- Hot switching failure. When a capacitive switch needs to interrupt high RF currents or sustain high transient RF voltages, abrupt chemical changes may occur at its surfaces leading to premature wear and failure. This is related to the dielectric charging phenomena.
- Shock-induced failure. High shocks ($>$30,000–100,000 g) may result in beam fracture particularly if contact is achieved. Such events are rare in most applications.

Failures related to cycling-induced fatigue, crack generation, and fracture are not typically observed under normal operating conditions. However, they may become important at extreme temperatures particularly for movable beams based on thin-film metals. Despite the aforementioned failure modes, the best switches today have achieved over 100 billion cycles under typical laboratory conditions when driven by 30 kHz bipolar bias waveforms with approximately 35 V peak amplitude. These devices were hot-switched at a power level of 20 dBm at 35 GHz [21]. However, these cannot be considered typical results. Early failures are found in several wafer samples. Additional research is required to increase the observed reliability and limit early failures that are commonly due to poor fabrication process control.

Cross-References

▶ Basic MEMS Actuators
▶ NEMS Piezoelectric Switches
▶ Piezoelectric MEMS Switch

References

1. Rebeiz, G.M.: RF MEMS Theory, Design, and Technology. Wiley, Hoboken (2003)
2. Webster, J.R., Dyck, C.W., Sullivan, J.P., Friedmann, T.A., Carton, A.J.: Performance of amorphous diamond RF MEMS capacitive switch. Electron. Lett. **40**(1), 43 (2004)
3. Goldsmith, C., Sumant, A., Auciello, O., Carlisle, J., Zeng, H., Hwang, J.C.M., Palego, C., Wang, W., Carpick, R., Adiga, V.P., Datta, A., Gudeman, C., O'Brien, S., Sampath, S.: Charging characteristics of ultra-nano-crystalline diamond in RF MEMS capacitive switches. In: Proceedings of the IEEE MTT-S International Microwave Digest, Anaheim, CA, pp. 1246–1249, May 2010
4. Goldsmith, C.L., Kanack, B.M., Lin, T., Norvell, B.R., Pang, L.Y., Powers, B., Rhoads, C., Seymour, D.: Micromechanical microwave switching. US Patent 5,619,061, 31 Oct 1994
5. Goldsmith, C.L., Yao, Z., Eshelman, S., Denniston, D.: Performance of low-loss RF MEMS capacitive switches. IEEE Microwave Wireless Compon. Lett. **8**(8), 269–271 (1998)
6. Peroulis, D.: RF MEMS devices for multifunctional integrated circuits and antennas. PhD Dissertation, The University of Michigan, Ann Arbor, (2003)
7. Peroulis, D., Pacheco, S.P., Sarabandi, K., Katehi, L.P.B.: Electromechanical considerations in developing low-voltage RF MEMS switches. IEEE T. Microw. Theory **51**(1), 259–270 (2003)
8. Peroulis, D., Pacheco, S.P., Katehi, L.P.B.: RF MEMS switches with enhanced power-handling capabilities. IEEE T. Microw. Theory **52**(1), 59–68 (2004)
9. Snow, M.: Comprehensive modeling of electrostatically actuated MEMS beams including uncertainty quantification. MSc Thesis, Purdue University, West Lafayette, (2010)
10. Younis, M.I., Abdel-Rahman, E.M., Nayfeh, A.: A reduced-order model for electrically actuated microbeam-based MEMS. J. Microelectromech. S. **12**(5), 672–680 (2003)
11. Guo, X., Alexeenko, A.: Compact model of squeeze-film damping based on rarefied flow simulations. J. Micromech. Microeng. **19**(4), 045026 (2009)

12. Parkos, D., Raghunathan, N., Venkattraman, A., Alexeenko, A., Peroulis, D.: Near-contact damping model and dynamic response of micro-beams under high-g loads. In: Proceedings of the IEEE International Conference on Micro Electro Mechanical Systems (MEMS), Cancun, Mexico, pp. 465–468, Jan 2011

13. Ayyaswamy, V., Alexeenko, A.: Coarse-grained model for RF MEMS device. MEMShub.org, http://memshub.org/resources/prismcg. Accessed on Mar 2011

14. Tousimis, http://www.tousimis.com/. Accessed on March 2011

15. Memtronics, http://www.memtronics.com/. Accessed on March 2011

16. Garg, A., Ayyaswamy, V., Kovacs, A., Alexeenko, A., Peroulis, D.: Direct measurement of field emission current in E-static MEMS structures. In: Proceedings of the 24th IEEE International Conference on Micro Electro Mechanical Systems (MEMS 2011), pp. 412–415, Jan 2011

17. Peng, Z., Yuan, X., Hwang, J.C.M., Forehand, D.I., Goldsmith, C.L.: Superposition model for dielectric charging of RF MEMS capacitive switches under bipolar control-voltage waveforms. IEEE T. Microw. Theory 55(12), 2911–2918 (2007)

18. Papaioannou, G., Exarchos, M.-N., Theonas, V., Wang, G., Papapolymerou, J.: Temperature study of the dielectric polarization effects of capacitive RF MEMS switches. IEEE T. Microw. Theory 53(11), 3467–3473 (2005)

19. Hsu, H.-H., Peroulis, D.: A viscoelastic-aware experimentally-derived model for analog RF MEMS varactors. In: Proceedings of the 23rd IEEE International Conference on Micro Electro Mechanical Systems (MEMS 2010), Wanchai, Hong Kong, pp. 783–786, Jan 2010

20. McLean, M., Brown, W.L., Vinci, R.P.: Temperature-dependent viscoelasticity in thin Au films and consequences for MEMS devices. IEEE/ASME J. Microelectromech. S. 19(6), 1299–1308 (2010)

21. Goldsmith, C., Maciel, J., McKillop, J.: Demonstrating reliability. IEEE Microwave Mag. 8(6), 56–60 (2007)

Capillarity Induced Folding

▶ Capillary Origami

Capillary Flow

Prashant R. Waghmare and Sushanta K. Mitra
Micro and Nano-scale Transport Laboratory,
Department of Mechanical Engineering,
University of Alberta, Edmonton, AB, Canada

Synonyms

Passive pumping; Surface tension–driven flow

Definition

The fluid flow in an enclosed conduit due to simultaneous changes in the inherent surface energies of fluid and solid surface of the conduit.

Introduction

Everything in the universe has its own state of energy, which is represented by possible combinations of 132 elements of periodic table. The simplest form of each element is an atom and each atom has three different components: proton, electron, and neutron. Further, each element has a fixed number of electrons, which arrange themselves in different shells, and the number of electrons in each shell can be determined by Bohr's theory. Several individual elements or atoms do not have sufficient number of electrons in their outer shell and this makes it unstable and further the element tries to become stable by searching for required electrons. The finding of sufficient electrons at the outer shell results in the formation of a molecule or in bulk cluster of molecules. The surface or interface formation with this cluster of molecules creates an imbalance in the arrangement and orientation of the molecules in the cluster, mainly across the interface. This imbalance represents the energy of system. Every system in the universe tries to attain the minimum state of energy and, therefore, in the case of liquid in the air, it has been observed that the raindrops always attain the spherical shape. Liquid molecules have capability to orient themselves. Hence, they can form the shape of minimum energy but in the case of solid molecules, they try to minimize the state of energy by covering up the liquid if it comes in contact. In this case, the solid surfaces are not in equilibrium with the saturated vapor, i.e., they are not in a state of minimum energy. The moment at which conduit or channel with high energy level surfaces comes in contact with lower energy level liquid interface, the solid surface tries to envelope itself with the liquid to attain the minimum energy. In the case of the higher energy liquid in comparison with the solid surfaces, the liquid interface tries to minimize its surface. The prior case can be illustrated with the water (72 dyn/cm) in glass channel (~200–300 dyn/cm) and the mercury (~700–735 dyn/cm) with the same glass channel is an example of the latter case.

The recent developments in the microfabrication technologies allowed to fabricate features of sizes from micro- to nanoscales. The surface to volume ratio of the feature increases as the scale of the feature decreases which in turn makes surface forces dominant over other forces. Because of high surface forces, very high pressure is required to pump the fluid in microchannels. Hence, researchers have developed different nonmechanical pumping mechanism with the help of electrokinetic and/or megnetohydrodynamic approaches. Generation and actuation of electric and/or magnetic field are an additional burden on the system which increases both the fabrication process and cost of the device. Therefore, attempts are being made to develop a flow without any external means. The fluid flow can be achieved by controlling surface chemistry, fluid properties like surface tension, or by changing the geometries. Such transport of fluid is called autonomous flow or autonomous pumping which is an ideal transport mechanism for microfluidic applications. The dominance of surface forces at microscale plays a significant role in deciding the pumping approach in microfluidic devices. Hence, nowadays surface tension–driven flow in microfluidic devices has widely attracted the attention of researchers. As explained earlier, the capillary action is the interplay between the surface energies or surface tension between the fluid and solid surface in contact. Moreover, at microscale, due to very high surface to volume ratio, the possibility of available surface area is very high. One can easily pump the fluid with capillarity provided the fluid has lower surface energy than the solid surfaces. For optimum design and function of any microfluidic device which works on capillary flow principle, it is essential to predict its behavior in advance and it is therefore necessary to perform the theoretical analysis of capillary flow within the microchannels. The prediction of the temporal variations in the flow front position along capillary length is the ultimate goal of the theoretical analysis. Therefore, generally the analysis is performed to predict the flow front position, i.e., penetration depth for given working and operating conditions.

Over the last century, the capillary phenomenon has become a topic of interest due to its importance in several areas. First time in literature, Washburn proposed a closed form solution for the penetration depth in a channel of millimeter dimension. The closed form solution is derived by balancing the surface tension force to viscous force and it is observed that the penetration depth is proportional to the square root of the time. As explained earlier, capillary phenomenon is the change in the surface energy process and to encompass the concept of change in the surface energy, one can use the thermodynamic approach for analysis. The surface energy of solid is the topic of ongoing debate. Moreover, there are several effects like dynamic contact angle, inlet effects, reservoir effects, suspension flow, etc., which require tedious and cumbersome analysis. Attempts are also being made to present analysis with microscopic energy balance approach where different forces are accounted in terms of different forms of energy. On the other hand, hydrodynamic models, based on the conventional fluid mechanics principle, are easy to implement. Such models are mainly developed by two distinct approaches in the literature, namely, differential and integral approach. The moving fluid-air interface with the differential approach becomes computationally costly, whereas the integral approach with moving control volume provides a simple form of ordinary differential equation. In such modeling or analysis, the governing equation for flow front transport is obtained by balancing different forces like viscous, inertial, gravity, pressure forces, etc. The velocity-dependent terms like inertial and viscous terms of the governing equation are determined with the velocity profile across the channel. In the literature, the steady state assumption is applied from the very entrance of the channel neglecting the entrance length effect. This assumption has been widely adopted till date as done by Washburn where the time and length scale used for the validation of the theoretical model was big. Hence, the assumption of a steady state velocity profile holds true. Whereas, in the case of microfluidic channels the length and timescale is very small; hence, the Washburn prediction does not follow the observation as demonstrated by Saha and Mitra [1, 2]. The later part of this entry is dedicated to emphasize the importance of such microscale effects in the analysis. Main emphasis is given to the integral approach based on modeling due to the ease of its adaptability. Therefore, in the next section, the overview of the modeling of the capillary transport is discussed in brief.

Mathematical Modeling

Figure 1 shows the microchannel of width 2B and depth of 2W is considered for the theoretical modeling.

Capillary Flow,
Fig. 1 Schematic of the
microchannel of width 2B,
depth 2W considered for
theoretical modeling [6]

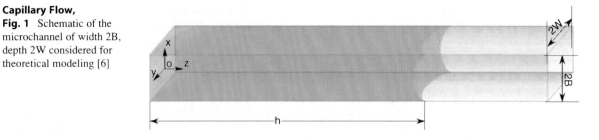

The momentum equation in integral form for homogeneous, incompressible, and Newtonian fluid can be written as [3]:

$$\sum F_z = \frac{\partial}{\partial t} \int_0^h \int_{-W}^W \int_{-B}^B \rho v_z dx dy dz + \int_{-W}^W$$
$$\times \int_{-B}^B v_z(-\rho v_z) dx dy \qquad (1)$$

where $\sum F_z$ refers to all forces present during the development of the fluid-air interface, ρ is the fluid density. Generally, forces present during the capillary transport are viscous (F_v), gravity$(F_g = 4\rho ghBW)$, pressure forces at the flow front(F_{pf}), and at the inlet(F_{pi}).

$$\sum F_z = \underbrace{F_v}_{\text{velocity dependent}} + \underbrace{F_g + F_{pf} + F_{pi}}_{\text{velocity independent}} \qquad (2)$$

The Eq. 1 contains three velocity-dependent terms, namely, transient, convective, and viscous force which is generally determined as velocity profile, v_z, across the channel. The fully developed flow assumption, i.e., Poiseuille flow assumption, is widely used in the literature neglecting the transience in the velocity profile [4, 5]. The consequence of such assumptions particularly at microscales will be discussed in detail in the later part.

The steady state velocity profile can be used to determine the velocity-dependent terms of the momentum equation. Moreover, remaining pressure force terms are calculated with available expressions (from the literature) for pressure fields at respective locations. The pressure force at the fluid-air interface can be determined by well-known Young-Laplace equation with fluid surface tension (σ) and equilibrium contact angle (θ_e). The approximated pressure field expression at the entrance of the microchannel is used widely to determine the pressure force at the entrance of the microchannel. Levin et al. [7], for the first time in the literature, claimed that atmospheric pressure cannot be used as entrance pressure at the inlet of the microchannel. The pressure field expression for circular capillary is derived by assuming a separate hemispherical control volume as fluid source other than the control volume considered within the microchannel. Further several researchers have used same expression for rectangular microchannels with an assumption of equivalent radius. In such analysis, with an equivalent radius assumption, the hemispherical control volume of equivalent radius of projected area of rectangular microchannel entrance is presented in Eq. 3.

$$p(o,t) = p_{atm} - \left\{ 1.11\rho\sqrt{BW}\frac{d^2h}{dt^2} + 1.58\rho\left(\frac{dh}{dt}\right)^2 + \frac{1.772\mu}{\sqrt{BW}}\frac{dh}{dt} \right\} \qquad (3)$$

The importance of an appropriate entrance pressure field expression for rectangular microchannel is also discussed in detail in later part of the study. Finally, determining all terms of Eq. 1 and rearranging as per order of differential operator, one can obtain the dimension form of the ordinary differential equation which governs the capillary transport in the microchannels. Further, nondimensional governing equation as shown in Eq. 4 can be obtained by performing nondimensional analysis with characteristic time, $t_0 = \frac{\rho(2B)^2}{12\mu}$ and characteristic length $h_0 = 2B$.

$$(h^* + C_1)\frac{d^2h^*}{dt^{*2}} + C_2\left(\frac{dh^*}{dt^*}\right)^2 + (C_3 + C_4h^*)\frac{dh^*}{dt^*}$$
$$+ C_5h^* + C_6 = 0 \qquad (4)$$

The coefficients of Eq. 4 are tabulated in Table 1. Two nondimensional numbers are obtained, i.e., Bond

Capillary Flow, Table 1 Constants of the generalized nondimensional governing equation for a capillary flow in a microchannel with fully developed velocity profile [3]

Constants	Expressions
C_1	$\frac{0.55}{\sqrt{\gamma}}$
C_2	0.958
C_3	1
C_4	$0.295\sqrt{\gamma}$
C_5	$\frac{Bo}{144Oh^2}$
C_6	$\frac{\gamma - cos\theta_e}{720Oh^2}$

number (Bo) and Ohnesorge number (Oh). The constants of this equation are functions of different nondimensional groups like Ohnesorge number (Oh), Bond number (Bo), and aspect ratio (γ). The Ohnesorge number represents the ratio of viscous to surface tension force, i.e., $Oh = \frac{\mu}{\sqrt{2B\rho\sigma}}$, the Bond number dictates the ratio of gravity to surface tension force, i.e., $\left(Bo = \frac{\rho g(2B^2)}{\sigma}\right)$.

The solution of Eq. 4 predicts the penetration depth with capillary flow. The numerical [8] and analytical [3] solutions of Eq. 4 are available in the literature.

Revisiting the Assumptions for Microscale Applications

As mentioned earlier, the velocity-dependent terms of momentum equation are determined with an assumption of a steady state. It is assumed that at the very entrance of the microchannel the flow is fully developed. However, in reality, three different flow regimes can be observed in the capillary flow: entry regime, Poiseuille regime, where the flow is fully developed, and the regime behind the fluid-air interface, i.e., surface tension regime. The steady state assumption, i.e., parabolic velocity profile assumption, is valid for steady state flow, whereas capillary flow is inherently a transient phenomenon. The parabolic velocity profile assumption is a reasonably good assumption for macroscale capillaries as shown by several researches [5, 7, 9]. Moreover, such assumptions are only valid in the case of a very high viscous fluid or very low Reynolds number flow which may not be true in every case [10]. Hence, it is important to consider and analyze such transience in the analysis at microscale. This can be tackled by considering the transient

or developing velocity profile instead of the steady state velocity profile as explained in the following section.

The transient momentum equation in the direction of the flow for pressure-driven flow is:

$$\rho\frac{\partial v_z}{\partial t} = \mu\frac{\partial^2 v_z}{\partial x^2} + \frac{dp}{dz} \tag{5}$$

The velocity in Eq. 5 is a combination of steady and transient part of velocity as depicted in Eq. 6 [11]:

$$v_z(x,t) = v_{z\infty}(x) + v_{zt}(x,t) \tag{6}$$

where $v_{z\infty}(x)$ is the fully developed or steady state velocity, i.e.,

$$v_{z\infty}(x) = \frac{B^2}{2\mu}\frac{dp}{dz}\left[1 - \left(\frac{x}{B}\right)^2\right] \tag{7}$$

The transient part of the velocity can be obtained by separation of the variable method which can be given as [6]:

$$v_{zt}(x,t) = 2\sum_{n=1}^{\infty}(-1)^n\left[\frac{1}{B\mu\lambda_n^3}\frac{dp}{dz}\right]cos(\lambda_n x)exp$$
$$\times\left(-v\lambda_n^2 t\right) \tag{8}$$

where v is the kinematic viscosity of the fluid and $\lambda_n = \frac{(2n-1)\pi}{2B}$. By combining Eqs. 5 and 7 the transient velocity profile $v_z(x,t)$ can be obtained by:

$$v_z(x,t) = \left\{\sum_{n=1}^{\infty}(-1)^n\frac{1}{B\mu}\left(\frac{2}{\lambda_n^3}\right)cos(\lambda_n x)exp\right.$$
$$\left.\times\left(-v\lambda_n^2 t\right) + \frac{1}{2\mu}(B^2 - x^2)\right\}\frac{dp}{dz} \tag{9}$$

Further, the average velocity across the channel can be represented as:

$$v_z(t)_{avg} = \frac{B^2}{3\mu}\left[1 - \sum_{n=1}^{\infty}\frac{96}{(2n-1)^4\pi^4}exp\right.$$
$$\left.\times\left(-\frac{(2n-1)^2\pi^2 vt}{4B^2}\right)\right]\frac{dp}{dz} \tag{10}$$

Capillary Flow,
Fig. 2 Transient response in
the difference in the
penetration depths with the
fully developed (steady state)
and developing (unsteady)
velocity profile under different
conditions [6]

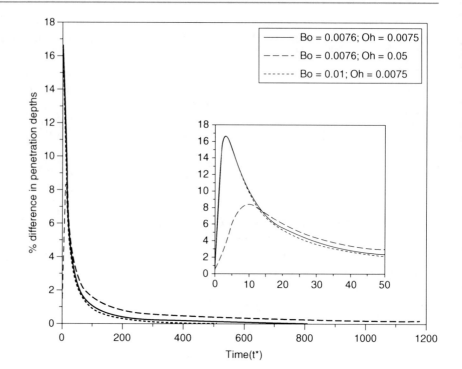

Finally, the velocity profile in terms of the penetration with transient velocity is:

$$v_z(x,t) = \frac{B^2}{2\mu}\alpha_1\left\{\sum_{n=1}^{\infty}(-1)^n\left[\frac{4}{(B\lambda_n)^3}\right]cos(\lambda_n x)exp\left(-v\lambda_n^2 t\right)\right.$$

$$\left.+\left(1-\frac{x^2}{B^2}\right)\right\} \times \left\{\frac{1}{\alpha_1\left[1-\sum_{n=1}^{\infty}\beta_1 exp\left(-\lambda_n^2 vt\right)\right]}\right\}\frac{dh}{dt}$$

$$(11)$$

where

$$\alpha_1 = \frac{\left[(\phi)^4 - 4exp - \frac{\phi^2 t}{3}\right]}{\left[(\phi)^4 - 6exp - \frac{\phi^2 t}{3}\right]}$$

and $\phi = \lambda_n B$.

A similar approach can be followed as explained in the previous section and further the governing equation for capillary transport can be derived with the transient velocity profile provided in Eq. 11. Moreover, the difference in the penetration depth with both approaches, i.e., with the steady state and transient velocity profile, under different operating conditions can be compared. Figure 2 shows the difference in the

penetration depth in such cases where the difference in the penetration depth is more at the beginning of the filling process, as shown in the inset of Fig. 2. This difference in the penetration depth decreases as the flow progresses along the microchannel where the flow becomes a fully developed flow. This can also be explained with the help of boundary layer theory which is the effect of fluid viscosity. The boundary layer thickness increases as the viscosity of fluid increases because of the retardation of flow due to increase in the viscosity, whereas in the case of the fluid density, the effect is opposite to viscosity. Therefore, the difference in penetration depth with the high density fluid (Bo = 0.01) is higher than the difference with the high viscous fluid. It is evident from the analysis that the transience effect in the analysis has a significant impact on the filling process prediction, particularly at the beginning of the filling process. At microscale, such difference needs to be accounted prior to the design.

As discussed earlier, the pressure force at the entrance of the microchannel is determined with the help of the pressure field at the microchannel entrance. Several researchers [3–8] have adopted the pressure field expression with an equivalent radius assumption. Levin et al. [6] developed an entrance pressure field expression for circular capillary, assuming

Capillary Flow, Fig. 3 The fluid volume from infinite reservoir considered as control volume for pressure field expression analysis in the case of rectangular microchannel. The *arrow* shows the direction of the fluid flow from the reservoir into the microchannel [6]

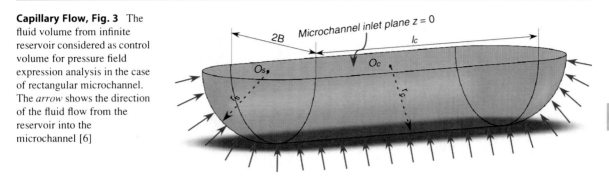

a hemispherical control volume as a separate control volume at the entrance which is responsible for a sink flow at the entrance of capillary and the pressure field. Moreover, a similar expression for rectangular capillaries is extended with an equivalent radius assumption. In such cases, the radius of circular capillary is replaced by the equivalent radius of projected area at the entrance of the channel. This is not a realistic representation for noncircular capillaries particularly for high aspect ratio microchannels where it is not appropriate to consider the hemispherical control volume for the sink flow or pressure field at the microchannel entrance. In the case of such geometries, the control volume needs to be considered as a combination of semicylinder and hemisphere as shown in Fig. 3.

The detailed derivation of the pressure field expression with this control volume can be seen in [6] which is:

$$p(0,t) = p_{atm} - \rho B \left\{ \left[\frac{4\gamma + 3(1-\gamma)}{24} \right] \right.$$

$$\left\{ \pi \left[\frac{1}{2\pi} + \frac{2}{\pi^2} + \frac{6}{10} \right] \right\} + \left[1 - \frac{2}{\pi} ln \frac{R_\infty}{B} \right] \right\} \frac{d^2 h}{dt^2}$$

$$+ \rho \left\{ \left[\frac{4(1-\gamma)}{\pi^2} - \frac{6}{5} \right] - \left[\frac{4\gamma + 3(1-\gamma)}{6} \right] \right.$$

$$\times \left[\frac{(2-\gamma)}{2\pi} - \frac{(1-\gamma)}{\pi^2} \right] \right\} \left(\frac{dh}{dt} \right)^2 - \frac{4\mu}{B}$$

$$\times \left[(2-\gamma) + \frac{(1-\gamma)}{\pi} \right] \frac{dh}{dt} \tag{12}$$

where R_∞ represents the radial distance far away from the control volume in the reservoir, where the sink action, i.e., entrance pressure force, disappears. One can re-derive the governing Eq. 4, using pressure field

expression presented in the Eq. 12, and determine the effect of such a pressure field on the analysis. Figure 4 shows the comparison of variations in the penetration depth with recently proposed pressure and with equivalent radius field expressions. The approximated pressure field overpredicts the penetration depth. The difference in the penetration depth with the proposed pressure field is significant, which shows that it is important to consider the proposed pressure field for a rectangular microchannel rather than an approximated pressure field.

The transport with a capillary action is the balance among surface, viscous, and other body forces which retard the flow as it progresses. Hence, the capillary flow always attains a steady state which is generally termed as an equilibrium penetration depth in the literature. If the length of the channel is longer than that of the equilibrium penetration depth, then flow front cannot reach the outlet and, therefore, the assistance to the capillary flow is attempted in such cases. Passive or nonmechanical pumping approaches combined with the capillary flow serve this enhancement. The scaling analysis suggests that the gravity force is less dominant at microscale [12], but several researchers have demonstrated that gravity can be used as an assistance to the capillary flow [13–15]. Generally, the capillary flow analysis is performed with an assumption of infinite reservoir. Hence, the reservoir effect and the gravitational force from the reservoir are generally neglected in the analysis. To accommodate the entrance effect of finite size reservoir at the inlet of the microchannel in the theoretical modeling, the entrance pressure field is developed for the arrangements shown in Fig. 5. The rectangular microchannel with rectangular reservoir on the top of the microchannel is considered and the pressure field with the gravity and reservoir effect is developed in the flow [16]. Moreover, this pressure field

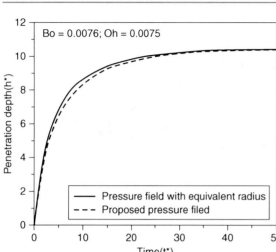

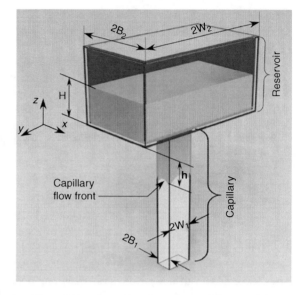

Capillary Flow, Fig. 4 The comparison of variations in the penetration depth with equivalent radius and recently proposed pressure field expressions. Figure 4 shows the comparison of penetration depth for $\gamma = 0.9$ with the corresponding difference in the penetration depth [6]

Capillary Flow, Fig. 5 Schematic of a gravity-assisted capillary flow in a vertically oriented capillary of width $2B_1$ and depth $2W_1$. The additional gravitational head from the fluid in a finite reservoir of size $(2B_2 \times 2W_2)$ is assisting the capillary flow [16]

is used to obtain the governing equation for capillary flow under the influence of gravity head from the reservoir.

The reservoir with three different levels of fluid in the reservoirs (H^*), namely, 10, 50, and 100, is considered for the analysis. Figure 6 shows the variations in the penetration depth (h^*) with different operating conditions. This analysis represents the interplay between the surface tension force and gravity head from the reservoir. The capillary flow takes place in the channel which remains same for all three cases, whereas the level of the fluid from the reservoir increases from case I to case III. Thus, for three cases, the capillary effect is the same but the gravity head is different. At the beginning of the transport, the fluid from the reservoir offers less inertia to the fluid transport and the capillary force dominates over the gravity from the reservoir. Therefore, at the beginning of the transport, the penetration depth with a lower reservoir fluid level ($H^* = 10.0$) is higher than the other two penetration depths as shown in the inset I. Similarly, the penetration depth with the highest reservoir fluid level ($H^* = 100.0$) has the lowest penetration depth as compared to others.

Moreover, as the fluid progresses in the microchannel, the momentum from the reservoir fluid assists the capillary flow and the gravitational force,

due to which the fluid from the reservoir becomes dominant over the capillary force within the microchannel. This results in transcendence among the penetration depths with a different gravity head. The penetration depth with the highest gravity head ($H^* = 100$) surpasses the penetration depth with gravity head $H^* = 50$ and $H = 10^*$ in inset I and II two, respectively. This can be attributed to as an interplay between the surface tension force, i.e., the capillarity and gravitational force from reservoir. In the case of microfluidic applications, the sizes of reservoir and microchannel are comparable to each other. Hence one cannot neglect the effect of the reservoir in such cases, particularly if it is surface tension–driven pumping. Further, one can assist the capillary flow with an appropriate arrangement of reservoir.

There are always certain limitations to the autonomous pumping which make them inadequate in long microchannels. Hence, it is important to enhance the pumping ability by other means. Further enhancement in the capillary flow can be achieved by coupling the capillary flow with the electroosmotic flow which is one of the electrokinetic pumping mechanisms. In most of the cases, the inner wall of a microchannel always has surface charges due to different mechanisms like ionization, dissociation of ions, isomorphic substitution, etc., [17]. These surface charges

Capillary Flow,
Fig. 6 Transient response of a flow front transport for different gravitational heads in the reservoir with Bo = 0.0055, Oh = 0.0084, $B_1/W_1 = 0.05$ $B_2/W_2 = 0.2$.
I. Flow front penetration rate for $H^* = 100$ surpasses the penetration rate for $H^* = 50$.
II: Flow front penetration rate for $H^* = 100$ surpasses the penetration rate for $H^* = 10$.
III: Flow front penetration rate for $H = 50$ surpasses the penetration rate for $H^* = 10$ [16]

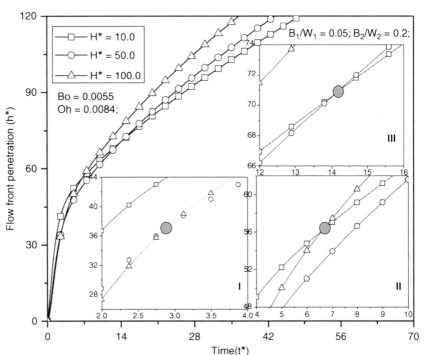

distribute ions of the electrolytes in a specific pattern when brought into contact with an electrolyte which is generally termed as the formation of electrical double layer (EDL). After applying the electric field across the channel, the movement of the ions takes place, which results in the movement of the fluid due to an electric field [18]. The electrolyte solution is transported with the capillary action; one can further assist the capillary flow. An additional body force due to electroosmotism is added to Eq. 2, which accommodates the additional effect of electroosmosis. Further one can analyze the interplay between the capillarity and electroosmotisms as presented in the recent studies [19]. Figure 7 shows the variation in the penetration depth of the capillary flow under the influence of electroosmotism. Through a nondimensional analysis, a new nondimensional number is proposed, i.e., Eo which represents the ratio between the surface tension force and electroosmotic force. The direction of the electroosmotic flow can be reversed by changing the electric field direction. Hence, negative and positive Eo numbers are observed in the analysis. The negative Eo numbers represent the change in the direction of the electric field as compared to positive Eo numbers. The pure capillary flow can be seen as Eo = 0.

The variation in the penetration depth under three different operating conditions is shown in Fig. 7. As observed in the pure capillary case (Eo = 0), the penetration depth attains the equilibrium penetration depth, whereas in the case of −Eo numbers, the equilibrium penetration depth increases with increment in the magnitude of −Eo numbers. This represents that in the case of electroosmotic flow with −Eo number, the capillary flow is assisted by electroosmotism. In this analysis, the nondimensional length of the microchannel (L^*) is considered as 300 and with −Eo = −0.01 the entire filling of the microchannel is observed. In the case of positive Eo numbers, it is observed that the electroosmotism acts in a opposite direction of the capillary flow. Hence, it retards the flow and this can be observed by the decrement in the equilibrium penetration depth with the increment in the positive Eo numbers.

For the enhancement in the capillary flow, a transport with additional gravity head and electroosmotic forces are considered. A generalized theoretical modeling for a gravity-assisted capillary flow with reservoir effects and electroosmotically assisted capillary flow is reported in brief. It is observed that even though the scaling among forces suggests that the

Capillary Flow,
Fig. 7 Variation in the penetration depth for vertically oriented channel with water as electrolyte, where B = 100 μm, W = 400 μm, L = 75 mm, ζ = −75 mV and constant contact angle is 27°. The inset shows the variation in the electric field within the electrolyte as the flow front progresses under different applied voltages [19]

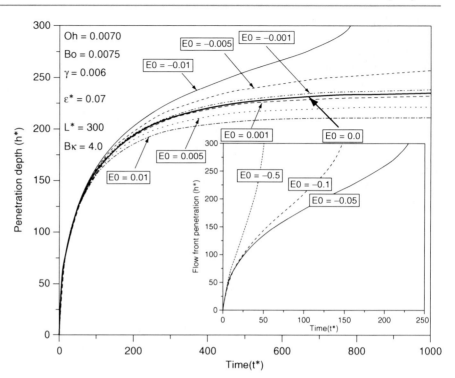

gravitation force is negligible at microscale, the reported analysis infers that with a finite reservoir, an added advantage due to gravity can be a useful tool to transport the fluid at microscale. This added force for the capillary transport can be utilized without any additional burden in the design of the LOC device. The electroosmotically assisted capillary flow model suggests that in a combined flow the electrokinetic parameters have an important influence on the capillary flow. Such electrokinetic flow approaches can be coupled to enhance the capillary flow transport in the microchannel.

The wetting properties of the fluid decide the capability of pumping with a capillary flow. Therefore, it is important to know the precise magnitude of wetting properties like contact angle and surface tension of the working fluid. The microfluidics has become a promising option for biomedical application and inclusion of biomolecules is an unavoidable part in such applications. In most cases, the biomolecules are attached with the microbeads and transported to the desired locations. It is evident from the experimental analysis that the inclusion of microbeads changes the wetting behavior drastically [20]. Therefore, it is necessary to consider the effect of microbeads in

the fluid for the analysis. This can be done by considering the following expressions for surface tension and contact angle: density and viscosity which are functions of volume fraction of microbeads. Such correlations of the surface tension and contact angle are provided in [20]. Such expressions for the variation in the contact angle and surface tension with the volume fraction can be readily used in modeling transport processes of microbead suspensions in micro-capillaries, used in the microfluidic devices.

In passive pumping, particularly with the capillary flow, different aspects due to microscale effects like aspect ratio–dependent velocity profile, contact angle at four walls, fluid-air interface dynamics in the case of suspension flow, etc., need to be investigated in detail. Theoretically the concept of electroosmotically assisted capillary flow has been presented but the experimental demonstration of such phenomena is also an interesting area of research. Moreover, wetting of biomolecule suspensions under transient effects instead of the steady state is also needed to be studied. The experimental study of the flow behind the front and at the entrance of the microchannels is also an interesting study to perform.

Cross-References

▶ AC Electroosmosis: Basics and Lab-on-a-Chip Applications
▶ Electrowetting
▶ Micro/Nano Flow Characterization Techniques
▶ Micropumps
▶ Surface Tension Effects of Nanostructures
▶ Wetting Transitions

References

1. Saha, A., Mitra, S.K.: Numerical study of capillary flow in microchannels with alternate hydrophilic-hydrophobic bottom wall. J. Fluid Eng. Trans. ASME **131**, 061202 (2009)
2. Saha, A., Mitra, S.: Effect of dynamic contact angle in a volume of fluid (VOF) model for a microfluidic capillary flow. J. Colloid Interface Sci. **339**, 461–480 (2009)
3. Xiao, Y., Yang, F., Pitchumani, R.: A generalized flow analysis of capillary flows in channels. J. Colloid Interface Sci. **298**, 880–888 (2006)
4. Washburn, E.: The dynamics of capillary flow. Phys. Rev. **17**, 273 (1921)
5. Chakraborty, S.: Electroosmotically driven capillary transport of typical non-Newtonian biofluid in rectangular microchannels. Anal. Chim. Acta **605**, 175–184 (2007)
6. Waghmare, P.R., Mitra, S.K.: A comprehensive theoretical model of capillary transport in rectangular microchannels. Microfluid. Nanofluid. (2011). doi:10.1007/s10404-011-0848-8
7. Levin, S., Reed, P., Watson, J.: A theory of the rate of rise a liquid in a capillary. In: Kerker, M. (ed.) Colloid and Interface Science, p. 403. Academic, New York (1976)
8. Marwadi, A., Xiao, Y., Pitchumani, R.: Theoretical analysis of capillary-driven nanoparticulate slurry flow during a micromold filling process. Int. J. Multiph. Flow **34**, 227 (2008)
9. Dreyer, M., Delgado, A., Rath, H.: Fluid motion in capillary vanes under reduced gravity. Microgravity Sci. Technol. **4**, 203 (1993)
10. Bhattacharya, S., Gurung, D.: Derivation of governing equation describing time-dependent penetration length in channel flows driven by non-mechanical forces. Anal. Chim. Acta **666**, 51–54 (2010)
11. Keh, H., Tseng, H.: Transient electrokinetic flow in fine capillaries. J. Colloid Interface Sci. **242**, 450 (2001)
12. Nguyen, N., Werely, S.: Fundamentals and Applications of Microfluidics. Artech House, New York (2003)
13. Yamada, H., Yoshida, Y., Terada, N., Hagihara, T., Teasawa, A.: Fabrication of gravity-driven microfluidic device. Rev. Sci. Instrum. **79**, 124301 (2008)
14. Jong, W.R., Kuo, T.H., Ho, S.W., Chiu, H.H., Peng, S.H.: Flows in rectangular microchannels driven by capillary force and gravity. Int. Commun. Heat Mass Transf. **34**, 186–196 (2007)
15. Kung, C., Chui, C., Chen, C., Chang, C., Chu, C.: Blood flow driven by surface tension in a microchannel. Microfluid Nanofluid **6**, 693 (2009)
16. Waghmare, P.R., Mitra, S.K.: Finite reservoir effect on capillary flow of microbead suspension in rectangular microchannels. J. Colloid Interface Sci. **351**(2), 561–569 (2010)
17. Hunter, R.: Zeat Potential in Colloid Science, Principle and Applications, Principle and Applications. Academic, London (1981)
18. Israelachvili, J.N.: Intermolecular and Surface Forces. Academic, London (1998)
19. Waghmare, P.R., Mitra, S.K.: Modeling of combined electroosmotic and capillary flow in microchannels. Anal. Chim. Acta **663**, 117–126 (2010)
20. Waghmare, P.R., Mitra, S.K.: Contact angle hysteresis of microbead suspensions. Langmuir **26**, 17082–17089 (2010)

Capillary Origami

Supone Manakasettharn, J. Ashley Taylor and Tom N. Krupenkin
Department of Mechanical Engineering, The University of Wisconsin-Madison, Madison, WI, USA

Synonyms

Capillarity induced folding; Elasto-capillary folding; Surface tension–powered self-assembly

Definition

Capillary origami is folding of an elastic planar structure into a three-dimensional (3D) structure by capillary action between a liquid droplet/bubble and a structure surface.

Why Capillary Origami?

The fabrication of 3D structures is one of the major challenges for micro- and nano-fabrication. Folding of an elastic planar structure after patterning and release is one technique to fabricate a 3D structure using self-assembly. The term *origami* is taken from the Japanese art of paper folding; while the actuation of the folding is accomplished by using capillary forces of a fluid droplet, hence the technique has been termed *capillary origami*. The combination of the folding process with capillary forces has resulted in a new technique for micro- and nano-fabrication.

History

The term *capillary origami* was first introduced in 2007 by Charlotte Py et al. to describe the folding of a polydimethylsiloxane (PDMS) sheet into a 3D structure by using capillary forces created by a water droplet [1]. As early as 1993, Syms and Yeatman demonstrated that 3D structures could be fabricated by folding surfaces using capillary forces produced by molten solder [2]. Later Richard R. A. Syms introduced the term *surface tension–powered self-assembly* to describe the technique [3, 4]. Both of these techniques are quite similar in that 3D structures can be produced by folding elastic thin films. Both use capillary forces for self-assembly. In the first example, various liquids such as water are used, while for the second study molten metals such as solder were used, which then solidified fixing the 3D microstructures.

Principles

At the macroscale level, the influence of capillary forces is negligible compared to other forces such as gravity, electrostatic, or magnetic. Because capillary forces scale linearly with the characteristic size of the system, at sub-millimeter dimensions capillary forces begin to dominate since the majority of other forces decrease much more rapidly than the first power of the length. For example, a human cannot walk on water because capillary forces produced at the water surface are much smaller than the gravitational force acting on a human, which scales as the cube of the length. On the other hand, the much smaller water strider can easily walk on water because capillary forces are large enough to balance the gravitational force produced by the water strider. For capillary origami, capillary forces need to be large enough to counteract the weight of the liquid droplet and the structural forces of the planar layer.

In terms of energy, for capillary origami one needs to consider the interplay of three different energies: capillary energy, bending energy, and gravitational potential energy. For a two-dimensional (2D) model, the capillary energy per unit length of the interface (2D analog of the surface energy) is defined as $E_c = L'\gamma$, where L' is the length of the interfacial surface of the fluid and γ is the surface tension [5]. The bending energy per unit length is approximately $E_b = \frac{LB}{2R^2}$,

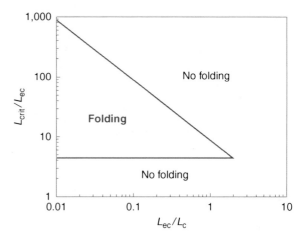

Capillary Origami, Fig. 1 Folding criteria plotted from (1) assuming complete circular folding and neglecting the effect of gravity

where L is the length of the structure and R is the radius of curvature [6]. $B = \frac{Eh^3}{12(1-v^2)}$ is the bending rigidity of the structure, where E is Young's modulus, h is thickness of the layer, and v is Poisson's ratio. If one only considers the mass of the fluid, assuming that it is much larger than the mass of the structure, then the gravitational potential energy per unit length is $E_g = \rho S g z$, where ρ is the density, S is the surface area, g is the constant of gravity, and z is the height of the center of mass. By neglecting the effect of gravity and assuming complete circular folding, de Langre et al. [7] derived simplified criteria for folding considering the interplay between capillary and bending energies, which are expressed as

$$\sqrt{2}\pi \; < \; \frac{L_{crit}}{L_{ec}} \; < \; 2\sqrt{2}\pi \frac{L_c}{L_{ec}} \qquad (1)$$

where L_{crit} is the critical length of a structure for folding to occur, $L_c = \sqrt{\frac{\gamma}{\rho g}}$ is the capillary length [5], and $L_{ec} = \sqrt{\frac{B}{\gamma}}$ is the elasto-capillary length [8].

The simplified criteria for folding derived from (1) can be plotted as shown in Fig. 1. To fold a structure requires $\frac{L_{ec}}{L_c} < 2$ or $L_c > \frac{L_{ec}}{2}$ or $\gamma > \frac{\sqrt{B\rho g}}{2}$ indicating that the capillary length must be larger than half of the elasto-capillary length so that the capillary effect can overcome bending rigidity of the structure. The other requirement for folding is $\frac{L_{crit}}{L_{ec}} > \sqrt{2}\pi \cong 4.44$ or $L_{crit} > 4.44 L_{ec}$ confirming that the length of the structure should also be long enough for a liquid droplet to

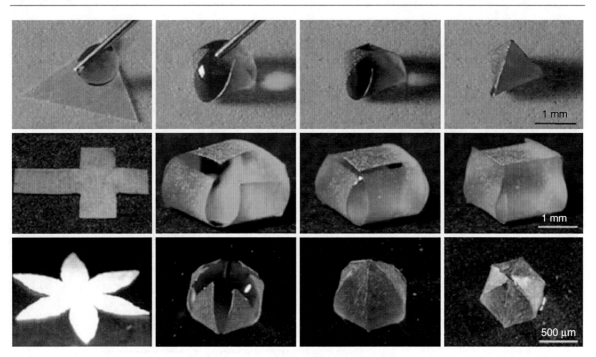

Capillary Origami, Fig. 2 Capillary origami 3D structures of a pyramid, a cube, and a quasi-sphere obtained by folding triangle-, cross-, and flower-shaped PDMS sheets, respectively, actuated with a water droplet (Reprinted with permission from [10]. Copyright 2007, American Institute of Physics)

Capillary Origami, Fig. 3 Capillary origami 3D structures formed from triangle- and flower-shaped templates using soap bubbles (scale bar: 2 cm) (Images reprinted from [11] with permission)

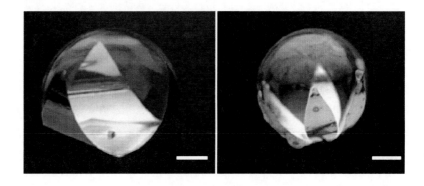

wet the surface to produce sufficient capillary forces to fold the structure. For the 3D structures the folding criteria become more complex. In particular in 3D the critical length also depends on the shape of the initial template such that $L_{crit} \cong 7L_{ec}$ for squares and $L_{crit} \cong 12L_{ec}$ for triangles [9]. Figure 2 shows examples of capillary origami structures of a pyramid, a cube, and a quasi-sphere obtained from folding triangle-, cross-, and flower-shaped PDMS sheets, respectively [10].

Besides using liquid droplets, capillary origami structures can be constructed by using soap bubbles as shown in Fig. 3. The weight of a soap bubble is much less than that of a liquid droplet especially for large droplets capable of covering centimeter size structures when gravitational forces become significant. A soap bubble was shown to fold a centimeter-size elastic structure, which cannot be accomplished using a liquid droplet [11].

Petals of a flower also can be folded into a structure similar to capillary origami when submerged in water as shown in Fig. 4. The folding of the flower in water is accomplished by the interplay of elastic, capillary, and hydrostatic forces. During submersion, hydrostatic pressure pushes against the back of petals, and surface tension prevents water from penetrating through the spacing between petals resulting in trapping an air

Capillary Origami, Fig. 4 The folding of an artificial flower when submerged in water (Reprinted with permission from [12]. Copyright 2009, American Institute of Physics)

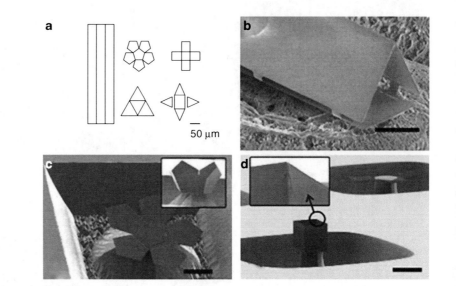

Capillary Origami, Fig. 5 (a) Schematics of initial templates. (b)–(d) SEM images of 3D microstructures after folding (scale bar: 50 μm) (Reprinted with permission from [13]. Copyright 2010, American Institute of Physics)

bubble inside a flower. The inside of the folded flower remains dry protected by the air bubble [12].

Applications

Capillary origami has been used to fabricate a number of 3D microstructures. Figure 5 illustrates the self-assembly of structures with various geometries. The initial planar templates are shown in Fig. 5(a), and the folded final 3D microstructures are shown in Fig. 5(b)–(d). The initial planar templates with lengths ranging from 50 to100 μm and a thickness of 1 μm were fabricated from silicon nitride thin films deposited and patterned by using standard micromachining processing typically used for integrated circuit and MEMS fabrication. Water droplets then were deposited on the templates to fold 3D microstructures [13]. Figure 6 shows another example of microfabrication of a quasi-spherical silicon solar cell based

on capillary origami. After fabrication by conventional micromachining processing, the initial flower-shaped silicon template was folded into a sphere using a water droplet. Unlike, conventional flat solar cells, this spherical solar cell enhanced light trapping and served as a passive tracking optical device, absorbing light from a wide range of incident angles [14].

Structures formed by capillary origami also can be actuated by using electrostatic fields to reversibly fold and unfold then. For this application we need to take into account the interplay of capillary, elastic, and electrostatic forces. As shown in Fig. 7, an electric field was applied between the droplet and the substrate. When the voltage was increased, the electrostatic force increased eventually overcoming capillary forces resulting in unfolding of the PDMS sheet. When the voltage was decreased below a certain threshold, the electrostatic force was no longer strong enough to prevent capillary forces from again folding the elastic sheet [15].

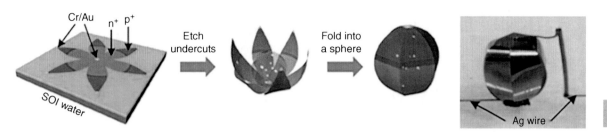

Capillary Origami, Fig. 6 Three schematics from left to right showing steps to fabricate a spherical-shaped silicon solar cell. The image at the far right shows the final spherical-shaped silicon solar cell (Images reprinted from [14] with permission)

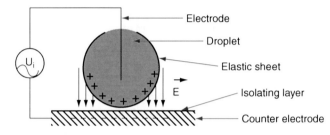

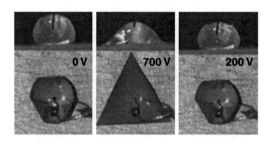

Capillary Origami, Fig. 7 Capillary origami controlled by an electric field. The schematic of the experimental setup is shown at the far left and three images to the right show the results of increasing voltage from 0 V to 700 V and decreasing voltage from 700 V to 200 V. Images from [15] – reproduced by permission of The Royal Society of Chemistry

Capillary origami is a simple and inexpensive method to fabricate 3D structures at the sub-millimeter scale. By using capillary forces, intricate and delicate 3D thin-film structures can easily be fabricated, which would be difficult to obtain by other means. More applications exploiting the advantages of capillary origami itself or in combination with electric fields can readily be envisioned. Ultimately one expects to see more commercial products based on this versatile technique.

Cross-References

▶ Self-assembly
▶ Surface Tension Effects of Nanostructures

References

1. Py, C., Reverdy, P., Doppler, L., Bico, J., Roman, B., Baroud, C.N.: Capillary origami: spontaneous wrapping of a droplet with an elastic sheet. Phys. Rev. Lett. **98**, 156103 (2007)
2. Syms, R.R.A., Yeatman, E.M.: Self-assembly of three-dimensional microstructures using rotation by surface tension forces. Electron. Lett. **29**, 662–664 (1993)
3. Syms, R.R.A.: Surface tension powered self-assembly of 3-D micro-optomechanical structures. J. Microelectromech. Syst. **8**, 448–455 (1999)
4. Syms, R.R.A., Yeatman, E.M., Bright, V.M., Whitesides, G.M.: Surface tension-powered self-assembly of microstructures – the state-of-the-art. J. Microelectromech. Syst. **12**, 387–417 (2003)
5. Berthier, J.: Microdrops and digital microfluids. William Andrew Pub, New York (2008)
6. Timoshenko, S., Woinowsky-Krieger, S.: Theory of plates and shells, 2nd edn. McGraw-Hill, New York (1959)
7. de Langre, E., Baroud, C.N., Reverdy, P.: Energy criteria for elasto-capillary wrapping. J. Fluids Struct. **26**, 205–217 (2010)
8. Bico, J., Roman, B., Moulin, L., Boudaoud, A.: Adhesion: elastocapillary coalescence in wet hair. Nature **432**, 690 (2004)
9. Py, C., Reverdy, P., Doppler, L., Bico, J., Roman, B., Baroud, C.N.: Capillarity induced folding of elastic sheets. Eur. Phys. J. Spec. Top. **166**, 67–71 (2009)
10. Py, C., Reverdy, P., Doppler, L., Bico, J., Roman, B., Baroud, C.: Capillary origami. Phys. Fluids **19**, 091104 (2007)
11. Roman, J., Bico, J.: Elasto-capillarity: deforming an elastic structure with a liquid droplet. J. Phys. Condens. Matter **22**, 493101 (2010)
12. Jung, S., Reis, P.M., James, J., Clanet, C., Bush, J.W.M.: Capillary origami in nature. Phys. Fluids **21**, 091110 (2009)
13. van Honschoten, J.W., Berenschot, J.W., Ondarcuhu, T., Sanders, R.G.P., Sundaram, J., Elwenspoek, M., Tas, N.R.: Elastocapillary fabrication of three-dimensional microstructures. Appl. Phys. Lett. **97**, 014103 (2010)

14. Guo, X., Li, H., Yeop Ahn, B., Duoss, E.B., Jimmy Hsia, K., Lewis, J.A., Nuzzo, R.G.: Two- and three-dimensional folding of thin film single-crystalline silicon for photovoltaic power applications. Proc. Natl Acad. Sci. U.S.A. **106**, 20149–20154 (2009)

15. Pineirua, M., Bico, J., Roman, B.: Capillary origami controlled by an electric field. Soft Matter. **6**, 4491–4496 (2010)

Carbon Nanotube Materials

▶ Computational Study of Nanomaterials: From Large-Scale Atomistic Simulations to Mesoscopic Modeling

Carbon Nanotube-Metal Contact

Wenguang Zhu
Department of Physics and Astronomy, The University of Tennessee, Knoxville, TN, USA

Synonyms

Carbon nanotube-metal interface

Definition

Carbon nanotube-metal contacts are widely present in many carbon nanotube-based nanodevices, and their electronic structures may significantly influence the operation and performance of carbon nanotube-based nanodevices.

Overview

Carbon nanotubes (CNTs) are quasi-one-dimensional materials with remarkable mechanical and electronic properties promising a wide range of applications from field-effect transistors (FETs) and chemical sensors to photodetectors and electroluminescent light emitters. In most of these CNT-based nanodevices, metals are present as electrodes in contact with the CNTs. Many factors including the CNT-metal contact geometry,

microscopic atomic details at the interface, and the resulting electronic structure can play a significant role in determining the functionality and performance of the devices. For instance, it has been demonstrated that an individual semiconducting CNT can operate either as a conventional FET or an unconventional Schottky barrier transistor, depending on the properties of the metal-CNT contact. In general, the electrical transport characteristics of the CNT-metal systems are sensitive to the choice of metal element as the electrode.

CNT-Metal Contact Geometry

There are two types of interface geometries of CNT-metal contacts, i.e., end contact and side contact [1, 2]. The end-contact geometry refers to the cases where metals are merely in contact with the open ends of one-dimensional CNTs, as illustrated in Fig. 1a. This contact geometry can be naturally achieved in the catalytic CVD growth of CNTs, where CNTs sprout from catalytic metal particles with the CNT axis normal to the metal surface. Figure 1b shows a sample experimental image of an end contact between a single-wall CNT and two Co tips in an in situ electron microscopy setup. The side-contact geometry refers to the cases where metals are in contact with the sidewall of CNTs, as illustrated in Fig. 1c. This contact geometry occurs when a CNT lays on the surface of a flat metal substrate. In most of CNT-based nanodevices, such as CNT FETs, metal strips are deposited from above to cover the CNTs laying on the surface as to build electrodes, fully covering sections of the CNTs, as shown in Fig. 1d. Among these two contact geometries, the side-contact geometry is more technologically relevant to CNT-based nanodevices.

Bonding and Wetting Properties of Metals on CNTs

In the end-contact geometry, metals form strong covalent bonds with carbon atoms at the open ends of CNTs [3]. The bonding energy can be as high as, e.g., 7.6 eV for a single bond at a CNT–Co contact, according to density functional calculations. Due to the strong covalent nature of the bonding, large mismatch-induced strains or high tensile strength can be built up at the interface.

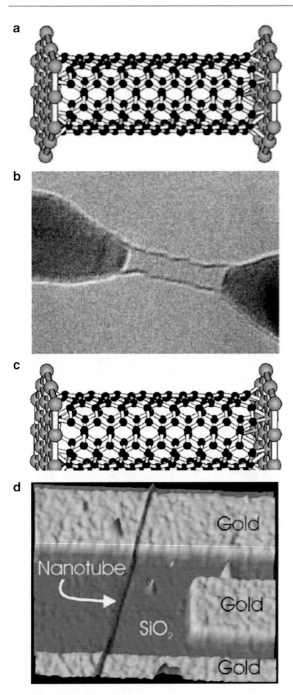

Carbon Nanotube-Metal Contact, Fig. 1 A schematic illustration of (**a**) an end contact and (**c**) a side contact between a CNT and a metal (From Palacios, J.J., Pérez-Jiménez, A.J., Louis, E., SanFabián, E., Vergés, J.A.: Phys. Rev. Lett. **90**, 106801 (2003), Fig. 1). (**b**) A CNT forming end contacts with Co tips (From Rodríguez-Manzo, J.A. et al.: Small, **5**, 2710–2715 (2009), Fig. 1). (**d**) A CNT forming side contacts on gold electrodes (From Anantram, M.P., Léonard F.: Rep. Prog. Phys. **69**, 507–561 (2006), Fig. 24)

In the side-contact geometry, metals and CNTs form much weaker bonds due to the nearly chemically inert side walls of CNTs [3]. Single-wall CNTs are built up of a cylindrically closed sheet of graphene, in which carbon atoms arranged in a honeycomb structure form very stable sp^2-hybridized covalent bonds with the p_z-orbitals of carbon extending normal to the sidewalls. The interaction between metals and CNTs in the side-contact geometry is determined by the hybridization between the carbon p_z-orbitals and the unbonded orbitals of the metals. Alkali and simple metals have binding energy around 1.5 eV per atom. Some transition metal atoms with unpaired d electrons, such as Sc, Ti, Co, Ni, Pd, Pt, form strong bonds with a binding energy around 2.0 eV per atom, whereas the transition metals with fully occupied d orbitals such as Cu, Au, Ag, and Zn have relatively weak binding with a binding energy less than 1.0 eV per atom. On the other hand, the binding energy of metal on CNTs also depends on the radius of CNTs. In general, the larger is the radius, the weaker the binding energy.

The wettability of metals on CNTs is critical to the electrical transport properties at CNT-metal contacts. In addition, CNTs can be used as templates to produce metallic nanowires with controllable radius by continuously coating the sidewalls of CNTs with metals. Experiments using different techniques such as electron beam evaporation, sputtering, and electrochemical approaches have achieved continuous coating of Ti and quasi-continuous coating of Ni and Pd on CNTs [3, 4]. Such metallic nanowires are ideal to be used as conducting interconnects in nanodevices. Metals such as Au, Al, Fe, Pb form isolated discrete clusters rather than a uniform coating layer on the surface of CNTs. Figure 2 shows sample TEM images of Ti, Ni, Pd, Au, and Fe coatings on CNTs [4]. The correlation between the wettability of these metals and their binding energies on CNTs is clear, i.e., metals with relatively strong binding energies with CNTs tend to form uniform coatings.

Electronic Structures of CNT-Metal Contacts

The electronic structure of CNT-metal contacts has a significant impact on the operation and performance of CNT-based nanodevices [2]. Due to the one-dimensional nature of CNTs and their special

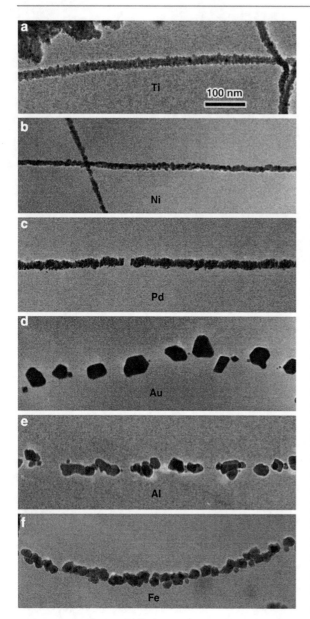

Carbon Nanotube-Metal Contact, Fig. 2 TEM images of (a) Ti, (b) Ni, (c) Pd, (d) Au, (e) Al, and (f) Fe coatings on carbon nanotubes

contact geometries, CNT-metal contacts exhibit some unusual features when compared to traditional planar contacts.

For metallic CNTs, as in contact with metals, ohmic contacts are normally formed at the interface, where no interface potential barrier exists, and the contact resistance is primarily determined by the wettability of the metal and the local atomic bonding and orbital hybridization at the interface [2]. Palladium is found to be optimal as electrodes to make ohmic contacts with metallic CNTs.

Semiconducting CNTs form either ohmic contacts or Schottky barriers at the interface with metals [2]. Figure 3 schematically illustrates the energy levels for an ohmic and a Schottky contact between a metal and a semiconducting CNT. A distinctive feature of CNT-metal contacts from traditional planar metal/semiconductor interfaces is that the height of the Schottky barrier formed at CNT-metal contacts strongly depends on the work function of the metal for a given semiconducting CNT [5, 6]. In general, at traditional planar metal/semiconductor interfaces, the Schottky barrier height shows very weak dependence on the metal work function due to the so-called Fermi-level pinning effect [7]. The strong dependence of the Schottky barrier on the metal work function in CNT-metal contacts is attributed to the reduced dimensionality of CNTs, which entirely changes the scaling of charge screening at the interface, making the depletion region decay rapidly in a direction normal to the interface and thus significantly weakening the Fermi-level pinning effect [5]. Experimental and theoretical work has shown that the interface Schottky barrier regions are much thinner in one dimension than those in three dimensions. In this case, charge carrier tunneling through the Schottky barriers becomes important. Because of the involvement of tunneling and thermionic emission in the carrier transport at the interfaces, the dependence of the on-current of CNT transistors on the Schottky barrier becomes very strong. Figure 4 shows experimental CNT-FET on-current and Schottky barrier height as a function of the CNT diameter for three different metal electrodes, Pd, Ti, and Al [6]. When the metal work functions are in the valence or conduction band of the semiconducting CNTs, ohmic contacts will likely be formed at the interface. Experimental measurements have shown that certain metals with high work functions, such as Pd, can produce nearly ohmic contacts with semiconducting CNTs. Ohmic contacts are more desirable in devices where contact resistance needs to be minimized. In addition to the metal work function, other factors, such as the contact geometry and the chemical bonding at the interface also play important roles in the transport properties of CNT-metal contacts.

Carbon Nanotube-Metal Contact, Fig. 3 A schematic illustration of the energy levels for an ohmic and a Schottky contact between a metal and a semiconducting CNT

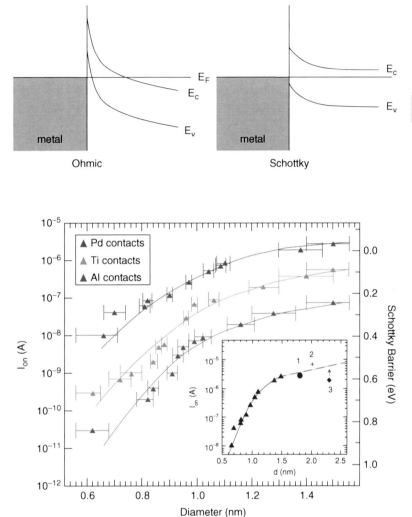

Carbon Nanotube-Metal Contact,

Fig. 4 Experimental CNT-FET on-current (*left axis*) and computed Schottky barrier height (*right axis*) as a function of the CNT diameter for three different metal electrodes, Pd, Ti, and Al

Cross-References

▶ Carbon Nanotubes for Chip Interconnections
▶ Carbon-Nanotubes
▶ CMOS-CNT Integration

References

1. Banhart, F.: Interactions between metals and carbon nanotubes: at the interface between old and new materials. Nanoscale **1**, 201–213 (2009)
2. Anantram, M.P., Léonard, F.: Physics of carbon nanotube electronic devices. Rep. Prog. Phys. **69**, 507–561 (2006)
3. Ciraci, S., Dag, S., Yildirim, T., Gülseren, O., Senger, R.T.: Functionalized carbon nanotubes and device applications. J. Phys. Condens. Matter **16**, R901–R960 (2004)
4. Zhang, Y., Franklin, N.W., Chen, R.J., Dai, H.J.: Metal coating on suspended carbon nanotubes and its implication to metal-tube interaction. Chem. Phys. Lett. **331**, 35–41 (2000)
5. Léonard, F., Tersoff, J.: Role of fermi-level pinning in nanotube schottky diodes. Phys. Rev. Lett. **84**, 4693–4696 (2000)
6. Chen, Z.H., Appenzeller, J., Knoch, J., Lin, Y.-M., Avouris, P.: The role of metal – nanotube contact in the performance of carbon nanotube field-effect transistors. Nano Lett. **5**, 1497–1502 (2005)
7. Tung, R.T.: Recent advances in Schottky barrier concepts. Mater. Sci. Eng. R **35**, 1–138 (2001)

Carbon Nanotube-Metal Interface

▶ Carbon Nanotube-Metal Contact

Carbon Nanotubes

▶ Ecotoxicology of Carbon Nanotubes Toward Amphibian Larvae
▶ Computational Study of Nanomaterials: From Large-Scale Atomistic Simulations to Mesoscopic Modeling

Carbon Nanotubes (CNTs)

▶ Chemical Vapor Deposition (CVD)
▶ Physical Vapor Deposition

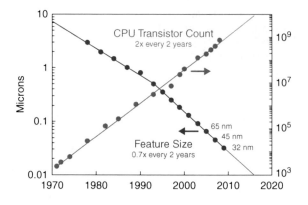

Carbon Nanotubes for Chip Interconnections, Fig. 1 Moore's Law: Transistor count has doubled while feature size has decreased by 0.7X every 2 years (Figure reprinted with permission from Kuhn [1])

Carbon Nanotubes for Chip Interconnections

Gilbert Daniel Nessim
Chemistry department, Bar-Ilan Institute of
Nanotechnology and Advanced Materials (BINA),
Bar-Ilan University, Ramat Gan, Israel

Synonyms

Carbon nanotubes for interconnects in integrated circuits; Carbon nanotubes for interconnects in microprocessors

Definition

Chip interconnections electrically connect various devices in a microprocessor. Today's established technology for interconnects is based on copper. However, it may be technically challenging to extend copper use to future interconnects in microprocessors with smaller lithographic dimensions due to materials properties limitations. Carbon nanotubes are currently investigated as a potential replacement for future integrated circuits (microprocessors). Although carbon nanotubes are a clear winner against copper in terms of materials properties, multiple fabrication challenges need to be overcome for carbon nanotubes to enter the semiconductor fab and replace copper for chip interconnections.

Motivation

Following over 40 years of successful fulfillment of Moore's law, stating that the number of transistors in a chip doubles every 2 years, we have already moved from microelectronics to nanoelectronics [1]. Although the *"end of scaling"* has been predicted many times in the past, enormous technical challenges, especially quantum mechanical issues and billion-dollar lithography investments, are a serious threat to further miniaturization (Fig. 1).

Today's latest processors are manufactured using the 32-nm technology. To move toward the 22-nm node and beyond, issues such as lithographic limitations, leaking currents in ultra-thin dielectrics (only a few monolayers thick), insufficient power and thermal dissipation, and interconnect reliability must be resolved [1]. At the transistor level, the performance is negatively affected by increased off-state currents due to short channel effects, increased gate leakage due to tunneling through nanometer-thin dielectric layers, and increased overall gate capacitance due to decreasing gate pitch.

Although quantum mechanical tunneling and leakage currents may eventually stop further scaling, efficient heat removal from a chip is currently the biggest obstacle. In this respect, the many kilometers of copper interconnects present in today's chips are the main culprit for heat generation. For instance, in 2004, Magen et al. [2] showed that for a microprocessor fabricated with the 0.13-μm-node technology consisting of 77 million transistors, interconnects consumed more than 50% of the total dynamic power.

Given the increased length of interconnects, their reduced cross section, and the increased current densities circulating into the interconnects of our latest chips, the problem has been further exacerbated.

Additionally, copper interconnects are a major contributor to the total resistance-capacitance (RC) delay of the chip, can fail by electromigration, and need a liner to avoid diffusion into the silicon. Bottom line, the interconnect issue is so serious that the International Semiconductor Roadmap [3] (ITRS, an expert team assessing the semiconductor industry's future technology requirements for the next 15 years) indicates copper interconnects as a possible dealbreaker to further miniaturization for IC nodes beyond 22 nm.

Many technology options are currently under investigation to replace copper for interconnects. Among them, we can mention other metals (mainly silicides), wireless, plasmonics, and optical interconnects. Most notably, there has been an intense research effort on new nanotechnology materials such as carbon nanotubes, which, at the theoretical level, could solve all the above technical issues suffered by copper.

The plan of this section is to first introduce the reader to copper interconnects' fabrication and limitations. Next, we will compare copper to carbon nanotubes (CNTs) and detail possible models for implementation. An important paragraph will focus on the state-of-the-art of CNT fabrication, prior to concluding on the outstanding issues and outlook for future CNT-based chip interconnections.

Background on Copper Interconnects and Dual-Damascene Process

In 1997, IBM introduced the revolutionary *"dual-damascene"* process to fabricate copper interconnects and to replace aluminum interconnects, the industry standard at the time. Compared to aluminum, copper presents two major advantages: (1) 50% lower resistivity (Cu ≈ 1.75 μm cm versus Al ≈ 3.3 μm cm) and (2) higher current densities before failure by electromigration (up to 5×10^6 A/cm^2) [4, 5]. Although, as a material, copper was a clear winner against aluminum, fabrication challenges delayed its introduction. Historically, we may be at a similar juncture with carbon nanotubes compared to copper as we were with copper compared to aluminum in 1997: in spite of their superior materials properties,

mainly fabrication issues are now preventing the introduction of nanotubes in the semiconductor industry to replace copper interconnects.

Copper diffuses into silicon, generating mid-gap states that significantly lower the minority carrier lifetime and which lead to leakage in diodes and bipolar transistors. Copper also diffuses through SiO$_2$ and low-k dielectrics, and therefore requires complete encapsulation in diffusion barriers. Since no dry etches were known for copper, IBM's bold innovation of polishing using chemical mechanical polishing (CMP), after electroplating the copper, was significantly at odds with the technological processes at that time in semiconductor fabrication.

The copper dual-damascene process consists of the following steps:
- Develop a pattern for wires or vias by patterned etching of the dielectric.
- Deposit a barrier layer (usually Ta) to prevent copper diffusion into silicon.
- Deposit a copper seed layer.
- Fill the vias with copper using electrodeposition.
- Remove excess copper using CMP.
- Repeat the process to lay the alternating layers of wires and vias which will form the complete wiring system of the chip (Fig. 2).

Typical microprocessor design follows a *"reverse scaling"* metallization scheme with multiple layers of interconnects labeled as local, intermediate, and global interconnects, with increasing width. Very thin local interconnects locally connect gates and transistors within a functional block and are usually found in the lower two metal layers. The wider and taller intermediate interconnects have lower resistance and provide clock and signal within a functional block up to 4 μm. Global interconnects are found at the top metal layers and provide power to all functions in addition to connecting functional blocks through clock and signal. They are usually longer than 4 μm (up to half of the chip perimeter) and exhibit very low resistance to minimize RC delay and voltage drop. Below are a typical cross section of an I.C. chip and a possible implementation using CNTs (Figs. 3 and 4).

Limitations of Copper Interconnections

Copper interconnects have efficiently scaled down to the current 32-nm-node microprocessors, although this

Carbon Nanotubes for Chip Interconnections,
Fig. 2 Dual-damascene process of copper filling an interconnect via (Figure reprinted with permission from Jackson et al. [21])

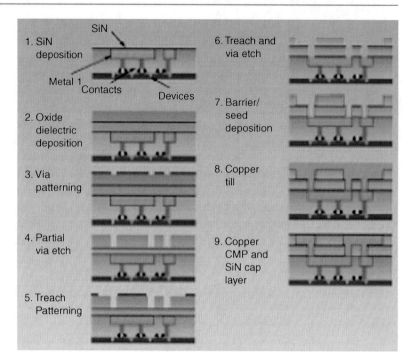

Carbon Nanotubes for Chip Interconnections,
Fig. 3 Typical cross sections of hierarchical scaling in current microprocessor (Figure reprinted with permission from the Semiconductor Industry Association [22])

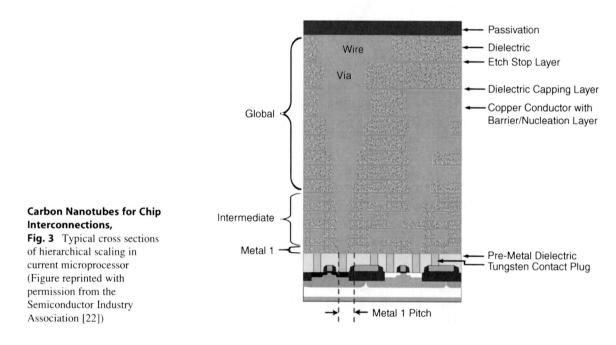

has required many technological advances to allow ever-shrinking copper cross sections to carry increasing currents without failure. However, we may be very close to smashing against a technical wall because of materials failure and related fabrication issues.

Alternative materials or technologies would require many changes in semiconductor fabrication and massive investments; thus the large semiconductor companies are doing the impossible to extend copper application to future nodes. It is clear that only when up against an insurmountable technical wall will the semiconductor industry switch to a new technology.

Electrical resistance is a major issue now that copper interconnect cross sections are comparable to

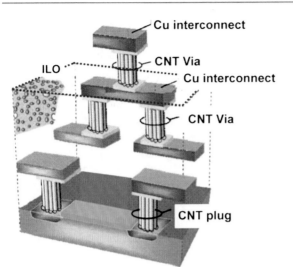

Cu interconnect
ILO
CNT Via
Cu interconnect
CNT Via
CNT plug

Carbon Nanotubes for Chip Interconnections, Fig. 4 Schematic view of possible implementation of carbon nanotube via interconnects in lieu of copper (Figure reprinted with permission from Awano et al. [7])

Carbon Nanotubes for Chip Interconnections, Table 1 Selected critical parameters for copper use as interconnects in future IC nodes (Data from the Semiconductor Industry Association [22])

Year of production (Estimated)	2010	2015	2020
MPU/ASIC metal 1 ½ pitch (nm) (contacted)	45	25	14
Total interconnect length (m/cm^2) – Metal 1 and 5 intermediate levels, active wiring only	2,222	4,000	7,143
Barrier/cladding thickness (for Cu intermediate wiring) (nm)	3.3	1.9	1.1
Interconnect RC delay [ps] for a 1-mm Cu intermediate wire, assumes no scattering and an effective ρ of 2.2 μΩ-cm	1,132	3,128	9,206

the mean free path of electrons in copper (∼40 nm in Cu at room temperature). Grain boundary and surface scattering are significant contributors to the increased resistance, especially now that we have reached nanoscale dimensions. At the microstructural level, the grain boundaries play an important role, hence, among other fabrication concerns, controlling the copper grain size during electrodeposition has allowed to limit the grain boundary scattering impact thus far.

The steep rise in interconnect resistance for smaller IC nodes is a major source of RC delays and directly affects the chip reliability by increasing the risk of electromigration failure, a major issue for further downscaling. Electromigration is the transport of material caused by the gradual movement of the copper ions due to the momentum transfer between conducting electrons and diffusing metal atoms, which occurs for high current densities, which can create voids leading to open circuits. Given that downscaling leads to a reduction of the interconnects' cross section, the problem is amplified at subsequently smaller nodes. To compound the issue, the need for a resistive diffusion barrier layer, also called a liner (usually Ta), to avoid copper diffusion into silicon, further reduces the available conductive copper cross section, thus increasing the risk of electromigration failure, especially as the operating temperature rises.

In addition to the increased resistance and the electromigration failure risk, many other aspects of the dual-damascene process are becoming potential sources of failure as the node shrinks. Among the many integration concerns, we can mention materials issues such as interface adhesion between the different materials (copper, low-k dielectrics, etc.), liner effectiveness, metal voids, CMP interface defects, etc. Concurrently, there is a long list of process-related issues such as the need for etch/strip/clean processes (to avoid damage to low-k dielectric materials), atomic layer deposition (ALD) processes to deposit liners, copper plating and CMP techniques, etc.

A few interesting numerical estimates taken from the 2009 projections from ITRS, [3] provide the reader with the magnitude of the technical challenge to extend copper interconnect technology (Table 1).

As already mentioned, many alternative technologies are currently being investigated for replacement of copper as interconnect material that would require significant chip redesigns and new fabrication technologies. Some examples include optical interconnects, radio frequency (RF) interconnects, plasmonics, and 3-D interconnects (probably still copper). The interested reader can find more details on these alternative technologies in the review paper from Havemann et al. [6] (now a little dated) or in the latest ITRS report on interconnects [3].

The Case for Carbon Nanotube Interconnects

An interesting solution, which has been the subject of intense research in recent years, is to replace copper

Carbon Nanotubes for Chip Interconnections, Fig. 5 Graphical representations of ideal graphene sheet, SWCNT, MWCNT (Figure reprinted with permission from Graham et al. [22])

with carbon nanotubes. If a reliable and repeatable fabrication process consistent with Complementary Metal Oxide Semiconductor (CMOS) technology requirements could be developed, integration into existing chip architectures may not require significant process redesign (Fig. 5).

Carbon nanotubes, which can be visualized as rolled sheets of graphene, have been widely investigated as a promising new material for many electrical device applications [5, 7] (e.g., transistor (CNT-FET), interconnects) as they exhibit exceptional electrical, thermal, and mechanical properties [8]. When comparing materials properties, CNTs are a clear winner against copper. Studies show that CNTs are stable for current densities up to 10^9 A/cm^2, two orders of magnitude higher than copper. CNTs can exhibit multichannel ballistic conduction over distances of microns. Because of their higher chemical stability relative to copper, diffusion barriers (liners) are not needed for CNTs, thus allowing a larger conductive cross section compared to copper for the same technology node. Additionally, their mechanical tensile strength (100 times that of steel) and their high thermal conductivity (comparable to diamond) give CNTs an edge compared to copper. Finally, growing CNTs in high aspect ratio vias could allow the design of chips with higher interlayer spacing to reduce overall RC losses and to decrease chip-layer energy dissipation [9].

Before examining possible models of CNT-based interconnect architectures, it is important to clearly understand CNTs' electrical properties, which represent the most critical material limitation to resolve with respect to copper. The electronic band structures of single-wall CNTs (SWCNTs) and of graphene are very similar. For graphene and metallic SWCNTs, the valence band and the conduction band

touch at specific points in the reciprocal space. For semiconducting SWCNTs, the conduction band and the valence band do not touch. Semiconducting SWCNTs have been extensively studied as channels in transistor devices while metallic SWCNTs have been considered for applications such as IC interconnects and field emission.

The resistance of a CNT contacted at both ends is the sum of three resistances [5, 10]:

$$R_{CNT} = R_Q + R_L + R_{CONTACT}$$

where R_Q is the quantum resistance, R_L is the scattering resistance, and $R_{CONTACT}$ is the contact resistance. We will now discuss these three resistances.

An ideal (defect-free) metallic SWCNT electrically contacted at both ends, in the absence of scattering or contact resistance, exhibits a resistance $R = 2R_Q \approx 13k\Omega$ as a SWCNT has two conduction channels. The quantum resistance $R_Q = 6.5k\Omega$ is due to the mismatch between the number of conduction channels in the nanotube and the macroscopic metallic contacts. The one-dimensional confinement of electrons, combined with the requirement for energy and momentum conservation, leads to ballistic conduction over distances in the order of a micron.

The scattering resistance is due to impurities or nanotube defects that reduce the electron mean free path, and depends on the length l of the SWCNT:

$$R_L = \frac{1}{2R_Q}\left(\frac{l}{l_0}\right)$$

For defect-free SWCNT lengths below a micron, we can neglect the scattering resistance

The contact resistance, which results from connecting the SWCNT to a contact (usually metallic), depends strongly on the material in contact with the nanotube, and on the difference between their work functions. The work functions of multiwall CNTs (MWCNTs) and SWCNTs have been estimated to be 4.95 and 5.10 eV, respectively [10]. Palladium has been found to be one of the materials minimizing the contact resistance, better than titanium or platinum contacts (which exhibited nonohmic behavior when in contact with CNTs) [10].

For interconnect applications, most often bundles of SWCNTs are considered. It is important to note that the coupling between adjacent SWCNTs is negligible

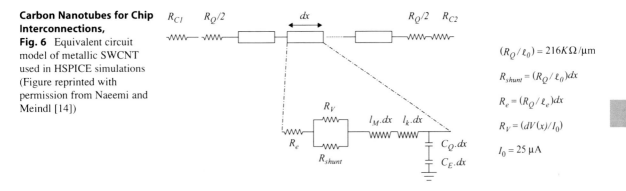

Carbon Nanotubes for Chip Interconnections, Fig. 6 Equivalent circuit model of metallic SWCNT used in HSPICE simulations (Figure reprinted with permission from Naeemi and Meindl [14])

$$(R_Q / \ell_0) = 216 K\Omega / \mu m$$

$$R_{shunt} = (R_Q / \ell_0) dx$$

$$R_e = (R_Q / \ell_e) dx$$

$$R_V = (dV(x)/I_0)$$

$$I_0 = 25 \mu A$$

since, for defect-free SWCNTs, the electrons would rather travel along the SWCNT axis (ballistic path) than across SWCNTs because of the large inter-CNT tunneling resistance (2–140 MΩ) [10]. Thus, the resistance of a bundle of SWCNTs can be viewed as a parallel circuit of the resistances of the individual SWCNTs. If we have n SWCNTs, the resistance of the bundle will be:

$$R_{SCWNT\ bundle} = \frac{R_{SCWNT}}{n}$$

The above overview related to SWCNTs. The electrical properties of MWCNTs have not been as extensively studied because of the additional complexities arising from their structure, as every shell has different electronic characteristics and chirality, in addition to interactions between the shells [11]. Geometrically, the interwall distance in a MWCNT is 0.34 nm, the same as the spacing between graphene sheets in graphite. What still has to be clarified is how the conductivity of a MWCNT varies with the number of walls.

Initially, it was thought that the conductance of a MWCNT occurred only through the most external wall, which seems to be the case at low bias and temperatures, where electronic transport is dominated by outer-shell conduction. However, theoretical models and experimental results indicate that shell-to-shell interactions can significantly lower the resistance of MWCNTs with many walls [5, 10].

One view is that the conductivity of a MWCNT with n walls is simply n times the conductivity of a SWCNT. Li et al. [12] experimentally measured an electrical resistance of only 34.4Ω for a large MWCNT with outer diameter of 100 nm and inner diameter of 50 nm (= > 74 walls). This value was much lower than

the one that could be calculated assuming all walls participated separately in the electrical conduction (i.e., calculated as the parallel of the resistances for each wall) showing that interwall coupling contributes to additional channels of conduction. Naeemi et al. [13] also assumed intercoupling between CNT walls in their models to increase the channels of conduction with increasing number of walls. However, their conductance was lower compared to that measured experimentally by Li's team. Bottom line: The conductivity of MWCNTs increases with the number of walls but the exact relationship has not yet been exactly clarified.

Models of CNTs as Interconnects

Various studies investigated replacing copper interconnects with bundles of CNTs (SWCNTs or MWCNTs) or with one large MWCNT. A first approach consists of using densely packed SWCNTs. Modeling SWCNTs as equivalent electrical circuits and using SPICE simulations, Naeemi et al. [14] showed that a target density of SWCNTs of at least 3.3×10^{13} CNTs/cm^2 was required. Currently achieved maximum densities for CNTs in vias barely reach 10^{12} CNTs/cm^2, which is still an order of magnitude smaller than required. Furthermore, since statistically only one-third of the SWCNTs grown are metallic (the other two-third are semiconducting), the conduction of the bundle will only occur in the metallic SWCNTs (Fig. 6).

Naeemi et al. [14] also compared SWCNT bundles to copper as local, intermediate, and global interconnects. They showed that, in SWCNT bundles, resistance and kinetic inductance decreased linearly with the number of nanotubes in the bundle, while magnetic inductance changed very slowly. The resistance of

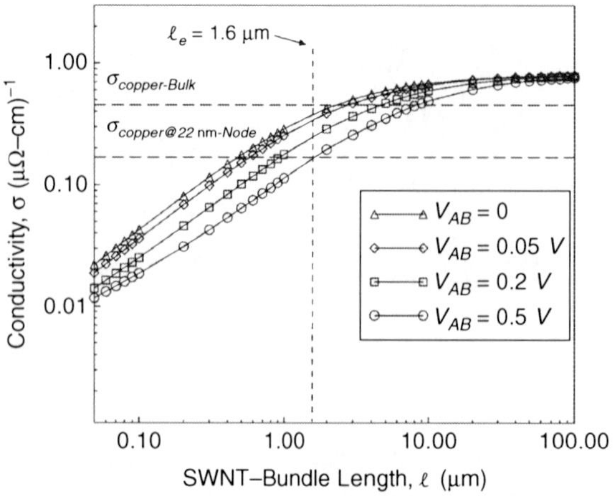

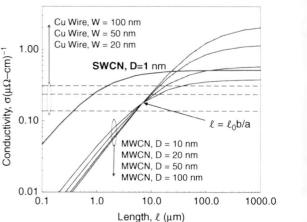

Carbon Nanotubes for Chip Interconnections,
Fig. 7 Conductivity of densely packed SWCNT bundles versus length for various bias voltages (Figure reproduced with permission from Naeemi and Meindl [14])

Carbon Nanotubes for Chip Interconnections,
Fig. 8 Conductivity of MWCNTs with various diameters compared to Cu wires and dense bundles of SWCNTs (Figure reproduced with permission from Naeemi and Meindl [13])

a bundle of SWCNTs with sufficient metallic nanotubes was smaller than the resistance of copper wires, while capacitance was comparable. SWCNT bundles also fared better compared to copper in reducing power dissipation, delay, and crosstalk. For local interconnects, they quantified the improvements as 50% reduction in capacitance, 48% reduction of capacitance coupling between adjacent lines, and 20% reduction in delay. For intermediate interconnects, the improvements were more marked, especially in terms of improved conductivities. For global interconnects, dense SWCNT bundles proved critical to improve bandwidth density (Fig. 7).

Using MWCNTs, which are all electrically conductive as they exhibit multiple channels of conduction (compared to only one-third metallic SWCNTs), could lower the resistivity of the bundle, although fewer of them can be packed in the same space because they usually have larger diameters (but also require a lower packing density compared to SWCNTs). In a different modeling study, Naeemi et al. [13] explored the suitability of MWCNTs as replacement for copper interconnects. They concluded that for long lengths (over 100 μm), MWCNTs have conductivities many times that of copper and even of SWCNT bundles. However, for short lengths (less than 10 μm), dense SWCNT bundles can exhibit a conductivity that is twice that of MWCNT bundles. Thus, for via applications, they recommended using dense bundles of SWCNTs or,

alternatively, bundles of MWCNTs with small diameter (i.e., with few walls) (Fig. 8).

Using an individual MWCNT with large diameter could offer high conductivity due to the participation of multiple walls to significantly increase the channels for conduction. However, as previously mentioned, the exact relationship between the number of walls and the conductance has yet to be clarified.

In conclusion, the choice of nanotubes may differ depending on the type of interconnect. For instance, dense bundles of SWCNTs or MWCNTs with few walls may be more suitable for small-section vertical vias, while dense bundles of larger MWCNTs may be more appropriate for long-range interconnects. The option of using a large MWCNT which fills all the space available needs to be further investigated. It is also plausible that hybrid systems of copper/SWCNTs/MWCNTs may be the best solution; for instance, small-section vertical vias may be replaced by dense SWCNT bundles, while larger long-range horizontal interconnects may still use copper, dense bundles of MWCNTs, or even metal-CNT composites.

Practical Implementation: Fabrication State of the Art and Outstanding Issues

Reliable and repeatable high-yield CNT fabrication compatible with CMOS standards is the main

Carbon Nanotubes for Chip Interconnections,
Fig. 9 Pictorials comparing the "grow-in-place" and "grow-then-place" techniques (Reproduced with permission from Professor Carl V. Thompson [18])

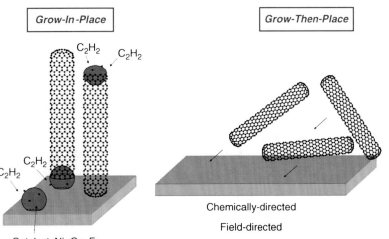

bottleneck in replacing copper in chip interconnections. Although hundreds of research teams have focused their efforts on nanotube growth and thousands of papers detailing growth recipes have been published, surprisingly, very few have focused on the growth on conductive substrates at CMOS-compatible processing temperature [7, 10, 15, 16]. It is still a challenge to reliably and consistently synthesize CNTs on conductive layers at temperatures below 400–450°C, the maximum temperature allowed in CMOS fabrication to avoid disrupting previous diffusion patterns. Furthermore, it is still difficult to precisely control CNT diameter and height, although chemical vapor deposition (CVD) from thin films of controlled thickness [17] or from nanoparticles [16] of controlled size has shown encouraging results.

To utilize carbon nanotubes in industrial applications, two main approaches have been considered: *"grow-in-place"* and *"grow-then-place"* [18] (Fig. 9).

Grow-then-place: This technique consists of first preparing nanotubes and subsequently transferring them to a substrate. Arc discharge and laser ablation are the main techniques used to synthesize free-standing nanotubes. The nanotubes may be subsequently selected (e.g., separating SWCNTs or metallic SWCNTs) and purified prior to use. To transfer them to another substrate, CNTs are usually functionalized in a way that they will attach to pre-patterned areas of the substrate which will attract functionalized CNTs. An interesting technique for interconnect vias is based on using electrophoresis to push CNTs dispersed in a liquid solution into a matrix with pits (e.g., porous alumina matrix).

The advantages of this method are that it places no restrictions on the process or temperature used for CNT synthesis and allows to pretreat the CNTs (e.g., select, purify, functionalize). The major drawback is that no successful and repeatable technique to transfer the CNTs to the substrate has been developed to date. The challenge of resolving this issue appears too high to make this technique a candidate for the CMOS industry. However, free-standing, purified, CNTs are manufactured by many companies and sold for other applications (e.g., CNT-polymer composites).

Grow-in-place: this technique usually consists of preparing the sample with a catalyst present in the locations where the nanotubes will be synthesized. For instance, a thin catalyst film can be deposited using e-beam evaporation or sputtering; alternatively, nanoparticles can be deposited on a substrate. Synthesis is usually performed using thermal or assisted (e.g., plasma) CVD.

This method has several advantages: (1) good control of nanotube position (CNTs will grow where there are catalysts), (2) proven recipes to obtain crystalline CNTs (at least on insulating substrates), (3) proven capabilities to obtain carpets of vertically aligned CNTs, (4) physical contact with the substrate, (5) electrical contact with the substrate, and (6) CVD techniques are commonplace in the CMOS industry.

The major drawbacks are that (1) the processing temperature should be below 400–450°C (CMOS-compatibility), thus putting serious limits on the synthesis method and (2) the CNTs should be directly synthesized on the substrate of choice, usually

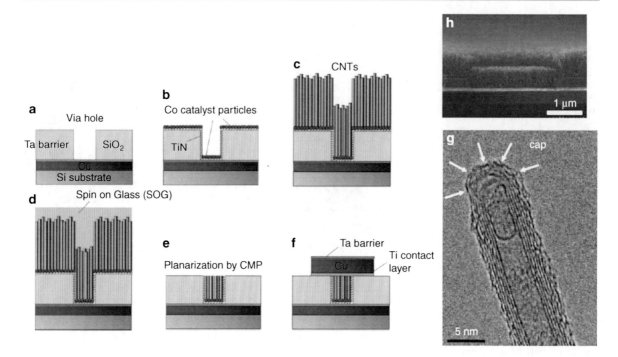

Carbon Nanotubes for Chip Interconnections, Fig. 10 Process to synthesize CNTs into pits, SEM cross section, and TEM showing crystalline MWCNT (Reprinted with permission from Yokoyama et al. [23])

a metallic layer to provide electrical contact. Although growing dense carpets of crystalline CNTs on insulating substrates such as alumina or silicon oxide has been achieved by many, CNT growth on metallic layers still remains a serious challenge. Interactions between the catalyst and the metallic substrate (e.g., alloying) are the major impediments for the successful growth of dense carpets of CNTs on metallic layers.

In addition to interesting results obtained from university research, good progress on the growth and characterization of carbon nanotubes for interconnects has been achieved by industrial laboratories, initially by Infineon and now by the Fujitsu laboratories. In 2002, Kreupl et al. [19] of Infineon showed that bundles of CNTs could be grown in pits of defined geometry.

More recently, Awano et al. [7] of Fujitsu grew bundles of MWCNTs in a 160 nm via at 450°C and measured an electrical resistance of 34Ω (the CNT density observed was 3×10^{11} CNTs/cm^2). This follows a previous result obtained earlier by the same team where they grew MWCNTs into a 2 μm via at temperatures close to 400°C with the lowest resistance measured of 0.6Ω after CMP and annealing in a hydrogen atmosphere [5] (Fig. 10).

Kreupl et al. [19] succeeded in growing a single MWCNT into a 25-nm hole and measured a high resistance of 20–30 kΩ. Given the difficulty of growing a single large MWCNT and the difficult task of making a precise electrical measurement, there may be room for further improvement if we could grow an individual, crystalline (defect-free) MWCNT with the maximum number of walls for a given external diameter, thus maximizing the number of channels of conduction.

Most of the effort on CNT synthesis to replace copper interconnects has focused on vertical growth of dense carpets of CNTs, which can be achieved by a high density of active catalyst dots. In contrast there have been fewer successful reports of horizontal growth, with less spectacular results. Many techniques have been used to achieve horizontal alignment among which we can mention high gas flow rates, electric fields, and epitaxial techniques to guide horizontal alignment of the nanotubes [10]. In 2010, Yan et al. [20] obtained an interesting horizontal growth of bundled CNTs with a density of 5×10^{10} CNTs/cm^2, which is approaching what has been achieved for vertical CNT growth (although still over an order of magnitude lower compared to the best result for vertical CNT growth) (Fig. 11).

**Carbon Nanotubes for Chip
Interconnections,**
Fig. 11 Scanning electron
microscope images of dense
carpets of horizontally aligned
CNTs grown using CVD
(Figure reproduced from Yan
et al. [20])

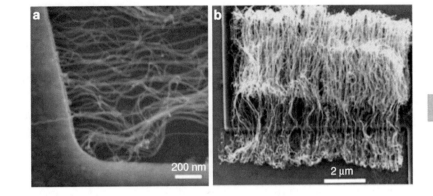

Although the experimental results obtained are encouraging, there are still numerous challenges that need to be resolved for CNTs to enter the semiconductor fab:

1. Increase the CNT areal density by one or two orders of magnitude. For SWCNTs, assuming all of them are metallic, a packing density of 10^{13}–10^{14} CNTs/cm^2 is required to compete with copper in terms of resistance, while for MWCNTs, the required packing density is lower and depends on the number of channels of conduction (i.e., number of walls). This will require, among other considerations, adequate catalyst and underlayer materials choice and deposition, possible surface pretreatment (e.g., plasma, reduction, and etching), maximum nucleation of active catalyst dots, and optimizing the CNT growth process. Although CNT areal density is an important issue, it may not be the dealbreaker.

2. Minimize the contact resistance between the CNTs and the substrate. To achieve maximum conductivity, the choice of the appropriate underlayer is critical; specific metals (e.g., Pd) and possibly silicides are good candidates. For MWCNTs, it is also important to ensure electrical contact with all the walls.

3. When using SWCNTs, synthesize only metallic SWCNTs (on average one-third of the SWCNTs grown) which are the ones participating in the electrical conduction. This is closely linked to the issue of chirality control, for which no solution has been proposed to date. Selective catalyst choice may provide an alternative avenue to synthesize a higher fraction of metallic SWCNTs.

4. Control growth direction of CNTs. This is especially challenging for horizontal interconnects where the directionality and the packing density achieved are still lagging compared to vertical

growth of CNTs, despite some interesting progress in this area [10, 20].

5. Synthesize crystalline, defect-free CNTs to ensure maximum electrical conductivity in the nanotube. This is challenging, especially when combined with the requirement of growing CNTs at low temperature to achieve CMOS compatibility.

6. Synthesize CNTs at temperatures below 400–450°C to ensure CMOS compatibility.

7. Repeatably yield the same CNT structures when the same process conditions are applied. This is a major issue since important variations in the structure and shape of the CNTs grown have been experimentally observed.

Conclusions and Outlook for CNTs as Chip Interconnections

In the past decade, the synthesis of CNTs and the understanding of their growth mechanisms have massively improved. However, for CNTs to enter the CMOS fab and replace copper, significant challenges still need to be resolved. In my opinion, the most significant challenge to overcome is developing a reliable and repeatable fabrication process consistent with CMOS conditions. To achieve that tall order, we need to improve our understanding of the CNT growth mechanisms. Although many simulation models and many experimentally-based insights have been achieved [10], there are still many questions related to CNT growth mechanisms that have not been fully answered. For instance:

– Which precursor gases favor CNT growth and which gases hinder CNT growth? What is the role of the gases in the resulting level of crystallinity of the CNTs grown? Could we pretreat the gases to improve the CNT yield or structure?

– What is the exact role of the catalyst? How does its materials properties and its lattice structure influence the resulting CNTs grown (in shape and structure)?

– What is the role of the underlayer (layer below the catalyst) and its interactions with the catalyst? Why is it so challenging to grow CNTs on metallic layers?

In addition to improving our mechanistic understanding of CNT growth, I believe that a parallel effort focused on developing better reactors is needed. Most researchers use standard CVD-based systems that were designed for a general purpose. A customized reactor, where the same growth conditions can be repeatably achieved with very small variations could provide the repeatability in results (CNT structures) that has eluded us so far.

Carbon nanotubes are already becoming a manufacturing reality in mechanical engineering applications (e.g., CNT-based composites) and many interesting results have been obtained to develop novel CNT structures for electrical applications. Although the jury is still out, if process repeatability could be achieved, we could hope, not only that CNTs will enter the CMOS fabs and replace copper for chip interconnections, but also that they will lead to innovative ventures requiring lower investments to develop integrated circuits with radically new architectural designs using carbon nanotubes as new building blocks.

Cross-References

▶ Carbon Nanotube-Metal Contact
▶ Carbon Nanotubes
▶ Chemical Vapor Deposition (CVD)
▶ CMOS-CNT Integration
▶ Nanotechnology
▶ Physical Vapor Deposition
▶ Synthesis of Carbon Nanotubes

References

1. Kuhn, K.J.: Moore's Law Past 32 nm: Future Challenges in Device Scaling. Intel Publication, Hillsboro (2009)
2. Magen, N., Kolodny, A., Weiser, U.: Interconnect-power dissipation in a microprocessor. In: Proceedings of the 2004 International Workshop, Paris, 1 Jan 2004
3. ITRS. International Technology Roadmap for Semiconductors - Interconnect 2009, International Sematech, Austin
4. Goel, A.K.: High-Speed VLSI Interconnections, 2nd edn. Wiley/IEEE, Hoboken (2007)
5. Nessim, G.D.: Carbon Nanotube Synthesis for Integrated Circuit Interconnects. Massachusetts Institute of Technology, Cambridge, MA (2009)
6. Havemann, R.H., Hutchby, J.A.: High-performance interconnects: an integration overview. Proc. IEEE 89(5), 586–601 (2001)
7. Awano, Y., Sato, S., Nihei, M., Sakai, T., Ohno, Y., Mizutani, T.: Carbon nanotubes for VLSI: interconnect and transistor applications. Proc. IEEE 98(12), 2015–2031 (2010)
8. Dresselhaus, M.S., Dresselhaus, G., Avouris, P. (eds.): Carbon Nanotubes: Synthesis, Structure, Properties, and Applications. Springer, Berlin (2001)
9. Chen, F., Joshi, A., Stojanović, V., Chandrakasan, A.: Scaling and evaluation of carbon nanotube interconnects for VLSI applications. In: Nanonets Symposium 07, Catania, 24–26 Sept 2007
10. Nessim, G.D.: Properties, synthesis, and growth mechanisms of carbon nanotubes with special focus on thermal chemical vapor deposition. Nanoscale 2(8), 1306–1323 (2010)
11. Collins, P.G., Avouris, P.: Multishell conduction in multiwalled carbon nanotubes. Appl. Phys 74(3), 329–332 (2002)
12. Li, H.J., Lu, W.G., Li, J.J., Bai, X.D., Gu, C.Z.: Multichannel ballistic transport in multiwall carbon nanotubes. Phys. Rev. Lett. 95(8), 086601 (2005)
13. Naeemi, A., Meindl, J.D.: Compact physical models for multiwall carbon-nanotube interconnects. IEEE Electr. Device Lett. 27(5), 338–340 (2006)
14. Naeemi, A., Meindl, J.D.: Design and performance modeling for single-walled carbon nanotubes as local, semiglobal, and global interconnects in gigascale integrated systems. IEEE T Electron Dev 54(1), 26–37 (2007)
15. Nessim, G.D., Seita, M., O'Brien, K.P., Hart, A.J., Bonaparte, R.K., Mitchell, R.R., Thompson, C.V.: Low temperature synthesis of vertically aligned carbon nanotubes with ohmic contact to metallic substrates enabled by thermal decomposition of the carbon feedstock. Nano Lett. 9(10), 3398–3405 (2009)
16. Awano, Y., Sato, S., Kondo, D., Ohfuti, M., Kawabata, A., Nihei, M., Yokoyama, N.: Carbon nanotube via interconnect technologies: size-classified catalyst nanoparticles and low-resistance ohmic contact formation. Phys. Status Solidi 203(14), 3611–3616 (2006)
17. Nessim, G.D., Hart, A.J., Kim, J.S., Acquaviva, D., Oh, J.H., Morgan, C.D., Seita, M., Leib, J.S., Thompson, C.V.: Tuning of vertically-aligned carbon nanotube diameter and areal density through catalyst pre-treatment. Nano Lett. 8(11), 3587–3593 (2008)
18. Thompson, C.V.: Carbon nanotubes as interconnects: emerging technology and potential reliability issues. In: 46th International Reliability Symposium; 2008: IEEE CFP08RPS-PRT, p. 368, 2008
19. Kreupl, F., Graham, A.P., Duesberg, G.S., Steinhogl, W., Liebau, M., Unger, E., Honlein, W.: Carbon nanotubes in interconnect applications. Microelectron. Eng. 64(1–4), 399–408 (2002)

20. Yan, F., Zhang, C., Cott, D., Zhong, G., Robertson, J.: High-density growth of horizontally aligned carbon nanotubes for interconnects. Phys Status Solidi. **247**(11–12), 2669–2672 (2010)

21. Jackson, R.L., Broadbent, E., Cacouris, T., Harrus, A., Biberger, M., Patton, E., Walsh, T.: Processing and integration of copper interconnects. In: Solid State Technology. Novellus Systems, San Jose (1998)

22. Graham, A.P., Duesberg, G.S., Hoenlein, W., Kreupl, F., Liebau, M., Martin, R., Rajasekharan, B., Pamler, W., Seidel, R., Steinhoegl, W., Unger, E.: How do carbon nanotubes fit into the semiconductor roadmap? Appl. Phys. **80**, 1141–1151 (2005). Copyright 2005, Springer Berlin/Heidelberg

23. Yokoyama, D., Iwasaki, T., Yoshida, T., Kawarada, H., Sato, S., Hyakushima, T., Nihei, M., Awano, Y.: Low temperature grown carbon nanotube interconnects using inner shells by chemical mechanical polishing. Appl. Phys. Lett. **91**, 263101 (2007). Copyright 2007, American Institute of Physics

Carbon Nanotubes for Interconnects in Integrated Circuits

▶ Carbon Nanotubes for Chip Interconnections

Carbon Nanotubes for Interconnects in Microprocessors

▶ Carbon Nanotubes for Chip Interconnections

Carbon Nanowalls

▶ Chemical Vapor Deposition (CVD)

Carbon-Nanotubes

▶ Robot-Based Automation on the Nanoscale

Catalyst

▶ Chemical Vapor Deposition (CVD)
▶ Physical Vapor Deposition

Catalytic Bimetallic Nanorods

▶ Molecular Modeling on Artificial Molecular Motors

Catalytic Chemical Vapor Deposition (CCVD)

▶ Chemical Vapor Deposition (CVD)

Catalytic Janus Particle

▶ Molecular Modeling on Artificial Molecular Motors

Cathodic Arc Deposition

▶ Physical Vapor Deposition

Cavity Optomechanics

▶ Optomechanical Resonators

Cell Adhesion

▶ Bioadhesion
▶ BioPatterning

Cell Manipulation Platform

▶ Biological Breadboard Platform for Studies of Cellular Dynamics

Cell Morphology

▶ BioPatterning

Cell Patterning

▶ BioPatterning

Cellular and Molecular Toxicity of Nanoparticles

▶ Cellular Mechanisms of Nanoparticle's Toxicity

Cellular Electronic Energy Transfer

▶ Micro/Nano Transport in Microbial Energy Harvesting

Cellular Imaging

▶ Electrical Impedance Tomography for Single Cell Imaging

Cellular Mechanisms of Nanoparticle's Toxicity

Francelyne Marano, Rina Guadagnini, Fernando Rodrigues-Lima, Jean-Marie Dupret, Armelle Baeza-Squiban and Sonja Boland
Laboratory of Molecular and Cellular Responses to Xenobiotics, University Paris Diderot - Paris 7, Unit of Functional and Adaptive Biology (BFA) CNRS EAC 4413, Paris cedex 13, France

Synonyms

Cellular and molecular toxicity of nanoparticles

Definition

The interaction between nanoparticles and cell triggers a cascade of molecular events which could induce toxicity and cell death. They are associated to the uptake of nanoparticles, their persistence at cellular level, their ability to release free radicals, and to induce an oxidative stress. The resulting activation of molecular pathways and transcription factors could lead to a pro-inflammatory response or, depending on the level of free radicals, apoptosis.

Background

The last five years have shown an increasing number of papers on the nanoparticle's mechanisms of cytotoxicity. What are the reasons? It is likely that the specific useful properties which appear at nanoscale can also lead to adverse effects. This hypothesis is strongly supported by in vivo and in vitro studies which compare the toxicity of NPs with their fine counterparts of the same chemical composition. These results have clearly demonstrated a higher toxicity of particles at nanoscale than at microscale. Moreover, it appears from experimental studies that solid nano-sized particles could be translocated beyond the respiratory tract and could induce a systemic response. The interstitial translocation of a same mass of particles is higher for ultrafine than fine particles after intratracheal instillation in rats [1]. Surface area, which is strongly increased for nanoparticles compared to microparticles of same chemical composition, and surface reactivity are considered as the principal indicators of NP's reactivity. It was shown that a toxic response could be observed even to apparently nontoxic substances when the exposure occurred in the nanometre size range. All these observations have lead to the development of a new field of toxicology, nanotoxicology [2]. However, the toxicological mechanisms which sustained the biological response are not yet clear and a matter of debates.

The concerns about the toxicity of engineered nanoparticles, which are increasingly used for industrial and medical applications, came also from the knowledge on the toxicity of non-intentional atmospheric particles. Short-term epidemiological studies in Europe and North America have shown an association between cardio-respiratory morbidity and mortality and an increased concentration of atmospheric fine particles [3]. Moreover, long-term epidemiological studies have also demonstrated an association between exposure to atmospheric particles

(Particulate Matter or PM10 and 2.5) and increased cancer risk [3]. In parallel, in vitro and in vivo studies on fine and ultrafine airborne particles such as Diesel exhaust particles and PM2.5, gave causal explanations to these adverse health effects (reviewed in [1]). They allow to define the molecular events induced by these particles in lung cells. The major event is a pro-inflammatory response which is characterized by various cytokine releases (pro-inflammatory mediators), associated to the activation of transcription factors, and signaling pathways. This was especially demonstrated for diesel exhaust particles (DEP), a major component of urban PM in Europe. These events are mostly induced by organic components of the DEP and are probably mediated by the generation of reactive oxygen species (ROS) during the metabolism of organic compounds (for a review see [4]). These findings were used as a background for the researches on biological mechanisms induced by NPs considering that fine and ultrafine atmospheric particles have great similarities with NPs, especially Diesel exhaust particles which are of nano-size and aggregate after their release in atmosphere.

It became rapidly obvious that the understanding of the cellular and molecular mechanisms leading to the biological effects of NPs is essential for the development of safe materials and accurate assays for risk assessment of engineered NPs [2] and several recent reviews were focused on demonstrated or hypothetic cellular mechanisms of these responses [4–6, 11].

The first event, when NPs enter in contact with the human body by inhalation, ingestion, dermal exposure, or intravenous application, is their interaction in the biological fluids and the cellular microenvironment with biological molecules such as proteins thus forming a protein corona [7]. Consequently, NPs do not directly interact with the cell membrane but through the protein and/or lipids of the corona. NPs-bound proteins may recognize and interact with the membraneous receptors or could bind nonspecifically cellular membranes. Whatever these interactions, they seem to play a central role which could determine further biological responses. In particular, these interactions may drive the uptake of NPs by the first target cells at the level of the biological barriers such as immune cells (macrophages, dendritic cells, and neutrophiles) or epithelial and endothelial cells. This uptake seems to be general for many NPs which are able to bind proteins at their surface and the paradigm

of "Trojan horse" was developed to explain this uptake and the further biological responses. One of the first responses is the direct or indirect production of ROS which is associated to the size, the chemical composition and the surface reactivity of the NPs. This common response occurs for a large number of NPs even with different chemical patterns and different abilities to form agglomerates, so that, the paradigm of the central role of oxidative stress was developed [5]. These authors suggested that "although not all materials have electronic configurations or surface properties to allow spontaneous ROS generation, particle interactions with cellular components are capable of generating oxidative stress." Further activation of nuclear factors and specific genetic programs are associated to the level of ROS production leading to cell death by necrosis and apoptosis or adaptive responses such as pro-inflammatory responses, antioxidant enzyme activation, repair processes, effects on cell cycle control and proliferation. Over the last years, numerous in vitro studies have confirmed this hypothesis leading to the development of assays using the detection of ROS or oxidative stress for the screening of NPs. However, new data during the last year have pointed out other specific effects of NPs which are not related to oxidative stress. For example, NPs can interact with membrane receptors, induce their aggregation and mimic sustained physiological responses through specific signaling pathways in the target cells. This type of mechanism may contribute to the development of diseases but could also be of use to develop therapeutic strategies whereby NPs activate or block specific receptors.

Cellular Uptake of Nanoparticles and Their Fate at Cellular Level

The uptake of particles by specialized immune cells in humans is a normal process which leads to their removal and contributes to the integrity of the body. However, depending on the level of the uptake, this process could induce an increasing release of inflammatory mediators and disturbance of phagocyte normal functions such as the clearance and the destruction of pathogens. One of the knowledge of the 50 last years on the effects of a sustained exposure to airborne particles, especially at occupational level, is the concept of overloading. If the mechanisms of

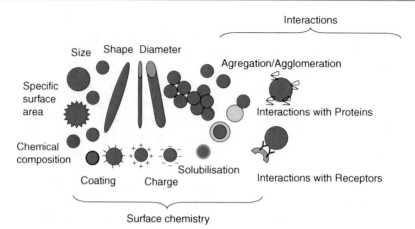

Cellular Mechanisms of Nanoparticle's Toxicity, Fig. 1 Different physicochemical characteristics of the nanomaterials involved in their biological activity: size, surface area, shape, bulk chemical composition, surface chemistry including solubility as well as surface charge or coatings, interactions between particles leading to agglomeration, and aggregation as well as with proteins leading to "corona" or with receptors in the cell membrane

clearance are not sufficient to eliminate the particles and if they are persistent, the particles could accumulate in the tissues, leading to a sustained inflammation and chronic pathologies. This was demonstrated for exposure to quartz, asbestos, coal, mineral dusts but also for long time exposure to heavy PM-polluted atmospheres such as in Mexico City [1]. These questions of uptake and persistence are fundamental for risk assessment evaluation of NPs. This may explain the number of papers published recently that analyze the mechanisms of uptake, the behavior and the translocation of various NPs. So far, it appears that the response depends on several parameters. Taken together, NP's surface and its specific chemical composition resulting from the engineering processes, the capacity of NPs to form aggregates (particle comprising strongly bonded or fused NPs) or agglomerates (collection of weakly bound NPs), the methods used for dispersion and experimental preparation determine in the NP's ability to adsorb or not specific biological compounds, such as proteins, to form the "corona" and to interact with biological membranes [7]. The amount and the structural/functional properties of the adsorbed proteins drive the interactions of these nanomaterials with the membranes and their uptake (Fig. 1). Recent studies have clearly identified a number of serum proteins such as albumin, IgG, IgM, IgA, apolipoprotein E, cytokines, or transferrin that bind to carbon black, titanium dioxide, acrylamide or polystyrene NPs [8]. Among the identified proteins, several are ligands for cellular receptors and may contribute to the biological

effects of NPs. For example, receptor aggregation induced by NPs could lead to cell signaling: coated gold NPs were able to bind and cross-link IgE-Fc epsilon receptors leading to degranulation and consequent release of chemical mediators [9].

On another hand, integrins such as $\alpha_5\beta_3$ are known to play a key role in cell signaling and their activation by extracellular ligands can modulate biological processes such as matrix remodeling, angiogenesis, tissue differentiation, and cell migration. These receptors were recently demonstrated as important membrane targets for carbon nanoparticles and their activation induced lung epithelial cell proliferation which was due at least in part to $\beta1$-integrin activation [6].

As far as, uptake process is concerned, it is likely that different cell types might have different uptake mechanisms, even for the same NPs. The possible pathways of cellular uptake were previously described by several authors (see [10]). It could occur through phagocytosis, macropinocytosis, clathrin-mediated endocytosis, non-clathrin, non-caveolae-mediated endocytosis, caveolae-mediated endocytosis, or diffusion (Fig. 2). These mechanisms have been described for different NPs and may occur for the same NP depending on the cell type, the medium, the level of aggregation. Therefore, uptake processes are considered as very complex and not easy to measure. Dawson et al. 2009 [12] have postulated that the uptake depends mostly on the size: NPs less than 100 nm can enter the cells and less than 40 nm in the nucleus. It was also suggested that the size of the NPs determine caveolin

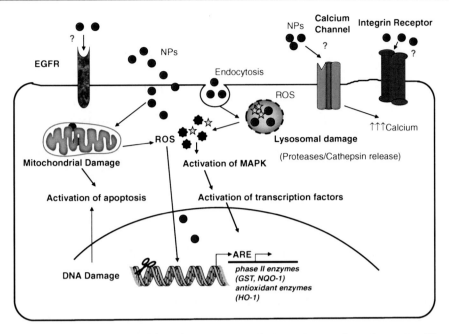

Cellular Mechanisms of Nanoparticle's Toxicity, Fig. 2 A schematic representation of NP's triggered cellular pathways through membrane receptors, ROS production and implication of oxidative stress in these responses. NPs could induce activation of EGF or integrin receptors can lead to apoptosis, inflammation or proliferation. ROS produced by NPs in immediate cellular environment or inside the cells lead to activation of redox-dependent signaling pathways like MAPK and the activation of transcription factors for example, AP-1, NF-kB, or Nrf2. They migrate to the nucleus and modify gene expression of cytobines, phase 2 enzymes (gluthation S transferase or GST, quinone oxydoréductase 1 or NQO-1), and antioxydant enzymes (hemeoxygenase 1 or HO-1). Oxidative stress could also results in the damage of different organelles like the mitochondria, lysosomes and nucleus resulting apoptosis. Accumulation of high intracellular calcium levels through a direct effect on calcium channel might also act as an alternative mechanism for the induction of these mechanisms (Adapted from [11])

versus clathrin dependent uptakes [13]. However, these oversimplified scenarios are refuted by obvious discrepancies in the recent literature about the optimal size, shape, and mechanisms of internalization of NPs.

The surface charge of the NPs could be an important factor for uptake since the negatively charged surface membrane could favor the positively charged NPs for higher internalization. However, negatively charged NPs were also shown to have enhanced uptake as compared to unfunctionalized NPs, perhaps by their possible interactions with proteins. Endocytosis of small NPs is energy-dependent and associated to lipid rafts, dynamin and F-actin mechanisms. Phagocytosis and macropinocytosis are mostly involved in endocytosis of large particles (more than 500 nm) and also in the uptake of the aggregates or agglomerates of NPs which could be promoted by their opsonisation in the biological fluids. Macropinocytosis (which is one kind of pinocytosis) is also an important mechanism for positively charged NPs and TiO_2 or carbon black aggregates internalization [14].

The behavior of the NPs after their uptake is another important question but, surprisingly, as far as now, little is known about the intracellular fate of NPs. Most of the transmission electron microscopy (TEM) observations have shown the NPs in cytoplasm vesicles limited by membranes. These vesicles could further be transported in the cytoplasm through the microtubule network. The biopersistence of nano-materials which are resistant to degradation in the endosomal compartment could be one of the factors of further toxicity and accumulation. However, several metal oxide NPs are toxic after dissolution in the cell. Indeed, the uptake of ZnO NPs into the lysosomal acidic medium accelerates their dissolution and the release of Zn^{2+} ions in the cytoplasm. Their excess could induce cytokine production and cytotoxicity and the initiation of acute inflammation at the level of the target organ such as the lung.

NPs such as TiO_2 or carbon black NPs were also observed only in the cytoplasm of cells [14]. Two explanations may be put forward. The first one is that

NPs could directly enter by diffusion through the lipid bilayer. It has been shown that cationic NPs could pass through cell membranes by generating transient holes without membrane disruption [15]. Another possible explanation could be the release of NPs after rupture of endosomal compartment. It was described that cationic NPs, after binding to lipid groups on the surface cell membrane, could be endocysed in vesicles and accumulated into the lysosomal compartment. Within, they are able to sequestrate protons which could lead to the activation of proton pump and further rupture of the ion homeostasis and lysosomal accumulation of water. The subsequent lysosomal swelling and membrane rupture leads to the cytoplasmic release of NPs [16]. In proliferating cells, these cytoplasmic NPs associated or not to microtubules, could enter in the nucleus during the mitosis, and could explain that non soluble NPs were observed in the nucleus [14]. More rarely, NPs were also observed within the mitochondrial matrix but, so far, no explanation was given to explain this organelle localisation.

The Cellular Stress Induced by Nanoparticles and Its Biological Consequences

Over the last 10 years of research conducted on the mechanisms of toxicity of non- intentional as well as engineered nanoparticles, has led to the establishment of a consensus within the scientific community of toxicologists to consider the central role of oxidative stress in cellular responses to NPs leading to inflammation or apoptosis [5, 17]. The concept of oxidative stress was developed for many years to explain dysfunctions leading to pathologies. Oxidative stress could occur when reactive oxygen species (ROS) are overproduced leading to an imbalance between ROS production and antioxidant defence capacity. It could also occur when the organism shows a deficiency in antioxidant systems and, especially in antioxidant enzymatic systems (superoxide dismutase, catalase, and glutathione peroxidase). An increased concentration of ROS, exceeding the antioxidant capacity of the cells, can lead to oxidative damage at molecular or cellular level.

ROS have important cellular roles either by acting as second messengers for the activation of specific pathways and gene expressions or by causing cell death. In the hierarchical oxidative stress model in response to NPs, Nel et al. [5] propose that a minor level of oxidative stress leads to the activation of the antioxidant protection whereas, at a higher level, cell membrane and organelles injuries could lead to cell death by apoptosis or necrosis, but specific signaling pathways and gene expression are involved at each step. The induction of oxidative stress by several NPs is due to their ability to produce ROS (TiO_2 for example) or to lead to their production. The surface properties of NPs modulate the production of ROS and the smaller they are; the higher is their surface area and their ability to react with biological components and to produce ROS. However, if this cellular induction appears to be general, all the NPs are not able to produce ROS and the cellular increase of the latter could be an indirect effect of the uptake.

ROS interact nonspecifically with biological compounds; however, some macromolecules are more sensitive such as the unsaturated lipids, the amino acids with a sulphydril group (SH) and guanine sites in nucleic acids. When lipid bilayer is attacked by ROS, cascade peroxidation occurs leading to the disorganization of the membranes and of their functions (exchange, barriers, and information). The most sensitive proteins contain methionine or cysteine residues, especially in their active site and their oxidation could lead to modify their activity and even to their inactivation.

The adaptive cellular responses to NPs are associated to the modulation of different redox-sensitive cellular pathways. Tyrosine kinases and serine/threonine kinase such as mitogen-activated protein kinases or MAP kinases were especially studied (ERK, p38, and JNK) in association with several transcription factors such as NFκB. The free radical can degrade the NFκB inhibitor IκB by the activation of the cascades leading to its proteolysis. The activation of NFκB induces its translocation within the nucleus and its link to consensus sequences in the promoter of numerous genes leading to their transcription. It is also true for other transcriptions factors such as AP1 and NrF2. The latter plays an essential role in the Antioxidant Response Element (ARE) mediated expression of phase 2 enzymes such as NQO1 (NADPH quinone oxidoreductase-1) and antioxidant enzymes such as heme-oxygenase-1 (HO-1). Indeed, HO-1 was found to be activated by CeO_2 NPs exposure of human bronchial cells via the p38-Nrf-2 signaling pathway. The ability of NPs to interact with these signaling pathways could partially explain their cytotoxicity. Recently, TiO_2 and SiO_2 NPs were demonstrated in vitro and in vivo to induce the release of $IL1\beta$ and $IL1\alpha$, two potent mediators of innate immunity, via the activation of inflammasome, a large multiprotein complex containing caspase 1 which cleaves pro $IL1\alpha$ and β in their active forms. These results lead to consider that these NPs could induce a potent inflammatory response. However, the mechanisms leading to this activation are not yet clear.

Another important target of ROS produced by NPs is DNA. Oxidative damage of DNA could generate intra-chain adducts and strand breakage. The bond between the base and desoxyribose could also be attacked leading to an abasic site and the attack on the sugar could create a single strand-break. The genotoxicity of NPs begins to be studied and recent reviews pointed out the possible genotoxic mechanisms.

However, oxidative stress appears now not sufficient to explain all the biological effects of NPs. The role of epidermal growth factor receptor (EGFR) was investigated by the group of K. Unfried with the demonstration that carbon black NPs induce apoptosis and proliferation via specific signaling pathways both using EGFR [18]. Carbon black NPs could also impair phagosome transport and cause cytoskeletal dysfunctions with a transient increase of intracellular calcium not associated with the induction of ROS since antioxidants did not suppress the response and which could be due to a direct effect on ion channels that control the calcium homeostasis in the cell [19]. Even if all the mechanisms are not completely demonstrated, it appears now that transmembrane receptors are implicated in NP-induced cell signaling and could lead to specific biological responses to NPs.

Nanoparticles and Cell Death

NPs have also been shown to induce either apoptotic or necrotic cell death in a variety of in vitro systems depending on the concentration and duration of exposure. This induction of cell death mechanism by NPs might act as basis of different pathologies and consequently it is important to understand NPs induced apoptosis pathways. Cells are able to undergo apoptosis through two major pathways, the extrinsic pathway with the activation of death receptors and the intrinsic

pathway with the central role of mitochondria, its permeabilization and the release of Cytochrome C leading to the activation of apoptosome. Recently, the permeabilization of lysosomal membrane was also shown to initiate apoptosis with the release of cathepsins and other hydrolases from the lysosomal lumen. The molecular pathways of apoptosis induction by carbon black and titanium dioxide NPs in human bronchial epithelial cells were recently studied. It was shown that the initial phase of apoptosis induction depends upon the chemical nature of the NPs. Carbon black NPs triggered the mitochondrial pathway, with the decrease of mitochondrial potential, the activation of bax (a pro-apoptotic protein of the Bcl2 family), and the release of cytochrome C, and the production of ROS is implicated in the downstream mitochondrial events. Whereas TiO_2 NPs induced lysosomal pathway with lipid peroxidation, lysosomal membrane destabilization, and catepsin B release [20], Lysosomal permeabilization has also been shown to be important in silica NPs induced apoptosis. These results point out the necessity of a careful characterization of the molecular mechanisms involved by NPs and not just describing the final outcome.

Future Directions of Research

The interactions between nanomaterials and their biological target are essential to explain their biological effect and the interest of the recent researches on the cellular mechanisms induced by NPs is to take in account the specificity of the cells and of their microenvironment. The first step is the formation of the corona in biological fluids whose composition and affinity kinetics strongly depend on the characteristics of NPs and, especially, their size and surface reactivity. This coating of proteins influences the aggregation, the final size and, finally, the uptake of NPs via the interaction with the membranes, their specific receptors or lipid rafts. It could determine if the nanomaterial is bioavailable and if it induces or not adverse interactions. The central mechanism proposed to explain the biological response is the oxidative stress. However, this paradigm is debated because very similar oxidative stress effects observed in cellular models and induced by different particles could lead in vivo to different pathological effects. It is now obvious that oxidative stress is a common and nonspecific mechanism in toxicology and that the responses at the level of the cell depend on the perturbation of the redox balance with a few number of induced signaling pathways. The different biological responses could depend on the tissue specificity which could lead to different diseases observed after occupational or environmental exposure to well known particles or fibers.

Recent studies have also shown that NPs could develop a response without a direct contact with the cells but after an induction of secreted factors, it is the "bystander effect." Small molecules such as purines could be increased at cytoplasmic level in response to NPs, transferred through the gap junctions within a tissue to activate specific receptors [10]. Moreover, NP-induced apoptosis was also demonstrated to be propagated through hydrogen peroxide mediated bystander killing in an in vitro model of human intestinal epithelium. These specific responses could explain the in vivo observed differences. Finally, the interactions of NPs with proteins, enzymes, cytokines, and growth factors, outside, or inside the cell lead to modify the functions of these proteins with a possible indirect pathological effect.

The large variety of engineered NPs in the market and under development makes these studies very complex. However, the development of safe nanomaterials depends on better knowledge of these specific interactions

Cross-References

► Ecotoxicity of Inorganic Nanoparticles: From Unicellular Organisms to Invertebrates
► Genotoxicity of Nanoparticles
► In Vivo Toxicity of Carbon Nanotubes
► In Vivo Toxicity of Titanium Dioxide and Gold Nanoparticles
► Quantum-Dot Toxicity
► In Vitro and In Vivo Toxicity of Silver Nanoparticles
► Toxicology: Plants and Nanoparticles

References

1. Donaldson, K., Borm, P. (eds.): Particle Toxicology, p. 434. CRC Press, Boca Raton (2007)
2. Oberdorster, G., Oberdorster, E., Oberdorster, J.: Nanotoxicology: an emerging discipline evolving from studies of ultrafine particles. Environ. Health Perspect. 113, 823–839 (2005)
3. Brunekreef, B., Holgate, S.T.: Air pollution and health. Lancet 360, 1233–1242 (2002)
4. Marano, F., Boland, S., Baeza-Squiban, A.: Particle-associated organics and proinflammatory signaling. In: Donaldson, K., Borm, P. (eds.) Particle Toxicology, pp. 211–226. CRC Press, Boca Raton (2007)
5. Nel, A., Xia, T., Madler, L., Li, N.: Toxic potential of materials at the nanolevel. Science 311, 622–627 (2006)
6. Unfried, K., Albrecht, C., Klotz, L.O., Mikecz, A.V., Grether-Beck, S., Schins, R.P.F.: Cellular responses to nanoparticles: target structures and mechanisms. Nanotoxicology 1, 52–71 (2007)
7. Nel, A.E., Madler, L., Velegol, D., Xia, T., Hoek, E.M., Somasundaran, P., Klaessig, F., Castranova, V., Thompson, M.: Understanding biophysicochemical interactions at the nano-bio interface. Nat. Mater. 8, 543–557 (2009)
8. Lynch, I., Salvati, A., Dawson, K.A.: Protein-nanoparticle interactions: what does the cell see? Nat. Nanotechnol. 4, 546–547 (2009)
9. Huang, Y.F., Liu, H., Xiong, X., Chen, Y., Tan, W.: Nanoparticle-mediated IgE-receptor aggregation and signaling in RBL mast cells. J. Am. Chem. Soc. 131, 17328–17334 (2009)
10. Bhabra, G., Sood, A., Fisher, B., Cartwright, L., Saunders, M., Evans, W.H., Surprenant, A., Lopez-Castejon, G., Mann, S., Davis, S.A., Hails, L.A., Ingham, E., Verkade, P., Lane, J., Heesom, K., Newson, R., Case, C.P.: Nanoparticles can cause DNA damage across a cellular barrier. Nat. Nanotechnol. 4, 876–883 (2009)
11. Marano, F., Hussain, S., Rodrigues-Lima, F., Baeza-Squiban, A., Boland, S.: Nanoparticles: molecular target and cell signaling. Arch. Toxicol. 85, 733–741 (2011). Online May 10
12. Dawson, K.A., Salvati, A., Lynch, I.: Nanotoxicology: nanoparticles reconstruct lipids. Nat. Nanotechnol. 4, 84–85 (2009)
13. Rejman, J., Oberle, V., Zuhorn, I.S., Hoekstra, D.: Size-dependent internalization of particles via the pathways of clathrin- and caveolae-mediated endocytosis. Biochem. J. 377, 159–169 (2004)
14. Hussain, S., Boland, S., Baeza-Squiban, A., Hamel, R., Thomassen, L.C., Martens, J.A., Billon-Galland, M.A., Fleury-Feith, J., Moisan, F., Pairon, J.C., Marano, F.: Oxidative stress and proinflammatory effects of carbon black and titanium dioxide nanoparticles: role of particle surface area and internalized amount. Toxicology 260, 142–149 (2009)
15. Gratton, S.E., Ropp, P.A., Pohlhaus, P.D., Luft, J.C., Madden, V.J., Napier, M.E., Desimone, J.M.: The effect of particle design on cellular internalization pathways. Proc. Natl. Acad. Sci. U.S.A. 105, 11613–11618 (2008)
16. Xia, T., Kovochich, M., Liong, M., Zink, J.I., Nel, A.E.: Cationic polystyrene nanosphere toxicity depends on cell-specific endocytic and mitochondrial injury pathways. ACS Nano 2, 85–96 (2008)
17. Ayres, J.G., Borm, P., Cassee, F.R., Castranova, V., Donaldson, K., Ghio, A., Harrison, R.M., Hider, R., Kelly, F., Kooter, I.M., Marano, F., Maynard, R.L., Mudway, I., Nel, A., Sioutas, C., Smith, S., Baeza-Squiban, A., Cho, A., Duggan, S., Froines, J.: Evaluating the toxicity of airborne particulate matter and nanoparticles by measuring oxidative stress

potential–a workshop report and consensus statement. Inhal. Toxicol. **20**, 75–99 (2008)

18. Sydlik, U., Bierhals, K., Soufi, M., Abel, J., Schins, R.P., Unfried, K.: Ultrafine carbon particles induce apoptosis and proliferation in rat lung epithelial cells via specific signaling pathways both using EGF-R. Am. J. Physiol. Lung Cell. Mol. Physiol. **291**, L725–L733 (2006)
19. Moller, W., Brown, D.M., Kreyling, W.G., Stone, V.: Ultrafine particles cause cytoskeletal dysfunctions in macrophages: role of intracellular calcium. Part. Fibre Toxicol. **2**, 7 (2005)
20. Hussain, S., Thomassen, L.C., Feracatu, I., Borot, M.C., Andreau, K., Fleury, J., Baeza-Squiban, A., Marano, F., Boland, S.: Carbon black and titanium oxide nanoparticles elicit distinct apoptosic pathways in bronchial epithelial cells. Part. Fibre Toxicol. **7**(10), 1–17 (2010). Online Apr.16

Cellular Toxicity

▶ Nanoparticle Cytotoxicity

Characterization of DNA Molecules by Mechanical Tweezers

▶ DNA Manipulation Based on Nanotweezers

Characterizations of Zinc Oxide Nanowires for Nanoelectronic Applications

▶ Fundamental Properties of Zinc Oxide Nanowires

Charge Transfer on Self-Assembled Monolayer Molecules

▶ Charge Transport in Self-Assembled Monolayers

Charge Transport in Carbon-Based Nanoscaled Materials

▶ Electronic Transport in Carbon Nanomaterials

Charge Transport in Self-Assembled Monolayers

Jeong Young Park
Graduate School of EEWS (WCU), Korea Advanced Institute of Science and Technology (KAIST), Daejeon, Republic of Korea

Synonyms

Charge transfer on self-assembled monolayer molecules

Definition

Charge transport in self-assembled monolayers (SAMs) is the transport of an electron or a hole through an organized molecule layer which is bound to a substrate.

Overview

Charge Transport Through Organic Molecules

Significant studies on charge transport properties through organic molecules have been carried out in the general area of molecule-based and molecule-controlled electronic devices, often termed "molecular electronics" [1, 2]. Self-assembled monolayers (SAMs) are composed of an organized layer of amphiphilic molecules in which one end of the molecule, the "head group," shows a special affinity for a substrate [3]. SAMs also consist of a tail with a functional group at the terminal end, as seen in Fig. 1.

Charge transport of organic molecules is usually limited by hopping processes and is therefore dominated by surface ordering. Self-assembled monolayers are a good model system of molecular electronics due to the ordered surface structure. In order to measure charge transport in a self-assembled monolayer, the substrate surface should be metallic. For example, a gold surface exhibits strong bonds with alkanethiol through S–H bonds. The other electrode should also be metallic for charge transport through the self-assembled monolayer. The measurement scheme of charge transport through a self-assembled monolayer

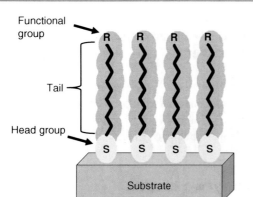

Charge Transport in Self-Assembled Monolayers, Fig. 1 Schematic of a self-assembled monolayer (SAMs) showing the head group that is bound to the substrate. SAMs consist of a tail with a functional group at the terminal

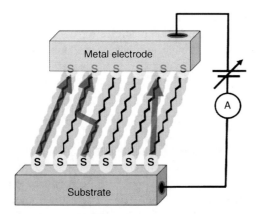

Charge Transport in Self-Assembled Monolayers, Fig. 2 Scheme of charge transport mechanisms through self-assembled monolayers. The dominant charge transport mechanism in a molecular junction involves "through-bond" (TB) tunneling, and "through space" (TS) as illustrated in the left and right transport channel, respectively

that represents a conductor-molecule-conductor junction is shown in Fig. 2.

Charge Transport Mechanism

For insulating molecules, such as alkane chains, electron transport occurs via tunneling mechanisms. When such molecules are placed between electrodes, the junction resistance changes exponentially: $R = R_0 \exp(\beta s)$, with electrode separation s, where R_0 is the contact resistance and β a decay parameter. In most experiments, the separation s is the length of the alkane chain. However, length is not the only important

parameter. Conformation and molecular orientation relative to the electrodes are also important. Other factors need to be considered as well, including energy positions of the highest occupied and lowest unoccupied molecular orbitals (HOMO, LUMO), electrode work function, and nature of the bonding to the electrodes.

Charge transport mechanisms through self-assembled monolayers consist mainly of three processes [4]. The dominant charge transport mechanism in a molecular junction is "through-bond" (TB) tunneling, where the current follows the bond overlaps along the molecules (as illustrated in the left transport channel of Fig. 2). Another contribution involves the charge transport from electrode to electrode, in which the molecule plays the role of a dielectric medium that is called "through space" (TS), as illustrated in the right transport channel of Fig. 2. The last contribution of charge transport pathway involves a chain-to-chain coupling as illustrated in the middle of Fig. 2. As the molecular chains tilt, the decrease of the electron tunneling distance leads to a lateral hop between the neighboring molecular chains.

Two Pathway Models

If electron transport was determined purely by tunneling through the alkane chains, one would expect the value of β to equal zero, since the tunneling distance is the same for all tilt angles. The nonzero value of β indicates the existence of either intermolecular charge transfer or variations in the S-Au bonding as a function of tilt that affect the conductivity in an exponential way with angle.

Slowinski et al. [4] proposed a two-pathway conductance model involving "through-bond" tunneling, and the "chain-to-chain" coupling. Assuming no effects due to changes in S-Au bonding, the first pathway is independent of tilt, while the second depends on the tilt angle. The tunneling current, thus, is given by

$$I_t = I_0 \exp(-\beta_{TB}d) + I_0 n_s \exp[-\beta_{TB}(d - d_{CC}\tan\Theta)] \times \exp(-\beta_{TS}d_{CC})$$

where I_t is the current at a specific tilt angle Θ, d is the length of the molecule, n_s is a statistical factor accounting for the number of pathways containing a single lateral hop as compared to those containing only through-bond hops, d is the diameter of the molecule chains, β_{TB} and β_{TS} are respectively through-bond and

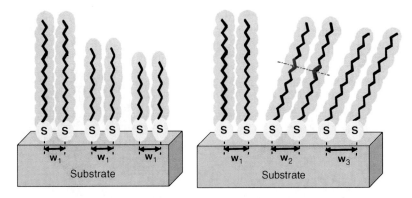

Charge Transport in Self-Assembled Monolayers, Fig. 3 Scheme of the measurement of the junction resistance for two different situations: (1) decreasing of the alkane chain (*left* part), and (2) the tilting of the alkane chain while maintaining the same number of carbon atoms (*right* part), which will yield the resistance (1) per unit length of molecule or (2) tilting angle of the molecules, respectively

through-space decay constants. For example, in case of C16 alkanethiol molecule chains, $d_{CC} = 4.3$ Å, $d = 24$ Å, and $n_s = 16$, i.e., the number of carbon atoms in the molecule.

Decay Constant upon Shortening and Tilting of Molecules

The junction resistance is dependent on electrode spacing for two different situations: (1) shortening of the alkane chain [5] and maintaining the same width (w) between chains and (2) tilting of the alkane chain but changing w [6, 7]. These measurements will yield the resistance per unit length of molecule or tilting angle of the molecules, respectively. The conductance decay constant β has already been measured using SAMs with different chain lengths when the separation between electrodes decreases as a function of the alkane chain length (the left image of Fig. 3). The decay constant, β, upon tilting of molecules can be measured using deformation with an AFM tip and simultaneous measurement of current (the right image of Fig. 3). This methodology will be described in the next section.

Basic Methodology

Preparation of Self-Assembled Monolayer

The organic molecular films on various types of substrates (conducting, semiconducting, or insulating substrates) have been prepared using techniques such as the Langmuir-Blodgett technique, dipping the substrates into solution with molecules, drop casting, or spin-coating [8].

As one example, details on the preparation of an alkanethiol SAM will be described below. Gold substrates (200–300 nm of gold coating over 1–4 nm of chromium layer n glass) are prepared by butane flame annealing in air after cleaning in acetone, chloroform, methanol, and a piranha solution (1:3; H_2O_2:H_2SO_4). The resulting surface consisted of large grains with flat terraces of (111) orientation (sizes up to 400 nm) separated by monatomic steps. Flatness and cleanness were tested by the quality of the lattice-resolved images of the gold substrate.

Two types of hexadecanethiol (C16) self-assembled monolayer can be formed on Au (111): complete monolayers of the molecules and islands of molecules covering only a fraction of the substrate. In the first case, the film was produced by immersing the substrate in 1 mM ethanolic solution of C16 for about 24 h, followed by rinsing with absolute ethanol and drying in a stream of nitrogen to remove weakly bound molecules. Incomplete monolayers in the form of islands were prepared by immersing the substrate in a 5 µM ethanolic solution of C16 for approximately 60 s, followed by rinsing. Samples consisting of islands facilitate the determination of the thickness of the molecular film relative to the surrounding exposed gold substrate. The molecular order of the islands improves with storage time at ambient conditions.

Techniques to Measure Charge Transport in Self-Assembled Monolayers

The current through a thiol SAM on a hanging Hg drop electrode can be measured in an electrochemical solution. The current was measured as a function of the monolayer thickness that can be tuned by two methods: by changing the number of carbons in the alkane chain and therefore its length; or, expansion of the Hg drop such that the monolayer surface coverage was reduced and the molecules increased their tilt angle with respect to the surface. Slowinski et al. determined the decay constants $\beta_{TB} = 0.91/\text{Å}$ and $\beta_{TS} = 1.31/\text{Å}$ by both a fit to their experimental data and by independent ab initio calculations. Mercury drop expansion experiments by Slowinsky et al. have shown a dependence of the current through the alkanethiol monolayers on surface concentration, prompting the authors to suggest the existence of additional pathways for charge transfer, like chain-to-chain tunneling.

Scanning tunneling microscopy and scanning tunneling spectroscopy have been used to reveal the atomic scale surface structure and charge transport properties of SAM layers [9, 10]. STM has been used to reveal various phases of surface structure and atomic scale defects, which could play a crucial role in the electrical transport.

Conductance measurements were performed with a conductive-probe atomic force microscopy (CP-AFM) system. The use of AFM with conducting tips provides the ability to vary the load on the nanocontact and also opens the way for exploring electron transfer as a function of molecular deformation. A junction is fabricated by placing a conducting AFM tip in contact with a metal-supported molecular film, such as a self-assembled monolayer (SAM) on Au, as shown in Fig. 4. The normal force feedback circuit of the AFM controls the mechanical load on the nanocontact while the current–voltage (I–V) characteristics are recorded. The possibility to control the load on the contact is an unusual characteristic of this kind of junction and provides the opportunity to establish a correlation between the mechanical deformation and electronic properties of organic molecules. The normal force exerted by the cantilever was kept constant during AFM imaging, while the current between tip and sample was recorded. It is crucial to carry out the experiment in the low load regime so that there is no damage to the surface. This can be confirmed by inspection of the images with Ångstrom depth sensitivity as well as by the reproducibility of the current and adhesion measurements. If the measured conductance did not change at constant load and did not show time-dependent behavior in the elastic regime, the tip

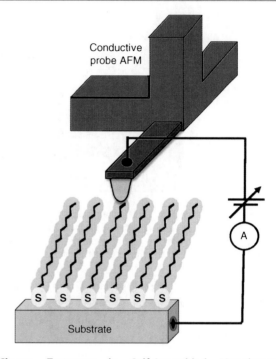

Charge Transport in Self-Assembled Monolayers, Fig. 4 Scheme of conductance measurements of SAM with a conductive-probe atomic force microscopy (CP-AFM) system

experiences minimal changes during subsequent contact measurements.

Key Research Findings

The molecular tilt induced by the pressure applied by the tip is one major factor that leads to increased film conductivity. By measuring the current between the conductive AFM tip and SAM as a function of the height of the molecules, the decay parameter (β) can be obtained [11]. Wold et al. studied the junction resistance as a function of load using AFM. The resistance was found to decrease with increasing load within two distinct power law scaling regimes [12]. Song et al. examined the dependence of the tunneling current through Au-alkanethiol-Au junctions on the tip-loading force [13]. It is found that the two-pathway model proposed by Slowinsky et al. can reasonably fit with the results, leading the authors to conclude that the tilt configuration of alkanethiol SAMs enhances the intermolecular charge transfer.

Charge Transport in Self-Assembled Monolayers, Fig. 5 AFM images (200 nm × 200 nm) of topography, and current images obtained simultaneously for a full monolayer of C16 on Au (111) surface. Lattice-resolved images of the film (inset in the *left* figure) reveal a lattice image of SAM (size: 2 nm × 2 nm)

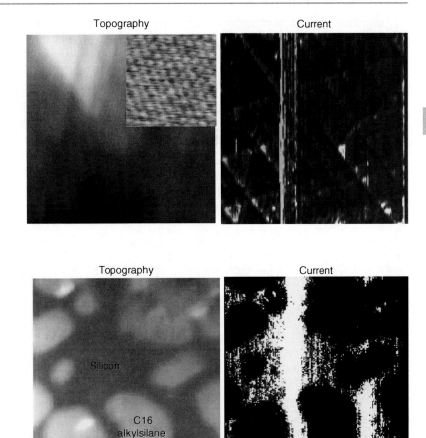

Charge Transport in Self-Assembled Monolayers, Fig. 6 AFM images (500 nm × 500 nm) of topographic, and current images, respectively, that were acquired simultaneously on hexadecylsilane SAM islands on silicon surface

Figure 5 shows topography and current images obtained simultaneously for a full monolayer of C16 on an Au (111) surface. The topographic image reveals the commonly found structure of the gold film substrate, composed of triangular-shaped terraces separated by atomic steps. Lattice-resolved images of the film (inset in the left figure) reveal a ($\sqrt{3} \times \sqrt{3}$)-R30° periodicity of the molecules relative to the gold substrate. Qi et al. measured current–voltage (I–V) characteristics on the C16 alkanethiol sample for loads varying between −20 and 120 nN, and found that the current changes in a stepwise manner and the plateaus are associated with the discrete tilt angle of the molecules. A stepwise response of the SAM film to pressure has been observed previously in other properties such as film height and friction of alkanesilanes on mica and alkanethiols on gold.

In order to measure the thickness of the self-assembled monolayer upon molecular deformation, the SAM islands that partially cover the substrate can be used. The heights of the islands can be obtained from topographical AFM images, while charge transport properties of alkanesilane SAMs on silicon surface are measured using AFM with a conducting tip. In this manner, the load applied to the tip-sample contact can be varied while simultaneously measuring electric conductance. Figure 6 shows the topographic and current images, respectively, that were acquired simultaneously on hexadecylsilane islands on a silicon surface. The image size is 500 × 500 nm. The hexadecylsilane islands are 100–200 nm in diameter and have a height of 1.6 nm at the applied load of 0 nN (or effective total load of 20 nN). It is also clear that the current measured on the alkanesilane island is much smaller than that measured on the silicon surface.

These changes were shown to correspond to the molecules adopting specific values of tilt angle relative to the surface, and explained as the result of methylene

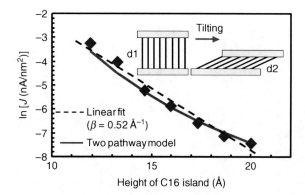

Charge Transport in Self-Assembled Monolayers, Fig. 7 Semilog plot of current density (nA/nm^2) as a function of the height of the hexadecylsilane SAM islands on a silicon surface. A decay constant (β) = 0.52 ± 0.04 Å^{-1} was found for the current passing through the film as a function of tip-substrate separation

groups interlocking with neighboring alkane chains. In the case of complete monolayers of alkanethiol SAM, the junction resistance (R) was measured as a function of the applied load [6]. These data were converted to current versus electrode separation by assigning each step in the current to a specific molecular tilt angle, following the sequence established in previous experiments. It was found that ln(R) increases approximately linearly with tip-surface separation, with an average slope β = 0.57 (±0.03) Å^{-1}. Similar measurement of the decay parameter upon the molecular tilts was carried out with a scanning tunneling microscope and simultaneous sensing of forces. By measuring the current as a function of applied load, a tunneling decay constant β = 0.53 (±0.02) Å^{-1} was obtained [14].

In the case of hexadecylsilane molecules, the local conductance of hexadecylsilane SAM islands on a silicon surface was measured with conductive-probe AFM. A semilog plot of current density (nA/nm^2) was obtained as a function of the height of the hexadecylsilane SAM islands on a silicon surface, as shown in Fig. 7. A decay constant (β) = 0.52 ± 0.04 Å^{-1} was found for the current passing through the film as a function of tip-substrate separation [7]. Figure 7 shows the best fit of the two-pathway model with the experimental current measurement as a function of the heights of molecule islands by using the fitting parameters of β_{TB} and β_{TS} that are 0.9 and 1.1 Å^{-1}, respectively. The good fit indicates that the

two-path tunneling model is a valid model to describe this observation.

While saturated hydrocarbon chains mainly interact with each other via weak van der Waals forces, much stronger intermolecular π-π interactions can be present in organic films comprised of conjugated/hybrid molecules. This influences charge transport significantly [5, 15]. In a conductance AFM study of two SAM systems, Fang et al. revealed the role of π-π stacking on charge transport and nanotribological properties of SAM consisting of aromatic molecules [16]. The two model molecules chosen in this study are (4-mercaptophenyl) anthrylacetylene (MPAA) and (4-mercaptophenyl)-phenylacetylene (MPPA). In MPPA, the end group is a single benzene ring, while in MPAA it is changed to a three fused benzene ring structure. This structural difference induces different degrees of lattice ordering in these two molecular SAM systems. Lattice resolution is readily achieved in the MPAA SAM, but it is not possible for the MPPA SAM under the same imaging conditions, indicating the MPAA is lacking long-range order. However, it is important to note that even without long-range order, the stronger intermolecular π-π stacking in the MPAA SAM greatly facilitates charge transport, resulting in approximately one order of magnitude higher conductivity than in the MPPA SAM.

Future Directions for Research

In this contribution, the basic concept of and recent progress on charge transport studies of organic SAM films formed by saturated hydrocarbon molecules and conjugated molecules has been outlined. Several techniques, including AFM, STM, and hanging Hg drop electrode, are used to elucidate the charge transport properties of SAM layers. A number of molecular scale factors such as packing density, lattice ordering, molecular deformation, grain boundaries, annealing induced morphological evolution, and phase separation play important roles in determining charge transport through SAM films. High resolution offered by scanning probe microscopy (SPM) is a key element in identifying and studying microstructures (e.g., molecular tilt, lattice ordering, defects, vacancies, grain boundaries) in organic films and their effects on electronic properties. Other advanced surface characterization techniques, such as SAM with nano-electrodes, in combination with conductive-probe atomic force microscopy, and spectroscopic techniques such as ultraviolet photoemission spectroscopy (UPS) and inverse photoemission spectroscopy (IPES), could be promising venues to explore the correlation between microstructures and electronic properties of organic films.

Cross-References

▶ Atomic Force Microscopy
▶ Conduction Mechanisms in Organic Semiconductors
▶ Electrode–Organic Interface Physics
▶ Scanning Tunneling Microscopy
▶ Self-Assembly

References

1. Aviram, A., Ratner, M.A.: Molecular Electronics: Science and Technology. New York Academy of Sciences, New York (1998)
2. Reed, M.A., Zhou, C., Muller, C.J., Burgin, T.P., Tour, J.M.: Conductance of a molecular junction. Science **278**, 252–254 (1997)
3. Ulman, A.: An Introduction to Ultrathin Organic Films from Langmuir-Blodgett to Self-Assembly. Academic, Boston (1991)
4. Slowinski, K., Chamberlain, R.V., Miller, C.J., Majda, M.: Through-bond and chain-to-chain coupling. Two pathways in electron tunneling through liquid alkanethiol monolayers on mercury electrodes. J. Am. Chem. Soc. **119**, 11910–11919 (1997)
5. Salomon, A., et al.: Comparison of electronic transport measurements on organic molecules. Adv. Mater. **15**, 1881–1890 (2003). doi:10.1002/adma.200306091
6. Qi, Y.B., et al.: Mechanical and charge transport properties of alkanethiol self-assembled monolayers on a Au(111) surface: The role of molecular tilt. Langmuir **24**, 2219–2223 (2008). doi:10.1021/la703147q
7. Park, J.Y., Qi, Y.B., Ashby, P.D., Hendriksen, B.L.M., Salmeron, M.: Electrical transport and mechanical properties of alkylsilane self-assembled monolayers on silicon surfaces probed by atomic force microscopy. J. Chem. Phys. **130**, 114705 (2009)
8. Barrena, E., Ocal, C., Salmeron, M.: Molecular packing changes of alkanethiols monolayers on Au(111) under applied pressure. J. Chem. Phys. **113**, 2413–2418 (2000)
9. Bumm, L.A., Arnold, J.J., Dunbar, T.D., Allara, D.L., Weiss, P.S.: Electron transfer through organic molecules. J. Phys. Chem. B **103**, 8122–8127 (1999)
10. Xu, B.Q., Tao, N.J.J.: Measurement of single-molecule resistance by repeated formation of molecular junctions. Science **301**, 1221–1223 (2003)
11. Wang, W.Y., Lee, T., Reed, M.A.: Electron tunnelling in self-assembled monolayers. Rep. Prog. Phys. **68**, 523–544 (2005)
12. Wold, D.J., Haag, R., Rampi, M.A., Frisbie, C.D.: Distance dependence of electron tunneling through self-assembled monolayers measured by conducting probe atomic force microscopy: Unsaturated versus saturated molecular junctions. J. Phys. Chem. B **106**, 2813–2816 (2002). doi:10.1021/jp013476t
13. Song, H., Lee, H., Lee, T.: Intermolecular chain-to-chain tunneling in metal-alkanethiol-metal junctions. J. Am. Chem. Soc. **129**, 3806 (2007)
14. Park, J.Y., Qi, Y.B., Ratera, I., Salmeron, M.: Noncontact to contact tunneling microscopy in self-assembled monolayers of alkylthiols on gold. J. Chem. Phys. **128**, 234701 (2008). doi:234701 10.1063/1.2938085
15. Yamamoto, S.I., Ogawa, K.: The electrical conduction of conjugated molecular CAMs studied by a conductive atomic force microscopy. Surf. Sci. **600**, 4294–4300 (2006)
16. Fang, L., Park, J.Y., Ma, H., Jen, A.K.Y., Salmeron, M.: Atomic force microscopy study of the mechanical and electrical properties of monolayer films of molecules with aromatic end groups. Langmuir **23**, 11522–11525 (2007)

Chem-FET

▶ Nanostructure Field Effect Transistor Biosensors

Chemical Beam Epitaxial (CBE)

▶ Physical Vapor Deposition

Chemical Blankening

▶ Chemical Milling and Photochemical Milling

Chemical Dry Etching

▶ Dry Etching

Chemical Etching

▶ Wet Etching

Chemical Milling and Photochemical Milling

Seajin Oh and Marc Madou
Department of Mechanical and Aerospace
Engineering & Biomedical Engineering, University of
California at Irvine, Irvine, CA, USA

Synonyms

Chemical blankening; Photoetching; Photofabrication;
Photomilling

Definition

Photochemical milling (PCM), also known as photochemical machining, is the process of fabricating high
precision metal workpieces using photographically
produced masks and etchants to corrosively remove
unwanted parts. This process is called wet etching in
MEMS fabrication techniques and can be also applied
to nonmetal materials. Wet etching, when combined
with nanolithography, is a useful process to fabricate
detailed nanostructures by extremely controlled
removal (Fig. 1).

Overview

Photochemical machining (PCM) produces three-
dimensional features by wet chemical etching
(Fig. 2). PCM yields burr-free and stress-free metal
products and allows for the machining of a wide range
of materials which would not be suitable for traditional
metal working techniques. PCM is also known as
photoetching, photomilling, photofabrication, or
chemical blankening [1]. There is a special type of
photochemical milling that uses light for initiating or
accelerating the wet etching process in metal or semiconductor materials.

The combination of photoresists and wet etching
enables the fabrication of very detailed structures
with complex geometry or large arrays of variable
etching profiles in thin (<2 mm) flat metal sheets.
Photoresists are made of synthetic polymers having
consistent properties. Liquid photoresist coats a thin

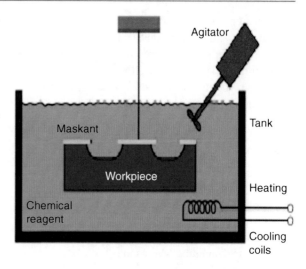

Chemical Milling and Photochemical Milling, Fig. 1
(**a**) Schematic illustration of photochemical milling process

film by dipping or spin casting which enables the
production of detailed patterns, but often creates pinholes in the thin layer. Thick dry photoresist films
applied by hot lamination have advantages of process
simplicity and reliability although the materials are
expensive.

The process of wet etching is based on the redox
chemistry of etchant reduction and metal oxidation,
which results in the formation of soluble metal-
containing ions that diffuse away from the reaction
metal surface. Many metals commonly used in
manufacturing industry are etched readily in aqueous
solutions comprising etchant (e.g., ferric chloride).
Metal oxides and virtually all materials can be etched
with a proper selection of etchant regardless of different etching rates.

Wet etching in the PCM process is isotropic where
the etchant attacks both downward into the material
and sideways under the edge of the resist layer and the
ratio of the depth to the undercut is termed the etch
factor (Fig. 3). In MEMS device fabrication, the undercut plays a key role in fabricating free-standing microstructure patterns (e.g., beams and cantilevers) that are
necessary where microstructures have to be flexible,
thermally isolated, small mass, double-sided contacts
with gases or liquid surroundings. In most cases, the
free-standing material is deposited as a film on
a substrate surface, termed a sacrificial material. The
etchant, possessing a lateral etching component, must
be sufficiently selective not to attach the free-standing

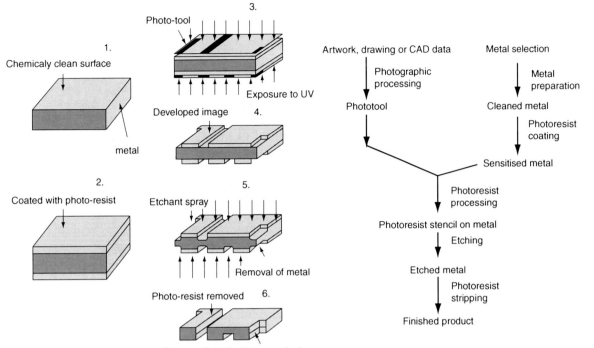

Chemical Milling and Photochemical Milling, Fig. 2 PCM process in metal sheet etching. 1. Chemically clean the metal surface. 2. Coat both sides of the plate with photoresist that adheres to the metal when exposed to UV light. 3. Expose plate and phototool to ensure image transfer. 4. Develop and create photomask image. 5. Spray metal with etchant or dip it in hot acidic solution to etch all material other than part covered with photoresist. 6. Rinse the plate to ensure photoresist and etchant removal [2].

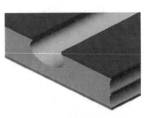

Chemical Milling and Photochemical Milling, Fig. 3 Schematic illustration of etch detail through line openings in the patterned photoresist (*blue*). Etch factor = Depth of etch (D)/ Undercut [½ (B−A)] [2]

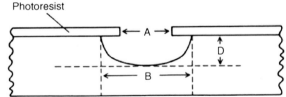

material. Further, novel nanolithography techniques enable to create submicrometer-scaled features.

Basic Methodology

Photolithography

An image of the profile of the flat feature is generated by computer-aided design (CAD) and electronically transferred onto a photographic film to produce a phototool, a photolithographic mask as known in MEMS. The photoresist is exposed through the phototool from ultraviolet source. There are two types of photoresist – negative and positive. UV lights soften the positive film, and the exposed area is released in the developing solution. The negative photoresist film has reversed pattern developing characteristics. Wet resist is applied to a metal sheet by a dipping process while spin coating is commonly used in micromachining (Fig. 4).

Wet Etching

In wet chemical etching, the components of the solid are changed into soluble chemical components which

Chemical Milling and Photochemical Milling, Fig. 4 Positive and negative photoresist [2]

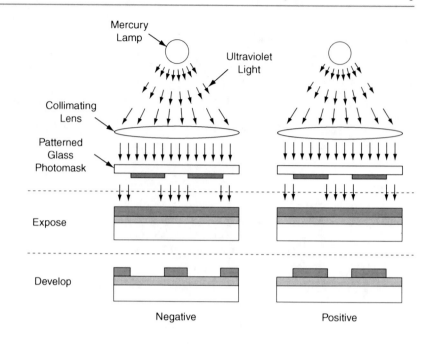

are transported by diffusion or convection away from the surface into the bulk of the solution. The solvent molecules form a shell around the dissolved solid particles that are mobile in the liquid phase. The specific interactions of the components of the liquid with the solid determine the reaction rate, which is attributed to a greater etching selectivity of the solid than dry etching methods. Water is used as the solvent in most wet etching processes [3].

1. In a metal or a semiconductor, the dissolution of metal or semiconductor is accompanied by an electron transfer and obeys the laws of electrochemistry. The oxidation reaction of metal or semiconductor M releases metal ions into solution and produce electrons.

$$M = M^{n+} + ne^- \quad \text{(anodic partial process)}$$

A secondary chemical redox process takes place to transfer the liberated electrons to an oxidizing agent OM in the solution.

$$OM + ne^- = OM^{n-} \quad \text{(cathodic partial process)}$$

The anodic and cathodic partial processes results in etching metal.

$$M_{\text{in solid}} + OM_{\text{in solution}} = M^{n+} + OM^{n-}$$

The etch rate corresponds to the number of metal ions produced at the solid surface per time unit, which is proportional to the interchanged anodic partial current I over the surface A, I/A. Metals and semiconductors often dissolve as complexes (e.g., $[MY_x]^+$) where smaller molecules or ions (ligands) form a chemically bound primary shell around the central atom. The etch rate can be changed by varying the concentration of reactants, temperature, viscosity, and convection of the solution.

2. In dielectric materials, acid–base reactions take place if the material to be etched reacts with hydrogen or hydroxyl ions. The cations are solvated by water as a strong polar solvent and they can diffuse rapidly into the bulk of the solution. Etching processes are applicable with oxides and hydroxides of metals and semiconductors at low pH-values,

$$M_xO_y + 2yH^+ = xM^{2y/x+} + yH_2O$$
$$M(OH)_x + xH^+ = M^{x+} + xH_2O$$

M = metal or semiconductor (e.g., Cu, Si)

Similarly, at very high pH-values some metals and semiconductors form stable water dissolvable complexes that are easily solvated by water.

Acidic anions, such as chloride and fluoride, or neutral molecules react as ligands, a typical example of which is the etching of SiO_2 in HF-containing etchants.

Chemical Milling and Photochemical Milling, Fig. 5 (*Left*) Schematic illustration of the process generating nanometer-scale lines by controlled undercutting. A pattern is produced in the photoresist by photolithography or soft lithography. An isotropic etch is applied to the substrate beneath the photoresist. Shallow undercutting of the base layer, followed by evaporation into the exposed areas, and lift-off generates $\sim 50 \pm 200$ nm gaps in the thin film at the edges of the photoresist pattern. The pattern is then used as an optical filter. (*Right*) SEM image of a cross section of linear trenches at the edges of a 100 nm line transferred into a Si <100> substrate. The trenches are 250 nm deep [4]

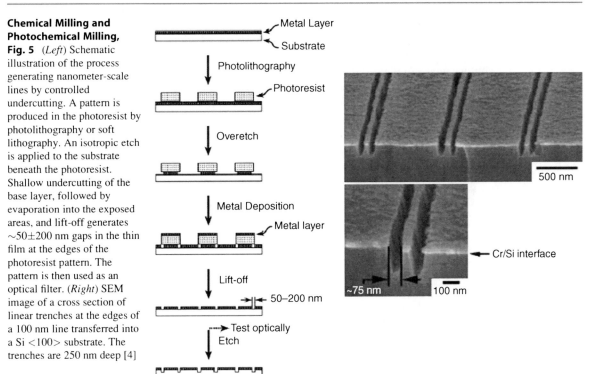

$$SiO_2 + 6HF = H_2SiF_6 + H_2O$$

Salt-like film is also dissolved by complexing agents. For example, copper chloride is etchable in neutral KCl solutions.

$$CuCl + 2Cl - = [CuCl_3]^{2-}$$

Key Research Findings for Nanotechnology

Edge Lithography

Undercutting by isotropic wet etching is applied to transfer the edges of a photoresist pattern into a feature of the final pattern, as illustrated in Fig. 5. The process generates 50 nm scale trenches by controlled undercutting. Currently, patterning features <100 nm are possible by advanced lithography techniques – deep ultraviolet, electron beam writing, extreme ultraviolet, and x-ray photolithography – but are prohibitively expensive. In contrast, edge lithography is a convenient, inexpensive technique for patterning features with nanometer-scale dimensions.

Wet Etching for Maskless Patterning

In the photochemical etching process, light exposure can increase the charge carrier density in near surface areas which enhances the anodical as well as the cathodical partial processes in wet etching. For example, defect electrons in the lower band left by light exposure support the release of cations and at the same time the released electrons in the conduction band are readily accepted by an oxidizing agent in the solution. In the same process, the intensive exposure with a focused beam enables direct pattern generation without a lithographic mask. This method is specified preferentially for patterning semiconductors deposited on a nonconducting substrate (e.g., GaN on sapphire) [5].

Future Directions for Research

Wet chemical etching is a simple, inexpensive, and well-understood process. The process plays a key role in the field of MEMS and nanotechnology. Continuous effort is being made to provide controllability, repeatability, and, most importantly, detail in fabricating microstructures. New etchant compositions have been developed to apply wet chemical etching to the materials constituting new devices. A representative example is selective removal of metal nitride films on sapphire for a new version of a light-emitting diode. At the same time, great efforts have been made to develop new lithography techniques for nanoscale patterns such as proximal probe lithography, very thin to monolayer lithography, and soft lithography. PCM equipped with the emerging lithography techniques can cause the paradigm shift in the creation of nanoscaled features [6].

Cross-References

▶ DUV Photolithography and Materials
▶ Dry Etching
▶ Electron Beam Lithography (EBL)
▶ EUV Lithography
▶ Nanoimprint Lithography
▶ Nanotechnology
▶ Stereolithography
▶ Wet Etching

References

1. Abate, K.: Photochemical etching of metals. Met. Finish. **100** (6A), 448–451 (2002)
2. Allen, D.M.: Photochemical machining: from manufacturing's best kept secret to a $6 billion rapid manufacturing process. CIRP J. Manuf. Syst. **53**, 559–572 (2005)
3. Köhler, J.M.: Wet chemical etching method. In: Etching in Microsystem Technology. Wiley, Weinheim (1999)
4. Love, J.C., Paul, K.E., Whitesides, G.M.: Fabrication of nanometer-scale features by controlled isotropic wet chemical etching. Adv. Mater. **13**(8), 604–607 (2001)
5. Bardwell, J.A., Webb, J.B., Tang, H., Fraser, J., Moisa, S.: Ultraviolet photoenhanced wet etching of GaN in K2S2O8 solution. J. Appl. Phys. **89**(7), 4142–4149 (2001)
6. Madou, M.: Fundamentals of Microfabrication, 2nd edn. CRC Press, Boca Raton (2002)

Chemical Modification

▶ Nanostructures for Surface Functionalization and Surface Properties

Chemical Solution Deposition

▶ Sol-Gel Method

Chemical Vapor Deposition (CVD)

Yoke Khin Yap
Department of Physics, Michigan Technological University, Houghton, MI, USA

Synonyms

Aerosol-assisted chemical vapor deposition (AACVD); Atmospheric pressure chemical vapor deposition (APCVD); Atomic layer chemical vapor deposition (ALCVD); Atomic layer deposition (ALD); Atomic layer epitaxial (ALE); Boron nitride nanotubes (BNNTs); Carbon nanotubes (CNTs); Carbon nanowalls; Catalyst; Catalytic chemical vapor deposition (CCVD); Cold-wall thermal chemical vapor deposition; Dissociated adsorption; Double-walled carbon nanotubes (DWCNTs); High-pressure carbon monoxide (HiPCO); Hot filament chemical vapor deposition (HFCVD); Hot-wall thermal chemical vapor deposition; Inductively coupled-plasma chemical vapor deposition (ICP-CVD); Low-pressure chemical vapor deposition (LPCVD); Metalorganic chemical vapor deposition (MOCVD); Multiwalled carbon nanotubes (MWCNTs); Nanobelts; Nanocombs; Nanoparticles; Nanotubes; Nanowires; Plasma-enhanced chemical vapor deposition (PECVD); Single-walled carbon nanotubes (SWCNTs); Thermal chemical vapor deposition; Ultrahigh vacuum chemical vapor deposition (UHVCVD); Vertically aligned carbon nanotubes

Definition

Chemical vapor deposition (CVD) is referred to as deposition process of thin films and nanostructures through chemical reactions of vapor phase precursors. Since CVD can be conducted using high purity precursors, it likely leads to thin film and nanostructures with high purity. The use of vapor phase precursors also enables better control on the composition and doping of thin films and nanostructures. For example, Si thin films can be deposited by decomposition of silane gas (SiH_4) by plasma or heat as follows: SiH_4 (g) → Si (s) + 2 H_2 (g). Doping of Si films with boron will lead to p-type Si films and can be achieved by the addition of B_2H_6 gas.

Classification

Classification by Operating Pressures

Chemical reactions involved in a CVD technique can be initiated by many ways leading to the classification of various types of CVD approaches. For example, CVD can be classified according to the operating pressures as follows:

- *Atmospheric pressure CVD* (APCVD) is referred to CVD processes that conducted at atmospheric pressure. The advantage of APCVD is simple experimental setup without the need of a vacuum system. The potential drawback will be the undesired contamination.
- *Low-pressure CVD* (LPCVD) is referred to CVD processes at pressures below and close to atmospheric pressure. One of the purposes of reducing the operation pressure is to avoid undesired reactions between precursors. LPCVD can also improve film uniformity.
- *Ultrahigh vacuum CVD* (UHVCVD) is LPCVD processes at a very low pressure, typically below $\sim 10^{-8}$ torr.

There are *CVD* processes that operate at high pressures. See details in section Classification by Excitation Techniques.

Classification by Excitation Techniques

In addition, CVD can be classified by the excitation techniques that initiate the chemical reactions, as follows.

- *Plasma-enhanced CVD* (PECVD) is referred to CVD processes that employed plasmas to initiate the needed chemical reactions. In general, PECVD could reduce the growth temperatures as the chemical reactions in PECVD are not ignited by heat. There are many subclassifications of PECVD that depend on the type of AC potential used for plasma generation. For example, RF-PECVD and MW-PECVD employed radio frequency (RF, typically at 13.56 MHz) and microwave (MW, 2.45 GHz) potential to dissociate gases and produce the needed plasmas. Some PECVD are classified by the configuration of plasma generation. For instance, inductively coupled plasma CVD (ICP-CVD) is actually RF-PECVD that uses an induction coil as the RF electrode outside the vacuum chamber and is sometimes called remote plasma CVD [1]. On the other hand, two RF plasmas can be used in a CVD system. For example, a dual-RF-PECVD approach was demonstrated for the growth of vertically aligned carbon nanofoils/nanowalls and carbon nanotubes (CNTs) [2, 3]. PECVD can also be obtained using DC plasma generated by a DC potential across a pair of electrodes and is simply called DC-PECVD.

- *Thermal CVD* is referred to CVD processes that employed heat to initiate the needed chemical reactions. The most common thermal CVD technique employed external furnace to control the growth temperatures of the entire reaction zone (e.g., vacuum quarts/glass chambers). This is sometimes called *hot-wall thermal CVD*. The advantages of this approach are including potential of large-scale synthesis. In contrast, there are several approaches to achieve the so-called *cold-wall thermal CVD*. For example, hot filaments (HF) are used to heat up the temperatures of the adjacent substrates while the needed chemical reactions are ignited with higher temperatures on the filaments. This approach is called *hot filament CVD* (HFCVD). Other heating approaches can also be used including IR lamps, laser beams, and passing current flows through a suspended Si chip [4].

Classification by the Precursor Type and Feeding Procedure

CVD can also be classified by the type of precursors or the procedures where precursors are introduced into the reaction chamber. Some of the examples are described as follows,

- *Aerosol-assisted CVD (AACVD)* involve the use of carrier gases (usually inert gases such as Ar, He, etc.) to transfer vapors of liquid-phase precursor into the reaction chamber in the form of aerosol. This approach enables the use of liquid precursors for the CVD process. The most well-known AACVD is *metalorganic CVD* (*MOCVD*) where metalorganic solids are dissolved in organic solvents. For example, Trimethyl-gallium [TMGa, $Ga(CH_3)_3$] is often used as the source of Ga to form GaN when it reacts with ammonia (NH_3) at high temperatures. Thus, *AACVD* can be viewed as a subclassification of *thermal CVD*.

- *Atomic layer CVD* (ALCVD) is better known as *atomic layer deposition* (ALD) or *atomic layer epitaxial* (ALE) [5]. ALD allow conformal deposition of monolayer of binary (or ternary) compounds by introducing the two (or three) reacting precursors in a pulsed-mode one after another. For example, monolayers of Zinc sulfide (ZnS) can be deposited by first exposing the growth surface with $ZnCl_2$. Once chemisorption of $ZnCl_2$ is completed, the reaction chamber will be purged with an inert gas. Thereafter, hydrogen sulfide (H_2S) will be introduced to the chamber so that the following reaction will take place on the growth surface: $ZnCl_2$ (g) + H_2S (g) → ZnS (s) + HCl (g). Since the chemical process is self-limiting, only one monolayer (or less) of ZnS will be formed in each cycle. The advantage of *ALD* is conformal coating on the full surface of the sample and the precision of film thickness control. The major drawback is the extreme low deposition rate. *ALD* is usually conducted at low temperatures (200–400°C) and may be viewed as a subclassification of *thermal CVD*. However, plasma and metalorganic precursors are sometime used in *ALD*.

Examples of CVD Approaches for Nanotechnology

Catalytic Chemical Vapor Deposition (CCVD)

Catalytic chemical vapor deposition (CCVD) is the simplest and most popular technique for the synthesis of carbon nanotubes (CNTs), graphene, and nanowires (NWs). *CCVD* is simply *thermal CVD* approach with the use of catalyst that induces chemical reactions for the formation of nanomaterials. A typical experimental layout for *CCVD* is shown in Fig. 1. As shown, the *CCVD* system consists of a tube furnace and a quartz tube chamber where chemical reactions are taking place.

Synthesis of Vertically Aligned Carbon Nanotubes by CCVD

A typical experimental setup of *CCVD* for the synthesis of carbon nanotubes (CNTs) is shown in Fig. 1a. In this case, the substrates (usually oxidized Si) are first deposited with catalyst films such as Fe, Ni, or Co by electron beam evaporation, sputtering, pulsed laser deposition, etc. [6–8]. These samples were then annealed at 600°C in Ar, N_2 or H_2 for about 30 min. During the annealing process, the catalyst films will be converted into nanoparticles that will serve as the growth sites for CNTs. After the annealing, the growth

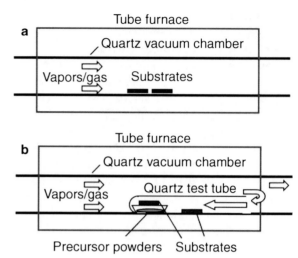

Chemical Vapor Deposition (CVD), Fig. 1 (a) Typical set up for CCVD and (b) modified double-tube configuration

temperatures ($\sim$600–800°C) and growth ambient (usually Ar, H_2, or their mixtures) will be set prior to the introduction of hydrocarbon source gas (methane, CH_4; ethylene C_2H_4, acetylene, C_2H_2). These hydrocarbon gases will be decomposed on the surface of the catalyst nanoparticles through the chemical process of dissociative adsorption [6].

For example, dissociated adsorption of C_2H_2 is summarized in Fig. 2 [6]. Figure 2a shows the adsorption of a C_2H_2 molecule (step *A*) on the surface of the Fe nanoparticle. This will lead to either the breaking of C–H bond (step *B1*) to form C_2H and H fragments or the breaking of C=C bond (step *B2*) to form two C–H fragments. The catalytic function of the Fe nanoparticle is to reduce the energy required for decomposition by a charge transfer from hydrocarbon molecules to Fe. According to a first principles calculation, the dissociation energy of the first hydrogen atom from an isolated C_2H_2 (step *A* to *B1*) in vacuum can be reduced from 5.58 eV to 0.96 eV. On the other hand, the energy barrier between *A* and *B2* is 1.25 eV. The C–H bond breaking (step *B1*) is followed by C=C bond breaking (step *C*) with a potential barrier of 1.02 eV. Whereas, C=C bond breaking (step *B2*) is followed by C–H bond breaking (step *C*) with an energy barrier of 0.61 eV. Both modes (*A* to *B1* to *C* or *A* to *B2* to *C*) are possible and give one C–H fragment, one C and one H. The decomposition of C_2H_2 is completed after the breaking of the last C–H

Chemical Vapor Deposition (CVD), Fig. 2 (a) Sequences of dissociative adsorption of C_2H_2 on Fe surface. See text for detailed description. The (**b**) decomposed carbon atoms (**c**) diffused into the solid-core Fe nanoparticle until (**d**) supersaturation and then (**e**) segregate as nanotubes

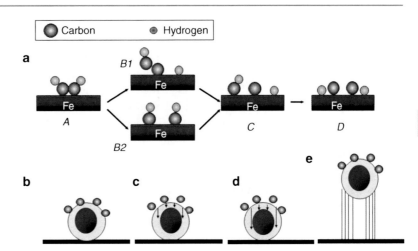

bond (step D) with the need of a potential energy of 0.61 eV.

The decomposed carbon atoms (Fig. 2b) will then diffuse into the subsurface of the Fe nanoparticle (Fig. 2c). At typical growth temperature (650–800°C), these nanoparticles are not melted even after considering the eutectic point of Fe-C phase. Since dissociative adsorption is an exothermic process, the near-surface temperatures of the catalytic nanoparticles will be higher than the growth temperatures. Therefore it is possible that the near surface region of the particles is melted. This will form the gas–liquid interface between carbon and Fe solid-core nanoparticles. Due to the high diffusion rate of carbon in Fe melt, a Fe-C alloy will start to form. When these nanoparticles become supersaturated with carbon (Fig. 2d) to a value critical for growth at the solid–liquid interface, the excess carbon will segregate as carbon nanotubes (Fig. 2e). A tip-growth mode is illustrated in this figure where the nanoparticles remained at the tips of CNTs. A base-growth mode is also possible where the nanoparticles remained at the bases of CNTs. Thus, the diameters of CNTs depend, to a certain extent, on the diameters of the nanoparticles. By controlling the thickness of the catalyst films and other parameters, *CCVD* enables the growth of vertically aligned (VA) single-, double-, and multiwalled CNTs [8].

In fact, one of the most impressive achievements in the growth of vertically aligned single-walled CNTs (VA-SWCNTs) is the so-called "*super growth*" [9]. This water-assisted approach was based on *CCVD* with the addition of 20–500 ppm of water vapors.

These water vapors were introduced into the growth chamber by flowing carrier gas (Ar, or He with 40% H_2) through a water bubbler. Ethylene was used as the carbon source along with various catalysts [Fe (1 nm), Al (10 nm)/Fe (1 nm), Al_2O_3 (10 nm)/Fe (1 nm), to Al_2O_3 (10 nm)/Co (1 nm)]. VA-SWCNTs with the length of several mm and diameter of 1–3 nm were reported. This approach also enabled the growth of double- and multiwalled CNTs [10].

The abovementioned *CCVD* approaches are achievable at relatively low temperatures and lead to VA-SWCNTs. Historically, SWCNTs were grown by *CCVD* at higher temperatures in a powder form. In 1996, Dai et. al, demonstrated the growth of SWCNTs at 1,200°C by using carbon monoxide (CO) as the carbon source gas and supported MoO_x powder as the catalyst [11]. In this case, powder form of SWCNTs was grown on the catalyst that was loaded on a quartz boat. Later, the growth temperature was reduced to 1,000°C by using CH_4 as the sources gas and results in the growth of randomly distributed SWCNTs on powders of supported metal oxide catalysts [12]. The diameters of these SWCNTs are 1–6 nm. Later, *CCVD* was modified into a high-pressure mode for large-scale synthesis of SWCNTs [13]. In this approach, liquid iron pentacarbonyl, $Fe(CO)_5$, was used as the catalyst and introduced into the CVD chamber by CO gas. At operational pressures of 1–10 atm. and temperatures of 800–1,200°C, $Fe(CO)_5$ thermally decomposed into iron clusters in gas phase and reacted with CO gas to produce SWCNTs. The yield of SWCNTs was found to increase with temperature and pressure. The average diameter of SWCNTs was decrease from ~1.0 nm at

1 atm. to 0.8 nm at 10 atm. This approach is now known as high-pressure carbon monoxide (HiPCO) technique, one of the major techniques for large-scale synthesis of SWCNTs with small diameters. Finally, alcohol was also used as the carbon source for the synthesis of SWCNTs by *CCVD* [14]. In this case, ethanol vapors (5 torr) were supplied to the reaction chamber that contained Fe/Co catalyst supported with zeolite at 700–800°C.

Synthesis of ZnO Nanostructures by CCVD

On the other hand, *CCVD* can be modified into a *double-tube* configuration as shown in Fig. 1b. This approach will enable the use of solid precursors to generate the needed growth vapors. For example, various ZnO nanostructures can be grown by such a setup by using ZnO and graphite powders as the precursors [15, 16]. As shown, these powders were loaded on a ceramic combustion boat which is contained at the end of a closed-end quartz tube. The substrates can be placed a distance away from the boat. The following reactions will occur at 1,100°C when oxygen gas (O_2) is introduced:

$$ZnO(s) + C(s) \rightarrow Zn(g) + CO(g)$$

$$2Zn(g) + O_2(g) \rightarrow 2ZnO(s)$$

Based on this approach, various ZnO nanostructures can be grown with and without the use of catalyst (gold films, Au), including nanotubes [15], nanowires, nanobelts, nanocombs [16], and nanosquids [17].

Synthesis of Boron Nitride Nanotubes by CCVD

The double-tube *CCVD* configuration in Fig. 1b was also used for the growth of boron nitride nanotubes (BNNTs) [18–20]. In this case, boron, magnesium oxide, and iron oxide were used as the precursors and loaded on the ceramic boat. The possible chemical reaction at 1,100–1,200°C is 4B (s) + MgO (s) + FeO (s) $\rightarrow$ 2B_2O_x (g) + MgO$_y$ (s) + FeO$_z$ (s), where x, y, and z are yet to be determined. When ammonia gas (NH_3) is introduced into the chamber, the generated boron oxide vapors (B_2O_x) will react with NH_3 to form BNNTs. The partially reduced MgO$_y$ and FeO$_z$ are the possible catalysts for the formation of BNNTs [18]. It is noted that the synthesis of BNNTs by these chemical processes usually require temperatures above 1,350°C. The key success for the low temperature

growth discussed here is due to the so-called "growth vapor trapping" approach obtained by placing the bare substrates directly on the ceramic boat. These substrates trapped the growth vapors from the precursors and enhanced the nucleation rate of BNNTs at low temperatures. In recent experiments, Fe, Ni, or MgO films were coated on the substrates used as the catalysts [19, 20]. Such an approach leads to patterned growth of BNNTs. The possible chemical reactions are:

$$B_2O_2(g) + MgO(s) + 2NH_3(g) \\ \rightarrow 2BN(BNNTs) + MgO(s) + 2H_2O(g) + H_2(g),$$

or

$$B_2O_2(g) + Ni(s) \text{ or } Fe(s) + 2NH_3(g) \\ \rightarrow 2BN(s, BNNTs) + Ni(s) \text{ or } Fe(s) + 2H_2O(g) \\ + H_2(g)$$

Cross-References

▶ Atomic Layer Deposition
▶ Carbon Nanotube-Metal Contact
▶ Carbon Nanotubes for Chip Interconnections
▶ Carbon-Nanotubes
▶ Focused-Ion-Beam Chemical-Vapor-Deposition (FIB-CVD)
▶ Synthesis of Carbon Nanotubes

References

1. Tsu, D.V., Lucovsky, G., Dvidson, B.N.: Effects of the nearest neighbors and the alloy matrix on SiH stretching vibrations in the amorphous SiO$_r$:H (0<r<2) alloy system. Phys. Rev. B **40**, 1795–1805 (1989)
2. Menda, J., et al.: Structural control of vertically aligned multiwalled carbon nanotubes by radio-frequency plasmas. Appl. Phys. Lett. **87**(173106), 3 (2005)
3. Hirao, T., et al.: Formation of vertically aligned carbon nanotubes by dual-RF-plasma chemical vapor deposition. Jpn. J. Appl. Phys. **40**, L631–L634 (2001)
4. van Laake, L., Hart, A.J., Slocum, A.H.: Suspended heated silicon platform for rapid thermal control of surface reactions with application to carbon nanotube synthesis. Rev. Sci. Instrum. **78**(083901), 9 (2007)
5. Leskelä, M., Ritala, M.: Atomic layer deposition chemistry: Recent developments and future challenges. Angew. Chem. Int. Ed. **42**, 5548–5554 (2003)

6. Kayastha, V.K., et al.: Controlling dissociative adsorption for effective growth of carbon nanotubes. Appl. Phys. Lett. **85**, 3265–3267 (2004)
7. Kayastha, V.K., et al.: High-density vertically aligned multiwalled carbon nanotubes with tubular structures. Appl. Phys. Lett. **86**(253105), 3 (2005)
8. Kayastha, V.K., et al.: Synthesis of vertically aligned single- and double- walled carbon nanotubes without etching agents. J. Phys. Chem. C **111**, 10158–10161 (2007)
9. Hata, K., et al.: Water-assisted highly efficient synthesis of impurity-free single-walled carbon nanotubes. Science **306**, 1362–1364 (2004)
10. Yamada, T., et al.: Size-selective growth of double-walled carbon nanotube forests from engineered iron catalysts. Nat. Nanotechnol. **1**, 131–136 (2006)
11. Dai, H., et al.: Single-wall nanotubes produced by metal-catalyzed disproportionation of carbon monoxide. Chem. Phys. Lett. **260**, 471–475 (1996)
12. Kong, J., Cassell, A.M., Dai, H.: Chemical vapor deposition of methane for single-walled carbon nanotubes. Chem. Phys. Lett. **292**, 567–574 (1998)
13. Nikolaev, P., et al.: Gas-phase catalytic growth of single-walled carbon nanotubes from carbon monoxide. Chem. Phys. Lett. **313**, 91–97 (1999)
14. Maruyama, S., et al.: Low-temperature synthesis of high-purity single-walled1 carbon nanotubes from alcohol. Chem. Phys. Lett. **360**, 229–234 (2002)
15. Mensah, S.L., et al.: Formation of single crystalline ZnO nanotubes without catalysts and templates. Appl. Phys. Lett. **90**, 113108 (2007)
16. Mensah, S.L., et al.: Selective growth of pure and long ZnO nanowires by controlled vapor concentration gradients. J. Phys. Chem. C **111**, 16092–16095 (2007)
17. Mensah, S.L., et al.: ZnO nnosquids: banching nnowires from nnotubes and nnorods. J. Nanosci. Nanotechnol. **8**, 233–236 (2008)
18. Lee, C.H., et al.: Effective growth of boron nitride nanotubes by thermal chemical vapor deposition. Nanotechnology **19**(455605), 5 (2008)
19. Lee, C.H., et al.: Patterned growth of boron nitride nanotubes by catalytic chemical vapor deposition. Chem. Mater. **22**, 1782–1787 (2010)
20. Wang, J., Lee, C.H., Yap, Y.K.: Recent advancements in boron nitride nanotubes. Nanoscale. **2**, 2028–2034 (2010)

Chemistry of Carbon Nanotubes

▶ Functionalization of Carbon Nanotubes

Chitosan

▶ Chitosan Nanoparticles

Chitosan Nanoparticles

Burcu Aslan[1], Hee Dong Han[2,4], Gabriel Lopez-Berestein[1,3,4,5] and Anil K. Sood[2,3,4,5]
[1]Department of Experimental Therapeutics, M.D. Anderson Cancer Center, The University of Texas, Houston, TX, USA
[2]Gynecologic Oncology, M.D. Anderson Cancer Center, The University of Texas, Houston, TX, USA
[3]Cancer Biology, M.D. Anderson Cancer Center, The University of Texas, Houston, TX, USA
[4]Center for RNA Interference and Non–coding RNA, M.D. Anderson Cancer Center, The University of Texas, Houston, TX, USA
[5]The Department of Nanomedicine and Bioengineering, UTHealth, Houston, TX, USA

Synonyms

Chitosan; Drug delivery system; Nanoparticles

Definition

Chitosan nanoparticles are biodegradable, nontoxic carriers for nucleotides and drugs with the potential for broad applications in human disease.

Overview

Characteristics of Chitosan

Chitosan is a natural cationic polysaccharide composed of randomly distributed N-acetyl-D-glucosamine and β-(1,4)-linked D-glucosamine. Chitosan can be chemically synthesized via alkaline deacetylation from chitin, which is the principal component of the protective cuticles of crustaceans [1]. Chitosan is biodegradable in vivo by enzymes such as lysozyme, which is endogenous and nontoxic [2]. In addition, biodegradation of chitosan is highly associated with the degree of deacetylation. These properties render chitosan particularly attractive for clinical and biological applications as a highly biocompatible material with low toxicity and immunogenicity.

Several techniques have been designed to assemble chitosan nanoparticles (CH-NPs) as a drug delivery system including emulsions, ionotropic gelation, micelles, and spray drying. A variety of therapeutic agents can be loaded into CH-NPs with high efficiency, which can then be injected intravenously, intraperitoneally, or intrathecally.

Chemical Modification of Chitosan

The abundant amine and hydroxyl groups present in chitosan offer a unique opportunity to attach targeting ligands or imaging agents. Numerous derivatives of chitosan have been designed and tailored to improve the physicochemical and adhesive properties of nanoparticles such as size, shape, charge, density, and solubility. Quaternized chitosan, N,N,N-trimethyl chitosan, thiolated chitosan, carboxyalkyl chitosan, sugar-bearing chitosan, bile acid-modified chitosan, and cyclodextrin-linked chitosan are among the modifications frequently utilized in chitosan-based drug delivery systems. Each modification offers unique properties and characteristics. For instance, trimethyl chitosan is soluble over a wide pH range and enhances the condensation capacity of plasmid DNA at neutral pH due to fixed positive charges on its backbone. Thiolation of chitosan provides free sulphydryl groups on its side chains and forms disulfide bonds with cysteine-rich subdomains of muco-glycoproteins on cell membranes and increases cellular uptake. In addition, both modifications have been used as nonviral carrier systems to combine the advantages of trimethyl chitosan and thiolated chitosan while minimizing their shortcomings [3].

Hydrophobic moieties can also be attached to chitosan to facilitate the incorporation of insoluble drugs, i.e., hydrophobic glycol chitosan nanoparticles. Chemical conjugation of hydrophobic 5β-cholanic acid to the hydrophilic glycol chitosan backbone allows for the incorporation of the water-insoluble drug camptothecin [4].

Key Research Findings

Preparation of Chitosan Nanoparticles

The amino groups of chitosan backbone can interact with anionic molecules such as tripolyphosphate (TPP). The ionic cross-linking of chitosan is advantageous since the method is easy and mostly performed under mild conditions without using organic solvents. Ionotropic gelation of chitosan using TPP for the

incorporation of low molecular drugs, proteins, DNA/ siRNA have been demonstrated [5]. CH-NPs are rapidly formed through ionic interactions between the negatively charged phosphates of TPP and positively charged amino groups of chitosan.

Drug Delivery

Ringsdorf first reported the concept of polymer–drug conjugates for delivering small molecule drugs [6]. The concept of polymer–drug conjugates allows chemical conjugation of a drug using a biodegradable spacer. The spacer is usually stable in the bloodstream, but cleaved at the target site by hydrolysis or enzymatic degradation. Based on this concept, several polymer–drug conjugates have been developed such as glycol chitosan conjugated with doxorubicin, which forms self-assembled nanoparticles in an aqueous condition. A paclitaxel-chitosan conjugate that can be cleaved at physiological conditions was developed for oral delivery of paclitaxel. N-succinyl-chitosan-mitomycin conjugates demonstrated high antitumor efficacy against a variety of murine tumor models of leukemia, melanoma, and primary and metastatic liver tumors.

Gene Delivery

Recently, plasmid DNA, siRNA, and oligonucleotide-loaded CH-NPs have been used for targeted gene silencing. Positively charged chitosan can easily form polyelectrolyte complexes with negatively charged nucleotides based on electrostatic interaction, incorporation, or adsorption, as illustrated in Fig. 1 [7]. The positive charge on the surface of CH-NPs is desirable to prevent aggregation due to electrostatic repulsion and increase binding efficiency with the negatively charged cell membrane by enhancing electrostatic interactions. However, other cellular uptake mechanisms such as clathrin-mediated endocytosis, caveolae-mediated endocytosis, and macropinocytosis may also be involved [8]. Moreover, the configuration of chitosan is modified under acidic pH by triggering the opening of tight junctions. Acidic pH results in an increase in the number of protonated amines on the chitosan leading to a further disruption on membrane organization. It has also been reported that chitosan can swirl across the membrane lipid bilayer and facilitate the cellular uptake of the polyplex due to increase in mole fraction of chitosan, leading to reduction in the polymeric chains which results in decreased molecular

Chitosan Nanoparticles,
Fig. 1 Preparation of
chitosan-based DNA/siRNA
nanoparticles based on
different mechanisms [21]

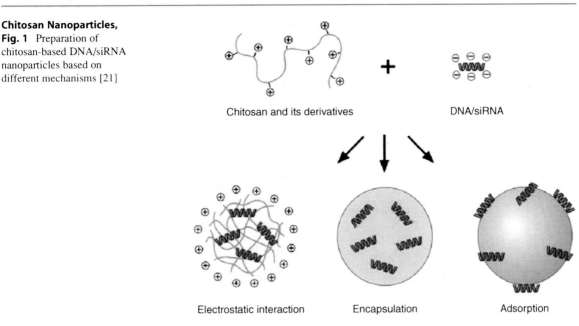

Chitosan and its derivatives   DNA/siRNA

Electrostatic interaction   Encapsulation   Adsorption

weight [7]. In addition, Lu et al. have recently reported the biodistribution of intravenously administered siRNA-CH-NPs in tumor-bearing mice. They demonstrated that CH-NPs allowed for a higher localization of siRNA in tumor tissues compared to other organs (Fig. 2) [9].

Electrostatic interactions between protonated amines of chitosan and negative charge of DNA or siRNA leads to spontaneous formation of highly compact encapsulation of either DNA or siRNA into CH-NPs [10]. Gel electrophoresis has been used to assess hydrogen bonding and hydrophobic interactions between chitosan and DNA [11]. However, a major limitation of siRNA delivery is its rapid degradation in plasma and cytoplasm. It has been reported that stable chitosan/siRNA complexes can protect siRNA degradation in circulation of CH-NPs in bloodstream to overcome extracellular and intracellular barriers. On the other hand, disassembly is also needed to allow release of siRNA. This emphasizes the importance of an appropriate balance between protection and release of siRNA for biological functionality [12].

Positively charged CH-NPs can bind to negatively charged cell surfaces with high affinity. CH-NPs are known to overcome endosomal escape via its "proton sponge effect." Once these nanoparticles penetrate into an acidifying lysosomal compartment, the unsaturated amino groups of chitosan distrain protons that are delivered by proton pumps (vATPase) which is called the "proton sponge mechanism." Subsequent lysosomal swelling and rupture leads to endo-lysosomal escape of nanoparticles [13].

Targeted Delivery

An ideal delivery system should lead to enhanced concentrations of therapeutic payloads at disease sites, and minimize potential non-desirable off-target effects, and ultimately raise the therapeutic index. Differences between tumor and normal tissue microenvironment and architecture, such as vascularization, overexpressed receptors, pH, temperature, ionic strength, and metabolites, can be exploited for selective targeting.

Targeted delivery systems have been designed to increase and/or facilitate uptake into target cells, and to protect therapeutic payloads. Recent work comparing non-targeted and targeted nanoparticles have shown that the primary role of the targeting ligands is to enhance selective binding efficiency to receptor in cell surface and cellular uptake into target cells, and to minimize accumulation in normal tissues. The addition of targeting ligands that provide specific ligand-receptor binding on nanoparticle-cell surface interactions can play a vital role in the ultimate location of nanoparticles. For example, nanoparticles decorated with specific moieties such as peptides, proteins, or antibodies can be targeted to cancer cells via cell-surface receptor proteins such as transferrin or folate

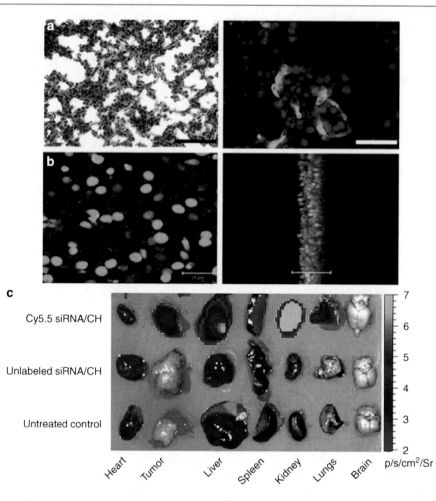

Chitosan Nanoparticles, Fig. 2 (**a**) Fluorescent siRNA distribution in tumor tissue. Hematoxylin and eosin, original magnification 2003 (*left*); tumor tissues were stained with anti-CD31 (*green*) antibody to detect endothelial cells (*right*). The scale bar represents 50 mm. (**b**) Fluorescent siRNA distribution in tumor tissue. Sections (8 mm thick) were stained with Sytox green and examined with confocal microscopy (scale bar represents 20 mm) (*left*); lateral view (*right*). Photographs taken every 1 mm were stacked and examined from the lateral view. Nuclei were labeled with Sytox green and fluorescent siRNA (*red*) was seen throughout the section. At all time points, punctated emissions of the siRNA were noted in the perinuclear regions of individual cells, and siRNA was seen in >80% of fields examined. (**c**) Shows fluorescence intensity overlaid on white light images of different mouse organs and tumor

receptors (known to be increased on a wide range of cancer cells). These targeting ligands enable nanoparticles to bind on to cell-surface receptors and penetrate cells by receptor-mediated endocytosis.

Target selective ligand-labeled CH-NPs can enhance receptor-mediated endocytosis. Various receptors on the tumor cell surface have been established as a target-binding site to achieve selective delivery. The overexpression of transferrin and folate in certain tumors has been exploited to deliver CH-NPs conjugated with these receptor's ligands [14, 15]. Another example is the $\alpha v \beta 3$ integrin, which is overexpressed in a wide range of tumors, and is largely absent in normal tissues. Han et al. [16] have recently reported that the administration of RGD peptide-labeled CH-NPs led to increased tumor delivery of siRNA-CH-NP and enhanced antitumor activity in ovarian carcinoma models (Figs. 3 and 4).

Chitosan-Based Environmental-Responsive Particles

Physiological alterations such as pH, temperature, ionic strength, and metabolites in the microenvironment of tumor have gained increased interest in terms

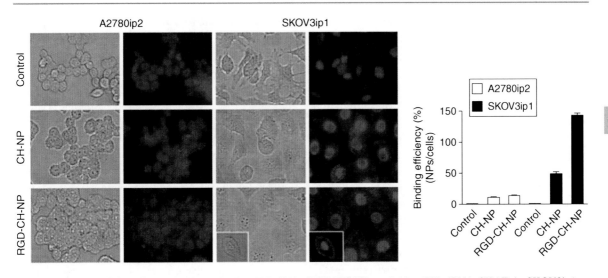

Chitosan Nanoparticles, Fig. 3 Binding of Alexa555 siRNA/RGD-CH-NPs and Alexa555 siRNA-CH-NP in SKOV3ip1 or A2780ip2 cells by fluorescence microscopy

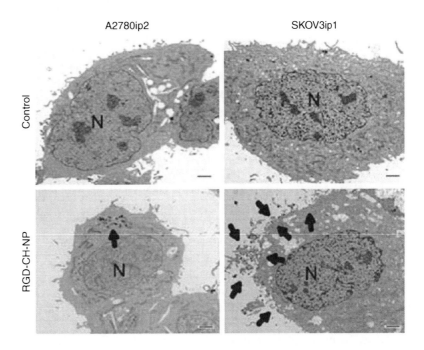

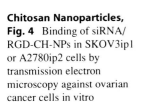

Chitosan Nanoparticles, Fig. 4 Binding of siRNA/RGD-CH-NPs in SKOV3ip1 or A2780ip2 cells by transmission electron microscopy against ovarian cancer cells in vitro

of targeted therapy. Potential differences in these parameters between tumor and normal tissue can be used for enhanced targeting. A novel, pH responsive NIPAAm/CH-NPs containing camptothecin and paclitaxel was successfully used to enhance tumor uptake and antitumor activity [17]. On the other hand, thermosensitive CH-g-poly(N-vinylcaprolactam) composite has been developed by an ionic cross-linking method and incorporated with 5-FU(5-

fluorouracil). This study showed that 40% of drug is released from the particles when the temperature is above a lower critical solution temperature of 38°C while only 5% of drug is released below 38°C, which confirms the drug release mechanism of this polymeric carrier system is based on temperature. These nanoparticles were more toxic to cancer cells while devoid of toxicity to normal cells, leading to enhance antitumor efficacy [18].

Chitosan-Based Magnetic Nanoparticles

Chitosan-based magnetic nanoparticles have been developed for magnetic resonance (MR) imaging via passive and tissue-specific targeting. The low-oxidizing ferromagnetic materials are the most commonly used compounds for nanoparticle formulations and provide a stable magnetic response. The accumulation of magnetic nanoparticles when injected intravenously can be induced when the tissue or organ is subjected to an external high-gradient magnetic field [19]. Drugs, DNA plasmids, or bioactive molecules are released into target tissues and effectively taken up by tumor cells after accumulation of these carriers. On the other hand, magnetic nanoparticles such as iron oxide nanoparticles (Fe_3O_4) are applied to oscillating magnetic fields, which results in the generation of heat and holds potential for rapid heating of tumor tissue.

When magnetic nanoparticles are administered systemically, they are rapidly coated with plasma proteins. These particles are taken up by the reticuloendothelial system, leading to decreased circulation time. Magnetic nanoparticles coated with hydrophilic materials such as chitosan could provide longer circulation time. In addition, chitosan-coated magnetic nanoparticles, which are used in magnetic resonance imaging, can also be functionalized and used as theranostic carriers due to amino and hydroxyl groups of chitosan.

Application of Chitosan Nanoparticles for SiRNA Delivery

Recently, siRNA targeted to EZH2 (a critical component of the polycomb repressive complex 2 [PRC2]), loaded CH-NPs were shown to enhance the delivery of siRNA to tumors, leading to downregulation of the target protein and subsequently enhanced antitumor activity. Lu et al. demonstrated target gene silencing using EZH2 siRNA loaded CH-NPs, leading to increased antitumor efficacy in animal tumor models. Moreover, Zhang et al. reported the use of intrathecal administration of siRNA targeted against specific muscarinic receptor subtypes loaded in CH-NPs in a rat model of pain. CH-NP/siRNA was distributed to the spinal cord and the dorsal root ganglion [20]. The administration of Chitosan-M_2-siRNA caused a large reduction in the inhibitory effect of muscarine on the rat paw withdrawal threshold from a heat stimulus. These studies support the use of CH-NPs as a delivery system for siRNA into neuronal tissues in vivo.

Future Directions for Research

Chitosan nanoparticles offer a unique potential for clinical and biological applications due to low immunogenicity, low toxicity, and high biocompatibility. In addition to its advantages such as protonated amine groups, chitosan can increase binding efficiency with cells because of electrostatic interactions. Therefore, CH-NPs may be used for broad applications in human disease. Moreover, chitosan allow modifications that will exploit the inherent physicochemical properties by conjugation of selective ligands. The highly desirable specific targeting of drugs has been elusive to date; however, CH-NPs can bring us closer to this goal.

Acknowledgments Portions of this work were supported by the NIH (CA 110793, 109298, P50 CA083639, P50 CA098258, CA128797, RC2GM092599, U54 CA151668), the Ovarian Cancer Research Fund, Inc. (Program Project Development Grant), the DOD (OC073399, W81XWH-10-1-0158, BC085265), the Zarrow Foundation, the Marcus Foundation, the Kim Medlin Fund, the Laura and John Arnold Foundation, the Estate of C. G. Johnson, Jr., the RGK Foundation, and the Betty Anne Asche Murray Distinguished Professorship.

Cross-References

▶ Effect of Surface Modification on Toxicity of Nanoparticles
▶ Nanomedicine
▶ Nanoparticle Cytotoxicity
▶ Nanoparticles

References

1. Muzzarelli, R.A.A.: Chitin (1–37). Pergamon, Elmsford (1977)
2. Khor, E.: Chitin: Fulfilling a Biomaterials Promise. Elsevier, Oxford, UK (2001)
3. Zhao, X., Yin, L., Ding, J., Tang, C., Gu, S., Yin, C., Mao, Y.: Thiolated trimethyl chitosan nanocomplexes as gene carriers with high in vitro and in vivo transfection efficiency. J. Control. Release **144**, 46–54 (2010)
4. Min, K.H., Park, K., Kim, Y., Bae, S.M., Lee, S., Jo, H.G., Park, R.W., In-San Kim, I.S., Jeong, S.Y., Kim, K., Kwon, I.C.: Hydrophobically modified glycol chitosan nanoparticles-encapsulated camptothecin enhance the drug stability and tumor targeting in cancer therapy. J. Control. Release **127**, 208–218 (2008)
5. Park, J.H., Saravanakumar, G., Kim, K., Kwon, I.C.: Targeted delivery of low molecular drugs using chitosan and its derivatives. Adv. Drug Deliv. Rev. **62**, 28–41 (2010)

6. Ringsdorf, H.: Structure and properties of pharmacologically active polymers. J. Polym. Sci. Polym. Symp. **51**, 135–153 (1975)

7. Lai, W.F., Lin, M.C.M.: Nucleic acid delivery with chitosan and its derivatives. J. Control. Release **134**, 158–168 (2009)

8. Nam, H.Y., Kwon, S.M., Chung, H., Lee, S.Y., Kwon, S.H., Jeon, H., Kim, Y., Park, J.H., Kim, J., Her, S., Oh, Y.K., Kwon, I.C., Kim, K., Jeong, S.Y.: Cellular uptake mechanism and intracellular fate of hydrophobically modified glycol chitosan nanoparticles. J. Control. Release **135**, 259–267 (2010)

9. Lu, C., Han, H.D., Mangala, L.S., Ali-Fehmi, R., Newton, C.S., Ozbun, L., Armaiz-Pena, G.N., Hu, W., Stone, R.L., Munkarah, A., Ravoori, M.K., Shahzad, M.M.K., Lee, J.W., Mora, E., Langley, R.R., Carroll, A.R., Matsuo, K., Spannuth, W.A., Schmandt, R., Jennings, N.J., Goodman, B.W., Jaffe, R.B., Nick, A.M., Kim, H.S., Guven, E.O., Chen, Y.H., Li, L.Y., Hsu, M.C., Coleman, R.L., Calin, G.A., Denkbas, E.B., Lim, J.Y., Lee, J.S., Kundra, V., Birrer, M.J., Hung, M.C., Lopez-Berestein, G., Sood, A.K.: Regulation of tumor angiogenesis by EZH2. Cancer Cell **18**, 185–197 (2010)

10. Mao, S., Sun, W., Kissel, T.: Chitosan-based formulations for delivery of DNA and siRNA. Adv. Drug Deliv. Rev. **62**, 12–27 (2010)

11. Messai, I., Lamalle, D., Munier, S., Verrier, B., Ataman-Onal, Y., Delair, T.: Poly(D, Llactic acid) and chitosan complexes: interactions with plasmid DNA. Colloids Surf. A Physicochem. Eng. Asp. **255**, 65–72 (2005)

12. Liu, X., Howard, K.A., Dong, M., Andersen, M.O., Rahbek, U.L., Johnsen, M.G., Hansen, O.C., Besenbacher, F., Kjems, J.: The influence of polymeric properties on chitosan/siRNA nanoparticle formulation and gene silencing. Biomaterials **28**, 1280–1288 (2007)

13. Nel, A.E., Mädler, L., Velegol, D., Xia, T., Hoek, E.M.V., Somasundaran, P., Klaessig, F., Castranova, C., Thompson, M.: Understanding biophysicochemical interactions at the nano–bio interface. Nat. Mater. **8**, 543–557 (2009)

14. Mao, H.Q., Roy, K., Troung-Le, V.L., Janes, K.A., Lin, K.Y., Wang, Y., August, J.T., Leong, K.W.: Chitosan–DNA nanoparticles as gene carriers: synthesis, characterization and transfection efficiency. J. Control. Release **70**(3), 399–421 (2001)

15. Fernandes, J.C., Wang, H., Jreyssaty, C., Benderdour, M., Lavigne, P., Qiu, X., Winnik, F.M., Zhang, X., Dai, K., Shi, Q.: Bone-protective effects of nonviral gene therapy with Folate–Chitosan DNA nanoparticle containing Interleukin-1 receptor antagonist gene in rats with adjuvant-induced arthritis. Mol. Ther. **16**(7), 1243–1251 (2008)

16. Han, H.D., Mangala, L.S., Lee, J.W., Shahzad, M.M.K., Kim, H.S., Shen, D., Nam, E.J., Mora, E.M., Stone, R.L., Lu, C., Lee, S.J., Roh, J.W., Nick, A.M., Lopez-Berestein, G., Sood, A.K.: Targeted gene silencing using RGD-Labeled chitosan nanoparticles. Clin. Cancer Res. **16**(15), 3910–3922 (2010)

17. Li, F., Wu, H., Zhang, H., Gu, C.H., Yang, Q.: Antitumor drug paclitaxel loaded pH-sensitive nanoparticles targeting tumor extracellular pH. Carbohydr. Polym. **77**(4), 773–778 (2009)

18. Rejinold, N.S., Chennazhi, K.P., Nair, S.V., Tamura, H., Jayakumar, R.: Carbohydr. Polym (2010). doi:10.1016/j.carbpol.2010.08.052

19. Veiseh, O., Gunn, J.W., Zhang, M.: Design and fabrication of magnetic nanoparticles for targeted drug delivery and imaging. Adv. Drug Deliv. Rev. **62**, 284–304 (2010)

20. Zhang, H.M., Chen, S.R., Cai, Y.Q., Richardson, T.E., Driver, L.C., Lopez-Berestein, G., Pan, H.L.: Signaling mechanisms mediating muscarinic enhancement of GABAergic synaptic transmission in the spinal cord. Neuroscience **158**, 1577–1588 (2009)

21. Lai, W.F., Lin, M.C.M.: Nucleic acid delivery with chitosan and its derivatives. J. Control. Release **134**, 158–168 (2009)

Clastogenicity or/and Aneugenicity

▶ Genotoxicity of Nanoparticles

Clinical Adhesives

▶ Bioadhesives

Cluster

▶ Synthesis of Subnanometric Metal Nanoparticles

CMOS (Complementary Metal-Oxide-Semiconductor)

▶ CMOS MEMS Fabrication Technologies

CMOS MEMS Biosensors

Michael S.-C. Lu
Department of Electrical Engineering, Institute of Electronics Engineering, and Institute of NanoEngineering and MicroSystems, National Tsing Hua University, Hsinchu, Taiwan, Republic of China

Synonyms

Integrated biosensors

Definition

CMOS MEMS biosensors are miniaturized biosensors fabricated on CMOS (complementary metal-oxide semiconductor) chips by the MEMS (microelectromechanical systems) technology.

Overview

A biosensor is a device designed to detect a biochemical molecule such as a particular deoxyribonucleic acid (DNA) sequence or particular protein. Many biosensors are affinity-based, meaning that they use an immobilized capture probe that selectively binds to the target molecule being sensed. Most biosensors require a label attached to the target, and the amount of detected label is assumed to correspond to the number of bound targets. Labels can be fluorophores, magnetic beads, gold nanoparticles, enzymes, or anything else allowing convenient binding and detection; however, labeling a biomolecule can change the associated binding properties, especially for protein targets.

An electrical biosensor is capable of detecting a binding event by producing an electrical current and/ or a voltage. Conventional optical detection methods require external instruments that are expensive and not amenable to miniaturization. Electrical bioassays hold great promise for numerous decentralized clinical applications ranging from emergency-room screening to point-of-care diagnostics due to their low cost, high sensitivity, specificity, speed, and portability. Miniaturization of an electrical biosensor can be achieved through microfabrication – widely known as the MEMS technology; in addition, the sensing circuits can be embedded on the same chip through integrated-circuit (IC) processes, among which the CMOS technology is the most popular choice for implementing various analog and digital circuits. A fully integrated CMOS MEMS biosensor array is capable of providing real-time high-throughput detection of multiple samples. CMOS biosensors can be implemented based on the electrochemical, impedimetric, ion-sensitive, magnetic, optical, and micromechanical approaches, which require different MEMS processes to construct the sensing interfaces. Some of the methods require labeling and some are label-free for bio-signal transduction. More details are provided in the following sections.

Sensing Principles and Key Research Findings

Electrochemical Biosensors

Electrochemical biosensors have been the subject of basic as well as applied research for many years. In 1970, Dr. Leland C. Clark demonstrated that glucose could be measured in whole blood with the presence of the glucose oxidase enzyme. Commercial glucose sensors based on electrochemical detection have been developed since then. Electrochemical biosensors typically depend on the presence of a suitable enzyme in the biorecognition layer to catalyze reaction of electroactive substances. Affinity-based electrochemical sensors use enzymes as labels that bind to antibodies, antigens, or oligonucleotides with a specific sequence.

Electrochemical biosensors can employ potentiometric, amperometric, and impedimetric sensing principles to convert the chemical information into a measurable electrical signal. In potentiometric devices, ion selective electrodes (ISE) are commonly used to transduce a biorecognition event into a potential signal between the working and the reference electrodes whose value, depending on the concentration of the analyte, can be predicted by the Nernst equation. The current flowing through the electrode is equal to or near zero.

Amperometric biosensors operate by applying a constant potential and monitoring the current associated with the reduction or oxidation of an electroactive species. The sensors can work in three- or four-electrode configurations. The former case consists of a reference, a working, and a counter electrode. The four-electrode setup has an additional working electrode such that oxidation and reduction take place at anode and cathode simultaneously. Working electrodes are normally made of noble metals which are critical to the operation of a sensor. Since these metals are not CMOS compatible, the electrodes must be formed after fabrication of CMOS circuits. Silver/silver chloride (Ag/AgCl) is commonly used as the reference electrode in many electrochemical biosensors. Since not all particles oxidized at the anode reach the cathode, a potentiostat, whose input and output are connected to a reference and to a counter electrode, respectively, is required to provide the difference current to the electrolyte and regulate the potential of the electrolyte to a constant value.

CMOS MEMS Biosensors, Fig. 1 Schematic of the CMOS interdigitated microelectrode for electrochemical DNA detection. On the *right*: schematic illustration of the redox-cycling process (Schienle et al. © 2004 IEEE)

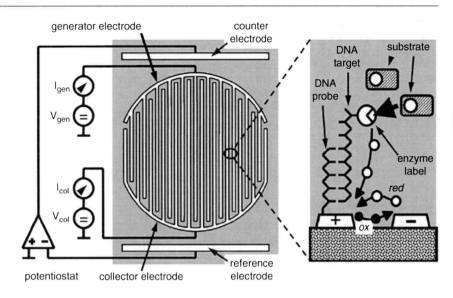

Interdigitated microelectrodes have been adopted in many electrochemical biosensors for quantitative analysis. The width and spacing of microelectrodes are reduced by microfabrication. The main advantage is the enhanced redox current due to fast redox recycling, as the chemical products produced at one side of the electrodes are readily collected at the other side of adjacent electrodes and regenerated to the original states. The relationship between the produced redox current and the analyte concentration has been derived by Aoki et al. [1].

Electrochemical DNA hybridization biosensors rely on the conversion of the DNA base-pair recognition event into an electrical signal. Schienle et al. [2] reported a CMOS electrochemical DNA sensor array with each sensing element consisting of interdigitated gold electrodes separated by 1 µm. As depicted in Fig. 1, probe molecules were immobilized on the gold surface through thiol coupling and the target molecules were tagged by an enzyme label (alkaline phosphatase). A chemical substrate (para-aminophenyl phosphate) was applied to the chip after hybridization. The enzyme label on the matched DNA strands cleaved the phosphate group and generated an electroactive compound (para-aminophenol), which was subsequently oxidized and reduced as the indicator of successful DNA hybridization. Levine et al. [3] also reported a CMOS electrochemical DNA sensor array which was operated based on conventional cyclic voltammetry (CV). During the measurement the electrode potential was scanned up and down in order to produce the redox reactions associated with the ferrocene labels attached to DNA strands. In addition to DNA detection, CMOS electrochemical sensors have been realized to allow high-throughput detection of dopamine and catecholamine release from adrenal chromaffin cells [4], since the release of neurotransmitters from secretory vesicles of biological cells is closely related to the function of a nervous system.

Impedimetric Biosensors

Changes in the electrical properties of a sensing interface (e.g., capacitance, resistance) can occur when a target biomolecule interacts with a probe-functionalized surface. A conventional impedimetric biosensor measures the electrical impedance of an electrode-solution interface in a.c. steady state with constant d.c. bias conditions. This approach, known as electrochemical impedance spectroscopy (EIS), is accomplished by imposing a small sinusoidal voltage over a range of frequencies and measuring the resulting current. The current–voltage ratio gives the impedance, which consists of both energy dissipation (resistor) and energy storage (capacitor) elements. Results obtained by EIS are often graphically represented by a Bode plot or a Nyquist plot. EIS reveals information about the reaction mechanism of an electrochemical process since different reaction steps can dominate at certain frequencies. Impedimetric biosensors can detect a variety of target analytes by simply varying the probe used. Changes in the impedance can be correlated to DNA hybridization, antigen–antibody reaction, or be used to detect biological cells.

CMOS MEMS Biosensors,
Fig. 2 Schematic of the
impedance extraction method
(Adopted by Lee et al. © 2010
Elsevier)

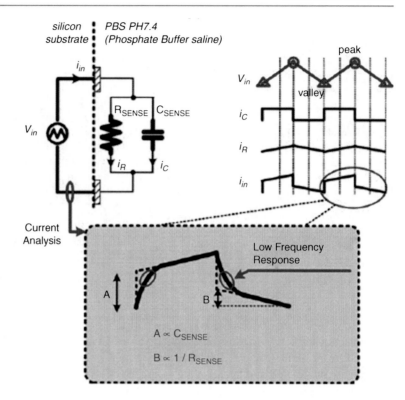

The interface impedance is commonly represented by an equivalent circuit model for analysis. E. Warburg [5] first proposed that the interface impedance can be represented by a polarization resistance in series with a polarization capacitor. The interface capacitance possesses a frequency dependency and is commonly represented by the Gouy-Chapman-Stern model as the series combination of the double-layer capacitance (Helmholtz capacitance) and the diffuse layer capacitance (Gouy-Chapman capacitance).

In addition to the frequency-domain measuring method, interface impedance changes can be measured by the potentiostatic step method where small potential steps are applied to the working electrode and the transient current responses, as determined by the time constant of the interface resistance and capacitance, are measured accordingly. Lee et al. [6] reported a fully integrated CMOS impedimetric sensor array for label-free detection of DNA hybridization. The changes in the reactive capacitance and the charge-transfer resistance on the gold sensing electrodes were extracted by applying a triangular voltage waveform and monitoring the produced currents. As illustrated in Fig. 2, the current flowing through the capacitor is associated with the slope of the applied

triangular wave, while the current flowing through the resistor is in proportion to the magnitude of the triangular wave. The reported detection limit was 10 nanomolar (nM).

It is important to distinguish the differences between faradaic and non-faradaic biosensors for impedance detection. In electrochemical terminology, a faradaic process involves charge transfer across an interface, while a transient current can flow without charge transfer in a non-faradaic process by charging a capacitor. The faradaic EIS requires the addition of a redox species which is alternately oxidized and reduced by the transfer of charges to and from the metal electrode. In contrast, no additional reagent is required for non-faradaic impedance spectroscopy. The associated impedance change is predominantly capacitive with the charge transfer resistance being omitted.

Miniaturized CMOS capacitive sensors have been developed for numerous biosensing applications. As the sensor size is small, monolithic integration provides the benefit of enhancing the signal-to-noise ratio by reducing the parasitic capacitance observed at the sensing node, which would otherwise negatively impact the detection limit during direct capacitance measurement. Stagni et al. [7] reported an 8 × 16

CMOS MEMS Biosensors, Fig. 3 (a) Schematic of the floating-gate ISFET structure. (b) Schematic of the open-gate ISFET structure

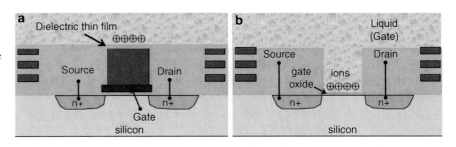

CMOS DNA sensor array where the bindings of complementary DNA strands on gold microelectrodes reduced the dielectric constant of the electrode-analyte impedance and the associated change was used to modify the charging and discharging transients of the detection circuit. Capacitive sensitivity can be enhanced by use of interdigitated microelectrodes with the minimum gap defined by the adopted CMOS process. Lu et al. [8] reported very sensitive detection of the neurotransmitter dopamine in the sub-femtomolar (fM) range. CMOS capacitive sensors have also been used for monitoring cellular activity since the morphological and physiological states of biological cells have correlations with their electrical properties [9].

The impedance change of a biosensing event can be made purely resistive with additional chemical modifications. Li et al. [10] reported a DNA array detection method in which the binding of immobilized DNA probes functionalized with gold nanoparticles produced a conductivity change between adjacent metal electrodes, which were made of gold metal in order to withstand the chemical in the clean processing. The detected signal can be further enhanced by additional silver deposition. Silicon dioxide was used as the underlying material under the gap for immobilization of the DNA probes. Detection limit of the target DNA concentration was 1 picomolar (pM).

Ion-Sensitive Field Effect Transistors (ISFET)

ISFETs were first developed in the early 1970s and have been utilized in various biosensing applications which depend on the type of receptor used for analyte recognition or how a signal is generated. Immunologically modified FETs and DNA-modified FETs detect the change of surface charges by monitoring the associated current–voltage relationship. Cell-based FETs detect the potential changes produced by biological cells due to the flow of ions across the cell membranes upon stimulations. For applications where it is required

to build an ISFET array to provide detection of multiple samples at different locations, monolithic integration is then preferred in order to reduce wiring complexity and noise interference.

CMOS ISFET sensors can be built by using a floating-gate or an open-gate structure as depicted in Fig. 3. The floating-gate structure is easier to fabricate as it requires minimal post-processing after completion of a conventional CMOS process. Silicon dioxide or silicon nitride in the CMOS passivation thin films can be used as the material for surface functionalization. The produced signal due to charges on sensor surface is capacitively coupled to the floating gate through a relatively thick dielectric layer which reduces the signal-coupling efficiency. The gate in an open-gate ISFET is removed and replaced by the aqueous solution whose potential is commonly set via a reference electrode. The exposed gate oxide is used for surface functionalization. Since the gate oxide thickness is only tens of angstroms in a conventional sub-μm CMOS process, sensitivity is thus significantly enhanced as compared to a floating-gate ISFET.

Accumulated charges on the ISFET surface modulate the threshold voltage of a MOS transistor, leading to a channel current change under fixed voltage biases of the drain, source, and gate terminals. The threshold voltage of an open-gate ISFET is expressed as:

$$V_{th} = E_{ref} - \Psi + \chi_{sol} - \frac{\phi_{si}}{q} - \frac{Q_{ox} + Q_{ss} + Q_B}{C_{ox}} + 2\phi_F$$

where E_{ref} is the reference electrode potential, Ψ is a chemical input parameter as a function of solution pH value, χ_{sol} is the surface dipole potential of the solvent, ϕ_{si} is the work function of silicon, q is the electron charge, C_{ox} is the gate oxide capacitance per unit area, Q_{ox} and Q_{ss} are the charges in the oxide and at the oxide-silicon interface, Q_B is the depletion charges in silicon and ϕ_F is the difference between the Fermi potential of the substrate and intrinsic silicon.

Kim et al. [11] reported the use of p-type ISFET fabricated in a standard CMOS process for detection of DNA immobilization and hybridization. Gold was deposited by post-processing as the gate material for immobilizing DNA due to its chemical affinity with thiols. The channel current increased during hybridization due to the negative charges present in the phosphate groups of DNA strands. Li et al. [12] presented a post-CMOS fabrication method to make open-gate ISFETs and demonstrated ultrasensitive dopamine detection in the fM range.

Magnetic Biosensors

Some of the magnetic biosensors require the use of magnetic beads as labels attached to the samples in order to induce a measurable electrical signal when specific binding on sensor surface occurs. Magnetic beads are made of small ferromagnetic or ferrimagnetic nanoparticles that exhibit a unique quality referred to as superparamagnetism in the presence of an externally applied magnetic field. This phenomenon, as discovered by Louis Néel (Nobel Physics Prize winner in 1970), has been used in numerous applications such as magnetic data storage and magnetic resonance imaging (MRI).

Several methods can be used to electronically detect the existence of magnetic beads, such as the GMR (giant magnetoresistance) effect discovered by the 2007 Nobel Physics Prize winners Albert Fert and Peter Grünberg. The effect appears in thin film structures composed of alternating ferromagnetic and nonmagnetic layers. A significant change in the electrical resistance is observed depending on whether the magnetization of adjacent ferromagnetic layers is in a parallel alignment (the low-resistance state) due to the applied magnetic field or an anti-parallel alignment (the high-resistance state) in the absence of the magnetic field.

Han et al. [13] reported a CMOS DNA sensor array that adopted the spin valve structure to observe the GMR effect. Magnetic thin films of nanometers in thickness were deposited and patterned after the conventional CMOS process. The biotinylated analyte DNA was captured by complementary probes immobilized on the sensor surface. Then streptavidin-coated magnetic labels were added and produced specific binding to the hybridized DNA. The stray magnetic field of the magnetic labels was detected as a resistance change in the sensor. In general, signal

modulation is required in the sensing scheme in order to separate the true bio-signals from the false ones caused by drifts or ionic solution interference. Other than the GMR principle, detection of magnetic beads can be achieved based on the Hall effect of a CMOS sensor. As discovered by Edwin Hall in 1879, the effect produces an electric field perpendicular to the magnetic induction vector and the original current direction. Detection of a single magnetic bead has been demonstrated [14].

To eliminate the needs of externally applied magnetic fields and post-CMOS fabrication for a magneto-resistive biosensor, Wang et al. [15] presented an inductive approach that can detect existence of a single magnetic bead on a CMOS chip. The sensing scheme used a highly stable integrated oscillator with an on-chip LC resonator. An a.c. electrical current through the on-chip inductor produced a magnetic field that polarized the magnetic particles present in its vicinity, leading to an increased effective inductance and therefore a reduced oscillation frequency. As the frequency shift due to a single micron-size magnetic bead is typically a few parts per million (ppm) of the resonant frequency, the sensing oscillator needs to have small phase noises at small offset frequencies to achieve a stable frequency behavior.

A miniaturized CMOS NMR (nuclear magnetic resonance) system has been reported by Sun et al. [16] for applications in biomolecular sensing. NMR was first discovered by Isidor Rabi who was awarded Nobel Prize in Physics in 1944. Magnetic nuclei, like ^1H and ^{31}P, could absorb radio frequency (RF) energy when placed in a magnetic field of a strength specific to the identity of the nuclei. The nucleus is described as being in resonance when the absorption occurs. Different atomic nuclei within a molecule exhibit different resonant frequencies for the same magnetic field strength. Essential chemical and structural information about a molecule can thus be studied by observing such magnetic resonant frequencies.

The reported CMOS NMR system consists of a magnet (static magnetic field), an RF coil surrounding a sample, and an RF transceiver linked to the coil as shown in Fig. 4. The RF magnetic field produced through the coil at the right frequency ω_o can excite nuclei spins within the sample. Once the RF excitation is stopped and the receiver is connected to the coil, the detected NMR signal displays an exponential relaxation in the precession of the net magnetic moment.

CMOS MEMS Biosensors, Fig. 4 Operation of the CMOS NMR system (Sun et al. © 2009 IEEE)

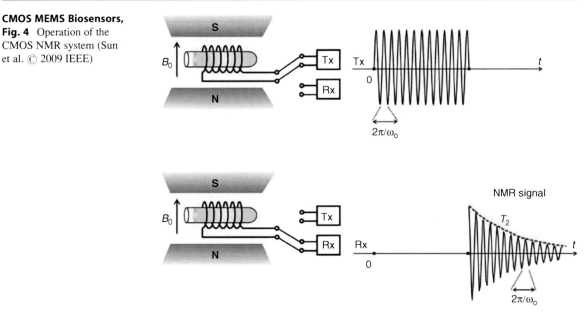

Both the resonant frequency and the relaxation's characteristic time are specific for the sample to be studied.

Optical Biosensors

Optical detection methods such as luminometry and fluorometry can be utilized to make CMOS-based biosensors. Luminometric methods, such as luciferase-based assays, involve the detection and quantification of light emission as a result of a chemical reaction. Luciferase is a generic term for the class of oxidative enzymes used in bioluminescence. Such methods have been used to detect pathogens and proteins, perform gene expression and regulation studies, and sequence DNA. Fluorometric methods require an excitation source to stimulate photoemission of the fluorescent-tagged species and optical filters to separate the generated photoemission from the high background interference. Luminometry is more amenable to miniaturization and integration than the fluorometric methods since no filters or excitation sources are required.

Eltoukhy et al. [17] reported a CMOS photodetector array which was directly integrated with a fiber-optic faceplate with immobilized luminescent reporters/probes for DNA synthesis by pyrosequencing. Pyrosequencing is a "sequencing by synthesis" technique which involves taking a single strand of the DNA to be sequenced and then synthesizing its complementary strand enzymatically. Synthesis of the complementary strand is achieved with one base pair at a time by monitoring the activity of

the DNA synthesizing enzyme with another chemiluminescent enzyme. The challenge of such a detection system is to achieve high sensitivity despite the presence of relatively high dark currents. The reported sensor array was able to detect emission rates below 10^{-6} lux over a long integration time (30 s).

Huang et al. [18] reported a CMOS microarray chip that leveraged low-cost integration of solid-state circuits for fluorescence-based diagnostics. The featured time-gated fluorescence detection was able to significantly reduce interferences from the external excitation source, eliminating the need for additional optical filters. Direct immobilization of DNA capture strands on the CMOS chip allows placement of optical detectors in close vicinity to the fluorescent labels, leading to improved collection efficiency and providing imaging resolution limited by pixel dimensions rather than diffraction optics.

In addition to the aforementioned approaches, a method based on detection of the incident light intensity using a normal light source has been developed [19]. The technique uses gold nanoparticles with silver enhancement to induce opacity on top of CMOS photodetectors when specific bindings occur. It does not require an external optical scanner and a specific light source as needed in the fluorescence-based method.

Micromechanical Cantilevers

Microcantilever-based biosensors have attracted considerable attention as a means of label-free detection of biomolecules. Intermolecular forces arising from adsorption of small molecules are known to induce surface stress. The specific binding between ligands and receptors on the surface of a microcantilever beam thus produces physical bending of the beam. Optical detection of the beam deflection due to hybridization of complementary oligonucleotides has been demonstrated [20]; however, the method requires external instruments that are not amenable to monolithic integration.

Piezoresistive detection is a viable alternative for CMOS integration because it is compatible with aqueous media. The piezoresistive effect is associated with the resistivity change of a semiconductor subject to a mechanical strain. By placing a MOS transistor into the base of a cantilever, modulation of the channel current underneath the gate region can be measured as a result of adsorption-induced surface stress [21]. Sensing resolution of the static beam deflection is limited by the flicker noise – the dominant source of noise in MOS transistors at low frequencies.

Summary

A miniaturized biosensing platform can be achieved through monolithic integration of sensing devices and detection circuits by the CMOS MEMS technology. Sensing devices that can be directly fabricated on a CMOS chip are introduced, including those based on electrochemical, impedimetric, ion-sensitive, magnetic, optical, and micromechanical approaches. Some of the methods require labeling (e.g., magnetic beads, nanoparticles) and some are label-free for bio-signal transduction. Some approaches use the devices (e.g., transistors) or the materials (e.g., metal electrodes) in a CMOS process for sensing, such that sensor performances can be enhanced in accordance with the scaling of CMOS technologies. Arrays of sensors can be conveniently fabricated on a CMOS chip such that sensing resolution and accuracy can be enhanced through statistical analysis of the collected data.

Cross-References

▶ Biosensors
▶ Nanogap Biosensors

References

1. Aoki, K., Morita, M., Niwa, O., Tabei, H.: Quantitative analysis of reversible diffusion-controlled currents of redox soluable species at interdigitated array electrodes under steady-state conditions. J. Electroanal. Chem. **256**, 269–282 (1988)
2. Schienle, M., Paulus, C., Frey, A., Hofmann, F., Holzapfl, B., Schindler-Bauer, P., Thewes, R.: A fully electronic DNA sensor with 128 positions and in-pixel A/D conversion. IEEE J. Solid State Circuits **39**, 2438–2445 (2004)
3. Levine, P.M., Gong, P., Levicky, R., Shepard, K.L.: Active CMOS sensor array for electrochemical biomolecular detection. IEEE J. Solid State Circuits **43**, 1859–1871 (2008)
4. Ayers, S., Berberian, K., Gillis, K.D., Lindau, M., Minch, B.A.: Post-CMOS fabrication of working electrodes for on-chip recordings of transmitter release. IEEE Trans. Biomed. Circuits Syst. **4**, 86–92 (2010)
5. Warburg, E.: Ueber das Verhalten sogenannter unpolarisbarer Elektroden gegen Wechselstrom. Ann. Phys. Chem. **67**, 493–499 (1899)
6. Lee, K., Lee, J., Sohn, M., Lee, B., Choi, S., Kim, S.K., Yoon, J., Cho, G.: One-chip electronic detection of DNA hybridization using precision impedance-based CMOS array sensor. Biosens. Bioelectron. **26**, 1373–1379 (2010)
7. Stagni, C., Guiducci, C., Benini, L., Riccò, B., Carrara, S., Samorí, B., Paulus, C., Schienle, M., Augustyniak, M., Thewes, R.: CMOS DNA sensor array with integrated A/D conversion based on label-free capacitance measurement. IEEE J. Solid State Circuits **41**, 2956–2963 (2006)
8. Lu, M.S.-C., Chen, Y.C., Huang, P.C.: 5 × 5 CMOS capacitive sensor array for detection of the neurotransmitter dopamine. Biosens. Bioelectron. **26**, 1093–1097 (2010)
9. Ghafar-Zadeh, E., Sawan, M., Chodavarapu, V.P., Hosseini-Nia, T.: Bacteria growth monitoring through a differential CMOS capacitive sensor. IEEE Trans. Biomed. Circuits Syst. **4**, 232–238 (2010)
10. Li, J., Xue, M., Lu, Z., Zhang, Z., Feng, C., Chan, M.: A high-density conduction-based micro-DNA identification array fabricated with a CMOS compatible process. IEEE Trans. Electron Devices **50**, 2165–2170 (2003)
11. Kim, D.S., Jeong, Y.T., Park, H.J., Shin, J.K., Choi, P., Lee, J.H., Lim, G.: An FET-type charge sensor for highly sensitive detection of DNA sequence. Biosens. Bioelectron. **20**, 69–74 (2004)
12. Li, D.C., Yang, P.H., Lu, M.S.-C.: CMOS open-gate ion-sensitive field-effect transistors for ultrasensitive dopamine detection. IEEE Trans. Electron Devices **57**, 2761–2767 (2010)
13. Han, S., Yu, H., Murmann, B., Pourmand, N., Wang, S.X.: A high-density magnetoresistive biosensor array with drift-compensation mechanism. In: IEEE International Solid-State Circuits Conference (ISSCC) Digest of Technical Papers, pp. 168–169, San Francisco, 11–15 Feb 2007
14. Besse, P., Boero, G., Demierre, M., Pott, V., Popovic, R.: Detection of a single magnetic microbead using a miniaturized silicon Hall sensor. Appl. Phys. Lett. **80**, 4199–4201 (2002)

15. Wang, H., Chen, Y., Hassibi, A., Scherer, A., Hajimiri, A.: A frequency-shift CMOS magnetic biosensor array with single-bead sensitivity and no external magnet. In: IEEE International Solid-State Circuits Conference (ISSCC) Digest of Technical Papers, pp. 438–439 (2009)
16. Sun, N., Liu, Y., Lee, H., Weissleder, R., Ham, D.: CMOS RF biosensor utilizing nuclear magnetic resonance. IEEE J. Solid State Circuits **44**, 1629–1643 (2009)
17. Eltoukhy, H., Salama, K., El Gamal, A.: A 0.18-µm CMOS bioluminescence detection lab-on-chip. IEEE J. Solid State Circuits **41**, 651–662 (2006)
18. Huang, T.D., Sorgenfrei, S., Gong, P., Levicky, R., Shepard, K.L.: A 0.18-µm CMOS array sensor for integrated time-resolved fluorescence detection. IEEE J. Solid State Circuits **44**, 1644–1654 (2009)
19. Xu, C., Li, J., Wang, Y., Cheng, L., Lu, Z., Chan, M.: A CMOS-compatible DNA microarray using optical detection together with a highly sensitive nanometallic particle protocol. IEEE Electron Device Lett. **26**, 240–242 (2005)
20. Fritz, J., Baller, M.K., Lang, H.P., Rothuizen, H., Vettiger, P., Meyer, E., Güntherodt, H.-J., Gerber, Ch., Gimzewski, J.K.: Translating biomolecular recognition into nanomechanics. Science **288**, 316–318 (2000)
21. Shekhawat, G., Tark, S.H., Dravid, V.P.: MOSFET-embedded microcantilevers for measuring deflection in biomolecular sensors. Science **311**, 1592–1595 (2006)

CMOS MEMS Fabrication Technologies

Hongwei Qu[1] and Huikai Xie[2]
[1]Department of Electrical and Computer Engineering, Oakland University, Rochester, MI, USA
[2]Department of Electrical and Computer Engineering, University of Florida, Gainesville, FL, USA

Synonyms

CMOS (complementary metal-oxide-semiconductor); CMOS-MEMS; Integration; MEMS (micro-electro-mechanical systems)

Definition

CMOS-MEMS are micromachined systems in which MEMS devices are integrated with CMOS circuitry on a single chip to enable miniaturization and performance improvement. CMOS-MEMS also refers to microfabrication technologies that are compatible with CMOS fabrication processes.

Overview

Microelectromechanical systems (MEMS) leverage semiconductor fabrication technologies to manufacture various miniature sensors and actuators. Due to their low cost and small size as well as their much improved reliability, MEMS devices have been widely used even in our daily life, e.g., MEMS accelerometers for automobiles' airbags, MEMS gyroscopes for electronic stability program (ESP) in automobile braking systems, MEMS tire pressure sensors; digital micromirror device (DMD)-enabled portable projectors, MEMS inkjet printers, MEMS resonators as frequency references, etc. Moreover, smart cell phones are now equipped with MEMS gyroscopes and accelerometers for motion actuated functions. They are also installed with surface-mounted MEMS microphones for even smaller size. The worldwide MEMS market reached 6.5 billion US dollars in 2010 [1].

Continuous miniaturization, expanded functionalities, lower cost, and improved performance are the ultimate goals of MEMS. The nature of MEMS strongly suggests direct integration of mechanical structures with electronics whose fabrication is dominated by CMOS technologies. In the last couple of decades, great efforts have been made in the integration of MEMS structures with ICs on a single CMOS substrate. The pioneering work for CMOS-MEMS transducers was done by H. Baltes and his coworkers at the Swiss Federal Institute of Technology Zurich (ETH) [2]. They employed both wet bulk silicon micromachining and surface micromachining techniques in the fabrication of integrated CMOS-MEMS devices. With the great advances in IC and MEMS technologies, the current focus of CMOS-MEMS integration technology is on the modification and standardization of CMOS technology to accommodate MEMS technology. One of the best-known commercial CMOS-MEMS devices is the digital micromirror device (DMD) manufactured by Texas Instruments. Recently, some CMOS foundries, such as TSMC, X-Fab and Global Foundries, have begun to offer CMOS-MEMS services for research and product developments.

This entry summarizes a variety of CMOS-MEMS technologies and devices that have been developed. Particular materials needed in associated CMOS-MEMS will also be introduced. Typical MEMS devices, including inertial sensors, resonators, and actuators, are exemplified in featuring the respective technologies.

Classification of CMOS-MEMS Technologies

MEMS can be integrated with CMOS electronics in many different ways. One common way to categorize CMOS-MEMS technologies is from the perspective of manufacturing processes. Based on the process sequence, CMOS-MEMS technologies can be classified into three categories: Pre-CMOS, Intra-CMOS, and Post-CMOS [3]. Due to its popularity and accessibility, post-CMOS will be described in more detail in this entry.

Pre-CMOS

It is widely accepted that pre-CMOS technologies are represented by the modular integration process originally developed at Sandia National Laboratories. As suggested by the name, in pre-CMOS technology, MEMS structures are pre-defined and embedded in a recess in silicon wafer; and the recess is then filled with oxide or other dielectrics. The wafer is then planarized prior to the following process steps for CMOS electronics [4]. In this "MEMS first" process, although MEMS structures are pre-defined, a wet etch after the completion of the standard CMOS processes is required to release the pre-defined MEMS structures. Due to the involvement of photolithography process needed for patterning the MEMS in the recess, the thickness of the MEMS structures is constrained by the lithographical limit.

Other methods for the formation of MEMS structures in pre-CMOS technologies, including wafer bonding and thinning for epitaxial and SOI wafers in which MEMS are prefabricated, have also been reported in the fabrication of a variety of MEMS devices.

Inter-CMOS

In early 1990s, Analog Devices, Inc (ADI) specifically developed a MEMS technology based on its BiCMOS process. This "iMEMS" technology, originally dedicated to manufacturing CMOS-MEMS accelerometers and gyroscopes, is an intermediate-CMOS-MEMS, or Inter-CMOS-MEMS technology in which the CMOS

process steps are mixed with additional polysilicon thin-film deposition and micromachining steps to form the sensor structures [5]. Infinion's pressure sensors are also fabricated using this kind of Inter-CMOS-MEMS technology. To reduce the residual stress in structural polysilicon, high temperature annealing is normally required in the Inter-CMOS-MEMS, which could pose a potential risk to CMOS interconnect and active layers. Thus, the thermal budget should be carefully designed. Moreover, it is almost impractical to perform intermediate CMOS and MEMS processed in separate foundries due to possible contaminations in the wafer transfer and processes. Therefore, a dedicated foundry used for both CMOS and MEMS is necessary for Inter-CMOS-MEMS technology, which may not take full advantages of mainstream technologies in either area.

Limitations of Pre- and Inter-CMOS-MEMS

Since surface micromachining and polysilicon are typically used, most of the Pre- and Inter-CMOS-MEMS technologies suffer from the limitations of thin-film structural materials. (1) Structural curling and cost associated with stress compensation: Due to the residual stress in the deposited thin films in the device, polysilicon structures often exhibit curling after release, resulting in reduced sensitivity, lower mechanical robustness, and increased temperature dependence. Although stress compensation can be realized via multiple controlled process steps, the associated cost is quite high. (2) Small size and/or mass: The curling of thin-film structures in turn limits the size of the overall microstructure. The mass is further reduced due to the small thickness of thin-film polysilicon. For inertial sensors, the smaller the structure mass, the lower performance of the device. (3) Parasitics: In a surface micromachined polysilicon accelerometer, depending on polysilicon wiring path, the parasitic impedance may considerably lower the static and dynamic performance of the device. (4) Cost and suboptimal processes of the dedicated foundry needed: The dedicated foundry combining CMOS and MEMS fabrications needed for pre- and inter-CMOS-MEMS is normally expensive and suboptimal for either fabrication. It is against the modern trends in which flexible accessibility of optimal and cost-effective processes are preferable.

CMOS MEMS Fabrication Technologies, Table 1   Representative thin-film deposition additive post-CMOS-MEMS technologies

Authors and references	Institute	Structural material	Sacrificial material	Interconnect material	Year
Hornbeck [7]	Texas instruments	Al	Photoresist	Al	1989 (invented in 1987)
Yun et al. [8]	UC-Berkeley	Polysilicon	SiO_2	W/TiN	1992
Franke et al. [9]	UC-Berkeley	Poly-SiGe	Ge or SiO_2	Al	1999
Sedky, Van Hoof et al. [10]	IMEC	Poly-SiGe	Ge	Al	1998

Post-CMOS

Post-CMOS-MEMS refer to the CMOS-MEMS processes in which all MEMS process steps are performed after the completion of the CMOS fabrication. The advantages of post-CMOS-MEMS over Pre- and Inter-CMOS-MEMS include process flexibility and accessibility and low cost. In contrast to both Pre-CMOS-MEMS and Inter-CMOS-MEMS, for Post-CMOS-MEMS technology, the fabrications of CMOS circuitry and MEMS structures are performed independently. The flexibility of foundry access makes it possible to take advantages of both advanced CMOS technologies and optimal MEMS fabrication. This is particularly attractive to research community in exploration of state-of-the-art in MEMS. Some design rules may need to be changed to accommodate MEMS structure design in the CMOS design stage. Meanwhile, post-CMOS microfabrication should be carefully designed, particularly considering the thermal budget, so as not to affect the on-chip CMOS electronics.

According to how MEMS structures are formed, post-CMOS-MEMS technologies fall into two categories: additive and subtractive. In additive post-CMOS-MEMS, structural materials are deposited on a CMOS substrate. In subtractive post-CMOS-MEMS, MEMS structures are created by selectively etching CMOS layers. Apparently, additive post-CMOS-MEMS methods require more stringent material compatibility with the CMOS technologies used. Thus, they are less utilized than subtractive post-CMOS-MEMS. The following introduction will focus more on subtractive post-CMOS-MEMS.

Additive MEMS Structures on CMOS Substrate

In additive post-CMOS-MEMS, metals, dielectrics, or polymers are deposited and patterned to form MEMS structures normally on top of the CMOS layers. Some commercial MEMS products are fabricated using additive post-CMOS-MEMS approaches. In this category, the best-known product is probably the digital mirror device (DMD), the core of the digital light processing (DLP) technology developed by Texas Instruments. In a DMD, tilting mirror plates and their driving electrodes are fabricated directly on top of CMOS circuits. Three sputtered aluminum layers are used to form the top mirror plate and the two parallel-plate electrodes for electrostatic actuation, respectively. The driving electrodes are addressed via CMOS memory cell. To release the mirror plate and top electrodes in the post-CMOS-MEMS fabrication of the mirrors, deep-UV hardened photoresist is used as the sacrificial layer.

In some circumstances where CMOS protection is well designed, electroplating can also be used to grow microstructures on top of CMOS electronics. Other structural and sacrificial materials, such as polycrystalline SiGe and Ge, have been used to create CMOS-MEMS as well [6]. Additive post-CMOS-MEMS processes, along with their respective materials, are summarized in the following table (Table 1).

In addition to the approaches of forming MEMS structures on top of the CMOS substrate by thin-film deposition, wafer bonding provides another method to directly integrate MEMS structures on CMOS substrate [11]. For instance, a prefabricated polysilicon capacitive acceleration sensor wafer is bonded to a CMOS wafer with read-out electronics. In a wafer-bonded piezoresistive accelerometer, the micromachined bulk silicon proof mass was sandwiched by a bottom glass cap and a top CMOS chip on which the conditioning circuit was integrated. SOI-CMOS-MEMS has also been attempted for monolithic integration of electronics with bulk MEMS structures. With 3-dimensional packaging enabled by technological breakthroughs such as through-silicon vias (TSVs), this integration method promises to be further developed in manufacturing complex microsystems. MEMS suppliers, including STMicroelectronics and InvenSense, have adopted wafer-to-wafer or chip-to-wafer bonding CMOS-MEMS integration.

Subtractive Post-CMOS-MEMS

In these devices, MEMS structures are formed from built-in CMOS thin-film stacks including metals and SiO_2, or from the silicon substrate. These materials are patterned and removed partially by wet or dry etching to release the MEMS structures. This section describes the thin-film and bulk CMOS-MEMS formed by such subtractive processes.

Subtractive CMOS-MEMS by Wet Etching

The first generation of CMOS-MEMS sensors was fabricated using a post-CMOS subtractive process in which silicon substrate was completely or partially removed using a wet etching method, leaving behind thin-film or bulk MEMS structures [2]. For thermal sensors in which beams or membranes consisting of dielectric layers, the substrate silicon is normally etched away completely to obtain thermally isolated structures. The silicon dioxide membrane can act as an intrinsic etch stop layer in backside silicon anisotropic wet etch using KOH, ethylene diamine-pyrocatechol (EDP), or Tetramethylammonium hydroxide (TMAH). A high-Q RF MEMS filter with an inter-metal dielectric layer as structural material was reported by IBM. A medical tactile sensor array was also reported in which the aluminum sacrificial layer was etched from the backside of the wafer after the CMOS substrate was etched through [12].

The silicon substrate can also be included in the MEMS structures using a wet etch process. The first method is to perform a time-controlled backside etch with a well-calibrated etching rate. A uniform single crystal silicon membrane with a desired thickness can be created. This method has been widely used in industry for fabrication of large volume products such as integrated pressure sensors. In cases where the silicon membrane thickness is not critical, even mechanical processing such as grinding can be used to create the backside cavity.

The second method involves the utilization of an automatic etch stop technique to create silicon membranes or MEMS structures. In this case, an anisotropic etch stops at the electrochemically biased p-n junction formed between the n-well and p-type substrate in CMOS [13]. Although the electrochemical electrode design and implementation are complicated, this process can be specifically used in the fabrication of highly sensitive pressure/force and thermal sensors. The anisotropic etch stop can also occur at highly doped p regions in the substrate. This method has been used in fabrication of many suspended structures including neural probes [14]. Note that the p++ doping process may not be available in a standard CMOS process. In the case where only a small portion of the silicon substrate needs to be removed to reduce the circuit-substrate coupling, a wet silicon etch can be performed from the front side. In wet silicon etching, either silicon nitride or additional polymers or both can be used to protect the front CMOS and pads.

Polymers sensitive to analytes can be coated on finished CMOS-MEMS structures for chemical and biological sensing. For example, the first CMOS-MEMS electronic nose was demonstrated by forming polymer-coated CMOS thin-film cantilevers on a CMOS chip [15].

Table 2 summarizes some representative devices that were fabricated using wet etching when this technology was dominant in post-CMOS micromachining. Bibliographies of these efforts can be found in the above citations in this section.

Subtractive Post-CMOS-MEMS by Dry Etching

Dry etching processes have quickly become popular in microfabrication for both MEMS research and industry. Particularly, the deep reactive ion etching (DRIE) technology, or Bosch process, has revolutionized subtractive post-CMOS microfabrication [23]. This section describes thin-film and bulk CMOS-MEMS devices fabricated using dry etching processes.

Most dry etching processes are based on plasma processes, such as reactive ion etch (RIE) and DRIE, while etchants in the vapor phase can also be used for dry etching. For example, vapor XeF_2 provides good isotropic etching of silicon, which has been used for releasing CMOS thin-film MEMS structures [24]. The combination of RIE and DRIE, performed from the front or back side, or both sides, has been used to fabricate a large variety of CMOS-MEMS devices. Depending on the structural materials and etching methods employed, subtractive post-CMOS can be divided into two types: thin-film processes and bulk processes.

Thin-Film Post-CMOS-MEMS Dry Processes In thin-film processes, structural materials are composed of CMOS thin films. Figure 1 depicts the process flow

CMOS MEMS Fabrication Technologies, Table 2 CMOS-MEMS devices enabled by subtractive process wet etching

Authors and references	Institutions	Year	Device	Device structure	Etching method
Wise et al. [16]	U. of Michigan	1979	Pressure sensor	Silicon diaphragm	Backside EDP etching
Wise et al. [14]	U. of Michigan	1985	Neuron probe array	CMOS thin films and Si substrate	EDP etching, p++ etching stop
Yoon and Wise [17]	U. of Michigan	1990	Mass flow sensor	CMOS thin films	Backside, SiO_2 etching stop
Baltes et al. [2]	ETH Zurich	1996	Thermal capacitor	CMOS thin films	Front side etching
Haberli et al. [18]	ETH Zurich	1996	Pressure sensor	CMOS thin films	Front side etching of aluminum as sacrificial layer
Schneider et al. [19]	ETH Zurich	1997	Thermal sensor	CMOS thin films and suspended substrate	PN junction electrochemical etch stop
Akiyama et al. [20]	U. of Neuchatel, ETH Zurich	2000	AFM probe	CMOS thin films and silicon substrate	N-well electrochemical etch stop
Schaufelbuhl et al. [21]	ETH Zurich	2001	Infrared imager	CMOS thin films	Backside KOH
Verd et al. [22]	U. of Barcelona	2006	RF MEMS	CMOS thin films	Front side SiO_2 etching

of a thin-film post-CMOS-MEMS process, which was originally developed at Carnegie Mellon University [25]. A sequenced process consisting of an isotropic SiO_2 etching, a silicon DRIE and an isotropic Si RIE releases the MEMS structure. In these process steps, the top metal layer acts as a mask to form the MEMS structures and to protect the CMOS circuitry, as seen in Fig. 1a, b. Anisotropic and isotropic silicon etching complete the process flow, as seen in Fig. 1c, d. Various inertial sensors have been fabricated using this thin-film technology. In all these inertial sensors, mechanical springs and proof masses are formed by the multiple-layer CMOS stacks consisting of SiO_2 and metals. The sensing capacitance is formed from sidewall capacitance between comb fingers. The multiple CMOS metal layers inside the comb fingers and other mechanical structures allow very flexible electrical wiring, facilitating different sensing schemes including vertical comb-drive sensing. Akustica, Inc. has commercialized digital microphones using a modified version of this process. Other sensors have also been demonstrated using similar thin-film technology, such as humidity sensors and chemical sensors.

All these thin-film post-CMOS dry etching processes have excellent CMOS compatibility and accessibility as well as design flexibility. However, a major issue is the large vertical curling and lateral buckling of suspended MEMS structures, which is caused by the residual stress in the stacked thin-film CMOS layers. Although structural curling can be tolerated for some small devices such as RF MEMS, for devices such as inertial sensors that need relatively large size, the impact of structural curling can be severe.

Bulk CMOS-MEMS Dry Process In order to overcome the structural curling and to increase the mass, flatness, and robustness of MEMS structures, single crystal silicon (SCS) may be included underneath the CMOS thin-film stacks. The SCS silicon structures are formed directly from the silicon substrate using DRIE. Figure 2 illustrates the process flow [26]. The process starts with the backside silicon DRIE to define the MEMS structure thickness by leaving a 10–100 μm-thick SCS membrane (Fig. 2a). Next, the same anisotropic SiO_2 etch as in the thin-film process is performed on the front side of wafer (chip) to expose the SCS to be removed (Fig. 2b). The following step differs from the thin-film process in that an anisotropic DRIE, instead of isotropic etch, finalizes the structure release by etching through the remaining SCS diaphragm, as shown in Fig. 2c. With the SCS underneath the CMOS interconnect layers included, large and flat MEMS microstructures can be obtained. If necessary, an optional time-controlled isotropic silicon etch can be added. This step will undercut the SCS underneath the designed narrow CMOS stacks to create thin-film structures (Fig. 2d). This step is particularly useful for fabricating capacitive inertial sensors. It can be used to form the electrical isolation structures between sensing fingers and silicon substrate.

The DRIE CMOS-MEMS technology has shown great advantages in the fabrication of relatively large MEMS devices such as micromirrors. Large flat mirror

CMOS MEMS Fabrication Technologies, Fig. 1 CMU post-CMOS fabrication process for MEMS structures made of CMOS thin films

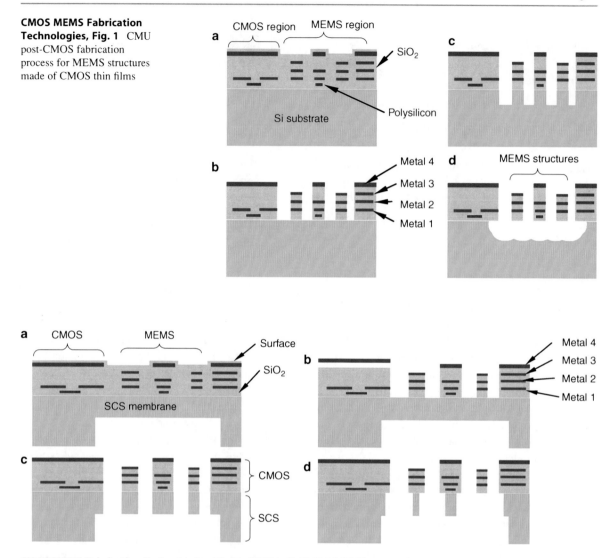

CMOS MEMS Fabrication Technologies, Fig. 2 DRIE bulk CMOS-MEMS process flow

can be obtained by including portion of silicon substrate underneath the aluminum mirror surface. A CMOS-MEMS gyroscope with a low noise floor of 0.02 degree/s/sqrtHZ has also been demonstrated using this technology [27].

By attaching SCS underneath the CMOS stack comb fingers, the sensing capacitance can be considerably increased for larger signal-to-noise ratio (SNR). Although CMOS thin films are still used in some microstructures for electrical isolation, the length of the thin-film portion is minimal to reduce the temperature effect. Compared to the thin-film dry CMOS-MEMS process, a backside silicon DRIE

step is added. This requires an additional backside lithography step to define the region for MEMS structures. The maximum thickness of the MEMS structures is limited by the aspect ratio that the silicon DRIE can achieve.

An Improved Bulk CMOS-MEMS Process The bulk CMOS-MEMS process depicted in Fig. 2 is useful in fabrication of many devices where SCS structures are desired to improve both mechanical and electrical performance of the devices. However, for some devices, very fine structures are formed in step (c) in Fig. 2; so the damage caused by the step (d) to these

CMOS MEMS Fabrication Technologies, Fig. 3 The modified bulk CMOS-MEMS process for separate etching of CMOS beams and SCS microstructures. Backside photoresist coating effectively reduces temperature in the device release, reducing deleterious non-uniform etching

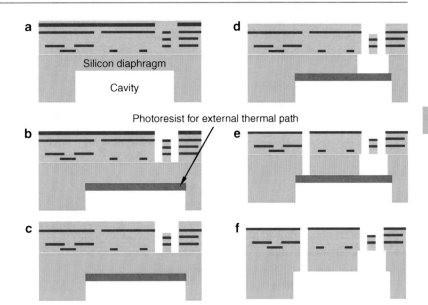

fine structures may be severe. This is particularly true for the fabrication of capacitive inertial sensors where narrow-gap sensing comb fingers are needed. For instance, in performing the isotropic silicon undercut to form the narrow CMOS beams for electrical isolation, the SCS in the comb fingers is also undercut. The sensing gap increases due to the undesired undercut greatly reduce sensitivity and signal-to-noise ratio (SNR). If the undercut occurs in mechanical structures such as suspension springs, the characteristics of the device will also be affected. Another issue is related to the thermal effect in the plasma etch for the SCS undercut. Upon completion of the silicon undercut, the greatly reduced thermal conductance from the isolated structure to the substrate can cause a temperature rise on the released structures. Slight over-etch is often necessary to accommodate process variations, but this will generate a large temperature rise on the suspended structures which in turn dramatically increases the SCS etching rate, resulting in uncontrollable and damaging results [28].

A modified dry bulk CMOS-MEMS process has been demonstrated to effectively address the issues caused by the undesired SCS undercut [28]. In the refined process illustrated in Fig. 3, the etching of the CMOS connection beams is performed separately from the etching of the microstructures where SCS is needed. The top metal layer is specifically used to define the connection beams. After their formation, the top metal layer is removed using a plasma or a wet etch. Then other microstructures are exposed after a SiO_2 etch. The direct etch-through of the remaining silicon on the microstructures will complete the release process. To reduce the thermal effect described above, a thick photoresist layer is patterned on the backside of the cavity. In the release step, the applied photoresist provides a thermal path that reduces the temperature rise on the etched-through structures. The removal of the photoresist using O_2 plasma etching completes the entire microfabrication process. Owing to the monolithic integration and large proof mass enabled by the inclusion of SCS, bulk CMOS-MEMS inertial sensors have demonstrated better performance than their thin-film counterparts [29]. The photoresist coating can also be replaced by sputtering a layer of metal such as aluminum.

Combined Wet/Dry Processes

In addition to the integration methods described above, efforts have been continuously made to integrate CMOS with MEMS using the combination of different microfabrication technologies. By combing silicon anisotropic wet etch with DRIE, some sophisticated surface and bulk MEMS structures such as bridges and cantilever arrays can be created. A multi-sensor system was demonstrated using a combined etch process [30]. In the accelerometers reported in [31], isotropic wet etching is used to remove metal layers in CMOS thin stacks to create parallel-plate-like vertical capacitors for gap-closing sensing. A silicon RIE follows to release the MEMS devices and break the coupling between the sensing thin films and the substrate. Sensitivities are largely increased with the gap-closing sensing compared to comb-finger sensing.

Summary

CMOS-MEMS technologies have been placed in pre-CMOS, intra-CMOS, and post-CMOS categories. Both pre-CMOS and intra-CMOS have issues such as dedicated foundries with suboptimal and less cost-effective processes. So it is normally impractical for academic research community to access these dedicated facilities. Post-CMOS provides excellent CMOS compatibility, foundry accessibility, and design flexibility, and the cost is also relatively low. While the process standardization and industrialization of CMOS-MEMS technologies are in continuous progress, innovative processing technologies have opened up new pathways for integration. Wafer bonding–based integration has blurred the boundary between pre- and post-CMOS-MEMS integrations. SOI-CMOS-MEMS have also been aggressively explored. The technologies involved in this new exploration have emerged as enabling means for three-dimensional and systems-in-package integrations. More recently, technologies to co-fabricate many subsystems including nano-systems are being pursued enthusiastically.

CMOS-MEMS integration will continually evolve with the emergence of new fabrication technologies and new materials. While some companies have demonstrated promising CMOS-MEMS products, more joint efforts from research community and industries are needed for new process transfer and standardization to allow large volume fabrication of new products.

Cross-References

▶ Integration
▶ MEMS
▶ Nanofabrication
▶ Sensors

References

1. Johnson, R.C.: MEMS market projected to hit double-digit growth, again. www.eetimes.com. Accessed 14 July 2011
2. Baltes, H., Paul, O., et al.: IC MEMS microtransducers. In: Proceedings of International Electronic Device Meeting IEDM '96, San Francisco, pp. 521–524 (1996)
3. Baltes, H., Brand, O., Fedder, G.K., Hierold, C., Korvink, J. G., Tabata, O.: CMOS-MEMS: Advanced Micro and Nanosys, 1st edn. Wiley, Weinheim (2005)
4. Smith, J.H., Montague, S., Sniegowski, J.J., Murray, J.R., McWhorter, P.J., Smith J.H.: Embedded micromechanical devices for the monolithic integration of MEMS with CMOS. In: Proceedings of International Electronic Device Meeting, IEDM '95, Washington D.C., pp. 609–612 (1995)
5. Kuehnel, W., Sherman, S.: A surface micromachined silicon accelerometer with on-chip detection circuitry. Sens. Actu. A Phys. 45, 7–16 (1994)
6. Franke, A.E., Heck, J.M., King, T.J., Howe, R.T.: Polycrystalline silicon-germanium films for integrated microsystems. J. Micromech. Sys. 12, 160–171 (2003)
7. Hornbeck, L.: Deformable-mirror spatial light modulators and applications. SPIE Crit. Rev. 1150, 86–102 (1989)
8. Yun, W., Howe, R.T., Gray, P.R.: Surface micromachined, digitally force-balanced accelerometer with integrated CMOS detection circuitry. Technical Digest of Solid State Sensors and Actuators Workshop, Hilton Head Island, pp. 126–131 (1992)
9. Franke, A.E., Bilic, D., Chang, D.T., Jones, P.T., King, T.J., Howe, R.T., Johnson, G.C.: Post-CMOS integration of germanium microstructures. In: The 12th IEEE International Conference on Micro Electro Mechanical Systems, Orlando, pp. 630–637 (1999)
10. Sedky, S., Fiorini, P., Caymax, M., Loreti, S., Baert, K., Hermans, L., Mertens, R.: Structural and mechanical properties of polycrystalline silicon germanium for micromachining applications. J. MEMS 7, 365–372 (1998)
11. Fedder, G.K., Howe, R.T., Liu, T.J., Quevy, E.P.: Technologies for cofabricating MEMS and electronics. Proc. IEEE 96, 306–322 (2008)
12. Salo, T., Vancura, T., Brand, O., Baltes, H.: CMOS-based sealed membranes for medical tactile sensor arrays. In: Proceedings of International Conference on Micro Electro Mechanical Systems, Kyoto, pp. 590–593 (2003)
13. Muller, T., Brandl, M., Brand, O., Baltes, H.T.: An industrial CMOS process family adapted for the fabrication of smart silicon sensors. Sen. Actu. A Phys. 84, 126–133 (2000)
14. Najafi, K., Wise, K.D., Mochizuki, T.K.: A high-yield IC-compatible multichannel recording array. IEEE T. Electron. Dev. 32, 1206–1211 (1985)
15. Baltes, H., Koll, A., Lange, D.H.: The CMOS MEMS nose-fact or fiction? In: Proceedings of IEEE International Symposium on Industrial Electronics ISIE '97, Guimaraes, vol. 1, pp. SS152–SS157 (1997)
16. Borky, J.M., Wise, K.D.: Integrated signal conditioning for silicon pressure sensors. IEEE T. Electron. Dev. ED-27, 927–930 (1979)

17. Yoon, E., Wise, K.D.: A multi-element monolithic mass flowmeter with on-chip CMOS readout electronics. Technical Digest of Solid State Sensors and Actuators Workshop, Hilton Head, pp. 161–164 (1990)

18. Haberli, A., Paul, O., Malcovati, P., Faccio, M., Maloberti, F., Baltes, H.: CMOS integration of a thermal pressure sensor system. In: IEEE International Symposium on Circuits and Systems, ISCAS '96, Atlanta, vol. 1, pp. 377–380 (1996)

19. Schneider, M., Muller, T., Haberli, A., Hornung, M., Baltes, H.: Integrated micromachined decoupled CMOS chip on chip. In: Proceedings of 10th IEEE International Workshop on MEMS, Nagoya, pp. 512–517 (1997)

20. Akiyama, T., Akiyama, T., Staufer, U., de Rooij, N.F., Lange, D., Hagleitner, C., Brand, O., Baltes, H., Tonin, A., Hidber, H.R.: Integrated atomic force microscopy array probe with metal-oxide-semiconductor field effect transistor stress sensor, thermal bimorph actuator, and on-chip complementary metal-oxide-semiconductor electronics. J. Vac. Sci. Technol. B **18**, 2669–2675 (2000)

21. Schaufelbuhl, A., Schneeberger, N., Munch, U., Waelti, M., Paul, O., Brand, O., Baltes, H., Menolfi, C., Huang, Q., Doering, E., Loepfe, M.: Uncooled low-cost thermal imager based on micromachined CMOS integrated sensor array. J. MEMS **10**, 503–510 (2001)

22. Verd, J., Uranga, A., Teva, J., Lopez, J.L., Torres, F., Esteve, J., Abadal, G., Perez-Murano, F., Barniol, N.: Integrated CMOS-MEMS with on-chip readout electronics for high-frequency applications. IEEE Electron. Dev. Lett. **27**, 495–497 (2006)

23. Laermer, F., Schilp, A.: Method of anisotropically etching silicon. US Patent 5,501,893, Robert Bosch Gmbh (1992)

24. Kruglick, E.J.J., Warneke, B.A., Pister, K.S.: CMOS 3-axis accelerometers with integrated amplifier. In: Proceedings of International Conference on Micro Electro Mechanical System, MEMS-98, Heidelberg, pp. 631–636 (1998)

25. Fedder, G.K., Santhanam, S., Reed, M.L., Eagle, S.C., Guillou, D.F., Lu, M.S.C., Carley, L.R.: Laminated high-aspect-ratio microstructures in a conventional CMOS process. Proceedings of International Conference on Micro Electro Mechanical Systems, MEMS-96, San Diego, CA, pp. 13–18 (1996)

26. Xie, H., Erdmann, L., Zhu, X., Gabriel, K.J., Fedder, G.K.: Post-CMOS processing for high-aspect-ratio integrated silicon microstructures. J. Microelectromech. Syst. **11**, 93–101 (2002)

27. Xie, H., Fedder, G.K.: Fabrication, characterization, and analysis of a DRIE CMOS-MEMS gyroscope. IEEE Sens. J. **3**, 622–631 (2003)

28. Qu, H., Xie, H.: Process development for CMOS-MEMS sensors with robust electrically isolated bulk silicon microstructures. J. Microelectromech. Syst. **16**, 1152–1161 (2007)

29. Qu, H., Fang, D., Xie, H.: A monolithic CMOS-MEMS 3-axis accelerometer with a low-noise, low-power dual-chopper amplifier. IEEE Sens. J. **8**, 1511–1518 (2008)

30. Hagleitner, C., Lange, D., Hierlemann, A., Brand, O., Baltes, H.: CMOS single-chip gas detection system comprising capacitive, calorimetric and mass-sensitive microsensors. IEEE J. Solid-St. Circc **37**, 1867–1878 (2002)

31. Tsai, M.H., Sun, C.M., Liu, Y.C., Wang, C.W., Fang, W.L.: Design and application of a metal wet-etching post-process for the improvement of CMOS-MEMS capacitive sensors. J. Micromech. Microeng. **19**, 105017 (2009)

CMOS-CNT Integration

Huikai Xie and Ying Zhou
Department of Electrical and Computer Engineering, University of Florida, Gainesville, FL, USA

Synonyms

Monolithic integration of carbon nanotubes

Definition

Carbon nanotubes can be directly grown on CMOS substrate without degrading the performance of CMOS electronics.

Introduction

With numerous outstanding electrical, mechanical, and chemical properties, carbon nanotubes (CNTs) have been explored for various applications with great success. As CMOS circuits possess powerful interfacing, signal amplification, conditioning, and processing capabilities, it is also highly desired to integrate CNTs with CMOS. CNTs may be either used as part of CMOS electronics or as sensing elements to form functioning nano-electromechanical systems (NEMS). Figure 1 shows a nanotube random-access memory (NRAM) which uses CNT ribbons as switches [1].

Many sensors based on CMOS-CNT hybrid systems have also been demonstrated, including mechanical, thermal, and chemical sensors [3, 4]. The integration of CMOS circuits with CNT sensors can increase signal-to-noise ratio and dynamic range, lower power consumption, and provide various controls and automations. Other efforts have been made to use multiwall carbon nanotubes (MWNTs) as CMOS interconnect for high frequency applications [5], or to apply CNT-based nano-electromechanical switches

CMOS-CNT Integration,
Fig. 1 The structure of
NRAM at (**a**) on and (**b**) off
states [2]

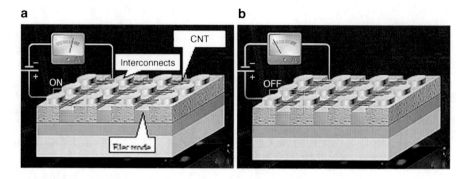

for leakage reduction in CMOS logic and memory circuits [6].

However, monolithic integration of CMOS and CNTs is still very challenging. Most CMOS-CNT systems have been realized either by a two-chip solution or low-throughput CNT manipulations. In this entry, CMOS-CNT integration approaches are reviewed, with a particular focus on a localized heating CNT synthesis method that can grow CNTs on foundry CMOS.

CNT Synthesis

There are three main methods for carbon nanotube synthesis: arc-discharge [7], laser ablation [8], and chemical vapor deposition (CVD) [9]. The first two methods involve evaporation of solid-state carbon precursors and condensation of carbon atoms to form nanotubes, where high annealing temperature, typically over 1,000°C, is required to remove defects and thus produce high-quality nanotubes. However, they tend to produce a mixture of nanotubes and other by-products such as catalytic metals, so the nanotubes must be selectively separated from the by-products. This requires post-growth purification and manipulation.

In   contrast,   the   CVD   method   employs a hydrocarbon gas as the carbon source and involves heating metal catalysts in a tube furnace to synthesize nanotubes. Nanotubes can grow either on the top (tip growth) or from the bottom (base growth). The diameters and locations of the grown CNTs can be controlled via catalyst size and catalyst patterning, and the orientation can be guided by an external electric field. Suitable catalysts that have been reported include Fe, Co, Mo, and Ni [10]. Compared to the arc-discharge and laser ablation methods, CVD uses much lower

growth temperature, but it is still too high for direct CNT growth on CMOS substrates.

In addition, during or after CNT growth, electrical contacts need to be formed for functional CNT-based devices. It is reported that Mo provides good ohmic contacts with nanotubes and shows excellent conductivity after growth, with resistance ranging from 20 kΩ to 1 MΩ per tube [11]. Several other metals, such as palladium, gold, titanium, tantalum, and tungsten, have also been investigated as possible electrode materials.

CMOS-CNT Integration

To integrate CNT on CMOS, there are several factors that must be taken into account: temperature budget, material compatibility, CNT type, CNT quality, and contamination. Depending on when CNTs are made, CMOS-CNT integration technology can be categorized as follows:
- Pre-CMOS: CMOS processes will be performed after CNTs are synthesized in place
- Intra-CMOS: CNT growth steps are inserted into CMOS fabrication steps
- Post-CMOS: CNTs are introduced after all CMOS processes have been done

For pre-CMOS, CNTs must go through standard CMOS process steps. This is very difficult to realize. There are temperature constraints, material compatibility, and contamination issues. There is no report about pre-CMOS CNT integration yet. For intra-CMOS, CNTs will be introduced at a later stage in the CMOS fabrication sequence, so it is easier to protect CNTs than in the pre-CMOS case. But temperature and contamination issues still must be considered. Post-CMOS, on the hand, completely eliminates

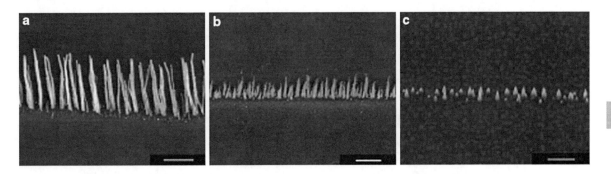

CMOS-CNT Integration, Fig. 2 SEM images of vertically aligned CNFs grown by PECVD deposition at (**a**) 500°C, (**b**) 270°C, and (**c**) 120°C (scale bars: (**a**) and (**b**) 1 μm and (**c**) 500 nm) [14]

CMOS contamination issues. It has potential to achieve mass production and low cost, but the temperature remains a limiting factor.

Intra-CMOS CNT Integration

Intra-CMOS (High Temperature CNTs)

Thermal CVD has been used to grow CNTs directly on CMOS substrate. For example, Tseng et al. demonstrated, for the first time, a process that monolithically integrates SWNTs with n-channel metal oxide semiconductor (NMOS) FET in a CVD furnace at 875°C [12]. However, the high synthesis temperature (typically 800–1,000°C for SWNT growth) may damage the aluminum metallization layers and change the characteristics of the on-chip transistors as well. Ghavanini et al. assessed the deterioration level of CMOS transistors with certain CNT CVD synthesis conditions applied, and they reported that one PMOS transistor lost its functions after the thermal CVD treatment (610°C, 22 min) [13]. As a result, the integrated circuits in Tseng's thermal CVD CNT synthesis can only consist of NMOS and use n+ polysilicon and molybdenum as interconnects, which make it incompatible with foundry CMOS processes.

Intra-CMOS (Low Temperature CNTs)

Some other attempts have been made to develop low temperature growth using various CVD methods. Hofmann et al. reported vertically aligned carbon nanotubes grown at temperature as low as 120°C by plasma-enhanced chemical vapor deposition (PECVD) [14]. However, the decrease in growth temperature jeopardizes both the quality and yield of the CNTs, as shown in Fig. 2. The synthesized products are actually defect-rich carbon nanofibers rather than MWNTs or SWNTs.

Intra-CMOS (Localized Heating)

Localized heating: To accommodate both the high temperature requirement (800–1,000°C) for high-quality SWNT synthesis and the temperature limitation of CMOS processing (<450°C), CNT synthesis based on localized heating has drawn great interest recently. Englander et al. demonstrated, for the first time, the localized synthesis of silicon nanowires and carbon nanotubes based on resistive heating [15]. The fabrication processes are shown in Fig. 3. Operated inside a room temperature chamber, the suspended micro-electromechanical system (MEMS) structures serve as resistive heaters to provide high temperature at predefined regions for optimal nanotube growth, leaving the rest of the chip area at low temperature. Using the localized heating concept, direct integration of nanotubes at specific areas can be potentially achieved in a CMOS compatible manner, and there is no need for additional assembly steps. However, the devices typically have large sizes and their fabrication processes are not fully compatible with the standard foundry CMOS processes. Although this concept has solved the temperature incompatibility problem between CNT synthesis and CMOS circuit protection, the fabrication processes of microheater structures still have to be well designed to fit into standard CMOS foundry processes and the resistor materials must be selected to meet the CMOS compatibility criteria.

Using the localized heating technique described above, on-chip growth using CMOS micro-hotplates was demonstrated by Haque et al. [16]. As shown in Fig. 4, tungsten was used for both the micro-hotplates (as the heating source) and interdigitated electrodes for nanotubes contacts. MWNTs have been successfully synthesized on the membrane, and simultaneously

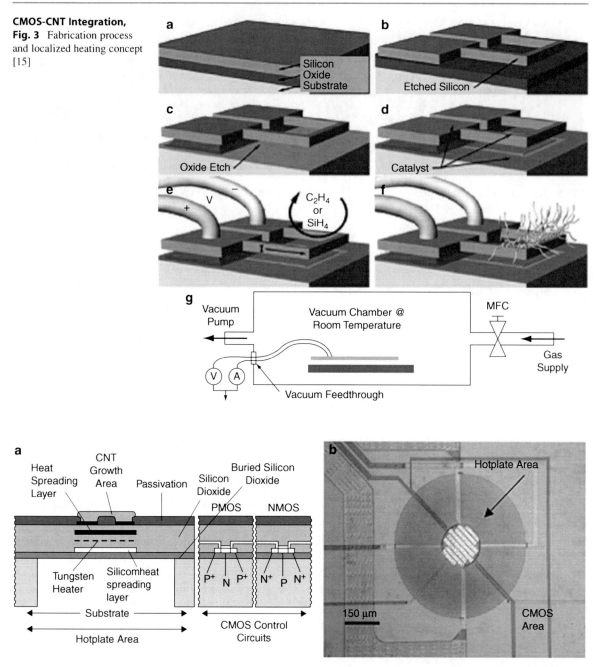

CMOS-CNT Integration, Fig. 3 Fabrication process and localized heating concept [15]

CMOS-CNT Integration, Fig. 4 (a) Schematic of the cross-sectional layout of the chip. (b) Optical image of the device top view showing the tungsten interdigitated electrodes on top of the membranes, heater radius = 75 μm, membrane radius = 280 μm [16]

connected to CMOS circuits through tungsten metallization. Although tungsten can survive the high temperature growth process, and has high connectivity and conductivity, Franklin et al. reported that no SWMTs were found to grow from catalyst particles on the tungsten electrodes, presumably due to the high catalytic activity of tungsten toward hydrocarbons [11]. Further, although the monolithic integration has been achieved, the utilization of tungsten, a refractory metal, as interconnect metal is limited in foundry CMOS, especially for mixed-signal CMOS processes. Moreover, this approach requires a backside bulk

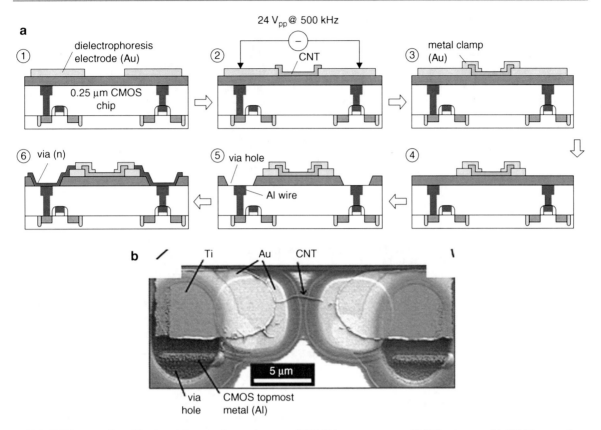

CMOS-CNT Integration, Fig. 5 (**a**) Process flow to integrate MWNT interconnects on CMOS substrate. (**b**) SEM image of one MWNT interconnect (wire and via) [5]

micromachining process and is limited to SOI CMOS substrates.

Post-CMOS CNT Integration

Post-CMOS (CNT Transfer and Assembly)

To overcome the temperature limitation, one possible solution is to grow nanotubes at high temperature first and then transfer them to the desired locations on CMOS substrates at low temperature. However, handling, maneuvering, and integrating these nanostructures with CMOS chips/wafers to form a complete system are very challenging. In the early stage, an atomic force microscope (AFM) tip was used to manipulate and position nanotubes into a predetermined location under the guide of scanning electron microscope (SEM) imaging [17]. Although this nanorobotic manipulation realized precise control over both the type and location of CNTs, its low throughput makes large scale assembly prohibitive.

Other post-growth CNT assembly methods include surface functionalization [18], liquid-crystalline processing [19], dielectrophoresis (DEP) [20], and large scale transfer of aligned nanotubes grown on quartz [21]. A 1 GHz CMOS circuit with CNT interconnects has been demonstrated using a DEP-assisted assembly technique [5]. The fabrication process flow and the assembled MWNT interconnect are shown in Fig. 5. The DEP process provides the capability of precisely positioning the nanotubes in a noncontact manner, which minimizes the parasitic capacitances and allows the circuits to operate at more than 1 GHz. However, to immobilize the DEP-trapped CNTs in place and to improve the electrical contact between CNTs and the electrodes, metal clamps must be selectively deposited at both ends of the CNTs (Fig. 5a, step 3). The process complexity and low yield ($\sim$ 8%, due to the MWNT DEP assembly limitation) are still the major concerns.

Post-CMOS (Localized Heating)

Monolithic CMOS-CNT integration is desirable to fully utilize the potentials of nanotubes for emerging

CMOS-CNT Integration, Fig. 6 (a) The 3D schematic showing the concept of the CMOS-integrated CNTs. The CVD chamber is kept at room temperature all the time. The *red* part represents the hot microheater that has been activated for high temperature nanotube synthesis. (b) Cross-sectional view of the device. (c) The schematic 3D microheater showing the local synthesis from the hotspot and self-assembly on the cold landing wall under the local electric field

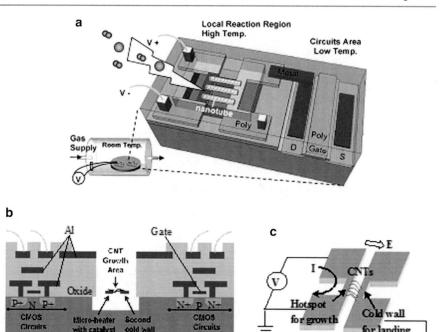

nanotechnology applications, but the approaches introduced above still cannot meet all the requirements and realize complete compatibility with CMOS processes. To solve the problem, a simple and scalable monolithic CMOS-CNT integration technique using a novel maskless post-CMOS surface micromachining processing has been proposed. This approach is fully compatible with commercial foundry CMOS processes and has no specific requirements on the type of metallization layers and substrates.

As illustrated in Fig. 6, the basic idea of the monolithic integration approach is to use maskless post-CMOS MEMS processing to form micro-cavities for thermal isolation and use the gate polysilicon to form resistors for localized heating as well as the nanotube-to-CMOS interconnect. The microheaters, made of the gate polysilicon, are deposited and patterned along with the gates of the transistors in the standard CMOS foundry processes. One of the top metal layers (i.e., the metal-3 layer as shown in Fig. 6b) is also patterned during the CMOS fabrication. It is used as an etching mask in the following post-CMOS microfabrication process for creating the micro-cavities. Finally, the polysilicon microheaters are exposed and suspended in a micro-cavity on a CMOS substrate. The circuits are covered under the metallization and passivation layers, as

illustrated in Fig. 6b. Unlike the traditional thermal CVD synthesis in which the whole chamber is heated to above 800°C, the CVD chamber is kept at room temperature all the time, with only the microheaters activated to provide the local high temperature for CNT growth (Fig. 6a, the red part represents the hot microheater).

The top view of a microheater design is shown in Fig. 6c. There are two polysilicon bridges: one as the microheater for generating high temperature to initiate CNT growth and the other for CNT landing. With the cold wall grounded, an E-field perpendicular to the surface of the two bridges will be induced during CNT growth. Activated by localized heating, the nanotubes will start to grow from the hotspot (i.e., the center of the microheater) and will eventually reach the secondary cold bridge under the guide of the local E-field. Since both the microheater bridge and the landing bridge are made of the gate polysilicon layer and have been interconnected with the metal layers in CMOS foundry process, the as-grown CNTs can be electrically connected to the CMOS circuitry on the same chip without any post-growth clamping or connection steps.

This technology has been verified at the chip level. The CMOS chips were fabricated in the AMI 0.5 μm

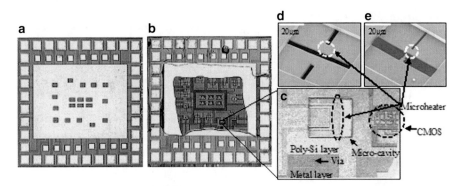

CMOS-CNT Integration, Fig. 7 (**a**) The CMOS chip photograph (1.5×1.5 mm^2) after foundry process; (**b**) The CMOS chip photograph after post-CMOS process (before final DRIE step); (**c**) Close-up optical image of one microheater and nearby circuit. CMOS circuit area, although visible, is protected under silicon dioxide layer. Only the microheater and secondary cold wall within the micro-cavity are exposed to synthesis gases. Polysilicon heater and metal wire are connected by via. (**d**) and (**e**) Closed-up SEM images of two microheaters

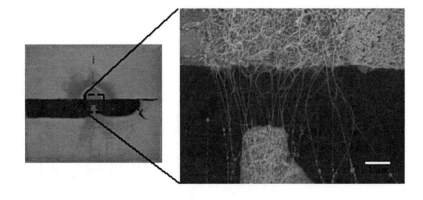

CMOS-CNT Integration, Fig. 8 Localized synthesis of carbon nanotubes grown from the 3×3 μm microheater, suspended across the trench and landed on the secondary polysilicon tip

3-metal CMOS process. Optical microscope images of a CMOS chip before and after MEMs fabrication are shown in Fig. 7a, b. The total chip area is 1.5×1.5 mm^2, including test circuits and 13 embedded microheaters. SEMs of two microheaters are shown in Fig. 7d, e, with resistances of 97 and 117 Ω, respectively. At about 2.5 V, red glowing was observed for the design in Fig. 7(e). This voltage was also used for the CNT growth.

Figure 8 shows one device with successful CNT growth, where individual suspended carbon nanotubes were grown from the 3×3 μm microheater shown in Fig. 7e and landed on the near polysilicon tip. The overall resistance of the CNTs is measured between the microheater and the cold polysilicon wall at room temperature. The typical resistances of in situ synthesized CNTs range from 5 to15 MΩ. The resistance variation from device to device is mainly due to the variation of the CNT quantity grown on each microheater. Junction effects of Schottky contacts were observed for self-assembled polysilicon/CNTs/ polysilicon heterojunctions.

After successful synthesis of carbon nanotubes, the influence of the localized heating on nearby CMOS circuits was evaluated. Simple circuits, such as inverters, were tested and proved working properly. There was no change to the rising and falling time after the CNT growth. The dc electrical characteristics of individual transistors had no considerable change after CNT growth, demonstrating the CMOS compatibility of this integration approach.

Summary

CMOS-CNT integration has been demonstrated by using both intra- and post- CMOS processes. Several methods have been developed to overcome the temperature conflict between CNT growth and CMOS, including using high temperature refractory metals for interconnect, low temperature CVD, transferring/ assembling CNTs prepared off site, and localized heating. Among these techniques, localized heating is very promising. Truly monolithic CNT-CMOS integration has been demonstrated on foundry CMOS substrate by employing MEMS and localized heating. This post-CMOS microfabrication is maskless, and the CNT growth does not affect the characteristics of the transistors on the same chip.

Cross-References

▶ Carbon Nanotube-Metal Contact
▶ Carbon Nanotubes for Chip Interconnections
▶ Carbon Nanotubes
▶ Synthesis of Carbon Nanotubes

References

1. Zhang, W., Jha, M., Shang, L.: NATURE: A hybrid nanotube/CMOS dynamically reconfigurable architecture. Design Automation Conference, 2006 43 rd ACM/IEEE, pp. 711–716 (2006)
2. Nantero, I.: "NRAM®," in http://www.nantero.com/mission.html, 2000–2009
3. Agarwal, V., Chen, C.-L., Dokmeci, M. R., Sonkusale, S.: A CMOS integrated thermal sensor based on single-walled carbon nanotubes. IEEE Sensors 2008 Conference, pp. 748–751 (2008)
4. Cho, T.S., Lee, K.-J., Kong, J., Chandrakasan, A.P.: A 32-uW 1.83-kS/s carbon nanotube chemical sensor system. IEEE J. Soild-State Circuits 44, 659–669 (2009)
5. Close, G.F., Yasuda, S., Paul, B., Fujita, S., Wong, H.-S.P.: A 1 GHz integrated circuit with carbon nanotube interconnects and silicon transistors. Nano Lett. 8, 706–709 (2008)
6. Chakraborty, R.S., Narasimhan, S., Bhunia, S.: Hybridization of CMOS with CNT-based nano-electromechanical switch for low leakage and Robust circuit design. IEEE Trans. Circuits Syst 54, 2480–2488 (2007)
7. Journet, C., Maser, W.K., Bernier, P., Loiseau, A., Lamy de la Chapelle, M., Lefrant, S., Deniard, P., Lee, R., Fischerk, J.E.: Large-scale production of single-walled carbon nanotubes by the electric-arc technique. Nature 388, 756–758 (1997)
8. Guo, T., Nikolaev, P., Thess, A., Colbert, D.T., Smalley, R.E.: Catalytic growth of single-walled nanotubes by laser vaporization. Chem. Phys. Lett. 243, 49–54 (1995)
9. Cassell, A.M., Raymakers, J.A., Kong, J., Dai, H.: Large Scale CVD Synthesis of Single-Walled Carbon Nanotubes. J. Phys. Chem. B 103, 6484–6492 (1999)
10. Meyyappan, M.: Carbon Nanotubes: Science and Applications. New York, CRC Press (2005)
11. Franklin, N.R., Wang, Q., Tombler, T.W., Javey, A., Shim, M., Dai, H.: Integration of suspended carbon nanotube arrays into electronic devices and electromechanical systems. Appl. Phys. Lett. 81, 913–915 (2002)
12. Tseng, Y.-C., Xuan, P., Javey, A., Malloy, R., Wang, Q., Bokor, J., Dai, H.: Monolithic integration of carbon nanotube devices with silicon MOS technology. Nano Lett. 4, 123–127 (2004)
13. Ghavanini, F.A., Poche, H.L., Berg, J., Saleem, A.M., Kabir, M.S., Lundgren, P., Enoksson, P.: Compatibility assessment of CVD growth of carbon nanofibers on bulk CMOS devices. Nano Lett. 8, 2437–2441 (2008)
14. Hofmann, S., Ducati, C., Robertson, J., Kleinsorge, B.: Low-temperature growth of carbon nanotubes by plasma-enhanced chemical vapor deposition. Appl. Phys. Lett. 83, 135–137 (2003)
15. Englander, O., Christensen, D., Lin, L.: Local synthesis of silicon nanowires and carbon nanotubes on microbridges. Appl. Phys. Lett. 82, 4797–4799 (2003)
16. Haque, M.S., Teo, K.B.K., Rupensinghe, N.L., Ali, S.Z., Haneef, I., Maeng, S., Park, J., Udrea, F., Milne, W.I.: On-chip deposition of carbon nanotubes using CMOS microhotplates. Nanotechnology 19, 025607 (2008)
17. Huang, X.M.H., Caldwell, R., Huang, L., Jun, S.C., Huang, M., Sfeir, M.Y., O'Brien, S.P., Hone, J.: Controlled placement of individual carbon nanotubes. Nano Lett. 5, 1515–1518 (2005)
18. Liu, J., Casavant, M.J., Cox, M., Walters, D.A., Boul, P., Lu, W., Rimberg, A.J., Smith, K.A., Colbert, D.T., Smalley, R.E.: Controlled deposition of individual single-walled carbon nanotubes on chemically functionalized templates. Chem. Phys. Lett. 303, 125–129 (1999)
19. Ko, H., Tsukruk, V.V.: Liquid-crystalline processing of highly oriented carbon nanotube arrays for thin-film transistors. Nano Lett. 6, 1443–1448 (2006)
20. Schwamb, T., Schirmer, N.C., Burg, B.R., Poulikakos, D.: Fountain-pen controlled dielectrophoresis for carbon nanotube-integration in device assembly. Appl. Phys. Lett. 93, 193104 (2008)
21. Ryu, K., Badmaev, A., Wang, C., Lin, A., Patil, N., Gomez, L., Kumar, A., Mitra, S., Wong, H.-S.P., Zhou, C.: CMOS-analogous wafer-scale nanotube-on-insulator approach for submicrometer devices and integrated circuits using aligned nanotubes. Nano Lett. 9, 189–197 (2009)

CMOS-MEMS

▶ CMOS MEMS Fabrication Technologies

CNT Arrays

▶ Vertically Aligned Carbon Nanotubes, Collective Mechanical Behavior

CNT Biosensor

▶ Nanostructure Field Effect Transistor Biosensors

CNT Brushes

▶ Vertically Aligned Carbon Nanotubes, Collective Mechanical Behavior

CNT Bundles

▶ Vertically Aligned Carbon Nanotubes, Collective Mechanical Behavior

CNT Foams

▶ Vertically Aligned Carbon Nanotubes, Collective Mechanical Behavior

CNT Forests

▶ Vertically Aligned Carbon Nanotubes, Collective Mechanical Behavior

CNT Mats

▶ Vertically Aligned Carbon Nanotubes, Collective Mechanical Behavior

CNT Turfs

▶ Vertically Aligned Carbon Nanotubes, Collective Mechanical Behavior

CNT-FET

▶ Nanostructure Field Effect Transistor Biosensors

Cob Web

▶ Spider Silk

Cochlea Implant

▶ Bio-inspired CMOS Cochlea

Cold-Wall Thermal Chemical Vapor Deposition

▶ Chemical Vapor Deposition (CVD)

Compliant Mechanisms

Larry L. Howell
Department of Mechanical Engineering,
Brigham Young University, Provo, UT, USA

Synonyms

Compliant systems; Flexures; Flexure mechanisms

Definition

Compliant mechanisms gain their motion from the deflection of elastic members.

Main Text

Compliant mechanisms offer an opportunity to achieve complex motions within the limitations of micro- and nano-fabrication. Because compliant mechanisms gain

Compliant Mechanisms, Fig. 1 A folded-beam suspension is an example of a widely used compliant mechanism in microelectromechanical systems (MEMS) applications

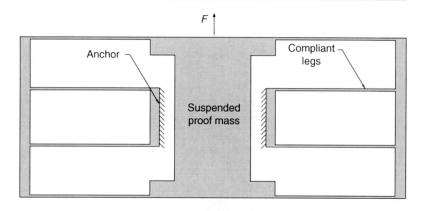

their motion from the constrained bending of flexible parts, they can achieve complex motion from simple topologies. Traditional mechanisms use rigid parts connected at articulating joints (such as hinges, axles, or bearings), which usually requires assembly of components and results in friction at the connecting surfaces [1–3]. Because traditional bearings are not practical and lubrication is problematic, friction and wear present major difficulties.

Nature provides an example of how to effectively address problems with motion at small scales. Most moving components in nature are flexible instead of stiff, and the motion comes from bending the flexible parts instead of rigid parts connected with hinges (for example, consider hearts, elephant trunks, and bee wings). The smaller the specimen, the more likely it is to use the deflection of flexible components to obtain its motion. And so it is with man-made systems as well, the smaller the device, the greater the advantages for using compliance [1].

Advantages of Compliant Mechanisms

Some of the advantages of compliant mechanisms at the micro- and nanoscales include the following:

Can be made from one layer of material. Compliant mechanisms can be fabricated from a single layer. This makes them compatible with many common microelectromechanical system (MEMS) fabrication methods, such as surface micromachining, bulk micromachining, and LIGA. For example, consider the folded beam suspension shown in Fig. 1. This device is often used as a suspension element in MEMS systems. It offers a simple approach for

Compliant Mechanisms, Fig. 2 This scanning electron micrograph shows a thermal actuator that uses multiple layers of compliant elements to achieve large amplification with a small footprint

constrained linear motion, and also integrates a return spring function. The device can achieve large deflections with reasonable off-axis stiffness. The compliant mechanism makes it possible to do these functions with a single layer of material.

No assembly required. Compliant mechanisms that gain all of their motion from the deflection of flexible components are "fully compliant mechanisms," where devices that combine both traditional and compliant elements are called "partially compliant mechanisms." Fully compliant mechanism can usually be fabricated without assembly of different components.

Small footprint. Some compliant mechanisms can also be designed to have a small footprint on the substrate on which they are built. Various strategies can be used to decrease the size of a mechanism. Figure 2 shows a thermal actuator that uses multiple layers to achieve a small footprint.

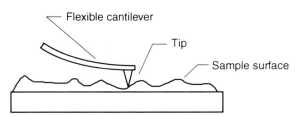

Compliant Mechanisms, Fig. 3 The cantilever of an atomic force microscope (AFM) is an example of compliance employed in high-precision instruments

Friction-free motion. Because compliant mechanisms gain their motion from deflection of flexible members rather than from traditional articulating joints, it is possible to reduce or eliminate the friction associated with rubbing surfaces. This results in reduced wear and eliminates the need for lubrication, as described next.

Wear-free motion. Wear can be particularly problematic at small scales, and the elimination of friction can result in the elimination of wear at the connecting surfaces of joints. For devices that are intended to undergo many cycles of motion, eliminating friction can dramatically increase the life of the system.

No need for lubrication. Another consequence of eliminating friction is that lubricants are not needed for the motion. This is particularly important at small scales where lubrication can be problematic.

High precision. Flexures have long been used in high precision instruments because of the repeatability of their motion. Some reasons for compliant mechanisms' precision are the backlash-free motion inherent in compliant mechanisms and the wear-free and friction-free motion described above. The cantilever associated with an atomic force microscope (Fig. 3) is an example application.

Integrated functions. Like similar systems in nature, compliant mechanisms have the ability to integrate multiple functions into few components. For example, compliant mechanisms often provide both the motion function and a return-spring function. Thermal actuators are another example of integration of functions, as described later.

High reliability. The combination of highly constrained motion of compliant mechanisms, the relative purity of materials used in micro/nanofabrication, and wear-free motion result in high reliability of compliant mechanisms at the micro/nanoscale.

Challenges of Compliant Mechanisms

Compliant mechanisms have many advantages, but they also have some significant challenges. A few of these are discussed below [1]:

Limited rotation. One clear drawback of compliant mechanisms is the general inability to undergo continuous rotation. Also, if a fully compliant mechanism is constructed from a single layer of material, then special care has to be taken to ensure that moving segments of the compliant mechanism do not collide with other segments of the same mechanism.

Dependence on material properties. The performance of compliant mechanisms is highly dependent on the material properties, which are not always well known.

Nonlinear motions. The deflections experienced by compliant mechanisms often extend beyond the range of linearized beam equations. This can make their analysis and design more complicated.

Fatigue analysis. Because most compliant mechanisms undergo repeated loading, it is important to consider the fatigue life of the device. Interestingly, because of the types of materials used and their purity, many MEMS compliant mechanisms will either fail on their first loading cycle or will have infinite fatigue life. Because of the low inertia of MEMS devices, it is often easy to quickly test a MEMS device to many millions of cycles. Factors such as stress concentrations, the operating temperature, and other environment conditions can affect the fatigue life.

Difficult design. Integration of functions into fewer components, nonlinear displacements, dependence on material properties, the need to avoid self-collisions during motion, and designing for appropriate fatigue life, all combine to make the design of compliant mechanisms nontrivial and often difficult.

Example Applications of Compliant Mechanisms

Examples of MEMS compliant mechanisms are shown here to further illustrate their properties and to demonstrate a few applications.

Digital Micromirrors. One of the most visible commercially available microelectromechanical systems is Texas Instruments' Digital Micromirror Device (DMD™), which is used in applications such as portable projectors. The DMD is a rectangular array of

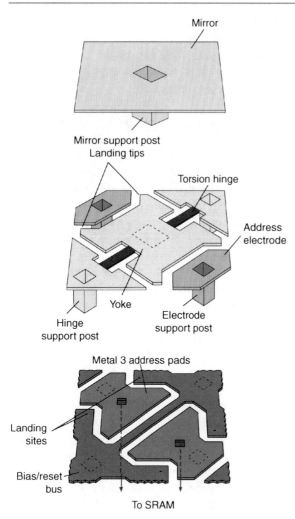

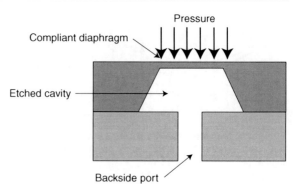

Compliant Mechanisms, Fig. 5 The strain on a compliant diaphragm of a piezoresistive pressure sensors results in a detectable change in resistance, which is correlated with the pressure

Compliant Mechanisms, Fig. 4 Texas Instrument's Digital Micromirror Device (DMD™) uses compliant torsion hinges to facilitate mirror motion (Illustration courtesy of Texas Instruments)

moving micromirrors that is combined with a light source, optics, and electronics to project high-quality color images. Figure 4 shows the architecture of a single DMD pixel. A 16-μm-square aluminum mirror is rigidly attached to a platform (the "yoke"). Flexible torsion hinges are used to connect the yoke to rigid posts. An applied voltage creates an electrostatic force that causes the mirror to rotate about the torsion hinges. When tilted in the on position, the mirror directs light from the light source to the projection optics and the pixel appears bright. When the mirror is tilted in the off position, the light is directed away from the projection optics and the pixel appears dark. The micromirrors can be combined in an array on a chip, and each micromirror is associated with the pixel of a projected

image. The torsion hinges use compliance to obtain motion while avoiding rubbing parts that cause friction and wear. The hinges can be deflected thousands of times per second and infinite fatigue life is essential.

Piezoresistive pressure sensors. A sensor is a device that responds to a physical input (such as motion, radiation, heat, pressure, magnetic field), and transmits a resulting signal that is usually used for detection, measurement, or control. Advantages of MEMS sensors are their size and their ability to be more closely integrated with their associated electronics. Piezoresistive sensing methods are among the most commonly employed sensing methods in MEMS. Piezoresistance is the change in resistivity caused by mechanical stresses applied to a material. Bulk micromachined pressure sensors have been commercially available since the 1970s. A typical design is illustrated in Fig. 5. A cavity is etched to create a compliant diaphragm that deflects under pressure. Piezoresistive elements on the diaphragm change resistance as the pressure increases; this change in resistance is measured and is correlated with the corresponding pressure.

Capacitive acceleration sensors. Accelerometers are another example of commercially successful MEMS sensors. Applications include automotive airbag safety systems, mobile electronics, hard drive protection, gaming, and others. Figure 6 illustrates an example of a surface micromachined capacitive accelerometer. Acceleration causes a displacement of the inertial mass connected to the compliant suspension, and the capacitance change between the comb fingers is detected.

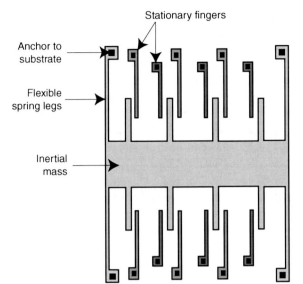

Compliant Mechanisms, Fig. 6 This accelerometer makes use of compliant legs that deflect under inertial loads. The deflection results in a detectable change in capacitance and is correlated with the corresponding acceleration

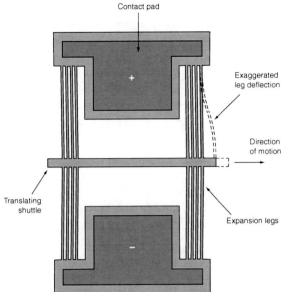

Compliant Mechanisms, Fig. 7 A schematic of a thermomechanical in-plane microactuor (TIM) that uses compliant expansion legs to amplify the motion caused by thermal expansion

Thermal actuators. A change in temperature causes an object to undergo a change in length, where the change is proportional to the material's coefficient of thermal expansion [4]. This length change is usually too small to be useful in most actuation purposes. Therefore, compliant mechanisms can be used to amplify the displacement of thermal actuators. Figure 7 illustrates an example of using compliant mechanisms to amplify thermal expansion in microactuators. Figure 8 shows a scanning electron micrograph of a thermomechanical in-plane microactuator (TIM) illustrated in Fig. 7. It consists of thin legs connecting both sides of a center shuttle. The leg ends not connected to the shuttle are anchored to bond pads on the substrate and are fabricated at a slight angle to bias motion in the desired direction. As voltage is applied across the bond pads, electric current flows through the thin legs. The legs have a small cross-sectional area and thus have a high electrical resistance, which causes the legs to heat up as the current passes through them. The shuttle moves forward to accommodate the resulting thermal expansion. Advantages of this device include its ability to obtain high deflections and large forces, as well as its ability to provide a wide range of output forces by changing the number of legs in the design.

Analysis and Design of Compliant Mechanisms

Multiple approaches are available for the analysis and design of compliant mechanisms. Three of the most developed approaches are described below.

Finite element analysis. Finite element methods are the most powerful and general methods available to analyze compliant mechanisms. Commercial software is currently available that has the capability of analyzing the large, nonlinear deflections often associated with compliant mechanisms. The general nature of the method makes it applicable for a wide range of geometries, materials, and applications. Increasingly powerful computational hardware has made it possible to analyze even very complex compliant mechanisms. It is also possible to use finite element methods in the design of compliant mechanisms, particularly once a preliminary design has been determined. But in the early phases of design, other methods (or hybrid methods) are often preferred so that many design iterations can be quickly analyzed.

Pseudo-rigid-body model. The pseudo-rigid-body model is used to model compliant mechanisms as traditional rigid-body mechanisms, which opens up the possibility of using the design and analysis methods developed for rigid-body mechanisms in the design of compliant mechanisms [1]. With the pseudo-rigid-body model approach, flexible parts are modeled as rigid links connected at appropriately placed pins, with springs to represent the compliant mechanism's resistance to motion. Extensive work has been done to develop pseudo-rigid-body models for a wide range of geometries and loading conditions. Consider a simple example. The micromechanism shown in Fig. 9 has a rigid shuttle that is guided by two flexible legs. (Note that the folded-beam suspension in Fig. 1 has four of these devices connected in series and parallel.) The pseudo-rigid-body model of the mechanism models the flexible legs as rigid links connected at pin joints with torsional springs. Using appropriately located joints and appropriately sized springs, this model is very accurate well into the nonlinear range. For example, if the flexible legs are single-walled carbon nanotubes, comparisons to molecular simulations have shown the pseudo-rigid-body model to provide accurate results [5]. The advantages of the pseudo-rigid-body model are realized during the early phases of design where many design iterations can be quickly evaluated, traditional mechanism design approaches can be employed, and motions can be easily visualized.

Topology optimization. Suppose that all that is known about a design is the desired performance and design domain. Topology optimization shows promise for designing compliant mechanisms under such conditions. The advantage is that very little prior knowledge about the resulting compliant mechanism is needed, and any biases of the designer are eliminated [6]. Topology optimization is often integrated with finite element methods to consider many possible ways of distributing material with the design domain. This has the potential to find designs that would not otherwise be discovered by other methods. Infinite possible topologies are possible and finite element methods can be employed to evaluate the different possibilities. The resolution of the design domain mesh can be a limiting factor, but once a desirable topology is identified, it can be further refined using other approaches.

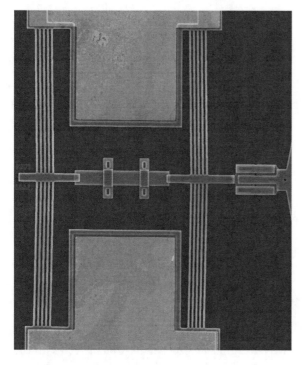

Compliant Mechanisms, Fig. 8 A scanning electron micrograph of a thermal actuator illustrated in Fig. 7

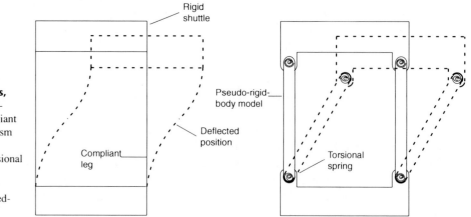

Compliant Mechanisms, Fig. 9 The pseudo-rigid-body model of the compliant parallel-guiding mechanism consists of appropriately located pin joints and torsional springs (This device is a building block of other devices, such as the folded-beam suspension)

Conclusion

Compliant mechanisms provide significant benefits for micro- and nano-motion applications. They can be compatible with many fabrication methods, do not require assembly, have friction-free and wear-free motion, provide high precision and high reliability, and they can integrate multiple functions into fewer components. The major challenges associated with compliant mechanisms come from the difficulty associated with their design, limited rotation, and the need to ensure adequate fatigue life. It is likely that compliant mechanisms will see increasing use in micro- and nano-mechanical systems as more people understand their advantages and have tools available for their development.

Cross-References

▶ AFM
▶ Basic MEMS Actuators
▶ Biomimetics
▶ Finite Element Methods for Computational Nano-optics
▶ Insect flight and Micro Air Vehicles (MAVs)
▶ MEMS on Flexible Substrates
▶ Nanogrippers
▶ Piezoresistivity
▶ Thermal Actuators

References

1. Howell, L.L.: Compliant Mechanisms. Wiley, New York (2001)
2. Lobontiu, N.: Compliant Mechanisms: Design of Flexure Hinges. CRC Press, Boca Raton (2003)
3. Smith, S.T.: Flexures: Elements of Elastic Mechanisms. Taylor & Francis, London (2000)
4. Howell, L.L., McLain, T.W., Baker, M.S., Lott, C.D.: Techniques in the design of thermomechanical microactuators. In: Leondes, C.T. (ed.) MEMS/NEMS Handbook, Techniques and Applications, pp. 187–200. Springer, New York (2006)
5. Howell, L.L., DiBiasio, C.M., Cullinan, M.A., Panas, R., Culpepper, M.L.: A pseudo-rigid-body model for large deflections of fixed-clamped carbon nanotubes. J. Mech. Robot. **2**, 034501 (2010)
6. Frecker, M.I., Ananthasuresh, G.K., Nishiwaki, S., Kikuchi, N., Kota, S.: Topological synthesis of compliant mechanisms using multi-criteria optimization. J. Mech. Des. **119**, 238–245 (1997)

Compliant Systems

▶ Compliant Mechanisms

Composite Materials

▶ Theory of Optical Metamaterials

Computational Micro/Nanofluidics: Unifier of Physical and Natural Sciences and Engineering

A. T. Conlisk
Department of Mechanical Engineering,
The Ohio State University, Columbus, OH, USA

Synonyms

Microscale fluid mechanics; Nanoscale fluid mechanics

Definition

Because of the small scale of the fluid conduits, electric fields must often be used to transport fluids especially at the nanoscale. This means that the fluids must be electrically conducting, and so microfluidics and nanofluidics require the user to be literate in fluid mechanics, heat and mass transfer, electrostatics, electrokinetics, electrochemistry, and if biomolecules are involved, molecular biology.

Introduction

The term microfluidics refers generally to internal flow in a tube or channel whose smallest dimension is under $100\,\mu m$. Nanofluidics refers to the same phenomenon in a conduit whose smallest dimension is less than $100\,nm$.

Microchannels and nanochannels have large surface-to-volume ratio, so that surface properties become enormously important. In fully developed channel flow, the pressure drop $\Delta p \sim \frac{1}{h^3}$, where h is the small dimension, and so the pressure drop is prohibitively large for a nanoscale channel. Thus, a solvent fluid such as water, proteins and other biomolecules, and other colloidal particles are most often transported electrokinetically. This means that the art of designing micro and nanodevices requires a significant amount of knowledge of fluid flow and mass transfer (biofluids are usually multicomponent mixtures) and often heat transfer, electrostatics, electrokinetics, electrochemistry, and molecular biology. Details of the character of these fields of study can be found in the recently published book by the author [1].

The study of micro/nanofluidics requires knowledge of all of the above-mentioned fields and so has a unifying effect. Moreover, nanofluidics opens the door to the discovery of the structure and conformation of biomaterials such as proteins and polysacchrides through molecular simulation.

In dealing with devices with small-scale features, microscale and below, there are three activities that normally comprise the design process; these are:
- Modeling: computational and theoretical
- Fabrication
- Experimental methods

Modeling is often done prior to the fabrication process as a guide as to what can be done. Experimental methods are usually used to assess the performance of a device, among other purposes.

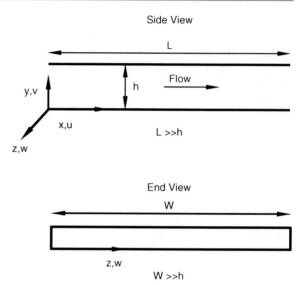

Computational Micro/Nanofluidics: Unifier of Physical and Natural Sciences and Engineering, Fig. 1 Geometry of a typical channel. In applications $h \ll W, L$ where W is the width of the channel and L its length in the primary flow direction. u, v, w are the fluid velocities in the x, y, z directions © A.T. Conlisk used with permission

This means that surface properties become very important at the microscale and nanoscale and surfaces are routinely engineered to achieve a desired objective. In most devices the nanoscale features interface directly with microscale features. A typical channel geometry is depicted on Fig. 1.

Surface-to-Volume Ratio

Consider a channel of rectangular cross-section having dimensions in the (x, y, z) coordinate system of (L, h, W) with the primary direction of fluid motion being in the x direction. Then the surface-to-volume ratio is given by

$$\frac{S}{V} = 2\left(\frac{1}{L} + \frac{1}{h} + \frac{1}{W}\right) = 6\,\mathrm{m}^{-1} \tag{1}$$

for a channel having all three dimensions $L = h = W = 1$ m. On the other hand, a channel having dimensions (10 μm, 10^{-2} μm, 10 μm),

$$\frac{S}{V} \sim 2 \times 10^8 \mathrm{m}^{-1} \tag{2}$$

Fluid Mechanics

Micro and nanofluidics generally involve the flow of electrically conducting fluids, *electrolyte solutions* that are assumed to be incompressible, having a constant density. Generally, the flows are internal, bounded on each side by walls, and the flows are assumed to be fully developed. In this case, referring to Fig. 1, the governing equation for the velocity u in a channel is given by

$$\mu \frac{\partial^2 u}{\partial y^2} = \frac{\partial p}{\partial x} - B_x \tag{3}$$

where p is the pressure and B_x is a body force. The no-slip condition is applied at each wall: $u = 0$ at $y = 0, h$.

Mass Transfer

The molar flux of species A for a dilute electrically conducting mixture is

$$\vec{N}_A = -D_{AB}\nabla c_A + m_A z_A c_A \vec{E} + c_A \vec{V} \qquad (4)$$

Here D_{AB} is the diffusion coefficient, R is the universal gas constant, T is the temperature, $z_A m_A$ is called the ionic mobility with $m_A = \frac{FD_{AB}}{RT}$, z_A is the valence, $F = 96500 \frac{Coul}{mole}$ is Faraday's constant, and $\vec{E}$ is the electric field. Equation 4 is called the Nernst-Planck equation and the electric field term in the flux equation is called *electrical migration*. The boundary condition of interest here is that the solid walls in Fig. 1 are impermeable to species A, or $N_{A_y} = 0$ at $y = 0, h$.

In one dimension, Eq. 4 can be integrated to give

$$c_A = c_{A0} e^{-z_A \phi} \qquad (5)$$

and this is termed the *Boltzmann distribution* for the concentration of species A.

The Electric Field

An electric field is set up around any charged body and is defined as the force per charge on a surface. Electrical charges are either positive or negative and like charges repel and opposite charges attract. For two bodies of charge q and q', the *electric field* is defined by

$$E = \frac{F}{q'} = \frac{q}{4\pi\varepsilon_e r^2} \frac{N}{C} \qquad (6)$$

and is directed outward from the body of charge q and toward the body having a charge q' if $q > 0$ and the electric field is in the opposite direction if $q < 0$. In general, the electric field is a vector. This formula is called *Coulomb's Law* and ε_e is called the *electrical permittivity*. The electrical permittivity is a transport property like the viscosity and thermal conductivity of a fluid.

The electric field due to a flat wall having a surface charge density σ in $\frac{Coulomb}{m^2}$ on one side is directed normal to the surface and has magnitude

$$E = \frac{\sigma}{2\varepsilon_e} \qquad (7)$$

A wire is characterized as having a line charge density and if charges are distributed over a volume, charge density is defined and called ρ_e in $\frac{Coulomb}{m^3}$.

The *electrical potential* is defined as the work done in moving a unit of charge and mathematically

$$\phi = -\int_a^b \vec{E} \cdot d\vec{s} \qquad (8)$$

The units of the electric potential are $\frac{Nm}{C} = 1\,Volt = 1\,V$. This formula is similar to the formula for mechanical work given by

$$W = -\int_a^b \vec{F} \cdot d\vec{s} \qquad (9)$$

In differential form, the electrical potential is given by

$$\vec{E} = -\nabla\phi \qquad (10)$$

For a single charge *Gauss's Law* is given by

$$\iint_S \varepsilon_e \vec{E} \cdot d\vec{A} = q \qquad (11)$$

For a volume that contains a continuous distribution of charge, ρ_e, summing over all the charges using the definition of the integral, Gauss's Law becomes

$$\iint_S \varepsilon_e \vec{E} \cdot d\vec{A} = \iint_V \rho_e dV \qquad (12)$$

Using Eq. 12 and the differential form of the definition of the electrical potential, it follows that

$$\nabla^2\phi = -\frac{\rho_e}{\varepsilon_e} \qquad (13)$$

This is a *Poisson equation* for the potential given the volume charge density. The combination of Eqs. 4 and 13 is called the *Poisson-Nernst-Planck* system of equations.

Electrochemistry

Electrochemistry may be broadly defined as the study of the electrical properties of chemical and biological

Computational Micro/Nanofluidics: Unifier of Physical and Natural Sciences and Engineering, Fig. 2 The electric double layer (EDL) consists of a layer of counter ions pinned to the wall, the Stern layer, and a diffuse layer of mobile ions outside that layer. The wall is shown as being negatively charged and the ζ–potential is defined as the electrical potential at the Stern plane. © A.T. Conlisk used with permission

where F is Faraday's constant, ε_e is the electrical permittivity of the medium, I is the *ionic strength* $I = \sum_i z_i^2 c_i$, c_i the concentrations of the electrolyte constituents at some reference location, R is the universal gas constant, z_i is the valence of species i and T is the temperature.

The ion distribution within the EDL can be described by using the number density, concentration, or mole fraction. Engineers usually prefer the dimensionless mole fraction whereas chemists usually use concentration or number density.

There are two views of the ion distribution within the electrical double layer that are generally thought to be valid and have been verified by numerical solutions of the governing equations (see the section on Electrokinetic Phenomena below). The Gouy–Chapman [3, 4] model of the electric double layer allows counterions to collect near the surface in much greater numbers than coions. This model as numerical solutions suggest [1] occurs at higher surface charge densities. The Debye–Hückel picture assumes that coions and counterions collect near the surface in roughly equal amounts, above and below a mean value. These pictures are depicted on Fig. 3.

material [2]. In particular much of electrochemistry pertinent to micro and nanofluidics involves the study of the behavior of *ionic solutions* and the *electrical double layer* (EDL). Electrochemistry of electrodes is important to understand the operation of a *battery*.

An ionic or electrolyte solution is a mixture of ions, or charged species immersed in a solvent, often water. It is the charged nature of ionic solutions that allows the fluid to move under the action of an electric field, provided by electrodes placed upstream and downstream of a channel in a *nanopore membrane*. The term *membrane* is used to mean a thin sheet of porous material that allows fluid to flow in channels that make up the porous part of the membrane. Those channels are often like those channels depicted on Fig. 1.

Because the surface-to-volume ratio is so large in a nanoscale channel, the properties of the surface are extremely important. Fluid can be moved by an electric field if the surfaces of a channel are charged. If the surface is negatively charged, a surplus of positive ions will arrange themselves near the wall. This is shown in Fig. 2. It is this excess charge that allows fluid to be transported by an externally applied electric field.

The nominal length scale associated with the EDL is the *Debye length* defined by

$$\lambda = \frac{\sqrt{\varepsilon_e RT}}{FI^{1/2}} \qquad (14)$$

Molecular Biology

The nanoscale is the scale of biology since many proteins and other biomolecules have nanoscale dimensions. Many of the applications of nanofluidics such as rapid molecular analysis, drug delivery, and biochemical sensing have a biological entity as an integral part of their operation. Moreover, using nanofluidic tools, DNA sequencing is now possible. The book by Alberts [6] is a useful tool for learning molecular biology.

Nucleic acids are polymers consisting of nucleotides. Those based on a sugar called *ribose* are called ribo nucleic acids (RNA) and those based on *deoxyribose* are called deoxyribonucleic acids (DNA). RNA is single stranded while DNA is usually double stranded although single-stranded DNA (ss-DNA) does exist. Nucleotides contain five-carbon sugars attached to one or more phosphate groups (a phosphorus central atom surrounded by four oxygens) and a base which can be either adenine (A), cytosine (C), guanine (G), or thymine (T). Two nucleotides connected by a hydrogen bond is called a base pair

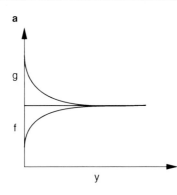

Computational Micro/Nanofluidics: Unifier of Physical and Natural Sciences and Engineering, Fig. 3 (**a**) Debye–Hückel [5] picture of the electric double layer. Here *g* denotes the cation mole fraction and *f* denotes the anion mole fraction. The Debye–Hückel model assumes the cation and anion wall mole fractions are symmetric about a mean value which occurs for low surface charge densities. (**b**) Gouy–Chapman model [3, 4] of the EDL allows many more counterions than coions to collect near the charged surface and is valid at higher surface charge densities. © A.T. Conlisk used with permission

(bp). *Protein synthesis* begins at a gene on a particular strand of a DNA molecule in a cell [6].

There are seven basic types of proteins classified according to their function although different authors use different terms to describe each class; see for example Alberts [6], panel 5–1. *Enzymes* are catalysts in biological reactions within the cell. For example, the immune system responds to foreign bacteria and viruses by producing *antibodies* that destroy or bind to the antigen, the foreign agent. The antigen is the catalyst, or *reaction enhancer* for inducing the immune response: the production of the antibodies. Proteins are responsible for many of the essential functions of the body, including moving material into and out of cells, regulating metabolism, managing temperature and *pH*, and muscle operation, among other functions.

Proteins are large and complex molecules, polymers made up of a total of 20 amino acids, and held together by peptide bonds. The 20 amino acids have *side chains* that can be *basic, acidic, polar* or *nonpolar.* Because they are so large they cannot described easily in a single chemical formula or picture. Thus, molecular biologists depict proteins and other macromolecules in distinct levels of structure. The *primary structure* is the amino acid sequence, the order in which the 20 amino acids appear. The *secondary structure* depicts the folding properties of a protein as depicted on Fig. 4. Proteins are further described by more complex folding of the secondary structure (*tertiary structure*) and a *quaternary structure* if the protein has more than one backbone.

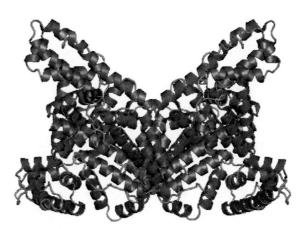

Computational Micro/Nanofluidics: Unifier of Physical and Natural Sciences and Engineering, Fig. 4 Ribbon view of the protein albumin depicting its folding pattern, the *secondary structure* of a protein. From the European Bioinformatics Institute, public domain, www.ebi.ac.uk

Proteins are usually negatively charged and thus nanopore membranes for rapid molecular analysis can be used to separate different types of proteins and other biomolecules based on different values of size and charge. Biomolecules are what is termed soft material in that they are porous and deform under stress. Indeed, recent measurements of the conformation of albumin show that it may take the shape of a wedge, looking like a piece of pie. With the explosive growth of computer capability, conformations of biomolecules are actually being computed using molecular simulation tools like molecular dynamics and Monte Carlo schemes.

Ion channels are natural conical nanopores whose walls are made of proteins that play a crucial role in the transport of biofluids to and from cells. The basic units of all living organisms are cells. In order to keep the cells functioning properly there needs to be a continuous flux of ions in and out of the cell and the cell components. The cell and many of its components are surrounded by a plasma membrane which provides selective transfer of ions. The membrane is made up of a double layer of lipid molecules (lipid bilayer) in which proteins are embedded. Ion channels are of two categories: carrier and pore. The *carrier* protein channel is based on the binding of the transport ion to a larger macromolecule, which brings it through the channel. A *pore* ion channel is a narrow, water-filled tunnel, permeable to the few ions and molecules small enough to fit through the tunnel (approximately 10 $\overset{\circ}{A}$ in diameter).

Electrokinetic Phenomena

As the scale of the channels in a nanopore membrane become smaller, pressure, the normal means for driving fluids through pipes and channels at macroscale (Fig. 1), becomes very difficult [1] since the pressure drop requires scales as h^{-3} where h is the (nanoscale) channel height. Since in many applications the fluids used are electrically conducting, electric fields can be used to effectively pump fluid. Moreover electrically charged particles can move relative to the bulk fluid motion and thus species of particles can be separated.

These *electrokinetic phenomena* are generally grouped into four classes [1]:

1. Electroosmosis (electroosmotic flow): the bulk motion of a fluid caused by an electric field
2. Electrophoresis: the motion of a charged particle in an otherwise motionless fluid or the motion of a charged particle relative to a bulk motion
3. Streaming potential or streaming current: the potential induced by a pressure gradient at zero current flow of an electrolyte mixture
4. Sedimentation potential: the electric field induced when charged particles move relative to a liquid under a gravitational or centrifugal or other force field

By far the two most important of these phenomena are electroosmosis and electrophoresis and for the purposes of the theme of this entry, electroosmosis is discussed exclusively.

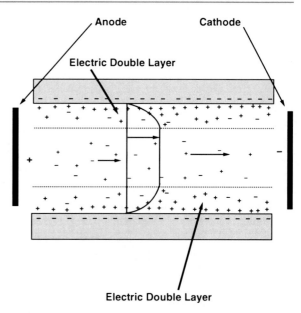

Computational Micro/Nanofluidics: Unifier of Physical and Natural Sciences and Engineering, Fig. 5 The combination of electrodes in the regions upstream and downstream of a charged channel or membrane, usually fluid reservoirs, causes electroosmotic flow. © A.T. Conlisk used with permission

The dimensionless form of the streamwise momentum equation in the fully developed flow region in the absence of a pressure gradient is

$$\varepsilon^2 \frac{\partial^2 u}{\partial y^2} = -\beta \sum_i z_i X_i \qquad (15)$$

and the Poisson equation for the potential in dimensionless form is

$$\varepsilon^2 \frac{\partial^2 \phi}{\partial y^2} = -\beta \sum_i z_i X_i \qquad (16)$$

where the partial derivatives in this one-dimensional fully developed analysis are really total derivatives, $\varepsilon = \frac{\lambda}{h}$ and $\beta = \frac{c}{I}$ where c is the total concentration including the solvent and I is the ionic strength. Here X_i is the mole fraction, but if the electrolyte concentrations are scaled on the ionic strength, $I = \frac{\lambda}{h} \sum_i z_i c_i$, $\beta = 1$. It is seen from Eq. 15 that the combination of the electrodes that create an electric field and the excess charge in the electrical double layers produces the electrical force that balances the viscous force causing the electrolyte to move and this is depicted on Fig. 5.

a

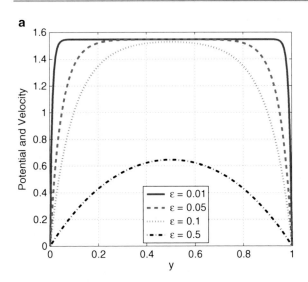

b

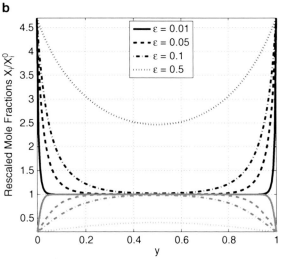

Computational Micro/Nanofluidics: Unifier of Physical and Natural Sciences and Engineering, Fig. 6 (a) Potential, velocity, and rescaled mole fractions for a 1:1 electrolyte for various values of ε. Here the dimensional potential on both walls is $\zeta^* = -40$ mV. In (b) the mole fractions are rescaled based on the upstream reservoir mole fractions as $\frac{X_i}{X_i^0}$. The cations are plotted in *black lines* and the anions are plotted in *gray lines*. © A.T. Conlisk used with permission

The fluid velocity satisfies the no-slip condition at the wall and the electric potential satisfies $\phi(0) = \phi(1) = 0$. Then both the equations and the boundary conditions are identical and on a dimensionless basis $u(y) = \phi(y)$. In reality, the potential does not vanish at the wall but if a Dirichlet boundary condition holds and the potential satisfies $\phi = \zeta$ at $y = 0,1$ and thus $u = \phi - \zeta$, where ζ is the *dimensionless* ζ-potential at the wall. Results for the potential and velocity and the concentrations scaled on the ionic strength are presented on Fig. 6 [1]. Note that for $\epsilon \ll 1$ the fluid velocity is constant and away from the walls of the channel, unlike the Poiseuille flow of pressure-driven flow.

Dimensional Analysis

The equivalence of the dimensionless velocity and the dimensionless potential is important because measurements of the velocity are equivalent to measurements of the electric potential. Moreover, it is noted that a Debye length of $\lambda = 1$ nm in a $h = 100$ nm channel gives the same value of $\varepsilon = 0.01$ as a $\lambda = 100$ nm Debye length in a $h = 10$ μm channel so that miscoscale measurements can be validated by nanoscale computations and conversely, nanoscale computations can be validated by miscoscale experiments. Note that this similarity analysis does not apply for unsteady flow since there is no time derivative in the potential equation.

Closure

Physics of fluids at the nanoscale is dominated by the large surface-to-volume ratio inherent at this length scale, and thus the surfaces of a channel or membrane become extremely important. Indeed, the pressure drop across a rectangular channel in a nanopore membrane, $\Delta p \sim \frac{1}{h^3}$, is prohibitively large for the efficient operation of a nanofluidic device if $h = O(\text{nm})$. Thus fluid, charged biomaterials such as proteins, and colloidal particles are most often transported electrokinetically at the nanoscale. The art of designing micro and nano fluidic devices therefore requires a significant amount of knowledge of fluid flow, mass transfer and often heat transfer, electrostatics, electrokinetics, electrochemistry, and molecular biology.

The common thread is micro/nanofluidics, which plays the role of unifying and integrating these fields. In particular, nanofluidics opens the door to reveal the structure and behavior of flows around nanoparticles and the conformation of proteins and other biomolecules using molecular simulations.

Cross-References

▶ Applications of Nanofluidics
▶ Electrokinetic Fluid Flow in Nanostructures
▶ Micro/Nano Flow Characterization Techniques
▶ Nanochannels for Nanofluidics: Fabrication Aspects
▶ Rapid Electrokinetic Patterning

References

1. Conlisk, A.T.: Essentials of Micro and Nanofluidics with Application to the Biological and Chemical Sciences. Cambridge University Press, Cambridge (2011)
2. Bockris, J.O.M, Reddy, A.K.N.: Modern Electrochemistry, vol. 1 Ionics, 2 edn. Plenum, New York/London (1998)
3. Gouy, G.: About the electric charge on the surface of an electrolyte. J. Phys. A **9**, 457–468 (1910)
4. Chapman, D.L.: A contribution to the theory of electrocapillarity. Phil. Mag. **25**, 475–481 (1913)
5. Debye, P., Huckel, E.: The interionic attraction theory of deviations from ideal behavior in solution. Z. Phys. **24**, 185 (1923)
6. Alberts, B., Bray, D., Hopkin, K., Johnson, A., Lewis, J., Raff, M., Roberts, K., Walter, P.: Essential Cell Biology. Garland Publishing, New York (1998)

Computational Study of Nanomaterials: From Large-Scale Atomistic Simulations to Mesoscopic Modeling

Leonid V. Zhigilei[1], Alexey N. Volkov[1] and Avinash M. Dongare[2]
[1]Department of Materials Science and Engineering, University of Virginia, Charlottesville, VA, USA
[2]Department of Materials Science and Engineering, North Carolina State University, Raleigh, NC, USA

Synonyms

Carbon nanotube materials; Carbon nanotubes; Computer modeling and simulation of materials; Dislocation dynamics; Kinetic Monte Carlo method; Mechanical properties of nanomaterials; Mesoscopic modeling; Metropolis Monte Carlo method; Molecular dynamics method; Multiscale modeling; Nanocrystalline materials; Nanofibrous materials and composites; Nanomaterials

Definitions

Nanomaterials (or nanostructured materials, nanocomposites) are materials with characteristic size of structural elements on the order of less than several hundreds of nanometers at least in one dimension. Examples of nanomaterials include nanocrystalline materials, nanofiber, nanotube, and nanoparticle-reinforced nanocomposites, and multilayered systems with submicron thickness of the layers.

Atomistic modeling is based on atoms as elementary units in the models, thus providing the atomic-level resolution in the computational studies of materials structure and properties. The main atomistic methods in material research are (1) molecular dynamics technique that yields "atomic movies" of the dynamic material behavior through the integration of the equations of motion of atoms and molecules, (2) Metropolis Monte Carlo method that enables evaluation of the equilibrium properties through the ensemble averaging over a sequence of random atomic configurations generated according to the desired statistical-mechanics distribution, and (3) kinetic Monte Carlo method that provides a computationally efficient way to study systems where the structural evolution is defined by a finite number of thermally activated elementary processes.

Mesoscopic modeling is a relatively new area of the computational materials science that considers material behavior at time- and length-scales intermediate between the atomistic and continuum levels. Mesoscopic models are system- /phenomenon-specific and adopt coarse-grained representations of the material structure, with elementary units in the models designed to provide a computationally efficient representation of individual crystal defects or other elements of micro/nanostructure. Examples of the mesoscopic models are coarse-grained models for molecular systems, discrete dislocation dynamics model for crystal plasticity, mesoscopic models for nanofibrous materials, cellular automata, and kinetic Monte Carlo Potts models for simulation of microstructural evolution in polycrystalline materials.

Computer Modeling of Nanomaterials

Rapid advances in synthesis of nanostructured materials combined with reports of their enhanced or unique properties have created, over the last decades, a new active area of materials research. Due to the nanoscopic size of the structural elements in nanomaterials, the interfacial regions, which represent an insignificant volume fraction in traditional materials with coarse microstructures, start to play the dominant role in defining the physical and mechanical properties of nanostructured materials. This implies that the behavior of nanomaterials cannot be understood and predicted by simply applying scaling arguments from the structure–property relationships developed for conventional polycrystalline, multiphase, and composite materials. New models and constitutive relations, therefore, are needed for an adequate description of the behavior and properties of nanomaterials.

Computer modeling is playing a prominent role in the development of the theoretical understanding of the connections between the atomic-level structure and the effective (macroscopic) properties of nanomaterials. Atomistic modeling has been at the forefront of computational investigation of nanomaterials and has revealed a wealth of information on structure and properties of individual structural elements (various nanolayers, nanoparticles, nanofibers, nanowires, and nanotubes) as well as the characteristics of the interfacial regions and modification of the material properties at the nanoscale. Due to the limitations on the time- and length-scales, inherent to atomistic models, it is often difficult to perform simulations for systems that include a number of structural elements that is sufficiently large to provide a reliable description of the macroscopic properties of the nanostructured materials. An emerging key component of the computer modeling of nanomaterials is, therefore, the development of novel mesoscopic simulation techniques capable of describing the collective behavior of large groups of the elements of the nanostructures and providing the missing link between the atomistic and continuum (macroscopic) descriptions. The capabilities and limitations of the atomistic and mesoscopic computational models used in investigations of the behavior and properties of nanomaterials are briefly discussed and illustrated by examples of recent applications below.

Atomistic Modeling

In atomistic models [1, 2], the individual atoms are considered as elementary units, thus providing the atomic-level resolution in the description of the material behavior and properties. In classical atomistic models, the electrons are not present explicitly but are introduced through the interatomic potential, $U(\vec{r}_1, \vec{r}_2, ..., \vec{r}_N)$, that describes the dependence of the potential energy of a system of N atoms on the positions $\vec{r}_i$ of the atoms. It is assumed that the electrons adjust to changes in atomic positions much faster than the atomic nuclei move (Born–Oppenheimer approximation), and the potential energy of a system of interacting atoms is uniquely defined by the atomic positions.

The interatomic potentials are commonly described by analytic functions designed and parameterized by fitting to available experimental data (e.g., equilibrium geometry of stable phases, density, cohesive energy, elastic moduli, vibrational frequencies, characteristics of the phase transitions, etc.). The interatomic potentials can also be evaluated through direct quantum mechanics–based electronic structure calculations in so-called first principles (ab initio) simulation techniques. The ab initio simulations, however, are computationally expensive and are largely limited to relatively small systems consisting of tens to thousands of atoms. The availability of reliable and easy-to-compute interatomic potential functions is one of the main conditions for the expansion of the area of applicability of atomistic techniques to realistic quantitative analysis of the behavior and properties of nanostructured materials.

The three atomistic computational techniques commonly used in materials research are:

1. Metropolis Monte Carlo method – the equilibrium properties of a system are obtained via ensemble averaging over a sequence of random atomic configurations, sampled with probability distribution characteristic for a given statistical mechanics ensemble. This is accomplished by setting up a random walk through the configurational space with specially designed choice of probabilities of going from one state to another. In the area of nanomaterials, the application of the method is largely limited to investigations of the equilibrium shapes of individual elements of nanostructures (e.g., nanoparticles) and surface structure/composition (e.g., surface reconstruction and compositional segregation [3]).

2. Kinetic Monte Carlo method – the evolution of a nanostructure can be obtained by performing atomic rearrangements governed by pre-defined

transition rates between the states, with time increments formulated so that they relate to the microscopic kinetics of the system. Kinetic Monte Carlo is effective when the structural and/or compositional changes in a nanostructure are defined by a relatively small number of thermally activated elementary processes, for example, when surface diffusion is responsible for the evolution of shapes of small crystallites [4] or growth of two-dimensional fractal-dendritic islands [5].

3. Molecular dynamics method – provides the complete information on the time evolution of a system of interacting atoms through the numerical integration of the equations of motion for all atoms in the system. This method is widely used in computational investigations of nanomaterials and is discussed in more detail below.

Molecular Dynamics Technique

Molecular dynamics (MD) is a computer simulation technique that allows one to follow the evolution of a system of N particles (atoms in the case of atomistic modeling) in time by solving classical equations of motion for all particles in the system,

$$m_i \frac{d^2 \vec{r}_i}{dt^2} = \vec{F}_i, \quad i = 1, 2, \ldots, N \quad (1)$$

where m_i and $\vec{r}_i$ are the mass and position of a particle i, and $\vec{F}_i$ is the force acting on this particle due to the interaction with other particles in the system. The force acting on the ith particle at a given time is defined by the gradient of the inter-particle interaction potential $U(\vec{r}_1, \vec{r}_2, \ldots, \vec{r}_N)$ that, in general, is a function of the positions of all the particles:

$$\vec{F}_i = -\vec{\nabla}_i U(\vec{r}_1, \vec{r}_2, \ldots, \vec{r}_N) \quad (2)$$

Once the initial conditions (initial positions and velocities of all particles in the system) and the interaction potential are defined, the equations of motion, (Eq. 1), can be solved numerically. The result of the solution is the trajectories (positions and velocities) of all the particles as a function of time, $\vec{r}_i(t), \vec{v}_i(t)$, which is the only direct output of an MD simulation. From the trajectories of all particles in the system, however, one can easily calculate the spatial and time evolution of structural and thermodynamic parameters of the system. For example, a detailed atomic-level analysis of

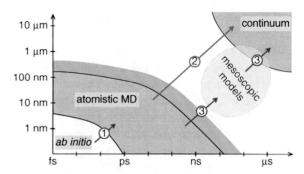

Computational Study of Nanomaterials: From Large-Scale Atomistic Simulations to Mesoscopic Modeling, Fig. 1 Schematic representation of the time- and length-scale domains of first-principles (ab initio) electronic structure calculations, classical atomistic MD, and continuum modeling of materials. The domain of continuum modeling can be different for different materials and corresponds to the time- and length-scales at which the effect of the micro/nanostructure can be averaged over to yield the effective material properties. The arrows show the connections between the computational methods used in multiscale modeling of materials: The *red arrow* #1 corresponds to the use of quantum mechanics–based electronic structure calculations to design interatomic potentials for classical MD simulations or to verify/correct the predictions of the classical atomistic simulations; the *green arrow* #2 corresponds to the direct use of the predictions of large-scale atomistic simulations of nanostructured materials for the design of continuum-level constitutive relations describing the material behavior and properties; and the two *blue arrows* #3 show a two-step path from atomistic to continuum material description through an intermediate mesoscopic modeling

the development of the defect structures or phase transformations can be performed and related to changes in temperature and pressure in the system (see examples below).

The main strength of the MD method is that only details of the interatomic interactions need to be specified, and no assumptions are made about the character of the processes under study. This is an important advantage that makes MD to be capable of discovering new physical phenomena or processes in the course of "computer experiments." Moreover, unlike in real experiments, the analysis of fast non-equilibrium processes in MD simulations can be performed with unlimited atomic-level resolution, providing complete information of the phenomena of interest.

The predictive power of the MD method, however, comes at a price of a high computational cost of the simulations, leading to severe limitations on time and length scales accessible for MD simulations, as shown schematically in Fig. 1. Although the record length-scale MD simulations have been demonstrated for systems containing more than 10^{12} atoms (corresponds to cubic samples on the order of 10 μm in size) with the use of hundreds of thousands of processors on one of the world-largest supercomputers [6], most of the systems studied in large-scale MD simulations do not exceed hundreds of nanometers even in simulations performed with computationally efficient parallel algorithms (shown by a green area extending the scales accessible for MD simulations in Fig. 1). Similarly, although the record long time-scales of up to hundreds of microseconds have been reported for simulations of protein folding performed through distributed computing [7], the duration of most of the simulations in the area of materials research does not exceed tens of nanoseconds.

Molecular Dynamics Simulations of Nanomaterials
Both the advantages and limitations of the MD method, briefly discussed above, have important implications for simulations of nanomaterials. The transition to the nanoscale size of the structural features can drastically change the material response to the external thermal, mechanical, or electromagnetic stimuli, making it necessary to develop new structure–properties relationships based on new mechanisms operating at the nanoscale. The MD method is in a unique position to provide a complete microscopic description of the atomic dynamics under various conditions without making any a priori assumptions on the mechanisms and processes defining the material behavior and properties.

On the other hand, the limitations on the time- and length-scales accessible to MD simulations make it difficult to directly predict the macroscopic material properties that are essentially the result of a homogenization of the processes occurring at the scale of the elements of the nanostructure. Most of the MD simulations have been aimed at investigation of the behavior of individual structural elements (nanofibers, nanoparticles, interfacial regions in multiphase systems, grain boundaries, etc.). The results of these simulations, while important for the mechanistic understanding of the elementary processes at the nanoscale, are often insufficient for making a direct connection to the macroscopic behavior and properties of nanomaterials.

With the fast growth of the available computing resources, however, there have been an increasing number of reports on MD simulations of systems that include multiple elements of nanostructures. A notable class of nanomaterials actively investigated in MD simulations is nanocrystalline materials – a new generation of advanced polycrystalline materials with submicron size of the grains. With a number of atoms on the order of several hundred thousands and more, it is possible to simulate a system consisting of tens of nanograins and to investigate the effective properties of the material (i.e., to make a direct link between the atomistic and continuum descriptions, as shown schematically by the green arrow #2 in Fig. 1). MD simulations of nanocrystalline materials addressing the mechanical [8, 9] and thermal transport [10] properties as well as the kinetics and mechanisms of phase transformations [11, 12] have been reported, with several examples illustrated in Fig. 2. In the first example, Fig. 2a, atomic-level analysis of the dislocation activity and grain-boundary processes occurring during mechanical deformation of an aluminum nanocrystalline system consisting of columnar grains is performed and the important role of mechanical twinning in the deformation behavior of the nanocrystalline material is revealed [9]. In the second example, Fig. 2b, the processes of void nucleation, growth and coalescence in the ductile failure of nanocrystalline copper subjected to an impact loading are investigated, providing important pieces of information necessary for the development of a predictive analytical model of the dynamic failure of nanocrystalline materials [8]. The third example, Fig. 2c, illustrates the effect of nanocrystalline structure on the mechanisms and kinetics of short pulse laser melting of thin gold films. It is shown that the initiation of melting at grain boundaries can steer the melting process along the path where the melting continues below the equilibrium melting temperature, and the crystalline regions shrink and disappear under conditions of substantial undercooling [11].

The brute force approach to the atomistic modeling of nanocrystalline materials (increase in the number of atoms in the system) has its limits in addressing the complex collective processes that involve many grains and may occur at a micrometer length scale and above. Further progress in this area may come through the development of concurrent multiscale approaches based on the use of different resolutions in the description of the intra-granular and grain boundary regions in a well-integrated computational model. An example of a multiscale approach is provided in Ref. [13], where

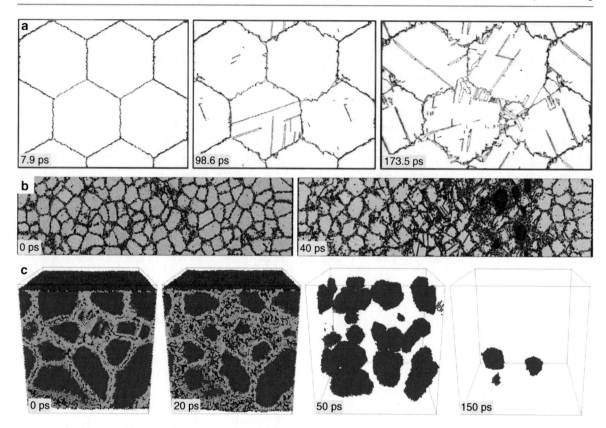

Computational Study of Nanomaterials: From Large-Scale Atomistic Simulations to Mesoscopic Modeling, Fig. 2 Snapshots from atomistic MD simulations of nanocrystalline materials: (**a**) mechanical deformation of nanocrystalline Al (only atoms in the twin boundaries left behind by partial dislocations and atoms in disordered regions are shown by red and blue colors, respectively) [9]; (**b**) spallation of nanocrystalline Cu due to the reflection of a shock wave from a surface of the sample (atoms that have local fcc, hcp, and disordered structure are shown by yellow, red, and green/blue colors, respectively) [8]; and (**c**) laser melting of a nanocrystalline Au film irradiated with a 200 fs laser pulse at a fluence close to the melting threshold (atoms that have local fcc surroundings are colored blue, atoms in the liquid regions are red and green, and in the snapshots for 50 and 150 ps the liquid regions are blanked to expose the remaining crystalline regions) [11]

scale-dependent constitutive equations are designed for a generalized finite element method (FEM) so that the atomistic MD equations of motion are reproduced in the regions where the FEM mesh is refined down to atomic level. This and other multiscale approaches can help to focus computational efforts on the important regions of the system where the critical atomic-scale processes take place. The practical applications of the multiscale methodology so far, however, have been largely limited to investigations of individual elements of material microstructure (crack tips, interfaces, and dislocation reactions), with the regions represented with coarse-grained resolution serving the purpose of adoptive boundary conditions. The perspective of the concurrent multiscale modeling of nanocrystalline materials remains unclear due to the close coupling

between the intra-granular and grain boundary processes. To enable the multiscale modeling of dynamic processes in nanocrystalline materials, the design of advanced computational descriptions of the coarse-grained parts of the model is needed so that the plastic deformation and thermal dissipation could be adequately described without switching to fully atomistic modeling.

Mesoscopic Modeling

A principal challenge in computer modeling of nanomaterials is presented by the gap between the atomistic description of individual structural elements and the macroscopic properties defined by the collective behavior of large groups of the structural elements. Apart from a small number of exceptions (e.g., simulations of nanocrystalline materials briefly discussed above), the direct analysis of the effective properties of nanostructured materials is still out of reach for atomistic simulations. Moreover, it is often difficult to translate the large amounts of data typically generated in atomistic simulations into key physical parameters that define the macroscopic material behavior. This difficulty can be approached through the development of mesoscopic computational models capable of representing the material behavior at time- and length-scales intermediate between the atomistic and continuum levels (prefix *meso* comes from the Greek word μέσος, which means middle or intermediate).

The mesoscopic models provide a "stepping stone" for bridging the gap between the atomistic and continuum descriptions of the material structure, as schematically shown by the blue arrows #3 in Fig. 1. Mesoscopic models are typically designed and parameterized based on the results of atomistic simulations or experimental measurements that provide information on the internal properties and interactions between the characteristic structural elements in the material of interest. The mesoscopic simulations can be performed for systems that include multiple elements of micro/nanostructure, thus enabling a reliable homogenization of the structural features to yield the effective macroscopic material properties. The general strategy in the development of a coarse-grained mesoscopic description of the material dynamics and properties includes the following steps:

1. Identifying the collective degrees of freedom *relevant for the phenomenon under study* (the focus on different properties of the same material may affect the choice of the structural elements of the model)
2. Designing, based on the results of atomic-level simulations and/or experimental data, a set of rules (or a mesoscopic force field) that governs the dynamics of the collective degrees of freedom
3. Adding a set of rules describing the changes in the properties of the dynamic elements in response to the local mechanical stresses and thermodynamic conditions

While the atomistic and continuum simulation techniques are well established and extensively used, the mesoscopic modeling is still in the early development stage. There is no universal mesoscopic technique or methodology, and the current state of the art in mesoscopic simulations is characterized by the development of system- /phenomenon-specific mesoscopic models. The mesoscopic models used in materials modeling can be roughly divided into two general categories: (1) the models based on lumping together groups of atoms into larger dynamic units or particles and (2) the models that represent the material microstructure and its evolution due to thermodynamic driving forces or mechanical loading at the level of individual crystal defects. The basic ideas underlying these two general classes of mesoscopic models are briefly discussed below.

The models where groups of atoms are combined into coarse-grained computational particles are practical for materials with well-defined structural hierarchy (that allows for a natural choice of the coarse-grained particles) and a relatively weak coupling between the *internal* atomic motions inside the coarse-grained particles and the *collective* motions of the particles. In contrast to atomic-level models, the atomic structure of the structural elements represented by the coarse-grained particles is not explicitly represented in this type of mesoscopic models. On the other hand, in contrast to continuum models, the coarse-grained particles allow one to explicitly reproduce the nanostructure of the material. Notable examples of mesoscopic models of this type are coarse-grained models for molecular systems [14–16] and mesoscopic models for carbon nanotubes and nanofibrous materials [17–19]. The individual molecules (or mers in polymer molecules) and nanotube/nanofiber segments are chosen as the dynamic units in these models. The collective dynamic degrees of freedom that correspond to the motion of the "mesoparticles" are explicitly accounted for in mesoscopic models, while the internal degrees of freedom are either neglected or described by a small number of internal state variables. The description of the internal states of the mesoparticles and the energy exchange between the dynamic degrees of freedom and the internal state variables becomes important for simulations of non-equilibrium phenomena that involve fast energy deposition from an external source, heat transfer, or dissipation of mechanical energy.

Another group of mesoscopic models is aimed at a computationally efficient description of the evolution of the defect structures in crystalline materials. The mesoscopic models from this group include the discrete dislocation dynamics model for simulation of crystal plasticity [20–22] and a broad class of methods designed for simulation of grain growth, recrystallization, and associated microstructural evolution (e.g., phase field models, cellular automata, and kinetic Monte Carlo Potts models) [21–23]. Despite the apparent diversity of the physical principles and computational algorithms adopted in different models listed above, the common characteristic of these models is the focus on a realistic description of the behavior and properties of individual crystal defects (grain boundaries and dislocations), their interactions with each other, and the collective evolution of the totality of crystal defects responsible for the changes in the microstructure.

Two examples of mesoscopic models (one for each of the two types of the models discussed above) and their relevance to the investigation of nanomaterials are considered in more detail next.

Discrete Dislocation Dynamics

The purpose of the discrete dislocation dynamics (DD) is to describe the plastic deformation in crystalline materials, which is largely defined by the motions, interactions, and multiplication of dislocations. Dislocations are linear crystal defects that generate long-range elastic strain fields in the surrounding elastic solid. The elastic strain field is accounting for ~90% of the dislocation energy and is responsible for the interactions of dislocations among themselves and with other crystal defects. The collective behavior of dislocations in the course of plastic deformation is defined by these long-range interactions as well as by a large number of local reactions (annihilation, formation of glissile junctions or sessile dislocation segments such as Lomer or Hirth locks) occurring when the anelastic core regions of the dislocation lines come into contact with each other. The basic idea of the DD model is to solve the dynamics of the dislocation lines in elastic continuum and to include information about the local reactions. The elementary unit in the discrete dislocation dynamics method is, therefore, a segment of a dislocation.

The continuous dislocation lines are discretized into segments, and the total force acting on each segment in the dislocation slip plane is calculated. The total force includes the contributions from the external force, the internal force due to the interaction with other dislocations and crystal defects that generate elastic fields, the

"self-force" that can be represented by a "line tension" force for small curvature of the dislocation, the Peierls force that acts like a friction resisting the dislocation motion, and the "image" force related to the stress relaxation in the vicinity of external or internal surfaces. Once the total forces and the associated resolved shear stresses, τ^*, acting on the dislocation segments are calculated, the segments can be displaced in a finite difference time integration algorithm applied to the equations connecting the dislocation velocity, v, and the resolved shear stress, for example, [21]

$$v = A \left(\frac{\tau^*}{\tau_0} \right)^m \exp \left(-\frac{\Delta U}{kT} \right) \qquad (3)$$

when the displacement of a dislocation segment is controlled by thermally activated events (ΔU is the activation energy for dislocation motion, m is the stress exponent, and τ_0 is the stress normalization constant) or

$$v = \tau^* b / B \qquad (4)$$

that corresponds to the Newtonian motion equation accounting for the atomic and electron drag force during the dislocation "free flight" between the obstacles (B is the effective drag coefficient and b is the Burgers vector).

Most of the applications of the DD model have been aimed at the investigation of the plastic deformation and hardening of single crystals (increase in dislocation density as a result of multiplication of dislocations present in the initial system). The extension of the DD modeling to nanomaterials is a challenging task as it requires an enhancement of the technique with a realistic description of the interactions between the dislocations and grain boundaries and/or interfaces as well as an incorporation of other mechanisms of plasticity (e.g., grain boundary sliding and twinning in nanocrystalline materials). There have only been several initial studies reporting the results of DD simulations of nanoscale metallic multilayered composites [24]. Due to the complexity of the plastic deformation mechanisms and the importance of anelastic short-range interactions among the crystal defects in nanomaterials, the development of novel hybrid computational methods combining the DD technique with other mesoscopic methods is likely to be required for realistic modeling of plastic deformation in this class of materials.

a
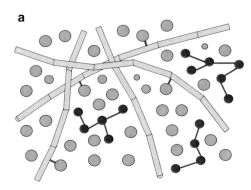

b
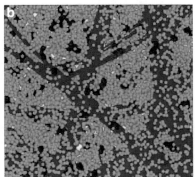

Computational Study of Nanomaterials: From Large-Scale Atomistic Simulations to Mesoscopic Modeling, Fig. 3 Schematic representation of the basic components of the dynamic mesoscopic model of a CNT-based nanocomposite material (**a**) and a corresponding molecular-level view of a part of the system where a network of CNT bundles (*blue color*) is embedded into an organic matrix (*green and red color*) (**b**)

Mesoscopic Model for Nanofibrous Materials

The design of new nanofibrous materials and composites is an area of materials research that currently experiences a rapid growth. The interest in this class of materials is fueled by a broad range of potential applications, ranging from fabrication of flexible/stretchable electronic and acoustic devices to the design of advanced nanocomposite materials with improved mechanical properties and thermal stability. The behavior and properties of nanofibrous materials are defined by the collective dynamics of the nanofibers and, in the case of nanocomposites, their interactions with the matrix. Depending on the structure of the material and the phenomenon of interest, the number of nanofibers that has to be included in the simulation in order to ensure a reliable prediction of the effective macroscopic properties can range from several hundreds to millions. The direct atomic-level simulation of systems consisting of large groups of nanofibers (the path shown by the green arrow #2 in Fig. 1) is beyond the capabilities of modern computing facilities. Thus, an alternative two-step path from atomistic investigation of individual structural elements and interfacial properties to the continuum material description through an intermediate mesoscopic modeling (blue arrows #3 in Fig. 1) appears to be the most viable approach to modeling of nanofibrous materials. An example of a mesoscopic computational model recently designed and parameterized for carbon nanotube (CNT)-based materials is briefly discussed below.

The mesoscopic model for fibrous materials and organic matrix nanocomposites adopts a coarse-grained description of the nanocomposite constituents (nano-fibers and matrix molecules), as schematically illustrated in Fig. 3. The individual CNTs are represented as chains of stretchable cylindrical segments [19], and the organic matrix is modeled by a combination of the conventional "bead-and-spring" model commonly used in polymer modeling [14, 15] and the "breathing sphere" model developed for simulation of simple molecular solids [16] and polymer solutions [25].

The degrees of freedom, for which equations of motion are solved in dynamic simulations or Metropolis Monte Carlo moves are performed in simulations aimed at finding the equilibrium structures, are the nodes defining the segments, the positions of the molecular units, and the radii of the spherical particles in the breathing sphere molecules. The potential energy of the system can be written as

$$U = U_{T(\text{int})} + U_{T-T} + U_{M-M} + U_{M(\text{int})} + U_{M-T} \quad (5)$$

where $U_{T(int)}$ is the potential that describes the internal strain energy associated with stretching and bending of individual CNTs, U_{T-T} is the energy of intertube interactions, U_{M-M} is the energy of chemical and nonbonding interactions in the molecular matrix, $U_{M(int)}$ is the internal breathing potential for the matrix units, and U_{M-T} is the energy of matrix – CNT interaction that can include both non-bonding van der Waals interactions and chemical bonding. The internal CNT potential $U_{T(int)}$ is parameterized based on the results of atomistic simulations [19] and accounts for the transition to the anharmonic regime of stretching (nonlinear

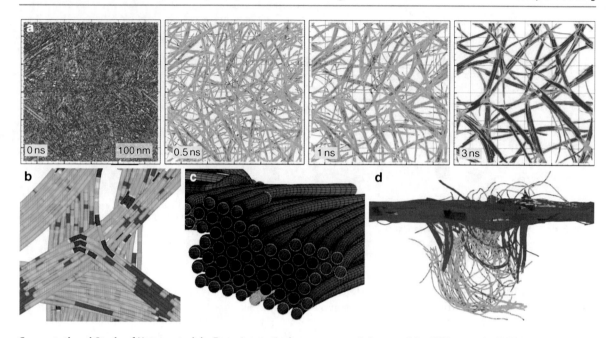

Computational Study of Nanomaterials: From Large-Scale Atomistic Simulations to Mesoscopic Modeling, Fig. 4 Snapshots from mesoscopic simulations of systems consisting of (10,10) single-walled carbon nanotubes: (**a**) spontaneous self-organization of CNTs into a continuous network of CNT bundles (CNT segments are colored according to the local intertube interaction energy) [18]; (**b**) an enlarged view of a structural element of the CNT network (CNT segments colored according to the local radii of curvature and the red color marks the segments adjacent to buckling kinks) [26]; (**c**) a cross-section of a typical bundle showing a hexagonal arrangement of CNTs in the bundle [18]; (**d**) snapshot from a simulation of a high-velocity impact of a spherical projectile on a free-standing thin CNT film

stress–strain dependence), fracture of nanotubes under tension, and bending buckling [26]. The intertube interaction term U_{T-T} is calculated based on the tubular potential method that allows for a computationally efficient and accurate representation of van der Waals interactions between CNT segments of arbitrary lengths and orientation [18]. The general procedure used in the formulation of the tubular potential is not limited to CNTs or graphitic structures. The tubular potential (and the mesoscopic model in general) can be parameterized for a diverse range of systems consisting of various types of nano- and micro-tubular elements, such as nanotubes, nanorodes, and microfibers.

First simulations performed with the mesoscopic model demonstrate that the model is capable of simulating the dynamic behavior of systems consisting of thousands of CNTs on a timescale extending up to tens of nanoseconds. In particular, simulations performed for systems composed of randomly distributed and oriented CNTs predict spontaneous self-assembly of CNTs into continuous networks of bundles with partial hexagonal ordering of CNTs in the bundles, Fig. 4a–c

[18, 26]. The bending buckling of CNTs (e.g., see Fig. 4b) is found to be an important factor responsible for the stability of the network structures formed by defect-free CNTs [26]. The structures produced in the simulations are similar to the structures of CNT films and buckypaper observed in experiments. Note that an atomic-level simulation of a system similar to the one shown in the left panel of Fig. 4 would require $\sim 2.5 \times 10^9$ atoms, making such simulation unfeasible.

Beyond the structural analysis of CNT materials, the development of the mesoscopic model opens up opportunities for investigation of a broad range of important phenomena. In particular, the dynamic nature of the model makes it possible to perform simulations of the processes occurring under conditions of fast mechanical loading (blast/impact resistance, response to the shock loading, etc.), as illustrated by a snapshot from a simulation of a high-velocity impact of a spherical projectile on a free-standing thin CNT film shown in Fig. 4d. With a proper parameterization, the mesoscopic model can also be adopted for calculation of electrical and thermal transport properties of complex nanofibrous materials [27].

Future Research Directions

The examples of application of the atomistic and mesoscopic computational techniques, briefly discussed above, demonstrate the ability of computer modeling to provide insights into the complex processes that define the behavior and properties of nanostructured materials. The fast advancement of experimental methods capable of probing nanostructured materials with high spatial and temporal resolution is an important factor that allows for verification of computational predictions and stimulates the improvement of the computational models. With further innovative development of computational methodology and the steady growth of the available computing resources, one can expect that both atomistic and mesoscopic modeling will continue to play an increasingly important role in nanomaterials research.

In the area of atomistic simulations, the development of new improved interatomic potentials (often with the help of ab initio electronic structure calculations, red arrow #1 in Fig. 1) makes material-specific computational predictions more accurate and enables simulations of complex multi-component and multiphase systems. Further progress can be expected in two directions that are already actively pursued: (1) large-scale MD simulations of the fast dynamic phenomena in nanocrystalline materials (high strain rate mechanical deformation, shock loading, impact resistance, response to fast heating, etc.) and (2) detailed investigation of the atomic structure and properties of individual structural elements in various nanomaterials (grain boundaries and interfaces, nanotubes, nanowires, and nanoparticles of various shapes). The information obtained in large-scale atomistic simulations of nanocrystalline materials can be used to formulate theoretical models translating the atomic-level picture of material behavior to the constitutive relations describing the dependence of the mechanical and thermal properties of these materials on the grain size distribution and characteristics of nanotexture (green arrow #2 in Fig. 1).

The results of the detailed analysis of the structural elements of the nanocomposite materials can be used in the design and parameterization of mesoscopic models, where the elementary units treated in the models correspond to building blocks of the nanostructure (elements of grain boundaries, segments of dislocations, etc.) or groups of atoms that have some distinct properties (belong to a molecule, a mer unit of a polymer chain, a nanotube, a nanoparticle in nanocomposite material, etc.). The design of novel system-specific mesoscopic models capable of bridging the gap between the atomistic modeling of structural elements of nanostructured materials and the continuum models (blue arrows #3 in Fig. 1) is likely to become an important trend in the computational investigation of nanomaterials. To achieve a realistic description of complex processes occurring in nanomaterials, the description of the elementary units of the mesoscopic models should become more flexible and sophisticated. In particular, an adequate description of the energy dissipation in nanomaterials can only be achieved if the energy exchange between the atomic degrees of freedom, excluded in the mesoscopic models, and the coarse-grained dynamic degrees of freedom is accounted for. A realistic representation of the dependence of the properties of the mesoscopic units of the models on local thermodynamic conditions can also be critical in modeling of a broad range of phenomena.

In general, the optimum strategy in investigation of nanomaterials is to use a well-integrated multiscale computational approach combining the ab initio and atomistic analysis of the constituents of nanostructure with mesoscopic modeling of the collective dynamics and kinetics of the structural evolution and properties, and leading to the improved theoretical understanding of the factors controlling the effective material properties. It is the improved understanding of the connections between the processes occurring at different time- and length-scales that is likely to be the key factor defining the pace of progress in the area of computational design of new nanocomposite materials.

Acknowledgment The authors acknowledge financial support provided by NSF through Grants No. CBET-1033919 and DMR-0907247, and AFOSR through Grant No. FA9550-10-1-0545. Computational support was provided by NCCS at ORNL (project No. MAT009).

Cross-References

▶ Ab initio DFT Simulations of Nanostructures
▶ Active Carbon Nanotube-Polymer Composites
▶ Carbon-Nanotubes
▶ Finite Element Methods for Computational Nano-optics

References

1. Allen, M.P., Tildesley, D.J.: Computer Simulation of Liquids. Clarendon, Oxford (1987)
2. Frenkel, D., Smit, B.: Understanding Molecular Simulation: From Algorithms to Applications. Academic, San Diego (1996)
3. Kelires, P.C., Tersoff, J.: Equilibrium alloy properties by direct simulation: Oscillatory segregation at the Si-Ge(100) 2×1 surface. Phys. Rev. Lett. **63**, 1164–1167 (1989)
4. Combe, N., Jensen, P., Pimpinelli, A.: Changing shapes in the nanoworld. Phys. Rev. Lett. **85**, 110–113 (2000)
5. Liu, H., Lin, Z., Zhigilei, L.V., Reinke, P.: Fractal structures in fullerene layers: simulation of the growth process. J. Phys. Chem. C **112**, 4687–4695 (2008)
6. Germann, T.C., Kadau, K.: Trillion-atom molecular dynamics becomes a reality. Int. J. Mod. Phys. C **19**, 1315–1319 (2008)
7. http://folding.stanford.edu/
8. Dongare, A.M., Rajendran, A.M., LaMattina, B., Zikry, M.A., Brenner, D.W.: Atomic scale studies of spall behavior in nanocrystalline Cu. J. Appl. Phys. **108**, 113518 (2010)
9. Yamakov, V., Wolf, D., Phillpot, S.R., Mukherjee, A.K., Gleiter, H.: Dislocation processes in the deformation of nanocrystalline aluminium by molecular-dynamics simulation. Nat. Mater. **1**, 45–49 (2002)
10. Ju, S., Liang, X.: Investigation of argon nanocrystalline thermal conductivity by molecular dynamics simulation. J. Appl. Phys. **108**, 104307 (2010)
11. Lin, Z., Bringa, E.M., Leveugle, E., Zhigilei, L.V.: Molecular dynamics simulation of laser melting of nanocrystalline Au. J. Phys. Chem. C **114**, 5686–5699 (2010)
12. Xiao, S., Hu, W., Yang, J.: Melting behaviors of nanocrystalline Ag. J. Phys. Chem. B **109**, 20339–20342 (2005)
13. Rudd, R.E., Broughton, J.Q.: Coarse-grained molecular dynamics and the atomic limit of finite elements. Phys. Rev. B **58**, R5893–R5896 (1998)
14. Colbourn, E.A. (ed.): Computer Simulation of Polymers. Longman Scientific and Technical, Harlow (1994)
15. Peter, C., Kremer, K.: Multiscale simulation of soft matter systems. Faraday Discuss. **144**, 9–24 (2010)
16. Zhigilei, L.V., Leveugle, E., Garrison, B.J., Yingling, Y.G., Zeifman, M.I.: Computer simulations of laser ablation of molecular substrates. Chem. Rev. **103**, 321–348 (2003)
17. Buehler, M.J.: Mesoscale modeling of mechanics of carbon nanotubes: self-assembly, self-folding, and fracture. J. Mater. Res. **21**, 2855–2869 (2006)
18. Volkov, A.N., Zhigilei, L.V.: Mesoscopic interaction potential for carbon nanotubes of arbitrary length and orientation. J. Phys. Chem. C **114**, 5513–5531 (2010)
19. Zhigilei, L.V., Wei, C., Srivastava, D.: Mesoscopic model for dynamic simulations of carbon nanotubes. Phys. Rev. B **71**, 165417 (2005)
20. Groh, S., Zbib, H.M.: Advances in discrete dislocations dynamics and multiscale modeling. J. Eng. Mater. Technol. **131**, 041209 (2009)
21. Kirchner, H.O., Kubin, L.P., Pontikis, V. (eds.): Computer simulation in materials science. Nano/meso/macroscopic space and time scales. Kluwer, Dordrecht (1996)
22. Raabe, D.: Computational materials science: the simulation of materials microstructures and properties. Wiley-VCH, Weinheim, New York (1998)
23. Holm, E.A., Battaile, C.C.: The computer simulation of microstructural evolution. JOM-J. Min. Met. Mat. S. **53**, 20–23 (2001)
24. Akasheh, F., Zbib, H.M., Hirth, J.P., Hoagland, R.G., Misra, A.: Dislocation dynamics analysis of dislocation intersections in nanoscale metallic multilayered composites. J. Appl. Phys. **101**, 084314 (2007)
25. Leveugle, E., Zhigilei, L.V.: Molecular dynamics simulation study of the ejection and transport of polymer molecules in matrix-assisted pulsed laser evaporation. J. Appl. Phys. **102**, 074914 (2007)
26. Volkov, A.N., Zhigilei, L.V.: Structural stability of carbon nanotube films: the role of bending buckling. ACS Nano **4**, 6187–6195 (2010)
27. Volkov, A.N., Zhigilei, L.V.: Scaling laws and mesoscopic modeling of thermal conductivity in carbon nanotube materials. Phys. Rev. Lett. **104**, 215902 (2010)

Computational Systems Bioinformatics for RNAi

Zheng Yin, Yubo Fan and Stephen TC Wong
Center for Bioengineering and Informatics,
Department of Systems Medicine and Bioengineering,
The Methodist Hospital Research Institute, Weill
Cornell Medical College, Houston, TX, USA

Synonyms

Automatic data analysis workflow for RNAi; Systems level data mining for RNAi

Definition

Computational systems bioinformatics for RNAi screening and therapeutics is defined as complete computational workflow applicable to the hypothesis generation from large-scale data from image-based RNAi screenings as well as the improvement of RNAi-based therapeutics; the workflow includes automatic image analysis compatible to large-scale cell image data, together with unbiased statistical analysis and gene function annotation.

Introduction

RNA interference (RNAi) defines the phenomenon of small RNA molecules binding to its complementary sequence in certain messenger RNA, recruiting a specific protein complex to dissect the whole mRNA and thus silencing the expression of the corresponding gene. It is a highly conserved system within living cells to quantitatively control the activity of genes. In 1998, Fire et al. first clarified the causality of this phenomenon and named it as RNAi [1], and the following decade saw RNAi evolving into a powerful tool for gene function study. In 2006, Fire and Mello were awarded Nobel Prize in Physiology or Medicine; and by 2007, scientific papers using high-throughput screening based on RNAi kept piling up while clinical trials of RNAi-based therapeutics on various diseases raised the expectation on a trend of soon-to-come "super drugs." Unfortunately, by late 2009 nearly all the first trend clinical trials have been terminated, meanwhile, researchers are struggling to effectively quantify the high-content information obtained from RNAi-based screening experiments.

Although facing obstacles, it is still believed that the combination of nanotechnology and systems biology would restore and amplify the glory of RNAi on both research and therapeutic areas. This entry will summarize the challenges facing RNAi-based therapeutics – especially the difficulty of delivery and some possible solution through chemoinformatics in nanometer scale; also difficulties facing RNAi-based high-content screening will be reviewed with suggestions on possible solutions.

RNAi-Based Therapeutics: How-to and What-to Deliver

Currently, various RNAi-based therapeutics are being tested in clinical trials but before the application becomes clinical, several problems have first to be solved. One of the critical challenges is how to specifically and effectively deliver the objects into the targeted cells. Serious side-effects in patients could be caused by off-target effects or immune response as reported in literatures [2, 3].

Apparently, the therapeutic goal is to achieve RNAi therapy, that is, systemically administered nucleic acids must survive in circulation long enough to reach their target tissue, enter the desired cells, "escape" their endosome or delivery packaging, and finally become incorporated into the RNA-induced silencing complex (RISC) – a towering task, and surprisingly, researchers have advanced a number of plausible solutions in recent years, including the use of specialized nanoparticle filled with an RNAi-based cancer therapy to target human cancer cells and silence the target gene. However, what making the task even more difficult for therapeutics is: everything now happens in vivo.

Chemoinformatics Solutions

Libraries have been built based on natural products and combinatorial chemistry with millions of compounds to date. The compound library is used to screen and locate small molecules to bind a particular protein, RNA or DNA. Virtual screening utilizing molecule fragments even can design unknown compounds with reasonable binding affinity to desired targets. However, these large-scale screening compounds tend to bind multiple targets (off-target effects) even after optimizations [4]. Also, it is virtually impossible to derive a solid algorithm or theory to screen and locate all bindings and inhibitions and their combinations every protein currently know. The lack of genome-scale coverage is a clear disadvantage of a compound based assay. Along with the lack of specificity comes the challenge of on-target potency. Even after a target has been identified the drug-like compounds have to be further engineered to increase efficiency and decrease off-target activities.

To a large extent, chemoinformatic approaches, where the target of a compound can be predicted by in silico alignment, modeling algorithms, and virtual screening, provide a more efficient way to screen compounds to the desired targets. A successful application has been reported for the inhibition of SARS protease

[5]. The chemoinformatics approach excels at optimization around a small number of well defined targets, while RNAi approaches can more readily identify unknown pathways and phenotypes with no prior knowledge of the target.

Data Analysis for RNAi HCS: Challenge from Millions of Cells

Large volumes of datasets generated from RNAi HCS of RNAi prohibits manual or even semi-manual analysis; thus, automated data analysis is desperately needed [6, 7]. In the context of genome-wide RNAi HCS using cultured cells from Drosophila, a series of automated methods on cell image processing [8], online phenotype discovery [9, 10], cell classification, and gene function annotation [11] have been developed. All these methods are integrated into an automated data analysis pipeline, G-CELLIQ (Genomic CELLular Image Quantitator), to support genome-wide RNAi HCS.

A lot of decisions need to be made en route to a genome-scale RNAi screens, including the selections of appropriate animal models, reagents for igniting RNAi, screening formats, and type of readouts (see [12] for a review). The focus here is arrayed high-content RNAi screening, a systematic screen with reagents spotted in 384- or 96-well plates and where each gene or gene group is knocked down individually. The readout of HCS is obtained through microscopy, which captures multiple phenotypic features simultaneously [13]. Compared with other screening approaches, RNAi HCS can offer broader insight into cellular physiology and provide informative and continuous phenotypic data generated by RNAi. However, the automated processing and analysis of the magnitude of image data presents great challenges to applying such findings at the genome level.

Image Processing: Cell Segmentation and Quantification

The scale of datasets has always been a huge obstacle. For example, in [6], approximately 17,000 overlapping cells were segmented semi-manually across 10 months from an HCS targeting around 200 genes or gene combinations. However, such low-dimensional screens are not genome-wide, as they can only cover 1–2% of the genome. For automatic data analysis, the following work needs to be accomplished.

1. Preprocessing: where empty images, incomplete cells, and other artifacts are identified and discarded
2. Cell segmentation: where dense or overlapping cells are segmented accurately
3. Feature extraction: where informative features are extracted to quantify cell morphology

Cell segmentation is the cornerstone for the whole data analysis workflow. While existing image analysis methods can handle the processing of standard images, they are limited in their scope and capability to handle genome-wide RNAi HCS analysis. Thresholding methods basically set a cutoff on the intensity of pixels and classify them into background and foreground (cell), and they may fail due to uneven background and illumination levels. Rule-based correction on over- and under-segmentation starts from relatively simple segmentation methods (like watershed), and use a series of heuristic rules, like distance between neighborhood nuclei and properties of putative cell boundaries, such methods suffer from difficulties in devising rules to merge the cell cytoplasm.

Phenotype Identification, Validation, and Classification

Given the quantified features describing cell morphology, the following work is essential to address the biological function of morphological profiles.

1. Define biologically meaningful phenotypes to compose cell populations based on single cell morphology.
2. Model existing phenotypes and identify novel phenotypes online to continuously generate new data.
3. Assign cellular phenotypes into different subphenotypes to address morphological changes caused by RNAi treatment.

Statistical tests, artificial neural networks [6], Support Vector Machine-Recursive Feature Elimination (SVM-RFE) [14], genetic algorithms, and various other methods have been used to select or extract informative subsets of features and model certain phenotypes. However, phenotypes are usually defined a priori from pilot datasets. Human intervention is currently necessary for image-based datasets of genetic or chemical perturbations where the dynamic range of cellular phenotypes cannot be predicted before data collection. Failing to accurately measure phenotypic variations will cause concomitant classification errors and mislead functional analysis; and it is impossible to perform manual analysis during the screening process where millions of images are acquired. Thus, the ability of these screens to identify new phenotypes is greatly limited [9].

Statistical Analysis and Gene Function Annotation

RNAi HCS inherits various statistical questions from traditional high-throughput screening, and the properties of image readouts raise specific challenges.

1. Summarizing gene function scores from quantified morphological change
2. Data triage and normalization based on readout from positive and negative control wells
3. Repeatability test and consolidation of scores from biological replicates
4. Cluster analysis, visualization, and biological interpretation of results

A comprehensive review [15] summarizes different dataset generated by RNAi screens and traditional small molecule screens. It also reviews statistical analysis methods applicable to most of problems outlined above. Quantitative morphological signatures (QMS) [6] represent the efforts on interpreting RNAi HCS datasets: use the similarity score to a panel of existing phenotypes to explore the broad phenotypic space. However, problems remain open when the discriminative ability of such scores is confounded by multiple phenotypes following a single RNAi treatment, so repeatability tests can become more complicated [6]. The use of publicly available databases on drugs and disease-related biological processes to interpret RNAi HCS datasets also remains unresolved.

G-CellIQ: An Integrated Automated Data Analysis Tool for RNAi High-Content Screening

Computational Architecture of G-CellIQ

G-CellIQ (Genomic CELLular Imaging Quantitator) is developed to process large volumes of digital images generated from large-scale HCS studies. The workflow for G-CellIQ can be simplified as "three modules handling three databases." Images generated from HCS are stored in a Raw image database; the Image processing/cell morphology quantification module segments each image into single cells and creates a quantified cell database; the Phenotype modeling and cell classification module compares each cell's

morphology to a panel of "reference" phenotypes (which can be defined both manually and automatically) and generates a morphology score; the annotation of gene function module then summarize single cell scores into scores for cell populations, images and wells, and the consolidated scores for involved genes form a single gene function profile database.

Image Processing and Cell Morphology Quantification

A segmentation method consisting of nuclear segmentation, cell body segmentation, and over-segmentation correction [8] is used in G-CELLIQ. Each cell body is described by 211 morphological features; automatic image quality control is applied to filter images where the signal from certain channels is extremely dark or bright or cells are located at the edge of an image.

Nuclear segmentation cell nuclei are first separated from background using a binarization method which implements an adaptive thresholding [8]. To segment clustered nuclei, the nuclei shape and intensity information are integrated into a combined image and processed with Gaussian filter and the nuclei centers are detected as local maxima in the gradient vector field (GVF); after that, nuclei are segmented using the marker-controlled seeded-watershed method.

Cell body segmentation Preliminary cell body segmentation is done using an adaptive thresholding algorithm. Due to the large size of HCS dataset, seeded-watershed method is used to segment the touching cell bodies. Results from nuclei segmentation are used as the seed information [16].

Over-segmentation correction is necessary when there are multiple nuclei existing within cells. For cell segments smaller than a given size threshold, its neighboring cell segments sharing the longest common boundary are determined and the intensity variation in a rectangular region across the common boundary of touching cell segments are calculated. If the intensity variation was smaller than a given threshold, the corresponding cell segments are merged.

Feature extraction The detailed shape and boundary information of nuclei and cell bodies is obtained through the proposed segmentation method. To capture the geometric and appearance properties, 211 morphology features belonging to five categories were extracted following [11].

Online Phenotype Discovery, Phenotype Modeling, and Cell Classification

Expert opinion is implemented to identify a panel of reference phenotypes while candidate groups of informative features are selected to describe typical phenotypes. SVM-RFE and GA-SVM method are used to select the feature sets with the best performance for cross validation. A series of SVM classifiers are trained to differentiate the reference phenotypes from all others, and a continuous value rather than binary class label is used as the output of SVM to indicate each cell's morphological similarity to typical phenotypes.

A novel method is designed to do online phenotype discovery [9, 10]; it is based on online phenotype modeling and iterative phenotype merging. For the modeling part, each existing phenotype is modeled through a Gaussian Mixture Model (GMM), and each model is continuously updated according to information of newly incorporated cells with a minimum classification error (MCE) method. For the merging part, the newly generated cell population is iteratively combined with each existing phenotype, one at a time. Then an improved gap statistics method is used to identify the number of possible phenotypes in the combination. Through cluster analysis, some of the cells in the new populations are assigned into the same cluster as samples from existing phenotypes, and those cells are merged by existing phenotypes to help update the phenotype models. On the other hand, some cells are never merged and remain as candidate novel phenotype for validation by statistical tests [9, 10].

Annotation of Gene Function

After assigning each cell a score vector corresponding to its similarity to reference phenotypes, all cells in control condition are pooled to model a baseline for cell morphology. All morphology scores are normalized to the Z-score relative to this control baseline. The scores for the qualified cells are then averaged to form scores for each well. In order to select repeatable wells from those undergoing the same dsRNA treatment, a series of repeatability tests are applied to scores for different well. The weighted average of the scores is then calculated for the repeatable wells and generated scores for each treatment condition (TC). Similar procedure consolidates the score from biological replicate TCs to form a score vector for each gene. Hierarchical

clustering is implemented group genes with similar function scores into the same pheno-cluster.

RNAi HCS Applying G-CellIQ

Since more than 75% of human disease genes have *Drosophila* orthologs, the pursuit of genes involved in normal fly morphogenesis and migration is expected to reveal mechanisms conserved in humans [17]. An example of using G-CellIQ in the context of RNAi HCS using cultured *Drosophila* cell lines is presented next.

Regulatory of cell shape change A genome-scale RNAi screening is carried out for regulators of *Drosophila* cell shape. Drosophila Kc187 cell lines are utilized in the screen and wild-type cells have hemocyte-like properties. Using dsRNA to target and inhibit the activity of specific genes/proteins, the role of individual genes in regulating morphology can be systematically determined.

Work in [9, 10] relies on part of the genome dataset to target the group of kinase-phosphatases in order to develop and validate online phenotype discovery methods. A panel of five existing phenotypes is set by expert labeling and online phenotype discovery. In order to address the level of penetrance in this dataset, the phenotypic scores for single cells are sorted according to similarity to wild-type cells, and if a certain well has significantly less (at least one standard-deviation) wild-type cells than control wells, wild-type cells from this well are removed to reveal the phenotypic change relating to RNAi treatment.

Roles of Rho family small GTPases in development and cancer Three automated and *quantitative* genome-wide screens for dsRNAs that induce the loss of the Rho-induced cytoskeletal structures (lamellipodia, filopodia, and stress fibers) are performed to identify putative Rho protein effectors. Candidate effectors identified in such image-based screens are readily validated in the context of the whole organism using the large number of mutant fly lines coupled with the vast arrays of in vivo techniques available to fly biologists [18].

Following image segmentation, feature extraction, classification of single cells, and scoring of individual wells, an image descriptor is assigned to each gene. Hierarchical clustering can then be used to cluster the image descriptor and identify groups of genes when targeted by RNAi result in quantitatively similar morphologies. Previous studies reported in [6] demonstrated that clustering results identifies groups of functionally related genes that operate in similar signaling pathways. These groups of genes are termed as "Pheno-clusters" [19]. To validate the hypothesis generation ability of G-CELLIQ, 32 dsRNAs/wells are randomly selected from a dataset where individual kinases and phosphatases were inhibited by RNAi. Each dsRNA/well was assigned an image descriptor, and hierarchical clustering was used to group genes/ wells. dsRNAs in this analysis clustered into two broad groups. One group of 19 conditions included 10/10 control conditions, as well as dsRNAs targeting the Insulin receptor (InR). Strikingly, the other large cluster of 13 conditions included 3/3 dsRNAs previously identified in a genome-wide screen for regulators of MAPK/ERK activation downstream of the EGF/EGFR activity [20]. These results demonstrate that automated high-throughput imaging can discriminate distinct morphologies and be used to model functional relationships between signaling molecules.

Cross-References

▶ RNAi in Biomedicine and Drug Delivery

References

1. Fire, A., et al.: Potent and specific genetic interference by double-stranded RNA in *Caenorhabditis elegans*. Nature **391**(6669), 806–811 (1998)
2. Hornung, V., et al.: Sequence-specific potent induction of IFN-[alpha] by short interfering RNA in plasmacytoid dendritic cells through TLR7. Nat. Med. **11**(3), 263–270 (2005)
3. Grimm, D., et al.: Fatality in mice due to oversaturation of cellular microRNA/short hairpin RNA pathways. Nature **441**(7092), 537–541 (2006)
4. Copeland, R.A., Pompliano, D.L., Meek, T.D.: Drug-target residence time and its implications for lead optimization. Nat. Rev. Drug Discov. **5**(9), 730–739 (2006)
5. Plewczynski, D., et al.: In silico prediction of SARS protease inhibitors by virtual high throughput screening. Chem. Biol. Drug Des. **69**(4), 269–279 (2007)
6. Bakal, C., et al.: Quantitative morphological signatures define local signaling networks regulating cell morphology. Science **316**, 1753–1756 (2007)
7. Zhou, X., Wong, S.T.C.: Computational systems bioinformatics and bioimaging for pathway analysis and drug screening. Proc. IEEE **96**(8), 1310–1331 (2008)
8. Li, F.H., et al.: High content image analysis for human H4 neuroglioma cells exposed to CuO nanoparticles. BMC Biotechnol. **7**, 66 (2007)
9. Yin, Z., et al.: Using iterative cluster merging with improved gap statistics to perform online phenotype discovery in the context of high-throughput RNAi screens. BMC Bioinformatics **9**(1), 264 (2008)
10. Yin, Z., et al.: Online phenotype discovery based on minimum classification error model. Pattern Recogn. **42**(4), 509–522 (2009)
11. Wang, J., et al.: Cellular phenotype recognition for high-content RNA interference genome-wide screening. J. Mol. Screen. **13**(1), 29–39 (2008)
12. Perrimon, N., Mathey-Prevot, B.: Applications of high-throughput RNAi screens to problems in cell and developmental biology. Genetics **175**, 7–16 (2007)
13. Carpenter, A.E., Sabatini, D.M.: Systematic genome-wide screens of gene function. Nat. Rev. Genet. **5**(1), 11–22 (2004)
14. Loo, L., Wu, L., Altshuler, S.: Image based multivariate profiling of drug responses from single cells. Nat. Methods **4**(5), 445–453 (2007)
15. Birmingham, A., et al.: Statistical methods for analysis of high-throughput RNA interference screens. Nat. Methods **6**(8), 569–575 (2009)
16. Yan, P., et al.: Automatic segmentation of RNAi fluorescent cellular images with interaction model. IEEE Trans. Inf. Technol. Biomed. **12**(1), 109–117 (2008)
17. Reiter, L.T., et al.: A systematic analysis of human disease-associated gene sequences in *Drosophila melanogaster*. Genome Res. **11**, 1114–1125 (2001)
18. Bier, E.: Drosophila, the golden bug, emerges as a tool for human genetics. Nat. Rev. Genet. **6**(1), 9–23 (2005)
19. Piano, F., et al.: Gene clustering based on RNAi phenotypes of ovary-enriched genes in *C. elegans*. Curr. Biol. **12**(22), 1959–1964 (2002)
20. Friedman, A., Perrimon, N.: Functional genomic RNAi screen for novel regulators of RTK/ERK signaling. Nature **444**, 230–234 (2006)

Computer Modeling and Simulation of Materials

▶ Computational Study of Nanomaterials: From Large-Scale Atomistic Simulations to Mesoscopic Modeling

Concentration Polarization

▶ Concentration Polarization at Micro/Nanofluidic Interfaces

Concentration Polarization at Micro/Nanofluidic Interfaces

Vishal V. R. Nandigana and N. R. Aluru
Department of Mechanical Science and Engineering, Beckman Institute for Advanced Science and Technology, University of Illinois at Urbana – Champaign, Urbana, IL, USA

Synonyms

Concentration polarization; Micro/nanofluidic devices; Nonlinear electrokinetic transport

Definition

Concentration polarization (CP) is a complex phenomenon observed at the interfaces of micro/nanofluidic devices due to the formation of significant concentration gradients in the electrolyte solution resulting in accumulation and depletion of ions near the interfaces.

This chapter provides an overview of the underlying theory and physics that is predominantly observed on the integration of microfluidic channels with nanofluidic devices. Nonlinear electrokinetic transport and concentration polarization phenomenon are discussed in detail along with recent advancements in utilizing this phenomenon for designing novel devices.

Electrical Double Layer (EDL) and Electroosmotic Flow

A solid in contact with an aqueous solution acquires a surface charge (σ_s) due to the dissociation of ionizable groups on the solid walls. The fixed surface charge on the solid surface in contact with the liquid develops a region of counterions (ions with charges opposite to the solid surface) in the liquid to maintain the electroneutrality at the solid–liquid interface. This screening region is denoted as the electrical double layer (EDL) or Debye length (DL). For instance, in the case of silica channel with KCl electrolyte solution, the dissociation of silanol groups would make the channel negatively charged and affect the distribution of K^+ counterions in the solution. The layer where the

ions get strongly attracted toward the channel surface due to the electrostatic force is called the inner layer with a typical thickness of one ion diameter. The outer Helmholtz plane separating the liquid and the diffusive layer constitute the liquid side part of the EDL. The ionic species in the diffusive layer are influenced by the local electrostatic potential, and the species distribution at equilibrium can be described by the Boltzmann equation. The thickness of the diffusive layer spans between 1 and 100 nm. The EDL thickness (λ_D) is given by [1]:

$$\lambda_D = \left(\frac{\varepsilon_0 \varepsilon_r RT}{F^2 \sum_{i=1}^{m} z_i^2 c_0} \right)^{1/2} \tag{1}$$

where F is Faraday's constant, z_i is the valence of ionic species i, c_0 is the bulk concentration of the electrolyte solution, ε_0 is the permittivity of free space, ε_r is the relative permittivity of the medium, m is the total number of ionic species, R is the universal gas constant, and T is the absolute temperature.

The thickness of the EDL plays a significant role in the transport of miniaturized devices. In microfluidic channels, the fluid transport is often controlled by electric fields, as it eliminates the use of external mechanical devices [1]. The electric field aids better control when compared to using pressure-driven techniques. Furthermore, the electric fields overcome the high pressures needed to transport the fluid at such length scales as the pressure follows a power–law relation with respect to the height (h) of the channel. The electric field acts on the charged counterions present at the interface of solution and stationary charged wall (i.e., at the EDL regions), resulting in the motion of the fluid which is referred as electroosmotic flow (EOF) [1]. As the EDL thickness in these channels is much smaller compared to their height $\left(\frac{h}{\lambda_D} \gg 1 \right)$, the fluid flow has a plug-like flow characteristic. However, recent advancements in the fabrication technology [2] have motivated researchers around the globe to investigate the transport phenomenon in channel sizes of the order of few hundreds of nanometers. Transport in these devices is referred to as "nanofluidics." The electrical double layer in these devices spans much of the diameter or channel height leading to many interesting transport phenomena compared to its microscopic counterpart. The electroosmotic velocity no longer follows a plug-like flow characteristic but follows a Poiseuille-like (parabolic) characteristic as the electrokinetic body force is not just confined to a thin layer adjacent to the channel surface. Along with the aforementioned difference, the micro and nanofluidic systems also exhibit a different ion transport characteristic which is discussed below. In nanofluidic systems, as the EDL thickness becomes comparable to the channel height $\left(\frac{h}{\lambda_D} \approx 1\right)$, there is a predominant transport of the counterions inside the channel, thus enabling the channel to be ion-selective [1]. These features are not observed in the microfluidic channels, as the counter-ionic space charge is confined to a very thin layer adjacent to the surface and the region away from the surface is essentially quasi-electroneutral (i.e., both co-ions and counterions are present away from the surface). Along with the EDL, the surface charge also plays a prominent role in controlling the transport inside the nanofluidic systems [2]. Similar ion-selective phenomenon was also observed in the intraparticle and intraskeleton mesopores of particulate and in membrane science [3].

Owing to the differences in the electrokinetic transport phenomena between the micro- and nanofluidic devices, the integration of these two devices paves way to complex physics. The models and the underlying theory developed to understand the electrokinetic transport in such systems are elaborated in section "Electrokinetic Theory for Micro/Nanochannels." A detailed discussion on the concentration polarization phenomenon and its applications are presented in section "Concentration Polarization." Finally, a brief summary is presented in section "Summary."

Electrokinetic Theory for Micro/Nanochannels

A complete set of equations for modeling the electrokinetic transport and to account for the EDL effects in micro/nanofluidic channels are presented. To understand the electrokinetic transport, space charge model developed by Gross et al. [4] is used extensively in the literature. The model solves the classical Poisson–Nernst–Planck (PNP) equations, which describe the electrochemical transport and the incompressible Navier–Stokes along with the continuity equations are solved to describe the movement of the fluid flow. These coupled systems of equations are more intensive, mathematically complicated, and computationally expensive. Though many linearized approximations

were proposed to this model to study the electrokinetic transport [2], the governing equations of the complete nonlinear space charge model is discussed in this chapter.

In electrokinetic flows, the total flux is contributed by three terms: a diffusive component resulting from the concentration gradient, an electrophoretic component arising due to the potential gradient, and a convective component originating from the fluid flow. The total flux of each species in the solution is given by

$$\boldsymbol{\Gamma}_i = -D_i \nabla c_i - \Omega_i z_i F c_i \nabla \phi + c_i \boldsymbol{u} \qquad (2)$$

where $\boldsymbol{\Gamma}_i$ is the flux vector, D_i is the diffusion coefficient, Ω_i is the ionic mobility, c_i is the concentration of the ith species, $\boldsymbol{u}$ is the velocity vector of the fluid flow, and ϕ is the electrical potential. Note that the ionic mobility is related to the diffusion coefficient by Einstein's relation, $\Omega_i = \frac{D_i}{RT}$ [5]. The electrical potential distribution is calculated by solving the Poisson equation,

$$\nabla \cdot (\epsilon_r \nabla \phi) = -\frac{\rho_e}{\epsilon_0} \qquad (3)$$

where ρ_e is the net space charge density of the ions defined as

$$\rho_e = F\left(\sum_{i=1}^{m} z_i c_i\right) \qquad (4)$$

The mass transfer of each buffer species is given by the Nernst–Planck equation,

$$\frac{\partial c_i}{\partial t} = -\nabla \cdot \boldsymbol{\Gamma}_i \qquad (5)$$

Equations 3, 5, and 2 are the classical Poisson–Nernst–Planck (PNP) equations, which describe the electrochemical transport. The incompressible Navier–Stokes and the continuity equations are considered to describe the movement of the fluid flow through the channel, i.e.,

$$\rho\left(\frac{\partial \boldsymbol{u}}{\partial t} + \boldsymbol{u} \cdot \nabla \boldsymbol{u}\right) = -\nabla p + \mu \nabla^2 \boldsymbol{u} + \rho_e \boldsymbol{E} \qquad (6)$$

$$\nabla \cdot \boldsymbol{u} = 0 \qquad (7)$$

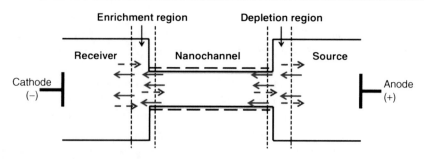

Concentration Polarization at Micro/Nanofluidic Interfaces, Fig. 1 Schematic illustration of ion-enrichment and ion-depletion effect in cation-selective micro/nanofluidic channel. The *solid arrows* indicate the flux of cations and the *dotted arrows* indicate the flux of anions. At the nanochannel–anodic junction, both the anions and cations are depleted, while there is an enhancement of both the ions at the cathode–nanochannel junction

where $\boldsymbol{u}$ is the velocity vector, p is the pressure, ρ and μ are the density and the viscosity of the fluid, respectively, and $\boldsymbol{E} = -\nabla\phi$ is the electric field. $\rho_e\boldsymbol{E}$ is the electrostatic body force acting on the fluid due to the space charge density and the applied electric field. Elaborate details on other simplified models are discussed in the review article of Schoch et al. [2].

Concentration Polarization

Concentration polarization (CP) is a complex phenomenon observed at the interface regions of micro/nanofluidic devices due to the formation of significant concentration gradients in the electrolyte solution near the interfaces causing accumulation of ions on the cathodic side and depletion of ions on the anodic side for a negatively charged nanochannel surface. This phenomenon was also observed in the field of colloid science and in membrane science which was extensively studied for over 40 years, and the early works of CP phenomenon is comprehensively reviewed by Rubinstein et al. [6]. The pioneering works from Rubinstein and his coworkers had revealed electrokinetic instabilities [7] in the concentration polarization regions leading to the breakdown of limiting current and resulting in the overlimiting conductance regimes in the ion exchange membranes. Such complex phenomenon could not be postulated using the classical equilibrium model of EDL [6]. All the underlying CP physics observed near the interfaces of micro/nanochannels are summarized in the following subsections.

Enrichment/Depletion Effects

In micro/nanofluidic devices, Pu et al. [8] first experimentally observed the CP effects near the interfaces and provided a simple model to explain the accumulation and depletion physics which is summarized below. For a negatively charged nanochannel, the EDL would be positively charged. For an overlapped EDL, as discussed in section "Electrical Double Layer (EDL) and Electroosmotic Flow," the nanochannel becomes ion-selective, resulting in higher cation concentration than anions. Thus, the flux of cations is higher compared to the anions in the nanochannel. With the application of positive potential at the source microchannel or reservoir (see Fig. 1), the cations move from the source (anode) reservoir to the receiving (cathode) reservoir end, while the anions move in the opposite direction through the nanochannel. At the cathodic side, the anion flux from the ends of reservoir to the nanochannel junction is higher compared to the anion flux from the junctions to the nanochannel as the anions are repelled by the negatively charged nanochannel. This difference in fluxes causes an accumulation of anions at the cathode–nanochannel junction. The cation flux from the nanochannel to the cathode junction is greater than from the cathode junction to the reservoir as the cations have to balance the anions present at this junction. This results in an accumulation of cations as well at the nanochannel–cathode junction. At the anodic side, the anion flux from the nanochannel to the anode junction cannot balance the anion flux from the anode–nanochannel junction to the reservoirs due to the limited anions passing through the nanochannel. This results in a depletion of anions at this junction. The cation flux from the reservoir to the anode–nanochannel junction is less than the cation flux entering the nanochannel as the cations are attracted by the positively charged nanochannel. This in turn leads to the depletion of cations at this junction. To summarize, for a negatively charged nanochannel, both the cations and anions accumulate at the cathodic interface and are depleted at the anodic interface. The phenomenon is reversed for a positively charged nanochannel surface. A schematic diagram highlighting the accumulation and the depletion physics for a negatively charged nanochannel is displayed in Fig. 1.

Nonlinear Electroosmosis

As discussed in the previous section, the integration of micro/nanofluidic devices leads to the accumulation and depletion of ions near the interfaces. Several numerical and experimental studies were carried out to gain a better understanding of the physics at these interfaces. The studies revealed complex and interesting physics at the depletion interface compared to the enrichment side. Rubinstein et al. [6] theoretically predicted the presence of space charges at the depletion region under large electric fields. The presence of the induced space charges near the depletion interface results in a nonequilibrium electrical double layer outside the nanochannel. The induced space charges under the action of the external electric field lead to nonlinear electroosmosis or otherwise known as electroosmosis of the second kind. The electroosmotic flow of the second kind was found to be directly proportional to the square of the applied electric field. Furthermore, the induced space charges also result in the generation of vortices at the depletion interface along with inducing large pressure and voltage gradients at this junction. Jin et al. [9] also reported similar physics from their extensive numerical study. In the case of a flat ion exchange membrane, Rubinstein et al. [7] derived a 2D nonequilibrium electroosmotic slip (u_s) for an applied voltage (V) using the linear stability analysis to impose strong vortex field near the membrane:

$$u_s = -\frac{1}{8}V^2 \frac{\frac{\partial^2 c}{\partial x \partial y}}{\frac{\partial c}{\partial y}} \qquad (8)$$

where x and y are the axes parallel and perpendicular to the ion-exchange membrane, respectively. Experiments performed by Kim et al. [10] also reveal the nonequilibrium EOF near the micro/nanofluidic junctions. The application of electric field on the surface of particles also results in such induced space charges which spread over a larger region than the primary EDL resulting in highly chaotic flow patterns. A recent review by Höltzel and Tallarek [3] provide a detailed discussion on the polarization effects around membranes, packed beds, and glass monoliths. Recent advancements by Rubinstein and Zaltzman [11], however, revealed that the extended space charge region was not a part of the EDL, but develop from the counterion concentration minimum zone with the co-ions expelled under the action of the electric field. They further claim that the space charges would be present in the system even without equilibrium EDL. Their analysis included the study of the space charge dynamics in concentration polarization regions using 1D and three-layer models of EDL. From the understanding of these extended space charge layers, another important phenomenon, namely, the nonlinear current characteristics in micro/nanochannels is addressed below.

Nonlinear Current–Voltage Characteristics

There has been growing interest in the development of micro/nanofluidic devices as ionic filters and nanofluidic batteries to control both ionic and molecular transport in aqueous solutions [2]. As discussed before, when the diameter/height of a charged channel scales comparable to the EDL, there is a predominant transport of the counterions inside the channel. Thus, the transport of electrical current inside the nanochannel is primarily due to the counterions. This feature enables the micro/nanochannel to be used as an ion exchange membrane. However, understanding the passage of ionic currents through such ion-selective solids is the most fundamental physical problem that has stimulated extensive research in this field for over a decade. Furthermore, the concentration polarization physics near the interfaces play a pivotal role in understanding the current–voltage characteristics. Figure 2a shows that the concentration gradients (CP regions) near the interfaces become steeper with the decrease in the ionic strength (c_s) and at higher electric fields (E_x). At low electric fields, the current increases linearly with the applied voltage following the Ohm's law (region I in Fig. 2b).

However, at higher electric fields, the ion concentration in the depleted CP zone (i.e., near the anodic

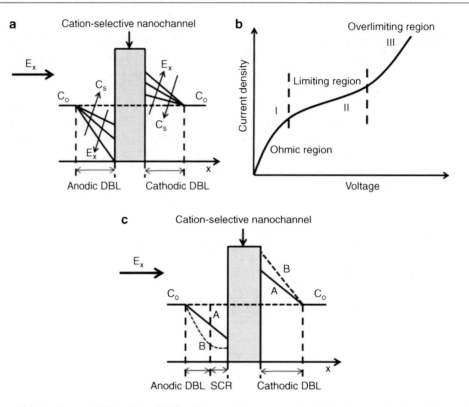

Concentration Polarization at Micro/Nanofluidic Interfaces, Fig. 2 (**a**) Schematic distribution of ionic concentration in equilibrium concentration polarization under axial electric field (E_x). The local electroneutrality is maintained at both enrichment (cathodic interface) and depletion (anodic interface) diffusion boundary layers (DBL). The concentration gradients become steeper with the decrease in the ionic strengths (c_s) and at higher electric fields, (**b**) displays the nonlinear current–voltage characteristics for an ion-selective micro/nanochannel, and (**c**) shows the nonequilibrium concentration distribution due to the induced space charge region (SCR) (shown as *dotted lines* (*B*)) in the depleted region under very large electric fields

interface region) reaches toward zero and the classical Levich analysis [5] predicts a diffusion-limited current saturation according to which a saturation of current density occurs at a constant level described as the "limiting-current density" (region II in Fig. 2b).

The ionic current can be calculated considering the Fick's first law:

$$I = nFAD \frac{dc}{dx} \qquad (9)$$

where n is the number of electrons transferred per molecule, D is the diffusion coefficient, and A is the electrode surface area. The concentration gradient is generally approximated by a linear variation (5),

$$I = nFAD \frac{c_0 - c(x = 0)}{\delta} \qquad (10)$$

where c_0 is the bulk electrolyte solution and δ is the diffusion boundary layer (DBL) thickness at the solid–liquid interface and $c(x = 0)$ represents the concentration at the anodic (depletion) solid–liquid interface. δ typically ranges between 10 and 400 μm in ion exchange membranes [6], while it depends on the microchannel length in the case of micro/nanochannels [10].

From Eq. 10 it is clear that the current I reaches a maximum value when the concentration at $c(x = 0) = 0$, resulting in a limiting/saturation current as predicted by the classical Levich theory.

$$I_{lim} = \frac{nFADc_0}{\delta} \qquad (11)$$

However, experimental studies in micro/nanofluidic devices (also in membrane science) revealed ionic currents larger than the limiting value and this regime was termed as the overlimiting current regime (region III in Fig. 2b). Further, the limiting current region is termed as the limiting resistance region (in micro/nanofluidic devices) due to the large but finite limiting differential resistance as the current does not saturate to a limiting value but has a slope which is smaller than the ohmic region. The nonlinear current characteristics in micro/nanochannels are a subject of intensive discussions in the literature [3]. Earlier studies indicated that water dissociation effects leading to the generation of H^+ and OH^- ions were responsible for such overlimiting currents. Later, Rubinstein et al. [7] postulated that some mechanism of mixing should be present that destroys the DBL as lower δ leads to higher currents (from Eq. 11). Maletzk et al. [12] coated the surface of cation-exchange membranes with a gel which does not allow mixing. From these experiments, they observed a saturation of current with no enhancements in the current, thereby confirming the earlier postulation of Rubinstein. Further, the fluid flow in the overlimiting regime revealed strong fluctuations indicating convection close to the surface. This convection was first attributed to the gravitational buoyant forces due to the concentration and temperature gradients. However, later theories have argued that the convection was not due to the gravitational instability in CP zones. Dukhin et al. [13] suggested the mechanism for mixing to be electroconvection. The type of electroconvection present in the overlimiting regime was revealed as the electroosmotic flow of the second kind. As discussed in the previous section, the induced space charges in the depletion zone (see Fig. 2c) under the action of the electric field results in the EOF of second kind and this convective instability tends to destroy the DBL leading to overlimiting currents as shown in Fig. 2b. Experiments by Kim et al. [10] also revealed the nonlinear currents due to the nonequilibrium EOF in the ion-selective micro/nanochannels. Similar physics was also observed in the perm-selective membranes and ion-selective particles [3]. In spite of all these postulations, the physics behind the extended space charge layer still remains largely unclear and there are a lot of potential research opportunities to fully understand this complex physics.

Propagation of Concentration Polarization

In this section, the conditions and the scenarios which can lead to the propagation of CP in micro/nanofluidic devices are highlighted. Zangle et al. [14] highlighted the phenomena of concentration polarization propagation using a simplified model of charged species transport and validated the same by conducting experiments and by comparing with other experimental results. The CP phenomenon was found to be governed by a type of Dukhin number, relating the bulk and the surface conductance. The inverse Dukhin number, for a symmetric electrolyte was specified as:

$$\frac{G_{bulk}}{G_\sigma} = \frac{Fhzc_0}{\sigma} \qquad (12)$$

where G_{bulk} is the bulk conductance, G_σ is the surface conductance, c_0 is the concentration outside the EDL, and σ is the wall surface charge density. Zangle et al. postulated that CP depends on Dukhin number and not on the ratio of channel height to the Debye length $\left(\frac{h}{\lambda_D}\right)$. Using their simplified model, they showed that both the enhancement and the depletion regimes at the interfaces of micro/nanofluidic channels propagate as shock waves under the following condition:

$$c_{o,r}^* h_n^* < \max(v_2^*, 2v_2^* - 1) \qquad (13)$$

where $c_{o,r}^* h_n^* = \frac{(v_1 z_1 - v_2 z_2) F h_n c_{o,r}}{-2v_1 \sigma}$ is an inverse Dukhin number describing the ratio of bulk to surface conductance as mentioned before. $v_2^* = \frac{v_2 z_2 F \eta}{\zeta_n \epsilon}$ is the mobility of the co-ion nondimensionalized by the electroosmotic mobility. $c_{o,r}$ is the reservoir electrolyte concentration, h_n is the nanochannel height, v_1 and v_2 are the mobilities, and z_1 and z_2 are the valences of the positive and negative ionic species, respectively. ζ_n is the nanochannel zeta potential, ϵ is the permittivity and η is the viscosity. Elaborate details of the model can be referred in [14]. From this model, they proposed a thumb rule to avoid propagation of CP and it was found that $c_{o,r}^* h_n^* \gg 1$. This condition for CP propagation was compared with 56 sets of experimental literature values and was found to give a sufficient first-hand prediction with regard to the concentration polarization propagation.

Though the model considers the effects of surface charge and the electrolyte concentration, the finite Pe effects which also play a critical role in the concentration polarization were not considered. Further, the experiments of Kim et al. [10] and the numerical studies performed by Jin et al. [9] also revealed that the applied potential also plays a pivotal role in the concentration polarization generation and propagation apart from the inverse Dukhin number. The shortcoming of the model was also highlighted by Zangle et al. in their work. Thus, still a clear and complete understanding of the CP regimes is yet to be reached and continuous efforts are being made to understand the physics at the micro/nanofluidic junctions to design advanced and novel devices.

Applications

In this section, various applications that have been developed utilizing the concentration polarization phenomenon are addressed. The applications range from preconcentrating biomolecules to fluid pumping and mixing and also in water desalination. A brief discussion of the aforementioned applications is presented below.

Preconcentration

Wang et al. [15] used the depletion region observed at the micro/nanojunction to preconcentrate proteins. The energy barrier created at the depletion region (due to the large voltage drop induced at this junction) prevents the entry of charged molecules into the nanochannel. This results in an increase in the concentration of the molecules near the depletion region. In their experiments, Wang et al. used two anodic microchannels which were independently controlled so that the direction of EOF can be aligned perpendicular to the axis of the nanopore. An increase of about 10^6–10^8 fold in the concentration of the protein was reported in their study. Over the past couple of years, similar preconcentration devices utilizing the CP effects were experimentally fabricated [3, 14]. Wang et al. [16] also presented an experimental approach to improve the binding kinetics and the immunoassay detection sensitivity using concentration polarization in micro/nanofluidic devices. The antigens were preconcentrated at the depletion region due to the strong electric field gradients resulting in the enhancement in the binding rates with the antibody beads.

Seawater Desalination

The phenomenon of concentration polarization witnessed in ion-selective membranes was successfully implemented to address the freshwater shortage issue by providing energy-efficient solution to water desalination. A microfluidic device was fabricated which provides 99% salt rejection at 50% recovery rate at a power consumption of less than 3.5Wh/L [17]. The CP depletion layer acts as a barrier for any charged species and these species were diverted away from the desalted water using suitable pressure and voltage fields. Their design also ensured salt ions and other debris to be driven away from the membrane thereby preventing any membrane fouling which is often observed in other desalination techniques.

Mixing, Pumping, and Other Applications

As discussed earlier, the induced space charges observed at the depletion region of micro/nanochannel along with the large electric field gradients at this region result in strong vortices. Kim et al. [18] enhanced the mixing efficiency of microfluidic devices using the vortices created at this interface. Further, as the electroosmotic flow of the second kind observed at the depletion region is directly proportional to the square of the applied electric field, Kim et al. [19] was able to pump fluids using the nonequilibrium EOF and observed a fivefold increase in the volumetric flow rates compared to similar devices utilizing equilibrium EOF. Yossifon et al. [20] used an asymmetric microchannel in conjunction with the nanochannels. The application of forward and reverse bias (at the overlimiting regime) voltage led to an asymmetric space charge polarization which resulted in the rectification of the current. Such membranes have potential applications in selective species separation.

Summary

The origin and the underlying physics that is present at the interfaces of micro/nanofluidic devices were discussed. The complex phenomenon of concentration polarization (CP) leading to the enrichment and depletion of ions near the micro–nano junctions and the concepts of induced space charges and nonlinear electrokinetic transport were briefly discussed in this chapter. Further, the various controversies surrounding the physical mechanism for nonlinear current characteristics have been highlighted. The criteria for concentration polarization propagation and the various applications that have been developed utilizing the concentration polarization phenomenon were discussed. Though, a lot of extensive work has been carried out to understand the CP physics, a clear and complete understanding of the CP regimes and the induced space charge dynamics is yet to be reached and efforts need to be directed in this area to understand the physics at the micro/nanofluidic junctions to design novel devices.

Cross-References

▶ Computational Micro/Nanofluidics: Unifier of Physical and Natural Sciences and Engineering
▶ Electrokinetic Fluid Flow in Nanostructures
▶ Integration of Nanostructures within Microfluidic Devices
▶ Surface-Modified Microfluidics and Nanofluidics

References

1. Karniadakis, G.E., Beskok, A., Aluru, N.R.: Microflows and Nanoflows: Fundamentals and Simulation. Springer, New York (2005)
2. Schoch, R.B., Han, J., Renaud, P.: Transport phenomena in nanofluidics. Rev. Mod. Phys. **80**, 839–883 (2008)
3. Höltzel, A., Tallarek, U.: Ionic conductance of nanopores in microscale analysis systems: where microfluidics meets nanofluidics. J. Sep. Sci. **30**, 1398–1419 (2007)
4. Gross, R.J., Osterle, J.F.: Membrane transport characteristics of ultrafine capillaries. J. Chem. Phys. **49**, 228–234 (1968)
5. Probstein, R.F.: Physiochemical Hydrodynamics: An Introduction. Wiley, New York (1994)
6. Rubinstein, I.: Electrodiffusion of Ions. SIAM, Philadelphia (1990)
7. Rubinstein, I., Zaltzman, B.: Electro–osmotic slip of the second kind and instability in concentration polarization at electrodialysis membranes. Math. Models Meth. Appl. Sci. **11**, 263–300 (2001)
8. Pu, Q., Yun, J., Temkin, H., Liu, S.: Ion–enrichment and ion–depletion effect of nanochannel structures. Nano Lett. **4**, 1099–1103 (2004)
9. Jin, X., Joseph, S., Gatimu, E.N., Bohn, P.W., Aluru, N.R.: Induced electrokinetic transport in micro – nanofluidic interconnect devices. Langmuir **23**, 13209–13222 (2007)
10. Kim, S.J., Wang, Y.-C., Lee, J.H., Jang, H., Han, J.: Concentration polarization and nonlinear electrokinetic flow near a nanofluidic channel. Phys. Rev. Lett. **99**, 044501 (2007)
11. Rubinstein, I., Zaltzman, B.: Dynamics of extended space charge in concentration polarization. Phys. Rev. E. **81**, 061502 (2010)
12. Maletzki, F., Rösler, H.-W., Staude, E.: Ion transfer across electrodialysis membranes in the overlimiting current range: stationary voltage current characteristics and current noise power spectra under different conditions of free convection. J. Membr. Sci. **71**, 105–116 (1992)
13. Dukhin, S.S.: Electrokinetic phenomena of the second kind and their applications. Adv. Colloid Interface Sci. **35**, 173–196 (1991)
14. Zangle, T.A., Mani, A., Santiago, J.G.: Theory and experiments of concentration polarization and ion focusing at microchannel and nanochannel interfaces. Chem. Soc. Rev. **39**, 1014–1035 (2010)
15. Wang, Y.-C., Stevens, A.L., Han, J.: Million–fold preconcentration of proteins and peptides by nanofluidic filter. Anal. Chem. **77**, 4293–4299 (2005)
16. Wang, Y.C., Han, J.: Pre-binding dynamic range and sensitivity enhancement for immuno-sensors using nanofluidic preconcentrator. Lab Chip **8**, 392–394 (2008)
17. Kim, S.J., Ko, S.H., Kang, K.H., Han, J.: Direct seawater desalination by ion concentration polarization. Nat. Nanotechnol. **5**, 297–301 (2010)
18. Kim, D., Raj, A., Zhu, L., Masel, R.I., Shannon, M.A.: Non-equilibrium electrokinetic micro/nano fluidic mixer. Lab Chip **8**, 625–628 (2008)
19. Kim, S.J., Li, L.D., Han, J.: Amplified electrokinetic response by concentration polarization near nanofluidic channel. Langmuir **25**, 7759–7765 (2009)
20. Yossifon, G., Chang, Y.-C., Chang, H.-C.: Rectification, gating voltage and interchannel communication of nanoslot arrays due to asymmetric entrance space charge polarization. Phys. Rev. Lett. **103**, 154502 (2009)

Conductance Injection

▶ Dynamic Clamp

Conduction Mechanisms in Organic Semiconductors

Weicong Li and Harry Kwok
Department of Electrical and Computer Engineering, University of Victoria, Victoria, Canada

Definition

Conduction mechanisms in organic semiconductors refer to the means by which electronic charges move through organic semiconductors under external stress particularly under the influence of an electrical field.

Overview

In order to understand the conduction mechanisms in organic semiconductors, it is necessary to first introduce the concept of *band theory*, which is well established in solid-state physics. Solids in general

Conduction Mechanisms in Organic Semiconductors, Fig. 1 Band structures of metal, semiconductor, and insulator

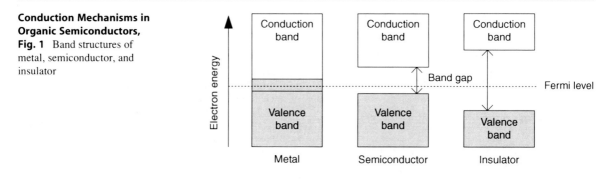

are made of atoms, each of which is composed of a positively charged nucleus surrounded by negatively charged electrons. In quantum mechanical terms, these electrons effectively reside in discrete energy states in orbits. When a large number of atoms (of order 10^{20} or more) are brought together to form a solid, the discrete energy states are so close together that energy bands begin to form. At the same time, there will be gaps between the energy bands which are known as the *band gaps*. Because of the presence of these energy gaps, there will be some energy bands that are almost fully occupied (known as the *valence bands*) and energy bands that are almost unoccupied (known as the *conduction bands*). Based on the band theory, solids are typically divided into the following three categories: *metals*, *semiconductors*, and *insulators*. In metals, there is an overlap between the energy bands so that the energy bands are partly filled by electrons at any temperature (even $T = 0$ K), while both the semiconductors and the insulators have fully filled valence bands and empty conduction bands at $T = 0$ K. To further study the distribution of electrons occupying the energy states in solids, it is also necessary to introduce the concept of Fermi level into the band theory, which represents the maximum energy of states that electrons can occupy at $T = 0$ K. Accordingly, all the allowed energy states below the Fermi level are occupied by electrons, and all the energy states above it are empty. When temperature is above 0 K, the probability that electrons occupy the state with energy E under thermodynamic equilibrium condition is given by Fermi-Dirac distribution function:

$$f(E) = \frac{1}{1 + \exp[(E - E_F)/kT]} \quad (1)$$

where E_F is the Fermi level, k is the Boltzmann constant, and T is the temperature in Kelvin. As mentioned earlier, in semiconductor and insulator, the valence band is fully occupied by electrons, and the conduction band is empty at $T = 0$ K. Therefore, one can infer that the Fermi level lies in the bandgap, between the valance and conduction bands. On the other hand, in metal, due to the fact that the energy bands are partly filled by electrons at any temperature, the Fermi level lies within the energy bands. The band structures of metal, semiconductor, and insulator, and the position of Fermi level in them are shown in Fig. 1. The distinction between the semiconductors and the insulators appears when temperature rises above 0 K. Because the band gap between conduction and valence bands in semiconductors is much narrower than that found in insulators, a fair amount of electrons can be thermally excited from the valence band to the conduction band in semiconductors at finite temperature, leading to measurable conductivity. This is not found in the insulators due to the larger band gaps even at room temperature, which lead to negligible probability of electrons occupying energy states in the conduction band, according to Eq. 1. The characteristic semiconductor band structure has allowed it to play an important role as the materials of choice in the prosperous electronic industry in the last few decades.

While inorganic semiconductors such as silicon dominated the electronic industry in the twentieth century, tremendous effort has been spent in the research and development of organic electronics in last decade due to the fact that organic semiconductors are usually easier and cheaper to form. Soluble organic materials, such as conjugated polymers, can be deposited in liquid phase (e.g., by printing and spin coating) onto large substrate areas at low processing temperature (below 100°C). Due to this advantage, organic electronics are particularly attractive in the making of displays, sensors, light sources, photovoltaic panels, radiofrequency identification detectors (RFID), and in devices used in optical communications. As a consequence, research on the charge conduction mechanisms in organic semiconductors and devices is of significant importance.

In general, conduction mechanisms primarily describe how electronic charges (referred to as carriers) move inside the solids under the influence of an external electrical field. The process produces a current. At the macroscopic level, the current density J in solids produced by external electrical field is given by

$$J = env = en\mu F \qquad (2)$$

where e is elementary charge of a single carrier, n the charge density, and v is the drift velocity. Furthermore, v can be expressed as the product of the charge mobility μ and the electrical field F. As can be seen in Eq. 2, a large current requires the presence of a substantial number of mobile charge carriers (electrons or holes). In organic semiconductors, mobile carriers are known to be produced from the distributed π-bonds, which are covalent chemical bonds resulting from the overlap of atomic orbitals. Thus, the limited current flow in many organic semiconductors are related to their irregular molecular structures which can result in low charge mobility in comparison to values found in silicon and other inorganic semiconductors. In addition, the more established conduction mechanisms based on band theory normally found in inorganic crystalline semiconductors are absent in the organic semiconductors. As mentioned earlier, band theory states that carriers could only exist and move in either the conduction bands or the valence bands because there are permitted energy states where carriers can reside and their movement between the energy states will produce a current. The use of "energy band" diagrams (see Fig. 1) to explain charge transport in organic semiconductors however is not possible. This is because of the presence of high densities of defects and trap states. Instead, charge transport in organic semiconductors is directly explained in terms of the energy (orbital) states which are termed either the *lowest unoccupied molecular orbital* (LUMO) or the *highest occupied molecular orbital* (HOMO). As such, LUMO and HOMO levels are not genuine energy bands and they are used merely to serve as references to demarcate ground state energy and the next activated state energy [1]. The distribution of these energy states known as the *density of states* (*DOS*) is usually considered to be

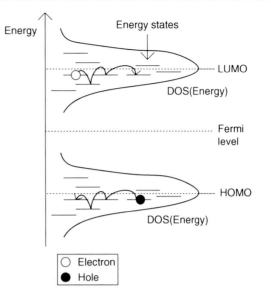

Conduction Mechanisms in Organic Semiconductors, Fig. 2 Density of (energy) states in an organic semiconductor

Gaussian centered at the LUMO and HOMO (see Fig. 2). The width of the Gaussian energy states depends on both the regularity of the molecular structure and the impurities present in the organic semiconductor.

As expected, in most organic semiconductors both the carrier density n and the charge mobility μ are low and the value of the latter often depends on the strength of the electrical field F in contrast to what is observed in inorganic semiconductors. In some organic semiconductors, the molecular structures can be highly disordered and different conduction mechanisms are found to predominate depending on the associated manufacturing process.

Basic Methodology

Many useful techniques have been proposed to study the conduction mechanisms in organic semiconductors including time-of-flight (TOF) experiment, space charge limited current (SCLC) measurement, and field-effect measurements using organic field-effect transistors (OFETs). These techniques, combined with the dependence on temperature, provide important information on mobility, trap concentration which are useful to assess the conduction mechanisms. Brief introductions to several different techniques commonly used are given here.

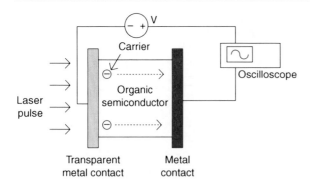

Conduction Mechanisms in Organic Semiconductors, Fig. 3 A schematic of the time-of-flight (TOF) experiment setup

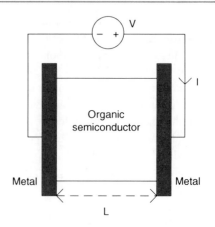

Conduction Mechanisms in Organic Semiconductors, Fig. 4 A simplified schematic of the setup used for space charge limited current (SCLC) measurement

Time-of-Flight (TOF) Experiment

As implied by its name, time-of-flight (TOF) experiment is the method of measuring the time it takes for one or a few carrier to travel a distance through the solid. When TOF experiment is used to study the conduction mechanisms in organic semiconductors (see Fig. 3), two metal electrodes (forming the contact) are deposited on the two ends of the organic semiconductor (one of the two is usually transparent). Initially, a few carriers are generated at one end near the transparent metal electrode using a short laser pulse with energy greater than the energy difference between HOMO and LUMO levels of organic semiconductor. The photo-generated carriers are then drifted toward the opposite end by an external electrical field generating a current pulse. By measuring the time delay of the current pulse, the velocity and the mobility of the charge carriers through the organic semiconductors can be computed.

Space Charge Limited Current (SCLC) Measurement

Due to the low mobility of carriers in organic semiconductors, the measured I–V characteristics usually deviate from Ohm's law (i.e., the linear relationship between current and voltage). This is illustrated in Fig. 4. In this case, if efficient charge injection from the metal electrode is achieved by choosing a suitable metal, the I–V characteristics will follow the space charge limited current equation as given by

$$I = \frac{9}{8}\theta\varepsilon_0\varepsilon_r\mu A \frac{V^2}{L^3} \qquad (3)$$

where θ is a parameter dependent on the traps present in the semiconductors, ε_0 the free space permittivity, ε_r the relative dielectric constant, μ drift mobility of injected charge carrier, A the cross section area of semiconductor, and L the distance between metal contacts. By measuring the I–V characteristics at different temperatures, one can determine the mobility and the trap density of traps. Thus, SCLC measurement is a very useful technique reflecting the conduction mechanisms in organic semiconductors.

Measurement Based on the Organic Field-Effect Transistor (OFET)

Field-effect transistor (FET) is an electronic device widely used in active circuits. It consists of a semiconductor with a conducting channel, an isolated gate controlling charge flow in the channel, a gate dielectric between the semiconductor and the gate, as well as source and drain regions forming the output terminals. Organic field-effect transistor (OFET) is a field-effect transistor formed on an organic semiconductor. The device configuration can have a top-gate or a bottom-gate as shown in Fig. 5.

The basic operation principle of the OFETs is very simple. When a bias voltage is applied between gate and source electrodes, carriers are injected from the source into the organic semiconductor forming an extremely thin accumulation layer (2 ~ 3 nm) at the interface between organic semiconductor and the dielectric. The carriers conduct a current across the source and the drain regions, and the current depends on the gate voltage as well as the charge mobility

Conduction Mechanisms in Organic Semiconductors, Fig. 5 Two device configurations for the OFETs: (a) top-gate, (b) bottom-gate

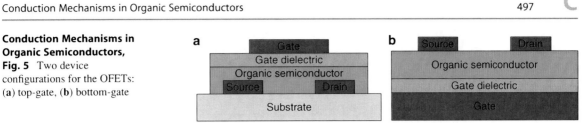

which is also dependent on the gate voltage and the drain-to-source voltage. The operation of the OFETs therefore relies on carrier accumulation in the field-effect structure in contrast to the case of the inorganic FETs which rely on either charge depletion or inversion. Therefore, OFETs are efficient tools to investigate the interfacial conduction mechanisms, while TOF and SCLC measurements are mainly used to study the bulk conduction mechanisms in the organic semiconductors. For example, *I–V* characteristics of OFETs are usually analyzed to determine parameters such as the charge mobility and the threshold voltage both of which are closed related to the density of traps at the interface. In addition, spectroscopic techniques are sometimes used to probe the morphology of the organic semiconductor interface, to look for the potential relationship between regularity of molecular structure and conduction performance.

Key Research Findings

Band-Like Transport

For highly purified and ordered organic molecular crystals, it is possible that band-like charge transport similar to that of the inorganic semiconductors may occur. The main feature found in band-like charge transport is the fact that the temperature dependence of the charge mobility has the following form:

$$\mu(T) \propto T^{-n}, \text{with } n = 1, 2, 3 \ldots \qquad (4)$$

In practice, n is usually positive which leads to increasing charge mobility when temperature decreases. In general, because the electrons are usually weakly delocalized even in the highly ordered organic crystals, the band widths of the HOMO and the LUMO are small compared to energy bands found in the inorganic semiconductors. As a result, room temperature charge mobilities observed in organic semiconductor crystals can only reach values in the range $1–20 \text{ cm}^2/\text{Vs}$ [2]. Band-like charge transport has been observed

in small-molecules and in single-crystal organic semiconductors (such as rubrene) formed by vapor deposition process. As a matter of fact, in the majority of organic semiconductors, traps and defects are formed during deposition which tends to destroy band-like properties.

Polaron Transport

A polaron is a quasiparticle composed of a charge carrier and its induced polarization field. In many organic materials, due to the low charge mobilities, carriers tend to polarize their surrounding lattice. As a result, polarization fields are formed around the carriers, which can no longer be considered as "naked." Instead, the carriers will be localized in potential minima created by the so-called molecular deformations [1]. In other words, a charge is trapped by the deformation it induces. Such an entity is known as a "polaron." Polarons can move between molecules similar to the carriers except that they also carry the deformations along. In many disordered molecular organic semiconductors, deformations associated with the trapped charges can be considerable and conduction mechanisms characterizing polaron transport are under intensive research by many research groups across the world.

Variable Range Hopping (VRH) Transport

For most organic semiconductors such as polymers and oligomers, their molecular structures are highly disordered and have considerable densities of defects and traps. The energy band diagrams are no longer suitable to describe the densities of states as these energy states are now localized. Furthermore, band-like charge transport can no longer explain the observed low charge mobilities and the fact that their values increase with temperature (as opposed to what is observed in band-like charge transport). One of the widely accepted theories, known as the variable range hopping (VRH) transport, is proved to give a reasonable explanation by describing charge transport in terms of hopping of the charge carriers between localized states as shown in Fig. 6.

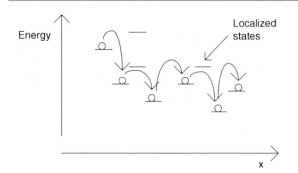

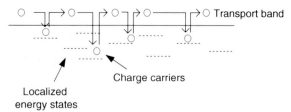

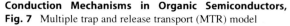

Conduction Mechanisms in Organic Semiconductors, Fig. 7 Multiple trap and release transport (MTR) model

Conduction Mechanisms in Organic Semiconductors, Fig. 6 Hopping transport in organic semiconductors

Hopping can be used to explain the lower mobility found in disordered organic semiconductors and instead of the power law dependence on temperature as in band-like charge transport, the temperature dependence in VRH charge transport exhibits temperature-dependent activation as well as dependence on the applied electric field as given by [3]

$$\mu(F, T) \propto \exp(-\Delta E/kT) * \exp(\beta\sqrt{F}/kT) \quad (5)$$

where μ is mobility, F is the electrical field, T is the temperature, ΔE is the activation energy, and β is a parameter related to disorder.

Multiple Trap and Release (MTR) Transport

Charge transport in OFETs is affected by defects and impurities which exist in the intrinsic part of the organic semiconductors per se and can also be linked to an inferior semiconductor/dielectric interface. As a result, the performance of OFETs is sample-dependent, which is one of the major difficulties in characterizing the properties of OFETs. As mentioned earlier, VRH transport is more suitable to account for charge transport in highly disordered organic semiconductors and, in contrast, another well-established and widely accepted charge transport model known as the "multiple trap and release" (MTR) model is frequently applied to the relatively well-ordered organic semiconductors, such as small molecules and molecular crystals. The basic principle of MTR model includes two important components: (1) a transport band containing delocalized energy states whereby carriers can move freely and (2) the presence of a high density of localized energy states located in the vicinity of the edge of transport band acting as traps. During charge transport, carriers move freely in the transport band

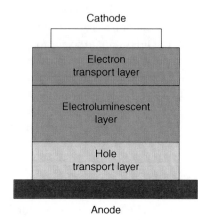

Conduction Mechanisms in Organic Semiconductors, Fig. 8 Schematic of a typical organic light-emitting diode (OLED)

with a high probability of being trapped at the localized energy states and then subsequently thermally released into transport band again. The basic illustration of MTR transport process is shown in Fig. 7.

The effective mobility (μ_{eff}) in the MTR model is actually smaller than the "real" mobility (μ_0) in the transport band in the absence of localized energy states and is given by [4]

$$\mu_{eff} = \mu_0\alpha\exp(-E_t/kT) \quad (6)$$

where α is the ratio of the effective density of energy states at the edge of the transport band to the density of traps in the localized energy states, and E_t is the energetic distance between the edge of the transport band and the localized energy states. If the localized energy states are energetically dispersive, α and E_t must be recalculated according to the trap distribution.

In the study of OFETs, the MTR model is widely used to account for charge transport due to the fact that it offers a reasonable explanation on the gate voltage

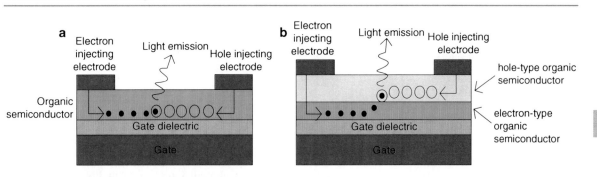

Conduction Mechanisms in Organic Semiconductors, Fig. 9 Schematic illustrations of organic light-emitting field-effect transistors (OLEFETs) with light emission in: (**a**) the single-layer configuration and (**b**) the multilayer configuration

dependent mobility usually observed in OFETs. As mentioned above, unlike inorganic semiconductors, organic semiconductors usually have Gaussian density of states (DOS). When a bias is applied to the gate of an OFET, the Fermi level at the dielectric-semiconductor interface will be shifted toward the transport band so that a fair amount of localized energy states near the edge of the transport band will be filled when the Fermi level is located closer to the transport band. As a result, the mobility of the carriers in the MTR model is actually improved because of the reduced density of traps leading to a reduced value of E_t. Therefore, the gate-voltage dependent effective mobility of the carriers in the MTR transport model is given by [5]

$$\mu_{eff} = \mu_0 \frac{N_c}{N_{t0}} \left[\frac{C_i(V_G - V_T)}{qN_{t0}} \right]^{\frac{T_0}{T} - 1} \quad (7)$$

where μ_0 is the mobility of the carriers in the transport band, N_c the effective density of states at the edge of the transport band, N_{t0} the total density of traps, C_i the capacitance of the insulator per unit area, and T_0 is a characteristic temperature related to the distribution of the DOS.

Devices

Organic Light-Emitting Diode (OLED)

Organic light-emitting diode (OLED) is an electroluminescent diode composed of organic materials serving as the electroluminescent layer and charge transport layer. The typical OLED structure is shown in Fig. 8. During operation of an OLED, a bias voltage is applied between anode and cathode. Holes

(electrons) are injected from the anode (cathode) into the electroluminescent layer through the hole (electron) transport layer. Because electrons and holes exist simultaneously in the same layer, there is a high probability that they recombine with each other due to electrostatic forces, leading to radiative emission. Therefore, efficient charge injection from both cathode and anode is requisite for the efficient operation of the OLED and current conduction is dominated by space charge limited current as introduced earlier.

Organic Light-Emitting Field-Effect Transistor (OLEFET)

Organic light-emitting field-effect transistor (OLEFET) is a novel organic device combining the function of current conduction of an OFET with electroluminescence in an OLED. The operation of the OLEFET is actually the same as that of the OFET. However, if proper materials are chosen as the source and the drain to give efficient charge injection and under favorable voltage bias condition, electrons and holes can be injected and transported separately in the OFET channel(s). This type of charge transport is known as ambipolar charge transport, which is unique and only found in an organic field-effect transistor. Furthermore, in ambipolar charge transport if the electrons and holes are allowed to recombine radiatively in an emitter layer to give out light, this type of OFET with electroluminescence functionality is usually called OLEFET. Various device structures have been proposed to realize ambipolar charge transport and light emission in OLEFETs, and, in most cases, the proposed structures fall into two main categories as far as the charge layers are concerned. These are the single-layer OLEFET and multilayer OLEFET as shown in Fig. 9.

Cross-References

▶ Electrode–Organic Interface Physics
▶ Flexible Electronics
▶ Optical and Electronic Properties
▶ Surface Electronic Structure

References

1. Kwok, H.L., Wu, Y.L., Sun, T.P.: Charge transport and optical effects in disordered organic semiconductors. In: Noginov, M.A., Dewar, G., McCall, M.W., Zheludev, N.I. (eds.) Tutorials in Complex Photonic Media, pp. 576–577. SPIE Press, Bellingham (2009)
2. Podzorov, V., Menard, E., Borissov, A., Kiryukhin, V., Rogers, J.A., Gershenson, M.E.: Intrinsic charge transport on the surface of organic semiconductors. Physical Review Letters **93**, 086602 (2004)
3. Brütting, W.: Physics of Organic Semiconductors. Wiley, Weinheim (2005)
4. Horowitz, G.: Organic field-effect transistors. Adv. Mat. **10**, 365–377 (1998)
5. Bao, Z., Locklin, J.: Organic Field-Effect Transistors. CRC Press, Boca Raton (2007)

Confocal Laser Scanning Microscopy

Reinhold Wannemacher
Madrid Institute for Advanced Studies, IMDEA
Nanociencia, Madrid, Spain

Synonyms

Confocal scanning optical microscopy (CSOM); Laser scanning confocal microscopy

Definition

A Confocal Laser Scanning Microscope (CLSM) images a point light source used for excitation onto the sample via the objective lens and images the excited focal volume onto a point detector using reflected, transmitted, emitted, or scattered light. In contrast to conventional microscopes, this scheme permits strong rejection of out-of-focus light and optical sectioning of the sample. In order to obtain an image, the focal volume must be scanned relative to the sample. Scanning can be performed in the lateral as well as in the axial directions and three-dimensional images of the sample can be generated in this way.

Operating Principle

Figure 1 illustrates the working principle of a fluorescence confocal microscope. A point light source, here the end of an optical fiber carrying the excitation light, is imaged onto the sample by the objective lens via a beam splitter. Emitted light from the focal spot is imaged through the beam splitter onto a pinhole via an auxiliary lens. The light is registered by the detector, because it passes through the pinhole (left-hand side of Fig. 1). Fluorescence from a fluorophore at an out-of-focus position in the excitation cone, on the other hand, arrives at the screen defocused. Therefore, only a small fraction of it passes through the pinhole and reaches the detector. The effect is understated by the schematic figure and actually much stronger in a real microscope because of the short focal length of the objective lens. The optical sectioning capability is the core of confocal microscopy and allows to render the object three-dimensionally under different angles by appropriate software once a stack of images at different depths has been acquired. In addition, the lateral resolution is slightly improved in confocal microscopy, compared to conventional microscopy, when the pinhole is small. On the other hand, an image can be acquired in this way only by serial scanning of the sample or of the excitation beam (for technical improvements in this respect see section Confocal Microscopy Involving Modified Illumination).

Because a CLSM operates with light, it may be used to image many different physical quantities. These may be simply reflected or transmitted intensity (brightfield confocal microscopy) or the intensity of fluorescence excited in the sample (fluorescence confocal microscopy). Other options include polarization and phase of reflected or transmitted light, as well as the intensity, wavelength, lifetime, time correlation, or recovery after photobleaching of fluorescence from the sample, or intensity, wavelength, and polarization of inelastically (Raman) scattered light. Moreover, a nonlinear response of the sample to the optical excitation near the laser focus, based, for example, on multiphoton excitation, second harmonic generation, or stimulated scattering processes may be used for

Confocal Laser Scanning Microscopy,
Fig. 1 Principle of confocal laser scanning microscopy, demonstrating the strong rejection of out-of-focus light

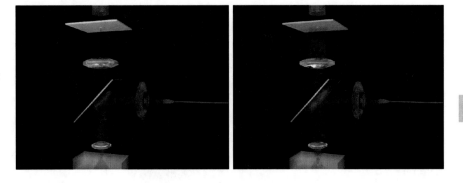

confocal microscopy. Some of these options will be discussed in section Variants of Confocal Laser Scanning Microscopy

Optical sectioning and three-dimensional image acquisition being the essential feature of confocal optical microscopy, it is worth mentioning here that an alternative (diffraction-limited) brightfield optical microscopy technique with similar capability is digital holographic microscopy, although this technique is far less widely known and used. Here, the object is reconstructed from an intensity camera image and no scanning is necessary. Moreover, a phase image is obtained in addition to an amplitude image. Significant improvements in object reconstruction algorithms have been made in recent years and lens-based versions with external reference beam as well as lensless versions have been demonstrated. Lens-based instruments with external reference beam are commercially available.

Basic Theory of the Confocal Microscope

The spatial resolution of modern high-quality conventional, as well as confocal microscopes is limited by diffraction. This means that within the design spectral range of the objective lens aberrations, such as spherical aberration, astigmatism, coma, field curvature, distortion, and chromatic aberration have a significantly smaller impact on the resolution of the microscope than diffraction. An important exception to this statement arises from aberrations due to refraction, when sample regions well inside refracting samples have to be imaged.

Most modern microscope objectives are now infinity-corrected, that means they are corrected for forming an image at infinity. A tube lens is in this case required to form a real image at finite distance.

The limitations by diffraction are, however, in any case dominated by the objective lens and not by the tube lens. This is due to the dependence of the diffraction limit on the opening angle of the rays contributing to the image, which is much smaller for the tube lens than for the objective lens.

Point Spread Function of the Confocal Microscope
The three-dimensional intensity distribution in the image space corresponding to a single-point object, demagnified by the magnification of the optical system, is called the (intensity) point spread function (PSF) of the lens. The PSF of a confocal microscope with an infinitesimally small pinhole is given by [1, 2]:

$$PS_{CF}(x, y, z) = PSF_{ill}(x, y, z) \cdot PSF_{det}(x, y, z) \quad (1)$$

Here, PS_{CF}, PSF_{ill}, and PSF_{det} represent the point spread functions for the confocal imaging and the illumination and detection paths respectively. Neglecting the contribution from the tube lens, as well as aberrations of the objective lens, the latter two functions are simply the point spread functions of a simple lens, which, in the paraxial and scalar approximation, can be calculated by means of the Huygens-Fresnel principle as [3]

$$PSF(x, y, z) = |h(x, y, z)|^2 \quad (2)$$

with

$$h(\vec{r}) = \frac{C'}{\lambda} \int \int_A \frac{e^{iks}}{s} dA \approx \frac{C}{\lambda} \int \int_\Omega e^{-ik\vec{q}\cdot\vec{r}} d\Omega \quad (3)$$

Here, $h(\vec{r})$ represents the scalar complex amplitude of the field in the image space at a position $\vec{r} = (x, y, z)$

Confocal Laser Scanning Microscopy, Fig. 2 Intensity point spread function of a conventional (**a**) and a confocal (**b**) microscope in the scalar and paraxial approximations. The numerical aperture of the objective lens is N.A.=0.5. z is the coordinate along the axis of the lens and r the lateral dimension, both measured in wavelengths

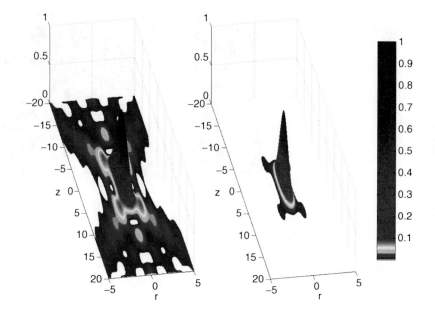

relative to the location of the geometric focus, λ is the wavelength, $k = 2\pi/\lambda$, s the distance between the point P at position $\bar{r}$ in the image space and a point Q at position $f\bar{q}$ in the pupil A of the lens of focal length f, $\bar{q}$ is a unit vector in the direction of Q, Ω is the solid angle subtended by the aperture of the lens as seen from the origin at the geometric focus, and C and C' are constants. Equation 3 assumes homogeneous illumination of the lens. Inhomogeneous illumination can be taken into account by multiplying the integrand with a corresponding pupil function (compare section Apodization). For a confocal microscope operating in reflection or transmission modes $\mathrm{PSF}_{\mathrm{ill}}(x, y, z) \approx \mathrm{PSF}_{\mathrm{det}}(x, y, z)$ and both functions are then identical to the one given in Eq. 2. This results in

$$\mathrm{PS}_{\mathrm{CF}}(x, y, z) = |h(x, y, z)|^4 \qquad (4)$$

Experimentally, this function would be observed when a point object is scanned through the focus of the instrument in both cases. It should be kept in mind here that the paraxial approximation is contrary to the actual typical situation in optical microscopy. The paraxial approximation, nevertheless, works surprisingly well even for N.A. ≈ 0.5 ($\theta_0 = 30°$ in the case of a dry lens, see below) and gives reasonable estimates of the diffraction limit even for higher numerical aperture objective lenses. Deficiencies of the scalar approximation will be discussed in section

Deficiencies of the Scalar and Paraxial Approximation. Effects of the Vector Character of Light

Figure 2 displays the three-dimensional intensity point spread function of a conventional (a) and a confocal (b) microscope calculated according to Eqs. 1 and 3. The numerical aperture of the objective lens is N.A. = 0.5. z is the coordinate along the axis of the lens and r the coordinate perpendicular to the optical axis, measured in wavelengths. The drastic reduction in side lobes for a confocal microscope is immediately evident. This also leads to drastic reduction in laser speckle in the case of coherent illumination.

Single-Point Resolution in the Focal Plane
In the focal plane ($z = 0$) the PSF of the conventional microscope, as calculated from Eq. 3, coincides with the well-known Airy pattern

$$\mathrm{PSF} = \left(\frac{2J_1(\mathrm{v})}{\mathrm{v}}\right)^2 \qquad (5)$$

with

$$\mathrm{v} = kr \cdot \mathrm{n} \sin \theta_0 = kr \cdot \mathrm{N.A.} \qquad (5a)$$

Here, r is the distance from the axis, θ_0 is the angle of a marginal ray passing through the aperture toward the geometric focus, relative to the optical axis, and n is the refractive index in the image space. From this expression, the lateral full width at half maximum (FWHM) of the PSF of the conventional microscope

$$\text{FWHM} = \frac{0.51\lambda}{\text{N.A.}} \quad (5b)$$

is easily calculated. Equation 4 yields

$$\text{PSF}_{\text{CF}} = \left(\frac{2J_1(v)}{v}\right)^4 \quad (5c)$$

for the PSF of the confocal microscope in reflection mode and, therefore

$$\text{FWHM}_{\text{CF}} = \frac{0.37\lambda}{\text{N.A.}} \quad (5d)$$

Equations 5b and 5d demonstrate the enhancement in lateral (single-point) resolution for a confocal microscope relative to a conventional one, in the case when the pinhole is closed completely. As an example, for $\lambda = 488$ nm, N.A. $= 0.9$, FWHM $= 277$ nm and $\text{FWHM}_{\text{CF}} = 201$ nm.

Single-Point Resolution on the Axis: Depth Response
Similarly, the PSF of the conventional microscope on the optical axis is calculated from Eq. 3 as

$$\text{PSF}(u) = \left(\frac{\sin(u/4)}{u/4}\right)^2 \quad (6)$$

with

$$u = nkz\sin^2\theta_0 \quad (6a)$$

It turns out that the definition

$$u = 4kz\sin^2(\theta_0/2) = 2nkz(1 - \cos\theta_0) \quad (6b)$$

which is equivalent to Eq. 6a in the paraxial approximation, is more appropriate at higher numerical apertures. The corresponding FWHM of the depth response is therefore

$$\text{FWHM} = \frac{0.89\lambda}{n(1 - \cos\theta_0)} \quad (6c)$$

for the conventional microscope. Correspondingly, for the confocal microscope

$$\text{PSF}_{\text{CF}}(u) = \left(\frac{\sin(u/4)}{u/4}\right)^4 \quad (6d)$$

and

$$\text{FWHM}_{\text{CF}} = \frac{0.64\lambda}{n(1 - \cos\theta_0)} \quad (6e)$$

For the parameters used in the example above, $\lambda = 488$ nm, N.A. $= 0.9$, FWHM $= 770$ nm and $\text{FWHM}_{\text{CF}} = 554$ nm. The FWHM of the point spread functions on the optical axis of the microscope is therefore about three times larger than the lateral FWHM for the conventional as well the confocal microscope, although the precise value depends on the numerical aperture and although the scalar theory used here is not applicable at large numerical aperture (see section Deficiencies of the Scalar and Paraxial Approximation. Effects of the Vector Character of Light.

V(z)
A clearer demonstration of the different optical sectioning capabilities of the conventional and confocal microscopes is obtained with planar objects instead of point objects. The depth response of a confocal microscope operating in reflection is often characterized by axially scanning a mirror through the focus position and registering the light intensity behind the pinhole during the scan. The corresponding amplitude function is called $V(z)$, an expression coined originally for the acoustic microscope, which is also a confocal instrument and to which the same scalar theory is applicable. Because the image of the illuminating infinitesimal pinhole is moving by a distance of $2z$ when the mirror moves by a distance z, the intensity detected behind the pinhole is derived from Eqs. 6, b, by replacing u by $2u$:

$$I(z) = |V(z)|^2 = \left|\frac{\sin(u/2)}{u/2}\right|^2$$
$$= \left|\frac{\sin(nkz(1 - \cos\theta_0))}{nkz(1 - \cos\theta_0)}\right|^2 \quad (7)$$

This equation predicts a central maximum of width

$$\text{FWHM}_{\text{CF}} = \frac{0.44\lambda}{n(1 - \cos\theta_0)} \quad (8)$$

which, in paraxial approximation $\theta_0 \ll 1$ for a dry lens becomes

$$\text{FWHM}_{\text{CF}} \approx \frac{0.89\lambda}{\text{N.A.}^2} \qquad (9)$$

Equation 7 also implies symmetric side lobes for negative and positive defocus. As an example, for a dry lens of N.A. $= 0.8$ and an operating wavelength of 488 nm Eq. 8 yields FWHM $= 537$ nm. In a conventional microscope, on the other hand, the signal received by a large area detector would be independent of the position of the mirror.

The simple theory presented so far predicts a symmetric $V(z)$ function. Whereas the width of the main peak of the $V(z)$ is typically very close to measurements performed with real lenses, the aberrations present in any real objective lens typically lead to deviations as far as the side maxima are concerned and in particular to asymmetry in $V(z)$ for positive and negative defocus. This may be used for quantitative characterization of, for example, the amount of spherical aberration present in the optical system. Interferometric versions of confocal microscopy, however, have been more traditionally used for this purpose.

Two-Point Resolution: Rayleigh and Sparrow Criteria

The single-point resolution of the confocal microscope, as given by the PSF discussed above, is in most cases not the relevant quantity to judge the resolution of the instrument, because what is really desired is the capability to resolve certain details of a microscopic object consisting of various parts. It is therefore important to quantify the two-point resolution of the instrument, which means the capability to resolve two point objects close to each other. There is some arbitrariness in this definition, because it depends on the subjective judgment, under which conditions two point objects are resolved in an image. Only the lateral two-point resolution will be discussed here.

The Rayleigh criterion defines two-point sources as resolved, if the image of the second point lies at the first zero of the image of the first one or at a larger distance. This leads to a lateral resolution

$$d_{\text{R}} = \frac{0.61\lambda}{\text{N.A.}} \qquad (10)$$

for the conventional microscope and

$$d_{\text{R,CF}} = \frac{0.56\lambda}{\text{N.A.}} \qquad (11)$$

for the confocal microscope, just 8% less than for the conventional microscope. In the case of coherent sources, d_{R} depends on the phase difference of the sources: whereas two out-of-phase coherent sources can be clearly resolved, because there will be a zero of intensity halfway between the images of the two sources, the sources cannot be resolved, if they are in phase with each other, because the maximum will lie in the middle between the images of the two individual sources.

There is another criterion for the two-point resolution, which is more generally applicable, because it does not refer to a zero of the response function. The Sparrow criterion states that two sources are considered to be resolved, when the intensity halfway between the two images is the same as the one at the individual image locations. For incoherent illumination, this results in

$$d_{\text{S}} = \frac{0.51\lambda}{\text{N.A.}} \qquad (12)$$

for *both* the conventional and confocal microscopes.

Coherence in Brightfield and Fluorescence Microscopy

The imaging of extended objects differs significantly for the conventional and confocal microscopes, respectively. For a CLSM operating in brightfield (reflection or transmission) mode the imaging is spatially coherent, because the illumination generates a spatially coherent field distribution in the focus of the objective lens, as given by the complex amplitude point spread function of the objective lens. For a CLSM in reflection mode this field distribution has to be multiplied by the reflectance R of the sample and this weighted field distribution is then imaged onto the pinhole implying convolution with the combined amplitude point spread function of the objective and pinhole relay lenses. Assuming imaging of the object onto the pinhole by the same objective lens that is used for excitation and neglecting contributions to the point spread function from the pinhole relay lens in the second imaging step the intensity behind the infinitesimally small pinhole can therefore be written as the convolution

$$I = \left| \int \int \int h(x,y,z)R(x,y,z)h(-x,-y,-z)\mathrm{d}x\mathrm{d}y\mathrm{d}z \right|^2 \tag{13}$$

Here, $h(x,y,z)$ is the amplitude PSF of the objective lens (compare Eq. 3). For the case of an even PSF Eq. 13 is equivalent to

$$I = \left| h^2 * R \right|^2 \tag{14}$$

that means the signal is given by the absolute square of the convolution of the amplitude PSF of the confocal microscope, $h^2(\vec{r})$, with the local amplitude reflectivity of the sample.

In the case of a conventional microscope, on the other hand, the illumination is approximately incoherent and therefore

$$I = |h|^2 * |R|^2 \tag{15}$$

In reality, for the conventional microscope, imaging is partially coherent, because the emission from each emitting point on the illumination source is imaged, due to diffraction at the condenser aperture, into a finite spatial region, which is occupied by a coherent field due to that point emitter, and the regions in the image space corresponding to neighboring, incoherently emitting points on the source partially overlap on the sample. This is true for critical illumination, where the spatial region would be given by the PSF of the condenser, as well as for Köhler illumination, where the spatial region is the whole illuminated region of the sample.

Fluorescence Confocal Microscopy

Fluorescence imaging is incoherent. Assuming that the fluorescence intensity is proportional to the excitation intensity the signal obtained in a confocal microscope with an infinitesimally small pinhole is

$$I = \left(|h(\lambda)|^2 |h(\beta\lambda)|^2 \right) * f \tag{16}$$

where it has again been assumed that, as in standard commercial confocal microscopes, the same objective lens is used for excitation and fluorescence imaging, respectively. Here, f represents the distribution of fluorescent centers in the sample, λ the excitation wavelength, $\beta\lambda$ the fluorescence wavelength, and β the ratio of both wavelengths, the Stokes ratio.

In the case of several different types of emitters, f would have to be weighted according to the spectral contribution of each emitter to the detected signal, which depends on the filters employed in detection for rejection of the excitation and also on the wavelength-dependent sensitivity of the detector. For a point emitter placed at the focus, the convolution with a δ function just yields the first two terms on the right-hand side of Eq. 16. For a conventional microscope, on the other hand, because the whole sample is illuminated, the single-point resolution is only determined by the intensity PSF of the objective lens at the fluorescence wavelength, and the excitation wavelength is irrelevant.

Effects of Finite Pinhole Size

A finite size of the pinhole is obviously required in order to obtain a measurable signal. This will reduce the lateral resolution as well as the optical sectioning capability. It can be shown [2] that for a single-point object the lateral resolution is almost unaffected by the size of the pinhole, if

$$v_P = \frac{2\pi}{\lambda} \frac{r_P}{M} \sin\alpha \leq 0.5 \tag{17}$$

where r_P is the radius of the pinhole and M the magnification of the lens. The maximum value allowed by Eq. 17 therefore sets a reasonable value for the pinhole size, if optimum lateral resolution and reasonable signal are desired. As an example, in the case of a 100x/0.8N.A. objective lens and $\lambda = 514$ nm a critical diameter of the pinhole of 10.2 μm is calculated from Eq. 17. The depth discrimination, as measured by moving a mirror through focus, on the other hand, is less affected and is essentially unaltered if

$$v_p \leq 2.5 \tag{18}$$

It is obvious from these equations that the pinhole size must be adapted when the objective lens is changed.

Apodization

Equation 3 assumed rectangular apodization, that is, homogeneous illumination of the lens pupil and neglects reflection losses at the lens. For a given objective lens rectangular apodization yields the smallest FWHM of the focus in the focal plane, at the expense of larger side maxima, compared to pupil functions falling off toward the edge of the objective lens. It can be achieved only approximately with Gaussian laser beams and is then equivalent to loss of a large fraction of the power of the excitation beam. Gaussian apodization, on the other hand, increases the FWHM, but reduces the side maxima.

Deconvolution

As described by Eq. 16 the image acquisition process in confocal fluorescence microscopy can be modeled as a convolution of the spatially dependent fluorescence of the sample with a point spread function (PSF) of the imaging system. This is true also for conventional non-confocal fluorescence microscopy. In addition, random noise is superimposed on the image. Deconvolution with the PSF would naturally seem the appropriate way to determine the true fluorescence distribution in both cases. Applied to confocal images the resolution may be improved. In the case of conventional micros-copy optical sectioning and removal of out-of-focus blur may be achieved by post-processing instead of employing hardware in the optical setup.

In general, the 3-D PSF, necessary for this proce-dure, can be obtained experimentally or analytically. In the experimental methods, images of one or more point-like objects are collected. The problem with this technique lies in the poor signal-to-noise ratio obtain-able with very small objects and the fact that the PSF may vary depending on the sample. In analytical cal-culations of the PSF aberrations of the optical system are often partially taken into account, whereas, on the other hand, the scalar approximation is most often used and the effects of the vector character of light (compare section Deficiencies of the Scalar and Paraxial Approximation: Effects of the Vector Char-acter of Light) are neglected.

Many 3D deconvolution methods are currently employed and some are available in commercial and non-commercial software packages [4]. The simplest class are *neighboring* methods, in which out-of-focus blur is removed by subtraction of neighboring (filtered) images within a stack. This method does not suffi-ciently remove noise. In contrast to that *linear* methods apply deconvolution to the whole stack of images at once. Examples are inverse filtering, Wiener filtering, the linear least squares, and the Tikhonov filtering techniques. The last three methods do not restore high frequency object components beyond the bandwidth of the PSF and inverse filtering suffers from noise amplification. All methods are very sensi-tive to error in the PSF. Therefore constrained iterative nonlinear algorithms are often employed, in which, starting from a guess for the true object, an error is minimized under certain constraints (positiveness of the image, finite support of the sample, etc.). In cases of strong noise in the image *statistical* iterative methods (like the maximum likelihood method) are favored. A computational alternative are *blind deconvolution* methods, in which the PSF of the optical system and the "true object" are simultaneously deter-mined in a converging iteration. These methods are, however, computationally demanding, sensitive to noise, and solutions may be non-unique.

Deficiencies of the Scalar and Paraxial Approximation: Effects of the Vector Character of Light

In view of many more recent developments in confocal microscopy, it appears useful to shortly discuss devia-tions from the simple scalar theory. These deviations become increasingly important with increasing numer-ical aperture of the objective lens. In the case of linear polarization of the excitation beam, cylindrical sym-metry is lost. The field in the focal plane and exactly on axis is then polarized in the direction of the excitation, but, away from the axis, it is elliptically polarized with a longitudinal component pointing in the direction of the optical axis. The intensity PSF becomes elongated and approximately elliptical, with the major axis of the ellipse in the direction of the excitation. Both effects increase with increasing numerical aperture. In the case of linearly polarized excitation, the two-point resolution of an ordinary optical microscope equipped with a well-corrected high numerical aperture objec-tive lens (as well as that of a similar confocal micro-scope) therefore depends on the orientation of the line connecting the two points relative to the incoming polarization (as well as on the orientation of the dipolar point reflectors or absorbers/emitters). Figure 3 shows the PSF for numerical aperture N.A. = 0.95 and the absolute squares of all electric field components in the focal plane. In the scalar approximation, the lines of constant intensity would, of course, be circles, which is clearly not the case in the figure appearing in the lower right corner of Fig. 3. Moreover, the maxima of the longitudinal field occur on both wings of the main maximum, along the direction of the incoming

Confocal Laser Scanning Microscopy, Fig. 3 PSF of a well-corrected microscope objective of a high-numerical aperture lens (N.A. = 0.95) for the case of linear polarization of the incoming beam, calculated using the Debye-Wolf integral. A constant pupil function has been assumed here and, correspondingly, reflection losses in the lens have been neglected. z is the coordinate along the axis of the lens and x, y are the lateral coordinates, all measured in wavelengths. The incoming polarization is along the x axis

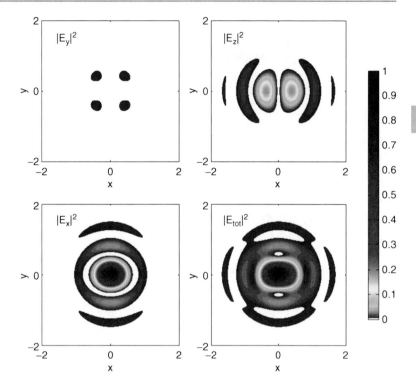

polarization (x direction) and the maximum absolute square of E_z is approximately 20% of that of E_x.

The vector diffraction problem was first solved by Richards and Wolf [5]. A somewhat more physical treatment employs expansion of the field in the image space into vector multipoles centered at the focus [6]. This latter approach is also of interest for matching the focal field distribution to the fields of a dipole via the amplitude distribution and the polarization in the pupil plane. In this way, the coupling of the field to single atoms or molecules can be significantly enhanced, which is of interest, for example, for quantum optical applications. The vectorial approach is in general also required to treat focusing of other distributions of intensity, phase, and polarization in the pupil plane of the lens. Examples relevant for applications include radially polarized excitation, producing a longitudinally polarized focus, azimuthally polarized excitation, which leads to a doughnut-shaped intensity distribution in the focal plane, or combinations of these distributions with scalar vortices, that means helical phase fronts. A longitudinal focus is essential for tip-enhanced Raman microscopy (TERS, compare section Confocal Raman Microscopy, ▶ Scanning Near-Field Optical Microscopy),

which combines a near field technique with confocal imaging. Other distributions are relevant for optical tweezers.

Instrumental Details

Scanning Techniques

Whereas scanning the sample is an option, particularly in laboratory setups, many commercial confocal microscopes employ lateral scanning of the laser beam and sample scanning in the vertical direction. Beam scanning is typically achieved using mirrors mounted on galvanometer motors, which allow to vary the angle at which the beam passes the rear focal plane of the objective lens. An example for a telecentric 4f system that allows to vary this angle without displacing the beam in the rear focal plane is shown in Fig. 4.

Recently, resonant galvanometer-based beam scanning systems have become commercially available, which employ torsion-spring based sinusoidal oscillations of the scan mirror with frequencies in the kilohertz range with open-loop operation for the fast scan axis. This permits frame rates on the order of

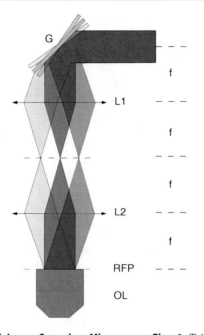

Confocal Laser Scanning Microscopy, Fig. 4 Telecentric lens system minimizing beam walk off. f: focal length of lenses L1 and L2, G: scan mirror mounted on galvanometer scanner, RFP: rear focal plane of objective lens OL

30 frames per second and in this way allows to study fast processes, such as diffusion in biological cells or to avoid blurring of the image due to movement of organs in *in vivo* studies. Alternative fast scanning confocal microscopes are discussed in section Variants of Confocal Laser Scanning Microscopy.

Excitation Sources and Beam Delivery

Ion lasers, HeNe lasers, diode-pumped solid state lasers, and diode lasers have all been used as continuous wave light sources in confocal laser scanning microscopy. Argon ion lasers provide a choice of several excitation wavelengths and are therefore still popular in spite of their low efficiency. In order to combine several laser beams dichroic beam splitters are employed. In many cases a software-controlled acousto-optic tunable filter (AOTF) selects the excitation wavelength(s) of choice from this combined beam. The AOTF is based on the diffraction of light from ultrasonic waves generated in a birefringent crystal by an ultrasonic transducer. Incident and diffracted waves propagate as ordinary and extraordinary wave in the crystal, respectively, or vice versa, and are therefore polarized perpendicular to each other. This

allows convenient rejection of the incident light by a polarizer. Because of momentum conservation the difference in the optical wave vectors of both beams must be equal to the acoustic wave vector. This means that the wavelength of the diffracted beam is controlled by the ultrasonic frequency. In a collinear AOTF both optical beams and the acoustic wave propagate in the same direction, independent of the optical wavelength.

The optical output of the AOTF is typically fed into an optical fiber, which delivers the beam to the input optics of the confocal scan head of the microscope. In brightfield reflection confocal microscopy, the beam splitter which directs the excitation light toward the objective lens (compare Fig. 1) induces a considerable loss for the excitation as well as for the detected light. This loss is minimized by a 50/50 beam splitter. In fiber-based confocal microscopes, instead of beam splitters 2 × 2 fiber couplers are typically employed. An improved version would make use of optical circulators, but these are presently not widely available for wavelengths in the visible range. In fluorescence confocal microscopy, the detected wavelength differs from the excitation wavelength and therefore dichroic dielectric beam splitters are used which are highly reflecting (transmitting) at the excitation (detection) wavelength and minimize losses in this way.

Detection

Standard detectors in confocal laser scanning microscopes are photomultipliers, which are in many cases operated in the analogue mode, which means by measuring the anode current and integrating over the pixel dwell time. Photon counting, on the other hand, is typically employed in fluorescence correlation and fluorescence lifetime microscopy (compare section Variants of Confocal Laser Scanning Microscopy), where usually avalanche photodiodes with high quantum efficiency and fast response times replace photomultipliers as detectors. Confocal microscopes which allow to spectrally disperse the light passing the confocal aperture often employ a charge-coupled camera (CCD) attached to a spectrograph to register the spectra. Back-illuminated Peltier or liquid nitrogen cooled CCD's provide high quantum efficiency (above 90% over a wide spectral range) and low background noise, which is important for single molecule detection or when working with less photostable fluorescent probes.

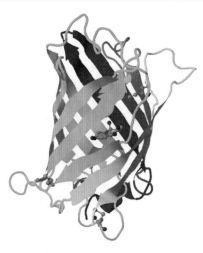

Confocal Laser Scanning Microscopy, Fig. 5 Tertiary structure of the Green Fluorescent Protein [8]. The fluorescent chromophore, composed of three amino acids is located in the center of the beta barrel protein cage, length about 4 nm, which prevents quenching of the fluorescence by water

Fluorescent Probes

Many samples are autofluorescent and therefore allow fluorescent imaging without having to introduce additional fluorescent probes. In biological samples, however, autofluorescence is typically weak and therefore the sample frequently had to be stained with appropriate dyes. The latter, however, are often highly phototoxic in living cells. An important development in light microscopy, including confocal laser scanning microscopy, started in the year 1994 when the green fluorescent protein (GFP, see Fig. 5) from the jellyfish *Aequorea victoria* was genetically expressed in bacteria making them fluorescent at room temperature. In the same way, it is now generally possible to label proteins of interest in biological cells with fluorescent proteins by genetic manipulation, which can be achieved, for example, by injection of a virus vector. The number of fluorescent proteins used in the field has exploded by now and they are widely used in optical microscopy because of their considerably reduced phototoxicity, brightness, and photostability [7]. Genetically modified fluorescent proteins from *Aequorea victoria* now cover the spectral range from the deep blue to yellow and others derived from Anthozoa species (corals and anemones), as well as other sources, span the entire visible spectrum. The tertiary structure and size of these fluorescent proteins is very similar to those derived from *Aequorea victoria*, although the amino acid sequences are quite different. Red emitting species with correspondingly longer excitation wavelengths are of particular interest because of reduced autofluorescence, deeper penetration, and better resistance toward high excitation density of biological tissue in this spectral region. Research is ongoing to improve brightness and photostability, reduce oligomerization and pH sensitivity, improve the appropriateness for fusion tagging, and reduce the time required for maturation of the protein in living organisms.

Variants of Confocal Laser Scanning Microscopy

Confocal Microscopy Involving Modified Illumination

Slit-Scanning Confocal Microscopes

Scanning a line focus, generated, for example, by a cylindrical lens, over the sample and imaging this line focus onto a slit aperture parallel to the image of the line focus still provides the optical sectioning capability of the confocal microscope, because the slit rejects out-of-focus light. At the same time, the frame rate is significantly increased, because scanning is necessary only in one direction. A spatially sensitive detector must be used to resolve light passing through different positions along the exit slit. This may be achieved by imaging the exit slit onto a one-dimensional detector array, read out synchronously with the scan, or by scanning an image of the exit slit, synchronously with the scan, across a two-dimensional detector, such as a CCD camera, forming a confocal image in this way. Disadvantages over the single-point scanning technique include reduced lateral resolution in the direction of the line focus, enhanced out-of-focus background, and, in the case of coherent illumination, increased laser speckle.

Spinning Disk Confocal Microscopes

Instead of scanning a single-point focus across the sample, multiple focal spots may be simultaneously generated and imaged each onto a confocal aperture. A white light version of such a confocal microscope based on a spinning Nipkow disk was introduced by Petran and Hadravsky already in the 1960s and later improved by Xiao, Corle, and Kino. The disk contains pinholes arranged in a spiral pattern, which are slightly displaced such that the whole sample is illuminated after the disk has rotated by a certain angle. In more recent versions, light from each focal spot on the sample passes the same pinhole in the disk that was used for excitation. A two-dimensional detector, such as the eye of the observer or a CCD camera, is used to register a confocal image while the disk is spinning. Nipkow disk based confocal microscopes are now commercially available from several manufacturers and provide fast confocal imaging, but at the cost of reduced flexibility, because the pinhole size cannot be varied and because the beams cannot be steered at will, as it is necessary, for example in some experiments involving photobleaching. Moreover, cross-talk between the different focal spots may occur. Another version of confocal microscopy with multiple focal spots, *swept field confocal microscopy*, leaves the pinhole array stationary and sweeps the image of this array over the sample. By switching between different pinhole arrays, the pinhole size can be varied.

The light efficiency of Nipkow disk based confocal microscopes may be significantly improved by adding a microlens array, mounted on a second disk, which is spinning on the same axis as the Nipkow disk and placed on top of the latter. Each microlens focuses incoming light onto one of the pinholes of the Nipkow disk. A dichroic beam splitter between both disks may be used to direct the detected fluorescence onto a camera. Another option is to use slit-shaped apertures on the Nipkow disk, which results in the same advantages and disadvantages as already described in section Slit-Scanning Confocal Microscopes. For a review of applications of spinning disk microscopes in life science, see reference [9].

Chromatic Brightfield Confocal Microscopy

A chromatic confocal microscope operating in reflection deliberately introduces chromatic aberrations into the imaging system. Scanning in the vertical direction is then replaced by simultaneous detection of different spectral components which encode the depth information, because the depth of the focus depends on the wavelength. A complete stack of images can be acquired in this way in a single two-dimensional mechanical scan of the sample or of the excitation beam. Broadband excitation may be provided by a white light lamp or by a femtosecond laser generated supercontinuum.

Structured Illumination Microscopy (SIM)

The SIM technique does not employ any pinhole and can be used with white light, but is related to

confocal microscopy in its optical sectioning capability [10]. Optical sectioning is achieved by acquiring a sequence of images of the sample with structured illumination. The simplest case of structured illumination is thereby produced by placing a grid of fully transparent and fully opaque stripes of equal width (one half the period L of the grid) into the illumination path and projecting this grid onto the sample. Only sample structures that are in focus will lead to significant variations of the image when the grid is displaced along the direction of periodicity, because only in focus the image of the grid within the sample is sharp. After acquiring three images with the grid displaced by 0, L/3, and 2L/3 the optical section can be calculated (in the simplest version of the SIM algorithm) as the root mean square of the three differences between the three images. More sophisticated deconvolution algorithms are available and many other structures for illumination can be used. Movement of a grid illumination pattern across the sample may be replaced by the generation of arbitrary patterns by digital mirror devices (DMD) based on microelectromechanical systems (MEMS), or on spatial light intensity modulators (SLM), based on liquid crystals.

Whereas optical sectioning can be achieved more easily in this way than in a standard confocal system with a single-point focus, there are also some problems related to this approach. SIM works badly in strongly scattering samples, because small differences on a large background have to be determined. This is related to a significant loss in bit resolution and, hence, dynamic range in the final image. Moreover, the optical sectioning capability of SIM is slightly worse than for the standard single focus confocal microscope.

Another version of SIM employs either a grid pattern or random aperture arrays on a spinning disk, which both allow a large throughput of the light from the illumination source of the order of 50%. Because of cross-talk between the light transmitted through neighboring apertures, the image acquired through the disk will contain a part that is not in focus. This part has to be subtracted from the image. A corresponding conventional image, to be subtracted from the partly confocal image, may be acquired by tilting the disk slightly, reflecting a second light source from the rear side of the disk toward the microscope objective and registering the corresponding image using a second camera. This procedure allows rapid optical sectioning and, hence, in vivo imaging of biological samples with very good signal-to-noise ratio with a comparatively simple instrument. A similar approach of subtracting the conventional image is based on a DMD, instead of a spinning disk and was termed *programmable array microscope*.

Versions of structured illumination microscopy providing a moderate degree of super-resolution are based on Moiré patterns produced by projecting a high spatial frequency grid onto the sample (high resolution SIM, HR-SIM). The Moiré pattern arises, because the observed signal is the product of the spatial distribution of the excitation with the concentration of the fluorophore and therefore contains spatial frequencies equal to differences between sample spatial frequencies and the one of the grid. The grid pattern must not only be shifted, but also be rotated, in order to be able to calculate the image. The method is able to increase the resolution by a factor of two beyond the diffraction limit.

Confocal Microscopy Beyond Brightfield and Standard Fluorescence

Confocal Raman Microscopy

In the Raman spectroscopy mode the inelastically scattered light from the sample is detected, where the frequency shift toward lower (Stokes signal) or higher photon energy (anti-Stokes signal) coincides with an internal vibration of the sample. Raman scattering is typically very weak and strong rejection of elastically scattered laser light is necessary. Historically, triple monochromators were often used for this purpose and still yield the highest spectral resolution, but in recent years dielectric long-pass filters with ultra-sharp transmission edges and high rejection factors have become available. This simplifies the instruments significantly, increases light efficiency, and in combination with slit-scanning techniques (see section Slit-Scanning Confocal Microscopes) allows rapid multispectral confocal Raman imaging with acquisition times per pixel in the millisecond range.

Coherent anti-Stokes Raman scattering (CARS) and *stimulated Raman scattering (SRS)* microscopies are nonlinear variants of Raman microscopy based on stimulated Raman scattering. Both techniques require two short pulse lasers operating at different frequencies v_1 and v_2. The overlapping beams are focused by the microscope objective into the sample. When the difference in the optical frequencies is tuned to the frequency of a characteristic vibrational frequency v_v of the sample ($v_1 - v_2 = v_v$), it will excite this vibration and at the same time generate anti-Stokes Raman scattered light at a frequency $2v_1 - v_2 = v_1 + v_v$ (CARS) or weakly deplete the pump and enhance the Stokes beams (SRS). Stimulated Raman scattering can be several orders of magnitude stronger than spontaneous Raman scattering and therefore allows rapid label-free imaging of a particular molecule in the sample. At least one of the two lasers necessary for CARS or SRS has to be tunable. This requirement maybe fulfilled, for example, by a Ti:sapphire laser or an optical parametric oscillator. Video rate *in vivo* SRS microscopy has been recently demonstrated and offers a number of advantages over CARS microscopy.

Another variant of confocal Raman microscopy, tip-enhanced Raman scattering (TERS) employs optical near fields of a sharp tip in order to increase the spatial resolution in Raman scattering over the diffraction limit. Confocal excitation and detection thereby reduces elastically scattered background in this setup.

Multiphoton Microscopy

Fluorescence excitation may in general be based on a linear process, in which a single photon excites the emitter into the excited state at energy E_1 from which it fluoresces, or on nonlinear processes, in which the emitter simultaneous absorbs n photons of energy E_1/n. Nonlinear processes are usually much less likely to occur than linear ones, with the probability strongly decreasing with the number of photons. Therefore, pulsed lasers are used for excitation, in which the optical power is concentrated in short pulses of widths in the femtosecond to picosecond range and the intensity during the pulse is very high. In addition, focusing the beam to a submicron spot leads, of course, to strong additional enhancement of the excitation probability.

Multiphoton microscopy has several important advantages over optical microscopy employing linear excitation. First, it is inherently confocal, in the sense that out-of-focus contributions to the detected signal are very small and therefore a confocal pinhole is not required (or the existing pinhole can be opened fully without loosing the optical sectioning capability). This is advantageous, in particular for strongly scattering samples, because it yields a higher detection efficiency. Second, the photon energy used for excitation is only one half (for two-photon excitation) or one third (for three-photon excitation) of that used in the linear case and the excitation wavelength therefore typically lies in the near infrared or even further in the infrared. Because the scattering in biological tissue and other inhomogeneous materials with inhomogeneities on a scale of the wavelength or below decreases strongly with the wavelength (proportional to λ^{-4} for very small inhomogeneities) the penetration depth is considerably larger for multiphoton excitation compared to single photon excitation of the same fluorophore. This means that three-dimensional imaging deep into tissue becomes possible. Third, photo-induced damage to the sample and the fluorophore is reduced, also because of the lower photon energy.

Instead of making use of multiple photon excitation of a fluorescent chromophore, it is also possible in some samples to detect light due to second harmonic generation (SHG) and to use that for label-free imaging. Other label-free and, in addition chemically specific, variants of multiphoton microscopy are CARS microscopy and SRS microscopy, as described in section Confocal Raman Microscopy.

Fluorescence Lifetime Imaging (FLIM)

The lifetime of a fluorophore may vary depending on local pH, oxygen or ion concentrations, or on intermolecular interactions, for example, fluorescence resonance energy transfer (FRET, see section Fluorescence Resonance Energy Transfer (FRET)). On the other hand, within some limits, it does not respond to the intensity of the excitation light, the fluorophore concentration, or photobleaching. The fluorescence lifetime is therefore a useful physical quantity that can be used for imaging and quantitative analysis of local pH, ion concentrations, or intermolecular interactions. Fluorescence lifetimes are typically in the range of a few picoseconds, when dominated by non-radiative processes, to several tens of nanoseconds, when limited by the radiative transition rate. In some cases it is useful to employ fluorophores with very long lifetimes, in particular when strong autofluorescence is present. Performing a lifetime measurement at each position of the excitation laser focus within the sample and representing the corresponding values as a gray scale or color value from a look-up table yields a confocal lifetime image. At the same time, a standard fluorescence intensity image can be obtained.

A common method of measuring fluorescence lifetime is time-correlated single photon counting (TCSPC). Here, a correlator, triggered by the short pulse of the exciting laser measures arrival times of fluorescence photons, typically detected by an avalanche photodiode with a short response time, and generates a histogram of delays. Fitting an exponential function to the histogram yields the fluorescence lifetime as the decay time of the exponential. Complications may arise when the decays are actually non-exponential. In the case of relatively long lifetimes in the nanosecond range, instead of TCSPC a gated image intensifier may be used to measure the number of photons falling into a time window defined by the gate. Both techniques operate in the time domain. When the lifetime is comparatively long, it is also possible to modulate the laser pulses in the MHz range and detect the phase shift of the corresponding modulation in detected fluorescence intensity, which depends in a simple way on the fluorescence lifetime. TCSPC, however, is the most flexible way of measurement, because it is not restricted to long lifetimes and allows to analyze non-exponential decays as well.

Fluorescence Resonance Energy Transfer (FRET)

FRET [11] is a resonant non-radiative energy transfer from a donor to an acceptor fluorophore due to the dipolar interaction. As the dipolar interaction energy is proportional to the third power of the donor acceptor distance R and because the probability for FRET to occur involves the square of an off-diagonal matrix element of the dipolar interaction, it falls off as R^6 and depends on the relative orientation between the molecules and the spectral overlap between the emission spectrum of the donor and the absorption spectrum of the acceptor. The probability is highest for parallel orientation of the donor and acceptor transition dipoles. Because of the steep fall off FRET can only occur if R is sufficiently small, that means, if the donor and the acceptor are sufficiently close to each other. Because the critical distance is only 1–10 nm, typically 4–6 nm, in all practical cases, the FRET mechanism provides a *molecular ruler* for measuring the donor-acceptor distance and in this way provides an indirect mechanism for studying structure on the nanometer scale, which cannot directly be resolved by diffraction-limited optical microscopy (super-resolution microscopy, compare section Super-Resolution, might, however, in the future partly supersede FRET studies). This is of particular interest for protein–protein and intra-protein interactions. For this purpose, the proteins of interest have to be labeled by fluorophores with overlapping emission and absorption spectra. In many cases fluorescent proteins (compare section Fluorescent Probes) are used for this purpose. A sensitive measure for FRET, which can be used for imaging, is the ratio of intensities of the donor and acceptor fluorescence peaks.

A problem with FRET confocal microscopy is that the signal-to-noise ratio is often very poor. Therefore, in many cases only the occurrence or absence of FRET is detected. The signal-to-noise ratio may be improved by measuring the donor lifetime, instead of the acceptor/donor fluorescence intensity ratio. This imaging option is usually known as FLIM-FRET (*fluorescence lifetime imaging – fluorescence resonance energy transfer*).

Fluorescence Recovery After Photobleaching (FRAP)

This microscopy technique, most often performed in a confocal set-up, bleaches the fluorescence of a certain sample region, often an intracellular organelle, and registers the recovery of fluorescence due to diffusion of the fluorescently labeled molecules. It may therefore be used to investigate the mobility of the target molecule within the surrounding structures.

Fluorescence Correlation Spectroscopy (FCS)

In contrast to the previous techniques, FCS is usually used not as an imaging technique, but with a fixed position of the confocal volume within the sample. The laser (usually a continuous wave laser) thereby excites fluorescent particles within this volume and particle movement in and out of the volume produces fluorescence intensity fluctuations. The autocorrelation function of these fluctuations provides information about the concentration, diffusion coefficient, and the mass of the particles. The diffusion coefficient depends on the viscosity of the medium via the Einstein–Smoluchowski relation. It may also depend on interactions of the particles with a micro-structured environment.

A variant of FCS is Fluorescence Cross Correlation Spectroscopy, in which the fluorescence intensity fluctuations of two fluorophores, labeling two different molecules, are measured simultaneously in two different channels. If the two molecules are bound in a dimer, the fluctuations will be highly correlated. Otherwise, no cross correlation is expected. The degree of cross correlation then is a measure of how many of the two different species of molecules are bound to each other.

Confocal Microscopy Sensitive to Phase

Interferometric versions of confocal laser scanning microscopy have been reported relatively early in the literature and were based on Mach-Zehnder or Michelson interferometers, for measurements in transmission or reflection, respectively [1]. In this way, it is possible to measure object topography or refractive index variations with interferometric precision. Interferometric confocal microscopes have the advantage over conventional interferometric microscopes that the shape of the wave front of the reference beam is irrelevant because only phase and amplitude at the pinhole is important. This means that the requirement for matching optics, otherwise necessary in interferometric microscopy, is strongly relaxed.

Often two beam splitters and two detectors, each with its own pinhole, are employed in the detection beam path of interferometric confocal microscopes based on Michelson interferometers. This allows to separate the conventional confocal signal and the pure interference signal as the sum and difference of the outputs of the two detectors. This is based on the fact that two beams at the inputs of a symmetric lossless beam splitter are combined with a $+\pi/2$, $-\pi/2$ phase shift relative to each other at the two outputs of the beam splitter, respectively. A non-interferometric variant of phase-sensitive confocal microscopy, *differential phase-contrast confocal microscopy*, is obtained by omitting the mirror generating the reference beam in the Michelson interferometer and obscuring one half of each of the relay lenses in a complimentary manner [2].

The sensitivity of interferometric confocal microscopes may be enhanced by using a spectrally shifted reference beam (*heterodyne interferometric confocal microscopy*). Topographic resolution of about 0.01 nm using such a heterodyne interferometric confocal microscope has been reported. In addition, the optical sectioning capability of the confocal microscope and the corresponding dependence of the signal on defocus can be used to avoid phase unwrapping ambiguities.

Another variant of interferometric confocal microscopy is called *4π microscopy*. Here, two opposing objective lenses are used for excitation and/or detection. This increases the available numerical aperture and therefore enhances the resolution, in particular in the axial direction. In addition, because of the coherent excitation the technique is inherently interferometric due to the interference of the two counterpropagating excitation beams in the focal region. This generates a standing wave interference pattern which modulates the main lobe of the focus in the axial direction and in this way allows to improve the axial resolution by a factor of about 4.5. The side lobes, due to interference, within the confocal main lobe may be effectively suppressed in the case of two-photon excitation. This suppression is due to the nonlinearity of the excitation process and can be additionally enhanced, if the sample fluorescence is also detected through both objective lenses in an interferometric setup. Because the position of maxima of the PSF for excitation and detection depends on the phase of both beams used for excitation and detection, precise control of the phase of both beams is required.

It should be mentioned here that a non-interferometric technique for phase measurement in optical microscopy is based on the so-called transport-of-intensity equation which can be derived from the paraxial time-dependent wave equation and relates the intensity and phase of a paraxial monochromatic wave to its longitudinal intensity derivative. The technique requires, however, the measurement of very small changes of intensity as a function of small defocus and therefore requires high bit resolution and long integration times.

Super-Resolution

Historically, the Abbe diffraction limit had been an insurmountable barrier in optical microscopy for many years. Many structures of interest, on the other hand, are significantly smaller than this limit and, particularly in biology, there has been a strong desire to significantly improve the resolution. In recent years, the diffraction limit has been broken in numerous ways, and this is based on some of the most exciting modern advancements in optics, which is still, although very old, a rapidly developing area of physics. The conceptually simplest way to achieve sub-Abbe resolution is by way of deconvolution (compare section Deconvolution). This does not, however, allow to achieve resolution enhancements of an order of magnitude, as urgently desired in many cases. One route to satisfy this demand employs non-propagating near fields. These near fields carry all optical information on length scales below the Abbe limit. The fact that these near fields are lost in far-field optical imaging can be viewed as the reason for the

limited resolution of standard (including confocal) optical microscopes. Near-field related techniques (compare → ► Scanning Near-Field Optical Microscopy) will not be discussed here (as an exception, compare TERS, section Confocal Raman Microscopy), although some of them may be combined with confocal imaging, for example *solid immersion lens microscopy* (SIL) or *total internal reflection microscopy* (TIRF). Other routes, however, have opened the way to far-field optical nanoscopy in recent years. They are based on prior knowledge about the sample (PALM/STORM) or on optical nonlinearity (STED, SSIM) [12]. STED uses a specially designed optical excitation and is a scanning microscopy technique. PALM/STORM is non-confocal, because parallel detection using a CCD camera is used. These techniques have reached lateral resolution in the range of 20 nm and ongoing research attempts to further push the resolution toward the molecular level (1–5 nm). Commercial instruments based on these techniques are increasingly becoming available [13]. The techniques will be shortly described, because of their potential as far field microscopy techniques to partly supersede standard confocal fluorescence laser scanning microscopy.

Super-Resolution by Prior Knowledge (Profilometry, PALM/STORM)

In cases where it is known that one and only one sharp interface between two homogeneous materials or one and only one point emitter or reflector in the focus is present, the position of the interface in the z direction or the three-dimensional position of the point emitter/reflector can be determined with a precision that is far beyond the Abbe limit and is limited only by the total amount of photons detected.

The simplest such technique consists in profiling a surface using a standard non-interferometric CLSM. By adjusting the z position to the wing of the $V(z)$ function, very slight changes in the topography of the sample surface can be measured, if the material and, hence, the reflectivity do not vary. Sub-nanometer depth resolution has been achieved in this way, averaged over the diffraction-limited lateral size of the confocal spot. As discussed in section Confocal Microscopy Sensitive to Phase the technique may be combined with interferometry to increase the depth resolution to about 0.01 nm.

In a principally similar way *photoactivated localization* microscopy (PALM) and *stochastic optical reconstruction microscopy* (STORM) determine the position of single-point emitters with nanometric precision. Because it must be avoided to have more than one molecule in the focal volume and such a sparse distribution of fluorescent emitters would not allow sub-diffraction limited resolution, it is necessary to separate the contributions from many individual molecules in some way. *Temporal* separation is key to current single molecule based super-resolution techniques. Other options, such as spectral separation, or more sophisticated schemes have also been discussed, however.

Current techniques use *photoactivable* molecules, which can be statistically turned on by another light source. Irreversible or reversible processes may be employed to turn off an activated subset. Determination of the positions of the activated molecules and multiple repetition then results in a software-generated image with nanometric resolution. Typically, thousands of images must be acquired and, correspondingly, imaging is slow, with a trade-off between spatial and temporal resolution. Effective frame rates of reconstructed images of about 1/(3 min) at a Nyquist-limited resolution of about 50 nm have been demonstrated for frames of 50×50 pixels. The obtainable resolution depends on the brightness of the fluorophores in the "on" state, as well as on the achievable contrast between "on" and "off" states. Fluorescent proteins as well as fluorescent dyes are currently in use [14].

PALM has hitherto mostly been used in total internal reflection configuration and is then restricted to essentially two-dimensional imaging. Lateral resolution down to ≈ 20 nm has been achieved. A three-dimensional STORM technique has achieved an image resolution of 20–30 nm in the lateral dimensions and 50–60 nm in the axial dimension in an illuminated sample volume of a few micrometers in thickness. An interferometric variant of PALM employing self-interference of fluorescent photons has improved the axial resolution to <20 nm for optically thin samples.

It may finally be worth noting that, whereas photoswitching is considered by some authors as a kind of optical nonlinearity, an optically nonlinear response is not essential to the breaking of the diffraction limit in the far field in the case of single molecule based techniques. For example, sub-diffraction imaging due to binding and detachment of diffusing probes has been demonstrated, which obviously does not involve any optical nonlinearity. Moreover, photoswitching and temporal selection of molecules in general represent conceptually only one, although currently the standard and most successful option to ensure that only one molecule is detected within the PSF of the objective lens. It may therefore be said that in this case the breaking of the diffraction barrier is based on the knowledge, however obtained, to have only one molecule within the PSF and on the facts that the center-of-mass position of a diffraction-limited spot can be determined to a much higher precision than given by the size of this spot and that this position corresponds to the position of the fluorescent molecule.

Super-Resolution due to Nonlinearity (SSIM, STED)

Some examples of confocal imaging methods employing optical nonlinearity have already been discussed in sections Confocal Raman Microscopy (CARS, SRS), and Multiphoton Microscopy. By reducing the size of the PSF for photons of the same wavelength optical nonlinearity is in principle able to enhance the resolution. In multiphoton microscopy, however, because long wavelength photons are used for excitation of fluorophores, the resolution is actually worse than that of its linear counterpart.

Saturated structured illumination microscopy, SSIM, on the other hand, provides resolution enhancement without fundamental limit. This nonlinear variant of SIM (compare section Structured Illumination Microscopy (SIM)) employs a nonlinear dependence of the fluorescence intensity on the excitation density, which is possible, for example by saturation of the excited state. In this case, the effective illumination generally contains higher spatial frequency components leading to increased resolution. Optical resolution beyond 50 nm has been reported using this technique.

Stimulated emission depletion microscopy (STED) selectively de-excites fluorescing molecules within the PSF of the confocal microscope by stimulated emission, using a second laser. This laser beam passes a spiral phase mask, which generates a zero of intensity in the center of the beam cross section. Stimulated emission and corresponding deexcitation of fluorophores then occur everywhere within the PSF, except close to the axis of the beam, and the size of the region, from which fluorescence can be collected depends on the intensity of the deexciting laser. Lateral resolution of about 20 nm has been reported using this setup. If STED is combined with a 4Pi setup, fluorescing spherical focal volumes of 40–45 nm diameter can be generated. STED is currently able to operate considerably faster than PALM. Video-rate STED microscopy with about 60 nm lateral resolution has been reported with a frame rate of 28 frames/s.

Cross-References

▶ Optical Techniques for Nanostructure Characterization
▶ Scanning Near-Field Optical Microscopy

References

1. Corle, R.C., Kino, G.S.: Confocal Scanning Optical Microscopy and Related Imaging Systems. Academic, San Diego (1996)
2. Wilson, T.: Confocal Microscopy. Academic, London (1990)
3. Born, M., Wolf, E.: Principles of Optics, 6th edn. Pergamon, Oxford (1980)
4. Sarder, P., Nehorai, A.: Deconvolution methods for 3-D fluorescence microscopy images. IEEE Signal Proc. Mag. **23**, 32–45 (2006)
5. Richards, B., Wolf, E.: Electromagnetic diffraction in optical systems. II. Structure of the image field in an aplanatic system. Proc. Roy. Soc. **A 253**, 358–379 (1959)
6. Sheppard, C.J.R., Török, P.: Efficient calculation of electromagnetic diffraction in optical systems using a multipole expansion. J. Mod. Opt. **44**, 803–818 (1997)
7. Chudakov, D.M., Matz, M.V., Lukyanov, S., Lukyanov, K.A.: Fluorescent proteins and their applications in imaging living cells and tissues. Physiol. Rev. **90**, 1103–1163 (2010)
8. Ormo, M., Cubitt, A.B., Kallio, K., Gross, L.A., Tsien, R.Y., Remington, S.J.: Crystal structure of the Aequorea victoria green fluorescent protein. Sci. **273**, 1392–1395 (1996). Image from the RCSB PDB (www.pdb.org) of PDB ID 1EMA (http://dx.doi.org/%2010.1021/ja1010652)
9. Gräf, R., Rietdorf, J., Zimmermann, T.: Live cell spinning disk microscopy. Adv. Biochem. Engin./Biotechnol. **95**, 57–75 (2005)
10. Langhorst, M.F., Schaffer, J., Goetze, B.: Structure brings clarity: structured illumination microscopy in cell biology. Biotechnol. J. **4**, 858–865 (2009)
11. Piston, D.W., Kremers, G.J.: Fluorescent protein FRET: the good, the bad and the ugly. Trends Biochem. Sci. **32**, 407–414 (2007)
12. Hell, S.W.: Far-field optical nanoscopy. Science **316**, 1153–1158 (2007)
13. Chi, K.R.: Ever-increasing resolution. Nature **462**, 675–678 (2009)
14. Heilemann, M., Dedecker, P., Hofkens, J., Sauer, M.: Photoswitches: key molecules for subdiffraction-resolution fluorescence imaging and molecular quantification. Laser Photon. Rev. **3**, 180–202 (2009)

Confocal Scanning Optical Microscopy (CSOM)

▶ Confocal Laser Scanning Microscopy

Conformal Electronics

▶ Flexible Electronics

Contour-Mode Resonators

▶ Laterally Vibrating Piezoelectric Resonators

Coupling Clamp

▶ Dynamic Clamp

Creep

▶ Nanomechanical Properties of Nanostructures

Cutaneous Delivery

▶ Dermal and Transdermal Delivery

Cuticle

▶ Arthropod Strain Sensors

Cylindrical Gold Nanoparticles

▶ Gold Nanorods

Decoration of Carbon Nanotubes

▶ Functionalization of Carbon Nanotubes

Density

▶ Interfacial Investigation of Protein Films Using Acoustic Waves

Dermal Absorption

▶ Dermal and Transdermal Delivery

Dermal and Transdermal Delivery

Biana Godin[1] and Elka Touitou[2]
[1]Department of Nanomedicine, The Methodist Hospital Research Institute, Houston, TX, USA
[2]Institute of Drug Research, School of Pharmacy, The Hebrew University of Jerusalem, Jerusalem, Israel

Synonyms

Dermal delivery – Cutaneous delivery; Dermal absorption; Skin delivery; Skin penetration
Transdermal delivery – Percutaneous absorption; Transcutaneous delivery; Transdermal permeation

Definition

Dermal delivery refers to the process of mass transport of active ingredients applied on the skin to various skin strata.

Transdermal delivery refers to the entire process of mass transport of substances applied on the skin surface and includes their absorption by each layer of the skin, their uptake by microcirculation of the skin, and distribution in the systemic circulation.

Introduction

History of Therapeutic Systems Applied on the Skin
The history of skin application of therapeutic substances goes back to ancient times. Numerous carriers, including conventional semisolid bases (creams, gels, ointments), matrix systems (clays, polymers), liquid systems (solutions, emulsions, suspensions), are being used for cutaneous application of therapeutics for centuries. Transdermal delivery, or delivery of active ingredients across the skin, was actually prototyped in ancient China and Egypt through the use of certain plasters and ointments. Several examples of medications topically applied on the skin to attain regional or systemic effect include the mustard plaster, used as a remedy for severe chest congestion, and Chinese balm or plaster-containing menthol and ginger (rich in salicylates) for regional pain relief. In the case of mustard plaster, which later was introduced to the US Pharmacopeia, powdered mustard seeds (*Brassica nigra*) mixed with warm water were applied to the patient's chest with a cloth. Following the

B. Bhushan (ed.), *Encyclopedia of Nanotechnology*, DOI 10.1007/978-90-481-9751-4,
© Springer Science+Business Media B.V. 2012

application, the change in the conditions to moist and warm activated some ingredients in mustard and particularly the enzyme myrosin. This enzyme is responsible for a hydrolysis of sinigrin and the release of active agent allyl isothiocyanate, which is transdermally absorbed, being a low molecular weight ($\sim$100 Da), mildly lipophilic substance. Other transdermal formulations recognized by the US pharmacopeia in the early twentieth century included Belladonna plaster (analgesic) and mercury ointment (for treatment of syphilis) [1].

Active development of the transdermal delivery systems started in the mid-1970s, exploring only the molecules which could intrinsically permeate the skin. The first transdermal delivery system, the scopolamine transdermal patch for treatment of motion sickness able to deliver the drug for 3 days, was granted with the FDA approval in 1979. The system was developed by the team of Alejandro Zaffaroni, who established ALZA Corporation. Currently, there are more than a dozen marketed transdermal delivery systems containing active ingredients such as nitroglycerin (cardiovascular conditions), nicotine (smoking cessation), various hormone combinations (contraceptives, hormone replacement therapy, etc.), lidocaine (local anesthesia), fentanyl (pain management), oxybutinin (urinary incontinence), methylphenidate (attention deficiency disorder), and others.

Throughout the history, the vast majority of medicaments applied to the skin surface were intended to cure superficial conditions. It can be explained by the intrinsic barrier function of the skin, which efficiently inhibits penetration of exogenous substances. To enable transdermal absorption or, on the contrary, enhanced accumulation of a therapeutic agent on the skin surface, it is important to thoroughly understand the skin physiology and barrier structure as well as the physicochemical properties of drug molecules and carrier systems.

Skin Barrier

The skin, or integument, is the largest body organ, comprising 15% of body weight and having a surface area of $\sim$2 m^2. Among the several functions of the skin, the main one is being a physical barrier to the environment, controlling the inward and outward passage of water, electrolytes, and xenobiotics and providing protection against pathogens, ultraviolet radiation, and mechanical insults. There are three structural layers to the skin: the epidermis, the dermis, and subcutis.

Hair follicles, sebaceous, sweat and apocrine glands are regarded as skin appendages. Epidermal layers of the skin have a dynamic turnover with cells of the outer strata continuously replenished by inner cells moving up to the surface from the basal layer. Epidermis and, particularly, its outermost layer stratum corneum (SC) represent the most important control elements in transport of molecules into and across the skin [2]. The actual investigation into the barrier properties of the skin started in the early 1920s [2]. It was initially proposed by Rein that an intermediate layer of cells between SC and viable epidermis is responsible for the resistance to transcutaneous transport of molecules. This hypothesis was further modified by the groups of Blank et al. and Sheuplein et al. who have shown for the first time that SC is the limiting step in the process of passive diffusion across the skin. In 1944, Winsor and Burch have shown that removal of SC by sandpaper dramatically increased permeation of water and solutes into the skin. Nevertheless, in later years, it was demonstrated by Michaels and colleagues that some of the exogenous substances will readily bypass the skin as a function of their physicochemical structure. Great contribution to the understanding of skin barrier structure was done by Kligman and colleagues, who in a series of studies separated the SC membrane and showed that it consists of well-defined polygonal corneocytes. The composition of intercorneocyte lipid domain of the SC was elucidated by Elias and coworkers [2]. Well along, the widely known model of SC was proposed. This model describes SC as a "brick-and-mortar" structure where the bricks are corneocytes, terminally differentiated metabolically inactive cells of the skin epithelium filled with insoluble keratins, and mortar is the continuous intercellular lipid phase, composed mostly of ceramides, cholesterol, free fatty acids, and cholesteryl esters (Fig. 1). This unique chemical composition of SC and very low water content ($\sim$15%w/w) impart a highly lipophilic nature to this protective layer. Thus, only low-molecular-weight xenobiotics with intermediate lipophilicity can cross this intrinsic barrier and be absorbed by the deeper layers of the skin or, through the network of blood capillaries located in the dermis, by the systemic circulation [2, 3]. Efficient administration of substances with distinct physicochemical characteristics (ionic, highly lipophilic, large molecular weight, etc.), aimed to act in the deep viable skin strata, regionally in the immediately underlying tissues

Dermal and Transdermal Delivery, Fig. 1 Schematic presentation of the skin structure. Stratum corneum, the outermost layer of the skin and the main barrier for absorption, consists of terminally differentiated anucleated corneocytes filled with proteins and intercellular stratum corneum lipids. The overall structure is frequently referred to as a "brick-and-mortar" wall

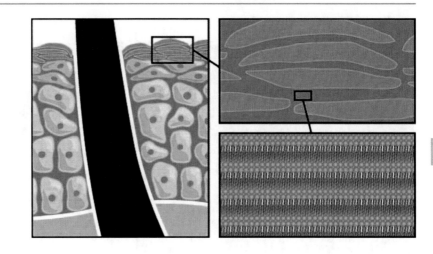

(muscles, ligaments, joints, etc.) or systemically, requires various techniques to enable their transport across the SC barrier [3, 4].

Percutaneous Absorption

The term percutaneous or transdermal absorption refers to the entire process of mass transport of substances applied topically and includes their absorption by each layer of the skin and finally their uptake by microcirculation of the skin. It can be broken down into a series of sequentially occurring transport events: (1) deposition of a penetrant molecule onto the SC, (2) diffusion through the SC, (3) diffusion through the viable epidermis, (4) passage through the upper part of papillary dermis, and finally (5) uptake into the microcirculation for subsequent systemic distribution. Since the viable tissue layers and the capillaries are quite permeable to solutes, diffusion across the SC is often regarded as the rate-limiting step [5].

The cells of the SC are anucleated and metabolically inactive; thus, passage through this layer is a passive process governed by classic diffusion laws and a simple equation for steady-state flux can be used to describe it. When the cumulative mass of a diffusant, m, passing per unit area, A, through a membrane is plotted, at some point, the graph approaches linearity and its slope yields the steady flux, J (dm/dt) (Eq. 1):

$$J = dm/dt = ADC_0K/h \qquad (1)$$

where C_0 represents the constant donor drug concentration; K, the partition coefficient of solute between membrane (SC) and the medium in which the molecule is applied; D, the diffusion coefficient; and h, the membrane thickness [5]. From Eq. 1, the ideal properties needed for a molecule to penetrate stratum corneum are: low molecular weight and high but balanced (optimal) partition coefficient, and a low melting point, correlating with good solubility. Permeation modifiers can affect various parameters in the equation to tailor the absorption into or across the skin. Skin permeation enhancers, for example, can increase the diffusion coefficient by affecting the partition coefficient to increase the solubility of the substance in SC, decrease the thickness of the membrane, or temporarily reduce/diminish the membrane properties of SC.

A molecule applied on the skin surface has three possible routes to reach the viable tissue: intracellular, through SC lipids; intercellular across the corneocytes; and via skin appendages (Fig. 2). The majority of transdermally absorbed molecules are believed to permeate the skin through the intercellular passage and, therefore, skin permeation enhancement methods disrupt or bypass the crystalline, semicrystalline, and liquid crystal domain formed by intercellular SC lipids. Because of the low fractional appendageal area (about 0.1%), this pathway usually adds little to steady-state drug flux, with the exclusion being highly polar molecules or macromolecules that struggle to cross intact SC. However, appendages may function as shunts important at short times prior to steady-state diffusion. Some nanovectors can enhance accumulation of bioactive materials in skin appendages, thus benefiting conditions such as acne and alopecia.

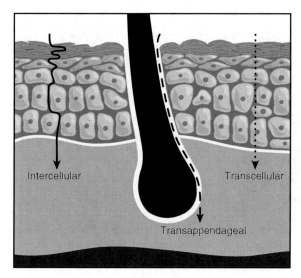

Dermal and Transdermal Delivery, Fig. 2 Schematic presentation of the routes for permeation of the agent applied to skin surface, to the viable layers of the skin and across the skin: intercellular, transappendageal, and transcellular routes

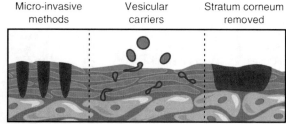

Dermal and Transdermal Delivery, Fig. 3 Nanotechnology- and microtechnology-based methods to enhance skin absorption of active materials: Microinvasive techniques (e.g., microfabricated needles, jet injectors); specially designed vesicular carriers (e.g., ethosomes, transfersomes); electrically assisted methods to temporarily remove SC barrier (e.g., radiofrequency ablation, electroporation)

Nano- and Microtechnologies in Dermal and Transdermal Delivery

The main goal of formulating the active agents in a carrier system is to achieve delivery of therapeutically relevant levels at the target tissue lowering the incidence of systemic adverse reactions. The desired site of action can vary from the skin surface to various body organs/systems. With a very few exceptions, the ultimate destination of the molecule applied on the skin will depend on the vehicle or vector system.

In the past three decades, the research focus in the area of cutaneous and transcutaneous delivery largely shifted toward the application of micro- and nanotechnologies to affect the permeation behavior of active agents applied on the skin surface. These technologies range from systems which enhance the transport of bioactives across the skin barrier (e.g., specially designed vesicles, microinvasive methods [6–9]) to those impeding skin penetration of small and mildly lipophilic molecules. There are several nanotechnology- and microtechnology-based mechanisms/carriers by which an enhanced dermal and transdermal systemic delivery can be achieved as schematically presented in Fig. 3. These techniques are used when the systemic effect is desired or when the applied agent should affect viable strata of the skin or beneath tissues and can be broadly categorized to specially designed nanovectors that temporarily affect

SC barrier properties and/or passively deliver drugs across the SC (e.g., ethosomes, transfersomes, surfactant-based elastic vesicles), minimally invasive techniques that actively birch SC and beneath strata (e.g., microneedles), and methods for removing portions of the SC (e.g., microporation, microabrasion). On the other hand, to maximize the effect of the drugs for alleviating superficial conditions and to minimize whole body distribution and related side effects, increased concentration and contact time of an active ingredient with skin surface/upper epidermis, while minimizing its absorption, are required. In this case, nanotechnology- and microtechnology-based approaches can be useful for preventing transcutaneous absorption of low molecular weight, mildly lipophilic agents which intrinsically cross the skin barrier (e.g., corticosteroids, sunscreens, retinoids). Large multilamellar liposomes, nonionic surfactant vesicles (niosomes), solid nano- and microparticles, dendrimers, and dendritic-core multi-shell nanotransporters are being investigated in this essence.

Further, in this entry, main techniques for skin application of drugs aiming at different purposes are described.

1. Nano- and microvectors: vesicles, microemulsions, and solid particles

There is a great versatility of nano- and microvectors suitable for topical, regional, and transdermal drug delivery of various active agents with distinct mechanisms of actions. These include phospholipid- and surfactant-based systems (liposomes, niosomes, ethosomes, transfersomes, elastic vesicles, micelles, etc.) and solid nano- and microparticles (solid lipid nanoparticles [SLN], polymeric nano- and microparticles, solgel particles, etc.). Several of these technologies were successfully tested in clinical trials and/or clinically used. Design considerations, structures and modes of actions of these nano- and microvectors are further discussed with examples of therapeutics/nutraceutics that were shown to benefit from skin application by means of a particular carrier.

Most early reports on traditional *liposomes*, containing phospholipids, water, and cholesterol, were published by Mezei and colleagues in the beginning of 1980s. These and further studies have shown that when applied to the skin, liposomes enable localized effect, depositing the encapsulated molecule in the SC. In vitro and in vivo studies in various animal models (from mice to rabbits) confirmed the local delivery behavior of liposomes [4, 6, 7]. While corticosteroids and retinoids were the most investigated drugs for liposome-based delivery, reports on xanthines, local anesthetics, proteins, enzymes, tissue growth factors, immunomodulators, and other biologically active substances have also been published [4, 10]. Results from a clinical study on a corticosteroid betamethasone dipropionate encapsulated in liposomes have shown that an antiinflammatory effect was more pronounced in patients with superficial skin inflammation (eczema) and not in patients with psoriasis, a condition involving deep skin strata [6]. A few in vivo studies and clinical trials with liposomally encapsulated retinoids confirmed enhanced topical localization with improved tolerability [4, 10]. The mode of action of traditional liposomes was investigated by a number of eminent scientists in the area of dermal and transdermal drug delivery. Starting from Junginger's and colleagues who provided valuable insights on the understanding of the interactions taking place on the liposomes-stratum corneum interface using freeze fracture electron microscopy, groups of Weiner, Touitou, Barry, Puglisi, Downing, Cevc and others investigated mechanisms involving a variety of microscopy techniques, radiolabeled vesicles distribution studies, and physicochemical analyses. It was shown that the ultrastructure of the intercellular lipid domains in SC is not affected by the liposomal application, no permeation of the liposome components to the deep layers of the skin is observed, and, in general, liposomes do not penetrate SC and appear as stacks on the skin surface [3, 6, 11]. This can explain the formation of drug reservoir in the SC and lack of penetration into the deeper layers of the skin. In general, liposomes are recognized as safe delivery system and approved for the last two decades by the regulatory agencies for skin application of pharmaceuticals (e.g., for acne and actinic keratosis) and cosmetic active ingredients as well as for other routes of drug administration (e.g., antifungal and anticancer intravenous medications) [4, 10].

More recent efforts on the front of the *phospholipid-based vesicles* shifted toward systems that can enhance the transport of active ingredients across the SC barrier. Ethosomes and transfersomes represent the two systems most investigated in this endeavor.

Ethosomes, mainly composed of phospholipids, relatively high concentrations of ethanol and water, were introduced by Touitou et al. [12]. The action of this vesicular system is generated by the presence of phospholipid vesicles and ethanol in the system. The presence of ethanol in the vesicle's structure enables malleability required to cross the "mortar" domain of the SC. On the other hand, ethanol also temporarily fluidizes the lipid phase in the SC, thus providing a background for permeation of the soft vesicles [6, 7, 12, 13]. The mechanism of action, proposed by Touitou et al. in 2000, can, thus, be described as three main simultaneously occurring processes: (1) fluidizing effect of ethanol on SC lipid domain and on phospholipid membrane in the ethosome; (2) entry and passage of the soft ethosomes across the SC lipid "mortar" phase; (3) fusion of the ethosomal vesicles with phospholipids in the viable skin strata and release of encapsulated components beneath the SC barrier. Moreover, the presence of alcohol enables encapsulation of molecules with a wide spectrum of physicochemical features. In vitro, in vivo, and clinical studies have shown that the composition of the system can tune their behavior in terms of delivery to various deep skin strata or transdermally. Numerous reports on ethosomal systems with various active ingredients were published by Touitou et al. and other research groups [6, 7, 10, 13, 14]. These include: antimicrobial and antiviral agents for deep skin infections, acne medications, antiinflammatory agents, cannabinoids, anti-alopecia therapeutics, anti-Parkinsonian trihexyphenydil HCl, antianxiety drug buspirone HCl, antipyretic paracetamol, large-molecular-weight compounds such as insulin and many others. Ethosomes are safe carriers and no skin adverse effects were reported with ethosomal products or in clinical studies. They are marketed for treatment of herpes simplex infection and used in cosmeceutic products.

Transfersomes were introduced by Cevc as deformable liposomes mainly composed of phospholipids, an "edge activator," such as sodium cholate, and water. The inventors argue that due to the presence of a polar substance in the lipid phase of the aggregate, the resulting vesicle can squeeze when the osmotic gradient is applied and enter through the SC pores with less than one tenth of its diameter. In the proposed mechanism of action, two features are claimed to be important: (1) nonoccluded conditions and (2) presence of the hydration gradient from the unhydrous SC (10–15% water) to the viable skin strata ($\sim$70% water). When applied to the intact skin, the ultradeformable transfersome vesicles can cross a nanoporous barrier essentially intact despite the extreme bilayer vesicle deformation and appreciable intra-pore friction. This is enabled by the self-controlled local bilayer composition adjustment to the shape and overall curvature imposed by the pore on a vesicle. Later, Barry and colleagues evaluated in vitro skin transport of molecules from transfersomes versus traditional vesicles. Both types raised maximum flux and skin deposition compared to saturated aqueous drug solution (maximum thermodynamic control) under a nonoccluded environment, but results were not as dramatic as detailed in earlier work. Transfersomes were proven in in vitro and in vivo studies transdermally to deliver steroids, NSAIDs, and large molecules (e.g., insulin, interferon-alpha, serum albumin) across human skin [6, 8]. A transfersome-based therapeutic, Diractin® or ketoprofen in Transfersome® gel developed by Idea-AG, was approved for clinical use. However, phase III clinical studies in Europe and USA comparing Diractin® with the locally applied ketoprofen-free vesicles (used as an "epicutaneous placebo"), for the treatment of osteoarthritis of the knee, were discouraging and revealed no statistically significant difference between Diractin® and the placebo.

Vesicles based on *synthetic surfactants* were also proposed for skin application of active ingredients. In this category, niosomes and elastic vesicles are further briefly described as carriers for topical or systemic drug delivery, respectively.

Niosomes, proposed by Handjani-Vila et al., are nonionic surfactant vesicles comprised of bilayer assemblies formed from certain synthetic amphiphilic molecules in aqueous medium. Niosomes can contain various surfactants generally used in pharmaceutical preparations, including polyoxyethylene alkyl ethers, sorbitan esters, polysorbates-cholesterol mixtures, alkyl glycerol ethers, and others. Similarly to liposomes, these carriers initially developed in L'Oreal for cosmetic applications enable enhanced retention of active agents within the superficial SC strata, as confirmed by various microscopy techniques. Various cosmetic products currently contain niosomes.

Synthetic surfactant-based *elastic vesicles*, introduced by Bouwstra's research group, were designed according to similar principles as phospholipid-based transfersomes. The most frequently reported "edge activator" in the formulation of these vesicles is octaoxyethylene laurate-ester (PEG-8-L). The molar concentration of PEG-8-L directly affects the bilayer's elasticity. Elastic vesicles were investigated for enhanced percutaneous transport of a number of drugs such as pergolide, rotigotine, estradiol, immune-modulators, and ketorolac [15]. As of today, no clinical studies with these systems have been reported.

Microemulsions typically contain two phases (aqueous and organic) and surfactant/cosurfactant and, due to a very low surface tension between the components, form spontaneously. Cyclodextrins were also used as a component in microemulsions. In general, the type of emulsion (w/o vs. o/w emulsion), the droplet size, the emollient, the emulsifier as well as the surfactant organization (micelles, lyotropic liquid crystals) in the emulsion may affect the cutaneous and percutaneous absorption. This system was used mainly for solubilization of lipophilic agents for skin application, thus increasing the concentration gradient and enabling skin deposition of higher concentrations of therapeutics. In some cases, this can allow for mild increase in percutaneous penetration as reported in several in vitro and in vivo studies [16].

Solid nano- and microparticles include such nanovectors as solid lipid nanoparticles (SLN), solgel particles, polymeric nano- and microparticles, metal (gold, silver, iron) and metal oxide (titanium oxide, zinc oxide, silica)-based particles, and microsponges. As it was mentioned above, efficient penetration of intact skin by non-deformable particles (organic or inorganic nanoparticles, conventional liposomes, micelles, etc.) is both elusive and unlikely. Best evidence for this comes from skin penetration studies with incompressible inorganic nanoparticles which lack any direct ability to affect the cutaneous barrier. Various sizes of inorganic and metallic particles were tested for skin delivery (30–200 nm) and all were found confined to the skin surface. Consistent with the notion, it is a general agreement in the field that solid particles with diameters of >30 nm accumulate on the skin surface and cannot transport across the skin, except through hair follicles [6, 8].

Solid Lipid Nanoparticles (SLN), initially designed to enable extended release of intravenous therapeutics, are relatively large on the nanoscale (<500 nm) dispersions of melted and solidified lipids [17]. Among the common SLN constituents are triglycerides, partial glycerides, fatty acids, steroids, and waxes which are formulated into a solid particles by methods such as high-pressure homogenization and microemulsion formation [17]. Diffusion of the active ingredients through the solid matrix is a slow process that enables prolonged release of the molecule on the superficial skin layers. Due to the low concentrations of the free active molecules present on the surface at each time point, the diffusion gradient and thus the driving force for passive diffusion is considerably low than when the actives are applied in liquid formulation. This prevents transdermal absorption of an active ingredient making SLN, and other solid particles, attractive carriers for substances that have to act on the skin surface but possess intrinsic physicochemical characteristics to be absorbed by the viable strata and from there by systemic circulation. Sunscreens, perfumes and insect repellents belong to this category [6, 17–19]. Being solid and opaque, SLN also possess an inherent ability to act as a physical sunscreen.

Polymeric carriers are another category of the solid nano- and microvectors for topical delivery. There are a plenty of structures and building blocks that can be used for the design of polymeric vectors. The polymers currently investigated and used for preparations of particles in biomedical applications range from biocompatible and biodegradable poly(lactic acid) (PLA), poly(glycolic acid) (PGA), and their copolymers (PLGA) to other representatives in the class, such as polymethyl methacrylate (PMMA), polycaprolactone, polyalkylcyanoacrylates, polyvinyl acetate (PVA), and polystyrene [4, 6, 20, 21]. Skin application of polymeric nano- and microscale systems represents a less explored venue compared to the

knowledge of their physicochemical properties and design variables accumulated in the drug delivery field. The first patent on polymeric particles related to skin delivery by Nuwayser was approved soon after the appearance of the first transdermal systems. The inventor proposed to use polymeric particles as a controlled-release component in the formulation of transdermal patches. Since then, several, mostly lipophilic drugs were incorporated into the polymeric particles applied on the skin, including steroids, minoxidil, and sunscreens such as octyl methoxycinnamate and benzophenone-3 [6, 8, 21]. Studies have shown that based on their shape and structure, polymeric particles can either accumulate in hair shafts enabling intrafollicular depot or retard skin permeation of active ingredients. A *microsponge delivery system* (MDS) is a recent technology based on porous nondegradable polymeric microspheres, typically 10–25 μ in diameter. MDS can entrap various active ingredients and then release them onto the skin over a period of time and in response to trigger (e.g., rubbing, temperature, pH change), while remaining on the skin surface. The technology is being used in cosmetics, over-the-counter skin care, sunscreens, and prescription products. MDS was formulated with benzoyl peroxide (acne medication), hydroquinone (bleaching cosmetic agent), fluconazole (antifungal), and retinoids (acne).

Solgel systems, introduced by Avnir in 1984, are highly porous, reactive organosilicates formed based on the hydrolytic polycondensation of silicon alkoxides with a general formula $Si(OR)_4$ or $R_0 nSi(OR)_4 \grave{A}n$ where R_0 is contains vinyl, methacryl, or epoxy groups. A flexible network of inorganic oxide or hybrid nanocomposites can be produced by reacting R_0 with additional groups. As a result, the characteristics of solgel nanocomposite are halfway between those of polymers and silica-based glasses. Solgel based systems were clinically approved for treatment of rosacea and acne and are used in sunscreen formulations.

2. Microinvasive techniques for enhanced skin absorption

Several microinvasive techniques are under investigation for transdermal delivery applications. Principles of ballistic high-velocity needle-free particles, microneedles, and skin microporation methods are described below.

Ballistic needle-free high-velocity particle system or PowderJect system was invented more than half a century ago for military use. Pulsed microjet injectors could be used to deliver drugs for local as well as systemic applications without using needles [8, 22]. The technology is based on transporting solid nanoparticles or liquid nanodroplets across the SC layer into the viable tissues, driven by a supersonic shock wave of helium. These systems are currently in use for delivery of proteins and vaccine. Jet-based noninvasive vaccination involves shooting suitably coated gold particles into the skin. The injectors in clinical use contain compressed gas that can discharge the contents across the skin barrier. Among the claimed advantages for these delivery systems are freedom from pain and needle phobia, improved efficacy and bioavailability, controlled release, accurate dosing, and safety. However, the reported problems with bruising and particles bouncing off skin surfaces are the main reasons for the PowderJect uncommon use. Another disadvantage is the use of extremely small volumes, lowering the application to very potent therapeutics. Nanoliter volume pulsed microjets comprised of 10–15 nL droplets were created to minimize the pain associated with this microinvasive transdermal drug delivery. To control the depth of jet penetration into the dermis, thus minimizing the pain, parameters such velocity and volume of the microinjections can be tuned: A high velocity (>100 m/s) of microjets allows their entry into the skin, whereas the extremely small volumes (from 2 nL) limit the penetration depth [22]. Due to the local adverse effects, the technology is generally proposed for acute usage, as in the case of skin immunization. Among currently used ballistic devices are Antares (insulin delivery), Mini-Jet, Biojector, Inject, and Penjet.

Microneedles (MN) are microinvasive tools to overcome the skin barrier and are currently one of the most vibrant domains of transdermal delivery research by active means [8, 9, 23].

MN were proposed by Prausnitz and coworkers in 1998 for transdermal delivery of high-molecular-weight bioactives. MN consist of an array of micro-projections, ranging from 20 to 2,000 µm in dimensions and smaller arrays are being developed. Application of MN to the SC surface creates transport pathways of micron dimensions that allow for transport of macromolecules, and aggregates such as supramolecular complexes and even nanoparticles. A variety of techniques and designs of MN have been proposed. MN can be produced by microfabrication, laser-based micromolding, saw-

diced and/or etched; can be biodegradable (polymer or nanoporous semiconductors based) or non-dissolving (metallic, silica), solid or hollow; can be designed for sampling or drug/vaccine delivery. The design of MN can involve various geometries, which have a direct impact on the diameter and the depth of the pores produced in the skin. As an example, MN can be blunt or possess a very high aspect ratio, can incorporate the active ingredient into the hollow structure, or be coated with the substance to deliver. Another key parameter in the design of efficient MN-based delivery system is the application and driving pressures. Generally, due to the skin elasticity, the average diameter of the transepidermal pores left in the skin after microneedles extraction equals to roughly only half of the needle diameter at the base [8, 9, 22]. While these systems were proven to be efficient in transcutaneous immunization and delivery of opiates and proteins, the long-term safety still remains an open question. It may explain the fact that despite the major investment of time and big hopes associated with percutaneous microneedles, only a few studies were performed on human subjects. Moreover, in these studies, mainly metallic MN have been evaluated. Safety of MN especially for chronic administration of therapeutics still holds a question mark. For the last decade the technology has been under development by several industrial companies mainly for the purpose of vaccination, including Zosano Pharma (Macroflux), 3 M Corporation (Microstructured transdermal system), Cornium International Inc (MicroCor), BD Technologies (Soluvia™ Prefillable Microinjection System) [24].

Generally, there is less data in the literature related to the methods of *upper skin layers' removal* for enhanced percutaneous absorption. Only the main principles of the techniques are given below. *Radiofrequency- or electroporation-*based removal of the superficial skin strata is based on microelectronic devices which use high-voltage currents [9, 23, 25]. These currents create a single or multiple (an array) microchannels across the stratum corneum deep into the viable epidermis. The efficacy of transport depends on the electrical parameters and the physicochemical properties of drugs. When applied in vivo, high-voltage pulses are reported to be well tolerated, inducing only muscle contractions; however, since the techniques may involve viable tissue, the safety for acute and chronic administration of actives yet remains the major question. A few studies reported in the literature show improved skin permeation of ionic molecules such as granisetron hydrochloride and diclofenac sodium, using the above techniques. Enhanced delivery of proteins has also been reported [26].

Summary

Extensive research during the past few decades has focused on various strategies utilizing nano-and microscaled delivery systems to enhance skin transport of dermal and systemic therapeutics or to retard cutaneous absorption of agents that should primarily act on the skin surface. This entry focused on suitable techniques for delivery of therapeutics either to the skin surface to the deep viable layers or to the systemic circulation. While currently available methods for administration of therapeutics onto the skin span over a wide range of micro- and nanotechnologies, new methods are being introduced in this dynamic research area. In many cases, the safety of chronic or acute use on the skin will determine the fate of the proposed method.

Cross-References

▶ Liposomes
▶ Nanomedicine
▶ Nanoparticles
▶ Radiofrequency
▶ Sol-Gel Method
▶ Solid Lipid Nanoparticles - SLN

References

1. Scheindlin, S.: Transdermal drug delivery: PAST, PRESENT, FUTURE. Mol. Interventions **4**, 308–312 (2004)
2. Elias, P.M., Ferngold, K.R. (eds.): Skin Barrier. Taylor & Francis, New York (2006)
3. Touitou, E.: Drug delivery across the skin. Expert Opin. Biol. Ther. **2**, 723–733 (2002)
4. Korting, H.C., Schafer-Korting, M.: Carriers in the topical treatment of skin disease. Handb. Exp. Pharmacol. **197**, 435–468 (2010)
5. Guy, R., Hadgraft, J.: Feasibility assessment in topical and transdermal delivery: mathematical models and in vitro studies. In: Guy R., Hadgraft J. (eds.) Transdermal Drug Delivery Systems. 2nd Edn. Revised and Expanded, pp. 1–23. Marcel Dekker Inc., New York- Basel (2003)
6. Godin, B., Touitou, E.: Nanoparticles aiming at specific targets – dermal and transdermal delivery. In: Domb, A.J., Tabata, Y., Ravi Kumar, M.N.V., Farber, S. (eds.) Nanoparticles for Pharmaceutical Applications. American Scientific, Valencia (2007)
7. Touitou, E., Godin, B.: Vesicular carriers for enhanced delivery through the skin. In: Touitou, E., Barry, B.W. (eds.) Enhancement in Drug Delivery, pp. 255–278. CRC Press, Boca Raton (2006)
8. Cevc, G., Vierl, U.: Nanotechnology and the transdermal route: a state of the art review and critical appraisal. J. Control. Release **141**, 277–299 (2010)
9. Prausnitz, M.R., Langer, R.: Transdermal drug delivery. Nat. Biotechnol. **26**, 1261–1268 (2008)
10. El Maghraby, G.M., Barry, B.W., Williams, A.C.: Liposomes and skin: from drug delivery to model membranes. Eur. J. Pharm. Sci. **34**, 203–222 (2008)
11. Honeywell-Nguyen, P.L., de Graaff, A.M., Groenink, H.W., Bouwstra, J.A.: The in vivo and in vitro interactions of elastic and rigid vesicles with human skin. Biochim. Biophys. Acta **1573**, 130–140 (2002)
12. Touitou, E., Dayan, N., Bergelson, L., Godin, B., Eliaz, M.: Ethosomes – novel vesicular carriers for enhanced delivery: characterization and skin penetration properties. J. Control. Release **65**, 403–418 (2000)
13. Godin, B., Touitou, E.: Ethosomes: new prospects in transdermal delivery. Crit. Rev. Ther. Drug Carrier Syst. **20**, 63–102 (2003)
14. Ainbinder, D., Paolino, D., Fresta, M., Touitou, E.: Drug delivery applications with ethosomes. J. Biomed. Nanotechnol. **6**, 558–568 (2010)
15. Loan Honeywell-Nguyen, P., Wouter Groenink, H.W., Bouwstra, J.A.: Elastic vesicles as a tool for dermal and transdermal delivery. J. Liposome Res. **16**, 273–280 (2006)
16. Date, A.A., Patravale, V.B.: Microemulsions: applications in transdermal and dermal delivery. Crit. Rev. Ther. Drug Carrier Syst. **24**, 547–596 (2007)
17. Shidhaye, S.S., Vaidya, R., Sutar, S., Patwardhan, A., Kadam, V.J.: Solid lipid nanoparticles and nanostructured lipid carriers–innovative generations of solid lipid carriers. Curr. Drug Deliv. **5**, 324–331 (2008)
18. Wissing, S.A., Müller, R.H.: Cosmetic applications for solid lipid nanoparticles (SLN). Int. J. Pharm. **254**, 65–68 (2003)
19. Schafer-Korting, M., Mehnert, W., Korting, H.C.: Lipid nanoparticles for improved topical application of drugs for skin diseases. Adv. Drug Deliv. Rev. **59**, 427–443 (2007)
20. Miyazaki, S.: Nanoparticles as carriers for enhanced skin penetration. In: Percutaneous Penetration Enhancers, 2nd edn, pp. 117–124. CRC Press, Taylor and Francis group, New York (2006)
21. Venuganti Venkata, V., Perumal Omathanu, P.: Nanosystems for dermal and transdermal drug delivery. In: Pathak Y., Thassu D (eds.) Drugs and the pharmaceutical sciences series, Drug Delivery Nanoparticles Formulation and Characterization, Vol 191. pp. 126–155. Informa Healthcare, New York. (2009)
22. Arora, A., Hakim, I., Baxter, J., Rathnasingham, R., Srinivasan, R., Fletcher, D.A., Mitragotri, S.: Needle-free delivery of macromolecules across the skin by nanoliter-volume pulsed microjets. Proc. Natl. Acad. Sci. U.S.A. **104**, 4255–4260 (2007)

23. Bal, S.M., Ding, Z., van Riet, E., Jiskoot, W., Bouwstra, J.A.: Advances in transcutaneous vaccine delivery: do all ways lead to Rome? J. Control. Release **148**, 266–282 (2010)
24. Kim, Y.C., Jarrahian, C., Zehrung, D., Mitragotri, S., Prausnitz, M.R.: Delivery systems for intradermal vaccination. Curr. Top. Microbiol. Immunol. **351**, 77–112 (2011)
25. Staples, M., Daniel, K., Cima, M.J., Langer, R.: Application of micro- and nano-electromechanical devices to drug delivery. Pharm. Res. **23**, 847–863 (2006)
26. Banga, A.K.: Microporation applications for enhancing drug delivery. Expert Opin. Drug Deliv. **6**, 343–354 (2009)

Detection of Nanoparticle Biomarkers in Blood

▶ Detection of Nanoparticulate Biomarkers in Blood

Detection of Nanoparticulate Biomarkers in Blood

Rajaram Krishnan[1] and Michael J. Heller[2,3]
[1]Biological Dynamics, Inc., University of California San Diego, San Diego, CA, USA
[2]Department of Nanoengineering, University of California San Diego, La Jolla, CA, USA
[3]Department of Bioengineering, University of California San Diego, La Jolla, CA, USA

Synonyms

Detection of nanoparticle biomarkers in blood

Definition

Detection of nanoparticulate biomarkers in blood is the ability to find nanoparticles in blood.

Overview

Challenges with Isolating and Detecting Nanoparticles in Blood

While the potential applications of nanotechnology in medicine are rapidly growing, a number of issues still need to be resolved before nanomedicine translates from the lab to the bedside [1]. Two important challenges will be the monitoring of drug delivery nanoparticles and the detection of cell-free-circulating (cfc) DNA nanoparticulate biomarkers [2, 3]. Presently, considerable research efforts are being carried out on the development of new drug delivery nanoparticle therapeutics [4]. For both research and clinical applications, it will be important to develop rapid, sensitive, and inexpensive monitoring techniques for determining levels of drug delivery nanoparticles directly into the patient's blood [2, 5, 6]. In another related area, the ability to detect cfc-DNA, cfc-RNA, and other cell-free-circulating nanoparticulate biomarkers directly into blood would represent a major advance for early cancer screening [7], residual disease detection [8], and chemotherapy monitoring [9].

Unfortunately, present methods for isolating nanoparticles and nanoparticulate biomarkers in the 10–500 nm range from blood are complex, expensive, and time consuming [5–9]. Adding even more challenge is that relatively large blood samples (1–10 ml) are needed when assaying for very low levels of nanoparticles or biomarkers. Thus, sample preparation is more often the weak link in monitoring and diagnostic assays than is the intrinsic sensitivity of the downstream detection technology. The sample preparation process for blood often involves centrifugation, filtration, washing, extractions, and a number of other steps before the analyte(s) can be identified [5–9]. The extended amount of time between blood drawing, cell separation, analyte extraction, and the final analysis also leads to significant loss of sensitivity and selectivity and to the degradation of the analytes [7–9]. Thus, there is a critical need for a novel robust technology, which will allow a variety of nanoscale entities to be manipulated, isolated, and rapidly detected *directly* from whole blood and other biological samples. Meeting this challenge will be required before therapeutic monitoring can be carried out in cost-effective point-of-care (POC) settings.

Dielectrophoresis as an Isolation Method

Dielectrophoresis (DEP) is an intrinsically powerful technique for separating cells and nanoparticles [10, 11]. DEP is the induced motion of particles produced by the dielectric differences between the particles and media in an asymmetric alternating current (AC) electric field [12]. For spherical particles, the DEP force equation is given by:

$$F_{DEP} = 2\pi\varepsilon_m r^3 Re[K(\omega)]\nabla E_{RMS}^2 \qquad (1)$$

where

$$Re[K(\omega)] = Re\left(\frac{\varepsilon_p^* - \varepsilon_m^*}{\varepsilon_p^* + 2\varepsilon_m^*}\right) \qquad (2)$$

where ε_p^* and ε_m^* are the complex dielectric permittivities of the particle and medium respectively, defined by $\varepsilon^* = \varepsilon\text{-}j\sigma/\omega$, where $j^2 = -1$, ε is the dielectric constant and σ is the conductivity. If a particle has a positive $Re[K(\omega)]$, it will migrate to the high-field regions, and if it has a negative $Re[K(\omega)]$, it will migrate to the low-field regions. Unfortunately, the use of DEP for practical applications has been limited by the need to carry out the process under low conductance (ionic strength) conditions [13, 14]. Thus, blood or any other high ionic strength sample has to be significantly diluted before DEP separations can be carried out [14]. Recently, a high conductance (HC) DEP method that allows both nanoparticles and cfc-DNA nanoparticulates to be manipulated, isolated, and detected under high ionic strength conditions has been developed [15–17]. HC-DEP sets the stage for new "seamless" sample to answer systems, which will allow a variety of drug delivery nanoparticles and other nanoscale biomarkers to be rapidly isolated and analyzed from clinically relevant amounts of complex undiluted biological samples.

Basic Methodology

Microelectrode Array Device Used for Dielectrophoresis

In these studies, all HC-DEP experiments were carried out using an electronic microarray device with 100 circular platinum microelectrodes 80 μm in diameter (Fig. 1a). The microarray is overcoated with a 10-μm thick porous polyacrylamide hydrogel layer and enclosed in a microfluidic cartridge, which

forms a 20 μL sample chamber. While only a 3 × 3 subset of microelectrodes was used for the DEP experiments, all 100 microelectrodes can be used if desired. AC electric fields were applied to the nine microelectrodes in a checkerboard-addressing pattern where each microelectrode has the opposite bias of its nearest neighbor. Figure 1b shows the asymmetric electric field distribution, which produces DEP positive (high-field) regions on the microelectrodes and the DEP negative (low-field) regions between the microelectrodes [15–17]. At AC frequencies in the 5,000–10,000 Hz range, nanoparticles, cfc-DNA, and other cellular nanoparticulates from about 10 to 500 nm in size will concentrate in the DEP high-field regions over the microelectrodes, while micron-size particles and cells will concentrate in the low-field regions. Figure 1c shows a composite half bright field/half fluorescence image for a prior DEP separation of 10 μm polystyrene particles and 40 nm red fluorescent nanoparticles carried out in 1× TBE buffer. The 10 μm particles are isolated into the low-field regions, and 40 nm red fluorescent nanoparticles are concentrated into the high-field regions. Detailed procedures for high conductance (HC) DEP separations were described earlier [15–17]. The general scheme for the HC-DEP separation of nanoparticles from whole blood (red and white cells) is shown in Fig. 1d–f. When the DEP field is applied, the nanoparticles (red dots) concentrate into the high-field regions where they are held firmly on the microelectrodes, and blood cells move into the low-field regions between the microelectrodes where they are held less firmly (Fig. 1e). A simple fluidic wash easily removes the blood cells while the nanoparticles remain in the high-field regions (Fig. 1f). The highly concentrated fluorescent nanoparticles can now be easily detected.

Preparation of Buffy Coat Blood and Whole Blood Experiments

For the HC-DEP experiments, concentrated 5× Tris Borate EDTA (TBE) buffer solution was obtained from USB Corporation (Cleveland, Ohio, USA) and diluted to 1× concentration and Dulbecco's phosphate buffer saline (1× PBS) solution was obtained from Invitrogen (Carlsbad, CA, USA) and diluted to 0.5× concentration. Human buffy coat blood was obtained

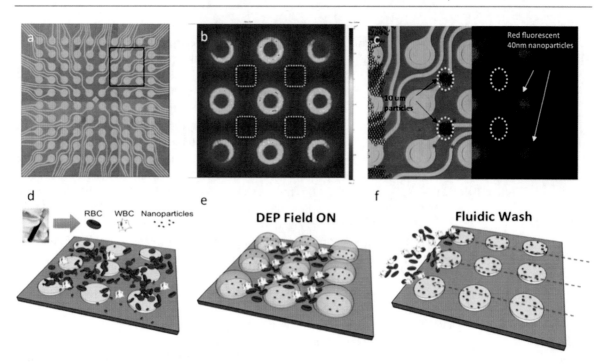

Detection of Nanoparticulate Biomarkers in Blood, Fig. 1 Microarray device and flow diagram for the separation of nanoparticles from blood. (**a**) Shows the electronic microarray device with 100 circular platinum microelectrodes 80 μm in diameter. (**b**) Alternating current (AC) electric fields were applied to a subset of nine microelectrodes in a checkerboard-addressing pattern. The electric field model for this geometry shows the DEP field maxima or high-field regions exist on the microelectrodes and the DEP field minima or low-field regions exist in the areas between the electrodes (*white dotted square*). (**c**) Shows a composite *half bright field/half red* fluorescence image for a prior DEP separation where 10 μm particles are isolated into the low-field regions and red fluorescent 40 nm nanoparticles are concentrated into the high-field regions (**d**) Shows the diagram of the microarray with *red* and *white* blood cells and *red* fluorescent nanoparticles before the DEP field is applied. (**e**) Shows the microarray after the DEP field is applied, with *red* fluorescent nanoparticles moving to the microelectrodes and the larger *red* and *white* blood cells moving between the microelectrodes. (**f**) Shows fluidic wash removing the blood cells, while the nanoparticles remain in the high-field regions

from San Diego Blood Bank (San Diego, CA). Buffy coat blood contains ~1/10 the number of red blood cells as whole blood. Whole blood samples were obtained from adult female Sprague Dawley rats using proper protocols. Conductivity measurements were made with an Accumet Research AR-50 Conductivity meter using 2 cell (range: 10–2,000 μS) and 4 cell (range: 1–200 mS) electrodes. Buffer conductivities were: 1× TBE: 1.09 mS/cm; 0.5× PBS: 7.6 mS/cm; buffy coat blood: 8.6 mS/cm, whole rat blood: 5.2 mS/cm (nanoparticle experiments). Fluorescent polystyrene nanoparticles (FluoSpheres) with NeutrAvidin were purchased from Invitrogen (Carlsbad, CA). The nanoparticles were 0.04 μm (40 nm) in diameter and red fluorescent (e × 585/em605). For human buffy coat blood experiments and rat blood experiments, 10 μL of the 40 nm red fluorescent nanoparticles was added from the stock solution to 300 μL of either buffy coat blood or whole rat blood. For comparison purposes, the number of nanoparticles in an Abraxane dosage [18] was calculated by using the recommended dosage of 260 mg/m^2 combined with an average estimated body surface area of 1.92 m^2 leading to a value of approximately 0.5 g of Abraxane per dosage. Since the mean diameter of an Abraxane nanoparticle is 130 nm, and estimating a density of ~1 g/mL, the number of circulating nanoparticles was calculated to be approximately 7.5×10^{10} nanoparticles per ml of blood (assuming 6 L of blood). The red fluorescent nanoparticles were thus serially diluted to the following amounts: 9.5×10^9, 9.5×10^{10}, and 9.5×10^{11} per mL in human buffy coat blood. The microarrays were controlled using a custom-made switching system that allows individual control over the voltage applied to each microelectrode.

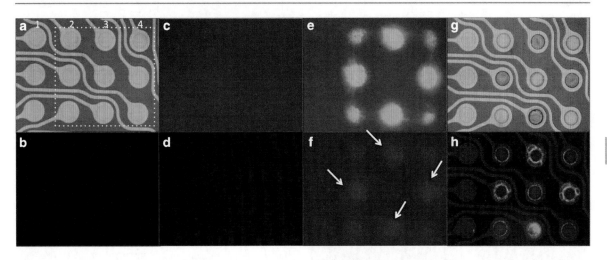

Detection of Nanoparticulate Biomarkers in Blood, Fig. 2 Separation and detection of 40 nm *red* fluorescent nanoparticles in buffy coat blood. (a) Shows the microarray in bright field and *red* fluorescence before the buffy coat blood sample was applied to the microarray. (b) Same as (a) in *red* fluorescence. (c) Shows the microarray in bright field after 20 μL of buffy coat blood containing 300 ng/μl of 40 nm *red* fluorescent nanoparticles were added to the microarray. (d) Same as (c) in *red* fluorescence. (e) Shows the microarray after the DEP field has been applied to a set of nine electrodes (columns 2, 3, and 4), while the three electrodes in the column 1 remained un-activated. The blood cells move away from the microelectrodes. (f) Same as (e) in *red* fluorescence where the *red* fluorescent nanoparticles can be seen concentrating on the microelectrodes. (g) Bright field image shows the microarray after it was washed with to remove the buffy coat blood cells (h) *Red* fluorescent image now shows the fluorescent nanoparticles concentrated on the microelectrodes

The microelectrodes were set to proper AC frequency and voltages using an Agilent 33120A Arbitrary Function Generator. AC frequencies ranged from 1,000 to 10,000 Hz, at 20 V peak to peak (pk-pk). The DEP field was generally applied for about 10–20 min. The DEP separations were visualized using a JenaLumar epifluorescent microscope (red fluorescence nanoparticles E × 585 nm, Em 605 nm). Both bright field and red fluorescent images were captured using an Optronics 24-bit RGB CCD camera. The image data was processed using a Canopus ADVC-55 video capture card connected to a laptop computer with Adobe Premiere Pro and Windows Movie Maker. The final fluorescence intensity data images were created by inputting fluorescent image frames of the video into MATLAB.

Key Research Findings

The initial HC-DEP experiment demonstrates the separation and detection of 40 nm red fluorescent nanoparticles in human buffy coat blood (0.86 S/m). Buffy coat blood was used in this experiment because it allows the complete DEP separation process to be observed. The DEP experiment was carried out at 10,000 Hz and 10 V pk-pk. Figure 2a, b shows the microarray in bright field and red fluorescence before the buffy coat blood sample was applied. Twelve microelectrodes are seen in the bright field image (Fig. 2a), and the red fluorescence image is dark (Fig. 2b). *All images show 12 microelectrodes, with the three control microelectrodes in column 1 not activated, and nine microelectrodes in columns 2, 3, and 4, which are activated.* Figure 2c, d shows images of the microarray in bright field and red fluorescence after 20 μL of buffy coat blood containing 300 ng/μL of 40 nm red fluorescent nanoparticles was added. The microelectrodes are not visible in the bright field image because of the high cell density (Fig. 2c), but a red fluorescent background is now seen in the fluorescent image (Fig. 2d). Figure 2e, f shows the microarray after the DEP field has been applied for 12 min to the nine microelectrodes. The bright field image shows the microelectrodes becoming more visible as the blood cells move into the low-field regions (Fig. 2e), and the fluorescent image shows the red fluorescent nanoparticles concentrating in the high-field regions around the microelectrodes (Fig. 2f). Finally, Fig. 2g, h shows the microarray after washing with 0.5 × PBS buffer to remove the blood cells. The bright field image shows the microarray is clear of cells (Fig. 2g), while the red fluorescent image shows intense fluorescence from the concentrated nanoparticles (Fig. 2h). The overall time from sample application to detection was less than 30 min.

The second HC-DEP experiment demonstrates the separation and detection of 40 nm red fluorescent nanoparticles in undiluted whole rat blood (0.52 S/m). This experiment was carried out at 10,000 Hz and 10 V pk-pk. Figure 3a, b shows the microarray before the whole blood sample was added. Figure 3c, d shows images of the microarray after 20 μL of whole blood containing 300 ng/μL of 40 nm red fluorescent nanoparticles was added. The DEP field was then applied for 11 min. Because of the very high cell density of whole blood, the movement of cells into the low-field regions and concentration of fluorescent nanoparticles into the high-field regions was not observable (while blood cells are present). After one fluidic wash with 0.5× PBS, the bright field image shows the microelectrodes with only a few cells present (Fig. 3e), and the red fluorescent image shows intense fluorescence from the nanoparticles concentrated around the microelectrodes (Fig. 3f). The microarray was then washed five times with 0.5× PBS, and the final images show no cells present (Fig. 3g), with significant red fluorescence remaining on the microelectrodes (Fig. 3h). Figure 3i is a 3D fluorescent intensity image (after the first wash) showing the relative levels of fluorescence on the activated microelectrodes versus the un-activated microelectrodes. The overall time from sample application to detection was less than 30 min.

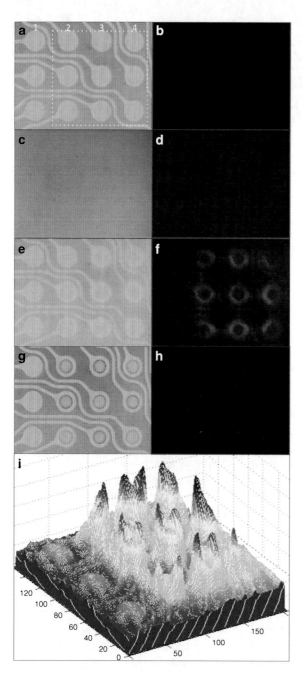

Detection of Nanoparticulate Biomarkers in Blood, Fig. 3
(continued)

Detection of Nanoparticulate Biomarkers in Blood, Fig. 3 Separation and detection of 40 nm *red* fluorescent nanoparticles in whole blood. (**a**) Bright field image before the whole blood sample was added. (**b**) Same as (**a**) in *red* fluorescence. (**c**) Bright field image after 20 μL of whole blood containing 300 ng/μL of 40 nm *red* fluorescent nanoparticles was added. (**d**) Same as (**c**) in *red* fluorescence. (**e**) Bright field image after DEP field was applied to a set of nine microelectrodes (columns 2, 3, and 4), and the blood was removed. The bright field image shows the underlying microelectrodes with some blood cells scattered across the surface. (**f**) Same as (**e**) where *red* fluorescent image shows the fluorescent nanoparticles concentrated over the microelectrode structures (columns 2, 3, and 4). (**g**) Bright field image after washing five times with 0.5× PBS – no cells remain. (**h**) Same as (**g**) – Red fluorescence from nanoparticles still remains on the microelectrodes. (**i**) 3D fluorescent intensity image after the first 0.5× PBS buffer wash showing the relative levels of fluorescence on all nine microelectrodes (produced using MATLAB)

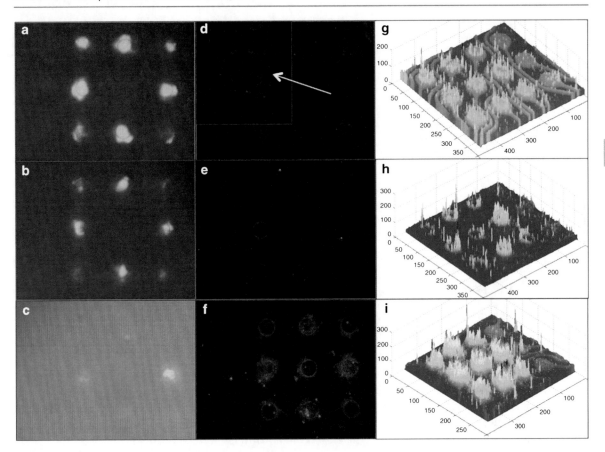

Detection of Nanoparticulate Biomarkers in Blood, Fig. 4 HC-DEP was carried out on samples of red fluorescent 40 nm nanoparticles at 9.5×10^9, 9.5×10^{10}, and 9.5×10^{11} particles/mL inserted in buffy coat blood. Blood cells are seen moving away from the activated microelectrodes. Figures (**a**) Bright field image for 9.5×10^9 nanoparticles/mL concentration, (**b**) Bright field image for 9.5×10^{10} nanoparticles/mL concentration, (**c**) Bright field image for 9.5×10^9 nanoparticles/mL concentration. (**d**) *Red* Fluorescent Image of (**a**) after the microarray was washed with 0.5× PBS. (**e**) *Red* Fluorescent Image of (**b**) after the microarray was washed with 0.5× PBS. (**f**) *Red* Fluorescent Image of (**c**) after the microarray was washed with 0.5× PBS. (**g**) 3D images showing the relative fluorescence intensity levels on the activated microelectrodes versus the un-activated microelectrodes for (**d**). (**h**) 3D images showing the relative fluorescence intensity levels for (**e**). (**i**) 3D images showing the relative fluorescence intensity levels for (**f**)

HC-DEP experiments were now carried out in order to determine if nanoparticles could be detected in blood at the basic dosage range used for present drug delivery nanoparticles. By way of example, the dosage for Abraxane drug delivery nanoparticles is approximately 7.5×10^{10} particles/mL blood [18]. Red fluorescent 40 nm nanoparticles at 9.5×10^9, 9.5×10^{10}, and 9.5×10^{11} particles/mL were made up in buffy coat blood, and HC-DEP was carried out at 10,000 Hz and 10 V pk-pk for 20 min. Figure 4a–c shows the bright field image for the three different concentrations of nanoparticles, where the blood cells can be seen moving away from the activated microelectrodes into the low-field regions. Figure 4d–f

now shows the red fluorescent images of the three different concentrations of nanoparticles after the microarray was washed with 0.5× PBS. Figure 4d shows the nanoparticles at the lowest concentration (9.5×10^9 particles/mL), with the enlargement of the microelectrode clearly showing fluorescence from the concentrated nanoparticles. Figure 4g–i shows the 3D fluorescent intensity images showing the relative levels of fluorescence on the activated microelectrodes versus the un-activated microelectrodes. These results demonstrate the intrinsic ability of HC-DEP to detect nanoparticles in blood at dosage levels now used for drug delivery nanoparticles.

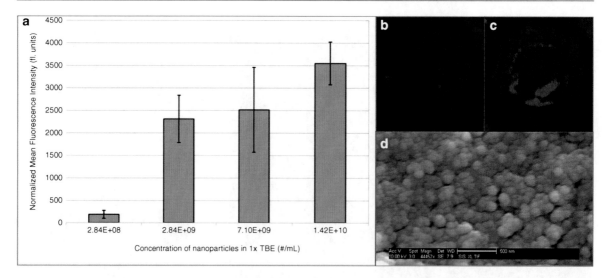

Detection of Nanoparticulate Biomarkers in Blood, Fig. 5 Detection levels for the *red* fluorescent 40 nm nanoparticles in $1\times$ TBE buffer. (**a**) Graph shows results after HC-DEP for a dilution series of nanoparticles, ranging from 2.8×10^8 to 2.8×10^{10} particles/mL. Nanoparticles could be detected down to the 2.8×10^8 particles/mL level. (**b**) *Red* fluorescent image before DEP field is applied. (**c**) *Red* Fluorescent image after the DEP field is applied. (**d**) SEM image of the microelectrode with clusters of the 40 nm nanoparticles clearly visible

Further experiments were carried out to determine the detection levels for the red fluorescent 40 nm nanoparticles in $1\times$ TBE using dilution series of nanoparticles, which ranged from 2.8×10^8 to 2.8×10^{10} particles/mL. HC-DEP was carried out at 10,000 Hz and 10 V pk-pk for 15 min. The results are presented in Fig. 5a (graph) and show that the nanoparticles can be detected down to the 2.8×10^8 particles/mL level. Because only a 1 µL volume of solution is actually being affected by the DEP field (on nine microelectrodes), the true number of nanoparticles being detected on each microelectrode is $<3 \times 10^5$ particles per microelectrode. A final experiment was now carried out to verify by scanning electron microscopy (SEM) that the 40 nm nanoparticles are truly being deposited into the high-field regions on the microelectrodes. For these experiments, a microelectrode array without a hydrogel layer was used. Figure 5d shows a red fluorescent image of one of the microelectrodes before the DEP field is applied. Figure 5e shows an image of the microelectrode after the DEP field was applied at 10,000 Hz and 10 V pk-pk for 4 min. Red fluorescence can be clearly seen concentrated around and on the microelectrode. Finally, Fig. 5f shows the SEM image of the same

microelectrode with clusters of the 40 nm nanoparticles clearly visible. In previous work, similar SEM results for HC-DEP using 200 nm nanoparticles were shown [17].

Future Directions for Research

The ability to rapidly detect and monitor the concentration of drug delivery nanoparticles will be important for future nanomedicine applications. HC-DEP has now been shown to be able to rapidly isolate and detect 40 nm nanoparticles from whole blood and buffy coat blood at clinically relevant levels, $\sim 9 \times 10^9$ particles/mL. Further studies in buffer show that even lower detection limits can easily be achieved, 2.8×10^8 particles/mL. The fact that the isolation of nanoparticles can be carried out directly in whole blood, with no sample processing, means that HC-DEP holds considerable promise for being a viable and cost-effective method for point-of-care applications. At the AC frequencies being used (5,000–10,000 kHz), HC-DEP provides a powerful nanoscopic tool that will allow a variety of important nanoscale entities in the 10–500 nm range to be manipulated and isolated relative to larger cells (red and white blood cells) and numerous smaller proteins and metabolic biomolecules. In addition to having high conductance (0.5–0.9 S/m) [19], whole blood is also one of the most complex of biological samples. Other important nanoscale entities that could be isolated using HC-DEP include: nanoparticle imaging agents, virus, prions, chylomicrons, large antibody and immunoglobulin complexes, and other cellular nanoparticulate biomarkers (nuclei, mitochondria, ribosomes, lysosomes, and various storage vacuoles). It should also be kept in mind that at higher AC frequencies (>10,000 Hz), DEP has the intrinsic ability for the higher resolution separation of different nanoparticles based on their AC crossover frequency and dielectric properties [20]. A final advantage for HC-DEP is the fact that the entities being isolated do not have to be labeled. The ability to post-label a specific biomarker or analyte will prove to be a significant advantage for future drug monitoring and diagnostic applications. Thus, numerous molecular biological detection techniques including PCR, immunochemistry, specific fluorescent dyes, and in situ hybridization can be easily carried out in the same sample chamber. Overall, the results of this study have shown that HC-DEP has enormous potential for a number of new "seamless" sample to answer nanomedicine applications.

Cross-References

▶ Dielectrophoresis
▶ Dielectrophoresis of Nucleic Acids

References

1. Editorial: Healthy challenges. Nat. Nanotechnol. **2**, 451 (2007)
2. Nishiyama, N.: Nanomedicine: nanocarriers shape up for long life. Nat. Nanotechnol. **2**, 203–204 (2007)
3. Ferrari, M.: Cancer nanotechnology: opportunities and challenges. Nat. Rev. Cancer **5**, 161–171 (2005)
4. Mu, L., Feng, S.S.: A novel controlled release formulation for the anticancer drug paclitaxel (Taxol(R)): PLGA nanoparticles containing vitamin E TPGS. J. Control. Release **86**, 33–48 (2003)
5. Duncan, R.: The dawning era of polymer therapeutics. Nat. Rev. Drug Discov. **2**, 347–360 (2003)
6. Nishiyama, N., Kataoka, K.: Current state, achievements, and future prospects of polymeric micelles as nanocarriers for drug and gene delivery. Pharmacol. Therapeut. **112**, 630–648 (2006)
7. Sozzi, G., et al.: Quantification of free circulating DNA as a diagnostic marker in lung cancer. J. Clin. Oncol. **21**, 3902–3908 (2003)
8. Board, R.E., et al.: DNA methylation in circulating tumor DNA as a biomarker for cancer. Biomarker Insights **2**, 307–319 (2007)
9. Gautschi, O., et al.: Circulating deoxyribonucleic acid as prognostic marker in non-small-cell lung cancer patients undergoing chemotherapy. J. Clin. Oncol. **22**, 4157–4164 (2004)
10. Albrecht, D.R., Underhill, G.H., Wassermann, T.B., Sah, R.L., Bhatia, S.N.: Probing the role of multicellular organization in three-dimensional microenvironments. Nat. Methods **3**, 369–375 (2006)
11. Morgan, H., Hughes, M.P., Green, N.G.: Separation of submicron bioparticles by dielectrophoresis. Biophys. J. **77**, 516–525 (1999)
12. Ramos, A., Morgan, H., Green, N.G., Castellanos, A.: Ac electrokinetics: a review of forces in microelectrode structures. J. Phys. D: Appl. Phys. **31**, 2338–2353 (1998)
13. Hughes, M.P.: Nanoparticle manipulation by electrostatic forces. In: Goodard, W.A., Brenner, D.W., Lyshevski, S.E. (eds.) Handbook of Nanoscience, Engineering, and Technology, 2nd edn, pp. 16-1–16-32. CRC Press/Taylor and Francis Group, Boca Raton (2007)
14. Cheng, J., et al.: Preparation and hybridization analysis of DNA/RNA from *E. coli* on microfabricated bioelectronic chips. Nat. Biotechnol. **16**, 541–546 (1998)
15. Krishnan, R., Sullivan, B.D., Mifflin, R.L., Esener, S.C., Heller, M.J.: Alternating current electrokinetic separation and detection of DNA nanoparticles in high-conductance solutions. Electrophoresis **29**, 1765–1774 (2008)
16. Krishnan, R., Heller, M.J.: An AC electrokinetic method for enhanced detection of DNA nanoparticles. J. Biophotonics **2**, 253–261 (2009)
17. Krishnan, R., Dehlinger, D.A., Gemmen, G.J., Mifflin, R.L., Esener, S.C., Heller, M.J.: Interaction of nanoparticles at the DEP microelectrode interface under high conductance conditions. Electrochem. Commun. **11**, 1661–1666 (2009)
18. Green, M.R., et al.: Abraxane, a novel Cremophor-free, albumin-bound particle form of paclitaxel for the treatment of advanced non-small-cell lung cancer. Ann. Oncol. **17**, 1263–1268 (2006)
19. Hirsch, F.G., et al.: The electrical conductivity of blood: I. relationship to erythrocyte conductivity. Blood **5**, 1017–1035 (1950)
20. Green, N.G., Ramos, A., Morgan, H.: Ac electrokinetics: a survey of sub-micrometre particle dynamics. J. Phys. D: Appl. Phys. **33**, 632–641 (2000)

Dielectric Force

▶ Dielectrophoresis

Dielectrophoresis

Nicolas G. Green[1] and Hossein Nili[1,2]
[1]School of Electronics and Computer Science, University of Southampton, Highfield, Southampton, UK
[2]Nano Research Group, University of Southampton, Highfield, Southampton, UK

Synonyms

AC electrokinetics; Dielectric force

Definition

Dielectrophoresis (DEP) is defined as the motion of polarizable particles under the influence of an applied *nonuniform* electric field, with the force arising from the interaction of the field and the dipole moment induced in the particle.

Introduction

Dielectrophoresis is a second-order effect arising from the net effect of the dielectric force acting on interfaces between different dielectrics. It is second order in that the field generates a charge on the interface due to the different polarizability of the dielectrics and then acts on that charge to produce a force. The second-order nature also means that the time average of the force in an AC field is nonzero. Where the interface is a closed surface surrounding a volume such as a particle or a droplet which is free to move, the net force results in motion where the field is nonuniform. This dielectrophoretic motion is due to the fact that the charge on the interface forms a dipole across the particle and the force on the two poles of the dipole is different since the electric field is nonuniform.

The phenomenon of dielectrophoresis was first recorded in the late nineteenth century as an effect arising in the behavior of soil samples. The term "dielectrophoresis" was coined by Herbert Pohl in 1951 to describe the motion of biological particles under the influence of nonuniform electric fields [1]. Pohl went on to study the effect in detail and in 1978 published the seminal work on the subject [2]. With the application of semiconductor fabrication methods allowing researchers to easily build field-generating electrodes on the scale of biological cells, dielectrophoresis gained traction in the late twentieth century. Applications followed for manipulation, characterization, and separation of micrometer-sized particles such as cells and bacteria down to nanometer-scale particles such as viruses and proteins.

Interfacial Polarization of Dielectric Materials

The forces which give rise to dielectrophoresis are general in that a force is generated on any surface between materials or any internal discontinuity in a material. Understanding of the force starts with an understanding of dielectric materials: materials which possess bound charges which *polarize* under the influence of an applied electric field. The bound charges can only move minute distances under the influence of an externally applied electric field.

Polarization
In electrostatics, the force acting between charges or the force on a charge in an electric field can be written as $\mathbf{F}_c = q\mathbf{E}$, where q is the charge and $\mathbf{E}$ is the electric field. When positive and negative charges are placed in an electric field, they are therefore displaced in opposite directions. If those charges are bound to atoms or molecules, they will move a short distance and then be held in displaced positions, producing a dipole moment in the atom/molecule.

The dipole moment is a fundamental object in electrostatics consisting of two equal and opposite charges, $+q$ and $-q$ for example, displaced by a vector distance $\mathbf{d}$. The dipole moment in this case is given by $\mathbf{p} = q\mathbf{d}$. Many molecules possess fixed dipole moments and in the case of dipoles arising from an applied electric field, the molecule then has an *induced* dipole moment. The process of this happening in a dielectric material, with all the molecules developing dipole moments, is referred to as *polarization*. In a dielectric, consisting of N molecules per cubic meter, the electric polarization $\mathbf{P}$, or the average vector dipole moment per molecule, is given by $\mathbf{P} = N\mathbf{p}$.

There are three basic polarization processes. Under an applied electric field, the center of charge of the electron cloud in a molecule moves slightly with respect to the center of charge of the nucleus; this is referred to as electronic polarization. In a solid such as crystalline potassium chloride, the ions of different sign move in different directions when subjected to an electric field; this is referred to as atomic polarization. If the molecules possess a permanent dipole moment, the material is described as polar, and the molecules both align with and become further polarized in an applied electric field; this is referred to as orientational polarization.

Charge Movement in Materials

Polarization causes charges to gather, either at points within the dielectric or at the surface [3, 4]. Differences in polarization on either side of an interface mean that there is a net charge at that interface. This charge can be generated by bound charges from polarization as described above or in real rather than ideal dielectrics by free charge associated with electrical conduction. In water-based biological systems where dielectrophoresis is typically applied, there is generally a fairly high conductivity arising from the ions and salts suspended in the solutions.

The important relationships can be derived simply by looking at charge densities and the relationship with polarization and the electric field. In a material, the bound volume charge density is related to the polarization by

$$\rho_b = -\nabla \cdot \mathbf{P} \tag{1}$$

and the total charge, which is the sum of the free charge density ρ_f and the bound charge density, is related to the electric field by Gauss's Law:

$$\nabla \cdot \mathbf{E} = \frac{\rho_f + \rho_b}{\varepsilon_o} \tag{2}$$

where ε_o is the permittivity of free space. Substituting Eq. 1 into Eq. 2 gives

$$\nabla \cdot (\varepsilon_o \mathbf{E} + \mathbf{P}) = \rho_f \tag{3}$$

This is generally written as

$$\nabla \cdot \mathbf{D} = \rho_f \tag{4}$$

where $\mathbf{D} = \varepsilon_o \mathbf{E} + \mathbf{P}$ is termed the electric flux density. For a linear isotropic dielectric $\mathbf{P}$ is proportional to $\mathbf{E}$

such that $\mathbf{P} = \varepsilon_o \chi \, \mathbf{E}$ where χ is the electric susceptibility of the dielectric. Therefore, for a linear and isotropic dielectric, from Eq. 3:

$$\mathbf{D} = \varepsilon_o \mathbf{E} + \mathbf{P} = \varepsilon_o(1 + \chi)\mathbf{E} = \varepsilon_o \varepsilon_r \mathbf{E} \tag{5}$$

where $\varepsilon_r = (1 + \chi)$ is the relative permittivity of the dielectric.

Frequency-Dependent Charge Processes

The extension of the simple electrostatic model to incorporate polarization effects and processes which depend on frequency can be illustrated using some simple models [3]. The most basic aspect of charge processes which involve frequency is that of conduction from the movement of free charges. This can be illustrated using a model parallel plate capacitor of area A and plate separation d, connected to an alternating potential ϕ of angular frequency ω, as shown in Fig. 1a. With an ideal dielectric filling the capacitor, the impedance is

$$Z = \frac{1}{i\omega C} \tag{6}$$

where $i^2 = -1$ and the capacitance C is given by

$$C = \varepsilon_o \varepsilon_r \frac{A}{d} \tag{7}$$

Most real dielectrics have a nonzero conductivity and the permittivity is dependent on the frequency of the applied electric field. With ideal dielectric replaced with a real dielectric of conductivity σ, the current in the circuit is the same as if the circuit were replaced with a capacitor (capacitance C from Eq. 8) representing the ideal dielectric and resistance in parallel given by

$$R = \frac{d}{\sigma A} \tag{8}$$

The total impedance of the circuit is then

$$Z = \frac{1}{1/R + i\omega C} = \frac{1}{\sigma \frac{A}{d} + i\omega \varepsilon_o \varepsilon_r \frac{A}{d}}$$

$$= \frac{1}{i\omega\left(\varepsilon_o \varepsilon_r - i\frac{\sigma}{\omega}\right)\frac{A}{d}} \tag{9}$$

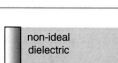 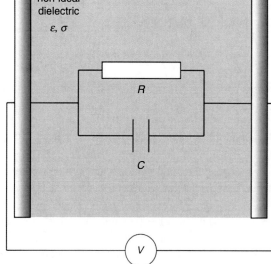 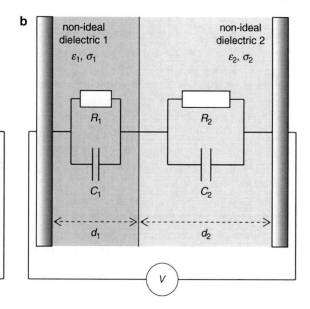

Dielectrophoresis, Fig. 1 (**a**) Parallel plate capacitor of plate separation d and area A containing a dielectric of permittivity $\varepsilon = \varepsilon_r\varepsilon_o$ and conductivity σ has the same impedance as the parallel circuit shown. (**b**) The same parallel plate capacitor completely filled by two dielectrics with permittivities ε_1, ε_2, conductivities σ_1, σ_2, and widths d_1 and d_2 has the equivalent circuit shown comprising that of the individual dielectrics in series

The real dielectric therefore behaves as a capacitor:

$$C = \tilde{\varepsilon}\frac{A}{d} \tag{10}$$

with a *complex polarizability* given by

$$\tilde{\varepsilon} = \varepsilon_o\varepsilon_r - i\frac{\sigma}{\omega} \tag{11}$$

This gives the frequency-dependent behavior of a basic homogeneous dielectric material. In general, the behavior of a dielectric is more complicated. Polarization mechanisms have a frequency-dependent nature and the material has additional frequency dependencies due to relaxation mechanisms.

Since polarization mechanisms involve the movement of charge to create dipoles and the rate of this movement is finite, each mechanism therefore has an associated characteristic time constant which is the time necessary for the polarization to reach its maximum value. As the frequency of the applied alternating field increases, the period over which the dipole moment is able to "relax" and then follow the field decreases. Therefore, there is a frequency at which these two periods are the same and beyond which the polarization no longer reaches its maximum. At much higher frequencies, no polarization occurs and

a relaxation or dispersion has occurred in the polarizability.

Of the three basic polarization mechanisms outlined previously, orientational polarization has the longest relaxation time. Atomic and electronic polarization will align with the field up to frequencies of the order of 10^{14} Hz and from the perspective of dielectrophoresis, which typically operates only up to 10^8 Hz, can be considered constant i.e., $\mathbf{P}_a = \varepsilon_o\chi_a\mathbf{E}$. The orientational polarization is given by a complex expression which accounts for the frequency variation:

$$\mathbf{P}_{or} = \frac{\varepsilon_o\chi_{or}\mathbf{E}}{1 + i\omega\tau_{or}} \tag{12}$$

where τ_{or} is the *relaxation time* and χ_{or} is the low-frequency static limit for the orientational polarization mechanism. The total frequency-dependent polarization is then

$$\mathbf{P} = \varepsilon_o\left(\chi_a + \frac{\chi_{or}}{1 + i\omega\tau_{or}}\right)\mathbf{E} \tag{13}$$

At the low-frequency limit, $\chi = \chi_a + \chi_{or} = \varepsilon_s - 1$, where ε_s is the relative permittivity measured in a static electric field. At the high-frequency limit, $\chi = \chi_a = \varepsilon_\infty - 1$, where ε_∞ is the relative permittivity at sufficiently high frequency that there is no orientational polarization. χ_{or} is therefore given by $\varepsilon_s - \varepsilon_\infty$ and the total polarization given by

$$\mathbf{P} = \varepsilon_o(\varepsilon^* - 1)\mathbf{E} \tag{14}$$

where

$$\varepsilon^* = \varepsilon_\infty + \frac{\varepsilon_s - \varepsilon_\infty}{1 + i\omega\tau_{or}} \tag{15}$$

This description gives the frequency-dependent polarization of a dielectric material and is normally written in separate real and imaginary components as the Debye formulations:

$$\varepsilon^* = \varepsilon' - i\varepsilon''$$

where $\varepsilon' = \varepsilon_\infty + \dfrac{\varepsilon_s - \varepsilon_\infty}{1 + \omega^2\tau_{or}^2}$ and $\varepsilon'' = \dfrac{\omega\tau_{or}(\varepsilon_s - \varepsilon_\infty)}{1 + \omega^2\tau_{or}^2}$

$$\tag{16}$$

Interfacial Polarization

As stated previously, experimental systems are further complicated by the presence of many dielectrics with differing properties, and the difference in polarizability causes charges to build up at the discontinuities or interfaces between the dielectrics [3]. As the polarizabilities of each dielectric are frequency dependent, so the surface charge density at the interface is also frequency dependent. As a result, the permittivity and conductivity of the whole system exhibit additional dispersions due solely to the polarization of the interfaces. The study of such heterogeneous dielectrics was first discussed by Maxwell and extended by Wagner and the dispersion in permittivity arising because of an interface is referred to as Maxwell-Wagner interfacial polarization. This can be illustrated by reference to a simple parallel plate capacitor as shown in Fig. 1b where there are two dielectrics with different properties. The two dielectrics have permittivities ε_1 and ε_2 and conductivities σ_1 and σ_2, respectively, and they can be represented as two resistor/capacitor parallel circuits in series. The impedance of the equivalent circuit is given by

$$Z = Z_1 + Z_2 = \frac{R_1}{1 + i\omega R_1 C_1} + \frac{R_1}{1 + i\omega R_2 C_2} \tag{17}$$

This can be reworked as Eq. 10, with complex permittivity given by

$$\tilde{\varepsilon} = \varepsilon_o\varepsilon' - i\varepsilon_o\varepsilon'' - i\frac{\sigma}{\omega} \tag{18}$$

ε' and ε'' are given by the Debye formulations (Eq. 16) with

$$\varepsilon_\infty = \frac{d\varepsilon_1\varepsilon_2}{d_1\varepsilon_2 + d_2\varepsilon_1} \quad \varepsilon_s = \frac{d(d_1\varepsilon_1\sigma_2^2 + d_2\varepsilon_2\sigma_1^2)}{(d_1\sigma_2 + d_2\sigma_1)^2}$$
$$\tau = \varepsilon_o\frac{d_1\varepsilon_2 + d_2\varepsilon_1}{d_1\sigma_2 + d_2\sigma_1} \tag{19}$$

σ is the conductivity of the whole system, given by

$$\sigma = \frac{d\sigma_1\sigma_2}{d_1\sigma_2 + d_2\sigma_1} \tag{20}$$

The resulting permittivity of the whole material in the capacitor has a Debye type relaxation solely due to interfacial polarization. The frequencies for interfacial polarization occur at lower frequencies than for orientational polarization and a plot of the complex permittivity of the system as a whole would look something like Fig. 2.

Dielectrophoresis

Interfacial Polarization of Particles

A particle suspended in a liquid is a special case of a multi-dielectric system with a closed surface surrounding a volume of a dielectric material (the particle) [3]. Some particles have a more complicated internal structure, which introduce additional interfaces to the model. Under the influence of an applied electric field, the particle/liquid interface is charged by Maxwell-Wagner interfacial polarization.

A schematic diagram of the resulting effect on the particle is shown in Fig. 3. The difference in polarizability results in a surface charge density around the interface at the surface of the particle. This charge has an opposite sign on either side of the particle in line with the electric field, therefore producing an induced dipole in the particle. The direction of the dipole is either with the electric field if the particle polarizability is greater than the fluid or against the field if the particle polarizability is smaller. The dipole magnitude and direction in general depend, by inference, not only on the particle internal properties but also on the properties of the suspending fluid medium.

Dielectrophoresis,
Fig. 2 The frequency
variation of the complex
permittivity of a dielectric,
taking into account the typical
relaxation mechanisms

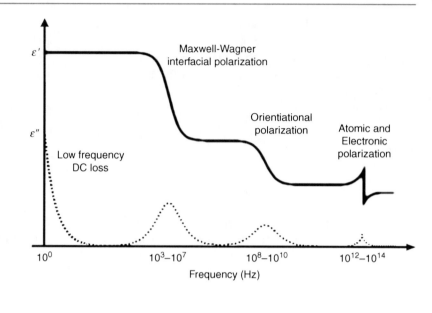

Dielectrophoresis,
Fig. 3 Electric field lines,
charges (distribution and net)
at dielectric interfaces, electric
field and effective dipole
moment vectors for when
a dielectric particle in
suspension is subjected to
a uniform electric field. The
effective dipole moment will
be aligned (**a**) with or
(**b**) against the applied electric
field depending on whether the
particle is more polarizable or
less polarizable than the
suspending medium

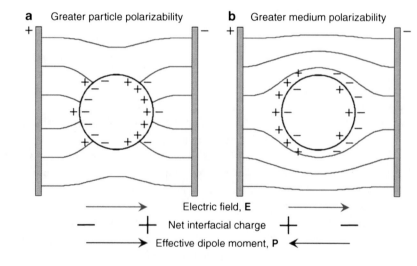

Determining an expression for the dipole moment of a spherical particle is an exercise in solving Poisson's equation using spherical harmonics which can be found in reference [5]. The solution of the electrical potential in spherical polar coordinates is

$$\phi_m = Er\cos\theta \left[\frac{a^3}{r^3} \left(\frac{\tilde{\varepsilon}_p - \tilde{\varepsilon}_m}{\tilde{\varepsilon}_p + 2\tilde{\varepsilon}_m} \right) - 1 \right] \qquad (21)$$

in the fluid medium (indicated by the subscript m) and

$$\phi_p = -\frac{3\tilde{\varepsilon}_m}{\tilde{\varepsilon}_p + 2\tilde{\varepsilon}_m} Er\cos\theta \qquad (22)$$

inside the particle (indicated by subscript p). Here, a is the radius of the particle. Comparison of Eq. 21 with the potential of a dipole:

$$\phi_{dipole} = \frac{qd\cos\theta}{4\pi\varepsilon_m r^2} \qquad (23)$$

shows that the electrical potential in the medium is the sum of the original uniform field and a dipole moment **p** where

$$\mathbf{p} = 4\pi\varepsilon_m \left(\frac{\tilde{\varepsilon}_p - \tilde{\varepsilon}_m}{\tilde{\varepsilon}_p + 2\tilde{\varepsilon}_m} \right) a^3 \mathbf{E} \qquad (24)$$

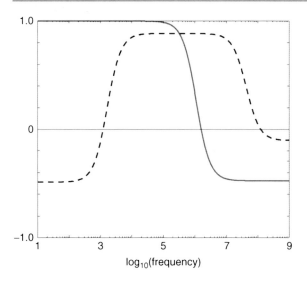

Dielectrophoresis, Fig. 4 Variations with frequency of the Clausius-Mossotti factor for two different particle types. Dispersions are due to Maxwell-Wagner interfacial polarization

This is the effective dipole moment of the sphere, with a size that is frequency dependent and undergoes a dispersion because of Maxwell-Wagner interfacial polarization. The effective polarizability of a spherical dielectric particle is

$$\tilde{\alpha} = 3\varepsilon_m \left(\frac{\tilde{\varepsilon}_p - \tilde{\varepsilon}_m}{\tilde{\varepsilon}_p + 2\tilde{\varepsilon}_m} \right) \qquad (25)$$

and the frequency dependency of the dipole and the effective polarizability is described by the Clausius-Mossotti factor:

$$\tilde{f}_{CM} = \frac{\tilde{\varepsilon}_p - \tilde{\varepsilon}_m}{\tilde{\varepsilon}_p + 2\tilde{\varepsilon}_m} \qquad (26)$$

The variation with frequency of the Clausius-Mossotti factor for two different particle types is shown in Fig. 4.

Translational Force on a Dipole in a Nonuniform Field

The simplest way to assign a force to a particle in a field is to consider the force exerted on a dipole by a nonuniform electric field [3, 4]. The dipole, shown in Fig. 5 (where the two equal and opposite charges $+q$ and $-q$ are separated by the vector $\mathbf{d}$), experiences a net force due to the nonuniform electric field $\mathbf{E}$ due to the fact that the field at, and therefore the force on,

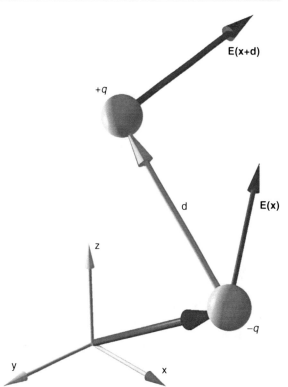

Dielectrophoresis, Fig. 5 The charges of the dipole $\mathbf{p} = Q\mathbf{d}$ in a nonuniform field $\mathbf{E}$ experience a different value of the electric field and as a result a different Coulomb force

the two poles of the dipole are not equal. The force can be represented as a sum of the forces on the two charges:

$$\mathbf{F} = q\mathbf{E}(\mathbf{x} + \mathbf{d}) - q\mathbf{E}(\mathbf{x}) \qquad (27)$$

where $\mathbf{x}$ can be taken to be the position vector of $-q$ without loss of generalization. This equation can be simplified if $|\mathbf{d}|$ is much smaller than $|\mathbf{x}|$ by expanding the electric field using a Taylor Series about $\mathbf{x}$, to give

$$\mathbf{E}(\mathbf{x} + \mathbf{d}) = \mathbf{E}(\mathbf{x}) + \mathbf{d} \cdot \nabla \mathbf{E}(\mathbf{x}) + \dots \qquad (28)$$

where all terms of order d^2 or higher have been neglected. If the dipole moment $\mathbf{p} = q\mathbf{d}$ remains finite in the limit $|\mathbf{d}| \rightarrow 0$, then substituting Eq. 28 into Eq. 27 gives the first-order force on an infinitesimal dipole as

$$\mathbf{F} = \mathbf{p} \cdot \nabla \mathbf{E} \qquad (29)$$

Only in a nonuniform field does this force have a nonzero net value.

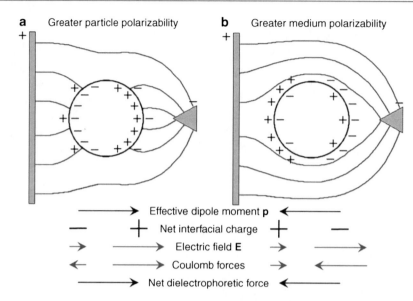

Dielectrophoresis, Fig. 6 Electric field lines, charges (distribution and net) at dielectric interfaces, directions of the applied electric field, effective dipole moment, Coulomb forces, and net dielectrophoretic force for when a dielectric particle in suspension is subjected to a nonuniform electric field. (**a**) When the particle is more polarizable than the suspending medium, the effective dipole moment is aligned with the applied electric field

and the DEP force acts to move the particle toward the region of highest electric field intensity. (**b**) When the particle is less polarizable than the suspending medium, the effective dipole moment is aligned against the applied electric field and the DEP force acts to move the particle away from the region of highest electric field strength

Dielectrophoretic Force

The resulting movement of particles is referred to as dielectrophoresis (DEP) and the force referred to as the dielectrophoretic force [3, 4]. Equation 29 is restricted to cases where the gradient of the field across the particle can be assumed to be constant, often referred to as the dipole approximation. In cases where the field is very nonuniform across a distance comparable in size to the particle, the higher order terms can no longer be ignored.

In the dipole approximation, the dielectrophoretic force on a particle is found by substituting the effective dipole moment, Eq. 24 into Eq. 29 to give

$$\mathbf{F}_{DEP} = \upsilon\alpha\mathbf{E}\cdot\nabla\mathbf{E} \qquad (30)$$

where υ is the volume of the particle and α is the effective polarizability (in this initial case a static polarizability). This can be rewritten using a vector identity as

$$\mathbf{F}_{DEP} = -\frac{1}{2}\upsilon\alpha\nabla(\mathbf{E}\cdot\mathbf{E}) = -\frac{1}{2}\upsilon\alpha\nabla|\mathbf{E}|^2 \qquad (31)$$

Since the effective polarizability of the particle can be either positive or negative, the dielectrophoretic force can act either in the direction of increasing field strength or in the opposite direction. Where particles are attracted toward regions of high field strength, the movement is referred to as *positive dielectrophoresis*. The opposite situation, where particles are repelled from high field regions, is called *negative dielectrophoresis*. The difference in the physical mechanism is illustrated in Fig. 6, where Coulomb forces on the two poles of the induced dipole are different owing to the nonuniform electric field. The force on the pole in the strongest electric field is the strongest but is toward high fields when the particle polarizability is greater (positive DEP) and away from high fields when the medium polarizability is greater (negative DEP).

Dielectrophoresis in an AC Field

Equation 31 is derived for a static electric field [3]. Deriving the force for an AC field requires an extension to account for the frequency and phase aspects associated with the time-varying charging. This extension also exposes additional force terms that produce related movement in particles.

Assuming an applied potential of a single frequency ω, then the time-dependent values in the system can be represented using phasors. An arbitrary, harmonic potential can be defined as $\phi(\mathbf{x}, t) = \text{Re}\left[\tilde{\phi}(\mathbf{x})e^{i\omega t}\right]$ where the tilde indicates the complex phasor ($\tilde{\phi} = \phi_R + i\phi_I$) and Re[. . .] indicates the real part of. The electric field is then given by $\mathbf{E}(\mathbf{x}, t) = \text{Re}\left[\tilde{\mathbf{E}}(\mathbf{x})e^{i\omega t}\right]$ where the vector $\tilde{\mathbf{E}} = -\nabla\tilde{\phi} = -(\nabla\phi_R + i\nabla\phi_I)$ is the corresponding phasor. The dipole moment of the particle is therefore $\tilde{\mathbf{p}} = v\tilde{\alpha}\mathbf{E}e^{i\omega t}$ and the time-averaged force on the particle is

$$\langle \mathbf{F}_{DEP} \rangle = \frac{1}{2}\text{Re}[(\tilde{\mathbf{p}} \cdot \nabla)\mathbf{E}^*] \qquad (32)$$

where * indicates complex conjugate.

If the phase is constant across the system, the field phasor can be assumed to be real without loss of generality (i.e., $\tilde{\mathbf{E}} = \mathbf{E} = -\nabla\phi_R$). The time-averaged force is then

$$\langle \mathbf{F}_{DEP} \rangle = \frac{1}{2}v\text{Re}[\tilde{\alpha}](\mathbf{E} \cdot \nabla)\mathbf{E} \qquad (33)$$

Using vector identities and the fact that the field is irrotational gives

$$\langle \mathbf{F}_{DEP} \rangle = \frac{1}{4}v\text{Re}[\tilde{\alpha}]\nabla(\mathbf{E} \cdot \mathbf{E}) \qquad (34)$$

This is the dielectrophoretic force, which is commonly written as

$$\langle \mathbf{F}_{DEP} \rangle = \frac{1}{4}v\text{Re}[\tilde{\alpha}]\nabla|\mathbf{E}|^2 \qquad (35)$$

Often the field is quoted using *root-mean-square* (*rms*) values so that Eq. 35 becomes

$$\langle \mathbf{F}_{DEP} \rangle = \frac{1}{2}v\text{Re}[\tilde{\alpha}]\nabla|\mathbf{E}_{rms}|^2 \qquad (36)$$

This is also the expression for the force in a DC field. The field term in Eq. 35 is the magnitude of the phasor and the difference between the two equations is the factor of a half. When using AC fields it is therefore extremely important to understand whether you are using amplitude or RMS values for the electric field. Inspection of these equations shows that the dielectrophoretic force depends on the volume of the particle and the gradient of the field magnitude squared. The force also depends on the real part of the effective polarizability.

This can be quantified for a spherical particle, where again the variation in the magnitude of the force with frequency is given by the real part of the Clausius-Mossotti factor. By substitution of the effective polarizability of a sphere (Eq. 25), the full expression for the time-averaged DEP force on a spherical particle is

$$\langle \mathbf{F}_{DEP} \rangle = \pi\varepsilon_m a^3 \text{Re}\left[\frac{\tilde{\varepsilon}_p - \tilde{\varepsilon}_m}{\tilde{\varepsilon}_p + 2\tilde{\varepsilon}_m}\right]\nabla|\mathbf{E}|^2 \qquad (37)$$

Examination of this equation shows that the magnitude of the force depends on the particle volume (a^3), the permittivity of the suspending medium, and the gradient of the field strength squared. The real part of the Clausius-Mossotti factor defines the frequency dependence and direction of the force. Expanding on the previous description, positive DEP occurs if the polarizability of the particle is greater than the suspending medium ($\text{Re}[\tilde{f}_{CM}] > 1$) and the particle moves toward regions of high electric field strength. Negative DEP occurs if the polarizability of the particle is less than the suspending medium ($\text{Re}[\tilde{f}_{CM}] < 1$) and the particles are repelled from regions of high field strength.

Dielectrophoresis in a Field with a Spatially Dependent Phase

For a general AC field, such as that generated by the application of multiple potentials of different phase, the derivation of the dielectrophoretic force is more involved [3]. The electric field in this case is $\mathbf{E}(\mathbf{x}, t) = \text{Re}\left[\tilde{\mathbf{E}}(\mathbf{x})e^{i\omega t}\right]$, where the vector $\tilde{\mathbf{E}} = -\nabla\tilde{\phi} = -(\nabla\phi_R + i\nabla\phi_I)$ is the corresponding *complex* phasor. The expression for the time-averaged force on the particle can then be derived from Eq. 32 with the vectors now consisting of complex components. The equation for the force is

$$\langle \mathbf{F}_{DEP} \rangle = \frac{1}{2}\text{Re}[(\tilde{\mathbf{p}} \cdot \nabla)\tilde{\mathbf{E}}^*] = \frac{1}{2}v\text{Re}[\tilde{\alpha}(\tilde{\mathbf{E}} \cdot \nabla)\tilde{\mathbf{E}}^*] \quad (38)$$

Again, using vector identities and the facts that the electric field is irrotational and has zero divergence (Gauss's law), the force expression becomes

$$\langle \mathbf{F}_{DEP} \rangle = \frac{1}{4} v \text{Re}[\tilde{\alpha}] \nabla (\widetilde{\mathbf{E}} \cdot \widetilde{\mathbf{E}}^*) - \frac{1}{2} v \text{Im}[\tilde{\alpha}] (\nabla \times (\widetilde{\mathbf{E}} \times \widetilde{\mathbf{E}}^*))$$

(39)

where Im [...] is the imaginary part of the function. This can be rewritten as

$$\langle \mathbf{F}_{DEP} \rangle = \frac{1}{4} v \text{Re}[\tilde{\alpha}] \nabla |\widetilde{\mathbf{E}}|^2 - \frac{1}{2} v \text{Im}[\tilde{\alpha}] (\nabla \times (\text{Re}[\widetilde{\mathbf{E}}] \times \text{Im}[\widetilde{\mathbf{E}}]))$$

(40)

where $|\widetilde{\mathbf{E}}|^2 = |\text{Re}[\widetilde{\mathbf{E}}]|^2 + |\text{Im}[\widetilde{\mathbf{E}}]|^2$. If there is no spatially varying phase, the phasor of the electric field can be taken to be real (i.e., $\text{Im}[\widetilde{\mathbf{E}}] = 0$), the second term on the right-hand side of equation is then zero and the expression for the force becomes that of Eq. 35. However, if there is a spatially varying phase, as in the case of traveling wave dielectrophoresis, the complete force expression must be used. The first term in the force expression depends on the frequency in the same manner as Eq. 35. The second term in the force equation depends, however, on the imaginary part of the effective polarizability, or rather the imaginary part of the Clausius-Mossotti factor. This force is zero at high and low frequencies, rising to a maximum value at the Maxwell-Wagner interfacial relaxation frequency.

A general statement would then be that the *general* dielectrophoretic force is divided into two components, one of which is the DEP force and another that depends on the phase of the field:

$$\langle \mathbf{F}_{DEP} \rangle = \underbrace{\frac{1}{4} v \text{Re}[\tilde{\alpha}] \nabla \left| \widetilde{\mathbf{E}} \right|^2}_{\text{Dielectrophoresis}}$$
$$\underbrace{- \frac{1}{2} v \text{Im}[\tilde{\alpha}] (\nabla \times (\text{Re}[\widetilde{\mathbf{E}}] \times \text{Im}[\widetilde{\mathbf{E}}]))}_{\text{Travelling Wave Dielectrophoresis}}$$

(41)

Generally, for traveling wave dielectrophoresis to be effective, a frequency must be chosen where both the real part is substantially high **and** the imaginary part, which only has a value close to the relaxation frequency, is nonzero.

Applications of Dielectrophoresis

The critical feature of dielectrophoresis is that it requires field nonuniformities in order to create movement. The rapid development of dielectrophoresis as a method was made capable by the production of microelectrodes using semiconductor fabrication methods [3, 4].

In addition to this, as most biological particles are dielectrics, dielectrophoresis is particularly fitted to lab-on-a-chip applications where multiple processes involving characterization, manipulation, and separation of bioparticles are integrated onto a single chip. Dielectrophoresis is advantageous over other means of inducing particle motion in its noninvasive nature and easy integration onto microdevices, not relying on any moving parts. Early biomedical applications of dielectrophoresis were limited to studying the response of cells to electric fields for characterization purposes. Nowadays, the technique has found widespread use for a variety of biotechnological and diagnostic applications. Examples include separation of human breast cancer cells from blood, trapping, and manipulation of DNA – of importance to the development of point-of-care devices for detection and identification of pathogenic microorganisms, aggregation of hemispherical cells as a first step toward the creation of artificial stem cell microniches in vitro, and creating engineered skin with artificial placodes of different sizes and shapes in different spatial patterns [6].

Dielectrophoresis has established its status as a versatile technique for separation of dielectric – biological and otherwise – particles in suspension. Through the frequency-dependent characteristics of dielectric particles, DEP has access to a wide range of particle properties and can be modified by the choice of suspending medium. Factors such as double layer and electro-convective effects can complicate DEP separation of dielectric particles [7]. Yet it has been shown that once these phenomena are correctly identified and incorporated in the design process, they can take their part in the toolbox of electrokinetic techniques for separation applications [8, 9].

Advances in fabrication technology have broadened the applicability of dielectrophoresis through realization of micro- and nano-electrode geometries that are capable of generating electric fields strong enough to move nanoparticles such as viruses and chromosomes [10]. At the nanoscale, effects such as thermal and hydrodynamic forces find increased significance. Yet it has been shown that the "nuisances" are controllable and not significant enough to hinder nanoparticle motion due to dielectrophoresis [11].

Applications of dielectrophoresis are not limited to the realm of diagnostics involving biological particles. As early as 1924, Hatfield achieved separation of minerals using DEP; interestingly, the work has been pursued by other research groups up until very recently. Other examples where dielectrophoresis has been used for industrial applications include construction of a current-limiting fuse using DEP collection of conductive particles between two electrodes, depositing a patterned coating of a nanostructured material onto a substrate using positive DEP, and the fabrication of nanoscale devices composed of movable components brought together in a fluid medium by exertion of the dielectrophoretic force – among other interactions [4].

With further developments in the design of suitable electrode geometry and in the modeling of phenomena that occur alongside dielectrophoretic motion of particles, DEP applications are expected to advance further, enabling manipulation and characterization of a yet wider range of particles.

Cross-References

- ▶ AC Electrokinetics of Nanoparticles
- ▶ Dielectrophoresis of Nucleic Acids
- ▶ Dielectrophoretic Nanoassembly of Nanotubes onto Nanoelectrodes
- ▶ Electric-Field-Assisted Deterministic Nanowire Assembly
- ▶ Electric Field–Directed Assembly of Bioderivatized Nanoparticles

References

1. Pohl, H.A.: The motion and precipitation of suspensoids in divergent electric fields. J. Appl. Phys. **22**, 869–871 (1951)
2. Pohl, H.A.: Dielectrophoresis: The Behavior of Neutral Matter in Nonuniform Electric Fields. Cambridge University Press, Cambridge (1978)
3. Morgan, H., Green, N.G.: AC Electrokinetics of Colloids and Nanoparticles. Research Studies Press, Philadelphia (2003)
4. Jones, T.B.: Electromechanics of Particles. Cambridge University Press, Cambridge (1995)
5. Lorrain, P., Corson, D.R., Lorrain, F.: Electromagnetic Fields and Waves. W.H. Freeman and Company, New York (1988)
6. Pethig, R.: Review article – dielectrophoresis: status of the theory, technology, and applications. Biomicrofluidics **4**, 022811 (2010)
7. Ramos, A., Morgan, H., Green, N.G., Castellanos, A.: AC electric-field-induced fluid flow in microelectrodes. J. Colloid Interface Sci. **217**, 420–422 (1999)
8. Gascoyne, P.R.C., Vykoukal, J.: Particle separation by dielectrophoresis. Electrophoresis **23**, 1973–1983 (2002)
9. Morgan, H., Hughes, M.P., Green, N.G.: Separation of submicron bioparticles by dielectrophoresis. Biophys. J. **77**, 516–525 (1999)
10. Hughes, M.P.: AC electrokinetics: applications for nanotechnology. Nanotechnology **11**, 124–132 (2000)
11. Ramos, A., Morgan, H., Green, N.G., Castellanos, A.: AC electrokinetics: a review of forces in microelectrode structures. J. Phys. D: Appl. Phys. **31**, 2338–2353 (1998)

Dielectrophoresis of Nucleic Acids

David J. Bakewell
Department of Electrical Engineering and Electronics, University of Liverpool, Liverpool, UK

Synonyms

Movement of polynucleotides in nonuniform electric field

Definitions

1. *Dielectrophoresis* (DEP) is the translational motion of an electrically polarizable body by the action of an externally applied, spatially nonuniform, electric field. DEP can occur with an electrically neutral body in a spatially nonuniform field and depends on the dielectric properties of the body relative to the surrounding medium [1]. The word *dielectrophoresis* can be said to be the conjunction of two parts: *phoresis* – meaning motion (from the Greek language) and *dielectro* – referring to the dependence of the motion on the dielectric properties of the body. DEP is distinguished from *electrophoresis* that requires a net charge on the body for Coulombic movement to occur in an externally applied, spatially uniform, electric field. DEP is

also distinguished from *electrorotation* (EROT) that refers to *rotational* motion of a body in a spatially inhomogeneous electric field.

2. *Nucleic acids* (NAs) are responsible for information storage, distribution, translation, and control throughout a living system, such as, a cell. There are two types of NAs central to life: deoxyribonucleic acid (DNA) and ribonucleic acid (RNA). DNA, discovered more than a century ago by Friedrich Miescher, is called a nucleic acid because it was extracted from the nuclei of cells (originally called nuclein) and showed acidic chemical properties [2, 3]. Nowadays, it is known that DNA can also be extracted from organelles in a cell, e.g., mitochondria, or other life forms, e.g., viruses, nonetheless, the term "nucleic" remains. Both DNA and RNA are comprised of nucleotides that consist of a nitrogenous base, a pentose sugar, and a phosphate link. A third type of NA is that which is not naturally occurring and is synthetically made, is artificial nucleic acid, e.g., peptide nucleic acid (PNA).

Introduction

The first, main exposition of dielectrophoresis (DEP) as an important method for moving an electrically neutral body was by Herbert Pohl [4], although the phenomenon was known as far back as 600 BC. DEP of a body suspended in a medium is usually achieved by applying an electrical potential between electrodes immersed in low conductivity electrolyte medium. Application of an electrical potential can yield an electric field that induces dipoles in the body. If the field is sufficiently nonuniform, the interaction of the dipole charges yields a net force on the body and causes it to move. Dependence on the dielectric polarizability of the body, rather than reliance on net electrical charge, is a key reason for considerable interest in using DEP for noncontact movement of micro- and nano-sized bodies. A corollary of the motion based on dielectric properties is that radio frequency (RF) electric fields can be used to move bodies by DEP and hence avoid problematic hydrolysis at the electrode-solvent interface. Developments in microfabrication methods that allow the generation of high electric field nonuniformities using RFs and low voltages has enabled DEP to become a popular method for moving

bodies in microenvironments, e.g., lab-on-chip (LOC) and micro-total analysis system (μTAS).

There is considerable interest in using DEP for moving nucleic acids (NAs) which are large molecules that play a central role in life. Both types of naturally occurring NAs, DNA and RNA, possess net negative electrical charge and electrophoresis is a commonly used tool in molecular biology laboratories today. DEP is less well known in current biology laboratories so that the focus in this essay lies with the DEP motion of these NAs in their native suspension at room temperature (25°C), i.e., water. Research into DEP in general, and in particular, DEP of DNA, has dramatically increased in recent years – as shown by world scientific journal publications quantified in Fig. 1. The plots show that in the last 4 years there are more than 30 publications per annum involving DEP of DNA. The topic, DEP of DNA, is of interest to nanotechnologists due to the central role DNA plays in life sciences and its properties as a nanoconstruction material. Progress in this field has been discussed in recent and highly informative reviews [1, 5, 6].

Conversely, there is comparatively little published on DEP of RNA or of PNA. This is perhaps because these NAs are more expensive and more complicated to prepare and handle than DNA; for instance, RNA is easily degraded (e.g., by an enzyme, RNAase) without extra laboratory precautions and facilities. This is unfortunate because RNA, for many scientists and technologists, is more important than DNA – see also "RNA world" hypothesis [3]. The following sections discuss fundamental structural, chemical, and physical properties of NAs for DEP. This is followed by a DEP of DNA application example and a summary of broader contributions to this field. Due to a need for concise referencing, many of the names of individual authors and groups are referred to as listed in reviews and books, rather than individual citations.

Fundamentals

This section describes the fundamentals of NAs, their structural and electrical properties in water, inducible polarizability, and the basic principles of DEP. The focus is on DNA since there are far fewer DEP studies on RNA and PNA and it is likely similar electrokinetic principles apply.

Dielectrophoresis of Nucleic Acids,
Fig. 1 Number of journal publications for years from 1995 to 2010 inclusive (Source: ISI Web of Knowledge v.4.10 © 2010 Thomson Reuters)

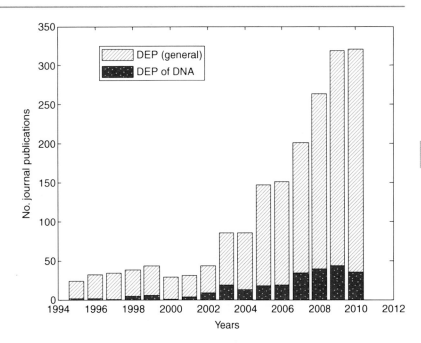

Nucleic Acids

The basic chemical building blocks of natural NAs are nucleotides as outlined in Fig. 2 and consist of three characteristic components:

– A nitrogen containing base that is either a
  – Purine with two possibilities: adenine (A) or guanine (G)
  – Pyrimidine with three possibilities: cytosine (C), thymine (T), or uracil (U)
– A pentose sugar ring (denoted S)
  – For DNA it is a 2′-deoxy-D-ribose
  – For RNA the ring is a D-ribose
– A phosphate (denoted P)

In DNA and RNA, the nucleotides are covalently linked with the phosphates acting as bridges between the pentose sugars. The 5′-phosphate group of one nucleotide is linked to the 3′-hydroxyl group of the next nucleotide. The linkages are often referred to as the "sugar-phosphate backbone." The spatial arrangement of the three constituents for each the four nucleotides that comprise DNA are shown in Fig. 2. A key property of NAs is that a single strand of linked nucleotides can form complementary pairs with another strand. That is, each base of a nucleotide on one strand can form hydrogen (H) bonds with the base of a nucleotide on the opposite strand. This occurs for all bases on each strand with purines pairing with pyramidines. Generally the following rules apply:

– DNA contains A, C, G, and T and they pair as A with T, C with G
– RNA contains A, C, G, and U and they can pair A with U, C with G

DNA and RNA are distinguished by their pentose sugars and they also differ by one of the bases (T rather than U). Only very rarely do exceptions to these rules occur. There are a number of different structures of DNA, two of the most common are single-stranded DNA (ssDNA) and double-stranded DNA (dsDNA). Figure 2 shows base pairing where A and C on the left ssDNA strand forms H-bonds with T and G on the right strand. There are different forms of dsDNA, namely, A, B, and Z [7]. The standard double-helical DNA usually referred to is (Crick and Watson) B-form.

Figure 3 shows a very short length of dsDNA as it occurs in aqueous (or water) solution, i.e., in the energetically favorable state as a double helix. The sugar-phosphate double-helical backbones are shown as thick lines. Structurally "backbone" is somewhat of a misnomer as it is the stacking of the nitrogen bases that gives the molecule its mechanical stiffness [7]. The sugar-phosphate backbones are hydrophilic and lie on the exterior of the macromolecule. The hydrophobic nucleotide base-pairs (bp) lie within the macromolecule, away from the polar water molecules. They are represented symbolically (A–T, C–G pairs) with respective double and triple H-bonds. The

Dielectrophoresis of Nucleic Acids, Fig. 2 Ladder sketch of dsDNA nucleic acid showing the constituents of each nucleotide containing a base, sugar (S), and phosphate (P), and pairing of left and right strands. Not to scale; for twist and writhe angles between molecules in 3D, see text for references

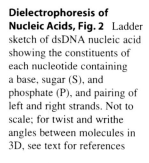
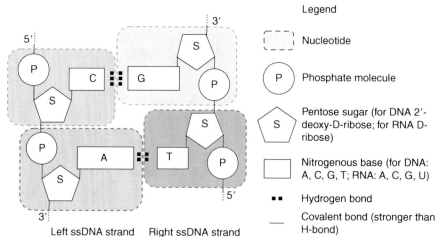

Left ssDNA strand   Right ssDNA strand

Legend

- - - Nucleotide

(P) Phosphate molecule

⬠ S Pentose sugar (for DNA 2'-deoxy-D-ribose; for RNA D-ribose)

▭ Nitrogenous base (for DNA: A, C, G, T; RNA: A, C, G, U)

▪▪ Hydrogen bond

— Covalent bond (stronger than H-bond)

displacement along the major helical axis between base-pairs is shown as 0.34 nm, so that the length between repetitive positions of the double helix is about 3.4 nm (10 bp). The diameter is about 2 nm.

Importantly, the H-bonds shown in Figs. 2 and 3 are weaker than the covalent bonds. This means that by moderate heating, or by using a different solvent, each dsDNA macromolecule can be reversibly separated into two complementary ssDNA strands. These ssDNA strands, in a solution with available nucleotides bases, can act as templates, thus enabling *replication*. They also enable detection: a ssDNA with a particular sequence will pair up with another ssDNA with the complementary sequence. In situations where a particular base does not match (e.g., A–A) and the others do, it is called a *mismatch*. These fundamental properties underpin the emergence of new technologies, such as, nanospheres and microarrays, where applications of DEP are being explored for enhancing concentration and other developments – as described in section Dielectrophoresis of DNA: General.

RNA is mainly involved with the reading of DNA, called *transcription*, and its *translation* into proteins and other macromolecules associated with cellular information. There are a number of different types of RNA that perform different functions inside a cell:

– Messenger RNA (mRNA): concerns with transfer of genetic "blueprint" information that codes for proteins. There are 20 different amino acids (AAs) found in nature, and DNA codes for them using the A, C, G, T base alphabet. This means that it requires a triplet of bases to sufficiently code for 20 AAs. That is, four bases to choose independently for first

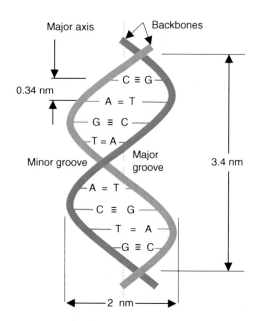

Dielectrophoresis of Nucleic Acids, Fig. 3 A short fragment of dsDNA as it appears in aqueous solution, double-helical B-form. See text for details

(base) position, four bases for the second, and so on, yields a total of $4^3 = 64$ possible AAs. This is more than enough to code for the actual 20 AAs needed to make proteins so the genetic code has redundancy. The triplet of bases that codes for each AA (and also DNA reading instructions) is called a *codon*.

– Transfer RNA (tRNA): reads or decodes each mRNA codon and transfers an appropriate AA onto a polypeptide chain that is being synthesized during production of a particular protein. tRNAs are small RNAs (about 80 nucleotides) with a characteristic clover leaf shape and act as adapters.

- Ribosomal RNA (rRNA): ribosomal RNA constitutes part of the ribosome. These are located in the cytoplasm of a cell and are responsible for catalyzing the synthesis of AA peptides and hence, protein, by decoding mRNA.
- Noncoding RNA is that RNA that is *not* mRNA (does not code for protein) or rRNA or tRNA, and are short ~22 nucleotide single strands. These perform a variety of functions that are being currently discovered. These include, e.g., small interfering RNA (siRNA) that can silence reading (gene expression) of DNA code, micro RNA (miRNA) that has been linked with cell cycle regulation, cardiac pathology, cancer, etc.

If RNA is extracted, e.g., from fish liver cells, the dominant amount of RNA is rRNA with the other amounts of RNA being much smaller. This means that mRNA needed for understanding protein production, e.g., for each gene, has to be carefully separated from the total amount of RNA.

Electrical Properties of Nucleic Acids

DNA is classified as a biological polyelectrolyte. These are biomolecules with a large number of charged chemical groups. An electrolyte or aqueous environment with salts acts as a supply for dissociated ions, e.g., Na^+, Mg^{2+}, OH^-, and Cl^-. DNA and RNA possess a net *negative* charge so they attract ions of the opposite charge, or *counterions*, by Coulombic forces to restore charge neutrality. The positive counterions, or cations, form a cloud around the NA and electrostatically screen the negative charge. Further away from the macromolecule, the counterions are more diffuse and the screening occurs in such a way as to attract other negative ions, or anions, with the same charge as the NA, and they are called *coions*.

The double-helical structure for DNA is shown to be straight in Fig. 3 but this only applies for relatively short lengths. The mechanical properties of DNA are partly determined by its chemical composition and the composition of the solvent. In a solvent with low counterion concentration there is little electrostatic screening so the polyelectrolyte repels itself and it tends to be straight. Conversely, in solvents with high counterion concentrations, there is considerable electrostatic screening so the DNA repels itself less and it

is more flexible. A measure of straightness is known as the persistence length, L_p, as sketched in Fig. 4. Typically for standard biological conditions for dsDNA, $L_p = 50$ nm [5]. The behavior of dsDNA in water is modeled as a worm-like chain with each straight link equal to the Kuhn length that is twice the persistence length, $L_K = 2L_p$ [8]. In this chain, due to thermal motion, the next link is independently and randomly oriented from the previous link. Each link is about 100 nm or about 300 bp. The persistence lengths of ssDNA and of ssRNA are much shorter than dsDNA, about 1 nm. The following discussion on dielectric polarization mechanisms includes both of the NAs but focuses on DNA since the literature, reviewed in [5], is much more extensive. In addition, dielectric spectroscopic measurements of the electrical properties of rRNA are often confounded by the presence of ribosomal proteins [9].

Dielectric Polarization Parameters

Polarization is a term that describes how charges, within a dielectric, respond to an externally applied electric field. Charges that are free to move over a long range reveal their movement on a macroscopic scale as conduction. If the movement of the charges is restricted they are said to be "polarized" [4]. In this respect, polarization is the "intention" of the charges to move in response to an applied electric field. The polarization can be expressed, in terms of an equivalent circuit with permittivity parameters, as the real (in-phase) permittivity ε' or *dielectric constant* and the free movement of charges as the imaginary (out-of phase) permittivity ε'', or *dielectric loss*. The permittivity parameters are frequency dependent: $\varepsilon \equiv \varepsilon(\omega)$. In addition the conductivity (or low-frequency Ohmic loss) of the electrolyte solution, σ, is included – see [10] and references therein, e.g., [19–21]. The complex permittivity ε is the combination of these:

$$\varepsilon(\omega) = \varepsilon'(\omega) - j\left[\varepsilon''(\omega) + \sigma/\omega\right] \quad (1)$$

where $j = \sqrt{-1}$, the angular frequency is $\omega = 2\pi f$ (rad s^{-1}) with frequency f (Hz). Often $\varepsilon'' << \sigma/\omega$, the superscript $'$ notation is omitted in the literature and the complex permittivity is written

$$\underline{\varepsilon}^* = \varepsilon - j\sigma/\omega \quad (2)$$

Dielectrophoresis of Nucleic Acids, Fig. 4 dsDNA (or DNA) as a negatively charged polyelectrolyte in solution attracts counterions (mostly cations). These form around the polyelectrolyte to electrostatically screen the charge of the DNA. Worm-like chain model of dsDNA, with close-up inset, in aqueous suspension, showing relatively straight segments of Kuhn length L_K. Not to scale

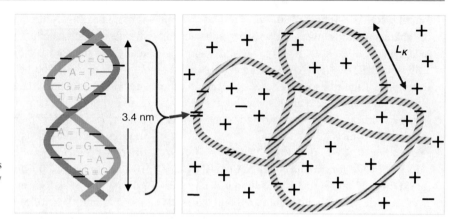

where the underscore denotes a complex quantity, "*" denotes conjugate, and ε is the permittivity (real part of $\underline{\varepsilon}^*$) that is assumed constant for the frequencies of interest.

A quantitative measure of the responsiveness of a body to an applied external electric field is the polarizability, α. A highly polarizable body, e.g., features many charges that are responsive to an electric field and their movement is in some way restricted. An important parameter for experimentally determining the value of the polarizability of a NA, such as, DNA, is the dielectric increment, or decrement – see Takashima (1989) cited in [10]:

$$\Delta\varepsilon' = \varepsilon'_{rl} - \varepsilon'_{rh} \tag{3}$$

where ε_{rl} and ε_{rh} are the low- and high-frequency relative permittivities, or limiting dielectric constants. The predicted polarizability α_m for a macromolecule determined using the experimentally measured dielectric decrement $\Delta\varepsilon'$ is given by:

$$\alpha_m = \frac{3\varepsilon_o \Delta\varepsilon'}{C_m} \tag{4}$$

where ε_o is the permittivity of free space, C_m is the number density (m^{-3}) of macromolecules, and the factor "3" accounts for random macromolecular orientation. Using dielectric mixture theory, the polarizability of each DNA or RNA macromolecule can be determined from the suspension by calculating the volume fraction of the NA. Another parameter is the relaxation time constant, τ_R (s), i.e. the time duration for charges to redistribute, or "relax" after being perturbed by an

electric field. The charges resonate at a dispersion frequency, f_R, or angular frequency ω_R:

$$\tau_R = \frac{1}{f_R} = \frac{2\pi}{\omega_R} \tag{5}$$

Dispersions arise when the oscillating cloud of charges can no longer follow the alternating electric field and this effect is observed in both real and imaginary parts of the complex permittivity. NAs can have more than one dielectric dispersion characterized by a decrement and associated time constant. Commercial time-domain dielectric spectrometers can be used to estimate the dispersion decrement $\Delta\varepsilon'$ and time constant τ_R using a sample, e.g., 150 microliters (μl), of macromolecules suspended in a solvent, such as, water. Characteristics of dielectric permittivities and losses for calf-thymus DNA, a macromolecule used in many dielectric spectroscopy studies, versus frequency are compiled and shown in [5]. Although there are variations in the experimentally measured permittivity ε', they all tend to decrease with increase in frequency and the polarizability follows the same trend.

The types of polarization include electronic, atomic, molecular, interfacial (or space-charge), and counterion polarization. The first three are attributed to the displacement, or orientation, of bound charges; the latter two concern movement on a larger scale. Of the first three types of polarization, only molecular dipole polarization tends to feature in the literature concerning dielectric properties of NAs. The asymmetric distribution of electrons in molecules gives rise to *permanent* dipoles. The interaction of such a dipole with an externally applied electric field causes a torque that attempts to orient the molecule in the field direction. The polarization is appropriately called *orientation*, or *dipole*, polarization. Water is an example of a molecule with a permanent dipole and manifests permittivity values, via ε and σ, that are practically frequency independent up to 17 GHz – see Grant et al. (1978) cited in [5].

The last two kinds of polarization, interfacial and counterion, involve large-scale charge movement. Currently, there is no universal consensus in the literature on the high-frequency polarization mechanisms for macromolecules, such as, DNA or RNA. At present, the emphasis tends to favor the Maxwell–Wagner interfacial polarization for molecular structures with well-defined interfaces with water molecules, and counterion fluctuation polarization, along the longitudinal axis, for DNA and similarly for RNA.

Counterion Fluctuation Polarization for DNA

It is the response of the counterions to an externally applied AC electric field that results in counterion polarization. Since there is Coulombic attraction between the charged body and the counterion layer, the counterions attempt to pull the charged body along with them as they follow the electric field. The counterion polarization mechanism is due to solution counterions (such as Na^+, Mg^{2+}, etc.) interacting with negatively charged phosphate groups along the DNA or RNA polyion backbone. The counterions move freely along lengths of the macromolecule in response to the component of the external electric field parallel to the major axis or "backbone"(Fig. 5). Their migration results in an induced dipole moment, and hence, an identifiable polarizability.

A popular model of counterion polarization is the "Mandel–Manning–Oosawa" model developed in various stages by these authors cited in [5, 10, 11]. In this model, the counterions move freely along macromolecular "subunit lengths" and are permitted to cross from one subunit to a neighboring subunit only by overcoming "potential barriers." The *subunit length* L_s is described as the length along the average macromolecular conformation between "breaks," or "potential barriers," resulting from perturbations in the equipotentials (due to conformational processes, folding, etc.). Counterions moving along these *subunit* lengths, under the influence of an external electric field, manifest *high*-frequency dispersion that is molecular weight *independent*. This polarization contrasts

with *molecular weight-dependent* counterion movement along the entire macromolecular contour length that results in *low*-frequency dispersion and may account for the static permittivity. A proportion of the counterions are so strongly attracted to the polyelectrolyte that they are said to "condense" onto the polyion backbone. Essentially there are three distinct phases:

1. *Condensed* counterions – these are sufficiently, but non-locally, bound (or "delocalized") to the phosphate groups of the DNA, and thereby neutralize a fraction of the DNA charge. In many respects the condensed counterion layer is similar to the Stern layer of the double layer model for colloids.
2. *Diffuse* counterions which are responsible for neutralizing the remainder of the DNA charge, with a density which decreases exponentially with distance from the axis.
3. *Bulk* ions or "added salt" ordinary aqueous solution ions.

In terms of their contribution to polarizability, the condensed counterion phase is the most important. A feature of the condensed state is that the local concentration of counterions around the DNA does not tend to zero when the solvent or bulk electrolyte concentration does. Condensation occurs when the condition $\xi \geq 1/|z|$ is satisfied, where z is the valence of each counterion and the charge density parameter, ξ, is given by:

$$\xi = q^2/(4\pi\varepsilon_m k_B T b) \tag{6}$$

where q is the elementary charge, $\varepsilon_m = \varepsilon_0\varepsilon_{rm}$ is the permittivity of the bulk electrolyte medium, ε_r is the relative permittivity, $k_B T$ is the Boltzmann temperature, and b is the average distance between charged sites; for B-DNA double helix, $b = 1.73$ Å. The "Manning–Mandel–Oosawa" model yields a generalized expression for scalar longitudinal polarizability α_s per subunit length L_s:

$$\alpha_s = \frac{z^2 q^2 L_s^2 n_{cc} A_{st}}{12 k_B T} \tag{7}$$

where A_{st} is the stability factor of the ionic phase and includes mutual repulsion between fixed charges on the backbone and the effect of Debye screening:

$$A_{st} = [1 - 2(|z|\xi - 1)\ln(\kappa_s b)]^{-1} \tag{8}$$

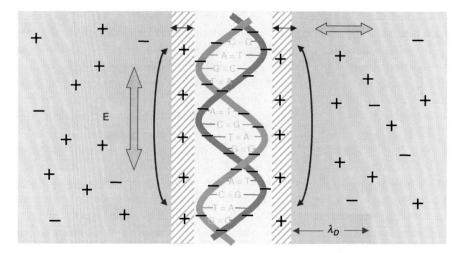

Dielectrophoresis of Nucleic Acids, Fig. 5 Counterion polarization on a short segment of DNA: a negatively charged sugarphosphate double helix attracts counterions. Components of an applied AC electric field (*red filled bidirectional arrows*) causes counterion movement in the longitudinal direction – along the DNA backbone, and transverse direction (*brown bi-directional line arrows*). The counterion movement tends to be within the "cylindrical" condensed layer (*green shaded /////*). The diffuse layer has characteristic length, λ_D, as shown and is referred to as the Debye screening length – see text for details

and the reciprocal of the Debye screening length, shown in Fig. 5, $\lambda_D = \kappa_s^{-1}$ (m) is given by:

$$\kappa_s = \left[\left(\frac{N_{Av}q^2}{\varepsilon_m k_B T} \right) \left(\sum_i C_i z^2 + \frac{C_p}{\xi} \right) \right]^{0.5} \quad (9)$$

where N_{Av} is Avogadro's number and C_i and C_p are the respective molar concentrations of ions in the bulk and diffuse phase, and phosphate groups. In the simplifying case where $C_p = 0$, (Eq. 9) reduces to the standard formula for the Debye reciprocal length of a balanced electrolyte [4, 10]. In (Eq. 7), n_{cc} is the number of condensed counterions that can be predicted theoretically:

$$n_{cc} = \phi_c L_s / |z| b \quad (10)$$

where the fraction of condensed counterions is $\phi_c = 1 - |z^{-1}| \xi^{-1}$. Equations 6–10 are equated with (Eq. 4) that determines the polarizability from a dielectric spectroscopy experiment. The relations $\alpha_m C_m = \alpha_s C_s$, where the number density of subunits is $C_s = N_{Av} C_p b / L_s$, are useful to bridge the theory and experiment. An expression for L_s based on the measured dielectric decrement $\Delta \varepsilon'$ is thus derived:

$$L_s = \sqrt{\frac{9 \Delta \varepsilon'}{\pi \varepsilon_{rm}(|z| \xi - 1) A_{st} N_{Av} C_p b}} \quad (11)$$

Assuming, e.g., $T = 298.2$ K $(25.0 \, °C)$, $\varepsilon_{rm} = 78.4$, $z = 1.00$ for monovalent cations, $C_p = 2.72$ mol/m^3, and $C_i = 1.10$ mol/m^3, $\xi = 4.132$, $\kappa_s = 9.754 \times 10^7$ (m^{-1}), $A_{st} = 3.764 \times 10^{-2}$, and $L_s = 3.308 \times 10^{-8} \sqrt{\Delta \varepsilon'}$. An expression for the relaxation time in terms of the subunit length was also derived:

$$\tau = \frac{L_s^2 q}{\pi^2 \mu k_B T} \quad (12)$$

Assuming the mobility value $\mu = 8.00 \times 10^{-8}$ at 25°C temperature, $L_s = 1.424 \times 10^{-4} \sqrt{\tau}$. A sample of dsDNA with measured values, $\Delta \varepsilon' = 9$ and $\tau = 500$ ns has predicted subunit lengths from (Eq. 11) and (Eq. 12) that approximately concur with each other, $L_s = 100$ nm. Physically, this value is very close to the worm-like chain Kuhn length for dsDNA. In the framework of the counterion model, the result would support the notion that equipotentials arise from natural curvature of the dsDNA suspended in solution.

Maxwell–Wagner Polarization for DNA

Interfacial, or *space-charge*, polarization arises from electrical charges being restricted in their movement at the interfaces between layers of different dielectric materials. Interfacial polarization results in dielectric dispersions when the aggregate of dissimilar materials is exposed to AC electric fields [4]. Maxwell–Wagner

(MW) polarization models for DNA suspensions have been undertaken by a number of researchers, e.g., Grosse (1989) and Saif et al. (1991) cited in [11].

MW interfacial polarization models the DNA macromolecule with the amino acid sugar-phosphate double helix as a 2 nm diameter, long cylindrical insulating core with low conductivity and permittivity. The insulator core is surrounded by a highly conducting sheath that represents the bound counterions on the negatively charged sugar-phosphate backbone. The conductive sheath interfaces with an electrolyte of low conductivity. The dielectric decrement for Maxwell–Wagner dispersion for DNA is predicted to be

$$\Delta\varepsilon' \cong 8\varepsilon_{rm}v_f/3 \qquad (13)$$

where v_f is the volume fraction. Similarly, the high-frequency relaxation is approximated in terms of the DNA *suspension* conductivity σ_s (or conductivity of the electrolyte medium, σ_m) and fraction of condensed counterions, ϕ_c:

$$f_R = \sigma_m\phi_c/(4\pi\varepsilon_o\varepsilon_{rm}v_f) \qquad (14)$$

where $\sigma_m = (1 - v_f)\sigma_s/(1 - \phi_c)$. Using $\sigma_s = 8.5 \times 10^{-3}$ S/m at 25°C, $v_f = 0.20\%$, $\varepsilon_{rm} = 78.4$, and the fraction of condensed counterions, $\phi_c = 0.25$, Eq. 13 and 14 yield $\Delta\varepsilon' \cong 0.42$ and $f_R \cong 162$ MHz.

An alternative approach that mimics the rod-shaped DNA macromolecule with conducting sheath by a randomly oriented shelled, prolate ellipsoid with a very short minor axis, and an extremely long major axis predicted a very high relaxation frequency of hundreds of MHz – see Bone et al. (1995) cited in [5, 11]. The MW interfacial polarization model tends to predict high relaxation frequencies in the order of hundreds of MHz and appears to be a less successful model for explaining the dielectric behavior of DNA than the counterion fluctuation model. This is perhaps because relatively dilute suspensions of DNA exist as worm-like entanglements with a poorly defined interface compared with the well-defined spherical, or rod-like, boundaries of colloids or some viruses.

Other Polarization Mechanisms

A variety of other polarization mechanisms include counterion polarization transverse to the DNA axis, rotation of water bound molecules along and across the grooves of the DNA, and relaxation of DNA polar groups. These polarization mechanisms tend to predict very high frequency dispersions up to hundreds of MHz. Also the relaxation times have temperature dependence that predicts much higher activation energies than measured and tend to have received less significance in the literature – see works by Xammara Oro, Mashimo, and Takashima referenced in Bone et al. (1995) as cited in [5, 11].

Dielectrophoresis

The principles of DEP are often described for an electrically neutral body although in practice the body, such as DNA, is often charged. Application of an electric field external to a neutral body induces an uneven spatial distribution of electrical charge, or *dipoles*, as shown in Fig. 6.

The external electric field also acts on the displaced charges themselves, so that the electric field can be said to act *twice* on the body. If the electric field is uniform, the field flux acting on electrical charges within each localized region yields Coulombic forces that cancel each other and the net force is zero. On the other hand, in a *nonuniform* electric field, Fig. 6, the density of the electric flux varies spatially between regions within the body, so that there is an imbalance of Coulombic forces and they sum to a non-zero net force. Assuming the body is more polarizable than the surrounding medium, it moves toward the region of the highest field nonuniformity (to the right). The imbalance of forces on the neutral body can be modeled by considering two elementary charges $+q$ and $-q$ at A and B, distance d apart, shown in Fig. 6 inset. The electric field at B is stronger than at A, $E_B > E_A$. Assuming the forces are acting in the horizontal direction, the sum of the Coulombic forces approximates with a Taylor series:

$$F_{total} = F_A + F_B = qE_B - qE_A$$
$$= q\left[E(x + d) - E(x)\right] \cong qd\frac{dE}{dx} \cong p\frac{dE}{dx} \qquad (15)$$

where $p = qd$ (C m) is the induced (or effective) dipole moment. This can be written as:

$$p = \alpha vE \qquad (16)$$

where α is the induced polarizability, or effective dipole moment, per unit volume, v, in unit electric field and has units Farad per meter (Fm^{-1}). It is

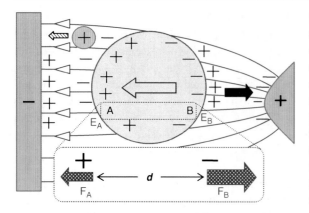

Dielectrophoresis of Nucleic Acids, Fig. 6 Dielectrophoresis: a neutral body (*yellow circle*) in a nonuniform electric field (*red unfilled arrows*) experiences a net force depending on the spatial distribution of electric field strength. In this example, where the polarizability of the body is greater than the surrounding medium, it moves to the right (*black arrow*). A positive test charge (\\\ *arrow*) moves to the left according to the direction of the electric field flux. *Inset*: charges at A and B displaced *d* apart interact with the electric field, E_A and E_B, generating Coulombic forces, F_A and F_B act in opposite directions (*blue-filled white-dot patterned arrows*), as shown. The force at B is greater than at A, $F_B > F_A$, so the neutral body moves to the right (*black filled arrow*)

assumed, for the force to be proportional to the derivative of E in (Eq. 15), that the dimension of the dipole p is small compared with the characteristic length of the electric field nonuniformity. Any spatial electric field phase variation is considered to be negligible, and so that consideration is confined solely to the in-phase component of the DEP force and that the polarizability is real, $\alpha \in R$. Combining (Eq. 15) and (Eq. 16):

$$F_{DEP}|_{DC} = \alpha v E \frac{dE}{dx} = \frac{1}{2}\alpha v \frac{dE^2}{dx} \qquad (17)$$

where the subscript signifies that this equation is applicable for a constant (DC) signal applied to electrodes and is positive, (i.e., acting toward the right). Expressions for the net force acting on a neutral bioparticle in a nonuniform electric field have also been developed using the energy variation principle and Maxwell stress tensor approach – see discussion and reference to Wang et al. (1997), Chap. 4, cited in [10]. The effective moment method is the most straightforward, and includes situations where there is dielectric loss.

Since the electric field acts twice, *inducing* and *acting on* dipoles within the body, the DEP force predicted by (Eq. 15) is proportional to the *square* of

the magnitude of the electric field. The body moves in accordance with the strength and inhomogeneity of the electric field, the volume and dielectric polarizability of the body, and also the dielectric properties of the suspending medium. Importantly, the net movement of the body does *not* depend on the direction of the electric field or on the electrode polarity. Consequently, time varying alternating current (AC) electrode voltages can be used to move the body rather than constant time direct current (DC) potentials. This feature, in turn, avoids problems of hydrolysis reactions occurring between electrode surfaces and bulk medium. AC radio frequencies above a few kHz applied to electrodes in moderately salty electrolytes, e.g., do not incur hydrolysis. This enables DEP to be implemented using microelectrodes with micron to submicron features for miniaturized applications.

Since AC signals are more useful than DC, the response observable of macromolecules moving in liquids implies that the small-time averaged DEP force is considered. It is evaluated as half the DC value. Generalizing (Eq. 17) to three dimensions:

$$\left\langle \vec{F}_{DEP}(\underline{x},t) \right\rangle_t = \vec{F}_{DEP}(\underline{x}) = \frac{1}{4}\alpha(f)v\vec{\nabla}\left|\vec{E}(\underline{x})\right|^2$$
$$= \frac{1}{2}\alpha(f)v\vec{\nabla}\left|\vec{E}_{rms}(\underline{x})\right|^2 \qquad (18)$$

where $\vec{\nabla}$ is the gradient operator, "$\rightarrow$," and $\langle \cdots \rangle_t$ denotes vector quantity and "small-time" average over a RF electric field oscillation period, the frequency dependence of the polarizability is made explicit, and "rms" is the root-mean-square. The DEP force is understood to be "almost instantaneous." In terms of understanding and computing the DEP force, the electric field gradient for realistic geometric electrode designs can be determined, analytically for simple cases, or by electromagnetic simulation software. The frequency-dependent polarizability, α, is inferred from dielectric measurements using (Eq. 4) or predicted from (Eq. 7) and the volume is usually known or can be approximated.

Note that for a sphere with radius r with MW polarization, $\alpha = 3\varepsilon_m \text{Re}\{f_{CM}\}$, where f_{CM} is the frequency-dependent Clausius–Mossotti factor and "Re" denotes real part. The time averaged force given by (Eq. 18) becomes:

$$\vec{F}_{DEP}(\underline{x}) = 2\pi\varepsilon_m r^3 \ \text{Re}\{f_{CM}\}\vec{\nabla}\left|\vec{E}_{rms}(\underline{x})\right|^2 \qquad (19)$$

which is the starting formula in many scientific publications, e.g., [1, 4–6]. In terms of scaling relations used in nanotechnology, a simple and effective way of considering the DEP force is to recall that E is determined by Poisson's or Laplace's equation, so:

$$F_{DEP} \propto \frac{r^3}{d^3} V^2 \qquad (20)$$

where d represents the characteristic length of the structure or feature size generating the nonuniform electric field. This means that if r and d decrease by the same amount, i.e., from μm to nm in body and feature size, the DEP force remains the same. This principle of invariance to scale enables DEP to be attractive by means of moving biomolecules on the nanoscale even if earlier work was on the micron-scale or larger. It also remains proportional to the square of the voltage, V, across the electrodes.

Dielectrophoretic Transport

NAs have little interial force compared with viscous drag force so that the former can be neglected. The viscous drag force can be approximated in the first instance if a charged macromolecule and associated counterions and coions, moving in a solvent (e.g., DNA in water), behave hydrodynamically as if they are a semirigid body, such as, a sphere or ellipsoid. As discussed in section Electrical Properties of Nucleic Acids, the rationale for this that DNA behaves as a worm-like chain retaining a roughly spherical (or ellipsoidal) shape due to ions and water molecules being attracted and dragged along with it. The complex of DNA, ions, and water can be made equivalent to a body with a hydrodynamic radius, r_h, and an effective drag force is approximated. As a consequence, the concentration flux of a population of single, non-interacting NAs, or an ensemble of experiments of single NAs, arising from a DEP force can be expressed in a very simple way:

$$\vec{J}_{DEP}(\underline{x}, t) = c(\underline{x}, t) \vec{F}_{DEP}(\underline{x}) / \zeta \qquad (21)$$

where the drag coefficient for a sphere is $\zeta = 6\pi\eta r_h$, where η is the dynamic viscosity of the solvent. The incessant collisions of water molecules on the NA gives rise to Brownian motion and hence, to a thermally driven diffusion flux:

$$\vec{J}_{Diff}(\underline{x}, t) = -D \vec{\nabla} c(\underline{x}, t) \qquad (22)$$

where the Einstein diffusion coefficient is $D = k_B T / \zeta$ with $k_B T$ being the Boltzmann temperature. The contributions from DEP and diffusion can be combined into a continuity equation, also known as the modified diffusion (MDE) or Fokker Planck equation:

$$\frac{\partial c(\underline{x}, t)}{\partial t} = -\nabla \cdot (\vec{J}_{DEP}(\underline{x}, t) + \vec{J}_{Diff}(\underline{x})) \qquad (23)$$

The evolution of (Eq. 23) is a starting point for very simple analytical and numerical simulations of DEP-driven transport of NAs.

Applications

There is considerable interest in controlling the movement of DNA in microfabricated environments for two general application areas: (1) DNA base sequence and gene expression analyses in LOC and μTAS microdevices and related biotechnological and biomedical applications and (2) using DNA as nanoscaffolds, or constructing material, in molecular electronics. A key motive for this application is that "top-down" fabrication approaches are limited on the nanoscale so there is strong technological interest in alternative "bottom-up" approaches. Although there is considerable research on DEP of DNA, there is very little work reported so far on DEP of RNA or PNA. This section starts with an introductory example of quantitative DEP of DNA, and continues to address wider contributions from the scientific community to this field.

Dielectrophoresis of DNA: An Introductory Example

DEP of DNA is typically accomplished with a standard RF signal generator, signal monitoring oscilloscope, and measurement apparatus. A wider range of RFs and conductivities can be used but a benchmark RF is 1 MHz with reverse osmosis (RO) water or buffer with conductivity less than 1 milliSiemen per meter (mS/m). The measurement process often involves imaging using fluorescence microscopy with modifications that enable recording and data processing. An experimental arrangement for observing DNA transport is shown in Fig. 7a. Tracking of DNA macromolecules as they move in real time is achieved by labeling them with a fluorescent dye and observing their

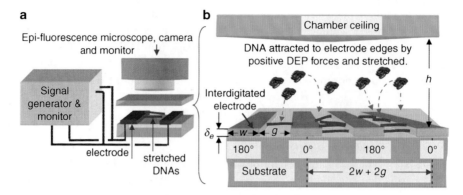

Dielectrophoresis of Nucleic Acids, Fig. 7 Scheme of DEP experimental apparatus (**a**) AC signal generator, monitoring oscilloscope, epi-fluorescence microscope, camera, monitor, DEP of DNA collection experiment. (**b**) Detailed inset of DEP collection showing interdigitated electrodes (fabricated with width w, interelectrode gap g, and thickness δ_e – not to scale) with electrical potential phases 180° apart. DNA molecules are attracted to the interelectrode gaps by positive DEP and stretched, as shown

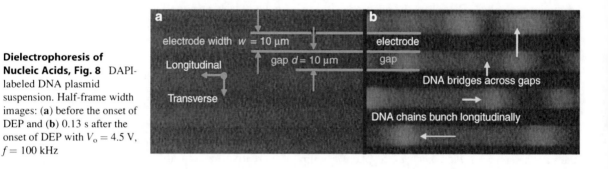

Dielectrophoresis of Nucleic Acids, Fig. 8 DAPI-labeled DNA plasmid suspension. Half-frame width images: (**a**) before the onset of DEP and (**b**) 0.13 s after the onset of DEP with $V_o = 4.5$ V, $f = 100$ kHz

micron-scale movement with an epi-fluorescence microscope with typically 200× or above magnification, as shown. Often a camera is used to record and display DNA movement on a PC/TV screen.

A perspective close-up view of the microelectrode interdigitated "finger" electrodes that are responsible for generating electric fields that drive DEP macromolecule transport is shown in Fig. 7b. The thickness of the electrodes, δ_e, has been exaggerated for illustration and is typically much less than the transverse width, w, and interelectrode gap, g, $\delta_e << w, g$. Standard microfabrication methods are used: UV patterned resist on glass or other suitable substrate, evaporated with metal, developed, and lift-off. The AC potentials to the electrodes are 180° apart so the potential pattern is repeated in the transverse direction every $2w + 2g$ so from (Eq. 18) the DEP force repeats and is symmetric about each center of the electrode. The regions of the planar array with the highest nonuniformity are the electrode edges so switching on the DEP attracts the DNA to that region. The DNA is shown to be coiled in

the suspension; plasmid DNA, e.g., has a shape similar to a rubber band that has been cut, twisted, and rejoined. If the DEP force is sufficiently strong, the coiled DNA collected by the DEP forces gets stretched by the electric field and builds up to form bridges that span between adjacent electrodes, as shown.

A part video-frame microscope image of DAPI stained 12 kilo-bp (kbp) plasmid DNA, taken from a similar study to [11], is shown in Fig. 8a before DEP was applied. The electrodes comprised 10 nm titanium onto glass substrate, 10 nm palladium, overlayed with 100 nm gold with dimensions $w = g = 10$ μm. The negatively supercoiled plasmid DNA strands cannot be seen individually but they were known to be about 4 μm in contour length or 2 μm looped end-to-end. Figure 8b shows the DNA collected into the interelectrode gap 0.13 s after DEP was applied. One plasmid alone cannot bridge the interelectrode gap, so it is thought that the strands elongate and link to each other, similar to a pearl chain, as they position between the electrodes. The solvent has started to evaporate so

**Dielectrophoresis of
Nucleic Acids,**
Fig. 9 Particle collection
under the action of positive
DEP force and release after the
DEP force is switched off
(**a**) cartoon showing particle
concentration (*side view*)
(**b**) concentration at the array
as a function of time, $c(t)$,
showing particle collection
and release

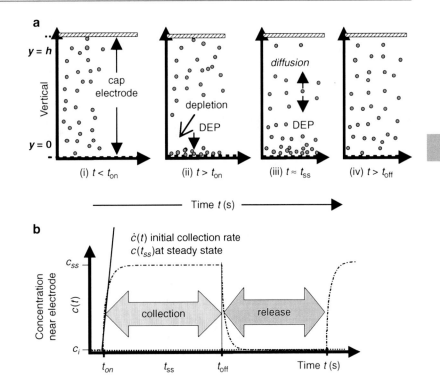

the transversally stretched DNA strands have tended to bundle longitudinally in places, as shown. After the DEP force is switched off, Brownian motion is responsible for the DNA strands being released (image not shown) and moving away from the planar array into the bulk solution.

A fluorescence spectroscopy technique was developed to quantify the DEP collection and for the release of nanoparticles and DNA (see [1, 11] and references therein to Bakewell and Morgan (2004)). Since the longitudinal length, l, of the electrodes $l >> w + g$, symmetry and periodicity patterns of the DEP pattern enabled image processing and quantification of video frames comprising a DEP experiment, and resulted in collection and release time profiles. A cartoon of DEP of plasmid DNA is shown in Fig. 9a and accompanying collection and release profiles in Fig. 9b as the DEP is switched on and off, or time modulated. The DNA plasmids initially uniformly suspended in a chamber, Fig. 9a (i), with concentration, c_i, as shown in the corresponding plot in Fig. 9b. After DEP is switched on at $t = t_{on}$ these are attracted toward a planar inter-digitated electrode array by positive DEP. As the plasmids rapidly collect near the array, the region above becomes depleted, and over time progresses toward the

microdevice cap, Fig. 9a (ii). Eventually the DEP flux becomes balanced by thermally driven diffusion and the concentration reaches steady-state (SS) at $t = t_{ss}$, shown in Fig. 9b (iii). Switching off DEP at $t = t_{off}$ releases the DNA that diffuses from the array and returns to the bulk solution, Fig. 9a (iv). Two quantification summary parameters measured for a designated volume around the electrode, shown in Fig. 9b, are (i) the initial collection rate, and (ii) the initial to SS transition. The former assumes, initially, that diffusion is negligible in (Eq. 23) so that, to a first approximation, the amount of fluorescence or DNA collected is proportional to the DEP force over a designated volume, $\dot{n}(0) \propto F_{DEP}$ where "." denotes time derivative. This enabled a comparison of DNA collections and DEP forces for different experimental parameter values, e.g., voltage, frequency-dependent polarizability. The other parameter uses the relation at SS, $\dot{n}(t_{ss}) = 0$, so the fluxes balance and again the DEP forces can be compared for different experimental parameters. The results indicated that the strength of collections also decreased with increasing frequency from 100 kHz to 5 MHz, and these trends concurred with dielectric polarizability measurements for the same DNA plasmid. The results also showed that for these electrodes the predicted DEP force is lower than expected from theory. A number of reasons were presented including the possible effects of AC electro-osmosis, polarization saturation, and distortive effect of the DNA on the electric field itself [11].

Dielectrophoresis of DNA: General

There have been many variations and advancements on the basic arrangement in the previous section. The literature indicates that Washizu's group in Japan pioneered most of the early and novel experiments on DEP of DNA, starting in the 1990s with their Fluid Integrated Circuit (FIC). The details of their work, e.g., Washizu et al. (1990, 1994, 1995, 2005), can be found in several recent and very comprehensive reviews [1, 5, 6] – along with many important and recent contributions from newly emergent groups around the world. The state-of-the-art and advancement for nanotechnology can be summarized into a couple of general categories:

1. Generation of DEP force, microfabricated environment, and DNA preparation:
   (a) The length of DNA that can be moved by DEP depends on the smallest feature size of the electrodes, DNA availability in the volume of solution that can attract DNA, etc., and ranges from 20 bases to tens of kilobases and upward – limited by the practicalities of sample preparation, e.g., a 1,000 kbp or 1 Mega-bp (Mbp) of DNA has a contour length of 340 μm, long enough to be sheared in a micropipette tip during preparation.
      • Many experiments have used 48.5 kbp (contour length 18.5 μm) λ-phage dsDNA, and more recently short sequences (e.g., 20 bp) of ssDNA that are commercially available.
      • Other types of DNA include 3D DNA origami which entails folded long ssDNA with short (oligonucleotide) ssDNA.
   (b) Methods for measuring DEP movement:
      • Fluorescence microscopy, as described in the introductory example, has been the standard method for visualizing and measuring DEP of DNA. Suitable dyes for visualization with fluorescence microscopy include DAPI, YOYO, TOTO, acridine orange, and PicoGreen.
      • Other label-free methods include impedance spectroscopy where the presence of the DNA alters the capacitance between the measurement electrodes. An ingenious arrangement is to use the same electrodes for DEP as for measurement and quantification, and was recently reported by Holzel and coworkers [12].
   (c) Voltage and frequency range:
      • Most DEP experiments use standard RF signal generators that range from 0 to 20 MHz and up to 30 V peak-to-peak. Investigators have tended to use a low-frequency range from 30 to 1 kHz or high frequency, e.g., 20 kHz–20 MHz. In principle there are no restrictions except to avoid hydrolysis at low RFs and this, in turn, depends on experimental factors, e.g., solution conductivity. The high-frequency response is influenced by impedance characteristics of the electrodes and supply buses.
   (d) Direct DEP force
      • "Traditional" DEP uses with metal electrodes that connect directly with buses to the AC supply. The electrodes have often been microfabricated using standard photo-lithographic methods as described for Fig. 7b. Consequently, planar metal electrodes are typically up to a few hundred nanometers in thickness and are essentially quasi-two-dimensional (~2D) with designs that are castellated, triangular, sinusoidal, polynomial, etc., to optimize DEP response. The metals used are gold, aluminum, titanium, platinum, and can be single or multilayered.
      • Some researchers have also used floating-potential electrodes (FPEs) between the main electrodes in an FIC-type device that do not require connections, e.g., authors Asbury (1998) and Washizu (2005), cited in [6, 11].
      • DEP of DNA, enhanced by nano-sized scale of the carbon nanotube (CNT) end features, has been demonstrated where the diameters of the nanotubes were as small as 1 nm. Single- or multi-walled carbon nanotubes (SWCNT or MWCNT) are attractive candidate electrodes since their nanoscale features yield high DEP forces – see Tuukkanen et al. (2007) cited in [13, 14].

(e) Other (indirect and new) methods for generating a DEP force:

- Electrode-less DEP (EDEP) or insulating DEP (iDEP) is a powerful method developed less than a decade ago for generating a DEP force by confining an existing uniform electric field using insulating boundaries. The boundaries are made from 3D microfabricated insulating posts made from quartz, polydimethylsiloxane (PDMS), polymethylmethacrylate (PMMA), or the end of a glass nanopipette – see Chou et al. (2001), Swami et al. (2009), and Ying et al. (2004) of Klenerman's group, cited in [1, 6].

- Optically induced DEP (ODEP) or optoelectronic tweezers (OETs) – as distinguished from "optical laser tweezers" – have been recently developed where an optical pattern, e.g., from a commercial photo-projector, illuminated onto a photoconductive layer of amorphous silicon, gives rise to a localized DEP force. The optical image has the benefit of acting as an electrode and being able to change position and shape in real time. It has recently been reported to have manipulated a single DNA macromolecule tethered to a micro-nanosphere [15].

2. DEP of DNA purpose or outcome:

(a) Electrostatic stretching and trapping of DNA to electrodes using DEP

- Temporary positioning: Since DNA is flexible on the micro scale, most reported DEP experiments, in one way or another, involve trapping and positioning DNA for the duration the DEP force is switched on.

- Semipermanent positioning or immobilization: use of chemical modification to bind the DNA to the electrode or substrate so that it remains tethered after the DEP force is switched off, e.g., avidin binding to aluminum (complementary biotin labeling of DNA) and thiol-labeling of gold electrodes (complementary thiol-labeling of DNA strands) – see works by Germishuizen et al. (2003, 2005) cited in [14], and Burns (2002, 2006) that modified DNA stretching with polymer-enhanced media [1, 5, 6].

- Applications of stretching and positioning of DNA include:

  - Measuring the lengths of stretched and oriented dsDNA prepared using restriction enzymes and simple cutting of stretched DNA using a UV laser – see works by Washizu (1990, 1995) cited in [6, 11].

  - Measuring the conductivity of DNA for molecular electronics – see the work by Burke (2003, 2004) cited in [6].

  - Probing the mechanical elastic properties of DNA – see Dalir et al. (2009) cited in [1].

  - Molecular surgery: optical tweezers control of a nanosphere coupled to an enzyme, e.g., DNase I, with cutting the bases of electrostatic stretched DNA was demonstrated by Yamamoto et al. (2000) in Washizu's group as cited in [6, 11]. These examples highlight that DNA base information can be used for sub-nanometer position-dependent function and underscore the importance of DEP.

  - DNA scaffolds: there is a large corpus of literature on DEP of CNTs, metals, and their oxides – see [14]. In addition to the current protocols on attachment of metal colloids to DNA, e.g., by Mirkin and coworkers, or in situ growth on substrates, DEP has been proposed for a solution-based method of metalized DNA [16].

(b) Separation and purification of DNA:

- DEP chromatography has been used to separate large (e.g., 48.5 kbp λ-phage dsDNA) and small (22 base ssDNA oligonucleotide) and also medium-large (2 kbp DNA) – see Kawabata and Washizu (2001) cited in [1, 5, 6]. Described simply and referring to the introductory example in Fig. 7b and descriptions [10, 14], the chromatograph is a type of field flow fractionation (FFF) where the DNA solution, driven externally by a pump flow, flows with the fluid lateral to the array (in the transverse direction) left to right and interacts with the vertically DEP-driven force generated by the electrode array. Large DNA macromolecules with strong polarizability and DEP force are pulled down onto the array, whereas small DNA fragments with weak polarizability continue to flow though the chamber, hence, separating the mixture.

- The separation of similar sized DNA by DEP has been made possible with nanosphere technology. Called "DEP enhancement," Kawabata and Washizu (2001), cited in [1, 5, 6], described the mixing and separation of 35 kbp T7-DNA with 48.5 kbp λ-DNA that had been linked to latex nanospheres. The T7-DNA nanosphere complexes underwent negative DEP whereas the λ-DNA underwent positive DEP, thus, separating the mixture.

(c) DEP of DNA has been combined with other biological, chemical, and electrokinetic methods, e.g., DC electrophoresis (EP), electro-osmosis (EO), and electrostatic liquid actuation. Three recent examples are:

- Development of DNA concentrators – see Yokokawa et al. (2010) cited in [12]. In that work, the DEP of 26 base ssDNA was compared with low level DC EP (i.e., electrode voltage was sufficiently low, less than a few volts, so that electrolysis did not occur). It was found that low level DC EP concentrated the ssDNA better than DEP but the solvent conductivity was low (2 mS/m) and possible degradation to the electrodes by using a DC voltage over time is unclear.
- EDEP or iDEP, achieved by using insulating constrictions, has resulted in a tenfold enhancement of DNA hybridization kinetics down to 10 picomolar concentration sensitivity using relatively high ionic buffer strengths – see Swami et al. (2009) cited in [1].
- DEP has also been recently been combined with nanosphere technology to significantly improve the hybridization kinetics of DNA with savings in the use of reagents and with lower ionic strength and simpler protocols than current microarray and Southern blot methods – see Cheng et al. (2010) cited in [1]. This offers a molecular scale DEP

enhancement and offers promising new advances in detection sensitivity, speed, portability, and ease-of-use for genomics research and diagnostics.

(d) DEP of DNA in "packages" or as "linked":

- Viruses contain DNA or RNA encapsulated with a protein coat. Viruses have been intensively studied for DEP response, particularly at RFs where the DEP changes from being positive to negative, thus optimizing virus separation – see works by separate groups, e.g., Morgan and Fuhr et al. reviewed in [1, 5, 6].
- Chromatin, that constitutes chromosomes and contains DNA wrapped around histones and other protein, has been lysed from bacteria (*E. coli*) and trapped by DEP at 500 Hz, as described by Prinz et al. (2002) cited in [1, 5, 6].
- DNA ligated with PNA and nanospheres showed that DNA could be orientated by DEP unidirectional and moreover imaged using a fluorescence microscope without the need for intercalating dyes – see Holzel and coworkers cited in [6].

Dielectrophoresis of RNA

DEP has been recently used to rapidly collect RNA from the nucleus of a living cancer cell using an AFM [17]. The AFM cone-shaped probe tip had been structurally modified to form nanometer scale concentric electrodes that enabled mRNA to be attracted and subsequently extracted from the cell nucleus. This novel and promising application of DEP may pave the way for enhanced analysis of single cell extracts with many applications; particularly, as mentioned in section Nucleic Acids, the mRNA forms only a small proportion of the total RNA in a cell. Time dependent DEP collections on of 16S and 23S subunit rRNA, extracted from *E. Coli*, onto planar interdigitated over a wide RF range, 3 kHz - 50 MHz, has been recently reported [18]. The in-depth study used an aqueous suspension with conductivity 13 mS/m and revealed positive to negative DEP transition above 9 MHz and a quadratic dependence for low voltages ($\sim 3V_{rms}$) - as predicted by Equations 17 – 20. Being the first demonstration of fast capture and release, DEP shows considerable promise for application in rRNA-based biosensing devices. Other reports at present have involved DEP of RNA in context, i.e., not as the main focus of study but in conjunction with DEP of DNA, or of cells, e.g., the lysing of cell contents Cheng et al. (1998) cited in [5, 11] – where cells undergo DEP positioning, or investigations on the effect of DEP on gene expression in cells, etc.

Outlook and Concluding Remarks

There have been many achievements and DEP of NAs, particularly DNA, is an active and promising research field with a scope that has advanced with new materials and technologies, e.g., microfluidics, nanospheres, microarrays, nanotubes, electronic nanodevices, and scanning probe microscopes. There is substantial opportunity for advancing theoretical understanding of polarization mechanisms that underpin DEP-induced motion of NAs; at present, current models favor counterion fluctuation. There is also considerable opportunity for advancement of nanoscale DEP applications, whether it is directed toward metallization of DNA for nanoconstruction or for advancing DNA hybridization kinetics that is needed in genomics and sensor research. DEP of RNA and indeed PNA appears to be research field open for investigation, particularly with the worldwide drive for single cell analysis and quantitative systems biology. There is also much to be accomplished in terms of integrating DEP-based methods with other electrokinetic, LOC, and μTAS miniaturization technologies and establishing reliable and quantifiable measures of functionality and performance.

Cross-References

► AC Electrokinetics of Nanoparticles
► Dielectrophoresis
► Dielectrophoretic Nanoassembly of Nanotubes onto Nanoelectrodes
► DNA Manipulation Based on Nanotweezers
► Electric-Field-Assisted Deterministic Nanowire Assembly
► RNAi in Biomedicine and Drug Delivery

References

1. Pethig, R.: Review article-dielectrophoresis: status of the theory, technology, and applications. Biomicrofluidics 4(2), 35 (2010)
2. Dahm, R.: Discovering DNA: Friedrich Miescher and the early years of nucleic acid research. Hum. Genet. 122(6), 565–581 (2008)
3. Nelson, D.L., Cox, M.M.: Lehninger Principles of Biochemistry, 5th edn. WH Freeman, New York (2008)
4. Pohl, H.A.: Dielectrophoresis. Cambridge University Press, Cambridge (1978)
5. Holzel, R.: Dielectric and dielectrophoretic properties of DNA. IET Nanobiotechnol. 3(2), 28–45 (2009)
6. Lapizco-Encinas, B.H., Rito-Palomares, M.: Dielectrophoresis for the manipulation of nanobioparticles. Electrophoresis 28(24), 4521–4538 (2007)
7. Calladine, C.R., Drew, H.R.: Understanding DNA – the Molecule and How it Works, 3rd edn. Academic, London (2004)
8. Bloomfield, V., Crothers, D.M., Tinoco, Jr., I.: Nucleic Acids: Structures, Properties, and Functions. University Science, Sausalito (2000)
9. Bonincontro, A., et al.: Radiofrequency dielectric spectroscopy of ribosome suspensions. Biochim. Biophys. Acta 1115(1), 49–53 (1991)
10. Morgan, H., Green, N.G.: AC Electrokinetics: colloids and nanoparticles. Research Studies Press, Baldock, England and IoP Publishing, Philadelphia, USA (2003)
11. Bakewell, D.J., Morgan, H.: Dielectrophoresis of DNA: time- and frequency-dependent collections on microelectrodes (vol 5, pg 1, 2006). IEEE Trans. Nanobioscience 5(2), 139–146 (2006)
12. Henning, A., Bier, F.F., Holzel, R.: Dielectrophoresis of DNA: quantification by impedance measurements. Biomicrofluidics 4(2), 9 (2010)
13. Regtmeier, J., et al.: Dielectrophoretic trapping and polarizability of DNA: the role of spatial conformation. Anal. Chem. 82(17), 7141–7149 (2010)
14. Zhang, C., et al.: Dielectrophoresis for manipulation of micro/nano particles in microfluidic systems. Anal. Bioanal. Chem. 396(1), 401–420 (2010)
15. Lin, Y.H., Chang, C.M., Lee, G.B.: Manipulation of single DNA molecules by using optically projected images. Opt. Express 17(17), 15318–15329 (2009)
16. Swami, A.S., Brun, N., Langevin, D.: Phase transfer of gold metallized DNA. J. Clust. Sci. 20(2), 281–290 (2009)
17. Nawarathna, D., et al.: Targeted messenger RNA profiling of transfected breast cancer gene in a living cell. Anal. Biochem. 408(2), 342–344 (2011)
18. Giraud, G., Pethig, R. et al.: Dielectrophoretic manipulation of ribosomal RNA, Biomicrofluidics 5(2), 024116 (2011).
19. Grant, E.H., Sheppard, R.J., and South, G.P.: Dielectric behaviour of biological molecules in solution. In: Monographs on physical biochemistry. Eds. Harrington, W. F. and Peacocke, A. R., OUP (1978)
20. Pethig, R.: Dielectric and electronic properties of biological materials. John Wiley, New York, USA (1979)
21. Takashima, S.: Electrical properties of biopolymers and membranes. Adam Hilger, Philadelphia, USA (1989)

Dielectrophoretic Assembly

▶ Electric-Field-Assisted Deterministic Nanowire Assembly

Dielectrophoretic Nanoassembly of Nanotubes onto Nanoelectrodes

Didi Xu[1], Arunkumar Subramanian[2], Lixin Dong[3] and Bradley J. Nelson[1]
[1]Institute of Robotics and Intelligent Systems, ETH Zurich, Zurich, Switzerland
[2]Department of Mechanical and Nuclear Engineering, Virginia Commonwealth University, Richmond, VA, USA
[3]Electrical and Computer Engineering, Michigan State University, East Lansing, MI, USA

Synonyms

Dielectrophoresis; Electrokinetic manipulation; Nanoassembly; Nanoelectrodes

Definition

Dielectrophoreis is an electrokinetic manipulation process where the interactions of a nonuniform electric field with a polarizable particle are employed to localize the particle in space with nanoscale precision. The nonuniform field strength around the particle exerts a dielectrophoretic (DEP) force on the dipole induced within the polarizable particle, which drives the particle toward the electric field maximum or minimum. In addition, the resulting torque aligns the particle along the direction of the applied electric field. Dielectrophoretic assembly has been widely used for high precision assembly of carbon nanotubes (CNTs) in order to overcome disadvantages such as the need for a complex setup and poor controllability, which are associated with other techniques to integrate CNTs onto functional nanostructures. Since this technique is performed at room-temperature, in a noncorrosive fluidic environment, and at low voltage, the electric field-based method is compatible with further postintegration nanostructuring steps such as those used in the Integrated Circuit (IC) industry.

Overview

Functional nanostructures, which incorporate an individual multiwalled carbon nanotube (MWNT) as the fundamental electromechanical transduction element, can be used to create tubular switches, memories, sensors, and resonators/oscillators with improved performance. This entry highlights two key unit processes that enable the fabrication of such constructs and provides an overview of potential applications using recently demonstrated nanomechanical switches. The first unit process relates to dielectrophoretic nanoassembly of single MWNTs onto arrays of nanoelectrode pairs fabricated on silicon chips [1, 2]. Appropriately optimized electrode designs insure that the assembled MWNT bridges the nanoelectrodes at its distal ends while remaining fully suspended and flat in the region between them (Fig. 1).

The second unit process involves the modification of the nested-shell, native MWNT geometry using Joule-heating-induced vaporization of carbon atoms at specific locations on the nanotube [2–4]. By engineering heat sinks along the NT length to modulate thermal dissipation and Joule-heating-induced temperatures, the vaporization of carbon atoms within the NT has been controlled with atomic-to-nanoscale [5]. This technique to selectively modify the NT geometry with control over location, length, and shells has provided us with the capability to create structures where some shells can axially translate or rotate with respect to others. Nanoelectromechanical systems (NEMS) devices based on such intershell displacement mechanisms have been demonstrated to be ultrafast, fatigue-free constructs with very low friction.

Basic Methodology

DEP Assembly Technique
Based on Pohl's theory [6], dielectrophoretic forces F_{DEP} and torques T_{DEP} can be calculated as

$$F_{DEP} = (p \cdot \nabla)E \qquad (1)$$

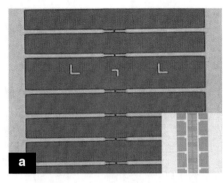

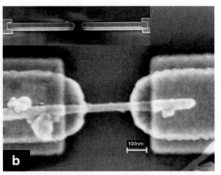

Dielectrophoretic Nanoassembly of Nanotubes onto Nanoelectrodes, Fig. 1 Dielectrophoretic assembly of MWNTs. (**a**) Nanoelectrode array design with a lower magnification image in inset. (**b**) SEM images of an assembled MWNT nanostructure at high and low magnification (Reproduced with permission from ref. 2, Copyright 2007 IOP)

$$T_{DEP} = p \times E \tag{2}$$

where p is the induced dipole moment of the particle and E is the electric field.

Because both DEP forces and torques are dependent on multiple parameters, such as particle and suspending medium properties and the applied electric field, the nanotube (NT) is modeled as a prolate ellipsoid with homogeneous dielectric properties. The large aspect ratio of NTs results in a needle-shaped particle with $l \gg r$, so the governing equations in dielectrophoresis could be simplified as [7, 8]

$$F_{DEP} = \frac{1}{2} \pi r^2 l \varepsilon_m \mathrm{Re}\left(\frac{\tilde{\varepsilon}_p - \tilde{\varepsilon}_m}{\tilde{\varepsilon}_m}\right) \nabla |E|^2 \tag{3}$$

$$\langle T_3 \rangle = \frac{1}{2} \pi r^2 l \varepsilon_m E^2 \sin\theta \cos\theta \mathrm{Re}\left(\frac{(\tilde{\varepsilon}_p - \tilde{\varepsilon}_m)^2}{\tilde{\varepsilon}_m(\tilde{\varepsilon}_m + \tilde{\varepsilon}_p)}\right) \tag{4}$$

where $\tilde{\varepsilon}_p$ and $\tilde{\varepsilon}_m$ are complex dielectric permittivities of the particle and the suspending medium, r and l are the radius and length of the CNT, and the angle between the electric field and the longest axis 1 is θ.

Another important aspect influencing the motion of CNTs is the viscous drag force, which is proportional to the relative velocity, retards the CNT in all directions. If gravity, Brownian motion, electro-osmosis, etc., are all negligible during DEP process, the instantaneous translation velocity is proportional to the instantaneous dielectrophoretic force [3, 4]. Similarly, the rotational friction factor of the CNT rotating about its center in x–y plane can also

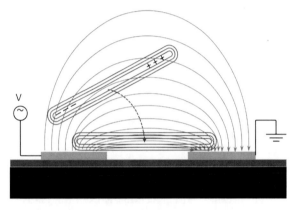

Dielectrophoretic Nanoassembly of Nanotubes onto Nanoelectrodes, Fig. 2 Dielectrophoresis concept

be calculated by hydrodynamic equations [9]. The schematic is shown in Fig. 2.

$$v_T \approx \frac{F_{DEP}}{f} \tag{5}$$

The constant f is the translation friction factor and depends on a range of parameters such as size, shape, and fluid viscosity.

The standard operation procedure of DEP assembly is as follows. First, the MWNTs are suspended and sonicated in ethanol (at a concentration of 5 μg/ml) for 10 h to insure homogeneity. The chip is then immersed in a reservoir containing this suspension and a nonuniform electric field is applied with a high frequency function generator. After deposition, the chip is removed from the reservoir, rinsed in clean ethanol, and brown-dried with nitrogen. The assembly

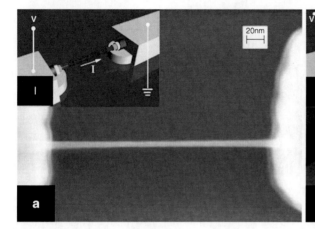

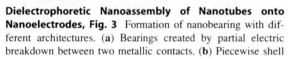

Dielectrophoretic Nanoassembly of Nanotubes onto Nanoelectrodes, Fig. 3 Formation of nanobearing with different architectures. (**a**) Bearings created by partial electric breakdown between two metallic contacts. (**b**) Piecewise shell engineering of single NTs between metallic contacts and floating contacts (Reproduced with permission from ref. 2, Copyright 2007 IOP)

parameters related to the target nanomaterial and nanoelectrode structure such as the AC/DC voltage components, polarizing frequency, and deposition time are modulated to optimize device deposition yields.

Current-Induced Shell Engineering of MWNT

Current driven shell engineering is useful in cases to employ Joule heating to vaporize carbon atoms and thin the NT between device contacts in order to tune its electrical resistance, mechanical stiffness, and clamping configuration [2]. A typical DEP assembled MWNT is sandwiched between two layers of metal at each end to improve electrical contact. The metallic contacts serve as heat sinks, and shell removal due to thermal stress occurs in the suspended segment, starting from the outermost and proceeding towards the inner ones. During current-induced vaporization, NT temperatures are estimated to be as high as 2,000 K. It is proven that at these elevated temperatures, defects such as kinks and holes are initiated at MWNT mid-lengths and result in individual shell removal in the region between contacting electrodes [3, 4]. An SEM image of a shell-structured nanotube is shown in Fig. 3a.

Modulation of Joule heat dissipation along NT lengths using additional heat sinks can be employed to achieve atomic-to-nanoscale control over shell etching locations on a MWNT [5]. It is found that the electric breakdown position can be controlled in

MWNTs by introducing additional metallic contacts located at different segments of a NT. A nanotube assembled onto three contacts is shown in Fig. 3b, and a schematic is shown in its inset. By applying a bias to the electrodes at distal ends and grounding the central metal contact (Fig. 3b), currents are driven through both suspended segments of the NT simultaneously and remove its outer shells in the respective segments. Detailed information on this technique can be found in references [2] and [5].

Key Research Findings

Nanotube Bearings

High bearing densities can be achieved through Joule-heat-driven vaporization of different segments of a NT, that is, a NT is assembled onto multiple metallic contacts using floating electrode DEP [2, 10]. Figure 4a shows independent, bidirectional nanobearings engineered at different locations along the length of a single MWNT bridging five electrodes. With a 220 nm device length, these constructs represent the smallest bearings reported to date and are expected to exhibit lower friction forces and faster response times during telescoping core movements. Each of the three NT segments in the central part of the fabricated device is anchored to the metal at only its outer shells, with inner shells capable of sliding inside the outer housing in both directions. An illustration of this shell structure

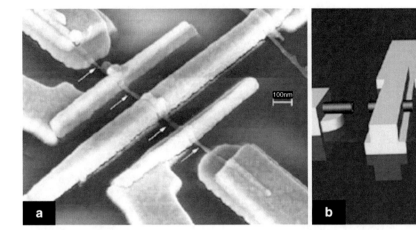

D

Dielectrophoretic Nanoassembly of Nanotubes onto Nanoelectrodes, Fig. 4 High-density NT bearings. (**a**) Telescoping segments formed with a 220 nm pitch and separated by 6–10 nm gaps. The *arrows* point to the inter-segment gaps. (**b**) Schematic illustration of the core shell mechanisms formed with the inter-segment gaps exaggerated to reveal the shell structure (Reproduced with permission from ref. 2, Copyright 2007 IOP)

is shown in Fig. 4b. This structure has a number of potential applications. One possible application involves resonators based on shell displacements, where neighboring nanotube segments electrically excite oscillations. The configurational stability and frequency spectrum of these constructs have been investigated. The most significant conclusion that emerges from the analysis of these open-capped structures is that, for a given NT (in terms of the number of shells and outer diameter), there exist only a finite number of configurations that are energetically stable. For an 11 shell 8 nm outer diameter CNT, resonant frequencies from 1.4 to 4.9 GHz for various stable inner shells exist [11]. Using these constructs, ultralow power electromechanical switches can be realized by electrostatically controlling the movement of NT cores and thereby, modulating the intersegment conductivity[10].

Nanotube Switch

Another energy-efficient, three-state NEMS switch which has been realized using the DEP assembled nanostructures is shown in Fig. 5 [12]. In this device, the contacts at the NT's distal ends are held at a common, ground potential and form the first terminal, T1, of the NEMS switch. The gate electrodes, which are located on either side of the NT in close proximity, represent terminals "T2" and "T3," respectively. In the device "off-state," the NT is not in contact with the gate electrodes on either side. When a voltage is applied to one of the gate electrodes, it induces an attractive electrostatic force on the nanotube and deforms it elastostatically. Beyond a threshold bias, which is characteristic of the NT Young's modulus, diameter, number of shells, length and device actuation gap, the NT comes in the contact with the gate electrode and the switch is turned "ON." Thus, the circuit between terminals T1_T2 and T1_T3 can be turned ON by applying the pull-in voltage at the respective actuation electrodes.

An SEM image of a representative device is shown in Fig. 6a, and its operational characteristics are shown in Fig. 6(b–d). The *I-V* curves during the switching transition initiated by a bias applied to terminal T2 is shown in Fig. 6b. One important characteristic that emerges from these curves is the hysteresis in device operation. This is due to the impact of van der Waals forces between the NT and gate electrode, which is onset at pull-in and opposes the subsequent pull-off event. The switching characteristics of the device constructed between terminals T3 and T1 can be seen in Fig. 6c. The pull-in voltage for this terminal was found to be higher as compared to that at terminal "T2" due to the larger NT-electrode gap at this terminal. The repeatability of these devices is highlighted in Fig. 6d using results from four cycles of switching induced by terminal T2. It can be observed from this plot that the turn-on bias remains repeatable.

Dielectrophoretic Nanoassembly of Nanotubes onto Nanoelectrodes,
Fig. 5 Three states of the device. The "OFF state" in the absence of external bias is shown on *top*. The "T1-T2 ON" state when a voltage is applied to terminal "T2" is shown in *bottom left*. The *bottom-right* image illustrates the "T1-T3 ON" state when a bias is applied to terminal "T3" (Reproduced with permission from ref. 11, Copyright 2009 ACS)

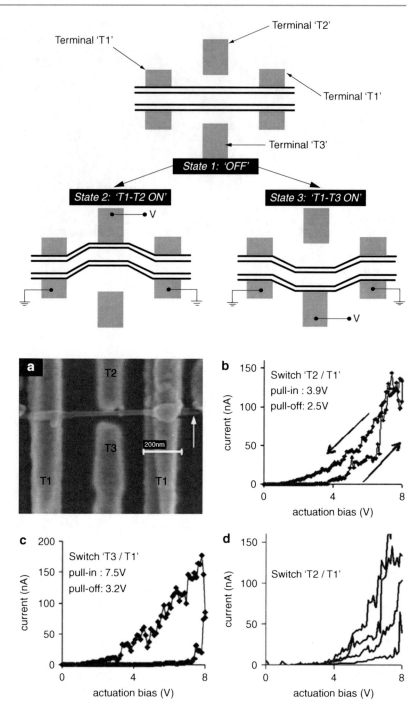

Dielectrophoretic Nanoassembly of Nanotubes onto Nanoelectrodes,
Fig. 6 Device characterization. (**a**) SEM image of a device with terminal labeling shown. The arrow points to a current-induced break in the NT to decouple it from the array. (**b**) I-V characteristics of the switch between terminals "T1-T2." The *arrows* point to the increasing and decreasing voltage parts of the curve. (**c**) I-V characteristics of the switch between terminals "T3-T1." (**d**) Device operation over four cycles shown to demonstrate the repeatability in device behaviors (Reproduced with permission from ref. 11, Copyright 2009 ACS)

Future Directions for Research

Results highlighted in this entry provide interesting pointers toward future nanodevices based on single MWNTs with novel performance regimes. Though significant progress has been made, future avenues

for research and development exist in each of the following thrust areas: nanofabrication, device engineering, and characterization. For instance, though DEP-based assembly has been demonstrated to be a robust method to assemble diverse materials with nanoscale precision, further improvements in assembly yields are required before these could be used within industrial processes. This is an important challenge that requires the identification of optimum deposition parameters within the multi-variable process.

In the area of device development, a number of different engineering embodiments can be realized using nanofabrication processes outlined in this report. These include microwave resonators based on elastostatic deformations and inter-shell displacements within MWNTs. These devices have the potential for use in near-term industrial applications such as mass, force, and resonant-pressure sensors. Another possible area of future research relates to developing integrated MEMS/NEMS platforms for probing the electromechanical properties of nanomechanical elements. For instance, appropriately designed constructs realized using a combination of DEP assembly and shell engineering can be used to understand the electromechanical coupling between individual shells within a MWNT and to provide insights into its suitability for use as piezoresistors in nanotransducers.

Cross-References

▶ AC Electrokinetics
▶ AC Electroosmosis: Basics and Lab-on-a-Chip Applications
▶ Dielectrophoresis
▶ Self-assembly

References

1. Xu, D., Subramanian, A., Dong, L.X., Nelson, B.J.: Shaping nanoelectrodes for ultrahigh precision dielectrophoretic assembly of carbon nanotubes. IEEE Trans. Nanotechnol. **8**(4), 449–456 (2009)
2. Subramanian, A., Dong, L.X., Tharian, J., Sennhauser, U., Nelson, B.J.: Batch fabrication of carbon nanotube bearings. Nanotechnology **18**(7), 075703 (2007)
3. Huang, J.Y., Chen, S., Jo, S.H., Wang, Z., Han, D.X., Chen, G., Dresselhaus, M.S., Ren, Z.F.: Atomic-scale imaging of wall-by-wall breakdown and concurrent transport measurements in multiwall carbon nanotubes. Physical Review Letters **94**(23), 236802 (2005)
4. Molhave, K., Gudnason, S.B., Pedersen, A.T., Clausen, C.H., Horsewell, A., Boggild, P.: Transmission electron microscopy study of individual carbon nanotube breakdown caused by Joule heating in air. Nano Lett. **6**(8), 1663–1668 (2006)
5. Subramanian, A., Choi, T.-Y., Dong, L.X., Tharian, J., Sennhauser, U., Poulikakos, D., Nelson, B.J.: Local control of elctric current driven shell etching of multiwalled carbon nanotubes. Appl. Phys. Mater. Sci. Process. **89**(1), 133–139 (2007)
6. Pohl, H.A.: Dielectrophoresis the Behavior of Neutral Matter in Nonuniform Electric Fields. Cambridge University Press, Cambridge (1978)
7. Jones, T.B.: Electromechanics of Particles. Cambridge University Press, Cambridge (1995)
8. Morgan, H., Green, N.G.: AC Electrokinetics: colloids and Nanoparticles. Research Studies, Hertfordshire (2003)
9. Koenig, S.H.: Brownian motion of an ellipsoid. A correction to Perrin's results. Biopolymers **14**, 2421–2423 (1975)
10. Subramanian, A., Dong, L.X., Nelson, B.J., Ferriera, A.: Supermolecular switches based on multiwalled carbon nanotubes. Appl. Phys. Lett. **96**(7), 073116 (2010)
11. Subramanian, A., Dong, L.X., Nelson, B.J.: Stability and analysis of configuration-tunable, bi-directional MWNT bearings. Nanotechnology **20**(49), 495704 (2009)
12. Subramanian, A., Alt, A.R., Dong, L.X., Kratochvil, B.E., Bolognesi, C.R., Nelson, B.J.: Electrostatic actuation and electromechanical switching behavior of one-dimensional nanostructures. ACS Nano **3**(10), 2953–2964 (2009)

Differential Scanning Calorimetry

▶ Nanocalorimetry

Digital Microfluidics

▶ Electrowetting

Dip-Pen Nanolithography

Yi Zhang[1], Rüdiger Berger[2] and Hans-Jürgen Butt[2]
[1]Shanghai Institute of Applied Physics, Chinese Academy of Sciences, Shanghai, China
[2]Max Planck Institute for Polymer Research, Mainz, Germany

Synonyms

Hard-tip soft-spring lithography; Polymer pen lithography; Scanning-probe lithography

Definition

Dip-pen nanolithography (DPN) is an atomic force microscope (AFM) [1]) based patterning technique in which molecules are transported from a sharp tip to a surface [2, 3]. With DPN, the molecules to be deposited are first loaded onto the tip by dipping the tip into a melt or solution of the substance to be deposited. Similar to a pen which people use to write notes on a paper, the atomically sharp tip is scanned in contact with a sample surface. The ink molecules are transported to the surface driven by capillary forces normally generated by the condensed water or melted ink when tip contacts surface. Typically, DPN allows surface patterning with a feature size on scales of less than 100 nm [4]. DPN is a quite unique method for patterning because it depends on the delivery of molecules to surfaces rather than energy (in the form of light, force, heat, or electricity) to make patterns. The DPN principle can be called a positive lithography technique in the sense that material is added rather than removed.

The History of DPN

The research on the deposition of organic materials by the AFM tip to a solid surface can be dated back to 1995 [2]. However, the first use of the term "dip-pen nanolithography" and the fast development of the DPN technique started in 1999 [3] in Prof. Chad Mirkin's lab in Northwestern University. Since then the excellent attributes of the DPN, such as direct-write capability, ultrahigh resolution, molecular ink and substrate general, and programmable abilities, have attracted researchers all over the world. Currently, several tens of groups are employing DPN technique for their researches and contributing on the development of the DPN technique at all aspects. In the last 5 years, a constant value of around 60–70 articles per year was published under the term "DPN." This value characterizes DPN to be a method of scientific importance and not to be a scientific niche. Meanwhile, commercialized instruments for parallel patterning of ink materials have been developed in early 2000s.

Description of the Method

The pens. Conventional DPN utilizes an AFM probe to transport ink materials from an ink-coated tip to a surface when the tip is engaged (Fig. 1). The size of AFM tip is on the scale of tens of nanometers to several micrometers. Accordingly, the feature size fabricated by the tip can be controlled in the range from about 10 nm to several micrometers by selecting suitable tips and depositing parameters. The motion of the AFM tip on the surface can be precisely controlled by the AFM system, which makes it convenient to fabricate programmed patterns. The same AFM tip also serves as an imaging tool for characterizing written nanoscopic features on a surface.

In order to control the deposition of molecules, specialized AFM tips were fabricated. For example, AFM cantilevers terminated with poly(methyl methacrylate) (PMMA) colloids have been used to pattern molecules by allowing the polymer on the tip to swell under different humidity conditions [6]. Even hollow tips connected with a reservoir were used for controlling the droplet size for liquid nano dispensing applications [7]. New procedures based on novel probes boost the potential of DPN into different research areas.

As a patterning technology, throughput is the biggest limitation of the single-probe DPN. Thus, parallelization capabilities must be developed in order to be competitive with standard lithographic techniques [8]. A possible solution to DPN parallelization is to develop, to fabricate, and to use two-dimensional (2D) tip arrays [9]. Linear and 2D arrays carrying more than 10,000 tips have been developed [8] and are commercially available. These tools allow one to fabricate molecule-based structures over long distances (centimeters) or large areas (square centimeters) with high efficiency. Moreover, the pens in a tip array can be independently actuated and used to make a different nanostructure, which provides more flexible control in DPN. The material that makes up the parallel tip arrays can be soft polymers [10]. The most recent development in this field indicated that a hard silicon tip array on a soft baking layer can overcome the throughput problems of conventional DPN and the resolution limits imposed by the use of soft polymer tips (Fig. 2) [11].

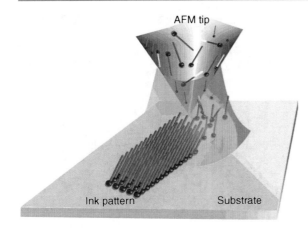

Dip-Pen Nanolithography, Fig. 1 Schematic drawing indicating ink transporting from an AFM tip onto a solid substrate. Image was reprinted from [5] with permission

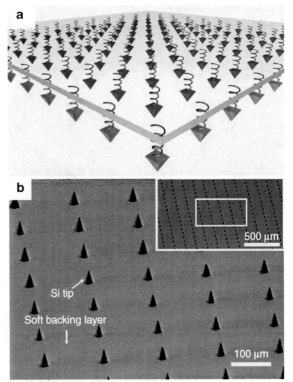

Dip-Pen Nanolithography, Fig. 2 (**a**) Schematic illustration of a hard tip array on a soft baking layer (the springs). (**b**) An SEM image of a real silicon tip array on a SiO$_2$/PDMS/glass substrate. The inset shows a large area of the array. Image was reprinted from [11] with permission

Ink materials. With DPN, "ink" substances are patterned to endow surfaces with predesignated functions at specific locations. Small molecules such as alkanethiols were first used as ink materials to be patterned on gold surface. Alkanethiol molecules deposited on gold surfaces normally undergo a self-assembly process and form a closely packed monolayer. Then DPN has been developed to pattern a variety of "ink–substrate" combinations. Inks studied to date include many types of small molecules, metal salts, nanoparticles, polymers, oligonucleotides, peptides, and proteins, while the substrates used include gold, glass, quartz, silicon, mica, germanium, and so on. In addition, researchers also have developed DPN techniques to pattern different chemical [12] or biological ink molecules on one substrate, so that multiple chemical or biochemical functionalities are integrated [4].

Alkanethiols were the first type of chemicals [1, 12] that were directly patterned on gold surface with DPN (Fig. 3). Alkanethiol molecules are important because they can be used to conveniently tailor the local chemical properties of the substrate by selecting appropriate tail group of the molecule. In addition, alkanethiols are easy to handle during tip coating and can be repeatable patterned on gold surface. Since then, alkanethiols have been widely explored as a model system for DPN to investigate the ink transport through AFM tip, molecular diffusion on the substrates, and phase separation of mixed ink molecules.

Patterning of biomolecules on various surfaces with DPN is particularly attractive [13]. The biomolecules can be patterned in a direct-write manner in either ambient or inert environments without risking cross-contamination. Since DPN has a very high spatial resolution, it is capable of fabricating high-density micro- or nano-arrays (Fig. 4) for biomolecular assays, including DNA and protein detection. The high spatial resolution that DPN brings to this field would allow an assay to screen for a correspondingly larger number of targets, or allow a fixed number of targets to be screened with a correspondingly smaller sample volume in a shorter period of time.

Tip coating. Tip coating is an important step of a DPN process. Researchers have developed a series of methods to coat AFM tips with molecular inks. For soluble ink materials, one can dip the tips into a dilute solution containing the ink molecules. When the tip is

Dip-Pen Nanolithography,
Fig. 3 Lateral force
microscopy images of two dot
arrays made by two
alkanethiols,
1-octadecanethiol (*left*) and
16-mercaptohexadecanoic
acid (*right*), respectively, on
thermally evaporated gold
films on silicon oxidized
substrates

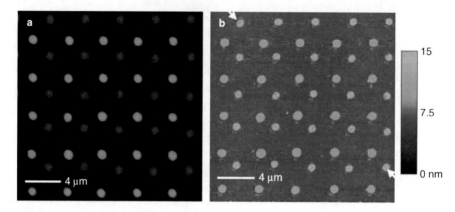

Dip-Pen Nanolithography, Fig. 4 Direct patterning of multiple-DNA inks by DPN. (**a**) Fluorescence image of two different DNA arrays. The arrays were made in two steps, each with an AFM tip that was coated with one DNA ink. After the first array was completed, the AFM tip was retracted and replaced with the second one coated with another DNA ink. The second tip was then relocated to the region of the first DNA array and fabricated the second array in-between the dots of the first array. (**b**) AFM image of the two DNA arrays shown in (**a**) decorated with gold nanoparticles. The DPN arrays were treated with a solution containing two different particles that were modified with DNA complementary to the first and second arrays, respectively. The two particles selectively assembled on the correct arrays. Image was reprinted from [14] with permission

taken out from the solution, the solvent should be removed, normally with a gas flow, prior to deposition. Coating of the tip may be conducted by placing the tips in the ink vapor and by cooling the tip. For liquid inks, tip coating is relatively simple, and dip-coating is normally adopted.

To facilitate tip inking and to increase the quantity of adsorbed ink, researchers have developed novel tip modifications methods. For example, AFM tips made of or coated with poly(dimethylsiloxane) (PDMS) can load more ink than bare AFM tips. Nowadays, researchers have developed microscopic inkwells as well as inkjet printing (Fig. 5) techniques that could be used to ink the different pens in a tip array for simultaneous DPN patterning of multiple inks [15].

Ink transport. Ink molecules transport from coated tip to solid substrate driven by a capillary force. It is a complicated process and is influenced by a lot of parameters. These include the chemical makeup of the ink, the substrate, and the tip, which affect the distribution and mobility of the ink on the tip and substrate, and the environment conditions under which the experiment is carried out. Since the ink-coated tip contact the substrate during DPN, a water meniscus would normally condense between a hydrophilic tip and a hydrophilic surface under room temperature and ambient conditions. It has been found that relative humidity of the environment affects ink transport [16], and today most DPN is conducted in a humidity-controlled chamber that helps to regulate ink

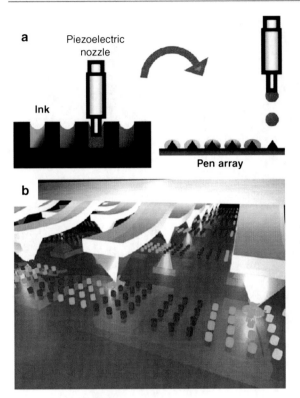

Dip-Pen Nanolithography, Fig. 5 Schematic drawings for (**a**) inking of pen arrays by inkjet printing and (**b**) multiplexed dip-pen nanolithography. After the nozzle of the ink jet printer was inked, the arrayed pens were independently addressed with different inks using the inkjet printer. The inked pens then make patterns containing multiplexed compositions on the surface. Image was reprinted from [15] with permission

transport. In a water-free or solvent-free environment, molecule transfer is hampered but not impossible. It is evident that different molecules are transported to the surface by fundamentally different mechanisms, which may be attributed to the different size, shape, solubility, and diffusion constant of the ink molecules.

To a better control over the ink transport, researchers have developed approaches that use ink chaperones to overcome the solubility and surface interactions that previously inhibited uniform pattern formation by DPN. One approach uses poly(ethylene glycol) (PEG) as a matrix to facilitate the transport of inks with different sizes and solubilities to an underlying substrate. Some others use viscous liquids such as glycerol, dimethylformamide, and so on, to assist ink transport. In some cases, the use of a liquid will also benefit biomolecule inks to keep their biological functions during DPN process.

Applications of DPN

By depositing ink molecules to surfaces, DPN has the ability to create architectures with controlled chemical composition and physical structure. With significant advances in the parallel printing capabilities in the context of DPN with arrayed tips, it has shown great potential to increase the throughput of DPN. Moreover, state-of-the-art DPN allows the individual control of the movement of single pens within a pen array. Therefore, further development of parallel DPN would dramatically increase the complexity and sophistication of the structures one can fabricate with DPN. These capabilities make DPN a unique and highly desirable tool for depositing ink materials on a variety of surfaces, which is important for fundamental studies and direct applications ranging from molecular electronics to materials assembly, and for investigating biological recognition [13].

Generating hard nanostructures. One approach is to write inorganic ink materials that are dispersed in a solvent by DPN. Then the solvent will be removed from the patterned features by post-DPN treatments such as heating or light curing, resulting in well-defined hard inorganic nanostructures [13]. The DPN fabricated hard nanostructures can serve as catalysts, nanoscale conductive circuits, and so on. For example, carbon nanotubes were synthesized on DPN-patterned nanoparticle catalysts [17].

Another approach is the combination of DPN with wet chemical etching. Using such an approach both metal and semiconductor nanostructures were fabricated (Fig. 6). DPN-generated alkanethiol monolayers can be employed as resists for creating solid-state structures by standard wet-etching techniques. By using this method, nanometer-sized complex features can be generated.

Generating templates for hierarchical assembly of materials. DPN-generated chemical templates of various sizes, shapes, and compositions have been used for the directed assembly of a lot of nanoscale building blocks such as carbon nanotubes (Fig. 7) and nanoparticles. In addition, DPN was found useful in bridging the nano-gaps between nanoscale building blocks, which should find broad applications, such as integration of nanocircuits, nanosoldering, and repair of failed nanoelectronics.

Fabricating biological nanoarrays. One can either directly fabricate nanoarrays of biological molecules on surfaces via DPN, or generate biological nanoarrays

Dip-Pen Nanolithography,
Fig. 6 AFM images (**a, b, d, e**)
and SEM image (**c**) of Au
nanostructures on a silicon
substrate. Image was reprinted
from [18] with permission

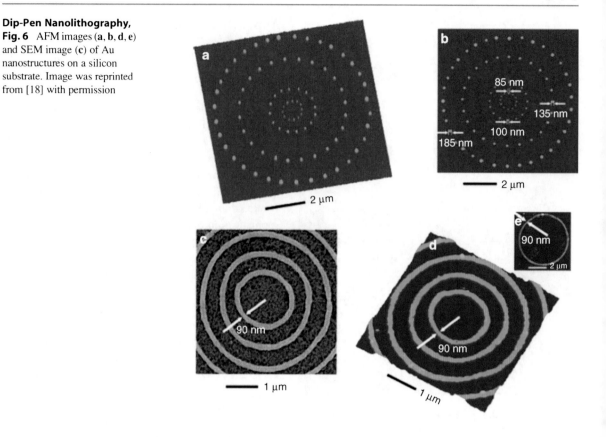

on a template fabricated with DPN methods. A variety of biological molecules, including DNA, peptides, proteins, viruses, and bacteria, have been patterned using direct-write or indirect adsorption approaches. With biological nanoarrays, a larger number of targets can be screened rapidly in an individual experiment with small amount of sample. For example, DPN-generated nanoarrays of the antibody have been used to screen for the HIV-1 virus antigen in serum samples [20].

Variants of DPN

Researchers constantly extend the basic capabilities of DPN beyond the range of patterning via a coated AFM tips (reviewed by Salaita et al. [13]). For example, researchers have employed heatable AFM tips to achieve melting and afterward patterning of some ink materials that are solids at room temperature and do not have appreciable solubility in a carrier solvent. For another example, by combining DPN with electrochemistry, a surface can also be electrochemically modified and

then rapidly coupled to ink material transported from the tip. To facilitate tip inking and ink transporting, several research groups have developed so-called nanofountain or nanopipet as AFM tips to deliver reagents directly to a substrate. These innovative variations of DPN have significantly expanded its capabilities.

Future Directions

Comparing with other lithographic techniques, DPN has unique properties and advantages. DPN is a positive lithography method. Therefore DPN is complementary to other scanning-probe-based, negative patterning methods [9]. DPN is superior over printing techniques in terms of patterning resolution, while still holds the capability of massive parallelization. The development on the independently actuated pen-array techniques, including software and hardware, is still underway. Further development in this direction would not only provide more flexible control and higher parallelization in DPN process, but also benefit other

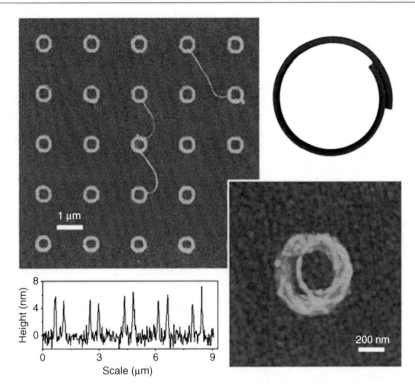

Dip-Pen Nanolithography, Fig. 7 (*Upper left*) AFM image and (*lower left*) height profiles of single-walled carbon nanotube (SWNT) rings assembled on a DPN-generated nanoarray. (*Upper right*) A model and (*lower right*) a zoom-in AFM image of a coiled SWNT ring. To achieve the assembly of SWNT, a gold substrate was first patterned with rings of 16-mercaptohexadecanonic acid (MHA). The bare gold regions of the substrate were then passivated with 1-octadecanethiol (ODT). The substrate was treated with a drop of 1,2-dichlorobenzene containing SWNTs. Because 1,2-dichlorobenzene wets the MHA rings but not the ODT passivated regions, SWNTs were selectively assembled on the MHA rings of the substrate. Image was reprinted from [19] with permission

scanning-probe-based lithography techniques. Then DPN should also find broader applications especially in industry of biotechnology.

In addition, DPN is very attractive for basic research. DPN is an appealing method to deposit a variety of organic molecules on surfaces. The zoo of organic molecules covers a variety of special functionalities, e.g., electrical conductivity, optical properties, and wetting properties. Thus, the behavior of small entities of molecules in respect to these properties can be studied on a nanometer lengths scale with ultrasmall amounts of material.

Cross-References

► Atomic Force Microscopy
► Nanoscale Printing
► Scanning Probe Microscopy
► Self-Assembled Monolayers

References

1. Binnig, G., Quate, C.F., Gerber, C.: Atomic force microscope. Phys. Rev. Lett. **56**, 930–933 (1986)
2. Jaschke, M., Butt, H.-J.: Deposition of organic material by the tip of a scanning force microscope. Langmuir **11**, 1061–1064 (1995)
3. Piner, R.D., Zhu, J., Xu, F., Hong, S.H., Mirkin, C.A.: "Dip-pen" nanolithography. Science **283**, 661–663 (1999)
4. Ginger, D.S., Zhang, H., Mirkin, C.A.: The evolution of dip-pen nanolithography. Angew. Chem. Int. Ed. **43**, 30–45 (2004)
5. Smith, J.C., Lee, K.B., Wang, Q., Finn, M.G., Johnson, J.E., Mrksich, M., Mirkin, C.A.: Nanopatterning the chemospecific immobilization of cowpea mosaic virus capsid. Nano. Lett. **3**, 883–886 (2003)

6. Kramer, M.A., Jaganathan, H., Ivanisevic, A.: Serial and parallel dip-pen nano lithography using a colloidal probe tip. J. Am. Chem. Soc. **132**, 4532–4533 (2010)
7. Meister, A., Krishnamoorthy, S., Hinderling, C., Pugin, R., Heinzelmann, H.: Local modification of micellar layers using nanoscale dispensing. Microelectron. Eng. **83**, 1509–1512 (2006)
8. Mirkin, C.A.: The power of the pen: development of massively parallel dip-pen nanolithography. ACS Nano **1**, 79–83 (2007)
9. Vettiger, P., Despont, M., Drechsler, U., Durig, U., Haberle, W., Lutwyche, M.I., Rothuizen, H.E., Stutz, R., Widmer, R., Binnig, G.K.: The "Millipede" – More than one thousand tips for future AFM data storage. IBM J. Res. Develop. **44**, 323–340 (2000)
10. Huo, F.W., Zheng, Z.J., Zheng, G.F., Giam, L.R., Zhang, H., Mirkin, C.A.: Polymer pen lithography. Science **321**, 1658–1660 (2008)
11. Shim, W., Braunschweig, A.B., Liao, X., Chai, J., Lim, J.K., Zheng, G., Mirkin, C.A.: Hard-tip, soft-spring lithography. Nature **469**, 516–520 (2011)
12. Hong, S.H., Zhu, J., Mirkin, C.A.: Multiple ink nanolithography: toward a multiple-pen nano-plotter. Science **286**, 523–525 (1999)
13. Salaita, K., Wang, Y.H., Mirkin, C.A.: Applications of dip-pen nanolithography. Nat. Nanotechnol. **2**, 145–155 (2007)
14. Demers, L.M., Ginger, D.S., Park, S.J., Li, Z., Chung, S.W., Mirkin, C.A.: Direct patterning of modified oligonucleotides on metals and insulators by dip-pen nanolithography. Science **296**, 1836–1838 (2002)
15. Wang, C.: Self-correcting inking strategy for cantilever arrays addressed by an inkjet printer and used for dip-pen nanolithography. Small **2**, 1666–1670 (2008)
16. Rozhok, S., Piner, R., Mirkin, C.A.: Dip-pen nanolithography: what controls ink transport? J. Phys. Chem. B **107**, 751–757 (2003)
17. Kang, S.W., Banerjee, D., Kaul, A.B., Megerian, K.G.: Nanopatterning of catalyst by dip pen nanolithography (DPN) for synthesis of carbon nanotubes (CNT). Scanning **32**, 42–48 (2010)
18. Zhang, H., Mirkin, C.A.: DPN-generated nanostructures made of gold, silver, and palladium. Chem. Mater. **16**, 1480–1484 (2004)
19. Wang, Y.: Controlling the shape, orientation, and linkage of carbon nanotube features with nano affinity templates. Proc. Natl Acad. Sci. U.S.A. **103**, 2026–2031 (2006)
20. Lee, K.B., Kim, E.Y., Mirkin, C.A., Wolinsky, S.M.: The use of nanoarrays for highly sensitive and selective detection of human immunodeficiency virus type 1 in plasma. Nano Lett. **4**, 1869–1872 (2004)

Directed Assembly

▶ Electric-Field-Assisted Deterministic Nanowire Assembly

Directed Self-Assembly

▶ Electric Field–Directed Assembly of Bioderivatized Nanoparticles

Disjoining Pressure and Capillary Adhesion

Seong H. Kim
Department of Chemical Engineering, Pennsylvania State University, University Park, PA, USA

Synonyms

Adsorbate adhesion; Liquid condensation; Liquid film; Liquid meniscus; Meniscus adhesion

Definition

Disjoining pressure, introduced by Derjaguin in 1936, may be defined as the difference between the thermodynamic equilibrium state pressure applied to surfaces separated by a thin film and the pressure in the bulk phase with which the film is in equilibrium [1]. Capillary adhesion refers to the adhesion due to a liquid meniscus bridging two solid surfaces which are wetted partially or completely by the liquid film. A curvature of the liquid meniscus induces a pressure difference across the meniscus, which depends on the liquid surface tension and the shape of the meniscus. The pressure difference as well as the surface tension of the liquid result in a net attractive force if the liquid wets the solid surface.

Occurrence

If you have been at a beach, you would have experienced capillary adhesion whether you recognized it or not. Walking on wet sand is easier than walking on dry sand. The wet sand can support your body weight, while the dry sand cannot and your feet sink into the sand. To build a sand castle the sand has to be wet. Wet sand can be shaped because particles adhere to each other. The strong adhesion is caused by liquid menisci, which form around the contact areas of two neighboring particles [2]. The force caused by such a liquid meniscus is called "capillary force" also termed meniscus force. Capillary adhesion phenomena are found and involved in various technical and biological systems. Capillary forces must be taken into account in studies of the adhesion between particles or particles to surfaces which are involved in processing powders, soils, and granular materials [3] and friction and wear of two contacting solid surfaces. They are also important at technological interfaces such as heads on magnetic hard disc memory and microelectromechanical system (MEMS) devices where they cause stiction [4]. Capillary forces play important roles in adhesion of insects, spiders, frogs, and geckos [5].

A liquid meniscus at the solid contact surface is readily formed if the solid surface can be wet with liquid. For example, if the solid surface is covered with a thin liquid film such as nonvolatile lubricants used for magnetic hard discs, the meniscus can be formed by flowing the liquid from the thin film into the contact region when an opposing surface comes into contact. Even if the solid surface is not covered with preloaded liquid films, the meniscus can be formed around the solid contacting point if the vapor in the surrounding medium is readily adsorbed on the solid surfaces. This phenomenon is called capillary condensation and it is quantified by Kelvin's equation (see below).

Key Research Findings

Disjoining Pressure

The disjoining pressure, Π, is the difference between the pressure of the interlayer on the surfaces confining it and the pressure in the bulk phase, the interlayer being part of this phase and/or in equilibrium with it [1]. Π can be attractive or repulsive. For example, in the case of two plates immersed in a liquid, Π is the pressure that must be applied to the liquid between the plates, in excess of the external pressure to the liquid, to maintain a given separation. The concept can be applied to soap films. In the case of the adsorbed layer at the solid–vapor interface, Π can be thought of as a mechanical pressure that must be applied to a bulk substance to bring it into equilibrium with a given film of the same substance.

Various methods exist to measure the disjoining pressure of liquid films. These include thin film balance [6], contact angle studies [7], and atomic force microscopy (AFM) [8]. The disjoining pressure can be regarded as a net of several components including dispersion interactions, overlapping of diffuse double layers in the case of charged surfaces, overlapping of adsorbed layers of neutral molecules, and alteration of solvent structure in the discrete region (such as salvation or hydration) [1]. The dispersion (van der Waals) interaction contribution can be estimated using the Derjaguin approximation, $\approx A/6\pi h^3$, where A is the Hamaker constant (typically $\sim 10^{-20}$ J) and h is the film thickness. Any deviation from this approximation must be due to specific chemical interactions such as ionic and covalent bonding of liquid molecules or structuring of liquid layers in the vicinity of solid surfaces.

Young–Laplace Equation

In the continuum model, the governing equation for the surface curvature of a liquid meniscus bridging two solid surfaces is the Young–Laplace equation [1, 3]. A curved liquid–vapor interface supports a pressure difference called Laplace pressure:

$$\Delta P_L = \gamma \left(\frac{1}{r_m} + \frac{1}{r_a} \right) = \frac{\gamma}{r_e} \qquad (1)$$

Here, γ is the surface tension of the liquid, r_m and r_a are the meridional and azimuthal radii of curvature of the surface, respectively, as shown in Fig. 1, and r_e is the effective radius of curvature. The curvature defined by r_m is in the plane of the paper and that defined by r_a is normal to the plane. Since the radius center of r_m is outside the meniscus, it has a negative value. The liquid surface is isobaric, meaning that r_e is constant along the meniscus surface, while r_m and r_a vary along the surface. Figure 1 also shows the contact angles of liquid with the sphere and the substrate surfaces (θ_1 and θ_2), which arise from the solid–liquid equilibrium (Young equation). The exact solution of the differential Young–Laplace equation for this geometry can be found in references [3] and [9].

Liquid Meniscus Formation Due to Vapor Adsorption

Thermodynamic equilibrium requires the chemical potential of the liquid meniscus to be equal to that of

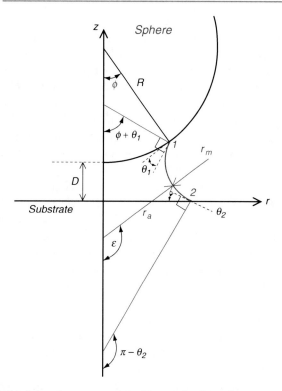

Disjoining Pressure and Capillary Adhesion, Fig. 1 Geometrical presentation of a liquid bridge between a sphere and a flat substrate [13]

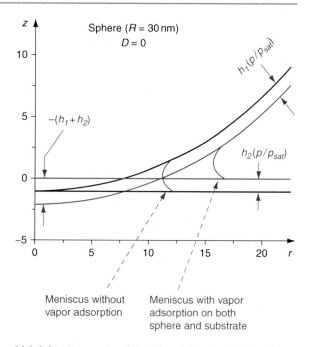

Disjoining Pressure and Capillary Adhesion, Fig. 2 Meniscus profiles for a sphere with a radius of 30 nm without (*blue line*) and with (*red line*) considering the equilibrium thickness of the adsorbate layer on the sphere, $h_1(p/p_{sat})$, and the thickness on the flat substrate, $h_2(p/p_{sat})$. The meniscus contact angle is set to $\theta_1 = \theta_2 = 0°$ [13]

the vapor. Assuming the molar volume (V_m) of the liquid substance to be constant and applying the Young–Laplace equation, (1), the Gibbs free energy of the liquid meniscus at a constant temperature is expressed as $\Delta G_{liq} = \int V_m dP = V_m \Delta P_L = V_m \gamma / r_e$. The Gibbs free energy of a vapor with respect to its saturation state is $\Delta G_{vap} = R_g T \cdot \ln(p/p_{sat})$. Setting ΔG_{liq} equal to ΔG_{vap} and replacing r_e with r_K gives the Kelvin equation [1, 3]:

$$V_m \Delta P_L = V_m \frac{\gamma}{r_K} = R_g T \ln(p/p_{sat}) \qquad (2)$$

$$r_K = \frac{\gamma V_m}{R_g T} \frac{1}{\ln(p/p_{sat})} = \frac{\lambda_K}{\ln(p/p_{sat})} \qquad (3)$$

Consider that the solid surface is covered with the adsorbate film which is in equilibrium with the vapor phase (Fig. 2). Then, the isothermal Gibbs energy relationship between the adsorbate and the vapor is as follows [10]:

$$V_m \Pi(h) = R_g T \ln(p/p_{sat}) \qquad (4)$$

where $\Pi(h)$ is the disjoining pressure of the adsorbed layer with a thickness h. The $V_m \Pi(h)$ term represents the molecular interactions of the adsorbate molecules with the solid surface [11]. From (2) and (4), it is obvious that the meniscus should be in equilibrium with the adsorbate film.

Capillary Force Calculation

The capillary force from the liquid meniscus depends on two components – the axial components (along the z-axis) of the surface tension force F_{st} and the Laplace pressure force F_{pl} at the point where the meniscus and the sphere meet (at the upper boundary point 1). For the geometry shown in 1, the two components can be calculated with the following equations:

$$F_{st} = 2\pi R \sin(\phi) \times \gamma \sin(\theta_1 + \phi) \qquad (5)$$

$$F_{pl} = -\pi R^2 \sin^2(\phi) \times \frac{\gamma}{r_K}$$

$$= -\pi R^2 \sin^2(\phi) \times \frac{R_g T}{V_m} \ln(p/p_{sat}) \qquad (6)$$

Here the positive sign is chosen for the attractive force. The term ϕ is often called a filling angle. The total capillary force is then the sum of these two components:

$$F_{cap} = F_{st} + F_{pl} = 2\pi\gamma R \\ \times \lfloor \sin(\phi)\sin(\theta_1 + \phi) - R\sin^2(\phi)/(2r_K) \rfloor \qquad (7)$$

Since the meniscus is in equilibrium, the force at the boundary point 1 is the same as the force at any point on the meniscus.

Simple Approximation
One of the widely used equations to describe the capillary force is $F_{cap} = 4\pi R\gamma \cos\theta$ [10]. This equation is derived from the Young–Laplace equation which contains two unknowns (principal radii of the meniscus) through several assumptions. One assumption is that the meniscus shape can be approximated to be circular. It also assumes that the radius of the tip is much larger than the cross-sectional radius of the liquid meniscus which, in turn, is much larger than the meridional curvature (external curvature) of the meniscus surface. This approximate form describes many macroscopic capillary phenomena; however, it does not work well as the contacting solid system gets smaller.

More Accurate Calculations
Figure 3 shows the capillary force (F_{cap}) and its two components (F_{pl} and F_{st}) for R = 100 nm calculated using the exact solution of the Young–Laplace equation, (1) [12, 13]. The force scale is normalized with $4\pi R\gamma$. The calculation shows a drastic difference in the p/p_{sat} dependence of F_{cap} due to the vapor adsorption. When the solid surface is assumed to be free of adsorbates (Fig. 3a), the capillary force decreases monotonically but insignificantly with p/p_{sat} until p/p_{sat} approaches the saturation point [14]. In contrast, the calculation including a type-II adsorption isotherm (inset in Fig. 2a) shows a large partial pressure dependence (Fig. 3b) [12, 13]. F_{cap} reaches a maximum value at $p/p_{sat} \sim 0.15$ and then decreases

upon further increase of p/p_{sat}. Note that the maximum p/p_{sat} range coincides with the knee point of the type-II adsorption isotherm which corresponds to the completion of the monolayer thickness. The result calculated with the adsorption isotherm is in good agreement with experimental measurement data [15]. The capillary force exerted by the meniscus formed by vapor adsorption and condensation is strongly affected by the structure of the adsorbate layer [16] as well as the geometry of the contacting solid surfaces [17].

Meniscus Stretch and Liquid Film on Solid
When the vapor adsorption from the environment is negligible (e.g., water does not adsorb readily on hydrophobic surfaces), the Laplace pressure of the meniscus ($P(h)$) under equilibrium conditions is equal to the disjoining pressure ($\Pi(h)$) of the liquid film deposited on the solid surface. Mate et al. demonstrated the use of AFM in calculating $r_{eff}(h)$ for a liquid lubricant film [18]. By slowly retracting an AFM probe in contact with a liquid film, the meniscus is stretched in a quasi-equilibrium state (the stretch rate is slow enough to allow for the change in the meridional and azimuthal radii in a manner that $r_{eff}(h)$ remains constant throughout the meniscus stretching). Then, $r_e(h)$ is obtained by fitting the force-distance curve data to the following equation:

$$\frac{F(d)}{4\pi R\gamma} = \left(1 + \frac{d}{2r_e(h)}\right) \qquad (8)$$

where $F(d)$ is the force measured by AFM, R is the radius of the AFM probe tip, and d is the tip-sample separation distance. The $r_e(h)$ is extracted from the slope of the linear portion of the retraction curve where the meniscus is stretched from the AFM tip.

Bowles and White later proposed a numerical solution to calculate the shape of the retraction curve based on the shape of the meniscus with a given $r_e(h)$ [19]. The White method analyzes this retraction curve based on a given $r_e(h)$ and the cantilever spring constant k. The White method fits the experimental data with the following equations:

$$\frac{F(d)}{4\pi R\gamma} = \frac{\lambda}{4\alpha^2} \qquad (9)$$

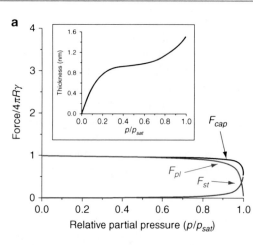
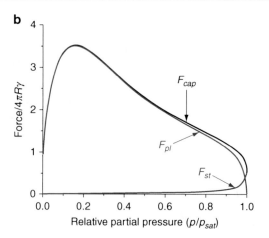

Disjoining Pressure and Capillary Adhesion, Fig. 3 Total (*black*), Laplace pressure component (*red*), and surface tension component (*blue*) of the capillary force for a sphere with R = 100 nm on a flat surface (**a**) without and (**b**) with taking into account the adsorption isotherm of ethanol. The forces are normalized with $4\pi R\gamma$, where γ = surface tension of liquid [13]. The inset in (**a**) shows the model adsorption isotherm used in this calculation. The isotherm curve is produced by fitting the experimental data reported in reference [15]

where $\lambda = \frac{r_e(h)}{R}$ and $\alpha = \frac{r_e(h)}{r_0(d)}$ are two parameters for the numerical solution of the shape of the meniscus on the AFM probe and $r_0(d)$ is a moving boundary where the lower edge of the meniscus meets the unperturbed liquid film of thickness h. Once $r_e(h)$ is determined, the disjoining pressure of the liquid film and the Laplace pressure of the liquid meniscus can be calculated using the surface tension of the liquid film.

Cross-References

▶ Nanotribology

References

1. Adamson, A.W.: Physical Chemistry of Surfaces, 5th edn. Wiley, New York (1990)
2. Bocquet, L., Charlaix, E., Ciliberto, S., Crassous, J.: Moisture-induced ageing in granular media and the kinetics of capillary condensation. Nature **396**, 735–737 (1998)
3. Kralchevsky, P.A., Nagayama, K.: Particles at Fluid Interfaces and Membranes. Elsevier, Amsterdam (2001)
4. Kim, S.H., Asay, D.B., Dugger, M.T.: Nanotribology and MEMS. NanoToday **2**, 22–29 (2007)
5. Autumn, K., Peattie, A.M.: Mechanisms of adhesion in Geckos. Integr. Comp. Biol. **42**, 1081–1090 (2002)
6. Claesson, P.M., Ederth, T., Bergeron, V., Rutland, M.W.: Techniques for measuring surface forces. Adv. Colloid Interface Sci. **67**, 119–183 (1996)
7. Leger, L., Joanny, J.F.: Liquid spreading. Rep. Prog. Phys. **55**, 431–486 (1992)
8. Saramago, B.: Thin liquid wetting films. Curr. Opin. Colloid Interface Sci. **15**, 330–340 (2010)
9. Orr, F.M., Scriven, L.E., Rivas, A.P.: Pendular rings between solids: meniscus properties and capillary force. J. Fluid Mech. **67**, 723–742 (1975)
10. Israelachvili, J.N.: Intermolecular and Surface Forces, 2nd edn. Academic, San Diego (1992)
11. Mate, C.M.: Application of disjoining and capillary pressure to liquid lubricant films in magnetic recording. J. Appl. Phys. **72**, 3084–3090 (1992)
12. Asay, D.B., de Boer, M.P., Kim, S.H.: Equilibrium vapor adsorption and capillary force: exact Laplace-Young equation solution and circular approximation approaches. J. Adhes. Sci. Technol. **24**, 2363–2382 (2010)
13. Hsiao, E., Marino, M.J., Kim, S.H.: Effects of gas adsorption isotherm and liquid contact angle on capillary force for sphere-on-flat and cone-on-flat geometries. J. Colloid Interface Sci. **352**, 549–557 (2010)
14. Butt, H.J., Kappl, M.: Normal capillary forces. Adv. Colloid Interface Sci. **146**, 48–60 (2009)
15. Asay, D.B., Kim, S.H.: Molar volume and adsorption isotherm dependence of capillary forces in nanoasperity contacts. Langmuir **23**, 12174–12178 (2007)
16. Asay, D.B., Kim, S.H.: Effects of adsorbed water layer structure on adhesion force of silicon oxide nanoasperity contact in humid ambient. J. Chem. Phys. **124**, 174712 (2006)
17. Köber, M., Sahagún, E., García-Mochales, P., Briones, F., Luna, M., José Sáenz, J.: Nanogeometry matters:

unexpected decrease of capillary adhesion forces with increasing relative humidity. Small **6**, 2725–2730 (2010)

18. Mate, C.M., Lorenz, M.R., Novotny, V.J.: Atomic force microscopy of polymeric liquid films. J. Chem. Phys. **90**, 7550–7555 (1989)
19. Bowles, A.P., Hsia, Y.T., Jones, P.M., Schneider, J.W., White, L.R.: Quasi- equilibrium AFM measurement of disjoining pressure in lubricant nano-films I: Fomblin Z03 on silica. Langmuir **22**, 11436–11446 (2006)

Dislocation Dynamics

▶ Computational Study of Nanomaterials: From Large-Scale Atomistic Simulations to Mesoscopic Modeling

Dispersion

▶ Fate of Manufactured Nanoparticles in Aqueous Environment

Dissociated Adsorption

▶ Chemical Vapor Deposition (CVD)

DNA Computing

▶ Molecular Computing

DNA FET

▶ Nanostructure Field Effect Transistor Biosensors

DNA Manipulation

▶ Robot-Based Automation on the Nanoscale

DNA Manipulation Based on Nanotweezers

Nicolas Lafitte[1], Yassine Haddab[2], Yann Le Gorrec[2], Momoko Kumemura[1], Laurent Jalabert[1], Christophe Yamahata[4], Nicolas Chaillet[2], Dominique Collard[1] and Hiroyuki Fujita[3]
[1]LIMMS/CNRS-IIS (UMI 2820), Institute of Industrial Science, The University of Tokyo, Tokyo, Japan
[2]FEMTO-ST/UFC-ENSMM-UTBM-CNRS, Besançon, France
[3]Center for International Research on MicroMechatronics (CIRMM), Institute of Industrial Science, The University of Tokyo, Meguro-ku, Tokyo, Japan
[4]Microsystem Lab., Ecole Polytechnique Fédérale de Lausanne, Lausanne, Switzerland

Synonyms

Characterization of DNA molecules by mechanical tweezers

Definition

DNA manipulation based on nanotweezers is the handling of DNA molecules using very small size tweezers for characterization purpose.

Introduction

Over the past two decades, studies on single-molecule biophysics have brought a large amount of information on DNA properties. The development of manipulation techniques and especially the emergence of new tools have allowed direct measurement of physical and chemical properties of DNA molecules. These properties turn out important for cell machinery understanding as structural modifications in the molecule induce changes in the interaction properties with proteins, and conversely. For example, topoisomerase enzymes unwind and wind DNA in order to facilitate DNA replication. A helicase protein moves along the

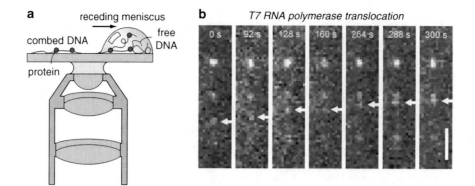

a receding meniscus

combed DNA — free DNA

protein

b *T7 RNA polymerase translocation*

0 s   92 s   128 s   160 s   264 s   288 s   300 s

DNA Manipulation Based on Nanotweezers, Fig. 1 Attachment of single DNA molecules on a glass surface. (**a**) DNA is stretched and immobilized using hydrophobic glass substrate and a receding air-water interface. After rehydration of the sample, the DNA stays firmly attached to the glass slide [5].

(**b**) Real-time visualization of the motion of an RNA polymerase along combed DNA strands. The directional movement of the RNA polymerase along a DNA molecule is observed using the incorporation of fluorescent into RNA strand (scale bar = 2.5 μm) [7]

unwound DNA, separating the two annealed nucleic acid strands of the double-stranded DNA. Polymerase enzymes reproduce new strands against the single-strand DNA templates. During all the replication process, the physical properties and the biological functions of the DNA change according to the interactions with the proteins allowing (or proscribing) the sequence of events [1–4].

Large advances in fluorescence microscopy have been achieved, enabling single-molecule visualization. Scientists developed powerful manipulation techniques where the DNA is attached to a substrate and observed by fluorescence [5, 6]. For instance, polymerase enzymes were fluorescently labeled to monitor their activity on a single molecule of DNA previously combed onto a surface (Fig. 1) [7]. Meanwhile, new approaches were developed to directly interact with the molecules, e.g., optical tweezers, magnetic tweezers (Fig. 2), atomic force cantilevers (known as AFM), and microfibers [8–10]. The response of single DNA molecules to a stretching or a twisting stress gives direct access to the physical properties of the molecule. Remarkable mechanical properties of DNA were discovered, and protein interactions with DNA were measured by the changes in the mechanical properties of the strands [11, 12].

However, all these new techniques, providing remarkable access to the biophysical properties of the molecules, still have a low throughput since the setup preparation and the experimentations are quite long and done one at a time. To move toward systematic biological or medical analysis, micro- and nanoelectromechanical systems (known as MNEMS for micro-nano-electromechanical systems) are more appropriate tools. They can be easily integrated into biological manipulation systems. Furthermore, they are produced at low cost by an efficient microfabrication process. In [13], DNA molecule bundle is trapped, stretched, and characterized in solution by silicon nanotweezers (Fig. 3).

Working Principle

Figure 3 shows a 3D illustration of the nanotweezers. Two sharp tips are used as electrodes for either dielectrophoresis trapping (cf. Trapping of DNA Molecules section) or conductivity measurements on the molecules. One tip is fixed when the other one can be moved through an electrostatic actuation. The gap between the electrodes can be sensed through a differential capacitor that measures the relative displacement of the moving electrode. Thus, the device consists of three parts:

• Two sharp tips
• A series of electrostatic comb-drive actuators
• A differential capacitive sensor

Laterally driven electrostatic comb-drive architectures are commonly used in MEMS devices since [14] with many industrial applications (e.g., microresonators, switches, micromirrors, accelerometers). When a voltage is applied between the two electrodes, the mobile parts move along the electric field paths tending to reduce the gap between the electrodes.

A differential capacitance sensor measures the tip displacement. This sensor is well adapted for bulk

a

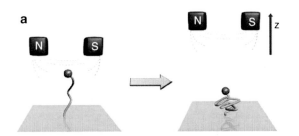

b

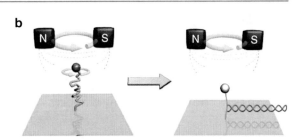

DNA Manipulation Based on Nanotweezers, Fig. 2 Schematic sketch of the magnetic tweezers technique. (**a**) The setup consists of a DNA molecule to a magnetic bead and to a glass surface for instance. A pair of magnets above the bead produces a magnetic field gradient (*dashed lines*) along the axial direction, which results in a force on the bead directed up toward the magnets. The magnets can be lowered (raised) to increase (decrease) the stretching force acting on the DNA. Magnet north and south poles are labeled N and S, respectively.

(**b**) Thanks to the magnetic properties of the bead, torque forces can also be applied to the sample. Starting from a relaxed configuration, the DNA molecule becomes more and more tangled as the magnets rotate, and eventually, loops of helices (plectonemes) are formed. The formation of plectonemes induces a progressive linear decrease in the DNA extension, which is analogous to the decrease in extension of a twisted and tangled telephone cord [20]

DNA Manipulation Based on Nanotweezers, Fig. 3 MNEMS nanotweezers for DNA manipulation and characterization [13]. (**a**) 3D schematic view of the design. The mobile tip is electrostatically actuated by V_{act}, and its displacement changes the sensor capacitance values C1 and C2 which are measured with an adequate electronic readout. AC electrodes allow applying an electric field in between the tips for molecule trapping. (dimensions: 5 mm × 6 mm). (**b**) Camera photograph of the device. (**c**) Close-up view (scanning electron microscope image) of the electrostatic comb-drive actuator

micromachining and compact integration [15]. The sensor consists of a series of central plates that moves relatively to fixed external plates, forming two variable capacitances C_1 and C_2. The values of the

capacitances change inversely with the gap between the electrodes. Hereafter, the capacitance variations are measured with an electronic readout allowing static or dynamic displacement measurements.

The micromachining of the device and of the sharp tips is explained in the next section, and the role of the sharp tips is discussed in the DNA trapping section.

Microfabrication of the Nanotweezers

The fabrication of the MEMS tweezers is based on standard microfabrication processes [16], i.e., reactive ion etching (RIE), local oxidation, and anisotropic etching of silicon. The starting material is a silicon-on-insulator (SOI) substrate having the following characteristics: $<100>$ −oriented crystallography Miller indices, 30-μm-thick silicon active layer, 2-μm-thick buried oxide insulator, and 400-μm-thick silicon handling substrate. The process flow is summarized below:

1. A thin Si3N4 layer is first deposited by low-pressure chemical vapor deposition (LPCVD) and patterned to form rectangles aligned along the $<100>$ directions (mask #1).
2. The Si3N4 and the Si over layer are etched by reactive ion etching (mask #2).
3. Next, a local oxidation of silicon (LOCOS) process is used to grow SiO2 on the top and sidewalls of the structured Si.
4. The Si3N4 layer is then removed.
5. A KOH wet anisotropic etching of Si is performed to obtain {111} facets, which make sharp opposing tips.
6. The buried oxide is removed by hydrofluoric acid (HF), and the handling silicon is structured by deep reactive ion etching (DRIE), using an aluminum mask on the backside (mask #3).

In the final step, a thin aluminum film is evaporated on the front side. Indeed, aluminum acts as an anchoring material for DNA molecules (cf. Trapping of DNA Molecules section). One should note that the process only requires three lithographic masks: one for defining the area of the sharp tips with silicon nitride (#1), one for microstructuring the silicon over layer (#2), and one for the backside etching of the handling substrate (#3).

Trapping of DNA Molecules

Figure 3b shows a scanning electron microscope (SEM) image of a DNA bundle anchored between the tips of nanotweezers. The trapping of molecule bundle with silicon tweezers is achieved when an AC electric field causes positive dielectrophoresis (called in the literature DEP): The DNA molecules elongate along the electric field lines and move toward the regions where the field is higher [17]. Sharp aluminum-coated electrodes create high electric gradients that cause the ends of the DNA strands to be attracted to the edges of the electrodes and become attached, resulting in the permanent formation of a bridging structure. Indeed, the adhesion of DNA molecules is strong on the metal as the molecules can be grafted (covalent bounds) on the metal and the metal oxide by its biphosphate function, present on the phosphate-deoxyribose backbone of the DNA, creating P-O-metal link by hydrolysis reaction.

The method can be used to trap single DNA molecules by tuning the parameters appropriately. Two approaches for single-molecule studies are currently investigated:

- A double-stranded λ-DNA molecule can be isolated by DC electrophoresis and consecutively trapped by AC DEP in a microfluidic device.
- Using pulsed DEP, silicon tweezers can be used to trap a single DNA molecule in an aqueous solution.

For all the biological experiments, a solution of double-stranded λ-DNA (48,502 base pairs, 16 μm long) was used. Double-stranded λ-DNA is the DNA of a bacteriophage known as Enterobacteria phage λ or λ-phage. After dilution in deionized water, a small droplet of the solution was deposited on a glass slide. The tweezers were mounted on an optical microscope and brought in contact with the surface of the droplet with a 3D precision micromanipulator (Fig. 4). Then, a high AC electric field (2 MV/m or 2 V/μm at 1 MHz frequency) was applied for few seconds in order to capture DNA molecules. After the experiments, the bundle could be broken and removed by blowing air and rinsing the tips with water. The DEP experiment could be repeated as long as the aluminum coating remained on the silicon tips.

Biocharacterization on DNA Molecules

The characterization of biological samples is done through the mechanical response of the system {tweezers + DNA bundle}. The frequency response of the tweezers being well identified, the changes in

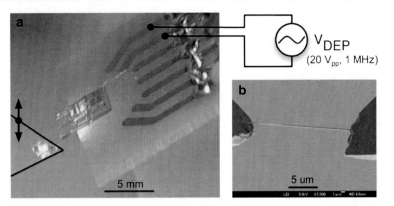

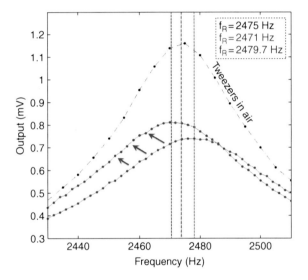

DNA Manipulation Based on Nanotweezers, Fig. 4 Trapping of DNA molecules with silicon nanotweezers by dielectrophoresis (DEP). (**a**) An AC voltage (V_{DEP}) is applied between the two tips of the tweezers in order to create a nonuniform electric field in between. A cover glass on which the DNA solution droplet is deposited is moved in z direction to come in contact with the tips. (**b**) Scanning electron microscope (SEM) image of a trapped molecule bundle

the frequency response are attributed to mechanical properties of the molecules. Recently, the real-time monitoring of the number of trapped DNA molecules during the dielectrophoresis was proven. During the trapping, the resonant frequency of the system increases resulting from the addition of the bundle stiffness to that of the sole tweezers. Simultaneously, the quality factor tends to decrease as the viscous losses increases with the bundle formation. In Fig. 5, a clear difference can be observed between frequency responses of the nanotweezers with and without trapped DNA molecules.

The DNA physical properties change with protein interactions, inducing changes of the whole system frequency response. Consequently, nanotweezers emerge as a relevant tool even for characterization of DNA-protein interactions. In a recent study, the digestion of the bundle by restriction enzymes was monitored in real time through the frequency response of the nanotweezers. HindIII proteins are known as restriction enzymes for DNA. Indeed, HindIII recognizes a specific sequence in the DNA code where to cleave the molecule [18].

As for the trapping of molecules, the resonance of the system {tweezers + DNA bundle} is monitored. Conversely, the resonant frequency decreases down to the initial resonant frequency of the sole tweezers when the digestion of the bundle is finished. The quality factor increases (Fig. 6). The evolution of the number of the molecules in the bundle, deduced from the single-molecule stiffness ($k_{\lambda\text{-DNA}} = 30\ \mu N/m$), shows an exponential decay with time, the extracted time constant depending inversely on the enzyme concentration [19].

DNA Manipulation Based on Nanotweezers, Fig. 5 Tweezers resonance evolution during digestion of the DNA bundle. Blue dots are the frequency response of the tweezers with a bundle (in solution). *Red dots* are the response of the tweezers after digestion of the bundle (without any bundle in solution). For comparison, the resonance of the sole tweezers in air is plotted in *black* [19]

The minimum number of molecules that can be sensed using such silicon nanotweezers (about 20 molecules) depends on the minimum frequency variation that can be experimentally measured. In recent works, dynamic control strategies have been developed to improve the overall system sensitivity to the variation of the number of DNA molecules. The objective of such control strategy is the detection of a variation on a single molecule of DNA.

DNA Manipulation Based on Nanotweezers,
Fig. 6 Time evolution of the resonance properties of the system {tweezers + DNA bundle}. Resonant frequency (in *blue dots*) decreases with the digestion of the bundle by the restriction enzymes when the quality factor (in *red dots*) increases. For comparison, resonant frequency of the sole tweezers is shown in deionized water (in *black dots*) and in a buffer without restriction enzymes (in *gray dots*) [19]

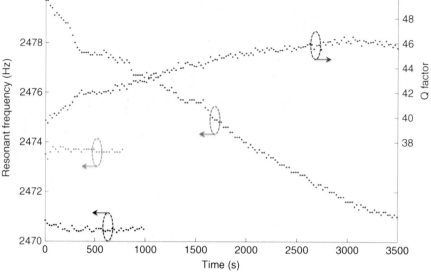

Outlook

Over the past two decades, remarkable techniques and tools were developed in order to perform single-molecule manipulation. Although these recent technological advances had a significant impact on the understanding of the interactions between DNA and the molecular machinery, they present some drawbacks that open new research issues.

Indeed, in most of the biological experiments on DNA, manipulations rely on complex systems as optical or magnetic tweezers. These techniques have a low throughput since their implementation is quite long and difficult. Furthermore, the molecule preparation is done one at a time. Here the direct sensing of biomolecular reactions is demonstrated using silicon nanotweezers. The experimental results obtained with the integrated MEMS nanotweezers showed the efficiency of such device for DNA characterization. Nevertheless, the main drawbacks of MEMS stem from the large size and high stiffness of the system relative to the biological molecule properties. Control strategies that are currently under development aim at reducing the tweezers stiffness using feedback to improve the sensitivity of the whole process toward single DNA molecule variation.

The improvement of resolution in DNA sensing brought new perspectives and questions to address.

The complexity of the DNA machinery is still far from being elucidated. More complex techniques or technologies are under development with, for instance, combination of different tools with fluorescent visualization, in order to follow multiprotein multienzyme complexes. Lastly, new developments on magnetic tweezers have recently achieved first results inside the nucleus of living cells. Performing such in vivo experiments inside cells will provide inestimable knowledge of molecules functions in their inherent environment.

Cross-References

▶ 3D Micro/Nanomanipulation with Force Spectroscopy
▶ Basic MEMS Actuators
▶ Dielectrophoresis of Nucleic Acids
▶ Nanogrippers
▶ Optical Tweezers

Bibliography

1. Bruce, A., Alexander, J., Julian, L., Martin, R., Keith, R., Peter, W.: The Molecular Biology of the Cell. Garland Science, New York (2002)
2. Neuman, K.C.: Single-molecule measurements of DNA topology and topoisomerases. J. Biol. Chem. **285**(25), 18967–18971 (2010). doi:10.1074/jbc.R109.092437

3. Dumont, S., Cheng, W., Serebrov, V., Beran, R.K., Tinoco, I., Pyle, A.M., Bustamante, C.: RNA translocation and unwinding mechanism of HCV NS3 helicase and its coordination by ATP. Nature **439**(7072), 105–108 (2006). doi:10.1038/nature04331

4. Hamdan, S.M., Johnson, D.E., Tanner, N.A., Lee, J.-B., Qimron, U., Tabor, S., van Oijen, A.M., Richardson, C.C.: Dynamic DNA helicase-DNA polymerase interactions assure processive replication fork movement. Mol. Cell **27**(4), 539–549 (2007). doi:10.1016/j.molcel.2007.06.020

5. Van Mameren, J., Peterman, E., Wuite, G.: See me, feel me: methods to concurrently visualize and manipulate single DNA molecules and associated proteins. Nucl. Acids Res. **36**, 4381–4389 (2008)

6. Haustein, E., Schwille, P.: Single-molecule spectroscopic methods. Curr. Opin. Struct. Biol. **14**(5), 531–540 (2004). doi:10.1016/j.sbi.2004.09.004

7. Kim, J.H., Larson, R.G.: Single-molecule analysis of 1D diffusion and transcription elongation of T7 RNA polymerase along individual stretched DNA molecules. Nucleic Acids Res. **35**(11), 3848–3858 (2007). doi:10.1093/nar/gkm332

8. Neuman, K.C., Nagy, A.: Single-molecule force spectroscopy: optical tweezers, magnetic tweezers and atomic force microscopy. Nat. Methods **5**(6), 491–505 (2008). doi:10.1038/nmeth.1218

9. Moffitt, J., Chemla, Y., Smith, S., Bustamante, C.: Recent advances in optical tweezers. Annu. Rev. Biochem. **77**, 205–228 (2008)

10. Cluzel, P., Lebrun, A., Heller, C., Lavery, R., Viovy, J.L., Chatenay, D., Caron, F.: DNA: an extensible molecule. Science **271**(5250), 792–794 (1996)

11. Bustamante, C., Bryant, Z., Smith, S.B.: Ten years of tension: single-molecule DNA mechanics. Nature **421**(6921), 423–427 (2003). doi:10.1038/nature01405

12. Strick, T., Allemand, J., Croquette, V., Bensimon, D.: Twisting and stretching single DNA molecules. Prog. Biophys. Mol. Biol. **74**(1–2), 115–140 (2000)

13. Yamahata, C., Collard, D., Legrand, B., Takekawa, T., Kumemura, M., Hashiguchi, G., Fujita, H.: Silicon nanotweezers with subnanometer resolution for the micromanipulation of biomolecules. J. Microelectromech. Syst. **17**(3), 623–631 (2008)

14. Tang, W., Nguyen, T.-C.H., Howe, R.: Laterally driven resonant microstructures. Sensors Actuators **20**, 25–32 (1989)

15. Sun, Y., Fry, S.N., Potasek, D.P., Bell, D.J., Nelson, B.J.: Characterizing fruit fly flight behavior using a microforce sensor with a new comb-drive configuration. J. Micromech. Syst. **14**, 4–11 (2005)

16. Senturia, S.: Microsystem Design. Kluwer, Boston (2000)

17. Washizu, M., Kurosawa, O.: Electrostatic manipulation of DNA in microfabricated structures. IEEE Trans. Ind. Appl. **26**(6), 1165–1172 (1990)

18. Roberts, R.J.: How restriction enzymes became the workhorses of molecular biology. Proc. Natl. Acad. Sci. U. S.A. **102**(17), 5905 (2005)

19. Kumemura, M., Collard, D., Yoshizawa, S., Fourmy, D., Lafitte, N., Jalabert, L., Takeuchi, S., Fujii, T., Fujita, H.: Direct bio-mechanical sensing of enzymatic reaction On DNA by silicon nanotweezers. In: 2010 IEEE 23rd International Conference on Micro Electro Mechanical Systems (MEMS), 915–918. Wanchai, Hong Kong (2010)

20. Salerno, D., Brogioli, D., Cassina, V., Turchi, D., Beretta, G. L., Seruggia, D., Ziano, R., Zunino, F., Mantegazza, F.: Magnetic tweezers measurements of the nanomechanical properties of DNA in the presence of drugs. Nucl. Acids Res. **38**(20), 7089–7099 (2010). doi:10.1093/nar/gkq597

DOLLOP

▶ Prenucleation Clusters

Doping in Organic Semiconductors

Yabing Qi
Energy Materials and Surface Sciences (EMSS) Unit, Okinawa Institute of Science and Technology, Kunigami-gun, Okinawa, Japan

Synonyms

Doping of organic semiconductors

Definition

There are two major ways to dope an organic semiconductor: chemical doping and electrochemical doping. Because of the additional complexity due to the presence of an electrolyte solution in the case of electrochemical doping, the focus is placed on chemical doping. In chemical doping, dopant molecules (typically a few weight percent) are introduced into an organic semiconducting host material, either by co-evaporation in vacuum or spin casting from solution.

Operation Principles

Doping in organic semiconductors can be rationalized based on energetics, i.e., charge carriers (electrons or holes) "flow" from the higher energy levels of host molecules to the lower energy levels of dopant molecules to minimize the total energy. It is known that the

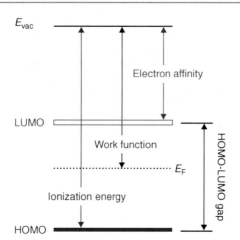

Doping in Organic Semiconductors, Fig. 1 Schematic drawing of energy levels in molecules

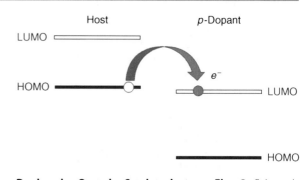

Doping in Organic Semiconductors, Fig. 2 Schematic drawing of charge transfer for p-type doping

energy levels in atoms are discrete and narrow in width. On the other hand, as a large number of atoms are brought to proximity to form solids with an ordered crystal structure, the energy levels evolve to continuous bands. Energy levels in molecules are somewhere between these two extremes, not as discrete as those in atoms, but also not as continuous as those in crystals. Figure 1 illustrates the energy levels in molecules. The highest energy levels that are occupied by electrons are the so-called highest occupied molecular orbitals (HOMO). Any energy level below HOMO is occupied by electrons. The lowest energy levels that are not occupied are the so-called lowest unoccupied orbitals (LUMO). Any energy level above LUMO is unoccupied. The energy required to excite an electron in HOMO of a molecule to the vacuum level is defined as the ionization energy (IE). The energy gain for an electron falling from the vacuum level to fill LUMO is electron affinity (EA). The energy difference between the Fermi level and the vacuum level is work function (WF).

In order to efficiently p-dope a host material, the EA of the dopant molecules needs be comparable to or even greater than the IE of the host molecules (see Fig. 2). Vice versa, an efficient n-type dopant is expected when the dopant molecules have an IE comparable to even lower than the EA of the host material. As an example, α-NPD (*N,N′*-diphenyl-*N,N′*-bis (1-naphthyl)-1,1′-biphenyl-4,4′-diamine) is a widely used hole transport material in organic light emitting diodes (OLEDs). The ionization energy and electron

affinity of α-NPD are 5.4 and 1.4 eV, respectively [3, 6]. An efficient p-type dopant for α-NPD should have an EA comparable to 5.4 eV. F_4-TCNQ (2,3,5,6-tetrafluoro-7,7,8,8-tetracyano-quinodimethane) has an EA of 5.24 eV, which has been shown to efficiently p-dope α-NPD [2]. It is worthwhile to note that in Fig. 2 an organic molecule dopant is assumed, although materials other than organic molecules have been used as dopants. For example, alkali metals have been used as p-type dopants and halogens have been used as n-type dopants.

Basic Methodology and Key Research Findings

Experimental techniques that have been commonly used in doping studies of organic semiconductors include ultraviolet photoemission spectroscopy (UPS), inverse photoemission spectroscopy (IPES), x-ray photoelectron spectroscopy (XPS), Rutherford backscattering (RBS), charge transport measurements, and scanning tunneling microscopy (STM).

The first set of measurements for a typical doping study are usually UPS and IPES measurements on neat films of host and dopant molecules, which provide fundamental information such as IE and EA. With such information for both materials, a crude judgment can be made whether doping can take place between these two materials. While electrochemical measurements combined with optical absorption measurements are also often cited in literature to yield values of IE and EA, it should be noted that (1) electrochemical measurements are usually performed in solution where intermolecular interaction is rather weak and (2) optical gap values need to be corrected for exciton

Doping in Organic Semiconductors,
Fig. 3 Position of the Fermi level in the HOMO–LUMO gap of α-NPD as a function of doping concentration of F$_4$-TCNQ. The values are taken from Gao and Kahn [2]. Reprinted with permission. Copyright 2003, American Institute of Physics

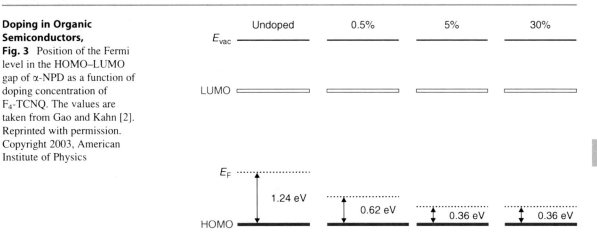

binding energy in order to be compared with the HOMO–LUMO gap (= IE − EA) measured by UPS and IPES.

In order to accurately assess effectiveness of doping, more direct measurements are required, one of which is to study the Fermi level position change by UPS as a function of doping concentration. When organic materials are not intentionally doped (so-called intrinsic), the Fermi level is often close to the mid-gap position. A Fermi level shift from the near mid-gap position to HOMO is expected for the p-type doping. For the n-type doping, on the other hand, the Fermi level shifts toward LUMO. Figure 3 schematically depicts the Fermi level shift toward HOMO as the concentration of F$_4$-TCNQ in α-NPD increases from 0% to 0.5%, and then to 5% [2]. It is worthwhile noting that as the doping concentration increases further to 30%, the Fermi level remains at approximately the same position, and is pinned at ~0.4 eV above HOMO. This is a result of the tail of density of states extended into the HOMO–LUMO gap. It has been proposed that various types of static and/or dynamic disorder in organic films are responsible for these states.

The second direct way to visualize the doping effect is to measure charge transport properties of undoped and doped films. Bulk conductivity enhancement and charge injection improvement at organic–electrode interfaces are two major doping-induced advantages, both of which can be verified via charge transport measurements. These two features are demonstrated in Fig. 4, where the current-voltage (I-V) characteristics are compared for undoped α-NPD, 0.5% F$_4$-TCNQ interface–doped α-NPD, and 0.5% F$_4$-TCNQ

bulk–doped α-NPD samples [2]. Detailed structures of the three types of samples are schematically shown on the right in Fig. 4. Due to the amorphous nature of α-NPD films, charge carrier hopping is believed to be the main conduction mechanism in α-NPD. The first sample is an undoped sample (represented by solid circles in Fig. 4). For this sample, the I-V curve basically follows a straight line with the slope approximately equal to 1 at low applied voltages (between 0.9 and 4 V). The electrical current in this region is dominated by ohmic conduction due to free charge carriers. As the applied bias increases to 6 V, the current begins to climb up with a much steeper slope. This rapid increase is called injection onset, caused by charge carrier injection at the organic/electrode interfaces. The steep slope is most likely caused by the filling-up of trap states in α-NPD, which in turn facilitate charge carrier hopping in α-NPD. Doping can be used to improve injection, which is demonstrated by the interface-doped α-NPD sample (represented by stars in Fig. 4). I-V characteristics at low applied voltages (between 0.9 and 2 V) share the same features as those of the undoped sample, i.e., linear on log-log scale with a slope of 1. The current in this regime is within a factor of 5 of that of the undoped one, which is expected from the fact that only a small portion of the film is doped. On the other hand, injection properties are significantly improved, and the injection onset occurs at a much lower bias (~3 V). For the bulk-doped sample (represented by open circles in Fig. 4), the low bias conductivity (i.e., in the ohmic conduction region) is 5 orders of magnitude higher than that of the undoped sample. Along the same line, doping has been implemented in organic field-effect transistors (OFETs) to tackle the contact resistance issue, which is one of the

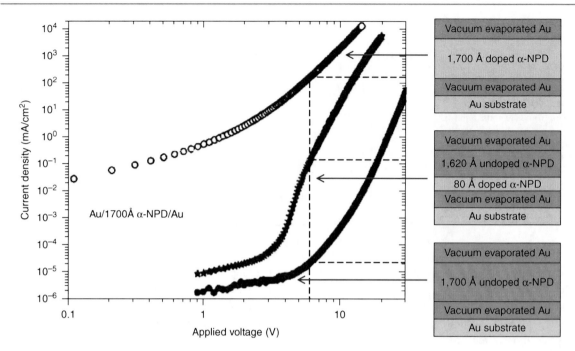

Doping in Organic Semiconductors, Fig. 4 I-V curves of undoped α-NPD, 0.5% F₄-TCNQ interface–doped α-NPD, and 0.5% F₄-TCNQ bulk–doped α-NPD samples. Detailed structures of the three types of samples are schematically shown on the right. Reprinted with permission from Gao and Kahn [2]. Copyright 2003, American Institute of Physics

common problems. Studies have shown that contact resistance in pentacene OFETs has been reduced by sevenfold by inserting a 10-nm-thick doped interface layer [4]. It is important to note that the stability of the devices was not affected significantly by doping of the contacts [4].

In some cases, as dopant molecules donate (or accept) electrons from the host material, oxidation states of some specific atoms in the dopant molecules change. This change can be used as an indicator for doping. Oxidation states can be determined by XPS. For example, such a change is revealed in an XPS study on cobaltocene and cobaltocene-doped tris(thieno)hexaazatriphenylene derivative [1].

Future Directions for Research

In the past decade or so, applications involving organic semiconductor (especially OLED-based products such as OLED lighting, OLED TV, OLED cell phone, etc.) have been commercialized successfully. Meanwhile, there is a rapidly growing interest in using doping to improve performance of these organic based devices [5]. Despite increased interest

and research efforts along this direction to make doping a process readily applicable in organic based devices, several key issues need to be addressed appropriately. First of all, air stability of dopants and doped films are crucial. This is particularly true for n-type dopants, because with a high-lying HOMO n-dopants are prone to oxidation when exposed to air. Secondly, to cut down fabrication cost, it is highly desirable to be able to introduce dopants via ambient pressure, low-temperature, and high-throughput techniques such as spin coating, spary pyrolysis, inkjet printing, etc. Thirdly, to ensure reliable operation of organic-based devices, dopants need to be spatially confined within the targeted region. Therefore, diffusional stability of dopants in the host matrix is of great value. Last but not least, systematic research is necessary to search for new dopants that are of low cost and high performance.

Cross-References

▶ Charge Transport in Self-assembled Monolayers
▶ Conduction Mechanisms in Organic Semiconductors

▶ Electrode–Organic Interface Physics
▶ Flexible Electronics
▶ Hybrid Solar Cells
▶ Maxwell–Wagner Effect
▶ Organic Actuators
▶ Organic Bioelectronics
▶ Organic Photovoltaics: Basic Concepts and Device Physics
▶ Organic Sensors
▶ Surface Electronic Structure

References

1. Chan, C.K., Amy, F., Zhang, Q., Barlow, S., Marder, S.R., Kahn, A.: Chem. Phys. Lett. **431**, 67 (2006)
2. Gao, W., Kahn, A.: J. Appl. Phys. **94**, 359 (2003)
3. Hill, I., Kahn, A.: J. Appl. Phys. **84**, 5583 (1998)
4. Tiwaria, S.P., Potscavage Jr., W.J., Sajoto, T., Barlow, S., Marder, S.R., Kippelen, B.: Org. Electron. **11**, 860 (2010)
5. Walzer, K., Maennig, B., Pfeiffer, M., Leo, K.: Chem. Rev. **107**, 1233 (2007)
6. Wan, A., Hwang, J., Amy, F., Kahn, A.: Org. Electron. **6**, 47 (2005)

Doping of Organic Semiconductors

▶ Doping in Organic Semiconductors

Double-Walled Carbon Nanotubes (DWCNTs)

▶ Chemical Vapor Deposition (CVD)

Droplet Microfluidics

▶ Electrowetting

Drug Delivery and Encapsulation

▶ Acoustic Nanoparticle Synthesis for Applications in Nanomedicine

Drug Delivery System

▶ Chitosan Nanoparticles

Dry Adhesion

▶ Gecko Adhesion

Dry Etching

Avinash P. Nayak, M. Saif Islam and
V. J. Logeeswaran
Electrical and Computer Engineering, University of California - Davis Integrated Nanodevices & Nanosystems Lab, Davis, CA, USA

Synonyms

Chemical dry etching; Gas etching; Physical dry etching; Physical-chemical etching; Plasma etching

Definition

In dry etching, plasmas or etchant gasses remove the substrate material. The reaction that takes place can be done utilizing high kinetic energy of particle beams, chemical reaction, or a combination of both.

Physical Dry Etching

Physical dry etching requires high energy kinetic energy (ion, electron, or photon) beams to etch off the substrate atoms. When the high energy particles knock out the atoms from the substrate surface, the material evaporates after leaving the substrate. There is no chemical reaction taking place and therefore only the material that is unmasked will be removed. The physical reaction taking place is illustrated in Fig. 1.

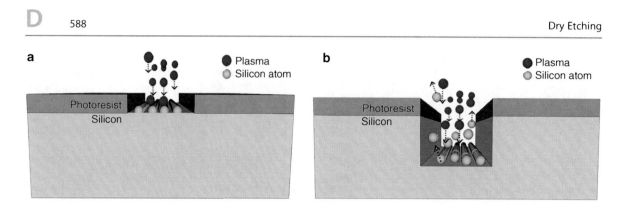

Dry Etching, Fig. 1 The plasma hits the silicon wafer with high energy to knock off the Si atoms on the surface. (**a**) The plasma atoms hitting the surface. (**b**) The silicon atoms being evaporated off from the surface

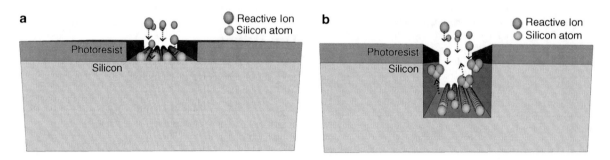

Dry Etching, Fig. 2 Process of a reactive ion interacting with the silicon surface. (**a**) The interaction between the reactive ion and the silicon atom. (**b**) A bond between the reactive ion and the silicon atom then chemically removes the silicon atoms from the surface

Chemical Dry Etching

Chemical dry etching (also called vapor phase etching) does not use liquid chemicals or etchants. This process involves a chemical reaction between etchant gases to attack the silicon surface. The chemical dry etching process is usually isotropic and exhibits high selectively. Anisotropic dry etching has the ability to etch with finer resolution and higher aspect ratio than isotropic etching. Due to the directional nature of dry etching, undercutting can be avoided. Figure 2 shows a rendition of the reaction that takes place in chemical dry etching. Some of the ions that are used in chemical dry etching is tetrafluoromethane (CH_4), sulfur hexafluoride (SF_6), nitrogen trifluoride (NF_3), chlorine gas (Cl_2), or fluorine (F_2).

Reactive Ion Etching

Reactive ion etching (RIE) uses both physical and chemical mechanisms to achieve high levels of resolution. The process is one of the most diverse and most widely used processes in industry and research. Since the process combines both physical and chemical interactions, the process is much faster. The high energy collision from the ionization helps to dissociate the etchant molecules into more reactive species.

In the RIE-process, cations are produced from reactive gases which are accelerated with high energy to the substrate and chemically react with the silicon. The typical RIE gasses for Si are CF_4, SF_6 and $BCl_2 + Cl_2$. As seen in Fig. 3, both physical and chemical reaction is taking place.

Applications

To fabricate highly efficient portable and flexible electronic devices for future applications, high-aspect-ratio vertically oriented silicon micropillars were fabricated using deep reactive ion etching (DRIE) by Logeeswaran et al. [1]. As shown in Fig. 4, a highly doped p-type Si(100) substrate with

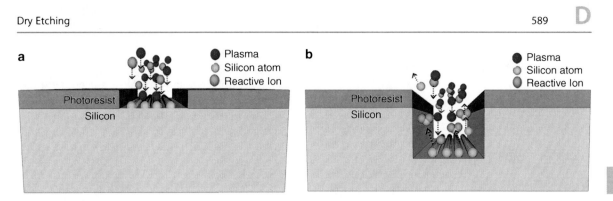

Dry Etching, Fig. 3 The RIE process. This process involves both physical and chemical reactions to etch off the silicon

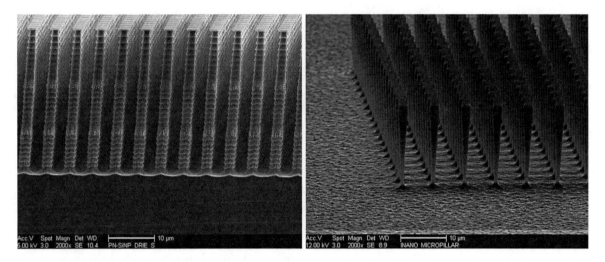

Dry Etching, Fig. 4 Silicon nanopillars fabricated using deep reactive ion etching

a doping concentration of $\sim 10^{19}$ cm^{-3} was patterned with a 2-μm dot etch mask. The substrate was kept at 10°C with SF$_6$ and C$_4$F$_8$ flows of 300 and 150 sccm, respectively, a source RF power at 1,800 W, and a substrate power at 20 W for a total etching time of $\sim$ 6 min. These nanowires were made using a Bosch process. The Bosch process is a pulsed-multiplexed etching technique which alternates between two modes to achieve extremely long and vertical micron-scaled nanowires.

An extensive study was done to determine the dependence of the etching performance by Chen et al. [2] Factors such as applied coil or electrode power, reactant gas flow rates, duty cycles, and chamber pressures were considered. These process parameters were modeled and experimentally determined to see the effect on the surface morphology and the mechanical performance of the silicon structures.

Cross-References

► Dry Etching
► EUV Lithography
► Nanofluidics
► Wet Etching

References

1. Logeeswaran, V.J., et al.: Harvesting and transferring vertical pillar arrays of single-crystal semiconductor devices to arbitrary substrates. IEEE T. Electron Dev **57**, 1856–1864 (2010)
2. Chen K.S., et al.: Effect of process parameters on the surface morphology and mechanical performance of silicon structures after deep reactive ion etching (DRIE), vol. 11. ETATS-UNIS: Institute of Electrical and Electronics Engineers, New York (2002)

Dual-Beam Piezoelectric Switches

▶ NEMS Piezoelectric Switches

DUV Lithography

▶ DUV Photolithography and Materials

DUV Photolithography and Materials

Garry J. Bordonaro
Cornell NanoScale Science and Technology Facility,
Cornell University, Ithaca, NY, USA

Synonyms

193-nm lithography; 248-nm lithography; DUV lithography; Photolithography

Definition

Deep Ultraviolet (DUV) Photolithography is the process of defining a pattern in a thin photosensitive polymer layer (photoresist) using controlled 254–193-nm light such that the resulting polymer pattern can be transferred into or onto the underlying substrate by etching, deposition, or implantation. The exposing light is passed through a chrome-on-quartz photomask, whose opaque areas act as a stencil of the desired pattern. The exposed polymer is then subjected to a chemical development process where the unwanted areas of polymer are removed, leaving the target areas unprotected from subsequent processing.

In semiconductor manufacturing, the light sources used are typically excimer lasers and produce either 248-nm or 193-nm light. The light is usually passed through the photomask and then through a reduction lens which reduces the pattern size by a factor of 5 or 4 times.

Introduction

The process of photolithography has been performed since at least the 1950s in the manufacturing of circuit boards and discrete electronic devices, initially using cyclized rubber-based photoresists as the patterning medium using contact lithography, where the photomask is brought into contact or near-contact with the substrate for direct 1:1 patterning, using mask aligning exposure tools. Modern DNQ-Novolak (Diazonaphthoquinone) photoresists were developed in the early 1970s allowing features smaller than 2 μm to be printed, and opened the door to the age of mass-produced silicon-based integrated circuits. The resulting economies of scale eventually led to the development of the personal computer for the mass market. The growth of this market, in turn, drove the development of advanced photoresist materials and exposure tools to satisfy the demands of the market for faster, smaller, and cheaper technology products.

The process of UV (Ultraviolet, 436–365 nm) photolithography as performed for the last 30 years remains basically unchanged in that time. A design pattern is generated using computer-aided design (CAD) software. The pattern data is converted into a standardized file format, usually GDSII or, more recently, OASIS. This data is then converted by specialized software into the format needed for the specific mask writing tool to be used, and then sent to the mask writer. The mask writer exposes the pattern on a thin resist film (100–500 nm) coated on top of a thin chrome layer (1,000 Å typ.), which has been deposited on top of the photomask substrate, which is most often fused silica or quartz, but can be other types of glass. The exposed pattern is then chemically developed, revealing the chrome layer underneath in the shape of the desired pattern. Next, the uncovered chrome is etched away with an acid, removing the designed pattern from the chrome layer. The end result is an opaque stencil made from thin chrome on a transparent plate, which is suitable for use in either a mask aligner for contact photolithography, or a projection photolithography tool. The photomask can then be used many times, particularly in projection photolithography, where the pattern area of the mask never contacts the wafer or the tool.

The device substrate, usually a silicon wafer, is spin-coated with a thin photoresist layer and baked at a temperature near the glass transition temperature of the resist to remove most of the solvent. Exposure of the mask pattern into the resist is then performed, either by bringing the wafer and mask together before exposure (contact or proximity photolithography, 1:1), or by mounting the mask in front of a (usually reduction) lens and projecting the image onto the wafer (projection photolithography, 4:1 typ.). Production tools currently are virtually all reduction projection printers, and they are what the remainder of this discussion will focus on. The DNQ-Novolak chemistries used in UV photolithography today are substantially the same as when they were first formulated, with some tweaking of wavelength absorption properties and sensitivities. The exposure light causes a chemical reaction within the photoresist, forming acid (for DNQ resist, indene carboxylic acid) in the exposed areas. An aqueous base solution (developer) removes these acidic areas of the resist, creating a three-dimensional replica of the pattern seen on the photomask. Positive-tone photoresists such as this are dominant in usage, but negative-tone versions, where the exposed areas *remain* after development, also are available. The high-pressure mercury (Hg) arc lamp light sources used still possess the same output spectrum characteristics as before, and the exposure tools using them, while more sophisticated and automated than before, are essentially the same (Fig. 1).

The emergence of modern-day DUV (Deep Ultraviolet, 248–193 nm) exposure tools and chemistry has profoundly changed this established process. What was once a comparatively forgiving series of process steps has now become a critically controlled time and temperature ballet with incredibly complex machinery and computerized systems. The development of DUV photolithography has caused engineers to push the boundaries of our understanding of the behavior of light, as well as our concepts regarding mechanical and chemical systems, beyond what we once considered the limits of our abilities. Most remarkably, it is done on a daily basis in High-Volume Manufacturing (HVM), and with great success.

History

Modern DUV lithography had its beginning in the early 1980s at IBM with Frechet, Willson, and Ito,

and their development of chemically amplified resists. The drive toward ever-smaller features in the structures that make up the electronic devices in the marketplace required the use of ever-shorter wavelength exposure light, which in turn required new photoresists with appropriate sensitivity to these new wavelengths. Unfortunately, the existing photoresist materials were too absorptive and insensitive in the DUV wavelength regions to be practical. New polymer systems were developed with sufficient transparency at these wavelengths, but their sensitivity to exposure was still far too low. Chemically Amplified (CA) resists overcame this by using a complex heat-activated de-protection method that multiplied the acid formed by exposure, which increased the material's DUV sensitivity by some orders of magnitude, leading to today's superior performers with resolution capabilities far better than DNQ-sensitized materials could ever achieve.

However, this was not an overnight success story. Although these resists began to appear in the early 1980s, and IBM had a small production line in Burlington, Vermont in the late 1980s, HVM using 248-nm photolithography did not begin until the late 1990s. There were two primary reasons for this: a lack of adequate photoresists, and delayed tool development. The CA resists described previously would take another 10 years to reach the stage of development where they were robust enough to survive the rigors of mass production with reasonable cost. Problems with airborne contamination and etch resistance also took years to overcome, but in the end DUV resist performance surpassed expectations, with etch resistance nearly equal to DNQ materials, and with much better sensitivity and resolution. The exposure tooling also encountered many roadblocks during development, including difficulty in identifying suitable fused silica material for lenses and masks, developing suitable 248-nm light sources, and, perhaps most importantly, finding funds to research all of these issues. The SEMATECH consortium played a vital role in this by helping to organize and direct efforts and funding in these matters. Industry-University collaborations also played key roles in areas like materials and chemical engineering. They reprised these roles during the similar progression in the development of 193-nm materials and tools in the 1990s. Thankfully, lessons learned in resist development and material qualification during the run-up to 248-nm HVM shortened the

DUV Photolithography and Materials, Fig. 1 Basic photolithography process

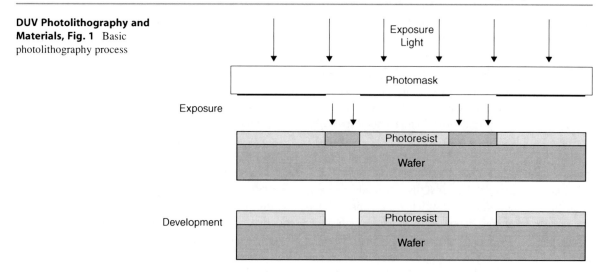

DUV Photolithography and Materials, Fig. 1 Basic photolithography process

193-nm development timeline by about 50%. This learning was also important in later investigations into the viability of 157-nm (F_2) photolithography, a technology that was ultimately rejected.

DUV Processing

DUV photolithography is at its core the same as UV photolithography. The two methods share many of the same basic processing steps: photoresist is spin-coated onto wafers, baked, exposed, post-baked, and developed. Similar tools are used for both technologies in HVM: spin coaters, hot plates, step-and-repeat systems (steppers), and aqueous developers. However, there are important differences in these processes besides the minimum value of the resulting pattern's Critical Dimensions (CD). The devil is, as usual, in the details (Fig. 2).

Wafer Preparation

Photolithographic processing always begins with cleaning of the wafer surface. Depending on the state of the wafer (new or in-process) the cleaning process can take many forms, and be either dry or wet. Final results depend heavily on the cleanliness of the wafer surface, so the importance of this step should not be minimalized. Typical new wafer cleaning may include the "standard cleans," SC1 and SC2, which derive from the well-known "RCA clean" originally used in vacuum tube manufacturing. SC1 is a 5:1:1 solution of water, ammonium hydroxide, and hydrogen peroxide

heated to ~75°C. SC2 is a 6:1:1 solution of water, hydrochloric acid, and hydrogen peroxide heated to ~75°C. Both baths are usually performed for ~10 min each, and an additional step consisting of a dip in dilute hydrofluoric acid is sometimes included as well. Cleaning of wafers which are "in-process" (partially completed) can take many forms, each dependent on the previous processing steps performed on the wafer, as specific residue will need to be addressed.

Once a clean surface is achieved, the paths of the two processes differ: UV wafers typically undergo a "priming" step, where the wafer surface is dehydrated using heat and sometimes vacuum, and then coated with a chemical primer, Hexamethyldisilazane (HMDS), to maintain the dehydrated state and optimize adhesion of the photoresist to the wafer surface [1]. Dehydration is necessary as photoresist will preferentially attach to water on the surface, causing serious degradation of resist adhesion. HMDS application will contain any remaining water and greatly increase the wetting angle of the surface, rendering it hydrophobic. DUV wafers, on the other hand, typically do not undergo this priming step.

In DUV lithography, a Bottom Antireflective Coating (BARC) is usually applied to the surface of the wafer by spin-coating. During exposure, light penetrates progressively deeper into the photoresist until complete exposure is achieved. At that point, the light reaches the bottom of the resist and is reflected by the wafer surface. BARC is a thin polymer layer which is designed, when applied at the proper thickness, to

Exposure Light

Reticle

Reduction Lens

X

Stepping Motion of Wafer

Y

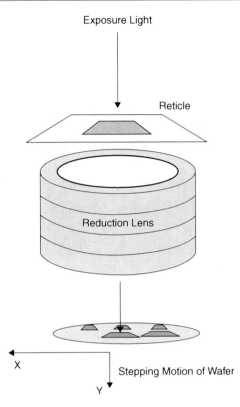

DUV Photolithography and Materials, Fig. 2 Wafer stepper schematic

absorb the light which would otherwise be reflected back into the resist by the wafer. BARC is also used in some UV processing where topography of the wafer surface can cause "notching" or unwanted exposure due to reflected light, particularly in the case of metal lines on the wafer. Light can be reflected randomly by irregularities in the surface, especially by features that have been etched into the wafer, or by deposited material on top of it, creating a background dose which can affect feature sizes. In some cases, the reflected light can be focused onto an area of resist, exposing it and creating unexpected patterns. BARC is necessary in most DUV processing due to the high reflectivity of silicon (or almost any material) at DUV wavelengths. This reflectivity, in combination with additional internal reflections caused by the refractive index mismatch at the air-resist interface, forms a standing wave within the bulk of the resist. The resulting constructive or destructive total effect of this wave on the exposure dose makes CD control increasingly difficult as feature sizes shrink, and the impact on the dose becomes more significant. Therefore, BARC is used

in most 248-nm processing and in all 193-nm processing. In fact, in advanced 193-nm processing multi-layer BARCs are common (Fig. 3).

Antireflective coatings are comprised of various polymer materials, usually linked to dyes to give them the desired wavelength absorbance. The polymers can be acrylic, vinyl, polyether, polyurethane, polysaccharide, or polystyrene, among others. The polymer is dissolved in a photoresist bowl-compatible solvent to allow easy integration into the processing line. There must be no intermixing with the photoresist applied after bake, and no chemical interactions that might cause footing or undercut at the base of the resist feature. Some BARCs can be developed away with the resist (D-BARC) to simplify processing or for applications such as metal lift-off.

There are two important benefits of this step: the internal reflective interference is greatly reduced or eliminated, allowing better process control, and the adhesion of the photoresist to the surface (now BARC) is greatly improved, negating the need for any priming [2]. The inevitable trade-off is that the process requires an additional spin and bake step at elevated temperature (~180–220°C), and sometimes an additional dry etch step after development to open the ARC at the bottom of the resist before the actual process step is performed. The etch chemistry can be O_2 plasma, or the same process that will be used in the following wafer etch step, but characterization of the etch rate is necessary. The thickness of the BARC must be adjusted to optimize performance with the existing film stack on the wafer at the time of exposure due to the interaction of the light with each of the layers, and must be re-optimized for any changes in the films during each step in the process. Photolithographic computer modeling can be of great help here, narrowing the empirical testing parameters to acceptable size.

Another type of antireflective coating, Top layer ARC, or TARC, is also used in some situations. This layer works in much the same manner as BARC, but is applied to the top of the resist layer. This layer can serve two purposes: it can act as an absorber for light reflected back from the wafer surface, reducing the standing wave intensity, and it can serve as a protective barrier layer for the photoresist, keeping any harmful chemicals from reaching it. In many immersion lithography applications (see DUV Exposure), it is used for this second function, preventing the

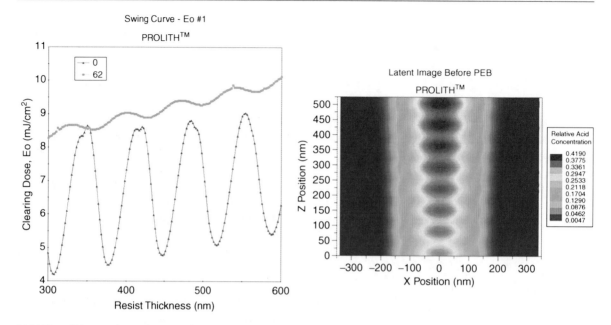

DUV Photolithography and Materials, Fig. 3 Standing wave dose swing of 63% reduced to 1% by BARC; 248 nm on Si

immersion fluid, usually water, from leaching chemicals from the surface of the resist and degrading the resulting image.

Wafer Flatness

One important difference between UV and DUV processing is the flatness requirement of the wafer substrate. Silicon wafers start as slices from a long single-crystal ingot, called a boule. These slices are polished to the required thickness and flatness on either one side or both sides. The surface polish is usually required to be <1 nm RMS, with <0.5 nm being more desirable. Wafers polished on both sides (Double-polished) typically have the flattest surfaces and least thickness variation. As will be shown, flatness is important for successful DUV lithography.

The most stringent requirement for wafer flatness in UV lithography is 2 μm TTV (Total Thickness Variation). The *least* stringent requirement for projection DUV lithography wafer flatness is 2 μm TTV, and is 0.5 μm TTV for many critical processes. This more demanding requirement is due to the properties of projection optics outlined below.

The resolution of a reduction lens is calculated based on the Rayleigh Criterion:

$$R = \frac{k_1 \lambda}{NA}$$

where

R = resolution of the system (1/2 of the smallest printable pitch)

λ = wavelength of exposure light

NA = numerical aperture of the lens

k_1 = process-dependent factor (≥ 0.8 typ. DNQ, ≤ 0.6 typ. DUV).

However, the Depth of Focus of the same system is similarly calculated:

$$DOF = \frac{k_2 \lambda}{NA^2}$$

where

DOF = depth of focus (usable focus range on-axis)

λ = wavelength of exposure light

NA = numerical aperture of the lens

k_2 = process-dependent factor (≤ 1 typ.).

The first equation shows that smaller wavelength light sources result in smaller features being printed (λ is directly proportional to R). It also shows that larger NA lenses result in smaller features being printed (NA is *inversely* proportional to R).

The second equation shows that smaller wavelength light sources result in a smaller depth of focus (λ is directly proportional to DOF). It also shows that larger NA lenses result in a much smaller depth of focus (NA is *inversely exponentially* proportional to DOF) [3].

From the information above it is clear that the same mechanisms that produce smaller features also decrease the depth of focus, which is typically the most important limiter of process latitude. The DOF is the tolerance of the system to vertical variations in the target surface, which can include resist non-uniformity, chuck non-flatness or contamination, focus system noise, and physical vibrations, as well as wafer thickness excursions. In other words, the *difficulty* of printing features increases *exponentially* as the minimum pitch decreases linearly. If the parameters for the first generation of HVM 248-nm exposure tools ($\lambda = 248$ nm, NA $= 0.7$, $k_1 = 0.6$, $k_2 = 1$) are substituted into these equations, the following values result: Resolution $= 213$ nm, Depth of Focus $= 506$ nm. Since the DOF figure represents the total available focus latitude, this might also be stated as ± 253 nm. At the time of these tools' introduction, the typical *minimum* photoresist thickness was ~ 600 nm, which already exceeded the focus capabilities of the new DUV systems.

This loss of latitude in processing caused a paradigm shift in the lithographic process flow: from then on, photoresists would have to be increasingly thinner, and all surfaces requiring DUV exposure would have to be very flat regardless of their current state of processing. The industry would depend on Chemical Mechanical Polishing (CMP) to provide the flat surfaces required for successful patterning of critical layers. CMP is a wet wafer polishing technique, where a slurry containing etch chemistries specific to the wafers to be polished is introduced to large rotating polishing pads. The polishing is performed using controlled pressure, continuous slurry flow, and constant motion of the wafer and pads to yield a uniformly flat surface across the wafer. Contamination is a constant issue in CMP processing, as any material left behind, such as copper, can contaminate an entire processing line. Post-CMP cleaning is, therefore, a critical step in the process flow. One example of innovative processing fostered by this flatness requirement is Damascene Metallization, where insulating material is deposited, patterned, and etched; then metal is deposited to fill the etched features, and finally CMP polished back down to the insulator, leaving a flat surface with metal in the previously etched areas. Many other variations exist, but the goal is that every process performed must result in a flat surface for the next lithography step.

Another important observation to be made from the above calculation is the powerful resolving ability of the photoresists used in DUV processing. The factor k_1 is an empirically derived number based on actual processing results. It can be influenced by many elements of wafer processing, such as uniformity, resist contrast, and development. The k_1 factor has a direct linear relationship to the system resolution. Note that the k_1 value is much smaller for DUV resists than for typical UV resists. This is due to the chemical amplification process described earlier. The above k_1 value is higher than typical values found in HVM today, which are often less than 0.4. This remarkable property of DUV photoresist enabled the printing of features *smaller* than the wavelength of the light used for exposure. In fact, 193 nm exposure tools were first introduced to HVM at around the 110 nm node, almost *one half* the exposure wavelength, a feat deemed impossible only a few years earlier (Table 1).

Photoresist Coating and Development

Spin-coating of photoresist is universally performed due to the efficiency and uniformity achieved with this process. Other methods exist, such as spray coating or meniscus, but they are reserved for special applications and do not appear in DUV processing. Wafer coating is performed by large, complex machines using precise control of dispense volume, spin speed, acceleration, and temperature parameters, to name only a few. While the tools used in UV processing are similar, the DUV versions are more tightly controlled in all parameters, have airborne chemical filtration (CA resists are sensitive to ammonia and NMP), and require higher uniformity of hot plate surface temperatures.

DUV photoresists are comprised of various materials including polyhydroxystyrene (PHS), acrylic polymer, and phenolic resin. The three general types of 248 nm photoresists are t-BOC PHS, ESCAP (PHS t-butyl acrylate), and acetal resin [4]. 193 nm resists can be polyacrylate, polymethylmethacrylate, or silsesquioxane resin, to name a few. The solvents used are primarily propylene glycol monomethyl ether acetate (PGMEA) and ethyl lactate, along with anisole and butyl acetate. Other materials are added to act as acid generators, acid inhibitors, and base quenchers, as well as surfactants to eliminate streaking during spin coating.

DUV Photolithography and Materials, Table 1 Various exposure wavelengths and their production capabilities at the time of the next wavelength technology introduction to HVM

λ (nm)	NA	Resolution (µm)	Total depth of focus (µm)	Overlay (nm)
436	0.38	1.000	3.020	±0.120
365	0.60	0.350	1.010	±0.050
248	0.75	0.130	0.340	±0.025
193	0.93	0.065	0.220	±0.012
193i[a]	1.35	0.038	0.106	±0.005

[a]Immersion, currently the leading-edge technology

A typical resist coating process for DUV lithography has two primary steps: BARC coating and resist coating. Beginning with a properly cleaned surface, the wafer is spin-coated with BARC, usually using a dynamic (wafer spinning) dispense process: a nozzle is moved out over the rotating wafer, and resist is released onto the wafer surface at a precise position and at a carefully controlled rate. The wafer's rotation is then ramped up at a high rate of acceleration (5–30 k rpm/s) to the final casting speed of 2–6 k rpm, although usually the maximum is 3 k rpm for 200 mm or larger wafers. The volume of exhaust air from the spin bowl is adjusted to obtain the maximum film thickness uniformity across the wafer and to simultaneously obtain the fewest number of defects. The wafer is then baked on a hotplate at 170–210°C for 60 s to remove solvent from the film, and is then moved to a chill plate for 20–30 s (Fig. 4).

A hotplate bake is analogous to baking in a convection oven, but has some important benefits. The hotplate can bring the wafer into intimate contact for uniform thermal transfer across the wafer, often using vacuum for better contact between the wafer and the plate, and the process drives the solvent out of the film from the inside at a high rate. Oven baking causes the resist to form an outer skin, which thickens as the drying process continues at a decreasing rate due to the increasing skin thickness. A typical hotplate bake time is 60 s, while an equivalent oven bake process may take 30 min or more, and cannot maintain similar uniformity across all wafers in a batch.

The BARC thickness can be anywhere from 40 to 120 nm depending on the application. After cooling, the wafer is coated with photoresist using the same methods as before, but this time it is baked at 110–140°C for 45–90 s to remove most of the casting solvent from the film, and then moved to a chill plate for 20–30 s to stop the bake process in a controlled

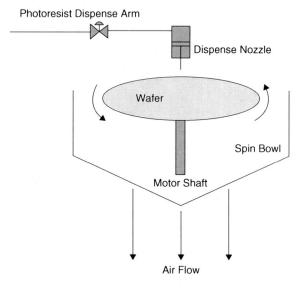

DUV Photolithography and Materials, Fig. 4 Resist coating schematic

way. All wafers must be baked for exactly the same length of time at exactly the same temperature for wafer-to-wafer repeatability to be maintained. The final thickness of the resist can be anywhere from <100 nm to >1 µm depending on the application. All of these process steps are performed by automated modules utilizing tight control of dispense volume and bake time and temperature, as well as the temperatures of the chemicals and the individual parts of the tool.

After exposure, DUV wafers must be Post-exposure Baked (PEB) to activate the acid amplification process. The PEB is a critical step, as the time and temperature both have a strong impact on the final CD and sidewall profile of the resist. Exposure of the resist by the tool creates a small amount of acid in the film, but not enough to cause well-developed features to form. The PEB activates the acid amplification process, which greatly multiplies the volume of acid in the exposed areas, making the resist soluble in the developer solution. The acid profile in the resist is sensitive to both the time and temperature of the PEB, as they determine the final acid diffusion length as well as the final acid concentration in the resist. Once the wafer cools it is placed on a spin chuck, and a developer solution, usually Tetramethylammonium Hydroxide (TMAH) or recently Tetrabutylammonium hydroxide (TBAH), is applied to the spinning surface by stream, spray, or other methods to form a puddle on the entire (now stationary) wafer. The developer remains in place for 30–60 s and is then rinsed off with deionized water as the wafer ramps up to speed, and continues to spin until dry. There have been recent reports of using alcohols as the developer to produce a negative-tone image in the resist, which is very useful in 193 nm processing as there are no negative-tone 193 nm resists currently available. The amount of water and the length of the rinse time are crucial to minimizing the creation of defects on the wafer from this process. Uniformity and repeatability are also key components of a successful development process (Fig. 5).

Because of the remarkable sensitivity of CA photoresists, any variation in temperature at any position on the wafer can cause a change in the final CD at that location. It is not uncommon for a DUV resist to have a CD change of 1 nm/°C. That may not seem like much, but it can make up more than 2% of current device CDs, which have a total error budget of 10% or less. Such CD variations can wreak havoc with the performance of multi-gigahertz devices, and significantly reduce yields in the fab (fabrication facility, or semiconductor factory). The uniformity of DUV hotplates is usually specified to be $\leq$0.5% variation across the surface, which can be larger than a 300 mm square. DNQ resists used for UV processing are not nearly as susceptible to these variations as engineers typically only perform PEB to reduce standing wave profiles in the resist sidewall.

DUV resist processing spurred the introduction of proximity hotplate baking, a method where instead of pulling the wafer into close contact with the surface using vacuum, the wafer is suspended a few tens of microns above the hotplate surface using (for example) ceramic beads at the wafer edge. The purpose of this spacing is to eliminate Across-Chip Line width Variations (ACLV) due to local changes in the bake temperature of the wafer that might be caused by

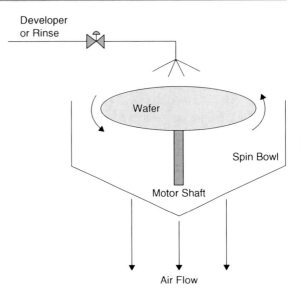

DUV Photolithography and Materials, Fig. 5 Development schematic

contamination or other nonuniformity of the hotplate surface. Longer bake times are required for this, but the resulting increase in across-chip and across-wafer CD control makes it worthwhile. Any variations in the CD from one area of a chip to another can decrease the maximum working speed of the device by skewing clock timings, or even cause the device to fail completely.

The uniformities of the spin-coated film thicknesses are very carefully controlled. With 193 nm resist film thickness on the order of 100 nm, and feature sizes reaching below 30 nm, there is no room for error. Any variation in the resist thickness (or the BARC thickness) will cause a measureable change in the CD at that location. For this reason, thickness variation across a 300 mm wafer is targeted to be <5 nm. Perhaps an even more difficult control issue is the level of defects. Defects are irregularities in the resist film which act to distort the imaging. Some examples are particles on top of, underneath, or imbedded into the resist film. If defects are in the same location as device features, they can cause incorrect imaging of the pattern, leading to poor device performance or failure. They can be caused by many things, including the resist itself. Defect control is important in all photolithography, but because the CDs are so small in DUV processing the problem is exacerbated. The allowable defect level for $\geq$30 nm defects on a 300 mm wafer is about 0.01/cm^2. The introduction of immersion lithography (see DUV Exposure), where water is used between the exposure lens and the wafer surface, has created entirely new types of defects. Some of these have been solved by hardware engineering, while others have required material and chemical solutions. Some solutions include modification of the wafer coating system and the use of protective top coating layers.

Considering that many wafer fabs start more than 10,000 new wafers each week, and that each wafer will undergo as many as 30 or more photolithography steps during the approximately 45-day process, it should be apparent that complete control of the coating process is essential to keeping CDs in spec and production profitable.

DUV Exposure

The tools used for DUV exposure are by and large the same as those used for g-line (436 nm) and i-line (365 nm) UV exposure. Economical HVM is only made possible by using the ubiquitous wafer stepper. This high-throughput, high-yield, high-uptime platform enables the incredible economies of scale that today supply the marketplace with billions of transistors at a cost to the end user of <$0.0000002 each. These aforementioned qualities are indeed required as the tools can cost anywhere from $7 M to more than $40 M each. Steppers use computer-controlled automated systems to load and pre-align wafers, compensate for environmental conditions, load and align reticles (photomasks used in projection exposure tools), step the wafers to each programmed location for alignment and focus measurements, expose the wafers, and unload the wafers to start the process again. The pattern image on the reticle is projected through the reduction lens and focused onto the wafer to print each image, or die. Each die will eventually be cut from the wafer and packaged into a chip for use in an electronic system. The placement accuracy of each die is measured by a laser interferometry system which monitors the wafer stage position. The wafer is leveled at the start to maximize the available focus range across the surface. Each die location on the wafer is measured for focus before each exposure using grazing angle light beams, and the individual die locations can be measured for accurate overlay of layer-to-layer patterns using through-the-lens grating comparison or off-axis diffraction systems. The location of each stepper is

scrutinized with regard to environmental factors such as electrical noise and vibration, as these can have a detrimental impact on the printed images. All of these systems must work together at high speeds and with high accuracy, while remaining operational >95% of the time. This is, of course, a simplification of what have been described as the most complex machines ever built.

DUV steppers utilize additional systems to level the wafer for *each die location* to maximize available DOF, correct focus error to a few tens of nanometers, and correct individual exposure lens element positions to reduce distortion within the field. Air filtration systems remove airborne contaminants from the atmosphere within the tool to prevent degradation of the resist profile and damage to the optical elements. Motors are used to generate motion opposite to the motions of the tool caused by the stage jumping from die to die. Alignment tolerances for both the wafer and the reticle are tighter due to shrinking feature sizes. At the same time, stage speeds and field sizes are increased to enhance productivity (Fig. 6).

The ever-increasing demands on DUV lithography for better alignment and lower distortion led to the introduction of the stepper-scanner systems now widely used in HVM. Scanners use a limited slit of the reduction lens for exposure, in order to optimize the best parts of the lens assembly. The reticle is scanned across this slit from end-to-end, while the wafer die location is scanned in the opposite direction under the lens. The result is a larger die than would be possible with a static system, with better distortion control and better overlay performance. The stage steps to the next die position and the scanning repeats in the reverse direction. Wafer stages can move at up to 700 mm/s, requiring counterbalances to move opposite the wafer and reticle stages to offset recoil, reduce stress on the factory floor, and dampen vibrations. Scanner throughput has reached over 200 aligned and exposed 300 mm wafers per hour using two wafer stages in tandem: one performs wafer focus and alignment mapping, while the other steps and scans the wafer under the exposure lens. Rates of 250 wafers per hour are predicted in the coming months.

Light Sources

Illumination systems for DUV exposure no longer use mercury arc lamps as the source, as the output intensity at DUV wavelengths is too low. Instead,

Exposure Light

Scan Direction of Reticle

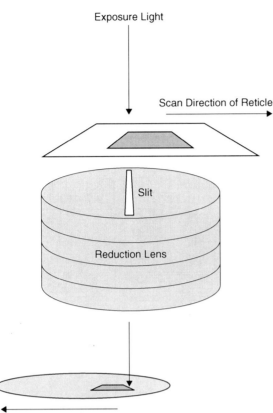

Slit

Reduction Lens

Scan Direction of Wafer

DUV Photolithography and Materials, Fig. 6 Wafer stepper-scanner schematic

excimer lasers provide high power illumination with up to 90 W at 6 KHz pulse frequency and sub-half picometer bandwidth. Excimer lasers have proven to be very reliable, and constant improvements have made them increasingly powerful and cost-effective. The laser can be located remotely from the tool, as far away as 20 m. The beam delivery systems utilize automated aiming mirrors in the Nitrogen-purged beam line to maintain optimum positioning at the tool. A complex system of lenses, expanders, and homogenizers shape the beam and ensure uniformity across the reticle, and from exposure to exposure. The laser uses a Fluorine gas–filled chamber (KrF for 248 nm, ArF for 193 nm) with high-voltage electrodes to create the illumination source. The chamber gas depletes over time, and new gas can be injected to replenish the mix, until the chamber must eventually be flushed and refilled. These laser gas injections and refills are monitored and performed automatically by the stepper software to keep laser power and

bandwidth within specifications. Laser maintenance is usually one of the less troublesome aspects of DUV lithography (Fig. 7).

The ramp-up of DUV photolithography also marked the acceptance of shaped illumination into HVM. Though proven for sometime, this Resolution Enhancement Technique (RET) was not often implemented in production environments. With the advent of sub-wavelength printing, shaped sources became a cost-effective enhancement to the wafer stepper. By using annular (off-axis) illumination, smaller features could be printed with greater depth of focus, albeit at the expense of (sometimes) throughput and printability of certain other feature sizes. Even greater enhancement was possible using quadropole or, in the extreme, dipole illumination. One-dimensional gratings could be printed at the diffraction limit of the system using binary masks, simply by inserting a shaped source. Programmability also was introduced, allowing any size annulus to be formed from keyboard commands, and source shapes to be called up automatically. Today these mechanisms have been developed to the point of "free form" source shape capability. Arrays of thousands of mirrors are used to create any random shape calculated to optimize the process with any particular mask. The forms of both the source and the mask are calculated together in a technique called Source Mask Optimization (SMO). Using free form sources, the engineer is allowed almost limitless flexibility in process optimization (Fig. 8).

Polarization of the illumination is now one of the variables the operator can manipulate to optimize tool performance with regard to particular pattern orientations. Because of the high angles of incidence of the exposure light to the resist surface, light is transmitted or reflected by the photoresist depending on its polarization. Polarization can be set in orthogonal directions, or in circular ones. Controlling the polarization of the exposure light can assist the engineer in controlling the printability of the resulting pattern.

Immersion Systems

The NA of exposure lenses has become so high, 0.93 in air, that the angles of light exiting the lens are as high as 70° from normal, leading to reflection of the light at the resist surface. The two polarizations of the light (TE and TM) will behave differently with one polarization reflecting away more than the other, and the

Standard DUV Excimer Laser Module Layout

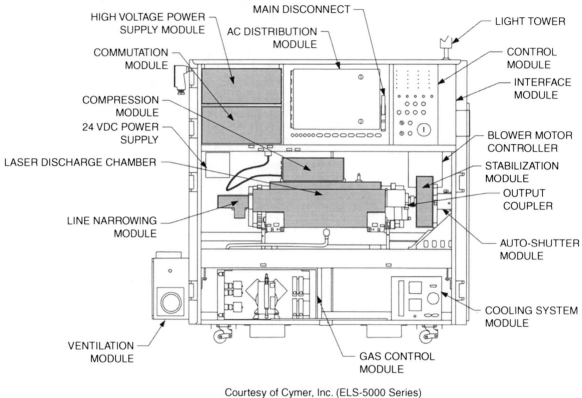

HIGH VOLTAGE POWER SUPPLY MODULE

COMMUTATION MODULE

COMPRESSION MODULE

24 VDC POWER SUPPLY

LASER DISCHARGE CHAMBER

LINE NARROWING MODULE

VENTILATION MODULE

MAIN DISCONNECT

AC DISTRIBUTION MODULE

LIGHT TOWER

CONTROL MODULE

INTERFACE MODULE

BLOWER MOTOR CONTROLLER

STABILIZATION MODULE

OUTPUT COUPLER

AUTO-SHUTTER MODULE

COOLING SYSTEM MODULE

GAS CONTROL MODULE

Courtesy of Cymer, Inc. (ELS-5000 Series)
www.cymer.com

DUV Photolithography and Materials, Fig. 7 248 nm DUV excimer laser diagram

penetrating light polarizations focusing at different depths within the resist. To overcome this, current leading-edge system vendors have implemented immersion systems at 193 nm, where purified water is used as the transport medium between the exposure lens and the resist. The water fills the space between the lens and the wafer surface, creating a higher index medium for the light to pass through. The higher index of the water is a better match to the indexes of both the lens glass and the photoresist, reducing both the exit angle of the light leaving the lens and the angle of incidence of the light entering the resist, thus reducing the reflectivity and polarization losses. The water must be injected into the space without wetting any other parts of the system, and must be kept only in the lens-wafer space throughout the exposure process. At the end of the exposures, it must be removed from the wafer without leaving any residue. To prevent defects the water must be bubble-free and leave no droplets

behind on the surface, even though the wafer moves at very high speeds. Elaborate systems consisting of de-gassers, showerheads, and pickup rings are used to inject and remove the water from the moving wafer surface (Fig. 9).

This additional complexity, introduced by immersion and polarization, drives the cost of tooling up but also allows features to be printed on a half-pitch as small as 38 nm in a single exposure. No smaller pitch can be resolved using these wavelengths unless a significantly higher refractive index medium can be found to replace the water currently in use, a doubtful prospect.

Mask Making

During the 1980s, mask making for the newly introduced reduction UV steppers initially relaxed the requirements for mask making compared to contact lithography, leading to what has been called the

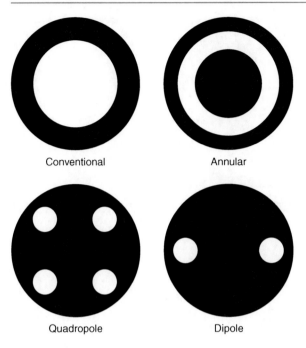

Conventional   Annular

Quadropole   Dipole

DUV Photolithography and Materials, Fig. 8 Illumination source shape examples

"mask makers' holiday," which lasted for some years. With features five times the size of the wafer CD, and only one field to make instead of the entire wafer, making acceptable masks presented a relatively simple task. With the introduction of DUV exposure at 4× reduction, this holiday was abruptly ended. The industry struggled to meet increasingly demanding specifications for feature size and placement, line-edge roughness, and defects. Shortly thereafter, even greater pressure was brought to bear by the inclusion of Optical Proximity Correction (OPC) features, which are assist features on the reticle sized below the resolution of the system, and Phase Shift Masks (PSM), which have some clear areas of the mask etched to a depth that causes the light passing through them to exit the mask 180° out of phase with the rest of the light. OPC may include assist features such as scattering bars or serifs, which are sub-resolution shapes that act to force the diffracted light into desired shapes on the wafer. Phase shifting areas can be used either to define features or to improve the fidelity of the printed pattern. By placing out-of-phase light at the edge of a feature, the location in the resist can be precisely defined. Critical level masks can today have features nearly as small as the wafer CDs, making masks very difficult and expensive to manufacture.

The addition of OPC and PSM to a mask can cause the data necessary to write the masks to explode, making file size and data rates problematic. The tools used to write the masks, laser and e-beam writers, can cost $10 M or more, and take many hours to complete each mask. Inspection and repair add more time and effort to the task. Using these RETs in manufacturing, one single mask can cost $60 K or more. Given that a mask set for a device may require 30 or more masks, with perhaps 10 critical levels, mask costs can surpass the $1 M per set mark. Frequent revisions in production, sometimes known as "re-spins," can obviously get very expensive. This in turn has driven the use of modeling in manufacturing.

Computer Modeling

Simulation of photolithography has been ongoing for around 30 years, initially in an effort to better understand and explain the process, then later as a way to optimize processing parameters. It has today evolved to the point where models can quickly and accurately predict not only the resulting resist pattern, but also the optimum illumination shape and mask pattern corrections. By modeling the process, expensive hours of DUV tool time are saved by greatly reducing the on-wafer testing required to explore the process parameter space. Dose, focus, and illumination settings can all be determined before even making the mask. Simulation of the process over the entire chip can be done to locate problems and optimize yield before "tape-out" of the mask design (submission of final data to the mask vendor), saving enormous time and effort, and thus cost. Current applications of computer power leverage the enormous data volumes generated in manufacturing to model systems and extract error contributions from individual system components such as the reticle, the lens, and even the wafer chuck. Once these values are measured, corrections can be applied to reduce or eliminate them (Fig. 10).

The resolution of DUV exposure systems has now been pushed so far into previously unknown, nonlinear regimes that the accepted concepts regarding refractive imaging systems cannot predict their behavior. The patterns that need to be exposed on the wafer cannot be defined in traditional ways due to previously unseen phenomena. Modeling and simulation have now

DUV Photolithography and Materials, Fig. 9 Immersion scanner schematic

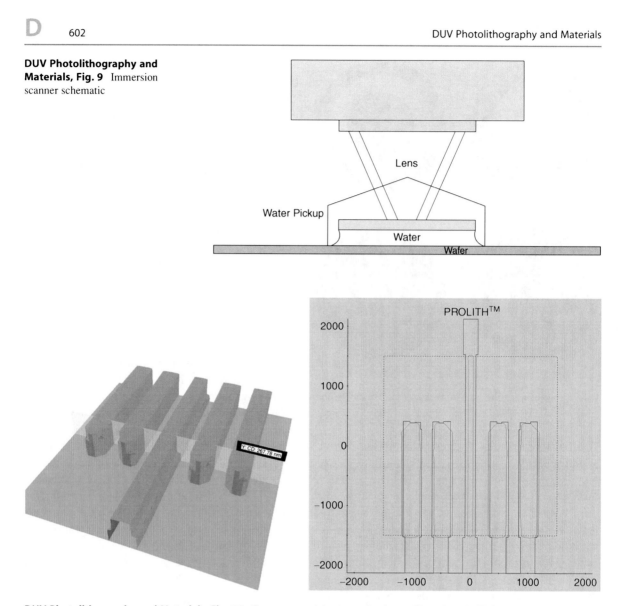

DUV Photolithography and Materials, Fig. 10 Computer model output of resist profile and mask OPC

allowed lithographers to reverse engineer the optical image, creating a new approach called Inverse Lithography Technology (ILT). A mask made using ILT may look nothing like the final printed pattern, but it is a collection of shapes which will, when used in conjunction with the complementary illumination source shape, cause a diffracted pattern that will ultimately form the desired image in the photoresist. The SMO approach combines ILT with variable shaping of the light source, in some cases with single pixel addressing. These techniques have allowed engineers to push the resolution of exposure systems to the limits of diffraction with acceptable control.

One of the major mask making issues that came to light in DUV lithography has to do with an unexpected effect that occurs when printing feature sizes below the wavelength of the exposure light. During exposure processes with low k_1 values, where the process becomes nonlinear, defects on the mask are printed on the wafer at dimensions larger than would normally be expected. This Mask Error Enhancement Factor (MEEF) can in some cases exceed $4\times$ or more, magnifying any error into a potential "killer" defect. Modeling these potential errors has led to better mask inspection methods focusing on the conditions particular to creating MEEF in order to prevent unexpected pattern failures. Simulations are also used to compute full-chip OPC to maximize the process window and increase yield. Without this computational modeling power, the advancement necessary to maintain "Moore's Law" would have lost momentum long ago.

Inspection

One of the most difficult aspects of DUV processing is the inspection process. Trying to determine exactly what is on the wafer after lithography is much more difficult than in typical UV patterning, due to the much smaller feature sizes and overlay requirements. DUV lithography has driven the rapid advancement of optical inspection tools utilizing actinic (248 nm, 193 nm) illumination for both wafers and masks. Scanning Electron Microscopes (SEM) are routinely used in HVM to check in-line wafers, which can then feedback and feed-forward data to other line processing tools. The information is then used to adjust processing parameters to compensate for measured results and improve yield. Atomic Force Microscopy (AFM) is also useful for determining surface characteristics and resist profiles. In recent years, a new method, called scatterometry, has gained popularity for measurement of dimensions which are difficult to resolve using standard tools. This method has the advantage of not requiring the formation of an image of the target to calculate its dimensions, something which can be impossible to do with light and is increasingly challenging with any tool at the scale of leading-edge devices. SEM inspection also has an unfortunate side effect on DUV resist: shrinkage. The area under inspection by the electron beam will actually change size through the process of inspection, making any resulting measurements questionable.

Another unique issue has also appeared in 193 nm DUV: the formation of a haze on the chrome side of reticles is causing new inspection techniques to be investigated. This haze can, over time, reduce transmission of the reticle affecting CD and even causing placement errors. New cleaning techniques are under investigation in order to counter this problem. Reticle inspection in general has undergone rapid development as 248 and 193 nm tools have advanced, in the quest to keep pace with shrinking CDs and exploding data sets.

Future Issues

One recurring theme in the decent to molecular-scale devices has been the battle against Line-edge Roughness (LER). LER may become the ultimate limiter of photolithography as the actual feature sizes approach the magnitude of the roughness. It manifests as small variations in sidewall linearity or smoothness, either vertically or horizontally. These variations affect the line width of device structures after pattern transfer, making control of the performance of the device difficult or impossible. When the printed line width is beyond the control of the lithographer, working devices become unreliable or unattainable. The operational speed of the devices, and in particular the timing of events within the device, depends on the uniform size of its structures. Once this uniformity is compromised, delicate timings are changed and device operations may fail.

The root cause of these variations in the sidewall plane has not definitively been explained by modeling, and ongoing work includes both investigations into the causes of LER, and the development of new materials and techniques to reduce or eliminate it. Materials such as molecular glass are intended to reduce the size of the polymer molecules in an effort to reduce the width variations. New methods and materials for development of the resist after exposure are also possible solutions. LER can be reduced by decreasing the sensitivity of the resist, thus increasing the required dose, but this would result in an unacceptable decrease in throughput.

High-index immersion materials have been investigated as a way to extend the usefulness of 193 nm tools. Replacement of the water used in immersion lithography systems with higher index fluid could extend 193 nm exposure resolution for another node or two at most. Possible candidates have usually included fluorinated liquids, but most have been dismissed as not having high enough index to be worth the investment to make them viable.

Current work is focused on Double-Patterning scenarios, where two mask patterns are exposed in the same layer. The use of two separate patterns solves the problem of shrinking pitch values, but requires either a resist that "forgets" the unwanted effects of the first exposure by "freezing" the image, or that an etch step be performed before the second mask is exposed. The latter method places demanding overlay requirements on the scanner as well as a heavy burden on the mask makers. This level of precision is unprecedented, with error tolerances in the range of 2–3 nm for layer-to-layer registration. Another effort is concentrating on "spacer" patterning. This is a method of self-aligned patterning where the wafer is patterned, a new layer of material is deposited, and then the material is etched back to form the new structures. This produces a pattern with about ½ the original pitch. This technique is already used by some companies in memory chip manufacturing.

Conclusion

DUV Photolithography is a dynamic and rapidly advancing technology. The race between manufacturers to make the smallest, fastest, and cheapest devices is seemingly endless. However, they may be approaching the end of their abilities to make things smaller. EUV Lithography (13.5 nm) is preparing to enter the production phase by the 11 nm node, but some uncertainty still exists. Even if it does come into the fab, how long will it be useful? Current predictions are that it could last until the 8 nm node, if it is ever reached. At the same time, Multiple E-beam and Imprint technologies are still in the running as well. Will anyone be able to afford these technologies?

One thing is certain: as soon as this entry goes to press it will become outdated. The latest advancements will change the direction of the industry, probably in unforeseen ways. In order to ascertain the current implementations of DUV Lithography, the reader should investigate the SPIE Advanced Lithography conference proceedings published since this entry was completed (2011), as well as search the publications of SEMATECH, ITRS, IEEE, BACUS, and IEDM. DUV Photolithography will likely remain in favor longer than many would imagine, but the leading edge will eventually pass it by.

Cross-References

▶ Atomic Force Microscopy
▶ Immersion Lithography Materials
▶ Nanoimprint Lithography

References

1. Dammel, R.: Diazonaphthoquinone-Based Resists. SPIE, Bellingham (1993)
2. Mack, C.A.: Field Guide to Optical Lithography. SPIE, Bellingham (2006)
3. Sheats, J.R., Smith, B.W.: Microlithography Science and Technology. Marcel Dekker, New York (1998)
4. Helbert, J.N.: Handbook of VLSI Microlithography. Noyes, Park Ridge (2001)

Dye Sensitized Solar Cells

▶ Nanomaterials for Excitonic Solar Cells

Dye-Doped Nanoparticles in Biomedical Diagnostics

Vladimir Gubala
Biomedical Diagnostics Institute, Dublin City University, Glasnevin, Dublin, Ireland

Synonyms

Dye-doped nanospheres in biomedical diagnostics

Definition

Dye-doped nanoparticles are new class of fluorescent labels. They are typically made of silica or polystyrene with tens of thousands of fluorescent dyes entrapped in the pores of its polymer matrix.

Overview and General Concept

When designing devices for biomedical diagnostics, it is essential to maximize the signal to noise ratio to achieve clinically relevant sensitivity and limits of detection. Simple assay designs involve first capturing an antigen onto a surface using one antibody, then measuring the surface concentration by visualizing the captured antigen through its reaction with a second, labeled antibody. There is increasing interest

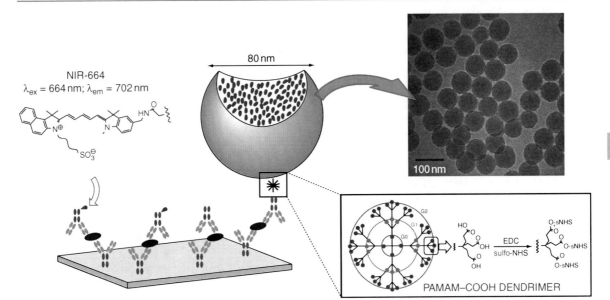

Dye-Doped Nanoparticles in Biomedical Diagnostics, Fig. 1 A cartoon illustrating the concept of fluorescence-linked immunosorbent assay using the detection antibody labeled with molecular fluorophore (NIR-664 in this particular case) and 80 nm nanoparticles, doped with tens of thousands of molecular dye molecules, thus significantly increasing the brightness of the label. The TEM image shows the high monodispersity and size-uniformity of 80-nm silica nanoparticles doped with NIR-664, prepared by reverse microemulsion method. Inset with PAMAM dendrimer highlights a unique modification of the NP surface that ensures high colloidal stability and efficient bioconjugation to biomolecules of interest

in fluorescence-based array sensors (biochips) for biomedical applications. While fluorescence detection offers high sensitivity, there is generally a low level of fluorescence signal from the biochip due to a monolayer of fluorescent labels, hence the importance of enhancing the fluorescence. In antibody-based assays, the measured signal can be amplified through the replacement of molecular luminophore with doped nanoparticles (NP) (Fig. 1).

Fluorescent labels are used for a range of applications including immunosorbent assays [1], immunocytochemistry [2], flow cytometry [3], and DNA/protein microarray analysis [4]. Current research in biomedical diagnostics is moving to inexpensive devices, using biochips that often have to measure biomolecule concentrations at the picomolar level in blood sample volumes on the microliter scale, without the possibility of any user manipulation of the sample. This is exceptionally challenging to achieve in a miniature bioassay device. First, the active biomolecules (antibody or nucleic acid) must be immobilized at the appropriate surface density while maintaining its reactivity. Second, the measured signal should be maximized relative to noise

and background contributions. In antibody-based assays, nonspecific binding (NSB) controls the background response and often results in a detection limit that is much higher than that defined by the equilibrium constants for the binding events.

To meet these new demands, fluorescent labels with improved physical and chemical properties are required: for example, high photostability and fluorescence intensity, with a reproducible signal under a variety of chemical and biological conditions. Excellent photostability and significant reductions in photobleaching is required for applications where high intensity of prolonged excitations are applied, such as intracellular optical imaging. Dye-doped nanoparticles (typically silica or polystyrene) stand out as excellent candidates as it is possible to dope silica NPs with a large number of fluorophores, increasing the total fluorescence of the label significantly. Despite the effect of fluorescence quenching phenomena within an NP with a large amount of dye incorporated in a small volume, the goal of producing a particle with brighter fluorophores protected inside an organic matrix, thereby increasing photostability and quantum efficiency is largely successful. Owing to their intense signal, fluorescent nanoparticles (NP) are useful as labels since they can be measured directly, without the need for any amplification step.

How to Select Appropriate Dye-Doped NPs for Your Application?

Dye-doped NPs are currently offered in a wide range of sizes, with entrapped fluorophores covering most of the useful spectrum of UV/V radiation and with a surface modification to contain chemically reactive groups, usually –COOH and –NH$_2$. Most of the scientists using dye-doped NPs will make their decision based on the characteristics of the instrument (e.g., excitations source and emission filters) and the type of application.

Typically, at least three parameters should be defined when selecting the appropriate dye-doped NP:

(a) What is the desired fluorescence property of the NP for your application?

The choice of the luminophore, particularly its excitation and emission wavelengths, obviously depends on the capacity of the instrument that is used to excite the fluorophores and collect the emitted light. However, it also depends on the application. At near infrared wavelengths there is low background interference from the fluorescence of biological molecules, solvent, and substrates. For example, whole blood has a weak absorption in the NIR region, thus reducing the need for whole-blood filtering for assays using whole blood. NIR light can also penetrate skin and tissue to several millimeters and this can enable fluorescence detection in dermatological or in vivo diagnostic devices. When preparing dye-doped NPs, it is also important to remember that not every readily available dye can be doped into the NP matrix. This will be further discussed later in this chapter.

(b) What is the desired size of the NP for your application?

In case of bioassays, the answer to this question is specific to the design of the assay. While it is tempting to think that the brightness will increase with increased size of the NP, one must also consider the increased surface area and different monolayer packing arrangement of larger NP when compared to smaller ones.

It is not necessarily true that by increasing the diameter of NP say by an order of 2, the fluorescence will also increase by the same order (Fig. 2). In general, dye-doped NPs in the range of 20–200 nm were successfully used in bioassays in the wider research community. As for optical imaging, the rate of uptake of nanoparticles by mammal cells depends on the particle size and shape. A number of articles emerged recently reporting that the optimum size of nanoparticles to enter the cells is around 50 nm. The cellular uptake of ∼50 nm NP is typically the highest and the NPs were shown to be trapped inside vesicles in the cytoplasm; however they did not enter the cell nucleus. Shape also plays an important role, with spherical nanoparticles being taken in larger numbers than rod-shaped nanoparticles. This is presumably due to the difference in curvature affecting the contact area with the cell membrane receptors.

(c) What is the desired surface modification for efficient conjugation with biomolecules?

It is relatively easy to functionalize silica or polystyrene NPs with reactive groups that enable bioconjugation. However, each step in the conjugation process modifies the zeta potential of the NP, thus affecting colloidal stability. Proteins are usually immobilized on the surface of polystyrene beads by physical adsorption. It is a relatively fast method for formation of reasonably homogeneous protein layer. For better results, it is often desirable to block the non-active surface area with bovine serum albumin or similar blocking agents. In case of silica particles, the colloidal stability and the density of the active molecules on the surface can be fine-tuned by carefully executed surface chemistry. Some aspects of the surface modification strategies of dye-doped silica NP will be discussed later in this chapter.

Basic Methodology

Synthesis of Dye-Doped Silica Nanoparticles

One obvious advantage of using silica particles as opposed to polystyrene is the significantly higher density of fused silica (2.2–2.6 g/cm^3), which facilitates the manipulation and purification of samples (centrifugation at 15,000 rpm for 1–2 min is usually sufficient to separate the beads from supernatant). Also, silica NPs are relatively nontoxic and chemically inert. They can be prepared in a range of sizes and the post-synthetic surface modification can be achieved with good reproducibility.

**Dye-Doped Nanoparticles
in Biomedical Diagnostics,
Fig. 2** A computer model
showing the correlation
between the brightness ratio
and the radius and number of
dye molecules entrapped
inside NP

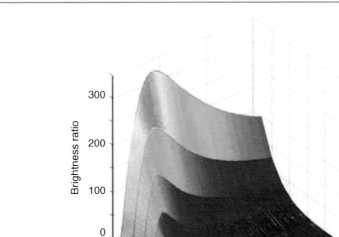

There are two main methods for synthesis of dye-doped silica particles.

1. Sol-gel technique via Stober method, which consists of the hydrolysis of a silica alkoxide precursor (such as tetraethylorthosilicate, TEOS) in an ethanol and aqueous ammonium hydroxide mixture.
2. *Reverse microemulsion* method based on a water-in-oil reverse microemulsion system.

Both methods have certain advantages and disadvantages. For example, the former method is comparatively simple and both organic and inorganic dyes can be incorporated in the silica matrix. A modified Stober method has been used to produce highly fluorescent dye-doped core–shell silica NPs with narrow size distribution by the Wiesner group at Cornell University (NY, USA) [5]. The latter technique is using stabilized water nanodroplets formed in the oil solution. Silane hydrolysis and the formation of NPs with the dye trapped inside occur inside such small microreactors. The size of the NP is determined by the nature of surfactant, the hydrolysis agent, and other parameters, such as the reaction time, oil/water ratio, etc. The resulting NPs usually show a high degree of uniformity and water dispersity. One of the limitations of this method is that, in most cases, it works well with inorganic dyes, some of which have lower quantum yields compared to organic fluorophores. By both techniques,

the dye molecule can be either physically adsorbed inside the pores of NP or covalently linked to its matrix. The second option is preferred, as the non-covalently attached fluorophores can eventually diffuse or "leach out" into a solution over time, thus reducing the overall brightness of the NP.

Surface Functionalization of Dye-Doped NPs

In general, the benefits of using dye-doped NPs can only be realized if they are efficiently coated with biorecognition elements (in most cases antibody, oligonucleotides, etc.), have good colloidal stability, and the ratio of specific to nonspecific binding (NSB) is sufficiently great. The colloidal stability of NPs can be modulated by carefully executed surface modification. Sufficient separation of NPs can be achieved by increasing either steric or electrostatic repulsions between two or more adjacent NPs. Current approaches to prepare colloidal solutions with low polydispersity index are mostly based on electrostatic interactions. In polystyrene particles, the introduction of charge on the surface can be done by copolymerization reaction of polystyrene monomer and its charged analogue. And even then, a longer-term colloidal stability is guaranteed only after addition of polar surfactants into the sample. Quite often, this surfactant must be removed prior the immobilization

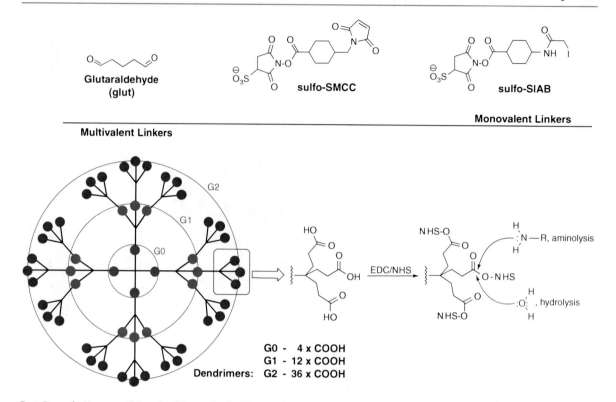

G0 - 4 x COOH
G1 - 12 x COOH
Dendrimers: G2 - 36 x COOH

Dye-Doped Nanoparticles in Biomedical Diagnostics, Fig. 3 Structures of the most popular monovalent linkers and a cartoon illustrating the multivalent feature of a Newkome-type dendrimer, activated by a dehydrating agent. (This figure was reprinted with permission of Elsevier)

of the antibody or DNA. On the other hand, the flexible silica chemistry provides versatile routes for surface modification. Different types of functional groups can be easily introduced onto the NP surface enabling conjugation with biomolecules of interest. The post-synthetic modification of the particle surface with charged species (e.g., phosphates, carboxylic acid, etc.) is not only aimed to reduce particle aggregation but also to provide some sort of reactive handle, where the reaction with antibodies takes place. Most of the biomolecules used as molecular recognition elements are charged species (proteins, DNA, etc.) and contribute positively toward the decrease of the polydispersity index of the sample.

One of the most elegant and efficient methods to achieve good colloidal stability of silica NPs is to introduce a mixture of charged precursors on the particle surface. The first one is usually chemically inert with low pKa value, assuring that the charge is present when working at physiologically relevant pH (7.0–7.4). The second one, typically at much lower quantity provides reactive group for the reaction with biomolecules or other cross-linking agents. In 2006, Tan and colleagues [6] proposed a way of improving the colloidal stability of silica NPs through the addition of a negatively charged, nonreactive alkoxysilane with phosphate group along with its counterpart carrying a reactive amine. Other very common groups that can be used as reactive "handles" include carboxylic acids, isothiocyanates, and epoxy group. Normally, the ratio between the nonreactive and reactive groups is relatively high (10:1) in order to maintain the suspension stability. As a consequence, the number of attachment sites on the NP surface is therefore limited. For this reason, it might be quite difficult to achieve high protein coupling ratios. Gubala et al. [7] overcame this disadvantage by using multivalent linkers such as dendrimers as antibody scaffolds (Fig. 3).

Bioconjugation Strategies

As mentioned previously in this chapter, the most straightforward method to immobilize large biomolecules like antibodies is through physical adsorption [8]. This method is very popular with polystyrene beads; however, it does not offer direct possibility of modulating the density of the captured protein on the NP surface. The control over the active surface area, which is the area of the NP surface covered with antibodies, can be achieved by using cross-linking reagent.

There is a great choice of commercial hetero- or homo-functional linkers available; however, their effect on NP stability, aggregation, solubility, and efficiency of bioconjugation was until recently poorly documented. This was striking, particularly when considering that this is the actual layer "sensed" by the molecule to be labeled.

In a sensitive and functional bioassay, one obvious issue in the use of NPs as labels is the fraction of the coupled antibody that is in fact active or available for reaction with antigen. This fraction can be rather small, which in turn can lead to diminished sensitivity and increased nonspecific binding. The minimization of NSB is essential for sensitive detection in an assay. Thus, it is clear that the strategy used to attach an antibody to NP surface is a key element that affects the activity of the bound antibody, the nonspecific binding, and the surface binding of particles.

Improved colloidal stability of silica NPs has been previously achieved by the addition of negatively charged nonreactive organosilanes in addition to organosilanes with functional groups available for bio-immobilization. The number of attachment sites on the nanoparticle surface is limited in order to maintain the suspension stability of the NP system. For this reason it is difficult to achieve high protein coupling ratios. Gubala [7] and colleagues at Biomedical Diagnostics Institute have recently demonstrated some advantages of using multivalent molecules such as dendrimers as antibody coupling scaffolds as illustrated in Fig. 3. Dendrimers are monodisperse, nanosized, hyperbranched "starburst polymers," with growing numbers of terminal functional groups with increasing generation number that have been extensively investigated for a variety of biomedical applications. By changing the generation of dendrimers, the authors were very efficiently able to control the surface area populated by biomolecules. The authors demonstrated that antibody-sensitized NPs prepared by using a multivalent linker showed a significantly lower limit of detection and higher sensitivity than with the homo- and hetero-functional cross-linkers (Fig. 4). The difference in performance was remarkable, particularly when compared with glutaraldehyde, one of the most common cross-linker still employed by many research laboratories. They reasoned that the multivalency of the dendrimers is one of the most significant factors behind the increase in the detection sensitivity. Dendrimers had a positive effect on NP stability and aggregation, were more reactive with biological materials, and were capable of immobilizing antibody at the appropriate surface density while maintaining its activity to the analyte of interest. Moreover, the multivalency of the dendrimer is a significant factor responsible for the improvement in the reaction yields even at lower protein concentration. This is very important, particularly when considering the cost of the bioorganic material that is used in bioconjugation reactions.

Dye-Doped Nanoparticles and the Implications on Assay Kinetics

A surface capture immunoassay can be considered in two ways: either incubation of antibody-sensitized particles with antigen followed by capture onto an antibody-sensitized surface of those particles that have bound antigen; or capture of antigen onto an antibody-sensitized surface followed by capture of antibody-sensitized particles onto the surface-bound antigen. In either case, the kinetics of reaction of a sensitized particle with a sensitized surface is a key element of the description of the process and hence of the assay design. The size of the particles and the fraction of the particle surface that is active for the antibody–antigen reaction are the important parameters. A simple kinetic model was developed for particle capture onto an antigen-loaded assay surface by means of an antibody–antigen reaction. This is similar to kinetic formulations treating single molecules, but also accounts for the fact that each particle has a multiplicity of antibodies present on its surface. The NPs-surface-bound antibodies can be considered as active binding sites. Depending on the coverage of such active sites on the particle, a collision of the particle with an antigen may or may not lead to reaction, with a probability proportional to the surface coverage of active sites on the particle and the surface coverage of antigen in the reaction well.

The capture reaction, of antibody-functionalized NPs onto the assay surface, can be written as:

$$P + S \rightleftharpoons PS \qquad (1)$$

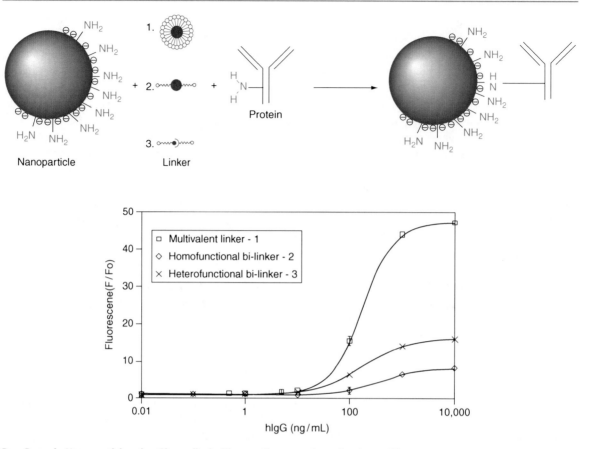

Dye-Doped Nanoparticles in Biomedical Diagnostics, Fig. 4 Bioconjugation strategies that rely on the use of cross-linking agent. The content of $-NH_2$ (or $-NH_3^+$) groups on the surface of the NP is typically limited to $\sim$10%. Higher amount of amines can cause aggregation. The lower density of reactive amines is offset by the multivalency of dendrimers. Their effect is threefold: (**a**) the use of dendrimers positively affects the colloidal stability of the NPs, (**b**) the reactivity of each amine on the surface is amplified by the number of dendrimers' surface groups, hence significantly improving reaction efficiency with usually expensive biomolecules, (**c**) dendrimers, due to their size, not only serve as linkers but also as spacers between the surface of the NP and the captured recognition elements, thus maintaining the activity of antibodies, oligonucleotides, etc. The net result is more sensitive assays when compared to the monovalent linkers and also improved assay kinetics

with forward rate constant k_{on} and reverse rate constant k_{off}. Here, S denotes an active antigen site on the assay capture surface. If the NP is bound to the surface by a single antibody–antigen interaction, then k_{off} will simply be the dissociation rate constant for the antigen–antibody complex. The association rate, k_{on}, will be determined by the probability of a reactive collision between an antibody-functionalized NP and an unoccupied reactive antigen site on the assay surface. A reactive collision, leading to coupling of the NP to the assay surface, would occur when a reactive part of the NP surface (an antibody, oriented with the binding site exposed) collides with a reactive part of the assay surface (an antigen, oriented with the epitope exposed). NPs are in a continuous state of collision with the assay surface, at a rate determined by the diffusion coefficient of the NPs (hence by their radius and the viscosity of the reaction medium) and by their concentration. As the NP concentration on the assay surface builds up, particles not only occupy sites but also physically block the assay surface area, on account of their size. The probability of obtaining a successful reactive collision between particle and capture surface also depends on the fraction of the capture surface that is active for reaction. Therefore, the probability that the collision will be with an active

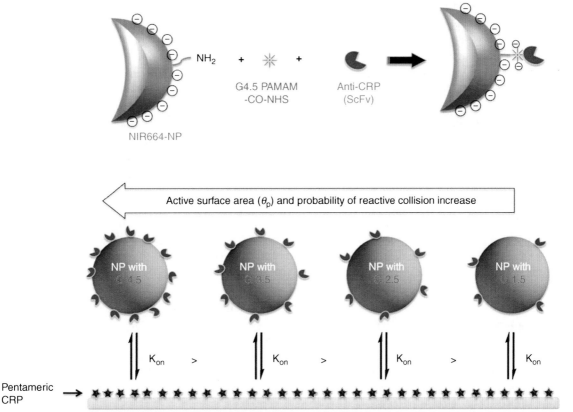

Dye-Doped Nanoparticles in Biomedical Diagnostics, Fig. 5 Top – A cartoon illustrating a bioconjugation reaction between single-chain fragments of C-reactive protein antibody (ScFv) and dye-doped silica NP. By varying the dendrimer generation, it is possible to vary the density of ScFv fragments on the surface of NP, hence modulate the active surface area of NPs. The net outcome is increased probability of reactive collision and hence association K_{on} constant. The effect of increased active antibody loading significantly outweighed any effect of a decreased NP diffusion coefficient compared to that of a molecular dye-labeled antibody

site on the assay capture surface will be a product of the following:

- The fraction of the total assay capture surface that is covered by accessible antigen with the binding epitope correctly exposed to facilitate antibody binding
- The fraction of the total assay surface that is unblocked by bound NPs
- A coverage-dependent factor that expresses the requirement that a NP requires a space whose smallest dimension is at least as large as the NP diameter, in order that the NP can fit into the space

The considerations of reaction probability are embedded in the capture rate constant, k_{on}, which depends on both the fraction of the surface of the antigen-loaded capture plate that is in fact active for the capture reaction (i.e., it depends on the surface state of the adsorbed antigen) and on the fraction of the surface of the antibody-sensitized NP that is active for reaction. It can be written as:

$$k_{on} = k_D \varepsilon \theta_S \theta_P \qquad (2)$$

where k_D is the diffusion-limit rate constant for NP consumption by the antigen-surface, dependent on the NP radius and the viscosity of the medium; ε is the reaction efficiency for a reactive collision; θ_S is the fraction of the capture surface area (here the antigen-loaded capture plate) that is active for the capture reaction, which in this case is dependent on the surface coverage of active adsorbed antigen on the prepared capture surface; and θ_P is the active fraction of the NP surface, in this case dependent on the surface coverage of active antibody attached to the NP. Therefore, the central objective of surface functionalization with is to increase θ_P, which can be achieved by using dendrimer as a linker (Fig. 5).

Gubala et al. have shown how the use of dendrimers as multivalent linkers effectively controls and maximizes the active fraction of the particle surface. Dendrimers have been presented as efficient coupling agents to conjugate antibodies to the surface of dye-doped silica nanoparticles. The protein-binding capacity of the dendrimer-sensitized NPs increased by the same factor as the number of surface carboxylate groups on the dendrimer used. The highest generation of the dendrimers (G4.5) used in their study showed the highest surface binding rate and the highest signal in the direct-binding FLISA. The effect of increased active antibody loading significantly outweighed any effect of a decreased NP diffusion coefficient compared to that of a molecular dye-labeled antibody. The practically important parameter, the ratio of the equilibrium fluorescence (offset corrected) to the nonspecific offset, or signal to background ratio, increased by a factor of $\sim$4 for the G4.5 dendrimer-conjugated NPs compared to the molecular dye-labeled antibody.

Future Directions

Successful detection devices for biomedical diagnostics frequently require high sensitivity and low LOD. Moreover, a device for point-of-care diagnostics must be both inexpensive and reliable under a variety of experimental conditions. In this chapter the advantages of employing dielectric, dye-doped NPs have been highlighted for the purpose of enhancing the performance of fluorescence-based assays. The objective of this chapter was to discuss the versatility of silica (or eventually polystyrene) as a host material for fluorescent dyes for many applications, specifically in the fields of nanobiotechnology and the life sciences. Silica as a host material represents an ideal scaffold for encapsulation of organic dyes in a silica matrix thus enhancing their stability and performance.

It is obvious from the growing literature on this topic that this area is still in a growth phase and that significant new developments are likely to emerge in the coming decades. Although co-doping of NPs with multiple dyes of distinguishable spectral properties has been demonstrated, the application of such barcoding NPs in multiplexed assays is expected to grow significantly. Similarly, the use of high-brightness NPs as labels in flow cytometry applications is attracting considerable interest. The versatility of silica as a host material and the feasibility to modulate the core–shell architecture provide a unique environment for innovative research toward creating integrated nanomaterials with highly fine-tuned functionality in the field of nanotechnology and beyond.

Cross-References

► Biosensors
► Nanoparticle Cytotoxicity
► Nanoparticles

References

1. McDonagh, C., Stranik, O., Nooney, R., MacCraith, B.D.: Nanoparticle strategies for enhancing the sensitivity of fluorescence-based biochips. Nanomedicine 4, 645–656 (2009)
2. Oertel, J., Huhn, D.: Immunocytochemical methods in haematology and oncology. J. Cancer Res. Clin. Oncol. 126, 425–440 (2000)
3. Brelje, T.C., Wessendorf, M.W., Sorenson, R.L.: Multicolor laser scanning confocal immunofluorescence microscopy: practical application and limitations, cell biological applications of confocal microscopy, 2nd edn, p. 165. Academic, San Diego (2002)
4. Schaferling, M., Nagl, S.: Optical technologies for the read out and quality control of DNA and protein microarrays. Anal. Bioanal. Chem. 385, 500–517 (2006)
5. Burns, A., Ow, H., Wiesner, U.: Fluorescent core-shell silica nanoparticles: towards "Lab on a Particle" architectures for nanobiotechnology. Chem. Soc. Rev. 35, 1028–1042 (2006)
6. Yan, J., Estévez, M.C., Smith, J.E., Wang, K., He, X., Wang, L., Tan, W.: Dye-doped nanoparticles for bioanalysis. Nano Today 2, 44–50 (2007)
7. Gubala, V., Le Guevel, X., Nooney, R., Williams, D.E., MacCraith, B.: A comparison of mono and multivalent linkers and their effect on the colloidal stability of nanoparticle and immunoassays performance. Talanta 81, 1833–1839 (2010)
8. Hermanson, G.T.: Bioconjugate techniques, 2nd edn. Academic/Elsevier, London (2008)

Dye-Doped Nanospheres in Biomedical Diagnostics

▶ Dye-Doped Nanoparticles in Biomedical Diagnostics

Dynamic Clamp

Thomas Nowotny[1] and Pablo Varona[2]
[1]School of Informatics, University of Sussex, Falmer, Brighton, UK
[2]Dpto. de Ingenieria Informatica, Universidad Autónoma de Madrid, Madrid, Spain

Synonyms

Artificial synapse; Conductance injection; Coupling clamp; Electronic expression; Electronic pharmacology; Hybrid network method; Neural activity-dependent closed-loop; Reactive current clamp; Synthesized conductance injection; Synthesized ionic conductance; Synthesized synaptic conductance

Definition

Dynamic clamp is a method of introducing simulated electrical components into electrically active biological cells using a fast real-time feedback loop between the cells and a computer.

The Dynamic Clamp Concept

Traditionally, neurophysiologists have used current and voltage clamp protocols to assess the electrical properties of neurons. In the current clamp technique, a current (typically a pulse) is injected into the neuron while the membrane potential is being recorded. In voltage clamp, the membrane potential is kept at a controlled value while the transmembrane current is being recorded. Both techniques have contributed to the understanding of the biophysical properties of excitable cells and allowed the design of conductance-based models of neurons.

The dynamic-clamp technique operates in a cyclic way (see Fig. 1). An electrode is inserted into a neuron, and the membrane potential is recorded, amplified, and digitalized into a computer that calculates a current to inject into a postsynaptic neuron. This postsynaptic cell can be the same or a different one from which the membrane potential is being recorded. The injected current is calculated after solving a model that describes ionic or synaptic conductances, or simulates the dynamics of individual neurons or neural networks. This process is repeated indefinitely with a fixed frequency. The time between two consecutive membrane potential acquisitions is known as the update rate. The correct selection of the update rate is critical for the proper function of the application. The maximum update rate is usually determined by the data acquisition board. A slow update rate can prevent the correct simulation of conductances and a realistic stimulation. On the other hand, a realistic update rate may require a fast computer, depending on the computational load of the models which can consist of several differential equations that need to be solved online. Depending on the goals of the experiment, hardware or real-time software implementation may be required to achieve a realistic closed-loop.

Through the closed-loop interaction, the dynamic-clamp technique can simulate the addition of new ionic and synaptic conductances into a living neuron. Modelers and experimentalists have used dynamic clamp to design new experiments on excitable cell properties which were impossible or difficult to carry out with classical techniques. For example, dynamic clamp has been used to simulate the effects of introducing or removing conductances, simulating drug effects, increasing or decreasing neural activity, simulating in vivo conditions, and modifying or building neural dynamics with artificial synapses and artificial neurons in hybrid circuits (see below).

The dynamic clamp allows for a large variety of activity-dependent experiments to address complex information processing and learning in neural systems [1, 2]. The closed-loop interaction can reveal dynamics hidden under traditional electrophysiological protocols. Dynamic clamp can also be used to control neural dynamics under normal or pathological conditions and to induce specific types of neural activity in circuits whose targets are under investigation.

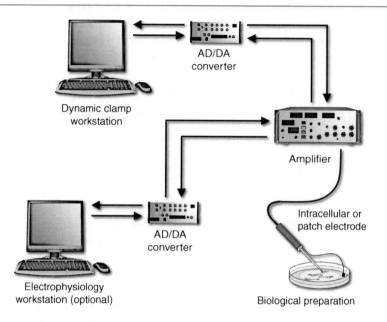

Dynamic Clamp, Fig. 1 The general dynamic clamp setup. The dynamic clamp system forms a closed observation-stimulus-loop in which the measured and amplified membrane potential (Vout) is input to the dynamic clamp hardware/software (dynamic clamp workstation), which calculates a corresponding transmembrane current according to a model specified by the user. The software then issues an appropriate current command, which is converted into a physical current injection by the amplifier. A second, independent, analog to digital (A/D) converter and PC with standard electrophysiology recording software monitors and saves all aspects of the experiment (electrophysiology workstation). The measurement-injection loop is repeated at 10–20 kHz, making the interaction essentially instantaneous for the target neuron

History of Dynamic Clamp

Dynamic clamp was independently invented by Sharp et al. [3] at Brandeis University and Robinson and Kawai [4] at Cambridge. The term "Dynamic clamp" was coined by the former group, while Robinson and Kawai originally referred to the technique as "synthesized synaptic conductance transients." The two systems also differed in their technical implementation.

Sharp et al. used a generic data acquisition board and developed the Microsoft DOS-based dclamp software to emulate artificial ionic and synaptic conductances.

Robinson and Kawai followed the "hardware route" and built analog circuitry (the SM-1 amplifier) to inject dynamic clamp conductances into neurons.

The MS DOS-based dclamp was later replaced by other solutions, the main developments being Real-Time Linux-based (RTLDC, RTlab, MRCI) solutions [5–8] and the Windows-based Dynclamp 2/4 [9].

Dynclamp 2/4 was further developed by Nowotny et al. since 2003 and under the name

StdpC [10, 11], including for the first time plastic synapses, in particular spike timing–dependent plasticity (STDP). Other additions included experimental automation through a scripting mechanism, a modern GUI-based on QT (Trolltech, now Nokia), and most recently active electrode compensation [12].

The real-time Linux-based projects RTLDC, RTLAB, and MRCI merged in 2004 into the common RTXI (real-time experimentation interface) project [13]. A major step forward for the usability of this system was the recent release on live CDs.

Besides these longer-term projects, independent other implementations of dynamic clamp have been developed over the years. These include QuB [14], g-Clamp [15], and the dynamic clamp subsystem of the commercial CED Signal software.

One of the more recent systems (RTBiomanager; [16]) now promises another step change in dynamic clamp research in systematically extending the concept of closed-loop real-time experimentation beyond the realm of pure electrophysiology.

Dynamic Clamp, Table 1 Common dynamic clamp implementations

Name	Technology	URL/Publication	Comments
RTXI (Real-Time eXperiment Interface)	Real-Time Linux, Comedi	http://www.rtxi.org/ [13]	Based on RTLDC and MRCI and RTlab
StdpC	Windows-based, custom DigiData driver, NIDAQmx	http://sourceforge.net/projects/stdpc/ [11]	Partially based on DynClamp2/4
G-clamp	Labview RT, NIDAQmx, embedded RT hardware subsystem	http://www.hornlab.neurobio.pitt.edu/ [15]	
SM-1, SM-2 amplifiers by Cambridge Conductance1,	Custom analog hardware and DSP board.	[4, 17]	
Dynamic Clamp in CED signal	Uses DSP capabilities of CED DAQs	http://www.ced.co.uk/	Offers a subset of models in StdpC; released in Fall 2010
QuB	Windows based, NIDAQmx	http://www.qub.buffalo.edu/wiki/index.php/Dynamic_Clamp [14]	
RTBiomanager	Real-Time Linux, Comedi	http://www.ii.uam.es/~gnb [16]	

The currently available dynamic clamp systems and their characteristic features are summarized in Table 1. They can roughly be categorized into hardware- (SM-1, SM-2) and software-based solutions. The latter category roughly divides into windows-based (StdpC, QuB, g-Clamp) and real-time Linux-based (RTXI, RTBiomanager) solutions. The hardware-based and real-time Linux-based systems guarantee hard real-time constraints, whereas the windows-based solutions work in a soft real-time framework.

Dynamic Clamp Applications/Configurations

Since its inception, dynamic clamp has seen several different extensions and modifications. The most common types of experiments are

Simulated Chemical Synapses

The simulation of chemical synapses reaches back to the original publications of Sharp et al. [3] and Robinson and Kawai [4]. It involves measuring the membrane potential of two neurons and injecting current into one of them – the designated postsynaptic cell – in order to simulate the action of a non-existing synapse or increasing the conductance of existing ones. Some researchers have also compensated existing synapses by using a negative conductance value. Further extensions of this protocol include
(a) Short-term depression of the simulated synapse
(b) Spike timing–dependent plasticity of the simulated synapse

(c) Using a simulated model neuron as the presynaptic and/or postsynaptic neuron
(d) Using an electronic circuit for the presynaptic neuron

A typical description of a chemical synapse is given by

$$I_{syn} = g_{syn}S(V_{post} - V_{rev})$$

$$\frac{dS}{dt} = \alpha(V_{pre})(1 - s) - \beta_0 s$$

$$\alpha(V_{pre}) = \left\{ \begin{array}{ll} \alpha_0 & 0 \leq t - t_{spike} \leq t_{release} \\ 0 & otherwise \end{array} \right\}$$

where I_{syn} denotes the postsynaptic current, g_{syn} the maximal synaptic conductance, S the fraction of active transmitter, V_{rev} the reversal potential of the synapse, and α_0 and β_0 are rate constants of synaptic transmitter activation and removal. $t_{release}$ denotes the duration of transmitter release in response to a presynaptic spike, and t_{spike} the time of the last occurrence of such a spike.

Simulating Electrotonic Connections (Gap Junctions)

A gap junction is essentially an Ohmic conductance between two cells through which current can flow in two directions. In dynamic clamp, a simulated gap junction is achieved by monitoring the membrane potential of two cells and injecting the appropriate currents into both cells according to Ohm's law and the observed membrane potential difference.

$$I_1 = g_{\text{gap}}(V_2 - V_1)$$
$$I_2 = g_{\text{gap}}(V_1 - V_2)$$

where I_1 and I_2 are the currents injected into the two neurons, and V_1 and V_2 are their respective membrane potentials. g_{gap} denotes the constant conductance of the gap junction.

A common extension are asymmetric or rectifying gap junctions, in which current can preferentially or exclusively flow in one direction.

Simulating Voltage-Dependent Ion Channels (Hodgkin–Huxley Conductances)

Voltage-dependent ion channels allow current flows into and out of a cell depending on its own membrane potential. For a simulated voltage-gated ion channel, the membrane potential of a cell is monitored, and the current of the simulated ionic conductance is injected into the same cell. Commonly, the conductance is described in the Hodgkin–Huxley formalism, for example, the sodium current is given by

$$I_{\text{Na}} = g_{\text{Na}} m h^3 (V - V_{\text{Na}})$$
$$\frac{dm}{dt} = \alpha_m(1-m) - \beta_m m \qquad \frac{dh}{dt} = \alpha_h(1-h) - \beta_h h$$
$$\alpha_m = \frac{(-35-V)/10}{\exp((-35-V)/10) - 1} \qquad \alpha_h = 0.07\exp(-60-V)/20$$
$$\beta_m = 4\exp((-60-V)/18) \qquad \beta_h = \frac{1}{\exp((-30-V)/10) + 1}$$

where I_{Na} denotes the sodium current, m and h are so-called activation and inactivation variables, V denotes the membrane potential of the neuron, and V_{Na} is the sodium reversal potential.

Pattern Clamp

Pattern clamp is the simple combination of simulating a presynaptic neuron and a strong unidirectional gap junction. It nevertheless deserves a separate discussion because if the gap junction conductance is large, the electrical activity of the target neuron will be almost identical to the simulated presynaptic activity pattern, allowing the experimenter to virtually control, "clamp" the neuron to a particular activity pattern (hence the name "pattern clamp"). While quite simple from a theoretical point of view, this mode of dynamic clamp is very powerful as it allows the experimenter to interact with the rest of the neuronal system through

the clamped neuron's natural synapses. This can include the release of neuromodulators and synaptic plasticity of the neuron's synapses.

Hybrid Circuits of Real and Artificial Neurons

Hybrid circuits of multiple living and artificial neurons implemented with dynamic clamp protocols provide a unique way to test model neuron and network dynamics within a realistic cellular and synaptic environment [18]. A hybrid circuit can be used to bridge between different description levels and allows precise changes in single cellular or synaptic parameters in order to assess their direct participation in the collective output. In this way, hybrid methods can result in more physiological ways of stimulating networks, creating direct links between experiments and modelling, and providing new tools and validations methods for neural network analysis.

Examples of Dynamic Clamp Results

Typical results from standard dynamic clamp applications are illustrated in Figs. 2–4. They were obtained using the StdpC dynamic clamp software on Windows XP and using a National Instruments M-series PCI-6229 data acquisition board (Figs. 2 and 3) and the real-time Linux system RTLDC on Fedora Core using a National Instruments PCI-MIO-16E-4 data acquisition board (Fig. 4).

Figure 2 illustrates results from simulating an excitatory chemical synapse between two cells of the mollusc *Lymnaea stagnalis* in an in vitro whole brain preparation. The cerebral giant cell (GCG) and the motoneuron B1 were impaled with sharp microelectrodes. A second sharp electrode was inserted into the B1 neuron for the purpose of current injection. The dynamic clamp was configured to use the tonically firing CGC as the presynaptic cell and the quiescent B1 neuron as the postsynaptic cell. The synapse was configured with reversal potential 0 mV, activation threshold -20 mV, activation slope of 25 mV, and time scale of 10 ms. The maximal synaptic conductance was varied, taking values 10, 20, 30, 50, 80, 120, 170, 230, 300, 390, 490, and 600 nS (encoded by line color in Fig. 2). Note, how the simulation of a chemical synapse in dynamic clamp is different from injecting a predefined EPSP, in particular if a postsynaptic spike is evoked (current reversal during the spike).

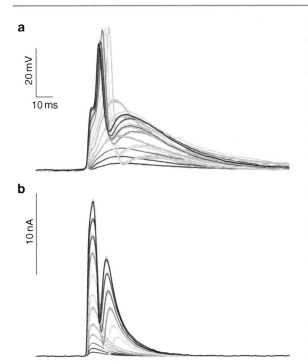

a

20 mV

10 ms

b

10 nA

Dynamic Clamp, Fig. 2 Excitatory postsynaptic potentials (EPSPs) and excitatory postsynaptic currents (EPSCs) of a simulated, excitatory, chemical conductance introduced into a molluscan cell. (**a**) EPSPs for simulated maximal synaptic conductance of 10, 20, 30, 50, 80, 120, 170, 230, 300, 390, 490, and 600 nS aligned to the occurrence of the presynaptic spike. For each maximal conductance, five individual EPSPs (*thin grey lines*) and their average thick lines (*colored online*) are shown. (**b**) Corresponding EPSCs. The *colors* of **a** and **b** are matched to mark the corresponding data sets. Note how the introduction of a simulated synaptic conductance is different from injecting a fixed current waveform: When a postsynaptic spike is elicited (*yellow to red traces (color online)*), the current through the simulated ion channel shows a marked dip and almost reverses (Figure modified from Kemenes, I., Marra, V., Crossley, M., Samu, D., Staras, K., Kemenes, G., Nowotny T. Dynamic Clamp with StdpC Software, Nature Prot. 6(3):405–417 (2011)

Figure 3 illustrates the results of performing pattern clamp on the CGC neuron in the mollusc *Lymnaea stagnalis*. An accelerating-decelerating burst pattern was defined using the StdpC spike generator facility. The pattern was imposed onto the CGC neuron using a high conductance-simulated gap junction. Figure 3 shows the results for gap junction conductances of 300 nS (B) and 2,000 nS (C). For the lesser conductance, pattern clamp is only partially successful, and some spikes of the pattern are displaced or missing. At 2,000 nS, the pattern clamp is almost flawless, including exact timing of all spikes in the burst pattern.

Interestingly, even though the gap junction conductance is almost 7 times higher in C than in B, the injected currents are only 2–3 times stronger due to the better control achieved. It is also important to note that the spikes in the CGC are of higher amplitude than the spikes of the target activity pattern (Fig. 3d, e). This indicates that the CGC spikes are genuine and will travel down the axon rather than being merely strong deflections of the membrane potential at the soma.

A somewhat more unusual example of dynamic clamp is shown in Fig. 4. Here, dynamic clamp is used to synchronize a model neuron and a real cell in order to fit the properties of the cell online and in real time. Within the dynamic clamp software, two copies of a neuron model with slightly different parameters are simulated together with a simulated coupling to a real cell impaled with two electrodes. Through the simulated coupling, the cell and the model cells synchronize (partially) which allows to assess how much the dynamics of each of the model cells differs from the real cells'. After a fixed testing period, the model cell with the more appropriate dynamics is retained and further compared to a new model. Through this iterative process, the model cells become increasingly similar to the real cell allowing indirect deduction of cell properties.

Limitations

Like any other electrophysiological technique, dynamic clamp has practical and principal limitations. The main principal limitation is that dynamic clamp simulates the electrical effect of simulated channels in the cell membrane, not the flow of specific ions. Where the presence of the ions is important as, for example, with calcium ions that act as second messengers, simulating an ion channel will have a different effect than the actual channel would have. A second principal limitation is related to the position of the electrode(s) used. If electrodes are far from the locus in the cell where real channels would be located, the effect of the injected current can be quite different from the current that flows through the channels directly. If, for example, the current injecting electrode is inserted into the soma, a Na current is simulated, that normally would be located at the axon hillock, and the axon hillock is electrically far removed from the soma, the dynamic clamp experiment may be difficult to interpret.

Practical limitations depend on the dynamic clamp implementation and the used hardware.

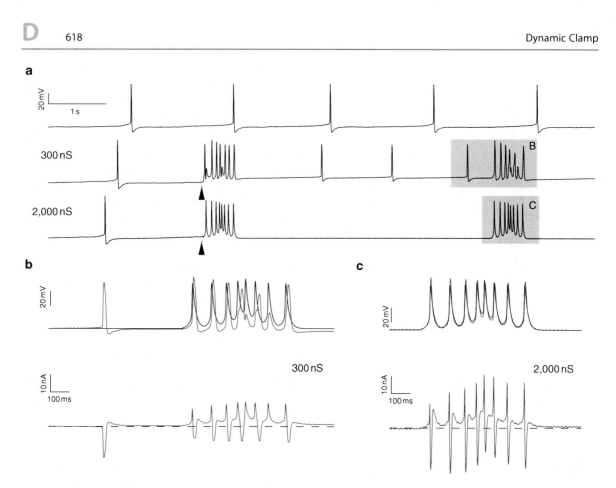

Dynamic Clamp, Fig. 3 Results of a pattern clamp experiment. The target cell spikes tonically when the pattern clamp is off (*top*). When the gap junction coupling in pattern clamp is not strong enough (gap junction conductance of 300 nS), the target neuron will only be clamped partially (*middle*) but sufficient conductance (gap junction conductance of 2,000 nS) leads to a complete clamp (*bottom*). Pattern clamp is switched on at the black arrowheads. Panels D and E are detailed views of the episodes shaded in grey in panel C. The grey (*red online*) trace was recorded from the target neuron, the black trace shows the desired pattern. Currents injected are shown in the bottom panels. With 300 nS conductance, the pattern clamp is not able to prevent intrinsic spiking (C, black arrowhead). This is readily achieved at 2 μS maximal conductance, even though the currents injected for this >6 times larger conductance are only 2–3 times larger owing to the better match achieved (From Kemenes, I., Marra, V., Crossley, M., Samu, D., Staras, K., Kemenes, G., Nowotny T. Dynamic Clamp with StdpC Software, Nature Prot. 6(3):405–417 (2011)

Dynamic-clamp protocols rely critically on the update rate used for the closed-loop interaction. In many applications, this update rate must be strictly accomplished, requiring precise timing and no jitter. Specific hardware, fast DSP systems, or real-time software technology can overcome these problems. Software implementations are more suitable than hardware approaches as they are easier to program, generally inexpensive, and more flexible to modify and customize to a particular experiment. A common problem in electrophysiology, and more so in dynamic clamp, is the series resistance and capacitance of electrodes. Common techniques such as bridge series resistance

compensation and capacitance compensation are generally not suitable for dynamic clamp due to the rapidly changing current injections. Most dynamic clamp experiments have, therefore, been conducted with two separate electrodes in the same cell, one for membrane potential measurements and one for current injection. Others have used low-resistance patch clamp electrodes to avoid the problem.

A recent alternative solution to the electrode artifact problem was introduced by Brette et al. [12], who used a sophisticated calibration method to estimate series resistance and capacitance artifacts with a digital model of the electrode and subtract them digitally

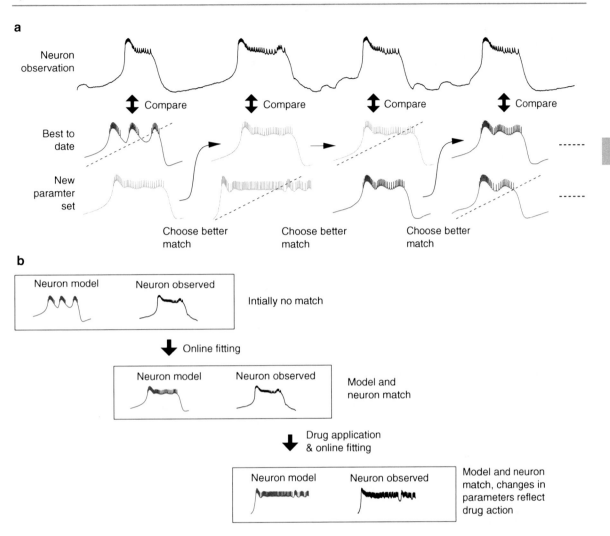

Dynamic Clamp, Fig. 4 Dynamic clamp used for online fitting. (**a**) two model neurons with slightly differing parameters are simulated alongside a real cell (a lateral pyloric cell (LP) of the lobster) to which they are coupled through a simulated electrical coupling. After a test period, the better matching model cell is retained and compared to a model with a new, slightly different, parameter set. (**b**) Initially, the model cells' dynamics can be quite different from the dynamics of the real cell. After some time, the models fit the cell increasingly better and eventually should be able to track slow changes within the cell. The data shown in this figure are illustrative excerpts from experiments with a prototype system programmed in RTLDC and used with a lobster LP cell (Levi and Nowotny 2005, unpublished)

and in real time. Using this new active electrode compensation (AEC) allows experimenters to conduct dynamic clamp experiments with a single high-resistance electrode.

Future Directions

The dynamic-clamp concept has an enormous potential to precisely control the spatiotemporal aspects of a stimulus and to build activity-dependent stimulus-response loops to interact with neural systems beyond the realm of electrophysiology. Establishing these loops can be an essential step toward understanding the dynamics of many neural processes and can bridge between traditionally disparate levels of analysis (e.g., subcellular, network, and system dynamics).

The same principles used in the dynamic-clamp technology can be generalized to develop new techniques of activity-dependent stimulation by enhancing or going beyond the electrophysiological realm in a broad spectrum of research in nervous systems (see Fig. 5).

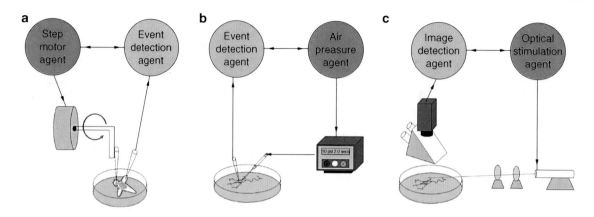

Dynamic Clamp, Fig. 5 Generalization of the dynamic-clamp concept for different stimulation and recording techniques. Computer algorithms implemented as intelligent agents monitor neural activity to exert goal-driven closed-loop stimulation. Panel A: activity-dependent mechanical stimulation. Neural activity is measured with intra- or extracellular recordings. Specific events detected in this activity are used to control a step motor for mechanical stimulation of sensory systems. Close-loop interaction can be used to automatically find receptive fields of cells that react to the sensory stimulation throughout the nervous system [15]. Panel B: activity-dependent drug microinjection. A microinjector delivers a timely and precise amount of neurotransmitter or neuromodulator to achieve a desired type and level of neural activity. Panel C: closed-loop optical recording with real-time event detection used to drive laser stimulation which can go down to nanoscale precision

For example, a generalization of the dynamic-clamp concept has been used to find receptive fields throughout the sensorymotor transformation by implementing activity-dependent mechanostimulation [19]. In this protocol, a step motor is controlled to exert mechanic stimulation as a function of neural events recorded with microelectrodes throughout this transformation. The protocol can explore different motor angles and speed until neural activity correlated with the stimuli is found. Another dynamic-clamp extension has implemented activity-dependent real-time drug microinjection to stimulate neural circuits [20] as a function of specific patterns recorded in the electrical activity. This protocol is used to constrain neural activity within specified limits and to produce robust temporal patterns. Other protocols can include closed-loop optical recording and stimulation, for example, in two-photon microscope setups which can lead to highly realistic experimental conditions that overcome the dynamic clamp limitations discussed above. Further extensions of the closed-loop interaction with neural systems will lead to novel brain-machine interfaces and prosthetic devices.

Cross-References

▶ Biomimetic Muscles and Actuators
▶ Biomimetics
▶ Biosensors
▶ Computational Systems Bioinformatics for RNAi
▶ Electrode–Organic Interface Physics
▶ Lab-on-a-Chip for Studies in C. elegans
▶ Nanorobotic Manipulation of Biological Cells
▶ Nanorobotics for Bioengineering
▶ Synthetic Biology

References

1. Prinz, A.A., Abbott, L.F., Marder, E.: The dynamic clamp comes of age. Trends Neurosci. **27**(4), 218–224 (2004)
2. Destexhe, A., Bal, T. (eds): Dynamic clamp: from principles to applications. Springer Series in Computational Neuroscience. Springer, New York (2009)
3. Sharp, A.A., O'Neil, M.B., Abbott, L.F., Marder, E.: The dynamic clamp: artificial conductances in biological neurons. Trends Neurosci. **16**(10), 389–394 (1993)
4. Robinson, H.P., Kawai, N.: Injection of digitally synthesized synaptic conductance transients to measure the integrative properties of neurons. J. Neurosci. Method. **49**(3), 157–165 (1993)
5. Dorval, A.D., Christini, D.J., White, J.A.: Real-time linux dynamic clamp: a fast and flexible way to construct virtual ion channels in living cells. Ann. Biomed. Eng. **29**, 897–907 (2001)
6. Culianu, C.A., Christini, D.J.: Real-time Linux experiment interface system: RTLab. In: Proceedings of the IEEE 29th Annual Bioengineering Conference, Newark, pp. 51–52 (2003)

7. Raikov, I., Preyer, A.J., Butera, R.: MRCI: a flexible real-time dynamic clamp system for electrophysiology experiments. J. Neurosci. Method. **132**, 109–123 (2004)

8. Butera Jr., R.J., Wilson, C.G., DelNegro, C.A., Smith, J.C.: A methodology for achieving high-speed rates for artificial conductance injection in electrically excitable biological cells. IEEE Trans. Biomed. Eng. **48**(12), 1460–1470 (2001)

9. Pinto, R.D., Elson, R.C., Szücs, A., Rabinovich, M.I., Selverston, A.I., Abarbanel, H.D.I.: Extended dynamic clamp: controlling up to four neurons using a single desktop computer interface. J. Neurosci. Method. **108**, 39–48 (2001)

10. Nowotny, T., Zhigulin, V.P., Selverston, A.I., Abarbanel, H.D., Rabinovich, M.I.: Enhancement of synchronization in a hybrid neural circuit by spike-timing dependant plasticity. J. Neurosci. **23**(30), 9776–9785 (2003)

11. Nowotny, T., Szücs, A., Pinto, R.D., Selverston, A.I.: Stdpc: a modern dynamic clamp. J Neurosci. Method. **158**(2), 287–299 (2006)

12. Brette, R., et al.: High-resolution intracellular recordings using a real-time computational model of the electrode. Neuron **59**(3), 379–391 (2008)

13. Lin, R.J., Bettencourt, J., White, J., Christini, D.J., Butera, R.J.: Real-time experiment Interface for biological control applications. Conf. Proc. IEEE Eng. Med. Biol. Soc. **1**, 4160–4163 (2010)

14. Milescu, L.S., Yamanishi, T., Ptak, K., Mogri, M.Z., Smith, J.C.: Real-time kinetic modelling of voltage-gated ion channels using dynamic clamp. Biophys. J. **95**, 66–87 (2008)

15. Kullmann, P.H.M., Wheeler, D.W., Beacom, J., Horn, J.P.: Implementation of a fast 16-bit dynamic clamp using LabVIEW-RT. J. Neurophysiol. **91**, 542–554 (2004)

16. Muñiz, C., Rodriguez, F.B., Varona, P.: RTBiomanager: a software platform to expand the applications of real-time technology in neuroscience. BMC Neurosci. **10**, 49 (2009)

17. Robinson, H.P.C.: A scriptable DSP-based system for dynamic conductance injection. J. Neurosci. Method. **169**(2), 271–281 (2008)

18. Le Masson, G., Renaud-Le Masson, S., Debay, D., Bal, T.: Feedback inhibition controls spike transfer in hybrid thalamic circuits. Nature **417**, 854–858 (2002)

19. Muñiz, C., Levi, R., Benkrid, M., Rodríguez, F.B., Varona, P.: Real-time control of stepper motors for mechanosensory stimulation. J. Neurosci. Meth. **172**(1), 105–111 (2008)

20. Chamorro, P., Levi, R., Rodriguez, F.B., Pinto, R.D., Varona, P.: Real-time activity-dependent drug microinjection. BMC Neurosci. **10**, 296 (2009)

E

e-Beam Lithography, EBL

▶ Electron Beam Lithography (EBL)

ECM

▶ Electrochemical Machining (ECM)

Ecotoxicity

▶ Ecotoxicology of Carbon Nanotubes Toward Amphibian Larvae
▶ Toxicology: Plants and Nanoparticles

Ecotoxicity of Inorganic Nanoparticles: From Unicellular Organisms to Invertebrates

Mélanie Auffan[1,2], Catherine Santaella[2,4],
Alain Thiéry[2,5], Christine Paillès[1,2], Jérôme Rose[7,8],
Wafa Achouak[2,4], Antoine Thill[2,6], Armand Masion[1,2],
Mark Wiesner[2,3] and Jean-Yves Bottero[1,2]
[1]CEREGE, UMR 6635 CNRS/Aix–Marseille Université, Aix–en–Provence, France
[2]iCEINT, International Consortium for the Environmental Implications of Nanotechnology, Center for the Environmental Implications of NanoTechnology, Aix–En–Provence, Cedex 4, France
[3]CEINT, Center for the Environmental Implications of NanoTechnology, Duke University, Durham, NC, USA
[4]Laboratoire d'Ecologie Microbienne de la Rhizosphère et d'Environnements Extrême,
UMR 6191 CNRS–CEA–Aix–Marseille Université de la Méditerranée, St Paul lez Durance, France
[5]IMEP, UMR 6116 CNRS/IRD, Aix–Marseille Université, Marseille, Cedex 03, France
[6]Laboratoire Interdisciplinaire sur l'Organisation Nanométrique et Supramoléculaire, UMR 3299 CEA/CNRS SIS2M, Gif–surYvette, France
[7]CEREGE UMR 6635– CNRS–Université Paul Cézanne Aix–Marseille III, Aix–Marseille Université, Europôle de l'Arbois BP 80, Aix–en–Provence Cedex 4, France
[8]GDRI ICEINT: International Center for the Environmental Implications of Nanotechnology, CNRS–CEA, Europôle de l'Arbois BP 80, Aix–en–Provence Cedex 4, France

Synonyms

Environmental toxicology; Nano-ecotoxicology; Toxicity of metal and metal oxide nanoparticles on Prokaryotes (bacteria), unicellular and invertebrate Eukaryotes

Definition

A toxic product is a chemical compound that harms the environment by affecting the biological organisms. Due to their novel properties, nanoparticles (NPs) cannot be considered as other organic and inorganic xenobiotics in the environment, e.g., pesticides, HAPs and PCBs dissolved metals, or also medicines. NPs have a mass, a charge, and above all a surface area. They are subject to phenomena of classical and

B. Bhushan (ed.), *Encyclopedia of Nanotechnology*, DOI 10.1007/978-90-481-9751-4,
© Springer Science+Business Media B.V. 2012

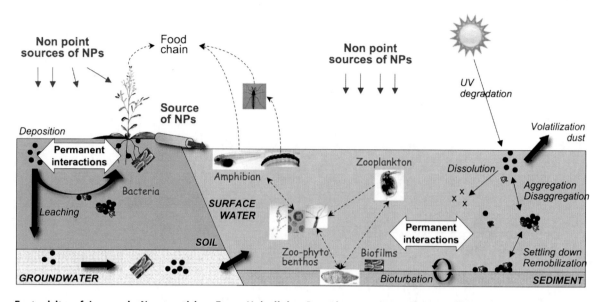

Ecotoxicity of Inorganic Nanoparticles: From Unicellular Organisms to Invertebrates, Fig. 1 Fate and transport of nanomaterials within aquatic and terrestrial ecosystems and the relationships with the living organisms

quantum physics. Their reactivity means that their surface atoms are labile, easily change their redox state, and highly reactive with respect to compounds in the environment. Considering the huge range of applications using NPs, it seems reasonable to expect their dissemination in the environment at each step in their life cycle, from design through production to use and disposal of finished products. To date, the data available show that NPs can cross biological membranes and distribute themselves within different compartments of living organisms, or can also induce a remote toxicity. Consequently, there is a need to elucidate how the NPs can lead to adverse effects on organisms in their natural environment, considering not only their effects on target organs and life traits, but also their fate and transfers within the food webs (Fig. 1) (Box 1).

Overview

The number of available publications (Fig. 2) dealing with the environmental impacts of NPs (called nano-ecotoxicology [1]) is presented in Fig. 2. Studies regarding algae, protozoa, and invertebrates are more recent (since 2006) than the ones regarding bacteria (which started in the 1980s) (*ISI Web of Knowledge source*) (Box 2). As a consequence, researches on bactericidal effects are more advanced (already at the

Box 1: Ecosystem and Food Chains

An ecosystem is the basic functional unit in ecology, as it includes both organisms and their abiotic environments (Fig. 1). It represents the highest level of ecological integration based on energy and biomass transfers. It is defined as a specific unit of all the organisms occupying a given area that interacts with the physical environment producing distinct trophic structure, biotic diversity, and material cycling.

scale of biofilms and the microbial community structures), while for the other models nano-ecotoxicology is still in its infancy (at the individual scale).

About 30 species of invertebrates are studied regarding the three main ecosystemic domains: freshwater, marine, and terrestrial. For each domain, the Crustacean Cladocera Daphniidae (freshwater), the Mollusks Bivalvia Mytilidae, and Annelids Polychaetes (marine) are the most studied invertebrates, while the terrestrial (earthworms and isopods) examples are scarce.

Considering the three-domain system that divides cellular life into archaea, bacteria, and eukaryote domains (based on differences in 16 S rRNA genes), the studied models are mostly Eukaryotes belonging to Opisthokonts Animals, Alveolata (Ciliata), Plants Chlorophyta (green algae), or Prokaryotes with

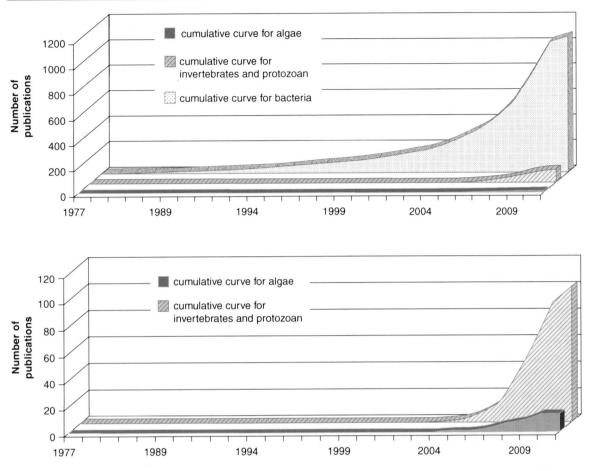

Ecotoxicity of Inorganic Nanoparticles: From Unicellular Organisms to Invertebrates, Fig. 2 Evolution of the nanoecotoxicology studies focused on the algae, invertebrates and protozoan, and bacteria since the 1980s (based on the ISI Web of Knowledge database)

planktonic bacteria. Following the Whittaker's model distinguishing the differences in nutrition, most of the studied invertebrates belong to the ingestion nutrition group (active suspensory feeders (Daphniidae), suspension feeders (Mytilidae), or carnivorous (Cnidaria, Polychaeta)), others are detritivorous (*Blaberus*), deposit feeders (*Eisenia*), or fluid suckers (*Caenorhabditis elegans*), and most of the autotrophic organisms are the eukaryotic photosynthetic green algae (Chlorophyta). Protozoa are unicellular heterotrophic organisms, belonging to the Ciliophora group, feeding by phagotroph.

Key Principles

Not intended to be an exhaustive list of results, this article discusses the relationships between the potential ecotoxicity of NPs with respect to their physicochemical properties, the organism functions, and the relationships between species. It has been recently shown that the surface reactivity and the chemical destabilization of metallic and metal oxide NPs were involved in their potential toxicity. NPs reactivity can be spontaneous (e.g., adsorption, oxidation, or reduction) or induced by an external event (e.g., the photocatalytic activity of TiO_2) [2]. Chemical destabilization can lead to redox changes within the lattice structure, the generation of reactive oxygen species (ROS), and to the release of ions in solution (e.g., Zn ions). The dissolution can be either reductive (e.g., the reduction of $Ce^{4+}O_2$ NPs into Ce^{3+}) or oxidative (e.g., Ag^0 NPs oxidation into Ag^+, Cu^0 NPs oxidation into Cu^{2+}) [3]. All these modifications can be critical events if the protective mechanisms of unicellular or pluricellular organisms are overwhelmed

Box 2: Bacteria as a Good Model for Ecotoxicology
Life on earth evolved from bacteria and life on earth depends on bacteria, as they make the essential elements oxygen, carbon, nitrogen, and sulfur available for other life forms on our planet. They play important roles in the global ecosystem. They are key players of the breakdown of dead organisms, and the release of nutrients back into the environment. Each human body hosts ten microorganisms for every human cell. Microbes inhabiting humans have been recently identified and shown to play a beneficial role in health by contributing to human metabolism while in turn they benefit from the nutrient-rich niche in the intestine. Since bacteria constitute the largest biological surface, nanoparticles (NPs) may encounter in different ecosystems from air, soil, water to insect and animal bodies. Their contribution to NPs fate in different ecosystems has to be taken into account, as they may be involved in NPs transformation, sequestration, solubilization, or aggregation.

The biology and functional ecology of bacteria in different ecosystems make them good models for NPs toxicity studies. NPs may show deleterious effect on microbial physiology and survival, altering hence ecosystems functioning. It is hence worth investigating NPs effects of exposure and subsequent bioaccumulation to determine their toxicity on microbial communities and model bacteria.

by a strong affinity with NPs. To highlight the relationship between surface reactivity and the ecotoxicological effects, this article is focused on metallic and metal oxide NPs.

Key Research Findings

Photocatalytic NPs: The Example of TiO$_2$

TiO$_2$ NPs (anatase and rutile) are the most used photocatalytic NPs as self-cleaning, opacifying, antimicrobial, antifouling agents, or UV-absorber in cosmetics, paints, etc., (see [4] and references therein). Recent studies have shown that they can be disseminated in the environment. For instance, the leaching out of TiO$_2$ NPs from exterior facades releases few $\mu g.L^{-1}$ of TiO$_2$ into stream water [5].

Bacteria

The ecotoxicity of TiO$_2$ NPs has been studied on rod-shaped *Escherichia coli*, *Pseudomonas fluorescens*, *Vibrio fischeri*, *Salmonella typhimurium*, bacillus-shaped *Nitrosomonas europaea* Gram-negative bacteria, and *Bacillus subtilis* Gram-positive bacteria [6, 7].

The mechanisms of ecotoxicity of TiO$_2$ mainly rely on a photocatalytic process at the surface of NPs generating ROS. The toxicity requires a contact between bacteria and TiO$_2$ NPs leading to the loss of membrane integrity and an increase in outer membrane permeability. The concentration of TiO$_2$ affecting bacterial survival in photocatalytic conditions varies between 10^{-3} and 5×10^{3} mg.L^{-1} [8]. For *E. coli*, the antimicrobial activity depends on biological parameters such as physiological state, generation, and initial concentration of bacteria [9] as well as the chemical structure and the shape of the TiO$_2$ NPs. Surprisingly, membrane damages were also observed in the absence of light, e.g., [10].

The uptake of well-dispersed TiO$_2$ NPs was observed by transmission electron microscopy in *S. typhimurium* [7] treated with 8 and 80 ng.mL^{-1} of TiO$_2$ for 60 min. This induced a weak mutagenic potential in *S. typhimurium* strains (TA98, TA1537). Mutagenic effects were also observed toward *E. coli* (WP2uvrA) [7, 11].

Size-dependent toxicity of TiO$_2$ NPs is controversial. Studies have shown that particle size did not affect antibacterial activity [12, 13]. However, Ivask et al. have shown that whereas nano and bulk TiO$_2$ both decrease viability of bacteria at high concentrations ($> 4,000$ mg.L^{-1}), the TiO$_2$ NPs were the only ones inducing bioluminescence of the superoxide anion sensing bacteria from 100 mg.L^{-1} [14]. Another size effect regards the size-dependent point of zero charge (PZC) of TiO$_2$, with a PZC of 4.8 for the smallest particles ($\varnothing$ 3.6 nm) and higher PZC values (6.2, 6.5, 6.8) for the largest ones ($\varnothing$ 8, 25, 200 nm) [6]. Since many experiments are performed at pH 5.5–6.8, exposure to bacteria and the adsorption onto their walls will be sensitive to this change in PZC [13].

Moreover, initially agglomerated TiO_2 NPs can be dispersed by bacteria as *P. fluorescens*. Such phenomenon can enhance the transport of NPs in the environment [15].

Algae

Since 2006, less than 20 papers illustrate the effects of NPs on marine and freshwater primary producers (algae). Overall, these studies demonstrate that primary producers (algae) are among the more sensitive forms of aquatic life to metallic NPs. The most studied freshwater green algae are *Pseudokirchneriella subcapitata*, but the results are controversial. Warheit et al. [16] observed acute effects with EC_{50} values of 21 mg.L^{-1} for TiO_2 NPs ($\varnothing$ 140 nm), and 72 h EC_{50} of 5.83 mg.L^{-1} for TiO_2 NPs ($\varnothing$ 25–70 nm). However, no toxicity was found at concentrations up to 100 mg.L^{-1} for TiO_2 NPs ($\varnothing$ <40 nm) [17]. The absence of ecotoxicity is attributed to the rapid aggregation/coagulation [17], whereas the toxicity is related to entrapping of algal cells by aggregated TiO_2 NPs [18]. Such controversy highlights that many studies address the question of ecotoxicity but very few concern the mechanisms of toxicity or internalization.

The effects of particle size and UV photoactivation were assessed using TiO_2 (anatase, $\varnothing$ 25 and 100 nm) on freshwater green algae *Desmodesmus subspicatus* [19]. The EC_{50} value was 44 mg.L^{-1} for the 25 nm NPs and greater than 50 mg L^{-1} for the 100 nm-TiO_2. UV photoactivation caused no additional effects. In some cases, the presence of TiO_2 NPs can also be beneficial. It increases the dry weight, chlorophyll synthesis and metabolism in photosynthetic organisms.

Finally, in a recent study [20], the toxicity of TiO_2 NPs was assessed for the first time on marine diatoms, chlorophyceae, and prymnesiophyceae. Ninety six hours NOEC were observed on all species in presence of TiO_2 NPs (anatase, $\varnothing$ 15–20 nm, 10–1,000 μg.L^{-1}).

Invertebrates

Terrestrial Invertebrates Ecotoxicological studies on terrestrial invertebrates are limited. They mostly regard the crustacean isopod *Porcellio scaber* [21, 22], the model *C. elegans*, and the earthworm *Eisenia fetida* [23, 24]. TiO_2 NPs ($\varnothing$ from 15 to < 50 nm) spread over the surface of dry leaf (1 mg TiO_2.g^{-1} of food for 14 days) did not induce effect on mortality,

weight change, or feeding behavior toward *Porcellio*. However, *E. fetida* exposed to artificial soils containing more than 1 g.kg^{-1} of TiO_2 NPs for 7 days bioaccumulates and exhibit significant mitochondria damage. Moreover, a strong inhibition of the cellulase activity of *E. fetida* occurred that directly impacted the digestion. However, it has not been clarified whether cellulase originates from the earthworms or the symbionts.

Freshwater Invertebrates The adverse effects of TiO_2 NPs on aquatic invertebrates have been more studied than other inorganic NPs. In freshwater crustaceans, only branchiopods have been tested: the cladocerans *Ceriodaphnia dubia*, *Chydorus sphaericus*, and the anostracans *Thamnocephalus platyurus*, as well as the brine shrimp *Artemia*.

Daphniid *Daphnia magna* are the most tested animals. TiO_2 NPs were mostly detected in their gut, but no clear conclusions can be drawn on the size-dependent toxicity of NPs. However, the aggregation state is important to consider [16]. Variable 48 h EC_{50} and LC_{50} ranging from 5.5 to 20,000 mg.L^{-1} have been reviewed [1]. Others studies regard *Hydra attenuata*, the Gastropod *Physa acuta*, and midge larvae *Chironomus riparius*. For instance, sublethal doses of alumina coated-TiO_2 NPs reduced embryo growth rate and hatchability, and induced developmental deformities of the embryos on the freshwater snail *Physa acuta* [25]. Embryonic growth and hatchability tests are useful endpoints to assess chronic toxicity and toxic thresholds of NPs in sediment environment. Although no snail mortality was observed significant change in antioxidant levels were observed at 0.05 or 0.5 g.kg^{-1}. In this study, the exposure duration appeared as a significant factor. Consequently, long-term exposure has to be taken into account since alterations of ecological populations and ecological functionality are possible.

Marine Invertebrates A significant decrease in casting rate, increase in cellular and DNA damage in coelomocytes was measured toward the sediment dwelling marine Polychaetes *Arenicola marina* [26]. Using Raman Scattering Microscopy, TiO_2 aggregates of diameter > 200 nm were located in the lumen of the gut and attached to the outer epithelium. On the bivalve *Mytilus galloprovincialis*, 24 h in vivo tests highlight that TiO_2 NPs ($\varnothing$ 22 nm, 51 m^2.g^{-1}) induced lysosomal membrane destabilization in the hemocytes and digestive gland [27]. TiO_2 also increased the catalase and glutathione transferase activity in the digestive glands (not in the gills) which implies an oxidative stress mechanism.

Chemically Unstable NPs

Bacteria

Metallic Ag NPs have antimicrobial properties, and are the most commonly used nanomaterials in consumer products (see [28, 29] and references therein). The widespread use of Ag NPs will increase the load up of silver into wastewater treatment plants. Predicted environmental concentration (PEC) for Ag NPs are 0.5–2 ngL^{-1} in surface waters, 32–111 ngL^{-1} in sewage treatment plant effluents, and 1.3–4.4 $mg.kg^{-1}$ in sewage sludge. Recently, silver sulfide NPs have been evidenced in the reducing and sulfur-rich environment of sewage sludge.

A wide range of Ag NPs sizes, shapes, crystal structure, and surface chemistry were exposed to bacteria at different pH and ionic strength. The toxicity proceeds through release of free silver ions associated with specific effects of Ag NPs. Ag ions inhibit the oxidation of endogenous substrates by *E. coli* and cause efflux of accumulated phosphates and metabolites. Ag^+ interferes with the respiratory chain, induces the collapse of the proton motive force, the de-energization of the cell, the disruption of DNA replication, and disturbs phosphate uptake. Silver binds to proteins or enzymes and also competes for cell surface binding sites of bacterial cells with metal ions, such as copper. Ag NPs attach to the surface of the Gram-negative membranes, can penetrate inside the bacteria, and drastically disturb the cell permeability and respiration, dissipate the proton motive force, collapse the plasma membrane potential, and deplete intracellular ATP levels. The mode of action of Ag NPs ($\varnothing$ 9.3 nm) 0.4 nM is similar to that of Ag^+ ions 6 µM. Free silver and Ag NPs generate ROS and cell membrane damage. Using superoxide anion sensor *E. coli* [14] showed that generation of ROS by Ag NPs contributed to the toxicity. Interestingly, extracellular polymeric substances produced by bacteria, neutralized free silver ions, and decreased the toxicity of Ag NPs [30]. River humic acids also mitigated the bactericidal action on *P. fluorescens*.

Size-, crystallinity-, and shape-dependent effects were observed in terms of antibacterial activity and ROS production. Direct interaction with the bacteria preferentially occurred for NPs ranging within 1–10 nm. Among tested Ag NPs sizes (10, 30–40, and 100 nm), only the 10 nm NPs inhibited methicillin-resistant *S. aureus* (the main cause of nosocomial infections worldwide) [31].

Quantum dots (QDs) are colloidal semiconductors with composition- and size-dependent absorption and emission. They are used as biological probes cellular imaging, tumor targeting, and diagnostics. The minimum inhibitory concentrations of ZnO NPs against *E. coli*, *Pseudomonas aeruginosa*, and *Staphylococcus aureus* were found between 125 and 500 $µg.mL^{-1}$ [32]. ZnO NPs induce toxicity via the dissolution and release of Zn^{2+} ions [6, 33]. The toxicity depends on the media composition (pH, ionic strength) and the illumination conditions. Both Gram-negative and Gram-positive membranes are disorganized after interactions with ZnO NPs [6, 34]. The increase in membrane permeability leads to the internalization of NPs [34]. A study using mutant strains has shown that CdSe and CdSe/ZnS QDs ($\varnothing$ <5 nm) can penetrate into *B. subtilis* and *E. coli* upon purine-processing mechanisms [35]. Extracellular polymeric substances excreted by bacteria prevent NPs internalization and decrease the toxicity observed [36]. Interestingly, ZnO NPs inside the cell are less crystallized than the initial NPs [34]. This raises the question of whether there is a direct internalization of the NPs, or a dissolution/precipitation within the cells.

CdTe QDs also induce toxicity toward Gram-positive methicillin-resistant *Shewanella aureus* and *B. subtilis*, with EC_{50} concentrations tenfold higher than toward Gram-negative strains *E. coli* and *P. aeruginosa* [37]. CdTe NPs bind to bacteria and impair the antioxidant system by photogeneration of ROS [37]. In that case, the release of Cd^{2+} has little or no role in toxicity of CdTe NPs [37], but the formation of TeO_2 and CdO by surface oxidation may cause cytotoxicity [38].

CeO_2 NPs are used as diesel fuel additives, polishing surface agents, as well as in coating and microelectronics. CeO_2 are nonstoichiometric NPs with cerium atoms in +4 and +3 oxidation states, and oxygen vacancies [39]. EC_{50} was near 5 mg L^{-1} for *E. coli* and ranged from 0.27 to 6.3 $mg.L^{-1}$ for *Anabaena* in pure water [40]. Under exposure to CeO_2 NPs, *N. europaea* cells have shown larger sedimentation coefficient than the control [6]. The toxicity of CeO_2 NPs is exerted by direct contact with cells [40], membrane damages [6], cell disruption [40], and release of free Ce^{3+} [41]. Even if free Ce^{3+} is toxic for *Anabaena* in pure water, it could not explain by itself the cytotoxicity observed. However, the reduction of the Ce^{4+} into Ce^{3+} at the surface of the CeO_2 NPs correlates with the cytotoxicity toward *E. coli*. No oxidative stress response was detected with *E. coli* or *B. subtilis*, but CeO_2 NPs and $CeCl_3$ alter the electron flow, and the respiration of the bacteria [42]. Pelettier et al. [42] also observed the disturbance of genes involved in sulfur metabolism, and an increase in the levels of cytochrome terminal oxidase (*cydAB*) transcripts known to be induced by iron limitation.

Algae

The short-term toxicity of Ag NPs ($\varnothing$ 25 nm) and $AgNO_3$ was tested toward the freshwater algae *Chlamydomonas reinhardtii* [43]. The toxicity measured by EC_{50} for Ag^+ was 18 times larger than for Ag NPs and much higher than those determined to inhibit algal growth in other studies. The toxic effects observed here reflect toxicity to photosynthesis. TEM showed no aggregation of Ag NPs. However, the measured Ag^+ concentration in the suspension of Ag NPs by itself cannot account for the observed toxicity since only 1% of the Ag had been oxidized [43]. Interestingly, once normalized by the concentration of dissolved Ag^+, the toxicity of Ag NPs was more important than that of $AgNO_3$. This suggests the importance of interactions between the surfaces of Ag NPs and algae cells.

In contrast, Ag^+ released from the oxidative dissolution of Ag NPs ($\varnothing$ 60–70 nm) severely suppressed cell growth, photosynthesis, and chlorophyll production in marine diatom *T. weissflogii* [44]. In both studies [43, 44] Ag NPs were not found to be toxic to either freshwater or marine unicellular algae at concentrations that could be expected in contaminated water bodies.

Regarding ZnO NPs, their toxicity toward the freshwater algae *P. subcapitata* was solely attributed to rapid dissolution at pH 7.6 with a saturation solubility in the range of 1 mg/L [45]. Toxicity of ZnO NPS ($\varnothing$ 20–30 nm) was assessed on marine phytoplankton (diatoms, chlorophytes, and prymnesiophytes) [20]. Even in saline water, the toxicity was related to dissolution, release, and uptake of free Zn ions.

Protozoa

Studies on adverse effects of NPs against unicellular eukaryotes are scarce. The toxicity of Ag NPs ($\varnothing$ 30–40 nm) on the Ciliate Protozoa *Paramecium caudatum* was reported by Kvitek et al. [46]. No toxicity was observed below 39 mg.L^{-1}, whereas ionic silver induced toxicity above 0.4 mg.L^{-1}. The authors noticed that the surfactant/polymer modification strongly affected the toxic behavior of Ag NPs.

The toxicity of CuO NPs (30) was investigated on Protozoans *Tetrahymena thermophila* [47]. A clear accumulation of the CuO NPs was observed in their food vacuoles. In all cases toxic effects of Ag and CuO NPs to protozoa were caused by their solubilized fraction. Interestingly, experiment performed in natural water versus artificial water remarkably (up to 140-fold) decreased the toxicity of CuO NPs depending mainly on the concentration of dissolved organic carbon.

Invertebrates

Terrestrial Invertebrates The effects of Cu NPs ($\varnothing$ 20–40 nm, and < 100 nm) and Cu ions were assessed on *Eisenia* in artificial soil [48]. Both Cu ions and NPs are bioaccumulated suggesting a route of entry into terrestrial food chains. No subchronic toxicity of Cu NPs was observed at concentrations up to 65 mgCu.kg^{-1} soils. Increases in metallothionein expression occurred from 20 mg kg^{-1} of Cu NPs and 10 mg kg^{-1} of $CuSO_4$. An oxidized and amorphous layer at the surface of the Cu NPs appears immediately in the soil and depends on particle size (XANES-based experiment). This highlights the necessity to consider the surface chemistry of Cu NPs during nano-ecotoxicology studies since the toxicant is likely to be more reflective of CuO than metallic Cu. *Eisenia* was also exposed to $AgNO_3$ and Ag NPs (PVP-coated) [49]. *Eisenia* consistently avoid soils contaminated with Ag NPs or $AgNO_3$ with faster avoidance for $AgNO_3$. In natural soil, *Eisenia* accumulated significantly higher concentrations of Ag than in artificial soil. Ag NPs incubated in the soil were 10–17% Ag (I), suggesting that Ag ions may be responsible for the effects observed with Ag NPs (EXAFS-based experiment). The earthworm *Lumbricus terrestris* was exposed to Ag NPs. Increased apoptotic activity was detected in a range of tissues both at acute and sublethal concentrations (down to 4 mg.kg^{-1} soil).

Comparing exposure in water and soil showed reduced bioavailability in soil reflected in the apoptotic response [50]. Toxicity of Ag NPs was also observed on *Drosophila melanogaster* in terms of egg development and pupae hatching. Ahamed et al. [51] show that coated-Ag NPs induce heat shock stress, oxidative stress, DNA damage, and apoptosis toward *D. melanogaster*.

C. elegans is a good model to assess the toxicity of chemically instable NPs at low doses. For Ag NPs, growth inhibition occurs at concentrations in the low mg.L^{-1} levels [52], and the metallothionein-deficient strain exhibited greater sensitivity than wild-strain. Both Ag and ZnO NPs release Zn and Ag ions that are responsible for most of the toxicity, but the NPs themselves did also cause toxicity on the nematode [52, 53]. CeO$_2$ NPs ($\varnothing$ 8.5 nm) also affect *C. elegans* at low doses inducing a decrease in the lifespan (-12% at 1 nM), ROS accumulation, and oxidative damage [54]. The best developed paradigm to explain these toxic effects are an oxidative stress and the ROS generation when the CeO$_2$ NPs reversibly alternate between Ce$_{(3+)}$ and Ce$_{(4+)}$ [54]. To date, further studies are needed to demonstrate whether the surface reactivity of CeO$_2$ NPs could trigger the generation of ROS or not.

Freshwater Invertebrates Heinlaan et al. [55] compared the toxicity of Cu and ZnO NPs between two branchiopods, *Daphnia magna* and an anostraca *Thamnocephalus platyurus*. The L(E)C$_{50}$ values of Cu NPs (90–224 mg.Cu.L^{-1}) are greatly higher than ZnO NPs (1.1–16 mg.L^{-1}). The toxicity of Cu NPs was also compared to Ag NPs [56] on *Ceriodaphnia dubia*. Whatever the water matrix tested, Ag and CuO NPs were highly toxic to the water flea. The lower toxicity of Ag NPs is explained by the silver complexation with dissolved organic compounds naturally present in the tested waters [57].

A TEM histological study on *Daphnia magna* observed the dispersion of Cu NPs in the midgut [58]. Questions raised on the respective localization of NPs between the peritrophic membrane and the midgut epithelium microvilli. This was completed by a study on the biokinetic uptake and efflux [59] showing that more than 70% of Ag NPs accumulate in the daphnids through ingestion. This highlights the importance of NPs transport along the aquatic food chain.

One study focused on *Hydra attenuata* has shown that CdSe/Cd QDs induced nonsynchronous tentacle retraction. The tentacle-writhing behavior is related to the position of metal core inside the QDs [60]. The organic coated QDs were able to enter into the head after 1 h, and the entire body within 24 h. The freshwater mussels, *Elliptio complanata*, are also disturbed by CdTe QDs: decrease in viability, lipid peroxidation of the gills and gut, and fall down of the immune activity of hemocytes [61]. The toxic effects are linked to the toxicity of their metal core, the coating, and dissolution rate. However, Velzeboer et al. [17] do not observe clear effects at concentrations up to 100 mg.L^{-1} toward the cladoceran *Chydorus sphaericus* exposed to CdSe/ZnS QDs. This lack of toxicity is explained by the aggregation of the NPs that decreases the bioavailability.

Ag NPs are efficient larvicides for mosquitoes involved in malaria fever. Sap-Iam et al. [62] found that 90% of *Aedes aegypti* larvae died after 3 h contact with 5 mg.L^{-1} of Ag NPS ($\varnothing$ 10 nm, PMA-coated). The involved mechanism is the penetration of Ag NPs through the larval membrane. Ag NPs in the intracellular space can bind to sulfur-containing proteins or phosphorous-containing compounds like DNA, leading to the denaturation of organelles and enzymes. However, Ag NPs had no effects on the hatchability of mosquito eggs due to the chitinized serosal cuticle that prevents the penetration within the eggs.

Marine Invertebrates A comparison between CuO NPs and the soluble form of Cu was done on *Scrobicularia plana* and *Nereis diversicolor* [63]. The annelid and bivalve organisms showed different behaviors and biochemical responses, but oxidative stress was observed for both species (increase in catalase and glutathione transferase activities).

Toxicity of QDs was also assessed on a coastal marine and salt lake rotifer, *Brachionus manjavacas* [64]. Population exposed to 0.3 μg.mL^{-1} of QDs ($\varnothing$ 37 nm) exhibited 50% of decrease in the growth rate while larger particles ($\varnothing$ up to 3,000 nm) caused no change. These larger particles remained confined in the gut, indicating NPs sizes as a critical factor in the internalization by tissue. Moreover, transfer of the F$_1$ offspring from NPs-exposed maternal females into NPs-free media demonstrated that NPs are cleared from the animals without residual adverse effects [64].

Chemically Stable NPs: The Example of Gold

Metallic Au NPs are used as contrast agents in electron microscopy, optical sensors, catalysts, and for therapeutic uses. Au metal is extremely resistant to oxidation, and essentially insoluble under ambient conditions. Although Au is resistant to oxidative dissolution, excess Au ions can remain in the suspension at the end of the synthesis process. Therefore, the possibility that Au associated with organisms was taken up as Au ions cannot be discounted without a determination of the speciation of Au in the exposure media [65].

Bacteria

Toxicity of Au NPs has been studied on *S. typhimurium*, *Salmonella enteritidis*, *Streptococcus mutans*, *E. coli*, and *Listeria monocytogenes*. Ammonium functionalized Au NPs ($\varnothing$ 2 nm) showed 3.1 ± 0.6 μM LC_{50} on *E. coli*. The toxicity is related to the strong electrostatic attraction between the negatively charged bilayer and the positively charged NPs [66]. This affinity for the surface was also observed with *S. typhimurium* strain TA 102 [67]. No toxic or mutagenic effects were observed. However, Au NPs demonstrated a Minimum Inhibitory Concentration (MIC) at 97 μg.mL^{-1} on *S. mutans*, which is a possible infectious etiology in dental caries. As a comparison, silver showed an average MIC of 4.86 μg.mL^{-1}.

Algae

One study focused on the impacts of Au NPs ($\varnothing$ 10 nm, amine-coated) toward the marine algae *Scenedesmus subspicatus*. LD_{50} value was reached at $1.6 \ 10^5$ Au NPs.cells^{-1} ($\sim$ 2.4 mg.L^{-1}). Au NPs were strongly adsorbed onto the cell wall leading to progressive intracellular and wall disturbances [68].

Invertebrates

Terrestrial Invertebrates Few in vivo studies document the interactions between Au NPs and terrestrial organisms. One recent paper evidences that Au NPs are bioavailable from soil to the model detritivore *Eisenia fetida* and have the potential to enter terrestrial food webs [65]. They are distributed among tissues, and affect the earthworm reproduction. Using several sizes of Au NPs, it was found that primary particle size ($\varnothing$ 20 or 55 nm) did not influence accumulated concentrations on a mass basis, while on a particle number basis the 20 nm NPs were more bioavailable. Differences in bioavailability can be related to the aggregation behavior in porous media [65].

The impacts of Au NPs ($\varnothing$ 50 nm) on the central nervous system (CNS) of male cockroach *Blaberus discoidalis* was assessed [69]. Negatively charged Au NPs interacted with the roach, transferred inside the nerve cord within 17 days, and encapsulated by the proteins of the CNS. No major impact on the life expectancy was observed after 2 months, whereas the locomotion was affected.

The effects toward *Drosophila melanogaster* wild-type (Oregon R+) were also studied using Au NPs (citrate-capped) [70]. Ingested Au NPs induced a strong reduction of the lifespan and fertility, DNA fragmentation, and overexpression of the stress proteins HSP_{70}. Au NPs were homogeneously distributed through the deep layers of the enteric tissue and in the reproductive organs. Most of them were in the endosomes near the lamellar structures of the rough endoplasmic reticulum. In the germinal tissues, Au NPs were homogeneously distributed in the spermatocytes and flagella, but only in the cytoplasm of the ovarioles. No data are available on the motility of spermatozoids [70].

Freshwater Invertebrates The impacts of Au NPs (amine-coated, $\varnothing$ 10 nm) was investigated on the freshwater benthic bivalve *Corbicula fluminea* [68]. Contamination via trophic exposure (7 days with *Scenedesmus subspicatus*) induced internalization through the bronchial and digestive barriers. The histological penetration of Au NPs in the stomach cells (cytoplasm and nucleus), or in epithelial cells of digestive gland (lysosomal vesicles) depended on the speed and intensity of the digestion. Strong bioconcentration factors were obtained in gills and visceral mass, respectively. Interestingly, dose-dependent genetic expressions were observed on the gills of *Corbicula* but not in the visceral mass [68]. At low doses [$1.6 \ 10^2$ Au NPs.cells^{-1}], there was an overexpression of genes associated to oxidative stress, while for [$1.6 \ 10^4$ Au NP.cells^{-1}], there was also decrease in the mitochondrial activity and respiratory rate. For the highest concentration [$1.6 \ 10^5$ Au NPs.cells^{-1}], biochemical mechanisms of detoxification were activated (changes in the glutathione transferase expression and metallothionein concentrations).

Marine Invertebrates Relatively little is known about how Au NPs nanoparticles interact with marine organisms. Au NPs ($\varnothing$ 15 nm) incubated with *Mytilus edulis* were detected in the digestive gland due to impaired feeding [71]. This suggests that Au NPs cause a modest level of oxidative stress sufficient to oxidize thiols in glutathione and proteins but without causing lipid peroxidation or induction of thioredoxin reductase activity. However, the mechanisms by which Au NPs might cause oxidative stress are still poorly understood.

Interspecies Relationships

Effects on Biofilms and Bacterial Communities
Effects of NPs were assessed toward biofilms and communities. In these complex systems, the microbial responses do not always correlate with results from studies performed on monocultures in vitro. For instance, TiO_2 NPs from wastewater and seawater sludge have low toxicity on *V. fischeri* (see [4] and references therein). However, low concentration (5.3 mg.L^{-1}) of TiO_2 NPs added to a stream microcosm under ambient UV radiation [72] damaged the cell walls of planktonic microorganisms due to intercellular action of ROS and accumulated in the benthic biofilms. Cell membrane damages were more pronounced in free-living cells than in biofilm cells, indicating the protective role of cell encapsulation against TiO_2 NPs. In biofilms, the surface properties of NPs govern their mobility, diffusion, and distribution, and structural parameters such as roughness and cell distribution of biofilm can be affected. Ag NPs also induced sloughing of *P. putida* biofilms. The presence of fulvic acid (which could play the role of a natural surfactant) reduced sloughing but increased uptake and bioaccumulation of the NPs [28]. Au NPs were also aggregated within flagella or biofilm network made from *Salmonella enteritidis* and *Listeria monocytogenes*. While bacterial cell walls and cytoplasm membrane were disintegrated, Au NPs did not penetrate into the bacteria [73].

Nitrifying bacteria are main players in wastewater treatment plants and in soil. Ag NPs concentrations lower than 1 mg.L^{-1} inhibit nitrifying organisms taken from a nitrifying reactor. This decrease correlates with the fraction of Ag NPs less than 5 nm, the intracellular ROS concentrations, but not with the photocatalytic ROS fractions (see [29] and references therein).

Soil bacterial communities were also studied after exposure to ZnO NPs (0.05, 0.1, and 0.5 mg.g^{-1} soil) in microcosms over 60 days. An alteration of the community structure and a decrease in the microbial biomass were observed [74]. The effects were higher than that of TiO_2 NPs, as reflected by lower DNA and stronger shifts in bacterial community composition, and modification of the soil enzyme activities [75]. However, while Ag NPs are bactericidal agents, bacterial community of estuarine sediments exposed to Ag NPs (for 20 days, 1 mg.L^{-1}, $\varnothing$ 58 nm) did not exhibit significant changes. This is related to the chloride amount in estuary water that alters the physicochemical behavior of Ag NPs [76] and the speciation of ionic silver.

Moreover, engineered NPs can be transferred from prey to predator. One study focused on bare CdSe QD accumulated in the *P. aeruginosa* bacteria, and transferred to and biomagnified in the *Tetrahymena thermophila* protozoa that prey on the bacteria. Cadmium concentrations in the protozoa predator were five times higher than their bacterial prey. However, QDs-treated bacteria affect the protozoa without lysing them. Consequently, intact QD remain available to higher trophic levels [77].

Effects of Invertebrate Assemblages
Land application of biosolids from wastewater treatment will be a major pathway for the introduction of manufactured NPs to the environment. The model organisms *Nicotiana tabacum* L. cv *xanthi* and *Manduca sexta* (tobacco hornworm) was used to investigate plant uptake and the potential for trophic transfer of Au NPs ($\varnothing$ 5, 10, 15 nm) [78]. This study presents the first evidence of trophic transfer of NPs from a terrestrial primary producer to a primary consumer as well as the first evidence of biomagnification of NPs within a terrestrial food web by mean factors of 6.2, 11.6, and 9.6 for the 5, 10, and 15 nm treatments, respectively.

Recent studies focused on freshwater simplified food webs. The first one regards carboxylated and biotinylated QDs transferred to rotifers *Brachionus calyciflorus* through dietary uptake of ciliated protozoans *Tetrahymena* interacting with *E. coli* [79]. QDs accumulation was limited in the ciliates and no enrichment was observed in the rotifers. The contaminated *E. coli* formed aggregates that were not eaten by *Tetrahymena*. No toxicity was observed in single cells *Tetrahymena* nor in *Brachionus calyciflorus*. Another study regards the trophic transfer of TiO_2 NPs in the food chain including *Daphnia magna* and its predator, the zebra fish *Danio rerio* [80]. Eight-day-old *D. magna* were exposed to TiO_2 ($\varnothing$ 21 nm, 0.1 and 1 mg.L^{-1}) and transferred to the 3-month-old *D. rerio* tanks. No biomagnification of TiO_2 NPs from *Daphnia* to zebra fish was found, but the dietary intake was found to be one of the main NPs exposure route for higher trophic level aquatic organisms.

However these studies regarding short trophic links do not take into account important parameters such as the colloidal destabilization of the NPs, the interaction between the surface of the NPs and (in)organic molecules naturally occurring or bio-excreted, or the flux between compartments of the ecosystems (aqueous phase, sediments, biota). To work under more realistic scenarios of exposure, few studies are performed on mesocosms. Ferry et al. [81] were the first to study the fate of positively charged Au NPs (65 nm × 15 nm) in estuarine mesocosms containing sediments, biofilms, primary producers, filter feeders, grazers, and omnivores. While water and sediments represented 99% of the total mass of the system, on a per mass basis the filter feeders and biofilms were the most effective sink of Au NPs. The estimated mass of biofilms was less than 0.5% but they recovered 60% of NPs. This high affinity is related to the negative surface charge of the biofilm interacting with the positively charged Au NPs and to the large surface area developed by the biofilms. These results are interesting since biofilms offer (1) a route into the food web through grazing by detritivores and (2) a route for mineralization through biofilm calcification.

Conclusion: Future Directions

Past experience with chemicals (e.g., methyl mercury, DDT, and PCBs) revealed dietary uptake at lower trophic levels and accumulation up the *food chain* to be an important route of contaminant exposure. All the observations that NPs can biomagnify highlight the importance of considering *dietary uptake* as a pathway for NPs exposure and raises questions about potential *ecoreceptor and human exposure* to NPs from long-term land and chronical exposure.

Ecotoxicity studies carried out at concentrations much higher than would ever be expected in the environment provide necessary data regarding the mechanisms of NPs toxicity. However complementary studies on the impact of NPs at concentrations comparable to those encountered in an environmental situation are required. To date, one challenge is to combine data from bioindicator organisms reflecting environmental health, and suitable biomarkers enlightening physiological processes [82].

Acknowledgements The authors would like to thank the CNRS and CEA for funding the International Consortium for the Environmental Implications of NanoTechnology and also the NSF and the US-EPA for funding the Center for the Environmental Implications of NanoTechnology. They also acknowledge financial support from the French National Agency (ANR) in the frame of the PNANO/NANAN, and P2N/MESONNET projects. We also apologize to the authors whose work was not cited. Due to the limitation of the number of references, we favored the citation of reviews and the most recent articles.

Cross-References

- ► Effect of Surface Modification on Toxicity of Nanoparticles
- ► Fate of Manufactured Nanoparticles in Aqueous Environment
- ► Nanoparticle Cytotoxicity
- ► Nanostructures for Surface Functionalization and Surface Properties
- ► Physicochemical Properties of Nanoparticles in Relation with Toxicity

Glossary

EC$_{50}$ The half maximal effective concentration referring to the concentration that induces a response halfway between the baseline and maximum after some specified exposure time.

EXAFS Extended X-ray Absorption Fine Structure.

IC$_{25}$ The inhibition concentration 25% is the statistical analysis used in chronic whole effluent toxicity tests to estimate the sublethal effects of the effluent sample. IC$_{25}$ estimates of the concentration causing 25% reduction in growth or reproduction of test organisms.

LC$_{50}$ The lethal concentration 50% is the dose required to kill half the members of a tested population for a test duration. It is a general indicator of a substance's acute toxicity.

LD$_{50}$ The median lethal dose 50%.

MIC Minimum inhibitory concentration.

NOEC No observed effect concentration, also called a no observable effect level (NOEL).

PEC Predicted environmental concentration.

ROS Reactive oxygen species.

TEM Transmission Electron Microscope.

XANES Xray Absorption Near Edge Structure.

References

1. Kahru, A., Dubourguier, H.-C.: From ecotoxicology to nanoecotoxicology. Toxicology **269**(2–3), 105–119 (2010)
2. Auffan, M., Rose, J., Bottero, J.Y., Lowry, G., Jolivet, J.P., Wiesner, M.R.: Towards a definition of inorganic nanoparticles from an environmental, health, and safety perspective. Nat. Nanotechnol. **4**, 634–641 (2009)
3. Auffan, M., Rose, J., Wiesner, M.R., Bottero, J.Y.: Chemical stability of metallic nanoparticles: a parameter controlling their potential toxicity in vitro. Environ. Pollut. **157**, 1127–1133 (2009)
4. Sharma, V.K.: Aggregation and toxicity of titanium dioxide nanoparticles in aquatic environment – a review. J. Environ. Sci. Health A Tox. Hazard. Subst. Environ. Eng. **44**(14), 1485–1495 (2009)
5. Kaegi, R., Sinnet, B., Zuleeg, S., Hagendorfer, H., Mueller, E., Vonbank, R., Boller, M., Burkhardt, M.: Release of silver nanoparticles from outdoor facades. Environ. Pollut. **158**(9), 2900–2905 (2010)
6. Fang, X.H., Yu, R., Li, B.Q., Somasundaran, P., Chandran, K.: Stresses exerted by ZnO, CeO$_2$ and anatase TiO$_2$ nanoparticles on the Nitrosomonas europaea. J. Colloid Interface Sci. **348**(2), 329–334 (2010)
7. Kumar, A., Pandey, A.K., Singh, S.S., Shanker, R., Dhawan, A.: Cellular uptake and mutagenic potential of metal oxide nanoparticles in bacterial cells. Chemosphere (2011). doi:10.1002/cyto.a.21085
8. Matsunaga, T., Okochi, M.: TiO$_2$-mediated photochemical disinfection of Escherichia coli using optical fibers. Environ. Sci. Technol. **29**, 501–505 (1995)
9. Rincón, A.-G., Pulgarin, C.: Bactericidal action of illuminated TiO$_2$ on pure Escherichia coli and natural bacterial consortia: post-irradiation events in the dark and assessment of the effective disinfection time. Appl. Catal. B Environ. **49**, 99–112 (2004)
10. Adams, L.K., Lyon, D.Y., Alvarez, P.J.J.: Comparative ecotoxicity of nanoscale TiO$_2$, SiO$_2$, and ZnO water suspensions. Water Res. **40**(19), 3527–3532 (2006)
11. Pan, X., Redding, J.E., Wiley, P.A., Wen, L., McConnell, J.S., Zhang, B.: Mutagenicity evaluation of metal oxide nanoparticles by the bacterial reverse mutation assay. Chemosphere **79**(1), 113–116 (2010)
12. Paoli, G.D.: A rapid GC-MS determination of gamma-hydroxybutyrate in saliva. J. Anal. Toxicol. **32**, 298–302 (2007)
13. Jiang, W., Mashayekhi, H., Xing, B.S.: Bacterial toxicity comparison between nano- and micro-scaled oxide particles. Environ. Pollut. **157**(5), 1619–1625 (2009)
14. Ivask, A., Bondarenko, O., Jepihhina, N., Kahru, A.: Profiling of the reactive oxygen species-related ecotoxicity of CuO, ZnO, TiO$_2$, silver and fullerene nanoparticles using a set of recombinant luminescent Escherichia coli strains. Anal. Bioanal. Chem. **398**(2), 701–716 (2010)
15. Horst, A.M., Neal, A.C., Mielke, R.E., Sislian, P.R., Suh, W.H., Madler, L., Stucky, G.D., Holden, P.A.: Dispersion of TiO nanoparticle agglomerates by Pseudomonas aeruginosa. Appl. Environ. Microbiol. **76**(21), 7292–7298 (2010)
16. Warheit, D.B., Hoke, R.A., Finlay, C., Donner, E.M., Reed, K.L., Sayes, C.M.: Development of a base set of toxicity tests using ultrafine TiO$_2$ particles as a component of nanoparticle risk management. Toxicol. Lett. **171**, 99–110 (2007)
17. Velzeboer, I., Hendriks, A.J., Ragas, A.M., van de Meent, D.: Aquatic ecotoxicity tests of some nanomaterials. Environ. Toxicol. Chem. **27**(9), 1942–1947 (2008)
18. Aruoja, V., Dubourguier, H.C., Kasemets, K., Kahru, A.: Toxicity of nanoparticles of CuO, ZnO and TiO$_2$ to microalgae Pseudokirchneriella subcapitata. Sci. Total Environ. **407**(4), 141–146 (2009)
19. Hund-Rinke, K., Simon, M.: Ecotoxic effect of photocatalytic active nanoparticles (TiO$_2$) on algae and daphnids. Environ. Sci. Poll. Res. **13**, 1–8 (2006)
20. Miller, R.J., Lenihan, H.S., Muller, E.B., Tseng, N., Hanna, S.K., Keller, A.A.: Impacts of metal oxide nanoparticles on marine phytoplankton. Environ. Sci. Technol. **44**, 7329–7334 (2010)
21. Jemec, A., Tisler, T., Drobne, D., Sepcic, K., Jamnik, P., Ros, M.: Biochemical biomarkers in chronically metal-stressed daphnids. Comp. Biochem. Physiol. **147C**, 61–68 (2008)
22. Drobne, D., Jemec, A., Tkalec, Z.P.: In vivo screening to determine hazards of nanoparticles: nanosized TiO$_2$. Environ. Poll. **157**(1157–1164) (2009)
23. Hu, C.W., Li, M., Cui, Y.B., Li, D.S., Chen, J., Yang, L.Y.: Toxicological effects of TiO$_2$ and ZnO nanoparticles in soil on eartworm Eisenia fetida. Soil Biol. Biochem. **42**, 586–591 (2010)
24. Heckman, L.-H., Hovgaard, M.B., Sutherland, D.S., Autrup, H., Besenbacher, F., Scott-Fordsmand, J.J.: Limit-test toxicity screening of selected inorganic nanoparticles to the earthworm Eisenia fetida. Ecotoxicology **20**, 226–233 (2011)
25. Musee, N., Oberholster, P.J., Sikhwivhilu, L., Botha, A.-M.: The effects of engineered nanoparticles on survival, reproduction, and behavior of freshwater snail, Physa acuta (Draparnaud, 1805). Chemosphere **81**, 1196–1203 (2010)
26. Galloway, T., Lewis, C., Dolciotti, I., Johnston, B.D., Moger, J., Regoli, F.: Sublethal toxicity of nano-titanium dioxide and carbon nanotubes in a sediment dwelling marine polychaete. Environ. Poll. **158**, 1748–1755 (2010)
27. Canesi, L., Fabbri, R., Gallo, G., Vallotto, D., Marcomini, A., Pojana, G.: Biomarkers in Mytilus galloprovincialis exposed to suspensions of selected nanoparticles (Nano

carbon black, C60 fullerene, Nano-TiO$_2$, Nano-SiO$_2$). Aquat. Toxicol. **100**, 168–177 (2010)

28. Fabrega, J., Fawcett, S.R., Renshaw, J.C., Lead, J.R.: Silver nanoparticle impact on bacterial growth: effect of pH, concentration, and organic matter. Environ. Sci. Technol. **43**(19), 7285–7290 (2009)

29. Marambio-Jones, C., Hoek, E.: A review of the antibacterial effects of silver nanomaterials and potential implications for human health and the environment. J. Nanopart. Res. **12**(5), 1531–1551 (2010)

30. Dimkpa, C.O., Calder, A., Gajjar, P., Merugu, S., Huang, W., Britt, D.B., McLean, J.E., Johnson, W.P., Anderson, A.J.: Interaction of silver nanoparticles with an environmentally beneficial bacterium, *Pseudomonas chlororaphis*. J. Hazard. Mater. **188**, 428–435 (2011)

31. Ayala-Núñez, N., Lara Villegas, H., del Carmen Ixtepan Turrent, L., Rodríguez Padilla, C.: Silver nanoparticles toxicity and bactericidal effect against methicillin-resistant *Staphylococcus aureus*: nanoscale does matter. Nanobiotechnol. **5**(1), 2–9 (2009)

32. Premanathan, M., Karthikeyan, K., Jeyasubramanian, K., Manivannan, G.: Selective toxicity of ZnO nanoparticles toward gram-positive bacteria and cancer cells by apoptosis through lipid peroxidation. Nanomed. Nanotech. Biol Med **7**(2), 184–192 (2011)

33. Li, M., Zhu, L.Z., Lin, D.H.: Toxicity of ZnO nanoparticles to *Escherichia coil*: mechanism and the influence of medium components. Environ. Sci. Technol. **45**(5), 1977–1983 (2011)

34. Huang, Z., Zheng, X., Yan, D., Yin, G., Liao, X., Kang, Y., Yao, Y., Huang, D., Hao, B.: Toxicological effect of ZnO nanoparticles based on bacteria. Langmuir **24**, 4140–4144 (2008)

35. Kloepfer, J.A., Mielke, R.E., Nadeau, J.L.: Uptake of CdSe and CdSe/ZnS quantum dots into bacteria *via* purine-dependent mechanisms. Appl. Environ. Microbiol. **71**(5), 2548–2557 (2005)

36. Brayner, R., Dahoumane, S.A., Yepremian, C., Djediat, C., Meyer, M., Coute, A., Fievet, F.: ZnO nanoparticles: synthesis, characterization, and ecotoxicological studies. Langmuir **26**(9), 6522–6528 (2010)

37. Dumas, E., Gao, C., Suffern, D., Bradforth, S.E., Dimitrijevic, N.M., Nadeau, J.L.: Interfacial charge transfer between CdTe quantum dots and gram negative *vs* gram positive bacteria. Environ. Sci. Technol. **44**(4), 1464–1470 (2010)

38. Schneider, R., Wolpert, C., Guilloteau, H., Balan, L., Lambert, J., Merlin, C.: The exposure of bacteria to CdTe-core quantum dots: the importance of surface chemistry on cytotoxicity. Nanotechnology **20**(22), 225101 (2009)

39. Schubert, D., Dargusch, R., Raitano, J., Chan, S.-W.: Cerium and yttrium oxide nanoparticles are neuroprotective. Biochem. Biophys. Res. Commun. **342**(1), 86–91 (2006)

40. Rodea-Palomares, I., Boltes, K., Fernandez-Pias, F., Leganas, F., Garcia-Calvo, E., Santiago, J., Rosal, R.: Physicochemical characterization and ecotoxicological assessment of CeO$_2$ nanoparticles using two aquatic microorganisms. Toxicol. Sci. **119**(1), 135–145 (2011)

41. Zeyons, O., Thill, A., Chauvat, F., Menguy, N., Cassier-Chauvat, C., Orear, C., Daraspe, J., Auffan, M., Rose, J.,

Spalla, O.: Direct and indirect CeO$_2$ nanoparticles toxicity for *E. coli* and *Synechocystis*. Nanotoxicology **3**(4), 284–295 (2009)

42. Pelletier, D.A., Suresh, A.K., Holton, G.A., McKeown, C.K., Wang, W., Gu, B.H., Mortensen, N.P., Allison, D.P., Joy, D.C., Allison, M.R., Brown, S.D., Phelps, T.J., Doktycz, M.J.: Effects of engineered cerium oxide nanoparticles on bacterial growth and viability. Appl. Environ. Microbiol. **76**(24), 7981–7989 (2010)

43. Navarro, E., Piccapietra, F., Wagner, B., Marconi, F., Kaegi, R., Odzak, N., Sigg, L., Behra, R.: Toxicity of silver nanoparticles to *Chlamydomonas reinhardtii*. Environ. Sci. Technol. **42**(23), 8959–8964 (2008)

44. Miao, A.-J., Schwehr, K.A., Xu, C., Zhang, S.-J., Luo, Z., Quigg, A., Santschi, P.H.: The algal toxicity of silver engineered nanoparticles and detoxification by exopolymeric substances. Environ. Poll. **157**, 3034–3041 (2009)

45. Franklin, N.M., Rogers, N.J., Apte, S.C., Batley, G.E., Gadd, G.E., Casey, P.S.: Comparative toxicity of nanoparticulate ZnO, bulk ZnO, and ZnCl$_2$ to a freshwater microalga (*Pseudokirchneriella subcapitata*): the importance of particle solubility. Environ. Sci. Technol. **41**(24), 8484–8490 (2007)

46. Kvitek, L., Vanickova, M., Panacek, A., Soukupova, J., Dittrich, M., Valentova, E., Prucek, R., Bancirova, M., Milde, D., Zboril, R.: Initial study on the toxicity of silver nanoparticles (NPs) against *Paramecium caudatum*. J. Phys. Chem. **C113**, 4296–4300 (2009)

47. Blinova, I., Ivask, A., Heinlaan, M., Mortimer, M., Kahru, A.: Ecotoxicity of nanoparticles of CuO and ZnO in natural water. Environ. Poll. **158**, 41–47 (2010)

48. Unrine, J.M., Tsyusko, O.V., Hunyadi, S.E., Judy, J.D., Bertsch, P.M.: Effects of particle size on chemical speciation and bioavailability of copper to earthworms exposed to copper nanoparticles. J. Environ. Qual. **39**(6), 1942–1953 (2010)

49. Shoults-Wilson, W.A., Reinsch, B.C., Tsyusko, O.V., Bertsch, P.M., Lowry, G.V., Unrine, J.M.: Role of particle size and soil type in toxicity of silver nanoparticles to earthworms. Soil. Sci. Soc. Am. J. **75**(2), 365–377 (2011)

50. Lapied, E., Moudilou, E., Exbrayat, J.-M., Oughton, D.H., Joner, E.J.: Silver nanoparticle exposure causes apoptotic response in the earthworm *Lumbricus terrestris* (Oligochaeta). Nanomedicine **5**(6), 975–984 (2010)

51. Ahamed, M., Posgai, R., Gorey, T.J., Nielsen, M., Hussain, S.M., Rowe, J.J.: Silver nanoparticles induced heat shock protein 70, oxidative stress and apoptosis in *Drosophila melanogaster*. Toxicol. Appl. Pharmacol. **242**, 263–269 (2010)

52. Meyer, J., Lord, C.A., Yang, X.Y., Turner, E.A., Badireddy, A.R., Marinakos, S.M., Chilkoti, A., Wiesner, M., Auffan, M.: Intracellular uptake and associated toxicity of silver nanoparticles in *Caenorhabditis elegans*. Aquat. Toxicol. **100**, 140–150 (2010)

53. Wang, H., Wick, R.L., Xing, B.: Toxicity of nanoparticulate and bulk ZnO, Al2O3 and TiO2 to the nematode *Caenorhabditis elegans*. Environ. Poll. **157**, 1171–1177 (2009)

54. Zhang, H., He, X., Zhang, Z., Zhang, P., Li, Y., Ma, Y., Kuang, Y., Zhao, Y., Chai, Z.: Nano-CeO$_2$ exhibits adverse

effects at environmental relevant concentrations. Environ. Sci. Technol. **45**, 3725–3730 (2011)

55. Heinlaan, M., Ivask, A., Blinova, I., Dubourguier, H.-C., Kahru, A.: Toxicity of nanosized and bulk ZnO, CuO and TiO$_2$ to bacteria *Vibrio fischeri* and crustaceans *Daphnia magna and Thamnocephalus platyurus*. Chemosphere **71**(7), 1308–1316 (2008)

56. Gao, J., Youn, S., Hovsepyan, A., Llaneza, V., Wang, Y., Bitton, G.: Dispersion and toxicity of selected manufactured nanomaterials in natural river water samples: effects of water chemical composition. Environ. Sci. Technol. **43**, 3322–3328 (2009)

57. Rodgers, J.H., Deaver, E., Rogers, P.L.: Partitioning and effects of silver in amended freshwater sediments. Ecotox. Environ. Safe. **37**(1), 1–9 (1997)

58. Heinlaan, M., Kahru, A., Kasemets, K., Arbeille, B., Prensier, G., Dubourguier, H.-C.: Changes in the *Daphnia magna* midgut upon ingestion of copper oxide nanoparticles: a transmission electron microscopy study. Water Res. **45**, 179–190 (2011)

59. Zhao, C.-M., Wang, W.-X.: Biokinetic uptake and efflux of silver nanoparticles in *Daphnia magna*. Environ. Sci. Technol. **44**, 7699–7704 (2010)

60. Malvindi, M.A., Carbone, L., Quarta, A., Tino, A., Manna, L., Pellegrino, T.: Rod-shaped nanocrystals elicit neuronal activity in vivo. Small **4**(10), 1747 (2008)

61. Gagné, F., Auclair, J., Turcotte, P., Fournier, M., Gagnon, C., Sauvé, S., Blaise, C.: Ecotoxicity of CdTe quantum dots to freshwater mussels: impacts on immune system, oxidative stress and genotoxicity. Aquat. Toxicol. **86**(3), 333–340 (2008)

62. Sap-Iam, N., Homklinchan, C., Larpudomlert, R., Warisnoicharoen, W., Sereemaspun, A., Dubas, S.T.: UV irradiation-induced silver nanoparticles as mosquito larvicides. J. Appl. Sci. **10**(23), 3132–3136 (2010)

63. Griffitt, R.J., Weil, R., Hyndman, K.A., Denslow, N.D., Powers, K., Taylor, D., Barber, D.S.: Exposure to copper nanoparticles causes gill injury and acute lethality in zebrafish (*Danio rerio*). Environ. Sci. Technol. **41**(23), 8178–8186 (2007)

64. Snell, T.W., Hicks, D.G.: Assessing toxicity of nanoparticles using *Brachionus manjavacas* (Rotifera). Environ. Toxicol. **26**(2), 146–152 (2011)

65. Unrine, J.M., Hunyadi, S.E., Tsyusko, O.V., Rao, W., Shoults-Wilson, W.A., Bertsch, P.M.: Evidence for bioavailability of Au nanoparticles from soil and biodistribution within earthworms (*Eisenia fetida*). Environ. Sci. Technol. **44**(21), 8308–8313 (2010)

66. Goodman, C.M., McCusker, C.D., Yilmaz, T., Rotello, V.M.: Toxicity of gold nanoparticles functionalized with cationic and anionic side chains. Bioconjug. Chem. **15**, 897–900 (2004)

67. Wang, S., Lawson, R., Ray, P.C., Yu, H.: Toxic effects of gold nanoparticles on *Salmonella typhimurium* bacteria. Toxicol. Ind. Health **27**(6), 547–554 (2011)

68. Renault, S., Baudrimont, M., Mesmer-Dudons, N., Gonzalez, P., Mornet, S., Brisson, A.: Impacts of gold nanoparticle exposure on two freshwater species: a phytoplanktonic alga (*Scenedesmus subspicatus*) and a benthic bivalve (*Corbicula fluminea*). Gold Bull. **41**(2), 116–126 (2008)

69. Rocha, A., Zhou, Y., Kundu, S., Gonzalez, J.M., Vinson, S.B., Liang, H.: In vivo observation of gold nanoparticles in the central nervous system of *Blaberus discoidalis*. J. Nanobiotechnol. **9**, 1–9 (2011)

70. Pompa, P., Vecchio, G., Galeone, A., Brunetti, V., Sabella, S., Maiorano, G., Falqui, A., Bertoni, G., Cingolani, R.: In vivo toxicity assessment of gold nanoparticles in *Drosophila melanogaster*. Nano. Res. **4**, 405–413 (2011)

71. Tedesco, S., Doyle, H., Blasco, J., Redmond, G., Sheehan, D.: Exposure of the blue mussel, *Mytilus edulis*, to gold nanoparticles and the pro-oxidant menadione. Comp. Biochem. Physiol C Toxicol. Pharmacol. **151**(2), 167–174 (2010)

72. Battin, T.J., Fvd, K., Weilhartner, A., Ottofuelling, S., Hofmann, T.: Nanostructured TiO$_2$: transport behavior and effects on aquatic microbial communities under environmental conditions. Environ. Sci. Technol. **43**(21), 8098–8104 (2009)

73. Sawosz, E., Chwalibog, A., Szeliga, J., et al.: Visualization of gold and platinum nanoparticles interacting with *Salmonella enteritidis and Listeria monocytogenes*. Int. J. Nanomedicine **5**, 631–637 (2010)

74. Ge, Y., Schimel, J.P., Holden, P.A.: Evidence for negative effects of TiO$_2$ and ZnO nanoparticles on soil bacterial communities. Environ. Sci. Technol. **45**(4), 1659–1664 (2011)

75. Kim, S., Kim, J., Lee, I.: Effects of Zn and ZnO nanoparticles and Zn^{2+} on soil enzyme activity and bioaccumulation of Zn in *Cucumis sativus*. Chem. Ecol. **27**(1), 49–55 (2011)

76. Bradford, A., Handy, R.D., Readman, J.W., Atfield, A., Muhling, M.: Impact of silver nanoparticle contamination on the genetic diversity of natural bacterial assemblages in estuarine sediments. Environ. Sci. Technol. **43**(12), 4530–4536 (2009)

77. Werlin, R., Priester, J.H., Mielke, R.E., Krämer, S., Jackson, S., Stoimenov, P.K., Stucky, G.D., Cherr, G.N., Orias, E., Holden, P.A.: Biomagnification of cadmium selenide quantum dots in a simple experimental microbial food chain. Nat. Nanotechnol. **6**, 65–71 (2011)

78. Judy, J.D., Unrine, J.M., Bertsch, P.M.: Evidence for biomagnification of gold nanoparticles within a terrestrial food chain. Environ. Sci. Technol. **45**(2), 776–781 (2010)

79. Holbrook, R.D., Murphy, K.E., Morrow, J.B., Cole, K.D.: Trophic transfer of nanoparticles in a simplified invertebrate food web. Nat. Nanotechnol. **3**, 352–355 (2008)

80. Zhu, X., Wang, J., Zhang, X., Chang, Y., Chen, Y.: Trophic transfer of TiO$_2$ nanoparticles from Daphnia to zebrafish in a simplified freshwater food chain. Chemosphere **79**, 928–933 (2010)

81. Ferry, J.L., Craig, P., Hexel, C., Sisco, P., Frey, R., Pennington, P.L., Fulton, M.H., Scott, I.G., Decho, A.W., Kashiwada, S., Murphy, C.J., Shaw, T.J.: Transfer of gold nanoparticles from the water column to the estuarine food web. Nat. Nanotechnol. **4**(7), 441–444 (2009)

82. Saez, G., Moreau, X., De Jong, L., Thiéry, A., Dolain, C., Bestel, I., Di Giorgio, C., De Meo, M., Bartélémy, P.: Development of new nano-tools: towards an integrative approach to address the societal question of nanotechnology? Nano Today **5**, 251–253 (2010)

Ecotoxicology of Carbon Nanotubes Toward Amphibian Larvae

Florence Mouchet[1,2], Périne Landois[3], Floriane Bourdiol[1,2,3], Isabelle Fourquaux[4], Pascal Puech[5], Emmanuel Flahaut[3,6] and Laury Gauthier[1,2]

[1]EcoLab – Laboratoire d'écologie fonctionnelle et environnement, Université de Toulouse, INP, UPS, Castanet Tolosan, France

[2]CNRS UMR 5245, EcoLab, Castanet Tolosan, France

[3]Institut Carnot Cirimat, Université de Toulouse, UPS, INP, Toulouse cedex 9, France

[4]CMEAB, Centre de Microscopie Electronique Appliquée à la Biologie, Université Paul Sabatier, Faculté de Médecine Rangueil, Toulouse cedex 4, France

[5]CEMES, Toulouse Cedex 4, France

[6]CNRS, Institut Carnot Cirimat, Toulouse, France

Synonyms

Amphibian larvae; Carbon nanotubes; Ecotoxicity

Definition

According to Truhaut [1], ecotoxicology is defined as the branch of toxicology concerned with the study of toxic effects, caused by natural or synthetic pollutants, to the constituents of ecosystems, animals (including human), vegetables, and microbial, in an integral context. Carbon NanoTubes (CNT), a man-made allotrope of carbon, are considered as emergent potential contaminants into the environment. The aim of the ecotoxicology is then to evaluate the potential toxicity of CNT toward living organisms. Amphibian's larvae are used as biological models to study the potential toxicity of CNT.

Carbon NanoTubes (CNT) are a man-made allotrope of carbon (1D). They were discovered in 1991 by Iijima [2] and are intensively studied since 20 years by the scientific community. Their structure can be described as a graphene sheet rolled up to form a hollow cylinder. CNT properties depend on their structure. Two main types of CNT are distinguished [3, 4]: Single-Walled CNT (SWNT) consisting of a single sheet and Multi-Walled CNT (MWNT) consisting of several concentric layers, having a larger diameter (Fig. 1). CNT may be closed by half-fullerene-like molecules. Among the MWNT, Double-Walled CNT (DWNT) are at the frontier between SWNT and MWNT, with morphology and properties very close to SWNT. CNT have a diameter from 1 to 100 nm and a length from less than one μm up to tens of μm or more (Fig. 2). Due to their exceptional physical (mechanic, electronic, thermal) and chemical properties particularly because of their nano-size, interest in CNT has grown rapidly and application areas are very wide: nano-composites, electron transmitters, electromagnetic shielding, batteries, super-capacitors, gas storage (including hydrogen), and nano-medicine [5]. Their current applications in the daily life of developed society are numerous as prototypes (flat-screens, tyres, sport equipments, etc.) and some others are emerging (paints, technical clothes, pharmaceutical products, etc.). Global CNT production capacity was estimated in 2002 over 2.5 t per day, and the correspondent market was estimated at $12 millions for 2002 and was expected to grow up to $700 millions by 2005 [6]. The recent analysis of John Oliver [7] reports that the global market for CNT grades based on committed production reached $103 millions in 2009. This market is projected to reach $167.2 millions in 2010 and $1 billion in 2014 at a compound annual growth rate of 58.9%. This same analyst states that MWNT grade market was nearly $103 million in 2009 and projected to reach $161 million in 2010 and $865.5 million in 2014, for a 5-year compound annual growth rate (CAGR) of 53.1%.

In respect to this economical context and the exponential production of CNT, it is likely that some of them will get into the environment by numerous ways (Fig. 3), during each step of their life cycle, from their production until the end of their life cycle, especially in the aquatic compartment which concentrates most pollutions. Indeed, CNT releases into the environment might come from different sources. Point sources may be the release of CNT at the production site, during manufacturing (via accident of release, landfills or wastewater effluents, etc.) or during their transport from the production site to industrial users. Nonpoint sources may be the dissemination of CNT during the use of products containing CNT, via numerous ways such as wear of products, until the end of their life (wet deposition from the atmosphere, storm-water runoff, surface water leakage with the waste disposal in

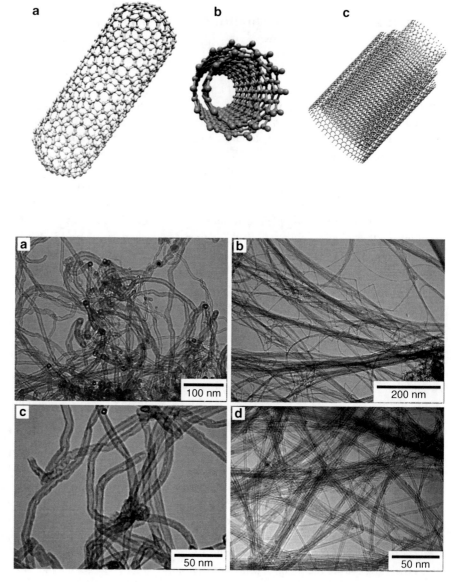

Ecotoxicology of Carbon Nanotubes Toward Amphibian Larvae, Fig. 1 Schematic representation of (**a**) Single-Walled Carbon Nanotube (SWNT), closed ends, (**b**) Double-Walled Carbon Nanotube, open ends, and (**c**) Multi-Walled Carbon Nanotube (MWNT), open ends (not to scale)

Ecotoxicology of Carbon Nanotubes Toward Amphibian Larvae, Fig. 2 TEM images of MWNT (**a, c**) and DWNT (**b, d**)

dumps, etc.) Contamination by CNT, as for the case of all chemical substances, may involve biogeochemical cycles, i.e., transport, dispersion, transformation, and/or concentration into compartments of the environment, but also may involve biological systems due to the potential ingestion, uptake, and excretion mechanisms of living organisms.

In this context, CNT could be classified as emergent contaminants and must receive considerable attention as new, unknown, and potentially hazardous materials. To date, several main questions are established by the scientific community about the quantities of CNT disposed

into the environment, their distribution into the environmental compartments, and their preferential areas of accumulation in these compartments, and finally their persistence or potential degradation in relation with some particular environmental conditions. On top of these geochemical concerns, the biological effects at molecular, sub-cellular, and cellular level from organism to ecosystem are of great concern and constitute the ecotoxicity area.

One strategy is to evaluate the potential ecotoxicological risk of CNT in the same way that risk assessment method is carried out during the assessment of the

Ecotoxicology of Carbon Nanotubes Toward Amphibian Larvae, Fig. 3 Environmental context of the CNT contamination toward the environment

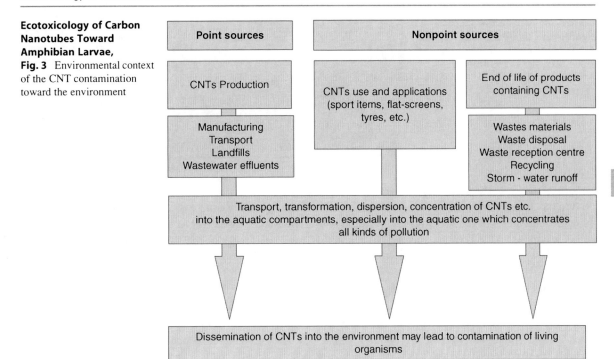

potential risk of a contaminant (e.g., chemical substances as pesticides or heavy metals) into the environment. This approach uses standardized methods in laboratory studies to determine biological effects using biological models, for example, amphibian larvae. Indeed, amphibians are well-known environmental health warning organisms due to their biphasic life cycle, permeable and sensitive eggs, skin, and gills [8]. Their specific physiology makes them particularly sensitive to the presence of contaminants in the water, which influences their behavior [9, 10]. In the same way, the huge quantity of DNA shared in a small number of chromosomes makes them good models for cytogenetic studies.

Among amphibian species, *Xenopus* and axolotl models (Fig. 4) were used to evaluate the potential ecotoxicity of CNT in laboratory conditions [11–14] consist in exposing amphibian larvae to different concentrations of CNT in water using standardized methods or adapted ones if necessary. The nature of CNT depends on the synthesis method used (Catalytic Chemical Vapor Deposition (CCVD), Electric–arc discharge, and Laser ablation are the most commonly used), and on their physical–chemical state (e.g., functionalization, dispersion). Three biological endpoints were investigated in amphibian larvae: the acute toxicity (mortality),

the chronic one according to the growth of larvae, and the genetic toxicity according to the induction of micronuclei in erythrocytes in the running blood as the expression of the clastogenic and/or aneugenic effects. Among toxic effects, genotoxicity may durably affect the aquatic ecosystems. The interaction between genotoxic compounds and DNA initially may cause structural changes in the DNA molecule. Unrepaired damage can generate other cell lesions and thus lead to tumor formation [15]. The evaluation of genotoxicity is particularly relevant because its assessment constitutes a warning system at short term to measure long-term effects at infra-lethal concentrations. Genotoxicity may be a useful biomarker as an intermediate endpoint in carcinogenesis, for example. This genetic toxicity represents the hidden risk because of its expression at long and very long term in the environment, contrary to the acute toxicity, for instance, which constitutes the visible risk. In the case of genetic toxicity, measured criteria will be DNA damage whereas in the case of acute toxicity, criteria will be mortality. In amphibian larvae, genome mutations may result in the formation of micronuclei, which are a consequence of chromosome fragmentation or malfunction of the mitotic apparatus.

**Ecotoxicology of Carbon
Nanotubes Toward
Amphibian Larvae,**
Fig. 4 *Xenopus laevis* (**a**) and
Ambystoma mexicanum
(**b**) larvae

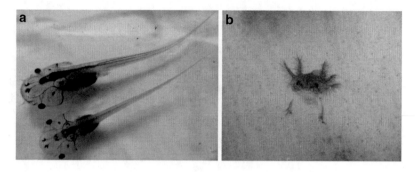

Acute, chronic, and genetic toxicities are the endpoints which were being investigated in amphibian larvae according to the French national [16] and international [17] standardized methods of the MicroNucleus Test (MNT). The MNT has been widely used with many amphibian species: *Pleurodeles waltl*, *Ambystoma mexicanum*, and *Xenopus laevis* [8, 9, 18–20]. The sensitivity and reliability of the MNT to detect chromosomal and /or genomic mutations make it a good method to analyze the potential cytogenetic damage caused by pure substances [21–23]. The use of MNT may provide an important tool for the prediction of the potential long-term effects on amphibians in the environment. One of the key functions of such biomarkers (micronucleus) is providing an "early warning" signal of significant biological effects (changes at the genetic/molecular level) with suborganism (molecular, biochemical, and physiological) responses preceding those occurring at higher levels of biological organization such as cellular, tissue, organ, whole-body levels and *in fine* at population level.

In this objective, two amphibian species were exposed to different kinds of CNT synthesized by different CCVD synthesis processes: DWNT [24] and MWNT. The three endpoints, described before, were investigated and analyzed. Moreover, the presence of CNT was investigated in the larvae using traditional microscopy methods, but also transmission electron microscopy (TEM) and Raman spectroscopy. This theme around CNT localization in biological matrix is a delicate point in this kind of work. Raman spectroscopy is an optical technique. Monochromatic light (laser) is focused on the sample. The scattered light, composed of the elastic scattering (Rayleigh) and the inelastic scattering due to an interaction with some quanta of energy like phonons (Raman), is analyzed through a spectrometer. With SWNT or DWNT, one-dimensional resonance phenomenon increases

strongly the Raman signal, while with large MWNT, the Raman signal is broadened and weaker.

In axolotl, Mouchet et al. [11] exposed the amphibian larvae under adapted standardized conditions to raw DWNT. The results obtained demonstrated that raw DWNT are neither acutely toxic nor genotoxic to larvae whatever their concentration in the water (ranging from 1 to 1,000 mg.L^{-1}). Nevertheless, visual observation under binocular showed that inside the gut there were large black masses of ingested CNT even at the lowest concentration tested.

Mouchet et al. [12–14] exposed *Xenopus* larvae under standardized conditions to raw DWNT or MWNT in water (ranging from 1 to 50 mg.L^{-1}). The results showed no genotoxicity in erythrocytes of larvae. In contrast, results showed chronic toxicity at 50 mgL^{-1} of MWNT (Fig. 5a) and from 10 mg/L of DWNT (Fig. 5b). A visual inspection of the larvae under binocular after exposure showed that internal gills were black-colored and seemed to be blocked up in the case of DWNT (Fig. 6c) but not in the case of MWNT (Fig. 6a), compared to larvae exposed without CNT (Fig. 6b). In the same way, the visual inspection of the digestive tract in larvae exposed to DWNT or MWNT revealed the presence of black masses into the intestine lumen suggesting that the larvae may have absorbed CNT present in the exposure medium (Fig. 6). In amphibian larvae, xenobiotics could enter in the body of larvae by two principal ways: the dermal route and the digestive route leading to potentially high ingestion rates of particles, especially particles of food susceptible to vehicle agglutinated CNT bundles and water containing suspended particles.

Using microscopic observations of intestine cross-sections in control and exposed larvae (Fig. 7), black masses were clearly identified within food intake in the intestine lumen in exposed larvae, but black masses were not observed in epithelial or in chorionic tissues.

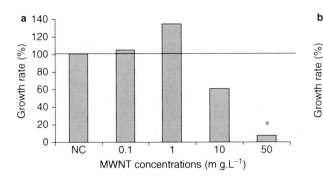

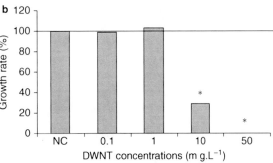

Ecotoxicology of Carbon Nanotubes Toward Amphibian Larvae, Fig. 5 Growth rate of *Xenopus larvae* exposed to 0.1, 1, 10 and 50 mg/L of raw CNT. * corresponds to a significant different size of larvae compared to the negative control group (mean value). *NC* Negative control

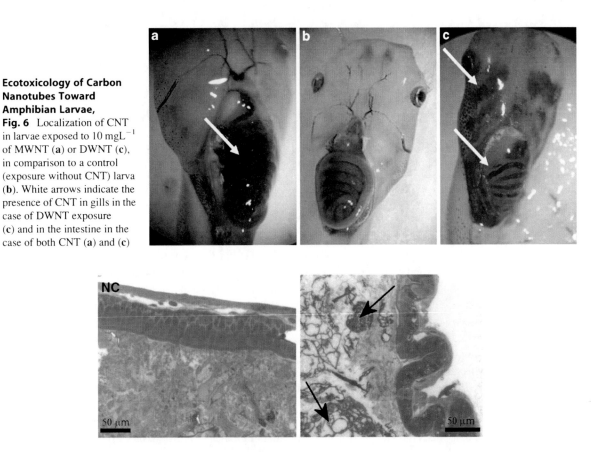

Ecotoxicology of Carbon Nanotubes Toward Amphibian Larvae, Fig. 6 Localization of CNT in larvae exposed to 10 mgL^{-1} of MWNT (**a**) or DWNT (**c**), in comparison to a control (exposure without CNT) larva (**b**). White arrows indicate the presence of CNT in gills in the case of DWNT exposure (**c**) and in the intestine in the case of both CNT (**a**) and (**c**)

Ecotoxicology of Carbon Nanotubes Toward Amphibian Larvae, Fig. 7 Optical microscopy photographs of histological thins of intestine of *Xenopus* larvae exposed to 10 mg/L of DWNT (on the right) compared to the control group (on the left). *NC* Negative control. Black arrows indicate the presence of black fibrous structures in the intestine lumen of CNT-exposed larvae. Sections showed were stained with methylene blue

It is not possible to identify CNT in the black masses of the intestine lumen by optical microscopy. Using TEM, the intestinal lumen, villi, and intestinal cells can be observed in detail. Figure 8 shows fibers in the intestinal lumen of larvae exposed to DWNT, near the villi but some of them are also observed in the control (not showed). These observations did not lead to the confirmation of the presence of CNT in the digestive

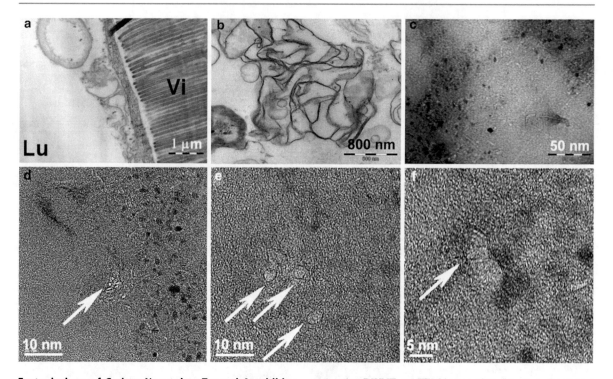

Ecotoxicology of Carbon Nanotubes Toward Amphibian Larvae, Fig. 8 Observation of area containing black fibrous structures in the intestine lumen of CNT-exposed larvae using Transmission Electron Microscopy (TEM) and 50 nm High-Resolution Transmission Electron Microscopy Electron micrographs (HRTEM) of *Xenopus* larvae intestine in animals exposed to DWNT. (**a**) TEM image of the intestine of a control larva, showing the lumen (Lu) from where nutrients are absorbed, as well as the intestinal wall (Vi: villi); (**b-f**) successive magnifications of (**a**) showing metal particles associated with CNT (**c, d**) and structures which could correspond to CNT (*white arrows*)

tissue at cellular level. Fibers were then observed at higher magnification (Fig. 8b and c). Discrimination of these fibers was realized under High-Resolution TEM. Figure 8d shows isolated CNT. At this magnification, (Fig. 8e and f) it was possible to show the presence of CNT in exposed larvae. Nevertheless, TEM is a local technique, also leading to possible ambiguity in terms of identification between CNT and biological compounds (ribosomes, organelles, etc.). Another method such as Raman spectroscopy is thus necessary to discriminate between them.

Two kinds of Raman analysis were performed on *Xenopus* intestine. The G or G'2D band can be used to identify and localize CNT in a biological matrix [12]. (1) Raman mapping is presented in Fig. 9 showing the optical microscopic image (slide) of an intestine sample of animal exposed to DWNT (Fig. 9a) and the corresponding Raman image from the same area (Fig. 9b). The superposition (Fig. 9c) clearly shows that, on the one hand, some of the black structures visible in optical microscopy do not contain any

DWNT, but that, on the other hand, CNT are present in some other places although optical microscopy does not show any clear evidence. (2) Figure 10 presents Raman line analysis in intestine, evidencing the presence of both kinds of CNT (DWNT and MWNT) in the intestine lumen, whereas no CNT were localized in the intestinal cells. Indeed, a sharp drop in intensity of the specific CNT band was observed. Raman spectroscopy analysis of different tissues of the larvae did not evidence the presence of CNT neither in blood (results not shown), nor in liver of amphibian suggesting that CNT do not cross the intestinal barrier.

To summarize, results showed different responses between the two amphibian species, *Xenopus* (anouran) and axolotl (urodeles) in relation with their different biology, larval growth, behavior, *etc*. A difference in sensitivity was generally observed in response to chemical pollutants, and *Xenopus* larvae are often found more sensitive than urodelian species to environmental substances [8]. Results highlight the potential risk of the studied CNT since acute and

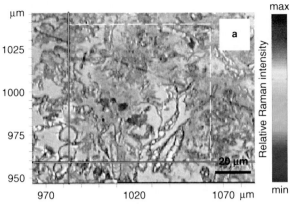
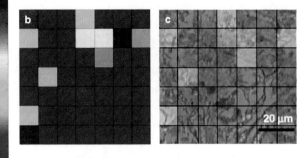
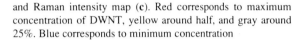

Ecotoxicology of Carbon Nanotubes Toward Amphibian Larvae, Fig. 9 Optical microscopy image of the studied intestine area (**a**); Raman intensity map (**b**) corresponding to the selected observation area of (**a**); Superposition of optical image and Raman intensity map (**c**). Red corresponds to maximum concentration of DWNT, yellow around half, and gray around 25%. Blue corresponds to minimum concentration

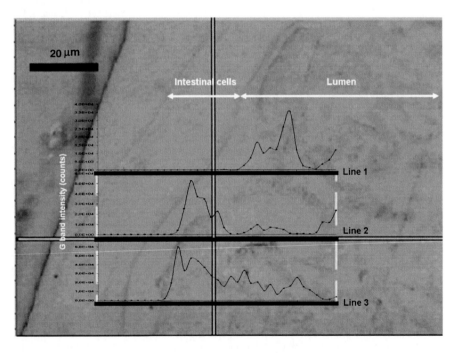

Ecotoxicology of Carbon Nanotubes Toward Amphibian Larvae, Fig. 10 Location of DWNT by Raman spectroscopy. The G band intensity extracted from Raman spectra is characteristic of the CNT. By acquiring spectroscopic lines, we follow the DWNT profile. Disappearance of the DWNT signal at the intestinal cell barrier is observed

chronic toxicities were observed in larvae exposed to raw CNT. In *Xenopus*, exposure to CNT leads to toxicity usually above 10 mg.L^{-1}, which corresponds to a high concentration. This level of contamination seems to be probably unrealistic in the global environmental context, except in the case of an accident at the production site, for example. Since CNT are ingested by larvae, one cannot exclude the possibility that CNT may be found later in the food chain, once released into the environment. In contrast, no genotoxicity was observed in larvae exposed to raw CNT in these conditions. Acute toxicity observed in *Xenopus* larvae may be mediated by several biological mechanisms: (1) branchial obstruction and /or gas exchange perturbation and/or anoxia, (2) and /or intestinal obstruction, (3) and/or competition between CNT and food. The presence of CNT was evidenced in the larvae using both electron microscopy and Raman analysis. The Raman imaging technique is very sensitive and thus allows assessing without doubt that the intestine of CNT exposed larvae contains CNT, as already suggested by the simple visual inspection of the larvae and the TEM observation. However, the image superposition analysis using Raman spectroscopy (Figs. 9, 10) clearly sharpened the first interpretation obtained by microscopy. Indeed, some of the black spots observed by optical microscopy do not contain any CNT, whereas CNT are clearly identified in some other places, although this technique does not show any clear evidence. Moreover, the abrupt vanishing of the intensity of the specific CNT bands in Raman spectroscopy when crossing the intestinal cell wall strongly suggests that raw CNT do not cross the intestinal barrier of amphibians.

Cross-References

▶ Carbon Nanotubes
▶ Electron Microscopy of Interactions Between Engineered Nanomaterials and Cells
▶ Genotoxicity of Nanoparticles
▶ In Vivo Toxicity of Carbon Nanotubes
▶ Synthesis of Carbon Nanotubes
▶ Transmission Electron Microscopy

References

1. Truhaut, R.: Eco-toxicology-objectives, principles and perspectives. Ecotoxicol. Environ. Saf. 1(2), 151–173 (1977)
2. Iijima, S.: Helical microtubules of graphitic carbon. Nature 354(6348), 5658 (1991)
3. Bethune, D.S., Kiang, C.H., Devries, M.S., Gorman, G., Savoy, R., Vazquez, J.: Cobalt-catalyzed growth of carbon nanotubes with single-atomic-layerwalls. Nature 363(6430), 605–607 (1993)
4. Iijima, S., Ichihashi, T.: Single-shell carbon nanotubes of 1-nm diameter. Nature 363(6430), 603–605 (1993)
5. Eklund, P., Ajayan, P., Blackmon, R., Hart, A.J., Kong, J., Pradhan, B., Rao, A., Rinzler, A.: International Assessment of Research and Development of Carbon Nanotube Manufacturing and Applications, World Technology Evaluation Center, Inc., 4800 Roland Avenue, Baltimore, MD, pp. 138. (2007)
6. Fuji-Keizai USA: Inc. CNT Wordwide. Carbon Nanotubes – Wordwide Status and Outlook: Applications, Applied Industries, Production, R&D and Commercial Implications. pp. 153 (2002)
7. NANO24D. BBC Research. Nanotechnologies. Carbon Nanotubes: Technologies and Global Markets. Published January 2000. Analyst: John Oliver. http://www.bccresearch.com/report/NAN024D.html. Accessed 15 May 2010 (2010)
8. Gauthier, L.: The amphibian micronucleus test, a model for in vivo monitoring of genotoxic aquatic pollution. Review paper. Int. J. Batrachol. 14(2), 53–84 (1996)
9. Gauthier, L., Tardy, E., Mouchet, F., Marty, J.: Biomonitoring of the genotoxic potential (micronucleus assay) and detoxifying activity (EROD induction) in the river Dadou (France), using the amphibian Xenopus laevis. Sci. Total Environ. 323, 47–61 (2004)
10. Bridges, C.M., Dwyer, F.J., Hardesty, D.K., Whites, D.W.: Comparative contaminant toxicity: are amphibian larvae more sensitive than fish? Bull. Env. Contam. Toxicol. 69, 562–569 (2002)
11. Mouchet, F., Landois, P., Flahaut, E., Pinelli, E., Gauthier, L.: Assessment of the potential in vivo ecotoxicity of Double-Walled Carbon Nanotubes (DWNT) in water, using the amphibian Ambystoma mexicanum. Nanotoxicology 1(2), 149–156 (2007)
12. Mouchet, F., Landois, P., Sarreméjean, E., Bernard, G., Puech, P., Pinelli, E., Flahaut, E., Gauthier, L.: Characterisation and in vivo ecotoxicity evaluation of double-wall carbon nanotubes in larvae of the amphibian Xenopus laevis. Aquat Toxicol. 87, 127–137 (2008)
13. Mouchet, F., Landois, P., Puech, P., Pinelli, E., Flahaut, E., Gauthier, L.: CNT ecotoxicity in amphibians: assessement of Multi Walled Carbon Nanotubes (MWNT) and Comparison with Double Walled Carbon Nanotubes (DWNT). Special focus environmental toxicity of nanoparticles. Nanomedecine 5(6), 963–974 (2010)
14. Mouchet, F., Landois, P., Datsyuk, V., Puech, P., Pinelli, E., Flahaut, E., Gauthier, L.: Use of the international amphibian micronuclei standardized procedure (ISO 21427–1) for in vivo evaluation of double-walled toxicity and genotoxicity in water. Environ. Toxicol. 26(2), 136–145 (2011)
15. Vuillaume, M.: Reduced oxygen species, mutation, induction and cancer initiation. Mutat. Res. 186(1), 43–72 (1987)
16. AFNOR: Association française de normalisation (the French National Organization for quality regulations). Norme NFT 90–325. Qualité de l'Eau. Evaluation de la génotoxicité au moyen de larves d'amphibien (Xenopus laevis, Pleurodeles waltl). AFNOR – Paris, pp. 17. (2000)
17. ISO: ISO International Standard. Water quality – Evaluation of genotoxicity by measurement of the induction of micronuclei – Part 1: Evaluation of genotoxicity using amphibian larvae. ISO 21427–1, ICS: 13.060.70, GENOVA – CH (2006)
18. Ferrier, V., Gauthier, L., Zoll-Moreux, C., L'Haridon, J.: Genotoxicity tests in amphibians – A review. In: Microscale Testing in Aquatic Toxicology: Advances, Techniques and Practice, vol. 35, pp. 507–519. CRC, Boca Raton, FL (1998)
19. Mouchet, F., Gauthier, L., Mailhes, C., Ferrier, V., Devaux, A.: Comparative Study of the Comet Assay and the Micronucleus Test in Amphibian Larvae (Xenopus laevis) Using Benzo(a)pyrene, Ethyl Methanesulfonate, and Methyl Methanesulfonate: Establishment of a Positive Control in the Amphibian Comet Assay. Environ. Toxicol. 20(1), 74–84 (2005)
20. Mouchet, F., Gauthier, L., Mailhes, C., Ferrier, V., Devaux, A.: Comparative evaluation of the genotoxicity of captan in amphibian larvae (Xenopus laevis and Pleurodeles waltl) using the comet assay and the micronucleus test. Environ. Toxicol. 21(3), 264–277 (2006)

21. Mouchet, F., Cren, S., Cunienq, C., Deydier, E., Guilet, R., Gauthier, L.: Assessment of lead ecotoxicity in water using the amphibian larvae (*Xenopus laevis*) and preliminary study of its immobilization in meat and bone meal combustion residues. Biometals **20**(2), 113–127 (2007)

22. Jaylet, A., Gauthier, L., Zoll, C.: Micronucleus test using peripheral red blood cells of amphibian larvae for detection of genotoxic agents in freshwater pollution. In: Sandhu, S.S., et al. (eds.) *In Situ* Evaluations of Biological Hazards of Environmental Pollutants, pp. 71–80. Plenum Press, New York (1990)

23. Gauthier, L., Van Der Gaag, M.A., L'Haridon, J., Ferrier, V., Fernandez, M.: In vivo detection of waste water and industrial effluent genotoxicity: use of the newt micronucleus test (Jalyet test). Sci. Total Environ. **138**, 249–269 (1993)

24. Flahaut, E., Bacsa, R., Peigney, A., Laurent, C.: Gram-scale CCVD synthesis of double-walled carbon nanotubes. Chem. Commun. **12**, 1442–1443 (2003)

Ecotribology

▶ Green Tribology and Nanoscience

EC-STM

▶ Electrochemical Scanning Tunneling Microscopy

Effect of Surface Modification on Toxicity of Nanoparticles

Malgorzata J. Rybak-Smith
Department of Pharmacology, University of Oxford, Oxford, UK

Synonyms

Influence of surface engineering on toxicity of nanoparticles; Influence of surface functionalization on toxicity of nanoparticles

Definitions

Nanoparticles – particles having at least one dimension between 1 and 100 nm.

Toxicity of nanoparticles – undesirable consequence of interactions of nanoparticles with human or mammalian cells (cytotoxicity), organs, such as lungs (pulmonary toxicity), spleen (splenic toxicity), liver (hepatotoxicity), kidneys (nephrotoxicity), immune system (immunocytotoxicity) and the nervous system (neurotoxicity).

Surface modification of nanoparticles – changing of physicochemical or biological characteristics of surface of nanoparticles by introducing chemical functional groups to the surface via creation of covalent bonds (covalent modification) or adsorption of biologically active molecules, such as proteins, surfactants, enzymes, antibodies, nucleic acids (non-covalent modification). Surface modification of nanoparticles can change their physicochemical properties, charge, size, hydrophobicity and make them more biocompatible for biomedical applications.

Nanoparticles are interesting novel materials which have a potential to be effectively used in nanomedicine and biotechnology. Their nano-sized character and exceptional physicochemical properties attract a lot of scientific attention. However, the size of nanoparticles allows them to penetrate various biological barriers within the body and cause toxicity. The potential toxicity of nanoparticles may also arise from interactions of various nanoparticles with biological systems and is highly undesirable. Toxicity of nanoparticles is a major obstacle for their use as potential drug or gene carriers, vaccines, or as biomedical devices. Medina et al. overviewed different classes of nanoparticles, including liposomes, carbon nanotubes (CNTs), gold shell nanoparticles, polymers, metallic nanoparticles, quantum dots (QDs), and carbon nanomaterials [1]. Nanoparticles can be used as biomarkers in molecular diagnosis or in drug delivery [1]. In the same entry, the pharmacological and diagnostic applications and potential toxicity of different classes of nanoparticles have been presented [1].

To minimize the toxicity of nanoparticles, surface engineering and functionalization can be applied. Surface of nanoparticles, also surface charge, can be modified in numerous ways. Non-covalent modification of the surface of nanoparticles can be achieved by coating or wrapping with biological molecules which may diminish the toxicity of nanoparticles and therefore make them more biocompatible. Coating with proteins, peptides, surfactants, polymers or wrapping with nucleic acids are commonly used approaches. Another approach involves creation of chemical bonds between functional groups present on the surface of nanoparticles and other biological molecules attached to the surface, and is called covalent modification. In this case, the covalent attachment of polyethylene glycol, PEG (PEGylation), carbohydrate or peptide conjugation, and acetylation may be applied. Functionalization of surface of nanoparticles can significantly improve their utility in biomedical applications as long as the functionalization is not toxic for cells and tissues by itself.

The unique physicochemical properties of CNTs make them of great interest for many biomedical applications. However, the presence of structural defects on their surface, metal impurities, length, hydrophobicity, and surface roughness can significantly affect CNT utility in biomedicine. Moreover, CNTs have a large surface area to mass ratio. The greater surface allows the greater interaction of CNTs with human plasma blood proteins, cellular membranes, and adsorption of biological molecules [2]. Smart et al. reviewed biocompatibility aspects of CNTs interactions with fibroblasts, osteoblasts, neuronal cells, ion channels, cellular membranes, antibodies, and the immune system [3].

Yu et al. presented that chemical modifications of surface of nanoparticles may improve their biocompatibility and can change their human plasma protein binding patterns, cytotoxicity, influence Reactive Oxygen Species (ROS) generation and immune responses [4]. Numerous modified MWCNTs showed lower immune responses than their precursor-carboxylated MWCNTs (MWCNT-COOH) [4]. Magrez et al. presented in vitro toxicity studies of carbon-based materials, like multiwalled carbon nanotubes (MWCNTs), carbon nanofibers and carbon nanoparticles on lung tumor cells [5]. All of the tested nanomaterials were toxic. Furthermore, enhanced cytotoxicity was observed after functionalization of the nanoparticles after an acid treatment [5]. The effect of various chemical functionalizations of CNTs on their biocompatibility has been reviewed by Foldvari et al. [6]. Covalent functionalization using 1,3-dipolar cycloaddition is commonly used approach and results in formation of CNT intermediates with amine groups present on the surface which can be further used to attach bioactive molecules such as streptavidin,

G-family proteins, or various peptides [6]. Moreover, the CNT intermediates can be used to attach small molecules, like amphotericin B (AmB), an effective antibiotic for treatment of chronic fungal infections [6]. Diminished toxicity and increased antibiotic activity of CNT-AmB conjugate was observed in Jurkat cells [6]. PEGylation is another commonly used approach which can improve the solubility and biocompatibility of CNTs [6]. Some bioactive molecules, for example, Cys-DNA or short interfering RNA (siRNA), can be non-covalently bound to single-walled carbon nanotubes (SWCNTs) [6]. The effect of functionalization of carbon nanoparticles such as SWCNTs, MWCNTs, carbon nanofibers, carbon graphite, carbon black, on cell toxicity was also discussed in the same review [6]. In general, the toxic effect of CNTs depends on their concentration, dose, purity, and surface properties of the CNTs (surface area, topology, presence of functional groups, presence of residues and impurities). SWCNTs and other carbon-based materials (e.g., carbon black, active carbon, carbon nanofibers) seem to be more toxic than MWCNTs [6]. The functionalization of CNTs can increase their dispersibility in biological media and therefore diminish their toxicity, compared with non-functionalized ("pristine") CNTs. "Soluble" functionalized CNTs may either enhance or decrease the toxicity compared with "pristine" CNTs. Possible enhanced toxicity of functionalized DWCNTs (oxidized) has been recently reported by de Gabory et al. [7]. SWCNTS and MWCNTs can cause cellular toxicity (apoptosis and necrosis were observed) [6]. The toxic effect depends on concentration of CNTs and time of incubation. It was concluded that the smaller the diameter of CNTs and the greater the aspect ratio, the greater toxic effect is observed [6]. In vivo oral, dermal, and pulmonary toxicity of CNTs were also described in this review [6].

Currently, little is known about interactions of nanoparticles with human lungs [8]. Engineered nanoparticles can cause pulmonary toxicity in humans, experimental animals and cell culture [8]. The increased ratio of surface area to mass of nanoparticles means that more potentially reactive groups present on their surface may interact with the lung cells and therefore impact their pulmonary toxicity [8]. Clearance of inhaled nanoparticles from the lungs depends mostly on particle size and surface charge [9]. Charged nanoparticles bind to lung tissues more effectively than non-charged ones [9]. Various surface coatings (albumin, lecithin, polysorbate 80, peptide attachments, PEG) were also found to influence nanoparticle cellular uptake and fate [9]. Nanoparticles associate with proteins or biopolymers present in human plasma or other biological fluids and form a dynamic protein corona [10].A long-lived protein corona formed from human plasma was studied on 50 and 100 nm sized, unmodified, carboxyl- and amine-modified polystyrene nanoparticles (PS) [10]. Surprisingly, both size and surface properties of the nanoparticles had a very considerable role in determining the nanoparticle coronas on the different particles of the same materials [10]. The size and surface modification of nanoparticles can impact the reactivity of biologically active proteins in the corona and likely have biological consequences [10].

Toxicity of various engineered nanoparticles, including CNTs, fullerenes, dendrimers, QDs, gold nanoparticles, silica, in vivo and in vitro, was summarized by De Jong et al. [11]. Some of the nanoparticles can cause lung inflammation, progression of plaque formation in vivo and oxidative stress in vitro and in vivo. Studies in vitro revealed that some of nanoparticles can cause platelet aggregation, inhibition of macrophage phagocytosis, and affect function of mitochondria [11]. Nevertheless, surface modification of QDs with N-acetylcysteine resulted in reduction of toxicity, compared to non-modified quantum dots, which can cause oxidative stress to plasma membranes and cellular compartments [11]. The effect of PEGylation on gold nanorod cytotoxicity was also presented in the same review [11]. Composition of nanoparticle formulation may influence interactions of the nanoparticles with tissues or cells and their potential toxicity [11]. Functionalization of nanoparticles may make them more specific for drug delivery purposes, diminish their toxicity to non-targeted organs, change their therapeutic effect, and make them safer and more biocompatible [11]. Coating of nanoparticles may prevent their agglomeration. For example, the use of polymers, like PEG, poly (vinylpyrrolidone) (PVP), chitosan, dextran, or surfactants as coatings, can keep nanoparticles separated in colloidal suspensions [11]. Crossing of the blood–brain barrier (BBB) by nanoparticles is especially challenging in drug delivery. Nanoparticle surface charges can affect the BBB crossing and must be considered for the brain toxicity. Neutral nanoparticles were found to

be not toxic for the BBB unlike cationic nanoparticles [11]. The toxicity of anionic nanoparticles was concentration dependent [11].

In vitro and in vivo toxicity studies on engineered gold nanoparticles (GNPs), including atomic clusters, colloidal particles, nanorods, gold nanoshells, and nanowires, were reviewed in great detail by Khlebtsov and Dykman [12]. Surface functionalization of GNPs can significantly influence their biodistribution in blood, liver, kidney, lung, spleen, muscle, brain, and bone [12]. Some of tested GNPs were coated with PEG, albumin, maltrodextrin, tumor necrosis factor (TNF), PEG-SH, IgG, maltose among many, and biodistribution studies on various animal models were summarized [12]. Interestingly, the adsorption of apolipoprotein on GNP surface may make possible for them to cross the BBB [12]. Also, cytotoxicity studies in vitro with cell cultures like human leukemia (K562), human breast carcinoma (SK-BR-3), human hepatocarcinoma (HepG2) and many more were presented in the same review [12]. The influence of a variety of surface coatings, such as DNA/transferrin, triphenylphosphine, cyclodextrin, cysteine, folic acid polyvinylpyrrolidone, glycolipid on GNP cytotoxicity was summarized [12].

Chemical modification of nanoparticles can impact their cellular uptake and cellular localization. The functionalization can change the binding of biological molecules, antibodies, etc., to the surface of nanoparticles and may have impact on drug delivery. Polymer-based nanoscale sized materials such as polymer micelles, nanogels, liposomes, polymer-DNA complexes can be of use in pharmacology. Batrakova and Kabanov broadly reviewed the biological activity of Pluronic block copolymers consisting of PEO and poly(propylene oxide) (PPO) blocks [13]. The amphiphilic character of these copolymers allows the interactions with hydrophobic surfaces and biological membranes [13]. These copolymers can self-assemble into pluronic micelles in aqueous solutions [13]. The core of these micelles is made up of hydrophobic PPO blocks while the shell consists of hydrophilic PEO chains [13]. The PEO shell can not only provide increased solubility, but also prevent the unwanted interactions with cells or plasma components. Pluronics have an ability to incorporate into plasma membranes followed by translocation into the cells which has an effect on various cellular functions [13]. Interestingly, the composition of the block copolymers can affect their biological functions. Block copolymers with intermediate lengths of PPO block and quite hydrophobic structure (such as Pluronic P85 or Pluronic L61), were the most efficiently transported into the cells [13]. On the other hand, hydrophilic block copolymers did not incorporate into the lipid bilayers and were not transported into the cells [13].

Fluorescent nanocrystals, also called quantum dots (QDs), can be utilized in drug delivery or cellular imaging [14]. The core of the QDs is usually a semiconductor like cadmium-selenium (CdSe), cadmium-tellurium (CdTe), indium-phosphate (InP), or indium-arsenate (InAs). Zinc sulfide (ZnS) is commonly used as the shell to stabilize the core of the QDs [14]. The structure of the QDs allows of the coating of the QDs and conjugation of biologically active molecules, such as antibodies, peptide ligands, which may reduce their toxicity and increase bioactivity [14]. Toxicity of cadmium and core material was widely discussed in this review [14]. Encapsulation of QD core with ZnS can not only stabilize the core of QD, but also improve their optical properties and bioavailability [14]. Toxicity studies of various QDs in vitro and in vivo were summarized by Ghaderi et al. [14]. In most of studies, CdSe and CdFe were the cores and ZnS was the shell [14]. The majority of tested QDs were not coated and were found cytotoxic for some of cell lines [14]. The size of QD plays an essential role in the observed cytotoxicity. In general, the smaller size the higher toxicity [14]. In one of the studies, preincubation of cells with N-acetylcysteine and bovine serum albumine (BSA) considerably reduced the cell death caused by QDs [14]. In some studies, generation of ROS caused the QDs toxicity [14]. Results of QD toxicity studies on mice and rats were also presented in the review [14]. CdSe/ZnS were coated with PEG and BSA and the latter were shown to be distributed to the liver, spleen, and other tissues [14]. Little is currently known about crossing of QDs through the BBB [14].

Jain et al. reviewed dendrimers toxicity in detail [15]. Positively charged dendrimers can interact with negatively charged biological membranes and cause the membrane destruction via nanohole formation and further cell lysis. Cationic dendrimers were shown to be less cytotoxic than anionic or PEGylated dendrimers [15]. Jain et al. summarized the cytotoxicity of positively charged dendrimers, polypropyleneimine (PPI), and polyamido amine (PAMAM)

against several cancer cell lines, including B16F10, CCRF, HepG2, COS-7, and Caco-2 [15]. In the same review, results of cytotoxicity and membrane disruption studies were presented for PPI dendrimers, PEG conjugated PPI dendrimers and PPI dendrimers with neutral acetamide groups on cultured Human Umbilical Vein Endothelial Cells (HUVEC) [15]. Moreover, hemolytic and hematological toxicity of dendrimers, including PPI and PAMAM, were presented [15]. The effect of various types of dendrimers, including glycine-, lactose-, mannose-, phenylalanine-coated PPI dendrimers on different blood parameters like white and red blood cells, hemoglobin, hematocrit, were summarized in the review [15]. It was concluded that the cationic dendrimers might show significant destruction of the hematological parameters [15]. However, immunogenicity has not been yet confirmed for dendrimers [15]. Results of biocompatible and surface engineered dendrimers, such as polyether imine dendrimers, SN-38 complexed G4-PAMAM dendrimers, glycine-coated PPI dendrimers, phenylalanine-coated PPI dendrimers, lactose-coated PPI dendrimers, PPI dendrimers with peripheral neutral acetamide groups, carboxylic acid-terminated PAMAM dendrimers, anionic PAMAM dendrimers, lauroyl- and PEGylated PAMAM dendrimers, and PEGylated PPI dendrimers, on various cell lines were summarized [15]. It was concluded that modification of the surface of dendrimers can diminish their toxicity [15]. Coating of surface of dendrimers with glycine, lactose, mannose, and galactose, PEG can also improve their hemolytic toxicity profile [15]. Surface engineering makes dendrimers more biocompatible for clinical applications. It allows them to mask the cationic charge on their surface which can be achieved by PEGylation, acetylation, peptide conjugation, and make them neutral. Also, introducing of negative charge on the surface of dendrimers makes them more biologically compatible [15]. Chemical modification of the surface of dendrimers can also allow them to be used as drug carriers or DNA and gene conjugates [15].

So far little is known about the interactions of nanoparticles with the central nervous system [16]. Hu et al. summarized results of recent in vitro studies on various cell lines, including cultured neuronal phenotype (PC12), brain cultures of immortalized mouse microglia (BV2), rat dopaminergic neurons (NR7), primary cultures of embryonic rat (Sprague–Dawley) striatum, and mouse neural stem cells (NSCs) [16]. Zinc oxide (ZnO) and anionic magnetic nanoparticles, for example, were neurotoxic to NSCs and PC12 cells, respectively [16]. Results of in vivo studies showed that the presence of charge on nanoparticles can have an influence on their neurotoxicity [16]. Changing the physicochemical properties of nanoparticles by conjugation of, for example, wheat germ agglutinin or coating with polysorbate surfactants, can impact the BBB transport mechanism [16]. Coating of nanoparticles with surfactants (Tween) may influence brain toxicity [16]. Cationic nanoparticles (gold, polystyrene) were shown to cause hemolysis and blood clotting whilst anionic particles were usually non-toxic [16].

Nanomaterials when placed in contact with human or mammalian body fluids and tissues are recognized by the immune system. Modification of nanoparticle surface with hydrophilic agents can influence their potential toxicity in vivo and in vitro. Coating or wrapping of nanoparticles can modify their interaction with blood proteins and cells. Dobrovolskaia et al. reviewed on how surface modification of various nanoparticles can change plasma protein binding and influence the nanoparticles interactions with red blood cells and platelets [17].

Functionalized nanoparticles can be unrecognizable by the immune system or can enhance the immune responses. Zolnik et al. presented how manipulation of physicochemical properties of nanoparticles can change their interactions with the components of the immune system [18]. The interactions may be undesirable and cause immunostimulation (interaction of nanoparticles with the complement system, cytokine secretion, immunogenicity) or immunosuppression. Nanoparticles can be recognized as foreign material entering the body. This may lead to toxicity. If nanoparticles (like carbon nanotubes) are too long they cannot be digested by macrophages and can cause granuloma formation. Modification of the surface properties of nanoparticles by changing their hydrophobicity and/or surface charge, may have a positive influence on compatibility of nanoparticles with the immune system. Immunosuppression may lower the body's defense against infection or may improve the therapeutic benefits of treatments for allergies and autoimmune diseases [18].

Nanoparticles can bind to coagulation proteins, blood plasma components, and cell-surface proteins. Binding of immune system complement proteins

activates the complement system and results in strong binding of several complement proteins to the CNTs [2]. Such complement activation may influence subsequent interaction of the CNTs with cells and tissues and induce potential toxicity. Protein binding occurs when CNTs are placed in contact with human blood or lung fluids and can have biological consequences [2]. Interactions of various CNTs with human plasma proteins and pulmonary surfactant proteins were reported [2]. Pre-coating of CNTs with bovine serum albumin (BSA), human serum albumin (HSA), human fibrinogen (FBG) or the detergent Tween 20 did not prevent the complement activation [2]. Frequently the surface of CNTs is not completely covered by the coating molecules, so some of potential binding sites on the surface of CNTs stay unobstructed and available for binding of various complement proteins [2].

The major limitation of nanoparticles in drug delivery is their rapid uptake by the mononuclear phagocyte system (MPS). However, the presence of PEG chains on the surface of nanoparticles can significantly change their interactions with blood proteins. Prolonged circulation of nanoparticles in the bloodstream is crucial for drug delivery purposes (e.g., particles should not be cleared by the liver and spleen before reaching their target). Moghimi et al. presented protein-binding processes to poly(ethylene glycol)–phospholipid (PEG-PL) engineered nanoparticles and the influence of surface PEGylation on complement activation and the fate of these engineered nanoparticles [19]. The effect of PEGylated liposome composition and size on complement activation in human serum was presented [19]. Anionic liposomes were shown to activate complement more than the methylated conjugates without charge [19].

Stable dispersion of nanoparticles in biological media is a necessary step toward their biomedical applications. It can be achieved by functionalization of CNT's hydrophobic surface with various chemical groups or adsorption of biologically active molecules. Although coating of CNTs with PEG_{5000}-phospholipid conjugates improves their stability in a biological buffer, the PEG-PL-coated CNTs can still activate the human complement system via the lectin pathway [19]. Complement activation by SWNTs in human serum with addition to N-acetylglucosamine, D-galactose, non-specific antibody, and Anti-MASP2 antibody in C1q-depleted serum was investigated by

Moghimi et al. [19]. Interestingly, PEG-phospholipid conjugates which form stable micellar structures with the size about 30 nm did not activate complement [19]. PEGylated liposomes with the size from 100 to 120 nm activated complement via classical and alternative pathway [19]. The lectin pathway was shown to be activated by 2–5 nm diameter CNTs with a length of 250 nm [19].

Effect of surface PEGylation of various liposomes on complement activation was presented by Moghimi et al. [20]. Anionic liposomes were shown to activate complement in vitro. However, incorporation of phosphatidylethanolamine-methoxypoly(ethylene glycol) (PE-mPEG) into the liposomal bilayer abolished the complement activation and was shown to be liposome-concentration dependent [20]. The drug Doxil®, commonly used in cancer therapy, is a negatively charged PEGylated liposome with doxorubicin enclosed. The drug is a strong activator of the human complement system and even its surface modification with PEG does not suppress complement activation [20]. However, the length of PEG can be an important factor in this as shown in some in vivo studies. Longer PEG chains were much more effective in preventing of complement activation than the shorter chains [20]. Other nanoparticles, very similar in structure to Doxil®, like hydrogenated soy phosphatidylcholine (HSPC) and cholesterol with the size of 100 nm, were shown not to activate the complement system in vitro. However, the presence of PEG_{2000}-conjugated distearoyl phosphatidylethanolamine, can cause significant complement activation in vitro [20]. Also, introduction of hydrogenated soy phosphatidylglycerol causes significant complement activation in vitro measured by an increase of SC5b-9 levels in human serum [20]. The negative charge plays a crucial role in the activation and the surface density of methoxypoly(ethylene glycol), mPEG, cannot prevent complement activation by liposomes as they cannot mask the negatively charged phosphodiester moieties [20].

Assessment of nanoparticles toxicity in vivo is a necessary step if the nanomaterials are to be used in biomedicine as efficient drugs or gene carriers. Functionalization of their surface by changing the surface charge and chemical composition can have a considerable impact on their toxic effect in vivo. Some nanomaterials can initiate oxidative stress in cells and release of ROS. This can initiate inflammation processes and cellular damage. Moreover, organs like the liver, spleen, kidney, and lungs can be affected by the oxidative stress [21]. Nanomaterials can also interact with mitochondria and nucleus, and such interactions are thought to be the main source of their toxicity [21]. Primary causes of various nanomaterials toxicity in vivo were summarized by Aillon et al. [21]. The authors discussed the influence of physicochemical properties of dendrimers, CNTs, QDs, and gold nanoparticles on their toxicity [21]. Examination of nanoparticle toxicity in vivo is still very limited. For most nanoparticles, their nano-size can cause their toxicity. Additionally, the surface charge on dendrimers, the length of CNTs, and the shape of gold nanoparticles can enhance their toxicity in vivo. Functionalization of the surface of the nanomaterials can diminish their toxicity. Chemical modification of the dendrimers surface can mask their charge, CNT hydrophobic surface can become more hydrophilic, and they can also become shorter by cutting using sonication. Hydrophobicity of "pristine" CNTs is a main obstacle in their biomedical applications. Moreover, they have a tendency to aggregate which can cause slow clearance from the body and oxidative stress in organs like liver, spleen, or lungs. Their length causes inefficient phagocytosis and damage of macrophages [21].

Studies on mice revealed that PEG-polyester dendritic hybrids did not cause any pathological changes in the liver, kidney, heart, lungs, and intestine of animals [21]. Similarly, PEGylated melanine dendrimer administration into mice did not cause liver and kidney toxicity [21].

Cadmium and selenium, heavy metals mostly used in QD cores, can be toxic in vivo [21]. Cadmium not only can accumulate in human tissues but also cross the BBB. However, electrostatic charge modification, adsorption, covalent modification, or chelation can have a considerable impact on the QD biocompatibility in vivo. As mentioned earlier, ZnS is commonly used as stabilizer of the QD surface, but also secondary coating of the QD can be applied to diminish its toxicity [21]. However, there is still a debate between researchers if the secondary coating of QD can prevent the cadmium leakage from their core [21]. Stability of the QD complex which consists of core, shell, and secondary coating is a very important factor in their toxicity [21].

Various covalent and non-covalent modifications of the surface of nanoparticles are essential approaches that allow to overcome the toxicity problems associated with nanomaterials. Surface functionalization gives an excellent opportunity to design novel nanoparticles with desirable physicochemical properties and biological activity. Nevertheless, it is essential that the functionalization is not damaging for cells or tissues by itself. Ideally, the functionalization should diminish nanoparticles toxicity and make them more biocompatible. This can significantly enhance their chances to be used in pharmacology and biomedical fields. The modification of surface properties of the nanoparticles can change their reactivity and have considerable impact on their cytotoxicity, ROS generation, immunocytotoxicity, pulmonary toxicity, and neurotoxicity. It can also influence their interactions with human and mammalian blood proteins and their adsorption on the surface. However, as the interactions with biological environments are quite complex, achieving the desirable biological or therapeutical responses of nanoparticles may be time consuming and not always work out. Sometimes quite opposite outcomes can be observed and the functionalized nanoparticles can be even more toxic either in vivo or in vitro than non-functionalized ones [2, 5, 19, 20]. Nevertheless, functionalization of nanoparticles surface can be considered as a useful tool for researchers and may reduce nanoparticles toxicity by making them more suitable for various biomedical applications.

Cross-References

▶ Carbon Nanotubes
▶ Cellular Mechanisms of Nanoparticle's Toxicity
▶ Functionalization of Carbon Nanotubes
▶ Gold Nanorods
▶ In Vivo Toxicity of Carbon Nanotubes
▶ In Vivo Toxicity of Titanium Dioxide and Gold Nanoparticles
▶ Nanoparticle Cytotoxicity
▶ Nanoparticles
▶ Quantum-Dot Toxicity

References

1. Medina, C., Santos-Martinez, M.J., Radomski, A., Corrigan, O.I.: Nanoparticles: pharmacological and toxicological significance. Br. J. Pharmacol. **150**, 552–558 (2007)
2. Rybak-Smith, M.J., Pondman, K., Salvador-Morales, C., Flahaut, E., Sim, R.B.: Recognition of carbon nanotubes by the human innate immune system. In: Klingeler, R., Sim, R.B. (eds.) Multi-Functional Carbon Nanotubes for Biomedical Applications, pp. 183–210. Springer, Heidelberg (2011)
3. Smart, S.K., Cassady, A.I., Lu, G.Q., Martin, D.J.: The biocompatibility of carbon nanotubes. Carbon **44**, 1034–1047 (2006)
4. Yu, Y., Zhang, Q., Mu, Q., Zhang, B., Yan, B.: Exploring the immunotoxicity of carbon nanotubes. Nanoscale Res. Lett. **3**, 271–277 (2008)
5. Magrez, A., Kasas, S., Salicio, V., Pasquier, N., Seo, J.W., Celio, M., Catsicas, S., Schwaller, B., Forró, L.: Cellular toxicity of carbon-based nanomaterials. Nano Lett. **6**, 1121–1125 (2006)
6. Foldvari, M., Bagonluri, M.: Carbon nanotubes as functional excipients for nanomedicines: II. Drug delivery and biocompatibility issues. Nanomed. Nanotechnol. **4**, 183–200 (2008)
7. de Gabory, L., Bareille, R., Daculsi, R., L'Azou, B., Flahaut, E., Bordenave, L.: Carbon nanotubes have a deleterious effect on the nose: the first in vitro data. Rhinology. **49**(4), 445–452 (2011)
8. Card, J.W., Zeldin, D.C., Bonner, J.C., Nestmann, E.R.: Pulmonary applications and toxicity of engineered nanoparticles. Am. J. Physiol. Lung Cell. Mol. Physiol. **295**, L400–L411 (2008)
9. Yang, W., Peters, J.I., Williams III, R.O.: Inhaled nanoparticles – a current review. Int. J. Pharm. **356**, 239–247 (2008)
10. Lundqvist, M., Stigler, J., Elia, G., Lynch, I., Cedervall, T., Dawson, K.A.: Nanoparticle size and surface properties determine the protein corona with possible implications for biological impacts. Proc. Natl. Acad. Sci. U.S.A. **105**(18), 14265–14270 (2008)
11. De Jong, W.H., Borm, P.J.A.: Drug delivery and nanoparticles: applications and hazards. Int. J. Nanomed. **3**, 133–149 (2008)
12. Khlebtsov, N., Dykman, L.: Biodistribution and toxicity of engineered gold nanoparticles: a review of *in vitro* and *in vivo* studies. Chem. Soc. Rev. (2010). doi:10.1039/c0cs00018c
13. Batrakova, E., Kabanov, A.V.: Pluronic block copolymers: evolution of drug delivery concept from inert nanocarriers to biological response modifiers. J. Control. Release **130**, 98–106 (2008)
14. Ghaderi, S., Ramesh, B., Seifalian, A.M.: Fluorescence nanoparticles "quantum dots" as drug delivery system and their toxicity: a review. J. Drug. Target. (2010). doi:10.3109/1061186X.2010.526227
15. Jain, K., Kesharwani, P., Gupta, U., Jain, N.K.: Dendrimer toxicity: let's meet the challenge. Int. J. Pharm. **394**, 122–142 (2010)

16. Hu, Y.-L., Gao, J.-Q.: Potential neurotoxicity of nanoparticles. Int. J. Pharm. **394**, 115–121 (2010)
17. Dobrovolskaia, M.A., McNeil, S.E.: Immunological properties of engineered nanomaterials. Nat. Nanotechnol. **2**, 469–478 (2007)
18. Zolnik, B.S., Gonzàlez-Fernàndez, Á., Sadrieh, N., Dobrovolskaia, M.A.: Minireview: nanoparticles and the immune system. Endocrinology **151**(2), 458–465 (2010)
19. Moghimi, S.M., Andersen, A.J., Hashemi, S.H., Lettiero, B., Ahmadvand, D., Hunter, A.C., Andersen, T.L., Hamad, I., Szebeni, J.: Complement activation cascade triggered by PEG-PL engineered nanomedicines and carbon nanotubes: The challenges ahead. J. Control. Release **146**, 175–181 (2010)
20. Moghimi, S.M., Szebeni, J.: Stealth liposomes and long circulating nanoparticles: critical issues in pharmacokinetics, opsonization and protein-binding properties. Prog. Lipid Res. **42**, 463–478 (2003)
21. Aillon, K.L., Xie, Y., El-Gendy, N., Berklan, C.J., Forrest, M.D.: Effects of nanomaterial physicochemical properties on *in vivo* toxicity. Adv. Drug. Deliver. Rev. **61**, 457–466 (2009)

Effective Media

▶ Theory of Optical Metamaterials

Elastic Modulus Tester

▶ Nanoindentation

Elasto-capillary Folding

▶ Capillary Origami

Electric Cooler

▶ Thermoelectric Heat Convertors

Electric Double Layer Capacitor

▶ Nanomaterials for Electrical Energy Storage Devices

Electric Field–Directed Assembly of Bioderivatized Nanoparticles

Youngjun Song[1] and Michael J. Heller[2,3]
[1]Department of Electrical and Computer Engineering, University of California San Diego, San Diego, CA, USA
[2]Department of Nanoengineering, University of California San Diego, La Jolla, CA, USA
[3]Department of Bioengineering, University of California San Diego, La Jolla, CA, USA

Synonyms

Bottom-up nanofabrication; Directed self-assembly; Nanoparticle self-assembly

Definition

Use of DC electric field microarray devices and electronic techniques to carry out the directed self-assembly of bioderivatized nanoparticles into high order structures.

Overview

Electronic microarray devices which produce reconfigurable electric field patterns can be used to control the self-assembly of nanocomponents into higher order structures. These devices allow the hierarchical assembly of nanocomponents to be carried within microscale geometries of electronic circuitry. Such nanofabrication devices and technologies have the potential for producing highly integrated nano/micro/macrostructures including nanophotonic-based large-scale arrays and displays; high-efficiency multiple band gap photovoltaics; fuel cells with nanostructured intermetallic catalysts and integrated fluidic channeling; smart-morphing nanocomposite materials; in vivo biosensor and mother-ship drug delivery devices. Electronic microarray devices that were originally developed for molecular diagnostic applications allow DNA, proteins, and almost any charged nanostructure to be rapidly transported and specifically bound at any site on the array surface. Microarray devices that have onboard integrated CMOS circuitry also allow rapid switching and precise control of the currents/voltages sourced to the microarray. Such microarray devices have been used to carry out the rapid highly parallel assisted self-assembly of biotin- and streptavidin-derivatized fluorescent nanoparticles and DNA-derivatized nanoparticles into multilayer structures. Electric field–directed assembly has major advantages over more classical passive layer-by-layer (LBL) self-assembly processes, which are generally one order of magnitude slower, nonparallel, and not reconfigurable in real time. The use of electronic microarray devices for assisted self-assembly represents a unique synergy of combining the best aspects of "top-down" and "bottom-up" technologies into a potentially viable nanomanufacturing process. It also enables a more bio-inspired logic to be followed for the hierarchal self-assembly of nanostructures into higher order structures, materials, and devices.

Nanotechnology Top-Down and Bottom-Up Processes

Nanotechnology includes a wide range of new ideas and concepts that are likely to enable novel applications in electronics, photonics, materials, energy conversion processes, biosensors, and biomedical devices. Some of the many challenges and opportunities in nanotechnology were discussed in the early National Nanotechnology Initiative meeting [1, 2]. One of the more important challenges is the area of nanofabrication, in particular the development of technologies which will lead to cost-effective nanomanufacturing processes. Enormous efforts are now being carried out on refining the more classical *"top-down"* photolithography processes to produce silicon-integrated (CMOS) electronic devices with nanometer scale features. While this goal has been achieved, the process requires billion dollar fabrication facilities and it does appear to be reaching some fundamental limits. So-called *bottom-up* self-assembly processes are also being studied and developed as possible new ways for producing nanodevices and nanomaterials. While there are now numerous examples of promising nanocomponents such as organic electron transfer molecules, quantum dots, carbon nanotubes, and nanowires, their assembly into higher order structures has been limited [3, 4]. Thus, the development of

a viable cost-effective bottom-up self-assembly nanofabrication process that allows billions of nanocomponents to be assembled into a higher order structure still remains a considerable challenge.

Living System and DNA Self-assembly – Living systems represent the best example of self-assembly processes that can be used for developing strategies for bottom-up nanofabrication. Living systems contain many types of biomolecules which have high fidelity recognition properties including deoxyribonucleic acid (DNA), ribonucleic acid (RNA), and proteins. Proteins serve as structural elements (collagen, keratin, silk, etc.), as binding recognition entities (antibodies) and as efficient catalytic macromolecules (enzymes). The nucleic acids DNA and RNA are involved in the storage and transfer of genetic information. All these biomolecules are able to interact and organize into higher order nanostructures which translate genetic information and perform biomolecular syntheses and energy conversion metabolic processes. Ultimately, all these biomolecules and bionanostructures are integrated and organized within membrane structures to form the entities called cells. Cells in turn can then replicate and differentiate (via these nanoscale processes) to form and maintain living organisms. Thus, biological systems have developed the ultimate "bottom-up" nanofabrication processes that allow component biomolecules and nanostructures with intrinsic self-assembly and catalytic properties to be organized into highly intricate living organisms. Of all the different biomolecules that could be useful for nanofabrication, the nucleic acids (DNA and RNA) with their high fidelity recognition and intrinsic self-assembly properties represent a most promising material for creating nanoelectronic, nanophotonic as well as many other types of organized nanostructures [5–7]. DNA, RNA, and other synthetic DNA analogues are programmable molecules which through their base sequence have intrinsic molecular recognition and self-assembly properties. Short DNA sequences, called oligonucleotides, are readily synthesized by automated techniques and can be modified with a variety of functional groups such as amines, biotin moieties, fluorescent or chromophore groups, and with charge transfer molecules. Additionally, synthetic DNA molecules can be attached to quantum dots, metallic nanoparticles, carbon nanotubes; as well as to surfaces like glass, silicon, gold, and semiconductor materials. Synthetic DNA molecules (oligonucleotides) represent an ideal type of "Molecular Legos" for the self-assembly of nanocomponents into more complex two- and three-dimensional higher order structures. At a first level, DNA sequences can be used as a kind of template for assembly on solid surfaces. The technique involves taking complementary DNA sequences and using them as a kind of selective glue to bind other DNA-modified macromolecules or nanostructures together. The base pairing property of DNA allows one single strand of DNA with a unique base sequence to recognize and bind together with its complementary DNA strand to form a stable double-stranded DNA structures. While high fidelity recognition molecules like DNA allows one to "self-assemble" higher order structures, the process has some significant limitations. First, for in vitro applications (i.e., in a test tube), DNA and other high fidelity recognition molecules like antibodies, streptavidins, and lectins work most efficiently when the complexity of the system is relatively low. That is as the complexity of the system increases (more unrelated DNA sequences, proteins, and other biomolecules), the high fidelity recognition properties of DNA molecules are overcome by nonspecific binding and other entropy-related factors, and the specific hybridization efficiency for the DNA molecules is considerably reduced. Under in vivo conditions (inside living cells), the binding interactions of high fidelity recognition molecules like DNA are much more controlled and compartmentalized, and the DNA hybridization process is assisted by structural protein elements and active dynamic enzyme molecules. Thus, new bottom-up nanofabrication processes based on self-assembly using high fidelity recognition molecules like DNA should also incorporate strategies for directing and controlling the overall process.

Electronic Microarray DNA Hybridization Technology – A variety of electronic microarrays have been designed and fabricated for DNA genotyping diagnostic applications [8–11]. An early 100 test-site electronic microarray which has been commercialized (Nanogen) has an inner set of 80-μ diameter platinum microelectrodes (Fig. 1a, b). Each microelectrode has an individual wire interconnect through which current and voltage are applied and regulated. The microarrays are fabricated from silicon wafers, with insulating layers of silicon dioxide, platinum microelectrodes, and gold connecting wires. Silicon dioxide/silicon nitride is used to cover and insulate the conducting wires, but not the surface of the platinum

microelectrodes. The whole surface of the microarray is covered with several microns of hydrogel (agarose or polyacrylamide) which forms a permeation layer. The permeation layer is impregnated with streptavidin which allows attachment of biotinylated DNA probes or other entities. The ability to use silicon and microlithography for fabrication of the DNA microarray allows a wide variety of devices to be designed. More sophisticated higher density electronic microarrays with 400 test-sites have on-chip CMOS control elements for regulating the current and voltages to the microelectrode at each test-site (Fig. 1c, d). These control elements are located in the underlying silicon structure and are not exposed to the aqueous samples that are applied to the chip surface when carrying out the DNA hybridization reactions. DNA hybridization reactions are carried out on these electronic microarrays by first applying a positive DC bias of ∼2–4 V to address or spot specific DNA molecules to the selected test-sites on the array. These DNA molecules are usually oligonucleotide "capture" probes or target DNA sequences that have been functionalized with biotin molecules. These biotinylated DNA molecules become strongly bound to streptavidin molecules which are cross-linked to the hydrogel layer covering the underlying microelectrode. In the next step, a positive DC electric field is applied to specific microelectrodes directing the hybridization of the other DNA molecules to the DNA sequences attached to the selected test-sites. (During the addressing and hybridization processes, other microelectrodes on the microarray are biased negative.) Electronic addressing also allows DNA probes to be spotted onto the array in a highly reproducible manner. Electronic microarrays can be formatted in a variety of ways that include reverse dot blot format (capture/identity sequences bound to test-sites), sandwich format (capture sequences bound to test-sites), and dot blot format (target sequences bound to test-sites). DNA hybridization assays involve the use of fluorescent DNA reporter probes designed to hybridize to specific target DNA sequences. The reporter groups are usually organic fluorophores that have been either attached to oligonucleotide probes or to the target/sample DNA/RNA sequences. After electronic addressing and hybridization are carried out, the microarray is analyzed using the fluorescent detection system. Reference [12] provides a good example of single nucleotide polymorphism (SNP) genotyping analysis that has been carried out on using electronic microarrays.

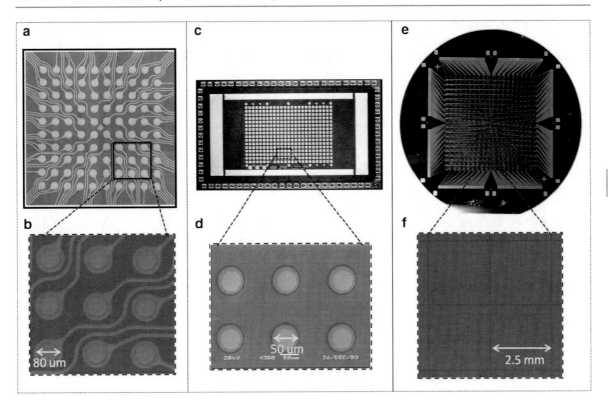

Electric Field–Directed Assembly of Bioderivatized Nanoparticles, Fig. 1 *Three different array devices for electric field–directed nanoparticle assembly.* (**a**) View of a 100 test-site microarray device with 80-μ diameter platinum microelectrodes, the device is approximately 5×5 mm in size. (**b**) A close-up of the 80 μ microelectrodes. (**c**) A 400 test-site CMOS electronic microarray device with 50-μ diameter platinum microelectrodes and four perimeter counterelectrodes, the device is approximately 5×7 mm in size. (**d**) A close-up of the 50-μ diameter microelectrodes. (**e**) A 4″ silicon wafer size electric field–directed assembler device. The total active deposition area is 5 cm² with 200 platinum electrodes which are 2.5×2.5 mm with 8 μm gaps. (**f**) Shows a close-up of four of the electrodes

Electric Field–Directed Nanofabrication

While electronic microarray devices were originally developed for DNA genotyping diagnostic applications, the resulting electric fields are able to transport any type of charged molecule or structure including proteins, antibodies, enzymes, nanostructures, cells, or micron-scale devices to or from any of the sites on the array surface. In principle, these active devices serve as a "mother-board" or "host-board" for the assisted assembly of DNA molecules or other entities into higher order or more complex structures. Since the DNA molecules have intrinsic programmable self-assembly properties, and can be derivatized with electronic or photonic groups, or attached to larger nanostructures (quantum dots, metallic nanoparticles, nanotubes), or microstructures, this provides the basis for a unique bottom-up nanofabrication process. Thus, electronic microarray devices can serve as motherboards that allow one to carry out a highly parallel electric field "pick and place" process for the heterogeneous integration of molecular, nanoscale, and micron-scale components into complex three-dimensional structures. Electric field–directed self-assembly technology is based on three key physical principles: (1) the use of functionalized DNA or other high fidelity recognition components as "Molecular Lego" blocks for nanofabrication; (2) the use of DNA or other high fidelity recognition components as a "selective glue" that provides intrinsic self-assembly properties to other molecular, nanoscale, or micron-scale components (metallic nanoparticles, quantum dots, carbon nanotubes, organic molecular electronic switches, micron and submicron silicon lift-off devices and components); and (3) the use of an active microelectronic array devices to provide electric field assistance or control to the intrinsic self-assembly of any modified electronic/photonic components and structures [13, 14].

Electric Field–Directed Self-assembly of Bioderivatized Nanoparticles – In addition to the more classical top-down processes such as photolithography, so-called bottom-up processes are being developed for carrying out self-assembly of nanostructures into higher order structures, materials, and devices [15]. To this end, considerable efforts have been carried out on both passive and active types of layer-by-layer (LBL) self-assembly processes as a way to make three-dimensional layered structures, which can have macroscopic x-y dimensions [16]. In cases where patterned structures are desired, the substrate material is generally pre-patterned using masking and a photolithographic process [17]. Other approaches to patterning include the use of optically patterned ITO films and active deposition of the nanoparticles. Nevertheless, limitations of passive LBL as well as active assembly processes provide considerable incentive to continue the development of better paradigms for nanofabrication and heterogeneous integration. Of particular importance to self-assembly processes involving biomolecular-derivatized nanoparticles is the prevention of nonspecific binding. Nonspecific binding due to cooperative hydrophobic, hydrogen bonding and ionic interactions between nanoparticles and the assembly substrate may often be as strong as the specific ligand binding interactions. Thus, once a nanoparticle has been bound nonspecifically, it is often difficult to remove it without perturbing or damaging the more specific nanoparticle interactions and structures. Nonspecific binding is often a concentration-dependent problem, whose probability increases when large amounts of derivatized nanoparticles interact with a physically and/or chemically complex substrate. Alternatively, trying to carry out stringent self-assembly at very low concentrations of biomolecular-derivatized nanoparticles is a slow process, and may still result in nonspecific binding to unprotected areas of the substrate. Therefore, there is a clear need for a rapid, directed assembly process which utilizes low concentrations of nanocomponents and reduces nonspecific interactions throughout the repeated multistep processes that will be required for fabricating complex higher order structures.

Electronic Microarrays and Techniques – Electronic microarrays and electric field techniques have now been used to carry out the highly parallel directed self-assembly of biomolecular-derivatized nanoparticles into multilayered higher order structures. The rapid and highly parallel assembly of biotin-, streptavidin-, and DNA-derivatized nanoparticles into structures with over 40 layers has been achieved [18–20]. The optimized electric field process allows nanoparticles at very low concentration in the bulk solution to be rapidly concentrated and specifically bound to the activated sites with minimal nonspecific binding. The electric field device used to obtain such results was an ACV 400 CMOS electronic microarray (Nanogen). This CMOS-controlled electronic microarray contains 400 50-μm diameter platinum microelectrodes in a 16×25 grid arrangement (Fig. 1c). The device has four additional large rectangular counterelectrodes on its perimeter. Each of the 400 microelectrodes is capable of independently producing from 0 to 5 V at up to 1 μA per microelectrode. The 400 microelectrodes on the array can be biased positive, negative, or left neutral in basically any combination; however, all four of the perimeter electrodes must be biased at the same polarity (positive or negative) or left neutral. The electronic microarray device can be set to source either current or voltage, and the activation time for each microelectrode can be controlled independently. The microarray device uses platinum electrodes as they present an inert interface to the solution. Additionally, the platinum electrodes do not degrade or disintegrate from the electrolysis reactions which occur at the voltages (>1.2 V) necessary to produce rapid electrophoretic transport of nanoparticles in the bulk solution. The ACV 400 site microarray comes coated with a 10-μm thick (hydrated) porous permeation layer composed of polyacrylamide impregnated with streptavidin (Fig. 2). This permeation layer is designed to prevent analyte molecules (DNA, proteins, etc.) and nanoparticles from contacting the microelectrodes, and to ameliorate the adverse effects of the electrolysis products on the binding reactions. In order to improve nanoparticle binding density, the microarray surface was overlayered with a biotin–dextran layer and a high density streptavidin layer (Fig. 2). The basic scheme for the electric field–directed self-assembly of biotin and streptavidin nanoparticles is shown in Fig. 3, and a full description of the actual procedure is given in Ref. [19].

Electric Field–Directed Assembly of Bioderivatized Nanoparticles, Fig. 2 *The 400 test-site CMOS electronic microarray device.* The 400 test-site device, with a cross-sectional diagram showing the silicon base, the CMOS control circuitry, the platinum microelectrodes, the silicon dioxide insulating structures between the microelectrodes, the overlaying polyacrylamide/streptavidin permeation gel layer, the biotin–dextran layer, and the final layer of streptavidin (the drawings are not to scale)

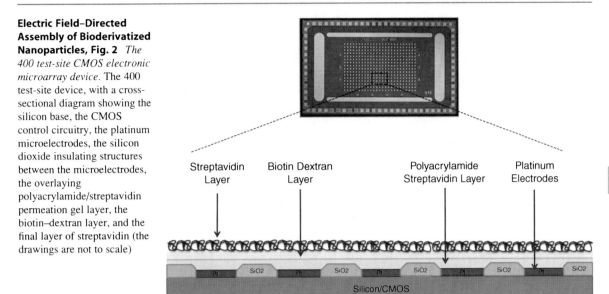

Procedure for Deposition of First Nanoparticle Layers – The microarray is first washed with a 100 mM solution of L-histidine to remove any excess streptavidin. The bulk of the liquid on the chip is then removed, and 20 μL of a 4 nM solution of biotinylated 40 nm red fluorescent nanoparticles in 100 mM L-histidine buffer is deposited onto the microarray surface. Because most of the derivatized nanoparticles used in the experiments have a net negative charge, microelectrodes used for deposition are biased positive, with the outer perimeter counterelectrodes being biased negative. A number of microelectrodes on the array are often left unbiased (not activated) and served as overall controls for nonspecific binding. Immediately after addition of the nanoparticle containing solution, the directed electrophoretic transport and deposition of nanoparticles to the specific array sites is carried out. In initial experiments, the microarray can be set up in a combinatorial fashion to test a variety of currents, voltages, and activation times to determine optimal deposition conditions. Generally, a current ramp from 0 to 0.4 μA in 0.025 μA increments is programmed along rows of microelectrodes on the array. Activation times from 0 to 20 s in 5 s intervals are programmed across groups of columns on the microarray. Several columns of microelectrodes are also left inactivated to serve as controls. Following directed electrophoretic transport of the biotinylated 40 nm nanoparticles, the microarray is washed six times with 100 mM L-histidine, and then

all liquid is removed from the microarray to prevent crystallization of the histidine upon drying [19].

Procedure Multilayer Nanoparticle Depositions – The basic combinatorial directed deposition procedure is repeated in order to determine optimal conditions (current, voltage, and activation time) for producing best quality alternating layers of biotin and streptavidin nanoparticles. Template experiments can be used for determining nanoparticle layering conditions (which could be run on the same microarray in highly parallel manner). The first template experiment involves alternating the deposition of biotin 40 nm red fluorescent nanoparticles and streptavidin 40 nm yellow-green fluorescent nanoparticles in a grid type format. In this format, groups of columns of microelectrodes are set to different activation times (typically between 0 and 30 s), and the rows of microelectrodes are set to different current levels (0.025–0.4 μA in 0.0.025 μA increments). Several repeats of the combinatorial pattern are done across the whole microarray, and intervening columns of microelectrodes are left inactivated to serve as controls. This combinatorial approach allows for a relatively quick determination of the optimal time and current conditions necessary to achieve the best deposition and layering. A second type of template experiment is used to determine the level of self-adhesion (nonspecific binding) between similar nanoparticles (biotin to biotin and streptavidin to streptavidin). This is carried out by having regions on

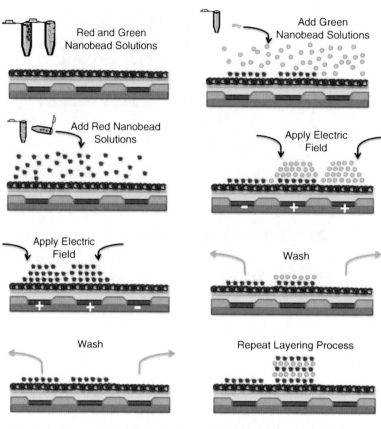

Electric Field–Directed Assembly of Bioderivatized Nanoparticles, Fig. 3 *The electric field–directed nanoparticle layering process.* The initial step involves adding a solution containing the biotinylated 40 nm *red* fluorescent nanoparticles onto the microarray device. The next step involves applying a positive DC biased to the desired test-sites on the microarray which attracts and concentrates the negatively charged biotinylated nanoparticles to the surface. The biotinylated nanoparticles become bound to the streptavidin layer, and the unattached nanoparticles are then washed away. In the next step, a solution containing streptavidin 40 nm *yellow-green* fluorescent nanoparticles is placed onto the microarray; the positively biased test-sites on the microarray attract and concentrate the negatively charged streptavidin nanoparticles which now become bound to the biotinylated nanoparticle monolayer. The unattached nanoparticles are then washed away. The alternate addressing, binding, and washing of biotin and streptavidin nanoparticles are repeated to form the multilayer nanoparticles structures

the microarray only activated on alternate deposition cycles, when the same type of nanoparticle solution is exposed to the microarray. This experiment tests if the same type of nanoparticle would accumulate without the presence of its specific ligand binding partner. In these experiments, a single column per template is activated during every deposition to serve as a positive control for biotin-to-biotin nanoparticle binding and streptavidin-to-streptavidin nanoparticle binding. Additional columns of microelectrodes are not activated throughout the experiment, and serve as overall controls for more general nonspecific binding to the microarray. All of these experiments are monitored in real time and imaged by an epifluorescent microscope set at 540 nm emission for the yellow-green fluorescent nanoparticles and 610 nm emission for the red fluorescent nanoparticles. Imaging was carried out using a Hamamatsu Orca-ER camera [19].

Multilayer Biotin–Streptavidin Nanoparticle Structures – One of the main advantages of the electric field–directed self-assembly fabrication process is the ability to direct the streptavidin–biotin ligand binding to only occur on the specific microelectrodes which have been positively biased. Figure 4a shows a section of the 400 site CMOS microarray device where alternate rows of microelectrodes were used to carry out

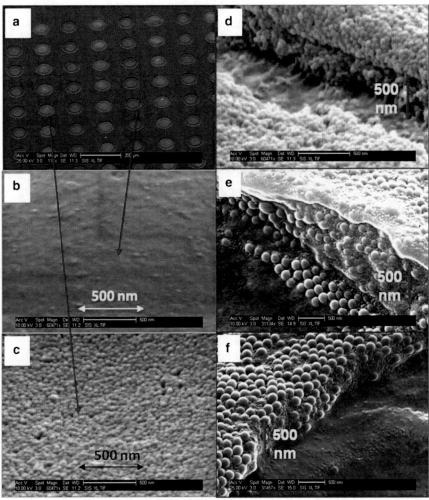

Electric Field–Directed Assembly of Bioderivatized Nanoparticles, Fig. 4 *Results of electric field–directed nanoparticle assembly/layering on a microarray device.* (**a**) Shows a section of the 400 site CMOS microarray device where alternate rows of microelectrodes were used to carry out nanoparticle layering. Rows with *bright circular spots* had 40 layers of alternating 40 nm *red* fluorescent biotin nanoparticles and 40 nm *green* fluorescent streptavidin nanoparticles deposited by applying a DC electric field. Rows with *lighter circular spots* had no electric field applied. (**b**) Shows an SEM image of a microelectrode which had no DC field applied. No nanoparticles have been deposited. (**c**) Shows a SEM image of a microelectrode to which the DC electric field was applied. Nanoparticles have been deposited on this surface. (**e**) Shows an SEM image of a cross section of a 40 layer structure of alternating 40 nm *red* fluorescent biotin nanoparticles and 40 nm *green* fluorescent streptavidin nanoparticles. (**f**) Shows an SEM image of a cross section of a 40 layer structure of alternating 40 nm biotin nanoparticles and 200 nm streptavidin nanoparticles. (**g**) Shows another SEM image of a cross section of a 40 layer structure of alternating 40 nm biotin nanoparticles and 200 nm streptavidin nanoparticles

biotin–streptavidin nanoparticle layering. Rows with bright circular spots had 40 layers of alternating 40 nm red fluorescent biotin nanoparticles and 40 nm yellow-green fluorescent streptavidin nanoparticles deposited by applying a DC electric field. Rows with lighter circular spots had no electric field applied. Figure 4b shows a scanning electron microscope (SEM) image of a microelectrode which had no DC field applied. It can clearly be seen that no biotin nanoparticles have bound to this surface, despite the fact that it contains a streptavidin binding layer and has been exposed to the biotin nanoparticles 20 times. Figure 4c shows an

SEM image of a microelectrode to which the DC electric field was applied. Nanoparticles have clearly been deposited on this surface. Multilayer nanostructures were verified by using SEM to image fractured cross sections of the 40 layer structures of alternating 40 nm red fluorescent biotin nanoparticles and 40 nm green fluorescent streptavidin nanoparticles (Fig. 4d). In other work, multilayer layers structures containing alternating 40 nm biotin nanoparticles and 200 nm streptavidin nanoparticles were fabricated. Figure 4e, f show the SEM image of cross sections of 40 layer structure of alternating 40 nm biotin nanoparticles and 200 nm streptavidin nanoparticles.

Multilayer DNA Nanoparticle Structures – In other work, electric field–directed hybridization was used to produce 20 layer nanostructures composed of DNA-derivatized nanoparticles [20]. Using the same CMOS electronic microarray device, DNA nanoparticles could be directed and concentrated such that rapid and specific hybridization occurs only on the activated sites. Nanoparticles layers were formed within 30 s of activation, and 20 layer structures were completed in under an hour. The results of this work carry significant implications for the future use of DNA nanoparticles for directed self-assembly nanofabrication of higher order structures. Advantages for using electric field directed fabrication of DNA nanoparticles include the following: First, the electric field process for the directed hybridization of DNA nanoparticles was significantly faster than the passive process; second, the electric field–directed process requires the use of only minimal concentrations of DNA nanoparticles, which would not be viable for passive hybridization process which requires much higher concentration of nanoparticles; third, the integrity of the biomolecular binding reactions (hybridization) and the ionic flux was maintained throughout the large number of process steps (20 depositions). Thus, high fidelity DNA hybridization and a large number/variety of unique DNA binding sequences can be used for nanofabrication; fourth, the higher order layered DNA nanoparticle structures can ultimately be removed from the microarray by a relatively simple lift-off procedure. Thus, an electric field array device can serve as a manufacturing platform for making higher order 3D structures; and fifth, the overall process demonstrated the nanofabrication of higher order structures from a relatively heterogeneous group of materials, that is, biotin–dextran polymers, streptavidin, and two different sets of DNA nanoparticles.

Future Work

Some important potential future applications for the electric field–directed nanofabrication process include miniaturized chemical and biosensor devices, "micron-sized" dispersible chemo/biosensors for environmental and bioagent detection, lab-on a-chip devices, and in vivo diagnostic/drug delivery systems. Figure 5 shows the scheme for a future glucose micro-biosensor device fabricated using electric field–directed nanoparticle assembly. The basic glucose micro-biosensor structure would include a bottom (lift-off) layer of biotin–streptavidin nanoparticles; a layer of glucose oxidase–streptavidin nanoparticles; a layer of luminol–dextran–biotin; a layer of streptavidin, a layer of peroxidase–biotin nanoparticles, a layer of red fluorescent streptavidin quantum dots, and a final upper capping layer of biotin–streptavidin nanoparticles. When placed in a sample containing glucose, the glucose would diffuse into the micro-biosensor device and react with the glucose oxidase to produce hydrogen peroxide. The hydrogen peroxide and luminol diffusing to the peroxidase layer would catalyze the chemiluminescence (blue light). The blue light via fluorescent energy transfer to red fluorescent quantum dots would now produce detectable red fluorescence signal. This demonstrates that different fluorescent color responses could be created. Electronic array devices have other advantages including that they can be designed in a wide variety of shapes and sizes. To this end, larger "macroarray" devices have now been fabricated which can produce layered nanostructures with macroscopic dimensions. Figure 1e, f show a 4 in. silicon wafer size electric field array device. The total active deposition area for this device is 5 cm^2 with 200 platinum electrodes which are 2.5×2.5 mm with 8μm gaps. Such "macroarray" devices will have applications for producing advanced fuel cells and photovoltaic materials. Overall electric field–directed nanofabrication using electronic array devices represents a unique synergy of combining the best aspects of "top-down" and "bottom-up" technologies into a viable nanofabrication process. It also represents a bio-inspired logic for self-assembly and hierarchal scaling of nanocomponents into integrated microscopic structures.

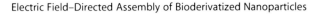

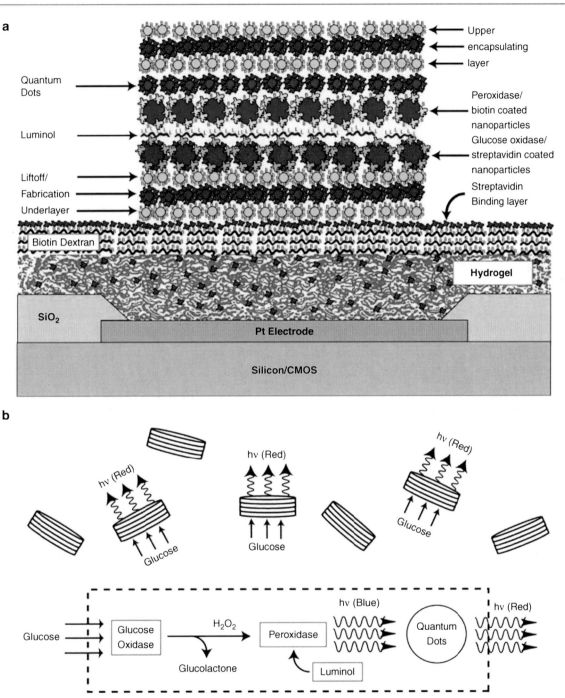

Electric Field–Directed Assembly of Bioderivatized Nanoparticles, Fig. 5 *Future glucose micro-biosensor device fabricated by electric field–directed assembly.* (**a**) The basic glucose micro-biosensor structure would include a bottom (lift-off) layer of biotin–streptavidin nanoparticles; a layer of glucose oxidase–streptavidin nanoparticles; a layer of luminol–dextran–biotin; a layer of streptavidin, a layer of peroxidase–biotin nanoparticles, a layer of *red* fluorescent streptavidin quantum dots, and a final upper capping layer of biotin–streptavidin nanoparticles. (**b**) When placed in a sample containing glucose, the glucose would diffuse into the micro-biosensor device and react with the glucose oxidase to produce hydrogen peroxide. The hydrogen peroxide and luminol diffusing to the peroxidase layer would catalyze the chemiluminescence (*blue light*). The *blue light* via fluorescent energy transfer to *red* fluorescent quantum dots would now produce detectable *red* fluorescence signal

Cross-References

► Bioinspired Synthesis of Nanomaterials
► Biomimetic Synthesis of Nanomaterials
► Electric-Field-Assisted Deterministic Nanowire Assembly
► Quantum Dot Nanophotonic Integrated Circuits
► Self-assembly
► Self-Assembly for Heterogeneous Integration of Microsystems
► Self-assembly of Nanostructures
► Self-repairing Photoelectrochemical Complexes Based on Nanoscale Synthetic and Biological Components

References

1. National Research Council: Small Wonders, Endless Frontiers: Review of the National Nanotechnology Initiative. National Academy Press, Washington, DC (2002)
2. National Science and Technology Council: The National Nanotechnology Initiative – Strategic Plan. National Science and Technology Council, Washington, DC (2004)
3. Hughes, M.P. (ed.): Nanoelectromechanics in Engineering and Biology. CRC Press, Boca Raton (2003)
4. Goddard, W.A., Brenner, D.W., Lyashevski, S.E., Lafrate, G.J. (eds.): Handbook of Nanoscience, Engineering and Technology. CRC Press, Boca Raton (2003)
5. Bashir, R.: Biological mediated assembly of artificial nanostructures and microstructures. In: Goddard, W.A., Brenner, D.W., Lyashevski, S.E., Lafrate, G.J. (eds.) Handbook of Nanoscience, Engineering and Technology, pp. 15-1–15-31. CRC Press, Boca Raton (2003)
6. Heller, M.J., Tullis, R.H.: Self-organizing molecular photonic structures based on functionalized synthetic DNA polymers. Nanotechnology 2, 165–171 (1991)
7. Hartmann, D.M., Schwartz, D., Tu, G., Heller, M.J., Esener, S.C.: Selective DNA attachment of particles to substrates. J. Mater. Res. 17(2), 473–478 (2002)
8. Sosnowski, R.G., Tu, E., Butler, W.F., O'Connell, J.P., Heller, M.J.: Rapid determination of single base mismatch in DNA hybrids by direct electric field control. Proc. Nat. Acad. Sci. USA 94, 1119–1123 (1997)
9. Heller, M.J., Tu, E., Holmsen, A., Sosnowski, R.G., O'Connell, J.P.: Active microelectronic arrays for DNA hybridization analysis. In: Schena, M. (ed.) DNA Microarrays: A Practical Approach, pp. 167–185. Oxford University Press, New York (1999)
10. Heller, M.J.: DNA microarray technology: devices, systems and applications. Annu. Rev. Biomed. Eng. 4, 129–153 (2002)
11. Heller, M.J., Tu, E., Martinsons, R., Anderson, R.R., Gurtner, C., Forster, A., Sosnowski, R.: Active microelectronic array systems for DNA hybridization, genotyping, pharmacogenomics and Nanofabrication Applications. In: Dekker, M., (eds.) Integrated Microfabricated Devices, Heller and Guttman, New York, pp. 223–270 (2002)
12. Gilles, P.N., Wu, D.J., Foster, C.B., Dillion, P.J., Channock, S.J.: Single nucleotide polymorphic discrimination by an electronic dot blot assay on semiconductor microchips'. Nat. Biotechnol. 17(4), 365–370 (1999)
13. Edman, C.F., Swint, R.B., Gurthner, C., Formosa, R.E., Roh, S.D., Lee, K.E., Swanson, P.D., Ackley, D.E., Colman, J.J., Heller, M.J.: Electric field directed assembly of an InGaAs LED onto silicon circuitry. IEEE Photonic. Tech. L. 12(9), 1198–1200 (2000)
14. US # 6,569,382, Methods and apparatus for the electronic homogeneous assembly and fabrication of devices, issued 27 May 2003
15. Mirkin, C.A., Letsinger, R.L., Mucic, R.C., Storhoff, J.J.: A DNA-based method for rationally assembling nanoparticles into macroscopic materials. Nature 382, 607–609 (1996)
16. Mardilovich, P., Kornilovitch, P.: Electrochemical fabrication of nanodimensional multilayer films. Nano Lett. 5(10), 1899–1904 (2005)
17. Hua, F., Shi, J., Lvov, Y., Cui, T.: Patterning of layer-by-layer self-assembled multiple types of nanoparticle thin films by lithographic technique. Nano Lett. 2(11), 1219–1222 (2002)
18. Dehlinger, D.A., Sullivan, B., Esener, S., Hodko, D., Swanson, P., Heller, M.J.: Automated combinatorial process for nanofabrication of structures using bioderivatized nanoparticles. J. Assoc. Lab Automation 12(5), 267–276 (2007)
19. Dehlinger, D.A., Sullivan, B.D., Esener, S., Heller, M.J.: Electric field directed assembly of biomolecular derivatized nanoparticles into higher order structures. Small 3(7), 1237–1244 (2007)
20. Dehlinger, D.A., Sullivan, B., Esener, S., Heller, M.J.: Directed hybridization of DNA nanoparticles into higher order structures. Nano Lett. 8, 4053–4060 (2008)

Electrical Impedance Cytometry

David Holmes[1] and Benjamin L. J. Webb[2]
[1]London Centre for Nanotechnology, University College London, London, UK
[2]Division of Infection & Immunity, University College London, London, UK

Synonyms

Impedance cytometry; Single cell impedance spectroscopy

Definition

Electrical impedance cytometry is a technique whereby the dielectric or impedance properties of biological cells are measured. An externally applied electric field is used to probe the cell or sample of cells. This can be achieved either through the application of one or more discrete excitation frequencies or via broadband frequency measurement techniques. Typically, a potential is applied between a pair of electrodes and the resulting current flowing through the system is measured. The impedance of the system is the ratio of the voltage to the current passing through the system. The dielectric properties of the cells can be derived from this measurement through the use of appropriate models. The development of microfluidics and lab-on-a-chip type devices has allowed single cell impedance measurements to be performed with high sensitivity and high throughput.

Overview

The electrical measurements of biological samples has a long history dating back over a century to the pioneering work in the 1910s by Höber, who measured the low and high frequency conductivity of suspensions of erythrocytes [1]. A decade later Fricke described the electrical properties of dispersive systems building on the work of Maxwell. Fricke was the first to measure the capacitive properties (8.1 mF/m^2) and thickness (3.3 nm) of the cell membrane – measurements which have subsequently been shown to be remarkably accurate [2]. The first single cell measurements were performed in the 1930s by Curtis and Cole [3].

Schwan [4] pioneered the field of biological impedance analysis; and was responsible for identifying the three main dielectric dispersions (α, β, γ) for suspensions of cells. These dispersions occur across a range of frequencies. The α-dispersion occurs at low-frequencies ($\sim$1 kHz) and arises from the polarization of the double layer around the particle. The β-dispersion occurs at intermediate frequencies (MHz) and arises from interfacial charging of the cell membrane. The γ-dispersion arises from the dipolar relaxation of water and occurs in the GHz range.

Experimental instrumentation and theoretical understanding have progressed in tandem, with highly sensitive techniques being developed, which allow the measurement of the dielectric properties of biological particles in fluid suspension. The majority of these techniques are however limited by the fact that they only provide an average value for the dielectric properties of a collection of particles. More recently, with the advent of micro- and nanofabrication techniques and the lab-on-a-chip, it has been possible to perform dielectric spectroscopic experiments on single biological particles suspended in physiological media.

This entry presents theory and models, which can be used to calculate the dielectric properties of cellular structures from impedance measurement. A practical realization of a single cell impedance measurement system is also presented along with data describing the electrical properties of single cells flowing in fluid suspension. Electrical measurement of cells growing on substrates and other more traditional bulk dielectric spectroscopy techniques (which measure the averaged properties of 100,000s of cells) is not covered; much theoretical and experimental data is available for these techniques. Such techniques are robust and well characterized with a number of commercial systems available.

Theoretical Background

Maxwell Mixture Theory

The single-shelled spherical model can be used to describe a cell in suspension. Figure 1a shows the spherical model of a cell comprising a single shell; the dielectric and geometrical parameters are shown.

The impedance of a dilute suspension of particles in a uniform electric field can be calculated in terms of individual particle dielectric properties by Maxwell's mixture theory; where the equivalent complex permittivity $\tilde{\varepsilon}_{mix}$ of a mixture of particles and suspending medium is given by:

$$\tilde{\varepsilon}_{mix} = \tilde{\varepsilon}_m \frac{1 + 2\Phi\tilde{f}_{CM}}{1 - \Phi\tilde{f}_{CM}} \quad (1)$$

where

$$\tilde{f}_{CM} = \frac{\tilde{\varepsilon}_p - \tilde{\varepsilon}_m}{\tilde{\varepsilon}_p + 2\tilde{\varepsilon}_m} \quad (2)$$

and $\tilde{\varepsilon} = \varepsilon - j\sigma/\omega$ is the complex permittivity, the subscripts m and p refer to the medium and particle respectively, $i^2 = -1$, ω the angular frequency, $\tilde{f}_{CM}$ is the Clausius-Mossotti and Φ is the volume fraction.

a

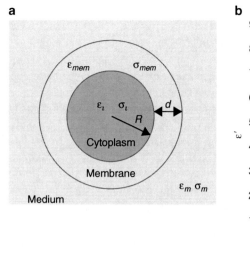

b

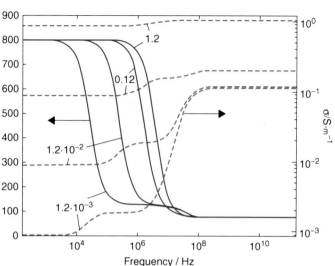

Electrical Impedance Cytometry, Fig. 1 (a) Single-shell electrical model of the spherical cell in suspension. (b) Plot showing the real and imaginary parts of the CM. The following parameters were used: $\varepsilon_o = 8.854 \times 10^{-12}$ F.m^{-2}, $R = 3 \times 10^{-6}$ m, $d = 5 \times 10^{-9}$ m, $\varepsilon_m = 80 \times \varepsilon_o$, $\varepsilon_{mem} = 5 \times \varepsilon_o$, $\sigma_{mem} = 10^{-8}$ S.m^{-1}, $\varepsilon_{\iota} = 60 \times \varepsilon_o$, $\sigma_{\iota} = 0.4$ S.m^{-1} and the medium conductivity varied between $\sigma_m = 1.2 \times 10^{-3}$ S.m^{-1} and 1.2 S.m^{-1}

The complex permittivity of the cell, $\tilde{\varepsilon}_p$ is a function of the dielectric properties of membrane and cytoplasm, cell membrane $\tilde{\varepsilon}_{mem}$ and internal properties $\tilde{\varepsilon}_{cyto}$, and cell geometry (where R is the inner radius and d the membrane thickness) given by:

$$\tilde{\varepsilon}_p = \tilde{\varepsilon}_{mem} \frac{\gamma^3 + 2\left(\dfrac{\tilde{\varepsilon}_{cyto} - \tilde{\varepsilon}_{mem}}{\tilde{\varepsilon}_{cyto} + 2\tilde{\varepsilon}_{mem}}\right)}{\gamma^3 - \left(\dfrac{\tilde{\varepsilon}_{cyto} - \tilde{\varepsilon}_{mem}}{\tilde{\varepsilon}_{cyto} + 2\tilde{\varepsilon}_{mem}}\right)} \quad (3)$$

where

$$\gamma = \frac{R + d}{R} \quad (4)$$

This is only valid for dilute suspensions ($\Phi < 10\%$). Interaction between the particles must be taken into account for higher particle concentrations. The mixture theory can be extended to high volume fractions:

$$1 - \Phi = \left(\frac{\tilde{\varepsilon}_{mix} - \tilde{\varepsilon}_p}{\tilde{\varepsilon}_m - \tilde{\varepsilon}_p}\right)\left(\frac{\tilde{\varepsilon}_m}{\tilde{\varepsilon}_{mix}}\right)^{1/3} \quad (5)$$

The impedance of the system is related to the complex permittivity as follows:

$$\tilde{Z}_{mix} = \frac{1}{j\omega\tilde{\varepsilon}_{mix}G_f} \quad (6)$$

where G_f is a geometric constant, which for an ideal parallel plate electrode system is simply the ratio of electrode area to gap A/g (m). Combining Eq. 6 with Eq. 1 allows the complex permittivity of a suspension of cells to be determined from the frequency-dependent impedance measurements. Typically, a cell suspension is measured using macroscopic planar or cylindrical electrode geometries where G_f can be easily defined. This is not the case for single cell impedance analysis, where the cell is located between a pair of microelectrodes of comparable size to that of the cell (ca. 10–20 μm). In this case the electric field is not uniform and the effect of the divergent field (fringing field) must be considered to correctly model the impedance. This involves detailed analysis of the field geometry and a modification to G_f [5, 6]. It is possible to use more complex electrode geometries, incorporating guard electrodes to confine the electric field and simplify the analysis; however, this requires careful electrode design and complicates the fabrication process.

Equivalent Circuit Model

For simplified analysis of the system, an electrical circuit analog is often used. Such an approach was

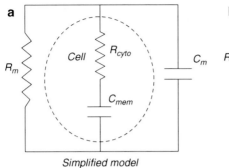
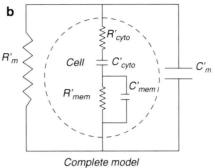

Electrical Impedance Cytometry, Fig. 2 Foster and Schwan's simplified circuit model for a single cell in suspension. (**a**) The cell is modeled as a resistor R_{cyto} (cytoplasm) and a capacitor C_{mem} (membrane) in series, with the suspending medium modeled as a resistor R_m and capacitor C_m. (**b**) The complete circuit model for a single-shelled particle in suspension. The particle is modeled as a resistor R'_{cyto} and a capacitor C'_{cyto} in series (cytoplasm) in combination with a resistor R'_{mem} and a capacitor C'_{mem} in parallel (membrane) (Adapted with permission from Sun et al. 2012, © 2010, Springer) [12]

first developed by Foster and Schwan [7]. The cell can be described in terms of an equivalent circuit model, as shown in Fig. 2.

Simplified Circuit Model

To a first level of approximation, the cell can be modeled as a resistor that describes the cytoplasm in series with a capacitor for the membrane as shown in Fig. 2a. The cell membrane resistance is generally much greater than the reactance of the membrane and can therefore be ignored. Likewise, the capacitance of the cell cytoplasm can be ignored when its reactance is small compared to the cell cytoplasm resistance. The values of the electrical components in the simple model circuit are given below. The suspending medium is described by:

$$R_m = \frac{1}{\sigma_m(1 - 3\Phi/2)G_f} \quad (7)$$

$$C_m = \varepsilon_\infty G_f \quad (8)$$

The simplified cell components are given by:

$$C_{mem} = \frac{9\Phi R C_{mem,0}}{4} G_f \quad (9)$$

$$R_{cyto} = \frac{4\left(\frac{1}{2\sigma_m} + \frac{1}{\sigma_{cyto}}\right)}{9\Phi G_f} \quad (10)$$

with specific membrane capacitance (per unit area) $C_{mem,0} = \varepsilon_{mem}/d$. The limiting high frequency permittivity of the suspension is related to the suspending medium permittivity according to:

$$\varepsilon_\infty \simeq \varepsilon_m\left[1 - 3\Phi\frac{\varepsilon_m - \varepsilon_{cyto}}{2\varepsilon_m + \varepsilon_{cyto}}\right] \quad (11)$$

This simplified circuit model has been used to interpret single cell impedance measurements providing good agreement with experiments (e.g., see Fig. 4).

Complete Circuit Model

For cases where the cell membrane conductance and cytoplasm capacitance values vary widely and cannot be ignored (e.g., during electroporation or cell lysis), a complete equivalent circuit model is required. This includes the resistance of the membrane and the capacitance of the cytoplasm. The complete circuit model is shown in Fig. 2b and the equations are given below. The suspending medium is described by:

$$R'_m = \frac{1}{\sigma_0 G_f} \quad (12)$$

$$C'_m = \varepsilon'_\infty G_f \quad (13)$$

The complete model cell components are described by:

$$R'_{mem} = \frac{1}{G_f}\left[\frac{\tau_1 + \tau_2}{\Delta\varepsilon_1 + \Delta\varepsilon_2} - \frac{1}{k_2 + k_3} - \frac{\tau_1\tau_2(k_2 + k_3)}{(\Delta\varepsilon_1 + \Delta\varepsilon_2)^2}\right] \quad (14)$$

$$C'_{mem} = \frac{\tau_1 \tau_2 (k_2 + k_3)}{(\Delta \varepsilon_1 + \Delta \varepsilon_2) R'_{mem}} \quad (15)$$

$$R'_{cyto} = \frac{1}{(k_2 + k_3) G_f} \quad (16)$$

$$C'_{cyto} = (\Delta \varepsilon_1 + \Delta \varepsilon_2) G_f \quad (17)$$

where R'_{mem} and C'_{mem} are the resistance and capacitance of the cell membrane, respectively. R'_{cyto} and C'_{cyto} are the resistance and capacitance of the cell cytoplasm, respectively. Full details of the above equations can be found in [8, 9].

Coulter Counter

Resistive pulse sensors are a specific case of the impedance cytometer. More commonly known as the Coulter counters, the technique measures the DC resistance between two electrically isolated fluid-filled chambers and the change in this current as a particle passes through a small connecting orifice [10]. Two large electrodes are positioned at either side of the orifice. As a particle passes through the orifice it displaces the conductive fluid and alters the resistance, this change results in a current pulse, the magnitude of which is proportional to the cell size. A simplified analysis of the Coulter counter was presented by Deblois and Bean [11]. The resistance of a tube, diameter D_t, length L_t, filled with electrolyte of resistivity ρ_m is:

$$R_t = \frac{4\rho_m L_t}{\pi D_t^2} \quad (18)$$

Sun et al. [12] showed that Maxwell's approximation can be used to evaluate the resistance of the tube containing a spherical particle (diameter d_p), giving an expression for the resistance, R_{mix} of a particle suspension:

$$R_{mix} = \frac{4\rho_{mix} L_t}{\pi D_t^2} = \frac{4\rho_m L_t}{\pi D_t^2} \left(1 + \frac{d_p^3}{D_t^2 L_t} \right) \quad (19)$$

The resistance change ΔR due to the particle is given by:

$$\Delta R = R_{mix} - R_t = \frac{4\rho_m d_p^3}{\pi D_t^4} \quad (20)$$

This expression is based on the assumption that the particle diameter is small compared with the tube (i.e., $\Phi \ll 1$), but ignores other issues such as access resistance to the tube. A solution with applicability over a broader range was given by Deblois and Bean [11]:

$$\Delta R_{(D_t/L_t \ll 1)} = \left(\frac{4\rho_m d_p^3}{\pi D_t^4} \right) \cdot F \left(\frac{d_p^3}{D_t^3} \right) \quad (21)$$

where $F(d_p^3/D_t^3)$ is a correction term that accounts for the nonuniformity of the current density within the tube.

Basic Methodology

The most successful implementation of a microfluidic impedance cytometry chip, in terms of measurement sensitivity and accuracy, is that of the facing electrode geometry as proposed by Gawad [13]. Other geometries have been implemented such as planar electrode geometries [14] and the so-called liquid electrodes [15]. The facing electrode implementation of a single cell impedance measurement system is shown in Fig. 3a. Electrodes on the upper and lower walls of the microfluidic channel are connected to an AC voltage source, which provides excitation signals at one or more frequencies. To minimize signal noise and drift a differential measurement is performed. The current flowing in the system is measured as a cell passes from one arm of the sensing circuit (two electrode pairs) to the other. The transit time between peaks, t_{tr} allows the particle velocity to be monitored, giving information about particle position within the channel and sample flow rates. An example of a microfluidic impedance cytometry chip is shown in Fig. 3b; the chip is constructed from glass substrates with platinum microelectrodes and channel sidewalls formed in polymer. The upper and lower channel walls (glass) allow for microscope observation of cells flowing though the device.

A simplified schematic of the impedance detection system is shown in Fig. 4. A differential detection circuit is used. The electrode pair on the left of Fig. 4a illustrates the equivalent circuit model for the detection region with no cell between the electrodes, the right electrode pair has the cell between the

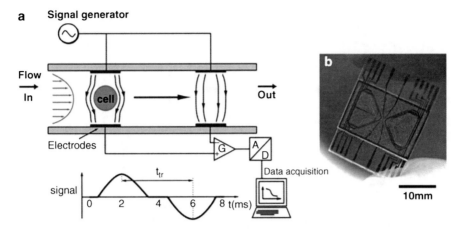

Electrical Impedance Cytometry, Fig. 3 (**a**) Schematic diagram showing the implementation of a single cell impedance measurement system. Electrodes on the upper and lower walls of the microfluidic channel are connected to an AC voltage source (providing excitation at one or more frequencies). The differential current flowing in the system is measured as a cell passes from one arm of the sensing circuit to the other. The transit time between peaks, t_{tr} allows the particle velocity to be monitored. (**b**) Photograph of a microfluidic impedance chip

electrodes. As described above, R_m and C_m are the equivalent resistance and capacitance of the medium, respectively, C_{mem} is the equivalent capacitance of the cell membrane, and R_{cyto} is equivalent resistance of the cell cytoplasm. C_{dl} represents the electrical double layer capacitance at the electrode–liquid interface. The impedance magnitude from a polymer bead (i.e., an insulating particle) and a cell of the same diameter are shown in Fig. 4b including a graphical indication of the frequency ranges over which the different structures of the cell have an influence on the impedance signal [16].

Key Research Findings

Single Cell Impedance Cytometer for Blood Analysis

One of the most commonly ordered tests in the doctor's office is complete blood count (CBC). Although simple and relatively inexpensive, this test is a very effective screening tool for a wide range of conditions. For example, a lower number of red blood cells could indicate anemia. A higher number of neutrophils could indicate an infection in progress, while changes in the number of lymphocytes could reveal diseases. Current standards for CBC follow stringent requirements for quality control and other regulations and rely on automated machines located either in hospitals or centralized laboratories. Impedance-based counting of

white blood cells and differentiating the major subtypes has been demonstrated and has the potential to replace the bulky hematology analyzers, currently only available in hospitals and centralized research labs.

Figure 5a shows the impedance magnitude versus frequency for purified leukocyte subpopulations (T-lymphocytes, monocytes, and neutrophils). The mean and standard deviation of $\sim$1,500 cells is plotted at each frequency point. The dashed lines in the plot show the best-fit circuit simulation (using PSPICE) incorporating the equivalent circuit model, electrical double layer, and differential amplification electronics of the system. Figure 5b shows the impedance-based analysis of clinical samples; the lymphocyte, neutrophil, and monocyte populations are clearly visible. Tests with patient samples are presented in Fig. 5c and show 95% correlation against commercial (optical/Coulter) blood analysis equipment, demonstrating the potential clinical utility of the impedance microcytometer for a point-of-care blood analysis system [16].

Impedance Labeling for Impedance Discrimination Based on Antigenic Expression

Electrical impedance cannot detect different antigen expressing cell subtypes with similar morphology; however, a method analogous to fluorescent antibody labeling was recently demonstrated which allows the discrimination of antigenically specific white blood cells [17]. This was demonstrated through the

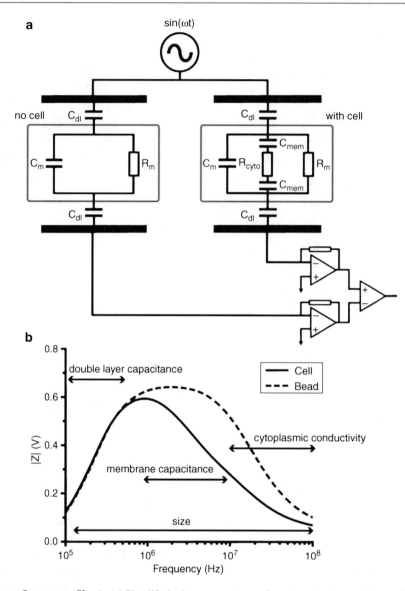

Electrical Impedance Cytometry, Fig. 4 (**a**) Simplified schematic of the impedance detection system shows the differential detection system. The electrode pair on the *left* illustrates the equivalent circuit model with no cell between the electrodes; the *right* electrode pair has the cell between the electrodes. R_m and C_m are the equivalent resistance and capacitance of the medium, respectively, C_{mem} is the equivalent capacitance of the cell membrane, R_{cyto} is equivalent resistance of the cell cytoplasm. C_{dl} represents the electrical double layer capacitance at the electrode–liquid interface. (**b**) The frequency-dependent impedance magnitude from a polymer bead (i.e., an insulating particle) and a cell of the same diameter (Adapted with permission from Holmes et al. 2009, copyright © 2009, RSC) [16]

impedance identification of lymphocyte subset expressing the CD4 surface protein. A suspension of blood cells was mixed with micron sized polymer beads coated with an antibody that specifically recognizes the CD4 antigen. The beads bind to the CD4 expressing cells (T-lymphocytes and monocytes), and in doing so change the impedance properties of these cells. Figure 6 shows the same blood sample with and without the addition of CD4 impedance labels. The bead complexed cells can be clearly seen as separate population on the scatter plot. The T-lymphocyte and monocyte populations are easily discriminated from one and other due to the larger size of the monocytes. The use of labels to selectively alter the impedance

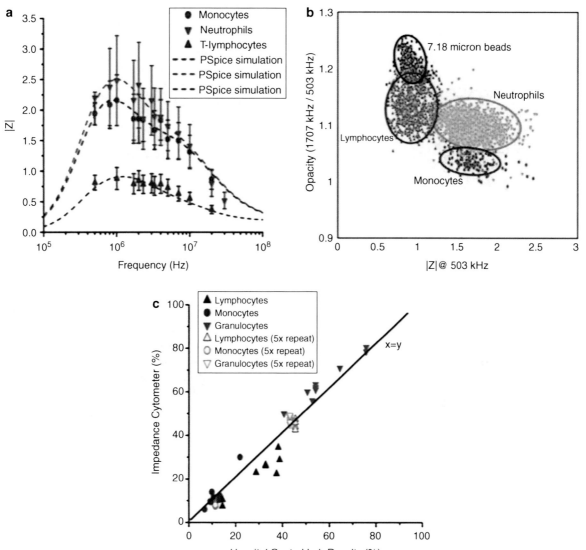

Electrical Impedance Cytometry, Fig. 5 (a) Data showing the impedance magnitude versus frequency for purified leukocyte subpopulations (T-lymphocytes, monocytes, and neutrophils). The mean and standard deviation of ~1,500 cells is plotted at each frequency point. The *dashed lines* show the best fit circuit simulation incorporating the equivalent circuit model, electrical double layer, and differential amplification electronics. (b) Data showing discrimination of leukocyte subpopulations from clinical blood sample. (b) Comparison of impedance-based blood analysis with hospital central lab analysis using a traditional hematology analyzer (Adapted with permission from Holmes et al. 2009, copyright © 2009, RSC) [16]

properties of certain cell subpopulations brings the molecular specificity of antibody labeling techniques to electrical cell discrimination.

Future Directions of Research

Impedance measurements represent a powerful noncontact method of analysis, with the potential to extract a wealth of information relating to the physiological and morphological properties of cells and other microscopic particles. The fact that the technique can probe cellular structures and extract information without the need for labeling makes it appealing for a range of diagnostic and other metrology applications where labeling can be expensive and may modify the cells under investigation (e.g., cell labeling can lead to activation).

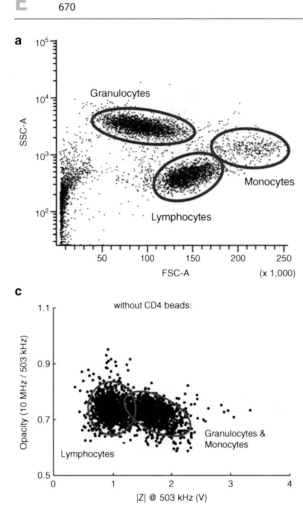

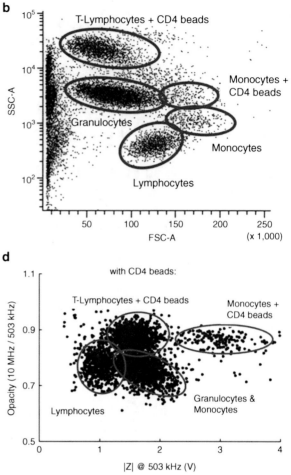

Electrical Impedance Cytometry, Fig. 6 (a) FACS data for saponin-lysed whole blood showing the three dominant leukocyte subpopulations. Some debris from the *red* cells is also visible, but the granulocyte and lymphocyte populations can be clearly identified, together with the smaller monocyte population (note axis on log scale). (b) The same blood sample complexed with CD4 beads, showing two new populations, the CD4+ T-lymphocytes and the subpopulation of monocytes that bind the beads. (c) and (d) show scatter plots from the impedance cytometer. The x-axis shows impedance measured at 503 kHz, which is a measure of cell size, and the y-axis shows the opacity (ratio of impedance at 10 MHz and 503 kHz). The CD4+ cell subpopulations are clearly visible in the impedance plot shown in (d) and are comparable with (b) (Adapted with permission from Holmes et al. 2010, copyright © 2010, ACS) [17]

Extension of the frequency range of the current impedance cytometry techniques to both higher and lower frequencies will allow probing of surface and internal properties of cells. Low-frequency (~1 kHz) measurement of the α-relaxation of cells is an important area of research, which has been poorly investigated to date due to the lack of experimental tools. High frequency (<50 MHz) measurements will allow probing of the intracellular compartment; current methods rely upon suspending cells in buffer of low conductivity to enhance the signal response and shift the membrane relaxation to lower frequencies (i.e.,

within the frequency response of the experimental apparatus). Impedance measurement of the cytoplasmic compartment will allow label-free monitoring of cell cycle, intracellular parasite detection (e.g., malaria), and a host of other biophysical measurements. Broadband frequency measurements have been implemented allowing full impedance spectra to be obtained rapidly and accurately [18].

Scaling down of impedance detection systems to allow measurements to be performed on viruses and other nanoparticles may be possible. A number of nonmammalian cell types have been measured using impedance cytometry (e.g., bacteria [19], marine algae [20]), and this combined with the integration of powerful sample preparation techniques and particle focusing will lead to the development of a range of powerful, robust, and portable devices, which will have relevance to a broad range of environmental, clinical, and life science applications.

Cross-References

▶ Electrical Impedance Tomography for Single Cell Imaging
▶ Micro- and Nanofluidic Devices for Medical Diagnostics
▶ Microfludic Whole-cell Biosensor
▶ On-Chip Flow Cytometry
▶ Single Cell Impedance Spectroscopy

References

1. Höber, R.: Eine Methode die elektrische Leitfaehigkeit im Innern von Zellenzu messen. Arch. Ges. Physiol. **133**, 237–259 (1910)
2. Fricke, H.: The electric capacity of suspensions of a red corpuscles of a dog. Phys. Rev. **26**, 682–687 (1925)
3. Curtis, H.J., Cole, K.S.: Transverse electric impedance of Nitella. J. Gen. Physiol. **21**, 189–201 (1937)
4. Schwan, H.P.: Electrical properties of tissue and cell suspensions. Adv. Biol. Med. Phys. **5**, 147–209 (1957)
5. Sun, T., Green, N.G., Gawad, S., Morgan, H.: Analytical electric field and sensitivity analysis for two microfluidic impedance cytometer designs. IET Nanobiotechnol. **1**, 69–79 (2007)
6. Linderholm, P., Seger, U., Renaud, Ph: Analytical expression for electric field between two facing strip electrodes in microchannel. Electron. Lett. IEE **42**, 145–147 (2006)
7. Foster, K.R., Schwan, H.P.: Dielectric properties of tissues and biological materials: a critical review. Crit. Rev. Biomed. Eng. **17**, 25–104 (1989)
8. Sun, T., Gawad, S., Green, N.G., Morgan, H.: Dielectric spectroscopy of single cells: time domain analysis using Maxwell's mixture equation. J. Phys. D: Appl. Phys. **40**, 1–8 (2007)
9. Sun, T., Bernabini, C., Morgan, H.: Single-colloidal particle impedance spectroscopy: complete equivalent circuit analysis of polyelectrolyte microcapsules. Langmuir **26**(6), 3821–3828 (2009)
10. Bayley, H., Martin, C.R.: Resistive-pulse sensing: from microbes to molecules. Chem. Rev. **100**, 2575–2594 (2000)
11. Deblois, R.W., Bean, C.P.: Counting and sizing of submicron particles by the resistive pulse technique. Rev. Sci. Instrum. **41**, 909–916 (1970)
12. Sun, T., Morgan, H.: Single-cell microfluidic impedance cytometry: a review. Microfluid. Nanofluid. **8**, 423–443 (2010)
13. Gawad, S., Cheung, K., Seger, U., Bertsch, A., Renaud, Ph: Dielectric spectroscopy in a micromachined flow cytometer: theoretical and practical considerations. Lab Chip **4**, 241–251 (2004)
14. Gawad, S., Schild, L., Renaud, Ph: Micromachined impedance spectroscopy flow cytometer for cell analysis and particle sizing. Lab Chip **1**, 76–82 (2001)
15. Demierre, N., Braschler, T., Linderholm, P., Seger, U., van Lintel, H., Renaud, Ph: Characterization and optimization of liquid electrodes for lateral dielectrophoresis. Lab Chip **7**, 355–365 (2007)
16. Holmes, D., Pettigrew, D., Reccius, C.H., Gwyer, J.D., Berkel, C.V., Holloway, J., Davie, D.E., Morgan, H.: Leukocyte analysis and differentiation using high speed microfluidic single cell impedance cytometry. Lab Chip **9**, 2881–2889 (2009)
17. Holmes, D., Morgan, H.: Single cell impedance cytometry for identification and counting of CD4 T-cells in human blood using impedance labels. Anal. Chem. **82**, 1455–1461 (2010)
18. Sun, T., Holmes, D., Gawad, S., Green, N.G., Morgan, H.: High speed multi-frequency impedance analysis of single particles in a microfluidic cytometer using maximum length sequences. Lab Chip **7**, 1034–1040 (2007)
19. Bernabini, C., Holmes, D., Morgan, H.: Micro-impedance cytometry for detection and analysis of micron-sized particles and bacteria. Lab Chip **11**(3), 407–412 (2011)
20. Benazzi, G., Holmes, D., Sun, T., Mowlem, M.C., Morgan, H.: Discrimination and analysis of phytoplankton using a microfabricated flow-cytometer. IET Nanobiotechnol. **1**(6), 94–101 (2007)

Electrical Impedance Spectroscopy

▶ Electrical Impedance Tomography for Single Cell Imaging

Electrical Impedance Tomography for Single Cell Imaging

Tao Sun
Research Laboratory of Electronics, Department of Electrical Engineering and Computer Science, Massachusetts Institute of Technology, Cambridge, MA, USA

Synonyms

Cellular imaging; Electrical impedance spectroscopy; Lab-on-a-chip; Rolled-up nanotechnology; Single cell analysis

Definition

Electrical impedance tomography (EIT) is an imaging technology that spatially characterizes the electrical properties of an object. It is an inverse problem: Given an object with unknown electrical properties, reconstruction of an image requires determination of the internal conductivity (or permittivity) profile of the object from a finite number of boundary measurements (current/voltage).

Introduction

The integration of mechanical elements, sensors, actuators, and electronics on a common substrate is known as Micro-Electro-Mechanical-Systems (MEMS). BioMEMS is a multidisciplinary subject across science and engineering and includes mathematics, biophysics, biochemistry, electrochemistry, microfluidics, micro/nano fabrication, integration, materials and surfaces, synthesis and detection technologies, accepted as a new concept – "Lab-On-a-Chip." For quantitatively understanding the functioning of the tissue and organisms, working at the cellular level is of prime interest for biomedical and clinical applications. Cellular analysis requires a combination of biophysical and biochemical approaches. Traditional analysis is performed with bulk techniques. However, measurements on large populations of cells can only provide average information. Individual cells, which look identical in appearance, generally have heterogeneous behavior. Moreover, the limitations of bulk techniques for studying the cell-to-cell and cell-to-surface interactions are apparent. Therefore, considerable attention is now paid to single cell analysis using microfluidic technologies [1].

Electrical impedance spectroscopy [2, 3] is a high-speed, noninvasive technique that is used to characterize the dielectric properties of biological cells. In the simplest sense, a biological cell consists of a conducting cytoplasm (with nucleus) covered by an ultrathin insulating cell membrane. When exposed to an ac electric field, at low frequencies the cell behaves as an insulating object. The cell membrane is equivalent to a capacitor and prevents the electric field lines from penetrating the cell. Therefore, information concerning cell size (and shape) can be obtained by measuring the electrical properties of the cell at low frequencies. At higher frequencies, the cell membrane

electrically becomes short-circuited. Thus, the internal properties of the cell (i.e., cytoplasmic resistance) can be probed. For adherent cells, the kinetics during culture, such as cell growth, differentiation, and the effect of drugs can be monitored using electrical impedance. Cells are grown on the surface of an array of electrodes and the impedance is monitored as a function of time. This technique is termed electric cell-substrate impedance sensing [4].

In electrical impedance tomography (EIT) [5], measurements of a sample are made between multiple electrodes. The technique generates images of the conductivity distribution inside an object. Although, the spatial resolution of EIT cannot be compared with magnetic resonance imaging, the technique is harmless, less costly, and has good temporal resolution.

Electrical Impedance Tomography

EIT is an inverse problem: Given an object with unknown electrical properties, reconstruction of an image requires determination of the internal conductivity (or permittivity) profile of the object from a finite number of boundary measurements (current/voltage). The theory of EIT is composed of two parts: the forward problem and the inverse problem. The forward problem corresponds to the calculation of the potentials on the electrodes for a known injected current and conductivity distribution. Solving the forward problem involves constructing the global Jacobian (sensitivity) matrix of the system, which is required for image reconstruction and the solution of the inverse problem. For EIT, the electric field distribution in considered quasi-electrostatic. If only low frequency ac is used, the analysis is restricted to mapping the conductivity distribution of the object. Therefore, the electrical potential ϕ, satisfies Laplace's equation:

$$\nabla \bullet \sigma \nabla \phi = 0 \tag{1}$$

Equation 1 gives a unique solution for the electric potential distribution, if the boundary conditions are defined. The forward problem is well-posed in terms of existence, uniqueness and stability. However, in the case of EIT, since the potential distribution is dependent on the conductivity distribution, it cannot be solved analytically for arbitrary values of σ. Therefore, a general approach is to set up a numerical model that can simulate the potential values on the electrodes as close as possible to the measured voltage data in an experiment. The numerical model discretizes the domain into small elements so that an approximate solution can be obtained. Using the finite element method (FEM) with the defined boundary conditions, the governing equation of the system is:

$$A \begin{bmatrix} \phi_p \\ V_l \end{bmatrix} = \begin{bmatrix} 0 \\ I \end{bmatrix} \quad \text{with} \quad A = \begin{bmatrix} A_C & A_E \\ A_E^T & A_D \end{bmatrix} \quad (2)$$

where A is the global admittance matrix of the system. The expressions of the compartments A_C, A_E, and A_D can be found in [6]. The potential on every node ϕ_p in the finite element model and the voltage on every electrode V_l can be obtained by solving Eq. 2.

Image reconstruction requires the knowledge of the conductivity distribution. In EIT, the changes in conductivity σ_q within the domain, and the changes in the potential distribution ϕ_p are linked by the Jacobian (sensitivity) matrix $J_{p,q}$:

$$J_{p,q} = \frac{\partial \phi_p}{\partial \sigma_q} \quad (3)$$

The inverse problem corresponds to calculation of the conductivity distribution when the injected current is known and the voltages are measured. It is "ill-posed" since the solutions of the inverse problem are not unique. The process of solving the inverse problem requires finding a stable value of σ, so that the difference between the numerical simulations $\phi(\sigma)$ and the measured voltage V is minimized.

$$f(\sigma) = \frac{1}{2}(\phi(\sigma) - V)^*(\phi(\sigma) - V)$$
$$= \frac{1}{2}\|\phi(\sigma) - V\|^2 \quad (4)$$

To find the minima of the objective function $f(\sigma)$, a Taylor series is used to expand $f(\sigma)$, neglecting higher-order terms in Eq. 6:

$$f(\sigma + \Delta\sigma) = f(\sigma) + f'(\sigma)\Delta\sigma + \frac{1}{2}f''(\sigma)\Delta\sigma^2 \quad (5)$$

Using the Newton–Raphson method to search for a step $\Delta\sigma$, such that $f(\sigma + \Delta\sigma) \cong 0$, the iterative solution of Eq 5 is:

$$\sigma_{n+1} = \sigma_n + \Delta\sigma \quad n > 0 \quad (6)$$

$$\Delta\sigma = -[Hf(\sigma_n)]^{-1}\nabla f(\sigma_n) \quad (7)$$

where $Hf(\sigma_n)$ is the Hessian matrix and $\nabla f(\sigma_n)$ is the gradient, given by:

$$\nabla f(\sigma) = f'(\sigma) = \phi'(\sigma)^*(\phi(\sigma) - V) \quad (8)$$

$$Hf(\sigma) = \phi'(\sigma)^*\phi'(\sigma) + \phi''(\sigma)(\phi(\sigma) - V) \quad (9)$$

For simplicity, the second-order derivative term in Eq. 9 can be neglected. Therefore, the expression for the adjusting step $\Delta\sigma$ becomes Eq. 6:

$$\Delta\sigma = -[\phi'(\sigma)^*\phi'(\sigma)]^{-1}\phi'(\sigma)^*(\phi(\sigma) - V)$$
$$= \phi'(\sigma)^\dagger(V - \phi(\sigma)) \quad (10)$$

where $\phi'(\sigma)^\dagger$ is the Moore–Penrose pseudo-inverse of $\phi'(\sigma)$, which is the inverse of the Jacobian matrix (Eq 3).

The solutions given by Eqs. 6 and 10 are based on the Gauss–Newton algorithm. Other algorithms can be used to minimize Eq. 4. For example, the linear generalized Tikhonov regularization can be encapsulated into the Gauss–Newton algorithm, which gives calculation of $\Delta\sigma$ as:

$$\Delta\sigma = [\phi'(\sigma)^*\phi'(\sigma) + \lambda^2 R^* R]^{-1}(\phi'(\sigma)^* V + \lambda^2 R^* R\sigma) \quad (11)$$

where R is either the identity matrix or other regularization matrix including certain *image prior* assumptions about σ for reconstruction. λ is a positive scalar regularization parameter.

Over the past 20 years, a large number of reconstruction algorithms have been developed to improve the computational efficiency, reconstruction sensitivity, and image fidelity. Examples include the modified Newton–Raphson method, back-projection algorithm, variational method, maximum a posteriori approach, Monte Carlo sampling method, and direct reconstruction algorithms. Most biomedical and clinical uses of EIT are in vivo and used to detect disease such as breast cancer [7] or to monitor brain function [8]. Interest has centered on in vitro applications, e.g., Davalos et al. [9] proposed using EIT for detecting and imaging electroporation of cells in tissue in real time. Recently, there has been an interest in miniaturizing the technology. Hou et al. [10] patterned 32 electrodes in a rectangular configuration to provide a two-dimensional map of the conductivity of carbon nanotube thin films. Chai et al. [11] fabricated a set of 4 × 4 complementary metal oxide semiconductor microelectrode array to image a cell culture environment. Linderholm et al. [12] used an array of interdigitated electrodes to monitor the migration, stratification, and influence of detergent on cells in culture.

EIT for Single Cell Imaging

Individual cells can be identified from differences in size and dielectric properties using electrical techniques. Recently, an on-chip two-dimensional cell imaging technology using EIT was developed [13]. The cell can be cultured at any position within the circular array for imaging (Fig. 1). To perform EIT, a stimulating signal is injected through one pair of electrodes and the voltage across all sequential electrode-pair combinations measured. The electrical stimulations and measurements are noncontact to the cell. The perturbation in the conductivity distribution in the system is characterized by the voltage data, and images the cell are reconstructed using software termed "Electrical Impedance Tomography and Diffuse Optical Tomography Reconstruction Software," EIDORS [14]. The first step toward single cell impedance tomography was demonstrated by imaging a large multinuclear single-cell organism – the amoeboid plasmodium of the slime mold *Physarum polycephalum*, Fig. 1b.

EIT Chip Fabrication

The EIT chip contains 16 electrodes positioned in a circular pattern, and made from a printed circuit board (PCB). The electrode length and spacing are identical, the electrodes are 35μm thick and the area of an individual electrode is 2.06 mm². A thin gold layer was galvanostatically plated onto the surface of the copper electrodes using a three-electrode potential station (PGSTAT128N, Autolab electrochemical instruments, the Netherlands) with a platinum-coated metal mesh counter electrode and Ag/AgCl electrode (REF321, Radiometer Analytical S. A., France) reference electrode. The thickness of the gold layers was 500 nm and controlled by adjusting the deposition

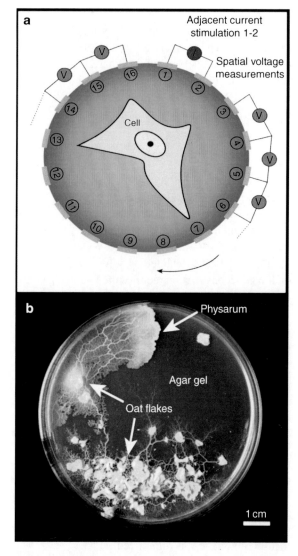

Electrical Impedance Tomography for Single Cell Imaging, Fig. 1 (**a**) Diagram showing the principle of electrical impedance tomography (EIT) for imaging single cell in culture in a circular electrode pattern. (**b**) Photograph showing *Physarum polycephalum* cultured (fed with oat flakes) on the surface of agar gel

time. To culture the plasmodium, a polydimethylsiloxane (PDMS) sheet (1 mm thick) with a 6 mm diameter hole was clamped over the EIT chip. A thin layer of agar gel was deposited over the surface of the electrodes, within the PDMS chamber, over which, the *Physarum polycephalum* was cultured. This avoids direct contact between the electrodes and the *Physarum polycephalum*. To eliminate evaporation, a second PDMS sheet and a polymethylmethacrylate (PMMA) plexiglass lid was used, as shown in Fig. 2.

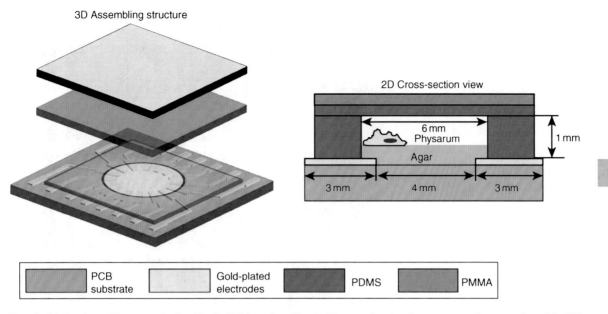

Electrical Impedance Tomography for Single Cell Imaging, Fig. 2 Diagram showing the structure and cross section of the EIT chip used to image *Physarum* in culture

System Instrumentation

Signal stimulations and measurements were performed using an Alpha-A impedance analyzer (Novocontrol Technologies, Germany) operating in four-electrode impedance measurement mode. A custom-designed electronic platform was built for multiplexed measurements, switching the pattern of stimulating signal and measuring voltages on different combinations of electrode pairs. Analogue chips, consisting of 16 × 8 crosspoint switches (Intersil CD22M3494) were used as multiplexers. The impedance analyzer and the multiplexing board were both controlled by a Linux PC through a GPIB-PCI interface (NI PCI-GPIB, National Instruments, USA) and a USB controller (FT2232D, FTDI Ltd. UK), respectively. Custom software was used for automated real-time measurements. The software sets the amplitude and frequency of the stimulating signal, the digital signals to the address lines of the multiplexers and controls the data acquisition. The highest temporal resolution of individual voltage measurement is 100 ms, which is limited by the data rate of the impedance analyzer.

EIT Measurement Scheme

The adjacent method was used for signal stimulation and voltage measurements. This method is more sensitive to variations in the impedance of objects near the electrodes than in the center of the chip. The stimulating signal is first applied through electrodes 1 and 2 (Fig. 1a) and the voltage measured from successive electrode pairs: 3-4, 4-5, ..., 15-16, giving 13 voltage measurements. Note that voltages across electrode pair 16-1, 1-2, and 2-3 cannot be measured since the stimulating signal is injected from electrode pair 1-2. Then a second set of measurements is made with the current applied through electrodes 2 and 3 giving a further 13 sets of voltage data. This process is repeated, generating a total of 16 × 13 = 208 voltage measurements, which requires approximately 20 s for a full scan. Owing to reciprocity, only half of the measurements are linearly independent, all the data are used in the image reconstruction.

Physarum polycephalum in Culture

The plasmodium of the slime mold *Physarum polycephalum* is an amoeboid multinucleated single cellular organism. It consists of fanlike advancing fronts connected by tubular structures found in the posterior region (Fig. 1b). This organism has been long used as a model organism for the study of mitosis, membrane potential, and protoplasmic streaming. A *Physarum* plasmodium can extend to over several centimeters in diameter; nevertheless, it still maintains itself as a single cell by protoplasmic flow. Being

Electrical Impedance Tomography for Single Cell Imaging, Fig. 3 (a) Optical images of *Physarum polycephalum* on the agar gel in the EIT chips, (b) Corresponding reconstructed images. The *Physarum* cell is more conductive than the agar gel, exhibiting as bright (*red*) region in the reconstructed images. (c) Filtered images, the ghost images are eliminated by filtering the low conductivity regions

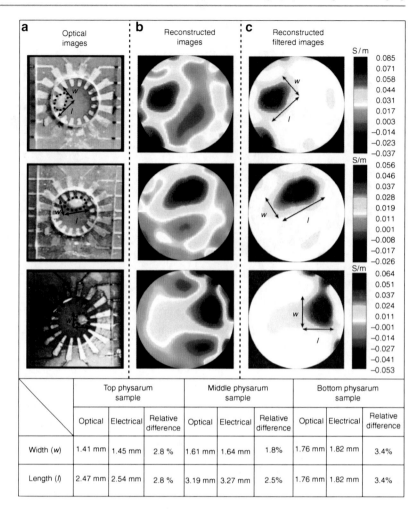

	Top physarum sample			Middle physarum sample			Bottom physarum sample		
	Optical	Electrical	Relative difference	Optical	Electrical	Relative difference	Optical	Electrical	Relative difference
Width (*w*)	1.41 mm	1.45 mm	2.8 %	1.61 mm	1.64 mm	1.8%	1.76 mm	1.82 mm	3.4%
Length (*l*)	2.47 mm	2.54 mm	2.8 %	3.19 mm	3.27 mm	2.5%	1.76 mm	1.82 mm	3.4%

a multi-nucleated cell, even if a cell is split into pieces, each piece can keep alive and act as single cell. It can be regarded as an autonomous distributed system and recently, its oscillatory behavior has been utilized directly for robotic control applications.

Results and Discussions

Different shapes of *Physarum polycephalum* were grown on the surface of the agar in different positions within the chip as shown in Fig. 3a. Two sets of voltage data were measured using the adjacent method (stimulating signal = 100 mV @ 200 kHz). One set of data was taken for pure agar gel in a closed EIT chip, providing a homogeneous background. The second set of data was measured after *Physarum* was grown

on the agar. Images were reconstructed using 5,184 elements in the forward model. The initial value of *hyper-parameter* was set to 0.1 and adjusted in the range between 0.05 and 0.2 for smooth boundaries between *Physarum polycephalum* and the background. Figure 3b shows the corresponding reconstructed images from the voltage data. Brighter (red) corresponds to higher conductivity and darker (blue) to lower conductivity. The brighter region where the *Physarum* cell locates (positive values in the reconstructed differential conductivity) can be clearly seen. The mirror ghost images are equivalent to regions of lower conductivity (negative values in the reconstructed differential conductivity), and produce variations in the reconstructed conductivity values. Because the ghost images are low conductivity regions, and the difference between *Physarum* and

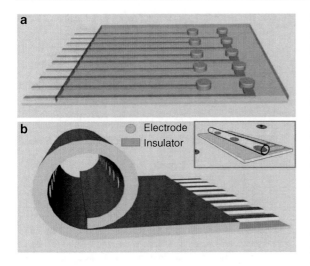

Electrical Impedance Tomography for Single Cell Imaging, Fig. 4 Schematic illustration of the proposed structure of rolled-up microspectroscope, integrated with pairs of microelectrodes for single cell imaging

the background is positive, the ghost images were eliminated by setting a threshold in the reconstructed conductivity, as shown in Fig. 3c. There are other artifacts, exhibiting both high and low conductivity regions, distributed along the edges of the reconstructed images, which may come from the incompleteness of the one-dimensional segment model of the electrodes in the forward model.

To analyze the resolution of the reconstructed images, the dimensions (the width and the length) of the *Physarum* cell from the optical and electrical images were compared. Dimensions of three samples (from top to bottom) are listed below the images in Fig. 3. The absolute difference in the dimensions between the optical and the reconstructed images ranges from 30μm to 80 μm. The relative difference, calculated as the ratio of the absolute difference to the dimension in the optical image, shows that the minimum is 1.8% and the maximum is 3.4%. The average size of a single *Physarum* cell is approximately 2 mm in one dimension.

Conclusion and Outlook

The developed system would be useful in diagnostic and biomedical applications, such as cancer and cancer treatment monitoring for characterizing the spatial electrical properties of tumor cell colonies. Understanding the tumor growth is essential for screening programs, clinical trials, and epidemiological investigations. Skourou et al. [15] demonstrated the feasibility of using electrical impedance technique to detect small tumors (<3 mm) and tumor-associated changes. The tumor growth varies but is generally from a few mm to tens of mm [16], which falls into the imaging scale and resolution of the developed system. It can be expected that electrically mapping the electrical properties of tumor spatially will benefit the study in the dynamics in tumor cells spreading and proliferation.

Although impedance-based microfluidic cytometry [2, 3] has been widely used for cell manipulation and characterization using various approaches, the microfabrication of the devices leads to the difficulties in bonding the top and bottom of the microchannels, aligning the micro-features (i.e., microelectrodes) exactly where they are required, fabricating the microelectrodes on the sidewalls of the microchannels, etc. A neat and smart technology is required to effectively address these problems in order to develop novel applications and push microfluidic cytometry into commercial markets, advocating point-of-care and on-side analysis.

A promising route is to use rolled-up microfabrication technology [17] to compensate the conventional microfabrication process. By using this technology, microchannels can be fabricated with a smooth inner/outer surface and a near to perfect circular cross section. Also, the microelectrodes can be located where they are needed with good control in position and alignment, thus offering the opportunities to address the above problems. Furthermore, the rolled-up technology can be applied to many materials and material combinations and thus it is easy to fabricate microchannels with customized surface chemical properties provided by the appropriate rolled-up materials. However, up to now, only few reports can be found to apply this technology toward bio-applications. Very recently, researchers fabricated rolled-up transparent microtubes as confined culture scaffolds for individual yeast cells [18]. Good biocompatibility of rolled-up microtubes has been demonstrated by monitoring the growth of yeast cells inside the microtubes. It would be interesting to explore the rolled-up microtubes from microfluidic approaches in single cell analysis applications, where different patterns of microelectrodes will be integrated into the inner surface of the rolled-up microtubes for single cell manipulation and characterization under laminar microflow. Cells will be flown and trapped inside the microtube. Rolled-up multiple layers of circular electrode-arrays allows performing 3D single cell imaging [19], as shown in Fig. 4. The dimensions (diameter and length) of the microspectroscopes can be easily controlled by lithography, etching, and deposition techniques. Therefore, the tubular microspectroscopes can be designed for analyzing different sizes of biological cells to obtain the optimal signal. The microelectrodes can be fabricated exactly at the desired positions and also desired shapes; hence, by using conventional photolithography, different cell analyzing functions can be fabricated on a single chip that integrates multifunctions for cell manipulation, identification, and imaging. Particularly, compared to the conventional microfluidic cytometry, the microelectrodes can be fabricated in a circular pattern inside the tubular microchannels. This allows many cell analysis applications to be extended from 2D to 3D, such as 3D single cell imaging using EIT by fabricating multiple sets of electrode-arrays inside the microtubes. Very recently, Caselli et al. [20] reported the simulation work of single cell imaging in a microfluidic cytometry using EIT. The authors reported detailed finite element modeling and numerical simulation results to demonstrate the feasibility of the device and method to extract information on cell morphology and to discriminate normal biconcave erythrocytes, spherocytes, and echynocytes. However, no device fabrication and experimental result has been reported yet. In summary, high-throughput and high-resolution microfluidic single cell impedance tomography is an important area that will continue to develop. It would be expected that 3D single cell imaging can lead to integrated point-of-care systems for biomedical and diagnostic applications.

Cross-References

▶ Electrical Impedance Cytometry
▶ Microfludic Whole-cell Biosensor

References

1. Whitesides, G.M.: The origins and the future of microfluidics. Nature **442**, 368–373 (2006)
2. Sun, T., Holmes, D., Gawad, S., Green, N.G., Morgan, H.: High speed multi-frequency impedance analysis for single particles in a microfluidic cytometer using maximum length sequences. Lab Chip **7**, 1034–1040 (2007)
3. Sun, T., Morgan, H.: Single-cell microfluidic impedance cytometry: a review. Microfluid. Nanofluid. **8**, 423–443 (2010)
4. Keese, C.R., Wegener, J., Walker, S.R., Giaever, I.: Electrical wound-healing assay for cells in vitro. PNAS **101**, 1554–1559 (2004)
5. Bayford, R.H.: Bioimpedance tomography (electrical impedance tomography). Ann. Rev. Biomed. Eng. **8**, 63–91 (2006)
6. Polydorides, N., Lionheart, W.R.B.: A matlab toolkit for three-dimensional electrical impedance tomography: a contribution to the electrical impedance and diffuse optical reconstruction software project. Meas. Sci. Technol. **13**, 1871–1883 (2002)
7. Cherepenin, V., Karpov, A., Korjenevsky, A., Kornienko, V., Mazaletskaya, A., Mazourov, D., Meister, D.: A 3D electrical impedance tomography (EIT) system for breast cancer detection. Physiol. Meas. **22**, 9–18 (2001)
8. Bagshaw, A.P., Liston, A.D., Bayford, R.H., Tizzard, A., Gibson, A.P., Tidswell, A.T., Sparkes, M.K., Dehghani, H., Binnie, C.D., Holder, D.S.: Validation of reconstruction algorithms for electrical impedance tomography of human brain function. Neuroimage **20**, 752–764 (2003)
9. Davalos, R., Otten, D.M., Mir, L.M., Runbinsky, B.: Electrical impedance tomography for imaging tissue electroporation. IEEE Trans. Biomed. Eng. **51**, 761–767 (2004)
10. Hou, T.-C., Loh, K.J., Lynch, J.P.: Spatial conductivity mapping of carbon nanotube composite thin films by electrical impedance tomography for sensing applications. Nanotechnology **18**, 315501 (2007)
11. Chai, K.T.C., Davies, J.H., Cumming, D.R.S.: Electrical impedance tomography for sensing with integrated microelectrodes on a CMOS microchip. Sensors Actuators B **127**, 97–101 (2007)
12. Linderholm, P., Marescot, L., Loke, M.H., Renaud, Ph: Cell culture imaging using microimpedance tomography. IEEE Trans. Biomed. Eng. **55**, 138–146 (2008)
13. Sun, T., Tsuda, S., Zauner, K.-P., Morgan, H.: On-chip electrical impedance tomography for imaging biological cells. Biosens. Bioelectron. **25**, 1109–1115 (2010)
14. Adler, A., Lionheart, W.R.B.: Uses and abuses of EIDORS: an extensible software base for EIT. Physiol. Meas. **27**, S25–S42 (2006)
15. Skourou, C., Hoopes, P.J., Strawbridge, R.R., Paulsen, K.: Feasibility studies of electrical impedance spectroscopy of early tumor detection in rats. Physiol. Meas. **25**, 335–346 (2004)
16. Weedon-Fekjær, H., Lindqvist, B.H., Vatten, L.J., Aalen, O. O., Tretli, S.: Breast cancer tumor growth estimated through mammography screening data. Breast Cancer Res. **10**, R41 (2008)
17. Schmidt, O.G., Eberl, K.: Thin solid films roll up into nanotubes. Nature **410**, 168 (2001)
18. Huang, G.S., Mei, Y.F., Thurmer, D.J., Coric, E., Schmidt, O.G.: Rolled-up transparent microtubes as two-dimensionally confined culture scaffolds of individual yeast cells. Lab Chip **9**, 263–268 (2009)
19. Sun, T.: Microfluidic-based rolled-up 3D electrical impedance spectroscope for single cell analysis, Research Grant

Proposal for Alexander von Humboldt Fellowship (unpublished material, 2009)

20. Caselli, F., Bisegna, P., Maceri, F.: EIT-inspired microfluidic cytometer for single-cell dielectric spectroscopy. J. Microelectromech. Syst. **19**, 1029–1040 (2010)

Electric-Field-Assisted Deterministic Nanowire Assembly

Theresa S. Mayer[1], Jeffrey S. Mayer[2] and Christine D. Keating[3]
[1]Department of Electrical Engineering and Materials Science and Engineering, The Pennsylvania State University, University Park, PA, USA
[2]Department of Electrical Engineering, Penn State University, University Park, PA, USA
[3]Department of Chemistry, Penn State University, University Park, PA, USA

Synonyms

Dielectrophoretic assembly; Directed assembly

Definition

Electric-field-assisted deterministic assembly uses the force induced on individual nanowires by a nonuniform electric field to position each nanowire with submicron registration accuracy relative to previously defined nanodevices on a substrate.

Overview

The electric-field force induced on solution-suspended, polarizable nanowires by a nonuniform alternating electric field can be used to attract, align, and position individual nanowires at predefined locations on a patterned substrate (e.g., Si integrated circuit) [1–4]. The magnitude and direction of the spatially varying electric-field force, which together control the final position and orientation of the nanowires, are determined by electrode structures on the substrate, bias voltages applied to the electrodes, the geometry and material properties of the nanowire, and material properties of the solution (Fig. 1a).

The force field can be tailored to provide submicron registration (overlay) accuracy between each assembled nanowire and a specific nanoscale feature on the substrate at integration densities $>10^6$ nanowires/cm^2 and with yields $>95\%$ [5]. Additionally, different nanowire populations (e.g., materials, coatings, etc.) can be directed to particular regions of the substrate by synchronizing the sequential delivery of each population with a unique electric-field force pattern [6]. Once assembled, the nanowires can be converted into one of many different device types – nanomechanical, electronic, optoelectronic – by adding mechanical and/or electrical contacts using conventional lithographic, deposition, and etching processes (Fig. 1b) [5, 6].

Basic Methodology

The electric-field-assisted deterministic nanowire assembly process can be divided into two phases: (1) attraction of the solution-suspended nanowires to a predefined area on the substrate and (2) alignment and positioning of the nanowires relative to other assembled wires or lithographic features on the substrate. In the first phase, the nanowires are suspended in the fluid far above the dielectric-coated substrate in a region where the gradient of the alternating electric field produced by the biased electrodes is relatively small but sufficient to polarize the individual nanowires. Each of the polarized nanowires experiences a force that is well modeled using the dielectrophoretic (DEP) force equation [7]:

$$\vec{F}_{DEP} = (\vec{p} \cdot \nabla)\vec{E},$$

where $\vec{p}$ is the effective dipole moment of the particle resulting from its polarization by the electric field $\vec{E}$. More specifically, numerical solutions of the force using the Maxwell stress tensor (MST) method show that it is proportional to $\nabla|\vec{E}|^2$ [8], which is the same field dependence as $\vec{F}_{DEP}$ for a spherical dielectric particle analyzed by Pohl [7]. In general, the force is significantly larger than the other hydrodynamic, gravitational, and Brownian forces acting on the nanowires, and causes the wires to align tangent to the electric field lines and to move in the direction of the highest field gradient. For a simple interdigitated electrode structure, the nanowires are attracted from solution

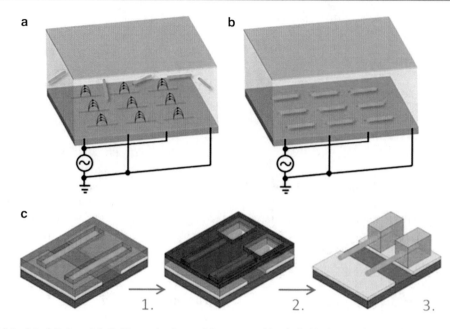

Electric-Field-Assisted Deterministic Nanowire Assembly, Fig. 1 *Deterministic Nanowire Assembly.* (**a**) Nonuniform electric fields originating from electrically biased electrode structures on the substrate extend into the fluid medium. (**b**) Electric-field forces acting on the solution-suspended nanowires are used to position individual nanowires within lithographic wells on the substrate. (**c**) Nanowires assembled in the wells are converted into cantilever devices by patterning windows to contact one end of each wire (step 1), electroplating metal clamps (step 2), and dissolving the sacrificial photoresist (step 3). After Ref. [5]

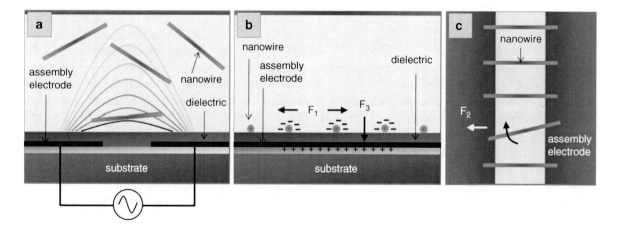

Electric-Field-Assisted Deterministic Nanowire Assembly, Fig. 2 *Deterministic Nanowire Assembly.* (**a**) Cross section of interdigitated electrode structure showing electric field lines. *Dark gray* designates the highest and light gray the lowest electric field magnitude. The nanowires align tangent to the field lines and are attracted to the high field region in the gap between the electrodes. (**b, c**) Electric-field forces due to like charge on the wire tips (F_1), opposite charge on the wires and on the biased metal electrode (F_2, F_3). Together these forces form uniformly spaced, aligned nanowire arrays

and are aligned such that they span the gap separating the biased electrodes (Fig. 2a).

In the second phase of the assembly process, the nanowires are on the substrate, and the induced charge distribution and electric field are more complex, resulting in forces between neighboring wires as well as between these same wires and the buried electrodes (Fig. 2b, c). In the first case, the like charge on the

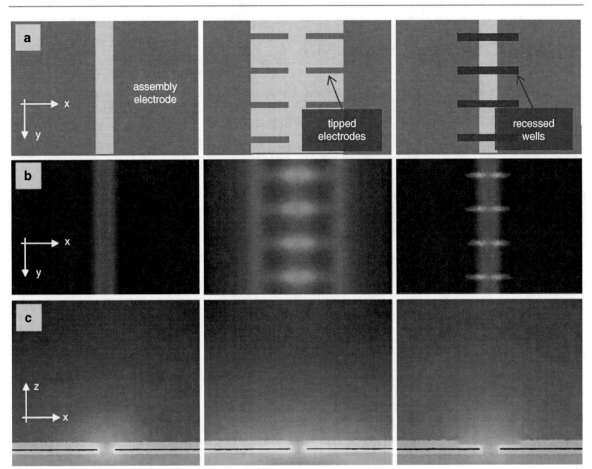

Electric-Field-Assisted Deterministic Nanowire Assembly, Fig. 3 *Electrode Structures.* (**a**) *Top and cross-sectional views* of the three different electrode structures all having a gap width of 5 μm. The sharp tipped electrodes and recessed wells are 1 μm wide and 10 μm long. (**b**) *Top view* of the electric field magnitude measured at the dielectric coating/solution interface. (**c**) *Cross-sectional view* of the electric field magnitude (plotted on a log scale) measured along the *center* of the tipped electrodes and recessed wells. *Dark gray* designates the highest and *light gray* the lowest electric field magnitude

nanowire tips gives rise to a net repulsive force between neighboring wires. This results in the formation of uniformly spaced arrays at high wire concentration. In the second case, if an aligned nanowire is offset from the center of the gap, the opposite charge on the wire and on the metal electrodes produces a lateral restoring force and torque. This causes the wires to align, centered across the gap, with end-to-end registration. It also fixes the position of the wires on the substrate when the solution is removed. As described in the next section "Key Research Findings," by tailoring the electrode structure design, the electric-field profile and forces the can be optimized to position individual wires with submicron registration accuracy relative to previously defined nanodevices on the substrate.

Key Research Findings

Assembly Electrodes and Electric-Field Profiles

The electric-field forces used for nanowire assembly are defined by the nonuniform field originating from the electrically biased electrode structure. Three different structures are presented in Fig. 3 to illustrate the range of field profiles that can be realized: (1) electrodes with a uniform-width gap and a planar dielectric coating, (2) electrodes with sharp tips and a planar dielectric coating, and (3) electrodes with a uniform-width gap and a dielectric coating patterned with wells that span the gap. Figure 3 also shows simulations of the electric field magnitude (COMSOL Multiphysics) measured at the interface between the dielectric coating and the solution (x–y plane, Fig. 3b) and through

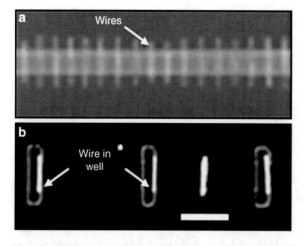

Electric-Field-Assisted Deterministic Nanowire Assembly, Fig. 4 *Nanowire Assembly.* Dark field optical microscope images of 200 nm diameter, ~7 μm long Rh nanowires assembled on electrodes with (**a**) a uniform-width gap, and (**b**) recessed wells that span a uniform-width gap. The wires were suspended in isopropanol for assembly, and a 10 kHz, 7 V_{rms} sinusoidal voltage was applied across the interdigitated assembly electrodes. Uniformly spaced, well-aligned nanowire arrays were assembled using the uniform-gap electrode structure. Deterministic positioning of individual nanowires was achieved using recessed wells. Scale bar = 7 μm

the cross section of the structure (x–z plane, Fig. 3c). For the first structure, the field strength is highest at the edge of the electrodes, and it is uniform along the length of the gap. This gives field lines that radiate outward from the gap, with a field gradient that is the largest near the substrate surface. Optical microscope images taken after electric-field-assisted assembly of template-synthesized Rh nanowires (200-nm diameter, 7-μm long wires; Fig. 4a) demonstrate that this field profile leads to the formation of parallel wire arrays along each gap. The spacing between adjacent nanowires is uniform, but the absolute position of each wire in the array is not well controlled. This limits the registration accuracy of the assembled wires relative to existing device features on the substrate.

The next two electrode structures overcome this limitation by producing spatially confined regions of high field strength using lithographic features. For the sharp-tipped electrode structure, the electric field magnitude is the strongest in the region surrounding the tips, with a field gradient that is the largest in the gap between the tips. This results in individual nanowires preferentially assembling across the gap, with additional wires aligning in the fringing field at an angle relative to the gap. A similar field profile can be

obtained by defining an array of recessed rectangular wells in the dielectric coating that separates the metal electrodes from the solution [5]. This structure has the advantage that it suppresses the lateral field between adjacent high field regions, which ensures that individual wires align to span the gap within the well (Fig. 4b). It is also possible to minimize the assembly of multiple wires on the sharp-tipped electrodes by adjusting the fluid flow to control the hydrodynamic forces acting on the wires during assembly [4]. Importantly, submicron registration accuracy between the assembled nanowire and features on the substrate can be achieved with both of these electrode structures.

Programmed Assembly of Different Nanowire Populations

Electric-field-assisted assembly can be used to deterministically position more than one type of nanowire on the same substrate. As illustrated in Fig. 5a, this is accomplished by synchronizing the sequential delivery of each nanowire population with a unique electric-field profile obtained by biasing a specific set of assembly electrodes. As a proof of concept, this programmed assembly approach was applied to create nanowire arrays containing three populations of wires, each functionalized with different nucleic acid probe molecules and positioned in different columns in the array [6]. Following assembly, the three types of wires were converted into nanomechanical cantilevers using conventional lithography and metal deposition processes. The probe-coated nanocantilever devices were incubated with a mixture of all three fluorescently labeled nucleic acid targets. Fluorescence microscopy images of the three column array (Fig. 5b) confirmed that the target molecules selectively hybridized to their complementary targets and that the assembly process positioned each nanowire population in the correct array column.

Examples of Applications

Adding heterogeneous populations of nanowire devices onto Si integrated circuits (ICs) will enable new and diverse functions ranging from energy harvesting to chemical and biological sensing [9]. Deterministic assembly offers a flexible and customizable alternative to traditional top-down nanofabrication because it completely decouples the nanowire synthesis process from the IC manufacturing process. This allows the use

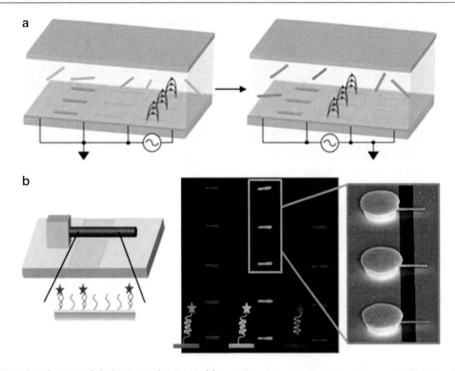

Electric-Field-Assisted Deterministic Nanowire Assembly, Fig. 5 *Programmed Nanowire Assembly*. (**a**) Scheme illustrating programmed deterministic assembly of different populations of bioprobe-coated nanowire. Each population is directed to the recessed wells in the electrically biased regions of the substrate. (**b**) Following programmed assembly of three populations of bioprobe-coated wires, nanomechanical cantilever devices (*left*) were fabricated as described in Fig. 1. Fluorescence images (*center*) show that the fluorescently labeled targets hybridized selectively to their complementary probes. The field emission scanning electron microscope image (*right*) shows the fabricated nanocantilever devices. After Ref. [6]

of emerging template, solution, and catalyst-based protocols for inexpensive batch synthesis of diverse materials such as metals, semiconductors, metal oxides, conducting polymers, and molecule-coated wires [10]. Off-chip wire synthesis also removes the constraints of thermal budget and chemical compatibility imposed by top-down manufacturing methods. This permits the integration of nanowires with properties optimized for a specific function.

Future Directions for Research

Advancing the electric-field-assisted deterministic assembly technique for use in practical applications will require ongoing research to develop new and scalable nanomanufacturing strategies that can provide high-yield integration of nanowire devices with the required registration accuracy over large substrate areas. For Si-based systems-in-a-package (SiP), the nanowires should be assembled at the back end of the Si IC process. This involves a highly integrated,

interdisciplinary research effort. For example, the availability of batch-synthesized nanowires with uniform properties (i.e., dimensions, microstructure, etc.) is essential to the success of this approach. Similarly, scalable methods should be developed to deliver aggregate-free, solution-suspended nanowires to large-area substrates with controlled and reproducible concentration. Additional research aimed at developing a comprehensive theoretical and experimental understanding of the electric-field forces will enable the design of new and more effective assembly electrode structures with patterns customized for different circuit applications. Finally, innovative nanofabrication methods should be applied to convert the assembled nanowires into high-performance electronic, optical, and sensing devices that can be monolithically integrated with the Si circuitry.

Cross-References

▶ Dielectrophoresis

References

1. Smith, P., Nordquist, C., Jackson, T.N., Mayer, T.S., Martin, B., Mbindyo, J., Mallouk, T.E.: Electric-field assisted assembly and alignment of metallic nanowires. Appl. Phys. Lett. **77**, 1399–1401 (2000)
2. Duan, X., Huang, Y., Cui, Y., Wang, J., Lieber, C.M.: Indium phosphide nanowires as building blocks for nanoscale electronic and optoelectronic devices. Nature **409**, 66–69 (2001)
3. Raychaudhuri, S., Dayeh, S., Wang, D., Yu, E.: Precise semiconductor nanowire placement through dielectrophoresis. Nano Lett. **9**, 2260–2266 (2009)
4. Freer, E., Grachev, O., Duan, X., Martin, S., Stumbo, D.: High-yield self-limiting single-nanowire assembly with dielectrophoresis. Nat. Nanotechnol. **5**, 525–530 (2010)
5. Li, M., Bhiladvala, R., Morrow, T., Sioss, J., Lew, K.K., Redwing, J.M., Keating, C.D., Mayer, T.S.: Bottom-up assembly of large-area nanowire resonator arrays. Nat. Nanotechnol. **3**(2), 88–92 (2008)
6. Morrow, T., Li, M., Kim, J., Mayer, T.S., Keating, C.D.: Programmed assembly of DNA-coated nanowire devices. Science **323**(5912), 352 (2009)
7. Pohl, H.: Dielectrophoresis: The Behavior of Neutral Matter in Nonuniform Electric Fields. Cambridge University Press, Cambridge/New York (1978)
8. Liu, Y., Chung, J., Liu, W., Ruoff, R.: Dielectrophoretic assembly of nanowires. J. Phys. Chem. B **110**, 14098–14106 (2006)
9. International Technology Roadmap for Semiconductors: Emerging research devices. In: International Technology Roadmap for Semiconductors, 2009 edn. International Technology Roadmap for Semiconductors (2009)
10. Xia, Y., Yang, P., Sun, Y., Wu, Y., Mayers, B., Gates, B., Yin, Y., Kim, F., Yan, H.: One-dimensional nanostructures: synthesis, characterization, and applications. Adv. Mater. **15**(5), 353–389 (2003)

Electrocapillarity

▶ Electrowetting

Electrochemical Machining (ECM)

Amitabha Ghosh
Bengal Engineering & Science University,
Howrah, India

Synonyms

ECM

Definition

Electrochemical machining is a metal removal–based manufacturing process that depends upon electrochemical dissolution for metal removal mechanism. It is one of the advanced manufacturing processes for producing parts with complex shapes using metals and alloys with low machinability.

Overview

Though electrochemical machining process is of relatively recent origin, the reverse process – electroplating – has been in use since a long time. The basic process is indicated in Fig. 1.

A metallic tool is fed at a constant rate toward the work, and the gap between the tool face and the work surface (in the machining area) is supplied with copious flow of an electrolyte (often brine) at a reasonably high rate. A large direct current at a low voltage (typically 10–15 V) is supplied to the work–tool system, the positive terminal being the work. Electrochemical dissolution causes material to be removed from the work piece and the electrolyte–material combination being such that no deposition takes place on the tool electrode (and the dissolved material and the ohmic heat generated in the tool–work gap are carried away by the electrolyte). The shape of the machined portion of the work conforms to the shape of the tool face as indicated in Fig. 1. Once the desired shape is achieved the tool is withdrawn.

Process Mechanism

According to Faraday's laws, the amount of material dissolved during an electrochemical process is proportional to the quantity of electricity passed; besides, the amounts of different substances dissolved by the same amount of electricity are proportional to their chemical equivalent weights. Thus,

$$m \propto \in It \tag{1}$$

where m is the weight in gram of a substance dissolved, I is the current in amperes, t is the time in seconds and $\in$ is the gram equivalent weight of the substance

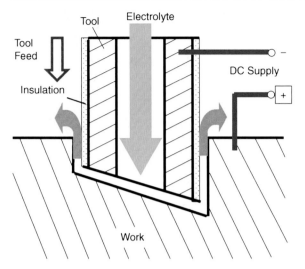

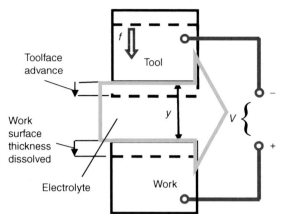

Electrochemical Machining (ECM), Fig. 2 Mechanics of ECM

Electrochemical Machining (ECM), Fig. 1 Basic scheme of electrochemical machining

(i.e., the work material). Equation 1 can be written in the form of the following equation:

$$m = \frac{\in It}{F} \tag{2}$$

Where F represents the constant of proportionality and is equal to 96,500 C. For ferrous work material using sodium chloride solution (brine) as the electrolyte, the following electrochemical reactions occur at the anode (work) and the cathode (tool); respectively:

$$Fe \rightarrow Fe^{++} + 2e^-$$

$$2H_2O + 2e^- \rightarrow H_2 \uparrow + 2(OH)^-$$

Thus, Fe dissolves at the anode and H_2 is released at the cathode. The reaction in the electrolyte is as shown below:

$$Fe^{++} + 2(OH)^- \rightarrow Fe(OH)_2 \downarrow$$

The insoluble $Fe(OH)_2$ is removed with the electrolyte as a precipitate and no deposition at the cathode occurs.

The gram equivalent weight of the work metal, $\in$ is given by A/Z, where Z is the valence of the produced ions and A is the atomic weight. If ρ be the density of the work material in gm/cm^3, Eq. 2 yields the following relation for the volume rate of material removal from the work piece:

$$Q = \frac{AI}{96500\rho Z} \, \text{cm}^3/\text{s} \tag{3}$$

The magnitude of the current I is given by

$$I = \frac{V - \Delta V}{R} \tag{4}$$

where V is the applied voltage in volts, ΔV represents the combined over voltage due to various effects and should be subtracted from the impressed voltage before Ohm's law can be applied, and R is the overall resistance (but most of it due to the electrolyte layer in the interelectrode gap). Typical electrolytic conductivity is $0.1–1.0 \, \Omega^{-1} cm^{-1}$ whereas that of iron is about $10^5 \Omega^{-1} cm^{-1}$.

Figure 2 shows symbolically the plane-parallel work–tool electrolyte combination with tool being advanced at a constant rate of f toward the work.

If the applied voltage is V and specific conductivity of the electrolyte is k, Eq. 2 yields the following relation for the current density at the work surface:

$$J = \frac{k(V - \Delta V)}{y} \tag{5}$$

Where, y is the gap thickness at the instant under consideration. So, after a short time interval Δt the tool face advances by $f \times \Delta t$. The work surface also recedes because of electrochemical dissolution. If δ be the depth of recession then using Eq. 3

$$\delta = Q \times \Delta t = \frac{AJ\Delta t}{96500\rho Z}$$

Replacing J in the above equation from Eq. 5

$$\delta = \frac{Ak(V - \Delta V)}{96500\rho Zy} \times \Delta t$$

But when the gap is stable the rate of depth of dissolution ($\delta/\Delta t$) and the rate of tool advance (f) must be equal. Hence, under equilibrium condition (i.e., $y = y_e$)

$$\frac{\delta}{\Delta t} = f$$

or,

$$\frac{Ak(V - \Delta V)}{96500\rho Zy_e} = f$$

or,

$$y_e = \frac{Ak(V - \Delta V)}{96500\rho Zf} \qquad (6)$$

It can be easily seen that this equilibrium condition is stable and for a given electrolyte–electrode combination, the equilibrium work–tool gap, y_e, depends upon the applied voltage and the feed rate.

Since an equilibrium gap is ultimately reached it is possible to arrive at a definite relationship between the tool and work geometries (i.e., a tool can be shaped to generate a desired work shape). This phenomenon makes the process suitable for machining objects with definite shapes.

Machining Force and Electrolyte Flow

At a first glance, it may appear that there is no machining force involved as there is no contact between the work and the tool. But, in reality, there exists large force acting on the electrodes due to a large pressure exerted by the electrolyte flowing through the work–tool gap.

Figure 3 indicates symbolically an ECM operation with an equilibrium work–tool gap of y_e. The flow velocity, v, is determined predominantly from the thermal conditions.

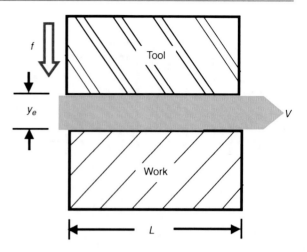

Electrochemical Machining (ECM), Fig. 3 Electrolyte flow through work–tool gap

If the variation in conductivity of the electrolyte with temperature is ignored, then the following equation for the required flow velocity, v, of the electrolyte can be obtained [1]:

$$v = \frac{J^2L}{k\Delta\theta\rho_e c_e} \qquad (7)$$

where, J is the current density, L is the length of the flow path, k is the specific electrical conductivity of the electrolyte, $\Delta\theta$ is the allowed temperature rise in the electrolyte, ρ_e and c_e are the density and specific heat of the electrolyte. If the gap is y_e and the viscosity of the electrolyte is η, the total pressure required (assuming turbulent flow) to achieve a flow velocity of v is given by the relation [1].

$$p = \frac{0.3164\rho_e v^2 L}{4y_e(R_e)^{1/4}} + \frac{\rho_e v^2}{2} \qquad (8)$$

where, R_e is the Reynolds's number. It can be easily demonstrated that to prevent boiling of the electrolyte the pressure for the electrolyte in a typical ECM operation is a few hundred kN/m^2. This can exert a reasonably large force on the tool face depending on the size.

Tool Design

The three primary aspects associated with the tool design for ECM operation are (a) the determination

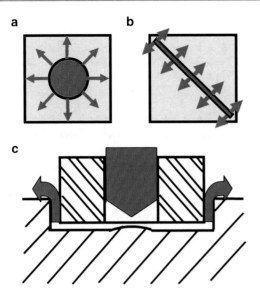

Electrochemical Machining (ECM), Fig. 4 Tool face showing the flow exit holes (**a**) normal shape, (**b**) narrow and long passage, (**c**) residual bulge with a flow path as shown in (**a**)

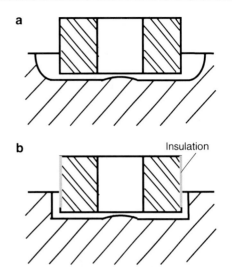

Electrochemical Machining (ECM), Fig. 5 Machined shape (**a**) without insulation, (**b**) with insulation

of the geometric shape of the tool that generates the desired work shape, (b) electrolyte flow path design that ensures smooth and ample flow of electrolyte all over the working area, and (c) design from other considerations like strength, insulation of the areas of tool surface which should not take part in the electrochemical process and fixing arrangements.

Since the equilibrium gap y_e can be accurately predicted it is possible to decide upon the required tool geometry that will generate the desired work shape. Design of electrolyte flow path is extremely important as the quality of the surface generated depends very significantly on the uniformity and smoothness of electrolyte flow. A large flow requires a large cross-sectional area of the orifice through which the flow emerges as indicated in Fig. 4a. This leads to a residual bulge on the machined surface of the work as shown in Fig. 4c. To avoid the residual bulge without reducing the flow path opening area, the passage can be in the form of a narrow but long slot as shown in Fig. 4b. Since the flow will be in a direction normal to the slot as shown, the slot geometry should be carefully decided so as to keep the whole machining area with ample and smooth flow of electrolyte.

To restrict machining only at the desired locations, it is essential to insulate some portions of the tool surface. Figure 5a, b indicates the result of ECM without and with insulation.

It is obvious that the tool material should be electrically conducting. As the desired shape has to be machined, tool material should have good machinability and reasonable mechanical strength.

Electrolytes

The functions accomplished by the electrolyte during ECM are: (a) completing the electrical path allowing passage of large current, (b) sustaining the electrochemical dissolution process, and (c) carrying away the heat and waste product from the machining area. So a good electrolyte should have large electrical specific conductivity. It should be also of such composition that deposition does not occur on the cathode (i.e., the tool) as otherwise the tool geometry can change. Besides, an electrolyte should be inexpensive, safe, and as noncorrosive as possible. Usually, an aqueous solution of inorganic compounds is used as shown below in the list:

Work material	Electrolyte
Ferrous	Chloride solution in water, usually NaCl
Nickel based	HCl or mixture of H_2SO_4 and brine
Titanium based	10% HF + 10% HCl + 10% HNO_3
Co–Cr–W based	NaCl
WC based	Strong alkaline solutions

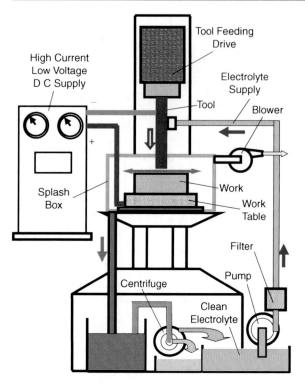

Electrochemical Machining (ECM), Fig. 6 Basic components of a typical ECM unit

ECM Unit

The basic components of a typical ECM unit are shown in Fig. 6. It consists of a work table holding the job.

The tool is fed by a low-feed-rate tool feeding drive with a facility to quickly retrieve the tool at a fast rate. The machining zone is enclosed in a box and the collected electrolyte and the precipitate are accumulated in a tank. A centrifuge separates the precipitate as sludge and the clean electrolyte is stoned in another tank from where it is recirculated by a pump through a filter as shown. The H_2 generated by the process is removed using an exhaust blower to prevent the chance of any explosion. A low voltage high amperage DC supply provides the required large current at a low voltage as indicated. All the units must be made of corrosion-free material like stainless steel, polymeric materials, etc. The whole system is supported by a rigid and strong structure as very large electrolyte pressure can generate a large force between the tool and the electrode; and any deformation can jeopardize the operation. All these requirements result in high cost of such machines.

References

1. Ghosh, A., Mallik, A.K.: Manufacturing Science, 2nd edn. Affiliated East West Press, New Delhi (2010)

Electrochemical Scanning Tunneling Microscopy

Ilya V. Pobelov, Chen Li and Thomas Wandlowski
Department of Chemistry and Biochemistry, University of Bern, Bern, Switzerland

Synonyms

EC-STM; In-situ STM

Definition

Electrochemical Scanning Tunneling Microscopy (EC-STM) is a conceptional and instrumental extension of Scanning Tunneling Microscopy toward its application to charged solid-liquid interfaces (electrochemical systems).

Introduction

With the invention of Scanning Tunneling Microscopy (STM) by Binning and Rohrer [1] a unique new tool for high-resolution imaging of structures and process (dynamics) at conducting surfaces became available [2]. Soon after its application in surface studies under ultrahigh vacuum (UHV) conditions this method was also adopted to electrochemical systems under potentiostatic or galvanostatic control and "in-situ" EC-STM was born [3–5]. This technique advanced tremendously the knowledge on surface structure and reactivity in electrochemistry because imaging of electrode surfaces was now possible in-situ, in real space, and with atomic-scale resolution [6–11].

Examples range from fundamental studies of reconstruction of metal single-crystal surfaces [12, 13], ordered adsorption of anions [14], and organic molecules [15, 16], the initial stages of metal deposition [17, 18] up to applications in electrocatalysis [19], plating [20], and corrosion [21]. During the last decade in-situ STM investigations were combined with in-situ Scanning Tunneling Spectroscopy (STS) to obtain electronic structure information [22] and to explore charge transport characteristics in single atom and molecular junctions [23, 24]. Besides imaging with atomic-scale resolution, the STM tip has been also employed as a tool in nanostructuring of electrode surfaces with high spatial and chemical resolution. Further details are given in a series of excellent reviews [6–8, 14].

The chapter starts with a brief introduction into Electrochemistry and EC-STM. Subsequently, the potential of STM in Electrochemistry is demonstrated in selected structure studies with bare and adlayer (anions, molecules, metal atoms) modified single-crystal surfaces, and by monitoring simple reactions, such as metal deposition. Finally, single-molecule transport experiments and nanostructuring of surfaces by the STM tip will be addressed.

Overview and Key Principles

Fundamentals of Electrochemistry

Electrochemistry represents chemical reactions that are connected with the transfer of electric charges [25–28]. Many of these processes occur at electrified solid/liquid interfaces. Typically, the solid phase acts as an electron conductor (metal or semiconductor electrode) and the liquid phase as ionic conductor (for example an aqueous electrolyte). The interface is composed of electronic and ionic charge carriers and characterized by an electric field as high as 10^9 Vm^{-1}. This region is called "electrical double layer" (EDL). In the absence of electrochemical reactions, no charge transfer across the interface occurs. The latter then resembles a capacitor that can be charged or discharged upon application of an external voltage ("ideally polarizable interface"). In the presence of an electrochemical (Faraday) reaction, charge or, in other words, an electrochemical current, passes the interface ("ideally non-polarizable interface"). The classical route to the study of electrochemical reactions rests on current and voltage measurements. Both quantities can be determined with high precision and are easily accessible, although they lack direct chemical or structural specificity.

Electrochemical experiments are typically carried out under potentiostatic conditions (i.e., under

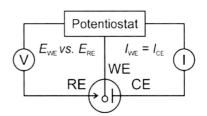

Electrochemical Scanning Tunneling Microscopy, Fig. 1 Schematics of a potentiostat and three-electrode electrochemical cell

potential control) in a three-electrode configuration composed of a working electrode (WE), the electrode of interest, a reference electrode (RE), and a counter electrode (CE), respectively (Fig. 1). For the choice of the RE and the conversion of the electrochemical potential scale into an absolute potential scale, we refer to the book by Ives and Janz [29] and the work of Heller et al. [30]. All electrodes are immersed into the electrolyte solution. The setting up of the potential differences and the measurement of the current are controlled by a potentiostat (Fig. 1).

The potentiostat measures current and potential in the CE and RE circuits. Due to voltage drops and polarization effects, the potential difference between WE and RE can be measured precisely only if there is negligible current flowing through the RE. The potential difference between the WE and the CE is adjusted via a feedback loop in such a way, that the desired value of the electrode potential E (E_{WE} vs. E_{RE}) is reached. As a consequence, the current flowing in the CE circuit (between WE and CE) is the current flowing at the WE at this electrode potential [26, 28].

Basics of Scanning Tunneling Microscopy

The STM technique [2] is based on electron tunneling, which is well-known from quantum mechanics. When a freely moving electron incidents an energy barrier of constant relative height Φ and width z it may penetrate through it with the probability

$$P(z) = \exp\left(-\frac{4\pi}{h}\sqrt{2m_e\Phi} \cdot z\right)$$

$$= \exp\left(-\kappa\sqrt{\Phi} \cdot z\right) \qquad (1)$$

Here $h = 6.626 \times 10^{-34}$ J·s is Planck's constant, $m_e = 9.11 \times 10^{-31}$ kg is the electron mass, and $\kappa = 10.12$ $eV^{-1/2} \cdot nm^{-1}$ is a combination of constants.

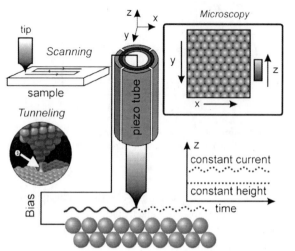

Electrochemical Scanning Tunneling Microscopy, Fig. 2 Schematics of an STM. *Upper-left*: tip scanning on the sample; *lower-left*: electron tunneling between tip and sample; *upper-right*: obtained STM image; *lower-right*: two operation modes

In STM, a sharp conductive tip (typically metal [2] or, more recently, also carbon nanotubes [31]) is brought up to few nanometers close to a conductive surface (Fig. 2). The potential difference E_b (bias voltage) applied between an atomically sharp tip and the sample causes electron tunneling via the insulating gap, which could be ultrahigh vacuum (UHV), gas, or liquid. In the low bias approximation [32], the tunneling current is proportional to E_b and to the electron tunneling probability:

$$I_T(z) \propto E_b \exp\left(-\kappa \sqrt{\Phi} \cdot z\right) \qquad (2)$$

The tunneling current I_T is measured by a highly sensitive preamplifier with a typical operational range of 1 pA to 100 nA. The x,y,z-movement of the tip in the nanometer scale is performed by a piezoelectric tube based on ceramics (often lead-zirconium-titanate) exhibiting a piezo-effect, that is, expanding or shrinking upon application of a voltage. One may distinguish between two operational modes of the STM (Fig. 2). In constant current mode a preset tunneling current I_T is kept constant by adjusting the z position of the tip at every measured point via a feedback circuit. The information on vertical displacement of the tip as a function of the lateral position is recorded as an array of (x,y,z) data and presented as a three-dimensional (3D) or two-dimensional (2D) image of the local density of states,

which reflects, in case of metals, the topography of the substrate surface directly.

In constant height mode the vertical position of the tip is kept constant during the lateral scanning while the value of I_T is recorded as a function of the tip position. The array of (x,y,I_T) data provides information on the surface topography and can be presented in a similar way as results of the constant-current mode STM.

In an STM experiment, one may achieve lateral x,y resolution of 0.1 nm and a vertical resolution up to 0.01 nm. This unprecedented capability of STM is due to the strong dependence of the tunneling current on the distance z between tip and sample (cf. Eq. 2). With a typical barrier height of one to a few eV, an increase of z by the size of a single metal atom (0.2–0.5 nm) decreases the tunnel current up to three orders of magnitude. Atomically sharp STM tips are required to avoid artifacts, and in this case, only the most protruding part of the tip acts as a probe. Finally, it is noted that STM per se is not providing "chemical information" on surface (and tip) composition!

Electrochemical (In-Situ) Scanning Tunneling Microscopy

Tunneling in an electrochemical experiment takes place in an electrically conducting environment, the electrolyte (aqueous, organic or ionic liquid as examples). Electrochemical reactions may occur at the tip as well as at the substrate. The resulting electrochemical current I_F is superimposed on the tunneling current I_T. To minimize I_F at the tip relative to I_T, the tip is coated with an inert and insulating layer made for example of glass [33], apiezon [34], polyethylene [35], or electrophoretic paint [36]. The remaining electrochemically active area of well-insulated tips amounts to 10^{-8}–10^{-7} cm^2 and yields residual electrochemical ("leakage") currents of typically less than 5 pA, which do not interfere with tunnel currents of 0.1–10 nA.

From an electrochemical point of view, substrate and tip represent two working electrodes (WE1 and WE2 in Fig. 3a), which potentials E_S and E_T are independently controlled with respect to the RE, and the bias voltage is given by the potential difference $E_b = E_T - E_S$. This functionality is controlled by a bipotentiostat, which represents a potentiostat with two independently adjustable working electrodes (4-electrode potentiostat).

Electrochemical Scanning Tunneling Microscopy, Fig. 3 (a) Schematics of an EC-STM setup; (b) assembled cell

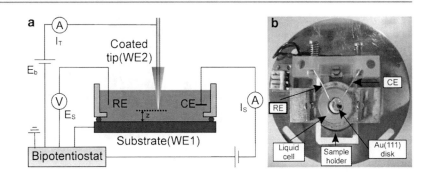

The type of substrate and its potential E_S are determined by the electrochemical problem to be studied. The choice of tip material and potential E_T are dictated by the requests of high mechanical and, in addition, chemical stability to minimize the Faradaic current I_F at the tip. Atomically sharp tips are typically prepared by electrochemical etching of thin metal wires, such as tungsten [37], platinum-iridium [38], or gold [39].

The choice of the RE depends on electrolyte and electrochemical problem under study. Examples are a Pt wire in acidic electrolytes, reference electrodes of the first kind (Me in Me^{z+} – containing electrolyte such as Cu/Cu^{2+} or Ag/Ag^{+}) or of the second kind ($Ag/AgCl$, Au/Au_xO_y, etc.). The materials of the CE are less critical. Often Pt or Au wires are chosen. Attempts to remove oxygen and to control the environment during an STM experiment led to the development of various custom-build environmental chambers. Figure 3b illustrates, as an example, a typical EC-STM cell.

Besides high resolution atomic-scale imaging, even up to video acquisition rates [40], most EC-STM setups are also capable to perform scanning tunneling spectroscopy (STS) measurements. The latter are carried out at fixed lateral (x,y) position of the tip above the substrate. In *current-distance STS experiments*, the tunneling current is measured at constant E_S and E_T as a function of the vertical tip displacement Δz. The independent control of the potentials of both tip and substrate with respect to the RE allows various modes of *current-voltage spectroscopy* in an electrochemical environment: In *variable bias STS mode*, the potential of one electrode (typically of the substrate WE1, E_S) is fixed, while that of the other (WE2, E_T) is swept over a wide range of potentials. This regime is equivalent to an *ex situ* current-voltage STS experiment with two electrodes. The *constant bias mode of current-voltage STS* is unique for the electrochemical environment.

Here, both E_S and E_T are swept simultaneously, so that their difference E_b is kept constant. Tunneling spectroscopy is an important technique for studies of electronic properties, and in particular of electron transfer at electrochemical solid-liquid interfaces, currently already approaching the single-molecule level [24].

Key Research Findings and Selected Applications

In the following section, the potential of EC-STM is demonstrated in structure studies on bare and adsorbate-modified single crystals, and by monitoring simple reactions such as metal deposition. Finally, single-molecule electron transport experiments and nanostructuring of surfaces are described.

Structure of Single-Crystalline Surfaces

For a detailed understanding of surface processes, the use of single crystals with structurally well-defined surfaces is highly desirable. After establishing high quality and lab-accessible preparation techniques, such as flame annealing [41], inductive annealing [42], and various chemical etching protocols [43, 44] STM revealed unprecedented microscopic details of electrode surfaces under electrochemical conditions since the early 1990 up to present [6–8, 10, 12, 13, 15]. The majority of atomic resolution studies was carried out with metals forming face-centered cubic (fcc) crystal lattices, such as Au, Ag, Cu, Ir, Ni, Pd, Pt, and Rh (c.f. reviews [6–8, 13, 44] and original literature cited therein). Selected studies were also reported for Fe-type [21] and semiconductor surfaces [45]. As an example, Fig. 4 shows in-situ STM images of the three low-index faces of silver single crystal electrodes recorded in 0.05 M H_2SO_4 [46], together

Electrochemical Scanning Tunneling Microscopy, Fig. 4 *Top row*: STM images of three low-index faces of silver single-crystal electrodes in 0.05 M H_2SO_4, 7 × 7 nm. *Bottom row*: models of the low-index faces of a FCC metal. The surface unit cells are indicated (Adapted from Ref. [46])

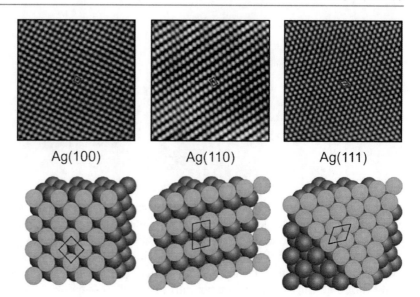

with schematic models of the respective surface structures.

The high spatial resolution of EC-STM provides not only access to steady state surface structures, such as the lattice arangement of surface atoms and defects (holes, steps, kinks, and/or islands) under potential control, but also to dynamic phenomena with a time-resolution as high as of several tens of milliseconds. Examples are step and island mobility [47] and surface reconstruction [12, 13].

Surface Reconstruction

Experiments on the formation and lifting of surface reconstruction represent classical examples of surface processes driven by electrode potential and solution composition [12, 13]. Surface reconstruction is the rearrangement of surface atoms into a different structure than the one dictated by the crystal plane, as driven by the minimization of the energy for surface atoms with an unsaturated coordination sphere. This phenomenon was first extensively studied in UHV [48]. Conclusive evidence about the existence of reconstructed surfaces under electrochemical conditions was obtained after electrochemists gained access to high-resolution structure-sensitive techniques, such as in-situ STM [12, 13]. Examples are the low-index phases of Au (Au(111)-(p × √3); Au(100)-(hex), Au(110)-(1 × 2) and Au(110)-(1 × 3)) and Pt (Pt(100)-(hex), Pt(110)-(1 × 2)) as prepared by sputtering and annealing cycles in UHV or annealing in a Bunsen burner or hydrogen flame [12, 44]. In particular, in-

situ STM and surface X-ray scattering experiments revealed that the reconstructed surfaces of the three low index Au single crystals are stable at negative surface charges, and are lifted upon positive polarization. The process is reversible. The adsorption of anions was found to assist the lifting of the reconstruction [12].

Figure 5 shows large-scale and atomic-scale resolution images of the thermally-induced Au(111)-(p × √3) reconstruction, as obtained after flame annealing in 0.1 M H_2SO_4 at $E_S = -0.20$ V, where the electrode is charged negatively. Changing the electrode potential toward positive charges causes the lifting of the reconstruction into the Au(111)-(1 × 1) structure. The extra amount of Au atoms from the ca. 4% more densely packed (p × √3) structure forms monatomic high gold islands, which are rather mobile. At sufficiently high potentials, the adsorbed sulfate ions form a 2D-ordered (√3 × √7) adlayer on top of the Au(111)-(1 × 1) surface. Changing the potential back to negative charge densities causes the reconstruction to reappear. However, this so-called electrochemically induced reconstruction is less ordered as compared to the thermally induced reconstruction. For further details the reader is referred to Refs. [12, 13, 16].

Adsorption of Anions

The study of adsorbates is of importance for electrochemistry, as many reactions proceed through intermediate adsorption stages [14, 28]. Anions exhibit a strong tendency to adsorb specifically at metal

Electrochemical Scanning Tunneling Microscopy, Fig. 5 Cyclic voltammogram and STM images of Au(111) in 0.1 M H_2SO_4 (Adapted from Ref. [16])

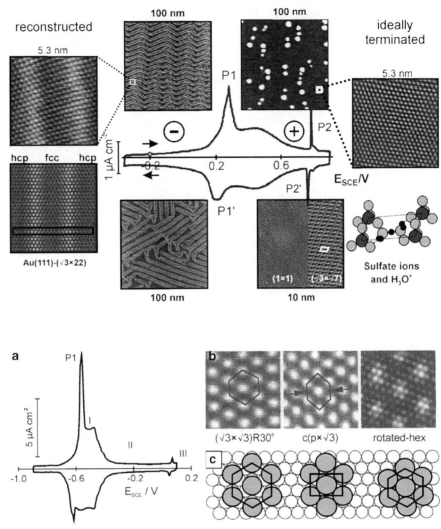

Electrochemical Scanning Tunneling Microscopy, Fig. 6 Adlayers of iodide on Au(111)-(1 × 1). (**a**) Cyclic voltammogram of Au (111) in 50 mM NaI, scan rate 50 mV·s^{-1}. The potential ranges of stability for the different monolayers of I on Au (111) are indicated, (**b**) EC-STM images, and (**c**) models of the I monolayers on Au(111) (Adapted from Refs. [8, 49])

surfaces, that is, to establish a direct bond with the metal by partial loss of their solvation shell. The direct contact to the electrode reduces the ionic character markedly, which allows a much higher surface concentration than in case of nonspecific (purely electrostatic) adsorption. Cyclic voltammograms often reveal sharp current spikes (Fig. 6a), indicative of phase transitions within the anionic adlayers and hence, of the existence of 2D-ordered phases. Structural aspects of these adlayers have been studied in detail by EC-STM in combination with other structure-sensitive in-situ techniques [6, 8, 10, 14, 15].

As an example, Fig. 6 demonstrates interfacial structures formed by iodide adsorbed on Au(111) [8, 49]. The cyclic voltammogram was recorded in 50 mM NaI aqueous electrolyte. The current peak P1

represents the lifting of the reconstruction Au(111)-$(p \times \sqrt{3}) \rightarrow$ Au(111)-(1 × 1), while the potential regions I, II, and III, which are separated by characteristic current peaks, could be identified as the following 2D-ordered iodide adlayers: $(\sqrt{3} \times \sqrt{3})R30°$, a compressed $c(p \times \sqrt{3})$ phase and a rotated hexagonal phase at the highest potential and iodide coverage, just before the onset of adlayer oxidation.

Adsorption and Self-Assembly of Organic Molecules

The adsorption and self-assembly of organic molecules at metal surfaces comprise a growing field of interest for fundamental research as well as technological applications. This is due to the immense importance of organic molecules in applied areas such as additives in plating baths (e.g., for superconformal growth in the Damascene process of Cu deposition [20, 50]) or as corrosion inhibitors, and to the growing field of molecular nanotechnology for designing materials and functionalized surfaces with novel electronic, magnetic, photonic, mechanical, and sensing/diagnostic properties [51].

Employing defined substrate surfaces, such as metal single crystals, and tailored molecules as coded information, molecular recognition pattern and self-organization as well as local functionalities have been studied by EC-STM under a wide range of conditions [6, 8, 15, 16, 24]. These investigations benefited from the application of concepts of supramolecular chemistry to create modified surfaces in the submonolayer, monolayer and multilayer regime. Adlayer structure motifs are based, for instance, on hydrogen-bonding, hydrophobic and electrostatic interactions, such as dipole, π-stacking, and metal-ion-ligand coordination [16]. In case of an electrochemical system, the substrate electrode WE1 acts as an electron source or sink and, eventually, as an additional coordination center. The electrode potential may be considered as a universal tuning source to fabricate and to address these molecular nanostructures. Structure motifs and processes of self-organization at surfaces can be investigated with atomic/molecular resolution employing EC-STM and STS.

A particular fascinating example of tailored molecular self-assembly at an electrified solid/liquid interface is benzene-1,3,5-tricarboxylic acid (trimesic acid (TMA), Fig. 7), a prototype material for engineering molecular nanostructures, on Au(111) in 0.05 M H$_2$SO$_4$ [16, 52]. Depending on the assembly conditions and the applied electrode potential, five 2D-ordered molecular adlayers were observed by EC-STM (Fig. 7). At a negatively charged or uncharged Au (111) surface, the molecules are adsorbed flat and assemble with increasing coverage in four distinctly different hydrogen-bonded patterns: the low-coverage honeycomb phase (IIa), sometimes even with guest molecules incorporated, a ribbon (IIb) and a herringbone (IIc) pattern as well as rows of hydrogen-bonded dimers (III). At a positively charged surface, the TMA molecules change their orientation from flat to upright and one carboxyl group deprotonates. The resulting carboxylate group coordinates with the positively charged surface and rows of a new stacking structure are formed (IV). These structure transitions are reversible upon changing the electrode polarization. The detailed mechanisms were explored in a combined EC-STM and infrared spectroscopy study [16, 52, 53].

Concluding this paragraph one should emphasize that the interpretation of EC-STM images of molecular adlayers requires great caution because the observed contrast pattern provides information on the electronic structure of the entire tunneling junction, and not a straightforward adlayer topography, as often naively assumed in the description of molecular adlayers.

Metal UPD and OPD

The deposition of metals from aqueous solutions has attracted the attention of electrochemists for more than hundred years. This is related to the importance in many industrial-scale processes such as plating and metal refining. Contemporary frontiers in this field include, for example, the optimization of the electrochemical copper deposition in the fabrication of high quality nm-wide integrated circuit interconnects (Damascene process) [20, 50]. A primary characteristics of the metal deposition process is the equilibrium potential E_{eq}, at which the rates of metal deposition and dissolution are equal. The dependence of E_{eq} on the concentration of the metal ions Me^{n+} in a solution, $C_{Me^{n+}}$, is described by the Nernst equation:

$$E_{eq} = E^0 + \frac{RT}{nF} \ln(a_{Me^{n+}}) \cong E^0 + \frac{RT}{nF} \ln(C_{Me^{n+}}) \quad (3)$$

with E^0 being the standard potential of the redox pair Me^{n+}/Me and $(a_{Me^{n+}})$ as the thermodynamic activity of the metal ions in solution. When an external potential $E < E_{eq}$ is applied to the metal electrode, which is in contact with a metal ion Me^{n+}- containing solution, bulk deposition (or overpotential deposition, OPD) of the metal Me takes place. At $E > E_{eq}$ the bulk dissolution of the metal phase is promoted. However, a (sub) monolayer of metal ions is deposited at $E > E_{eq}$ when the energy of the metal-substrate interaction is higher than that between the atoms of the deposited metal. This process is referred to as underpotential deposition (UPD) [17, 54, 55]. Metal deposition onto a foreign substrate (e.g., Cu on Au) often starts with the formation of a monolayer at underpotentials, that is, at potentials positive to the Nernst potential for the respective bulk phase. UPD of metals is important as the first stage of many metal bulk deposition processes, to

Electrochemical Scanning Tunneling Microscopy, Fig. 7 Steady state voltammogram for Au(111)-(1 × 1) in 0.05 M H₂SO₄ in the presence of 3 mM TMA, scan rate 10 mV·s⁻¹, together with typical EC-STM images of the TMA adlayers (size 10 × 10 nm): (IIa) hexagonal honeycomb phase, $E = -0.18$ V; (IIb) ribbon-type motif, $E = 0.03$ V; (IIc) herringbone motif, $E = 0.21$ V; (III) hydrogen-bonded linear dimers, $E = 0.34$ V; (IV) ordered chemisorbed striped phase of TMA at $E = 0.80$ V (Adapted from Ref. [52])

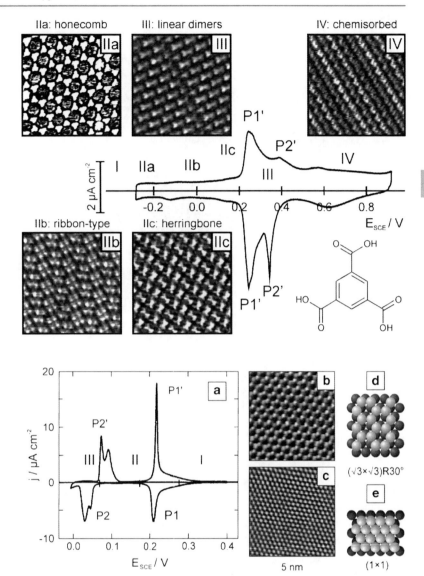

Electrochemical Scanning Tunneling Microscopy, Fig. 8 (a) Cyclic voltammogram of a Au(111)-(1 × 1) electrode in 0.1 M H₂SO₄ and 1 mM CuSO₄, scan rate 1 mV·s⁻¹ (Adapted from Ref. [44]). (b) ($\sqrt{3} \times \sqrt{3}$) R30° and (1 × 1) Cu UPD adlayers on Au(111). The models of both phases are depicted in panels (d) and (e)

tailor electrocatalytic properties of metal surfaces as well as in fundamental electrochemical surface science. The structures and process involved were studied in detail by EC-STM, in combination with classical electrochemical and other structure-sensitive techniques for a wide range of systems and conditions [6–8, 15, 54–56].

As an example, Fig. 8 represents the archetype UPD system Cu on Au(111) in 0.05 M H₂SO₄ [15, 57]. The cyclic voltammogram shows two well-defined pairs of current peaks P1/P1′ and P2/P2′ corresponding to energetically different adsorption/desorption processes in regions I to III. In the first step (P1), the transition

between randomly adsorbed copper ions and (hydrogen) sulfate ions (region I) into an ordered layer of copper ions and co-adsorbed sulfate ions takes place. Among other structure-sensitive techniques EC-STM revealed a ($\sqrt{3} \times \sqrt{3}$)R30° pattern of bright features (region II) [18]. The correct interpretation of this contrast pattern is based on a surface X-ray scattering study of Toney et al. [58]. These authors demonstrated that Cu ions adsorb in hollow positions of the underlying Au(111) lattice and form a honeycomb-like hexagonal network with a 2/3 Cu coverage. Sulfate ions are positioned in the center of the hexagons with one S-O bond facing toward the electrolyte, which

Electrochemical Scanning Tunneling Microscopy, Fig. 9 (**a**) STM images of Au (111)-(1 × 1) in 0.05 M H_2SO_4 + 5·10^{-5} M $CuSO_4$ at 0.1 V and −0.185 V versus Cu^{2+}/Cu; (**b**) Au(111)-(1 × 1) in 0.1 M H_2SO_4 + 10^{-3} M $CuSO_4$ at 0.15 V and −0.3 V versus Cu^{2+}/Cu (Adapted from Ref. [57])

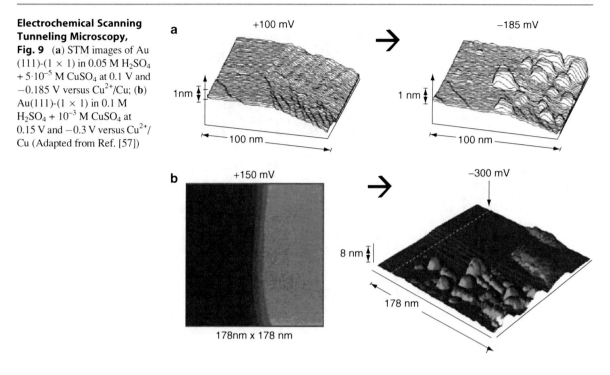

corresponds to a 1/3 sulfate coverage. The pattern resolved by EC-STM represents only the co-adsorbed anion structure. Further decrease of the potential leads to a second interfacial phase transition in this system, as indicated by the peak P2, and a pseudomorphic monolayer of Cu with the same structure as the underlying Au(111) surface is formed (region III). The potential range III is very narrow, and limited by the onset of copper overpotential (bulk) deposition (OPD), which proceeds at $E < E_{eq}$ with $E_{eq} = 0.012$ V versus SCE.

OPD of Cu on Au(hkl) was among the first systems studied by EC-STM [18]. The research on Cu-OPD and many other technologically important metals can be divided in three main directions addressing (1) the initial stages of bulk deposition, (2) the growth of the metal phase, and (3) the influence of additives on deposit morphology and deposition rate. Structure and dynamic studies by EC-STM contributed substantially to the advancement of knowledge in all three target areas [6, 7, 9, 17, 18].

The morphology of the deposit during the initial stages of the OPD process depends on the nature and distribution of surface defects. For example, the deposition of Cu on Au(111) proceeds first by nucleation on surface defects, such as steps, kinks, and screw dislocation, and only at higher overpotentials also on terrace

sites (Fig. 9). The nucleation-and-growth behavior leads to inhomogeneous film thicknesses and hence, to rough metal overlayers. The presence of organic additives such a benzotriazole, thiourea, crystal violet, or polyethylene-glycol, which act as so-called levelers and brighteners, modify the deposition process in such a way that smooth and uniform overlayers emerge. As an example, the addition of crystal violet induces a quasi-two-dimensional growth, which causes the substrate to be quickly covered by a rather uniform Cu-film on Au [18]. This particular study gives an illustration on the importance of additives for the optimization of deposition processes, such as that of Cu in the Damascene Technology [50].

Current-Distance Spectroscopy: Measurements of the Tunneling Barrier

Current-distance STS represents a complementary approach for the characterization of the normal distribution of electric charges and adsorbates at electrochemical interfaces, as compared to STM imaging. Particularly interesting insights on the electronic structure of adlayers were obtained by comparison of measured and simulated tunneling responses [22]. Figure 10a shows current-distance curves recorded on Au(111) in 0.1 M H_2SO_4 at different potentials. Unlike expected from a simple constant tunneling barrier

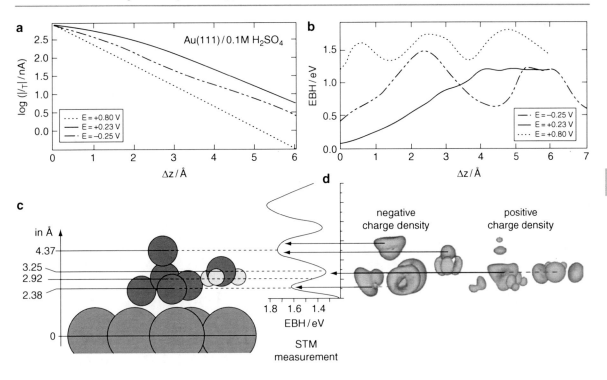

Electrochemical Scanning Tunneling Microscopy, Fig. 10 (a) Experimental current-distance STS traces (b) corresponding effective barrier height profiles measured at the Au(111)/0.1 M H$_2$SO$_4$ interface at different potentials (c) DFT-optimized structure of the sulfate adlayer adsorbed on Au(111), $E = 0.8$ V. (d) Comparison of calculated charge distribution for the structure shown in (c) and measured effective barrier height profiles (Adapted from Ref. [22])

model (cf. Eq. 1), the log (I)-Δz curves demonstrate pronounced deviations from linear behavior. They were also represented in terms of an effective barrier height (EBH), calculated as

$$\Phi_{eff} = \kappa^{-2} \left(\frac{\partial \ln I}{\partial z} \right)^2 \qquad (4)$$

The profiles of the EBH obtained from curves in Fig. 10a are displayed in Fig. 10b. They exhibit minima and maxima with height and position depending on the electrode potential. The distance scale in Fig. 10a and b is relative due to the experimental uncertainty of the absolute tip separation Δz in the STM experiment. To understand the origin of the variation of the EBH, structure and electron density distribution of the experimentally studied interfaces were modeled in DFT calculations [22]. As an example, Fig. 10c demonstrates the calculated optimal structure of the sulfate adlayer on Au(111) with the well-known ($\sqrt{3} \times \sqrt{7}$) R19.1° geometry (Fig. 5) and the corresponding distribution of positive and negative charge densities (Fig. 10d). Electron tunneling is facilitated by the

positive local charge density and hindered by the negative ones. Therefore, the positions of the maxima and minima in the EBH profile were attributed to the maxima and minima of charge density (Fig. 10d). The distance scale of the experimental curves was shifted so that the features of the calculated charge distribution normal to the surface coincided with the features of the experimental curve. The report published in Ref. [22] demonstrates the power of complementary experimental and theoretical approaches to the study of charged interfaces by EC-STM/STS.

Single-Molecule Charge Transport and Switching

In this paragraph, an advanced application of EC-STM for studies of electrochemically controlled electron transport in single-molecule junctions is described [59]. The latter are formed by a molecule with two chemical anchoring groups trapped to two metal electrodes. The STM tip is used as an active tool for contacting one anchoring group of a molecule bound to the substrate electrode by the second anchoring group. The method is demonstrated for the redox-active molecule perylene-3,4:9,10-tetra-carboxylic

Electrochemical Scanning Tunneling Microscopy, Fig. 11 Single-molecule conductance experiments of T-PBI immobilized on Au(111)-(1 × 1) in 0.05 M LiClO$_4$, pH ≈ 12. (**a**) Chemical structure of T-PBI. (**b**) Cyclic voltammogram of a T-PBI monolayer, scan rate 0.2 V·s^{-1}, and single-molecule conductance of T-PBI as a function of the substrate potential. (**c**) Typical current-distance retraction traces for T-PBI molecular junctions at $E_b = 0.10$ V, $E_S = -0.82$ V versus SCE. (**d**) Conductance histogram for T-PBI constructed from curves such as presented in panel (**c**). The inset shows an STM image of a disordered T-PBI adlayer on Au(111) (200 × 200 nm, $I_T = 0.1$ nA; $E_S = -0.02$ V versus SCE, $E_{bias} = 0.1$ V) [59]

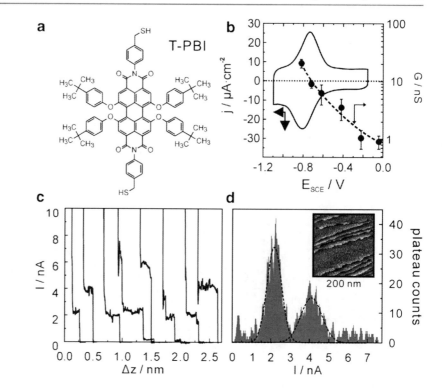

acid bisimide (T-PBI, Fig. 11a). The cyclic voltammogram of a T-PBI modified Au(111) electrode (Fig. 11b) shows a broad current peak at $E^0 = 0.79$ V, which represents a reversible two-electron reduction/ oxidation of the immobilized organic adlayer. A typical EC-STM image of a disordered densely packed T-PBI adlayer on an Au(111)-(1 × 1) surface is shown as inset in Fig. 11d.

Single Au |molecule| Au junction conductance measurements were carried out according to a "break junction" or "stretching" technique [60]: A sharp Au STM tip, capable of imaging experiments with atomic resolution, was brought to a preset tunneling position. Subsequently, the STM feedback was switched off, and the tip approached the adsorbate-modified substrate surface at constant x-y position. The approach was stopped at a preset tunneling current of 5–20 μA, which corresponds to a strong tip-adlayer interaction. After a dwelling time of ~100 ms, sufficient to create molecular junctions between tip and substrate, the tip was retracted and the corresponding current-distance traces were recorded. Examples, acquired at $E_S = -0.82$ V and $E_b = 0.10$ V, are shown in Fig. 11c. The traces exhibit characteristic plateaus, which are separated by abrupt steps. These observations are attributed to the stochastic formation and breaking

of individual respective multiple Au |T-PBI| Au junctions. Unlike in case of a tunneling current, the current flowing through a molecular junction is rather constant during the separation of tip and sample electrodes, and quickly drops by orders of magnitude when the junction is broken (Fig. 11c). The statistical analysis of several thousands of individual traces leads to conductance histograms, such as plotted in Fig. 11d. The histogram shows two pronounced peaks with positions close to 2 nA and 4 nA, with the second peak being broader and lower than the first one. The latter is a typical indication of molecular junctions corresponding to one and two molecules bound between electrodes. The single junction conductance is estimated for this particular experiment at $E_s = -0.82$ V and $E_b = 0.10$ V as $G = 22 \pm 4$ nS. These experiments were repeated at different substrate potentials. Figure 11b illustrates the corresponding single junction conductance as a function of substrate potential. The reduction of T-PBI upon decrease of the potential leads to an increase of the junction conductance by factor 20 over the whole measured potential range.

These experiments illustrate one of the frontier areas in electrochemical nanotechnology. They represent a direct approach toward exploring device-analog functions at electrified solid/liquid interfaces.

Electrochemical Scanning Tunneling Microscopy, Fig. 12 Various approaches to nanostructuring of metal electrodes by the EC-STM tip (Adapted from Ref. [61])

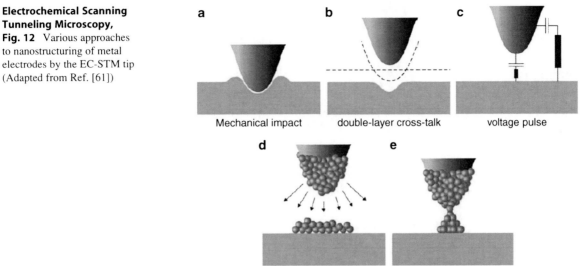

a   Mechanical impact

b   double-layer cross-talk

c   voltage pulse

d   local oversaturation

e   jump-to-contact

EC-STM Based Nanostructuring

The STM tip can also be used as a tool to actively structure electrode surfaces [7, 61]. Several existing strategies are illustrated in Fig. 12: (a) formation of active surface sites by mechanical impact of the tip [62], (b) local surface dissolution by double layer cross talk upon application of a steady tip potential [63], (c) dissolution initiated by short voltage pulses [64], (d) deposition of metal upon the local oversaturation of solution with metal ions during a "burst-like" dissolution of the metal from the STM tip [65], and (e) transfer of metal clusters from the tip to the surface [66]. While surface or monolayer defects as created by the mechanical impact of the tip are in general poorly defined and lead to non-reproducible surface features, the more controlled noncontact regimes allow fabrication of much more uniform, stable surface nanostructures with unique electric, magnetic, and catalytic properties. Fast dissolution of predeposited metal from the tip (Fig. 12d) can be used to create a local increase of the concentration of metal ions leading to metal deposition in the close vicinity of the tip (cf. Eq. 3). If the substrate potential E_S is only slightly more positive than E_{eq} as defined by the bulk concentration of metal ions, local metal deposition may occur [65]. The generated metal clusters are metastable, as their dissolution at the substrate potential used is kinetically hindered [17].

In the "jump-to-contact" technique (Fig. 12e), metal predeposited on the tip is transferred to the substrate upon approaching the tip toward the surface

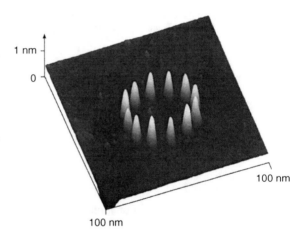

Electrochemical Scanning Tunneling Microscopy, Fig. 13 Cu clusters formed by the "jump-to-contact" technique on a Au(111) electrode in a solution of 0.05 M H_2SO_4 and 1 mM $CuSO_4$. The clusters are 0.8 nm high, the diameter of the circle is about 40 nm (Adapted from Ref. [7])

without reaching mechanical contact [7, 61, 66]. Jump-to-contact occurs, and a metal bridge is formed between tip and the substrate. When the tip is retracted from the surface, the metal bridge breaks and leaves a small metal [66] or alloy [67] cluster on the surface (Fig. 13). Using this technique, small and uniform clusters of various metals can be generated on metal and semiconducting surfaces with high reproducibility and lateral precision having a wide range of unique properties. The tip potential is always held in the OPD range and, provided that the metal deposition is sufficiently fast, the cluster transfer process can be repeated at kHz rates [61, 66].

Summary

In this article we briefly described the theoretical and experimental background of electrochemical scanning tunneling microscopy. We presented selected case studies covering classical and state-of-the-art applications of EC-STM and STS, illustrating the high potential of this technique in modern research and technology.

Acknowledgements This work was supported by the Swiss National Science Foundation (grant 200021_124643), the Volkswagen Foundation, FP7 through FUNMOLS, and the German Science Foundation within the priority program 1243.

Note This article is dedicated to Prof. Dieter M. Kolb ([†] 2011), who contributed with pioneering research to the development of Electrochemical Scanning Tunneling Microscopy.

Cross-References

▶ AFM in Liquids
▶ Atomic Force Microscopy
▶ Charge Transport in Self-Assembled Monolayers
▶ Force Modulation in Atomic Force Microscopy
▶ Nanoindentation
▶ Nanoscale Properties of Solid–Liquid Interfaces
▶ Nanostructures for Surface Functionalization and Surface Properties
▶ Plating
▶ Scanning Tunneling Microscopy
▶ Scanning Tunneling Spectroscopy
▶ Self-Assembled Monolayers

References

1. Binning, G., Rohrer, H., Gerber, C.H., Weibel, E.: Surface studies by scanning tunneling microscopy. Phys. Rev. Lett. **49**, 57–61 (1982)
2. Chen, C.J.: Introduction to Scanning Tunneling Microscopy. Oxford Science, New York (2008)
3. Liu, H.-Y., Fan, F.-R.F., Lin, C.W., Bard, A.J.: Scanning electrochemical and tunneling ultramicroelectrode microscope for high-resolution examination of electrode surfaces in solution. J. Am. Chem. Soc. **108**, 3838–3839 (1986)
4. Lustenberger, P., Rohrer, H., Christoph, R., Siegenthaler, H.: Scanning tunneling microscopy at potential controlled electrode surfaces in electrolytic environment. J. Electroanal. Chem. **243**, 225–235 (1988)
5. Wiechers, J., Twomey, T., Kolb, D.M., Behm, R.J.: An in-situ scanning tunneling microscopy study of Au(111) with atomic scale resolution. J. Electroanal. Chem. **248**, 451–460 (1988)
6. Gewirth, A.A., Niece, B.K.: Electrochemical applications of in situ scanning probe microscopy. Chem. Rev. **97**, 1129–1162 (1997)
7. Kolb, D.M.: An atomistic view of electrochemistry. Surf. Sci. **500**, 722–740 (2002)
8. Itaya, K.: In situ scanning tunneling microscopy in electrolyte solutions. Prog. Surf. Sci. **58**, 121–248 (1998)
9. Kolb, D.: Structure studies of metal electrodes by in-situ scanning tunneling microscopy. Electrochim. Acta **45**, 2387–2402 (2000)
10. Wandelt, K., Thurgate, S. (eds.).: Solid-Liquid Interfaces: Macroscopic Phenomena – Microscopic Understanding. Topics in Applied Physics, vol. 85. Springer, New York (2003)
11. Bowker, M., Davies, P.R. (eds.).: Scanning Tunneling Microscopy in Surface Science, Nanoscience and Catalysis. Wiley-VCH, Weinheim (2010)
12. Kolb, D.M.: Reconstruction phenomena at metal-electrolyte interfaces. Prog. Surf. Sci. **51**, 109–173 (1996)
13. Dakkouri, A.S., Kolb, D.M.: Reconstruction of gold surfaces. In: Wieckowski, A. (ed.) Interfacial Electrochemistry: Theory, Experiment and Applications, pp. 151–173. Marcel Dekker, New York (1999)
14. Magnussen, O.M.: Ordered anion adlayers on metal electrode surfaces. Chem. Rev. **102**, 679–725 (2002)
15. Wandlowski, Th.: Phase transitions in two-dimensional adlayers at electrode surfaces: thermodynamics, kinetics and structural aspects. In: Urbakh, M., Gileadi, M. (eds.) Encyclopedia of Electrochemistry, vol. 1, pp. 383–467. Wiley-VCH, Weinheim (2002)
16. Han, B., Li, Z., Li, C., Pobelov, I., Su, G., Aguilar-Sanchez, R., Wandlowski, Th.: From self-assembly to charge transport with single molecules – an electrochemical approach. Top. Curr. Chem. **287**, 181–255 (2009)
17. Budevski, E., Staikov, G., Lorenz, W.J.: Electrochemical Phase Formation and Growth. Wiley-VCH, Weinheim (1996)
18. Batina, N., Will, T., Kolb, D.M.: Study of the initial stages of copper deposition by in situ scanning tunnelling microscopy. Faraday Discuss. **94**, 93–106 (1992)
19. Meier, J., Friedrich, K.A., Stimming, U.: Novel method for the investigation of single nanoparticle reactivity. Faraday Discuss. **121**, 365–372 (2002)
20. Moffat, T.P., Wheeler, D., Josell, D.: Superconformal Film Growth. In: Staikov, G. (ed.) Electrocrystallization in Nanotechnology, pp. 96–114. Wiley-VCH, Weinheim
21. Maurice, V., Marcus, P.: Scanning Tunneling Microscopy and Atomic Force Microscopy. In: Marcus, P. (eds.) Analytical Methods in Corrosion Science and Enginering, pp. 133–168. CRC Press, Boca Raton (2006)
22. Simeone, F.C., Kolb, D.M., Venkatachalam, S., Jacob, T.: The Au(111)/Electrolyte interface: a tunnel-spectroscopic and DFT investigation. Angew. Chem. Int. Ed. **46**, 8903–8906 (2007)

23. Xu, B., He, H., Boussaad, S., Tao, N.J.: Electrochemical properties of atomic-scale metal wires. Electrochim. Acta. **48**, 3085–3091 (2003)

24. Li, C., Mishchenko, A., Wandlowski, Th.: Charge transport in single molecule junctions at the solid-liquid interface. In: Metzger, R.M. (eds.) Unimolecular and Supramolecular Electronics Springer (2011), (in press)

25. Sato, N.: Electrochemistry at Metal and Semiconductor Electrodes. Elsevier, Amsterdam (1998)

26. Bard, A.J., Faulkner, L.: Electrochemical Methods: Fundamentals and Applications. Wiley, New York (2001)

27. Hamann, C.H., Hamnett, A., Vielstich, W.: Electrochemistry. Wiley-VCH, Weinheim (2007)

28. Schmickler, W., Santos, E.: Interfacial Electrochemistry. Springer, Heidelberg (2010)

29. Ives, D.J.G., Janz, G.J.: Reference Electrode – Theory and Practice. Academic, New York (1961)

30. Reiss, H., Heller, A.: The absolute potential of the standard hydrogen electrode: a new estimate. J. Phys. Chem. **89**, 4207–4213 (1985)

31. Dai, H., Hafner, J.H., Rinzler, A.G., Colbert, D.T., Smalley, R.E.: Nanotubes as nanoprobes in scanning probe microscopy. Nature **384**, 147–150 (1996)

32. Tersoff, J., Hamann, D.R.: Theory and application for the scanning tunneling microscope. Phys. Rev. Lett. **50**, 1998–2001 (1983)

33. Siegenthaler, H.: STM in Electrochemistry. In: Wiesendanger, R., Güntherodt, H.-J. (eds.) Scanning Tunneling Microscopy II, Surface Sciences, vol. 28, pp. 7–49. Springer, Berlin/Heidelberg (1995)

34. Nagahara, L.A., Thundat, T., Lindsay, S.M.: Preparation and characterization of STM tips for electrochemical studies. Rev. Sci. Instrum. **60**, 3128–3130 (1989)

35. Dretschkow, Th., Lampner, D., Wandlowski, Th.: Structural transitions in 2, 2′-bipyridine adlayers on Au(111) – an in-situ STM study. J. Electroanal. Chem. **458**, 121–138 (1998)

36. Bach, C.E., Nichols, R.J., Beckmann, W., Meyer, H., Schulte, A., Besenhard, J.O., Jannakoudakis, P.D.: Effective insulation of scanning tunneling microscopy tips for electrochemical studies using an electropainting method. J. Electrochem. Soc. **140**, 1281–1284 (1993)

37. Ibe, J.P., Bey Jr., P.P., Brandow, S.L., Brizzolara, R.A., Burnham, N.A., DiLella, D.P., Lee, K.P., Marrian, C.R.K., Colton, R.J.: On the electrochemical etching of tips for scanning tunneling microscopy. J. Vac. Sci. Technol. A **8**, 3570–3575 (1990)

38. Zhang, B., Wang, E.: Fabrication of STM tips with controlled geometry by electrochemical etching and ECSTM tips coated with paraffin. Electrochim. Acta. **39**, 103–106 (1994)

39. Ren, B., Picardi, G., Pettinger, B.: Preparation of gold tips suitable for tip-enhanced Raman spectroscopy and light emission by electrochemical etching. Rev. Sci. Instrum. **75**, 837–841 (2004)

40. Tansel, T., Magnussen, O.M.: Video STM studies of adsorbate diffusion at electrochemical interfaces. Phys. Rev. Lett. **96**, 026101 (2006)

41. Clavilier, J.P., Faure, R., Guinet, G., Durand, R.: Preparation of monocrystalline Pt microelectrodes and electrochemical study of the plane surfaces cut in the direction of the {111} and {110} planes. J. Electroanal. Chem. **107**, 205–209 (1980)

42. Schweizer, M., Kolb, D.M.: First observation of an ordered sulfate adlayer on Ag single crystal electrodes. Surf. Sci. **544**, 93–102 (2003)

43. Villegas, I., Ehlers, C.B., Stickney, J.L.: Ordering of copper single-crystal surfaces in solution. J. Electrochem. Soc. **137**, 3143–3148 (1990)

44. Kibler, L.A.: Preparation and Characterization of Noble Metal Single Crystal Electrode Surfaces, ISE, Barcelona (2003)

45. Allongue, P., Kieling, V., Gerischer, H.: Etching mechanism and atomic structure of H-Si(111) surfaces prepared in NH₄F. Electrochim. Acta **40**, 1353–1360 (1995)

46. Dietterle, M.: Untersuchungen zur elektrolytischen Cu-Abscheidung und zur Dynamik von Stufenkanten auf niedrigindizierten Ag-Elektroden: Eine in-situ STM Studie, Ph. D. thesis, Universität Ulm (1996)

47. Giesen, M.: Step and Island dynamics at solid/vacuum and solid/liquid interfaces. Prog. Surf. Sci. **68**, 1–153 (2001)

48. Somorjai, G.A., van Hove, M.V.: Adsorbate-Induced Restructuring of Surfaces. Progr. Surf. Sci. **30**, 201–231 (1989)

49. Yamada, T., Batina, N., Itaya, K.: Structure of electrochemically deposited iodine adlayer on Au(111) studied by ultrahigh-vacuum instrumentation and in situ STM. J. Phys. Chem. **99**, 8817–8823 (1995)

50. Vereecken, P.M., Binstead, R.A., Deligianni, H., Andricacos, P.C.: The chemistry of additives in damascene copper plating. IBM J. Res. Dev. **49**, 3–18 (2005)

51. Lindsay, S.M.: Introduction to Nanoscience. Oxford University Press, Oxford (2010)

52. Li, Z., Han, B., Wan, L.J., Wandlowski, Th.: Supramolecular nanostructures of 1,3,5-benzenetricarboxylic acid at electrified Au(111)/0. 05 M H₂SO₄ interfaces: an in situ scanning tunneling microscopy study. Langmuir **21**, 6915–6928 (2005)

53. Han, B., Li, Z., Pronkin, S., Wandlowski, Th.: In situ ATR-SEIRAS study of adsorption and phase formation of trimesic Acid on Au(111–25 nm) film electrodes. Can. J. Chem. **82**, 1481–1494 (2004)

54. Kolb, D. M.: Physical and Electrochemical Properties of Metal Monolayers on Metallic Substrates. In: Gerischer, H., Tobias, C. W. (eds.) Advances in Electrochemistry and Electrochemical Engineering, John Wiley & Sons, New York, vol. 11, p. 125 (1978)

55. Herrero, E., Buller, L.J., Abruña, H.D.: Underpotential deposition at single crystal surfaces of Au, Pt, Ag and other materials. Chem. Rev. **101**, 1897–1930 (2001)

56. Allongue, P., Maroun, F.: Metal electrodeposition on single crystal metal surfaces: mechanisms structure and applications. Curr. Opin. Solid State Mater. Sci. **10**, 173–181 (2006)

57. Schneeweiss, M., Kolb, D.: The initial stages of copper deposition on bare and chemically modified gold electrodes. Phys. Stat. Sol. A **173**, 51–71 (1999)

58. Toney, M.F., Howard, J.N., Richer, J.: Electrochemical deposition of copper on gold electrode in sulfuric acid: resolution of interfacial structure. Phys. Rev. Lett. **75**, 4472–4475 (1995)

59. Li, C., Mishchenko, A., Li, Z., Pobelov, I., Wandlowski, Th., Li, X.Q., Würthner, F., Bagrets, A., Evers, F.:

Electrochemical gate-controlled electron transport of redox-active single perylene bisimide molecular junctions. J. Phys. Condens. Matter **20**, 374122 (2008)

60. Xu, B.Q., Tao, N.J.: Measurement of single-molecule resistance by repeated formation of molecular junctions. Science **301**, 1221–1223 (2003)

61. Kolb, D.M., Simeone, F.C.: Electrochemical nanostructuring with an STM: a status report. Electrochim. Acta **50**, 2989–2996 (2005)

62. Jaklevic, R.C., Elie, L.: Scanning-tunneling-microscope observation of surface diffusion on an atomic scale: Au on Au(111). Phys. Rev. Lett. **60**, 120–123 (1988)

63. Xie, Z.-X., Kolb, D.M.: Spatially confined copper dissolution by an STM tip: a new type of electrochemical reaction? J. Electroanal. Chem. **481**, 177–182 (2000)

64. Schuster, R., Kirchner, V., Allongue, P., Ertl, G.: Electrochemical Micromachining. Science **289**, 98–101 (2000)

65. Schindler, W., Hofmann, D., Kirschner, J.: Nanoscale Electrodeposition: A New Route to Magnetic Nanostructures? J. Appl. Phys. **87**, 7007–7009 (2000)

66. Kolb, D.M., Ullmann, R., Will, T.: Nanofabrication of small copper clusters on gold(111) electrodes by a scanning tunneling microscope. Science **275**, 1097–1099 (1997)

67. Del Pópolo, M.G., Leiva, E.P.M., Schmickler, W.: On the stability of electrochemically generated nanoclusters – a computer simulation. Angew. Chem. Int. Ed. **40**, 4674–4676 (2001)

68. Simeone, F.C., Kolb, D.M., Venkatachalam, S., Jacob, T.: The Au(111)/Electrolyte interface: a tunnel-spectroscopic and DFT investigation. Surf. Sci. **602**, 1401–1407 (2008)

Electrode–Organic Interface Physics

Michael G. Helander[1], Zhibin Wang[1] and Zheng-Hong Lu[1,2]
[1]Department of Materials Science and Engineering, University of Toronto, Toronto, Ontario, Canada
[2]Department of Physics, Yunnan University, Yunnan, Kunming, PR China

Synonyms

Band alignment; Energy-level alignment; Metal–organic interfaces

Definition

Electrode–organic interface physics describes the formation of an electric contact between an organic semiconductor material and an electrode, typically a metal. The alignment of the energy-levels of the organic material with the energy-levels of the electrode dictates the energetic barrier which must be overcome to move electrical charges between the electrode and organic material. Depending on the magnitude of the energetic barrier at the interface, the injection of charge across the interface will be different.

Introduction

Organic electronic devices are electrical devices, such as light emitting diodes, photovoltaics, or thin film transistors, in which the active materials consist entirely of organic materials with semiconducting properties. Since organic materials consist of discrete molecular units, there is no need for long range order between adjacent molecules or polymer chains. Organic semiconductors therefore have the advantage over traditional inorganic semiconductors, such as Si or GaAs, in that they do not require perfect single crystal films to operate in real devices. Complicated multilayer structures with nanometer-scale thicknesses can thus be easily fabricated from organic materials using low-cost roll-to-roll printing or vacuum coating techniques. However, the discrete nature of organic semiconductors also implies that they typically contain almost no *intrinsic* charge carriers (i.e., electrons or holes), and thus act as insulators until electrical charges are injected into them. In electrical device applications this means that all of the holes and electrons within a device must be injected from the anode and cathode respectively. As a result, device stability, performance, and lifetime are greatly influenced by the interface between the organic materials and the electrode contacts.

Energy-Level Alignment

Of particular interest for device applications is the energy-level alignment at electrode–organic interfaces, or in other words the alignment of the molecular orbitals of the organic to the energy-levels of the electrode (typically the Fermi level). For traditional inorganic semiconductors the energy-level alignment at semiconductor–metal contacts is typically described by the Schottky barrier height, which is the potential barrier between the Fermi level of the metal and the conduction band or valence band edge of the semiconductor. Analogous to the Schottky barrier height at semiconductor–metal interfaces the energy-level alignment at electrode–organic interfaces is also typically represented by an injection barrier height ϕ_B. This injection barrier represents the potential barrier which must be overcome to move an electrical charge from the Fermi level of the electrode into the highest occupied molecule orbital (HOMO) or lowest unoccupied molecular orbital (LUMO) of the organic (or vice versa). Metal–organic interfaces were the first contacts to be extensively characterized in organic electronics, owing to their similarity to Schottky contacts in traditional semiconductors.

Work Function

The work function of a uniform surface of a conductor is defined as the minimum energy required to remove an electron from the interior of the conductor to just outside the surface, where "just outside" refers to a distance that is large enough that the image force is negligible, but small compared to the physical dimensions of the crystal (typically $\sim 10^{-4}$ cm). In other words, the work function is the difference between the electrochemical potential $\bar{\mu}$ of electrons in the bulk and the electrostatic potential energy $-e\Phi_{vac}$ of an electron in the vacuum just outside the surface [1]:

$$e\phi_m = -e\Phi_{vac} - \bar{\mu}. \tag{1}$$

The energy-level corresponding to $-e\Phi_{vac}$ is referred to as the *local* vacuum level E_{vac} and is not to be confused with the vacuum level at infinity E_0, which represents an electron at rest at infinity (i.e., zero potential energy) [2]. Since the Fermi level E_F relative to the vacuum level at infinity E_0 is the electrochemical potential of electrons in the conductor [3],

$$E_0 - E_F = -\bar{\mu}, \tag{2}$$

the work function given by (1) is equivalent to the difference in potential energy of an electron between the local vacuum level E_{vac} and the Fermi level:

$$e\phi_m = E_{vac} - E_F. \tag{3}$$

From (3) it is clear that if the local vacuum level E_{vac} varies across a sample, as is the case for the different faces of a single crystal, the work function will also vary.

Schottky Contacts

The energy-level alignment at semiconductor–metal contacts has been extensively studied owing to their importance in microelectronics. In the Schottky–Mott limit the vacuum level of the semiconductor and metal align, forming a region of net space charge at the interface. No charge is transferred across the interface and hence the semiconductor bands are forced to bend to accommodate the potential difference. The Schottky barrier height for electrons is then given by the difference between the work function of the metal ϕ_m and the electron affinity of the semiconductor χ (i.e., the conduction band edge). However, the Schottky–Mott limit is rarely observed at most metal–semiconductor interfaces due to charge transfer across the interface (i.e., an interfacial dipole). As a result, the Fermi level cannot move freely in the bandgap of the organic and thus tends to be pinned by the dipole layer. The interface slope parameter is a convenient parameter that quantifies this phenomenon, and is commonly defined in terms of the Schottky barrier height for an electron or hole [4]:

$$S_\phi \equiv \frac{d\phi_{Bn}}{d\phi_m} = -\frac{d\phi_{Bp}}{d\phi_m}, \tag{4}$$

where S_ϕ is the interface slope and ϕ_{Bn} and ϕ_{Bp} are the barrier heights for electrons and holes respectively. In the Schottky–Mott limit $S_\phi = 1$ (vacuum level alignment); and for Fermi level pinning $S_\phi = 0$ (interfacial dipole).

In the case of organic semiconductors many studies have experimentally demonstrated a strong correlation between the metal work function and the barrier height for holes or electrons. In analogy to inorganic semiconductors, in the Schottky–Mott limit the vacuum level of the organic and metal align, forming a region of net space charge at the interface. The barrier height for holes is then given by the offset between the HOMO of the organic E_{HOMO} and the work function of the metal:

$$e\phi_{Bp} = E_{HOMO} - e\phi_m. \tag{5}$$

However, since organic semiconductors contain almost no free charge carriers, band bending in the organic cannot fully accommodate the potential difference. Hence, the Schottky–Mott limit is rarely observed at metal–organic interfaces due to the

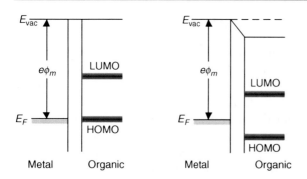

Electrode–Organic Interface Physics, Fig. 1 Schematic diagram of energy-level alignment at the metal–organic interface in the Schottky–Mott limit (*left*) and with an interfacial dipole (*right*)

formation of a strong interfacial dipole between the metal surface and organic molecules. Accounting for the effect of an interfacial dipole, the injection barrier for holes is then given by the following:

$$e\phi_{Bp} = E_{HOMO} - e(\phi_m - \Delta).\qquad(6)$$

From the above equation it is clear that a negative dipole reduces the hole injection barrier, while a positive dipole enhances electron injection. From (6) an effective metal work function can be defined as the difference between the pristine metal work function and the dipole:

$$\phi_{m,eff} = \phi_m - \Delta,\qquad(7)$$

where $\phi_{m,eff}$ is the effective metal work function (Fig. 1).

Interface Dipole Theory

Deviation from the Schottky–Mott limit was first described by Bardeen [5] as the result of a large density of surface states; charge transfer between the metal and these surface states acts to pin the Fermi level. In fact surface states were first postulated by Tamm and Shockley prior to the Schottky model, but were largely ignored. Heine [6] later demonstrated that surface states do not exist in the gap of most metal–semiconductor interfaces, or at least not for modern devices produced by the high vacuum deposition of metal onto a clean semiconductor surface. In Bardeen's time semiconductor–metal contacts were formed by laminating a piece of metal and semiconductor, which inherently results in a "dirty" interface. Heine

postulated that for a "clean" interface, gap states are induced in the semiconductor as a result of the rapidly decaying tail of the electronic wave function from the metal. These gap states would also tend to pin the Fermi level same as for Bardeen's surface states. Tersoff [7] later identified the gap states as metal-induced gap states (MIGs), which derive from the virtual gap states (VIGs) of the semiconductor complex band structure.

Regardless of their specific origin, interface states are independent of the energy-levels in the semiconductor, and hence can either be donor like or acceptor like when close to the valence or conduction bands respectively. A charge neutrality level ϕ_{CNL} is defined as the point at which the interface states are equally donor- and acceptor-like (i.e., the transition point from donor to acceptor states). Cowley and Sze [8] applied this notion to derive the dependence of the Schottky barrier height upon the metal work function. They argued that at semiconductor–metal interfaces the charge neutrality level of the semiconductor will tend to align with the Fermi level of the metal as a result of charge transfer between the metal and the interface states, forming an interfacial dipole. In other words, the redistribution of charge at the interface (dipole) offsets the vacuum level of the metal and semiconductor, resulting in an effective metal work function that is different from the vacuum metal work function:

$$\phi_{m,eff} = \phi_{CNL} + S_\phi(\phi_m - \phi_{CNL}).\qquad(8)$$

The interfacial dipole is then the difference between $\phi_{m,eff}$ and ϕ_m:

$$\Delta = (1 - S_\phi)(\phi_m - \phi_{CNL}).\qquad(9)$$

Interface Slope Parameter

In their analysis, Cowley and Sze assumed a constant continuum of interface states across the semiconductor gap. Based on this assumption, the interface slope parameter, which in this context represents the dielectric screening strength of the semiconductor, is given by:

$$S_\phi = \frac{\varepsilon_i}{\varepsilon_i + e\delta D_s},\qquad(10)$$

where ε_i is the interface permittivity, δ is the characteristic thickness of the interface (atomic length), and

D_s is the density of interface states. However, since these parameters can only be measured indirectly, accurate determination of the interface slope parameter using (10) is problematic at best. Luckily, Mönch [9] empirically found that the interface slope parameter (for weakly interacting interfaces) is dependent on the optical dielectric constant:

$$S_\phi = \frac{1}{1 + 0.1(\varepsilon_\infty - 1)^2}, \qquad (11)$$

where ε_∞ is the optical dielectric constant (high frequency limit of the dielectric function). One important caveat to consider for (11) is that it only accurately describes the interface slope parameter for weakly interacting and laterally homogenous Schottky contacts. If interfacial chemical reactions occur, or if the interface is patchy, then S_ϕ often deviates from the ideal theoretical dielectric value.

Recently it was shown that organic semiconductors also follow the same empirical relationship between S_ϕ and ε_∞ described by (11), same as for Schottky contacts with solid Xenon. However, many organic molecules either interact fairly strongly with clean metal surfaces or do not wet the surface of metals uniformly. The experimentally determined value of S_ϕ for clean metal surfaces is, therefore, often different than the ideal theoretical value described by (11). For example, the experimental value of S_ϕ for C_{60} is nearly 0, while the theoretical value is 0.53 assuming $\varepsilon_\infty = 4$. In this case it is well known that C_{60} interacts strongly with most clean metal surfaces, which results in complete pinning of the Fermi level (i.e., $S_\phi = 0$). For terminated surfaces, such as an oxidized metal surface (e.g., Al/Al_2O_3), the interaction between the metal and organic is screened by the adsorbed surface termination layer. In this case, S_ϕ approaches the ideal theoretical value for a weakly interacting interface. Metal oxides in particular form ideal contacts with most organics since the first layer of adsorbed molecules tends to wet the oxide surface, forming a laterally homogeneous contact. One important consequence of this is that a clean metal and oxide-terminated metal will behave differently in contact with an organic semiconductor, even if the effective metal work function of the oxide-terminated metal is the *same* as the work function of the clean metal surface. In general, the clean metal will pin the Fermi level stronger than the oxide-terminated metal.

Since the interfacial dipole is indicative of charge transfer between the metal and semiconductor, electronegativity is often used in place of work function to describe the dependence of barrier height on the contacting metal. In this case a linear dependence of the injection barrier on the electronegativity *difference* of the metal and semiconductor is observed:

$$\phi_{m,\text{eff}} = \phi_{\text{CNL}} + S_x(X_m - X_s), \qquad (12)$$

where X_m and X_s are the electronegativity of the metal and semiconductor respectively and S_x is the electronegativity-based interface slope parameter given by:

$$S_x = \frac{d\phi_{\text{Bn}}}{dX_m} = -\frac{d\phi_{\text{Bp}}}{dX_m}. \qquad (13)$$

One important distinction between the two different interface slope parameters is that S_ϕ is a dimensionless quantity with an upper limit of 1, while S_x on the other hand is *not* dimensionless and has an arbitrary upper limit depending on the units of electronegativity used. There have been several attempts to link the two interface slope parameters via a constant of proportionality:

$$S_\phi = A_x S_x, \qquad (14)$$

where the value of A_x varies depending on the electronegativity scale used. In the case of organic semiconductors, there is no clear definition of electronegativity and hence previous attempts to relate injection barriers at metal–organic interfaces to electronegativity values are in general not very useful.

Metal–Dielectric–Organic Interfaces

Although interface dipole theory can explain the energy-level alignment at metal–organic interfaces in most device applications, the metal contacts are often modified with various dielectric buffer layers to improve charge injection. Dielectric buffer layers, often referred to as electron or hole injection layers (EIL or HIL) since they dramatically enhance carrier injection, typically fall into one of three categories: (1) wide bandgap insulators such as LiF, (2) semiconducting transition metal oxides such as V_2O_5, or (3) organic molecules, such as fullerene (C_{60}). Despite the appreciable difference in material properties for these three different types of dielectrics, one important similarity has been consistently reported in the literature. For all three types of dielectrics, strong interfacial dipoles (i.e., a rapid change in vacuum level across the interface) have been reported at the interfaces formed with metals and organics.

The energy-level alignment at real electrode–organic interfaces will therefore typically depend on the properties of the underlying metal as well as the dielectric buffer layer. A similar situation occurs at metal–oxide–semiconductor (MOS) interfaces used in advanced high-κ dielectric metal gate MOS field effect transistors (MOSFETs). It was discovered that the effective work function of the metal gate was strongly influenced by the properties of the oxide dielectric. Robertson [10] proposed that interface dipole theory could be applied to each of the interfaces (i.e., metal–oxide and oxide–semiconductor) to describe the trend in effective work function, or in other words that (12) could be applied to both interfaces to describe the net effective work function. A similar approach can thus be applied to metal–dielectric–organic interfaces to describe the energy-level alignment.

In order to calculate the net effect of the oxide layer on the injection barrier, the two interfaces can be considered in series. At the metal–dielectric interface the interaction between the dielectric and the metal creates an interfacial dipole layer that modifies the work function of the pristine metal surface. This results in the organic molecules deposited on top of the dielectric modified metal experiencing an effective metal work function as given by (12). Treating the metal–oxide as a layered anode structure, the effective metal work function can be plugged back into (12) to extract the effective work function of the metal–oxide after consideration of the effect of the organic layer:

$$
\begin{aligned}
\phi_{m,\text{eff}} = \left(1 - S_\phi^{\text{org}}\right)\phi_{\text{CNL}}^{\text{org}} \\
+ S_\phi^{\text{org}}\left[\phi_{\text{CNL}}^{\text{ox}} + S_\phi^{\text{ox}}\left(\phi_m - \phi_{\text{CNL}}^{\text{ox}}\right)\right],
\end{aligned}
\tag{15}
$$

where S_ϕ^{org}, S_ϕ^{ox}, $\phi_{\text{CNL}}^{\text{org}}$, and $\phi_{\text{CNL}}^{\text{ox}}$ are the interface slope parameters and charge neutrality levels of the organic and oxide respectively. The total combined dipole across the interface can then be calculated using (9). The barrier height for holes is then taken as the

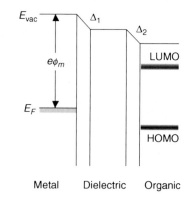

Electrode–Organic Interface Physics, Fig. 2 Schematic diagram of energy-level alignment at a metal–dielectric–organic interface

difference between this combined effective work function and the HOMO of the organic (Fig. 2).

Dielectrics typically have $S_\phi < 1$, which implies that the Fermi level is not completely pinned at most metal–dielectric–organic interfaces. As a result, the barrier height at the interface is partially dependent on the work function of the underlying metal. However, in a few special cases S_ϕ can approach unity, implying that the Fermi level is pinned and hence the energy-level alignment at the interface is independent of the underlying metal. Buckminsterfullerene (C_{60}) is one example of such a dielectric layer. In contact with clean metal surfaces C_{60} can accept up to six electrons from the metal, becoming itself partially metallic. As a result, the optical dielectric constant (ε_∞) of the C_{60} approaches infinity, such that S_ϕ goes to 1 as given by (11). Therefore C_{60} can be used as a "universal" charge injection layer as it pins the Fermi level of any metal to its charge neutrality level.

Charge Injection and Transport

Up to this point only the barrier height at electrode–organic interfaces has been discussed. However, in real devices what is measured is the flow of current through the device as a function of applied bias. Therefore how the barrier height discussed above affects the injection and transport of charge in real devices is of interest. At traditional semiconductor–metal interfaces the process of charge injection is typically divided into two different regimes, space charge limited current (SCLC) and injected limited current (ILC).

Space Charge Limited Current

The first regime, which represents the maximum current that a semiconductor can sustain in the bulk (i.e., the amount of carriers in thermal equilibrium), is called the SCLC. A semiconductor–metal contact capable of injecting enough charges to sustain the SCLC is called an Ohmic contact. In most traditional semiconductors an Ohmic contact is assumed to represent a zero barrier height. One significant feature of SCLC is that the spatial distribution of electric field is described by $F(x) \propto x^{1/2}$, where x is the distance from the charge-injecting contact. The electric field at the interface of an Ohmic contact is therefore equal to zero. For unipolar transport (i.e., transport of only one type of charge carrier) in a perfect insulator (no intrinsic carriers) without traps the SCLC is given by the Mott-Gurney law [11]:

$$J = \frac{9}{8}\varepsilon_0\varepsilon\mu\frac{V^2}{d^3}, \qquad (16)$$

where V is the applied voltage, d is the thickness of the film, and μ is the field-independent mobility.

With further consideration of an exponential distribution of trap states, the trap-charged limited current (TCLC) based on (16) follows:

$$J \propto \mu\frac{V^{l+1}}{d^{2l+1}}, \qquad (17)$$

where l is a parameter derived from the trap distribution. Notice that the injection of charge depends only on the mobility, thickness of the semiconductor, and applied voltage as the barrier height is assumed to be zero.

In the case of organic semiconductors, however, the mobility typically depends on electric field. In disordered organic materials, it is believed that all electronic states are localized and participate in conduction through thermally activated hopping, which yields a Poole–Frenkel-like field dependence of the mobility:

$$\mu(F) = \mu_0\exp(\beta\sqrt{F}). \qquad (18)$$

Under the assumption of a Poole–Frenkel dependence, an approximation to the SCLC for a field-dependent mobility is given by [12]:

$$J_{\text{SCLC}} = \frac{9}{8}\varepsilon\varepsilon_0\mu_0\exp\left(0.89\beta\sqrt{\frac{V}{d}}\right)\frac{V^2}{d^3}. \qquad (19)$$

There has been significant experimental evidence that this model does indeed describe the SCLC in many organic semiconductors. One interesting caveat however due to the hopping-based transport in organic materials is that the barrier height for an Ohmic contact does not have to be zero in the typically operating electric field range of most device applications. Specific criteria for determining this critical threshold barrier height will be discussed in detail below.

Injection-Limited Current

The second regime of charge injection in semiconductors represents the situation where a significant barrier height exists at the interface, such that the transport of charge in the semiconductor is limited by the injection of charge at the semiconductor–metal contact. The ILC at Schottky contacts has traditionally been described by one of two processes. For high electric field strength and large barrier heights, Fowler–Nordheim (FN) tunneling, also known as field emission, dominates the injection current [4]:

$$J_{\text{FN}} = \frac{q^3F^2}{8\pi h\phi_B}\exp\left[-\frac{8\pi\sqrt{2m^*}\phi_B^{3/2}}{3qhF}\right] \qquad (20)$$

where F is the electric field strength, ϕ_B is the barrier height, m^* is the effective mass of an electron or hole in the semiconductor, q is the electronic charge, and h is the Planck constant. For thermally activated processes at lower electric field, Richardson–Schottky (RS) emission, also known as thermionic emission, dominates the injection current [4]:

$$J_{\text{RS}} = A^*T^2\exp\left[-\frac{\left(\phi_B - \sqrt{qF/4\pi\varepsilon}\right)}{k_BT}\right] \qquad (21)$$

where A^* is the modified Richardson constant, ε is the permittivity of the semiconductor, k_B is the Boltzmann constant, and T is the temperature.

In the case of organic semiconductors, the ILC was early on modeled as diffusion-limited thermionic emission. However, both FN tunneling and RS emission are derived for delocalized Bloch waves in the semiconductor, whereas the electronic states in most organic semiconductors are highly localized due to their discrete molecular nature. In addition, at room temperature FN tunneling typically requires electric field strengths several orders of magnitude greater than what are observed in real organic devices ($\sim 10^5$ V/cm).

To overcome these deficiencies Scott and Malliaras [13] have proposed a modified thermionic emission model for the ILC in organic semiconductors based on the original solution to the drift-diffusion equation for injection into a wide bandgap intrinsic semiconductor solved by Emtage and O'Dwyer [14]. This model determines the equilibrium contributions to the current density for charge carriers recombining with their own image analogous to Langevin recombination of an electron-hole pair in the bulk:

$$J_{\text{ILC}} = 4N_0\psi^2 e\mu F \exp\left(\frac{-e\phi_B}{k_B T}\right)\exp(f^{1/2}), \qquad (22)$$

where N_0 is the density of chargeable sites in the organic film, ϕ_B is the barrier height, F is the electric field at the charge-injecting contact, μ is the electric field-dependent mobility, k_B is the Boltzmann constant, T is temperature, e is the electron charge, and ψ is a function of the reduced electric field ($f = e^3 F/4\pi\varepsilon k_B^2 T^2$):

$$\psi = f^{-1} + f^{-1/2} - f^{-1}(1 + 2f^{1/2})^{1/2} \qquad (23)$$

There has been significant experimental evidence that this model does indeed describe the ILC in many organic semiconductors. Also, under ILC conditions the spatial electric field distribution is assumed to be uniform such that $F(x) = V/d$; whereas for SCLC the value of V/d only gives the average value of the electric field. This suggests that SCLC can be distinguished from ILC by the electric field at the electrode–organic interface.

Quasi-Ohmic Contact

The two conditions discussed above, SCLC and ILC, correspond to the upper and lower limits respectively. What about the case in between SCLC and ILC, when the electric field at the electrode contact is $0 < F(x = 0) < V/d$ (see Fig. 3)? This indicates that there is in fact no clear boundary between SCLC and ILC, but rather an intermediate regime in between referred to as "quasi-Ohmic." Since the quasi-Ohmic

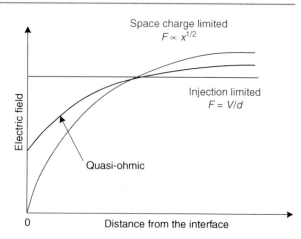

Electrode–Organic Interface Physics, Fig. 3 Calculated electric field at the metal–organic interface as a function of the barrier height for a typical organic semiconductor with a mobility of 10^{-4}–10^{-3} cm^2 V^{-1} s^{-1} and a thickness of 1,000 nm

regime exhibits characteristics of both SCLC and ILC, it is incorrect to apply the models for either case. Moreover, it is important to note that the boundaries of the quasi-Ohmic regime are dependent on the field-dependent carrier mobility μ, the barrier height ϕ_B, the applied voltage V, and the device thickness d. As a result the same electrode–organic contact may display characteristics of ILC, quasi-Ohmic, and SCLC in different ranges of applied bias. A new approach is thus required to properly describe the quasi-Ohmic regime.

As discussed above, the quasi-Ohmic regime in between SCLC and ILC cannot be described using traditional models. As a result a new approach that includes both the injection at the interfaces and the transport in the bulk of the organic is required to deal with this special case. One approach to deal with this situation is to use a time domain simulation under the transport models developed for inorganic semiconductors (i.e., the drift-diffusion and Poisson equations) to describe the distribution of electric field at steady state.

In the simulation, space x and time t are discrete. The injection current density described by (22) and (23) serves as a boundary condition at $x = 0$, $t > 0$ (i.e., the charge-injecting contact). The time-dependent continuity equation follows:

$$\frac{\partial p(x,t)}{\partial t} = -\frac{1}{e}\frac{\partial J(x,t)}{\partial x}, \qquad (24)$$

where p is the total density of holes and J is the conduction current density. The relation between the electric field and the charge density can be expressed by the Poisson equation as:

$$\frac{\varepsilon_r \varepsilon_0}{e} \frac{\partial F(x,t)}{\partial x} = p(x,t). \qquad (25)$$

In (25), ε_r and ε_0 are the dielectric constants of the organic and the dielectric permittivity in vacuum respectively. The conduction current density can be calculated through the drift-diffusion equation:

$$J(x,t) = ep(x,t)\mu(F)F - eD\frac{\partial p(x,t)}{\partial x}, \qquad (26)$$

where D is the diffusion coefficient, which can be obtained from Einstein's relation (as a function of the field-dependent mobility):

$$D = \frac{\mu k_B T}{e}. \qquad (27)$$

The other boundary conditions are:

$$\begin{cases} V = \int_0^d F dx \\ F(x,t=0) = V/d, \\ p(x,t=0) = 0 \\ J(x,t=0) = 0 \end{cases} \qquad (28)$$

where d is the thickness of the film and V is the applied voltage. It is noted that the transient current density J_t is contributed by the displacement current and the response of the charge carrier density as:

$$J_t(x,t) = \varepsilon_r \varepsilon_0 \frac{\partial F(x,t)}{\partial t} + J(x,t). \qquad (29)$$

From (29) the total transient current at $x = d$ can be calculated until the steady state is reached. Also, the spatial distribution of electric field at steady state can be obtained from the simulation.

From these simulation results the boundaries of the quasi-Ohmic regime (i.e., the lower limit of SCLC and the upper limit of ILC) can then be defined as the convergence of the simulation results with the SCLC from (19) and the ILC from (22). Also, these boundaries represent the strict limits for SCLC and ILC in terms of the electric field at the charge-injecting

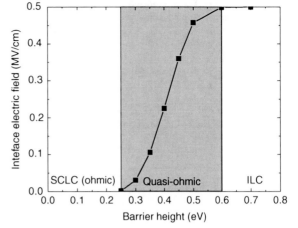

Electrode–Organic Interface Physics, Fig. 4 Calculated electric field at the metal–organic interface as a function of the barrier height for a typical organic semiconductor with a mobility of 10^{-4}–10^{-3} cm^2 V^{-1} s^{-1} and a thickness of 1,000 nm

contact. Figure 4 shows an example of the calculated electric field at the electrode–organic interface for a typical organic semiconductor with a mobility of 10^{-4}–10^{-3} cm^2 V^{-1} s^{-1}. The calculation shows that the interfacial electric field depends strongly on the barrier height in the quasi-Ohmic regime. The interfacial electric field approaches zero and the average values (i.e., $F = V/d$) at low ($< \sim 0.25$ eV) and high ($> \sim 0.55$ eV) barrier height which respectively correspond to SCLC regime and ILC regime.

Future Directions

The energy-level alignment and charge injection at electrode–organic interfaces is based on the classical device physics developed for inorganic semiconductors. It can be applied to a wide variety of different materials and interfaces, even newly developed organic materials.

Cross-References

▶ Organic Photovoltaics: Basic Concepts and Device Physics

References

1. Herring, C., Nichols, M.H.: Thermionic emission. Rev. Mod. Phys. **21**, 185 (1949)

2. Hagstrum, H.D.: The determination of energy-level shifts which accompany chemisorption. Surf. Sci. **54**, 197 (1976)
3. Van Rysselberghe, P.: A note on work functions and chemical potentials. J. Chem. Phys. **21**, 1550 (1953)
4. Sze, S.M., Ng, K.K.: Physics of Semicondcutor Devices. Wiley, Hoboken (2007)
5. Bardeen, J.: Surface states and rectification at a metal semi-conductor contact. Phys. Rev. **71**, 717 (1947)
6. Heine, V.: Theory of surface states. Phys. Rev. **138**, A1689 (1965)
7. Tersoff, J.: Schottky barrier heights and the continuum of gap states. Phys. Rev. Lett. **52**, 465 (1984)
8. Cowley, A.M., Sze, S.M.: Surface states and barrier height of metal-semiconductor systems. J. Appl. Phys. **36**, 3212 (1965)
9. Mönch, W.: Role of virtual gap states and defects in metal-semiconductor contacts. Phys. Rev. Lett. **58**, 1260 (1987)
10. Robertson, J.: Band offsets of wide-band-gap oxides and implications for future electronic devices. J. Vac. Sci. Technol. B **18**, 1785 (2000)
11. Lampert, M.A., Mark, P.: Current Injection in Solids. Academic, New York (1970)
12. Murgatroyd, P.N.: Theory of space-charge-limited current enhanced by Frenkel effect. J. Phys. D Appl. Phys. **3**, 151 (1970)
13. Scott, J.C., Malliaras, G.G.: Charge injection and recombination at the metal-organic interface. Chem. Phys. Lett. **299**, 115 (1999)
14. Emtage, P.R., O'Dwyer, J.J.: Richardson–Schottky effect in insulators. Phys. Rev. Lett. **16**, 356 (1966)

Electrohydrodynamic Forming

▶ Electrospinning

Electrokinetic Fluid Flow in Nanostructures

Francesca Carpino[1], Larry R. Gibson II[1], Dane A. Grismer[1] and Paul W. Bohn[1,2]
[1]Department of Chemical and Biomolecular Engineering, University of Notre Dame, Notre Dame, IN, USA
[2]Department of Chemistry and Biochemistry, University of Notre Dame, Notre Dame, IN, USA

Synonyms

Nanofluidics

Definition

Electrokinetic flow in nanostructures involves the study of the behavior, manipulation, and control of fluids that are confined to structures characterized by nanometer scale (typically ≤ 100 nm) dimensions. Fluids confined in these structures exhibit behaviors not observed on longer length scales, because characteristic physical scaling lengths, e.g., Debye length, in the fluid are commensurate with the dimensions of the nanostructure itself. A dominant factor in determining the electrokinetic (EK) flow behavior in physical nanostructures is the size of the electrical double layer (EDL) associated with the physical boundaries of the nanostructure and its relation to the channel dimensions.

Overview

Motivation. Although the engineering drivers for reducing the size of laboratory devices are clear (reduced materials costs, reduced power consumption, less waste, lower operating costs, etc.), the case for doing science at reduced dimensions needs to be made more carefully. Factors which justify working at the nanometer scale include: (1) new transport phenomena, (2) exploiting the enhanced surface-to-volume ratio, (3) using diffusion as a viable transport mechanism, and (4) integrating large molecules or molecular complexes with very small (1–10 nm) physical structures. In particular:

- The similarity of the Debye length, κ^{-1}, and the channel diameter, a, accesses a new electrokinetic flow regime ($\kappa a \sim 1$) that is not available at longer length scales, even in μm-scale capillaries. By changing κ^{-1} it is possible to move from a regime, $\kappa a \gg 1$, where flow is dominated by electrophoresis to a regime, $\kappa a \sim 1$, where electroosmotic flow is the dominant transport mechanism. This is possible, because in the nanochannels there is a preponderance of counterions over co-ions; in fact it is easy to achieve conditions where every mobile counterion in the pore is of one polarity, i.e., there are no co-ions.

- The huge increase in surface-to-volume ratio dramatically increases the efficiency of physical partitioning and chemical reactions and reduces Joule heating effects. Comparing a 20 μm open tubular pore featuring a 10-nm thick coating to a 200-nm nanopore with the same coating, the capacity factor, k', for molecular separations increases by a factor of 117.

- Due to their small size, adjacent nanostructures can be sufficiently close to one another to allow solution emerging from neighboring pores to rapidly mix through diffusion alone. Using the Stokes–Einstein infinite dilution diffusivity for a 1-nm particle in H_2O at 300 K and a structure supporting a pore density of $\sim 5 \times 10^8$ cm^{-2} yields a diffusion time of $\tau_D \sim 450$ μs to achieve full lateral mixing.

Early Studies. All investigations of nanoscale electrokinetic transport ultimately trace their origin to a seminal 1965 study of EK transport in nanometer scale cylindrical capillaries by Rice and Whitehead [1], in which the behavior of electroosmosis, streaming potential, current density distributions, and the electroviscous effect in cylindrical capillaries of nanometer dimensions were described within the Debye–Hückel approximation. Rice and Whitehead began with the Poisson–Boltzmann equation for a narrow cylindrical capillary at small, $\zeta < 25$ mV, zeta potential,

$$\frac{1}{r}\frac{d}{dr}\left(r\frac{d\varphi}{dr}\right) = \kappa^2 \varphi \tag{1}$$

where

$$\kappa = \sqrt{8\pi n e^2 / \varepsilon k T} \tag{2}$$

is the inverse Debye length. The solution of (1) is

$$\varphi = \varphi_0 \frac{I_0(\kappa r)}{I_0(\kappa a)} \tag{3}$$

where I_0 is the modified Bessel function of the first kind, which allows the charge density to be recovered directly using the Poisson equation. For an infinite cylindrical tube, the equation of motion under a combination of electrical and pressure-driven flow can then be written as

$$\frac{1}{r}\frac{d}{dr}\left(r\frac{dv_z}{dr}\right) = \frac{1}{\eta}\frac{dp}{dz} - \frac{F_z}{\eta} \tag{4}$$

where the body force, F_z, is driven by the action of the applied field on the net charge density in the double layer, $\rho(r)$. The important result from this analysis is

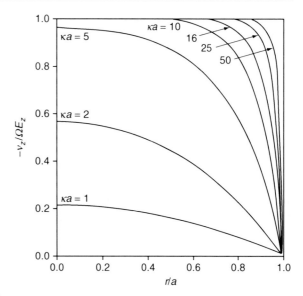

Electrokinetic Fluid Flow in Nanostructures, Fig. 1 EOF velocity profiles in infinite cylindrical capillary (Adapted with permission from Ref. [1])

that two flow regimes are established. When $\kappa a \gg 1$, the radial velocity profile, $v_z(r)$, reduces to the classical result, except very near the cylinder wall, i.e., the "plug flow" familiar from studies of capillary electrophoresis in μm-scale capillaries is recovered. When $\kappa a \sim 1$, $v_z(r)$ is proportional to $(a^2 - r^2)$, a behavior which is equivalent to the Poiseuille flow at nanometer dimensions, as shown in Fig. 1 and has been verified experimentally [2].

Fundamentals. When $\kappa a \sim 1$ there is significant interaction between the electrical double layers associated with opposing surfaces of the nanostructure. In general, transport across nanofluidic features at $\kappa a \sim 1$ depends on the sum of interactions among charge, electrophoretic mobility, molecular size, nanochannel cross section, ionic strength, and surface charge. A 100 nm $\times$ 1 μm nanochannel might exhibit a typical surface charge density ($\sigma \sim 2 \cdot 10^{-3}$ C m^{-2}), resulting in $\sim$ 4,000 elementary charges on the interior wall of the nanochannel, requiring ≥ 1 mM counterion concentration in order to balance the surface charge. The ionic strength of the solution can then be adjusted to control the relative populations of counter- and co-ions in the nanochannel. At low ionic strength, EDL overlap is significant, and counterion concentration and nanochannel conductance are determined purely by surface charge density. For common buffer concentrations (1–100 mM), κ^{-1} can range from 10 to

1 nm. Small molecule diffusion times across such nanochannels are of the order of 10^{-5}–10^{-4} s, which is advantageous if one wishes to implement heterogeneous reactions within nanochannels (*vide infra*). Adsorption (or desorption) of surface monolayers ($\Gamma = 10^{-10}$–10^{-11} mol/cm^2) within nanopores can cause the concentration to change $\sim$ 1–5 mM, which has important implications for selective capture and subsequent release in purification and concentration applications.

Flow rate normalized to cross-sectional area is less in nanochannels than in microchannels and limits translocation velocities. In addition to applied field, molecules are affected by the electric field due to the electrical double layer (EDL). As a result, counterions with larger charge are more attracted to the surface of the nanochannel and their migration is impeded most. Co-ions are repelled from edges of the channel and are transported faster than counterions. Neutral species are not significantly affected by the EDL and exhibit intermediate velocities. This complex interplay between charge and molecular size can have interesting consequences for differential transport and molecular separations, especially of macromolecules. For example, molecules which occupy or sweep out a significant portion ($> a/10$) of the nanochannel can exhibit altered molecular transport characteristics. Smaller molecules, which approach the channel walls more closely, show slower translocation velocities than larger molecules, which present a larger cross section to the higher flow velocities near the center of the nanochannel. This phenomenon mediates Ogston sieving, which has been exploited to effect DNA separations in nanochannels [3]. In addition, macromolecules that are larger than the cross section of the nanochannel undergo entropically disfavored conformational changes upon entering the confined space represented by the nanochannel. Longer DNA strands surmount such entropic barriers more rapidly, because they have a larger contact area at the entrance and larger increase in entropy facilitates faster escape from the constriction, an observation which has been explored to fabricate nanoscale sieving media.

Applications

Electrokinetic Flow Switching. Clearly the special characteristics of nanoscale EK flow present rich possibilities for studying the fundamental physics of fluid transport at this scale. Beyond the interesting fluid physics, nanofluidic architectures present great opportunities for construction of integrated microfluidic architectures exhibiting digital fluidic transfer by exploiting the unique properties of nanocapillary array membranes (NCAMs) as the fluidic analogs of transistors in integrated electronic circuitry, (Fig. 2). NCAMs can be physically realized in a variety of forms, including spatially random distributions of nuclear track-etched nanopores in polycarbonate membranes or focused ion beam milled arrays in poly (methylmethacrylate). Fluid volume elements can be transferred from one microfluidic channel to another by manipulating the bias potential applied across an NCAM. Kuo et al. demonstrated that this fluid transfer is linear in driving potential up to a threshold value, V_{thresh}, beyond which quantitative mass transfer occurs, i.e., at $V_{appl} > V_{thresh}$ all of the analyte species which enter the cross section of the NCAM and the microfluidic channel are transferred quantitatively across the NCAM [4]. In addition, the pore size determines the mass selectivity of fluid transfer; surprisingly, hindered transport is observed for molecular sizes as little as $a/20$, where a is the pore diameter [5]. Integration of nanofluidic components also permits fluidic samples to be handled with discrimination based on molecular size and charge using valves, gates, and other fluidic components, similar to the way that integrated electronic components control information flow in VLSI circuits [6]. In addition, 3D integrated microfluidics constitute a powerful method to enhance fluid handling capabilities, because disparate fluidic manipulations such as preparation, concentration, tagging, separation, or affinity recognition may be accomplished in distinct physical locations.

Molecular Separations. Naturally occurring confined nanoscale geometries take advantage of several unique phenomena to produce ionic and molecular separations of remarkable power and utility. For example, transmembrane potassium channels routinely permit the passage of K$^+$, while excluding smaller, similarly charged Na$^+$ ions [7]. Such observations in nature have generated an interest in biomimetic nanostructures as tools to separate entities ranging from small molecules to proteins. Although synthetic replication of such complex ion channels has yet to be realized, significant progress in the field has been made.

The simplest form of a molecular separation utilizing nanostructures is filtration using a pressure gradient

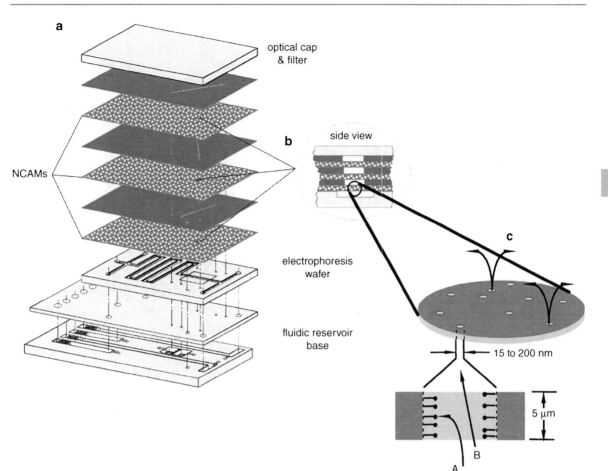

Electrokinetic Fluid Flow in Nanostructures, Fig. 2
(**a**) Exploded schematic view of a multilevel vertically integrated µTAS system displaying a fluid input layer (*bottom*), a pre-separation (*electrophoresis*) layer, and three microfluidic layers separated by nanocapillary array membrane-switching layers. (**b**) Side view of the vertically stacked microfluidic channels separated by NCAMs. (**c**) Schematic diagram of a single NCAM with an exploded view of an individual nanopore with chemically derivatized interior. (**a**) and (**b**) denote two species that have been introduced at the mouth of the pore, with (**a**) being sequestered by the molecular recognition motifs lining the pore wall and (**b**) transferred across the pore (Adapted with permission from Ref. [16])

across a membrane. Provided both the surface charge density, σ, of the enclosing boundary and the ionic strength, I, of the mobile phase are negligible, size-based separation is feasible. Under these conditions, establishing either a pressure or concentration gradient will drive the extraction of appropriately sized analytes across the separation device. However, these options offer limited control, and the separations can be very slow.

Fortunately, EK transport is a powerful alternative. In fact, with the appropriate conditions to ensure $\kappa a \sim 1$ (*vide supra*), nanochannels, nanotubes, and nanopores may become completely perm-selective for counterions (relative to the zeta potential, ζ, of the structural boundary). Garcia et al. successfully separated charged from uncharged dye molecules in nanochannels using EK transport [8]. In this experiment, the EDL was sufficiently large to localize the charged dye molecules (co-ions relative to ζ of the nanochannel walls) in the center of the nanochannel, while the uncharged molecules interacted with channel walls. The induced electroosmotic velocity (maximal at the center of channel) eluted the charged dye molecules significantly faster than the uncharged species, essentially inducing radially varying capillary electrophoresis within the nanochannel.

In addition to EK transport, molecular separations can be enhanced by modifying the nanostructure surface chemistry. For example, the perm-selectivity of

Electrokinetic Fluid Flow in Nanostructures, Fig. 3
(a) Orthogonal electric fields, E_x and E_y, serve as driving forces for the molecules and provide the energy needed to deform from their minimal energy states. (b) The longer molecule (*red*) has a greater rate of passage through confined regions due to a greater probability of deforming its shape than the shorter molecule (*blue*) (Reproduced with permission from Ref. [14])

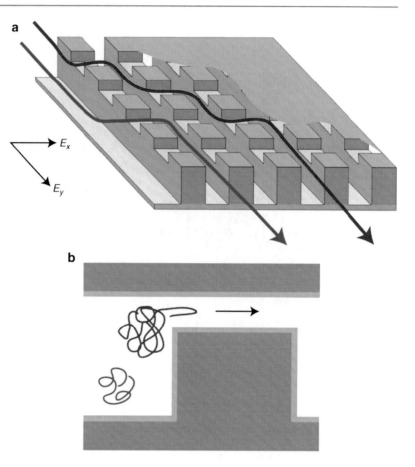

nanopores (due to $\kappa a \sim 1$) in polycarbonate NCAMs can be enhanced by uniformly functionalizing pores in order to control both the pore size and surface charge density, thereby enabling size- and charged-based separations of small molecules as well as the resolution of mixtures of proteins of similar size but distinct isoelectric points (pI) [9, 10]. Similarly, Lokuge et al. realized actively tuned size-dependent separations of labeled dextrans, by modulating the temperature of NCAMs decorated with poly(N-isopropylacrylamide), thus causing the temperature-responsive polymer to cycle through its lower critical solution temperature, contracting and expanding to open and close the pores, respectively [11].

In order to extend the separation capabilities offered by nature, the screening quality of hydrophobic molecules has been exploited. Unfortunately, the conditions required for EK-enhanced separations typically do not favor hydrophobic molecules. In fact, high charge density regions within nanopores, for example, may act as barriers to entry. Fortunately, surfactants can be added at concentrations above the critical micelle concentration (CMC) to form micelles, which can be used to sequester hydrophobic molecules and transport them across the nanopores. This strategy was used to observe the micelle-enhanced transport of a sparingly soluble local anesthetic dissolved in nonionic surfactant across polymeric nanoporous membranes [12].

The fact that nanostructures can be fabricated to be commensurate with biomolecular scaling lengths also creates new modes for single-molecule manipulation and separation. For example, in bulk solution DNA molecules equilibrate to a spherical shape with a characteristic radius of gyration, R_0 [13]. To enter structures with any cross-sectional dimension less than $2R_0$, the molecule has to deform from its minimal energy state, a process that leads to entropic trapping. Figure 3 shows a periodic nanofilter array used to separate molecules with different mobilities, which are determined by the ease with which they deform from the preferred conformational state, a property that is inversely proportional to molecular size [14].

Additionally, confinement within nanochannels enables study of physical properties of DNA molecules in an uncoiled, linear form [15].

Whether applied independently or coupled, size-, charge-, and hydrophobicity-based molecular separations within nanopores, nanochannels, and nanotubes offer a range of options, especially when they are incorporated into larger microscale geometries and utilized as electrically switchable molecular gates [16]. Such devices have the potential to enhance or augment well-established separation processes, including microdialysis, capillary electrophoresis, and micellar electrokinetic chromatography positioned either upstream or downstream of the nanostructure.

Chemical Interactions and Reactions. Large surface-to-volume ratios and small cross-sectional areas make it possible for molecules traversing nanopores to have numerous interactions with the channel surface [15]. This phenomenon can be exploited to useful effect, for example, to selectively pattern the interior of nanochannels and nanopores in a process known as diffusion-limited patterning. The same phenomenon can be used to manipulate complex mass-limited samples, particularly in microfluidic/nanofluidic hybrid devices. Kim et al. exploited the molecular recognition abilities of antibodies to selectively capture, purify, and release target analytes from a mixture [6]. Gold-thiol chemistry was used to immobilize Fab′ onto the electrolessly gold-plated interiors of NCAM nanopores. A mixture containing the target antigen, insulin, was transported through the NCAM by EK flow, after which, a releasing agent was applied to remove captured analytes. Mass spectra of the releasate showed strong specific retention of insulin – at 23-fold of the control intensity – but no presence of the other species, indicating the specific capture, concentration, and release of insulin. Thus, the high surface area within the NCAM and the increase in wall-collision frequency, Ω_W, render this immobilization strategy applicable to a broad range of analytes, especially those contained within mass-limited samples.

The effects of confinement in nanostructures are also apparent in altering the nature of fluidic transport. De Santo et al. observed sub-diffusive motion in nanochannels exhibiting high degrees of confinement [17]. Using fluorescence correlation spectroscopy (FCS), small, cationic rhodamine 6 G (Rh6G) molecules and charge-neutral labeled macromolecules, dextran and PEG, were observed under no-flow conditions in nanochannels of 30, 20, and 10 nm height. While Rh6G maintains its bulk diffusion coefficient in all cases, both dextran and PEG exhibit hindered diffusion – nearly an order of magnitude smaller at the highest degree of confinement. The decrease in diffusion for macromolecules is explained by anomalous diffusion, the degree of which increases with decreasing channel dimensions, until the theoretical limit for single-file diffusion is achieved. These findings emphasize the importance of surface interactions under confinement.

Concentration Polarization. Aside from direct effects on transport and reactivity inside nanofluidic channels, the presence of nanochannels can affect properties of liquid samples in adjacent microscale structures. Concentration polarization (CP), for example, occurs near ion-selective nanostructures. Surface-charged nanochannels/nanopores show selective permeability to ions due to EDL overlapping at $\kappa a \sim 1$ [18]. At low ionic strengths (<10 mM) the Debye length is comparable to the channel diameter, a, and the nanochannel is occupied entirely by counterions. Figure 4 illustrates CP at a cation perm-selective nanostructure [19]. When a dc bias is applied across the nanochannel, CP develops. On the anodic side, the anions move toward the anode migrating away from the anionic membrane interface where their concentration decreases, while only cations can enter and pass through the perm-selective nanostructure. On the cathodic side, anions cannot enter the cation-selective nanostructure, so cations exit the membrane, causing an increase in local concentration. This phenomenon leads to concentration gradients established by the ion-enriched and ion-depleted zones thus created.

When an analyte bears the correct charge to be segregated in the ion enrichment zone, the local concentration can be increased, sometimes by orders of magnitude, thereby enhancing the sensitivity in chemical analysis applications. Han and coworkers observed strong nonlinear electrokinetic flow at the micro-nanostructure interface giving rise to strong CP [20]. The capability to preconcentrate analytes using CP prior to analysis has obvious applications in biosensing, where analytes may be present at concentrations far below available limits of detection (LOD). Wang et al. [21] utilized a nanofluidic preconcentrator and a bead-based immunoassay to enhance binding kinetics, sensitivity, and the

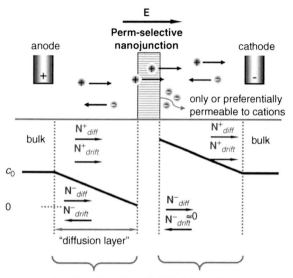

Electrokinetic Fluid Flow in Nanostructures, Fig. 4 Concentration polarization observed at the interfaces of a cation perm-selective nanostructure in the presence of current flow. Ion enrichment is observed at the cathodic side, while ions deplete at the anodic side (Reproduced with permission from Ref. [19])

dynamic range of detection in immuno-sensing of R-phycoerythrin, achieving a 500-fold sensitivity enhancement and lowering the LOD from 50 pM to < 100 fM. Ramsey and coworkers [22] developed a comparable preconcentration system using a highly ion-conductive charge-selective polymer [poly-AMPS (2-acrylamido-2-methyl-1-propanesulfonic acid)]. Such sample preconcentration devices could be potentially integrated to a wide range of detectors, such as mass spectrometers, to improve the detection of low-level proteins.

Resistive Pulse Sensing. Because nanochannels are commensurate in size with macromolecular scaling lengths, they can be used as chemical and biochemical sensors. The translocation of large molecules through nanopores containing high ionic strength electrolyte can produce a transient ionic current simultaneous with the temporary pore blockage [23], which can be used to count and size macromolecules, using the same principles as in commercial Coulter counters. The resulting current transients are termed resistive pulses. The majority of effort has targeted the use of biological ion channels to support DNA or RNA translocation, with α-Hemolysin being the most popular choice due to its well-studied structure, stability, and

ability to introduce specific structural modifications. α-Hemolysin is a 293-amino acid protein secreted by *Staphylococcus aureus*. Since the passage of each ssDNA or RNA molecule induces a current transient with a temporal duration proportional to its length, careful characterization of resistive pulse widths can be used to acquire information on molecular size. In favorable cases, these measurements can yield information capable of discriminating oligonucleotide composition [24].

Adding either chemical or geometric asymmetry to nanopores leads to structures with the ability to discriminate among chemical species. Conical nanopores can be produced from track-etched polymer membranes, which are etched in base, while neutralizing the incoming basic solution from the distal side. Conical nanopores display distinct advantages for resistive pulse sensing, including higher sensitivity due to ionic current being focused at the conical tip, larger ion currents at equivalent diameters, and being more resistant to fouling [25]. For example, protein biosensors have been fabricated from surface biofunctionalized nanotubes by electrolessly plating a layer of gold onto the interior surfaces of conical nanopores, followed by self-assembly of thiolated biotin derivatives [26]. Streptavidin was subsequently detected, and blockage time was related to streptavidin concentration. This strategy highlights one of the strengths of metal-coated nanostructures – by exploiting self-assembly chemistry a wide range of sensor chemistries can be realized within the same physical structure.

Single-Molecule Manipulations. When coupled with traditional optically based single-molecule detection methods, nanofluidic features enable detection to be performed at more biologically relevant concentrations. The diffraction-limited spot size in a confocal fluorescence correlation spectroscopy (FCS) experiment is on the order of 0.2 fL, requiring nanomolar or lower concentrations for single-molecule analysis [27]. As seen in Fig. 5, nanoscale channels use only a fraction of this volume, decreasing the detection volume by up to more than an order of magnitude, while at the same time allowing a corresponding increase in the fluorophore concentration [28].

At the cost of flow-through experimental capabilities, zero mode waveguides (ZMWs) enable a further reduction in excitation volume beyond that of

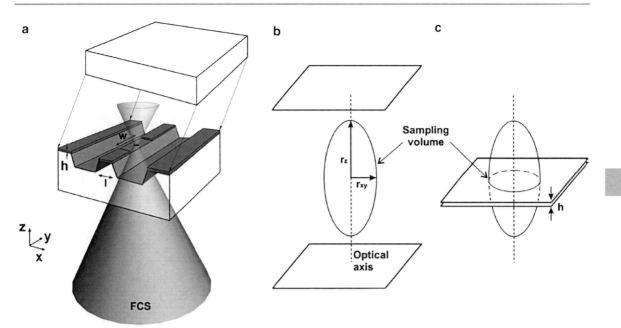

Electrokinetic Fluid Flow in Nanostructures, Fig. 5 FCS used for observation of sample inside a nanochannel: (**a**) schematic of hybrid microfluidic/nanofluidic system and FCS probe (not to scale); (**b**) FCS ellipsoidal detection volume with height $2r_z$ and width of $2r_{xy}$ in the case of molecules diffusing freely in 3D; and (**c**) FCS detection volume of molecules confined between two walls diffusing in 2D (when $h << 2r_z$) (Reproduced with permission from Ref. [28])

diffraction-limited techniques [15]. ZMWs are created by making a nanopore or nanopit in an opaque film such as Al or Au. The physical volume of these features confines the sample and excitation volumes to be on the order of 100 zL, ($1\ zL = 10^{-21}\ L$), more than three orders of magnitude smaller than the lower limit of the diffraction-limited focal volume in FCS experiments. Correspondingly, single-molecule observations can take place at physiologically relevant concentrations, $1\ \mu M < C < 1\ mM$, even for proteins with $\mu M\ K_D$ values. The signal-to-noise is significantly increased, because the electromagnetic field is effectively confined within the ZMW, which enhances the signal relative to the reduced background by eliminating light propagation into the far field.

Conclusions

Realizing EK flow in nanostructures opens up a range of new phenomena that support important new applications. Especially important are those applications that enable the space-time manipulations of molecular distributions in fluids, such as molecular separations, concentration polarization, and EK flow switching.

In addition, new approaches to chemical sensing that exploit novel nanofluidic properties are realized through the scaling of mass transport for specific chemical interactions that in turn supports resistive pulse sensing and single-molecule manipulations.

Acknowledgments Work described in this entry that was carried out in the authors' laboratories was supported by the National Science Foundation, the Department of Energy and the Army Corps of Engineers.

Cross-References

► Applications of Nanofluidics
► Micro/Nano Flow Characterization Techniques

References

1. Rice, C.L., Whitehead, R.: Electrokinetic flow in a narrow cylindrical capillary. J. Phys. Chem. **69**, 4017–4024 (1965)
2. Ramsey, J.M., Alarie, J.P., Jacobson, S.C.: Molecular transport through nanometer confined channels. In Sixth International Conference on Miniaturized Chemical and Biochemical Analysis Systems, p. 314. Nara (2002)

3. Stein, D., van der Heyden, F.H.J., Koopmans, W.J.A., Dekker, C.: Pressure-driven transport of confined DNA polymers in fluidic channels. Proc. Natl. Acad. Sci. U.S.A. **103**, 15853–15858 (2006)

4. Kuo, T.C., Cannon, D.M., Chen, Y.N., Tulock, J.J., Shannon, M.A., Sweedler, J.V., Bohn, P.W.: Gateable nanofluidic interconnects for multilayered microfluidic separation systems. Anal. Chem. **75**, 1861–1867 (2003)

5. Zhang, Y., Timperman, A.T.: Integration of nanocapillary arrays into microfluidic devices for use as analyte concentrators. Analyst **128**, 537–542 (2003)

6. Kim, B.-Y., Swearingen, C.B., Ho, J.A., Romanova, E.V., Bohn, P.W., Sweedler, J.V.: Direct immobilization of Fab' in nanocapillaries for manipulating mass limited samples. J. Am. Chem. Soc. **129**, 7620–7626 (2007)

7. Doyle, D.A., Cabral, J.M., Pfuetzner, R.A., Kuo, A.L., Gulbis, J.M., Cohen, S.L., Chait, B.T., MacKinnon, R.: The structure of the potassium channel: molecular basis of K + conduction and selectivity. Science **280**, 69–77 (1998)

8. Garcia, A.L., Ista, L.K., Petsev, D.N., O'Brien, M.J., Bisong, P., Mammoli, A.A., Brueck, S.R.J., Lopez, G.P.: Electrokinetic molecular separation in nanoscale fluidic channels. Lab Chip **5**, 1271–1276 (2005)

9. Savariar, E.N., Krishnamoorthy, K., Thayumanavan, S.: Molecular discrimination inside polymer nanotubules. Nat. Nanotechnol. **3**, 112–117 (2008)

10. Osmanbeyoglu, H.U., Hur, T.B., Kim, H.K.: Thin alumina nanoporous membranes for similar size biomolecule separation. J. Membr. Sci. **343**, 1–6 (2009)

11. Lokuge, I., Wang, X., Bohn, P.W.: Temperature controlled flow switching in nanocapillary array membranes mediated by poly(N-isopropylacrylamide) polymer brushes grafted by atom transfer radical polymerization. Langmuir **23**, 305–311 (2007)

12. Bhown, A.S., Stroeve, P.: Micelle-mediated transport of a sparingly soluble drug through nanoporous membranes. Ind. Eng. Chem. Res. **46**, 6118–6125 (2007)

13. Schoch, R.B., Han, J.Y., Renaud, P.: Transport phenomena in nanofluidics. Rev. Mod. Phys. **80**, 839–883 (2008)

14. Austin, R.: Nanofluidics - a fork in the nano-road. Nat. Nanotechnol. **2**, 79–80 (2007)

15. Piruska, A., Gong, M., Sweedler, J.V., Bohn, P.W.: Nanofluidics in chemical analysis. Chem. Soc. Rev. **39**, 1060–1072 (2010)

16. Gatimu, E.N., King, T.L., Sweedler, J.V., Bohna, P.W.: Three-dimensional integrated microfluidic architectures enabled through electrically switchable nanocapillary array membranes. Biomicrofluidics **021502**, 1–11 (2007)

17. De Santo, I., Causa, F., Netti, P.A.: Subdiffusive molecular motion in nanochannels observed by fluorescence correlation spectroscopy. Anal. Chem. **82**, 997–1005 (2010)

18. Holtzel, A., Tallarek, U.: Ionic conductance of nanopores in microscale analysis systems: where microfluidics meets nanofluidics. J. Sep. Sci. **30**, 1398–1419 (2007)

19. Kim, S.J., Song, Y.-A., Han, J.: Nanofluidic concentration devices for biomolecules utilizing ion concentration polarization: theory, fabrication, and applications. Chem. Soc. Rev. **39**, 912–922 (2010)

20. Kim, S.J., Wang, Y.-C., Lee, J.H., Jang, H., Han, J.: Concentration polarization and nonlinear electrokinetic flow near a nanofluidic channel. Phys. Rev. Lett. **99**, 044501 (2007)

21. Wang, Y.-C., Han, J.: Pre-binding dynamic range and sensitivity enhacement for immuno-sensors using nanofluidic preconcentrator. Lab Chip **8**, 392–394 (2008)

22. Chun, H., Chung, T.D., Ramsey, J.M.: High yield sample preconcentration using a highly ion-conductive charge-selective polymer. Anal. Chem. **82**, 6287–6292 (2010)

23. Heng, J.B., Ho, C., Kim, T., Timp, R., Aksimentiev, A., Grinkova, Y.V., Sligar, S., Schulten, K., Timp, G.: Sizing DNA using a nanometer-diameter pore. Biophys. J. **87**, 2905–2911 (2004)

24. Rhee, M., Burns, M.A.: Nanopore sequencing technology: research trends and applications. Trends Biotechnol. **24**, 580–586 (2006)

25. Vlassiouk, I., Kozel, T.R., Siwy, Z.S.: Biosensing with nanofluidic diodes. J. Am. Chem. Soc. **131**, 8211–8220 (2009)

26. Siwy, Z., Trofin, L., Kohli, P., Baker, L.A., Trautmann, C., Martin, C.R.: Protein biosensors based on biofunctionalized conical gold nanotubes. J. Am. Chem. Soc. **127**, 5000–5001 (2005)

27. Levene, M.J., Korlach, J., Turner, S.W., Foquet, M., Craighead, H.G., Webb, W.W.: Zero-mode waveguides for single-molecule analysis at high concentrations. Science **299**, 682–686 (2003)

28. Durand, N.F.Y., Dellagiacoma, C., Goetschmann, R., Bertsch, A., Marki, I., Lasser, T., Renaud, P.: Direct observation of transitions between surface-dominated and bulk diffusion regimes in nanochannels. Anal. Chem. **81**, 5407–5412 (2009)

Electrokinetic Manipulation

▶ Dielectrophoretic Nanoassembly of Nanotubes onto Nanoelectrodes

Electron Beam Evaporation

▶ Physical Vapor Deposition

Electron Beam Lithography (EBL)

Nezih Pala and Mustafa Karabiyik
Department of Electrical and Computer Engineering, Florida International University, Miami, FL, USA

Synonyms

e-beam lithography, EBL

Definition

Electron beam lithography (often abbreviated as e-beam lithography or EBL) is the process of transferring a pattern onto the surface of a substrate by first scanning a thin layer of organic film (called resist) on the surface by a tightly focused and precisely controlled electron beam (exposure) and then selectively removing the exposed or nonexposed regions of the resist in a solvent (developing). The process allows patterning of very small features, often with the dimensions of submicrometer down to a few nanometers, either covering the selected areas of the surface by the resist or exposing otherwise resist-covered areas. The exposed areas could be further processed for etching or thin-film deposition while the covered parts are protected during these processes. The advantage of e-beam lithography stems from the shorter wavelength of accelerated electrons compared to the wavelength of ultraviolet (UV) light used in photolithography, which allows defining much smaller diffraction-limited features. On the other hand, direct writing of patterns by scanning electron beam is a slow process and results in low throughput. Therefore, EBL is used for preparing photomasks for photolithography or for direct writing of small-area, low-volume patterns for research purposes.

Principles of Electron Beam Lithography

There are two driving forces for the development of electron beam lithography and other patterning technologies (e.g., X-ray lithography) as potential alternatives to the conventional UV-based photolithography: higher resolution (smaller feature size) and cost.

The maximum resolution of an optical system is limited with the wavelength of the beam used. Conventional photolithography systems, which rely on UV light, can therefore achieve a minimum "diffraction-limited" feature size of several hundreds of nanometers. Using resolution enhancement techniques such as phase shifting and immersion lithography, it is possible to improve the resolution to some extent. However, the fundamental laws of optics will continue to be a limitation. Moreover, resolution enhancement techniques for photolithography caused both cost and cycle time of masks to steadily increase with the rising level of complexity for every technology node. The cost of an optical mask set for 90-nm technology has surpassed the $1 million mark and exceeds $2 million for the 65-nm node. As a result, prototyping is becoming extremely expensive and time consuming. Mask amortization for low- and medium-volume chip production is becoming all but unaffordable. For this reason, various maskless lithography approaches are being pursued in attempts to solve the cost and cycle time dilemmas. Especially in the framework of nanotechnology, variations of EBL emerge as valid alternatives for lithographic patterning (see Fig. 1).

Electron Optics

The power of electron beam lithography to define nanometer-scale patterns comes from the short wavelength of electrons. In 1924, Louis de Broglie proposed that moving particles could show wave-like properties and therefore can be considered and modeled as waves in an appropriate framework. His theory was fully confirmed with the observation of electron diffraction in crystals in 1929. In his work, de Broglie developed a simple formula for the wavelength of a moving particle:

$$\lambda = \frac{h}{p} = \frac{h}{m \cdot v} \tag{1}$$

where h is Planck's constant (4.135×10^{-15} eV.s), p is the momentum, m is the mass of the electron (9.11×10^{-31} kg), and v is its speed. This neglects special relativity effects, because it is assumed that the electrons are traveling sufficiently below the speed of light. If the electron were traveling at the speed of light, c (2.9979×10^8 m/s) which is its maximum theoretical speed, it would have a wavelength of 2.43 pm. An electron's velocity can be determined from its kinetic energy, by the formula

$$E_{kin} = \frac{1}{2}mv^2 \tag{2}$$

Thus, the de Broglie wavelength for electrons with kinetic energy E_{kin} is:

Electron Beam Lithography (EBL), Fig. 1 Resolution of different lithographic techniques as a function of throughput

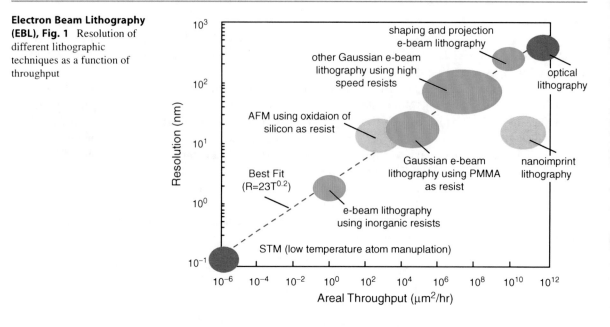

$$\lambda = \frac{h}{\sqrt{2mE_{kin}}} = \frac{1.23}{\sqrt{E_{kin}}}\,[nm] \qquad (3)$$

where E_{kin} is expressed in eV. So if an electron were accelerated to an energy of 10 keV, then it would have $\lambda = 0.12$ A. This clearly shows that if lithography could be done with electrons, it would have a huge advantage over current optical lithography systems, which are limited by their wavelength. This is just a theoretical limit, however. EBL systems use projection and with most forms of EBL, no mask is used at all. So except in high-resolution systems (with very low throughput) where electron diffraction can come into play, the resolution of the system is limited by the beam column rather than the wavelength of the electrons. The resist also places limitations on the resolution of the beam, which can be greater than those effects, and in most cases, scattering in the resist places the limiting factor for the resolution.

Equipment

Although modern electron beam lithography systems are quite sophisticated to achieve very high resolution, precision, and repeatability, they all have the same fundamental parts in common. Figure 2 shows the schematic of a typical EBL system with major equipment parts.

Electron Sources

Electron source is the most essential part of any electron beam system. This is where the electrons are generated, which will eventually reach the resist on the sample and expose it. An ideal electron source should have a high intensity (brightness), high uniformity, small spot size, good stability, and long life. Brightness is measured in units of amperes per unit volume per steradian. Higher brightness is desired to have increased reaction rate with the resist and thereby shorter exposure time and higher throughput. High-resolution patterning, on the other hand, requires lower beam intensities. Virtual size of the source is also important since it determines the demagnification needed to be applied to the column. The smaller virtual size provides smaller beam spot size on the wafer with minimum number of lenses. This makes the column less complicated, and allows a higher resolution to be achieved. Another very important feature of a good electron source is the low energy dispersion of the emitted electrons. An electron beam with a wide energy distribution is comparable to white light, while an energy beam with a narrow spectrum is comparable to a laser source. A narrow electron energy spread is desired as it reduces chromatic aberrations. Characteristics of different electron sources are summarized in Table 1 [1].

In an electron source, electrons are removed from cathode of the gun either by heating the cathode or by applying a large electric field. The first process is

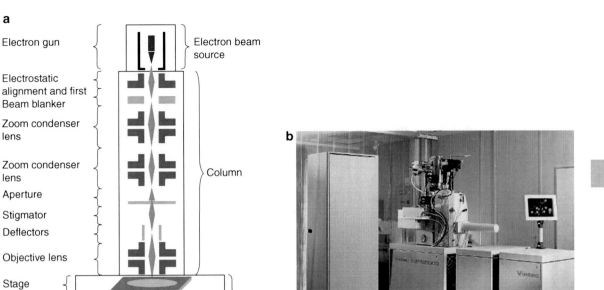

Electron Beam Lithography (EBL), Fig. 2 (a) Schematic of a typical electron beam lithography system. (b) A typical EBL system (Courtesy of Vistec Electron Beam Lithography Group)

Electron Beam Lithography (EBL), Table 1 Characteristics of different electron sources

Source type	Material	Brightness (A/cm²rad)	Source size	Energy dispersion (eV)	Vacuum level (Torr)	Emitter temp. (K)
Tungsten Thermionic	W	$\sim10^5$	25 μm	2–3	10^{-6}	$\sim$3,000
LaB$_6$ Thermionic	LaB$_6$	$\sim10^6$	10 μm	2–3	10^{-8}	$\sim$2,000–3,000
Thermal field aided emission	Zr/O/W	$\sim10^8$	20 nm	0.9	10^{-9}	$\sim$1,800
Cold field emission	W	$\sim10^9$	5 nm	0.22	10^{-10}	Ambience

called thermionic emission (Fig. 3a) and the latter is called field emission (Fig. 3b). A combination of two, which is called thermal field–aided emission, is also sometimes employed.

For thermionic emission sources, tungsten or tungsten thoriated filaments were commonly used as emitters. In such sources, the filament is heated by passing current through it and electrons are emitted thermionically from a sharp tip. Tungsten filaments allow operation at pressures as high as 0.1 mtorr, but their low current density ($\sim$0.5 A/cm²) results in very low brightness ($\sim$2 × 10⁴ A/cm² sr). Thoriated tungsten cathodes have somewhat lower brightness at the same filament current and require higher vacuum (0.01 mtorr), but their maximum current density can

be as large as 3 A/cm². Another disadvantage of tungsten sources is the high-energy spread due to the high operating temperature (2,700 K). Lanthanum hexaboride (LaB$_6$) is a newer thermionic source which can achieve very high current density over 20 A/cm² and good brightness of nearly 2 × 10⁴ A/cm² sr at a lower temperature of 1,800 K. It has a slightly smaller source size than tungsten; however, it requires a better vacuum of 10^{-8} Torr, which can be easily achieved in present systems and must be well protected against sudden vacuum loss.

A major concern with all thermionic sources is the finite source size or crossover diameter. Because a large volume of wire is heated, the thermionic sources produce broad beams. The energy distribution

Electron Beam Lithography (EBL), Fig. 3 Simplified cross-sectional schematics of (**a**) thermionic emission (**b**) field emission electron sources

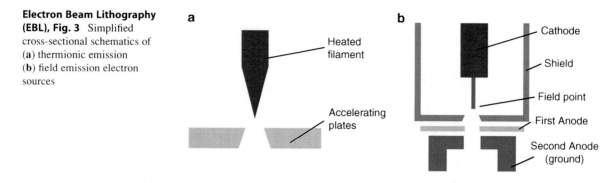

from such a source is also quite broad. This leads to focusing problems similar to those found in large optical partially incoherent sources. For typical LaB_6 sources, the crossover diameter is about 10 μm. Therefore, a demagnification of 100× is required to achieve 0.1-μm spot size. Even though the source has a large current density, the inherent source brightness is greatly reduced in order to achieve deep submicron resolution.

Field emission electron sources are based on application of an electric field high enough to extract the electrons through the surface potential barrier of the emitter (see Fig. 3b). In field emitters, cathode tip should be sharp, made of a material with reduced work function and be placed close to the extraction electrode to operate at a low voltage. Tungsten tips provide the extremely high fields necessary for electron extraction. An example of an electron source and the electrodes is shown in Fig. 3b. Two anodes are normally used. The first anode is the extraction electrode, and is used to extract the electrons from the cathode tip, and the second is used to accelerate the electrons to their full potential. The shield electrode is used to prevent thermally generated electrons from entering the beam. Thus, electrons are only emitted from the tungsten tip. Cold field emission sources have the disadvantage of high short-term noise, which is due to atoms adsorbing onto the tip. This causes fluctuations in the electron current in the short term as well as in the long term. Intensity fluctuations in the beam current caused by tip absorption can be reduced by ultrahigh vacuum environment.

Thermionic field emission sources are the best available electron sources. Thermal field emission sources combine the best of both field and thermionic emission sources. A tungsten needle (with approximately 1-μm tip) is used at only 1,800 K, thus increasing the lifetime of the source. The barrier to field

emission of electrons is lowered because the needle is heated. Its brightness, source size, and energy spread are all better than the standard thermionic sources. Recently Zr/W/O tips have become available with crossover diameters as small as 200 Å for thermionic field emission. Such emitters can provide very high current densities up to 1,000 A/cm² and beam diameter as small as 100 Å without compromising the brightness. They, however, require a constant vacuum of at least 1×10^{-8} Torr for stable operation.

Once an electron beam is generated, it is focused and steered in the column using lenses, apertures, and deflectors. In the following sections, main components of a typical EBL column will be described.

Electron Lenses

The physical principles of electromagnetic lenses can be described by using the basic laws of electromagnetism. Electron motion can be modeled as EM waves, which implies that it can be focused and manipulated analogous to the classical optics systems (geometric optics). At the same time, electrons maintain the characteristic properties of classical charged particles. Therefore, electrons are manipulated in two basic ways: through an external electric field or a magnetic field.

From the second law of Newton, $\mathbf{F} = m \cdot (d^2 x/dt^2)$, the trajectory and speed of electrons can be controlled by external forces, more precisely, by electromagnetic forces. According to the Coulomb's law, an electric field will exert a force on a charged particle. An electron will experience a force of $\mathbf{F}_C = -e\mathbf{E}$, where e is the elementary charge of an electron, and $\mathbf{E}$ is the electric field. According to Lorentz's Law, an electron will experience a force when traveling perpendicular to a magnetic field. This force will be perpendicular to the magnetic field and the electron's velocity, and is given by $\mathbf{F}_L = -e.\mathbf{v} \times \mathbf{B}$. The contributions could be

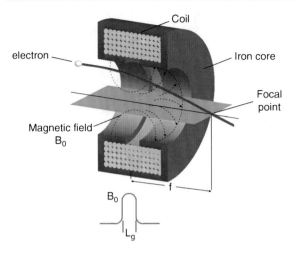

Electron Beam Lithography (EBL), Fig. 4 Cross-sectional schematic of a magnetic electron lensn [2]

summed together; however, normally in EBL systems, both magnetic and electric fields are not used simultaneously.

Magnetic lenses are used primarily because the aberrations produced by magnetic lenses are less severe than those produced by electrostatic lenses. A cross section of a magnetic lens is shown in Fig. 4. The radial magnetic fields cause the electrons to experience rotation about the optical axis upon entering the lens. The electrons now have a velocity component in the tangential direction about the optical axis. This component now interacts with the axial magnetic field, imparting a force on the electrons toward the optical axis, bringing them close to the axis.

Neglecting the actual aberrations that exist in the beam trajectories, electrons are focused to a certain distance, f, from the center of the lens, determined by the electron mass m, unit charge e, magnetic field B_0, the gap L_g, and V_0 the voltage representing the electron's velocity:

$$f \cong \frac{8mV_0}{eL_gB_0^2} \qquad (4)$$

Electrostatic lens operation, to force electrons to converge in some point of the optical axis, is similar to magnetic lens one. The realization is accomplished by three plates provided with a central aperture. Central plate has a variable potential and the first and third plates are connected to the ground. In general, electrostatic lenses are used as condenser lenses of the electron source since the distortions inherent to these lenses are less critical here. The deflection unit is in charge of deviating the beam through the sample surface, within what is called the scan field. Ideally, the minimum degradation of beam is desired, i.e., precise deflection, constant beam size, and no hysteresis.

Beam Deflectors
Beam deflectors are used to scan the electron beam over the target, across an area called the scan field. To have a predictable deflection for precise patterning of ultrasmall features, this process must be very linear, with minimal degradation to the electron beam spot size and with no hysteresis. Like in electron lenses, deflection can be done by an electric field or by a magnetic field. Electrostatic beam deflection results in more distortions in the beam, so magnetic deflection is normally used. Beam deflectors are one of the most important components; however, any nonlinearities or problems with beam deflectors can easily be fixed in the software, which drives them.

Apertures
There are a few different types of apertures, blanking apertures, and beam-limiting apertures. Blanking apertures deflect the beam by deflecting the beam away from the aperture hole. Beam-limiting apertures set the beam convergence angle, α, (the angle between the beam trajectory and the normal line to the target) which controls the effect of lens aberrations and resolution but also limits the beam current.

Beam Blankers
Beam blankers are used to turn off or "blank" the beam. This is necessary when doing vector scanning and the beam needs to be moved from one part of the wafer to another. The beam must be blanked in a time, which is very small compared to the time it takes to illuminate one pixel on the array, and the beam cannot move along the substrate while it is being blanked. A common blanker is shown in Fig. 5 and it works by applying a voltage to the upper plate, which deflects the electron beam away from the center of the column.

Stigmators
The imperfections in fabrication and assembling of the column cause astigmatism. Astigmatism results in different focusing conditions at different positions on the sample surface. This means that the ideal circular section of the beam becomes elliptic and, consequently

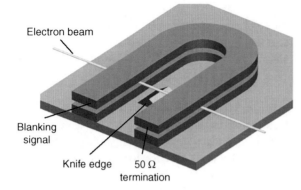

Electron Beam Lithography (EBL), Fig. 5 Schematics of a typical beam blanker

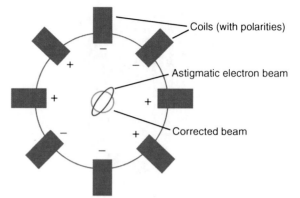

Electron Beam Lithography (EBL), Fig. 6 Schematic of a stigmator with eight poles. Alternate coils are connected in series but have different polarities. Astigmation of the electron beam can be corrected by changing the size of the current through the two sets of coils

distorts the image. Concerning to lithography, beam shape does not correspond to the model used for calculating the exposure dose; therefore, the pattern to be transferred is also distorted from the original design. The stigmator system is responsible for correcting beam shape to be circular again. It consists of four or eight poles that surround the optical axis. Adjustment is performed by the balance of electrical signal of the poles (Fig. 6).

Stage

EBL patterning is performed in a high vacuum chamber. The sample is fixed on a sample holder by screws, spring-loaded mechanisms, double-sided tapes, or silver paint. Wafer holders usually contain a set of test standards for calibration focusing and position. Calibration subroutines can be performed immediately before the actual writing or even during the writing in between different fields as often as desired.

The dimensions and mobility of the stage determine the size and accessibility of the sample that can be patterned. The equipment used can lodge up to 300 mm wafers and movement is possible for XYZ, rotation and tilt (Fig. 7a). The capabilities of the system benefit from precise and motorized controlled displacements. The stage moved by piezoelectric motors and the position is controlled by laser interferometry system to achieve nanometer-scale precision. The stage is also placed on vibration isolation system to reduce the mechanical vibrations, which might affect the spatial precision. At the same time, computer monitors, transformers, and vacuum pumps are kept separate or controlled and shielding is used to avoid interferences. In addition, the chamber may also be

equipped with a charge-coupled device (CCD) camera to visualize inside the chamber and to assess the control of sample positioning.

Limitations of the Equipment

Different from conventional optical lenses, electromagnetic lenses are only converging. In reference to aberrations, their quality is so poor that field size and convergence angle (numerical aperture) are limited. From electron source specifications, the beam diameter as a function of virtual source and column reduction is determined. The distortions caused by the column arise as spherical aberrations (astigmatism) and they are originated in both the lenses and the deflectors. Chromatic aberrations appear when electrons present a certain energetic spectra. This may be caused in the beam source itself, but also can be increased by the so-called Boersch and Loeffler effects, where energetic distribution increases as a result of electron collisions. Both phenomena contribute to decrease the precision of lenses and deflectors.

In general, the theoretical effective beam can be expressed as the quadratic sum of each contribution and optimal point is a compromise between all involved factors. For high resolution, the use of high magnification and beam energy is combined with low energetic dispersion and short focal distance. In consequence, write field is smaller and beam current should be reduced, which means that exposure process is slower, throughput is limited, and flexibility is constrained. The final spot diameter on the surface of the wafer is given by

a

Wafer clips   Faraday cup

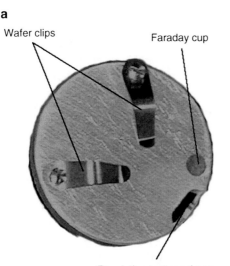

Resolution test specimen

b

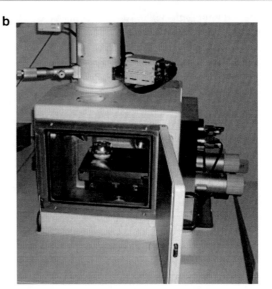

Electron Beam Lithography (EBL), Fig. 7 (**a**) A simple EBL sample holder with two wafer clips to hold the sample, a Faraday cup to measure the beam current, and gold on carbon resolution sample for resolution test. (**b**) Wafer chamber of an EBL system with the door open

$$d^2 = d_d^2 + d_o^2 + d_S^2 + d_C^2 \qquad (5)$$

where d_s is the spherical aberration, d_C is the chromatic aberration due to the nonzero energy distribution, and d_o is the perfect lens diameter, which is limited by the finite source size and space charge and d_d is the diffraction limit.

Another limitation is called stitching and stems from the limited deflection of the beam. Typically, for very small beam deflections electrostatic deflection "lenses" are used, larger beam deflections require electromagnetic scanning. Because of the inaccuracy and because of the finite number of steps in the exposure grid, the writing field is on the order of 100 μm–1 mm, larger patterns require stage movements. An accurate stage shift is critical for stitching (tiling writing fields exactly against each other) and pattern overlay (aligning a pattern to a previously made one). Large patterns spanning several write fields may suffer from *stitching errors* due to the finite inaccuracies in stage movement.

Resists for Electron Beam Lithography

Interactions between electromagnetic radiation (either UV or electron beam) and polymers, which are used as resists (either for optical or e-beam lithography) result in one of the following two chemical reactions: (1) *cross-linking* (2) *chain scission* (also known as fragmentation). The resists are usually long chains of carbon-based polymers. By interacting with incoming radiation, atoms in adjacent chains may be displaced and the carbon atoms will bond directly. This process is called cross-linking. Highly cross-linked molecules dissolve more slowly in an organic solvent (developer). A material in which cross-linking is the dominant reaction upon exposure is a *negative resist*. Radiation can also disrupt the polymer chains, breaking them up to smaller pieces. This reduces the molecular weight of the resist, which makes it more soluble in developer solution. A material in which chain scission is the dominant reaction upon exposure is a *positive resist*. The most important resist criteria are contrast and sensitivity for the exposure type and energy and etch resistance.

Tone determines the areas that will be removed in the development. For positive tone resists, patterned areas are removed with the developer, whereas for the negative ones, the result is the inverse of exposed features, resist is not dissolved where irradiated.

Sensitivity quantifies the minimum amount of delivered dose that is required to achieve the selective development, where dose is the number of electrons per unit area.

Resolution defines the minimum feature size or the smallest distance between two patterns that can be resolved.

Electron Beam Lithography (EBL), Table 2 Commonly used EBL resists

Resist	Tone	Sensitivity @20 keV ($\mu C/cm^2$)	Resolution (nm)	Contrast	Etch resistance
PMMA	+	100	10	2.0	Poor
P(MMA-MAA)	+	5	100	1.0	Poor
PBS	+	1	250	2.0	Good
NEB-31	−	30	50	3.0	Good
EBR-9	+	10	200	3.0	Poor
ZEP	+	30	10	2.0	Good
UV-5	+	10	150	3.0	Good
SAL-606	−	8.4	100	2.0	Moderate
COP	−	0.3	1,000	0.8	Poor

Contrast describes how abrupt is the dependence of thickness on dose, which also has implications upon resolution. It is determined from the linear slope of the sensitivity curves.

Etch resistance determines the resist integrity under chemical (wet) and physical (dry) etching processes.

Other resist variables are adhesion to the different substrates, compatibility with conventional techniques, or shelf life. Table 2 summarizes the properties of commonly used EBL resists.

Positive Resists

Polymethyl methacrylate (PMMA) was one of the first resists developed for EBL and remains the most commonly used positive resist. PMMA has extremely high resolution, and its ultimate resolution has been demonstrated to be less than 10 nm. In PMMA, both cross-linking and fragmenting of the polymeric chains occur (see Fig. 8), but the rate of scission is much larger than that of cross-linking. If PMMA is exposed to ten times the normal critical dose, chain-linking will occur and it will behave like a negative resist; however, its resolution will be degraded to about 50 nm. One of the major disadvantages of PMMA is low sensitivity. PMMA also has poor resistance to dry etching, which makes it limited to mask making where only a thin Cr wet etch on a flat substrate is required. When it is used for direct writing, PMMA is typically used as the imaging resist for a liftoff process. PMMA is available in low-molecular-weight form and high-molecular-weight form, of which the former is more sensitive than the latter. Thus, if two layers of PMMA are used, with the low-molecular-weight form on the bottom, this will cause the resist to develop with an undercut, thus making it suitable for a liftoff process.

The copolymer, methyl methacrylate and methacrylic Acid (P(MMA-MAA)), provides a three- to fourfold improvement in sensitivity relative to PMMA and image thermal stability of 160°C. Since P(MMA-MAA) can also be developed in MIBK:IPA solvent as used for PMMA, a single-step development of mixed layers of PMMA and P(MMA-MAA) is possible for enhancing the sensitivity and thermal stability of the resist.

A new resist called ZEP has an order of magnitude improvement in sensitivity versus PMMA and it has 2.5 times better etch resistance compared to PMMA although it is still not as resistant as novolak-based resists. ZEP has about the same resolution as PMMA of about 10 nm and it has demonstrated an ability to produce lines of width 10 nm with a pitch of 50 nm. Other important positive resists include PBS (polybutene-1-sulfone) and EBR-9 (a copolymer of trifluoroethyl a-chloroacrylate and tetrafluropropyl a-chloroacrylate) which have high sensitivity. It is noted that the desired properties of a resist are high resolution and high sensitivity (high speed). Unfortunately, the resist that has higher sensitivity, including those mentioned here, usually has lower resolution, especially compared to PMMA.

Negative Resists

A number of negative EBL resists are also available. These resists have components on the polymer chain that enhance the cross-linking. Typical cross-linking components include chloromethyl styrene, epoxies, and vinyl groups. Negative resists tend to have less bias but they have problems with scum and swelling during development and bridging between features.

Commonly used negative e-beam resists consist of the Shipley advanced lithography (SAL) product line, an epoxy copolymer of glycidyl methacrylate and ethylacrylate (P(GMA-EA)), also known as COP, and a partially chloromethylated polystyrene (CMS). While COP has high sensitivity, CMS possesses modest resolution at modest sensitivity. The SAL offers many new deep ultraviolet (DUV) resists through the use of chemically amplified resist materials. Fast versions of CAP have demonstrated high-resolution capability with 100-nm lines at sensitivity higher than PMMA. Also, as it was mentioned before, PMMA

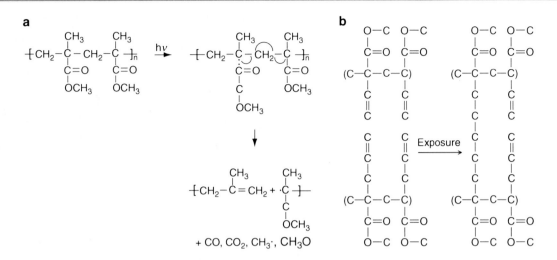

Electron Beam Lithography (EBL), Fig. 8 (**a**) Mechanism of radiation-induced chain scission in PMMA (**b**) Cross-linking of PMMA structure where it is modified through the addition of a C=C side chain

can exhibit negative tone when exposed to a dose one order of magnitude higher [3].

Electron-Substrate Interactions

Electron beam generation and control technologies allow achieving beam diameters below the nanometer range. However, electrons being massive charged particles undergo interactions with the resist and the substrate and result in larger minimum resolution. Therefore, electron-solid interaction must be carefully taken into account to obtain small feature size by EBL.

Typical electron beam lithography machines use electron beams with 10–100 keV energy per electron. Therefore, the free path of an electron is 10 μm or more, which is at least an order of magnitude more than the resist thickness. Thus, the electrons can easily penetrate the resist layer and reach the substrate. When an electron beam hits the surface of a resist deposited on a substrate, it experiences elastic and inelastic collisions with the resist and substrate atoms and molecules. Elastic collisions result in backward scattering whereas inelastic collisions create small angle forward scattering (see Fig. 9). The energetic transfer in the resist mainly causes the resist exposure, whereas interaction with the substrate underneath is manifested as heating and creates more backscattered electrons. Usually, elastic collisions are expressed in terms of the unscreened single electron scattering of Rutherford. It evaluates the process as a probability of interaction

that strongly depends on the solid atomic number and the electron energy. Inelastic collisions are the cause of the loss of electron energy and it is typically described by the Bethe's continuous slowing down model. As a result, electron trajectories are mostly dependent on electron energy and the substrate composition.

In the resist, forward scattering is the more probable process, i.e., main parameter is the energy, while the interaction with the substrate comes from backscattering, which is highly dependent on both charge energy and substrate nature. Thus, electron trajectories are typically distributed in a certain 3D interaction volume that reaches few microns in range of 10–20 keV for incoming electron energy. The usual pearl shape of electron distribution in the resists can be easily explained from the low atomic number and density of the resists (Fig. 7b). High-energy electrons initially are just slightly deviated from the incoming direction, since elastic contribution is weak ($\sim Z^2$), but once they slow down, lateral scattering dominates ($1/E^2$). Electron trajectories in resist and substrate are often modeled by Rutherford scattering and simulated using Monte Carlo techniques. However, resultant increase in the beam diameter can also be estimated by an empirical formula [5]:

$$\Delta d = 0.9 \left(\frac{t_R}{V_b} \right)^{1.5} \tag{6}$$

where Δd is the change in beam diameter after entering the resist, t_R is the resist thickness, and V_b is the

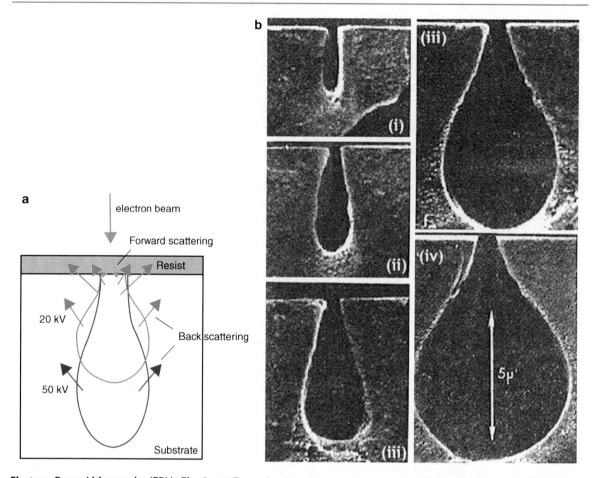

Electron Beam Lithography (EBL), Fig. 9 (a) Forward and backward scattering in resist and substrate. (b) Experimental evidence of pearl-shaped electron trajectories caused by electron-resist scattering. A 5-μm-thick layer of PMMA is developed with increasing times from (i) to (iv) [4]

accelerating voltage of the electron beam. Forward scattering is detrimental if very high resolution is required, and can be reduced by increasing the accelerating voltage. Forward scattering does have one advantage, however; it can be used to create a negative sidewall angle in the developed resist. This is advantageous when doing liftoff process with positive resist, because a slight negative sidewall angle is required. This eliminates the need for using two layers of different resists and simplifies the processing.

The electrons impinging on the resist and substrate have a large energy (typically 10–100 keV). Their forward scattering results in generation of *secondary electrons* with lower kinetic energies (typically 20–50 eV). The major part of the resist exposure is due to these electrons. Since they have low energies, their range is only a few nanometers. Therefore, they contribute little to the proximity effect. However, this

phenomenon, together with the forward scattering, effectively causes a widening of the exposure region. This is one of the main limiting factors in resolution of e-beam lithography machines. The distance a typical electron travels before losing all its energy depends on both the energy of the primary electrons and the type of material it is traveling in. The fraction of electrons that are backscattered is roughly independent of beam energy. It does, however, have a strong relation to the substrate material. Substrates with a low atomic number give less backscattering than substrates with high atomic number.

Proximity Effects

As it is mentioned above, as the electrons penetrate the resist and the substrate, they experience many scattering events. In backscattering, an electron collides with the much heavier nucleus, which results in an elastic scattering event. The electron retains (most of) its energy, but changes its direction. The scattering angle may be large in this case. After large angle scattering events in the substrate, electrons may return back through the resist at a significant distance from the incident beam, thereby causing additional resist exposure outside the desired area. This phenomenon caused by electron backscattering is called *proximity effect*. It is important to correct for the proximity effects to avoid erroneous pattern developing.

Proximity effects may be observed in the final pattern in two different forms: intrashape and intershape proximity effects. The intrashape proximity effect causes smaller shapes to be underexposed relative to larger shapes written with the same dose. In this case, each pixel in a smaller shape gets less exposure from scattered electrons, because there are fewer neighboring exposed pixels. In a positive resist, intrashape proximity effects may result in underdeveloping of narrow lines or small features. Intershape proximity effects are the result of the fact that exposed shapes are also affected by other nearby shapes, with electrons scattering through the intervening resist. Densely packed shapes will strongly interact, providing much "cooperative exposure" between the shapes, and will thus require a lower exposure dose than sparse, isolated shapes, which get fewer scattered electrons and will thus be underexposed relative to their cousins with closer neighboring shapes. In a positive resist, intershape proximity effects may result in accidental exposure of narrow fields in between large exposed areas and thereby developing completely. These effects are illustrated in Fig. 10.

The first step of the proximity-effect correction is an accurate knowledge of the energy density profile deposited in the electron-resist layer due to a point or pixel exposure (often called point spread function). In general, this profile is a function of the system setup. An important property of these profiles is that the shape is independent of dose as well as position, assuming a planar and homogeneous substrate. This profile is often approximated by the sum of two Gaussian distributions:

$$f(r) = \frac{1}{\pi(1+\eta)} \left\{ \frac{1}{\alpha^2} \exp\left[-\left(\frac{r}{\alpha^2}\right)^2\right] + \frac{\eta}{\beta^2} \exp\left[-\left(\frac{r}{\beta^2}\right)^2\right] \right\} \tag{7}$$

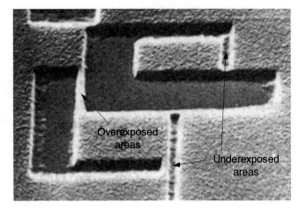

Electron Beam Lithography (EBL), Fig. 10 SEM picture of a positive resist pattern affected by the intrashape and intershape proximity effects [5]

where r is lateral distance from incident beam; α is standard deviation of forward scattered electron distribution; β is standard deviation of backscattered electron distribution; η is ratio of energy deposited by forward scattered electrons to the energy deposited by backscattered electrons. The first term represents the contribution of forward scattered electrons, whereas the second term is related with the backward scattered electrons. Equation 7 is normalized so that

$$\int_0^\infty f(r)2\pi dr = 1 \tag{8}$$

for some substrates and resists, a better match between the experimental results and the model is achieved with the addition of a third contribution. As a matter of fact, deposited energy distribution is not uniquely caused by proximity effects in terms of electron scattering, so, due to this, more accuracy is obtained when an exponential decay modulates proximity function:

$$f(r) = \frac{1}{\pi(1+\eta+v)} \left\{ \frac{1}{\alpha^2} \exp\left[-\left(\frac{r}{\alpha^2}\right)^2\right] + \frac{\eta}{\beta^2} \exp\left[-\left(\frac{r}{\beta^2}\right)^2\right] + \frac{v}{2\gamma^2} \exp\left[-\left(\frac{2}{\gamma^2}\right)\right] \right\} \tag{9}$$

where v is efficiency of exponential function and γ is decay of exponential function.

The models discussed above are two-dimensional versions of an essential three-dimensional phenomenon. In general, the energy profile depends upon depth

Electron Beam Lithography (EBL), Fig. 11 Electron trajectory simulations using Monte Carlo for 10 mm of resist on a silicon wafer for different electron energies [6]

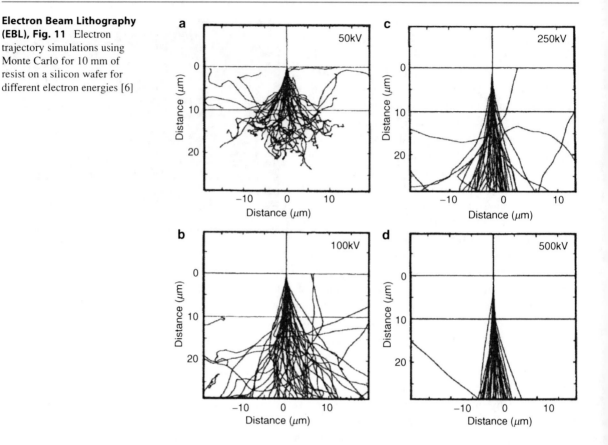

as well as radius. By averaging out the depth dependence, a two-dimensional profile can be obtained out of a three-dimensional profile. The simplified models allow great reduction in computation time for the exposure estimation and correction. For certain applications, it may be necessary to use a three-dimensional profile. In this case, a Monte Carlo simulation of electron scattering in the resist layer can be used. Electron scattering in resists and substrates can be modeled with reasonable accuracy by assuming that the electrons continuously slow down as described by the Bethe equation, while undergoing elastic scattering, as described by the screened Rutherford formula (Fig. 11).

Once the proximity function is correctly determined with the coefficients, proximity effects can be corrected for by several methods.

Dose modification and its variants are the most common correction. It is based on determination of the required dose for each pixel with a reasonable accuracy while being computationally practicable. Numerous variants of the "self-consistent dose

correction" have been developed. In its simplest form, it is basically the reverse of the exposure estimation. Let Q_j be the dose applied to pixel j and let N be the total amount of pixels. The total energy on pixel i will be:

$$E_i = \sum_{j=1}^{N=} R_{ij} Q_j \text{ with } R_{ij} = \frac{\Delta V_f}{t} \left(\frac{1}{\pi \alpha^2} \exp \left(-\frac{r_{ij}^2}{\alpha^2} \right) + \frac{\eta}{\beta^2} \exp \left(-\frac{r_{ij}^2}{\beta^2} \right) \right)$$

(10)

where r_{ij} is the distance between pixel centers of i and j. This equation can be written in matrix notation for all i:

$$[E_i] = [R_{ij}][Q_j]$$

(11)

Solving this set of equations with matrix operations will provide a proximity-effect corrected pattern. However, this "self-consistent" scheme will not be a perfect correction since only the exposed pixels are

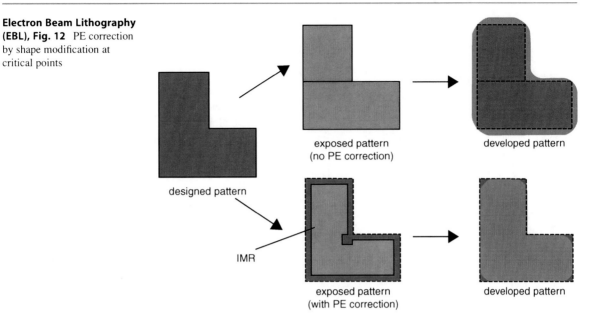

Electron Beam Lithography (EBL), Fig. 12 PE correction by shape modification at critical points

exposed pattern (no PE correction)

developed pattern

designed pattern

IMR

exposed pattern (with PE correction)

developed pattern

considered in this equation. The problem can be simplified, thereby reducing the calculation time, by splitting the dose modification into a problem of forward scattering and back scattering correction. With dose modification, it is possible to achieve superior proximity-effect correction. The main disadvantage is that with very large circuits, it may require large computation times.

In another correction method, the shapes found in the pattern image are modified in such a way that the developed image will resemble the intended image as close as possible, while a single dose is used for the entire circuit. A good example of a shape modification method is the correction scheme in PYRAMID [7]. PYRAMID takes a pattern with rectangular circuit elements. The circuit is then passed to a correction hierarchy, which adjusts each element via precalculated rule tables. This rule table is created using exposure estimation as described earlier. The first step is to replace each rectangle with its inner maximal rectangle (IMR). The second step is to correct the effect of interaction among the different circuit elements. Each edge facing other circuit elements will be adjusted so that the midpoint of the edge will be equal to the experimentally determined development threshold. Even better results can be obtained by bending the edge when appropriate. The final step is to modify the shapes at critical points, that is, junctions between adjacent rectangles (Fig. 12). The major advantage of

the shape modification method is that accurate results can be obtained without being computationally expensive and without losing throughput. However, it may not be as flexible as the other methods and experimental data are needed to obtain the necessary rule tables.

Background exposure correction, often referred to as GHOST, works by writing a second exposure, which is the inverse of the intended image (Fig. 13). This is done in such a way that the background dose is brought to a constant level. The main advantage of this method is that it is one of the easiest proximity-effect correction methods found in the literature and can be used with virtually all electron beam machines. However, there are several disadvantages with this method. One problem may be a contrast reduction (although this is true for many other correction types too). However, the main problem is that it only provides correction to the backscattering component, where the forward scattering remains uncorrected. There is also a loss of throughput because of the double exposure.

Electron Beam Lithography Systems

Although they all are based on the same basic principles, different types of electron beam lithography systems are available. These systems as well as some proposed future systems will be briefly discussed in this section.

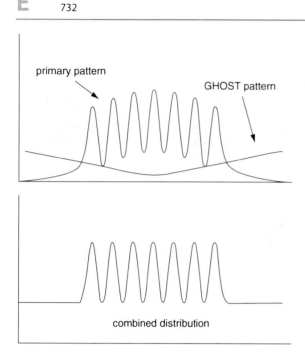

Electron Beam Lithography (EBL), Fig. 13 The figure on the top shows a pattern of lines, and the GHOST pattern superimposed. The figure on the bottom shows the summed contribution of the GHOST pattern and the line pattern [5]

Gaussian Beam EBL Systems

The most EBL systems today are based on scanning electron microscope (SEM) or scanning transmission electron microscope (STEM) systems since they share the same electron optics column. In fact, the first EBL systems were direct descendents of scanning electron microscopes and for research applications, it is very common to convert an electron microscope into an electron beam lithography system using a relatively low cost accessory (~ $100 k). This provides a huge saving especially for academic research facilities considering that dedicated e-beam writing systems are very expensive (~$4 M). SEMs normally have some lenses to focus the beam on the specimen, a beam-blanking mechanism, and some magnetic deflection coils used to scan the beam across the sample. Controlling these components by a computer allows transfer of any design in digital format to an exposure pattern. Digital-to-analog (DAC) converters are the key components for this transfer. The area to be scanned by the beam is divided into fields and subfields. This makes high speed (subfield) scanning achievable with DAC electronics when the address contains fewer bits (e.g., 12 bits). Then the large-field scanning is performed by tiling together subfields with

a high-precision DAC with larger addressing (e.g., 20 bits or larger). For the given DAC sizes and 1-nm address grid, it can be estimated that the maximum subfield size is 2^{12} nm (~4 μm) and the maximum main field is 2^{20} nm (~1 mm) [8]. The scan trajectory is defined as the path, which the beam would be deflected along if every pixel were to be exposed. The actual pattern is written by opening and closing a shutter. The scan method in which the beam scans the entire field by following the bit map data organized serially is called *raster scan* (Fig. 14a). Since every pixel must be scanned serially in raster scan, exposure time is independent of the pattern and longer than the alternative method *vector scan*.

Vector scan method is developed to improve the throughput by directing the electron beam to only those parts of the chip to be exposed (Fig. 14b). The digital address of each area to expose is sent to DAC and the beam scans over only the pixels, which must be exposed. Some systems use two sets of DACs: The first DAC drives a deflector, which directs the beam onto the corner of some primitive shape on the target wafer; the second DAC drives the same deflector or a different deflector in a raster pattern to fill in the shape. A moveable x−y substrate stage is often used, and is normally controlled using a laser interferometer to give accuracy down to the wavelength of the laser being used. Another advantage of the vector scan is the higher precision. Ignoring system aberrations and nonlinearities, the precision of the pattern address is determined by the width of the digital word. Therefore, pixels can be placed on an extremely fine grid if a high-speed, wide word DAC is used.

Variable-Shaped Beam and Mask Projection EBL Systems

As it was explained in detail before, direct write systems use a small electron beam spot that is moved with respect to the wafer to expose the pattern: one pixel at a time. This technique inherently suffers from low throughput to become a major process for economically manufacturing complex ICs. Inspired by the photolithography techniques, special EBL systems similarly have been developed to improve the throughput.

In one approach, the beam is partially intercepted by a square aperture. The electron beam is deflected and rotated electrostatically to adjust the width, length, and angle of the line exposed on the wafer. Since most

Electron Beam Lithography (EBL), Fig. 14 Schematic representation of (**a**) Raster scan (**b**) Vector scan

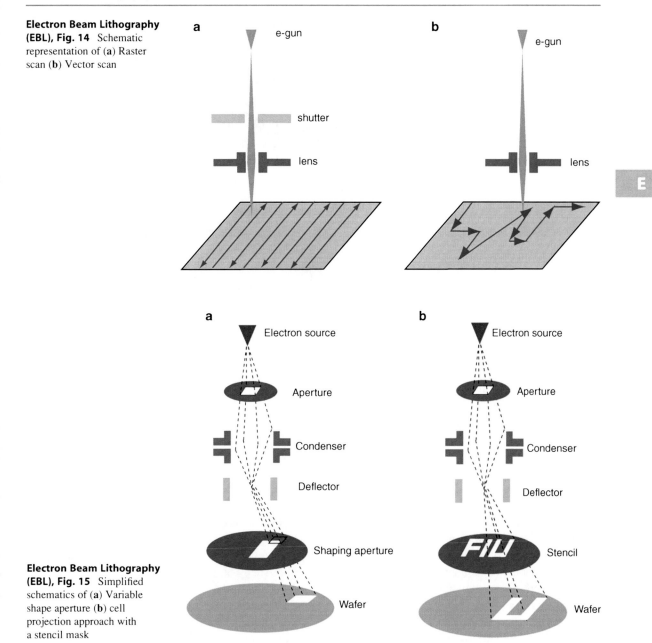

Electron Beam Lithography (EBL), Fig. 15 Simplified schematics of (**a**) Variable shape aperture (**b**) cell projection approach with a stencil mask

images are made up of rectangles, hundreds or even thousands of pixels can be exposed simultaneously. Line width is somewhat more difficult to control in this method but for many applications, the increase in throughput is sufficient to compensate for the reduction in line width control (Fig. 15a).

In the cell projection approach, a mask with predetermined geometries similar to the photomasks in photolithography is placed between the wafer and the beam source (Fig. 15b). Since the short penetration

length of electrons prohibits the use of a quartz mask, a stencil mask is used instead. Highly repetitive features such as DRAM cells can be transferred at a much higher throughput than conventional EBL. Cell projection has an advantage over shaped beam systems in that complex shapes can be created on a stencil mask beyond rectangles and triangles, at a modest increase in throughput.

To be able to stop highly energetic electrons, the stencil masks in cell projection method have to be large

Electron Beam Lithography (EBL), Fig. 16 Schematic diagram of the SCALPEL principle showing the function of the scattering mask in forming image contrast [9]

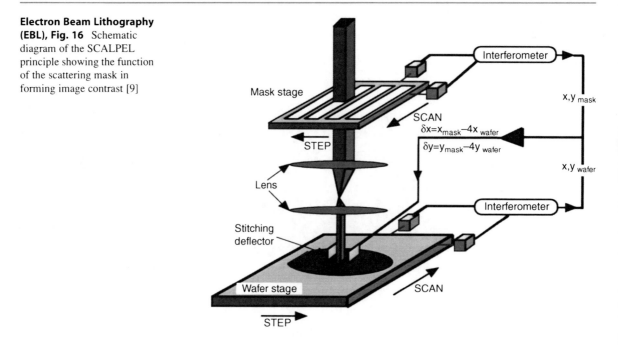

even to pattern a small area. The energetic electrons absorbed by the mask also heat up the mask, which can significantly alter the size of the features. New projection EBL systems have been developed to solve these problems. SCattering with Angular Limitation Projection Electron-beam Lithography (SCALPEL) system developed by Bell Labs is a noteworthy approach. SCALPEL approach combines the high resolution and wide process latitude inherent in electron beam lithography with the throughput of a parallel projection system and avoids the problems of stencil masks. In the SCALPEL system, a mask consisting of a low atomic number membrane and a high atomic number pattern layer is uniformly illuminated with high-energy (100 keV) electrons (Fig. 16). The entire mask structure is essentially transparent to the electron beam and little of its energy is deposited there. The portions of the beam, which pass through the high atomic number pattern layer, are scattered through angles of a few milliradians. An aperture in the back focal plane of the electron projection imaging lenses stops the scattered electrons and produces a high contrast image at the plane of the semiconductor wafer. The SCALPEL approach separates the pattern formation and energy absorption to maintain pattern fidelity. The pattern is defined by the mask although the energy from the unwanted parts of the beam is deposited on a mechanical aperture [9]. Throughput on the

SCALPEL system has been predicted to be approximately 45 wafers per hour. This is ten times higher than raster-scan EBL systems, and only 50% slower than optical lithography machines. It is estimated by SCALPEL's designers that SCALPEL will be cheaper than optical lithography systems, which use RET and hence, could be used for the sub-100-nm ULSI nodes. However, further development and commercialization of SCALPEL was stopped in January 2001.

IBM has also developed a projection EBL system, which is dubbed projection reduction exposure with variable-axis immersion lenses (PREVAIL). The PREVAIL system increases the throughput by shifting the main optical axis of the electron beam through the use of variable-axis lens designed at IBM. Deflections within each subfield can be superimposed onto the magnetic and electric fields used in the variable-axis lens. This variable-axis system allows the beam to be moved quicker from cell to cell on the mask, and reduces the dependence on slower mechanical components (Fig. 17).

Multi-beam EBL Systems

Throughput of EBL can also be increased by employing multiple beams. One way to obtain multiple beams is to use a series of electron sources rather than a single source. With the advancements of nanomanufacturing technologies, micromachined electron gun

Electron Beam Lithography (EBL), Fig. 17 Core technology of the IBM's PREVAIL EBL system: Curvilinear variable-axis lenses (CVAL)

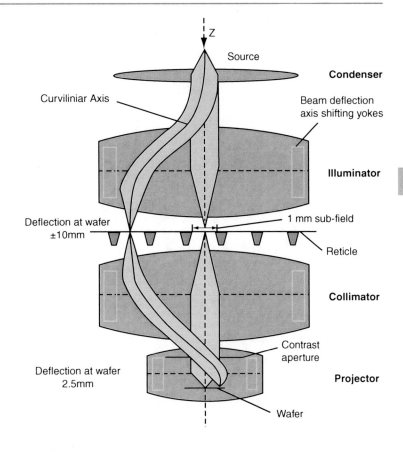

(MEG) devices have been proposed. A MEG device consists of monolithically integrated lenses feedback resistors and heaters. MEGs operate at much lower voltages (~200 V) than conventional electron guns and yield a spot size of 8.7 nm at a working distance of 20 μm. The lower voltage operation of these emitters mean that the electrons will have negligible backscatter and no proximity-effect corrections will be needed; however, exposure times may be longer due to the low narrow/low current beam. Since this technology is still at prototype stage, some details regarding the sample mounting and possible throughput are not clear. Very similar approach named sTM aligned field emission (SAFE) microcolumn arrays have also been proposed. SAFE technology uses a 2D array of miniaturized electron sources (Fig. 18). Although details of the manufacturing technology have not been fully disclosed, SAFE system is estimated to achieve a resolution of 25 nm and throughput of 60 count 200 mm diameter wafers per hour.

An alternative method proposed to obtain multiple beams is using one large electron beam and then dividing it into smaller beams by using miniature MEMS lenses and apertures. Smaller secondary beams are then focused on the target wafer [11]. This system has a 32×32 array of subelectron beams which cover a total of $3:6 \times 3:6$ mm area. It would require arrays of small lenses, blanking deflectors, blanking apertures, and beam deflectors, all implemented in a silicon substrate. The deflectors only need to deflect the beam in one direction, and a stage can move the sample in the other direction. No prototype of such system has been produced yet.

Recently, a novel maskless EBL system called the reflective electron beam lithography (REBL) has been proposed by KLA-Tencor Corporation [12]. REBL system incorporates six core technologies: reflective electron optics, the digital pattern generator, time domain integration, gray tone exposure, rotary stage, and optical wafer registration. Reflective electron optics enables the use of the digital pattern generator (DPG) chip to independently control 1×10^6 beams for massively parallel exposure. The electron optics ray diagram of the REBL nanowriter is shown in Fig. 19.

Electron Beam Lithography (EBL), Fig. 18 Conceptual drawing of microcolumn array for parallel e-beam lithography [10]

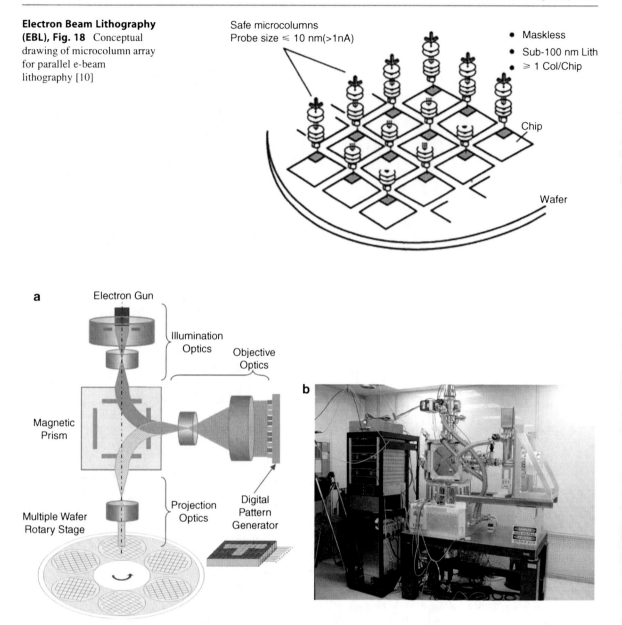

Electron Beam Lithography (EBL), Fig. 19 (a) Diagram of REBL nanowriter concept. (b) First REBL column on test stand for characterization [12]

A large area cathode of relatively low reduced brightness ($B_r < 10^4$ Am^{-2} sr^{-1} V^{-1}) generates an illuminating electron beam. The cathode lens forms a crossover at the center of the condenser lens. The condenser lens forms an image of the source at the virtual center of the magnetic prism. The prism is set up to bend the optical axis exactly 90° and establish the focal lengths in the x and y planes to be equal, with equal magnifications in the x and y axes. When these three conditions are satisfied, the prism behaves as a 1:1 imaging lens, imaging the crossover from the center of the condenser lens to the center of the transfer lens. The transfer lens in turn images the source image at the virtual prism center onto the DPG through the DPG lens. The electrostatic DPG lens serves a dual purpose: It forms a virtual image of the crossover at infinity, and it decelerates the electrons to within a few volts of the cathode potential. When a mirror is turned "on" at the DPG, it will reflect electrons from the illuminating beam. These reflected electrons are reaccelerated back toward the prism as they travel through the accelerating field of the electrostatic DPG lens. A crossover is formed at the center of the transfer lens by the DPG lens. The transfer lens images the DPG at the virtual center of the prism but along the diagonal 90° clockwise from the first image formed on the illuminating arm. The reflected electron beam, going in the opposite direction, is therefore bent in the opposing direction. The prism thus bends the reflected beam away from the gun to project it onto the wafer. The prism also forms a crossover at the back focal plane of the demagnification lens, while the demagnification lens focuses the DPG image from the virtual center of the prism onto the wafer plane. The prototype REBL system has demagnification of 50 and DPG pixel pitch of 1.5 μm which produce a pixel pitch at the wafer plane of 30 nm. The system targets five to seven wafers levels per hour on average at the 45-nm node with extendibility to the commercial mainstream lithography market for the 32-nm node and below.

Examples of Application

EBL is arguably the most versatile technique in nanofabrication technologies. Ability of direct writing nanoscale features allows researchers to fabricate experimental proof-of-concept and prototype devices, circuits and systems in diverse fields including electronics, photonics, microelectromechanical systems, and material sciences. Although it is impossible to cover all these applications, selected ones will be presented in this section to give a glimpse of the capabilities of EBL.

Transistors for High-Frequency Applications

With their high breakdown voltage and large sheet carrier concentrations, nitride-based wide band gap materials and devices attracted great deal of attention in the last decade. Particularly AlGaN/GaN HEMTs have been researched for their potential applications in wireless communication and radar replacing the incumbent GaAs devices. Fabrication of AlGaN/GaN single devices and monolithically integrated microwave integrated circuits (MMICs) involves several nanofabrication steps. Among them, fabrication of transistor gate is particularly important as the gate length is the smallest dimension in the transistors and is the most important parameter that determines frequency response of the transistors. Since the gate resistance is an important parasitic element affecting key parameters like f_{max}, it should be minimized. To minimize the gate resistance for a given gate width, it is essential to have a large cross section. However, a large cross section would mean a long gate footprint affecting frequency response. This problem can be tackled by making a mushroom like (T-gate) structures as shown in Fig 20. A T-gate is a structure that has a narrow footprint on the semiconductor and large head. The large head of the gate increases the cross-section area of the structure and thereby reduces the gate resistance. To fabricate T-gates one takes advantage of the difference in developing times for a given dose for the two commonly used e-beam resists, PMMA and copolymer (P(MAA-MMA)). This provides sufficient contrast to fabricate the three-dimensional shape required.

The process for fabrication of gates usually involves spinning three-layer stack, consisting of PMMA/copolymer/PMMA. Then, the resist is exposed with three passes: one strong centerline dose, which helps in clearing the resist to define the foot of the gate first and two light side area doses to help develop a big head. CAD layout is designed accordingly. Patterns are written by an e-beam lithography tool using low current (~1 nA). After developing the resists, a metal stack (e.g., Ni/Au) with adequate thickness is deposited. Liftoff is performed usually in methylene chloride. Figure 20 shows the SEM picture of a T-gate with a footprint of 100 nm.

Surface Plasmonic Devices

Surface plasmon polaritons (SPPs) originate from the collective oscillations of conduction electrons coupled with photons propagating at metal-dielectric interfaces. Localization of SPPs can find wide range of applications in technology like SPP-based Bragg reflectors, emitters, and filters. Their massive localized electric fields are used to enhance weak Raman signals. Using surface plasmon enhancement even a single molecule can be detected on plasmonic surfaces. SPP cavity structures can be used as a template in surface-enhanced Raman spectroscopy (SERS) applications.

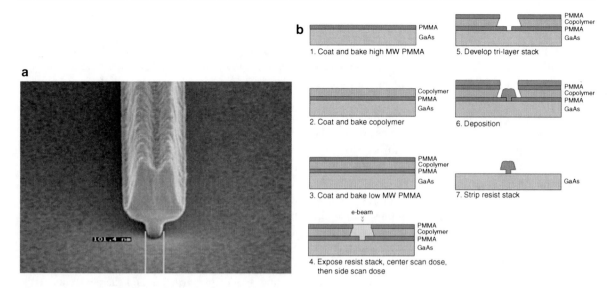

Electron Beam Lithography (EBL), Fig. 20 (a) SEM picture of a 100-nm "T-gate" of an AlGaN/GaN HEMT fabricated by 3-layer resist EBL. (b) Typical process flow for "T-gate" fabrication using tri-layer process (Courtesy of MicroChem Corp.)

Electron Beam Lithography (EBL), Fig. 21 Top view SEM image of a uniform grating fabricated by e-beam lithography with a period of 440 nm

Template supports a localized plasmonic mode that can be excited by the laser to be used in SERS. These templates are created by electron beam lithography. SPP-localized modes are adjusted by tuning the periodicity of the metallic grating. A sample image of metallic grating created by EBL is shown in Fig. 21.

PMMA is used as resist to create a uniform grating by EBL. The line width of the exposed regions depends on the following parameters: dose exposed per unit length, develop time, developer concentration, film thickness of PMAA, baking time and temperature, period of the grating (proximity affect), type of PMMA, silver thickness under PMMA, type of substrate. Correction of proximity effects is extremely important to obtain precise grating spacing, especially for small periods. In addition to using a PE correction software, an empirical "dose matrix" test method is also employed to determine the optimum exposure parameters. In a typical dose matrix test, the exposed dose on the resist is varied and other parameters are kept unchanged. Different fields are exposed with different doses. All fields are processed to the end. After checking the dose matrix with SEM, best exposure doses are selected to create gratings for measurements.

Light Trapping for Solar Cells

Light trapping with metal nanostructures on thin-film photovoltaic solar cells increases the solar cell efficiency. Light is coupled to metal nanoparticles and then coupled to the thin-film solar active layer over a wide range of angles. This procedure enhances the effective path length and absorption in the layer. So, metal nanoparticles can be used to increase efficiency of the devices based on the absorption of light. Enhancement effects of metallic nanoparticles are also demonstrated for dye-sensitized solar cells. Periodically structured metallic nanoparticles can trap the light onto the thin-film solar cell. Such structures can be created by EBL. First, an array of nanoholes is patterned on PMMA over an ITO-coated microscope slide Fig. 22a. After standard EBL fabrication process, the desired nanoparticle-coated thin-film solar cell

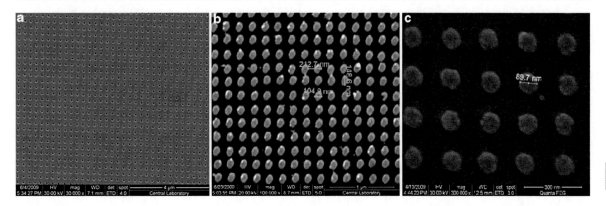

Electron Beam Lithography (EBL), Fig. 22 (a) Array of 100-nm diameter nanoholes with a period of 300 nm patterned on PMMA over an ITO-coated microscope slide. (b) SEM images of metal nanoparticles patterned over ITO-coated glass with different beam currents. (c) Enlarged image of the nanoparticles (Courtesy of Urcan Guler)

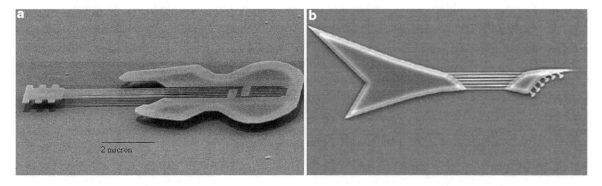

Electron Beam Lithography (EBL), Fig. 23 (a) World's smallest guitar, the "nano-guitar" fabricated on silicon using e-beam lithography [13] (b) Newer version of the "nano-guitar" which can actually be played by focused laser beams [14]

structure can be obtained (Fig. 21b, c). Period and size of the nanoparticles can be tuned for desired absorption spectrum of the incident light.

Micro and Nano Electromechanical Systems (MEMS and NEMS)

MEMS and NEMS were envisioned long time ago and put forward as a challenge by Richard Feynman in his seminal talk "There is plenty of room at the bottom" in 1959. Advancements in the nanofabrication technologies have allowed replication of many mechanical systems in nanometer scale and even creation of novel systems by taking advantage of physics at this scale. Electron beam lithography is one of the key technologies, which made possible the development of MEMS and NEMS. A striking example of the capabilities of MEMS fabrication by EBL is the *nano-guitar* In 1997, Cornell University researchers built the world's smallest guitar – about the size of a red blood cell – to demonstrate the possibility of manufacturing tiny mechanical devices using nanofabrication techniques [13]. It was 10 μm long, with six strings each about 50 nm, or 100 atoms, wide and made from crystalline silicon (Fig. 23a). Six years later the same group fabricated a new version of the nano-guitar, which is about five times larger than the original, but still so small that its shape can only be seen in a microscope [14]. Its strings are really silicon bars, 150 by 200 nm in cross section and ranging from 6 to 12 μm in length (Fig. 23b). The strings vibrate at frequencies 17 octaves higher than those of a real guitar, or about 130,000 times higher. What is unique about the second version is that it is possible to "play" it by sending focused laser beams. Since there is no practical microphone available for picking up the guitar sounds, however, the reflected laser light was computer processed to provide an equivalent acoustic trace at a much lower frequency. Both demonstrate the

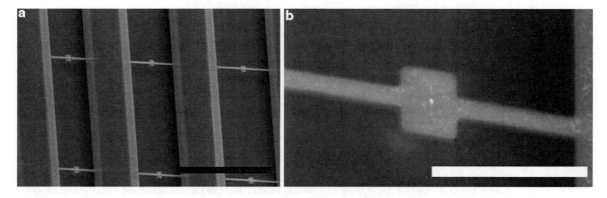

Electron Beam Lithography (EBL), Fig. 24 (**a**) SEM micrograph of arrays of bridge oscillators. (**b**) Oblique-angle SEM micrograph of oscillator. The diameter of the gold pad at the center is 50 nm. Scale bars correspond to 2 μm [15]

power of nanotechnology for a new generation of electromechanical devices.

One of the most widely explored application of MEMS and NEMS is sensitive detection of bound mass. With selective binding surfaces, the devices could be exquisitely sensitive to binding of selected chemical or biological species. In the selected case, EBL-supported nanofabrication capability was used to fabricate arrays of both polycrystalline silicon and silicon nitride resonators with evaporated gold contact pads. Circular gold pads 50–400 nm in diameter were deposited on the bridge resonators to serve as anchors for self-assembled thiol monolayer (SAM) for selective binding of target substances (Fig. 24). Detection of the resonant frequency shift allowed the determination of the mass of the adsorbed SAM. Nanoscale resonators presented quality factor of Q = 8,500 and allowed detection of 2.7 attogram ($= 2.7 \times 10^{-18}$ g) minimum resolvable mass [15].

Cross-References

▶ Scanning Electron Microscopy

References

1. Rius Sune, G: Electron lithography for nanofabrication. Ph.D. dissertation, Universitat Autonoma de Barcelona (2008)
2. Herriott, D.R., Brewer, G.R.: Electron-beam lithography machines, chapter 3. In: Brewer, G.R. (ed.) Electron-Beam Technology in Microelectronic Fabrication, pp. 141–216. Academic, New York (1980)
3. Tseng, A.A., Chen, K., Chen, C.D., Ma, K.J.: Electron beam lithography in nanoscale fabrication: recent development. IEEE Trans. Electron. Packag. Manuf. **26**(2), 141–149 (2003)
4. Goldstein, J., Newbmy, D., Joy, D., Lyman, C., Echlin, P., Lifshin, E., Sawyer, L., Michael, J.: Scanning Electron Microscopy and X-ray Microanalysis, 3rd edn. Springer, New York (2003). ISBN 0 306 47292 47299
5. Rai-Choudhury, P.: Handbook of Microlithography, Micromachining and Microfabrication, volume 1: Microlithography. SPIE Press Monograph, vol. PM39. SPIE Press, Bellingham (1997)
6. Jones, G., Blythe, S., Ahmed, H.: Very high voltage (500 kV) electron beam lithography for thick resists and high resolution. J. Vac. Sci. Technol. B **5**(1), 120–123 (1987)
7. Lee, S., Cook, B.: PYRAMID – a hierarchical, rule-based approach toward proximity effect correction – part II: correction. IEEE Trans. Semiconduct. Manuf. **11**(1), 117–128 (1998)
8. Tennant, D.M., Bleir, A.R.: Electron beam lithography of nanostructures. In: Wiederrecht, G. (ed.) Handbook of Nanofabrication. Elsevier, Amsterdam (2009)
9. Harriott, L.R.: Scattering with angular limitation projection electron beam lithography for suboptical lithography. J. Vac. Sci. Technol. B **15**(6), 2130–2135 (1997)
10. Chang, T.H.P., Kern, D.P.: Arrayed miniature electron beam columns for high throughput sub-100 nm lithography. J. Vac. Sci. Technol. **10**(6), 2743–2748 (1992)
11. Shimazu, N., Saito, K., Fujinami, M.: An approach to a high-throughput e-beam writer with a single-gun muliple-path system. Jpn. J. Appl. Phys. **35**(12B), 6689–6695 (1995)
12. Petric, P., Bevis, C., Carroll, A., Percy, H., Zywno, M., Standiford, K., Brodie, A., Bareket, N., Grella, L.: REBL: A novel approach to high speed maskless electron beam direct write lithography. J. Vac. Sci. Technol. **B27**(1), 161 (2009)
13. Photo by D. Carr and H. Craighead, Cornell Press Release, July 1997.
14. Sekaric, L.: The high and low notes of the Universe, Physics News Update, Number 659 (2003)
15. Ilic, B., Craighead, H.G., Krylov, S., Senaratne, W., Ober, C., Neuzil, P.: Attogram detection using nanoelectro-mechanical oscillators. J. Appl. Phys. **95**, 3694 (2004)

Electron Beam Physical Vapor Deposition (EBPVD)

▶ Physical Vapor Deposition

Electron Microscopy of Interactions Between Engineered Nanomaterials and Cells

Alexandra Porter and Eva McGuire
Department of Materials, Imperial College London, London, UK

Synonyms

FIB-SEM; SEM; TEM

Definition

The application of ion beam and transmission electron microscopy techniques to image, analyze, and study the 3-D structure and chemistry of nanomaterials inside cells.

Overview

Direct imaging of the structure and chemical analysis of tissues, cells, and biomaterials at the nanoscale is critical for understanding a wide variety nanoscale processes in biological systems. Whether in comprehending the role of amyloid fibrils in Alzheimer's disease or the toxicity of nanomaterials, there is a significant need for such studies to develop the potential for new forms of treatment and to advance the design of bio- and nano-materials [1–9]. Analytical electron microscopy is an indispensable technique for characterization of the material-biology interface because information can be obtained with very high spatial and energy resolution. However, this field remains wide open due to technical difficulties with sample preparation, particularly of hard-soft interfaces, and in achieving adequate spatial resolution and contrast between individual components within the cells and tissues while preserving them in their near native state and maintaining adequate stability of samples under the electron beam. Traditional preparation methods in electron microscopy use chemically intrusive stains and fixatives to preserve tissue and cell structure; however, these protocols will both distort the structure and modify the chemistry of the material. It is crucial to preserve the chemistry of the bio- or nanomaterial in the biological system as any slight modification in their chemistry or oxidation state caused during processing could make the results ambiguous and liable to misinterpretation which in turn could impact treatment strategies. Furthermore, the electron beam can damage the sample, altering the structure and changing the local chemistry. Methods to minimize this effect include the development of minimal dose methods as well as instruments that can operate at a range of accelerating voltages so that the experiments can be performed under optimum conditions. Once this has been achieved, the only option remaining is to cool the sample to reduce the damage rate. This entry aims to discuss development and application of analytical and 3-D microscopy techniques to analyze a range of bio-/nanomaterials interfaces at medium to high spatial resolution from ~ 1 μm to 1 nm using a range of techniques, namely dual-beam focused ion beam milling, energy-filtered TEM (EFTEM), high-angle annular dark-field scanning transmission electron microscopy (HAADF-STEM), and electron tomography. Traditional methods of sample preparation and using mass-thickness contrast for the imaging of biological samples will be discussed first as these still remain important techniques for both imaging and screening samples.

Sample Preparation

One of the greatest challenges for imaging nanoparticles within cells or biomaterials cell interfaces is in the sample preparation. When preparing samples for TEM, it is essential to induce minimal alteration to cell ultrastructure and also to the distribution and chemistry of nanoparticles within the cell. There are two main techniques available to prepare TEM samples at room temperature: fixation, dehydration, and embedding followed by ultramicrotomy; and focused ion beam milling (FIB). With ultramicrotomy, ultrathin resin-embedded sections containing the cell–nanomaterial interface are mechanically sectioned using a diamond knife. Here the challenge lies in matching the hardness of the embedding resin with the hardness of the cell/nanomaterial to ensure that the interface does not damage during sectioning [1]. For this reason, choice of embedding resin is crucial. FIB is also an excellent technique for preparing and analyzing interfaces for electron microscopy [6, 9]. A clear advantage of FIB milling is that chemical processing and also mechanical damage can be obviated, thus reducing damage to the specimen prior to examination. The FIB process will be described later in this entry.

Staining

To increase contrast from cell organelles, heavy metal stains are often employed. Stains such as Os, Pb, and U leave heavy metals in specific regions of the structure. For example, osmium tetroxide is a strong oxidant that cross-links lipids; it therefore acts not only as a secondary fixative of membranes, but also increases contrast in TEM images because of the high atomic number of the osmium atoms. These stains can however, obscure visualization and characterization of low contrast nanostructures. To overcome this problem, stained and unstained sections have been compared to correlate distributions of nanoparticles within cell structure [4, 5].

Section Thickness

In ultramicrotomy, standard sample thicknesses for TEM sections can vary from 70 nm to several hundred nanometers. This results in the acquisition of a two-dimensional image containing three-dimensional information which can, in turn, result in the "blurring" of data or loss of information. For electron energy loss (EEL)-based techniques or high-resolution imaging, optimum contrast can be achieved by sectioning ultrathin sections of around 20 nm in thickness. By using these ultrathin sections, two-dimensional slices through cells are obtained, therefore minimizing the three-dimensional information and allowing for high-resolution imaging of nanoparticles within the cell. Contrast from cell organelles can also be significantly improved by sectioning at 20–40 nm or by leaving the section under the electron beam in the TEM to thin. The limitation with ultrathin sections, however, is durability of the section under the beam, thus, when energy or spatial resolution is not critical, thicker sample thickness (~70 nm) will provide sufficient contrast.

Rapid Freezing Techniques

Such as freeze substitution and freeze drying are also available to prepare cell–nanomaterial interfaces. These methods involve rapidly freezing the tissues or cells, typically in liquid propane or liquid nitrogen, either by drastically decreasing the temperature and/or by applying a high pressure. The cells are rapidly frozen to avoid ice crystal formation which would cause mechanical injury to the cells. After the cells/tissues have been frozen, these materials are slowly heated to allow the frozen water in the material to sublime from the solid phase to the gas phase. This process is called freeze drying. Sublimation of the material is required to avoid shrinkage, which would occur when rapidly heating the sample. If ice is not sublimed into the gas phase, the surface tension of the material will increase, and the material could collapse under this tension as the water leaves the liquid specimen. Freeze substitution is based on rapid freezing of tissues followed by solution ("substitution") of ice at temperatures well below 0°C. For best morphological and histochemical preservation, substituting fluids should, in general, contain both chemical fixing agent and solvent for ice, e.g., 1% solutions of osmium tetroxide in acetone or mercuric chloride in ethanol. Preservation of structure is generally poorer after substitution in solvent alone. Osmium tetroxide can also act to enhance the contrast acquired from the cell organelles. These techniques have many advantages over traditional room temperature preparation methods, particularly the preservation of structure, chemistry, and the accurate localization of many soluble and labile substances. Both these methods have been used to study the process of the formation of nanoparticles of amorphous calcium phosphate inside osteoblast and also neuronal cells. In these studies, cryo-fixation methods are critical to preserve both the chemistry of the labile calcium phosphate particles and the structure of the cell [10].

Bright-Field TEM Imaging and Mass-Thickness Contrast

Mass-thickness contrast is the contrast mechanism that is traditionally used for imaging biological samples. As electrons pass through the sample, they are scattered off-axis by elastic scattering events. Electrons that interact with the electron clouds of atoms

Electron Microscopy of Interactions Between Engineered Nanomaterials and Cells, Fig. 1 (**a**) The insertion of an objective aperture to select the direct beam removes most scattered electrons and may greatly increase image contrast. (**b**) The mechanism by which differences in mass and thickness create image contrast in a bright-field TEM image (Figure adapted from [12])

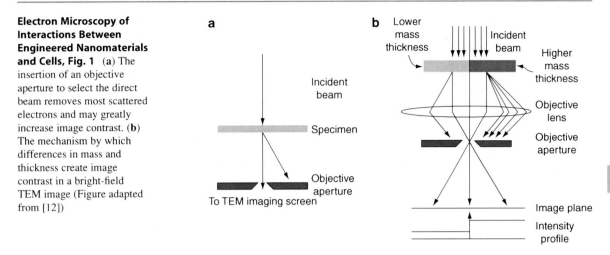

in the sample are scattered through relatively small angles and may still contribute to the image, but electrons that are scattered due to interactions with an atomic nucleus are scattered through a higher angle and are less likely to contribute to the image generated. The cross section of scattering depends on the atomic number, Z, and on the density and thickness of the sample. In bright-field imaging, the direct beam is used to generate bright-field TEM images. The insertion of an objective aperture selects the angular range of electrons contributing to the image and may greatly improve image contrast as demonstrated in Fig. 1a. The mechanism by which contrast in a bright-field TEM image is formed from differences in mass and thickness is demonstrated in Fig. 1b. The areas of the sample that have higher mass thickness (ρt) produce areas of lower intensity in the resulting TEM image than areas of lower mass thickness. Mass-thickness contrast from cell organelles can be enhanced by staining the cells with heavy metals such as osmium tetroxide.

Analytical Transmission Electron Microscopy

In a TEM, energy loss can either be measured in series, i.e., an image is acquired over a defined energy loss window, known as energy-filtered TEM (EFTEM); or in parallel, where an entire energy loss spectrum is acquired from a point in the specimen, known as electron energy loss spectroscopy (EELS). EELS is a technique where high-energy electrons incident on a sample scatter from, and therefore lose energy to, atoms in the sample. The energy lost in a particular scattering event is lost to an atom's electrons and only

takes discrete values (quanta) which are different for every element. The electrons are transmitted through a sample, and then the energy that they lose is recorded in the form of a loss spectrum. The spectrum of each element is comprised of peaks or edges positioned at characteristic energy-losses. The shape and position of each edge also provides information about the bonding environment of the sample being analyzed. Therefore, EELS spectra provide detailed information about the elemental composition and bonding state within a sample with sub-nanometer spatial resolution. EFTEM is a technique based on EELS which records images taken at different energies. The basis of this technique is that an energy-filter slit is used which only allows electrons through which have lost certain energies. By selecting different energies, elemental EFTEM maps can be produced. Using this technique, it is possible to acquire two-dimensional, elemental maps using electrons with an energy loss characteristic of a core level, interband transition or Plasmon resonance energy. With EFTEM, the EELS spectrum may be extracted by acquiring a series of energy-filtered images, using a narrow energy window (1–3 eV). Image processing software may then be employed to extract the EELS spectra over a given energy range.

EFTEM is an excellent technique for nanobioscience applications because it can detect low atomic number elements (such as biological constituents) with high sensitivity. Using this technique, maps of the distributions of the elements in the sample can be resolved and, thus, the chemistry of the cell or presence of nanoparticles and their locations in the cells determined [3, 4]. For example, low-loss EFTEM enables clear differentiation between *unlabelled* C_{60}, SWNTs, and the cell cytoplasmic contents [3, 4]. TEM and EELS can be also applied to study intracellular free ion concentrations, the oxidation state of engineered nanoparticles in the cellular environment [11], mineral distributions in osteoporotic bone, or the formation of and the oxidation state of iron storage protein ferritin in the brains of patients suffering from Alzheimer's disease [13].

In scanning transmission electron microscopy (STEM), the incident electron beam is focused to a probe and scanned across the sample. Detectors placed at different post-specimen locations can collect different scattered or unscattered beams. The scattered (or unscattered) electrons are collected serially, and the image is displayed pixel by pixel. Combined with EELS, it can be used to map the chemistry or oxidation state of materials nanoscale resolution. In recent work, Jantou et al. have shown that STEM-EELS can provide new insights into the chemistry and bonding at the collagen-hydroxyapatite interface in ivory dentin and observed fine structure on the carbon K-edge spectra, suggesting that functional groups at the interface between the collagen and mineral phase can be identified [9]. This work potentially provides information on the modification of the surface amino acid groups by the inorganic phase [9]. Such approaches offer enormous potential for improved characterization of the chemistry of nanoparticles within tissues to advance knowledge of how these structures interact with cell organelles (Fig. 2).

HAADF-STEM is an imaging mode that is highly sensitive to the atomic number of the sample. HAADF-STEM relies on collecting electrons which have been scattered to high angles by Rutherford scattering. The HAADF detector has a large central aperture and collects only electrons that are scattered to relatively high angles (e.g., >30 mrad, the exact range of scattering angles is determined by the camera length and the geometry of the detector). The intensity of a HAADF-STEM image is therefore dependent on Z^n ($n \sim 2$). As the detector collection angles are reduced, the contribution of mass thickness (ρt) increases. Thus, it is possible for the user to control the kinds of electron/specimen interactions that contribute to an image (Fig. 3). HAADF-STEM has been combined with energy dispersive x-ray spectroscopy (EDX) to image Amyloid-beta aggregates inside cells using a selenium enhancement technique [7]. The rationale for this study is that a number of challenges arise when attempting to image the interactions between amyloid or other protein aggregates and cells or tissue. The most significant of these is a lack of contrast between the carbon-rich protein aggregate and the carbonaceous cellular environment, making the aggregates extremely difficult to distinguish and identify with confidence. A variety of strategies have previously been employed to overcome this, but these all involve the application of chemically intrusive labeling techniques such as immunogold labeling or tagging the proteins with fluorescent tags [7]. Instead, peptides were assembled from selenium-analogues of the sulfur-containing methionine peptides, and then used HAADF-STEM to detect the selenium-doped species selectively within the carbon-rich background of the cell. The presence of selenium was confirmed by the use of energy dispersive X-ray spectroscopy (EDX). The energies of the X-rays generated by the interactions between the electron beam and the atoms in the sample can be examined for peaks at energies characteristic of specific elements.

Electron Tomography

TEM and STEM imaging creates two-dimensional projections of a specimen. One disadvantage of this is that structural details from different depths within the specimen are superimposed on top of one another. Electron tomography (ET) involves collecting a series of images while tilting the specimen in the TEM around a single axis at regular intervals. It is impossible from this projection alone to determine the structure of the original object. For this reason, electron tomography is carried out – the process of taking a large number of two-dimensional electron micrographs of a specimen at different tilt angles and then reconstructing these to generate a volume that represents the original object.

The collection of a tomographic dataset produces a large number of two-dimensional projections of the region of interest which can be reconstructed into a three-dimensional volume that closely resembles the original object. The correlation between the three-dimensional reconstructed volume and the original object is limited by the data collected – the quality of the images and number of images collected – and the technique used to reconstruct the dataset.

Data collection for electron tomography involves tilting the specimen around an axis perpendicular to

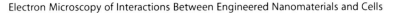

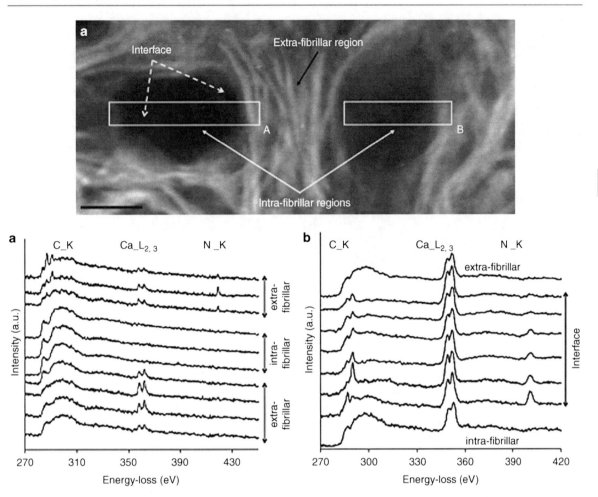

a C_K   Ca_L$_{2,3}$ N_K   extra-fibrillar / intra-fibrillar / extra-fibrillar

Intensity (a.u.)

270   310   350   390   430
Energy-loss (eV)

b C_K   Ca_L$_{2,3}$ N_K   extra-fibrillar ... Interface ... intra-fibrillar

Intensity (a.u.)

270   300   330   360   390   420
Energy-loss (eV)

Electron Microscopy of Interactions Between Engineered Nanomaterials and Cells, Fig. 2 STEM image of a transverse section of ivory dentine. Fine structure on the carbon K-edge can be seen at the interface between the intra- and extra-fibrillar regions which can be attributed to the molecular amino acid signatures. P L2,3-, O K- (not shown), and Ca L2,3- edges can be used to investigate local variations in mineral chemistry [9]

Electron Microscopy of Interactions Between Engineered Nanomaterials and Cells, Fig. 3 (a) HAADF-STEM image taken with inner and outer collection angles of 8.3 and 41.5 mrad, respectively. (b) HAADF-STEM image taken with inner and outer collection angles of 32.5 and 161.0 mrad, respectively. Bright contrast from the nulclear membrane is enhanced in (a). *Arrows* delineate the nuclear membrane

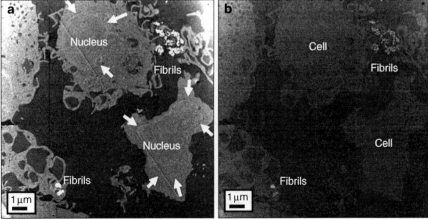

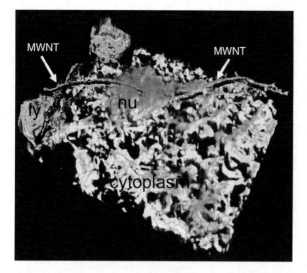

Electron Microscopy of Interactions Between Engineered Nanomaterials and Cells, Fig. 4 HAADF-STEM tomogram of a necrotic macrophage cell illustrating MWNTs penetrating the nucleus (nu) [15]

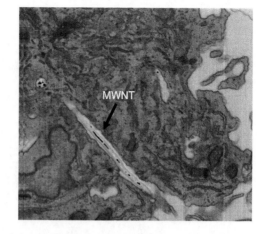

Electron Microscopy of Interactions Between Engineered Nanomaterials and Cells, Fig. 5 FEG-SEM image of a MWNT internalized inside a phagosome in a macrophage cell. Image taken from an osmicated cell in backscattered electron mode

the electron beam and collecting images at successive angles. Specimen shift must be minimized during data acquisition, in particular in the direction parallel to the tilt axis, as only the area that is in every image can be used for the reconstruction of the data. Images are collected up to as high tilt angles as possible to minimize the size of the missing "wedge" in the dataset. It is impossible to take images of the specimen at tilt angles up to $\pm 90°$; therefore, there will always be some information missing from the dataset.

A three-dimensional structure can then be reconstructed from the series of images collected to give high-resolution structural information about the material being analyzed. ET combined with EFTEM can also be used to yield three-dimensional compositional information [14]. Biological structures frequently need to be determined in all three dimensions because their chemical and physical properties are highly related to their topography. ET reconstructions are crucial to confirm that the nanostructures are localized within cell organelles and not just lying on the surface of the sample (Fig. 4).

Ion Beam Milling

A recent development in three-dimensional imaging is the "slice-and-view (S&V)" method in the dual-beam focused ion beam (DB-FIB). The DB-FIB combines an ultrahigh-resolution field emission gun scanning electron microscope (FEG-SEM) with a precise FIB system. In this system, the SEM uses electrons as a source for imaging and monitoring the milling process, whereas the FIB uses gallium positive ions as a source to ablate the surface of a specimen via the sputtering of substrate. This unique FIB-SEM combination allows the investigator to image the architecture of selected specimens by SEM and FIB in parallel. The focused ion beam can be used to mill away a precise volume to reveal the internal surfaces at a designated site in the specimen. The three-dimensional morphology of the sample can then be built up by milling and recording successive image slices (Fig. 5), the entire three-dimensional volume of the sample can then be reconstructed from the serial images to give information about the architecture of the sample. Milling can be performed with a precision of ~ 10 nm depth and up to 100 µm width allowing for analysis, in three dimensions, of the structure of materials from 0.1 mm to <100 nm. Approximately 500 serial images from a sample can be generated automatically overnight; therefore, both quantitative and qualitative information can be acquired. This technique has successfully been used to study the ultrastructure of mouse cortex tissue in three dimensions [16] and has great potential for mapping the distributions of nanoparticles inside tissues which can otherwise be analogous to the problem of "finding a needle in a haystack." The FEG-SEM method can also be used to prepare site specific lamella regions in the FIB for subsequent analysis in the TEM. More recently, this method has shown to be highly promising for preparing and studying frozen hydrated cells and tissue-material interfaces [17–19]. The advantage here is that this method involves no exposure to toxic or aggressive staining agents and the process induces little or no mechanical damage as would be experienced during cryo-ultramicrotomy.

Electron Beam Damage

Electron beam damage can be the limiting factor in the TEM study of biological materials. Many of the electron/specimen interactions that produce useful signals in the TEM result in damage of the specimen that can lead to structural and chemical changes in the sample [12]. This means that the region of the specimen under examination is no longer representative of the original sample and interpretation of the images and spectra is more difficult. Damage can result from both elastic and inelastic scattering of electrons. There are two main mechanisms of electron beam damage in the TEM: radiolysis and knock-on damage.

Radiolysis or ionization damage results in the breaking up of chemical bonds in the sample [12]. Organic solids, such as polymers and biological materials, are most prone to radiolysis [12]. Incident electrons can break down the main chain of a molecule, resulting in the loss of mass, or can break off side chains which may in turn react with other molecules in the sample and cross-link to form a new structure. Loss of mass results in a thinning of the sample or even the drilling of holes in the sample. Crystalline samples can lose their crystallinity due to ionization damage. The extent of damage due to radiolysis can be minimized by increasing the accelerating voltage as this decreases the interaction cross section. However, this leads to a reduction in image contrast, as less electron scattering takes place on interaction with the sample, and can result in an increase in knock-on damage.

Knock-on or atomic displacement damage occurs when electrons in the beam transfer enough energy to atoms in the sample to knock them out of their atomic sites leading to the creation of vacancies in the specimen [12]. This damage mechanism will generate

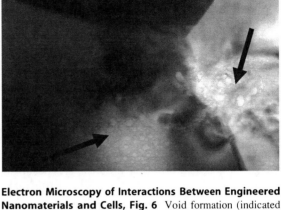

Electron Microscopy of Interactions Between Engineered Nanomaterials and Cells, Fig. 6 Void formation (indicated by black arrows) in an hydroxyaptite ceramic implant generated by knock-on damage to the sample under the electron beam

artifacts, such as amorphization of carbon nanotubes or void formation in calcium phosphate ceramics (Fig. 6). Electrons in the beam must possess enough kinetic energy to knock a specimen atom out of its site; this energy depends on the type of sample and is known as the threshold energy. This type of damage therefore increases with increasing accelerating voltage. Knock-on damage is not expected at an acceleration voltage of 100 kV (unless hydrogen atoms are present), but it is anticipated at higher voltages [20]. Cooling the specimen will not reduce the effects of knock-on damage. Keeping the energy of the electron beam low is the only way to avoid this type of damage. If energies above the threshold energy must be used, the total electron dose should be minimized by minimizing the time the electron beam is incident on the sample.

Heat is also a major cause of damage in the TEM. Much of the energy that is transferred to the sample when beam electrons collide with specimen electrons ends up as heat within the specimen [20]. Organic specimens are particularly susceptible to heat damage as they have low thermal conductivity, and so the region of interest cannot dissipate the thermal energy it gains under the beam. Heat damage can be minimized by using a higher accelerating voltage as this minimizes the interaction cross section.

Future Research Directions

Rapid developments are currently taking place in the field of electron microscopy, both to increase energy and spatial resolution, and also to increase automation of the equipment so that data can become significantly more quantitative and so it can be obtained with less work by the user, a feature that biological electron microscopy has traditionally suffered from. Exciting developments are occurring in the area of dual-beam FIB milling such that it will now be possible to correlate across length scales to link macroscale processes to structure and chemical changes occurring at the nanoscale. These methods will make major advancements in linking the physicochemical properties of nanoparticles to their effect on tissue structure and more generally will guide our understanding of tissue pathologies such as neurodegenerative diseases.

Acknowledgments The authors thank the contributions of Prof. David McComb, Imperial College, and Dr. Ben Lich at the FEI company, Eindhoven.

Cross-References

▶ Carbon Nanotubes
▶ Nanoparticle Cytotoxicity

References

1. Porter, A.E., Patel, N., Skepper, J.N., Best, S.M., Bonfield, W.: Comparison of in vivo dissolution processes in hydroxyapatite and silicon-substituted hydroxyapatite bioceramics. Biomaterials **24**, 4609–4620 (2003)
2. Porter, A.E., Gass, M.H., Muller, K., Skepper, J., Midgley, P., Welland, M.: Uptake of C_{60} by human monocyte macrophages, its localization and implications for toxicity: studied by high resolution electron microscopy and electron tomography. Acta Biomater. **2**(4), 409–419 (2006)
3. Porter, A.E., Gass, M.H., Muller, K., Skepper, J., Midgley, P., Welland, M.: Visualizing the uptake of C_{60} to the cytoplasm and nucleus of human monocyte-derived macrophage cells using energy-filtered transmission electron microscopy and electron tomography. Environ. Sci. Technol. **41**(8), 3012–3017 (2007)
4. Porter, A.E., Gass, M.H., Muller, K., Skepper, J., Midgley, P., Welland, M.: Direct imaging of single-walled carbon nanotubes in cells. Nat. Nanotechnol. **2**(11), 713–717 (2007)
5. Uchida, M., Willits, D.A., Muller, K., Willis, A.F., Jackiw, L., Jutila, M., Young, M.J., Porter, A.E., Douglas, T.: Intracellular distribution of macrophage targeting ferritin-iron oxide nanocomposite. Adv. Mater. **21**(4), 458–462 (2009)
6. Porter, A.E., Nalla, R.K., Minor, A., Jinshek, J.R., Kisielowski, C., Radmilovic, V., Kinney, J.H., Ritchie, R.O.: Changes in mineralization with aging-induced transparency in human root dentin. Biomaterials **26**(36), 7650–7660 (2005)
7. Porter, A.E., Knowles, T.P., Muller, K., Meehan, S., McGuire, E., Skepper, J., Welland, M.E., Dobson, C.M.: Imaging amyloid fibrils within cells using a Se-labelling strategy. J. Mol. Biol. **392**(4), 868–871 (2009)
8. Gentleman, E., Swain, R.J., Evans, N.D., Boonrungsiman, S., Jell, G., Ball, M.D., Shean, T.A., Oyen, M.L., Porter, A., Stevens, M.M.: Comparative materials differences revealed in engineered bone as a function of cell-specific differentiation. Nat. Mater. **8**, 763–770 (2009)
9. Jantou, V., Horton, M.H., McComb, D.W.: The nanomorphological relationships between apatite crystals and collagen fibrils in ivory dentine. Biomaterials **31**(19), 5275–5286 (2010)
10. Aronova, M.A., Kim, Y.C., Pivovarova, N.B., Andrews, S.B., Leapman, R.D.: Quantitative EFTEM mapping of near physiological calcium concentrations in biological specimens. Ultramicroscopy **109**, 201–212 (2009)
11. Bhabra, G., Sood, A., Fisher, B., Cartwright, L., Saunders, M., Evans, W.H., Surprenant, A., Lopez-Castejon, G., Mann, S., Davis, S.A., Hails, L.A., Ingham, E., Verkade, P., Lane, J., Heesom, K., Newson, R., Case, C.P.: Nanoparticles can cause DNA damage across a cellular barrier. Nat. Nanotechnol. **4**, 876–883 (2009)
12. Williams, D.B., Carter, C.B.: Transmission electron microscopy: a textbook for materials science. Springer, New York/London (2009)
13. Collingwood, J.F., Chong, R.K.K., Kasama, T., Cervera-Gontard, L., Dunin-Borkowski, R.E., Perry, G., P'osfaig, M., Siedlake, S.L., Simpson, E.T., Smith, M.A., Dobson, J.: Three-dimensional tomographic imaging and characterization of iron compounds within Alzheimer's plaque core material. J. Alzheimers Dis. **14**, 235–245 (2008)
14. Gass, M.H., Koziol, K.K.K., Windle, A.H., Midgley, P.A.: Four-dimensional spectral tomography of carbonaceous nanocomposites. Nano Lett. **6**(3), 376–379 (2006)
15. Cheng, C., Muller, K., Koziol, K., Skepper, J.N., Midgley, P.A., Welland, M.E., Porter, A.E.: Toxicity and imaging of multi-walled carbon nanotubes in human macrophage cells. Biomaterials **30**(25), 4152–4160 (2009)
16. Knott, G., Marchman, H., Wall, D., Lich, B.: Serial section scanning electron microscopy of adult brain tissue using focussed ion beam milling. J. Neurosci. **28**(12), 2959–2964 (2008)
17. Marko, M., Hsieh, C., Scalek, R., Frank, J., Mannela, C.: Focused-ion-beam thinning of frozen hydrated biological

specimens for cryo-electron microscopy. Nat. Meth. **4**, 215–217 (2007)

18. Rigort, A., Bauerlien, F.J.B., Leis, A., Gruska, M., Hoffmann, C., Laugks, T., Bohm, U., Eibauer, M., Gnaegi, H., Baumeister, W.: Micromaching tools and correlative approaches for cellular cryo-electron tomography. J. Struct. Biol. **172**(2), 169–179 (2010)

19. Hayles, M.F., de Winter, D.A.M., Schneijdenberg, C.T.W.M., Meeldijk, J.D., Lucken, U., Persoon, H., de Water, J., de Jong, F., Humbel, B.M., Verkleij, A.J.: The making of frozen-hydrated, vitreous lamellas from cells for cryo-electron microscopy. J. Struct. Biol. **172**(2), 180–190 (2010)

20. Egerton, R.F., Li, P., Malac, M.: Radiation damage in the TEM and SEM. Micron **35**(6), 399–409 (2004)

Electron Transport in Carbon Nanotubes and Graphene Nanoribbons

▶ Electronic Transport in Carbon Nanomaterials

Electron-Beam-Induced Chemical Vapor Deposition

▶ Electron-Beam-Induced Deposition

Electron-Beam-Induced Decomposition and Growth

▶ Electron-Beam-Induced Deposition

Electron-Beam-Induced Deposition

Guoqiang Xie
Institute for Materials Research, Tohoku University, Sendai, Japan

Synonyms

Electron-beam-induced chemical vapor deposition; Electron-beam-induced decomposition and growth

Definition

Electron-beam-induced deposition is a process for growing patterns with different sizes, shapes, and materials in the nanometer or submicron scale. In this process, an energetic electron beam in a vacuum chamber is focused on a substrate surface on which precursor molecules, containing the element to be deposited (e.g., organometallic compound or hydrocarbon), are adsorbed. As a result of complex beam-induced surface reactions, the precursor molecules adsorbed in and near to the irradiated area are dissociated into nonvolatile and volatile parts by the energetic electrons. The nonvolatile materials are deposited on the surface of the substrate to form a deposit, while the volatile parts are pumped away by the vacuum system. The advantage of electron-beam-induced deposition over conventional lithography methods is that two- and even three-dimensional (3D) structures are patterned and deposited simultaneously, making it a fast, one-step technique. Due to the controllability of the electron beam, the fabrication of the position-, size-, and morphology-controllable nanostructures can be easily achieved.

Overview

Originally, electron-beam-induced deposition was known as contamination growth in electron microscopy. Broers et al. [1] used contamination growth patterns as an etching mask to define 8-nm-wide metal lines. In the last decades, electron-beam-induced deposition for maskless nanofabrication has gained more importance as a tool, and it has been studied by many researchers because of its higher resolution due to the smaller beam spot. At an early stage, electron-beam-induced deposition has mainly been performed in scanning electron microscopes (SEMs) for various gases and conditions, and the technique was found to be very successful for a number of applications, making variously shaped nanostructures from metals such as W, Au, and Pt [2, 3]. However, the smallest structures fabricated in scanning electron microscopes have a typical width of 15–20 nm, no matter how small the primary electron beam is [4]. This is usually attributed to the fact that the distribution of secondary electrons, rather than primary electrons, determines the size of the deposits. The profile of the secondary electrons exhibits some spread, particularly for the energy range of scanning electron microscopes. Even for primary electrons of zero diameter, the profile of secondary electrons has a range of more than 15 nm [5]. Hence, it was believed that the resolution limit of electron-beam-induced deposition was at most 15 nm for the fabrication in scanning electron microscopes. Recently, several groups have tried to minimize the secondary electron effect by using transmission electron microscopes (TEMs) and scanning transmission electron microscopes (STEMs). Transmission electron microscopes utilize electrons of higher primary energy than scanning electron microscopes and can produce more focused beams with less secondary electron emission. Mitsuishi et al. [6] fabricated nanodots of 3.5 nm in diameter using a 200-keV field emission scanning transmission electron microscope. They found that the dot size does not depend on the substrate thickness; rather, the deposition period and speed are strongly dependent on the intensity of the electron beam. Tanaka et al. [7] have also fabricated nanodots containing W by electron-beam-induced deposition in a 200-keV ultrahigh vacuum field emission transmission electron microscope (UHV-FE-TEM). The size of the dots can be controlled by changing the pressure in the chamber. The smallest particle size was about 1.5 nm in diameter.

Physics of Material Supply and Deposition

Aiming at a rate optimization in the electron-beam-induced deposition, one has to know how the deposition rate depends on the various relevant process parameters. Some theoretical analyses about the deposition process have been developed by Scheuer et al. [8, 9] for flood-beam systems and adapted for scanned-beam systems by Petzold and Heard [10]. Following the theoretical approach of Scheuer [8], it is assumed that only those molecules adsorbed at the substrate surface undergo dissociation induced by electron beam irradiation. Thus, only a monolayer of adsorbed molecules can contribute to the layer growth. Figure 1 illustrates the different processes on the surface [8]. The density (N) of adsorbed molecules on the substrate surface varied in time (t) is given by the following equation:

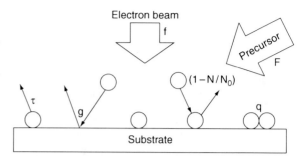

Electron-Beam-Induced Deposition, Fig. 1 Illustration of surface interaction of the deposition process

$$\frac{dN}{dt} = g \cdot F \left(1 - \frac{N}{N_0}\right) - \frac{N}{\tau} - q \cdot N \cdot f \qquad (1)$$

where f is the electron flux density, q is the cross section for dissociation of the adsorbed molecules under electron beam irradiation, F is the molecular flux density arriving on the substrate, g is the sticking coefficient, N_0 is the molecule density in a monolayer, and τ is the mean lifetime of an adsorbed molecule.

If $dN/dt = 0$, equilibrium is obtained, resulting in an equilibrium density (N_E):

$$N_E = N_0 \cdot \frac{(g \cdot F/N_0)}{(g \cdot F/N_0) + 1/\tau + q \cdot f} \qquad (2)$$

This means that $N_E \leq N_0$ for all supply and irradiation conditions, i.e., the maximum surface coverage at any time is one monolayer (ML). The growth rate (R) of the layer is given as follows:

$$R = v \cdot N \cdot q \cdot f \qquad (3)$$

where v is the volume occupied by a dissociated molecule or its fractions. In the equilibrium case, $N = N_E$. From Eqs. 1, 2, 3, the steady-state growth equation is derived:

$$R = v \cdot N_0 \cdot \frac{(g \cdot F/N_0) \cdot q \cdot f}{(g \cdot F/N_0) + 1/\tau + q \cdot f} \qquad (4)$$

In practice, the influence of the finite lifetime (τ) of the molecules can be made negligible, e.g., by cooling the sample. It is then worthwhile to note a few properties of the above equation. Obviously, the growth rate increases when the electron flux (f) and/or the molecular flux (F) is increased. In practice, these parameters are limited, e.g., by electron optical constraints, the vapor supply and differential pumping system, or simply the fact that only a limited vapor pressure can be tolerated above the sample before significant electron scattering starts. Under these conditions, the maximum rate obtainable for any given electron flux is

$$R_{max} = v \cdot N_0 \cdot q \cdot f \qquad (5)$$

and choosing

$$q \cdot f = \mathrm{g} \cdot F / N_0 \qquad (6)$$

yields 50% of the maximum obtainable growth rate at supply F. This condition can be considered a reasonable compromise of parameters for efficient use of both electron and molecule supply. Increasing either flux to infinity while keeping the other one constant only increases the growth rate by a factor of 2, while decreasing either flux by one order of magnitude reduces the growth rate by more than a factor of 5.

Key Research Findings

High-quality, intense, and focused electron beams have been used for the fabrication of nanomaterials with controlled position, size, and morphology. Electron-beam-induced deposition was used as an advanced method to fabricate nanostructures in situ by a maskless process. The reaction of precursor gas molecules with electrons near the material surface causes the formation of a solid deposit whose size and shape depend on the substrate materials and the profile of the electrons. Using the electron-beam-induced deposition technique, a variety of nanostructures, such as nanodots, nanowires, nanotubes, nanopatterns, two- or three-dimensional nano-objects, and so on, have been fabricated [2, 11–13].

When precursor gases containing tungsten (W), iron (Fe), and other metals are introduced into the microscope, noble-metal nanostructures can be fabricated in situ by electron-beam-induced deposition. The electron beam controls both the nucleation and growth processes. For the substrate materials, conventionally two kinds of materials, namely, conductive

and insulator materials, are used. When the conductive materials, such as silicon, were used as a substrate, two- or three-dimensional nanostructures of W, Pt, Fe compounds, and so on, were fabricated by the electron-beam-induced deposition with the beam-scanning technique [2, 11]. This process is effective for the direct fabrication of nanosized structures by taking advantage of its fine probe of 1–10 nm in diameter. The effect of a vacuum on the size and growth process of the fabricated nanodots containing W was investigated using an ultrahigh vacuum field emission transmission electron microscope (UHV-FE-TEM). The size of the dots can be controlled by changing the dose of electrons and the partial pressure of the precursor. The smallest particle size obtained was about 1.5 nm in diameter at 1.5×10^{-6} Pa pressure, a probe size of 1 nm, and with 5 s irradiation [7]. Furthermore, nanodots, nanorods, and freestanding square frames were also fabricated by the electron-beam-induced deposition in an ultrahigh vacuum field emission scanning electron microscope (UHV-FE-SEM) using the precursor gas of iron carbonyl gas, $Fe(CO)_5$. The typical line width and length of the iron freestanding square frames were 30–50 nm and about 300 nm, respectively [14]. The iron freestanding square frame structures were magnetized by introducing the specimen into a magnetic field of about 1 T under the objective lens of the transmission electron microscope. After that, the objective lens was turned off, and the residual magnetic field was estimated to be less than 100 mT. The application of the magnetic field to iron three-dimensional nanostructures produced an electron hologram from the nanostructures. It was found that the magnetic field leaked from the nanostructure body, which appeared to act as a "nanomagnet" because of the square and ring shapes.

When insulator materials were used as a substrate, inhomogeneous deposition occurred because the charges accumulated on the insulator substrate upon electron irradiation during electron-beam-induced deposition. It has been reported that carbon dendrite-like or filament-like structures were grown on insulator substrates under electron beam irradiation in an electron microscope at a relatively poor vacuum condition [15]. The charging up of the surface of the insulator substrate due to electron beam irradiation and the high concentration of hydrocarbon molecules in the vacuum chamber are considered to be the formed reasons. By using insulator substrates such as Al_2O_3 [16] and SiO_2 [17], characteristic morphologies of nanostructures, such as arrays of nanowhiskers (or nanowires), arrays of nanodendrites, and fractal-like nanotree structures, were fabricated in transmission electron microscopes (TEMs) by the electron-beam-induced deposition. The typical size of the diameter of a nanowhisker, the tip of a nanodendrite, and the tip of a nanotree was about 3 nm and was almost completely independent of the size of the electron beam. The nanostructures with different morphologies can be obtained by controlling the intensity of the electron beam during electron-beam-induced deposition. The process of growth of the nanostructures is based on mechanisms in which the formation of the nanostructures is related to the nanoscale unevenness of the charge distribution on the surface of the substrate, the movement of the charges to the convex surface of the substrate, and the accumulation of charges at the tip of the grown nanostructure. Furthermore, novel composite nanostructures of Pt nanoparticle/W nanodendrite or Au nanoparticle/W nanodendrite were also fabricated by the decoration of W nanodendrites with metallic elements [18]. The nanostructures have advanced features such as a large specific surface area, a freestanding structure on substrates, a typical size of several nanometers, and high purity. Because there are many possible combinations for the precursor and substrate, it is expected that various nanostructures may be fabricated on various substrates. Therefore, the nanostructures and their fabrication process may be applied in technology to realize various functional nanomaterials such as catalysts, sensor materials, and emitters.

Instead of using thin films of insulator materials, an anodic porous alumina membrane with the ordered nanohole arrays was used as a substrate; ordered array of nanostructures can be fabricated by the selective deposition of the precursor molecules into the ordered nanohole arrays using the electron-beam-induced deposition process. As an example, a porous alumina membrane on an aluminum substrate was prepared by a two-step anodization process, and then the aluminum was removed by etching. The diameter of the holes was about 40 nm, and the interval between the centers of the holes was about 100 nm. The total thickness of the membrane was about 1,000 nm, but there were electron transparent regions at the bottoms of the holes (the barrier-layer thickness of

the membrane was about 50 nm [19]). Tungsten hexacarbonyl, $W(CO)_6$, was used as a precursor gas source. The gas was supplied through a nozzle placed near the membrane on the side of the pores. The electron beam simultaneously irradiated the membrane from its reverse side. The deposition was carried out at room temperature. The diameter of electron beam during irradiation was about 1,900 nm. The irradiation period was 1 min. It was found that the size of the formed pattern was almost the same as that of the electron-beam-irradiated area. Also, the deposition was only conducted in the pores of the electron-beam-irradiated area. Almost all of the pores in the electron-beam-irradiated area were filled with the deposits, while there was no deposit on the places other than the pores. The as-deposited nanoparticles contained many nanocrystallites. The nanocrystallites were identified to be the equilibrium phase of bcc structure metal W at room temperature [20].

Future Directions for Research

The further development of electron-beam-induced deposition requires the application of state-of-the-art electron microscopy technology such as precise positioning of a nanometer-scale electron beam, a high current for the achievement of small structures with a much higher fabrication speed, and an ultrahigh vacuum environment for eliminating contamination of the deposits.

There are many possible combinations for precursor and substrate; therefore, various functional nanostructures or compound nanostructures can be produced by electron-beam-induced deposition. However, the growth conditions for each combination of precursor and substrate should be different, and it should be further investigated. The fundamental processes involved in the interaction of an electron beam with substrates, the adsorption of precursor molecules on a substrate, and the decomposition of molecules by energetic electrons during electron-beam-induced deposition should also be further clarified.

Cross-References

▶ Atomic Layer Deposition
▶ Chemical Vapor Deposition (CVD)

▶ Electron Beam Lithography (EBL)

▶ Focused-Ion-Beam Chemical-Vapor-Deposition (FIB-CVD)

▶ Self-assembly

▶ Self-assembly of Nanostructures

References

1. Broers, A.N., Molzen, W.W., Cuomo, J.J., Wittels, N.D.: Electron-beam fabrication of 80-Å metal structures. Appl. Phys. Lett. **29**, 596–598 (1976)

2. Koops, H.W.P., Kretz, J., Rudolph, M., Weber, M., Dahm, G., Lee, K.L.: Characterization and application of materials grown by electron-beam-induced deposition. Jpn. J. Appl. Phys. (Part 1) **33**, 7099–7107 (1994)

3. Koops, H.W.P., Schössler, C., Kaya, A., Weber, M.: Conductive dots, wires, and supertips for field electron emitters produced by electron-beam induced deposition on samples having increased temperature. J. Vac. Sci. Technol. B **14**, 4105–4109 (1996)

4. Kohlmann-von Platen, K.T., Chlebek, J., Weiss, M., Reimer, K., Oertel, H., Brünger, W.H.: Resolution limits in electron-beam induced tungsten deposition. J. Vac. Sci. Technol. B **11**, 2219–2223 (1993)

5. Silvis-Cividjian, N., Hagen, C.W., Leunissen, L.H.A., Kruit, P.: The role of secondary electrons in electron-beam-induced-deposition spatial resolution. Microelectron. Eng. **61–62**, 693–699 (2002)

6. Mitsuishi, K., Shimojo, M., Han, M., Furuya, K.: Electron-beam-induced deposition using a subnanometer-sized probe of high-energy electrons. Appl. Phys. Lett. **83**, 2064–2066 (2003)

7. Tanaka, M., Shimojo, M., Mitsuishi, K., Furuya, K.: The size dependence of the nano-dots formed by electron-beam-induced deposition on the partial pressure of the precursor. Appl. Phys. A. **78**, 543–546 (2004)

8. Scheuer, V., Koops, H., Tschudi, T.: Electron beam decomposition of carbonyls on silicon. Microelectron. Eng. **5**, 423–430 (1986)

9. Koops, H.W.P., Weiel, R., Kern, D.P., Baum, T.H.: High-resolution electron-beam induced deposition. J. Vac. Sci. Technol. B **6**, 477–481 (1988)

10. Petzold, H.C., Heard, P.J.: Ion-induced deposition for x-ray mask repair: rate optimization using a time-dependent model. J. Vac. Sci. Technol. B **9**, 2664–2669 (1991)

11. Furuya, K.: Nanofabrication by advanced electron microscopy using intense and focused beam. Sci. Technol. Adv. Mater. **9**, 014110 (2008)

12. Song, M., Furuya, K.: Fabrication and characterization of nanostructures on insulator substrates by electron-beam-induced deposition. Sci. Technol. Adv. Mater. **9**, 023002 (2008)

13. Dong, L., Arai, F., Fukuda, T.: Electron-beam-induced deposition with carbon nanotube emitters. Appl. Phys. Lett. **81**, 1919–1921 (2002)

14. Takeguchi, M., Shimojo, M., Furuya, K.: Fabrication of alpha-iron and iron carbide nanostructures by electron-beam induced chemical vapor deposition and postdeposition heat treatment. Jpn. J. Appl. Phys. **44**, 5631–5634 (2005)

15. Banhart, F.: Laplacian growth of amorphous carbon filaments in a non-diffusion-limited experiment. Phys. Rev. E **52**, 5156–5160 (1995)

16. Song, M., Mitsuishi, K., Tanaka, M., Takeguchi, M., Shimojo, M., Furuya, K.: Fabrication of self-standing nanowires, nanodendrites, and nanofractal-like trees on insulator substrates with an electron-beam-induced deposition. Appl. Phys. A **80**, 1431–1436 (2005)

17. Xie, G.Q., Song, M., Mitsuishi, K., Furuya, K.: Growth of tungsten nanodendrites on SiO$_2$ substrate using electron-beam-induced deposition. J. Nanosci. Nanotech. **5**, 615–619 (2005)

18. Xie, G.Q., Song, M., Furuya, K., Louzguine, D.V., Inoue, A.: Compound nanostructures formed by metal nanoparticles dispersed on nanodendrites grown on insulator substrates. Appl. Phys. Lett. **88**, 263120 (2006)

19. Xie, G.Q., Song, M., Mitsuishi, K., Furuya, K.: Selective tungsten deposition into ordered nanohole arrays of anodic porous alumina by electron-beam-induced deposition. Appl. Phys. A **79**, 1843–1846 (2004)

20. Xie, G.Q., Song, M., Mitsuishi, K., Furuya, K.: Fabrication of ordered array of tungsten nanoparticles on anodic porous alumina by electron-beam-induced selective deposition. J. Vac. Sci. Technol. B **22**, 2589–2593 (2004)

Electronic Contact Testing Cards

▶ MEMS High-Density Probe Cards

Electronic Expression

▶ Dynamic Clamp

Electronic Pharmacology

▶ Dynamic Clamp

Electronic Structure Calculations

▶ Ab Initio DFT Simulations of Nanostructures

Electronic Transport in Carbon Nanomaterials

Alejandro Lopez-Bezanilla[1], Stephan Roche[2,3],
Eduardo Cruz-Silva[4], Bobby G. Sumpter[5] and
Vincent Meunier[6]
[1]National Center for Computational Sciences, Oak
Ridge National Laboratory, Oak Ridge, TN, USA
[2]CIN2 (ICN–CSIC), Catalan Institute of
Nanotechnology, Universidad Autónoma de
Barcelona, Bellaterra (Barcelona), Spain
[3]Institució Catalana de Recerca i Estudis Avançats
(ICREA), Barcelona, Spain
[4]Department of Polymer Science and Engineering,
University of Massachusetts Amherst, Amherst,
MA, USA
[5]Computer Science and Mathematics Division and
Center for Nanophase Materials Sciences, Oak Ridge
National Laboratory, Oak Ridge, TN, USA
[6]Department of Physics, Applied Physics, and
Astronomy, Rensselaer Polytechnic Institute, Troy,
NY, USA

Synonyms

Charge transport in carbon-based nanoscaled
materials; Electron transport in carbon nanotubes and
graphene nanoribbons

Definition

Electronic transport is the transport of a charge carrier,
either an electron or a hole, through a system that is
coupled to two or more electrodes which serve as
a source and sink for the charge carriers.

Single-walled carbon nanomaterials like carbon
nanotubes and graphene nanoribbons are, respectively,
tubular and planar arrangements of a one-atom-thick
carbon sheet which possess a quasi one-dimensional
(1D) geometry.

Introduction

As predicted by the miniaturization roadmap of the
microelectronic industry, the gate length in metal-
oxide-silicon-based field effect transistors
(MOSFETs) has now been downscaled to a few tens
of nanometers. This length reduction allows for the
massive integration of billions of interconnected transistor devices, while preserving impressive current–
voltage characteristics of individual components [1].
However, despite the sustained development of
silicon-based nanoelectronics, the search for alternative materials for electronics has motivated the scientific community toward exploring the potential of
molecular, organic, and carbon-based electronics
such as carbon nanotubes or graphene [2]. The integration of carbon nanotubes (CNTs) and graphene
nanoribbons (GNRs) as electron channels in electronic
devices has become the focus of significant research
and is the topic of this work. It should be stressed at the
onset that, to date, carbon-based nanodevices have not
been found to outperform silicon devices in mainstream microelectronics. Therefore, other types of
applications need to be targeted. Among the wealth
of possibilities, (bio)sensing at the nanoscale for medical, optical switches, photovoltaic cells, or nanoelectromechanical devices (NEMS) constitute some of the
most promising alternatives [3, 4].

In spite of their promise, the use of such devices will
remain elusive without a precise understanding of the
intrinsic transport properties of carbon-based materials
[5, 6]. Here, the advances made in the use of controlled
chemical doping to fine-tune the transport properties of
the pristine material as a particularly promising
method of fundamental research are reviewed. Both
experimental and theoretical studies have considered
external agents like adsorbed atoms and molecules,
substitutional impurities, and functional groups as
possible candidates to introduce new functionalities
in pristine carbon nanosystems. For instance,
electroactive systems such as π-conjugated polymers
display electronic and optical properties that can be
tuned by appropriate molecular design. These systems
also adopt several conformations with varying physical properties when molecular groups are subjected to
oxidation or protonation through doping reactions.
These polymers can also form molecular composites
with other electroactive components of different
chemical composition. Interesting properties emerge
from the combination of CNTs and GNRs unique
properties with those of the polymers themselves and
open up many opportunities for potential applications,
such as data storage media, photovoltaic cells,

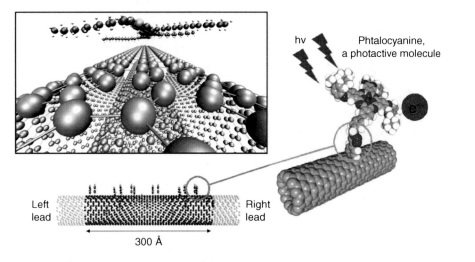

Electronic Transport in Carbon Nanomaterials, Fig. 1 Physisorption (*top left*) and chemisorption (*bottom*) of organic molecules on carbon nanotubes. A phthalocyanine (*bottom*) is attached to a carbon nanotube by using a chain of phenyl groups that serve as linkers. This chemical attachment allows the nanotube to evacuate the electrons that result from photoactivation of the phthalocyanine. Modeling the degradation of electron transport abilities of the nanotube upon grafting of the phenyl groups is a fundamentally important task. In contrast, physisorbtion (*top left*) represents a noninvasive type of functionalization. The organic molecule(s) sit(s) in the proximity of the nanotube such that chemical bonds are not established and the electronic structure is only partially affected

etc. π-conjugated polymers are also known to strongly interact with a carbon nanotube surface through π-stacking interactions to form supramolecular adducts (Fig. 1a). This type of noncovalent functionalization allows for nanotube modification without the introduction of defects in the original geometry. At the same time, this procedure facilitates the molecular mixing of both composite components [7].

Conjugated π-networks decorate the surface of quasi-one-dimensional structures of CNTs and GNRs and confer to them either semiconducting or metallic properties, in addition to quasi-ballistic transport on micrometer length scales. Some external agents that do not necessarily modify the system geometry can alter the electronic transport features. For example, hole and electron doping chemistry does not significantly modify the position of individual carbon atoms but can create strongly localized states, electrostatic fields, and conduction gaps or localization phenomena. Conversely, the chemical modification of nanotubes or nanoribbons by covalent attachment of functional groups causes simultaneous geometrical and electronic structure distortions. The alteration of the surface of the carbon nanostructures perturbs the crystal periodicity and modifies the electronic band structure and, in turn, the native quasi-ballistic nature of the electron transport properties [8, 9].

The experimental achievements in the field of nanotube functionalization have been spectacular. For instance, the "Click Chemistry" [10] approach succeeded in functionalizing and characterizing CNTs with zinc-phthalocyanine, with the aim of preparing a nanotube-based scaffold, which would make it easier to fabricate more complex functional structures. The attachment of phthalocyanines to nanotubes and fullerenes has emerged as a superb approach to carbon nanostructure phthalocyanine-based photovoltaic applications and other electronic devices [11]. Phthalocyanines are planar and electron-rich macrocycles characterized by remarkably high extinction coefficients in the red/near-infrared region (which is an important part of the solar radiation spectrum) and outstanding photostability and singular physical properties. These features make them exceptional building blocks for their incorporation in photovoltaic devices. Theoretical and computational studies support the understanding of how molecular states of phthalocyanines or other photoactive species mix with carbon nanotubes extended states (Fig. 1b).

Numerous questions regarding chemical functionalization still remain unsolved, especially since the goal is to develop novel applications. For instance, how does the interaction between nanotube and adsorbate affect their physical properties? What grafting process should be used to minimize the modification of the intrinsic and desirable electron transport capabilities of pristine nanotubes? In this regard, theoretical advances have allowed tremendous progress toward the understanding of some of the physical and chemical issues related to the modification of the electronic structure of carbon-based systems. However, the complexity of the calculations involving modified structures does not always permit a systematic comparison of theoretical modeling at size and timescales relevant to experiments. This complexity arises from the need to include a large number of atoms in the computer simulation, as well as from the accuracy required to treat systems in a realistic manner. Notwithstanding, DFT-based approaches provide a set of powerful tools for the quantum mechanical investigations of the physical and chemical properties of nanosystems. The self-consistent calculation of the electronic structure of any material is critical whenever charge transfer between two interacting systems plays a major role in the interaction. Examples of this effect include chemical doping [13] or structural alterations introduced by topological defects that force local change in aromaticity [14]. Variations of the charge distribution leads to a number of effects, such as the appearance of local polarization and modification of the local geometry. For instance, when a functional group such as an aryl ring is positioned close to the surface of a CNT or a GNR, a covalent bond between two carbon atoms of both objects is established [15]. The covalent bond corresponds to the archetypical sharing of electrons between the two moieties and creates a dipolar moment that polarizes the charge distribution in the surrounding of the new carbon–carbon σ-bond between the individual molecules. The polarized charge distribution gives rise to an electrical field that perturbs the charge distribution that initially caused the polarization, and so forth in a self-consistent cycle.

The present entry primarily focuses on some recent works toward understanding the surface modification of carbon nanotubes and graphene nanoribbons by chemical doping such as adsorbed functional groups and substitutional doping. Depending on the chemical interaction, the transport properties of the carbon materials can be significantly altered. All the results shown here are based on DFT modeling, as implemented in the SIESTA [16, 17] and NWChem [18] codes. The localized orbital basis set, which constitutes one of the essential features of both codes, along with their

reliable and fast calculations, are the main advantages of this approach. The tight-binding-like Hamiltonian obtained from the self-consistent calculation of an electronic structure presents the compactness required to study the transport property within the Green's function formalism.

In order to set the theoretical background to this entry, the principal concepts underlying the Landauer-Büttiker [19–21] theory of coherent electronic transport in low-dimensional systems are presented in section "The Landauer-Büttiker Quantum Transport Theory," along with a detailed description of the decimation technique in section "Green Function," which is particularly well suited for the calculation of long disordered systems. In section "Quantum Transport in Nanostructured Carbon Materials," a study of metallic and semiconducting functionalized carbon nanotubes by exohedral interactions with carbene and phenyl groups, endohedral amphoteric doping, and substitutional doping with N, P, B, S, and heteroatomics such as N–P is presented. In section "Summary and Conclusions," a summary and some conclusions are provided.

The Landauer-Büttiker Quantum Transport Theory

First, the basic ideas of the transport formalism are explored. The electronic transport calculations that will be presented are based on the Landauer-Büttiker formulation of conductance, which is particularly adequate for the study of electron motion along a device channel connected to two semi-infinite leads. The Landauer-Büttikker approach considers the transport properties of a phase coherent system formed by scattering region and leads, providing a simple and powerful framework for quantum transport in 1D mesoscopic systems. A device is defined as a material region where charge carriers spend a certain amount of time during their propagation. The electronic wave packets propagate quantum mechanically along a channel characterized by a cross-section and a longitudinal length. The channel is connected to left and right 1D semi-infinite leads that are in thermodynamical equilibrium with infinitely larger electron reservoirs. The Landauer-Büttiker approach is based on the classical concept of scattering: an incoming electron has a certain probability to be backscattered or transmitted when it crosses the device due to impurities. Material impurities are seen by charges as attractive or repulsive potential barriers and the mutual interaction leads to various material conductance regimes. This transport model treats electron interferences from a quantum mechanical point of view corresponding to the coherent dynamics of the electron.

In 1957, Rolf Landauer introduced the revolutionary concept that contacts play an important role in transport phenomena, and should therefore be properly treated. In addition to the molecule or device resistance, the intrinsic resistance of the whole system is determined by the coupling of the channel with the electrodes. Landauer's formulation for the quantum mechanical problem of electronic transport consists in establishing a one-to-one correspondence between the conductance of a 1D conductor and the probability of an electron to transmit without reflection.

Landauer's model deals with strictly 1D systems, the conductance of which can be expressed by:

$$G = \frac{e^2}{h} \frac{T}{R},$$

where e is the electronic charge, h is the Planck constant, and T and R are the transmission and reflection probabilities, respectively. This equation has an apparent algebraic limitation: when $T = 1$ (i.e., perfect conductor) the conductance of the device becomes infinite. This result caused controversy at the time of Landauer's seminal publication since it is clear that no channel has an infinite conductance. Büttiker's work put the controversy to rest by considering the contribution of all the electronic channels of transmission. The resulting Landauer-Büttiker approach states that, in a two-probe system, the conductance per channel is

$$G_n = \frac{e^2}{h} T_n,$$

where T_n is the transmission coefficient for the nth channel. In this formula, several assumptions are made: there is no electron–electron interaction and each charge propagates coherently through the conductor with only possible elastic scattering events, since all the inelastic processes take place in the reservoirs. The final expression that takes into account all channel contributions, in addition to spin degeneracy therefore reads

$$G = G_0 \sum_{n=1}^{N} T_n,$$

where $G_0 = \frac{2e^2}{h}$ is the quantum of conductance and the sum runs over all the different conducting channels at a given energy. This formula is usually known as the two-probe conductance formula. This description differs from Landauer's original formula in the conductance of a perfect 1D conductor is equal to G_0, which corresponds to maximum value of conductance. Reciprocally:

$$R = \frac{1}{G_0} = \frac{h}{2e^2} = 12.906 \text{ k}\Omega$$

is the quantum of resistance. It is striking that this theory predicts that a perfect conductor has a minimum resistance larger than zero. The actual experimental verification of this theory devised in the 1950s was only achieved in the late 1980s by van Wees et al. [22] and Wharam et al. [23] who unambiguously demonstrated the finite minimum resistance of each individual channel of conductance.

Green Function

In this section, a primer on the use of Green functions in transport calculations is provided. In this work, electronic transport of single-particle problems are solved, the energetic of which is described by a density functional Hamiltonian, H, used in Schrödinger equation as:

$$H\Psi_i = \varepsilon_i \Psi_i \Rightarrow \left[-\frac{1}{2}\nabla^2 + V(r) - \varepsilon_i \right] \Psi_i(r) = 0,$$

where V(r) is a potential and ε_i are the eigenvalues associated to the eigenfunctions $\Psi_i(r)$. The wavefunctions can be expressed by an expansion in a complete set of basis functions $\{\varphi_v(r)\}$:

$$\Psi_i(r) = \sum_{v=1}^{N} c_{iv} \varphi_v(r).$$

Truncating the number of basis functions leads to an operational approximation. A sufficient large number of basis functions can reduce the imprecision.

The Hamiltonian matrix elements of H in the $\{\varphi_v(r)\}$ basis are defined as:

$$H_{vv'} \equiv \int d3r\varphi_v^*(r)\left[-\frac{1}{2}\nabla^2 + V(r)\right]\varphi_{v'}(r)$$

and the overlap matrix element between two functions is defined as:

$$S_{vv'} \equiv \int d3r\varphi_v^*(r)\varphi_{v'}^*(r)$$

Inserting expression (2) into (1) a generalized eigenvalue problem is obtained:

$$\sum_{v'} H_{vv'}c_{iv'} = \varepsilon_i \sum_{v'} S_{vv'}c_{iv'} \quad \forall v.$$

From this expression, the Green's function is defined as:

$$\sum_{v''}(\varepsilon S_{vv''} - H_{vv'})G_{v''v'}(\varepsilon) \equiv \delta_{vv'}.$$

Within a tight-binding framework, the basis set can be formed by a set of atomic orbitals centered on each atom of the system that confers a compact form to the matrix. Inverting the matrix for a given value of ε is, in principle, the only calculation needed to determine the Green's function of the electronic system. In this work, the Hamiltonian and overlap matrices are provided by a DFT calculation of a system composed of hundred of atoms. Since the SIESTA and NWChem codes employ a basis set of atomic-like orbitals, the Hamiltonian matrix that result from a simulation meets the operational requirements of size and compactness to determine Green's function. Further, an appropriate choice of basis set size leads to a relatively compact set of matrices within the linear combination of atomic orbital model.

In a generalized consideration of the transport problem of a two-terminal geometry, the finite device is sandwiched between two semi-infinite electrodes, and the Hamiltonian of the entire system (i.e., an infinite size Hamiltonian) can be written as:

$$H = \begin{bmatrix} H_L & -\gamma_L & 0 \\ -\tilde{\gamma}_L & H_0 & -\tilde{\gamma}_R \\ 0 & -\gamma_R & H_R \end{bmatrix}$$

where H_0 is the device Hamiltonian, $H_{L,(R)}$ are the bulk left (right) electrode Hamiltonians, and $\gamma_{L,R}$ represent the tight-binding-like coupling between the electrodes and the device channel. The Green's function operator of the system can be expressed as:

$$[\varepsilon \cdot S - H]G(E) = I,$$

where I is the identity matrix and ε is the complex energy that imposes causality and is chosen to avoid the real axis poles of the Green's function by extending its domain to the complex plane. Depending on the positive or negative semi-infinite complex plane definition of ε, the Green's function can be labeled as $G_0^{r(a)}(\varepsilon)$, that is, the retarded (advanced) Green functions of the device channel (fulfilling the relation $G_0^r(\varepsilon) = \tilde{G}_0^a(\varepsilon)$) where $\sim$ denotes the complex conjugated. Physically, these two Green functions represent an incoming and an outgoing wave formed upon the same excitation. It is worth noting that, while the contact function subspaces are semi-infinite, the device subspace has a finite representation. To obtain a finite matrix representation of the electrodes, it is useful to resort to the self-energy representation $\Sigma_{L,R}$ of both the left (L) and the right (R) electrodes:

$$\Sigma_L = \tilde{\gamma}_L g_L \gamma_L$$

$$\Sigma_R = \tilde{\gamma}_R g_R \gamma_R$$

where

$$g_L = [E \cdot S - H_L]^{-1}$$

$$g_R = [\varepsilon \cdot S - H_R]^{-1}$$

are the (surface) Green functions of the left and right contact, respectively. The transformation of these two infinite matrices into two (N,N) matrices to obtain a finite expression is usually performed by taking advantage of the semi-infinite lattice symmetry of the two leads, either by analytical or iterative techniques. Once the self-energy terms are calculated, one can obtain the final expression of the Green's function matrix of the whole device as:

$$G_0 = [\varepsilon \cdot S - H_0 - \Sigma_L - \Sigma_R]^{-1}.$$

From this expression, the self-energies can be interpreted as additive potential terms of the device channel Hamiltonian whose final expression represents an effective Hamiltonian that accounts for the presence of the electrodes. The numerical representation within the chosen basis now corresponds to a matrix of finite rank. This is an undeniable advantage. Note that this standard procedure for the calculation of Green's function of a two-terminal system can be easily extended for multiterminal geometries, since each electrode can be included by a proper self-energy added to the expression of G_0.

Once the Green's function of the device channel is determined, the Caroli formula [24] is applied to calculate the transmission coefficients. In its compact form, the transmission between the right and left electrodes $T(\varepsilon)$ is:

$$T(\varepsilon) = tr\left[\Gamma_R(\varepsilon)G_0^r(\varepsilon)\Gamma_L(\varepsilon)G_0^a(\varepsilon)\right],$$

where the contact terms, Γ_L and Γ_R for left (L) and right (R) electrodes are:

$$\Gamma_{R(L)}(\varepsilon) = i\left[\Sigma_{R(L)}^r(\varepsilon) - \Sigma_{R(L)}^a(\varepsilon)\right].$$

This expression for $T(\varepsilon)$ is equivalent to that introduced in section "The Landauer-Büttiker Quantum Transport Theory" for the transmission coefficients.

Quantum Transport in Nanostructured Carbon Materials

Chemical Functionalization of Graphene Nanoribbons

Graphene is a zero-gap semiconductor and is therefore unsuitable for achieving efficient field-effect functionality with an acceptable ON/OFF states current ratio. One possibility to increase the zero-gap of two-dimensional (2D) graphene single layers is to shrink its lateral dimension. This dimensional confinement effectively opens an electronic bandgap that decreases in magnitude when the nanoribbon width increases. In spite of this, theoretical predictions and experimental observations have reported energy band-gaps far too small or very unstable in regard to edge reconstruction and defects, thus preventing envisioning these

graphene-based devices from achieving improved performance relative to the ultimate CMOS-FETs.

To further introduce suitable band-gaps, one can instead recourse to larger width graphene nanoribbons (above 10 nm in lateral sizes) and compensate the loss due to the bandgap shrinking by triggering mobility gaps through doping or surface modification. Mobility gaps are direct consequences of a wide distribution of quasi-bound states over the valence band (for acceptor-type impurities) or the conduction band (for donor-type impurities) in the first conductance plateau when dopants are randomly distributed across the ribbon width because of the strong dependence of the scattering potential with the dopant position in respect to the ribbon edges. The electronic transport properties of GNRs modified by means of substitutional doping with atoms such as boron, nitrogen, or phosphorus and chemical attachment of functional groups have been simulated and has shown promising features [12].

Graphene is an aromatic material: its electronic ground state corresponds to completely delocalized π-electrons. It follows that all conjugated chemical bonds are equivalent. This symmetry is broken when a functional group is grafted onto graphene's surface since it removes a π-bond, thereby inducing an sp^2 to sp^3 rehybridization and a net magnetic moment [25]. However, this situation is not stable and as soon as a second atom is chemisorbed on graphene's basal plane, the unpaired electron participates in a second sp^3 hybridization and the magnetic moment vanishes. This formation of radicals associated with the removal of one p_z orbital in the conjugated graphene network also takes place in nanotube networks. Chemical functionalization of graphene-based materials is a promising strategy to reversibly tune the electronic properties without aggressively reducing the ribbon width. For instance, pairs of hydrogen (H–) and hydroxyl (OH–) functional groups attached on GNRs at opposite positions in a hexagonal graphene ring introduce the simultaneous appearance of two bonds with different charge densities in the system. The oxygen atom of the OH– group is a good acceptor of electrons while H– adopts a donor behavior when it is located in the vicinity of a C atom. The two equivalent bonds can also be treated by considering two phenyl rings covalently attached at opposite sides of a hexagonal ring.

Several chemical groups can be simultaneously attached, which allows for a wide range of

Electronic Transport in Carbon Nanomaterials, Fig. 2 Conductance of narrow GNRs. Conductance of a17GNR as a function of defect position. The locations of the defects for (**a–d**) are illustrated at the *bottom*. The (**d**) case shows the conductance in the case of two pairs of functional groups

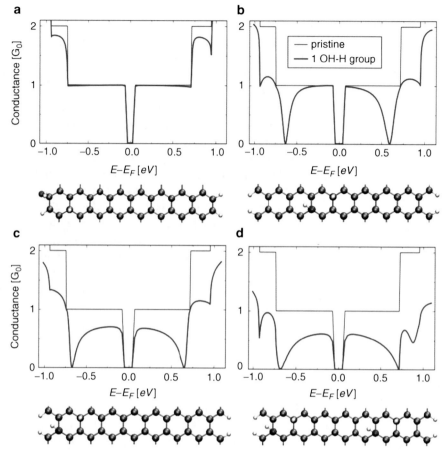

conductance and structural modulations depending on the relative positions of the addends. The particular configuration of a nanoribbon with functionalization at the edge is also of special interest. In this case, no localized states are introduced by the functional groups at energies close to the Fermi level and, therefore, the conductance ability of the ribbon is not significantly affected.

As explained in section "Green Function," a complete device can be set up once ab initio calculations are performed to obtain the Hamiltonian and overlap matrices associated with short functionalized nanoribbon sections. To compute the conductance of clean and defected GNRs, the multichannel Landauer-Büttiker technique described in sections "The Landauer-Büttiker Quantum Transport Theory" and "Green Function" is used. The channel is defined between two perfect GNR-based right (left) leads, which impose chemical potentials μ_1 (μ_2). This choice of leads also optimizes the coupling with the channel.

First the properties of an armchair-edged nanoribbon with 17 bonds along its length (hereafter called a17GNR) are discussed. The dependence of the a17GNR conductance as a function of the H– and OH–group position is shown in Fig. 2 for a set of different configurations. If the C–OH bond is located at the GNR edge, no backscattering is observed in the first conductance plateau (Fig. 2a), although some weak conductance decay is seen for higher energy sub-bands. Edge functionalization does not alter the electronic properties of the aGNR in the first sub-band since the impurity states are located far away from the Fermi level. In contrast, when the C–OH bond is shifted out of the edge line, the conductance is more significantly reduced as shown in Fig. 2b or in Fig. 2c. Figure 2b shows similarity with the situation for pair of grafted phenyls, whereas other locations (Fig. 2c) exhibit a larger suppression of conductance in the first plateau, although higher energy sub-bands remain largely unaffected. This dependence of the

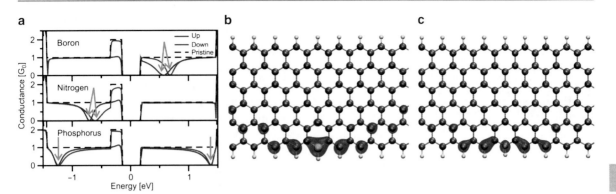

Electronic Transport in Carbon Nanomaterials, Fig. 3 (a) Electronic quantum conductance for a zigzag graphene nanoribbon (12zGNR) doped with boron, nitrogen, or phosphorus. *Green arrows* indicate the conductance dips caused by the localized states associated to the doping atoms. In panels (**b**) and (**c**) illustrate the charge distribution of such localized states for boron (**b**) and nitrogen (**c**). All energies are relative to the Fermi energy of pristine nanotube

conductance on the location of the OH/H pair can be rationalized by analyzing the variation of electrostatic screening effects that dictate the effective scattering potential. The conductance decay is driven by back-scattering on individual functional groups and by quantum interferences due to the proximity of the defects.

Zigzag-edged GNRs (zGNRs) display a number of different electronic features compared to aGNRs. As opposed to aGNRs, zGNRs present a magnetic ordering at the edges as a consequence of the strong localization of the unpaired π and π* electrons in the ground state. These edge states are spin-degenerate but both ribbon edge states exhibit different spin orientations as a result of edge–edge interactions, and hence the difference between ferromagnetic and antiferromagnetic spin states disappears gradually for wide enough ribbons. First-principles calculations show that zGNRs are metallic in the ferromagnetic configuration whereas they exhibit semiconducting behavior in the antiferromagnetic state. zGNRs' magnetic properties have interesting implications for the development of electronic and spintronic devices.

The case of GNR substitutionally doped with boron, nitrogen, and phosphorus atoms, which modify the electronic and transport properties of the ribbons, is now considered. The presence of a dopant typically induces localized states and a drop in conductance at some particular energies. The dependence of the conductance drop position in the transmission spectrum on the specific doping site was investigated within an ab initio approach. DFT is well suited to calculate the charge transfer between the doping atoms and the

GNRs and thus to find the exact position in energy of the bound states. Nitrogen and phosphorus atoms preferentially substitute at the edges of zigzag and armchair nanoribbons, whereas boron atoms preferentially substitute at the edges of zGNRs but next to the edge of aGNRs. Spin-dependent conductance in the most energetically favorable boron-doped and nitrogen-doped zGNRs leads to the possibility of utilizing such structures in spintronic devices. The most favorable configuration of phosphorus atoms does not present a significant degree of spin-dependent behavior, but it displays both a donor-like and an acceptor-like state. The conductance of doped armchair nanoribbon exhibits spin degeneracy for all cases, but in phosphorus-doped zigzag ribbons, spin-dependent behavior can emerge as the doping is placed at particular sites across the ribbon width (Fig. 3).

The large predicted variation of resonant energies with the dopant position with respect to the GNR edges [26] indicates that a random distribution of impurities along the ribbon surface should lead to a rather uniform reduction of the conductance over valence states on the first plateau of conductance. Turning to the quantum transport properties of boron doped GNRs up to 10 nm in width, first-principles calculations are combined with tight-binding models to investigate the effect of chemical doping on charge conduction for ribbons length up to 1 μm with randomly distributed substitutional boron impurities. Due to the large size of the corresponding unit cells, a statistical study of the mesoscopic transport using the fully ab initio Hamiltonian is not directly possible. Instead, a tight-binding model parameterized by accurate first-principles

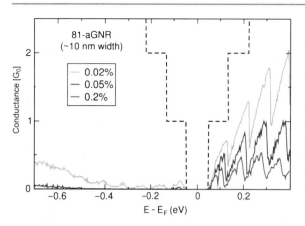

Electronic Transport in Carbon Nanomaterials, Fig. 4 Average conductance as a function of energy for a semiconducting 81-aGNR (*dashed line*) and three selected doping rates (0.02%, 0.05%, and 0.2%, from *top to bottom*)

calculations is used. In a first step, the scattering potential induced by the dopant is extracted from those obtained with DFT. The analysis of the ab initio onsite and hopping terms around the doping atom obtained from self-consistent DFT calculations within an atomic-like basis, allows a simple nearest-neighbor tight-binding model to be built. The substitutional impurity modifies mainly the onsite terms, creating a potential well on a typical length scale of about 10 Å.

Figure 4 shows the conductance of a 10 nm wide armchair nanoribbon with a low dose for boron doping. For a doping density of about 0.2%, the system presents a mobility gap of the order of 1 eV. When lowering the doping level to 0.05%, the mobility gap reduces to about 0.5 eV and finally becomes less than 0.1 eV for lower density. The 0.2% case is obtained for a fixed nanoribbon width and length, so that adjustments need to be performed for lateral or longitudinal size upscaling, but the method is straightforward once the transport length scales (mean free paths, localization length) have been computed [12]. One notes however that the existence of mobility gaps (with conductance several orders of magnitude lower than the quantum conductance) can hardly allow a straightforward quantitative estimation of resulting ON/OFF current ratio, since this would require computing the charge flow in a self-consistent manner (using a Schrödinger-Poisson solver). This turns out to be critical since accumulated charges inside the ribbon channel screen the impurity potential in an unpredictable way. These types of calculations would need to be performed using nonequilibrium Green

functions (NEGF) formalism and deserve further consideration [27].

To summarize the results presented thus far, it is clear that for any graphene-based application to be viable, it is critical to understand the effect of defects and disorder. In the following, junctions of nanoribbons based on an ordered array of structural defects [28] are treated. Such arrays of defects have been observed experimentally and could be visualized as grain boundaries in a graphene sheet [29]. In particular, an array of pentagons and heptagons can be used to join an armchair and a zigzag nanoribbon, as shown in Fig. 5. The spin resolved band structure for a (4,4) hybrid nanoribbon is shown in Fig. 5a. Note the presence of an energy gap of ~0.2 eV in the minority spin states. For ribbons with an even width, there is a gap in the spin down states, that is, the systems exhibit half-metallicity and these hybrid systems behave as a spin polarized conductor (Fig. 5b). The isosurface plots of the wave functions near the Fermi level are indicated by the arrows on Fig. 5a. Some of these bands (top) are closely related to the zigzag nanoribbon edge states although, unlike zGNRs, there is only one edge state yielding a net magnetic moment. Another interesting case arises when an armchair segment is sandwiched in between two zGNRs electrodes, as shown in Fig. 5c. In this case, the two different lattices can be joined while keeping the threefold connectivity of each inner carbon atoms by introducing a set of pentagon-heptagon defects. The results presented in Fig. 5d indicate that the conductance near the Fermi level is reduced as the armchair section is increased, denoting a tunneling driven transport across the junctions.

Chemical Functionalization in Carbon Nanotubes

Amphoteric doping: It has been demonstrated that DFT-based transport calculations can be used at the fundamental level to devise a novel nanoelectronic device by exploiting the interplay between the intrinsic properties of pristine nanotubes and those of chemical dopants [30]. Here, a bistability is built in to form a nanoswitch whose conductance state depends on the relative position of a guest molecule inside of a carbon nanotube (see Fig. 6). Depending on the mode of operation, the proposed device is capable of acting as an electrical switch or a nonvolatile memory element. The switching mechanism consists of a significant change in the flow of electrons in a circuit due to different orientations of a molecule

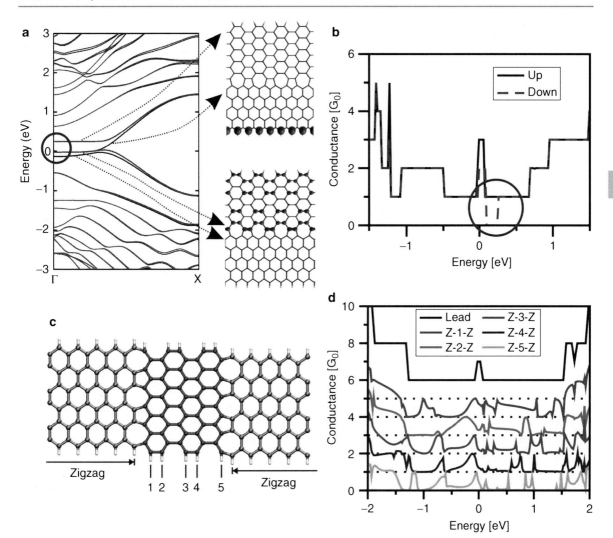

Electronic Transport in Carbon Nanomaterials, Fig. 5
(**a**) Computed spin resolved band structure for a (4,4) hybrid nanoribbon. The energy gap ($E_g = 0.27$ eV) on the minority spin (*red*) is indicated by a *blue circle*. The *arrows* show the wave function plot for the bands close to the Fermi level at the Γ-point. (**b**) Quantum conductance for a typical (4,4) hybrid graphene nanoribbon. The nanoribbon is spin degenerate in most of the energy spectrum, except for the region close to the Fermi level, where spin-polarized conductance is observed. (**c**) Example of a zigzag-armchair-zigzag device joined by 5–7 defects. (**d**) Quantum conductance for a set of these devices. It is observed that the conductance is reduced as the armchair section is increased, revealing a tunneling driven transport

relative to the host nanotube. As shown in Fig. 6, one stable position of the molecule yields a high current across the device ("ON" conducting state) while a change in orientation is associated with an important, measurable decrease of the transmission property ("OFF" nonconducting state). The information is therefore stored in the form of the orientation of the molecule (parallel for "OFF" and perpendicular for "ON"), which retains its position even when the external source of energy is switched off (nonvolatility). In addition, due to the intrinsic mechanical properties of the nanotube, the molecule's orientation can be reliably switched back and forth mechanically. The information storage mechanism remains valid for any molecule–metallic nanotube combination as long as it displays the desired properties of significant charge transfer, and the presence of at least two metastable molecule positions with respect to the nanotube core. In addition, these two positions, in order to present distinct transport properties, must be characterized by clearly different

Electronic Transport in Carbon Nanomaterials, Fig. 6 *Left panel*: total DFT energy of a "guest" F_4TCNQ molecule as a function of the angle between the main molecule axis and the "host" nanotube axis (inserts show the orientation). *Right panels*: conductance profile of the two different open systems where the active region is seamlessly connected to semi-infinite (10,10) carbon nanotubes

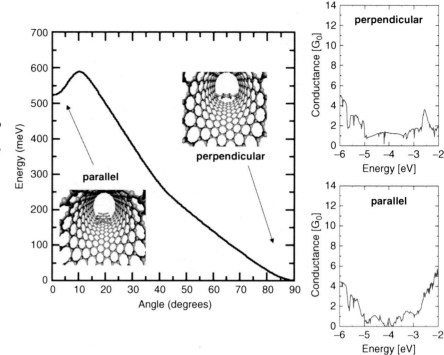

types of interaction (such as different amounts of charge transfer for instance). These are necessary conditions but they are not always simple to guarantee a priori, and computational exploration is needed to properly select the correct combination of systems.

In this regard, it was shown that computational approaches paired with density functional theory studies can be routinely used to identify relevant combinations of molecules and wires/CNTs for realistic applications. In particular, a class of devices where a noncovalently bound molecule that acts as a passive gate can be extended to a large number of molecule-nanotube pairs, giving great versatility in the choice of nanotube diameter and doping molecule. Specifically, for a given molecule, it is possible not only to identify a nanotube that displays the characteristics of a memory element device but also to develop a detailed understanding of the criteria governing the functioning of this device.

Nanonetworks: Carbon nanotubes can be combined to create 2D and 3D (three-dimensional) networks. These complex systems can be tailored from nanostructures as building blocks and are the foundations for constructing multifunctional nano- and microdevices. However, assembling nanostructures into ordered micronetworks remains a significant

challenge in nanotechnology. The most suitable building blocks for assembling such networks are nanoparticle clusters, nanotubes, and nanowires. Unfortunately, little is known regarding the different ways networks can be created and their physicochemical properties as a function of their architecture. It is expected that, when 1D nanostructures are connected covalently [31], the resulting assemblies will possess mechanical, electronic, and porosity properties that are strikingly different from those of the isolated 1D blocks. In extensive theoretical studies, it has been shown that the properties of 2D and 3D networks built from 1D units are dictated by the specific architecture of these arrays. Specifically, the hierarchy concept as a practical way to design complex nanostructures from basic nano-building blocks was introduced. It was also demonstrated that nanotubes could be joined to make super networks with new properties emerging from those of the individual building blocks (i.e., the nanotubes). In addition to the unusual mechanical and electronic properties, the porosity of these systems makes them good candidates for catalysts, sensors, filters, or molecular and energy storage properties. The crystalline 2D and 3D networks are also expected to present unusual optical properties, in particular when the pore periodicity approaches the wavelength of different light sources, such as optical photonic crystals. The power of theoretical calculations in predicting novel materials with enhanced electronic and mechanical properties, using single atoms as building blocks, has been demonstrated repeatedly. These systems were studied by examining how atomic level modification of the networks can be made to tailor the properties of the current flow, a key ingredient for smart and addressable nanoarrays. The electrical current could be efficiently guided in 2D nanotube networks by introducing specific topological defects within the periodic framework. Using semiempirical transport calculations coupled with the Landauer-Büttiker formalism of quantum transport in multiterminal nanoscale systems, a detailed analysis of the processes governing the atomic-scale design of nanotube circuits was possible. When defects are introduced as patches in specific sites, they act as reflecting centers that re-inject electrons along specific paths, via a wave reflection process (see Fig. 7 left panel). This type of defect can be incorporated while preserving the threefold connectivity of each carbon atom embedded within the graphitic lattice. These findings introduce a new way to explore bottom-up design, at the nanometer scale, of complex nanotube circuits that could be extended to 3D nanosystems and applied in the fabrication of nanoelectronic devices.

Using knowledge acquired during the nanonetwork characterization and the behavior of nanostructures relative to local changes in chemistry, it was shown that the actual realization of networks can be obtained by inserting minute amounts of sulfur in precursors used for nanotube formation. This method was found to lead to the creation of a large range of structures, including crossbars and multiterminal junctions (Fig. 7 right panel) [32]. It also clarified the key role of sulfur in nanotube growth. Most notably, it was demonstrated that sulfur not only promotes the C-H separation in the precursor but also triggers the emergence of negative and positive local curvature, which are essential in the development of multiterminal, covalent systems. Here, the dopant's main role is to create nanostructures whose properties emerge from those of the individual components. For instance, sulfur triggers branching and initiates network creation but does not play a major beneficial role in the electronic transport properties of the resulting object.

Heteroatom Doping: Substitutional doping of carbon nanotubes and graphene clearly offers a practical path to tailor their physical and chemical properties by creating new states that modify their electronic structure [33]. The presence of these states, as discussed above, originates from the different electronic configurations of the doping atoms. These modifications in the electronic structure and surface reactivity can help to achieve new improved materials. Nitrogen and boron are among the most studied substitutional dopants used in nanotube research but phosphorus and phosphorus-nitrogen heteroatomic doping of carbon nanotubes have been successfully realized in experiments by using thermolysis of mixtures of ferrocene and triphenyl phosphine dissolved in benzylamine and ethanol. In order to fully understand the structure and properties of these new types of doped nanotubes, extensive electronic structure and transport calculations were used.

DFT calculations revealed an electronic band structure displaying the presence of localized (P) and semilocalized (PN) states around the doping atoms. In contrast to nitrogen, these states are normal to the nanotube surface and do not modify the intrinsic nanotube metallicity, and therefore semiconducting nanotubes remain so regardless of the doping. However, these electronic states behave as scattering centers for carriers with energies close to the localized state energy. Electronic transport calculations on pristine, nitrogen, phosphorus, and P–N doped nanotubes clarified the different effects of the dopants on their conductance. The dips observed in the quantum conductance (Fig. 8a) correspond to the localized and semilocalized states located around the doping atoms (Fig. 8b). The calculations of the quantum conductance on semiconducting zigzag nantubes showed that phosphorus doping do not modify their intrinsic semiconducting behavior, as opposed to the effect observed for N-doped nanotubes. Phosphorus and PN doping in a (10,0) nanotube only creates bound and quasibound states around the phosphorus atom, which are dispersionless (flat bands in the band structure), and projected as sharp peaks in the density of states.

Due to the sp^3 hybridization, substitutional P atoms induce highly localized states which modify the chemical properties of the surface of carbon nanotubes. These sites have a strong affinity toward acceptor molecules. On the other hand, it was found that P–N co-dopants not only have a reduced affinity for acceptor molecules, but that the P–N bond can also accept charge, resulting in affinity toward donor molecules because of a partial positive

Electronic Transport in Carbon Nanomaterials,
Fig. 7 *Left panel*: An ordered carbon nanotube network where defects introduced at the intersection of the nanotubes act as electron scattering centers (insert). *Right panel*: some of the different types of carbon nanotube architectures possible by introducing small amounts of sulfur into the precursors for the synthesis

Electronic Transport in Carbon Nanomaterials,
Fig. 8 (**a**) Electronic quantum conductance of nitrogen, phosphorus, and phosphorus–nitrogen doped (6,6) SWCNTs, where the dips in conductance caused by scattering at the localized states can be observed.
(**b**) Real space representation of the localized wavefunctions for nitrogen and phosphorus doped nanotubes.
(**c**) Variations in the electronic conductance for a PN-doped nanotube after the physical/ chemical absorption of different molecular species at the phosphorus localized state.
(**d**) Variations of the charge density distribution around a NO_2 molecule absorbed at a PN site. *Black/white* clouds represent increase/decrease of the electronic charge density. All energies are relative to the Fermi energy of pristine nanotube

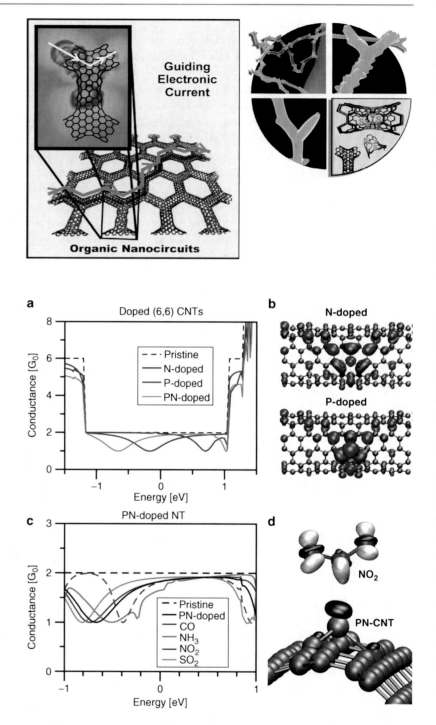

charge at the phosphorus atom. For example, CO and NH_3 were found to adsorb only on PN-doped nanotubes, O_2 was adsorbed only on P-doped nanotubes, while NO_2 and SO_2 were adsorbed on both P- and PN-doped nanotubes. A natural bond orbital (NBO) analysis showed that the P atom has a very different chemical

environment in P- and PN-doped nanotubes as a result of variations in charge and local bonding conditions.

The chemical changes at the localized electronic states affect its energy, resulting in different positions for dips in the conductance that are associated with them (Fig. 8c). SO_2 and CO cause only minor shifts in the conductance, while NH_3 causes the suppression of a conductance dip, and NO_2 and O_2 had similar effects in the conductance of P-doped carbon nanotubes. Since changes in the conductance at Fermi energy of about 10% can be detected experimentally and, in particular, since modifications of the conductance slope near the Fermi energy can be an alternative for experimental measurements, the results of this study demonstrated the potential for a molecular sensor with identifiable selectivity that is based on P- and PN-doped carbon nanotubes.

Summary and Conclusions

At the nanoscale a number of new phenomena resulting from quantum effects and confinement, begin to dominate the processes governing the electron flow across the active device. Understanding those quantum effects is critical in the quest to develop new device concepts or novel mechanisms tuned to desired functionality and performance. Given the inherent difficulty in the integration of nanoscale systems like carbon nanotubes and graphene nanoribbons (e.g., controlled manipulation without agglomeration), into precise arrangements for devices, and the complexity in the assessment of the properties of these devices, a tremendous effort has been devoted in past years toward the development of theoretical methods and computational tools that make it possible to design and test new systems for desired and tailored characteristics. Realistic treatment of the fundamental properties of novel devices at the quantum mechanical level is now becoming a more routine task mainly due to accurate approaches such as density functional theory, nonequilibrium Green function formalism for electron transport in open-systems, and the advent of larger and more powerful computers. Extensive applications of these approaches are beginning to uncover new behavior in cabon-based materials that are uniquely suited for nanoelectronic applications.

Cross-References

References

1. Narendra, S.G., Chandrakasan, A.P.: Leakage in Nanometer CMOS Technologies. Springer, New York (2006)
2. Avouris, P., Chen, Z., Perebeinos, V.: Carbon-based electronics. Nat. Nanotechnol. **2**(10), 605–615 (2007)
3. Allen, B.L., Kichambare, P.D., Star, A.: Carbon nanotube field-effect transistor-based biosensors. Adv. Mater. **19**(11), 1439–1451 (2007)
4. Vincent, D., Auvray, S., Borghetti, J., Chung, C.-L., Lefevre, R., Lopez-Bezanilla, A., Nguyen, K., Robert, G., Schmidt, G., Anghel, C., Chimot, N., Lyonnais, S., Streiff, S., Campidelli, S., Chenevier, P., Filoramo, A., Goffman, M.F., Goux-Capes, L., Latil, S., Blase, X., Triozon, F., Roche, S., Bourgoin, J.-P.: Carbon nanotube chemistry and assembly for electronic devices. C. R. Phys. **10**(4), 330–347 (2009)
5. Charlier, J.-C., Blase, X., Roche, S.: Electronic and transport properties of nanotubes. Rev. Mod. Phys. **79**(2), 677–732 (2007)
6. Castro Neto, A.H., Guinea, F., Peres, N.M.R., Novoselov, K.S., Geim, A.K.: The electronic properties of graphene. Rev. Mod. Phys. **81**(1), 109–162 (2009)
7. Sumpter, B.G., Meunier, V., Jiang, D.E.: New insight into carbon nanotube electronic structure selectivity. Small **4**, 2035 (2008); Linton, D., Driva, P., Sumpter, B., Inanov, I., Geohegan, D., Feigerle, C., Dadmun, M.D.: The importance of chain connectivity in the formation of non-covalent interactions between polymers and single-walled carbon nanotubes and its impact on dispersion. Soft Matter **6**, 2801 (2010)
8. Lopez-Bezanilla, A., Triozon, F., Latil, S., Blase, X., Roche, S.: Effect of the chemical functionalization on charge transport in carbon nanotubes at the mesoscopic scale. Nano Lett. **9**(3), 940–944 (2009)
9. Lopez-Bezanilla, A., Triozon, F., Roche, S.: Chemical functionalization effects on armchair graphene nanoribbon transport. Nano Lett. **9**(7), 2537–2541 (2009)
10. Kolb, H.C., Finn, M.G., Sharpless, K.B.: Click chemistry: diverse chemical function from a few good reactions. Angew. Chem. Ed. **40**(11), 2004+ (2001)
11. Campidelli, S., Ballesteros, B., Filoramo, A., Diaz, D., Torre, G., Torres, T., Aminur Rahman, G.M., Ehli, C., Kiessling, D., Werner, F., Sgobba, V., Guldi, D.M., Ciofi, C., Prato, M., Bourgoin, J.-P.: Facile decoration of functionalized single-wall carbon nanotubes with phthalocyanines via "click chemistry". J. Am. Chem. Soc. **130**(34), 11503–11509 (2008)
12. Biel, B., Triozon, F., Blase, X., Roche, S.: Nano Lett. **9**(7), 2725–2729 (2009)
13. Biel, B., Blase, X., Triozon, F., Roche, S.: Anomalous doping effects on charge transport in graphene nanoribbons. Phys. Rev. Lett. **102**(9), 096803 (2009)

14. Dubois, S.M.M., Lopez-Bezanilla, A., Cresti, A., Triozon, F., Biel, B., Charlier, J.-C., Roche, S.: Quantum transport in graphene nanoribbons: effects of edge reconstruction and chemical reactivity. ACS Nano **4**(4), 1971–1976 (2010); Valiev, M., Bylaska, E.J., Govind, N., Kowalski, K., Straatsma, T.P., van Dam, H.J.J., Wang, D., Nieplocha, J., Apra, E., Windus, T.L., de Jong, W.A.: NWChem: a comprehensive and scalable open-source solution for large scale molecular simulations. Comput. Phys. Commun. **181**, 1477 (2010)

15. Jiang, D.E., Sumpter, B.G., Dai, S.: How do aryl groups attach to a graphene sheet? J. Phys. Chem. B **110**, 23628 (2006)

16. Ordejon, P., Artacho, E., Soler, J.M.: Self-consistent order-N density-functional calculations for very large systems. Phy. Rev. B **53**(16), 10441–10444 (1996)

17. Soler, J.M., Artacho, E., Gale, J.D., Garcia, A., Junquera, J., Ordejon, P., Sanchez-Portal, D.: The SIESTA method for ab initio order-N materials simulation. J. Phys. Condens. Matter **14**(11), 2745–2779 (2002)

18. Valiev, M., Bylaska, E.J., Govind, N., Kowalski, K., Straatsma, T.P., van Dam, H.J.J., Wang, D., Nieplocha, J., Apra, E., Windus, T.L., de Jong, W.A.: NWChem: a comprehensive and scalable open-source solution for large scale molecular simulations. Comput. Phys. Commun. **181**, 1477 (2010)

19. Landauer, R.: IBM J. Res. Dev. **32**, 306 (1988)

20. Büttiker, M.: IBM J. Res. Dev. **32**, 317 (1988)

21. Büttiker, M., Imry, Y., Landauer, R., Pinhas, S.: Generalized many channel conductance formula with application to small rings. Phys. Rev. B **31**(10), 6207–6215 (1985)

22. van Wees, B.J., et al.: Phys. Rev. Lett. **60**, 848–850 (1988)

23. Wharam, D.A., et al.: J. Phys. C: Solid State Phys. **21**, L209 (1988)

24. Caroli, C., Combescot, R., Nozieres, P., Saint-James, D.: J. Phys. C **4**, 916 (1971)

25. Boukhvalov, D.W., Katsnelson, M.I.: Chemical functionalization of graphene with defects. Nano Lett. **8**(12), 4373 (2008)

26. Cruz-Silva, E., Barnett, Z.M., Sumpter, B.G., Meunier, V.: Structural, magnetic, and transport properties of substitutionally doped graphene nanoribbons from first principles. Phys. Rev. B **23**, 155445 (2011)

27. Brandbyge, M., Mozos, J.L., Ordejón, P., Taylor, J., Stokbro, K.: Density-functional method for non-equilibrium electron transport. Phy. Rev. B **65**, 165401 (2002)

28. Botello-Mendez, A.R., Cruz-Silva, E., Lopez-Urias, F., Sumpter, B.G., Meunier, V., Terrones, M., Terrones, H.: Spin polarized conductance in hybrid graphene nanoribbons using 5–7 defects. ACS Nano **3**, 3606–3612 (2009)

29. Simonis, P., Goffaux, C., Thiry, P.A., Biro, L.P., Lambin, P., Meunier, V.: STM study of a grain boundary in graphite. Surf. Sci. **511**, 319–322 (2002)

30. Wenchang, L., Meunier, V., Sumpter, B.G., Bernholc, J.: Density functional theory studies of quantum transport in molecular systems. Int. J. Quantum Chem. **106**, 3334 (2006); Meunier, V., Sumpter, B.G.: Amphoteric doping of carbon nanotubes by encapsulation of organic molecules: electronic properties and quantum conductance. J. Chem. Phys. **123**, 024705 (2005); Meunier, V., Sumpter, B.G.: Tuning the conductance of carbon nanotubes with encapsulated molecules. Nanotechnology **18**, 424032 (2007); Meunier, V., Kalinin, S.V., Sumpter, B.G.: Nonvolatile memory elements based on the intercalation of organic molecules inside carbon nanotubes. Phys. Rev. Lett. **98**, 056401 (2007)

31. Rodriguez-Manzo, J.A., Banhart, F., Terrones, M., Terrones, H., Gobert, N., Ajayan, P.M., Sumpter, B.G., Meunier, V.: Covalent metal-nanotube heterojunctions as ultimate nano-contacts. Proc. Natl. Acad. Sci. U. S. A. **106**, 4591 (2009)

32. Romo-Herrera, J.M., Sumpter, B.G., Cullen, D.A., Terrones, H., Cruz-Silva, E., Smith, D.J., Meunier, V., Terrones, M.: An atomistic branching mechanism for carbon nanotubes: sulfur as the triggering agent. Angew. Chem. Int. Ed. **47**, 2948 (2008); Romo-Herrera, J.M., Cullen, D.A., Cruz-Silva, E., Sumpter, B.G., Meunier, V., Terrones, H., Smith, D., Terrones, M.: The role of sulfur in the synthesis of novel carbon morphologies: from covalent Y-junctions to sea urchin-like structures. Adv. Funct. Mater. **19**, 1193 (2009); Sumpter, B.G., Huang, J., Meunier, V., Romo-Herrera, J.M., Cruz-Silva, E., Terrones, H., Terrones, M.: A theoretical and experimental study on manipulating the structure and properties of carbon nanotubes using substitutional dopants. Int. J. Quantum Chem. **109**, 97–118 (2009)

33. Cruz-Silva, E., López-Urías, F., Munoz-Sandoval, E., Sumpter, B.G., Terrones, H., Charlier, J.-C., Meunier, V., Terrones, M.: Electronic transport and mechanical properties of phosphorus and phosphorus-nitrogen doped carbon nanotubes. ACS Nano **3**, 1913–1921 (2009); Cruz-Silva, E., Cullen, D.A., Gu, L., Romo-Herrera, J.M., Munoz-Sandoval, E., Lopez-Urias, F., Sumpter, B.G., Meunier, V., Charlier, J.C., Smith, D.J., Terrones, H., Terrones, M.: Heterodoped nanotubes: theory, synthesis, and characterization of phosphorus-nitrogen doped multiwalled carbon nanotubes. ACS Nano **2**, 441 (2008); Sumpter, B.G., Meunier, V., Romo-Herrera, J.M., Cruz-Silva, E., Cullen, D.A., Terrones, H., Smith, D.J., Terrones, M.: Nitrogen-mediated carbon nanotube growth: diameter reduction, metallicity, bundle dispersibility, and bamboo-like structure formation. ACS Nano **1**, 369 (2007); Maciel, I.O., Campos-Delgado, J., Cruz-Silva, E., Pimenta, M.A., Sumpter, B.G., Meunier, V., Lopez-Urias, F., Muñoz-Sandoval, E., Terrones, M., Terrones, H., Jorio, A.: Synthesis, electronic structure and Raman scattering of phosphorous-doped single-wall carbon nanotubes. Nano Lett. **9**, 2267 (2009); Cruz-Silva, E., Lopez-Urias, F., Munoz-Sandoval, E., Sumpter, B.G., Terrones, H., Charlier, J.-C., Meunier, V., Terrones, M.: Phosphorus and phosphorus–nitrogen doped carbon nanotubes for ultrasensitive and selective molecular detection. Nanoscale **3**(3), 1008–1013 (2011)

Electronic Visual Prosthesis

▶ Artificial Retina: Focus on Clinical and Fabrication Considerations

Electroplating

▶ Nano-sized Nanocrystalline and Nano-twinned Metals

Electroresponsive Polymers

▶ Biomimetic Muscles and Actuators Using Electroactive Polymers (EAP)

Electrospinning

Michael J. Laudenslager[1] and
Wolfgang M. Sigmund[1,2]
[1]Department of Materials Science and Engineering, University of Florida, Gainesville, FL, USA
[2]Department of Energy Engineering, Hanyang University, Seoul, Republic of Korea

Synonyms

Electrohydrodynamic forming

Definition

Electrospinning is a process to obtain polymer, ceramic, metallic, and composite fibers from solutions, dispersions, or melts as a liquid jet accelerates through an electric field.

Introduction

Electrospinning produces long fibers with diameters that range from tens of nanometers to several microns. Despite the nanoscale diameters, the fiber lengths can reach several meters. Figure 1 shows a typical scanning electron microscope image of randomly oriented electrospun nanofibers. While initial developments in the technique primarily dealt with polymers, further advances have demonstrated a diverse range of materials including ceramic, metallic, and composite

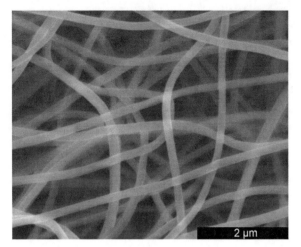

Electrospinning, Fig. 1 Electrospun fibers spun from a titanium butoxide sol-gel and polyvinylpyrrolidone solution

systems. Additionally, this technique also provides control over the fiber structure and orientation.

The study of electrospinning begins with the field of electrohydrodynamics, which describes a range of phenomena that occur as liquids interact with electric fields. A single droplet when placed in an electric field deforms from a spherical droplet into an elongated Taylor cone [1]. With higher voltages, the droplet begins to eject fluid. Figure 2 shows three stages in droplet deformation as the electric field is increased. The droplet was formed from a highly viscous polyvinylpyrrolidone solution dissolved in water. Depending on the solution properties, two related phenomena occur: electrospraying and electrospinning.

Electrospraying forms discrete particles from nanometers to microns in diameter, whereas electrospinning forms long, continuous strands of fibers. However, a range of behaviors exists between these two extremes. In electrospinning, chain entanglements from high molecular weight polymers prevent the fluid from breaking apart. One of the great advantages of this process is that virtually any soluble polymer can be electrospun into nanofibers. As the strand accelerates through the field, it initially travels along a straight path. Due to instabilities in the electric field, its trajectory changes and the fiber whips around, spiraling toward nearby grounded surfaces. Although the study of electrohydrodynamics has existed for

Electrospinning, Fig. 2 Three images showing droplet deformation and eventual fiber ejection as the electric field increases

some time, electrospinning has only recently garnered much attention.

Electrospinning has several distinct advantages over other fiber processing techniques. Conventional fiber processing techniques are not readily capable of producing fibers with the nanoscale diameters that are characteristic of electrospun materials. Conversely, nanomaterial synthesis techniques can produce structures with similar diameters, but only at significantly shorter length scales. Furthermore, the electrospinning technique is applicable to an extensive range of material systems, while other nanomaterial growth methods are often highly specific to each material. The major drawback of the electrospinning technique is that the production rate of electrospun fibers is significantly slower than conventional techniques, typically on the order of milligrams per hour. Overall, the nanoscale dimensions and high aspect ratio of electrospun fibers creates many interesting avenues for research. The dimensions of these fibers greatly improve several of their properties such as porosity, surface area, and grain size (in the case of ceramic fibers). With these enhanced properties, electrospun fibers are applied to a number of disparate fields including filtration, cell growth, catalysis, energy storage and conversion devices, and hydrophobic surfaces.

Historical Developments in the Electrospinning Process

The first published use of the word electrospinning appeared in the literature in 1995. However, the study of charged fluids has existed for far longer. William Gilbert first published on the topic in 1600, where he describes the deformation of liquid droplets in an electric field [2]. Further insights into the field were reported by Lord Rayleigh who modeled the ability of charged fluids to eject liquid in 1882 [3]. The next major publications occurred in the 1964 by Sir Geoffrey Taylor. Taylor was the first to provide detailed descriptions of the deformation of liquid droplets [1]. To his credit, the distinctive shape of the deformed charged droplets has been given the name Taylor cones (or sometimes Gilbert-Taylor cones).

During this time, the industry also took notice of the field. Several important commercial developments in the field began in 1902. At this time, the US patent office recognized two methods for the dispersion of droplets via electrical means [4, 5]. Furthermore, in 1934, Anton Formhals patented the electrospinning process. Two years later, in 1936, the first commercial development of electrospun fibers began. I.V. Petryanov-Sokolov developed filters from electrospun fibers. Today, numerous companies currently manufacture electrospinning equipment and sell electrospun products.

Electrospinning Setup

Figure 3 shows a typical electrospinning setup. An advancement pump controls the flow rate of the polymer inside the syringe. Next, a power supply generates a charge on the liquid droplet. The electric field causes the droplet to deform and eject fluid. As the fluid travels, instabilities in the electric field cause the fiber to undergo a spiraling motion. This motion serves two purposes: It helps to thin the fiber, and it allows the solvent to evaporate prior to collecting on the substrate. The substrate is a grounded, conductive plate typically made from aluminum foil. Although this is the most commonly used setup, numerous modifications exist that can alter many of the properties of electrospun fibers.

Electrospinning,
Fig. 3 Typical laboratory
scale electrospinning setup

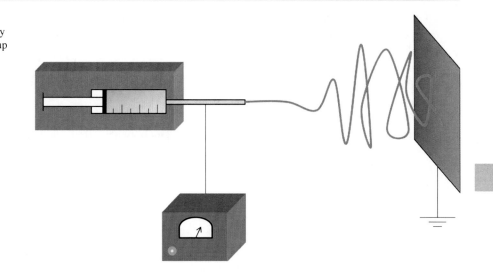

Morphology of Electrospun Fibers

Porosity. As the fibers collect in randomly oriented
meshes, the space left between fibers makes a highly
porous structure. This porosity is dependent on the
fiber diameter, and any alignment techniques that are
used. The high surface area exhibited by electrospun
mats is also due to the porosity. It is also possible to
achieve porosity within individual fibers. Two
methods are commonly used to produce porous
fibers. One method exploits phase separation in
a composite polymer system. Selectively removing
one of the polymers after phase separation leaves
behind a highly porous structure. Carefully adjusting
the solvent, humidity, and molecular weight can also
lead to highly porous surfaces with specific polymer
systems [6]. In both cases, the fibers themselves are
in the micron range, while the pores are on the
nanoscale.

Core-shell and hollow fibers. Sun et al. used a novel
electrospinning apparatus consisting of concentric
capillaries to produce core-shell fibers [7]. Figure 4
demonstrates a typical core-shell electrospinning
apparatus. Separate tubes feed material into each
compartment of the device. In this case, two different
polymers are used. The same device is also used to

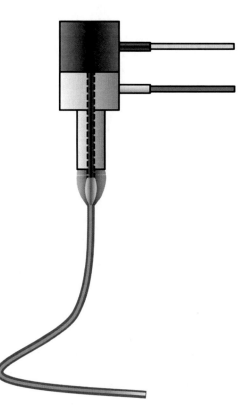

Electrospinning, Fig. 4 Device for core-shell and hollow fiber
production

form hollow fibers. Pumping an immiscible oil through the inner capillary and subsequently removing the oil leaves behind hollow fibers [8].

Beaded fibers. Not all electrospun nanofibers form flat, uniform surfaces. Under certain conditions, large droplets called beads form on the electrospun strands. Several factors influence bead formation: low viscosity, low electric fields, and high surface tension. Therefore, adjusting the solution concentration, solvent, and ions present in the system can greatly influence the presence of beads [9].

Ribbons. These fibers do not have the circular cross section that is characteristic of nanofibers. Under certain conditions, a volatile solvent can induce the formation of a skin on electrospun fibers. As the material dries, the tubular structure collapses in on itself forming flattened ribbon structures [10].

Branched fibers. Short branches protruding from the sides of the nanofibers are reported due to perturbations in the cross-sectional areas of the fibers in polycaprolactone dissolved in acetone. These structures occur more frequently with large diameter fibers produced from highly viscous solutions in strong electrical fields [10].

Aligned fibers. Several different techniques produce aligned electrospun fibers [11]. The techniques either use a novel collection device or modify the electric or magnetic field to prevent the whipping motion. In one method, a high-speed mandrel replaces the standard collection plate. The rotating motion aligns the fibers around the mandrel. Another modification to the collector is to create a long gap in the collection substrate. The conductive sides of the gap attract the fibers, and residual charges within the fibers repel other fibers leading to parallel arrays collecting across the gap. A final technique involves using a magnetic field to prevent the whipping motion of fibers.

Methods to upscale production. The greatest limitation of the electrospinning process is the slow production rate. Increasing the fiber production output from a single needle is not possible. However, it is possible to outscale the technique by assembling arrays of spinnerets in close proximity. This can result in numerous complications. Multiple spinnerets in close proximity cause electric field interference, which creates an uneven coating of fibers. Additionally, a single clogged needle in a large array could ruin the uniformity of the fiber mat. Several startup companies have begun developing other novel methods to increase fiber production. Some of these methods bypass the use of needles altogether. However, their efforts are beyond the scope of this entry.

Electrospun Materials

Polymers. The most frequently electrospun materials are polymers. Both synthetic and natural polymers are extensively reported for a variety of applications [12]. Due to the ability to use biocompatible materials and solvents, these materials are particularly interesting for biomedical and filtration applications. However, the electrospinning process is not limited to producing only polymeric fibers.

Ceramics. Several methods are able to produce ceramic fibers. The most common route to ceramic fibers is to incorporate a sol-gel material into a polymer. Heat treatments can burn out the polymer, leaving behind pure ceramic fibers. Similarly, incorporation of ceramic nanoparticles into electrospun polymer fibers followed by calcination is another pathway to ceramic fibers.

Metals. Only a few studies report purely metallic electrospun fibers. One method is to reduce ceramic nanofibers into metallic fibers. Another method uses electrospun fibers indirectly. Polymer nanofibers are coated in a thin metallic layer, and subsequently the polymer core is removed. This process leaves behind hollow metallic nanotubes. The potential advantages of metallic nanofibers for catalysis ensure that research into the area of metallic fibers will continue to expand.

Composites. As long as one of the materials is capable of being electrospun, the entire system will produce fibers. Due to this fact, a myriad of composite systems exist. These systems typically involve the incorporation of nanomaterials into the fibers. By carefully selecting the polymer and heat treatment conditions, the polymer can be turned into a carbon nanofiber support for the nanoparticles. Systems containing multiple polymers are also common. For the few polymers that prove intractable for electrospinning, a second polymer is often added creating a new composite system.

Analytical Models for Fiber Diameter

The unique properties of electrospun fibers stem from their nanoscale diameters. Therefore, it is of critical importance to understand the parameters that influence the fiber diameter. However, a complex relationship exists. The solution properties, Taylor cone shape and size, electric field, and whipping motion of the fiber all affect the fiber diameter. The solution properties (viscosity and surface tension) interact with gravity to form an equilibrium droplet shape. Appling an electric field deforms this shape and, at a sufficiently high voltage, ejects fluid from the droplet. Furthermore, adjusting the flow rate of the liquid through the capillary alters the size of the droplet. Finally, the radius of curvature as the jet whips toward the target further affects the final diameter.

Combining all of these terms, Fridrikh et al. proposed an analytical model to determine the final nanofiber diameter with an error of around 20% for polymer fibers [13]. Equation 1 shows the relationship between the electrospinning parameters and the terminal fiber radius. The terminal radius depends on surface tension (γ), dielectric constant of the outside medium (ϵ), total current directed toward the lower electrode (I), radius of curvature divided by the diameter of the jet (χ), and flow rate (Q). This equation assumes that all solutions and voltages produce electrospun fibers, which is not the case. However, one can use the equation to determine which parameters have the greatest influence over the terminal fiber diameter. From the equation, it is apparent that flow rate and current are the most influential terms.

$$r_{terminal} = \left(\gamma \epsilon \frac{Q^2}{I^2} \frac{2}{\pi (2\ln\chi - 3)} \right)^{\frac{1}{3}} \qquad (1)$$

Incorporation of ceramic precursors into polymer fibers greatly enhances the electrical conductivity by several orders of magnitude. For these highly conductive solutions, one must consider the conductivity due to the bulk charge and not limit the model to the surface charge alone. The high conductivity invalidates some of the critical assumptions in Eq. 1. Calcination further complicates the prediction of diameters in ceramic systems; fiber diameters shrink as calcination removes the polymeric material. However, the polymer loading and final fiber diameter is linearly related, and can be determined using a correction factor [14].

Properties of Electrospun Fibers

Mechanical properties. The mechanical properties of nanofibers can vary greatly depending on the materials

involved and their processing conditions. The most common method to test fiber properties is to generate stress strain curves from large fiber mats. Measuring the properties of individual fibers is more challenging. To this end, several studies report various methods using atomic force microscopy measurements to determine the mechanical properties of individual fibers. The results of these studies suggest that the electrospinning process induces a molecular orientation within the samples, which increases the Young's modulus. The mechanical properties of ceramic fibers introduce special challenges. While polymer nanofibers are typically flexible, ceramic fibers are often extremely brittle. However, reports also exist of flexible ceramic fibers.

Applications and Publication Trends

Electrospinning applications span a range of diverse fields. Figure 5 shows the breakdown of publications according to their field from a title search on ISI web of knowledge. The first publications appeared in 1995, and the number of active researchers in the field has rapidly increased. The first publications in electrospun ceramic fibers appeared in 2003. Along with this development, publications in energy materials began to appear. The list of publications was categorized based on key terms relevant to each field. Papers that did not contain any of these terms are labeled "Other." This category includes research into the fundamental concepts of electrospinning as well as modifications to the electrospinning setup to produce new structures and morphologies. The following sections outline how researchers utilize electrospun fibers in their respective fields.

Biomedical. Approximately half of electrospinning publications fall under the field of biomedical research. This is a broad area of research that comprises several subfields: scaffolds, wound healing, and drug delivery [15, 16]. Many biocompatible materials are electrospun for these applications. The porous nature of the fibers improves cell proliferation. Furthermore, by orienting the fibers, directed cell growth is possible. Drug delivery utilizes the high surface area to volume ratio of the fibers to increase the loading of drug molecules.

Energy materials. Recent developments in ceramic fiber processing have opened many new avenues in

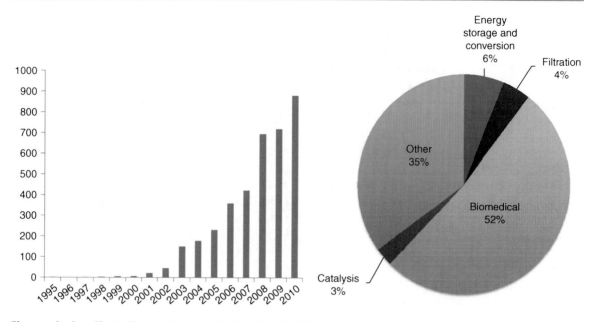

Electrospinning, Fig. 5 Electrospinning publications through 2010

energy harvesting and conversion [17]. The one-dimensional structure of nanofibers confines grain growth, which can improve electrical conductivity. Researchers have assembled batteries, photovoltaics, hydrogen storage materials, capacitors, thermoelectric, and fuel cells from electrospun nanofibers. Some of these devices also make use of polymer fibers such as battery and fuel cell membranes.

Filtration. The highly porous network created by electrospun fibers makes them interesting filtration materials. As air flows through filters, there is a change in pressure. Higher drops in pressure require more energy to force air through the filtration devices. Assembling filters from thinner, uniform fibers, reduces the pressure drop, which lowers the energy requirements to force air through the filters. While most nanofilters are assembled from polymer fibers, ceramic fibers are also investigated due to their chemical and thermal stability [18].

Catalysis. The interconnected structure of electrospun fibers coupled with their high surface area makes electrospun fibers interesting catalyst materials. The connected fiber structure allows facile recovery of the catalytic material. Researchers have explored several methods to use nanofibers in catalysis. The most straightforward method is to embed catalytic nanoparticles in a polymer support material. Other studies have explored photocatalysis using

ceramic fibers. Finally, a few researchers have heated polymer fibers embedded with metal salts to produce purely metallic nanofibers [19].

Super hydrophobic surfaces. Water repellent surfaces have many potential industrial applications, particularly due to the self-cleaning effect observed in super hydrophobic surfaces (surfaces with a water contact angle >150°). A combination of high surface roughness and chemistry is necessary to achieve super hydrophobicity. By carefully controlling the morphology of electrospun fibers, a variety of super hydrophobic materials are reported [20]. These fibers often make use of the multiple scale of surface roughness created from beaded fibers.

Outlook

Electrospinning is an incredibly versatile nanomaterial processing technique. Using this technique, researchers can generate a wide range of materials and nanostructures from relatively simple, inexpensive equipment. These materials have the potential to purify air, heal wounds, deliver drugs, catalyze reactions, and produce and store clean energy. The full potential of electrospinning is just starting to be realized. As novel techniques are developed to enhance the output of electrospun fibers, industrial usage of the technique will increase. Further developments in this field will continue to unlock the potential of this technique to solve important scientific challenges.

Acknowledgments This work was supported by WCU (World Class University) program through the Korea Science and Engineering Foundation (R31-2008-000-10092).

Cross-References

▶ Lotus Effect
▶ Nanomaterials for Electrical Energy Storage Devices
▶ Nanomaterials for Excitonic Solar Cells
▶ Nanomechanical Properties of Nanostructures
▶ Nanomedicine
▶ Nanostructures for Energy
▶ Scanning Electron Microscopy
▶ Sol-gel Method
▶ TiO$_2$ Nanotube Arrays: Growth and Application

References

1. Taylor, G.: Disintegration of water drops in an electric field. Proc. R. Soc. Lond. A Math. Phys. Sci. **280**, 383–397 (1964). doi:10.1098/rspa.1964.0151
2. Gilbert, W., Mottelay, P.: De Magnete. Dover, New York (1958)
3. Rayleigh, L.: On the equilibrium of liquid conducting masses charged with electricity. Phil. Mag. **14**, 184 (1882)
4. Cooley, J.: Apparatus for electrically dispersing fluids, 4 Feb 1902
5. Morton, W.: Method of dispersing fluids, Issued 29 July 1902
6. Casper, C.L., Stephens, J.S., Tassi, N.G., Chase, D.B., Rabolt, J.F.: controlling surface morphology of electrospun polystyrene fibers: effect of humidity and molecular weight in the electrospinning process. Macromolecules **37**, 573–578 (2003). doi:10.1021/ma0351975
7. Sun, Z., Zussman, E., Yarin, A., Wendorff, J., Greiner, A.: Compound core-shell polymer nanofibers by co-electrospinning. Adv. Mater. **15**, 1929–1932 (2003)
8. Li, D., Xia, Y.: direct fabrication of composite and ceramic hollow nanofibers by electrospinning. Nano Lett. **4**, 933–938 (2004). doi:10.1021/nl049590f
9. Fong, H., Chun, I., Reneker, D.H.: Beaded nanofibers formed during electrospinning. Polymer **40**, 4585–4592 (1999)
10. Reneker, D.H., Yarin, A.L.: Electrospinning jets and polymer nanofibers. Polymer **49**, 2387–2425 (2008)
11. Teo, W., Ramakrishna, S.: A review on electrospinning design and nanofibre assemblies. Nanotechnology **17**, 89 (2006)
12. Schiffman, J., Schauer, C.: A review: electrospinning of biopolymer nanofibers and their applications. Polym. Rev. **48**, 317–352 (2008)
13. Fridrikh, S., Yu, J., Brenner, M., Rutledge, G.: Controlling the fiber diameter during electrospinning. Phys. Rev. Lett. **90**, 144502 (2003)
14. Sigmund, W., et al.: Processing and structure relationships in electrospinning of ceramic fiber systems. J. Am. Ceram. Soc. **89**, 395–407 (2006)
15. Liang, D., Hsiao, B.S., Chu, B.: Functional electrospun nanofibrous scaffolds for biomedical applications. Adv. Drug Deliv. Rev. **59**, 1392–1412 (2007). doi:10.1016/j.addr.2007.04.021
16. Li, W.J., Mauck, R.L., Tuan, R.S.: Electrospun nanofibrous scaffolds: production, characterization, and applications for tissue engineering and drug delivery. J. Biomed. Nanotechnol. **1**, 259–275 (2005)
17. Laudenslager, M., Scheffler, R., Sigmund, W.: Electrospun materials for energy harvesting, conversion, and storage: A review. Pure Appl. Chem. **82**, 2137–2156 (2010). doi:10.1351/PAC-CON-09-11-49
18. Barhate, R., Ramakrishna, S.: Nanofibrous filtering media: filtration problems and solutions from tiny materials. J. Membr. Sci. **296**, 1–8 (2007)
19. Dersch, R., Steinhart, M., Boudriot, U., Greiner, A., Wendorff, J.: Nanoprocessing of polymers: applications in medicine, sensors, catalysis, photonics. Polym. Adv. Technol. **16**, 276–282 (2005)
20. Ma, M., Hill, R.M., Rutledge, G.C.: A review of recent results on superhydrophobic materials based on micro-and nanofibers. In: Superhydrophobic Surfaces, p. 241. Brill, Leiden (2009)

Electrostatic Actuation of Droplets

▶ Electrowetting

Electrostatic MEMS Microphones

Neal A. Hall
Electrical and Computer Engineering
University of Texas at Austin, Austin, TX, USA

Synonyms

MEMS capacitive microphone; MEMS condenser microphone; MEMS microphone; Silicon microphone

Definition

A microphone fabricated using surface and bulk silicon micromachining techniques and operating on the principle of a variable capacitance.

Electrostatic MEMS Microphones, Fig. 1 (*Top*) Schematic of a silicon MEMS die containing the sensing structure and (*bottom*) a schematic of a complete electrostatic MEMS microphone surface-mountable package. The package contains either a *top* or *bottom* sound inlet but typically not both

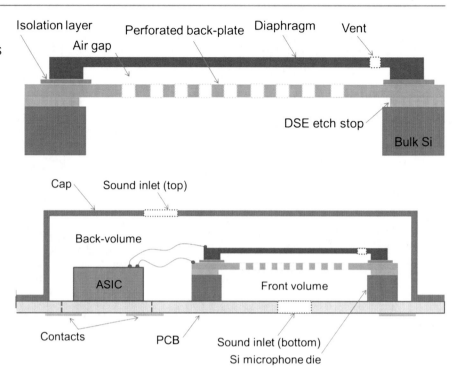

Applications and Background

Electrostatic MEMS microphones, like their larger counterparts, convert pressure waves in air into electrical signals. MEMS microphones were successfully commercialized in 2003 and began competing with electret condenser microphones (ECMs) in consumer electronic device markets, most notably in cellular phones. Compared to ECMs which use heat-sensitive electret foils, MEMS microphones have a critical advantage in that they can withstand high temperatures encountered in standard lead-free solder reflow cycles used in automated pick and place manufacturing, and this in turn results in significant cost savings to system integrators [1]. Since their entry into the market, electrostatic MEMS microphones have become one of the highest growth areas for MEMS, growing from less than 300 million units shipped in 2007 to over 1 billion units shipped in 2011 [2]. Today many major semiconductor companies manufacture and sell electrostatic MEMS microphones including Knowles, Analog Devices, Infineon, and ST Microelectronics. The top four buyers of MEMS microphone in 2011 were Apple, Samsung, LG, and Motorola [3]. Apple's iPhone 4 product alone contains three electrostatic MEMS microphones – two in the body of the phone and a third in the mobile headset.

Operation and Construction

Field variables used to characterize sound include both pressure and particle vibrations. Although some new techniques attempt to measure particle velocity directly [4], the vast majority of microphones including electrostatic MEMS microphones are constructed as dynamic pressure sensors. Generally speaking, the range of sound pressures of interest in most applications is as low as 200 μPa root mean squared (rms) (20 dB) and as high as 10 Pa (114 dB). Electrostatic MEMS microphones must detect these small pressure signals in the presence of a large "DC" background atmospheric pressure, approximately 100 kPa. The frequency range typically of interest is the audio bandwidth, 20 Hz–20 kHz, although only recently have commercial models of MEMS microphones approached these limits. The corresponding wavelength range is 17 m–17 mm. For most of this range, MEMS microphones packages are significantly smaller than the wavelength of sound being measured.

Figure 1 details important features of a silicon microphone die and also a complete surface mount electrostatic MEMS microphone package. Referring to the top image, a compliant, pressure-sensitive diaphragm is suspended over a rigid and perforated backplate to form a variable capacitor. When sound

Electrostatic MEMS Microphones, Fig. 2 Photographs of a commercial MEMS microphone. This device has a footprint of 4.72 × 3.76 mm and a profile of 1.25 mm. The image at *left* is with the cap removed and shows details of the PCB, ASIC, and MEMS die. The figure at far *right* shows the electrical contacts on the *bottom* of the package that interface with a customer's system (Images borrowed courtesy of Infineon and first appearing in Ref. [5])

Electrostatic MEMS Microphones,
Fig. 3 Scanning electron micrographs (SEMs) of a silicon die containing an electrostatic MEMS sensing structure (Images borrowed courtesy of Infineon and first appearing in Ref. [5])

perforated polysilicon back plate (2 μm)

1mm

flexible polysilicon membrane (0.3 μm)

pressure is applied to the diaphragm, the diaphragm moves vertically and the resulting change in the parallel-plate capacitance is detected. The capacitive structure must be electrically biased and loaded with charge, and the high output impedance of the signal must be buffered with an amplifier. These features are implemented by a small application specific integrated circuit (ASIC) as shown in Fig. 1 bottom. To sense sound pressure effectively, one side of the diaphragm must be exposed to the incoming sound pressure while the opposing side must be sealed off from the environment. A common package to accomplish this is also shown in Fig. 1. The MEMS die and ASIC are both mounted on a common printed circuit board (PCB) which contains a small opening on the bottom surface to let sound pressure in. A sealed back volume is created by a cap mounted directly to the PCB. Electrical vias through the PCB lead to contacts on the bottom side to create a complete surface-mountable package.

As a variation to the system shown in Fig. 1 (bottom), the sound inlet is sometimes placed on the cap rather than the PCB to form a "top-inlet" configuration. A second common variation found on the MEMS die is a backplate fabricated above rather than below the diaphragm. Photographs and SEMs of a commercial electrostatic MEMS microphone that

implements both of these variations are presented in Figs. 2 and 3, respectively. In the center image of Fig. 2, one can observe the sound inlet on the cap of the device. The SEMs in Fig. 3 highlight the perforated backplate suspended approximately 2 μm above the pressure sensitive membrane.

A final important feature to note in Fig. 1 is the presence of a small low-frequency vent, most commonly fabricated into the microphone diaphragm. The vent provides the device with a high flow resistance leakage path enabling sound pressure to bypass the diaphragm and reach the back volume. This feature creates a lower limiting frequency and prevents the diaphragm from deforming and responding to DC atmospheric pressure and low-frequency barometric pressure fluctuations, which are orders of magnitude larger than the audible sound pressures that the device is designed to sense.

Although Fig. 1 presents the most common construction of MEMS microphones, single-chip architectures have also been demonstrated combing the MEMS capacitive sensing structure with the CMOS electronics on the same die. At least one commercial embodiment of this approach exists. Figure 4 below presents a photograph of a microphone made by Akustica (acquired by Bosch in 2009) in which one

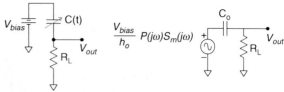

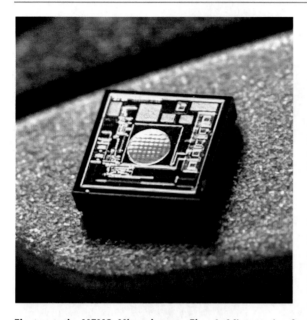

Electrostatic MEMS Microphones, Fig. 4 Micrograph of a single-chip electrostatic MEMS microphone made by Akustica. The sensing diaphragm can be seen in the center of the structure. This die measures approximately 1 × 1 mm (Image borrowed courtesy of Akustica, Inc. [6])

can observe the pressure-sensitive diaphragm structure along with electronics. The total die size is 1 × 1 mm. An obvious advantage of this approach is size, with space being a premium in small consumer electronic devices (e.g., cell phones and laptop computers). A second advantage cited is potentially lower per unit cost. The MEMS structure is fabricated in layers already required in the CMOS process, so only few additional processing steps are required for the MEMS. The peripheral packaging requirements are also reduced. A third advantage cited is electronic design flexibility owing to the proximity between the sensing capacitance and the electronics, which reduces parasitic capacitance associated with bond pads and wire bonds that are otherwise required to route signals to off chip electronics as shown in the two-chip embodiment of Fig. 1. This proximity may also be advantageous for shielding against electromagnetic interference created by neighboring electronics in consumer products.

Transduction Principle

The variable capacitance transduction principle is presented in Fig. 5. The capacitive sensing structure is biased through a high impedance resistance, R_L,

Electrostatic MEMS Microphones, Fig. 5 (*Left*) Circuit schematic for biasing an electrostatic MEMS microphone and measuring changes in capacitance, and (*right*) small signal equivalent circuit about the bias point

shown at left. Electrostatic forces pull the diaphragm in toward the backplate to create a standoff distance h_o typically between 1 and 2 μm and a nominal parallel-plate capacitance C_o. A phenomena known as electrostatic collapse of the diaphragm limits the bias voltage that can be applied [7], which is typically in the range of 2–11 V and depends on the design [1]. The instantaneous voltage across the parallel-plate capacitor is given by $v = q/C = qh/(\varepsilon_o A)$, where C is the instantaneous capacitance, q is the instantaneous charge, h is the instantaneous backplate-diaphragm standoff distance, and A is the area of the structure. Small diaphragm vibrations create small changes in h, and the resulting small signal voltage generated about the bias point is therefore $dv = q_o/(\varepsilon_o A) \, dh = (v_{bias}/h_o) \, dh$. The output impedance associated with the small signal voltage is the device capacitance itself, C_o, and the small signal equivalent circuit is that shown in Fig. 4 (right), with dh replaced by the product of the incoming acoustic pressure $P(j\omega)$ and the mechanical sensitivity of the device, $S_m(j\omega)$, defined below.

v_{bias}/h_o is commonly referred to as an electrical sensitivity and has units of volts per meter of diaphragm displacement. It is also the electric field in the diaphragm-backplate gap created by application of the bias voltage. $S_m(j\omega)$ is a transfer function relating the displacement of the diaphragm to incoming sound pressure. As shown in Fig. 5 right, the small signal voltage generated is the product of the incoming pressure phasor and the electrical and mechanical sensitivities. The small capacitance of the device creates a high output impedance, and v_{out} in Fig. 4 is therefore followed by an amplifier with high input impedance such as a JFET [8]. Digital microphones are also common, in which case the buffered analog output is followed by a sigma-delta analog to digital converter.

Microfabrication of Sensing Structure

The structure shown in Fig. 1 (top) can be fabricated using many different fabrication process flows with many options for materials. Almost all electrostatic MEMS microphones use thin surface micromachined layers for the backplate and diaphragm, and a deep silicon etch (DSE) bulk micromachining step to form the front volume. The surface machined layers are typically between 100 nm and 2 µm thick. Figure 1 (top) suggests a simple fab flow option, in which the first step is to deposit or grow a material with high selectivity against a silicon deep reactive ion etch (DRIE) process, such as silicon dioxide. The first structural layer is then deposited and patterned to form the backplate. Polysilicon is a common material choice. A thin isolation layer (commonly silicon nitride) is deposited and etched to provide electrical isolation between the diaphragm and backplate, followed by deposition and etching of the sacrificial layer to form what will become the air gap. Low temperature oxide (LTO) and TEOS oxide are common choices for the sacrificial layer. After deposition and patterning of the diaphragm, which is commonly formed using polysilicon, the surface steps are completed and the device is etched through the silicon wafer from the opposing side up to the DSE etch stop. Both isotropic etches (e.g., DRIE) and anisotropic wet etches (e.g., KOH) are used in commercial versions of the technology [1]. The sacrificial layer(s) at the surface are then removed and the wafers are diced to realize completed die. Some commercial processes use a silicon on insulator (SOI) wafer as a starting point and use the epitaxial Si layer for formation of the microphone backplate [9].

As electrostatic MEMS microphones continue to mature into commercial device applications, several innovations at the microfabrication level continue to be driven by reliability and device-to-device repeatability criteria. For thin (1–2 µm thick) clamped diaphragms employed in MEMS microphones, even small residual film stress resulting from the deposition process can have a major influence on the compliance of the sensing diaphragm which in turn directly affects the total sensitivity of the device measured in volts/Pa. Controlling film stress in microfabrication processes to ensure device-to-device repeatability has practical limitations. Further, changing environmental conditions (e.g., temperature) can affect sensitivity output of a single device during the course of day-to-day operation. This has motivated the development of innovative structures designed for immunity of output sensitivity to these variations. Figure 6 presents a device by Knowles in which the sensing diaphragm is not clamped, but rather is freely floating until pulled into contact with the support posts by electrostatic forces upon biasing. By allowing the diaphragm to float freely rather than clamping and anchoring it to the substrate, zero diaphragm tension is achieved regardless of film stresses introduced in the deposition process. The free plate design is achieved by completely surrounding the plate with sacrificial layers rather than anchoring it to the substrate during the processing sequence.

The sensing membrane in the single-chip microphone developed by Akustica is made using metal interconnect layers that are already part of standard CMOS processes [10]. Specifically, a metal layer is patterned to form a perforated screen which becomes buried in oxide layers during the subsequent CMOS processing steps. Upon completion of the CMOS process flow, the oxide on top of the wafer is etched down to the screen. The perforations in the screen then allow silicon etchant to reach the Si wafer beneath the screen to create a cavity behind the screen. This cavity is analogous and identical in function to the back volume labeled in Fig. 1. To create a sealed diaphragm, a conformal polymer is deposited to fill the screen perforations. Figure 7 shows an SEM of the metal screen layer forming the diaphragm. The interesting serpentine pattern of the screen in this design is aimed at removing bowing due to stresses inherent in the metal.

Dynamic Model of Package

More so than many other MEMS and Nano-systems, packaging has a profound effect on the overall dynamics of electrostatic MEMS microphones. The use of impedance-based network analogies such as the one presented in Fig. 8 are pervasive in the study of MEMS microphones [11]. Such models are based on the observation that acoustical and mechanical systems obey rules analogous to Kirchhoff's voltage and current laws. In acoustical impedance networks, pressure is a node variable and the sum of pressure drops across elements in a closed loop is equal to zero. Similarly, volumetric flow of air entering and leaving nodes in the network is conserved. In such models, acoustic pressure is a potential or effort variable and therefore analogous to voltage in electrical circuits. Acoustic volume velocity (with units of m^3/s) flows through

Electrostatic MEMS Microphones, Fig. 6 A schematic and SEMs highlighting a free diaphragm design by Knowles (Image borrowed courtesy of Knowles Acoustics and originally appearing in Ref. [1])

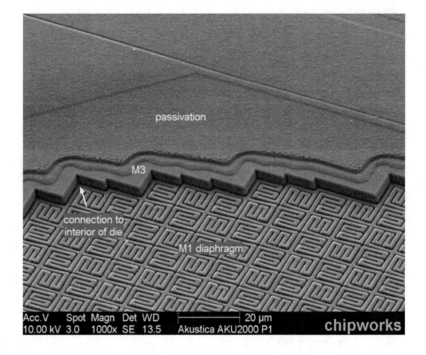

Electrostatic MEMS Microphones, Fig. 7 SEM showing the diaphragm layer and anchoring in Akustica's CMOS microphone (Image borrowed courtesy of MEMS Investor Journal, Inc. and Chipworks)

various elements in the circuit and is therefore analogous to current. Cavities or volumes in acoustical systems have the ability to store potential energy by storing compressed air and are analogous to capacitors in electrical networks, while small inlets and outlets force air to accelerate upon passage through and therefore store energy by virtue of momentum, analogous to an inductor in an electrical network. Similarly, the mechanical compliance and mass of the diaphragm are represented by capacitors and inductors, respectively, with appropriately computed values. The backplate resistance R_b plays a dominant role in shaping the frequency response of the system and also in the noise analysis as discussed below. R_b is a flow resistance generated by the movement of air through the backplate perforations as the diaphragm vibrates

**Electrostatic MEMS
Microphones,
Fig. 8** Network model
superimposed on the physical
package schematic to
emphasize the important
package dynamics in
electrostatic MEMS
microphones

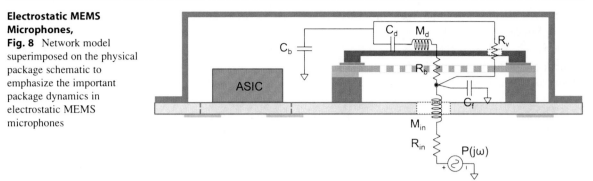

relative the stationary backplate. Network models provide a way to combine mechanical, acoustical, and electrical features of the device into a unified model that can be simulated with common circuit simulator tools such as SPICE.

Figure 8 presents a common network model superimposed on the physical device structure. Subscripts in, f, b, d and v refer to elements associated with the inlet, front cavity, back cavity, diaphragm, and vent, respectively. The $S_m(j\omega)$ transfer function referenced in Fig. 5 relating an incoming pressure phasor to motion of the diaphragm may be modeled and computed based on the network model in Fig. 8.

Insight into the dynamics of the complete package arises from analysis of the network model in the low-frequency limit, the pass band, and near the fundamental resonance of the system. The model makes clear how the vent feature represented by R_v creates a lower limiting frequency by creating a low-frequency bypass around the diaphragm compliance C_d. To see this further, the section of the model relating the front cavity pressure, P_f, to the drive pressure across the diaphragm $(P_f–P_b)$ is shown in Fig. 9. A low-frequency approximation is shown and M_d and R_b are therefore not included.

For microphones packaged as shown in Fig. 8, C_b is often significantly larger than C_d. The low-frequency response of the microphone is then characterized by a simple first-order high-pass filter, with a zero at $\omega = 0$, and a pole or break frequency at $\omega = 1/(R_vC_b)$, above which the microphone operates in a flat band with $S(j\omega)$ dominated by the series compliance $C_b + C_d$. A fundamental resonance of the system characterized by two complex conjugate poles is common of electrostatic MEMS microphones. This LRC type resonance arises from the series combination of the inertial elements $M_d + M_{in}$, the compliance elements, $C_d + C_b$, and the loss elements $R_b + R_{in}$. This

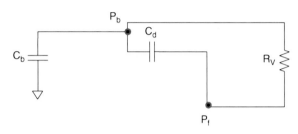

Electrostatic MEMS Microphones, Fig. 9 Section of the low-frequency equivalent circuit of Fig. 8 relating the front volume pressure to the back volume pressure

fundamental resonance is typically designed to be close to or above the desired upper limiting frequency of the microphone.

Dominant Noise Sources, dB(A) Weighting, SNR Definition

Signal-to-noise ratio is an important figure of merit for small acoustic transducers, defining the realm of applications in which a particular device technology is suited. Noise floors for electrostatic MEMS microphones are most meaningful when referred to an input pressure level. The input-pressure-referred-noise spectrum is commonly integrated using an A-weighted filter which takes into account the frequency response of a healthy human ear. The sound pressure level (SPL) of the A-weighted rms pressure level is computed and a dB(A) noise level is reported for the device. Common noise levels for MEMS microphones are around 34 dB(A). Often a signal-to-noise ratio (SNR) is reported, and in these cases the convention is to assume a 1 Pa rms, or 94 dB SPL, input. SNR is then defined by the following expression: SNR = 94 dB − A weighted noise. Additional performance-based figures of merit for MEMS electrostatic microphones include dynamic range, upper limiting frequency, and lower limiting frequency.

Scaling Challenges for Small-Scale Electrostatic Microphones

Electrostatic MEMS microphones face challenges in achieving low noise floors, and this has limited their ability to address applications such as hearing aids which require noise floors in the 20–25 dB(A) range. A fundamental noise limit in miniature microphones is thermal-mechanical noise due to the dissipative mechanisms in the acoustical and mechanical parts of the system [12]. In particular, the backplate resistance R_b labeled in Fig. 5 generates a thermal-mechanical pressure noise contribution that is seen at the input of the device across the pass band of operation and therefore impacts the input-pressure-referred noise directly. P_{tm}, the thermal-mechanical noise due to the mechanical flow resistance R_b, can be computed as:

$$P_{tm} = \frac{1}{A_d} \sqrt{4k_b T R_b} = \frac{1}{A_d} \sqrt{4k_b T R_{ba} A_d} \propto \sqrt{\frac{R_{ba}}{A_d}}$$

$$\propto \frac{\sqrt{R_{ba}}}{a}$$

where k_b is Boltzmann's constant, T is the ambient temperature, A_d and a are the area and radius of the diaphragm and backplate (assumed to the same), and R_{ba} is the mechanical flow resistance per unit area. R_{ba} is a function of the perforation ratio and perforation geometry [13] but not the size of the backplate itself since R_{ba} is defined on a per unit area basis. For a given backplate design, decreasing the diaphragm radius by a factor of 2 results in a doubling (i.e., 6 dB) in the input-referred thermal noise limit. While backplates can be redesigned to achieve lower R_{ba}, this path faces practical limitations as higher perforation ratios reduce the total active capacitance of the device. Further, the lack of damping in the system can result in an undesirably high resonance Q (i.e., small damping ratio) at the fundamental resonance of the system discussed above. In the pass band, the device behaves as a second-order system with a damping ratio:

$$\varsigma = \frac{R_{ab} A_d}{2M\omega_n} \propto \sqrt{\frac{R_{ba}}{\rho t \omega_n}}$$

where M is the combined inertia of the diaphragm and sound inlet, ω_n is the fundamental radial natural frequency of the system, and ρ and t are the density and thickness of the diaphragm. Within limits dictated by fabrication robustness, the latter two variables provide a means for achieving a desired damping ratio for an R_{ba} predetermined by noise considerations.

In addition to the above mechanical and acoustical scaling considerations, the lower cutoff frequency of the microphone may also be limited by the high-pass nature of the small signal electrical circuit shown in Fig. 5 (right). This electrical filter is independent of the lower limiting frequency inherent to $S_m(j\omega)$ discussed above due to the package dynamics. For a given bias resistor, the lower limiting frequency of the microphone will increase with reduced device capacitance. Although in principle GΩ bias resistors can be used, preventing leakage across such elements amid all required environmental operating conditions can prove challenging.

Other Techniques and Future Directions

Other types of MEMS microphones include piezoelectric and optical. The former have shown to be advantageous for special applications demanding high linearity and high dynamic range (e.g., aero-acoustic measurement arrays and sensors for wind tunnel testing) [14], while recent work with the later work aims to achieve significantly lower noise floors than that achievable with electrostatic MEMS microphones [15]. Active research is also aimed at demonstrating designs that are inherently directional. An innovation inspired by the directional hearing mechanism of a special parasitoid fly has recently been demonstrated [16]. Until recently, little attention was given to SNR as an important parameter for MEMS microphones in the consumer electronics commercial sector. Recently, however, manufacturers of electrostatic MEMS microphones have started introducing lower noise models, with the lowest noise models to date having input-referred noise levels equal to 29 dBA. This may be driven by a recent trend in smartphone applications: the use of multiple microphones in a single product for implementation of ambient noise cancelation algorithms. As a significant departure from traditional close-talking voice input, hands-free and headset-free application modalities may continue to drive performance increases and new innovations for electrostatic MEMS microphones.

Cross-References

▶ CMOS MEMS Fabrication Technologies
▶ Integrated Micro-Acoustic Devices
▶ MEMS Packaging

References

1. Loeppert, P.V., Lee, S.B.: SiSonicTM – the first commercialized MEMS microphone. Presented at the Solid-state sensors, actuators, and microsystems workshop, Hilton Head Island, South Carolina, 2006
2. Tekedia.com: MEMS microphones break the billion unit barrier. http://tekedia.com/34372/mems-microphones-break-billion-unit-barrier/ (2012, January). Accessed 1 Feb 2012
3. Lowensohn, J.: Apple bests Samsung as buyer of tiny mics. http://news.cnet.com/8301-13579_3-57362174-37/apple-bests-samsung-as-buyer-of-tiny-mics/ (2012, January 19). Accessed 1 Feb 2012
4. De Bree, H.-E., Leussink, P., Korthorst, T., Jansen, H., Lammerink, T.S.J., Elwenspoek, M.: The μ-flown: a novel device for measuring acoustic flows. Sens. Actuators A Phys. **54**, 552–557 (1996)
5. Fuldner, M., Dehe, A.: Challenges of high SNR (signal to noise) silicon micromachined microphones. Presented at the 19th international congress on acoustics, Madrid, 2007
6. Akustica, Inc. New microphones boost acoustic performance in consumer electronics. http://www.akustica.com (2011). Accessed 1 Feb 2012
7. Osterberg, P., Yie, H., Cai, X., White, J., Senturia, S.: Self-consistent simulation and modelling of electrostatically deformed diaphragms. In: Proceedings of the IEEE Micro Electro Mechanical Systems. An Investigation of Micro Structures, Sensors, Actuators, Machines and Robotic Systems. Oiso, Japan, pp. 28–32 (1994)
8. Brauer, M., Dehe, A., Fuldner, M., Barzen, S., Laur, R.: Improved signal-to-noise ratio of silicon microphones by a high-impedance resistor. J. Micromech. Microeng. **14**, S86 (2004)
9. Weigold, J.W., Brosnihan, T.J., Bergeron, J., Zhang, X.: A MEMS condenser microphone for consumer applications. Presented at the 19th IEEE international conference on micro electro mechanical systems. Istanbul, Turkey (2006)
10. Neumann, J.J.J., Kaighman, G.: CMOS-MEMS membrane for audio-frequency acoustic actuation. Sens. Actuators A Phys. **95**, 175–182 (2002)
11. Beranek, L.: Acoustics. McGraw-Hill, New York (1954)
12. Gabrielson, T.B.: Fundamental noise limits for miniature acoustic and vibration sensors. J. Vib. Acoust. **117**, 405–410 (1995)
13. Homentcovschi, D., Miles, R.N.: Modeling of viscous damping of perforated planar microstructures. Applications in acoustics. J. Acoust. Soc. Am **116**, 2939–2947 (2004)
14. Horowitz, S., Nishida, T., Cattafesta, L., Sheplak, M.: Development of a micromachined piezoelectric microphone for aeroacoustics applications. J. Acoust. Soc. Am **122**, 3428–3436 (2007)
15. Kuntzman, M., Garcia, C., Onaran, G., Avenson, B., Kirk, K., Hall, N.: Performance and modeling of a fully packaged micromachined optical microphone. J. Microelectromech. Syst. **20**, 828–833 (2011)
16. Miles, R.N., Su, Q., Cui, W., Shetye, M., Degertekin, F.L., Bicen, B., Garcia, C., Jones, S.A., Hall, N.A.: A low-noise differential microphone inspired by the ears of the parasitoid fly *Ormia ochracea*. J. Acoust. Soc. Am **125**, 2013–2026 (2009)

Electrostatic RF MEMS Switches

▶ Capacitive MEMS Switches

Electrothermomechanical Actuators

▶ Thermal Actuators

Electrowetting

CJ Kim
Mechanical and Aerospace Engineering Department,
University of California, Los Angeles (UCLA),
Los Angeles, CA, USA

Synonyms

Digital microfluidics; Droplet microfluidics; Electrocapillarity; Electrostatic actuation of droplets; Electrowetting-on-dielectric (EWOD)

Definition

Making a surface more wetting to a liquid by applying voltages.

Introduction

When a material (typically solid) and a liquid are in contact, the application of an electric potential between them may cause the wettability of the material to increase, which is exhibited by a decrease of the observed contact angle. This phenomenon is called electrowetting – a term reminiscent of the more traditional electrocapillarity. In recent years, the development of various electrode and material configurations for technological applications – mostly in the rapidly expanding field of microfluidics – has given rise to additional terms such as electrowetting-on-dielectric (EWOD). Because technologies relying on the actuation of liquids by applied electrical potentials are still new, the terms used to describe such systems can be difficult to delineate. It will therefore be helpful to introduce a few terms related to electrowetting with a little bit of the history and theoretical background behind each of them. While the basic approaches to utilizing the electrowetting phenomenon to obtain physical results are presented here, applications of electrowetting to manipulate droplets or to develop devices and systems are deferred to other related entries as well as other works referenced later in this entry.

Electrocapillarity and Electrowetting

A typical liquid (e.g., water) rises in a vertical glass capillary against gravity by the intermolecular attraction between the liquid and the inner wall of the capillary. A terminal height is reached when the gravity of the liquid column equals the attractive force or, more precisely, the vertical component of the liquid-air interfacial tension. The vertical component is larger if the liquid wets the glass surface better (e.g., the surface is clean). The degree of wetting is expressed by the contact angle θ of the liquid on the solid surface by the Young equation:

$$\gamma \cos \theta = \gamma_{sg} - \gamma_{sl} \tag{1}$$

where γ, γ_{sg}, and γ_{sl} are the interfacial energy (or interfacial tension, surface tension) between liquid and air (gas), solid and air, and solid and liquid, respectively. On the other hand, if the liquid does not wet the glass (e.g., mercury), i.e., the contact angle is greater than 90°, the liquid falls down in the capillary. A terminal falling depth against gravity is determined by the surface tension and the contact angle.

In 1875, Gabriel Lippmann showed that a mercury column in a glass capillary, which is dipped into an electrolyte bath, rises or falls when a voltage is applied

between the mercury and the electrolyte [1]. This experiment, which is called the Lippmann electrometer, demonstrated that the interfacial tension between the mercury and the electrolyte is a function of the electric potential across the interface. Perhaps termed originally to depict the initial experimental configuration involving a capillary tube, *electrocapillarity* nevertheless describes a general phenomenon of interfacial energy changing by electric fields.

In 1981, Beni and Hackwood [2] introduced the term *electrowetting* to describe a manifestation of electrocapillarity that occurs when the configuration of materials and electrodes is somewhat different than that of the classical Lippmann electrometer. Specifically, in the traditional electrocapillarity measurement, voltage is applied between a metallic *liquid* (i.e., mercury) and an aqueous liquid (i.e., electrolyte), while both materials are surrounded by an insulating *solid* (i.e., glass capillary). For electrowetting, on the other hand, voltage is applied between a metallic *solid* and an aqueous liquid surrounded by an insulating *fluid* (e.g., air or oil). Figure 1 schematically illustrates these two phenomena, which operate under the same general principle but are distinguishable by how the materials are configured.

Readers may realize that, despite their apparent differences, electrocapillarity and electrowetting describe essentially the same principle. One can do a mental exercise that involves dictating phase changes of a given material. Rotate Fig. 1a 90° clockwise; remove the right dielectric solid and convert the left dielectric solid into a fluid (i.e., vapor or liquid); and convert the liquid metal into a flat solid metal and duplicate it at the bottom. Now one has Fig. 1b. So, the two terms describe one phenomenon depending on the users' focus of interest. If the interest is a liquid column moving in a solid capillary, electrocapillarity describes the phenomenon well. If the interest is how well a liquid wets a solid surface, e.g., what is the contact angle, electrowetting is a better description.

Some readers may raise another question, however. There exists a contact angle of the liquid electrolyte on the solid surface in the electrocapillarity of Fig. 1a, and there exists a column of the liquid electrolyte in a metallic capillary in the electrowetting of Fig. 1b. The confusion may be avoided if one notes that the main interest should be in the interface across which the electric potential is applied. In Fig. 1a, the liquid-liquid interface, across which the voltage is applied,

Electrowetting,
Fig. 1 Comparison between
electrocapillarity (**a**) and
electrowetting (**b**), drawn to
maintain the original
configurations presented by
Beni and Hackwood [2].
A *grid pattern* implies a solid,
and a *wave pattern* implies
a fluid. A *white* background
implies a dielectric, and a *gray*
(*blue* if seen in color)
background implies
a conductor

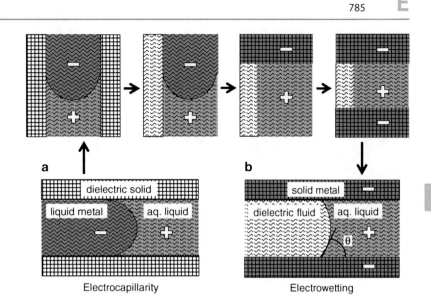

Electrocapillarity   Electrowetting

travels in a capillary, exhibiting capillary action. In
Fig. 1b, the solid-liquid interface, across which the
electric potential is applied, forms a contact angle,
exhibiting wetting.

Electrowetting and Electrowetting-On-Dielectric (EWOD)

Historically, for both the electrocapillarity and the
electrowetting phenomena, the main interest was the
interface between an aqueous liquid and a metal.
Whether the metal was a liquid (for electrocapillarity)
or a solid (for electrowetting), the electric double layer
(EDL) at the interface played a critical role. Since the
underlying mechanism for both phenomena was that
the free interfacial energy can be modulated by
increasing or decreasing the capacitive energy stored
at the EDL via an electric potential across it, the
mechanism would be valid only while the EDL functions as an insulating capacitor, i.e., below ~1 V range
in practice. An applied voltage above such a limit
would induce electrochemical reactions, such as electrolytic gas generation, and alter the surface. In the
case of electrowetting, it was difficult to induce contact
angle changes large enough to be useful without
degrading the surface. Promising utilities were limited
to the case of a liquid metal in an electrolyte-filled
capillary, e.g., [3, 4].

Building on the findings of Minnema et al. [5], in
1993 Berge [6] reported that it is possible to perform

electrowetting on a dielectric material when it is placed
(as a solid film) between the liquid and the electrode. In
a nutshell, the dielectric film functions as the primary
energy-storing capacitor, thereby serving the same function as the EDL in conventional electrowetting. Since it
required a very high voltage for electrowetting to be
noticeable on an insulator, it is fitting that the first such
observation was made during investigations of "water
trees" (degradation structure grown in a polymer due to
humidity and an electric field) in high-voltage cables
[5]. Several kV were commonly used in the early
reports. Since reducing the operating voltage range
was simply a matter of making the dielectric layer
thinner and of higher quality materials, the thin-film
techniques of microelectromechanical systems
(MEMS) played a timely role in helping experimenters
reduce the dielectric thickness from the millimeter
range [5, 6] to the micrometer range [7]. To distinguish
it from the conventional practice of electrowetting on
a metal surface and because it provided unprecedented
viability for practical applications, Lee [8] and Moon
et al. [7] named this new material configuration
electrowetting-on-dielectric (EWOD). Figure 2 compares EWOD with conventional electrowetting.

Although a significantly higher voltage is necessary
to charge the dielectric capacitor of EWOD than the
EDL of electrowetting in order to obtain the same
degree of wetting, much higher voltages can be applied
to the dielectric for additional wetting. In other words,
a larger contact angle change can be achieved with
EWOD by allowing high voltages, whereas

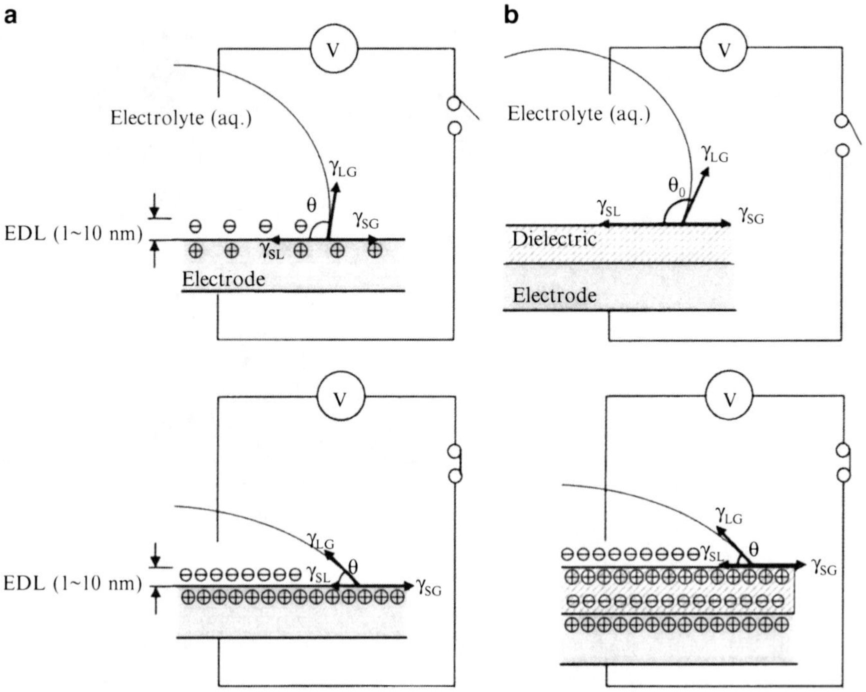

Electrowetting,
Fig. 2 Electrowetting and
EWOD [7]. (**a**) Illustration of
conventional electrowetting.
Top: With no external voltage
applied, charges are
distributed at the electrode-
electrolyte interface, forming
an EDL. *Bottom*: With an
external voltage applied,
charge density at EDL
changes, and the contact angle
decrease or increase. (**b**)
Illustration of EWOD. *Top*:
With no external voltage
applied, there is little charge
accumulation at the interface.
Bottom: With an external
voltage applied, charge
accumulates at the interfaces,
and the contact angle
θ decreases

conventional electrowetting is severely limited by the small breakdown voltage of the EDL. Figure 3 explains the trade-off between dielectric thickness and operating voltage in EWOD, and why it is advantageous to use EWOD (compared to conventional electrowetting) despite the relatively high voltages that are needed. The thick solid curve in the figure shows that the voltage required to induce a specified increase in wetting (i.e., a reduction of contact angle $\Delta\theta$) is proportional to the square root of the thickness of the dielectric layer for a given dielectric material (Teflon AF is assumed in the figure), while the thin solid straight line shows that the breakdown voltage of the dielectric layer is proportional to the thickness. For successful operation, the electrowetting voltage should be smaller than the breakdown voltage. Although the required voltage continues to increase with the dielectric thickness, the margin of safety increases as well.

Despite the higher operation voltages, e.g., 15–80 V in air or 10–50 V immersed in oil in present experimental reports, EWOD is preferred over electrowetting on a conductor (which uses less than 2 V) in most cases. One may lower the voltage requirements of EWOD without sacrificing the electrowetting effect by coating the dielectric with a very thin layer of a very hydrophobic material. This strategy works by reducing the contact angle hysteresis, thereby enabling

the contact lines (and thus droplets as well) to move more easily. A wide variety of droplet-medium combinations can be manipulated with EWOD, e.g., water in air, water in oil, oil-encapsulated water in air, oil in air, gas in water, etc. To learn more about the fundamentals of electrowetting and EWOD, readers are encouraged to consult [10].

Theories and Equations

The phenomenon of electrowetting can be interpreted (or understood) thermodynamically as well as electromechanically. In the thermodynamic interpretation, assuming conventional electrowetting, the electrical energy accumulated at the solid-liquid interface, $cV^2/2$ (where c is the capacitance per area of the EDL and V the voltage across it), decreases the interfacial energy. For EWOD, the interpretation is slightly modified: the electrical energy accumulated in between the electrode and the liquid, $cV^2/2$, where c is the capacitance per area and V the voltage applied between them, makes the interfacial energy between the solid surface and the liquid decrease. The interfacial energy is exhibited as the contact angle of the liquid on the solid surface θ, which is easily measurable; as the solid-liquid interfacial energy decreases, so

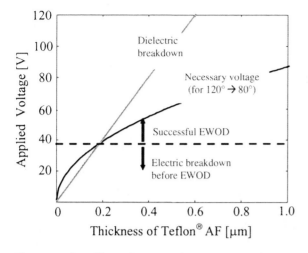

Electrowetting, Fig. 3 Electrowetting voltage and dielectric thickness [7]. The *thick solid line* represents the voltage required to obtain a certain degree of wetting (drawn for a contact angle reduction from 120° to 80° – enough reduction to slide a water droplet on many nonwetting solid surfaces in air) with Teflon AF as the sole dielectric, and the *thin solid line* represents the breakdown voltage for the same dielectric layer. A desired wettability increase is possible only where the electrowetting curve (the parabolic curve) stays safely below the *breakdown line* (the *straight line*). In other words, the dielectric should be thicker than where the two lines cross (~0.2 μ in the figure), as marked "Successful EWOD." If a smaller degree of wetting (e.g., moving a droplet on a solid surface coated with or immersed in oil) were acceptable, the corresponding electrowetting curve would be drawn lower, allowing lower voltages and thinner dielectrics. For conventional electrowetting, in contrast, the electrowetting curve would be drawn much lower than the EWOD curve in the figure due to the huge capacitance of the EDL (typically only several nanometers thick). Only ~1 V was used to induce enough wetting (70°–40° to rough estimations) on gold to draw water in air environment [9]. However, that 1 V, which was the minimum needed to move the interface, was the maximum that could be applied before severe electrolysis. With no margin of safety, the actuation was not reversible

does the contact angle. By combining this energy equation with the Young equation and following the thermodynamic principle of energy minimization, the electrowetting equation (often called Lippmann-Young Equation) can be derived and written as follows:

$$\cos\theta = \cos\theta_0 + \frac{cV^2}{2\gamma} \qquad (2)$$

where θ_0 is the contact angle when there are no charges at the solid-liquid interface, i.e., no EDL exists, and γ is liquid-fluid (air being a fluid in Eq. 1) interfacial

energy. The fluid may be a gas or a liquid that is immiscible with the liquid of interest, to which voltage is applied. For EWOD, where the specific capacitance can be obtained from the dielectric properties as $c = \varepsilon/d$ (ε is the permittivity, and d is the thickness of the dielectric), θ_0 coincides with the contact angle at no applied voltage, i.e., the Young angle, as the EDL has a negligible effect in this case. Considering the wide range of dielectric materials (i.e., representing a wide range of permitivities) and different liquid-fluid combinations used, it is convenient to define the second term on the right side of the equation as:

$$Ew = \frac{cV^2}{2\gamma} \qquad (3)$$

The dimensionless Electrowetting Number *Ew*, defined as Eq. 3, expresses the importance of electrical energy relative to the free energy at the liquid-fluid interface. The liquid-fluid interface can be liquid-gas, liquid-vapor, or liquid-liquid if immiscible liquids. Although a liquid-air system is considered a standard in this essay, liquid-liquid systems of aqueous solution and oil are found in many electrowetting devices.

The importance of EWOD over conventional electrowetting and the growing view of the phenomenon as a physical force have led some researchers to the electromechanical interpretation of electrowetting, as summarized in Fig. 4. Considering the interfaces close to the solid-liquid-fluid triple point as illustrated in Fig. 4b, one can see that the applied voltage produces normal electrostatic forces on the solid-liquid interface. Also, the fringe electric field on the liquid-fluid interface near the triple point exerts electrostatic traction normal to the liquid-fluid boundary right above the contact point (i.e., within χ in the figure), resulting in a net force parallel to the solid surface and causing droplet spreading. These electrostatic forces deform the liquid-fluid interface very close (submicron range in most practical cases) to the triple point, and this deformation reduces the slope of the interface away from the region. This reduced slope is the contact angle apparent to the observer. This pure electromechanical interpretation of electrowetting was made by theoretical derivations, e.g., [11, 12], and observed in experiments [13]. Under the electromechanical view, the electrowetting force is not infinitely concentrated at the contact point.

Electrowetting,
Fig. 4 Electromechanical interpretation of electrowetting, illustrated for EWOD [10]. (a) With no voltage applied, the contact angle of a liquid is the Young angle. (b) With voltage V applied, the contact angle decreases by electrowetting. In the figure magnifying the area of solid-liquid-fluid triple point, χ is on the order of the dielectric thickness d, which is on the order of 1 μm in many EWOD devices

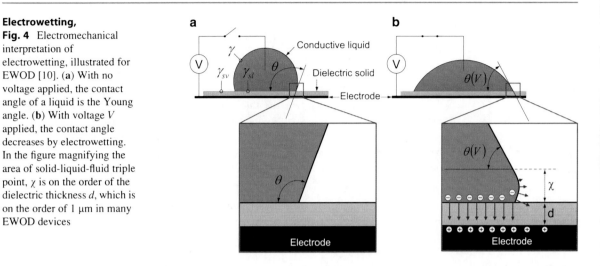

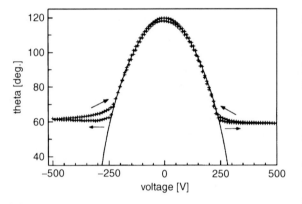

Electrowetting, Fig. 5 A typical electrowetting curve [14] follows the electrowetting equation (Eq. 2) until the electrowetting effect saturates at high voltages (Reprinted with permission from American Chemical Society)

Electrowetting Curve

Since the electrowetting equation (Eq. 2), derived by both the thermodynamic and the electromechanical approaches, matches the experiments quite well in most cases, it plays an important role in understanding the phenomenon and performing research and development for applications. An example is shown in Fig. 5, which also reveals an important limitation: contact angle saturation. When the voltage is applied above a certain limit, a further increase of voltage does not decrease the contact angle anymore, e.g., not below 60° in Fig. 5. Instead, excessively high voltages only bring about detrimental effects, such as the liquid-fluid interface becoming unstable and ejecting small

droplets. Contact angle saturation is one of the most common sources of frustration for those engaged in research and development involving electrowetting. Despite its importance, however, the origin of saturation is still not well understood.

Summary

While explaining electrowetting, other closely related terms (electrocapillarity and EWOD) have also been described. Based on the same fundamental concept of electric potential affecting the interfacial energy, each portrays the electrowetting phenomenon in a somewhat different material configuration. Depending on the user's interest, one configuration may be more convenient than others, although EWOD is by far the most widely used configuration today. Basic theories of electrowetting have been presented through the conventional thermodynamic interpretation as well as the relatively new electromechanical interpretation, both of which result in the same electrowetting equation. The electrowetting equation is very useful as experimental data usually match the electrowetting curve. Today electrowetting is used in a wide range of applications including optical, biomedical, and even electronic.

Cross-References

▶ Dielectrophoresis

References

1. Lippmann, G.: Relations entre les phenomenes electriques et capillaries. Ann. Chim. Phys. **5**, 494–549 (1875)
2. Beni, G., Hackwood, S.: Electro-wetting displays. Appl. Phys. Lett. **38**, 207–209 (1981)
3. Prins, M.W.J., Welters, W.J.J., Weekamp, J.W.: Fluid control in multichannel structures by electrocapillary pressure. Science **291**, 277–280 (2001)
4. Lee, J., Kim, C.-J.: Surface-tension-driven microactuation based on continuous electrowetting. J. Microelectrom. Syst. **9**, 171–180 (2000)
5. Minnema, L., Barneveld, H.A., Rinkel, P.D.: An investigation into the mechanism of water treeing in polyethylene high-voltage cables. IEEE Trans. Electr. Insul. **EI-15**(6), 461–472 (1980)
6. Berge, B.: Electrocapillarity and wetting of insulator films by water. Comptes Rendus de l'Academie des Sciences Serie II **317**, 157–163 (1993)
7. Moon, M., Cho, S.K., Garrell, R.L., Kim, C.-J.: Low voltage electrowetting-on-dielectric. J. Appl. Phys. **92**, 4080–4087 (2002)
8. Lee, J.: Microactuation by continuous electrowetting and electrowetting: theory, fabrication, and demonstration. Ph.D. Thesis, University of California, Los Angeles (2000)
9. Lee, J., Moon, H., Fowler, J., Schoellhammer, T., Kim, C.-J.: Electrowetting and electrowetting-on-dielectric for microscale liquid handling. Sensor. Actuator. **A95**, 259–268 (2002)
10. Nelson, W.C., Kim, C.-J.: Droplet actuation by electrowetting-on-dielectric (EWOD): a review. J. Adhes. Sci. Technol. (2012). (In print)
11. Jones, T.B.: On the relationship of dielectrophoresis and electrowetting. Langmuir **18**, 4437–4443 (2002)
12. Kang, K.H.: How electrostatic fields change contact angle in electrowetting. Langmuir **18**, 10318–10322 (2002)
13. Mugele, F., Buehrle, J.: Equilibrium drop surface profiles in electric fields. J. Phys. Condens. Matter **19**, 375112–375132 (2007)
14. Verheijen, H.J.J., Prins, M.W.J.: Reversible electrowetting and trapping of charge: model and experiments. Langmuir **15**, 6616–6620 (1999)

Electrowetting-on-Dielectric (EWOD)

▶ Electrowetting

Energy-Level Alignment

▶ Electrode–Organic Interface Physics

Engineered Nanoparticles

▶ Physicochemical Properties of Nanoparticles in Relation with Toxicity

Environmental Impact

▶ Fate of Manufactured Nanoparticles in Aqueous Environment

Environmental Toxicology

▶ Ecotoxicity of Inorganic Nanoparticles: From Unicellular Organisms to Invertebrates

Epiretinal Implant

▶ Artificial Retina: Focus on Clinical and Fabrication Considerations
▶ Epiretinal Prosthesis

Epiretinal Prosthesis

Guoxing Wang[1] and Robert J. Greenberg[2]
[1]School of Microelectronics, Shanghai Jiao Tong University (SJTU), Minhang, Shanghai P R, China
[2]Second Sight Medical Products (SSMP), Sylmar, CA, USA

Synonyms

Epiretinal implant; Retinal prosthesis

Definition

Epiretinal prosthesis refers to the device designed to restore partial vision for the blind people by stimulating the retina through electrodes on the retina surface.

Overview

Introduction to Retina

Vision is the main method for people obtaining information from the outside world. Unfortunately, it has been estimated that there are about 250 million blind patients in the world, and the number is increasing year by year. Researchers have tried many ways to address the causes of blindness. Electrical stimulation has been proposed as a means to restore some sight for blind patients. The idea behind this is similar to what has been done in cochlear implants, where electrical stimulation of hair cells in the inner ear has been used to successfully restore the hearing for the deaf people [1]. In retinal prosthesis, electrical pulses are being delivered to the retina, invoking a neural response that corresponds to the information of the image and creating a useful visual perception for the blind patients.

As is shown in Fig. 1, the retina is composed of several neuron cell layers, located in the back of the eye. The innermost layer in the retina that is closest to the lens and front of the eye is called the ganglion cell layer while the photosensors (the rods and cones) lie outermost in the retina against the pigment epithelium and choroid. The photosensors convert light into chemical/electrical signals which are conducted to the ganglion cells through intermediate layers of neurons. The ganglion cell axons form the optic nerve, carrying the image information to the brain. Interestingly, light has to travel through the thickness of the retina before striking and activating the rods and cones where the absorption of photons by the visual pigment of the photoreceptors is translated into a chemical/electrical message that can stimulate the succeeding neurons of the retina. The ganglion cells transmit all the information that has been processed by the retinal layers to the brain in terms of neuronal spikes.

Main Causes of Blindness

Some retinal diseases, unfortunately, disturb the above process and can prevent the patients from getting useful vision. Age-related macular degeneration (AMD) is one of those and the leading cause of blindness in the developed world [2]. It results from abnormal aging of retina. As the cones of the fovea die, the central vision is affected first and thus patients lose the capability of seeing objects in details. Retinitis pigmentosa (RP) is another disease that causes blindness [2]. This is a genetically related disease and refers to a collective name for a number of genetic defects that result in photoreceptor loss. RP affects the rods (used in night vision) first and then the cones (used in ambient daylight levels). There are 100,000 people in the USA affected by retinitis pigmentosa, according to the National Eye Institute.

In both RP and AMD, the vision is impaired due to the damage to the photoreceptors that transform photons to neural signals. Postmortem evaluations of retina with RP or AMD have shown that a large number of cells remain relatively healthy in the inner retina compared to the outer retina [3]. This has led to the idea of building a device to electrically stimulate the remaining viable cells in the retina to restore vision. In other words, the device can function as a replacement for the photoreceptors, converting the light information to electrical signals which can be interpreted by the remaining retinal neuron cells. The remaining ganglion cell neurons can pass the information to the brain, which perceives patterns of light and dark spots corresponding to the electrical signals generated by the device.

Types of Visual Prostheses

Retinal prostheses refer to the devices that mainly aim to fix the vision loss problems through electrically stimulating the retina. Electrodes are used to deliver electrical pulses to the retina. Depending on where the electrodes are placed, there are two leading approaches of retinal prostheses: epiretinal and subretinal. For epiretinal prosthesis, the electrodes are placed inside the vitreous, close to the inner layers of ganglion cells and nerve fibers. In contrast, in the sub-retinal prosthesis, the electrodes are placed under the retina where the rods and cones were. This entry will mainly explain the challenges and progresses associated with epiretinal prostheses. Other types of visual prostheses will be briefly discussed.

If the electrodes are implanted between the pigment epithelial layer and the outer layer of the retina, it is termed subretinal prosthesis. Compared with epiretinal approach, sub-retinal prosthesis has the potential advantage of closer proximity to the natural circuits of the retina and easier fixation of the electrodes since they are kind of sandwiched between the tissue layers. The disadvantages might include a more difficult surgery and more engineering challenges such as heat dissipation and power management. A German team

Epiretinal Prosthesis,
Fig. 1 Schematic of the human eye (**a**) and the retina-layered structure (**b**). (Adapted from Webvision, http://webvision.med.utah.edu/)

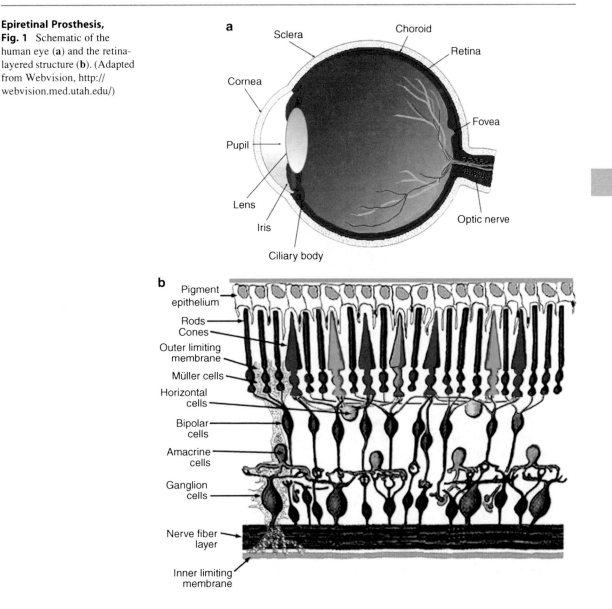

led by the University Eye Hospital in Tübingen was formed in 1995 by Eberhart Zrenner to develop a subretinal prosthesis. Recently, they reported very promising results with acute human studies [4], but creating a reliable long-term subretinal device has proven more challenging. Also, currently the subretinal surgery requires filling the eye with toxic silicone oil. Clearly there are many issues to be resolved with the subretinal approach before it can be a viable practical therapy.

Another group led by Joseph Rizzo at the Massachusetts Eye and Ear Infirmary and John Wyatt at MIT also worked on retinal prosthesis and recently reported

a subretinal stimulator with animal implantation results [5].

It is also worthwhile mentioning that similar to retinal stimulation, researchers worldwide have also proposed to electrical stimulation of optic nerve and visual cortex to invoke phosphenes to provide visional inputs to the patients. Among those are Dr. Claude Veraart from University of Catholique de Louvain at Belgium, Dr. Tohru YAGI from Tokyo Institute of Technology, Dr. Palankar from Stanford University, Dr. Nigel Lovell from University of New South Wales at Australia, and Dr. Mohammad Sawan from Ecole Polytechnique de Montreal at Canada. Despite all the

Epiretinal Prosthesis,
Fig. 2 Scheme of epiretinal prosthesis components (Source: Artificial Retina Project, Department of Energy, http://artificialretina.energy.gov, January 2011)

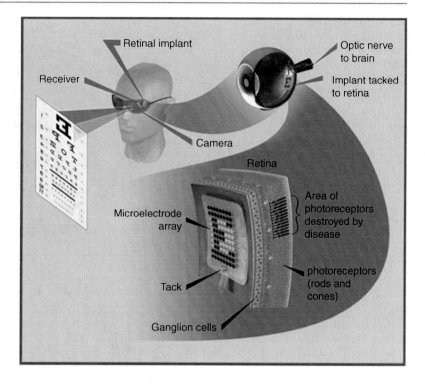

challenges ahead, with the great efforts by the scientists, engineers, and entrepreneurs, there is no doubt people will move forward in curing blindness.

Principles of Epiretinal Prosthesis

Figure 2 shows the concept of epiretinal prosthesis. The device can be partitioned into two components, the external unit and the implant unit. The external unit includes the camera, signal processor, and power and data transmitter. The implant unit includes the power and data receiver, stimulator, and electrodes.

The whole device works as follows. An external video camera captures the image in the field of vision. This image is processed by a video signal processing unit that converts this image into stimulation data for the electrical stimulus. This data is then wirelessly transmitted to and received by an integrated circuit chip in the receiver inside the eye. The chip converts the signal to actual electrical stimulus currents which are delivered to the retina through the microelectrode array as shown in the figure. The electrode array is fixated on the retina by a retinal tack.

Image Capture

A digital camera, most likely mounted on the patient's eyeglasses or head, captures the images in the field of vision. The raw data from the camera has to be processed to convert the video information to the format that can be understood by the implant chip to stimulate the retina. For example, the raw data needs to be downsampled, since most cameras today have orders of higher number of pixels than needed for a prosthetic device with only tens of electrodes. To date, Second Sight Medical Products, the leading company in retinal prosthesis, is carrying out a chronic clinical trial for a device with 60 electrodes (The Argus™ II). Devices with more electrodes are also being researched worldwide [6]. Besides downsampling, the video processing has to carry out complex procedures such as image enhancement.

Research is also being conducted to move the camera inside the eyeball. An intraocular camera used for image capturing may include an optical imaging system with an image sensor array, which may be enclosed in an implantable biocompatible housing together with other electronic components. The difficulty lies in the miniaturization of such an optical system in a way that is protected from the harsh corrosive environment of the body. Heat is also a challenge. Leaving the camera outside has the advantages of lower power consumption inside and thus less heat dissipation. In addition, it allows for convenient refinement and upgrades of signal processing algorithms and functionality in the exterior blocks.

Besides the imaging processing capability, the external unit should also provide options for personal customizations. The unit should have an easily accessible user interface that permits the patient to do some necessary image enhancement adjustments on his/her own. An error detection function should also be built in.

Wireless Telemetry

The implant needs power for the functions of electrical stimulation, image data recovery, and other circuits. The power required by the implant is provided by wireless telemetry. The overall power needed from telemetry is a function of the stimulation parameters such as stimulation voltage and stimulation current amplitude, which in turn depend on many other parameters, such as the electrodes shape, distance to the retina, and even disease conditions of patients [3]. It has been estimated that a peak power of 250 mW may be needed for 1,000 electrodes stimulation [7]. During stimulation, the actual power needed by the implant may change with the stimulation patterns, which are a function of the image "seen" by the patients.

On the other hand, the maximum power that the implant can consume is governed by the safety issue that the temperature of the implant contacting the tissue should not exceed the normal temperature by a few degrees. Heating effects of RF power and chip operation have been studied experimentally and through modeling [8]. In sum, these studies suggest that wireless data and power transmission will not create a significant thermal effect, while the chip operation may generate a significant amount of heat. However, the liquid environment of the eye acts as a heat sink that is capable of dissipating a significant amount of power. An electronic chip positioned away from the retina can run at considerably higher power than a chip positioned on the retinal surface [9].

The required data rate for transmitting the image data from the external unit to the implant unit depends on the stimulus parameters and the refreshing rate of the stimulation. Assuming a total of 20 bits for one electrode stimulus current definition [10] and a refreshing rate of 50 frames per second, for 100 electrodes, the minimum data rate is $20 \times 50 \times 100 = 100$ kbps (kilobits per second). Obviously higher data rate will be needed if more electrodes are being stimulated.

Besides relaying the image information from the external unit to the implant unit, it is also necessary to monitor the implant status for safety and also for optimal control purpose. For example, it is important to know the temperature of the implant and possibly shut down the device when the temperature exceeds a predetermined safety limit. Also, the change of pH value of the solution due to electrical stimulation could potentially be an indicator of tissue damage; thus it should be transmitted to the external unit for safety control [11]. The power level information inside the implant can also be detected and relayed back to the external unit to achieve an optimal power transfer system. The electrode impedance may be another parameter that needs to be transmitted back to the external unit for determining optimal stimulation patterns [12]. In sum, a communication channel is required to monitor the important implant parameters such as the temperature, pH value, power level, and electrode impedance, etc. These physiological signals generally change slowly and thus do not require a high data rate communication channel.

In summary, the telemetry system for retinal prosthesis needs to perform three functions, as shown in Fig. 3:

- Transmitting power to the implant
- Transmitting image data to the implant
- Transmitting data from implant to the external unit

To power the implant unit, the epiretinal prosthesis, like most other implantable applications, use inductive coupling as a means for wirelessly transmitting power. When an electrical current flows through a wire, a magnetic field is generated. A coil is formed when this conducting wire is wound into one or several loops. The magnetic field lines generated by each loop of coil combine with the lines generated by other loops to produce a concentrated field in the center of the coil. When another (receiving) coil is placed close to the transmitting coil, the magnetic field generated by the transmitting coil passes through the windings of the second coil and these two coils are magnetically (inductively) coupled. Further, if the current of the transmitting coil is alternating, by Faraday's law, the time-varying magnetic field generated by

**Epiretinal Prosthesis,
Fig. 3** Illustration of an
inductive power and data
telemetry for epiretinal
prosthesis

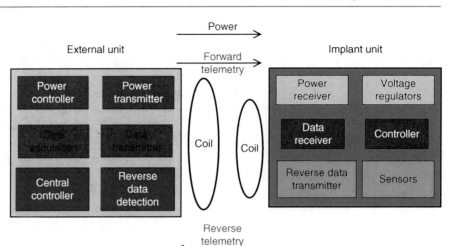

the alternating current induces an alternating voltage in the receiving coil, thus enabling energy transfer from the transmitting coil to the receiving coil.

The whole system is powered by a portable battery outside the body with power transmitted to the implant through the coupled coil pair. The power amplifier at the primary side converts the battery DC power into radio frequency (RF) power to be radiated by the primary coil and thus to be received by the secondary coil. This power is then converted back to DC by the power recovery circuits to operate the implant electronics.

The size and shape of the coils are highly critical for an efficient power transfer and deserves special attention. Firstly, to facilitate the energy transfer, the two coils should be constructed with a shape that can maximize the coupling and be placed as close as possible. Secondly, the wire diameter of the coil should be chosen according to the frequency of the electrical current to minimize the losses associated with electrical resistance. For example, one way to reduce the resistance is to construct the coil with several wires (strands) twisted together (for example, litz wire) to increase the equivalent cross-section conducting area and reduce the resistance. However, the number of strands is normally limited by the maximum allowable size of the coils. Another important parameter of the coil is the number of the loops of the conductor (called number of turns). With the same current in the transmitting coil, more turns can generate higher magnetic field thus inducing a higher voltage across the implanted receiving coil. However, as the number of turns increases, the equivalent resistance of the coil

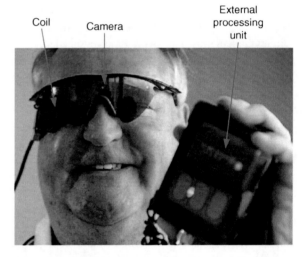

Epiretinal Prosthesis, Fig. 4 Photo of external unit developed by Second Sight Medical Products. Coil is on the side of the eyeglasses

increases too, which causes higher power dissipation on the coil. For a given size of coil, the number of strands trades off with the number of turns, and both should be optimized for a given application to achieve high power efficiency. The coil being implanted should be constructed as small as possible to minimize the patient discomfort and to ease surgery/injection procedure. The external coil, mounted on eyeglasses, however, is usually much larger for better coupling as shown in Fig. 4.

The magnetically coupled coil pair can also be used to transmit data wirelessly from the external unit to the implant unit. By modulating the power carrier (used to

Epiretinal Prosthesis,
Fig. 5 Scheme of typical
biphasic pulses for stimulating
retina

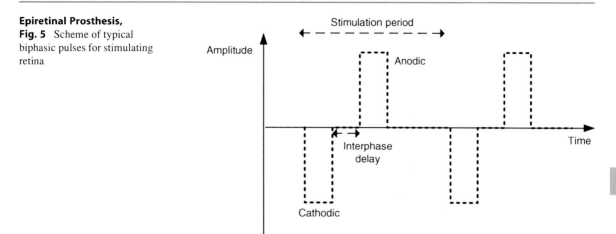

Stimulating the Retina

Electrical current is delivered from the electrodes to the retina, and presumably the retinal cells are stimulated, which generate neuronal spikes to be passed to brain. A high-density electrode array is composed of a substrate that provides mechanical support and the conducting material, for instance, a noble metal such as platinum, creating the electrical connection between the stimulator and the tissue. At the electrode/tissue interface, the electronic current is transformed into ionic current which sets up a voltage gradient that then activates the neurons.

It is desirable for the electrodes to be placed on to the macular region and close to the retina as this will create more efficient stimulation. The electrode array, as illustrated in Fig. 2, is fixated onto the retina by a tack. Optimizing this interface is an important area of research.

As shown in Fig. 5, a typical electrical stimulus is biphasic with the pulse of a cathodic phase and an anodic phase. The interphase interval (time between the two phase) separates the pulses slightly to make sure that the next pulse would not reverse the physiological effect of the previous one. The stimulus information contains the amplitude, width, interphase delay, and the repetition rate of the pulses. All this information highly correlates to the image information and the actual stimulation strategy may also be patient-specific. In other words, for the same picture being "seen" by the patient, the stimulation currents may be very different.

The integrated circuit chip is responsible for generating such current according to the commands it receives through data telemetry from the external unit. The stimulator is miniaturized to be accommodated within the constraints of the available space and designed to be controlled with the minimum power to cut down the possible heat dissipation inside the body.

Current Research Progress

In the early 1990s, Humayun and colleagues demonstrated that controlled electrical stimulation of the retina in individuals blind from RP and AMD results in visual percepts in the form of light dots. Experiments using pattern electrical stimulation of a human retina resulted in subjects perceiving simple forms of vision with the percepts corresponding in time and location to the electrical stimulus. These were short-term tests, with the stimulating devices temporarily positioned on the retina.

In 1998, Second Sight Medical Products was founded to build a retinal prosthesis. A permanent implant, developed by Second Sight Medical Products, with 16 electrodes (4 × 4 array) was implanted in patients starting in 2002. The implanted subjects were able to use a camera to detect the presence or absence of ambient light, to detect motion, read a large letter, count objects, and recognize and differentiate objects such as cup from plate. The longest implant still in use by a patient today is over 7 years to date.

transmit power), data can be sensed by the implant unit. Research is also being conducted to transmit data using a different carrier instead of the power carrier, mainly to increase the amount of data that can be transmitted without sacrificing the power transmission efficiency.

Starting in 2007, Second Sight went on to a clinical trial for its second-generation device, Argus II, which has 60 electrodes. So far there have been more than 30 patients implanted with centers in the USA, England, France, Geneva, Mexico, etc. The patients with such devices, for the first time, could see a series of letters, thus recognizing simple words and even sentences. Some patients could recognize numbers on cards held in their palm.

One natural assumption is that with more electrodes, the epiretinal prosthetic device could provide better vision inputs for the suffered blind patients and thus restore more meaningful vision to the patients, such as face recognition, unaided mobility.

Simulations of prosthetic vision suggest that 600–1,000 electrodes will be needed to restore visual function to a level that would allow blind to read, navigate a room unaided, and recognize faces [13, 14]. But, through the use of scanning, clearly far fewer can make a significant impact. As the number of stimulation outputs increases, the complexity of the stimulator electronics increases, requiring larger amounts of power to provide to the implant and a higher data rate, for transmitting the image information to the stimulating electronics. Protecting these high-channel-count devices from the body is also increasingly difficult as the numbers of electrodes increases. To date, only the Argus I and II from Second Sight have demonstrated long-term reliability in patients.

Future Direction for Research

With the advancement of technologies in the areas of microelectronics, mechanics, materials, there is no doubt that devices with better performance than current state of the art will emerge in the future. Needless to say, this involves a great deal of engineering challenges and trade-offs. Can 1,000 analog drivers be packed into a microchip that can be small enough to be implanted? Can enough data be transmitted into the device without consuming too much power to actually overheat the implant? Can a hermetic package be built to support 1,000 electrodes?

Another important area is to understand the neural response of the retina when stimulated by electrical current. Many experiments have been carried out to find out the relationship between stimulation current

(such as the magnitude, waveform, and frequency) and the perception, but more work is needed.

Cross-References

▶ Bio-Inspired CMOS Cochlea
▶ Biomimetics
▶ Biosensors
▶ Nanostructured Functionalized Surfaces
▶ Organic Bioelectronics

References

1. Sanders, R.S., Lee, M.T.: Implantable pacemakers. Proc. IEEE **84**(3), 480–486 (1996)
2. Margalit, E., Sadda, S.R.: Retinal and optic nerve diseases. Artif. Organs **27**(11), 963–974 (2003)
3. Humayun, M.S., Prince, M., de Juan, E.J., Barron, Y., Moskowitz, M., Klock, I.B., Milam, A.H.: Morphometric analysis of the extramacular retina from postmortem eyes with retinitis pigmentosa. Invest Ophthalmol Vis Sci **40**(1), 143–148 (1999)
4. Zrenner, J., et al.: Subretinal electronic chips allow blind patients to read letters and combine them to words. Proc. R. Soc. B. (2010). doi:10.1098/rspb.2010.1747
5. Shire, D.B., Kelly, S.K., Jinghua, C., Doyle, P., Gingerich, M.D., Cogan, S.F., Drohan, W.A., Mendoza, O., Theogarajan, L., Wyatt, J.L., Rizzo, J.F.: Development and implantation of a minimally invasive wireless subretinal neurostimulator. IEEE Trans. Biomed. Eng. **56**(10), 2502–2511 (2009)
6. Weiland, J.D., Humayun, M.S.: Visual prosthesis. Proc. IEEE **96**(7), 1076–1084 (2008)
7. Wang, G., Liu, W., Sivaprakasam, M., Kendir, G.A.: Design and analysis of an adaptive transcutaneous power telemetry for biomedical implants. IEEE Trans. Circ. Syst. I. **52**, 2109–2117 (2005)
8. Singh, V., Roy, A., Castro, R., McClure, K., Weiland, J., Humayun, M., Lazzi, G.: Specific Absorption Rate and Current Densities in the Human Eye and Head Induced by the Telemetry Link of a Dual-Unit Epiretinal Prosthesis. IEEE Trans. Antennas Propag. **57**(10), 3110–3118 (2009)
9. Piyathaisere, D.V., Margalit, E., Chen, S.J., Shyu, J.S., D'Anna, S.A., Weiland, J.D., Grebe, R.R., Grebe, L., Fujii, G., Kim, S.Y., Greenberg, R.J., De Juan Jr., E., Humayun, M.S.: Heat effects on the retina. Ophthalmic Surg. Lasers Imaging **34**(2), 114–120 (2003)
10. Sivaprakasam, M., Liu, W., Wang, G., Weiland, J.D., Humayun, M.S.: Architecture tradeoffs in high density microstimulators for retinal prosthesis. IEEE Trans. Circ. Syst. I Spec. Issue Biomed. Circ. Syst. **52**(12), 2629–2641 (2005)
11. Chu, A.P., Morris K., Greenberg R.J., Zhou D.M.: Stimulus induced pH changes in retinal implants. In: 26th annual international conference of the engineering in medicine

and biology society. IEEE, San Francisco, California, vol. 2, pp. 4160–4162 (2004)

12. Shah, S., Chu, A., Zhou, D., Greenberg, R., Guven, D., Humayun, M.S., Weiland, J.D.: Intraocular impedance as a function of the position in the eye, electrode material and electrode size. In: Conference proceedings, 26th annual international conference of the engineering in medicine and biology society. IEEE, San Francisco, California, vol. 2, pp. 4169–4171 (2004)

13. Cha, K., Horch, K.W., Normann, R.A.: Mobility performance with a pixelized vision system. Vis. Res. **32**(7), 1367–1372 (1992)

14. Hayes, J.S., Yin, J.T., Piyathaisere, D.V., Weiland, J., Humayun, M.S., Dagnelie, G.: Visually Guided Performance of Simple Tasks Using Simulated Prosthetic Vision. Artif. Organs **27**(11), 1016–1028 (2003)

EUV Lithography

Chimaobi Mbanaso and Gregory Denbeaux
College of Nanoscale Science and Engineering, University at Albany, Albany, NY, USA

Synonyms

Extreme ultraviolet lithography (EUVL); Soft x-ray lithography

Definition

EUV lithography is a technology that uses radiation near 13.5 nm wavelength to transfer patterns from a reflective mask to a substrate coated with a light-sensitive material called a photoresist. This patterning technique occurs under vacuum conditions in an all-reflective optics configuration and is used to fabricate integrated circuits with printed features that are smaller than 32 nm. The technology is considered to be the likely next-generation lithography method with the prospect of succeeding the current immersion lithography technique.

Overview

Semiconductor integrated circuits (ICs) which are embedded in various electronic devices are fabricated using a lithography process. As technology has advanced since the invention of the IC in 1958, lithography has played a vital role in patterning smaller features to produce faster and more powerful operating devices. Decreasing the exposure wavelength used for lithography is one of the ways in which the printing of ICs has steadily improved using optical lithography techniques. Extreme ultraviolet (EUV) lithography operating at 13.5 nm wavelength will possibly succeed the state-of-art lithography method as the next-generation technique of choice for continuing the shrinking of ICs [1–9]. Other next-generation lithography methods that have been considered include proximity x-ray lithography, electron projection lithography, ion projection lithography, and multi-e-beam direct-write [6, 10]. Before the adoption of the next lithography technology, important requirements such as throughput, cost of ownership, and image quality have to be considered [6].

In lithography, an optical projection printing system uses light to image a mask pattern onto a wafer substrate that has been coated with a light-sensitive material called a photoresist. The projection lens system typically reduces the mask pattern by a factor of four onto the photoresist [11, 12]. The lithography process defines the critical dimensions of the printed features while other process steps such as etching and metallization contribute toward the overall fabrication of the device. In EUV lithography, 13.5 nm wavelength of light enables the transfer of patterns from a reflective mask to a photoresist-coated substrate. The short EUV wavelength is easily absorbed by most materials; hence, the lithography process must be operated under vacuum conditions and by using special reflective materials in the optical system in contrast to conventional lithography techniques that use refractive optics [3, 4].

One of the drivers for implementing EUV lithography is the capacity to print high-resolution features by the following fundamental relationships describing the lithographic system with resolution (R) and depth of focus (DOF). Thus,

$$R = k_1 \frac{\lambda}{NA} \tag{1}$$

$$DOF = \pm k_2 \frac{\lambda}{NA^2} \tag{2}$$

where λ is the optical wavelength and NA is the numerical aperture of the projection system [3]. The parameters k_1 and k_2 are determined through empirical means and depend on the critical dimension (CD) tolerance and the size of an acceptable IC manufacturing process window [5]. The printing of smaller features in the semiconductor industry has followed a trend in which the number of transistors in an IC has doubled after approximately every 24 months (Moore's law) [13]. Strategies to improve the optical lithography technique to keep pace with Moore's law have led to modification of the lithography system according to Eq. 1; hence, decreasing the wavelength of light, fabricating high NA optics or decreasing k_1 have led to smaller patterned features [8]. The optical lithography technique that used the 436 nm g-line illumination wavelength during the early 1980s printed features with dimensions of over 2 μm by using a system with an NA of 0.15 and a k_1 of 0.8 [10]. At present, 193 nm lithography uses an immersion liquid along with various resolution enhancement techniques such as phase shifted masks, optical proximity correction, off-axis illumination, and double patterning lithography to print features below 32 nm [10]. Shortening the wavelength by using 13.5 nm is the next practical way to enable the printing of smaller features for future IC devices.

Initial concepts of EUV lithography were introduced in the 1980s from research efforts in Japan and the United States while using a broader radiation wavelength range of about 10–30 nm, referred to as soft x-rays [3, 4]. National laboratories such as the Lawrence Livermore National Laboratories (LLNL), Sandia National Laboratory (SNL), Lawrence Berkeley National Laboratory (LBNL), as well as the AT&T Bell Laboratories performed some of the early research in using EUV radiation in an all-reflective projection lithography system [4]. Since then, research efforts by industry consortia and semiconductor manufacturers have all been instrumental in leading developmental efforts in EUV lithography for high-volume manufacturing. Some of the early EUV development programs which commenced in the 1990s include Extreme Ultraviolet Limited Liability Company (EUV LLC), the European Extreme UV Concept Lithography Development System (EUCLIDES), and the Japanese Association of Super-Advanced Electronics Technologies (ASET). These programs and many others since then [4], have supported the development of light sources, optics, masks, photoresists, and vacuum equipments needed for EUV lithography. The

EUV Lithography, Fig. 1 Schematic showing the EUV lithography projection optical system based on reflective mirrors (Courtesy of Carl Zeiss SMT GmbH)

sections that follow discuss these essential components of an EUV exposure tool and their development for this technology (Fig. 1).

EUV Light Sources

Historically, progressively shorter wavelengths in the lithography tools have been used as one of ways to reduce the feature sizes of IC devices. The visible g-line at 436 nm wavelength was used in the early 1980s for the lithography process [11]. Its wavelength successor was the ultraviolet i-line at 365 nm which was used in the late 1980s [11]. Both the g-line and i-line radiation wavelengths were produced with a mercury arc lamp. With further decrease in feature sizes came the advent of deep ultraviolet (DUV) 248 nm wavelength lithography using a KrF (krypton fluoride) excimer laser in the late 1990s and then in 2001 [11], the ArF (argon fluoride) excimer laser was used for advanced lithography and has been modified to use an immersion liquid. ArF lithography uses light at 193 nm wavelength. The fluorine (F_2) excimer laser, which operates at 157 nm wavelength, was briefly considered to be the next lithography step toward shorter wavelengths but did not proceed to high-volume manufacturing [10, 11]. EUV lithography, operating at 13.5 nm, is considered the next step in shortening the wavelength of light used for lithography to continue the scaling of device features.

To generate high-power radiation at 13.5 nm for EUV lithography, high-temperature plasmas (approximately 30 eV) of a fuel material are used [3, 14]. For high-volume manufacturing applications, these hot plasmas of a target fuel such as xenon, tin, or lithium are generated from either an electrical discharge or by using an intense laser pulse. These two viable commercial solutions for generating EUV radiation are referred to as discharge produced plasma (DPP) and laser produced plasma (LPP). Other sources of EUV radiation such as synchrotrons were used for early demonstrations and pioneer studies of EUV lithography but requirements such as space restrictions limited their practical use for commercial applications [6, 7]. Tin is currently the leading target fuel for high-volume manufacturing EUV sources because of its higher conversion efficiency (CE) compared to xenon and lithium [3, 14]. CE, which is the ratio of the generated energy from the EUV source in the 2% bandwidth of 13.5 nm wavelength to the energy input to the EUV source, is used to estimate the utility requirements and fuel choice [3].

In order to focus the light generated by the EUV source, collector optics that are either grazing incidence or normal incidence are used. The point where the light is focused is called the intermediate focus (IF) and connects the source-collector module to the illuminator and projection mirrors of the EUV lithography scanner. Only a portion of the EUV radiation emitted by sources is collected and transmitted to the IF due to factors such as the geometrical restrictions in the collector design, the efficiency of the collectors to reflect the EUV radiation, absorption of EUV by background gases and spectral purity filters (SPFs) along the propagation path to the IF [3]. SPFs are needed to suppress out-of-band radiation produced outside the 2% bandwidth of the 13.5 nm wavelength by the plasma sources. The presence of these filters can further attenuate the in-band EUV power. Other filters include debris barriers to mitigate high-energy particles from the plasma source which can otherwise lead to degradation of reflective optics in close proximity to the source [3, 4].

EUV source specifications are determined based on consumer requirements which include throughput, cost of ownership, and image quality. For instance, for high-volume manufacturing, the EUV power generated must be sufficient to ensure that a high wafer throughput, which is estimated to be about 100 wafers per hour, is achieved. This equates to a power of 115–180 W depending on the sensitivity of the photoresist at 5–10 mJ/cm^2, respectively. Hence, sources that meet the power requirements, while maintaining energy stability and reliability during operation, are needed. EUV sources can also be used for metrology purposes and the specifications for such applications are somewhat different from lithography exposure tools [6].

EUV Multilayer Optics

In contrast to prior optical lithography technologies that use transmission optics for propagating the light to the photoresist, EUV technology needs vacuum conditions and reflective optics to avoid absorption of the short wavelength at 13.5 nm. The optics system used for EUV lithography consists of reflective illumination optics used to effectively allow the mask to be illuminated and reflective projection optics to image the pattern on the photoresist. The reflective optics consist of repeated bilayer pairs to make a multilayer mirror material [3, 15]. The mathematical relationship between the bilayer spacing, d, incidence angle, θ, and wavelength, λ, is shown in the equation below,

$$n\lambda = 2d \cos \theta \qquad (3)$$

where n is an integer. The bilayer stack to reflect EUV radiation from a multilayer consists of a metal (high Z material) and a spacer (low Z material) which correspond to materials with high and low atomic numbers (Z). Historically, two of the common multilayers considered for EUV lithography were Molybdenum/Beryllium (Mo/Be) and Molybedenum/Silicon (Mo/Si) corresponding to best reflectivity performance at wavelengths in the 11.3–11.6 nm and 13.3–13.6 nm range, respectively [3]. The Mo/Si for 13.5 nm is widely used as the metal/spacer material of choice for EUV technology with a typical stack consisting of 40

Mo/Si layer pairs. These mirrors have shown a peak reflectivity of close to 70% at 13.5 nm [15]. The quality of the multilayer coatings contributes to the reflectivity of the multilayer mirrors and is dependent on the deposition process. Magnetron sputtering [3, 15] is one of the common deposition techniques for EUV lithography mirrors because of the ability to coat large optics, the relatively fast sputtering rate and reproducibility from process to process. Some of the other deposition methods include ion-beam sputtering, electron-beam evaporation, and pulsed laser deposition [3].

Contaminants, which include water and hydrocarbons, in the presence of EUV photons can lead to oxidation or surface contamination resulting in a loss of reflectivity [3, 16]. The operational lifetime of the projection optics used in the EUV system must satisfy greater than 30,000 light-on hours [3, 5]. Thus, degradation at the surface of the optics must be kept at a minimum. Molecular contamination has been attributed to cracking of adsorbed molecules at the mirror surface by EUV-induced secondary electrons [16]. Typical species causing contamination are oxygen-containing molecules and carbon-containing molecules. A small increase in the layer of oxide (less than 1.5 nm) on the surface of these multilayer mirrors leads to a reflectance loss of greater than 1%, and is unacceptable [3]. Such optics contamination will decrease the throughput of the exposure tool and printing uniformity. Strategies to reduce contamination of the optics include using protective capping layers, such as ruthenium, that are resistant to oxidation [3, 16]. Other strategies to extend the lifetime of the optics and reduce carbon growth include improvement of the vacuum conditions and adopting various cleaning techniques that would efficiently remove the carbon but not damage the underlying mirror. Possible removal methods include the use of molecular oxygen and atomic hydrogen [3].

EUV Masks

The EUV mask, which contains the pattern to be imaged onto the wafer, has to be reflective to avoid absorption of the short wavelength of light; hence, it is different from conventional transmissive optical masks. The EUV mask undergoes two main fabrication processes. These are the development of the mask

blank and the patterning of the mask. For the mask blank fabrication, a flat and defect-free substrate with a low coefficient of thermal expansion material maintains the rigidity of the mask with minimum distortion [3, 6]. On this substrate, a reflective multilayer structure consisting of alternating layers of Mo/Si is deposited and covered by a capping layer to prevent oxidation. The multilayer on the EUV mask ensures high reflectivity at the 13.5 nm wavelength and consists of 40 pairs of Mo/Si bilayers as shown in Fig. 2. An absorber layer coating, antireflective coating on the topside of the mask as well as a conductive layer coating on the backside of the mask are all performed as part of the mask blank fabrication process. The absorber layer helps in the pattern-forming process by providing contrast to the reflective multilayer through absorption and minimum reflection, and materials such as tantalum nitride (TaN) are used for this function [3]. To minimize reflectivity during mask inspection using deep ultraviolet (DUV) light, an antireflective coating can be applied to the top of the absorber layer. A buffer layer material such as silicon dioxide (SiO_2) or ruthenium can be used to protect the multilayer structure during the etching or repair of the absorber layer [3]. The conductive backside coating is made to be compatible with the electrostatic chuck used to support and flatten the mask in the exposure tool as well as other process tools. Electrostatic chucking ensures that the clamping forces are evenly distributed over the mask area to maintain flatness. The next step in the fabrication of the EUV masks is the mask patterning by electron beam writing and dry etching. The writing process is similar to that used for conventional optical masks but with tighter restrictions on distortions and thermal specifications.

Mask defects can occur in the absorber layer, the multilayer reflective coating, or in the substrate below the multilayer coating. Substrate defects can propagate through the multilayer coating and cause a phase defect which is the perturbation of the reflected wave front [3]. Amplitude defects can be caused by particles near the top of the multilayer or if there is a flaw in the multilayer. Other defects can occur in the absorber layer where a portion of the absorber is missing from its desired location (clear defect) or if there is absorber material in an undesired location (opaque defect) [3]. Unlike optical lithography that uses a highly transparent membrane (pellicle) to be placed over the mask to

EUV Lithography,
Fig. 2 Schematic showing
a stacking structure of
a reflective mask used for
EUV lithography

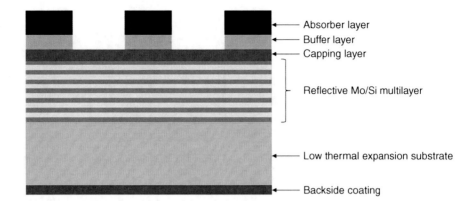

Absorber layer
Buffer layer
Capping layer
Reflective Mo/Si multilayer
Low thermal expansion substrate
Backside coating

provide protection from small particles that can fall and cause defects, the absorption of the short 13.5 nm wavelength makes it unusable in EUV lithography. Nevertheless, during shipping, storage, or handling, various protection schemes have been evaluated such as the dual pod concept [3]. The dual pod is a special mask carrier that gives double protection to the mask by using an inner and outer compartment to maintain a particle-free mask before use in an EUV exposure tool.

To avoid defects from being imaged on the wafer, the EUV mask must go through the manufacturing process defect-free and remain that way during use. Reliable and accurate EUV mask inspection is needed to capture and characterize printable mask defects at various levels of fabrication. Inspection tools that use EUV wavelength (actinic) and DUV wavelengths (non-actinic) have been used for this purpose [3]. However, to meet the requirements for high-volume manufacturing, actinic inspection and imaging tools will likely be needed due to the limits of DUV inspection techniques [3].

EUV Photoresists

Photoresists, which are the light-sensitive materials used in the lithography process, have undergone much progress from the Novolac® photoresists used in the i-line (365 nm) era to chemically amplified photoresists used for 248 and 193 nm lithography [12, 17]. The platform for EUV photoresists, also based on chemical amplification, is similar to those used for 248 and 193 nm optical lithography. These photoresists were first used in the mass production of semiconductor devices during the transition from the i-line of the mercury lamp to the KrF (248 nm) excimer laser [18]. These chemically amplified photoresists have been further employed from KrF (248 nm) to the current ArF (193 nm) immersion lithography and exhibit high sensitivity and resolution [6]. In chemically amplified photoresists, pattern formation occurs through deposition of energy in the photoresist upon exposure to the incident radiation leading to the generation of acids. The acids catalyze the pattern forming reaction at elevated temperatures referred to as the post exposure bake (PEB) step. The image in the photoresist is then developed in an aqueous base to reveal the pattern. The efficiency of these processes depends on the absorption efficiency of the incident radiation, the number of acids generated, and the efficiency of the catalytic chain reaction [12]. In EUV, the mechanism for interaction of radiation with the photoresist is different than exposures using 248 and 193 nm radiation technologies. The high energy of the EUV photon, 92 eV (13.5 nm), ionizes the components of the photoresist film to generate secondary electrons which contribute to the chemical reactions occurring during exposure [17, 18]. The properties of these photoresists depend on the photoacid generator (PAG) and the polymer used in the formulation and affects the specifications for resolution, line edge roughness (LER), and sensitivity [12]. The sensitivity of the photoresist required for EUV lithography is dependent on the power produced by the source. One of the concerns of using EUV photoresists in the vacuum environment with all the nearby reflective optics is the release of hydrocarbon contaminants during exposure, which could decrease the lifetime of the optics [19]. The outgassing of photoresist materials exposed to EUV wavelengths has been studied using various methods [19].

EUV Lithography,
Fig. 3 ASML NXE:3100
which is the pre-production
TWINSCAN system EUV
lithography scanners
(Courtesy of Bill Pierson,
ASML)

EUV Research Status

EUV technology has undergone rigorous research
efforts which began over two decades ago in order to
meet the requirements for high-volume manufacturing
and cost of ownership. One of the critical areas that
needs further development is the EUV source.
Currently, the collectable power of EUV is limited
and must be improved for EUV technology at
high-volume manufacturing throughput levels.
For EUV technology to be implemented, demonstrations of the reliability and stability for source power
have to be achieved. Defect-free masks with the
corresponding inspection infrastructure that can
find all the printable EUV mask defects must be developed. SEMATECH (Semiconductor Manufacturing
Technology, www.sematech.org) is contributing
actively to develop the infrastructure for EUV blank
and mask inspection tools. In addition, photoresists
that meet the requirements of resolution, sensitivity,
and line edge roughness (LER) need to be available as
well.

Currently, ASML (ASM lithography, www.asml.
com) has developed the semiconductor industry's first
EUV lithography production platform (TWINSCAN
NXE). The system platform is based on the learning
and achievements from the EUV prototype tools (Alpha
Demo Tools – ADT) located at IMEC (Interuniversity
MicroElectronics Center, www.imec.be) in Belgium
and CNSE (College of Nanoscale Science and Engineering, www.cnse.albany.edu) in Albany, New York

[20]. The system architecture is based on a six-mirror
lens, and the EUV radiation is generated from
a tin-based plasma source [20]. The preproduction
TWINSCAN NXE system is the NXE:3100 shown in
Fig. 3, with *NA* as 0.25 and resolution of 27 nm, and six
of them have been produced [20]. The NXE:3300B will
follow the NXE:3100 to target volume production
at a resolution of 22 nm and less at an *NA* of 0.32
in 2012 with throughput estimated to be 125 wafers
per hour [20].

Cross-References

▶ DUV Photolithography and Materials
▶ Electron Beam Lithography (EBL)
▶ Immersion Lithography
▶ Immersion Lithography Materials
▶ Nanoimprint Lithography
▶ Nanoimprinting

References

1. Wu, B., Kumar, A.: Extreme Ultraviolet Lithography.
   McGraw-Hill, San Francisco (2009)
2. Wurm, S.: EUV lithography development and research challenges for the 22 nm half-pitch. J. Photopolym. Sci.
   Technol. **22**(1), 31–42 (2009)
3. Bakshi, V.: EUV Lithography. SPIE Press, Bellingham
   (2009)
4. Kemp, K., Wurm, S.: EUV lithography. C. R. Phys. **7**, 875–
   886 (2006)

5. Wurm, S.: Extreme ultraviolet lithography development in the United States. Jpn. J. Appl. Phys. **46**(B9), 6105–6112 (2007)
6. Wu, B., Kumar, A.: Extreme ultraviolet lithography: a review. J. Vac. Sci. Technol. B **25**(6), 1743–1761 (2007)
7. Gwyn, C.W., Stulen, R., Sweeney, D., Attwood, D.: Extreme ultraviolet lithography. J. Vac. Sci. Technol. B **16**(6), 3142–3149 (1998)
8. Chkhalo, N.I., Salashchenko, N.N.: Projection XEUV-nanolithography. Nucl. Instrum. Method Phys. Res. Sect. A **603**, 147–149 (2009)
9. Fay, B.: Advanced optical lithography development, from UV to EUV. Microelectron. Eng. **61/62**, 11–24 (2002)
10. Lin, B.J.: The ending of optical lithography and the prospects of its successors. Microelectron. Eng. **83**, 604–613 (2006)
11. Rothschild, M., Bloomstein, T.M., Fedynshyn, T.H., Kunz, R.R., Liberman, V., Switkes, M., Efremow, N.N., Palmacci, S.T., Sedlacek, J.H.C., Hardy, D.E., Grenville, A.: Recent trends in optical lithography. Lincoln Lab. J. **14**(2), 221–236 (2003)
12. Ronse, K.: Optical lithography – a historical perspective. C.R. Phys. **7**, 844–857 (2006)
13. Moore, G.E.: Cramming more components onto integrated circuits. Electronics **38**, 114–117 (1965)
14. Bakshi, V.: EUV Sources for Lithography. SPIE Press, Bellingham (2006)
15. Feigl, T., Yulin, S., Benoit, N., Kaiser, N.: EUV multilayer optics. Microelectron. Eng. **83**, 703–706 (2006)
16. Mertens, B., Weiss, M., Meiling, H., Klein, R., Louis, E., Kurt, R., Wedowski, M., Trenkler, H., Wolschrijn, B., Jansen, R., Runstraat, A., Moors, R., Spee, K., Ploger, S., Kruijs, R.: Progress in EUV optics lifetime expectations. Microelectron. Eng. **73–74**, 16–22 (2004)
17. Brainard, R.L., Barclay, G.G., Anderson, E.H., Ocola, L.E.: Photoresists for next generation lithography. Microelectron. Eng. **61/62**, 707–715 (2002)
18. Kozawa, T., Tagawa, S.: Radiation chemistry in chemically amplified photoresists. Jpn. J. Appl. Phys. **49**, 030001 (2010)
19. Dentinger, P.M.: Outgassing of photoresist materials at extreme ultraviolet wavelengths. J. Vac. Sci. Technol. B **18**(6), 3364–3370 (2000)
20. ASML's customer magazine: ASML Images. Summer Edition. Veldhoven, The Netherlands (2010)

Excitonic Solar Cell

▶ Nanomaterials for Excitonic Solar Cells

Exoskeleton

▶ Arthropod Strain Sensors

Exposure and Toxicity of Metal and Oxide Nanoparticles to Earthworms

Claire Coutris[1] and Erik J. Joner[2]
[1]Department of Plant and Environmental Sciences, Norwegian University of Life Sciences, Ås, Norway
[2]Bioforsk Soil and Environment, Ås, Norway

Synonyms

Bioaccessibility/bioavailability; Lumbricidae/oligochaeta/earthworms; Soil/terrestrial ecosystem/terrestrial compartment; Toxicity/ecotoxicity; Uptake/internalization/sequestration/biodistribution

Definition

Exposure and toxicity are the two variables that are required to assess the risk that a substance poses to an organism or the environment. Exposure determines to which extent an organism enters in contact with a substance, while toxicity measures to which extent an organism is impaired by the substance.

Background

The increasing use of engineered nanomaterials (ENMs) in consumer products is leading to a new type of environmental pollution for which the knowledge related to environmental hazards and risks is very limited. Toxicity and even ecotoxicity have been examined for a range of ENMs, and a majority of the reports conclude that several ENMs have toxic properties under certain conditions. Yet, exposure conditions seem to govern the outcome of such studies which range from instillation into mammalian airways to exposure of ENMs mixed into soil.

Soil is a particular matrix. It is recognized as a major recipient and sink for environmental pollution with ENMs [1], particularly through application of sewage sludge which contains most of the ENMs entering wastewater [2]. But soil is also an environmental compartment that is essential for food production and a number of important ecological and biogeochemical functions. Soil is an extremely complex matrix, with high specific surface area, a large amount of charged surfaces, abundant amorphous colloidal material, and a wide range of dissolved ions and dissolved organic matter, all of which eludes precise characterization. From one soil to another, there may be a large variation with respect to composition and properties due to differences in parent material and the extent and stage of biological and climate-dependent soil-forming processes. The complexity and diversity of soils is a challenge in terrestrial ecotoxicity studies, as soil properties will influence bioavailability and thereby exposure of organisms and ultimately the apparent toxicity to ENMs. Which properties are most important for ENM bioavailability probably depend on the ENM in question, but remains to be examined in detail.

Differences in soil composition translate into differences in soil properties that can be quantified by physical and chemical analyses. Though no systematic examination exist of which soil properties affect the bioavailability of different ENMs, certain soil properties that affect the bioavailability of other environmental pollutants are likely to be important. These are, therefore, reported in soil-based ecotoxicity studies on ENMs and include, or should include, texture (particle size distribution), organic matter content, pH, cation exchange capacity (CEC), and base saturation (BS; quantity of exchangeable Na, K, Mg, and Ca as a percentage of CEC). Even though other parameters may be equally important or provide more detailed information that may be used to explain interactions between a soil and ENMs, these are often too specialized to be available in most soil science labs (e.g., equipment for characterization of specific surface area, clay mineralogy, or soil organic matter quality).

When ENMs are characterized, specific surface area has become a useful parameter believed to predict both surface reactivity, capacity to interact with dissolved ions and molecules, as well as propensity for dissolution. Small nanoparticles typically have specific surface areas (measured by BET; the Brunauer-Emmett-Teller method) in the order of 30–50 $m^2\ g^{-1}$ [3]. Though impressive it may seem, soil constituents like clay may have BET values an order of magnitude higher, ranging from 50 to 1,100 $m^2\ g^{-1}$. Also humic soil organic matter has BET values in the order of 500–800 $m^2\ g^{-1}$. As soils commonly contain at least 3–5% organic matter and often at least 10% clay, there are ample possibilities for interactions between ENMs

and the soil matrix, which may affect exposure and toxicity to soil organisms.

Standard OECD and ISO test guidelines for soil organisms like earthworms commonly exclude the use of soil, and those standardized tests that permit the use of soil prescribe the use of soil types that are low in clay and organic matter. Artificial "soils" are frequently used in which weakly transformed peat constitute the organic fraction and sand and low swelling clay represent the mineral fraction. Not only are these "soils" poor as proxies for real soil with respect to surface area and surface charge, but they are also void of organisms which constitute an integral part of natural soils.

To test the impact of environmental toxicants on soil organisms, earthworms have frequently been used due to the ecological importance of their activity related to soil organic matter turnover and their burrowing that ensures soil porosity [4]. Earthworms are also convenient as test animals, as several stress-related and toxicological endpoints may be used in tests of limited duration to assess both acute and chronic effects at different levels (behavioral, physiological, reproductive, and genetic).

Eisenia fetida is a model earthworm used in standardized soil toxicity tests issued by the US-EPA and the OECD. This species has a short generation time and reproduces easily under lab conditions. Part of its genome is also sequenced for use in transcriptomic studies. Yet, the extensive use of this species is often criticized as it is epigeic (living in organic surface layer or compost), rather than endogeic or anecic (living and burrowing in soil, *stricto sensu*).

Exposure

The question of possible risks of ENMs does not depend solely on their toxicity, but also requires information on exposure (Risk = Hazard × Exposure). The latter aspect has, however, received far less attention than toxicity studies, probably due to methodological difficulties, as ENMs are difficult to track and quantify in organisms and their tissues, and even more so in complex media like water, sediments, soil, and waste. Exposure is somewhat differently defined in toxicology and ecotoxicology. In higher organisms in vivo, exposure is often concerned with toxicokinetic aspects of absorption, distribution, metabolism, and excretion

(ADME), though the route of exposure is also given (oral, dermal, respiratory, etc.). In ecotoxicology, the perception of exposure is much closer to the colloquial understanding of the word and implies to which extent an organism gets in contact with a toxic compound, ingests it and excretes it. Two earthworm species living in the same soil but having different feeding habits (e.g., anecic vs. endogeic) may thus be exposed to ENMs in surface applied sewage sludge in very different ways. Also, a worm living in a clay-rich soil containing ENMs may be differently exposed compared to the same worm species living in a sandy soil containing the same type and amount of ENMs. This aspect of exposure is closely related to the concept of bioavailability, which is taking into account the interactions between a nutrient or a toxic compound and its surrounding matrix. Studies on earthworm exposure to ENMs have sought to vary both the way ENMs are provided (mixed into soil or feed) and matrix composition (soil vs. water, and soil composition). Also, pure measurements describing uptake and excretion have been made for some ENMs. The few earthworm exposure studies that exist are described under each category of ENM, together with data on ecotoxicity.

Experimental Evidence

Several types of metal and metal oxide nanomaterials have been studied for their potential bioavailability and toxicity to earthworms, namely, Ag, Au, Co, Cu, Fe, Ni, Al_2O_3, SiO_2, TiO_2, ZnO, and ZrO_2. Most of these studies used the earthworm *E. fetida*, while others used *Lumbricus terrestris*, *L. rubellus*, or *E. veneta*. Also tests using the closely related pot worm *Enchytraeus albidus* have provided relevant data in similar tests, since this species also belongs to the subclass Oligochaeta. A variety of endpoints are reported, from survival, body burden, and reproduction to apoptosis and expression of genes involved in metal homeostasis and oxidative stress. State-of-the-art results are presented for each metal and metal oxide type below.

Ag

A variety of silver nanoparticles (Ag NPs) with a range of sizes (9–50 nm diam.) and coatings (citrate, oleic acid, polyvinylpyrrolidone, or none) have been exposed to earthworms through water, food, or different types of soil at concentrations ranging from 0.3 to 1,000 mg kg^{-1}. Bioaccumulation has generally been higher for AgNO$_3$ than for Ag NPs, and the various sizes and coatings tested have affected bioaccumulation to no or only limited extent. What has made a difference is the soil type used in these tests, so that higher bioaccumulation of Ag NPs has been observed in soil with lower organic matter and higher sand content. In artificial OECD soil, bioaccumulation factors (BAFs, being the concentration ratio of accumulated ENMs/soil ENMs) have been very low and seem to indicate that Ag NPs measured in earthworms merely result from incomplete gut clearance after exposure. In soil with lower organic matter and higher sand content, Ag NPs have been more bioaccessible and BAFs higher (BAF = 0.05–0.08 in sandy loam while <0.03 in OECD soil) [5, 6]. Survival and growth of worms were not affected by Ag NP concentrations as high as 1,000 mg kg^{-1}, contrary to AgNO$_3$, which reduced survival to approx. 3% at the same concentration [7]. Reproduction measured as cocoon production has been affected at lower concentrations with significant reductions at ≥94 mg kg^{-1} for AgNO$_3$, and at ≥800 mg kg^{-1} Ag NPs in an experiment by Shoults-Wilson et al. [6]. In this experiment, no differences were found between particles of different size or with different coatings. A more sensitive endpoint yet is the measurement of avoidance, as demonstrated in an experiment by Shoults-Wilson et al. [8] where worms given the choice between uncontaminated soil and soil spiked with 7 mg kg^{-1} Ag NPs, avoided the latter when measured after 48 h exposure. A similar concentration of AgNO$_3$ induced the same response, but faster. This concentration is close to predicted environmental concentrations of Ag NPs in sludge from sewage treatment plants in Europe and the United States (1.29–6.24 mg kg^{-1}) [1]. Also enhanced frequency of apoptosis measured in the cuticle, the intestinal epithelium, and the chloragogenous tissue of earthworms (*L. terrestris*) exposed to concentrations as low as 4 mg kg^{-1} in soil indicate that earthworms may be negatively affected by environmentally relevant concentrations of Ag NPs [9]. The two latter reports do, however, not prove any lasting negative impact, but may simply be due to stress from which the worms can recover in time. The studies conducted so far have been based on freshly spiked media. It appears, however, that Ag NPs may become more bioavailable than AgNO$_3$ over time as Ag NPs can act as constant sources releasing ionic and bioavailable Ag. This is indicated, e.g., by the results of Shoults-Wilson et al. [10] who showed that 11–20% of Ag in 30–50 nm Ag NPs was oxidized after 28 days in OECD soil, and supported by results from sequential chemical extractions of soils spiked with different types of Ag NPs (Coutris et al., unpublished results). Until more sensitive endpoints are examined, e.g., related to transcriptomic analyses, that can confirm or preclude proper toxicity effects of environmentally relevant AgNP concentrations, conclusions must be drawn from the physiological and reproductive endpoints available. Based on such data it seems like Ag NPs have a low toxicity and low bioavailability to earthworms.

Au

Gold NPs (4 and 18 nm diam., coated with citrate) have been found in the gut and the skin of earthworms exposed for 1 week to Au NPs in artificial OECD soil spiked with 3 mg Au kg^{-1}. A similar body distribution was found in worms exposed 48 h on filter paper to a 10 mg L^{-1} Au NPs suspension [11]. In this study, Au uptake was particle size dependent, with uptake of Au ions being the highest followed by 4 and 18 nm Au NPs. In another experiment with Au NPs (20 and 55 nm diam., 5–50 mg kg^{-1} soil) Unrine et al. [12] confirmed that Au NPs were bioavailable to E. fetida through exposure in artificial OECD soil, and that Au ions were taken up to a larger extent than Au NPs. However, in this case, no difference in uptake was noticed between Au NPs on a mass basis (whereas on a particle number basis, more 20 nm Au NPs were taken up than 55 nm Au NPs). Although no effects on survival or growth were noticed, both ions and NPs had negative effects on earthworm reproduction. This response was not purely dose dependent, as e.g., 55 nm Au NPs had no negative effect on reproduction at 50 mg Au kg^{-1}, whereas at 20 mg kg^{-1} a negative effect was observed. Also here, ions were more toxic than NPs, but smaller NPs did not appear more toxic than larger ones, and the two size classes tested formed aggregates of similar size. Finally, none of the earthworm genes related to metal homeostasis and oxidative and cellular stress were expressed differently in control and Au NPs treatments, whereas $HAuCl_4$ at ≥ 2.5 mg kg^{-1} resulted in the upregulation of metallothionein [12].

Co

Cobalt NPs (Co core/Co_3O_4 shell, 4 nm diam.) have been found to be readily bioavailable to E. fetida

exposed though their diet to 1 mg Co kg^{-1} (BAF = 0.69 ± 0.04 for Co NPs and 0.88 ± 0.05 for $CoCl_2$). Co from both particulate and ionic sources was only slowly excreted from the worms (biological half-life 200 days) and showed a rather similar biodistribution within internal organs of the worms [5]. The highest accumulation was found in the blood and the intestinal tract and the least in the reproductive organs. Still, at high concentrations (87 mg kg^{-1} food), Co from Co NPs have been found in reproductive organs and cocoons produced after the worms had been transferred to a non contaminated environment [13].

Cu

Copper NPs (Cu/CuO, 80 nm diam.) added to artificial OECD soil at 1,000 mg kg^{-1} had no acute effect on the survival of E. fetida, but induced a dramatic decrease in reproduction, whereas a similar concentration of $CuCl_2$ had both an acute and chronic effect [7]. The same particles have been tested at 450 and 750 mg kg^{-1} on Enchytraeus albidus, and indicated that both $CuCl_2$ and Cu NPs caused oxidative stress damage, yet in different ways and with more pronounced effects in the case of $CuCl_2$ [14]. An experiment by Unrine et al. [15], where E. fetida was exposed to artificial OECD soil spiked with Cu NPs (20–40 nm and 100 nm diam.), has showed that the expression of metallothionein genes was upregulated at ≥ 20 mg kg^{-1} Cu NPs and ≥ 10 mg kg^{-1} $CuSO_4$, while no effect on the regulation of other genes related to oxidative and general stress was detected. Synchrotron-based X-ray analyses indicated that both types of Cu NPs had been oxidized within 4 weeks so that they contained a mixture of zero-valent Cu, Cu_2O, and CuO. Smaller particles were also more oxidized than larger ones. The authors concluded that oxidized Cu NPs were able to enter the food chain from soil, but that only concentrations >65 mg kg^{-1} would induce adverse effects.

Fe

Ecotoxicity data for nano zero-valent iron (nZVI) on terrestrial macro- and mesofauna are only starting to emerge. Preliminary studies have shown that earthworms are sensitive to nZVI concentrations of ≥ 100 mg kg^{-1}. Both the compost worm E. fetida, used in standardized OECD tests, and the epigeic earthworm Lumbricus rubellus showed signs of adverse effects during long-term exposure to 100 mg nZVI kg^{-1}, whereas acute (mortality after 14 days exposure in sandy soil) toxicity was observed only at concentrations ≥ 500 mg nZVI kg^{-1} for both species (El-Temsah and Joner, unpublished results).

Ni

The only published study assessing the potential toxicity of Nickel NPs (20 nm diam.) to E. fetida did not reveal any acute effects on the survival or reproduction. The test featured 28 days exposure in artificial OECD soil containing 1,000 mg kg^{-1} Ni NPs. On the other hand, a similar concentration of NiCl$_2$ induced a high mortality and impaired reproduction dramatically (98% decrease in cocoon production and none of the cocoons hatched) [7].

Al$_2$O$_3$

Al$_2$O$_3$ NPs (12–14 nm diam.) added to artificial OECD soil at 1,000 mg kg^{-1} did not show any acute effects on the survival or reproduction of E. fetida [7]. Another study with similarly sized Al$_2$O$_3$ NPs (11 nm) found that cocoon production was reduced only at concentrations $\geq 3,000$ mg kg^{-1}, and that avoidance was apparent when concentrations reached $\geq 5,000$ mg kg^{-1} [16]. Although no predicted environmental concentrations are available for this type of nanomaterials, it seems unlikely that such concentrations will be found in the environment.

TiO$_2$

TiO$_2$ NPs (21 nm diam., 73% anatase, 27% rutile) added to artificial OECD soil at 1,000 mg kg^{-1} had no acute effect on the survival of Eisenia fetida, but induced a weak but significant adverse effect on reproduction [7]. TiO$_2$ NPs (10–20 nm diam., rutile) added to artificial OECD soil induced lipid peroxidation, DNA damage, and mitochondria damage only at very high concentrations (5,000 mg kg^{-1}) [17]. Another experiment with Lumbricus terrestris exposed to an aged TiO$_2$ composite (100–300 nm, 1–100 mg kg^{-1}) through water, food, and soil showed neither effect on survival nor bioaccumulation, but enhanced apoptotic frequency in the cuticle, the intestinal epithelium, and chloragogenous tissue of worms exposed to 100 mg kg^{-1} in water [18]. Similar effects were not observed when exposed in soil or feed.

ZnO

No acute toxic effects of Zn on survival were found in E. veneta exposed to ZnCl$_2$ or ZnO NPs (≤ 100 nm

diam.) at 250 and 750 mg ZnO kg^{-1} soil for 21 days [19]. Growth was observed in all worms irrespective of Zn source and dose, but was slightly reduced for ZnCl$_2$ at the highest concentration. Cocoon production was reduced by 50% at 750 mg kg^{-1} in the case of ZnO NPs, whereas the same Zn concentration led to an almost complete inhibition in the case of ZnCl$_2$. No reduction in immune activity of coelomocytes was found in presence of ZnO NPs, even at the highest concentration tested, whereas a 20% reduction was observed at 750 mg kg^{-1} for ZnCl$_2$. In another study with 7 days exposure of E. fetida to ZnO NPs (10–20 nm diam.) Hu et al. [17] observed increased lipid peroxidation at ≥ 100 mg kg^{-1}, reduced activity of superoxide dismutase and biochemical metabolism (measured as catalase activity) at ≥ 500 mg kg^{-1}, induced DNA damage at 1,000 mg kg^{-1}, and mitochondria damage at 5,000 mg kg^{-1}. This study made no comparisons to ionic Zn. A third study where E. fetida was exposed to ZnO NPs in agar for 4 days found close to 100% mortality of ZnO NPs at 1,000 mg ZnO kg^{-1} [20]. The same authors also used a contact test where worms were exposed to ZnO NPs added as an aqueous suspension to filter paper, but this test showed no dose-dependent toxicity. Both the agar test and the filter paper test did, however, show that environmental conditions can influence toxicity, as lower negative effects were observed when ZnO NPs suspensions were prepared with water containing salts or dissolved humic material, as compared to exposure to ZnO NPs suspended in deionized water.

SiO$_2$ and ZrO$_2$

To date, the adverse effects of SiO$_2$ and ZrO$_2$ nanoparticles have only been examined one study, which did not reveal any detrimental effects on the survival or reproduction of E. fetida when exposed to 1,000 mg kg^{-1} SiO$_2$ NPs (5–15 nm diam.) or ZrO$_2$ NPs (20–30 nm diam.) in artificial OECD soil [7].

Concluding Remarks

Testing of metallic and oxide ENMs for adverse effects on soil invertebrates has only been attempted for a few years, and the limited amount of data available, with respect to tested ENMs, earthworm species, ecotoxicological endpoints, and exposure conditions, calls for cautious conclusions. Still, it seems that ENMs made of elements with known antimicrobial properties, like Ag and Cu, are among those that should be treated with most concern as they have so far shown the lowest adverse effect concentrations. Earthworms may not only be affected by direct toxicity of these ENMs, but will also suffer if microorganisms on which they depend are affected. Earthworms strongly depend on microorganisms living both endosymbiotically in the coelomic cavity and in the digestive tract, and any adverse effects on these organisms are bound to affect the worm's physiology and health. Regarding the specific effects of ENM versus the same elements in a soluble form, there are several indications that the effect of ENMs stems, at least partly, from dissolving ions released by from the surface of ENMs. While free ions often react rapidly with other ions or more complex molecules to form insoluble metal species (e.g., AgCl), ENMs may continue to act in a toxic manner by surface interaction or continued release of ions. In this way, ions may appear more acutely toxic, while corresponding ENM may eventually appear to have more chronic effects.

Cross-References

▶ Cellular Mechanisms of Nanoparticle's Toxicity
▶ Ecotoxicity of Inorganic Nanoparticles: From Unicellular Organisms to Invertebrates
▶ Ecotoxicology of Carbon Nanotubes Toward Amphibian Larvae
▶ Effect of Surface Modification on Toxicity of Nanoparticles
▶ Electron Microscopy of Interactions Between Engineered Nanomaterials and Cells
▶ Fate of Manufactured Nanoparticles in Aqueous Environment
▶ Genotoxicity of Nanoparticles
▶ In Vitro and In Vivo Toxicity of Silver Nanoparticles
▶ In Vivo Toxicity of Titanium Dioxide and Gold Nanoparticles
▶ Nanoparticle Cytotoxicity
▶ Nanoparticles
▶ Physicochemical Properties of Nanoparticles in Relation with Toxicity
▶ Quantum-Dot Toxicity
▶ Toxicology: Plants and Nanoparticles

References

1. Gottschalk, F., Sonderer, T., Scholz, R.W., Nowack, B.: Modeled environmental concentrations of engineered nanomaterials (TiO_2, ZnO, Ag, CNT, fullerenes) for different regions. Environ. Sci. Technol. **43**, 9216–9222 (2009)
2. Kiser, M.A., Westerhoff, P., Benn, T., Wang, Y., Perez-Rivera, J., Hristovski, K.: Titanium nanomaterial removal and release from wastewater treatment plants. Environ. Sci. Technol. **43**, 6757–6763 (2009)
3. Auffan, M., Rose, J., Bottero, J.-Y., Lowry, G.V., Jolivet, J.-P., Wiesner, M.R.: Towards a definition of inorganic nanoparticles from an environmental, health and safety perspective. Nat. Nanotechnol. **4**, 634–641 (2009)
4. Edwards, C.A., Bohlen, P.J.: Biology and Ecology of Earthworms. Chapman and Hall, London (1996)
5. Coutris, C., Hertel-Aas, T., Lapied, E., Joner, E.J., Oughton, D.H.: Bioavailability of cobalt and silver nanoparticles to the earthworm *Eisenia fetida*. Nanotoxicology doi:10.3109/17435390.2011.569094 (online Apr 2011)
6. Shoults-Wilson, W.A., Reinsch, B.C., Tsyusko, O.V., Bertsch, P.M., Lowry, G.V., Unrine, J.M.: Role of particle size and soil type in toxicity of silver nanoparticles to earthworms. Soil Sci. Soc. Am. J. **75**, 365–377 (2011)
7. Heckmann, L.H., Hovgaard, M.B., Sutherland, D.S., Autrup, H., Besenbacher, F., Scott-Fordsmand, J.J.: Limit-test toxicity screening of selected inorganic nanoparticles to the earthworm *Eisenia fetida*. Ecotoxicology **20**, 226–233 (2011)
8. Shoults-Wilson, W.A., Zhurbich, O.I., McNear, D.H., Tsyusko, O.V., Bertsch, P.M., Unrine, J.M.: Evidence for avoidance of Ag nanoparticles by earthworms (*Eisenia fetida*). Ecotoxicology **20**, 385–396 (2011)
9. Lapied, E., Moudilou, E., Exbrayat, J.M., Oughton, D.H., Joner, E.J.: Silver nanoparticle exposure causes apoptotic response in the earthworm *Lumbricus terrestris* (Oligochaeta). Nanomedicine **5**, 975–984 (2010)
10. Shoults-Wilson, W.A., Reinsch, B.C., Tsyusko, O.V., Bertsch, P.M., Lowry, G.V., Unrine, J.M.: Effect of silver nanoparticle surface coating on bioaccumulation and reproductive toxicity in earthworms (*Eisenia fetida*). Nanotoxicology doi:10.3109/17435390.2010.537382 (online Dec 2010)
11. Unrine, J., Bertsch, P., Hunyadi, S.E.: Bioavailability, trophic transfer, and toxicity of manufactured metal and metal oxide nanoparticles in terrestrial environments. In: Grassian, V.H. (ed.) Nanoscience and Nanotechnology, pp. 343–364. Wiley, Hoboken (2008)
12. Unrine, J.M., Hunyadi, S.E., Tsyusko, O.V., Rao, W., Shoults-Wilson, W.A., Bertsch, P.M.: Evidence for bioavailability of Au nanoparticles from soil and biodistribution within earthworms (*Eisenia fetida*). Environ. Sci. Technol. **44**, 8308–8313 (2010)
13. Oughton, D.H., Hertel-Aas, T., Pellicer, E., Mendoza, E., Joner, E.J.: Neutron activation of engineered nanoparticles as a tool in studies on their environmental fate and uptake in organisms. Environ. Toxicol. Chem. **27**, 1883–1887 (2008)
14. Gomes, S.I.L., Novais, S.C., Gravato, C. et al.: Effect of Cu-nanoparticles versus one Cu-salt: analysis of stress biomarkers response in *Enchytraeus albidus* (Oligochaeta).

Nanotoxicology   doi:10.3109/17435390.2011.562327
(online Apr 2011)

15. Unrine, J.M., Tsyusko, O.V., Hunyadi, S.E., Judy, J.D.,
    Bertsch, P.M.: Effects of particle size on chemical specia-
    tion and bioavailability of copper to earthworms (*Eisenia
    fetida*) exposed to copper nanoparticles. J. Environ. Qual.
    39, 1942–1953 (2010)

16. Coleman, J.G., Johnson, D.R., Stanley, J.K., et al.:
    Assessing the fate and effects of nano aluminum oxide in
    the terrestrial earthworm, *Eisenia fetida*. Environ. Toxicol.
    Chem. **29**, 1575–1580 (2010)

17. Hu, C.W., Li, M., Cui, Y.B., Li, D.S., Chen, J., Yang, L.Y.:
    Toxicological effects of TiO_2 and ZnO nanoparticles in soil
    on earthworm *Eisenia fetida*. Soil Biol. Biochem. **42**,
    586–591 (2010)

18. Lapied, E., Nahmani, J.Y., Moudilou, E. *et al.* Ecotoxico-
    logical effects of an aged TiO_2 nanocomposite measured as
    apoptosis in the anecic earthworm *Lumbricus terrestris* after
    exposure through water, food and soil. Environ. Int. **37**,
    1105–1110 (2011)

19. Hooper, H.L., Jurkschat, K., Morgan, A.J. et al.: Compara-
    tive chronic toxicity of nanoparticulate and ionic zinc to the
    earthworm *Eisenia veneta* in a soil matrix. Environ. Int. **37**,
    1111–1117 (2011)

20. Li, L.-Z., Zhou, D.-M., Peijnenburg, W.J.G.M. et al.: Tox-
    icity of zinc oxide nanoparticles in the earthworm *Eisenia
    fetida* and subcellular fractionation of Zn. Environ. Int. **37**,
    1098–1104 (2011)

Extreme Ultraviolet Lithography (EUVL)

▶ EUV Lithography

Extrusion Spinning

▶ Spider Silk